W9-BWQ-437

Turn the graphing calculator into a powerful tool for your success!

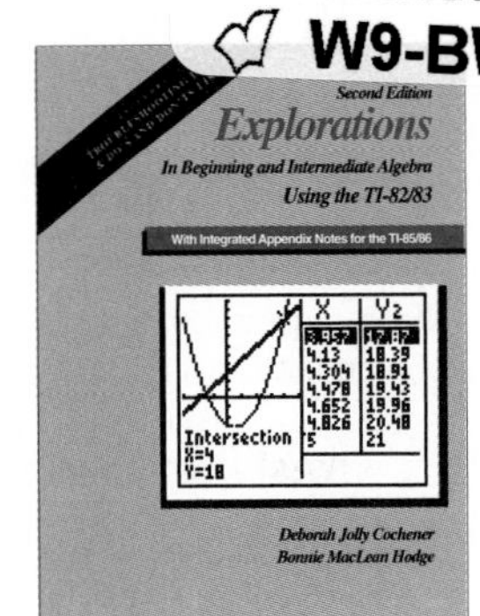

Explorations in Beginning and Intermediate Algebra Using the TI-82/83 with Integrated Appendix Notes for the TI-85/86, Second Edition

by Deborah J. Cochener and Bonnie M. Hodge,
both of Austin Peay State University

352 pages. Spiralbound. 8 1/2" x 11". ISBN: 0-534-36149-8.
©1999. Published by Brooks/Cole.

You can quickly learn to use the graphing calculator to develop problem-solving and critical-thinking skills that will improve your performance in beginning and intermediate algebra!

Designed to help you succeed in your algebra course, this unique and student-friendly workbook improves both your understanding *and* retention of algebra concepts—using the graphing calculator. By integrating technology into mathematics, the authors help you develop problem-solving and critical-thinking skills.

To guide you in your explorations, you'll find:

- hands-on applications with solutions
- correlation charts that relate course topics to the workbook units
- key charts (specific to the TI-82, TI-83, TI-85, and TI-86) that show which units introduce keys on the calculator
- a *Troubleshooting Section* to help you avoid common errors

Other primary features to help you succeed:

- *In Your Own Words*, a section at the end of each unit, gives you the opportunity to summarize main points from the unit and personalize the material for your own use
- Appendices that address the linking capabilities of the calculators and menu maps for each calculator that are designed to assist the you in locating a needed menu or function
- An optional unit that addresses the complex number operations available on the TI-83, TI-85, and TI-86
- *Extra for Experts* problems that extend the concepts within the unit
- *Accuracy Checks*—five-to-ten-question quizzes—that correspond to most of the units. The answers are posted at the Brooks/Cole web site.

Topics

This text contains 36 units divided into the following subsections:

Basic Calculator Operations
Graphically Solving Equations and Inequalities
Graphing and Applications of Equations in Two Variables
Stat Plots

Order your copy today!

To receive a sampler of ***Explorations in Beginning and Intermediate Algebra Using the TI-82/83 with Integrated Appendix Notes for the TI-85/86, Second Edition***, simply mail in the order form on the other side of this page, call (800) 354-9706, or visit us on the Internet: http://www.brookscole.com

ORDER FORM

Call our toll-free number (800) 354-9706 to purchase, or use the form below and mail or fax to (831) 375-6414.

No risk. All of our books and software are backed by our 30-day, money-back guarantee. Major credit cards accepted. We accept purchase orders from your company or institution. The cost of shipping will be added to your bill. To order, simply fill out this coupon and return it to Brooks/Cole along with your check, money order, or credit card information.

______ Yes! I would like to order ***Explorations in Beginning and Intermediate Algebra Using the TI-82/83 with Integrated Appendix Notes for the TI-85/86, Second Edition***, by Deborah J. Cochener and Bonnie M. Hodge, ISBN: 0-534-36149-8 for $26.95. *Note: Pricing is subject to change. Please call (800) 354-9706 to receive current pricing information.*

Residents of AL AZ CA CT CO, FL, GA, IL, IN, KS, KY, LA, MA, MD, MI, MN, MO, NC, NJ, NY, OH, PA, RI, SC, TN, TX, UT, VA, WA, WI must add appropriate state sales tax.

Subtotal	________
Tax	________
Handling	$4.00
TOTAL	________

Payment Options

______ Payment enclosed (check or money order)

______ Please charge the following credit card:

____ VISA ____ MasterCard ____ American Express

Card #:______________________________ Expiration Date:______________

Contact Name:_________________________ Phone:______________

Signature:______________________________

(Note: Credit card billing and shipping address must be the same.)

All orders under 250 pounds will be shipped via UPS, unless the customer requests a specific carrier. Average shipping charges are estimated at 10% of the order. The cost of shipping will be added to your bill. Customer's requested shipper:____________

Prices are subject to change without notice. We will refund any pre-payments for unshipped, out-of-stock titles after 150 days and for not-yet-published titles after 180 days, unless an earlier date is requested in writing by you.

Please ship my order to: *(please print)*

Name ____

Street Address ______________________________________

City ____________________ State __________ Zip ____________

Telephone (________)____________________ e-mail:______________

Mail to:

Brooks/Cole Publishing Company
Source Code 9BCMA
511 Forest Lodge Road
Pacific Grove, CA 93950-5098
Phone: (800) 354-9706 • Fax: (831) 375-6414

I(T)P **International Thomson Publishing Education Group**

Intermediate Algebra

To the mathematics instructors of Edgewood High School, the University of Redlands, and California State University at Los Angeles, whose love of their discipline and dedication to their students continue to be an inspiration to me.

AST

To Karl Jacobs, a fine mentor and college president.

RDG

Intermediate Algebra

Alan S. Tussy
Citrus College

R. David Gustafson
Rock Valley College

Brooks/Cole Publishing Company

I(T)P® *An International Thomson Publishing Company*

Pacific Grove • Albany • Bonn • Boston • Cincinnati • Detroit • Johannesburg • London • Madrid • Melbourne
Mexico City • New York • Paris • San Francisco • Singapore • Tokyo • Toronto • Washington

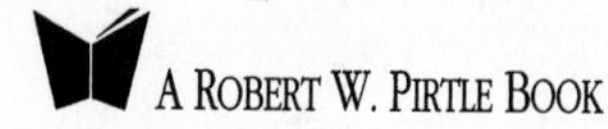

A ROBERT W. PIRTLE BOOK

Sponsoring Editor: *Robert W. Pirtle*
Marketing Team: *Jennifer Huber, Christine Davis, Debra Johnston*
Editorial Assistant: *Erin Wickersham*
Production Editor: *Ellen Brownstein*
Production Service: *Hoyt Publishing Services*
Manuscript Editor: *David Hoyt*
Photo Editor: *Terry Powell*

Interior & Cover Design: *Vernon T. Boes*
Interior Illustration: *Lori Heckelman*
Cover Illustration: *Joe Miles*
Art Coordinator: *David Hoyt*
Typesetting: *The Clarinda Company*
Cover Printing: *Phoenix Color Corp.*
Printing and Binding: *Banta Book Group*

For more information, contact:

BROOKS/COLE PUBLISHING COMPANY
511 Forest Lodge Road
Pacific Grove, CA 93950
USA

International Thomson Publishing Europe
Berkshire House 168-173
High Holborn
London WC1V 7AA
England

Thomas Nelson Australia
102 Dodds Street
South Melbourne, 3205
Victoria, Australia

Nelson Canada
1120 Birchmount Road
Scarborough, Ontario
Canada M1K 5G4

International Thomson Editores
Seneca 53
Col. Polanco
11560 México, D. F., México

International Thomson Publishing GmbH
Königswinterer Strasse 418
53227 Bonn
Germany

International Thomson Publishing Asia
60 Albert Street #15-01
Albert Complex
Singapore 189969

International Thomson Publishing Japan
Hirakawacho Kyowa Building, 3F
2-2-1 Hirakawacho
Chiyoda-ku, Tokyo 102
Japan

Printed in the United States of America

10 9 8 7 6 5 4 3 2

Library of Congress Cataloging-in-Publication Data

Tussy, Alan S., [date]
Intermediate algebra / Alan S. Tussy, R. David Gustafson.
p. cm.
Includes index.
ISBN 0-534-35581-1 (alk. paper)
1. Algebra. I. Gustafson, R. David (Roy David), [date].
II. Title.
QA154.2.T87 1998
512.9—dc21

98-31048
CIP

THIS BOOK IS PRINTED ON
ACID-FREE RECYCLED PAPER

PREFACE

For the Instructor

Algebra is used to describe numerical relationships. It is a language in its own right. The purpose of this textbook is to teach students how to read, write, speak, and think mathematically using the language of algebra. It presents all the topics associated with a second course in algebra. We have used a blend of the traditional and the reform instructional approaches to do this. In this book, you will find the vocabulary, practice, and well-defined pedagogy of a traditional approach. You will also find that we emphasize the reasoning, modeling, communicating, and technological skills that are such a big part of today's reform movement.

This textbook expands the students' mathematical reasoning abilities and gives them a set of mathematical survival skills that will help them succeed in a world that increasingly requires that every person become a better analytical thinker. We believe it will hold student attrition to a minimum while preparing students to succeed in the next course—whether that is college algebra, trigonometry, statistics, or finite mathematics.

Features of the text

A Review of Basic Algebra

Chapter 1 begins with a review of the fundamental algebraic concepts of variable, expression, and equation. Students translate English phrases to mathematical symbols and use a five-step problem-solving strategy to solve a wide variety of application problems.

An Innovative Table of Contents

We have made several changes to the usual table of contents found in most intermediate algebra books. An early discussion of graphing on the rectangular coordinate system and of functions appears in Chapter 2. That is followed by Chapter 3, Solving Systems of Equations. Chapter 4 deals exclusively with inequalities.

Interactivity

Most worked examples in the text are accompanied by Self Checks. This feature allows students to practice skills discussed in the example by working a similar problem. Because the Self Check problems are adjacent to the worked examples, students can easily refer to the solution and author's notes of the example as they solve the Self Check. Author's notes are used to explain the steps in the solutions of examples. The notes are extensive so as to increase the students' ability to read and write mathematics.

Most examples have Self Checks.

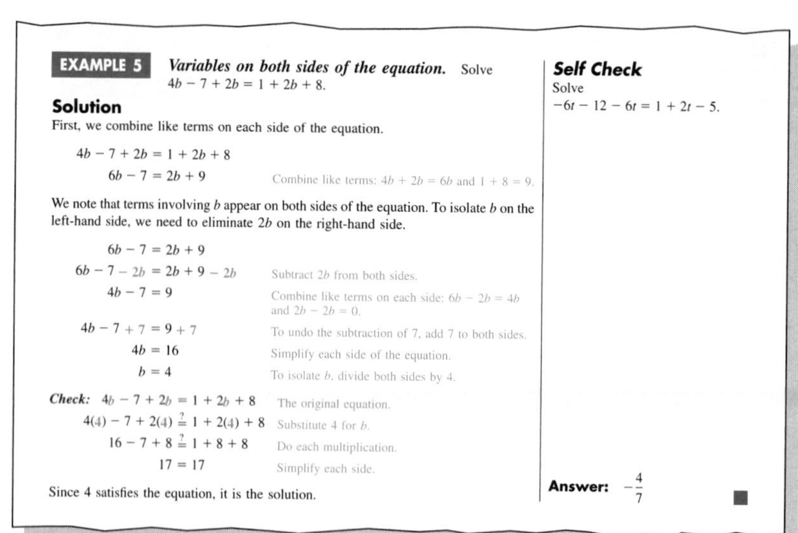

EXAMPLE 5 ***Variables on both sides of the equation.*** Solve $4b - 7 + 2b = 1 + 2b + 8$.

Solution

First, we combine like terms on each side of the equation.

$$4b - 7 + 2b = 1 + 2b + 8$$
$$6b - 7 = 2b + 9 \quad \text{Combine like terms: } 4b + 2b = 6b \text{ and } 1 + 8 = 9.$$

We note that terms involving b appear on both sides of the equation. To isolate b on the left-hand side, we need to eliminate $2b$ on the right-hand side.

$$6b - 7 = 2b + 9$$
$$6b - 7 - 2b = 2b + 9 - 2b \quad \text{Subtract } 2b \text{ from both sides.}$$
$$4b - 7 = 9 \quad \text{Combine like terms on each side: } 6b - 2b = 4b \text{ and } 2b - 2b = 0.$$
$$4b - 7 + 7 = 9 + 7 \quad \text{To undo the subtraction of 7, add 7 to both sides.}$$
$$4b = 16 \quad \text{Simplify each side of the equation.}$$
$$b = 4 \quad \text{To isolate } b \text{, divide both sides by 4.}$$

Check:
$$4b - 7 + 2b = 1 + 2b + 8 \quad \text{The original equation.}$$
$$4(4) - 7 + 2(4) \stackrel{?}{=} 1 + 2(4) + 8 \quad \text{Substitute 4 for } b.$$
$$16 - 7 + 8 \stackrel{?}{=} 1 + 8 + 8 \quad \text{Do each multiplication.}$$
$$17 = 17 \quad \text{Simplify each side.}$$

Since 4 satisfies the equation, it is the solution.

Self Check

Solve $-6t - 12 - 6t = 1 + 2t - 5$.

Answer: $-\frac{4}{7}$

Author's notes explain the steps in the solution process.

In-Depth Coverage of Geometry

Perimeter, area, and volume, as well as many other geometric concepts, are used in a variety of contexts throughout the book. We have included many drawings to help the students improve their ability to spot visual patterns in their everyday lives.

Geometric topics are presented in a practical setting.

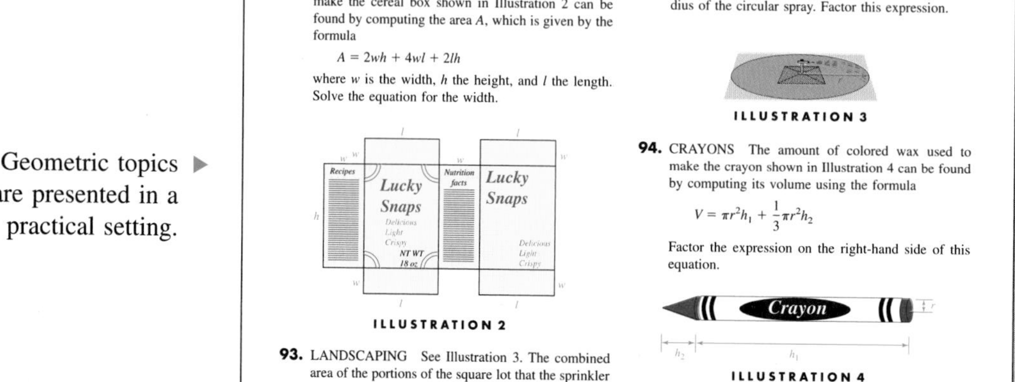

92. PACKAGING The amount of cardboard needed to make the cereal box shown in Illustration 2 can be found by computing the area A, which is given by the formula

$$A = 2wh + 4wl + 2lh$$

where w is the width, h the height, and l the length. Solve the equation for the width.

ILLUSTRATION 2

93. LANDSCAPING See Illustration 3. The combined area of the portions of the square lot that the sprinkler doesn't reach is given by $4r^2 - \pi r^2$, where r is the radius of the circular spray. Factor this expression.

ILLUSTRATION 3

94. CRAYONS The amount of colored wax used to make the crayon shown in Illustration 4 can be found by computing its volume using the formula

$$V = \pi r^2 h_1 + \frac{1}{3}\pi r^2 h_2$$

Factor the expression on the right-hand side of this equation.

ILLUSTRATION 4

Study Sets—More Than Just Exercises

The problems at the end of each section are called Study Sets. Each Study Set includes Vocabulary, Notation, and Writing problems designed to help students improve their ability to read, write, and communicate mathematical ideas. The problems in the Concepts section of the Study Sets encourage students to engage in independent thinking and reinforce major ideas through exploration. In the Practice section of the Study Sets, students get the drill necessary to master the material. In the Applications section, students deal with real-life situations that involve the topics being studied. Each Study Set concludes with a Review section consisting of problems similar to those in previous sections.

◀ Each Study Set contains Vocabulary, Concepts, Notation, Practice, Applications, Writing, and Review sections.

STUDY SET Section 5.1

VOCABULARY *In Exercises 1–4, fill in the blanks to make the statements true.*

1. In the exponential expression x^n, x is called the ____________, and n is called the ____________.

2. The expression $x \cdot x \cdot x \cdot x \cdot x$ contains five ____________ of x.

3. $\{1, 2, 3, 4, 5, 6, \ldots\}$ is the set of ____________ numbers.

4. $\{\ldots, -3, -2, -1, 0, 1, 2, 3, \ldots\}$ is the set of ____________.

CONCEPTS *In Exercises 5–12, complete the rules for exponents. Assume that $x \neq 0$ and $y \neq 0$.*

5. $x^m x^n = \square$ **6.** $(x^m)^n = \square$ **7.** $(xy)^n = \square$ **8.** $\left(\frac{x}{y}\right)^n = \square$

9. $x^0 = \square$ **10.** $x^{-n} =$

13. An expression with a negative expon... as an equivalent expression with a p... Explain.

15. A cube is shown in Illustration 1.
a. Find the area of its base.
b. Find its volume.

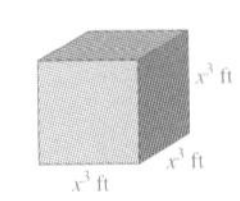

ILLUSTRATION 1

NOTATION *In Exercises 17–18, co...*

17. $\frac{x^5x^4}{x^{-2}} = \frac{x^\square}{x^{-2}}$ Keep the base and ...
$= x^{9-\square}$ Keep the base and ... exponents.
$= x^{11}$

NOTATION *In Exercises 17–18, complete each solution.*

17. Solve $-5x - 1 \geq -11$.
$-5x - 1 \geq -11$
$-5x \geq \square$ Add 1 to both sides.
$\frac{-5x}{-5} \square \frac{-10}{-5}$ Divide both sides by -5.
$x \leq 2$ Do the divisions.
Using interval notation, the result is $(\square, 2]$.

18. Solve $3 - 6x < 17 + x$.
$3 - 6x < 17 + x$
$3 - \square < 17$ Subtract x from both sides.
$-7x < \square$ Subtract 3 from both sides.
$\frac{-7x}{-7} \square \frac{14}{-7}$ Divide both sides by -7.
$x > -2$ Do the divisions.
Using interval notation, the result is $(-2, \square)$.

PRACTICE *In Exercises 19–42, solve each inequality. Give the result in interval notation and then graph the solution set.*

19. $3x > -9$
20. $4x < -36$
21. $-30y \leq -600$
23. $0.6x \geq 36$
25. $3 > -\frac{9}{10}x$
27. $x + 4 < 5$
29. $-5t + 3 \leq 5$
31. $7 < \frac{5}{3}a - 3$
33. $0.4x + 0.4 \leq 0.1x + 0.8$
35. $3(z - 2) \leq 2(z + 7)$
37. $-11(2 - b) < 4(2b + 2$
39. $\frac{1}{2}y + 2 \geq \frac{1}{3}y - 4$
41. $\frac{2}{3}x + \frac{3}{2}(x - 5) \leq x$

APPLICATIONS

43. REAL ESTATE Refer... For which regions of th... equality true in 1997?
Median sales price < U...

44. MUSIC INDUSTRY ... 2 on the next page. D... which the following ine...
CDs shipped > cassette...

APPLICATIONS

101. MICROSCOPES Illustration 3 shows the relative sizes of some chemical and biological structures, expressed as fractions of a meter (m). Express each fraction shown in the illustration as a power of 10, from the largest to the smallest.

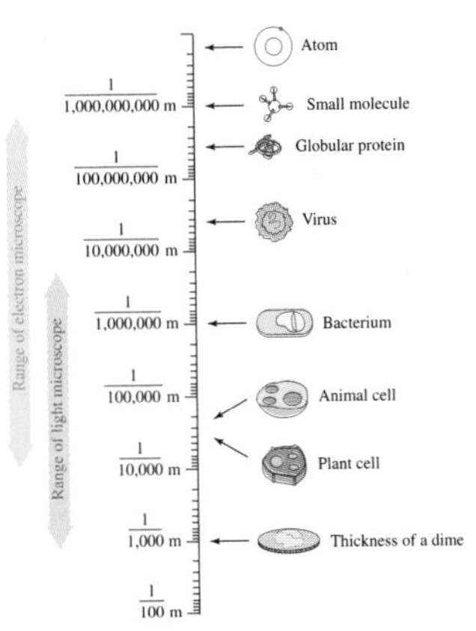

ILLUSTRATION 3

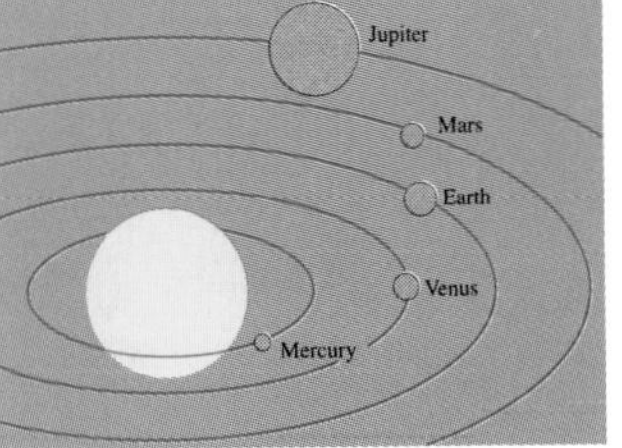

ILLUSTRATION 4

102. ASTRONOMY See Illustration 4. The distance d, in miles, of the nth planet from the sun is given by the formula

$$d = 9{,}275{,}200[3(2^{n-2}) + 4]$$

Find the distance of earth and Mars from the sun.

103. LICENSE PLATES The number of different license plates of the form three digits followed by three letters, as in Illustration 5, is $10 \cdot 10 \cdot 10 \cdot 26 \cdot 26 \cdot 26$. Write this expression using exponents. Then evaluate it.

ILLUSTRATION 5

104. PHYSICS Albert Einstein's work in the area of special relativity resulted in the observation that the total energy E of a body is equal to its total mass m times the square of the speed of light c. This relationship is given by the famous equation $E = mc^2$. Identify the base and exponent on the right-hand side.

WRITING *Write a paragraph using your own words.*

105. Explain how an exponential expression with a negative exponent can be expressed as an equivalent expression with a positive exponent. Give an example.

106. In the definition of x^{-n}, x cannot be 0. Why not?

REVIEW *In Exercises 107–110, solve each inequality. Give the result in interval notation and then graph the solution set.*

107. $a + 5 < 6$
108. $-9x + 5 \geq 15$
109. $6(t - 2) \leq 4(t + 7)$
110. $\frac{1}{4}p - \frac{1}{3} \leq p + 2$

5.1 Exponents **307**

Problem-Solving Strategy

One of the major objectives of this textbook is to make students better problem solvers. To this end, we use a five-step problem-solving strategy throughout the book. The five steps are: *Analyze the problem, Form an equation, Solve the equation, State the conclusion,* and *Check the result.*

Applications and Connections to Other Disciplines

A distinguishing feature of this book is its wealth of application problems. We have included numerous applications from disciplines such as science, economics, business, manufacturing, history, and entertainment, as well as mathematics.

EXAMPLE 4 ***Triathlon course.*** The Ironman Triathlon in Hawaii includes swimming, long-distance running, and cycling. The long-distance run is 11 times longer than the distance the competitors swim. The distance they cycle is 85.8 miles longer than the run. Overall, the competition covers 140.6 miles. Find the length of each part of the triathlon and round each length to the nearest tenth of a mile.

◀ Every example has a title.

Analyze the problem The entire triathlon course covers a distance of 140.6 miles. We note that the distance the competitors run is related to the distance they swim, and the distance they cycle is related to the distance they run.

Form an equation If x represents the distance the competitors swim, then $11x$ represents the length of the long-distance run, and $11x + 85.8$ represents the distance they cycle. From the diagram in Figure 1-15, we see that the sum of the individual parts of the triathlon must equal the total distance covered.

140.6 mi

Swimming x mi | Running $11x$ mi | Cycling

The distance they swim	plus	the distance they run	plu
x	$+$	$11x$	$+$

We now solve the equation.

Solve the equation

$$x + 11x + 11x + 85.8 = 140.6$$
$$23x + 85.8 = 140.6$$
$$23x = 54.8$$
$$x \approx 2.382608$$

State the conclusion To the nearest tenth, the distance the co run is $11x$, or approximately 11(2.3826 tenth, that is 26.2 miles. The distance 26.20869565 + 85.8 = 112.0086957 mi

Check the result If we add the lengths of the three parts we get 140.6 miles. The answers check

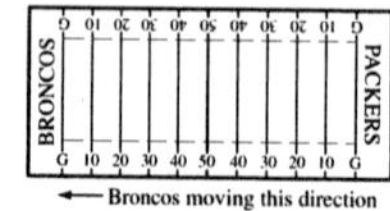

ILLUSTRATION 3

26. TRACK AND FIELD In the shot put, the solid metal ball must land in a marked sector for it to be a fair throw. In Illustration 4, graph the system of inequalities that describes the region in which a shot must land.

$$\begin{cases} y \le \frac{3}{8}x \\ y \ge -\frac{3}{8}x \\ x \ge 0 \end{cases}$$

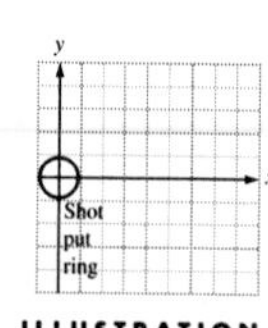

ILLUSTRATION 4

Every application problem has a title. ▶

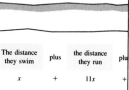

27. NO-FLY ZONES After the Gulf War, U.S. and Allied forces enforced northern and southern "no-fly" zones over Iraq. Iraqi aircraft was prohibited from flying in this air space. If x represents the north latitude parallel measurement, the no-fly zones can be described by

$x \ge 36$ or $x \le 33$

On the map in Illustration 5, shade the regions of Iraq over which there was a no-fly zone.

28. CARDIOVASCULAR FITNESS The graph in Illustration 6 shows the range of pulse rates that persons ages 20–90 should maintain during aerobic exercise to get the most benefit from the training. The shaded region "Effective Training Heart Rate Zone" can be described by a system of linear inequalities. Determine what inequality symbol should be inserted in each blank.

ILLUSTRATION 5

$$\begin{cases} x \,\square\, 20 \\ x \,\square\, 90 \\ y \,\square\, -0.87x + 191 \\ y \,\square\, -0.72x + 158 \end{cases}$$

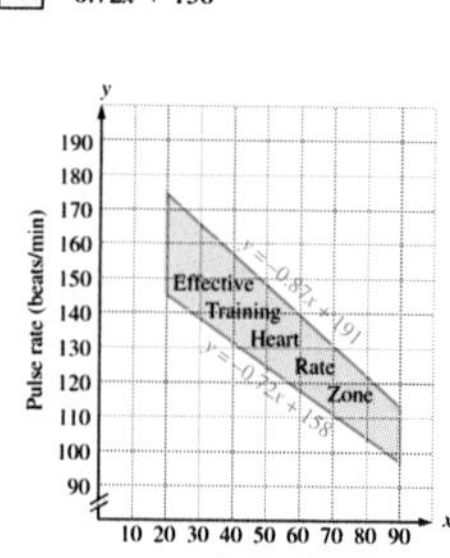

ILLUSTRATION 6

Increased Emphasis on Graphing

In response to recent trends in mathematics education, we have increased the emphasis on learning mathematics through graphing. Graphing on the number line is discussed in Section 1.3. Chapter 2 introduces graphing using the rectangular coordinate system. Throughout the text, students are asked to graph functions and interpret graphs.

Functions and Modeling

The concept of function is introduced in Chapter 2 and is stressed throughout the text. Students learn to use function notation, graph functions, and write functions that mathematically model many interesting real-life situations. By the end of the course, students will recognize families of functions, their graphs, and areas of application.

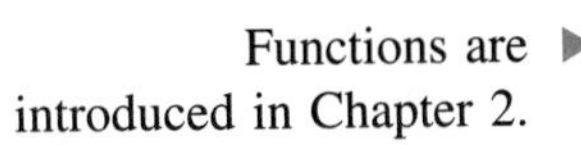
Functions are ▶ introduced in Chapter 2.

Linear functions

In Section 2.2, we graphed equations whose graphs were lines. Such equations define a basic type of function called a **linear function**.

Linear functions A **linear function** is a function defined by an equation that can be written in the form

$$f(x) = mx + b \quad \text{or} \quad y = mx + b$$

where m is the slope of the line graph and $(0, b)$ is the y-intercept.

EXAMPLE 8 ***Manicurist.*** A recent graduate of a cosmetology school rents a station from the owner of a beauty salon for \$18 a day. She expects to make \$12 profit from each customer she serves. Write a linear function describing her daily income if she serves c customers per day. Then graph the function.

Solution The manicurist makes a profit of \$12 per customer, so if she serves c customers a day, she will make \$12c. To find her income, we *btract* the \$18 rental fee she pays profit. Therefore the income func-) = 12c − 18.
raph of this linear function, shown 2-47, is a line with slope 12 and in- −18). Since the manicurist cannot gative number of customers, we do l the line into quadrant III.

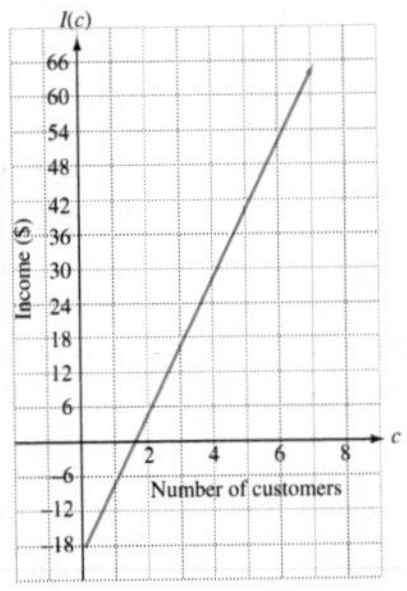

FIGURE 2-47

2.5 *Introduction to Functions* **145**

APPLICATIONS

49. FREEWAY DESIGN A Grand Avenue exit off the 210 Freeway is to be constructed. (See Illustration 5.) Sketch the off-ramp design on the graph if the sides of the pavement are defined by the functions $f(x) = (x - 1)^3$ and $f(x) = (x - 3)^3$.

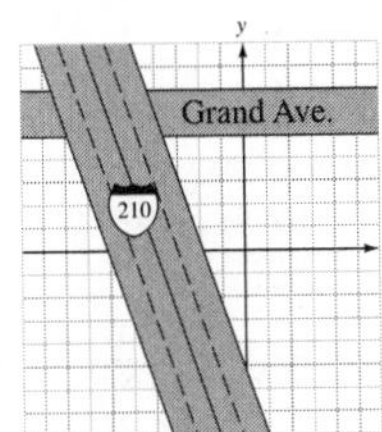

ILLUSTRATION 5

50. OPTICS See Illustration 6. The law of reflection states that the angle of reflection is equal to the angle of incidence. What function studied in this section mathematically models the path of the reflected light beam with an angle of incidence measuring 45°?

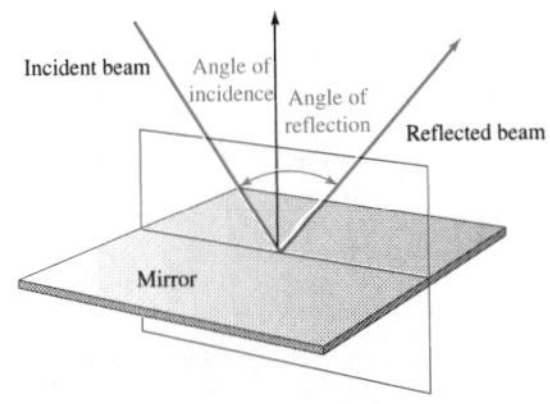

ILLUSTRATION 6

51. CENTER OF GRAVITY See Illustration 7. As a diver performs a $1\frac{1}{2}$-somersault in the tuck position, her center of gravity follows a path that can be described by a graph shape studied in this section. What graph shape is that?

ILLUSTRATION 7

52. FLASHLIGHTS Light beams coming from a flashlight bulb are reflected outward by a parabolic mirror as parallel rays.

a. The cross-sectional view of a parabolic mirror is given by the function $f(x) = x^2$ for the following values of x: −0.7, −0.6, −0.5, −0.4, −0.3, −0.2, −0.1, 0, 0.1, 0.2, 0.3, 0.4, 0.5, 0.6, 0.7. Sketch the parabolic mirror using the graph in Illustration 8.

b. From the light bulb filament at (0, 0.25), draw a line segment representing a beam of light that strikes the mirror at (−0.4, 0.16) and then reflects outward, parallel to the y-axis.

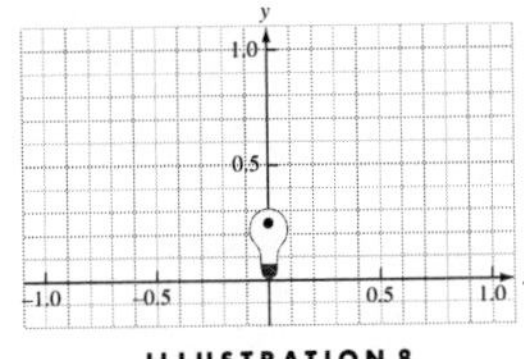

ILLUSTRATION 8

◀ Exercises demonstrate that functions model real-life situations.

Group Work

A one-page feature called Accent on Teamwork appears near the end of each chapter. It gives the instructor a set of problems that can be assigned as group work or to individual students as outside-of-class projects.

Key Concepts

Nine important algebraic concepts are highlighted in one-page Key Concept features, appearing near the end of each chapter. Each Key Concept page summarizes a concept and gives students the opportunity to review the role it plays in the overall picture.

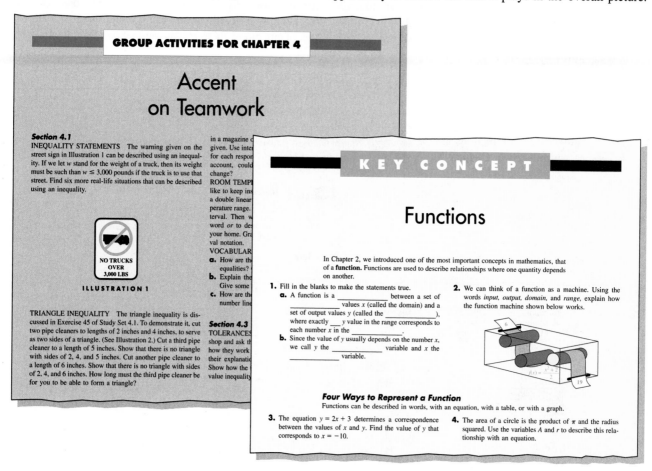

GROUP ACTIVITIES FOR CHAPTER 4

Accent on Teamwork

Section 4.1

INEQUALITY STATEMENTS The warning given on the street sign in Illustration 1 can be described using an inequality. If we let w stand for the weight of a truck, then its weight must be such than $w \leq 3{,}000$ pounds if the truck is to use that street. Find six more real-life situations that can be described using an inequality.

NO TRUCKS
OVER
3,000 LBS

ILLUSTRATION 1

TRIANGLE INEQUALITY The triangle inequality is discussed in Exercise 45 of Study Set 4.1. To demonstrate it, cut two pipe cleaners to lengths of 2 inches and 4 inches, to serve as two sides of a triangle. (See Illustration 2.) Cut a third pipe cleaner to a length of 5 inches. Show that there is no triangle with sides of 2, 4, and 5 inches. Cut another pipe cleaner to a length of 6 inches. Show that there is no triangle with sides of 2, 4, and 6 inches. How long must the third pipe cleaner be for you to be able to form a triangle?

in a magazine c
given. Use inte
for each respo
account, could
change?
ROOM TEMP
like to keep ins
a double linear
perature range.
terval. Then w
word *or* to de
your home. Gr
val notation.
VOCABULAR
a. How are th
equalities?
b. Explain th
Give some
c. How are th
number lin

Section 4.3

TOLERANCES
shop and ask t
how they work
their explanati
Show how the
value inequalit

KEY CONCEPT

Functions

In Chapter 2, we introduced one of the most important concepts in mathematics, that of a **function.** Functions are used to describe relationships where one quantity depends on another.

1. Fill in the blanks to make the statements true.

a. A function is a ____________ between a set of ____________ values x (called the domain) and a set of output values y (called the ____________), where exactly ___ y value in the range corresponds to each number x in the ____________.

b. Since the value of y usually depends on the number x, we call y the ____________ variable and x the ____________ variable.

2. We can think of a function as a machine. Using the words *input, output, domain,* and *range,* explain how the function machine shown below works.

6
$f(x) = \frac{x^2+2}{2}$
19

Four Ways to Represent a Function

Functions can be described in words, with an equation, with a table, or with a graph.

3. The equation $y = 2x + 3$ determines a correspondence between the values of x and y. Find the value of y that corresponds to $x = -10$.

4. The area of a circle is the product of π and the radius squared. Use the variables A and r to describe this relationship with an equation.

Calculators

For instructors who wish to use calculators as part of the instruction in this course, the text includes an Accent on Technology feature that introduces keystrokes and shows how scientific calculators and graphing calculators can be used to solve problems.

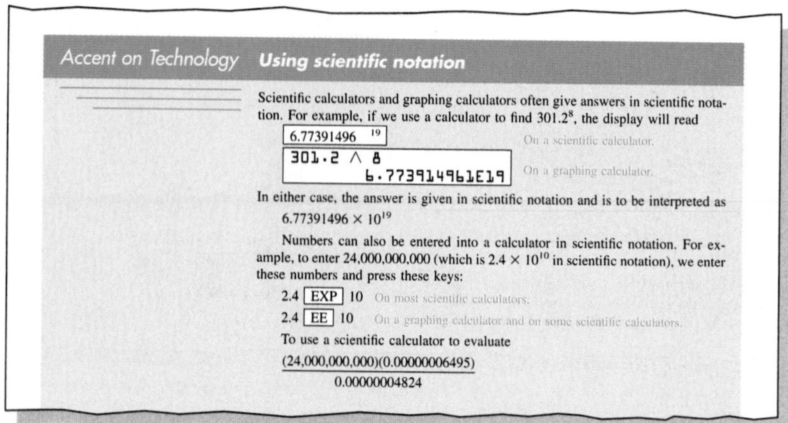

Accent on Technology **Using scientific notation**

Scientific calculators and graphing calculators often give answers in scientific notation. For example, if we use a calculator to find 301.2^8, the display will read

6.77391496 19 — On a scientific calculator.

301.2 ^ 8
6.77391496E19 — On a graphing calculator.

In either case, the answer is given in scientific notation and is to be interpreted as

$6.77391496 \times 10^{19}$

Numbers can also be entered into a calculator in scientific notation. For example, to enter 24,000,000,000 (which is 2.4×10^{10} in scientific notation), we enter these numbers and press these keys:

2.4 EXP 10 — On most scientific calculators.

2.4 EE 10 — On a graphing calculator and on some scientific calculators.

To use a scientific calculator to evaluate

$\frac{(24{,}000{,}000{,}000)(0.00000006495)}{0.00000004824}$

Systematic Review

Each Study Set ends with a Review section that contains problems similar to those in previous sections. Each chapter ends with a Chapter Review and a Chapter Test. The chapter reviews have been designed to be "user friendly." In a unique format, the reviews list the important concepts of each section of the chapter in one column, with appropriate review problems running parallel in a second column. In addition, Cumulative Review Exercises appear after Chapters 2, 4, 6, 8, and 9.

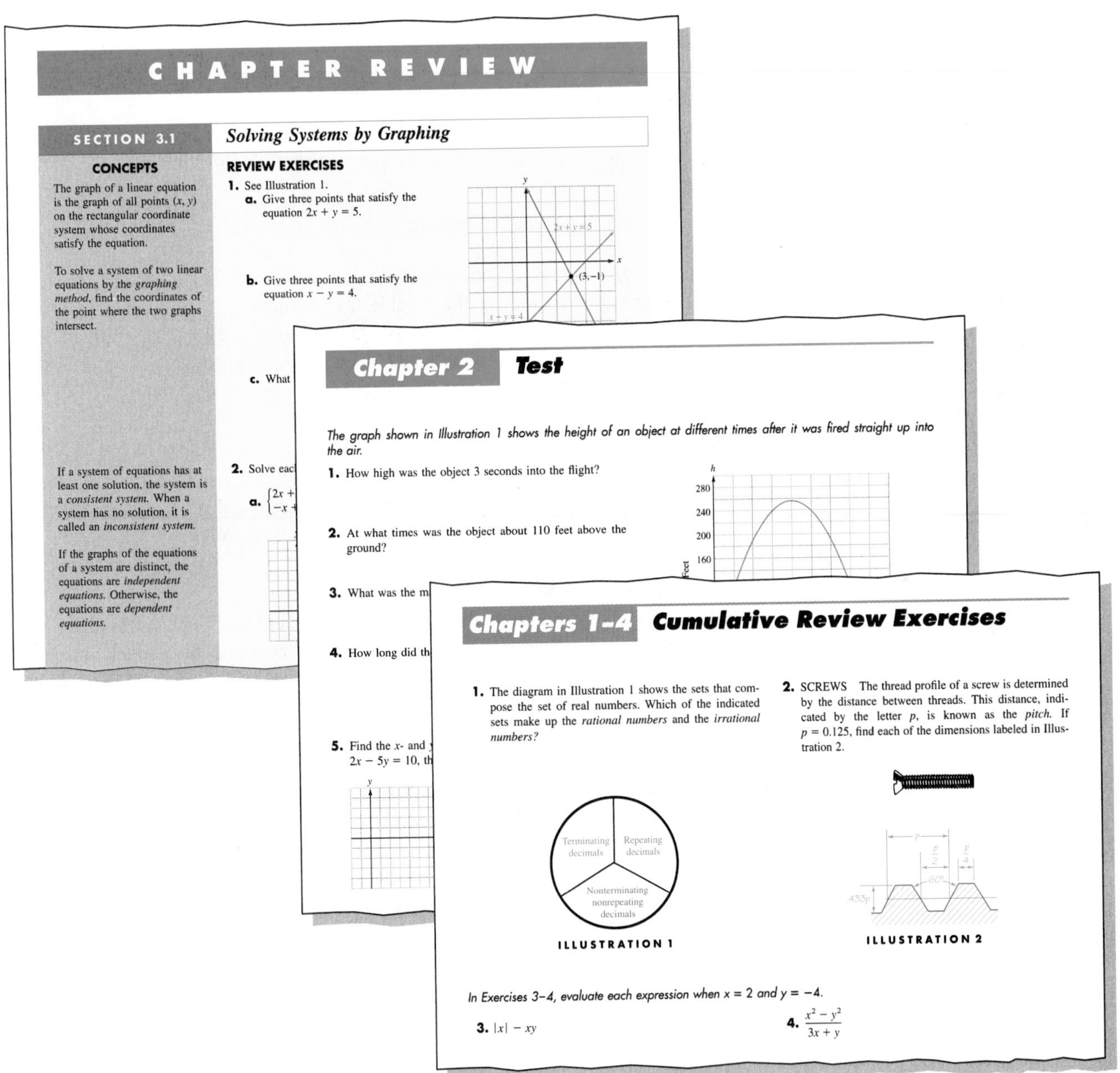

Appendixes

Material on conic sections is included in Appendix I. Appendix II covers the binomial theorem. For each, study sets are provided for student practice.

Student support

We have included many features that make *Intermediate Algebra* very accessible to students.

Worked Examples

The text contains hundreds of worked examples, many with several parts. Explanatory notes make the examples easy to follow.

Self Checks

Most of the examples are accompanied by Self Check problems, which allow students to practice the skills demonstrated in the worked examples.

Author's Notes

Author's notes, printed in red, are used to explain the steps in the solutions of examples. The notes are extensive so as to increase the students' ability to read and write mathematics.

Problems

The book includes numerous carefully graded exercises. In the Annotated Instructor's Edition, answers are printed in blue beside the problems. In the student edition, an appendix provides the answers to most of the odd-numbered exercises in the Study Sets as well as all the answers to the Key Concept, Chapter Review, Chapter Test, and Cumulative Review problems.

Functional Use of Color

For easy reference, definition boxes (in yellow), strategy boxes (in purple), and rule boxes (in blue) are color-coded. In addition, the book uses color to highlight terms and expressions that you would point to in a classroom discussion.

Study Skills and Math Anxiety

These two topics are discussed in detail in the section entitled "For the Student" at the end of this preface. In "Success in Intermediate Algebra," students are asked to design a personal strategy for studying and learning the material. "Taking a Math Test" helps students prepare for a test and then gives them suggestions for improving their performance.

Reading and Writing Mathematics

Also included (on pages xviii–xix) are two features to help students improve their ability to read and write mathematics. "Reading Mathematics" helps students get the most out of the examples in this book by showing them how to read the solutions properly. "Writing Mathematics" highlights the characteristics of a well-written solution.

Videotapes

The videotape series that accompanies this book shows students the steps in solving many examples in the text. A video logo placed next to an example indicates that the example is taught on tape.

Ancillaries for the instructor

Annotated Instructor's Edition

In the Annotated Instructor's Edition, the answers to all exercises are printed in blue next to the exercises.

Thomson World Class Learning and Testing Tools

This integrated testing software package features algorithmic test generation, online testing, and class management capabilities.

Tutorial Video Series

These videotapes contain worked-out solutions to problems from the text and new problems, with added instruction.

Printed Test Items

This manual contains test exercises and answers, arranged according to the organization of the text.

Complete Solutions Manual

This instructor's manual includes complete solutions to both the even- and the odd-numbered problems, arranged according to the organization of the main text.

Ancillaries for the student

Student Video

This videotape contains worked-out examples of the concepts most students have trouble understanding.

Student Solutions Manual

The manual provides the solutions for all the odd-numbered problems in the main text.

Web Site

The web site contains text-specific practice and study material, additional instruction, and online quizzes.

Acknowledgments

We are grateful to the following instructors, who have reviewed the text at various stages of its development. Their comments and suggestions have proven invaluable in making this a better book. We sincerely thank them for lending their time and talent to this project.

Mary Lou Hammond
Spokane Community College

Judith Jones
Valencia Community College

Theresa Jones
Amarillo College

Janice McFatter
Gulf Coast Community College

Betty Weissbecker
J. Sargeant Reynolds Community College

Cathleen Zucco
SUNY-New Paltz

We want to express our gratitude to Bob Billups, George Carlson, Robin Carter, Jim Cope, Terry Damron, Marion Hammond, Karl Hunsicker, Doug Keebaugh, Arnold Kondo, John McKeown, Kent Miller, Donna Neff, Steve Odrich, Eric Rabitoy, Dave Ryba, Chris Scott, Liz Tussy, and the Citrus College Library Staff for their help with some of the application problems in the textbook.

We would also like to thank Bob Pirtle, Jennifer Huber, Ellen Brownstein, David Hoyt, Lori Heckelman, Vernon Boes, Erin Wickersham, Melissa Duge, Diane Koenig, and The Clarinda Company for their help in creating this book.

Alan S. Tussy
R. David Gustafson

For the Student

Success in intermediate algebra

To be successful in mathematics, you need to know how to study it. The following checklist will help you develop your own personal strategy to study and learn the material. The suggestions listed below require some time and self-discipline on your part, but it will be worth the effort. This will help you get the most out of this course.

As you read each of the following statements, place a check mark in the box if you can truthfully answer Yes. If you can't answer Yes, think of what you might do to make the suggestion part of your personal study plan. You should go over this checklist several times during the semester to be sure you are following it.

Preparing for the Class

- ☐ I have made a commitment to myself to give this course my best effort.
- ☐ I have the proper materials: a pencil with an eraser, paper, a notebook, a ruler, a calculator, and a calendar or day planner.
- ☐ I am willing to spend a minimum of two hours doing homework for every hour of class.
- ☐ I will try to work on this subject every day.
- ☐ I have a copy of the class syllabus. I understand the requirements of the course and how I will be graded.
- ☐ I have tried to schedule a free hour after the class to give me time to review my notes and begin the homework assignment.

Class Participation

- ☐ I will regularly attend the class sessions and be on time.
- ☐ When I am absent, I will find out what the class studied, get a copy of any notes or handouts, and make up the work that was assigned when I was gone.
- ☐ I will sit where I can hear the instructor and see the chalkboard.
- ☐ I will pay attention in class and take careful notes.
- ☐ I will ask the instructor questions when I don't understand the material.
- ☐ When tests, quizzes, or homework papers are passed back and discussed in class, I will write down the correct solutions for the problems I missed so that I can learn from my mistakes.

Study Sessions

- ☐ I will find a comfortable and quiet place to study.
- ☐ I realize that reading a math book is different than reading from a newspaper or a novel. Quite often, it will take more than one reading to understand the material.
- ☐ After studying an example in the textbook, I will work the accompanying Self Check.
- ☐ I will begin the homework assignment only after reading the assigned section.
- ☐ I will try to use the mathematical vocabulary mentioned in the book and used by my instructor when I am writing or talking about the topics studied in the course.
- ☐ I will look for opportunities to explain the material to others.
- ☐ I will check all of my answers to the problems with those provided in the back of the book (or with the *Student Solutions Manual*) and reconcile any differences.
- ☐ My homework will be organized and neat. My solutions will show all the necessary steps.

- ☐ I will try to work some review problems every day.
- ☐ After completing the homework assignment, I will read the next section to prepare for the coming class session.
- ☐ I will keep a notebook containing my class notes, homework papers, quizzes, tests, and any handouts—all in order by date.

Special Help

- ☐ I know my instructor's office hours and am willing to go in to ask for help.
- ☐ I have formed a study group with classmates that meets regularly to discuss the material and work on problems.
- ☐ When I need additional explanation of a topic, I view the video and check the web site.
- ☐ I take advantage of extra tutorial assistance that my school offers for mathematics courses.
- ☐ I have purchased the *Student Solutions Manual* that accompanies the text, and I use it.

To follow each of these suggestions will take time. It takes a lot of practice to learn mathematics, just as with any other skill.

No doubt, you will sometimes become frustrated along the way. This is natural. When it occurs, take a break and come back to the material after you have had time to clear your thoughts. Keep in mind that the skills and discipline you learn in this course will help make for a brighter future. Good luck!

Taking a math test

The best way to relieve anxiety about taking a mathematics test is to know that you are well-prepared for it and that you have a plan. Before any test, ask yourself three questions. When? What? How?

When Will I Study?

1. When is the test?

2. When will I begin to review for the test?

3. What are the dates and times that I will reserve for studying for the test?

What Will I Study?

1. What sections will the test cover?
2. Has the instructor indicated any types of problems that are guaranteed to be on the test?

How Will I Prepare for the Test?

Put a check mark by each method you will use to prepare for the test.

- ☐ Review the class notes.
- ☐ Outline the chapter(s) on a piece of poster board to see the big picture and to see how the topics relate to one another.
- ☐ Recite the important formulas, definitions, vocabulary, and rules into a tape recorder.
- ☐ Make flash cards for the important formulas, definitions, vocabulary, and rules.

- ☐ Rework problems from the homework assignments.
- ☐ Rework each of the Self Check problems in the text.
- ☐ Form a study group to discuss and practice the topics to be tested.
- ☐ Complete the appropriate Chapter Review(s) and the Chapter Test(s).
- ☐ Review the Warnings given in the text.
- ☐ Work on improving speed in answering questions.
- ☐ Review the methods that can be used to check answers.
- ☐ Write a sample test, trying to think of questions the instructor will ask.
- ☐ Complete the appropriate Cumulative Review Exercises.
- ☐ Get organized the night before the test. Have materials ready to go so that the trip to school will not be hurried.
- ☐ Take some time to relax immediately before the test. Don't study right up to the last minute.

Taking the Test

Here are some tips that can help improve your performance on a mathematics test.

- When you receive the test, scan it, looking for the types of problems you had expected to see. Do them first.
- Read the instructions carefully.
- Write down any formulas or rules as soon as you receive the test.
- Don't spend too much time on any one problem until you have attempted all the problems.
- If your instructor gives partial credit, at least try to begin a solution.
- Save the most difficult problems for last.
- Don't be afraid to skip a problem and come back to it later.
- If you finish early, go back over your work and look for mistakes.

Reading mathematics

To get the most out of this book, you need to learn how to read it correctly. A mathematics textbook must be read differently than a novel or a newspaper. For one thing, you need to read it slowly and carefully. At times, you will have to reread a section to understand its content. You should also have pencil and paper with you when reading a mathematics book, so that you can work along with the text to understand the concepts presented.

Perhaps the most informative parts of a mathematics book are its examples. Each example in this textbook consists of a problem and its corresponding solution. One form of solution that is used many times in this book is shown in the diagram on the next page. It is important that you follow the "flow" of its steps if you are to understand the mathematics involved. For this solution form, the basic idea is this:

- A property, rule, or procedure is applied to the original expression to obtain an equivalent expression. We show that the two expressions are equivalent by writing an equals sign between them. The property, rule, or procedure that was used is then listed next to the equivalent expression in the form of an author's note, printed in red.
- The process of writing equivalent expressions and explaining the reasons behind them continues, step by step, until the final result is obtained.

The solution in the following diagram consists of three steps, but solutions have varying lengths.

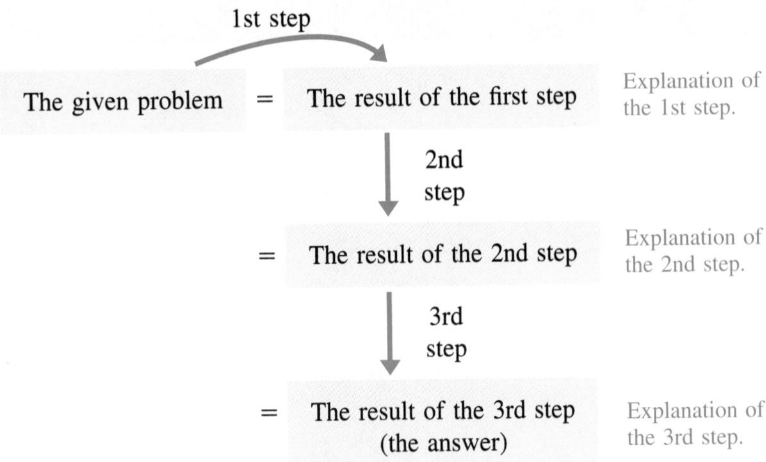

Writing mathematics

One of the major objectives of this course is for you to learn how to write solutions to problems properly. A written solution to a problem should explain your thinking in a series of neat and organized mathematical steps. Think of a solution as a mathematical essay—one that your instructor and other students should be able to read and understand. Some solutions will be longer than others, but they must all be in the proper format and use the correct notation. To learn how to do this will take time and practice.

To give you an idea of what will be expected, let's look at two samples of student work. In the first, we have highlighted some important characteristics of a well-written solution. The second sample is poorly done and would not be acceptable.

Evaluate: $35 - 2^2 \cdot 3$.

A well-written solution:

The problem was copied from the textbook. ▶ $35 - 2^2 \cdot 3 = 35 - 4 \cdot 3$ ◀ The first step of the solution is written here.

$= 35 - 12$ ◀ The steps are written under each other in a neat, organized manner.

$= 23$

▲ The equals signs are lined up vertically.

A poorly written solution:

The problem wasn't copied from the text. ▶ $2^2 = 4 = 35 - \underbrace{4 \cdot 3}_{12}$ ◀ An equals sign is improperly used.

SUB: 35
−12
23 → =(23) ◀ The work is disorganized and difficult to follow.

CONTENTS

1 *A Review of Basic Algebra 1*

2 *Graphs, Equations of Lines, and Functions 91*

3 *Systems of Equations 177*

4 *Inequalities 239*

5 *Exponents, Polynomials, and Polynomial Functions 298*

6 *Rational Expressions 381*

7 *Radicals and Rational Exponents 462*

8 *Quadratic Equations, Functions, and Inequalities 531*

Intermediate Algebra

A Review of Basic Algebra

1

Campus Connection

The *Chemistry* Department

In a chemistry class, students study many formulas. For example, in one chemistry lab experiment, students measure the temperature, pressure, and volume of a gas sample. Then they find the number of molecules in the sample by solving the formula $PV = nR(T + 273)$ for n and substituting data values for P, V, R, and T. In this chapter, we will work with formulas from a wide variety of disciplines, including some that you would see in a chemistry class.

ALGEBRA IS A MATHEMATICAL LANGUAGE THAT CAN BE USED TO SOLVE MANY TYPES OF PROBLEMS.

1.1 Describing Numerical Relationships

In this section, you will learn about

- **Variables, algebraic expressions, and equations**
- **Verbal models**
- **Constructing tables**
- **Graphical models**
- **Formulas**

INTRODUCTION. To solve problems using algebra, you will need to learn how to read it, write it, and speak it. In this section, we discuss several ways in which numerical relationships are described.

Variables, algebraic expressions, and equations

The rental agreement for a banquet hall is shown in Figure 1-1.

Rental Agreement
ROYAL VISTA BANQUET ROOM
Wedding Receptions•Dances•Reunions•Fashion Shows

Rented To____________________Date____________
Lessee's Address____________________________

Rental Charges
• $100 per hour
• Nonrefundable $200 cleanup fee

Terms and conditions
Lessor leases the undersigned lessee the above described property upon the terms and conditions set forth on this page and on the back of this page. Lessee promises to pay rental cost stated herein.

FIGURE 1-1

In examining the rental agreement, we see that two operations need to be performed to calculate the cost of renting the banquet hall.

- First, we must *multiply* the hourly rental cost of \$100 by the number of hours the hall is to be rented.
- To that result, we must then *add* the cleanup fee of \$200.

In words, we can describe the process as follows.

The cost of renting the hall	is	100	times	the number of hours it is rented	plus	200.

We can describe the procedure used to calculate this cost in a more concise way by using *variables* and mathematical symbols. A **variable** is a letter that is used to stand for a number. If we let C stand for the cost of renting the hall and h stand for the number of hours it is rented, the words can be translated to form a *mathematical model.*

The cost of renting the hall	is	100	times	the number of hours it is rented	plus	200.
C	$=$	100	$\cdot$	h	$+$	200

The statement $C = 100h + 200$ is called an **equation.**

Equation — An **equation** is a mathematical sentence that contains an = sign.

Here are some examples of equations.

$2 + 3 = 5 \qquad 3x - 2 = 4x + 10 \qquad P = a + b + c$

On the right-hand side of the equation $C = 100h + 200$, the notation $100h + 200$ is called an **algebraic expression.** Algebraic expressions are the building blocks of equations.

Algebraic expressions — Variables and numbers (called **constants**) can be combined with the operations of addition, subtraction, multiplication, division, raising to a power, and finding a root to create **algebraic expressions.**

Here are some examples of algebraic expressions.

$5a - 12$	This algebraic expression is a combination of the constants 5 and 12, the variable a, and the operations of multiplication and subtraction.
$\dfrac{50 - y}{3y^3}$	This algebraic expression is a combination of the constants 50 and 3, the variable y, and the operations of subtraction, multiplication, division, and raising to a power.
$c = \sqrt{a^2 + b^2}$	The right-hand side of this equation is the algebraic expression $\sqrt{a^2 + b^2}$.

Verbal models

Table 1-1 lists some words and phrases that are often used in mathematics to denote the operations of addition, subtraction, multiplication, and division.

Addition +	Subtraction −	Multiplication ·	Division ÷
added to	subtracted from	multiplied by	divided by
plus	difference	product	quotient
the sum of	less than	times	ratio
more than	decreased by	percent (or fraction) of	half
increased by	reduced by	twice	into
greater than	minus	triple	per

TABLE 1-1

In the banquet hall example, the equation $C = 100h + 200$ was used to describe a procedure to calculate the cost of renting the hall. Using vocabulary from Table 1-1, we can write a **verbal model** that also describes this procedure. One such model is:

The cost of renting the hall (in dollars) is the *product* of 100 and the number of hours it is rented, *increased by* 200.

EXAMPLE 1 ***Describing numerical relationships.*** The cost to have a dinner catered is \$6 per person. For groups of more than 200, a \$100 discount is given. Write a mathematical model and a verbal model that describe the relationship between the catering cost and the number of people being served, for groups larger than 200.

Solution

To find the catering cost C (in dollars) for groups larger than 200, we need to *multiply* the number of people served n by \$6 and then *subtract* the \$100 discount.

The catering cost	is	6	times	the number of people served	minus	100.

In symbols, the mathematical model is:

$$C = 6n - 100$$

A verbal model is:

The catering cost (in dollars) is the *product* of 6 and the number of people served, *decreased by 100.*

Self Check

After winning a lottery, three friends split the prize equally. Each person then had to pay \$2,000 in taxes on his or her share. Write a mathematical model and a verbal model that relate the amount of each person's share, after taxes, to the amount of the lottery prize.

Answers: $S = \frac{P}{3} - 2{,}000$; each person's share (in dollars), after taxes, is the quotient of the lottery prize and 3, decreased by 2,000

Constructing tables

In the banquet hall example, the equation $C = 100h + 200$ can be used to determine the cost of renting the banquet hall for *any* number of hours.

EXAMPLE 2 ***Using an equation.*** Find the cost of renting the banquet hall for 3 hours and for 4 hours. Write the results in a table.

Solution

We begin by constructing a table with the appropriate column headings: h for the number of hours the hall is rented and C for the cost (in dollars) to rent the hall. Then we enter the number of hours of each rental time in the left column.

h	C
3	
4	

Next, we use the equation $C = 100h + 200$ to find the total rental cost for 3 hours and for 4 hours.

$$\begin{aligned} C &= 100h + 200 \\ C &= 100(3) + 200 && \text{Replace } h \text{ with 3.} \\ &= 300 + 200 && \text{Do the multiplication.} \\ &= 500 \end{aligned}$$

$$\begin{aligned} C &= 100h + 200 \\ C &= 100(4) + 200 && \text{Replace } h \text{ with 4.} \\ &= 400 + 200 && \text{Do the multiplication.} \\ &= 600 \end{aligned}$$

Finally, we enter these results in the right-hand column of the table: \$500 for a 3-hour rental and \$600 for a 4-hour rental.

h	C
3	500
4	600

Self Check

Find the cost of renting the hall for 6 hours and for 7 hours. Write the results in the table.

h	C
6	
7	

Answer: For 6 hours, the cost is \$800. For 7 hours, the cost is \$900. ■

Graphical models

The cost of renting the banquet hall for various lengths of time can also be presented in a **bar graph.** The bar graph in Figure 1-2(a) has a **horizontal axis** labeled "Number of hours the hall is rented." The **vertical axis** of the graph, labeled "Cost to rent the hall (\$)," is scaled in units of 50 dollars. The bars directly over each of the times (1, 2, 3, 4, 5, 6, and 7 hours) extend to a height that gives the corresponding cost to rent the hall. For example, if the hall is rented for 5 hours, the height of the bar indicates that the cost is \$700.

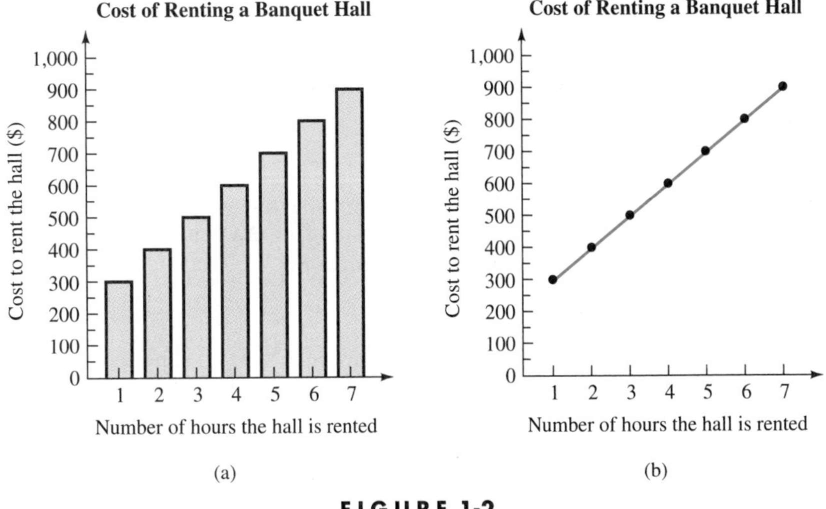

FIGURE 1-2

A **line graph** can also be used to show the rental costs. This type of graph consists of a series of dots drawn at the correct "height," connected with line segments. (See Figure 1-2(b).) Not only does the line graph present all the information contained in the bar graph, but it also provides additional information. We can use the line graph to find the cost of renting the banquet hall for a length of time not shown in the bar graph.

EXAMPLE 3 ***Reading a line graph.*** Use the line graph in Figure 1-2(b) to determine the cost of renting the hall for $4\frac{1}{2}$ hours.

Solution

In Figure 1-3, we locate $4\frac{1}{2}$ on the horizontal axis of the line graph and draw a line straight up to intersect the graph. From the point of intersection with the graph, we draw a horizontal line to the left that intersects the vertical axis. On the vertical axis, we can read that the rental cost is \$650 for $4\frac{1}{2}$ hours.

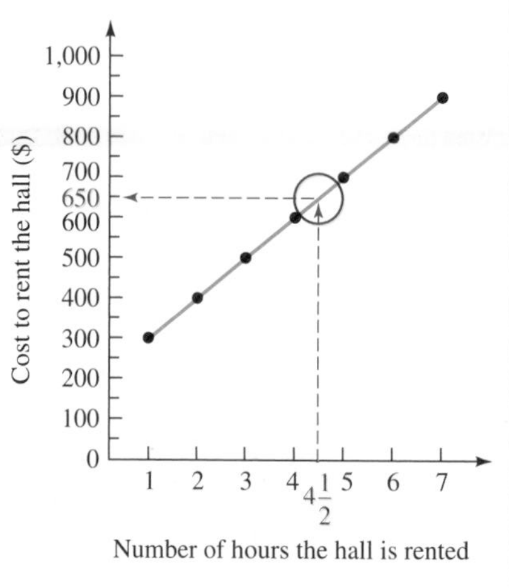

FIGURE 1-3

Self Check

Use Figure 1-3 to find the cost of renting the banquet hall for $6\frac{1}{2}$ hours.

Answer: \$850

Formulas

Equations that express a known relationship between two or more quantities, represented by variables, are called **formulas.** Formulas are used in many fields. Automotive technology, economics, medicine, retail sales, and banking are just a few examples.

EXAMPLE 4 ***Translating from words to symbols.*** Use variables to express each relationship as a formula.

a. The distance in miles traveled by a vehicle is the product of its average rate of speed in mph and the time in hours it travels at that rate.

b. The sale price of an item is the difference between the regular price and the discount.

Solution

a. The word *product* indicates multiplication. If we let d stand for the distance traveled in miles, r for the vehicle's average rate of speed in mph, and t for the length of time traveled in hours, we can write the formula as

$$d = rt$$

b. The word *difference* indicates subtraction. If we let s stand for the sale price of the item, p for the regular price, and d for the discount, we have

$$s = p - d$$

Self Check

Express each relationship as a formula:

a. The retail price of an item is the sum of its wholesale cost and the markup.

b. The simple interest earned by a deposit is the product of the principal, the annual rate of interest, and the time.

Answers: **a.** $r = c + m$, **b.** $I = Prt$

The *perimeter* of a geometric figure is a measure of the distance around it. Table 1-2 shows the formulas for the perimeters of several geometric figures. The perimeter of a circle is called its **circumference.**

Figure	Name	Perimeter/ circumference
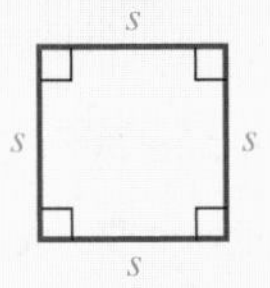	Square	$P = 4s$
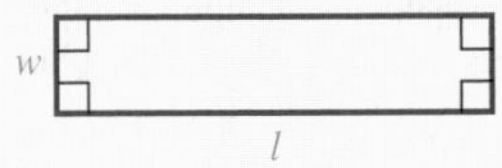	Rectangle	$P = 2l + 2w$
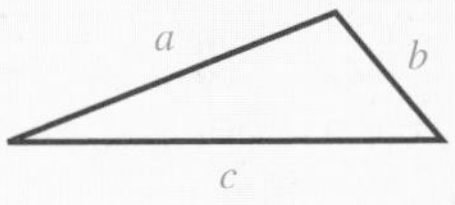	Triangle	$P = a + b + c$
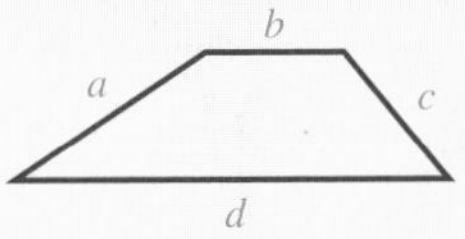	Trapezoid	$P = a + b + c + d$
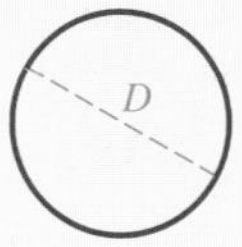	Circle	$C = \pi D$ (π is approximately 3.1416.)

TABLE 1-2

EXAMPLE 5 ***Translating from symbols to words.*** The formula for the perimeter of a rectangle is $P = 2l + 2w$. Express the formula in words.

Solution

In the formula, the length and width are each *multiplied by two.* Then we *add* those results.

The perimeter of a rectangle is the sum of twice the length and twice the width.

Self Check

Express the formula for the circumference of a circle, $C = \pi D$, in words.

Answer: The circumference of a circle is the product of pi and the length of the diameter. ■

EXAMPLE 6 ***Using perimeter formulas.*** Find the number of feet of redwood edging needed to outline a square flower bed having sides 6.5 feet long.

Solution

To find the amount of redwood edging needed, we need to find the perimeter of the square flower bed.

$P = 4s$ The formula for the perimeter of a square.

$P = 4(6.5)$ Substitute 6.5 for s, the length of one side of the square.

$= 26$ Do the multiplication.

26 feet of redwood edging is needed to outline the flower bed.

Self Check

Find the amount of fencing needed to enclose a triangular lot with sides 150 ft, 205.5 ft, and 165 ft long.

Answer: 520.5 ft ■

STUDY SET Section 1.1

VOCABULARY *In Exercises 1–8, fill in the blanks to make the statements true.*

1. An ______________ is a mathematical sentence that contains an = sign.

2. A ______________ is a letter that is used to stand for a number.

3. Variables and numbers can be combined with mathematical operations to create ______________.

4. Phrases such as *the sum, increased by,* and *more than* are used to indicate the operation of ______________.

5. A ______________ is an equation that expresses a known relationship between two or more quantities represented by variables.

6. Phrases such as *the difference, decreased by,* and *less than* are used to indicate the operation of ______________.

7. The distance around a geometric figure is its ______________.

8. The ______________ of a circle is the distance around it.

CONCEPTS *In Exercises 9–16, classify each expression as an algebraic expression or an equation.*

9. $x - 5$

10. $x - 5 = 5$

11. $P = a + b + c$

12. $\frac{x + y}{8}$

13. $d = rt$

14. Prt

15. $2x^2$

16. $T = 6n - 400$

17. a. What type of graph is shown in Illustration 1?

b. What units are used to scale the horizontal axis? The vertical axis?

c. Estimate the height of the candle after it has burned for $3\frac{1}{2}$ hours. For 8 hours.

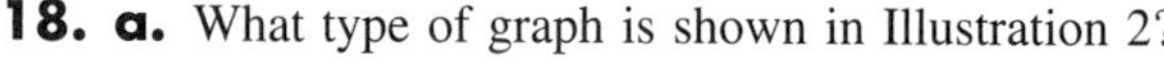

18. a. What type of graph is shown in Illustration 2?

b. What units are used to scale the horizontal axis? The vertical axis?

c. In what year was the sale of hunting and firearms equipment the greatest? The least?

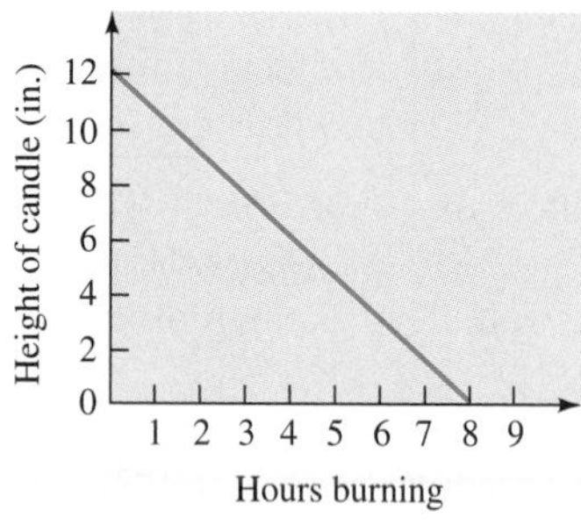

ILLUSTRATION 1

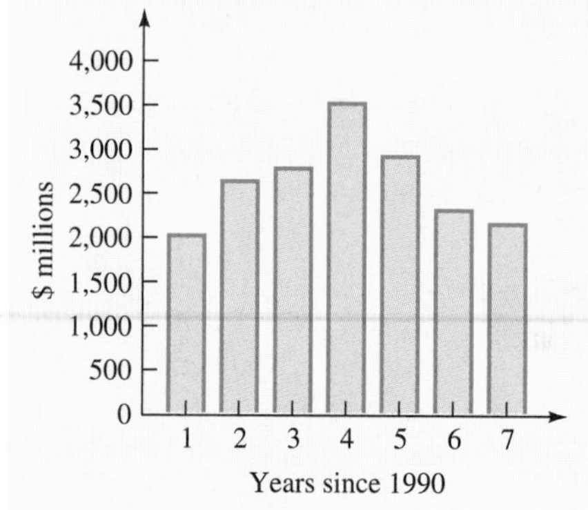

Based on data from the National Sporting Goods Assn.

ILLUSTRATION 2

In Exercises 19–26, translate each verbal model to a mathematical model.

19.

The cost each semester	is	\$13	times	the number of units taken	plus	the student services fee of \$24.

20.

The yearly salary	is	$25,000	plus	$75	times	the number of years of experience.

21. The quotient of the number of clients and seventy-five gives the number of social workers needed.

22. The difference between 500 and the number of people in a theater gives the number of unsold tickets.

23. Each test score was increased by 15 points to give a new adjusted test score.

24. The weight of a super-size order of French fries is twice that of a regular-size order.

25. The product of the number of boxes of crayons in a case and 12 gives the number of crayons in a case.

26. The perimeter of an equilateral triangle can be found by tripling the length of one of its sides.

In Exercises 27–28, use the data in each table to find a formula that mathematically describes the relationship between the two quantities. Then state the relationship in words. (Answers may vary.)

27.

Tower height (ft)	Height of base (ft)
15.5	5.5
22	12
25.25	15.25
45.125	35.125

28.

Seasonal employees	Employees
25	75
50	100
60	110
80	130

NOTATION

29. Algebraic expressions are combinations of variables, constants, and the operations of mathematics. For the algebraic expression $2D + 4$,

a. What variable is used?

b. What constants are used?

c. What operations are used?

30. Consider the equation $T = 3r - 15$.

a. How many variables does the equation contain?

b. What operations are indicated on the right-hand side of the equation?

c. Is the right-hand side of the equation an algebraic expression?

PRACTICE *In Exercises 31–34, use the given equation (formula) to complete each table.*

31. $c = \frac{p}{12}$

Number of packages	Cartons
24	
72	
180	

32. $y = 100c$

Number of centuries	Years
1	
6	
21	

33. $n = 22.44 - K$

K	n
0	
1.01	
22.44	

34. $y = x + 15$

x	y
0	
15	
30	

35. The lengths of the two parallel sides of a trapezoid are 10 inches and 15 inches. The other two sides are each 6 inches long. Find the perimeter of the trapezoid.

36. Find the distance the wheel of a bicycle rolls in one revolution if the diameter of the wheel is 20 inches. Round to the nearest tenth.

37. Find the perimeter of a square quilt that has sides 2 yards long.

38. Find the perimeter of a triangular postage stamp with sides 1.8, 1.8, and 1.5 centimeters long.

APPLICATIONS

39. CARPET CLEANING See the advertisement shown in Illustration 3.

Rent the in–home
Carpet Cleaning System
"Do it yourself and save!"
Safe, effective
Costs only $10 an hour
plus $20 for supplies

ILLUSTRATION 3

a. Write a verbal model that states the relationship between the cost C of renting the carpet-cleaning system and the number of hours h it is rented.

b. Translate the verbal model written in part a to a mathematical model

c. See Illustration 4. Use your result from part b to complete the table, and then draw a line graph.

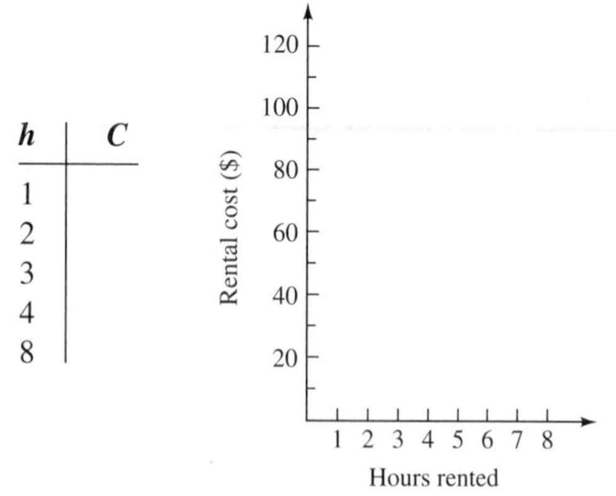

h	C
1	
2	
3	
4	
8	

ILLUSTRATION 4

40. UTILITY VEHICLES What geometric concept applies when finding the length of the plastic trim around the cargo area floor mat shown in Illustration 5? Estimate the amount of trim used.

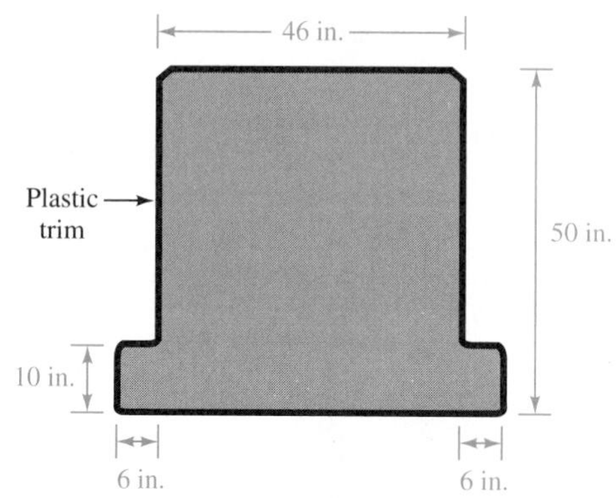

ILLUSTRATION 5

41. FINISH CARPENTRY The miter saw in Illustration 6 can pivot a total of 180° to make an angled cut on a piece of cabinet molding.

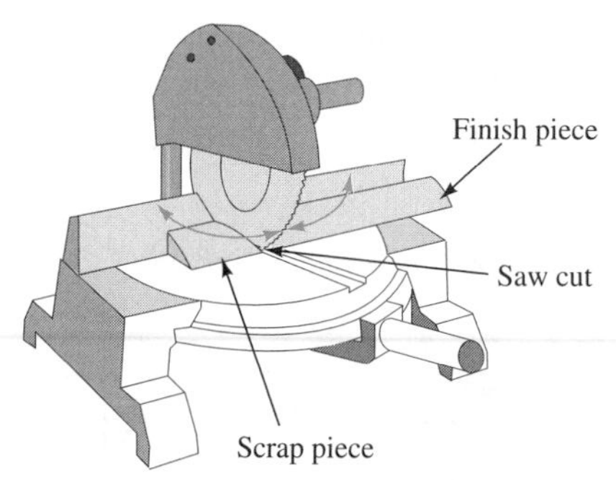

ILLUSTRATION 6

a. Write a verbal model that states the relationship between the angle measure on the scrap piece of molding and the angle measure on the finish piece of molding after a saw cut is made.

b. Translate the verbal model written in part a to a mathematical model.

c. Use your results from part b to complete the table and draw a line graph in Illustration 7.

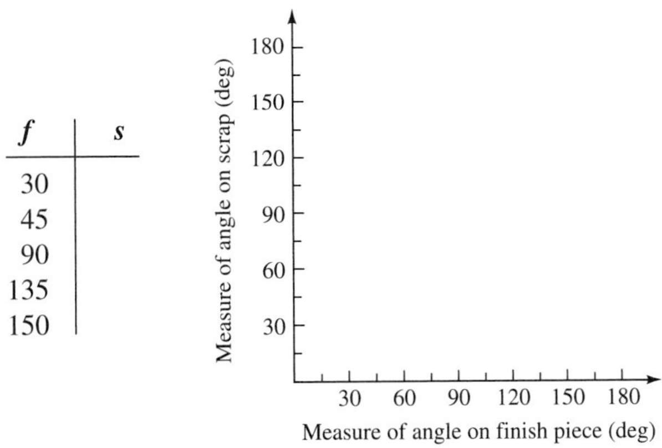

f	s
30	
45	
90	
135	
150	

ILLUSTRATION 7

42. PRODUCTION PLANNING Use the diagram in Illustration 8 to write four formulas involving r that could be used by planners to order the necessary number of oak mounting plates, bar holders, chrome bars, and mounting screws for a production run that will produce r towel racks.

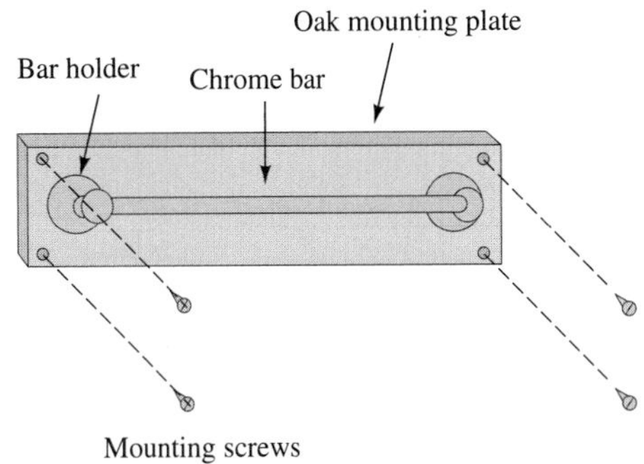

ILLUSTRATION 8

WRITING *Write a paragraph using your own words.*

43. Explain the difference between an expression and an equation. Give examples.

44. Name four ways in which numerical relationships were described in this section. Give examples.

45. Use each word below in a sentence so that a mathematical operation is indicated. If you are unsure of the meaning of a word, look it up in a dictionary.

quadrupled	deleted	bisected
confiscated	annexed	docked

46. Consider the formula $h = 7d$, where d is the age of a dog and h is the equivalent "human age" of the dog. Explain what information is given by the formula.

1.2 *The Real Number System*

In this section, you will learn about

- **Natural numbers, whole numbers, and integers**
- **Rational numbers**
- **Irrational numbers**
- **Real numbers**
- **The real number line**
- **Inequality symbols**
- **Absolute value**

INTRODUCTION. In this course, we will work with many different types of numbers. For example, we will use positive numbers to express distances. We will encounter some temperatures that will be negative. We will work with fractional parts of an hour, and we will use decimals to write numbers in scientific notation. The solutions of some geometry problems will contain π or a square root. In this section, we will define many types of numbers. We will show that together, they make up a collection of numbers called the *real numbers.*

Natural numbers, whole numbers, and integers

The graph in Figure 1-4 shows the daily low temperatures for Anchorage, Alaska for the first seven days of January, 1998. On the horizontal axis, the numbers 1, 2, 3, 4, 5, 6, and 7 have been used to denote the calendar days of the month. This collection of numbers is called a **set,** and the members, or **elements,** of the set can be listed within **braces** { }.

{1, 2, 3, 4, 5, 6, 7}

Each of the numbers 1, 2, 3, 4, 5, 6, and 7 is a member of a basic set called the **natural numbers.** The natural numbers are the numbers we count with.

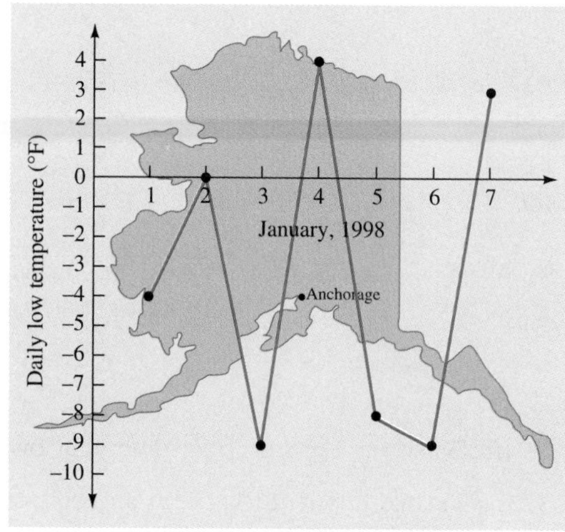

FIGURE 1-4

Natural numbers The set of **natural numbers** is {1, 2, 3, 4, 5, 6, 7, 8, 9, 10, . . .}.

The three dots used in this definition indicate that the list of natural numbers continues on forever.

Two important **subsets** of the set of natural numbers are the prime numbers and the composite numbers.

Prime numbers The **prime numbers** are the natural numbers greater than 1 that are divisible only by themselves and 1.

The first ten prime numbers are 2, 3, 5, 7, 11, 13, 17, 19, 23, and 29.

Composite numbers The **composite numbers** are the natural numbers greater than 1 that are not prime numbers.

The first ten composite numbers are 4, 6, 8, 9, 10, 12, 14, 15, 16, and 18.

The natural numbers together with 0 make up a set of numbers called the **whole numbers.**

Whole numbers The set of **whole numbers** is {0, 1, 2, 3, 4, 5, 6, 7, 8, 9, 10, . . .}.

Since every natural number is also a whole number, we say that the set of natural numbers is a **subset** of the set of whole numbers.

The graph in Figure 1-4 contains both **positive numbers,** numbers greater than 0, and **negative numbers,** numbers less than 0. For example, on January 7, the low temperature in Anchorage was 3°F (3 degrees above zero). On January 3, the low temperature was −9°F (9 degrees below zero). January 2, the temperature was 0°F. 0 is neither positive nor negative. The numbers used to represent the temperatures in the graph are members of a set of numbers called the **integers.**

Integers

The set of **integers** is $\{\ldots, -4, -3, -2, -1, 0, 1, 2, 3, 4, \ldots\}$.

Integers that are divisible by 2 are called **even integers,** and integers that are not divisible by 2 are called **odd integers.**

Even integers: $\ldots, -6, -4, -2, 0, 2, 4, 6, \ldots$

Odd integers: $\ldots, -5, -3, -1, 1, 3, 5, \ldots$

Since every whole number is also an integer, the set of whole numbers is a subset of the set of integers.

Rational numbers

In this course, we will work with positive and negative fractions. For example, if money is deposited in a savings account for 7 months, we say it is deposited for $\frac{7}{12}$ of a year. If a tank loses 40 gallons of water over a 3-minute period, we can describe the change per minute in the number of gallons of water in the tank using the fraction $-\frac{40}{3}$.

We will also work with mixed numbers. For instance, we might speak of $5\frac{7}{8}$ cups of flour or of a river that is $3\frac{1}{2}$ feet below flood stage $\left(-3\frac{1}{2}\right)$. Fractions and mixed numbers are part of the set of numbers called the **rational numbers.**

Rational numbers

A **rational number** is any number that can be written as a fraction with an integer in its numerator and a nonzero integer in its denominator.

Some examples of rational numbers are $\frac{7}{12}$, $-\frac{40}{3}$, and $\frac{25}{99}$.

The mixed numbers $5\frac{7}{8}$ and $-3\frac{1}{2}$ are also rational numbers, because they can be written as fractions with integer numerators and nonzero integer denominators:

$$5\frac{7}{8} = \frac{47}{8} \quad \text{and} \quad -3\frac{1}{2} = \frac{-7}{2}$$

The next example shows that every integer is a rational number.

EXAMPLE 1 ***Rational numbers.*** Show that each of the following numbers is a rational number by writing it as a fraction with an integer numerator and a nonzero integer denominator: **a.** −4 and **b.** 0.

Solution

a. −4 is a rational number, because −4 can be written as the fraction $\frac{-4}{1}$.

b. 0 is a rational number, because $0 = \frac{0}{1}$.

We see that all integers are rational numbers, because they can be written as fractions with a denominator of 1. Therefore, the set of integers is a subset of the set of rational numbers.

Self Check

Show that each number is a rational number: **a.** $10\frac{1}{2}$, **b.** 99, and **c.** −3.

Answers: **a.** $10\frac{1}{2} = \frac{21}{2}$, **b.** $99 = \frac{99}{1}$, **c.** $-3 = \frac{-3}{1}$

We will also work with decimals. Some examples of uses of decimals are as follows.

- The interest rate of a loan was $11\% = 0.11$.
- In baseball, the distance from home plate to second base is 127.279 feet.
- The third-quarter loss for a business was -2.7 million dollars.

Terminating decimals such as 0.11, 127.279, and -2.7 are rational numbers, since they can be written as fractions with integer numerators and nonzero integer denominators.

$$0.11 = \frac{11}{100} \qquad 127.279 = 127\frac{279}{1{,}000} = \frac{127{,}279}{1{,}000} \qquad -2.7 = -2\frac{7}{10} = \frac{-27}{10}$$

Examples of **repeating decimals** are 0.333. . . and 4.252525. . . . The three dots indicate that the digits continue in the pattern shown. Any repeating decimal can be expressed as a fraction with an integer numerator and a nonzero integer denominator. For example, $0.333\ldots = \frac{1}{3}$ and $4.252525\ldots = 4\frac{25}{99} = \frac{421}{99}$. Since every repeating decimal can be written as a fraction, repeating decimals are also rational numbers.

Rational numbers The set of **rational numbers** is the set of all terminating and all repeating decimals.

EXAMPLE 2 ***Terminating and repeating decimals.*** Change each fraction to a decimal and tell whether the decimal terminates or repeats: **a.** $\frac{4}{5}$ and **b.** $\frac{17}{6}$.

Solution

a. To change $\frac{4}{5}$ to a decimal, we divide the numerator by the denominator.

$$\begin{array}{r} .8 \\ 5\overline{)4.0} \\ \underline{40} \\ 0 \end{array}$$ Write a decimal point and a 0 to the right of 4.

In decimal form, $\frac{4}{5}$ is 0.8. This is a terminating decimal.

b. To change $\frac{17}{6}$ to a decimal, we use a calculator to do the division and obtain 2.8333. . . . This is a repeating decimal, because the digit 3 repeats forever. This decimal can be written as $2.8\overline{3}$, where the overbar indicates that the 3 repeats.

Self Check

Change each fraction to a decimal and tell whether it terminates or repeats: **a.** $\frac{25}{990}$ and **b.** $\frac{47}{50}$.

Answers: **a.** $\frac{25}{990} = 0.0\overline{25}$, repeating decimal, **b.** $\frac{47}{50} = 0.94$, terminating decimal ■

Irrational numbers

Some decimals cannot be written as fractions. As an example, let's consider the amount of concrete needed to anchor a basketball pole in the ground, as diagrammed in Figure 1-5. Using a formula from geometry, we will soon be able to show that the exact amount of concrete needed is π cubic feet. Expressed in decimal form,

$$\pi = 3.141592654\ldots$$

This **nonterminating, nonrepeating decimal** cannot be written as a fraction with an integer for its numerator and its denominator. Therefore, π is not a rational number. It is an example of an **irrational number.**

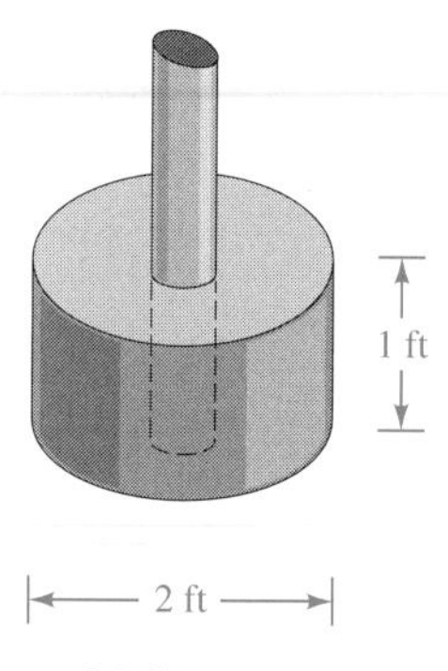

FIGURE 1-5

Irrational numbers The set of **irrational numbers** is the set of all real numbers that are not rational. Irrational numbers are nonterminating, nonrepeating decimals.

Accent on Technology Approximating irrational numbers

We can approximate the value of irrational numbers by using a scientific calculator. To find the approximate value of π, we simply press the $\boxed{\pi}$ key.

Keystrokes: $\boxed{\pi}$ (you may have to use a $\boxed{\text{2nd}}$ or $\boxed{\text{Shift}}$ key first) $\boxed{3.141592654}$

We see that $\pi \approx 3.141592654$. (Read $\approx$ as "is approximately equal to.") If a problem calls for an approximate answer, we can round this decimal. To the nearest thousandth, $\pi = 3.142$.

Another irrational number is $\sqrt{2}$. To approximate $\sqrt{2}$, we enter 2 and press the $\boxed{\sqrt{\ }}$ key (the square root key).

Keystrokes: 2 $\boxed{\sqrt{\ }}$ $\boxed{1.414213562}$

We see that $\sqrt{2} \approx 1.414213562$. To the nearest hundredth, $\sqrt{2} = 1.41$.

Some other examples of irrational numbers are

$-\sqrt{7} = -2.64575131\ldots$ This is a negative, nonterminating, nonrepeating decimal.

$2\pi = 6.283185307\ldots$ We evaluate 2π using a calculator: $2\pi = 2 \cdot \pi$.

WARNING! Students often *incorrectly* classify a number such as 4.12122122212222. . . as a repeating decimal. Although the decimal part of this number does exhibit a "pattern," no single block of digits repeats forever. Therefore, it is a nonterminating, nonrepeating decimal—an irrational number.

Real numbers

Together, the set of rational numbers and the set of irrational numbers form the set of **real numbers.** This means that every real number can be written as either a terminating, a repeating, or a nonterminating, nonrepeating decimal. Thus, the set of real numbers is the set of all decimals.

The real numbers A **real number** is any number that is either a rational or an irrational number.

Figure 1-6 shows how the sets of numbers discussed in this section are related. It also gives some examples of each type of number.

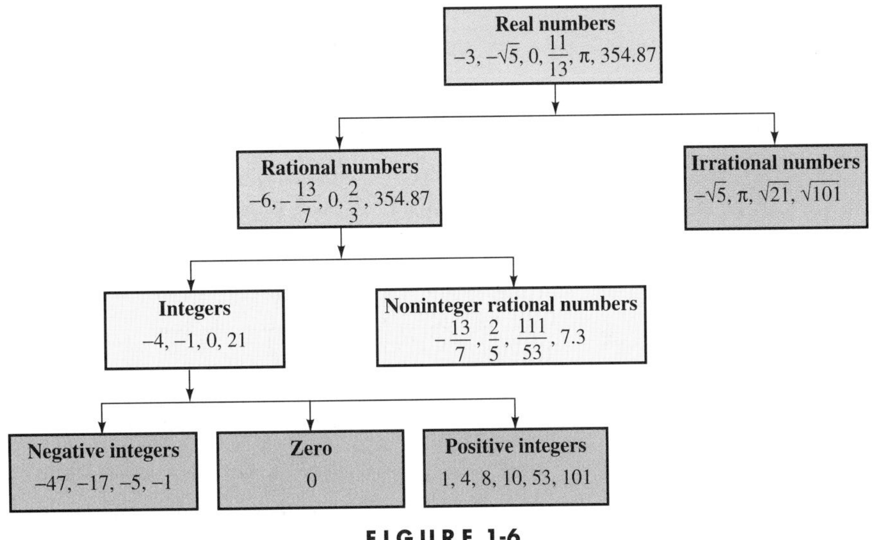

FIGURE 1-6

EXAMPLE 3 ***Classifying real numbers.*** Which numbers in the following set are natural numbers, whole numbers, integers, rational numbers, irrational numbers, and real numbers?

$$\left\{\tfrac{5}{8}, -0.02, 45, -9, \sqrt{7}, 5\tfrac{2}{3}, 0, 1.727227222\ldots\right\}$$

Solution

Natural numbers:	45
Whole numbers:	45, 0
Integers:	45, −9, 0
Rational numbers:	$\frac{5}{8}$, −0.02, 45, −9, $5\frac{2}{3}$, 0
Irrational numbers:	$\sqrt{7}$, 1.727227222. . .
Real numbers:	$\frac{5}{8}$, −0.02, 45, −9, $\sqrt{7}$, $5\frac{2}{3}$, 0, 1.727227222. . .

Self Check

Use the instructions for Example 3 with the following set:
$\left\{-\pi, -\frac{22}{7}, -5, 3.4, 1, \frac{16}{5}, 9.\overline{7}\right\}$

Answers: natural numbers: 1; whole numbers: 1; integers: −5, 1; rational numbers: $-\frac{22}{7}$, −5, 3.4, 1, $\frac{16}{5}$, $9.\overline{7}$; irrational numbers: $-\pi$; real numbers: all

The real number line

We can illustrate real numbers using a **number line.** To each real number, there corresponds a point on the line. Furthermore, to each point on the line, there corresponds a number, called its **coordinate.** The point labeled 0 is called the **origin.**

Figure 1-7 shows the **graph** of −3 and 4. As we move from left to right on the number line, the coordinates of the points get larger. Since the graph of 4 is to the right of the graph of −3, we know that 4 is greater than −3 and that −3 is less than 4.

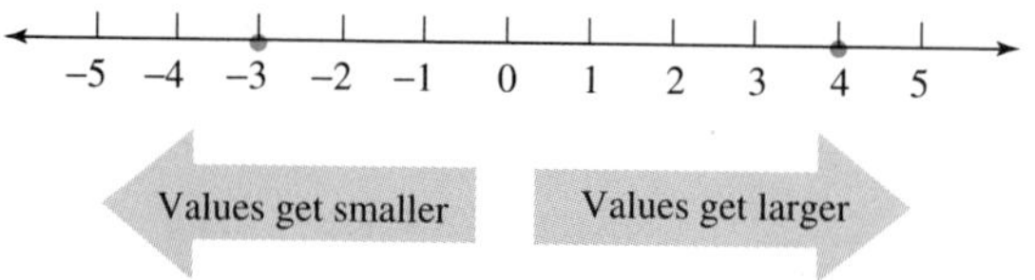

FIGURE 1-7

EXAMPLE 4 ***Graphing on the number line.*** Graph $-\frac{8}{3}$, −1.1, 0.565656. . . , $\frac{\pi}{2}$, $-\sqrt{15}$, and $2\sqrt{2}$.

Solution

To help locate the graph of each number, we make some observations.

- Expressed as a mixed number, $-\frac{8}{3} = -2\frac{2}{3}$.
- Since −1.1 is less than −1, its graph is to the *left* of −1.
- 0.565656. . . ≈ 0.6
- From a calculator, $\frac{\pi}{2} \approx 1.6$.
- From a calculator, $-\sqrt{15} \approx -3.9$.
- $2\sqrt{2}$ means $2 \cdot \sqrt{2}$. From a calculator, $2\sqrt{2} \approx 2.8$.

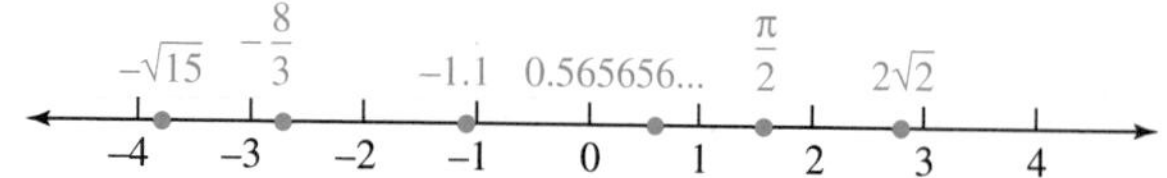

Self Check

Graph π, $-2.\overline{1}$, $\sqrt{3}$, $\frac{11}{4}$, and −0.9.

Answers:

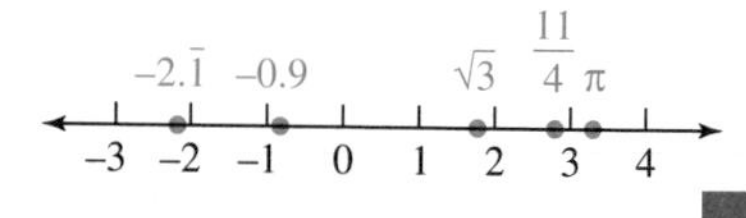

Inequality symbols

To show that two quantities are not equal, we use one of the **inequality symbols** shown in Table 1-3.

Symbol	Read as	Examples
$\neq$	"is not equal to"	$6 \neq 9$ and $0.33 \neq \frac{3}{5}$
$<$	"is less than"	$\frac{22}{3} < \frac{23}{3}$ and $-7 < -6$
$>$	"is greater than"	$19 > 5$ and $\frac{1}{2} > 0.3$
$\leq$	"is less than or equal to"	$3.5 \leq 3.\overline{5}$ and $1\frac{4}{5} \leq 1.8$
$\geq$	"is greater than or equal to"	$29 \geq 29$ and $-15.2 \geq -16.7$

TABLE 1-3

It is always possible to write an inequality with the inequality symbol pointing in the opposite direction. For example,

$-3 < 4$ is equivalent to $4 > -3$

$5.3 \geq -2.9$ is equivalent to $-2.9 \leq 5.3$

EXAMPLE 5 ***Inequality symbols.*** Use one of the symbols $>$ or $<$ to make each statement true: **a.** $-24 \ \square \ -25$ and **b.** $\frac{3}{4} \ \square \ 0.76$.

Solution

a. Since -24 is to the right of -25 on the number line, $-24 > -25$.

b. If we express the fraction $\frac{3}{4}$ as a decimal, we can easily compare it to 0.76. Since $\frac{3}{4} = 0.75$, $\frac{3}{4} < 0.76$.

Self Check

Use one of the symbols $\geq$ or $\leq$ to make each statement true:

a. $\frac{2}{3} \ \square \ \frac{4}{3}$ and **b.** $8\frac{1}{2} \ \square \ 8.4$.

Answers: **a.** $\leq$, **b.** $\geq$ ■

Absolute value

In Figure 1-8, we can see that -3 and 3 are both a distance of 3 units away from zero on the number line. Because of this, we say that -3 and 3 are **opposites** or **negatives.**

Parentheses are needed to express the opposite of a negative number. For example, the opposite, or negative, of -3 is written as $-(-3)$. Since -3 and 3 are the same distance from zero, the opposite of -3 is 3. Therefore, $-(-3) = 3$. This leads us to the following conclusion.

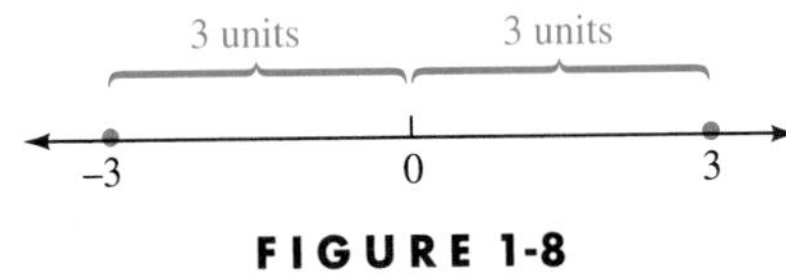

FIGURE 1-8

The double negative rule If x is any real number, then $-(-x) = x$.

The **absolute value** of any real number is the distance between the number and zero on a number line. To indicate absolute value, the number is inserted between two vertical bars. For example, the points shown in Figure 1-8 with coordinates of 3 and -3 both lie 3 units from zero. Thus, $|3| = |-3| = 3$.

The absolute value of a number can be defined more formally.

Absolute value For any real number x, $\begin{cases} \text{If } x \geq 0, \text{ then } |x| = x. \\ \text{If } x < 0, \text{ then } |x| = -x. \end{cases}$

WARNING! Since absolute value expresses distance, the absolute value of a number is always positive or zero, but never negative.

EXAMPLE 6 ***Evaluating absolute values.*** Find each absolute value: **a.** $|34|$, **b.** $\left|-\frac{4}{5}\right|$, **c.** $|0|$, **d.** $-|-1.8|$.

Solution

a. Since 34 is a distance of 34 from 0 on a number line, $|34| = 34$.

b. $-\frac{4}{5}$ is a distance of $\frac{4}{5}$ from 0 on a number line. Therefore, $\left|-\frac{4}{5}\right| = \frac{4}{5}$.

c. $|0| = 0$

d. $-|-1.8| = -(1.8)$ Find $|-1.8|$ first: $|-1.8| = 1.8$.

$= -1.8$ Rewrite without parentheses.

Self Check

Find each absolute value.

a. $|-9.6|$

b. $-|-12|$

c. $\left|\frac{3}{2}\right|$

Answers: **a.** 9.6, **b.** −12, **c.** $\frac{3}{2}$

STUDY SET Section 1.2

VOCABULARY *In Exercises 1–8, fill in the blanks to make the statements true.*

1. A ______ number is any number that can be written as a fraction with an integer in its numerator and a nonzero integer in its denominator.

2. The ______ numbers are the natural numbers greater than 1 that are divisible only by themselves and 1.

3. The ______ of any real number is the distance between the number and zero on a number line.

4. Together, the set of rational numbers and the set of irrational numbers form the set of ______ numbers.

5. ______ numbers are nonterminating, nonrepeating decimals.

6. ______ numbers are greater than 0, and ______ numbers are less than 0.

7. A ______ number is a natural number greater than 1 that is not prime.

8. ___ is neither positive nor negative.

CONCEPTS *In Exercises 9–20, list the elements in the set* $\left\{-3, -\frac{8}{5}, 0, \frac{2}{3}, 1, 2, \sqrt{3}, \pi, 4.75, 9, 16.\overline{6}\right\}$ *that satisfy the given condition.*

9. Natural number

10. Whole number

11. Integer

12. Rational number

13. Irrational number

14. Real number

15. Even natural number

16. Odd integer

17. Prime number

18. Composite number

19. Odd composite number

20. Odd prime number

In Exercises 21–24, tell whether each number is a repeating or a nonrepeating decimal, and whether it is a rational or an irrational number.

21. 0.090090009. . . .

22. $0.\overline{09}$

23. 5.41414141. . .

24. 1.414213562. . .

25. Show each of the following numbers is a rational number by expressing it as a fraction with an integer in its numerator and a nonzero integer in its denominator.

$7, -7\frac{3}{5}, 0.007, 700.1$

26. Tell whether each statement is true or false.
a. All prime numbers are odd numbers.
b. $6 \geq 6$
c. 0 is neither even nor odd.
d. If x is negative, $|x| = -x$.

27. Find x if $|x| = 3.5$.

28. Name two numbers that are 6 units away from −2 on the number line.

29. The diagram in Illustration 1 can be used to show how the natural numbers, whole numbers, integers, rational numbers, and irrational numbers make up the set of real numbers. If the natural numbers can be represented as shown, label each of the other sets.

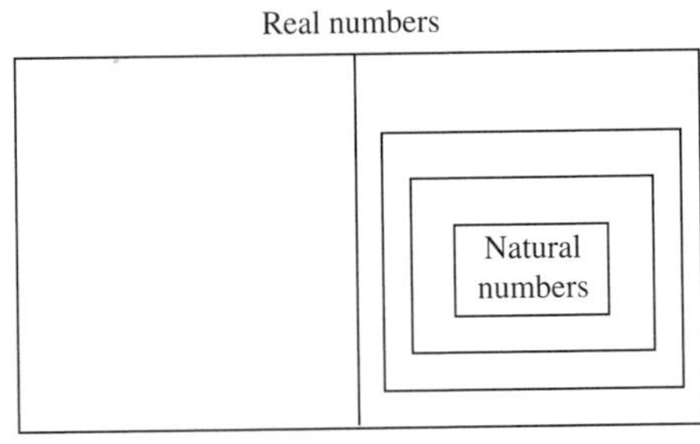

ILLUSTRATION 1

30. The formula $C = \pi D$ gives the circumference C of a circle, where D is the length of its diameter. Find the circumference of the man's wedding ring shown in Illustration 2. Give an *exact* answer and then an *approximate* answer, rounded to the nearest hundredth of an inch.

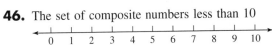

ILLUSTRATION 2

NOTATION *In Exercises 31–34, fill in the blanks to make the statements true.*

31. The symbol $<$ means "______________."

32. $|-2|$ is read as "the ______________ value ____ -2."

33. The symbols { } are called ______________.

34. The symbol $\geq$ means "______________."

PRACTICE *In Exercises 35–38, change each fraction into a decimal and classify the result as a terminating or a repeating decimal.*

35. $\frac{7}{8}$

36. $\frac{8}{3}$

37. $-\frac{11}{15}$

38. $-\frac{19}{16}$

In Exercises 39–48, graph each set on a number line.

39. $\left\{-\frac{5}{2}, -0.1, 2.142765. . . , \frac{\pi}{3}, -\sqrt{11}, 2\sqrt{3}\right\}$

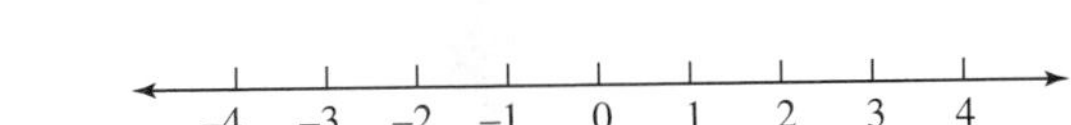

40. $\left\{2\frac{1}{9}, -3.821134. . . , -\frac{\pi}{2}, \sqrt{15}, -0.9, \frac{\sqrt{2}}{2}\right\}$

−4 −3 −2 −1 0 1 2 3 4

41. $\{3.\overline{15}, \frac{22}{7}, 3\frac{1}{8}, \pi, \sqrt{10}, 3.1\}$

3.1 3.15 3.2

42. $\{-0.\overline{331}, -0.331, -\frac{1}{3}, -\sqrt{0.11}\}$

−0.34 −0.335 −0.330

43. The set of prime numbers less than 8

0 1 2 3 4 5 6 7 8

44. The set of integers between -7 and 0

−7 −6 −5 −4 −3 −2 −1 0

45. The set of odd integers between 10 and 18

10 11 12 13 14 15 16 17 18

46. The set of composite numbers less than 10

0 1 2 3 4 5 6 7 8 9 10

47. The set of positive odd integers less than 12

0 1 2 3 4 5 6 7 8 9 10 11 12

48. The set of negative even integers greater than -7

−7 −6 −5 −4 −3 −2 −1 0

In Exercises 49–56, insert either a $<$ or a $>$ symbol to make a true statement.

49. 8 ☐ 9

50. 9 ☐ 0

51. $-(-5)$ ☐ -10

52. $|-3|$ ☐ $-(-6)$

53. -7.999 ☐ -7.1

54. $4\frac{1}{2}$ ☐ $\frac{7}{2}$

55. $6.\overline{1}$ ☐ 6

56. -6.07 ☐ $-\frac{17}{6}$

In Exercises 57–60, write each statement with the inequality symbol pointing in the opposite direction.

57. $19 > 12$ **58.** $-3 \geq -5$ **59.** $-6 \leq -5$ **60.** $-10 < 0$

In Exercises 61–68, write each expression without using absolute value symbols.

61. $|20|$ **62.** $|-20|$ **63.** $-|-6|$ **64.** $-|-8|$

65. $|-5.9|$ **66.** $-|1.\overline{27}|$ **67.** $\left|\frac{5}{4}\right|$ **68.** $\left|-\frac{5}{16}\right|$

APPLICATIONS

69. DRAFTING Express each dimension in the drawing of a bracket in Illustration 3 as a four-place decimal.

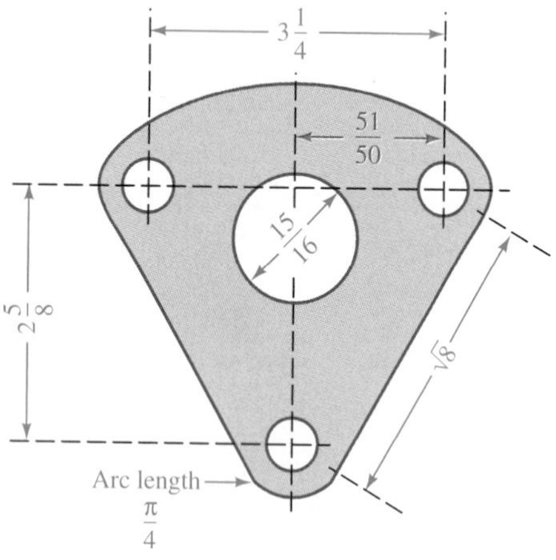

$3\frac{1}{4}$, $\frac{51}{50}$, $\frac{15}{16}$, $2\frac{5}{8}$, $\sqrt{8}$, Arc length $\frac{\pi}{4}$

ILLUSTRATION 3

70. pH SCALE The pH scale, shown in Illustration 4, is used to measure the strength of acids and bases (alkalines) in chemistry. It can be thought of as a number line. On the scale, graph and label each pH measurement given in the table.

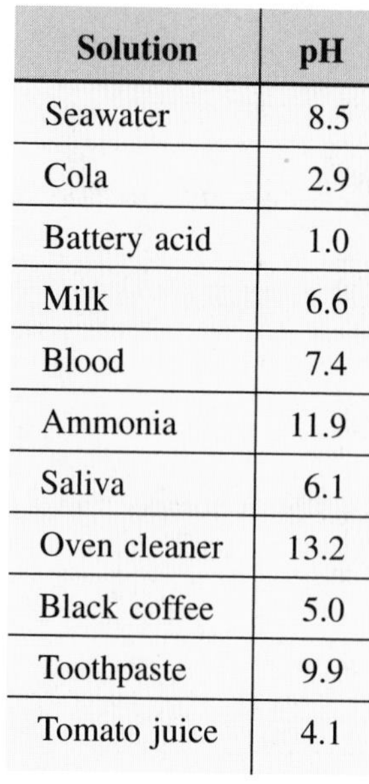

Solution	pH
Seawater	8.5
Cola	2.9
Battery acid	1.0
Milk	6.6
Blood	7.4
Ammonia	11.9
Saliva	6.1
Oven cleaner	13.2
Black coffee	5.0
Toothpaste	9.9
Tomato juice	4.1

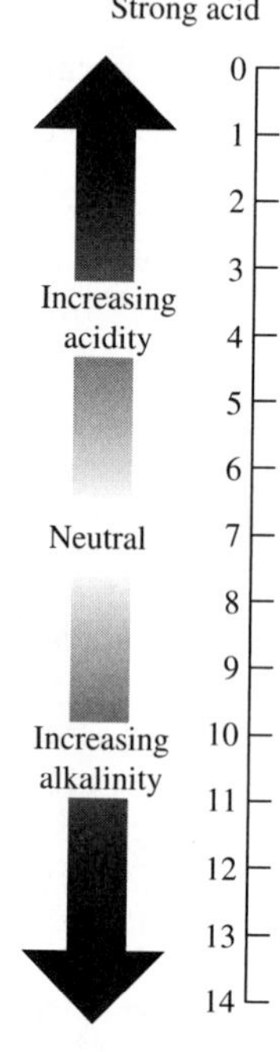

ILLUSTRATION 4

WRITING *Write a paragraph using your own words.*

71. Explain why the whole numbers are a subset of the integers.

72. What is a real number? Give examples.

73. Explain why there are no even prime numbers greater than 2.

74. Explain why every integer is a rational number, but not every rational number is an integer.

REVIEW

75. Is $\frac{3x-4}{2}$ an equation or an expression?

76. Translate into mathematical symbols: The weight of an object in ounces is 16 times its weight in pounds.

In Exercises 77 and 78, complete the table.

77. $T = x - 1.5$

x	T
3.7	
10	
30.6	

78. $j = 3m$

m	j
0	
15	
300	

1.3 Operations with Real Numbers

In this section, you will learn about

- **Rules for adding real numbers**
- **A rule for subtracting real numbers**
- **Rules for multiplying real numbers**
- **Rules for dividing real numbers**
- **Raising a real number to a power**
- **Finding a square root of a real number**
- **Order of operations**
- **Evaluating algebraic expressions**

INTRODUCTION. Six operations can be performed with real numbers: addition, subtraction, multiplication, division, raising to a power, and finding a root. In this section, we will review the rules for performing each of these operations. We will also discuss the procedure used to evaluate numerical expressions involving several operations.

Rules for adding real numbers

When two numbers are added, we call the result their **sum.** To find the sum of 2 and 3, we can represent the numbers with arrows, as shown in Figure 1-9(a). Since the endpoint of the second arrow is at 5, we have $2 + 3 = 5$.

To add -2 and -3, we can draw arrows as shown in Figure 1-9(b). Since the endpoint of the second arrow is at -5, we have $-2 + (-3) = -5$.

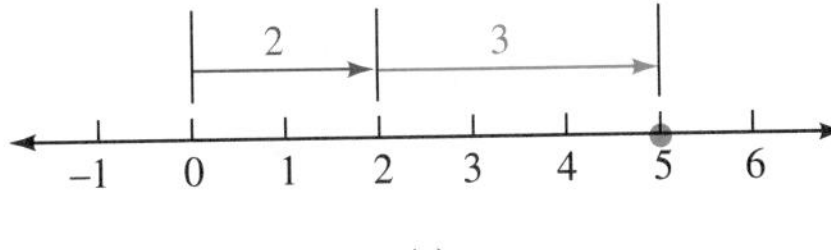

(a)

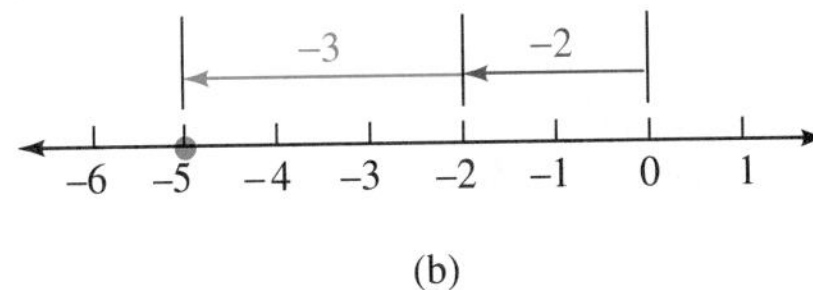

(b)

FIGURE 1-9

To add -6 and 2, we can draw arrows as shown in Figure 1-10(a). Since the endpoint of the second arrow is at -4, we have $-6 + 2 = -4$.

To add 7 and -4, we can draw arrows as shown in Figure 1-10(b). Since the endpoint of the second arrow is at 3, we have $7 + (-4) = 3$.

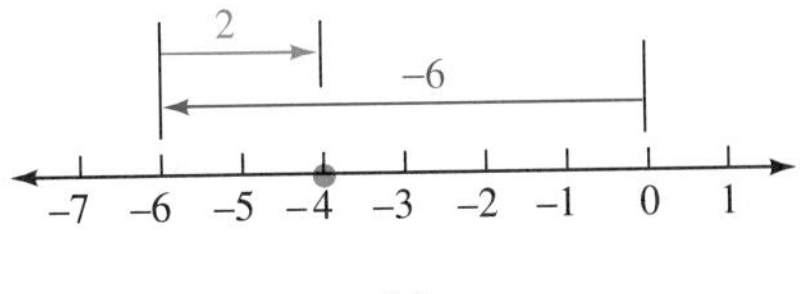

(a)

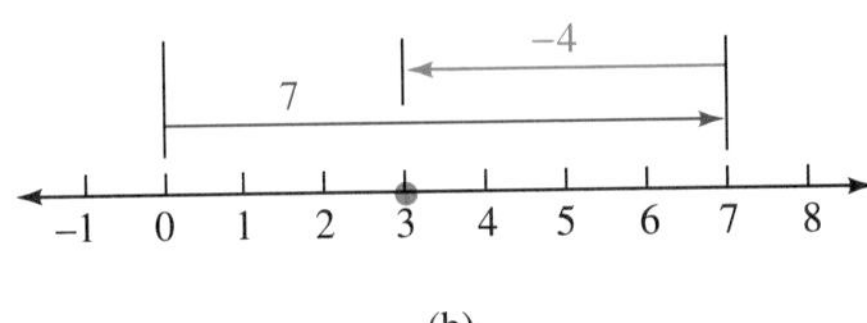

(b)

FIGURE 1-10

These examples suggest the following rules.

Adding real numbers	*With like signs:* Add the absolute values of the numbers and keep the common sign. *With unlike signs:* Subtract the absolute values of the numbers (the smaller from the larger) and keep the sign of the number with the larger absolute value.

EXAMPLE 1 ***Adding real numbers.*** Find each sum: **a.** $-5 + (-3)$, **b.** $8.9 + (-5.1)$, and **c.** $-\frac{13}{15} + \frac{3}{5}$.

Solution

a. $-5 + (-3) = -8$ — Because the numbers have like signs, add their absolute values and keep the common negative sign: $-(5 + 3) = -8$.

b. $8.9 + (-5.1) = 3.8$ — Since the numbers have unlike signs, subtract their absolute values. The sum is positive, because 8.9 has the larger absolute value: $+(8.9 - 5.1) = 3.8$.

c. $-\frac{13}{15} + \frac{3}{5} = -\frac{13}{15} + \frac{9}{15}$ — Express $\frac{3}{5}$ in terms of the lowest common denominator, 15: $\frac{3}{5} = \frac{3 \cdot 3}{5 \cdot 3} = \frac{9}{15}$.

$= -\frac{4}{15}$ — Subtract the absolute values of the fractions. Keep the − sign, because $-\frac{13}{15}$ has the larger absolute value: $-\left(\frac{13}{15} - \frac{9}{15}\right) = -\frac{4}{15}$.

Self Check

Add: **a.** $-70.4 + (-21.2)$, **b.** $-34 + 25$, and **c.** $\frac{7}{4} + \left(-\frac{3}{2}\right)$.

Answers: **a.** -91.6, **b.** -9, **c.** $\frac{1}{4}$

A rule for subtracting real numbers

When two numbers are subtracted, we call the result their **difference.** To find a difference, we can change the subtraction into an equivalent addition. For example, the subtraction $7 - 4$ is equivalent to the addition $7 + (-4)$, because they have the same answer:

$$7 - 4 = 3 \quad \text{and} \quad 7 + (-4) = 3$$

This suggests that to subtract two numbers, we can change the sign of the number being subtracted and add. In other words, subtraction is the same as adding the *opposite* of the number being subtracted.

Subtracting real numbers	If x and y are real numbers, then $x - y = x + (-y)$.

EXAMPLE 2 ***Subtracting real numbers.*** Find each difference: **a.** $12 - 4$, **b.** $-1.3 - 5.5$, and **c.** $-\frac{14}{3} - \left(-\frac{7}{3}\right)$.

Solution

In each case, we add the opposite of the number that is being subtracted.

a. $12 - 4 = 12 + (-4)$ — Here, 4 is being subtracted, so we change the sign of 4 and add.

$= 8$ — Use the rule for adding two numbers with unlike signs.

b. $-1.3 - 5.5 = -1.3 + (-5.5)$ — Change the sign of 5.5 and add.

$= -6.8$ — Use the rule for adding two numbers with like signs.

c. $-\frac{14}{3} - \left(-\frac{7}{3}\right) = -\frac{14}{3} + \frac{7}{3}$ — Change the sign of $-\frac{7}{3}$ and add.

$= -\frac{7}{3}$ — Use the rule for adding two numbers with unlike signs.

Self Check

Subtract: **a.** $-15 - 4$, **b.** $\frac{5}{9} - \frac{7}{9}$, and **c.** $-12.1 - (-7.6)$.

Answers: **a.** -19, **b.** $-\frac{2}{9}$, **c.** -4.5

EXAMPLE 3 ***The U.S. federal debt.*** Debt can be represented by a negative number. The line graph in Figure 1-11 shows the mounting federal debt, in trillions of dollars. Find the change in the federal debt for the years 1960–1996.

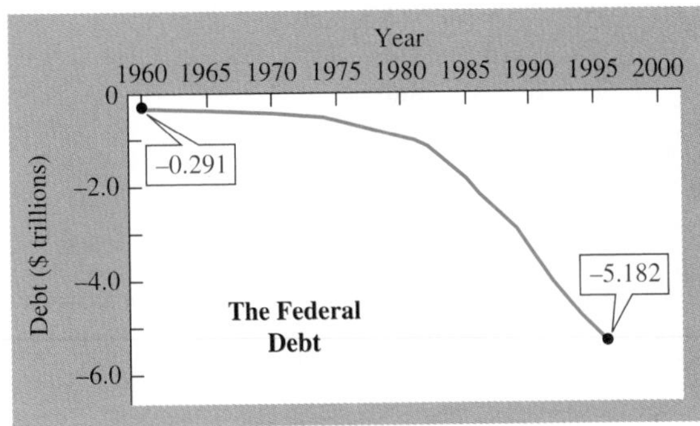

Based on data from the *Wall Street Journal Almanac* (1998), p. 132.

FIGURE 1-11

Solution To find the change, we subtract the earlier value from the later value.

$$\underbrace{-5.182}_{\text{Debt in 1996}} - \underbrace{(-0.291)}_{\text{Debt in 1960}} = -5.182 + 0.291$$
$$= -4.891$$

The negative result indicates that the debt grew over the years 1960–1996 by $4.891 trillion. ■

Rules for multiplying real numbers

When two numbers are multiplied, we call the numbers **factors** and the result their **product.** We can find the product of 5 and 4 by using 4 in an addition five times:

$$5(4) = 4 + 4 + 4 + 4 + 4 = 20$$

We can find the product of 5 and -4 by using -4 in an addition five times.

$$5(-4) = (-4) + (-4) + (-4) + (-4) + (-4) = -20$$

Since multiplication by a negative number can be defined as repeated subtraction, we can find the product of -5 and 4 by using 4 in a subtraction five times:

$$-5(4) = -4 - 4 - 4 - 4 - 4$$
$$= -4 + (-4) + (-4) + (-4) + (-4) \quad \text{Add the opposite of each 4.}$$
$$= -20 \quad \text{Do the additions.}$$

We can find the product of -5 and -4 by using -4 in a subtraction five times:

$$-5(-4) = -(-4) - (-4) - (-4) - (-4) - (-4)$$
$$= 4 + 4 + 4 + 4 + 4 \quad \text{Add the opposite of each } -4.$$
$$= 20 \quad \text{Do the additions.}$$

The products 5(4) and $-5(-4)$ both equal 20, and the products $5(-4)$ and $-5(4)$ both equal -20. These results suggest the following rules for multiplying real numbers.

Multiplying real numbers	*With like signs:* Multiply their absolute values. The product is positive. *With unlike signs:* Multiply their absolute values. The product is negative.

EXAMPLE 4 ***Multiplying real numbers.*** Find each product: **a.** $4(-7)$, **b.** $-5.2(-3)$, and **c.** $-\frac{7}{9}\left(\frac{3}{16}\right)$.

Solution

a. $4(-7) = -28$ Multiply the absolute values of 4 and -7: $4 \cdot 7 = 28$. Since the signs are unlike, the product is negative.

b. $-5.2(-3) = 15.6$ Multiply the absolute values: $5.2 \cdot 3 = 15.6$. Since the signs are alike, the product is positive.

c. $-\frac{7}{9}\left(\frac{3}{16}\right) = -\frac{7 \cdot 3}{9 \cdot 16}$ Multiply the numerators and multiply the denominators. Since the signs are unlike, the product is negative.

$= -\frac{7 \cdot \overset{1}{\cancel{3}}}{\underset{1}{\cancel{3}} \cdot 3 \cdot 16}$ Factor 9 as $3 \cdot 3$ and divide out the common factor of 3.

$= -\frac{7}{48}$ Multiply in the numerator and denominator.

Self Check

Multiply: **a.** $(-6)(5)$, **b.** $(-4.1)(-8)$, and **c.** $\left(\frac{4}{3}\right)\left(-\frac{1}{8}\right)$.

Answers: **a.** -30, **b.** 32.8, **c.** $-\frac{1}{6}$

Rules for dividing real numbers

When two numbers are divided, we call the result their **quotient.** In the division $\frac{x}{y} = q$ $(y \neq 0)$, the quotient q is a number such that $y \cdot q = x$. We can use this relationship to find rules for dividing real numbers. We consider four divisions:

$$\frac{10}{2} = 5, \text{ because } 2(5) = 10 \qquad \frac{-10}{-2} = 5, \text{ because } -2(5) = -10$$

$$\frac{-10}{2} = -5, \text{ because } 2(-5) = -10 \qquad \frac{10}{-2} = -5, \text{ because } -2(-5) = 10$$

These results suggest the following rules for dividing real numbers.

Dividing real numbers	*With like signs:* Divide their absolute values. The quotient is positive. *With unlike signs:* Divide their absolute values. The quotient is negative.

EXAMPLE 5 ***Dividing real numbers.*** Find each quotient: **a.** $\frac{-44}{11}$ and **b.** $\frac{2.7}{-9}$

Solution

a. $\frac{-44}{11} = -4$ Divide the absolute values of -44 and 11: $\frac{44}{11} = 4$. Since the signs are unlike, the quotient is negative.

b. $\frac{2.7}{-9} = -0.3$ Divide the absolute values: $\frac{2.7}{9} = 0.3$. Since the signs are unlike, the quotient is negative.

Self Check

Divide **a.** $\frac{55}{-5}$ and **b.** $\frac{-7.2}{-6}$.

Answers: **a.** -11, **b.** 1.2

To divide two fractions, we multiply the first fraction by the reciprocal of the second fraction. In symbols, if a, b, c, and d are real numbers, then

$$\frac{a}{b} \div \frac{c}{d} = \frac{a}{b} \cdot \frac{d}{c} \qquad (b \neq 0, c \neq 0, \text{ and } d \neq 0)$$

EXAMPLE 6 ***Dividing fractions.*** Find each quotient: **a.** $\frac{2}{3} \div \left(-\frac{3}{5}\right)$ and **b.** $-\frac{1}{2} \div (-6)$.

Solution

a. $\frac{2}{3} \div \left(-\frac{3}{5}\right) = \frac{2}{3} \cdot \left(-\frac{5}{3}\right)$ Multiply by the reciprocal of $-\frac{3}{5}$, which is $-\frac{5}{3}$.

$= -\frac{10}{9}$

b. $-\frac{1}{2} \div (-6) = -\frac{1}{2} \cdot \left(-\frac{1}{6}\right)$ Multiply by the reciprocal of -6, which is $-\frac{1}{6}$.

$= \frac{1}{12}$

Self Check

Divide: **a.** $-\frac{7}{8} \div \frac{2}{3}$ and **b.** $-\frac{1}{10} \div \left(-\frac{1}{5}\right)$.

Answers: **a.** $-\frac{21}{16}$, **b.** $\frac{1}{2}$ ■

WARNING! Students often confuse two similar-looking division problems. For example, consider $\frac{0}{4}$ and $\frac{4}{0}$. We know that $\frac{0}{4} = 0$, because $4 \cdot 0 = 0$. On the other hand, $\frac{4}{0}$ is undefined, because there is no real number q such that $0 \cdot q = 4$. In general, if $x \neq 0$, $\frac{0}{x} = 0$ and $\frac{x}{0}$ is undefined.

Raising a real number to a power

Exponents indicate repeated multiplication. For example,

$3^2 = 3 \cdot 3$ Read 3^2 as "3 to the second power" or "3 squared."

$(-9.1)^3 = (-9.1)(-9.1)(-9.1)$ Read $(-9.1)^3$ as "-9.1 to the third power" or "-9.1 cubed."

$\left(\frac{2}{3}\right)^4 = \left(\frac{2}{3}\right)\left(\frac{2}{3}\right)\left(\frac{2}{3}\right)\left(\frac{2}{3}\right)$ Read $\left(\frac{2}{3}\right)^4$ as "$\frac{2}{3}$ to the fourth power."

These examples suggest the following definition.

Natural-number exponents

If n is a natural number, then

$$x^n = \overbrace{x \cdot x \cdot x \cdot \cdots \cdot x}^{n \text{ factors of } x}$$

The exponential expression x^n is called a **power of x**, and we read it as "x to the nth power." In this expression, x is called the **base,** and n is called the **exponent.** A natural-number exponent tells how many times the base of an exponential expression is to be used as a factor in a product.

Base $\longrightarrow x^n \longleftarrow$ Exponent

EXAMPLE 7 ***Evaluating exponential expressions.*** Find each power: **a.** $(-2)^5$, **b.** $\left(\frac{3}{4}\right)^2$, and **c.** -0.1 cubed.

Solution

In each case, we use the fact that an exponent tells how many times the base is to be used as a factor in a product.

a. $(-2)^5 = (-2)(-2)(-2)(-2)(-2) = -32$

b. $\left(\dfrac{3}{4}\right)^2 = \dfrac{3}{4}\left(\dfrac{3}{4}\right) = \dfrac{9}{16}$

c. -0.1 cubed means $(-0.1)^3$.

$$(-0.1)^3 = (-0.1)(-0.1)(-0.1) = -0.001$$

Self Check

Find each power.

a. $(-3)^3$

b. $(0.8)^2$

c. 2^4

d. $\frac{7}{5}$ squared

Answers: **a.** -27, **b.** 0.64, **c.** 16, **d.** $\frac{49}{25}$

WARNING! Although the expressions $(-3)^2$ and -3^2 look alike, they are not. In $(-3)^2$, the base is -3. In -3^2, the base is 3. The $-$ sign in front of 3^2 means the opposite of 3^2. When we evaluate them, we see that the results are different:

$$(-3)^2 = (-3)(-3) = 9 \qquad -3^2 = -(3 \cdot 3) = -9$$

Accent on Technology — Remodeling a kitchen

A homeowner plans to install the cooking island shown in Figure 1-12. To find the number of square feet of kitchen floor space that will be lost, we substitute 3.25 for s in the formula for the area of a square, $A = s^2$. Using the squaring key $\boxed{x^2}$, we can evaluate $(3.25)^2$ by entering these numbers and pressing these keys:

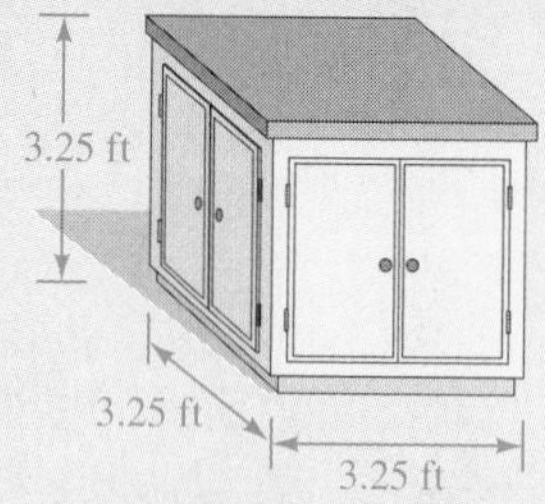

Keystrokes: 3.25 $\boxed{x^2}$ 10.5625

About 10.6 square feet of floor space will be lost.

FIGURE 1-12

The number of cubic feet of storage space that the cooking island will add can be found by substituting 3.25 for s in the formula for the volume of a cube, $V = s^3$. Using the exponential key $\boxed{y^x}$ ($\boxed{x^y}$ on some calculators), we can evaluate $(3.25)^3$ by entering these numbers and pressing these keys:

Keystrokes: 3.25 $\boxed{y^x}$ 3 $\boxed{=}$ 34.328125

The island will add about 34.3 cubic feet of storage space.

Finding a square root of a real number

Since the product $3 \cdot 3$ can be denoted by the exponential expression 3^2, we say that 3 is squared. The opposite of the squaring a number is called finding its **square root.**

All positive numbers have two square roots: one that is positive and one that is negative. For example, the two square roots of 9 are 3 and -3. The number 3 is a square root of 9, because $3^2 = 9$, and -3 is a square root of 9, because $(-3)^2 = 9$.

The symbol $\sqrt{}$, called a **radical sign,** is used to represent the positive (or *principal*) square root of a number.

Principal square root

A number b is a square root of a if $b^2 = a$.

If $a > 0$, the expression $\sqrt{a}$ represents the **principal** (or positive) **square root** of a. The principal square root of 0 is 0: $\sqrt{0} = 0$.

The principal square root of a positive number is always positive. Although 3 and -3 are both square roots of 9, only 3 is the principal square root. The symbol $\sqrt{9}$ represents 3. To represent -3, we place a $-$ sign in front of the radical:

$$\sqrt{9} = 3 \qquad \text{and} \qquad -\sqrt{9} = -3$$

EXAMPLE 8 ***Finding square roots.*** Evaluate **a.** $\sqrt{121}$, **b.** $-\sqrt{49}$, **c.** $\sqrt{\frac{1}{4}}$, and **d.** $\sqrt{0.09}$

Solution

a. $\sqrt{121} = 11$, because $11^2 = 121$.

b. Since $\sqrt{49} = 7$, $-\sqrt{49} = -7$.

c. $\sqrt{\frac{1}{4}} = \frac{1}{2}$, because $\left(\frac{1}{2}\right)^2 = \frac{1}{4}$.

d. $\sqrt{0.09} = 0.3$, because $(0.3)^2 = 0.09$.

Self Check

Find each square root:

a. $\sqrt{64}$ **b.** $-\sqrt{100}$

c. $\sqrt{\frac{4}{25}}$ **d.** $\sqrt{1}$

e. $\sqrt{0.81}$ **f.** $-\sqrt{400}$

Answers: **a.** 8, **b.** -10, **c.** $\frac{2}{5}$, **d.** 1, **e.** 0.9, **f.** -20 ■

Order of operations

Quite often, we will have to evaluate expressions involving more than one operation. For example, consider the expression $3 + 2 \cdot 5$. We can evaluate it in two different ways. We can do the addition first and then do the multiplication. Or we can do the multiplication first and then do the addition. However, we get different results.

Method 1: Do the addition first

$3 + 2 \cdot 5 = 5 \cdot 5$ Add 3 and 2 first.

$= 25$ Do the multiplication.

Method 2: Do the multiplication first

$3 + 2 \cdot 5 = 3 + 10$ Multiply 2 and 5 first.

$= 13$ Do the addition.

Different results

This example shows that we need to establish an order of operations. Otherwise, the same expression can have two different values. To guarantee that calculations will have one correct result, we will use the following rules.

Rules for the order of operations

Do all calculations within each pair of grouping symbols (parentheses, brackets, or braces), working from the innermost pair to the outermost pair.

1. Evaluate all powers and roots.
2. Do all multiplications and divisions, working from left to right.
3. Do all additions and subtractions, working from left to right.

When all grouping symbols have been removed, repeat the steps above to finish the calculation.

In a fraction, simplify the numerator and the denominator separately. Then simplify the fraction, whenever possible.

To evaluate $3 + 2 \cdot 5$ correctly, we must apply the rules for the order of operations. Since the expression does not contain grouping symbols, we follow steps 1, 2, and 3 of the previous list. The expression does not contain any powers or roots, so we do the multiplication first, followed by the addition.

$$3 + \mathbf{2 \cdot 5} = 3 + \mathbf{10} \quad \text{Ignore the addition for now and do the multiplication of 2 and 5.}$$
$$= 13 \quad \text{Next, do the addition.}$$

Using the rules for the order of operations, we see that the correct answer is 13.

EXAMPLE 9 ***Using the rules for the order of operations.*** Evaluate each expression: **a.** $-5 + 4(-3)^2$ and **b.** $-\frac{10}{5} - 5(3) + 6$.

Solution

a. Although the expression contains parentheses, there are no operations to perform within the parentheses. So we proceed with steps 1, 2, and 3 of the rules for the order of operations.

$$-5 + 4(-3)^2 = -5 + 4(9) \quad \text{First, evaluate the power: } (-3)^2 = 9.$$
$$= -5 + 36 \quad \text{Do the multiplication: } 4(9) = 36.$$
$$= 31 \quad \text{Do the addition.}$$

b. Since the expression does not contain any powers, we do the multiplications and divisions, working from left to right.

$$\frac{-10}{5} - 5(3) + 6 = -2 - 5(3) + 6 \quad \text{Do the division: } \tfrac{-10}{5} = -2.$$
$$= -2 - 15 + 6 \quad \text{Do the multiplication: } 5(3) = 15.$$
$$= -17 + 6 \quad \text{Working from left to right, do the subtraction: } -2 - 15 = -17.$$
$$= -11 \quad \text{Do the addition.}$$

Self Check

Evaluate each expression:

a. $-9 + 2(-4)^2$

b. $\frac{20}{-5} - (-6)(-5) + (-12)$

Answers: **a.** 23, **b.** -46 ■

EXAMPLE 10 ***Order of operations within grouping symbols.*** Evaluate each expression: **a.** $3 - (4 - 8)^2$ and **b.** $2 + 3[-2 - 8(4 - 3^2)]$.

Solution

a. We begin by doing the operation within the parentheses.

$$3 - (4 - 8)^2 = 3 - (-4)^2 \quad \text{Do the subtraction: } 4 - 8 = -4.$$
$$= 3 - 16 \quad \text{Evaluate the power: } (-4)^2 = 16.$$
$$= -13 \quad \text{Do the subtraction.}$$

b. First, we do the work within the innermost grouping symbols (the parentheses), using the rules for the order of operations.

$$2 + 3[-2 - 8(4 - 3^2)] = 2 + 3[-2 - 8(4 - 9)] \quad \text{Find the power: } 3^2 = 9.$$
$$= 2 + 3[-2 - 8(-5)] \quad \text{Do the subtraction: } 4 - 9 = -5.$$

Next, we work within the brackets.

Self Check

Evaluate each expression:

a. $(5 - 3)^3 - 40$

b. $-3[-2(5^3 - 3) + 4] - 1$

$= 2 + 3[-2 - (-40)]$ Do the multiplication: $8(-5) = -40$.

$= 2 + 3[-2 + 40]$ Write the subtraction as addition of the opposite.

$= 2 + 3[38]$ Do the addition: $-2 + 40 = 38$.

$= 2 + 114$ Do the multiplication: $3[38] = 114$.

$= 116$ Do the addition.

Answers: **a.** -32, **b.** 719 ■

EXAMPLE 11 ***Working with absolute values.*** Evaluate $|-45 + 30|(2 - 7)$.

Solution

Since the absolute value bars are grouping symbols, we do the operations within the absolute value bars and the parentheses first.

$|-45 + 30|(2 - 7) = |-15|(-5)$ Do the addition within the absolute value symbol and the subtraction within the parentheses.

$= 15(-5)$ Find the absolute value: $|-15| = 15$.

$= -75$ Do the multiplication.

Self Check

Evaluate $2|-25 - (-6)(3)|$.

Answer: 14 ■

Evaluating algebraic expressions

Recall that an algebraic expression is a combination of variables and numbers with the operations of arithmetic. To evaluate an algebraic expression, we substitute specific numbers for the variables and then apply the rules for the order of operations. When doing this, it is a good idea to write parentheses around a number when it is substituted into the expression in place of a variable.

EXAMPLE 12 ***Evaluating algebraic expressions.*** If $a = -2$, $b = 9$, and $c = -1$, evaluate

a. $-\frac{1}{2}a^2$ and **b.** $\frac{-a\sqrt{b} + 3c^3}{c(c - b)}$.

Solution

a. We substitute -2 for a and use the rules for the order of operations.

$-\frac{1}{2}a^2 = -\frac{1}{2}(-2)^2$ Substitute -2 for a. Write parentheses around -2 so that it is squared.

$= -\frac{1}{2}(4)$ Evaluate the power: $(-2)^2 = 4$.

$= -2$ Do the multiplication.

b. $\frac{-a\sqrt{b} + 3c^3}{c(c - b)} = \frac{-(-2)\sqrt{9} + 3(-1)^3}{-1(-1 - 9)}$ Substitute -2 for a, 9 for b, and -1 for c.

$= \frac{-(-2)(3) + 3(-1)}{-1(-10)}$ In the numerator, evaluate the square root and the power: $\sqrt{9} = 3$ and $(-1)^3 = -1$. In the denominator, do the subtraction.

$= \frac{2(3) + 3(-1)}{-1(-10)}$ In the numerator, simplify: $-(-2) = 2$.

$= \frac{6 + (-3)}{10}$ In the numerator, do the multiplications. In the denominator, do the multiplication.

$= \frac{3}{10}$ In the numerator, do the addition.

Self Check

If $r = 2$, $s = -5$, and $t = 3$, evaluate:

a. $-\frac{1}{3}rst$ and

b. $\frac{\sqrt{-5s}}{(s + t)r^2}$

Answers: **a.** 10, **b.** $-\frac{5}{8}$ ■

EXAMPLE 13 ***Surface area.*** Find the amount of material needed to make the lampshade shown in Figure 1-13 by finding the surface area of the shade using the formula

$$S = \pi(R_1 + R_2)\sqrt{(R_1 - R_2)^2 + h^2}$$

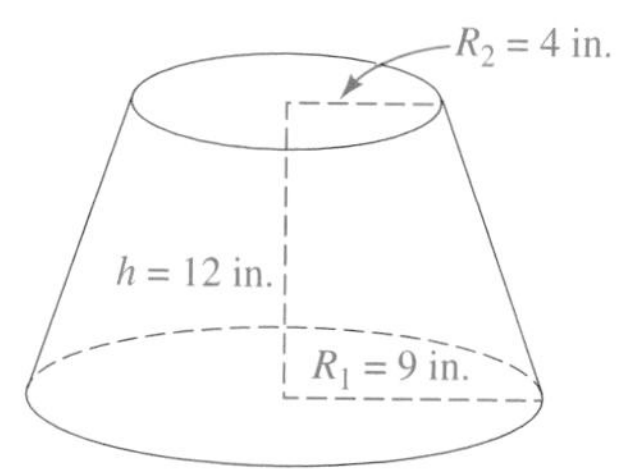

FIGURE 1-13

Solution To find the surface area of the shade, we evaluate the algebraic expression on the right-hand side of the formula using the values given in Figure 1-13.

In the formula, the 1 and 2 in the notation R_1 (read as "R sub-one") and R_2 (read as "R sub-two") are called **subscripts.** They distinguish between the lower radius and the upper radius of the shade.

$$S = \pi(R_1 + R_2)\sqrt{(R_1 - R_2)^2 + h^2}$$

$S = \pi(9 + 4)\sqrt{(9 - 4)^2 + (12)^2}$	Substitute 9 for R_1, 4 for R_2, and 12 for h.
$= \pi(13)\sqrt{(5)^2 + (12)^2}$	Do the operations inside the parentheses.
$= \pi(13)\sqrt{25 + 144}$	Evaluate the powers within the radical.
$= \pi(13)\sqrt{169}$	Do the addition within the radical.
$= \pi(13)(13)$	Evaluate the square root: $\sqrt{169} = 13$.
$= 169\pi$	Do the multiplication.
≈ 530.9291585	Use a calculator to approximate 169π.

The surface area of the shade is 169π square inches. Rounded to the nearest tenth, 530.9 square inches of material are needed to make the shade. ■

Recall that the area of a two-dimensional geometric figure is a measure of the surface it encloses. Table 1-4 shows the formulas for the areas of several geometric figures.

Figure	Name	Area	Figure	Name	Area
	Square	$A = s^2$		Triangle	$A = \frac{1}{2}bh$
	Rectangle	$A = lw$		Trapezoid	$A = \frac{1}{2}h(b_1 + b_2)$
	Circle	$A = \pi r^2$			

TABLE 1-4

The volume of a three-dimensional geometric figure is a measure of its capacity. Table 1-5 shows the formulas for the volumes of several geometric figures.

Figure	Name	Volume	Figure	Name	Volume
	Cube	$V = s^3$		Cylinder	$V = Bh$*
	Rectangular solid	$V = lwh$		Cone	$V = \frac{1}{3}Bh$*
	Sphere	$V = \frac{4}{3}\pi r^3$		Pyramid	$V = \frac{1}{3}Bh$*

*B represents the area of the base.

TABLE 1-5

EXAMPLE 14 ***Finding area and volume.*** Find the amount of skin covered by a rectangular band-aid $\frac{5}{8}$ inches wide and 3 inches long.

Solution

To find the amount of skin covered by the band-aid, we need to find the area of the band-aid.

$A = lw$ The formula for the area of a rectangle.

$A = 3 \cdot \frac{5}{8}$ Substitute 3 for l and $\frac{5}{8}$ for w.

$A = \frac{15}{8}$

The band-aid covers $\frac{15}{8}$ or $1\frac{7}{8}$ in.^{2} (square inches) of skin.

Self Check

Find the volume of a medicine cabinet that is 23 inches tall, $13\frac{1}{2}$ inches wide, and 4 inches deep.

Answer: 1,242 in.^{3}

STUDY SET Section 1.3

VOCABULARY *In Exercises 1–10, fill in the blanks to make the statements true.*

1. When adding two numbers, the result is called the ______________. When subtracting two numbers, the result is called the ______________.

2. When multiplying two numbers, the result is called the ______________. When dividing two numbers, the result is called the ______________.

3. To ______________ an algebraic expression, we substitute values for the variables and then apply the rules for the order of operations.

4. In the expression $9 + 6[22 - (6 - 1)]$, the ______________ are the innermost grouping symbols, and the brackets are the ______________ grouping symbols.

5. 6^2 can be read as "six ______________," and 6^3 can be read as "six ______________."

6. 4^5 is the fifth ______________ of four.

7. In the exponential expression x^2, x is the ______________, and 2 is the ______________.

8. An ______________ is used to represent repeated multiplication.

9. Subtraction is the same as adding the ______________ of the number being subtracted.

10. To find the ______________ in a quantity, we subtract the earlier value from the later value.

CONCEPTS

11. Consider the expression $6 + 3 \cdot 2$.
a. In what two different ways might we evaluate the given expression?
b. Which result from part a is correct and why?

12. **a.** What operations does the expression $60 - (-9)^2 + 5(-1)$ contain?
b. In what order should they be performed?

13. What are we finding when we calculate
a. the amount of surface a circle encloses?
b. the capacity of a cylinder?

14. **a.** What is the related multiplication statement for the division statement $\frac{0}{6} = 0$?
b. Why isn't there a related multiplication statement for $\frac{6}{0}$?

NOTATION *In Exercises 15–16, complete each solution.*

15. $60 - 20 \cdot 2^3 = 60 - 20(\square)$ Find the power.
$= 60 - \square$ Multiply.
$= -100$ Subtract.

16. $9 + 8[(1 + 4) \cdot 5] = 9 + 8[(\square) \cdot 5]$ Add in the parentheses.
$= 9 + 8(\square)$ Multiply in the brackets.
$= 9 + \square$ Multiply.
$= 209$ Add.

17. What is the name of this symbol $\sqrt{}$?

18. What is the one number that a fraction cannot have as its denominator?

PRACTICE *In Exercises 19–20, complete the table. Then draw a bar graph on the axes provided.*

19.

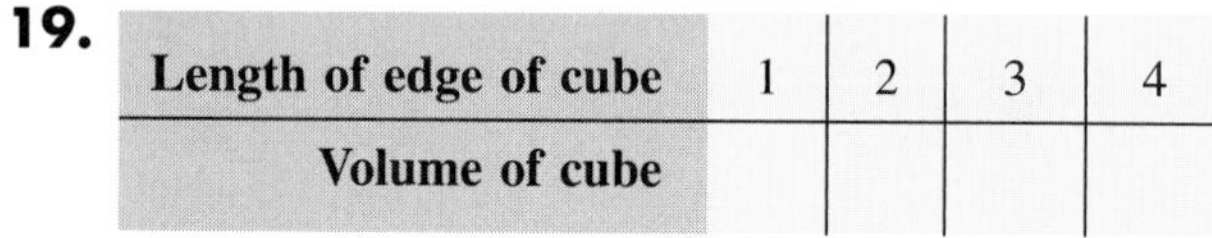

Length of edge of cube	1	2	3	4
Volume of cube				

20.

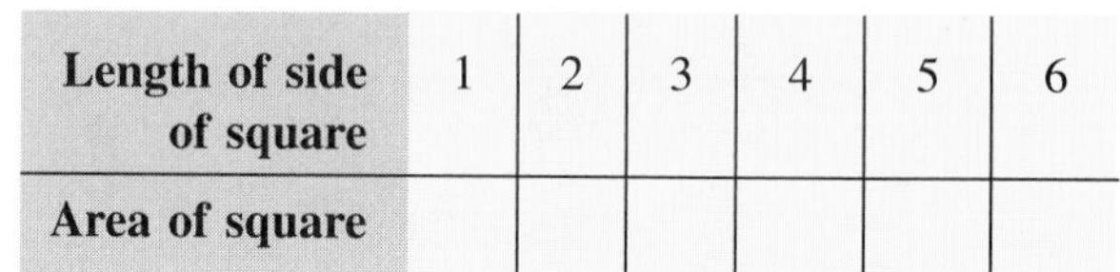

Length of side of square	1	2	3	4	5	6
Area of square						

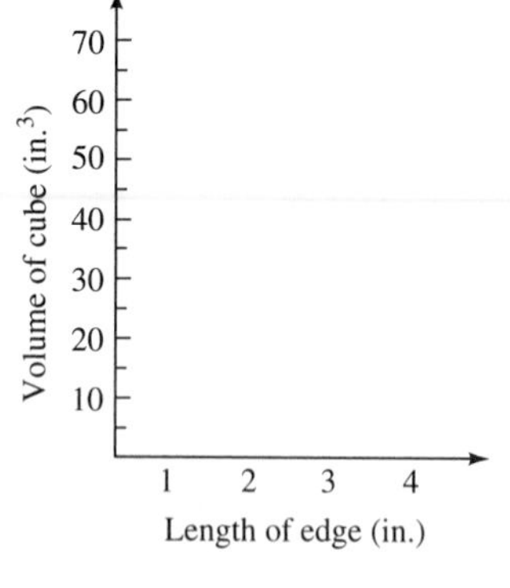

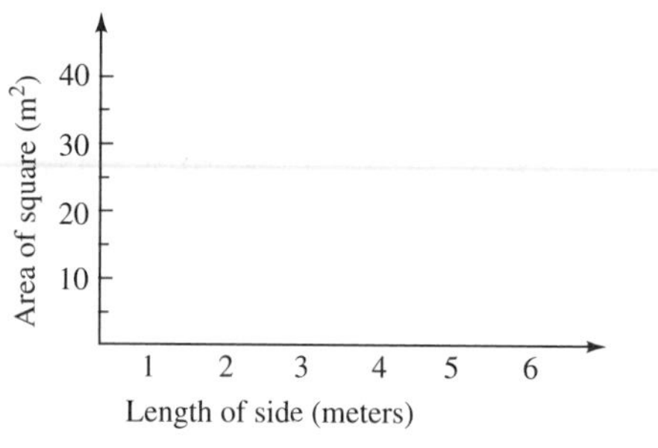

In Exercises 21–40, do the operations.

21. $-3 + (-5)$

22. $-2 + (-8)$

23. $-7.1 + 2.8$

24. $3.1 + (-5.2)$

25. $-3 - 4$

26. $-11 - (-17)$

27. $-3.3 - (-3.3)$

28. $0.14 - (-0.13)$

29. $-2(6)$

30. $-3(-7)$

31. $\dfrac{-8}{4}$

32. $\dfrac{-16}{-4}$

33. $\frac{1}{2} + \left(-\frac{1}{3}\right)$

34. $-\frac{3}{4} + \left(-\frac{1}{5}\right)$

35. $\frac{1}{2} - \left(-\frac{3}{5}\right)$

36. $\frac{1}{26} - \frac{11}{13}$

37. $\left(-\frac{3}{5}\right)\left(\frac{10}{7}\right)$

38. $\left(-\frac{6}{7}\right)\left(-\frac{5}{12}\right)$

39. $-\frac{16}{5} \div \left(-\frac{10}{3}\right)$

40. $-\frac{5}{24} \div \frac{10}{3}$

In Exercises 41–52, evaluate each expression.

41. 12^2

42. 9^2

43. -5^2

44. $(-5)^2$

45. $4 \cdot 2^3$

46. $(4 \cdot 2)^3$

47. $(1.3)^2$

48. $\left(\frac{3}{5}\right)^2$

49. $\sqrt{64}$

50. $\sqrt{121}$

51. $-\sqrt{\frac{9}{16}}$

52. $-\sqrt{0.16}$

In Exercises 53–64, evaluate each expression.

53. $3 - 5 \cdot 4$

54. $12 - 2 \cdot 3$

55. $\left(-3 - \sqrt{25}\right)^2$

56. $4^2 - (-2)^2$

57. $2 + 3\left(\frac{25}{5}\right) + (-4)$

58. $(-2)^3\left(\frac{-6}{-2}\right)(-1)$

59. $\dfrac{-\sqrt{49} - 3^2}{2 \cdot 4}$

60. $\frac{1}{2}\left(\frac{1}{8}\right) + \left(-\frac{1}{4}\right)^2$

61. $-2|4 - 8|$

62. $|\sqrt{49} - 8(4 - 7)|$

63. $(4 + 2 \cdot 3)^4$

64. $|9 - 5(1 - 8)|$

In Exercises 65–76, evaluate each expression.

65. $3 + 2[-1 - 4(5)]$

66. $-3[5^2 - (7 - 3)^2]$

67. $3 - [3^3 + (3 - 1)^3]$

68. $3|-(3 \cdot 5 - 2 \cdot 6)^2|$

69. $\dfrac{|-25| - 2(-5)}{2^4 - 9}$

70. $\dfrac{2[-4 - 2(3 - 1)]}{3(3)(2)}$

71. $\dfrac{3[-9 + 2(7 - 3)]}{(8 - 5)(9 - 7)}$

72. $\dfrac{5 \cdot 4 \cdot 3 \cdot 2 \cdot 1}{4(3 \cdot 2 \cdot 1)}$

73. $\dfrac{(6 - 5)^4 + 21}{27 - \left(\sqrt{16}\right)^2}$

74. $\dfrac{3(3{,}246 - 1{,}111)}{561 - 546}$

75. $54^3 - 16^4 + 19(3)$

76. $\dfrac{36^2 - 2(48)}{(25)^2 - \sqrt{105{,}625}}$

In Exercises 77–84, evaluate each expression for the given values of the variables.

77. $\dfrac{y_2 - y_1}{x_2 - x_1}$ for $x_1 = -3$, $x_2 = 5$, $y_1 = 12$, $y_2 = -4$

78. $P_0\left(1 + \frac{r}{k}\right)^{kt}$ for $P_0 = 500$, $r = 4$, $k = 2$, $t = 3$

79. $(x + y)(x^2 - xy + y^2)$ for $x = -4$, $y = 5$

80. $\dfrac{-b + \sqrt{b^2 - 4ac}}{2a}$ for $a = 1$, $b = 2$, $c = -3$

81. $\dfrac{x^2}{a^2} + \dfrac{y^2}{b^2}$ for $x = -3$, $y = -4$, $a = 5$, $b = -5$

82. $\frac{n}{2}[2a_1 + (n - 1)d]$ for $n = 50$, $a_1 = -4$, $d = 5$

83. $\sqrt{(x_2 - x_1)^2 + (y_2 - y_1)^2}$ for $x_1 = -2$, $x_2 = 4$, $y_1 = 4$, $y_2 = -4$

84. $\dfrac{|Ax_0 + By_0 + C|}{\sqrt{A^2 + B^2}}$ for $A = 3$, $B = 4$, $C = -5$, $x_0 = 2$, and $y_0 = -1$

APPLICATIONS

85. ALUMINUM FOIL Find the number of *square feet* of aluminum foil on a roll if the dimensions printed on the box are $8\frac{1}{3}$ yards × 12 inches.

86. HOCKEY A goal is scored in hockey when the puck, a vulcanized rubber disk 2.5 cm (1 in.) thick and 7.6 cm (3 in.) in diameter, is driven into the opponent's goal. Find the volume of a puck in cubic centimeters and cubic inches. Round to the nearest tenth.

87. PAPER PRODUCTS When folded, the paper sheet shown in Illustration 1 forms a rectangular-shaped envelope. The formula

$$A = \tfrac{1}{2}h_1(b_1 + b_2) + b_3h_3 + \tfrac{1}{2}b_1h_2 + b_1b_3$$

gives the amount of paper (in square units) used in the design. Explain what each of the four terms in the formula finds. Then evaluate the formula for $b_1 = 6$, $b_2 = 2$, $b_3 = 3$, $h_1 = 2$, $h_2 = 2.5$, and $h_3 = 3$. All dimensions are in inches.

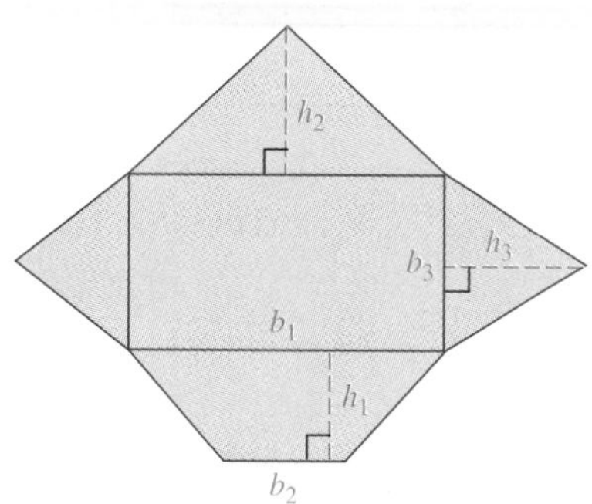

ILLUSTRATION 1

88. BOND MARKET The graph in Illustration 2 shows the quarterly investment in U.S. bond funds from 1995 through 1997.

a. In the graph, what does a negative net cash inflow indicate?

b. Find the net cash inflow for each of the years 1995, 1996, and 1997.

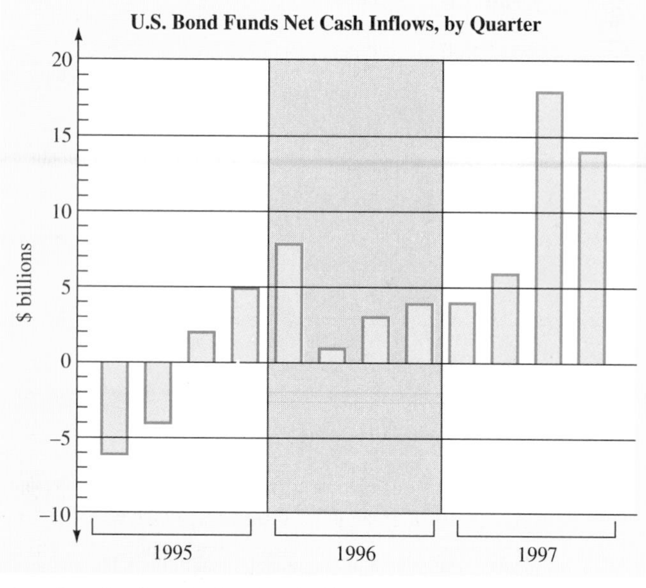

Based on data from the *Los Angeles Times* (Jan. 6, 1998)

ILLUSTRATION 2

c. What was the difference in the net cash inflow for the first quarter of 1995 compared to the first quarter of 1996 and the first quarter of 1996 compared to the first quarter of 1997?

89. ACCOUNTING On a financial balance sheet, debts (negative numbers) are denoted within parentheses. Assets (positive numbers) are written without parentheses. What is the 1998 fund balance for the preschool whose 1998 financial records are shown in Illustration 3?

Community Care Preschool Balance Sheet, June, 1998	
Fund balances	
Classroom supplies	$ 5,889
Emergency needs	927
Holiday program	(2,928)
Insurance	1,645
Janitorial	(894)
Licensing	715
Maintenance	(6,321)
BALANCE	?

ILLUSTRATION 3

90. TEMPERATURE EXTREMES The highest and lowest temperatures ever recorded in several cities are shown in Illustration 4. List the cities in order, from the smallest to the largest range in temperature extremes.

City	Extreme temperatures	
	Highest	Lowest
Atlanta, Georgia	105	−8
Boise, Idaho	111	−25
Helena, Montana	105	−42
New York, New York	107	−3
Omaha, Nebraska	114	−23

ILLUSTRATION 4

91. ICE CREAM See Illustration 5. If the two equal-sized scoops of ice cream melt completely into the cone, will they overflow the cone?

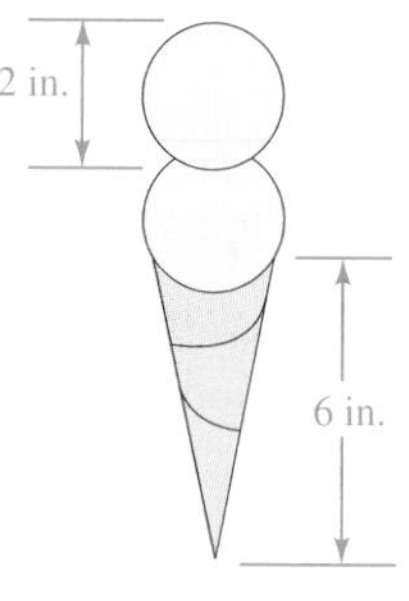

ILLUSTRATION 5

92. PHYSICS Waves are motions that carry energy from one place to another. Illustration 6 shows an example of a wave called a *standing wave.* What is the difference in the height of the crest of the wave and the depth of the trough of the wave?

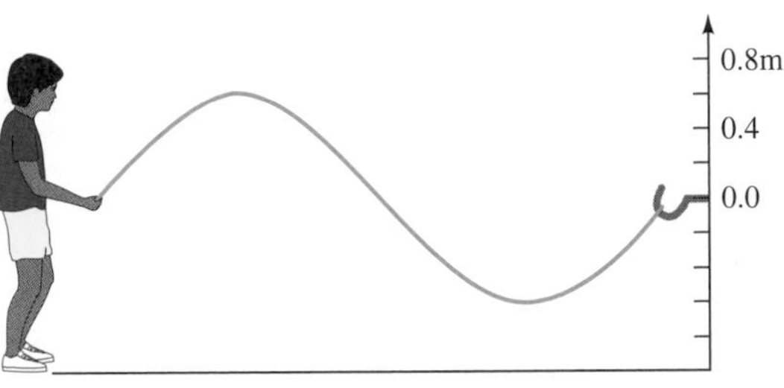

ILLUSTRATION 6

93. PEDIATRICS Young's rule, shown below, is used by some doctors to calculate dosage for infants and children.

$$\frac{\text{Age of child}}{\text{Age of child} + 12}\left(\begin{matrix}\text{average}\\ \text{adult dose}\end{matrix}\right) = \text{child's dose}$$

The syringe in Illustration 7 shows the adult dose of a certain medication. Use Young's rule to determine the dosage for a 6-year-old child. Then use an arrow to locate the dosage on the calibration.

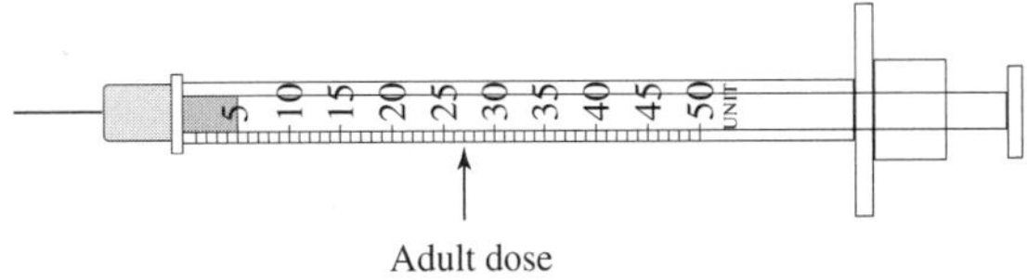

ILLUSTRATION 7

94. DOSAGES The adult dosage of procaine penicillin is 300,000 units daily. Calculate the dosage for a 12-year-old child using Young's rule. (See Exercise 93.)

WRITING *Write a paragraph using your own words.*

95. Explain what the statement $x - y = x + (-y)$ means.

96. Explain why rules for the order of operations are necessary.

REVIEW

97. What two numbers are a distance of 5 away from -2 on the number line?

98. Place the proper symbol ($>$ or $<$) in the blank: -4.6 ☐ -4.5.

99. List the set of integers.

100. Translate into mathematical symbols: ten less than twice x.

101. True or false: The real numbers is the set of all decimals.

102. True or false: Irrational numbers are nonterminating, nonrepeating decimals.

1.4 Simplifying Algebraic Expressions

In this section, you will learn about

- **Properties of real numbers**
- **Properties of 0 and 1**
- **More properties of real numbers**
- **Simplifying algebraic expressions**
- **The distributive property**
- **Combining like terms**

Introduction. Suppose we are asked to find the total dollar amount of the checks that have been recorded in the check register shown in Figure 1-14. As the notes in the figure point out, we can simplify the computation by adding the numbers in a different order than the way they are entered. The *commutative* and *associative* properties of addition guarantee that we will obtain the same result whether we add them in their original order or in this more convenient way. In this section, we will introduce some important properties of real numbers. Then, we will see how they can be used to simplify expressions containing variables.

Number	Date	Description of Transaction	Payment/Debt	
101	3/6	DR. OKAMOTO, DDS	$64	00
102	3/6	UNION OIL CO.	$25	00
103	3/8	STATER BROS.	$16	00
104	3/9	LITTLE LEAGUE	$75	00

Add $64.00 and $16.00 to get $80.00.

Add $25.00 and $75.00 to get $100.00.

Now add the two subtotals to get the total dollar amount of the checks: $80.00 + $100.00 = $180.00.

FIGURE 1-14

Properties of real numbers

When we work with real numbers, we will use the following properties.

Properties of real numbers

If a, b, and c are real numbers, then we have

The associative properties for addition and multiplication

$$(a + b) + c = a + (b + c) \qquad (ab)c = a(bc)$$

The commutative properties for addition and multiplication

$$a + b = b + a \qquad ab = ba$$

The distributive property of multiplication over addition

$$a(b + c) = ab + ac$$

The *associative properties* enable us to group the numbers in a sum or a product any way that we wish and get the same result.

EXAMPLE 1 ***Using the associative property.*** Evaluate $(14 + 94) + 6$ in two ways.

Solution

$$(14 + 94) + 6 = 108 + 6 \quad \text{Work within the parentheses first.}$$
$$= 114$$

To evaluate the expression in a second way, we apply the associative property of addition.

$$(14 + 94) + 6 = 14 + (94 + 6) \quad \text{Use parentheses to group 94 with 6.}$$
$$= 14 + 100 \quad \text{Do the addition within the parentheses.}$$
$$= 114$$

Notice that the results are the same.

Self Check

Evaluate $2 \cdot (50 \cdot 37)$ in two ways.

Answer: 3,700 ■

WARNING! Subtraction and division are not associative, because different groupings give different results. For example,

$$(8 - 4) - 2 = 4 - 2 = 2 \quad \text{but} \quad 8 - (4 - 2) = 8 - 2 = 6$$
$$(8 \div 4) \div 2 = 2 \div 2 = 1 \quad \text{but} \quad 8 \div (4 \div 2) = 8 \div 2 = 4$$

The *commutative properties* enable us to add or multiply two numbers in either order and obtain the same result. Here are two examples.

$$3 + (-5) = -2 \quad \text{and} \quad -5 + 3 = -2$$
$$-2.6(-8) = 20.8 \quad \text{and} \quad -8(-2.6) = 20.8$$

WARNING! Subtraction and division are not commutative, because doing these operations in different orders will give different results. For example,

$$8 - 4 = 4 \quad \text{but} \quad 4 - 8 = -4$$
$$8 \div 4 = 2 \quad \text{but} \quad 4 \div 8 = \frac{1}{2}$$

The *distributive property* enables us to evaluate certain types of expressions involving a multiplication and an addition. As an example, let's consider the expression $4(5 + 3)$, which can be evaluated in two ways.

Method 1: Rules for the Order of Operations

In this method, we compute the sum inside the parentheses first.

$4(5 + 3) = 4(8)$	Do the addition inside the parentheses first.
$= 32$	Do the multiplication.

Method 2: The Distributive Property

In this method, we distribute the 4 across the 5 and the 3, find each product separately, and add the results.

	First product		Second product	
$4(5 + 3) =$	$4 \cdot 5$	$+$	$4 \cdot 3$	To apply the distributive property, we multiply each term inside the parentheses by the factor outside the parentheses.
$=$	20	$+$	12	Do the multiplications first: $4 \cdot 5 = 20$ and $4 \cdot 3 = 12$.
$=$	32			Do the addition.

Notice that each method gives a result of 32.

Properties of 0 and 1

The real numbers 0 and 1 have important special properties.

Properties of 0 and 1

Additive identity: The sum of 0 and any number is the number itself.

$$0 + a = a + 0 = a$$

Multiplicative identity: The product of 1 and any number is the number itself.

$$1 \cdot a = a \cdot 1 = a$$

Multiplication property of 0: The product of any number and 0 is 0.

$$a \cdot 0 = 0 \cdot a = 0$$

For example,

$$7 + 0 = 7, \qquad 1(5.4) = 5.4, \qquad \left(-\frac{7}{3}\right)1 = -\frac{7}{3}, \qquad \text{and} \qquad -19(0) = 0$$

More properties of real numbers

If the sum of two numbers is 0, they are called **additive inverses, negatives,** or **opposites** of each other. For example, 6 and -6 are negatives, because $6 + (-6) = 0$.

The additive inverse property

For every real number a, there is a real number $-a$ such that

$$a + (-a) = -a + a = 0$$

If the product of two numbers is 1, the numbers are called **multiplicative inverses** or **reciprocals** of each other.

The multiplicative inverse property

For every nonzero real number a, there exists a real number $\frac{1}{a}$ such that

$$a \cdot \frac{1}{a} = \frac{1}{a} \cdot a = 1$$

Some example of reciprocals are

- 5 and $\frac{1}{5}$ are reciprocals, because $5\left(\frac{1}{5}\right) = 1$.
- $\frac{3}{2}$ and $\frac{2}{3}$ are reciprocals, because $\frac{3}{2}\left(\frac{2}{3}\right) = 1$.
- -0.25 and -4 are reciprocals, because $-0.25(-4) = 1$.

The reciprocal of 0 does not exist, because $\frac{1}{0}$ is undefined.

Recall that when a number is divided by 1, the result is the number itself, and when a number is divided by itself, the result is 1.

Division properties

Division by 1: If x represents any real number, then $\frac{x}{1} = x$.

Division of a number by itself: If x represents any real number, then $\frac{x}{x} = 1$ $(x \neq 0)$.

Examples of two types of division involving zero that we will encounter in this course are

$$\frac{0}{5} \quad \text{and} \quad \frac{5}{0}$$

In the first case, we have division *of* zero. In the second case, we have division *by* zero.

Division with 0

Division of 0: For any real number x, if $x \neq 0$, then $\frac{0}{x} = 0$.

Division by 0: For any real number x, $\frac{x}{0}$ is undefined.

Simplifying algebraic expressions

To **simplify algebraic expressions,** we use properties of the real numbers to write the expressions in a less complicated form. As an example, let's consider the expression $6(5x)$ and simplify it as follows:

$$6(5x) = 6 \cdot (5 \cdot x)$$ $6(5x) = 6 \cdot 5x$ and $5x = 5 \cdot x$.

$$= (6 \cdot 5) \cdot x$$ Apply the associative property of multiplication to group the 5 with the 6.

$$= 30x$$ Do the multiplication inside the parentheses: $6 \cdot 5 = 30$.

Since $6(5x) = 30x$, we say that $6(5x)$ simplifies to $30x$.

EXAMPLE 2 ***Simplifying algebraic expressions.*** Simplify each expression: **a.** $9(10t)$, **b.** $-21a(-4)$, and **c.** $-5.3r(-2s)$.

Solution

a. $9(10t) = (9 \cdot 10)t$ Use the associative property of multiplication to regroup the factors.

$= 90t$ Do the multiplication inside the parentheses: $9 \cdot 10 = 90$.

b. $-21a(-4) = -21(-4)a$ Use the commutative property of multiplication to change the order of the factors a and -4.

$= [-21(-4)]a$ Use the associative property of multiplication to group the numbers together.

$= 84a$ Do the multiplication within the brackets: $-21(-4) = 84$.

c. $-5.3r(-2s) = [-5.3(-2)](r \cdot s)$ Use the commutative and associative properties to group the numbers and group the variables.

$= 10.6rs$ Do the multiplications: $-5.3(-2) = 10.6$ and $r \cdot s = rs$.

Self Check

Simplify each expression: **a.** $14 \cdot 3s$, **b.** $-1.6b(3)$, and **c.** $-\frac{2}{3}x(-9y)$.

Answers: **a.** $42s$, **b.** $-4.8b$, **c.** $6xy$ ■

The distributive property

We can use the distributive property to *remove the parentheses* when an algebraic expression is multiplied by a quantity. For example, to remove the parentheses in the expression $5(x + 2)$, we proceed as follows:

$$5(x + 2) = 5 \cdot x + 5 \cdot 2$$ Distribute the 5. The arrows help us visualize the distribution.

$$= 5x + 10$$ Do the multiplications.

Since subtraction is the same as adding the opposite, the distributive property also holds for subtraction.

$$5(x - 2) = 5 \cdot x - 5 \cdot 2$$ Distribute the 5.

$$= 5x - 10$$ Do the multiplications.

EXAMPLE 3 ***Applying the distributive property.*** Use the distributive property to remove parentheses: **a.** $6(a + 9)$ and **b.** $-15(4b - 1)$.

Solution

a. $6(a + 9) = 6 \cdot a + 6 \cdot 9$ Distribute the 6.

$= 6a + 54$ Do the multiplications.

b. $-15(4b - 1) = -15(4b) - (-15)(1)$ Distribute the -15.

$= -60b - (-15)$ Do the multiplications.

$= -60b + 15$ Add the opposite of -15, which is 15.

Self Check

Remove parentheses: **a.** $9(r + 4)$ and **b.** $-11(-3x - 5)$.

Answers: **a.** $9r + 36$, **b.** $33x + 55$ ■

WARNING! If an expression contains parentheses, it does not necessarily mean that the distributive property can be applied. For example, the distributive property does not apply to

$5(12x)$ or $5(-4 \cdot y)$ Here, a product is multiplied by 5.

However, the distributive property does apply to

$5(12 + x)$ and $5(-4 - y)$ Here, a sum and a difference are multiplied by 5.

A more general form of the distributive property is the **extended distributive property.**

$$a(b + c + d + e + \cdots) = ab + ac + ad + ae + \cdots$$

EXAMPLE 4 ***Applying the extended distributive property.*** Remove parentheses: $-0.5(7 - 5y + 6z)$.

Solution

$-0.5(7 - 5y + 6z)$

$= -0.5(7) - (-0.5)(5y) + (-0.5)(6z)$ Distribute the -0.5.

$= -3.5 - (-2.5y) + (-3z)$ Do the multiplications.

$= -3.5 + 2.5y - 3z$

Self Check Remove parentheses: $\frac{1}{3}(-6t + 3s - 9)$.

Answer: $-2t + s - 3$ ■

Since multiplication is commutative, we can write the distributive property in the following forms.

$$(b + c)a = ba + ca, \quad (b - c)a = ba - ca, \quad (b + c + d)a = ba + ca + da$$

EXAMPLE 5 ***Applying the distributive property.*** Remove parentheses: $(-8 - 3y)(-30)$.

Solution

$(-8 - 3y)(-30) = -8(-30) - 3y(-30)$ Distribute the -30.

$= 240 - (-90y)$ Do the multiplications.

$= 240 + 90y$

Self Check Remove parentheses: $(-5s + 4t)(-10)$.

Answer: $50s - 40t$ ■

To use the distributive property to simplify $-(x + 3)$, we note that the negative sign in front of the parentheses represents the number -1.

The $-$ sign represents -1.

$-(x + 3) = -1(x + 3)$

$= -1(x) + (-1)(3)$ Distribute the -1.

$= -x + (-3)$ Do the multiplications.

$= -x - 3$

EXAMPLE 6 ***Distributing a negative sign.*** Simplify $-(-21 - 20m)$.

Solution

$-(-21 - 20m) = -1(-21 - 20m)$ Write the $-$ sign in front of the parentheses as -1.

$= -1(-21) - (-1)(20m)$ Distribute the -1.

$= 21 - (-20m)$ Do the multiplications.

$= 21 + 20m$

Self Check Simplify $-(-27k + 15)$.

Answer: $27k - 15$ ■

Combining like terms

An **algebraic term** is either a number or the product of numbers (called **constants**) and variables. Some examples of terms are $3x$, $-7y$, y^2, and 8. The **numerical coefficients** of these terms are 3, -7, 1, and 8, respectively.

In algebraic expressions, terms are separated by + and − signs. For example, the expression $3x^2 + 2x - 4$ has three terms, and the expression $3x + 7y$ has two terms.

Terms with the same variables and the same exponents are called **like terms** or **similar terms:**

$5x$ and $6x$ are like terms	$27x^2y^3$ and $-326x^2y^3$ are like terms
$4x$ and $-17y$ are unlike terms (because they have different variables)	$15x^2y$ and $6xy^2$ are unlike terms (because the variables have different exponents)

If we are to add (or subtract) objects, they must have the same units. For example, we can add dollars to dollars and inches to inches, but we cannot add dollars to inches. The same is true when working with terms of an algebraic expression. They can be added or subtracted only when they are like terms.

This expression can be simplified, because it contains like terms.	This expression cannot be simplified, because its terms are not like terms.
$5x + 6x$ (↑ ↑ Like terms.)	$5x + 6y$ (↑ ↑ Unlike terms.)
The variable parts are identical.	The variable parts are not identical.

To simplify expressions containing like terms, we use the distributive property. For example,

$$5x + 6x = (5 + 6)x = 11x \quad \text{and} \quad 32y - 16y = (32 - 16)y = 16y$$

These examples suggest the following rule.

Combining like terms

To combine like terms, add (or subtract) their coefficients and keep the same variables with the same exponents.

EXAMPLE 7 ***Combining like terms.*** Simplify each expression: **a.** $-8f + (-12f)$ and **b.** $0.56s - 0.2s$.

Solution

a. $-8f + (-12f) = -20f$ Add the coefficients of the like terms: $-8 + (-12) = -20$. Keep the variable f.

b. $0.56s^3 - 0.2s^3 = 0.36s^3$ Subtract the coefficients of the like terms: $0.56 - 0.2 = 0.36$. Keep s^3.

Self Check

Simplify by combining like terms:
a. $5k + 8k$ and
b. $-600a^2 - (-800a^2)$.

Answers: **a.** $13k$, **b.** $200a^2$ ■

EXAMPLE 8 ***Combining like terms.*** Simplify $9b - B - 14b + 34B$.

Solution

This expression has four terms. The uppercase B and the lowercase b are different variables. So the first and third terms are like terms, and the second and fourth terms are like terms.

$9b - B - 14b + 34B = -5b + 33B$ Combine like terms: $9b - 14b = -5b$ and $-B + 34B = 33B$.

Self Check

Simplify $8R + 7r - 14R - 21r$.

Answer: $-6R - 14r$ ■

EXAMPLE 9 ***Simplifying algebraic expressions.*** Simplify $9(x + 1) - 3(7x - 1)$.

Solution

We apply the distributive property and then combine like terms.

$$9(x + 1) - 3(7x - 1) = 9x + 9 - 21x + 3$$ Use the distributive property twice.

$$= -12x + 12$$ Combine like terms: $9x - 21x = -12x$ and $9 + 3 = 12$.

Self Check

Simplify $-5(y - 4) + 2(4y + 8)$.

Answer: $3y + 36$ ■

STUDY SET Section 1.4

VOCABULARY *In Exercises 1–8, fill in the blanks to make the statements true.*

1. ______________ terms are terms with the same variables and the same exponents.

2. To combine like terms, add (or subtract) their ______________ and keep the same variables and exponents.

3. A number or the product of numbers and variables is called an algebraic ______________.

4. $\frac{1}{3}$ and 3 are ______________, because $\frac{1}{3} \cdot 3 = 1$.

5. To ______________ algebraic expressions, we use properties of real numbers to write the expressions in a less complicated form.

6. The distributive property is used to ______________ parentheses when a sum or difference is multiplied by a quantity.

7. Division by 0 is ______________.

8. The numerical ______________ of the term $-8c$ is -8.

CONCEPTS

9. Using the variables x, y, and z, write the associative property of addition.

10. Using the variables x and y, write the commutative property of multiplication.

11. Using the variables r, s, and t, write the distributive property of multiplication over addition.

12. **a.** What is the additive identity?
b. What is the multiplicative identity?
c. Simplify $-(-10)$.

13. What number should be
a. subtracted from 5 to obtain 0?
b. added to 5 to obtain 0?

14. By what number should
a. 5 be divided to obtain 1?
b. 5 be multiplied to obtain 1?

15. Give the reciprocal.
a. $\frac{15}{16}$ **b.** -20
c. 0.5 **d.** x

16. Does the distributive property apply?
a. $2|x - 3|$ **b.** $2(3 \cdot 5)$
c. $2(3x)$ **d.** $2(x - 3)$

17. Consider the expression $2x^2 - x + 6$.
a. What are the terms of the expression?
b. Give the coefficient of each term.

18. Which properties of real numbers involve changing *order* and which involve changing *grouping?*

In Exercises 19–26, tell whether the terms are like terms. If they are, combine them.

19. $2x, 6x$ **20.** $-3x, 5y$ **21.** $-5xy, -7yz$ **22.** $-3t^2, 12t^2$
23. $3x^2, -5x^2$ **24.** $5y^2, 7xy$ **25.** $xy, 3xt$ **26.** $-4x, -5x$

NOTATION

27. In $-(x - 7)$, what does the negative sign in front of the parentheses represent?

28. Do the division, if possible.

a. $\frac{0}{8}$ **b.** $\frac{8}{0}$

PRACTICE *In Exercises 29–40, fill in the blanks to make the statements true by applying the given property of the real numbers.*

29. $3 + 7 =$ ____________
Commutative property of addition

30. $2(5 \cdot 97) =$ ____________
Associative property of multiplication

31. $3(2 + d) =$ ____________
Distributive property

32. $1 \cdot y =$ ____________
Commutative property of multiplication

33. $c + 0 =$ ___
Additive identity property

34. $-4(x - 2) =$ ____________
Distributive property and simplifying

35. $25 \cdot \frac{1}{25} =$ ___
Multiplicative inverse property

36. $z + (9 - 27) =$ ____________
Commutative property of addition

37. $8 + (7 + a) =$ ____________
Associative property of addition

38. ___ $\cdot 3 = 3$
Multiplicative identity property

39. $(x + y)2 =$ ____________
Commutative property of multiplication

40. $h + (-h) =$ ___
Additive inverse property

In Exercises 41–44, evaluate each side of the equation separately to show that the same result is obtained. Identify the property of real numbers that is being illustrated.

41. $(37.9 + 25.2) + 14.3 = 37.9 + (25.2 + 14.3)$

42. $7.1(3.9 + 8.8) = 7.1 \cdot 3.9 + 7.1 \cdot 8.8$

43. $2.73(4.534 + 57.12) = 2.73 \cdot 4.534 + 2.73 \cdot 57.12$

44. $(6.789 + 345.1) + 27.347 = (345.1 + 6.789) + 27.347$

In Exercises 45–54, remove the parentheses.

45. $-4(t - 3)$

46. $-4(-t + 3)$

47. $-(t - 3)$

48. $-(-t + 3)$

49. $\frac{2}{3}(3s - 9)$

50. $\frac{1}{5}(5s - 15)$

51. $0.7(s + 2)$

52. $2.5(6s - 8)$

53. $3\left(\frac{4}{3}x - \frac{5}{3}y + \frac{1}{3}\right)$

54. $6\left(-\frac{4}{3} + \frac{8}{6}s + \frac{16}{2}t\right)$

In Exercises 55–62, simplify each expression.

55. $9(8m)$

56. $12n(4)$

57. $5(-9q)$

58. $-7(2t)$

59. $(-5p)(-6b)$

60. $(-7d)(-7e)$

61. $-5(8r)(-2y)$

62. $-7s(-4t)(-1)$

In Exercises 63–88, simplify each expression by combining like terms.

63. $3x + 15x$

64. $12y - 17y$

65. $18x^2 - 5x^2$

66. $37x^2 + 3x^2$

67. $-9x + 9x$

68. $-26y + 26y$

69. $-b^2 + b^2$

70. $-3c^3 + 3c^3$

71. $8x + 5x - 7x$

72. $-y + 3y + 6y$

73. $3x^2 + 2x^2 - 5x^2$

74. $8x^3 - x^3 + 2x^3$

75. $3.8h - 0.7h$

76. $-5.7m + 5.3m$

77. $\frac{3}{5}t + \frac{1}{5}t$

78. $\frac{3}{16}x - \frac{5}{16}x$

79. $4(y + 9) - 8y$

80. $-(4 + z) + 2z$

81. $2z + 5(z - 4)$

82. $12(2m + 11) - 11$

83. $8(2c + 7) - 2(c - 3)$

84. $9(z + 3) - 5(3 - z)$

85. $2x^2 + 4(3x - x^2) + 3x$

86. $3p^2 - 6(5p^2 + p) + p^2$

87. $-(a + 2) - (a - b)$

88. $3z - 2(y - z) + y$

APPLICATIONS

89. PARKING AREA A restaurant parking lot is shown in Illustration 1.

a. Express the area of the entire parking lot as the product of its length and width.

b. Express the area of the entire lot as the sum of the areas of the self-parking space and the valet parking space.

c. Write an equation that shows that your answers to parts a and b are equal. What property of real numbers is illustrated by this example?

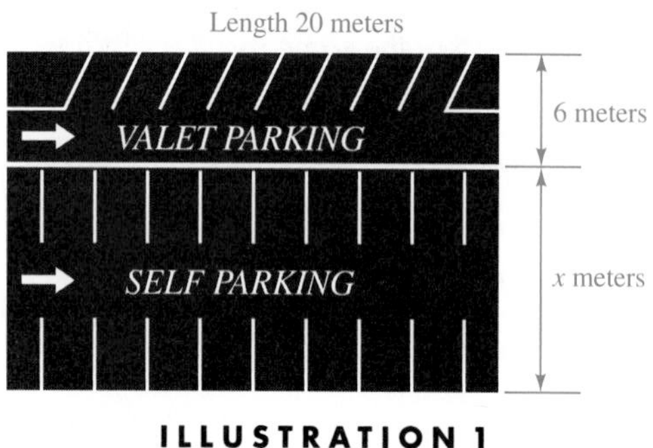

ILLUSTRATION 1

90. CROSS SECTION When the steel casting shown in Illustration 2 is cut down the middle, we see that it has a uniform cross section consisting of two identical trapezoids. What is the area of the cross section? (The measurements are in inches.)

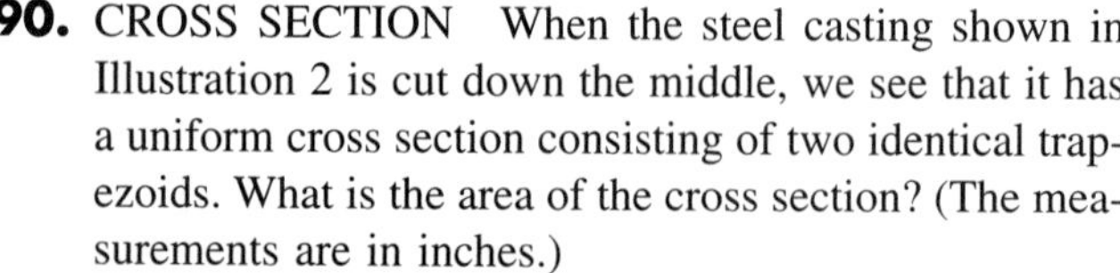

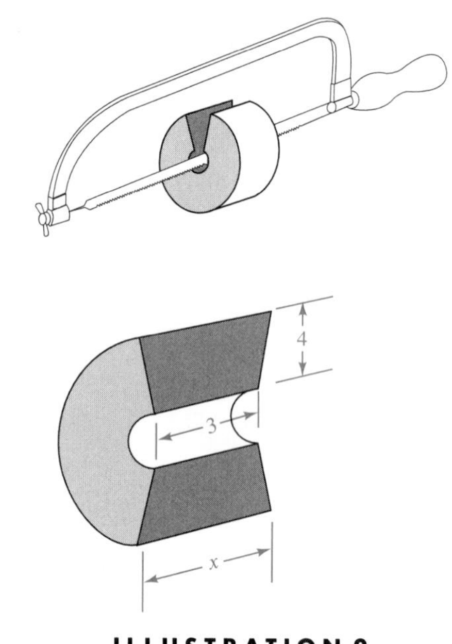

ILLUSTRATION 2

WRITING *Write a paragraph using your own words.*

91. Explain why the distributive property does not apply when simplifying $6(2 \cdot x)$.

92. Show that the operation of subtraction is not commutative.

93. What are like terms?

94. Use each of the words *commute, associate,* and *distribute* in a sentence in which the context is nonmathematical.

REVIEW *In Exercises 95–102, evaluate each expression.*

95. $-5.6 - (-5.6)$

96. $-(-4 \cdot 2)^3$

97. $\left(-\frac{3}{2}\right)\left(\frac{7}{12}\right)$

98. $-\sqrt{\frac{9}{16}}$

99. $(4 + 2 \cdot 3)^3$

100. $-3|4 - 8|$

101. $\frac{-\sqrt{64} - 5^2}{2 \cdot 4 + 3}$

102. $\frac{1}{2} - \left(-\frac{4}{5}\right)$

1.5 Solving Linear Equations and Formulas

In this section, you will learn about

- **Solutions of equations**
- **Linear equations**
- **Properties of equality**
- **Solving linear equations**
- **Simplifying expressions to solve equations**
- **Identities and impossible equations**
- **Solving formulas**

INTRODUCTION. In one method that we will often use to solve problems, we begin by letting a variable stand for an unknown quantity. Then we write an equation involving the variable to mathematically describe the situation. Finally, we perform a series of steps on the equation to find the value represented by the variable. The process of determining the value (or values) represented by a variable is called *solving the equation.* In this section, we will discuss an equation-solving strategy to solve one basic type of equation called a *linear equation* in one variable.

Solutions of equations

An **equation** is a statement indicating that two quantities are equal. The equation $2 + 4 = 6$ is true, and the equation $2 + 5 = 6$ is false. If an equation contains a variable (say, x), it can be either true or false, depending on the value of x. For example, if $x = 1$, the equation $7x - 3 = 4$ is true.

$7x - 3 = 4$	At this stage, we don't know whether the left- and right-hand sides of the equation are equal, so we use an "is possibly equal to" sign $\stackrel{?}{=}$.
$7(1) - 3 \stackrel{?}{=} 4$	Substitute 1 for x.
$7 - 3 \stackrel{?}{=} 4$	Do the multiplication.
$4 = 4$	We obtain a true statement.

Since 1 makes the equation true, we say that 1 **satisfies** the equation. However, the equation is false for all other values of x.

The set of numbers that satisfy an equation is called its **solution set.** The elements of the solution set are called **solutions** or **roots** of the equation. Finding the solution set of an equation is called **solving the equation.**

EXAMPLE 1 ***Checking a solution.*** Determine whether 2 is a solution of $3x + 2 = 2x + 4$.

Solution

We substitute 2 for x wherever it appears in the equation and see whether it satisfies the equation.

$3x + 2 = 2x + 4$	The original equation.
$3(2) + 2 \stackrel{?}{=} 2(2) + 4$	Substitute 2 for x.
$6 + 2 \stackrel{?}{=} 4 + 4$	Do the multiplication.
$8 = 8$	Do the addition.

Since $8 = 8$ is a true statement, the number 2 satisfies the equation. It is a solution of $3x + 2 = 2x + 4$.

Self Check

Is -5 a solution of $2x - 5 = 3x$?

Answer: yes

Linear equations

In practice, we are not told the solutions of an equation—we need to find the solutions (that is, solve the equation) ourselves. In this text, you will learn how to solve many different types of equations. The easiest to solve are **linear equations.**

Linear equations

A **linear equation in one variable** (say, x) is any equation that can be written in the form

$$ax - c = 0 \quad (a \text{ and } c \text{ are real numbers and } a \neq 0)$$

Some examples of linear equations are

$$-2x - 8 = 0 \qquad 3(a + 2) = 20 \qquad 4b - 7 + 2b = 1 + 2b + 8$$

Linear equations are also called **first-degree equations,** because the highest power on the variable is 1.

Properties of equality

When solving linear equations, the objective is to *isolate* the variable on one side of the equation. This is achieved by "undoing" the operations performed on the variable. As we undo the operations, we produce a series of simpler equations, all having the same solution set. Such equations are called **equivalent equations.**

Equivalent equations

Equations with the same solution set are called **equivalent equations.**

The following properties are used to isolate a variable on one side of an equation.

Properties of equality

If a, b, and c are real numbers and $a = b$, then

$a + c = b + c$ **(Addition property of equality)**

$a - c = b - c$ **(Subtraction property of equality)**

If $c \neq 0$, then

$ca = cb$ **(Multiplication property of equality)**

$\frac{a}{c} = \frac{b}{c}$ **(Division property of equality)**

In words, we have

If any quantity is added to (or subtracted from) both sides of an equation, a new equation is formed that is equivalent to the original equation.

If both sides of an equation are multiplied (or divided) by the same nonzero quantity, a new equation is formed that is equivalent to the original equation.

Solving linear equations

EXAMPLE 2 ***Isolating the variable by "undoing" operations.***
Solve $-2x - 8 = 0$ and give the solution set.

Solution

We note that x is multiplied by -2 and then 8 is subtracted from that product. To isolate x on the left-hand side of the equation, we undo these operations in the *reverse* order.

Self Check

Solve $-3a + 15 = 0$ and give the solution set.

- To undo the subtraction of 8, we add 8 to both sides.
- To undo the multiplication by -2, we divide both sides by -2.

$-2x - 8 + 8 = 0 + 8$ Add 8 to both sides.

$-2x = 8$ Simplify both sides of the equation.

$\frac{-2x}{-2} = \frac{8}{-2}$ Divide both sides by -2.

$x = -4$ Simplify both sides of the equation.

Check: We substitute -4 for x to verify that it satisfies the original equation.

$-2x - 8 = 0$

$-2(-4) - 8 \stackrel{?}{=} 0$ Substitute -4 for x.

$8 - 8 \stackrel{?}{=} 0$ Do the multiplication.

$0 = 0$ Do the subtraction.

Since -4 satisfies the original equation, it is the solution. The solution set is $\{-4\}$.

Answer: $\{5\}$

EXAMPLE 3 ***Multiplying by the reciprocal.*** Solve $\frac{3}{4}y = -7$.

Solution

On the left-hand side, y is multiplied by $\frac{3}{4}$. We can undo the multiplication by dividing both sides by $\frac{3}{4}$. Since division by $\frac{3}{4}$ is equivalent to multiplication by its **reciprocal,** we can isolate y by multiplying both sides by $\frac{4}{3}$.

$\frac{3}{4}y = -7$

$\frac{4}{3}\left(\frac{3}{4}y\right) = \frac{4}{3}(-7)$ Use the multiplication property of equality: Multiply both sides by the reciprocal of $\frac{3}{4}$, which is $\frac{4}{3}$.

$\left(\frac{4}{3} \cdot \frac{3}{4}\right)y = \frac{4}{3}(-7)$ Use the associative property of multiplication to regroup.

$1y = \frac{4}{3}(-7)$ The product of a number and its reciprocal is 1: $\frac{4}{3} \cdot \frac{3}{4} = 1$.

$y = -\frac{28}{3}$ On the right-hand side, do the multiplication.

Check: $\frac{3}{4}y = -7$ The original equation.

$\frac{3}{4}\left(-\frac{28}{3}\right) \stackrel{?}{=} -7$ Substitute $-\frac{28}{3}$ for y.

$-\frac{\overset{1}{\cancel{3}} \cdot \overset{1}{\cancel{4}} \cdot 7}{\underset{1}{\cancel{4}} \cdot \underset{1}{\cancel{3}}} \stackrel{?}{=} -7$ Multiply the numerators and the denominators. Factor 28 and divide out the common factors.

$-7 = -7$ Simplify the left-hand side.

Since $-\frac{28}{3}$ satisfies the original equation, it is the solution.

Self Check

Solve $\frac{2}{3}b - 3 = -15$.

Answer: -18

Simplifying expressions to solve equations

To solve more complicated equations, we often need to apply the distributive property and combine like terms.

EXAMPLE 4 ***Using the distributive property.*** Solve $3(a + 2) = 20$.

Solution

We begin by applying the distributive property to remove parentheses on the left-hand side.

$$\begin{aligned} 3(a + 2) &= 20 && \\ 3a + 6 &= 20 && \text{Distribute the 3.} \\ 3a + 6 - 6 &= 20 - 6 && \text{To undo the addition of 6, subtract 6 from both sides.} \\ 3a &= 14 && \text{Simplify each side of the equation.} \\ \frac{3a}{3} &= \frac{14}{3} && \text{To undo the multiplication by 3, divide both sides by 3.} \\ a &= \frac{14}{3} && \text{Simplify the left-hand side.} \end{aligned}$$

Check:

$$\begin{aligned} 3(a + 2) &= 20 && \text{The original equation.} \\ 3\left(\frac{14}{3} + 2\right) &\stackrel{?}{=} 20 && \text{Substitute } \tfrac{14}{3} \text{ for } a. \\ 3\left(\frac{14}{3} + \frac{6}{3}\right) &\stackrel{?}{=} 20 && \text{Get a common denominator: } 2 = \tfrac{6}{3}. \\ 3\left(\frac{20}{3}\right) &\stackrel{?}{=} 20 && \text{Add the fractions: } \tfrac{14}{3} + \tfrac{6}{3} = \tfrac{20}{3}. \\ 20 &= 20 && \text{Simplify the left-hand side.} \end{aligned}$$

Since $\frac{14}{3}$ satisfies the equation, it is the solution.

Self Check

Solve $-2(x + 3) = 18$.

Answer: -12

EXAMPLE 5 ***Variables on both sides of the equation.*** Solve $4b - 7 + 2b = 1 + 2b + 8$.

Solution

First, we combine like terms on each side of the equation.

$$\begin{aligned} 4b - 7 + 2b &= 1 + 2b + 8 && \\ 6b - 7 &= 2b + 9 && \text{Combine like terms: } 4b + 2b = 6b \text{ and } 1 + 8 = 9. \end{aligned}$$

We note that terms involving b appear on both sides of the equation. To isolate b on the left-hand side, we need to eliminate $2b$ on the right-hand side.

$$\begin{aligned} 6b - 7 &= 2b + 9 && \\ 6b - 7 - 2b &= 2b + 9 - 2b && \text{Subtract } 2b \text{ from both sides.} \\ 4b - 7 &= 9 && \text{Combine like terms on each side: } 6b - 2b = 4b \text{ and } 2b - 2b = 0. \\ 4b - 7 + 7 &= 9 + 7 && \text{To undo the subtraction of 7, add 7 to both sides.} \\ 4b &= 16 && \text{Simplify each side of the equation.} \\ b &= 4 && \text{To isolate } b \text{, divide both sides by 4.} \end{aligned}$$

Check:

$$\begin{aligned} 4b - 7 + 2b &= 1 + 2b + 8 && \text{The original equation.} \\ 4(4) - 7 + 2(4) &\stackrel{?}{=} 1 + 2(4) + 8 && \text{Substitute 4 for } b. \\ 16 - 7 + 8 &\stackrel{?}{=} 1 + 8 + 8 && \text{Do each multiplication.} \\ 17 &= 17 && \text{Simplify each side.} \end{aligned}$$

Since 4 satisfies the equation, it is the solution.

Self Check

Solve
$-6t - 12 - 6t = 1 + 2t - 5$.

Answer: $-\frac{4}{7}$

In general, to solve linear equations in one variable, we will follow these steps.

Solving linear equations

1. If the equation contains fractions, multiply both sides of the equation by a number that will eliminate the denominators.
2. Use the distributive property to remove all sets of parentheses and combine like terms.
3. Use the addition and subtraction properties to get all variables on one side of the equation and all numbers on the other side. Combine like terms, if necessary.
4. Use the multiplication and division properties to make the coefficient of the variable equal to 1.
5. Check the result by replacing the variable with the possible solution and verifying that the number satisfies the equation.

EXAMPLE 6 ***Using the equation-solving strategy.*** Solve $\frac{1}{3}(6x - 15) = \frac{3}{2}(x - 2) + 2$.

Solution *Step 1:* Since 6 is the smallest number that can be divided by both 2 and 3, we multiply both sides of the equation by 6 to clear it of fractions.

$$\frac{1}{3}(6x - 15) = \frac{3}{2}(x - 2) + 2$$

$$6\left[\frac{1}{3}(6x - 15)\right] = 6\left[\frac{3}{2}(x - 2) + 2\right]$$ To eliminate the fractions, multiply both sides by 6.

$$2(6x - 15) = 6 \cdot \frac{3}{2}(x - 2) + 6 \cdot 2$$ On the left-hand side, do the multiplication: $6 \cdot \frac{1}{3} = 2$. On the right-hand side, distribute the 6.

$$2(6x - 15) = 9(x - 2) + 12$$ Do the multiplications on the right-hand side.

Step 2: We use the distributive property to remove parentheses and then combine like terms.

$$12x - 30 = 9x - 18 + 12$$ On the left-hand side, distribute the 2. On the right-hand side, distribute the 9.

$$12x - 30 = 9x - 6$$ Combine like terms: $-18 + 12 = -6$.

Step 3: We use the addition and subtraction properties of equality by subtracting $9x$ from both sides and by adding 30 to both sides.

$$12x - 30 - 9x + 30 = 9x - 6 - 9x + 30$$

$$3x = 24$$ On each side, combine like terms.

Step 4: The coefficient of the variable x is 3. To undo the multiplication by 3, we divide both sides by 3.

$$\frac{3x}{3} = \frac{24}{3}$$ Divide both sides by 3.

$$x = 8$$ Do the divisions.

Step 5: We check by substituting 8 for x in the original equation and simplifying:

$$\frac{1}{3}(6x - 15) = \frac{3}{2}(x - 2) + 2$$
$$\frac{1}{3}[6(8) - 15] \stackrel{?}{=} \frac{3}{2}(8 - 2) + 2$$
$$\frac{1}{3}(48 - 15) \stackrel{?}{=} \frac{3}{2}(6) + 2$$
$$\frac{1}{3}(33) \stackrel{?}{=} 9 + 2$$
$$11 = 11$$

Since 8 satisfies the equation, it is the solution. ■

EXAMPLE 7 ***Using the equation-solving strategy.***

Solve $\frac{x+2}{5} - 4x = \frac{8}{5} - \frac{x+9}{2}$.

Solution

Some of the steps used to solve an equation can be done in your head, as you will see in this example.

$$\frac{x+2}{5} - 4x = \frac{8}{5} - \frac{x+9}{2}$$

$$10\left(\frac{x+2}{5} - 4x\right) = 10\left(\frac{8}{5} - \frac{x+9}{2}\right)$$ To eliminate the fractions, multiply both sides by 10.

$$10 \cdot \frac{x+2}{5} - 10 \cdot 4x = 10 \cdot \frac{8}{5} - 10 \cdot \frac{x+9}{2}$$ On each side, distribute the 10.

$$2(x + 2) - 40x = 2(8) - 5(x + 9)$$ Do each multiplication by 10.

$$2x + 4 - 40x = 16 - 5x - 45$$ On each side, remove parentheses.

$$-38x + 4 = -5x - 29$$ On each side, combine like terms.

$$-33x = -33$$ Add $5x$ to both sides. Subtract 4 from both sides. These steps are done in your head—we don't show them.

$$x = 1$$ Divide both sides by -33. This step is also done in your head.

Check:

$$\frac{x+2}{5} - 4x = \frac{8}{5} - \frac{x+9}{2}$$

$$\frac{1+2}{5} - 4(1) \stackrel{?}{=} \frac{8}{5} - \frac{1+9}{2}$$ Substitute 1 for x.

$$\frac{3}{5} - 4 \stackrel{?}{=} \frac{8}{5} - 5$$

$$\frac{3}{5} - \frac{20}{5} \stackrel{?}{=} \frac{8}{5} - \frac{25}{5}$$ Write 4 as $\frac{20}{5}$ and 5 as $\frac{25}{5}$.

$$-\frac{17}{5} = -\frac{17}{5}$$ Do each subtraction.

Since 1 satisfies the equation, it is the solution. ■

Self Check

Solve $\frac{a+3}{2} + 2a = \frac{3}{2} - \frac{a+27}{5}$.

Answer: -2

EXAMPLE 8 ***Solving equations involving decimals.*** Solve $-35.6 = 77.89 - x$.

Solution

$$-35.6 = 77.89 - x$$

$$-35.6 - 77.89 = 77.89 - x - 77.89 \quad \text{Subtract 77.89 from both sides.}$$

$$-113.49 = -x \quad \text{Simplify each side of the equation.}$$

$$-113.49 = -1x \quad -x = -1x.$$

$$\frac{-113.49}{-1} = \frac{-1x}{-1} \quad \text{To isolate } x\text{, divide both sides by } -1.$$

$$113.49 = x \quad \text{Simplify each side of the equation.}$$

$$x = 113.49$$

Check to see that $x = 113.49$ satisfies the equation.

Self Check

Solve $-1.3 = -2.6 - x$.

Answer: -1.3

For more complicated equations involving decimals, we can multiply both sides of the equation by a power of 10 to clear the equation of decimals. In the following example, multiplying both sides by 100 changes the decimals in the equation to integers, which are easier to work with.

EXAMPLE 9 ***Clearing an equation of decimals.*** Solve $0.04(12) + 0.01x = 0.02(12 + x)$.

$$0.04(12) + 0.01x = 0.02(12 + x)$$

$$100[0.04(12) + 0.01x] = 100[0.02(12 + x)] \quad \text{To make 0.04, 0.01, and 0.02 integers, we multiply both sides by 100.}$$

$$100 \cdot 0.04(12) + 100 \cdot 0.01x = 100 \cdot 0.02(12 + x) \quad \text{On the left-hand side, distribute the 100.}$$

$$4(12) + 1x = 2(12 + x) \quad \text{Do the three multiplications by 100.}$$

$$48 + x = 24 + 2x \quad \text{Remove parentheses.}$$

$$48 + x - 24 - x = 24 + 2x - 24 - x \quad \text{Subtract 24 and } x \text{ from both sides.}$$

$$24 = x \quad \text{Simplify each side.}$$

$$x = 24$$

Check by substituting 24 for x in the original equation.

Identities and impossible equations

The equations discussed so far are called **conditional equations.** For these equations, some numbers satisfy the equation and others do not. An **identity** is an equation that is satisfied by every number for which both sides of the equation are defined.

EXAMPLE 10 ***An equation that is an identity.*** Solve $-2(x - 1) - 4 = -4(1 + x) + (2x + 2)$.

Solution

$$-2(x - 1) - 4 = -4(1 + x) + (2x + 2)$$

$$-2x + 2 - 4 = -4 - 4x + 2x + 2 \quad \text{Use the distributive property.}$$

$$-2x - 2 = -2x - 2 \quad \text{On each side, combine like terms.}$$

$$-2 = -2 \quad \text{Add } 2x \text{ to both sides.}$$

The result $-2 = -2$ indicates that the equation will be true for every value of x. Since the equation is an identity, the solution set is the set of real numbers.

Self Check

Solve
$3(a + 4) + 5 = 2(a - 1) + a + 19$
and give the solution set.

Answer: all real numbers

An **impossible equation** or a **contradiction** is an equation that has no solution.

EXAMPLE 11 ***An impossible equation.*** Solve $-6.2(-x - 1) - 4 = 4.2x - (-2x)$.

Solution

$$-6.2(-x - 1) - 4 = 4.2x - (-2x)$$

$6.2x + 6.2 - 4 = 4.2x + 2x$ On the left-hand side, remove parentheses. On the right-hand side, write the subtraction as addition of the opposite.

$6.2x + 2.2 = 6.2x$ On each side, combine like terms.

$6.2x + 2.2 - 6.2x = 6.2x - 6.2x$ Subtract $6.2x$ from both sides.

$2.2 = 0$ Simplify each side.

The result $2.2 = 0$ (a false statement) indicates that there is no value for x that will make the equation true. The solution set is the empty set, denoted as $\varnothing$.

Self Check

Solve $3(a + 4) + 2 = 2(a - 1) + a + 19$ and give the solution set.

Answer: $\varnothing$ ■

Solving formulas

To solve a formula for a variable means to isolate that variable on one side of the equation and have all other quantities on the other side. We can use the skills discussed in this section to solve many types of formulas for a specified variable.

EXAMPLE 12 ***Solving a formula for a specified variable.*** Solve $A = \frac{1}{2}bh$ for h.

Solution

$$A = \frac{1}{2}bh$$

$2A = bh$ To eliminate the fraction, multiply both sides by 2.

$\frac{2A}{b} = h$ To isolate h, divide both sides by b.

$h = \frac{2A}{b}$ Write the equation with h on the left-hand side.

Self Check

Solve $A = \frac{1}{2}bh$ for b.

Answer: $b = \frac{2A}{h}$ ■

WARNING! When asked to solve a formula for a specific variable, we must isolate that variable on one side of the equation. For example, it would be incorrect to say that $t = \frac{4A - S}{t}$ is solved for t, because t appears on *both* sides of the equation.

EXAMPLE 13 ***Solving a formula for a specified variable.*** For simple interest, the formula $A = P + Prt$ gives the amount of money in an account at the end of a specific time. A represents the amount, P the principal, r the rate of interest, and t the time. We can solve the formula for t as follows:

Solution

$$A = P + Prt$$

$A - P = Prt$ To isolate the term involving t, subtract P from both sides.

$\frac{A - P}{Pr} = t$ To isolate t, divide both sides by Pr.

$t = \frac{A - P}{Pr}$ Write the equation with t on the left-hand side.

Self Check

Solve $A = P + Prt$ for r.

Answer: $r = \frac{A - P}{Pt}$ ■

EXAMPLE 14 ***Solving a formula for a specified variable.*** The formula $F = \frac{9}{5}C + 32$ converts degrees Celsius to degrees Fahrenheit. Solve the formula for C.

Solution

$$F = \frac{9}{5}C + 32$$

$$F - 32 = \frac{9}{5}C \quad \text{To isolate the term involving } C \text{, subtract 32 from both sides.}$$

$$\frac{5}{9}(F - 32) = \frac{5}{9}\left(\frac{9}{5}C\right) \quad \text{To isolate } C \text{, multiply both sides by } \tfrac{5}{9}.$$

$$\frac{5}{9}(F - 32) = C \quad \tfrac{5}{9} \cdot \tfrac{9}{5} = 1.$$

$$C = \frac{5}{9}(F - 32)$$

To convert degrees Fahrenheit to degrees Celsius, we can use the formula $C = \frac{5}{9}(F - 32)$.

Self Check

Solve $S = \dfrac{180(n - 2)}{5}$ for n.

Answer: $n = \dfrac{5S + 360}{180}$ or $n = \dfrac{5S}{180} + 2$ ■

STUDY SET Section 1.5

VOCABULARY *In Exercises 1–6, fill in the blanks to make the statements true.*

1. An __________ is a statement that two quantities are equal.

2. If two equations have the same solution set, they are called __________ equations.

3. If a number is substituted for a variable in an equation and the equation is true, we say that the number __________ the equation.

4. An __________ is an equation that is true for all values of its variable.

5. An impossible equation is true for no __________ of its variable.

6. $3x + 1 = 10$ is a __________ equation in one variable.

CONCEPTS *In Exercises 7–8, fill in the blanks to make the statements true.*

7. If a, b, and c are real numbers, and $a = b$, then $a + c = b + \square$ and $a - c = b - \square$.

8. If a, b, and c are real numbers, and $a = b$, then $c \cdot a = \square \cdot b$ and $\frac{a}{c} = \square$ $(c \neq 0)$.

9. a. Simplify $5y + 2 - 3y$.
b. Solve $5y + 2 - 3y = 8$.
c. Evaluate $5y + 2 - 3y$ for $y = 8$.

10. a. In $-5b + 3 = -18$, what can we do to both sides to undo the addition of 3?
b. In $-5b = -18$, what can we do to both sides to undo the multiplication by -5?

11. Is $a = \dfrac{3x + 2}{a}$ solved for a? Why or why not?

12. Is $2x = 3m + 2$ solved for x? Why or why not?

13. When solving $\dfrac{x + 1}{3} - \dfrac{2}{15} = \dfrac{x - 1}{5}$, why should we multiply both sides by 15?

14. When solving $1.45x - 0.5(1 - x) = 0.7x$, why should we multiply both sides by 100?

NOTATION *In Exercises 15–16, complete each solution.*

15. Solve the equation $-2(x + 7) = 20$.

$-2(x + 7) = 20$

$\square - 14 = 20$ Distribute the -2.

$-2x - 14 + \square = 20 + \square$ Add 14 to both sides.

$-2x = \square$ Do the additions.

$\frac{-2x}{\square} = \frac{34}{\square}$ Divide both sides by -2.

$x = -17$ Do the divisions.

16. Solve the equation $\frac{d}{3} + 4 = 1$.

$\frac{d}{3} + 4 = 1$

$\frac{d}{3} + 4 - \square = 1 - \square$ Subtract 4 from both sides.

$\frac{d}{3} = \square$ Do the subtractions.

$\square\left(\frac{d}{3}\right) = \square(-3)$ Multiply both sides by 3.

$d = -9$ Do the multiplications.

17. Fill in the blanks to make the statements true.

a. $-x = \square x$ **b.** $\frac{2t}{3} = \square t$

18. When checking a solution of an equation, the symbol $\stackrel{?}{=}$ is used. What does it mean?

PRACTICE *In Exercises 19–22, tell whether 5 is a solution of each equation.*

19. $3x + 2 = 17$

20. $7x - 2 = 53 - 5x$

21. $3(2m - 3) = 15$

22. $\frac{3}{5}p - 5 = -2$

In Exercises 23–56, solve each equation.

23. $\frac{x}{4} = 7$

24. $-\frac{x}{6} = 8$

25. $-\frac{4}{5}s = 16$

26. $-3 = -\frac{9}{8}s$

27. $200 = 34 - t$

28. $8 - x = -12$

29. $1.6a = 4.032$

30. $0.52 = 0.05y$

31. $3x + 1 = 3$

32. $8k - 2 = 13$

33. $3(k - 4) = -36$

34. $4(x + 6) = 84$

35. $4j + 12.54 = 18.12$

36. $9.8 - 15r = -15.7$

37. $4a - 22 - a = -2a - 7$

38. $a + 18 = 5a - 3 + a$

39. $2(2x + 1) = x + 15 + 2x$

40. $-2(x + 5) = x + 30 - 2x$

41. $2(a - 5) - (3a + 1) = 0$

42. $8(3a - 5) - 4(2a + 3) = 12$

43. $9(x + 2) = -6(4 - x) + 18$

44. $3(x + 2) - 2 = -(5 + x) + x$

45. $\frac{1}{2}x - 4 = -1 + 2x$

46. $2x + 3 = \frac{2}{3}x - 1$

47. $\frac{b}{2} - \frac{b}{3} = 4$

48. $\frac{w}{2} + \frac{w}{3} = 10$

49. $\frac{a + 1}{3} + \frac{a - 1}{5} = \frac{2}{15}$

50. $\frac{2z + 3}{3} + \frac{3z - 4}{6} = \frac{z - 2}{2}$

51. $\frac{5a}{2} - 12 = \frac{a}{3} + 1$

52. $5 - \frac{x + 2}{3} = 7 - x$

53. $0.45 = 16.95 - 0.25(75 - 3x)$

54. $0.02x + 0.0175(15{,}000 - x) = 277.5$

55. $0.04(12) + 0.01t - 0.02(12 + t) = 0$

56. $0.25(t + 32) = 3.2 + x$

In Exercises 57–62, solve each equation. If the equation is an identity or an impossible equation, so indicate.

57. $4(2 - 3t) + 6t = -6t + 8$

58. $2x - 6 = -2x + 4(x - 2)$

59. $3(x - 4) + 6 = -2(x + 4) + 5x$

60. $2(x - 3) = \frac{3}{2}(x - 4) + \frac{x}{2}$

61. $2y + 1 = 5(0.2y + 1) - (4 - y)$

62. $-3x = -2x + 1 - (5 + x)$

In Exercises 63–78, solve each formula for the indicated variable.

63. $V = \frac{1}{3}Bh$ for B

64. $A = \frac{1}{2}bh$ for b

65. $I = Prt$ for t

66. $E = mc^2$ for m

67. $P = 2l + 2w$ for w

68. $T - W = ma$ for W

69. $A = \frac{1}{2}h(B + b)$ for B

70. $l = a + (n - 1)d$ for n

71. $y = mx + b$ for x

72. $\lambda = Ax + AB$ for B

73. $\bar{v} = \frac{1}{2}(v + v_0)$ for v_0

74. $l = a + (n - 1)d$ for d

75. $S = \frac{a - lr}{1 - r}$ for l

76. $s = \frac{1}{2}gt^2 + vt$ for g

77. $S = \frac{n(a + l)}{2}$ for l

78. $K = \frac{Mv_0^2}{2} + \frac{Iw^2}{2}$ for I

APPLICATIONS

79. CONVERTING TEMPERATURES In preparing an American almanac for release in Europe, editors need to convert temperature ranges for the planets from degrees Fahrenheit to degrees Celsius. Solve the formula $F = \frac{9}{5}C + 32$ for C. Then use your result to make the conversions for the data shown in Illustration 1. Round to the nearest degree.

Planet	High °F	Low °F	High °C	Low °C
Mercury	810	−290		
Earth	136	−129		
Mars	63	−87		

ILLUSTRATION 1

80. THERMODYNAMICS In thermodynamics, the Gibbs free-energy function is given by the formula $G = U - TS + pV$. Solve for S.

81. WIPER DESIGN The area cleaned by the windshield wiper assembly shown in Illustration 2 is given by the formula

$$A = \frac{d\pi(r_1^2 - r_2^2)}{360}$$

Engineers have determined the amount of windshield area that needs to be cleaned by the wiper for two different vehicles. Solve the equation for d and use your result to find the number of degrees d the wiper arm must swing in each case. Round to the nearest degree.

Vehicle	Area cleared	d (deg)
Luxury car	513 in.^{2}	
Sport utility vehicle	586 in.^{2}	

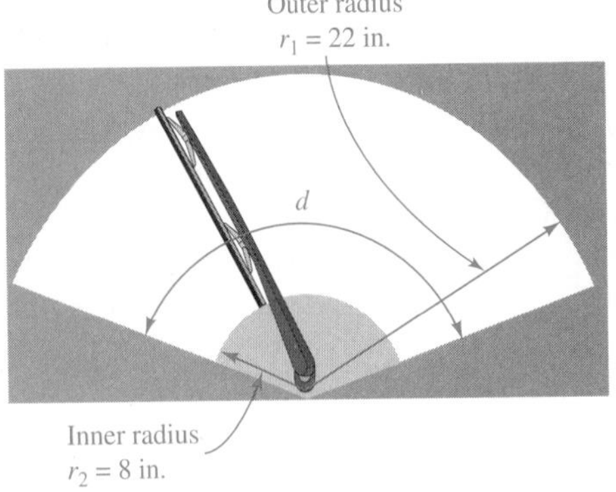

ILLUSTRATION 2

82. ELECTRONICS Illustration 3 is a schematic diagram of a resistor connected to a voltage source of 60 volts. As a result, the resistor dissipates power in the form of heat. The power P lost when a voltage E is placed across a resistance R (in ohms) is given by the formula

$$P = \frac{E^2}{R}$$

Solve for R. If P is 4.8 watts and E is 60 volts, find R.

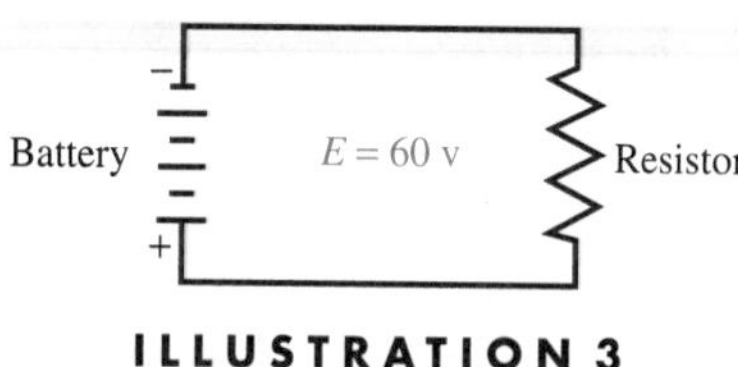

ILLUSTRATION 3

83. CHEMISTRY LAB In chemistry, the ideal gas law equation is $PV = nR(T + 273)$, where P is the pressure, V the volume, T the temperature, and n is the number of moles of a gas. R is a constant, 0.082. Solve the equation for n. Then use your result and the data from the student lab notebook in Illustration 4 to find the value of n to the nearest thousandth for trial 1 and trial 2.

Ideal gas law
Lab #1

Betsy Kinsell
Chem 1
Section A

Data:	Pressure (Atmosph.)	Volume (Liters)	Temp (°C)
Trial 1	0.900	0.250	90
Trial 2	1.250	1.560	−10

R = 0.082 (Constant)

ILLUSTRATION 4

84. INVESTMENT An amount P, invested at a simple interest rate r, will grow to an amount A in t years according to the formula $A = P(1 + rt)$. Solve for P. Suppose a man invested some money at 5.5%. If after 5 years, he had \$6,693.75 on deposit, what amount did he originally invest?

85. COST OF ELECTRICITY The cost of electricity in a certain city is given by the formula $C = 0.07n + 6.50$, where C is the cost and n is the number of kilowatt hours used. Solve for n. Then find the number of kilowatt hours used each month by the homeowner whose checks to pay the monthly electric bills are shown in Illustration 5.

86. COST OF WATER A monthly water bill in a certain city is calculated by using the formula $n = \frac{5{,}000C - 17{,}500}{6}$, where n is the number of gallons used and C is the

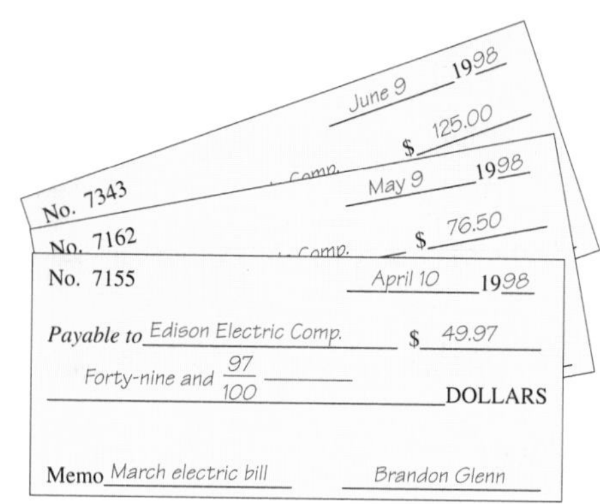

ILLUSTRATION 5

monthly cost. Solve for C and compute the bill for quantities of 500, 1,200, and 2,500 gallons.

87. SURFACE AREA To find the amount of tin needed to make a coffee can such as that shown in Illustration 6, we use the formula for the surface area of a right circular cylinder,

$$A = 2\pi r^2 + 2\pi rh$$

Solve the formula for h.

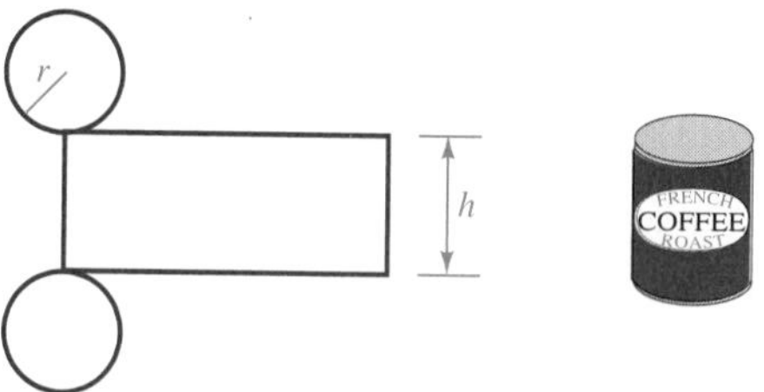

ILLUSTRATION 6

88. CARPENTRY A regular polygon has n equal sides and n equal angles. The measure a of an interior angle in degrees is given by $a = 180\left(1 - \frac{2}{n}\right)$. Solve for n. How many sides does the outdoor bandstand design shown in Illustration 7 have if the performance platform is a regular polygon with interior angles measuring 135°?

ILLUSTRATION 7

WRITING *Write a paragraph using your own words.*

89. What does it mean to *solve an equation?*

90. Why doesn't the equation $x = x + 1$ have a real-number solution?

91. If you are asked to solve the equation $2s = 7t - s$ for s, why is $s = \frac{7t - s}{2}$ an incorrect answer?

92. When solving a linear equation in one variable, the objective is to isolate the variable on one side of the equation. What does that mean?

REVIEW *In Exercises 93–100, simplify each expression.*

93. $-(4 + t) + 2t$

94. $12(2r + 11) - 11 - 3$

95. $4(b + 8) - 8b$

96. $-2(m - 3) + 8(2m + 7)$

97. $3.8b - 0.9b$

98. $-5.7p + 5.1p$

99. $\frac{3}{5}t + \frac{2}{5}t$

100. $-\frac{3}{16}x - \frac{5}{16}x$

1.6 Using Equations to Solve Problems

In this section, you will learn about

- **A problem-solving strategy**
- **Translating words to form an equation**
- **Analyzing a problem**
- **Number–value problems**
- **Drawing diagrams**
- **Geometry problems**
- **Using formulas to solve problems**

INTRODUCTION. One of the objectives of this course is to improve your problem-solving abilities. In the next two sections, you will have the opportunity to do that as we discuss how to use equations to solve many different types of problems.

A problem-solving strategy

The key to problem solving is understanding the problem and devising a plan for solving it. The following list provides a strategy for solving problems.

Problem solving

1. ***Analyze the problem*** by reading it carefully to understand the given facts. What information is given? What vocabulary is given? What are you asked to find? Often a diagram or table will help you visualize the facts of the problem.
2. ***Form an equation*** by picking a variable to represent the quantity to be found. Then express all other quantities mentioned as expressions involving the variable. Finally, write an equation expressing a quantity in two different ways.
3. ***Solve the equation.***
4. ***State the conclusion.***
5. ***Check the result*** in the words of the problem.

Translating words to form an equation

In order to solve problems, which are almost always given in words, we must translate those words into mathematical symbols. In the next example, we use translation to write an equation that mathematically models the situation.

EXAMPLE 1 ***Top U.S. employers.*** As of early 1998, the top two employers in the United States were the United States Postal Service and Wal-Mart Stores, employing a combined total of 1,375,000 people. If Wal-Mart employed 25,000 fewer people than the Postal Service, how many people worked for the Postal Service?

Analyze the problem We are to find the number of people working for the Postal Service. Translating two key phrases in the problem will help us write an equation.

- The phrase "employing a combined total of 1,375,000 people" indicates that if we add the number of Postal Service employees and the number of Wal-Mart employees, the result will be 1,375,000.
- The phrase "Wal-Mart employed 25,000 fewer people than the Postal Service" indicates that if the Postal Service employed x people, Wal-Mart employed $x - 25{,}000$ people.

Form an equation To form an equation, we translate the words into mathematical symbols. We let x represent the number of people employed by the Postal Service. Then $x - 25{,}000$ represents the number of people employed by Wal-Mart.

The number of people employed by the Postal Service	plus	the number of people employed by Wal-Mart	is	1,375,000.
x	$+$	$x - 25{,}000$	$=$	$1{,}375{,}000$

Now we solve the equation.

Solve the equation

$$x + x - 25{,}000 = 1{,}375{,}000$$

$$2x - 25{,}000 = 1{,}375{,}000 \quad \text{Combine like terms.}$$

$$2x = 1{,}400{,}000 \quad \text{Add 25,000 to both sides.}$$

$$x = 700{,}000 \quad \text{Divide both sides by 2.}$$

State the conclusion As of 1998, the Postal Service employed 700,000 people.

Check the result If the Postal Service employed 700,000, Wal-Mart employed 25,000 less, or 675,000 people. Adding 700,000 and 675,000, we get 1,375,000. The answer checks. ■

Analyzing a problem

Sometimes the wording of a problem doesn't contain key phrases that directly translate to an equation. In such cases, an analysis of the problem often gives clues that help us write an equation.

EXAMPLE 2 ***Travel promotion.*** The price of a 7-day Alaskan cruise, normally \$2,752 per person, is reduced by \$1.75 per person for large groups traveling together. How large a group is needed for the price to be \$2,500 per person?

Analyze the problem For each member of the group, the cost is reduced by \$1.75. For a group of 20 people, the \$2,752 price is reduced by 20(\$1.75) = \$35.

The price of the cruise = \$2,752 − 20(\$1.75)

For a group of 30 people, the \$2,752 cost is reduced by 30(\$1.75) = \$52.50.

The price of the cruise = \$2,752 − 30(\$1.75)

Form an equation We can describe this situation in words. If we let x represent the size of the group necessary for the price of the cruise to be \$2,500 per person, we can translate the words into an equation.

The price of the cruise	is	\$2,752	minus	the number of people in the group	times	\$1.75.
2,500	=	2,752	−	x	·	1.75

We now solve the equation.

Solve the equation

$$2{,}500 = 2{,}752 - 1.75x$$

$$2{,}500 - \mathbf{2{,}752} = 2{,}752 - 1.75x - \mathbf{2{,}752} \quad \text{Subtract 2,752 from both sides.}$$

$$-252 = -1.75x \quad \text{Simplify each side.}$$

$$144 = x \quad \text{Divide both sides by } -1.75.$$

State the conclusion If 144 people travel together, the price will be \$2,500 per person.

Check the result For 144 people, the cruise cost of \$2,752 will be reduced by 144(\$1.75) = \$252. If we subtract, \$2,752 − \$252 = \$2,500. The answer checks. ■

Number–value problems

Some problems deal with quantities that have a value. In these problems we must distinguish between the *number of* and the *value of* the unknown quantity. For problems such as these, we will use the relationship

Number · value = total value

EXAMPLE 3 ***Portfolio analysis.*** A college foundation owns stock in Kodak (selling at \$58 per share), Coca-Cola (selling at \$65 per share), and Microsoft (selling at \$125 per share). The foundation owns an equal number of shares of Kodak and Coca-Cola stock, but five times as many shares of Microsoft stock. If this portfolio is worth \$523,600, how many shares of each stock does the foundation own?

Analyze the problem The value of the Kodak stock plus the value of the Coca-Cola stock plus the value of the Microsoft stock must equal \$523,600. We need to find the number of shares of each of these stocks that the foundation has in its portfolio.

Form an equation If we let x represent the number of shares of Kodak stock, then x also represents the number of shares of Coca-Cola stock. Since the foundation owns five times as many shares of Microsoft stock as Kodak or Coca-Cola stock, $5x$ represents the number of shares of Microsoft. The value of the shares of each company's stock would be the *product* of the number of shares of that stock and its per-share value.

Stock	Number of shares ·	Value per share =	Total value of the stock
Kodak	x	\$58	\$58x
Coca-Cola	x	\$65	\$65x
Microsoft	$5x$	\$125	\$125(5x)

We can now form the equation.

The value of Kodak stock	plus	the value of Coca-Cola stock	plus	the value of Microsoft stock	is	the total value of all of the stock.
$58x$	$+$	$65x$	$+$	$125(5x)$	$=$	$523,600$

Next, we solve the equation.

Solve the equation

$$58x + 65x + 125(5x) = 523,600$$
$$58x + 65x + 625x = 523,600 \quad 125(5x) = 625x.$$
$$748x = 523,600 \quad \text{Combine like terms on the left-hand side.}$$
$$x = 700 \quad \text{Divide both sides by 748.}$$

State the conclusion The foundation owns 700 shares of Kodak, 700 shares of Coca-Cola, and $5(700) = 3,500$ shares of Microsoft.

Check the result The value of 700 shares of Kodak stock is $700(\$58) = \$40,600$. The value of 700 shares of Coca-Cola stock is $700(\$65) = \$45,500$. The value of 3,500 shares of Microsoft is $3,500(\$125) = \$437,500$. The sum is $\$40,600 + \$45,500 + \$437,500 = \$523,600$. The answers check. ■

Drawing diagrams

When solving problems, diagrams are often helpful, because they allow us to visualize the facts of the problem.

EXAMPLE 4 ***Triathlon course.*** The Ironman Triathlon in Hawaii includes swimming, long-distance running, and cycling. The long-distance run is 11 times longer than the distance the competitors swim. The distance they cycle is 85.8 miles longer than the run. Overall, the competition covers 140.6 miles. Find the length of each part of the triathlon and round each length to the nearest tenth of a mile.

Analyze the problem The entire triathlon course covers a distance of 140.6 miles. We note that the distance the competitors run is related to the distance they swim, and the distance they cycle is related to the distance they run.

Form an equation If x represents the distance the competitors swim, then $11x$ represents the length of the long-distance run, and $11x + 85.8$ represents the distance they cycle. From the diagram in Figure 1-15, we see that the sum of the individual parts of the triathlon must equal the total distance covered.

FIGURE 1-15

The distance they swim	plus	the distance they run	plus	the distance they cycle	is	the total length of the course.
x	$+$	$11x$	$+$	$11x + 85.8$	$=$	140.6

We now solve the equation.

Solve the equation

$$x + 11x + 11x + 85.8 = 140.6$$
$$23x + 85.8 = 140.6 \quad \text{Combine like terms.}$$
$$23x = 54.8 \quad \text{Subtract 85.8 from both sides.}$$
$$x \approx 2.382608696 \quad \text{Divide both sides by 23.}$$

State the conclusion To the nearest tenth, the distance the competitors swim is 2.4 miles. The distance they run is $11x$, or approximately $11(2.382608696) = 26.20869565$ miles. To the nearest tenth, that is 26.2 miles. The distance they cycle is $11x + 85.8$, or approximately $26.20869565 + 85.8 = 112.0086957$ miles. To the nearest tenth, that is 112.0 miles.

Check the result If we add the lengths of the three parts of the triathlon and round to the nearest tenth, we get 140.6 miles. The answers check. ■

Geometry problems

Sometimes a geometric fact or formula is helpful in solving a problem. Figure 1-16 shows several geometric figures. A **right angle** is an angle whose measure is 90°. A **straight angle** is an angle whose measure is 180°. An **acute angle** is an angle whose measure is greater than 0° and less than 90°. An angle whose measure is greater than 90° and less than 180° is called an **obtuse angle.**

If the sum of two angles equals 90°, the angles are called **complementary,** and each angle is called the **complement** of the other. If the sum of two angles equals 180°, the angles are called **supplementary,** and each angle is the **supplement** of the other.

A **right triangle** is a triangle with one right angle. An **isosceles triangle** is a triangle with two sides of equal measure that meet to form the **vertex angle.** The angles opposite the equal sides, called the **base angles,** are also equal. An **equilateral triangle** is a triangle with three equal sides and three equal angles.

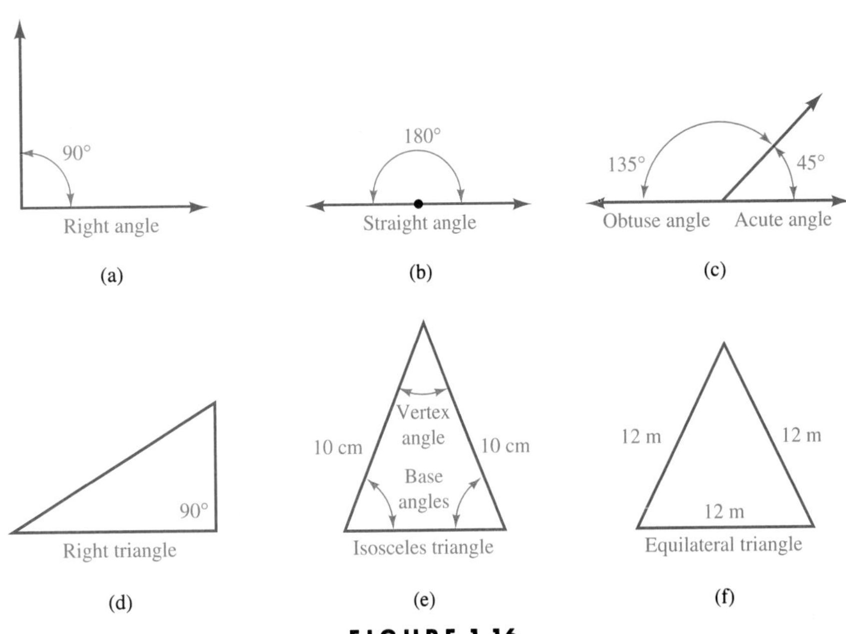

FIGURE 1-16

EXAMPLE 5 ***Flag design.*** The flag of Guyana, a republic on the northern coast of South America, is one isosceles triangle superimposed over another isosceles triangle on a field of green, as shown in Figure 1-17. The measure of a base angle of the larger triangle is 14° more than the measure of a base angle of the smaller triangle. The measure of the vertex angle of the larger triangle is 34°. Find the measure of each base angle of the smaller triangle.

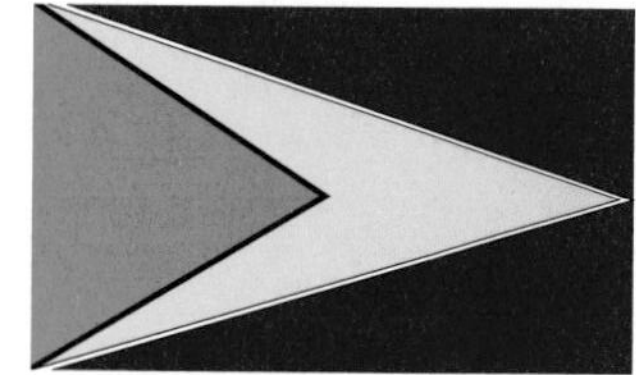

FIGURE 1-17

Analyze the problem We are working with isosceles triangles. Therefore, the base angles of the smaller triangle have the same measure, and the base angles of the larger triangle have the same measure.

Form an equation If we let x represent the measure in degrees of one base angle of the smaller triangle, then the measure of its other base angle is also x. (See Figure 1-18.) The measure of a base angle of the larger triangle is $(x + 14)°$, since its measure is 14° more than the measure of a base angle of the smaller triangle. We are given that the vertex angle of the larger triangle measures 34°.

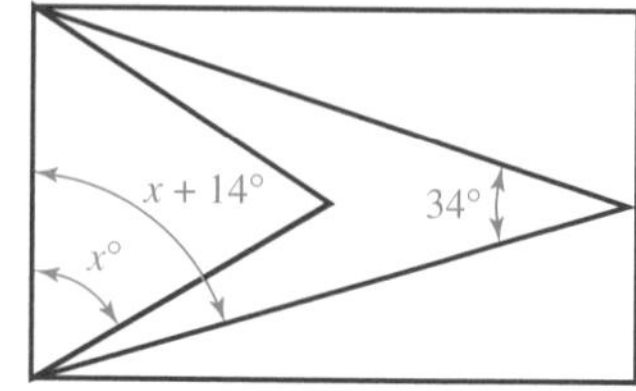

FIGURE 1-18

The sum of the measures of the angles of any triangle (in this case, the larger triangle) is 180°.

The measure of one base angle	plus	the measure of the other base angle	plus	the measure of the vertex angle	is	180°.
$x + 14$	$+$	$x + 14$	$+$	34	$=$	180

We now solve the equation to find x.

Solve the equation

$$
\begin{aligned}
x + 14 + x + 14 + 34 &= 180 \\
2x + 62 &= 180 && \text{Combine like terms.} \\
2x &= 118 && \text{Subtract 62 from both sides.} \\
x &= 59 && \text{Divide both sides by 2.}
\end{aligned}
$$

State the conclusion The measure of each base angle of the smaller triangle is 59°.

Check the result If $x = 59$, then $x + 14 = 73$. The sum of the measures of each base angle and the vertex angle of the *larger* triangle is $73° + 73° + 34° = 180°$. The answer checks. ■

Using formulas to solve problems

When preparing to write an equation to solve a problem, the given facts of the problem often suggest a formula that can be used to mathematically model the situation.

EXAMPLE 6 ***Designing a kennel.*** A man has a 50-foot roll of fencing to make a rectangular kennel. If he wants the kennel to be 6 feet longer than it is wide, find its dimensions.

Analyze the problem The perimeter, P, of the rectangular kennel is 50 feet. Recall that the formula for the perimeter of a rectangle is $P = 2l + 2w$. We need to find its length and width.

Form an equation We let w represent the width of the kennel. Then the length, which is 6 feet more than the width, is represented by the expression $w + 6$. See Figure 1-19.

FIGURE 1-19

We can now form the equation by substituting 50 for P, $w + 6$ for the length, and w for the width in the formula for the perimeter of a rectangle.

$$P = 2l + 2w$$
$$50 = 2(w + 6) + 2w$$

Now we solve the equation.

Solve the equation

$50 = 2(w + 6) + 2w$	
$50 = 2w + 12 + 2w$	Use the distributive property to remove parentheses.
$50 = 4w + 12$	Combine like terms.
$38 = 4w$	Subtract 12 from both sides.
$9.5 = w$	Divide both sides by 4.

State the conclusion The width of the kennel is 9.5 feet. The length would be 6 feet more than this, or 15.5 feet.

Check the result If a rectangle has a width of 9.5 feet and a length of 15.5 feet, its length is 6 feet more than its width, and the perimeter is 2(9.5) feet + 2(15.5) feet = 50 feet. ■

STUDY SET Section 1.6

VOCABULARY *In Exercises 1–6, fill in the blanks to make the statements true.*

1. An ______________ angle has a measure of more than 0° and less than 90°.

2. A ______________ angle is an angle whose measure is 90°.

3. If the sum of the measures of two angles equals 90°, the angles are called ______________ angles.

4. If the sum of the measures of two angles equals 180°, the angles are called ______________ angles.

5. If a triangle has a right angle, it is called a ______________ triangle.

6. If a triangle has two sides with equal measures, it is called an ______________ triangle.

CONCEPTS

7. The unit used to measure the intensity of sound is called the *decibel.* In Illustration 1, translate the comments in the right-hand column into mathematical symbols to complete the decibel column.

	Decibels	Compared to conversation
Conversation	d	—
Vacuum cleaner		15 decibels more
Circular saw		10 decibels less than twice
Jet takeoff		20 decibels more than twice
Whispering		10 decibels less than half
Rock band		Twice the decibel level

ILLUSTRATION 1

8. Illustration 2 shows the four types of problems an instructor put on a history test.
a. Complete the table.
b. Which type of question appears the most on the test?
c. Write an algebraic expression that represents the total number of points on the test.

Type of question	Number	Value	Total value
Multiple choice	x	5	
True/false	$3x$	2	
Essay	$x - 2$	10	
Fill-in	x	5	

ILLUSTRATION 2

9. Illustration 3 shows the three pieces that fit together to make a flute.
a. What is the length of the shortest piece?
b. What is the length of the longest piece?
c. Write an algebraic expression that represents the length of the flute.

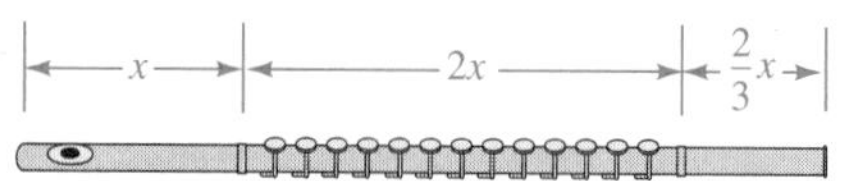

ILLUSTRATION 3

10. For each picture in Illustration 4, what geometric concept studied in this section is illustrated?

Radiation warning

(a) (b)

ILLUSTRATION 4

11. Write an algebraic expression that represents the length in inches of the head of the tennis racquet shown in Illustration 5.

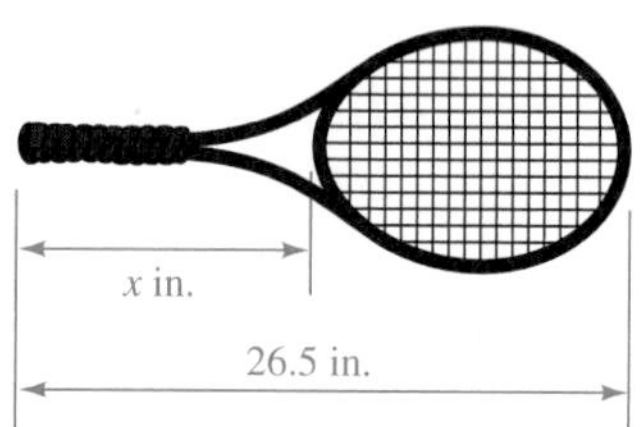

ILLUSTRATION 5

12. The surface area of the earth is 510,066,000 square kilometers (km^2). If we let x represent the number of km^2 covered by water, what would the algebraic expression $510{,}066{,}000 - x$ represent?

APPLICATIONS

13. CEREAL SALES In 1996, the two top-selling cereals were Kellogg's Frosted Flakes and General Mills' Cheerios, with combined sales of $582,200,000. Cheerios sales were $9,400,000 less than those of Frosted Flakes. What were the 1996 sales for each cereal?

14. MARKET SHARE In 1996, sales of the top three brands of toothpaste captured 57.8% of the market share. Crest, the top seller, had 8.4% more of the market than the number two seller, Colgate. Mentadent, the number three seller, had a market share that was 7.6% less than that of Colgate. What was the market share in 1996 for each of these brands?

15. SPRING TOUR A group of junior high students will be touring Washington, DC. Their chaperons will have the \$1,810 cost of the tour reduced by \$15.50 for each student they personally supervise. How many students will a chaperon have to supervise so that his or her cost to take the tour will be \$1,500?

16. MACHINING Each pass through a lumber plane shaves off 0.015 inch of thickness from a board. How many times must a board, originally 0.875 inch thick, be run through the planer if a board of thickness 0.74 inch is desired?

17. MOVING EXPENSES To help move his furniture, a man rents a truck for \$29.25 per day plus 15¢ per mile. If he has budgeted \$75 for transportation expenses, how many miles will he be able to drive the truck if the move takes 1 day?

18. COMPUTING SALARIES A student working for a delivery company earns \$47.50 per day plus \$1.75 for each package she delivers. How many deliveries must she make each day to earn \$100 a day?

19. VALUE OF AN IRA In an Individual Retirement Account (IRA) valued at \$53,900, a couple has 500 shares of stock, some in Big Bank Corporation and some in Safe Savings and Loan. If Big Bank sells for \$115 per share and Safe Savings sells for \$97 per share, how many shares of each does the couple own?

20. ASSETS OF A PENSION FUND A pension fund owns 2,000 more shares in mutual stock funds than mutual bond funds—12,000 shares overall. Currently, the stock funds sell for \$12 per share, and the bond funds sell for \$15 per share. How many shares of each does the pension fund own if the value of the securities is \$165,000?

21. SELLING CALCULATORS Last month, a bookstore ran the ad shown in Illustration 6. Sales of \$4,980 were generated, with 15 more graphing calculators sold than scientific calculators. How many of each type of calculator did the bookstore sell?

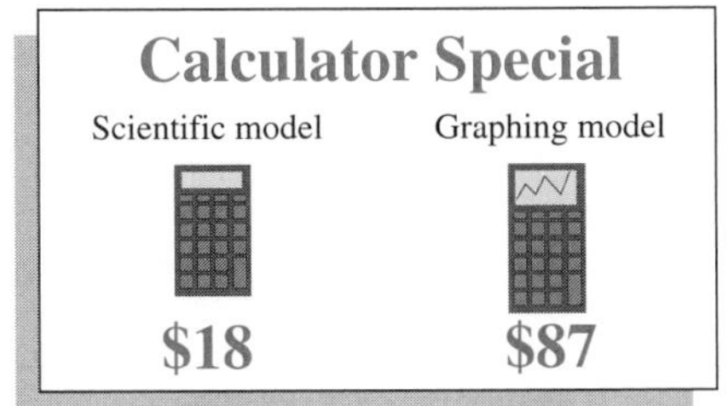

ILLUSTRATION 6

22. SELLING GRASS SEED A seed company sells two grades of grass seed. A 100-pound bag of a mixture of rye and Kentucky bluegrass sells for \$245, and a 100-pound bag of bluegrass sells for \$347. How many bags of each are sold in a week when the receipts for 19 bags are \$5,369?

23. WOODWORKING The carpenter in Illustration 7 saws a board into two pieces. He wants one piece to be 1 foot longer than twice the length of the shorter piece. Find the length of each piece.

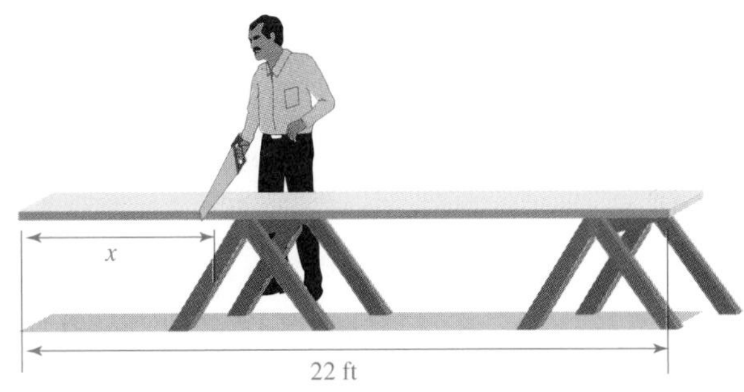

ILLUSTRATION 7

24. STATUE OF LIBERTY From the foundation of the large pedestal on which it sits, to the top of the torch, the Statue of Liberty National Monument measures 305 feet. The pedestal is 3 feet taller than the statue. Find the height of the pedestal and the height of the statue.

25. NURSING Illustration 8 shows the angle a needle should make with the skin when administering a certain type of intradermal injection. Find the measure of both angles labeled in the illustration.

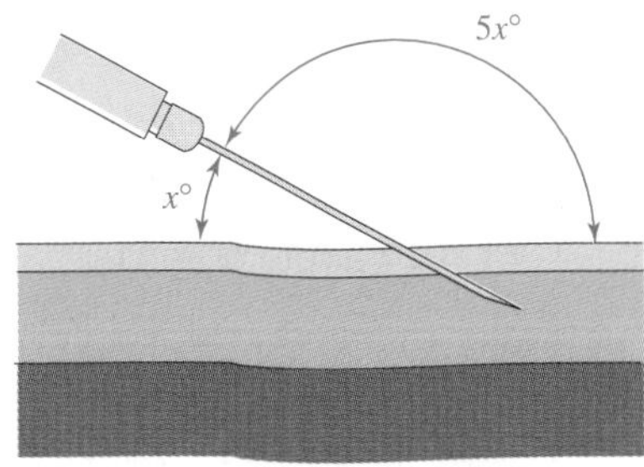

ILLUSTRATION 8

26. SUPPLEMENTARY ANGLES Refer to Illustration 9 and find x.

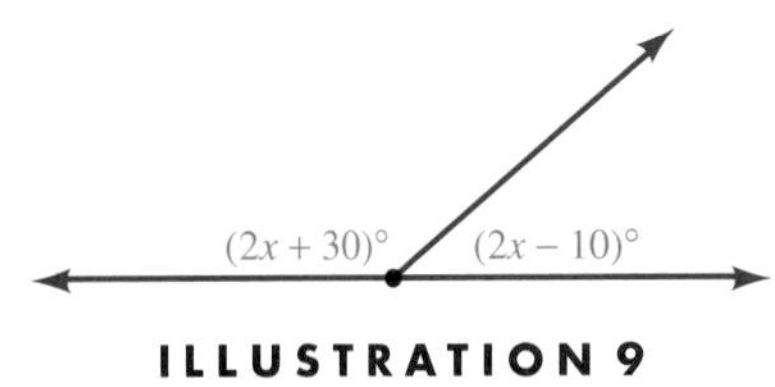

ILLUSTRATION 9

27. ARCHITECTURE Because of soft soil and a shallow foundation, the Leaning Tower of Pisa in Italy is not vertical, as shown in Illustration 10. How many degrees from vertical is the tower?

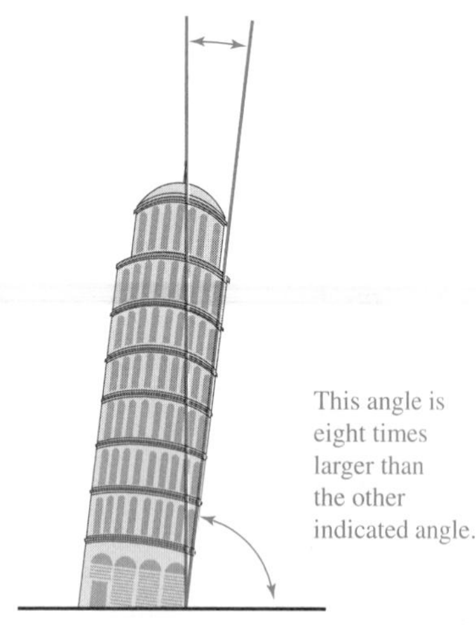

ILLUSTRATION 10

28. KITCHEN STEPSTOOL The sum of the measures of the three angles of any triangle is 180°. In Illustration 11, the measure of angle 2 is 10° larger than the measure of angle 1. The measure of angle 3 is 10° larger than the measure of angle 2. Find each angle measure.

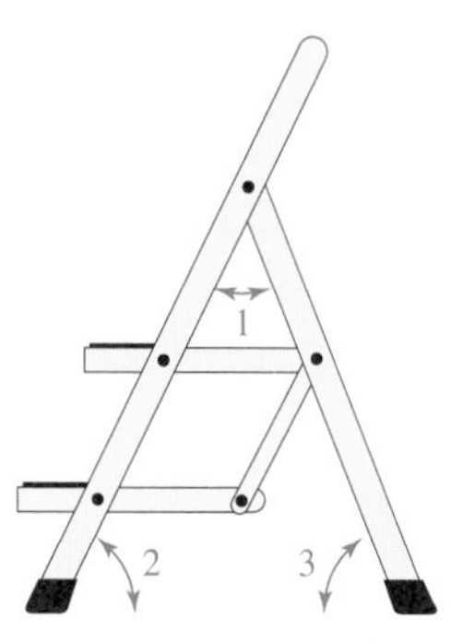

ILLUSTRATION 11

29. SUPPLEMENTARY ANGLES AND PARALLEL LINES In Illustration 12, lines r and s are cut by a third line l to form angles 1 and 2. When lines r and s are parallel, angles 1 and 2 are supplementary. If angle $1 = (x + 50)°$, angle $2 = (2x - 20)°$, and lines r and s are parallel, find x.

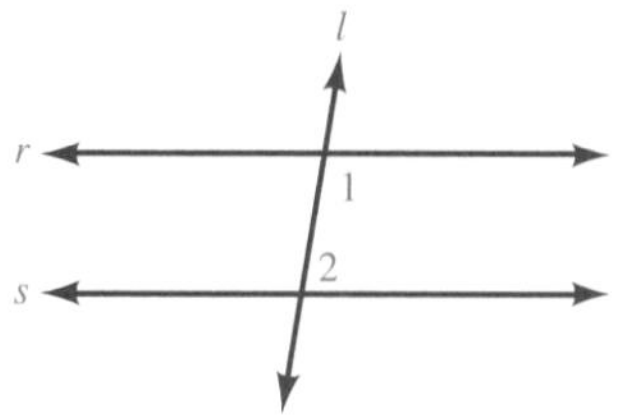

ILLUSTRATION 12

30. VERTICAL ANGLES When two lines intersect as in Illustration 13, four angles are formed. Angles that are side-by-side, such as $\angle 1$ (angle 1) and $\angle 2$, are called **adjacent angles.** Angles that are nonadjacent, such as $\angle 1$ and $\angle 3$ or $\angle 2$ and $\angle 4$, are called **vertical angles.** From geometry, we know that if two lines intersect, vertical angles have the same measure. If $m(\angle 1) = (3x + 10)°$ and $m(\angle 3) = (5x - 10)°$, find x. (Read $m(\angle 1)$ as "the measure of $\angle 1$.")

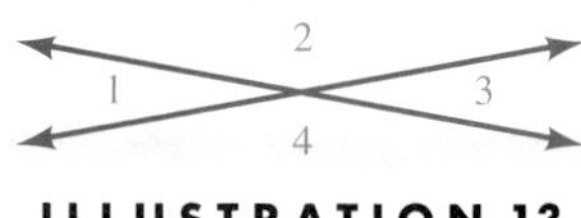

ILLUSTRATION 13

31. ANGLES OF A QUADRILATERAL The sum of the angles of any four-sided figure (called a *quadrilateral*) is 360°. The quadrilateral shown in Illustration 14 has two equal base angles. Find x.

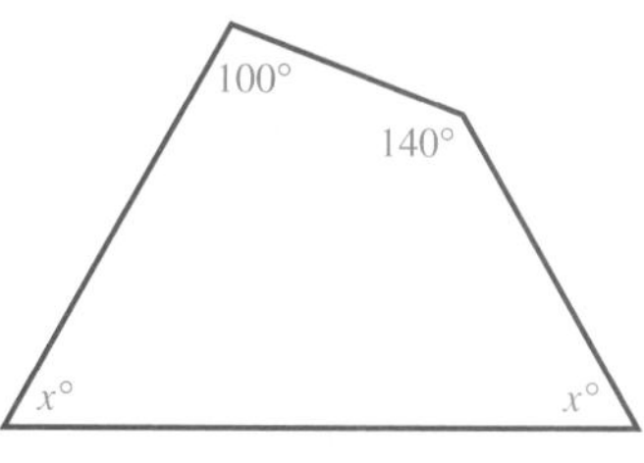

ILLUSTRATION 14

32. HEIGHT OF A TRIANGLE If the height of a triangle with a base of 8 inches is tripled, its area is increased by 96 square inches. Find the height of the triangle.

33. GOLDEN RECTANGLES Throughout history, most artists and designers have felt that rectangles having a length 1.618 times as long as their width have the most visually attractive shape. Such rectangles are known as *golden rectangles.* Measure the length and width of the rectangles in Illustration 15. Which of the rectangles is closest to being a golden rectangle? Do you agree that it is more visually attractive?

i.

ii.

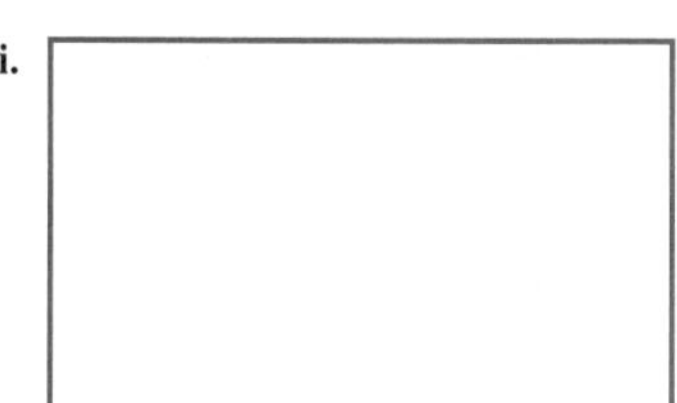

ILLUSTRATION 15

34. QUILTING A woman is planning to make a quilt in the shape of a golden rectangle. (See Exercise 33.) She has exactly 22 feet of a special lace that she plans to sew around the edge of the quilt. What should the length and width of the quilt be? Round both answers up to the nearest hundredth.

35. FENCING A PASTURE A farmer has 624 feet of fencing to enclose the pasture shown in Illustration 16. Because a river runs along one side, fencing will be needed on only three sides. Find the dimensions of the pasture if its length is double its width.

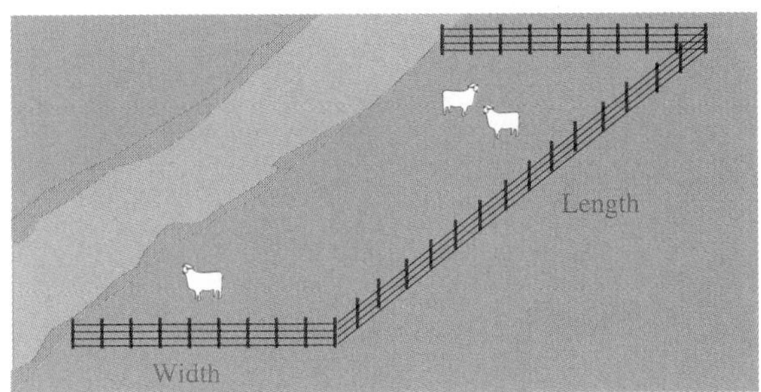

ILLUSTRATION 16

36. FENCING A PEN A man has 150 feet of fencing to build the pen shown in Illustration 17. If one end is a square, find the outside dimensions.

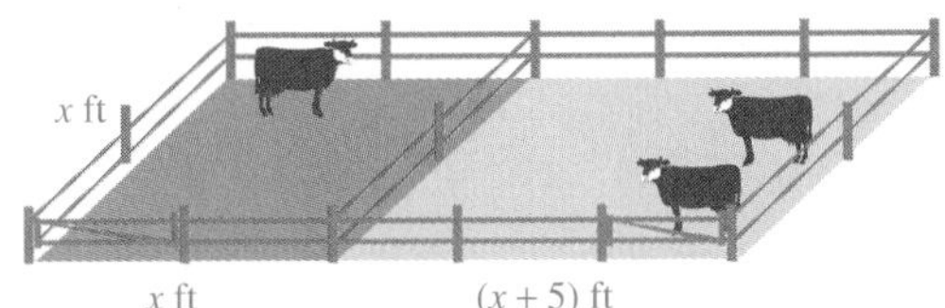

ILLUSTRATION 17

37. ENCLOSING A SWIMMING POOL A woman wants to enclose the pool shown in Illustration 18 and have a walkway of uniform width all the way around. How wide will the walkway be if the woman uses 180 feet of fencing?

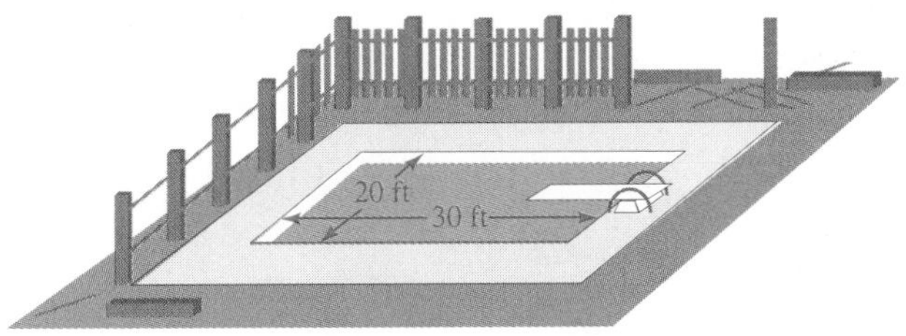

ILLUSTRATION 18

38. INSTALLING SOLAR HEATING One solar panel in Illustration 19 is to be 3 feet wider than the other. To be equally efficient, they must have the same area. Find the width of each.

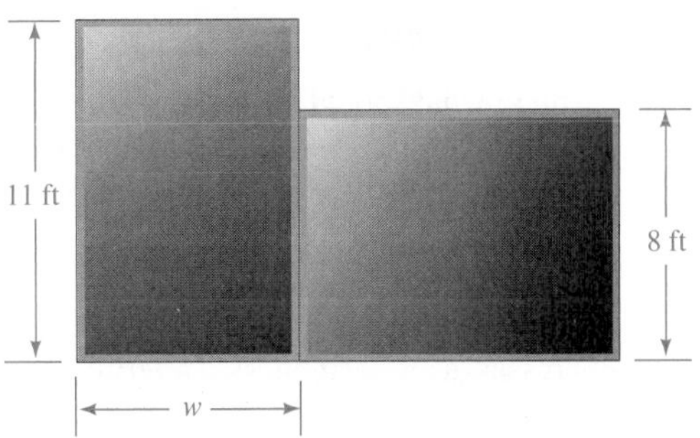

ILLUSTRATION 19

39. MAKING FURNITURE A woodworker wants to put two partitions crosswise in a drawer that is 28 inches deep, as shown in Illustration 20. He wants to place the partitions so that the spaces created increase by 3 inches from front to back. If the thickness of each partition is $\frac{1}{2}$ inch, how far from the front end should he place the first partition?

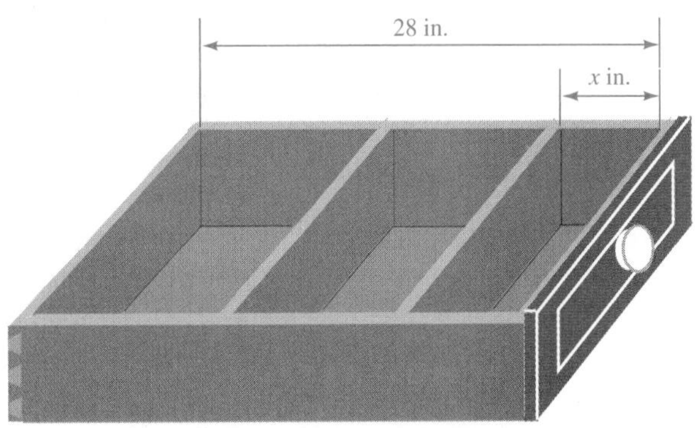

ILLUSTRATION 20

40. BUILDING SHELVES A carpenter wants to put four shelves on an 8-foot wall so that the five spaces created decrease by 6 inches as we move up the wall. (See Illustration 21.) If the thickness of each shelf is $\frac{3}{4}$ inch, how far will the bottom shelf be from the floor?

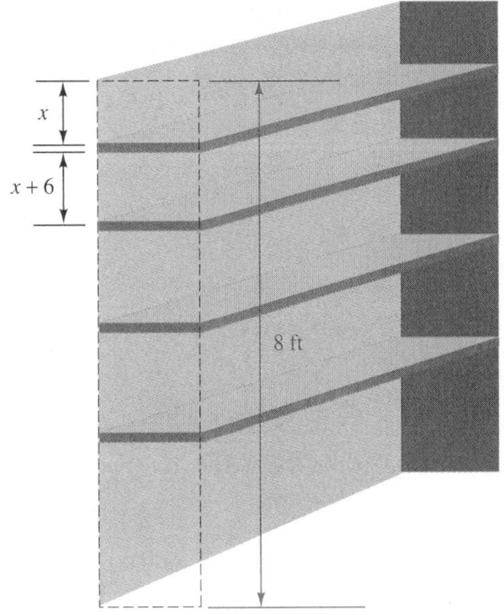

ILLUSTRATION 21

WRITING *Write a paragraph using your own words.*

41. Briefly explain what should be accomplished in each of the steps *(analyze, form, solve, state,* and *check)* of the problem-solving strategy used in this section.

42. What gave you the most difficulty when solving the application problems of this section?

REVIEW

43. When expressed as a decimal, is $\frac{7}{9}$ a terminating or repeating decimal?

44. Solve $x + 20 = 4x - 1 + 2x$.

45. List the integers.

46. Solve $2x + 2 = \frac{2}{3}x - 2$.

47. Evaluate $\sqrt{\frac{25}{16}}$.

48. Solve $T - R = ma$ for R.

1.7 More Applications of Equations

In this section, you will learn about

- **Percent problems**
- **Statistics problems**
- **Investment problems**
- **Uniform motion problems**
- **Mixture problems**

Introduction. In this section, we will again use equations as we solve a variety of problems.

Percent problems

Numeric information is often expressed using percents. Percent means per one hundred. We can use the **percent formula** (Amount = percent · base) to solve many types of percent problems.

EXAMPLE 1 ***Disposable diapers.*** In 1996, Huggies, the leading brand of disposable diapers, had sales of \$1,455,000,000. Use the information in the circle graph in Figure 1-20 to determine the amount of money spent in 1996 on disposable diapers.

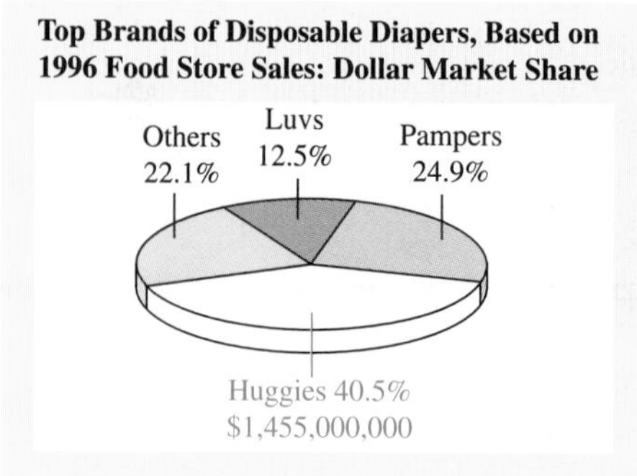

Based on data from *The Wall Street Journal Almanac* (1998)

FIGURE 1-20

Analyze the problem We are told that Huggies had sales of \$1,455,000,000. The circle graph indicates that \$1,455,000,000 was 40.5% of total disposable diaper sales in 1996.

Form an equation If we let x represent total disposable diaper sales (in dollars) in 1996, we can translate the words into an equation.

\$1,455,000,000	is	40.5%	of	what number?	
\$1,455,000,000	=	40.5%	·	x	In this context, "is" translates to = and "of" to multiplication.

We now solve the equation.

Solve the equation

$$1{,}455{,}000{,}000 = 40.5\% \cdot x$$

$$1{,}455{,}000{,}000 = 0.405x \quad \text{Write 40.5\% as a decimal: } 40.5\% = 0.405.$$

$$\frac{1{,}455{,}000{,}000}{\mathbf{0.405}} = \frac{0.405x}{\mathbf{0.405}} \quad \text{Divide both sides by 0.405.}$$

$$3{,}592{,}592{,}593 = x$$

State the conclusion In 1996, sales of disposable diapers totaled \$3,592,592,593.

Check the result If diaper sales totaled \$3,592,592,593 in 1996, then the \$1,455,000,000 spent on Huggies was about $\frac{1{,}500{,}000{,}000}{3{,}600{,}000{,}000} = \frac{15}{36} \approx 42\%$ of the total. The answer of \$3,592,592,593 seems reasonable. ■

When the regular price of merchandise is reduced, the amount of reduction is called **markdown** (or discount).

Sale price	=	regular price	−	markdown

Usually, the markdown is expressed as a percent of the regular price.

Markdown	=	percent of markdown	·	regular price

EXAMPLE 2 ***Wedding gowns.*** At a bridal shop, a wedding gown that normally sells for \$397.98 is on sale for \$265.32. Find the percent of markdown.

Analyze the problem In this case, \$265.32 is the sale price, \$397.98 is the regular price, and the markdown is the *product* of \$397.98 and the percent of markdown.

Form an equation We let r represent the percent of markdown, expressed as a decimal. We then substitute \$265.32 for the sale price and \$397.98 for the regular price in the formula

Sale price	is	regular price	minus	markdown.	
265.32	=	397.98	−	$r \cdot 397.98$	Markdown = percent of markdown · regular price

We now solve the equation.

Solve the equation

$$265.32 = 397.98 - r \cdot 397.98$$

$$265.32 = 397.98 - 397.98r \quad \text{Rewrite } r \cdot 397.98 \text{ as } 397.98r.$$

$$-132.66 = -397.98r \quad \text{Subtract 397.98 from both sides.}$$

$$\frac{-132.66}{-397.98} = r \quad \text{Divide both sides by } -397.98.$$

$$0.333333\ldots = r \quad \text{Do the division using a calculator.}$$

$$33.3333\ldots\% = r$$ To write the decimal as a percent, move the decimal point two places to the right and insert a % sign.

State the conclusion The percent of markdown on the wedding gown is 33.3333. . . % or $33\frac{1}{3}\%$.

Check the result The markdown is $33\frac{1}{3}\%$ of \$397.98, or \$132.66. The sale price is \$397.98 − \$132.66, or \$265.32. The answer checks. ■

Percents are often used to describe how a quantity has changed. To describe such changes, we use **percent increase** or **percent decrease.**

EXAMPLE 3 ***Home video rentals.*** The years 1985–1990 saw tremendous growth for the home video rental industry. Annual rental revenue increased from \$2.55 billion in 1985 to \$6.63 billion in 1990. What was the percent increase in rental revenue?

Analyze the problem To find the percent increase, we first find the *amount of increase* by subtracting the smaller number from the larger. Note that we are working in billions of dollars.

$$\$6.63 - \$2.55 = \$4.08 \quad \text{Subtract the 1985 revenue from the 1990 revenue.}$$

Form an equation Next, we find what percent of the original \$2.55 billion revenue the \$4.08 billion increase was. We will let x = the unknown percent and translate the words into an equation.

What percent	of	2.55	is	4.08?	
x	$\cdot$	2.55	=	4.08	x is the percent, 2.55 is the base, and 4.08 is amount.

Solve the equation

$$x \cdot 2.55 = 4.08$$

$$2.55x = 4.08$$

$$x = \frac{4.08}{2.55} \quad \text{Divide both sides by 2.55.}$$

$$x = 1.6 \quad \text{Use a calculator to do the division.}$$

$$x = 160\%$$

State the conclusion There was a 160% increase in annual video rental revenue between 1985 and 1990.

Check the result A 100% increase in the 1985 revenue would be \$2.55 billion. A 200% increase would be 2(\$2.55 billion) = \$5.1 billion. Our result of a 160% increase, which corresponds to a dollar increase of \$4.08 billion, seems reasonable. ■

Statistics problems

Statistics is a branch of mathematics that deals with the analysis of numerical data. Three types of averages are commonly used in statistics as measures of central tendency of a distribution of numbers: the **mean,** the **median,** and the **mode.**

Mean, median, and mode

The **mean** of several values is the sum of those values divided by the number of values.

$$\text{Mean} = \frac{\text{sum of the values}}{\text{number of values}}$$

The **median** of several values is the middle value. To find the median,

1. Arrange the values in increasing order.
2. If there is an odd number of values, choose the middle value.
3. If there is an even number of values, add the middle two values and divide by 2.

The **mode** of several values is the value that occurs most often.

EXAMPLE 4 ***Physiology experiment.*** As part of a project for a physiology class, a student measured ten people's reaction time to a visual stimulus. Their reaction times, in hundredths of a second, are listed below. Find **a.** the mean, **b.** the median, and **c.** the mode of the distribution.

0.29, 0.24, 0.21, 0.39, 0.28, 0.25, 0.17, 0.28, 0.33, 0.26

Solution **a.** To find the mean we add the values and divide by the number of values, which is 10.

$$\text{Mean} = \frac{0.29 + 0.24 + 0.21 + 0.39 + 0.28 + 0.25 + 0.17 + 0.28 + 0.33 + 0.26}{10} = 0.27 \text{ second}$$

b. To find the median, we first arrange the values in increasing order:

0.17, 0.21, 0.24, 0.25, 0.26, 0.28 , 0.28, 0.29, 0.33, 0.39

Because there is an even number of measurements, the median will be the sum of the middle two values, 0.26 and 0.28, divided by 2. Thus, the median is

$$\text{Median} = \frac{0.26 + 0.28}{2} = 0.27 \text{ second}$$

c. Since the time 0.28 second occurs most often, it is the mode. ■

EXAMPLE 5 ***Bank service charges.*** When the average (mean) daily balance of a customer's checking account falls below \$500 in any week, the bank assesses a \$15 service charge. What minimum balance will the account shown in Figure 1-21 need to have on Friday to avoid the service charge?

Security Savings

☐ Weekly Statement ☐

Acct: 201-234-002 Type: checking

Day	Date	Daily balance	Comments
Mon	3/11	\$730.70	
Tue	3/12	\$350.19	
Wed	3/13	–\$50.19	overdrawn
Thu	3/14	\$275.55	
Fri	3/15		

FIGURE 1-21

Analyze the problem We can find the average (mean) daily balance for the week by adding the daily balances and dividing by 5. We want the mean to be \$500 so that there is no service charge.

Form an equation We will let x = the minimum balance needed on Friday. Then we translate the words into mathematical symbols.

The sum of the five daily balances	divided by	5	is	\$500.

$$\frac{730.70 + 350.19 + (-50.19) + 275.55 + x}{5} = 500$$

Solve the equation

$$\frac{730.70 + 350.19 + (-50.19) + 275.55 + x}{5} = 500$$

$$\frac{1{,}306.25 + x}{5} = 500 \quad \text{Combine like terms in the numerator.}$$

$$5\left(\frac{1{,}306.25 + x}{5}\right) = 5(500) \quad \text{Multiply both sides by 5.}$$

$$1{,}306.25 + x = 2{,}500$$

$$x = 1{,}193.75 \quad \text{Subtract 1,306.25 from both sides.}$$

State the conclusion On Friday, the account balance needs to be \$1,193.75 to avoid a service charge.

Check the result Check the result by adding the five daily balances and dividing by 5. ■

Investment problems

The money an investment earns is called *interest.* **Simple interest** is computed by the formula $I = Prt$, where I is the interest earned, P is the principal (amount invested), r is the annual interest rate, and t is the length of time the principal is invested.

EXAMPLE 6 ***Interest income.*** To protect against a major loss, a financial analyst suggests a diversified plan for a client who has \$50,000 to invest for one year.

1. Alco Development, Inc. Builds mini-malls. High yield: 12% per year. Risky!
2. Certificate of Deposit (CD). Insured, safe. Low yield: 4.5% annual interest.

If the client puts some money in each investment and wants to earn \$3,600 in interest, how much should be invested at each rate?

Analyze the problem In this case, we are working with two investments made at two different rates for 1 year. If we add the interest from the two investments, the sum should equal \$3,600.

Form an equation If we let x represent the number of dollars invested at 12%, the interest earned is $I = Prt = \$x(12\%)(1) = \$0.12x$. If $\$x$ is invested at 12%, there is $\$(50{,}000 - x)$ to invest at 4.5%, which will earn $\$0.045(50{,}000 - x)$ in interest. These facts are listed in Figure 1-22.

	P ·	r ·	t =	I
Alco Development, Inc.	x	0.12	1	$0.12x$
Certificate of deposit	$50{,}000 - x$	0.045	1	$0.045(50{,}000 - x)$

FIGURE 1-22

The sum of the two amounts of interest should equal \$3,600. We now translate the verbal model into an equation.

The interest earned at 12%	plus	the interest earned at 4.5%	is	the total interest earned.
$0.12x$	$+$	$0.045(50{,}000 - x)$	$=$	$3{,}600$

Now we solve the equation.

Solve the equation

$$0.12x + 0.045(50{,}000 - x) = 3{,}600$$

$$1{,}000[0.12x + 0.045(50{,}000 - x)] = 1{,}000(3{,}600)$$ To eliminate the decimals, multiply both sides by 1,000.

$$120x + 45(50{,}000 - x) = 3{,}600{,}000$$ Distribute the 1,000 and simplify both sides.

$$120x + 2{,}250{,}000 - 45x = 3{,}600{,}000$$ Remove parentheses.

$$75x + 2{,}250{,}000 = 3{,}600{,}000$$ Combine like terms.

$$75x = 1{,}350{,}000$$ Subtract 2,250,000 from both sides.

$$x = 18{,}000$$ Divide both sides by 75.

State the conclusion \$18,000 should be invested at 12% and \$(50,000 − 18,000) = \$32,000 should be invested at 4.5%.

Check the result The annual interest on \$18,000 is 0.12(\$18,000) = \$2,160. The interest earned on \$32,000 is 0.045(\$32,000) = \$1,440. The total interest is \$2,160 + \$1,440 = \$3,600. The answers check. ■

Uniform motion problems

Problems that involve an object traveling at a constant rate for a specified period of time over a certain distance are called **uniform motion** problems. To solve these problems, we use the formula $d = rt$, where d is distance, r is rate, and t is time.

EXAMPLE 7 ***Travel time.*** After a weekend stay on her grandparents' farm, a girl is to return home, 385 miles away. To split up the drive, the parents and grandparents start at the same time and drive toward each other, planning to meet somewhere along the way. If the parents travel at an average rate of 60 mph and the grandparents at 50 mph, how long will it take them to reach each other so that the child can be reunited with her parents?

Analyze the problem The cars are traveling toward each other as shown in Figure 1-23(a). We know the rates the cars are traveling (60 mph and 50 mph). We also know that they will travel for the same amount of time.

Form an equation We can let t represent the time that each car travels. Then the distance traveled by the parents is $rt = 60t$, and the distance traveled by the grandparents is $50t$. This information is organized in the table in Figure 1-23(b). The sum of the distances traveled by the parents and grandparents is 385 miles, the distance between the child's home and the grandparents' farm.

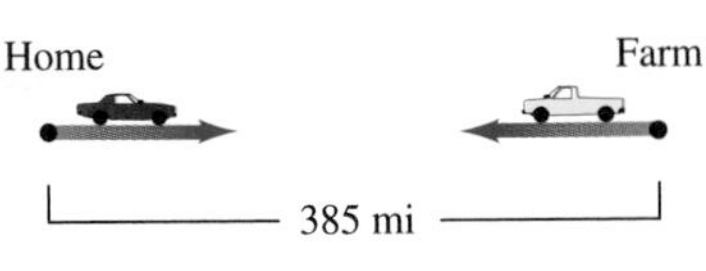

(a)

	r ·	t =	d
Parents	60	t	$60t$
Grandparents	50	t	$50t$

(b)

FIGURE 1-23

The distance the parents travel	plus	the distance the grandparents travel	is	the distance between the child's home and the grandparents' farm.
$60t$	$+$	$50t$	$=$	385

We now solve the equation.

Solve the equation

$$60t + 50t = 385$$
$$110t = 385 \quad \text{Combine like terms.}$$
$$t = 3.5 \quad \text{Divide both sides by 110.}$$

State the conclusion The parents and grandparents will meet in $3\frac{1}{2}$ hours.

Check the result The parents travel $3.5(60) = 210$ miles. The grandparents travel $3.5(50) = 175$ miles. The total distance traveled is $210 + 175 = 385$ miles. The answer checks. ■

WARNING! When using the formula $d = rt$, make sure that the units match. For example, if the rate is given in miles per *hour,* the time must be expressed in *hours* and the distance in miles.

Mixture problems

We now discuss two types of mixture problems. In the first example, a *dry mixture* of a specified value is created from two differently priced components.

EXAMPLE 8 ***Mixing nuts.*** The owner of a candy store notices that 20 pounds of gourmet cashews are getting stale. They did not sell because of their high price of \$12 per pound. The owner decides to mix peanuts with the cashews to lower the price per pound. If peanuts sell for \$3 per pound, how many pounds of peanuts must be mixed with the cashews to make a mixture that could be sold for \$6 per pound?

Analyze the problem To solve this problem, we will use the formula $v = pn$, where v represents value, p represents the price per pound, and n represents the number of pounds.

Form an equation We can let x represent the number of pounds of peanuts to be used. Then $20 + x$ represents the number of pounds in the mixture. We enter the known information in the table shown in Figure 1-24. The value of the cashews plus the value of the peanuts will be equal to the value of the mixture.

	p ·	n =	v
Cashews	12	20	240
Peanuts	3	x	$3x$
Mixture	6	$20 + x$	$6(20 + x)$

FIGURE 1-24

Next we translate the verbal model into mathematical symbols.

The value of the cashews	plus	the value of the peanuts	is	the value of the mixture.
240	$+$	$3x$	$=$	$6(20 + x)$

We now solve the equation.

Solve the equation

$$240 + 3x = 6(20 + x)$$

$$240 + 3x = 120 + 6x \quad \text{Use the distributive property to remove parentheses.}$$

$$120 = 3x \quad \text{Subtract } 3x \text{ and 120 from both sides.}$$

$$40 = x \quad \text{Divide both sides by 3.}$$

State the conclusion The owner should mix 40 pounds of peanuts with the 20 pounds of cashews.

Check the result The cashews are valued at \$12(20) = \$240, and the peanuts are valued at \$3(40) = \$120. The mixture is valued at \$6(60) = \$360. Since the value of the cashews plus the value of the peanuts equals the value of the mixture, the answer checks. ■

In the next example, a *liquid mixture* of a desired strength is to be made from two solutions with different concentrations.

EXAMPLE 9 ***Milk production.*** Owners of a dairy find that milk with a 2% butterfat content is their best seller. Suppose a stainless steel tank at the dairy contains 12 liters of milk containing 4% butterfat. How much 1% milk must be mixed with it to get a mixture that is 2% butterfat?

Analyze the problem In Figure 1-25(a), the first tank contains 12 liters of 4% milk. The second tank contains the amount of 1% milk that must be added to the first tank to obtain the desired concentration. The third tank, the 2% mixture, contains the contents of the first two tanks.

The amount of butterfat in a tank is the *product* of the percent butterfat and the amount of milk in the tank. So the first tank contains 4% of 12 liters = 0.04(12) liters of butterfat.

Form an equation If we let l represent the number of liters of 1% milk, then the second tank contains $0.01l$ liters of butterfat. Upon mixing, the third tank will contain $12 + l$ liters of milk that is 2% butterfat, or $0.02(12 + l)$ liters of butterfat. The sum of the amounts of butterfat in the first two tanks is $0.04(12) + 0.01l$. This should equal the amount of butterfat in the third tank, which is $0.02(12 + l)$ liters. This information is presented in table form in Figure 1-25(b).

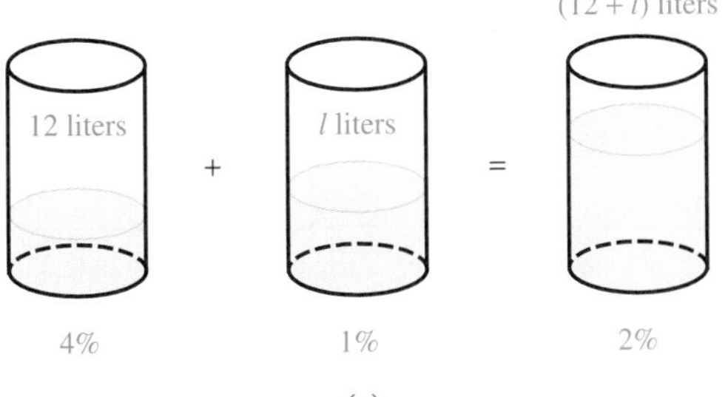

(a)

	Percent butterfat ·	Amount of milk =	Amount of butterfat
4% milk	0.04	12	0.04(12)
1% milk	0.01	l	$0.1l$
2% milk	0.02	$12 + l$	$0.02(12 + l)$

(b)

FIGURE 1-25

Next we translate the words into mathematical symbols.

The amount of butterfat in 12 liters of 4% milk	plus	the amount of butterfat in l liters of 1% milk	is	the amount of butterfat in $(12 + l)$ liters of mixture.
0.04(12)	+	$0.01l$	=	$0.02(12 + l)$

Solve the equation

$$0.04(12) + 0.01l = 0.02(12 + l)$$

$$4(12) + 1l = 2(12 + l) \quad \text{Multiply both sides by 100.}$$

$$48 + l = 24 + 2l \quad \text{Use the distributive property to remove parentheses.}$$

$$24 = l \quad \text{Subtract 24 and } l \text{ from both sides.}$$

State the conclusion Thus, 24 liters of 1% milk should be added to get a mixture that is 2% butterfat.

Check the result 12 liters of 4% milk contains 0.48 liters of butterfat, and 24 liters of 1% milk contains 0.24 liters of butterfat. This gives a total of 36 liters of a mixture that contains 0.72 liters of butterfat. Since this is a 2% solution, the answer checks. ■

STUDY SET Section 1.7

VOCABULARY *In Exercises 1–6, fill in the blanks to make the statements true.*

1. When an investment is made, the amount of money invested is called the _______________.

2. The value that occurs the most in a distribution of numbers is called the _______________.

3. The middle value of a distribution of numbers is called the _______________.

4. The _______________ of several values is the sum of those values divided by the number of values.

5. In the statement, "10 is 20% of 50," 10 is the _______________, and 50 is the _______________.

6. When the regular price of an item is reduced, the amount of reduction is called the _______________.

CONCEPTS

7. Complete the table in Illustration 1 for each 1-year investment.

Account	Principal	Rate	Interest earned
CD	$1,500	6%	
Bonds	$\$x$	5.65%	
Stocks	$\$(850 - x)$	7%	

ILLUSTRATION 1

8. Complete the table in Illustration 2.

	Price	Pounds	Value
M & M plain	$2.49	30	
M & M peanut	$2.79	p	
Mixture	$2.59	$p + 30$	

ILLUSTRATION 2

9. a. If $5,000 of $30,000 is invested at 5%, how much is left to be invested at another rate?

b. If $\$x$ of $30,000 is invested at 5%, how much is left to be invested at another rate?

10. Complete the table in Illustration 3, given the speed of light and sound through air and water.

	Rate	Time	Distance
Light (air)	180,693 mi/sec	60 sec	
Light (water)	136,711 mi/sec	60 sec	
Sound (air)	1,088 ft/sec	x sec	
Sound (water)	5,440 ft/sec	$(x - 3)$ sec	

ILLUSTRATION 3

11. The two coolers shown in Illustration 4 contain a mixture of punch concentrate and water.

a. What percent water is the punch in cooler 1?

b. Find the amount of punch concentrate in cooler 1 and cooler 2.

c. If the contents of each cooler are mixed together, how much punch will there be?

d. How much punch concentrate will the mixture contain?

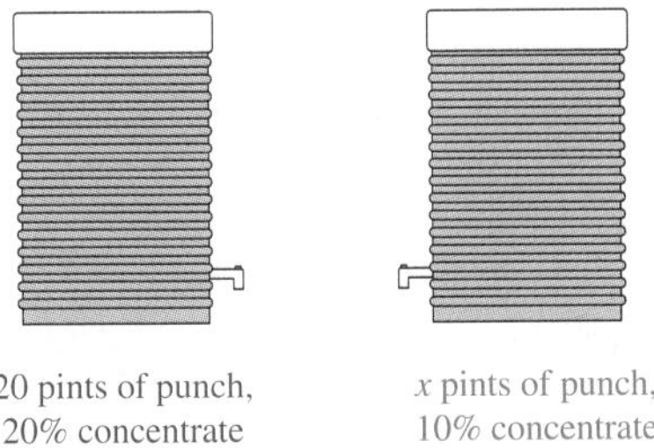

ILLUSTRATION 4

12. The top three grossing films for the weekend of November 7–9, 1997, are shown in Illustration 5. See the "% chg. from last" column in the table.

a. Why is there a blank for *Starship Troopers?*

b. Explain what +1,118% means for the movie *Bean.*

c. What does the negative entry for *I Know What You Did Last Summer* mean?

	Film	Wks. in release	Wknd. total millions	% chg. from last
1	Starship Troopers	1	$22.1	—
2	Bean	3	12.7	+1,118
3	I Know What You Did Last Summer	4	6.5	−31

Based on data from Entertainment Data Inc.

ILLUSTRATION 5

NOTATION *In Exercises 13–16, translate each statement into mathematical symbols.*

13. What number is 5% of 10.56?

14. 16 is what percent of 55?

15. 32.5 is 74% of what number?

16. What is 83.5% of 245?

In Exercises 17–18, complete each solution.

17. Solve $0.09x + 0.08(2{,}000 - x) = 400$.

$$\square[0.09x + 0.08(2{,}000 - x)] = \square(400)$$
$$\square + 8(2{,}000 - x) = \square$$
$$9x + \square - 8x = 40{,}000$$
$$x = 24{,}000$$

18. Solve $0.2(5) + 0.6l = 0.4(5 + l)$.

$$\square[0.2(5) + 0.6l) = \square[0.4(5 + l)]$$
$$\square(5) + 6l = \square(5 + l)$$
$$10 + 6l = 20 + \square$$
$$\square = 10$$
$$l = 5$$

19. What formula is used to solve simple interest problems?

20. What formula is used to solve uniform motion problems?

21. What formula is used to solve dry mixture problems.

22. Give the percent formula.

APPLICATIONS

23. ENERGY In 1994, the United States alone accounted for 24% of the world's total energy consumption, using 85.64 quadrillion British thermal units (Btu). A quadrillion is 1,000,000,000,000,000. What was the world's energy consumption in 1994? Round to the nearest hundredth of a quadrillion.

24. PERSONAL COMPUTERS In 1998, 47.8 million, or 47.3%, of U.S. households had a personal computer. How many U.S. households were there in 1998? Round to the nearest tenth of a million.

25. BUYING A WASHER AND A DRYER Find the percent of markdown of the sale in Illustration 6.

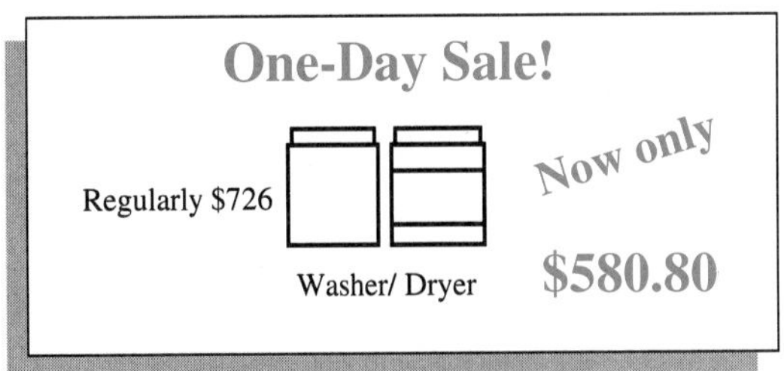

One-Day Sale!
Regularly $726
Now only
$580.80
Washer/ Dryer

ILLUSTRATION 6

26. BUYING FURNITURE A bedroom set regularly sells for $983. If it is on sale for $737.25, what is the percent of markdown?

27. FLEA MARKET A vendor sells tool chests at a flea market for $65. If he makes a profit of 30% on each unit sold, what does he pay the manufacturer for each tool chest? (*Hint:* The retail price = the wholesale price + the markup.)

28. MANAGING A BOOKSTORE A bookstore sells a textbook for $39.20. If the bookstore makes a profit of 40% on each sale, what does the bookstore pay the publisher for each book? (*Hint:* The retail price = the wholesale price + the markup.)

29. IMPROVING PERFORMANCE The graph in Illustration 7 shows how the installation of a special computer chip increases the horsepower of a truck. What is the percent increase in horsepower for the engine running at 4,000 revolutions per minute (rpm)? Round to the nearest percent.

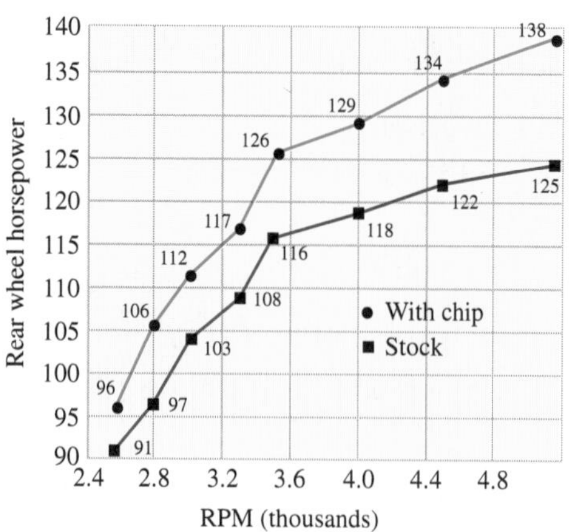

ILLUSTRATION 7

30. GREENHOUSE GASES Over the next fifteen years, the United States is striving to cut its carbon dioxide emissions from a high of 4,970.0 million tons per year to 4,671.8 million tons per year. This will help to lessen air pollution. What percent of decrease is this?

31. BROADWAY SHOWS Complete the table in Illustration 8 to find the percent increase in attendance at Broadway shows for each season compared to the previous season. Round to the nearest tenth.

Season	Broadway attendance	% increase
1994–95	9.04 million	—
1995–96	9.45 million	
1996–97	10.57 million	

Source: League of American Theater Producers Inc.

ILLUSTRATION 8

32. THE ART MARKET The line graph in Illustration 9 shows the annual auction sales at Christies, a major art auction house. Between what two years was there the greatest drop in sales? Approximate the percent of decrease for this period.

Christie's Annual Auction Sales ($ millions)

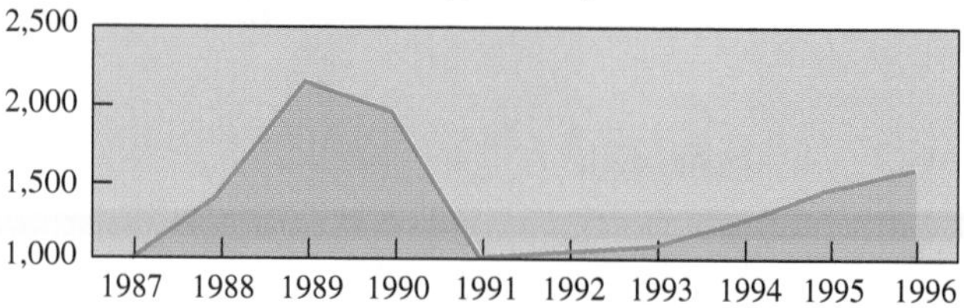

Based on data from *The Wall Street Journal Almanac* (1998)

ILLUSTRATION 9

33. FUEL EFFICIENCY The ten most fuel-efficient cars in 1997, based on manufacturer's estimated city and highway average miles per gallon (mpg), are shown in Illustration 10. Find the median and mode of both sets of data.

Model	mpg city/hwy
Geo Metro LSi	39/43
Honda Civic HX coupe	35/41
Honda Civic LX sedan	33/38
Mazda Protege	31/35
Nissan Sentra GXE	30/40
Toyota Paseo	29/37
Saturn SL1	29/40
Dodge Neon Sport Coupe	29/38
Hyundai Accent	29/38
Toyota Tercel DX	28/38

ILLUSTRATION 10

34. SPORT FISHING The report shown below lists the fishing conditions at Pyramid Lake for a Saturday in January. Find the median and the mode of the weights of the stripped bass caught at the lake.

Pyramid Lake—Some stripped bass are biting but are on the small side. Striking jigs and plastic worms. Water is cold: 38°. Weights of fish caught (lbs): 6, 9, 4, 7, 4, 3, 3, 5, 6, 9, 4, 5, 8, 13, 4, 5, 4, 6, 9

35. JOB TESTING To be accepted into a police training program, a recruit must have an average score of 85 on a battery of four tests. If a candidate scored 78 on the oral test, 91 on the physical fitness test, and 87 on the psychological test, what is the lowest score he can obtain on the written test and still be accepted into the training program?

36. GAS MILEAGE The city mileage estimates of Dodge models selling for $15,000–$20,000 are listed in Illustration 11. Suppose management wants to add a new model in this price range. What mileage should the new model get if the city mileage average for this group must then be 20.8 mpg?

Model	City mileage (mpg)
Avenger	20.3
Dakota SLT	14.1
Neon Highline	28.2
Stratus	16.9

ILLUSTRATION 11

37. INVESTING MONEY Lured by the ad in Illustration 12, a woman invested $12,000, some in a money market account and the rest in a 5-year CD. How much was invested in each account if the income from both investments is $1,060 per year?

First Republic Savings and Loan	
Account	**Rate**
NOW	5.5%
Savings	7.5%
Money Market	8.0%
Checking	4.0%
5-year CD	9.0%

ILLUSTRATION 12

38. ENTREPRENEURS Last year, a women's professional organization made two small-business loans totaling $28,000 to young women beginning their own businesses. The money was lent at 7% and 10% simple interest rates. If the annual income the organization received from these loans was $2,560, what was each loan amount?

39. INHERITANCE Paul split an inheritance between two investments, one a certificate of deposit paying 7% annual interest, and the other a promising biotech company offering an annual return on investment of 10%. He invested twice as much in the 10% investment as he did in the 7% investment. If his combined annual income from the two investments was $4,050, how much did he inherit?

40. INCOME TAX RETURN On the federal income tax form Schedule B in Illustration 13, a taxpayer forgot to write in the amount of interest income he earned for the year. From what is written on the form, determine the amount of interest earned from each investment and the amount he invested in stocks.

Schedule B-Interest and Dividend Income		
Part 1 Interest Income (See pages 12 and B1.)	**Note:** If you had over $400 in taxable income, use this form	
	1 List name of payer.	**Amount**
	① MONEY MARKET ACCT.	
	DEPOSITED $15,000 @ 3.3%	SAME
		AMOUNT
	② STOCKS	FROM
	EARNED 5%	EACH

ILLUSTRATION 13

41. TRAVEL TIME At lunchtime, a man called his wife from his office to tell her that they needed to switch vehicles so that he could use the family van to pick up some building materials after work. The wife then left their home, traveling towards his office in their van at 35 mph. At the same time, the husband left his office in his car, traveling towards their home at 45 mph. If his office is 20 miles from their home, how long will it take them to meet so they can switch vehicles?

42. CYCLING A cyclist leaves his training base for a morning workout, riding at the rate of 18 mph. One hour later, his support staff leaves the base in a car going 45 mph in the same direction. How long will it take the support staff to catch up with the cyclist?

43. RADIO COMMUNICATION At 2 P.M., two military convoys leave Eagle River, WI, one headed north and one headed south. The convoy headed north averages 50 mph, and the convoy headed south averages 40 mph. They will lose radio contact when the distance between them is more than 135 miles. When will this occur?

44. RUNNING A MARATHON RACE Two marathon runners leave the starting gate, one running 12 mph and the other 10 mph. If they maintain the pace, how long will it take for them to be one-quarter of a mile apart?

45. JET SKIING A jetski can go 12 mph in still water. If a rider goes upstream for 3 hours against a current of 4 mph, how long will it take the rider to return? (*Hint:* Upstream speed is (12 − 4) mph; how far can the rider go in 3 hours?)

46. PHYSICAL FITNESS For her workout, Sarah walks north at the rate of 3 mph and returns at the rate of 4 mph. How many miles does she walk if the round trip takes 3.5 hours?

47. SWEET AND SOUR The owner of a candy store wants to make a 30-pound mixture of two candies to sell for $1 per pound. If red licorice bits sell for 95¢ per pound and lemon gumdrops sell for $1.10 per pound, how many pounds of each should be used?

48. HEALTH FOOD A pound of dried pineapple bits sells for $6.19, a pound of dried banana chips sells for $4.19, and a pound of raisins sells for $2.39 a pound. Two pounds of raisins are to be mixed with equal amounts of pineapple and banana to create a trail mix that will sell for $4.19 a pound. How many pounds of pineapple and banana chips should be used?

49. DILUTING SOLUTIONS In Illustration 14, how much water should be added to 20 ounces of a 15% solution of alcohol to dilute it to a 10% solution?

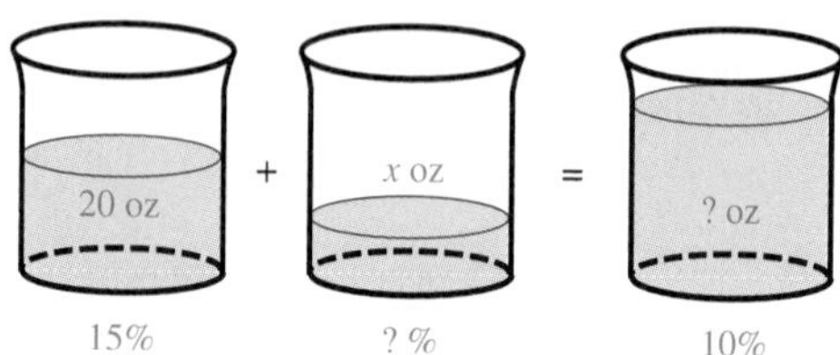

ILLUSTRATION 14

50. INCREASING CONCENTRATION The beaker shown in Illustration 15 contains a 2% saltwater solution.

a. How much water must be boiled away to increase the concentration of the salt solution from 2% to 3%?

b. Where on the beaker would the new water level be?

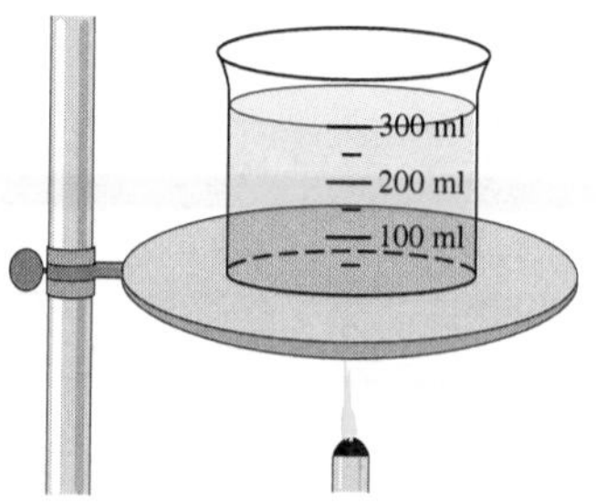

ILLUSTRATION 15

51. DAIRY FOODS Cream is approximately 22% butterfat. How many gallons of cream must be mixed with milk testing at 2% butterfat to get 20 gallons of milk containing 4% butterfat?

52. LOWERING FAT How many pounds of extra-lean hamburger that is 7% fat must be mixed with 30 pounds of lean hamburger that is 15% fat to obtain a mixture that is 10% fat?

WRITING *Write a paragraph using your own words.*

53. If a car travels at 60 mph for 30 minutes, explain why the distance traveled is *not* $60 \cdot 30 = 1{,}800$ miles.

54. If a mixture is to be made from solutions with concentrations of 12% and 30%, can the mixture have a concentration less than 12% or greater than 30%? Explain.

REVIEW *In Exercises 55–58, solve each equation.*

55. $9x - 3 = 6x$

56. $7a + 2 = 12 - 4(a - 3)$

57. $\frac{8(y-5)}{3} = 2(y - 4)$

58. $\frac{t-1}{3} = \frac{t+2}{6} + 2$

KEY CONCEPT

Let x =

In Chapter 1, we discussed one of the most important problem-solving techniques used in algebra. In this method, the first step is to let a variable represent the unknown quantity. Then we use the variable in writing an equation that mathematically describes the situation in question. Finally, we solve the equation to find the value represented by the variable and state the solution called for in the problem. Let's review some of the key parts of this problem-solving method.

Analyzing the Problem

Always begin problem solving by reading the problem carefully. What facts are given? What are we asked to find? Can we infer anything? In Exercises 1 and 2, what do we know and what are we asked to find?

1. GEOGRAPHY Of the 48 contiguous states, 4 more lie east of the Mississippi River than lie west of the Mississippi. How many states are west of the Mississippi River?

2. GEOMETRY In a right triangle, the measure of one acute angle is 5° more than twice that of the other angle. Find the measure of the smallest angle.

Letting a Variable Represent an Unknown Quantity

Most of the time, we let the variable represent what we are asked to find. In Exercises 3 and 4, what quantity should the variable represent? State your response in the form "Let $x =$ . . .".

3. CAMPING To make anchor lines for a tent, a 60-foot rope is cut into four pieces, each successive piece twice as long as the previous one. Find the length of each anchor line.

4. INSURANCE COVERAGE While waiting for his van to be repaired, a man rents a car for $12 per day and 10 cents per mile. His insurance company will pay up to $100 of the rental fee. If he needs the car for two days, how many miles of driving will his policy cover?

Forming an Equation

As Exercises 5–8 show, several methods can be used to help form an equation.

5. TRANSLATION For each phrase, what operation is indicated?

a. less than **b.** of
c. increased by **d.** ratio

6. FORMULAS What formula is suggested by each type of problem?

a. Uniform motion **b.** Simple interest
c. Dry mixture **d.** Perimeter of a rectangle

7. TABLES Complete the table. What equation is suggested?

	% alcohol	Amount	Amount alcohol
Spray 1	0.15	15 oz	
Spray 2	0.50	x oz	
Mixture	0.40	$(15 + x)$ oz	

8. DIAGRAMS What equation is suggested by the diagram below?

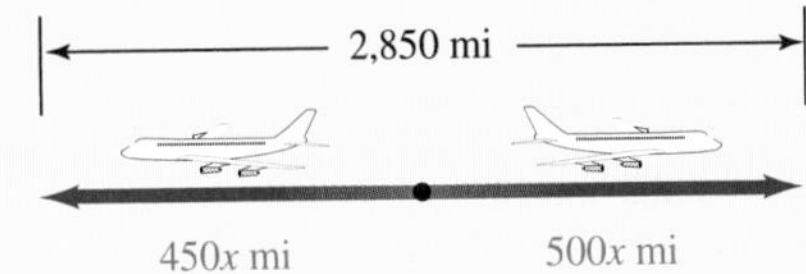

Accent on Teamwork

Section 1.1

PRODUCTION PLANNING In the Study Set for Section 1.1, Exercise 42 asks for a series of equations to be written to help a production planner order the correct number of parts for a production run of r towel racks. See Illustration 1.

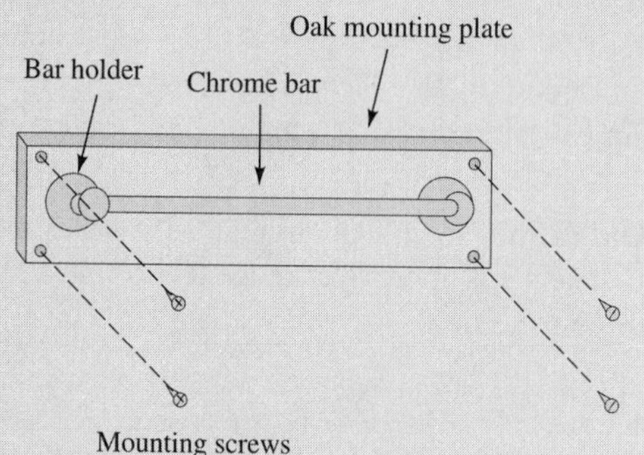

ILLUSTRATION 1

a. Decide on a product for which you will be the production planner. Make a detailed drawing of it, like the one in Illustration 1.
b. For one component of your product, create a table that gives the number that should be ordered for production runs of 50, 100, 150, and 200 units. Do the same for a second component of your product, using a bar graph, and for a third component, using a line graph.
c. For each of the three components in part b, write an equation that will give the number of components to be ordered for a production run of u units.

Section 1.2

NUMBER LINES Exercise 70 in Study Set 1.2 shows how a number line is used in chemistry to illustrate pH. Go to the library and look in some textbooks and encyclopedias to find examples of number lines that are used in other disciplines, such as history, medicine, psychology, and cosmetology.

Section 1.3

ORDER OF OPERATIONS In making a cake from a mix, the instructions must be followed carefully. Otherwise, the results can be disastrous. Think of two other multistep processes. Explain why the steps must be performed in the proper order or the outcome is greatly affected. Think of two processes for which the outcome is not affected by the order in which the steps are performed.

Section 1.4

PROPERTIES OF REAL NUMBERS Use the numbers 100 and 10 to show that there is no commutative property of subtraction or of division. Use 100, 10, and 2 to show that there is no associative property of subtraction or of division.

Section 1.5

SUBTRACTION PROPERTY OF EQUALITY Borrow a scale and some weights from your school's science department and use them as part of a class presentation to model how the subtraction property of equality is used to solve the equation $x + 2 = 5$.

Section 1.6

GOLDEN RECTANGLE See Exercise 33 in Study Set 1.6. Cut out a set of six rectangles, each with the same width but with different lengths. One of the rectangles should have a length 1.618 times its width; this is the golden rectangle. Paste the rectangles on a piece of poster board randomly. (See Illustration 2.) Survey 50 people. Ask them to pick the rectangle that is "the most visually appealing." See if the golden rectangle is picked most often. Make a table showing the results.

ILLUSTRATION 2

Section 1.7

MIXTURE PROBLEM Make a 10% solution of lemonade by mixing 1 paper cup of lemonade concentrate with 9 paper cups of water in a pitcher. In a similar manner, using a 4-to-6 ratio, make a 40% solution in another pitcher. Then use algebra to determine how many ounces of the 40% lemonade must be mixed with 8 ounces of the 10% lemonade to get a 20% solution. Measure out the appropriate amounts and make the 20% mixture. Taste each of the three solutions. Can you tell the difference in the concentration?

CHAPTER REVIEW

SECTION 1.1 Describing Numerical Relationships

CONCEPT

A *variable* is a letter that stands for a number.

An *equation* is a mathematical sentence that contains an = sign.

Formulas are equations that express a known relationship between two or more quantities.

Bar graphs and *line graphs* are used to describe numerical relationships.

The *perimeter* of a geometric figure is the distance around it.

REVIEW EXERCISES

1. Translate each verbal model into a mathematical model.

a. The cost C (in dollars) to rent t tables is \$15 more than the product of \$2 and t.

b. A rectangle has an area of 25 in.^{2}. The length of the rectangle is the quotient of its area and its width.

c. The waiting period for a business license is now 3 weeks less than it used to be.

2. To determine the proper cooking time for prime rib, a cookbook suggests using the formula $T = 30p$, where T is the cooking time in minutes and p is the weight of the prime rib in pounds. Use this formula to complete the table.

p	T
6.0	
6.5	
7.0	
7.5	
8.0	

3. Use the data from the table in Exercise 2 to draw each type of graph.

a. Bar graph

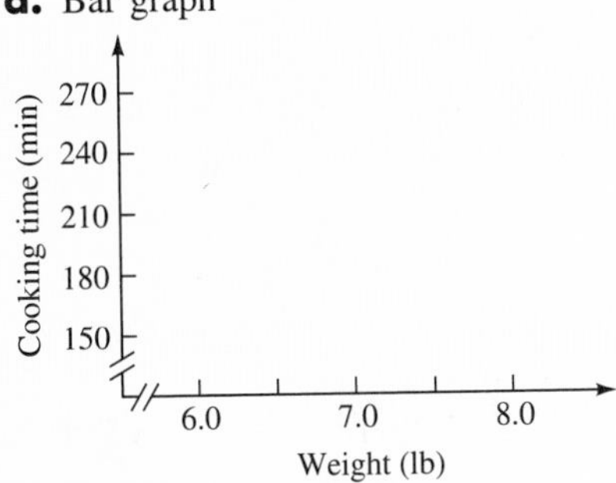

b. Line graph

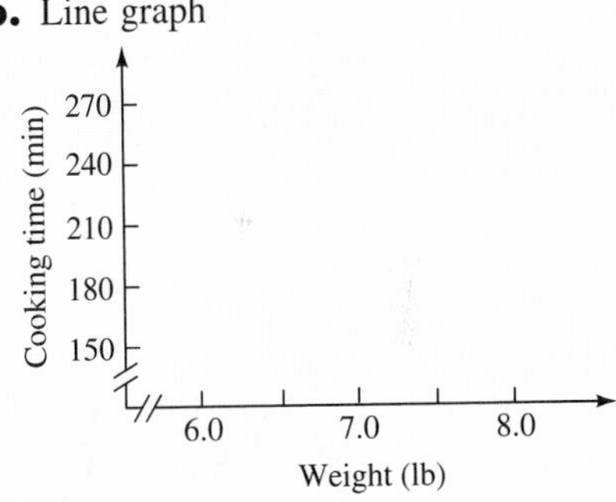

4. The owner of a new business wants to frame the first dollar bill her business ever received. How long a piece of molding will she need if a dollar is 2.625 inches wide and 6.125 inches long?

SECTION 1.2 The Real Number System

Natural numbers:
$\{1, 2, 3, \ldots\}$

Whole numbers:
$\{0, 1, 2, 3, \ldots\}$

Integers:
$\{\ldots, -3, -2, -1, 0, 1, 2, 3, \ldots\}$

5. List the numbers in $\left\{-5, 0, -\sqrt{3}, 2.4, 7, -\frac{2}{3}, -3.\overline{6}, \pi, \frac{15}{4}, 0.13242368\ldots\right\}$ that satisfy the given condition.

a. Natural numbers

b. Whole numbers

c. Integers

d. Rational numbers

e. Irrational numbers

f. Real numbers

g. Negative numbers

h. Positive numbers

Prime numbers:
2, 3, 5, 7, 11, 13, . . .

Composite numbers:
4, 6, 8, 9, 10, 12, . . .

Integers divisible by 2 are *even integers.* Integers not divisible by 2 are *odd integers.*

i. Prime numbers **j.** Composite numbers

k. Even integers **l.** Odd integers

Rational numbers are numbers that can be written as $\frac{a}{b}$ $(b \neq 0)$, where a and b are integers. They form the set of all terminating and all repeating decimals.

6. Use one of the symbols > or < to make each statement true.
a. $-16 \, \square \, -17$ **b.** $-(-1.8) \, \square \, 2\frac{1}{2}$

7. Tell whether each statement is true or false.
a. $23.000001 \geq 23.1$ **b.** $-11 \leq -11$

Irrational numbers are numbers that can be written as nonterminating, nonrepeating decimals.

8. Graph the prime numbers between 20 and 30 on the number line.

20 21 22 23 24 25 26 27 28 29 30

A *real number* is any number that is either a rational or an irrational number.

9. Graph the set $\left\{2.75, 2.\overline{3}, \sqrt{7}, \frac{8}{3}, \frac{3\pi}{4}\right\}$ on the number line.

2 2.5 3

For any real number x:

$$\begin{cases} \text{If } x \geq 0, \text{ then } |x| = x. \\ \text{If } x < 0, \text{ then } |x| = -x. \end{cases}$$

10. Write each expression without using absolute value symbols.
a. $|-18|$ **b.** $-|6.26|$

SECTION 1.3 *Operations with Real Numbers*

Adding real numbers: With like signs, add the absolute values and keep the common sign. With unlike signs, subtract the absolute values and keep the sign of the number that has the greatest absolute value.

Subtracting real numbers:

$$x - y = x + (-y)$$

Multiplying and dividing real numbers: With like signs, multiply (or divide) their absolute values. The sign is positive. With unlike signs, multiply (or divide) their absolute values. The sign is negative.

11. Do the operations.
a. $-3 + (-4)$ **b.** $-70.5 + 80.6$

c. $-\frac{1}{2} - \frac{1}{4}$ **d.** $-6 - (-8)$

e. $(-4.2)(-3.0)$ **f.** $-\frac{1}{10} \cdot \frac{5}{16}$

g. $\frac{-2.2}{-11}$ **h.** $-\frac{9}{8} \div 21$

i. $15 - 25 - 23$ **j.** $-3.5 + (-7.1) + 4.9$

k. $-3(-5)(-8)$ **l.** $-1(-1)(-1)(-1)$

x^n is a *power of x*. x is the base, and n is the *exponent*.

$$x^n = \underbrace{x \cdot x \cdot x \cdot \cdots \cdot x}_{n \text{ factors of } x}$$

A number b is a *square root* of a if $b^2 = a$. $\sqrt{a}$ represents the *principal* (positive) *square root* of a.

Order of operations: Work from the innermost pair to the outermost pair of grouping symbols in the following order.

1. Evaluate all powers and roots.
2. Do all multiplications and divisions, working from left to right.
3. Do all additions and subtractions, working from left to right.

When the grouping symbols have been removed, repeat the steps above to finish the calculation. In a fraction, simplify the numerator and denominator separately, and then simplify the fraction.

To *evaluate an algebraic expression,* substitute the values for the variables and then apply the rules for the order of operations.

The *area* of a figure is the amount of surface it encloses. The *volume* of a figure is its capacity.

12. Evaluate each expression.

a. $(-3)^5$ **b.** $\left(-\frac{2}{3}\right)^2$

c. 0.3 cubed **d.** -5^2

13. Evaluate each expression.

a. $\sqrt{4}$ **b.** $-\sqrt{100}$

c. $\sqrt{\frac{9}{25}}$ **d.** $\sqrt{0.64}$

14. Evaluate each expression.

a. $-6 + 2(-5)^2$ **b.** $\frac{-20}{4} - (-3)(-2)\left(-\sqrt{1}\right)$

c. $4 - (5 - 9)^2$ **d.** $4 + 6[-1 - 5(25 - 3^3)]$

e. $2|-1.3 + (-2.7)|$ **f.** $\frac{(7-6)^4 + 32}{36 - \left(\sqrt{16} + 1\right)^2}$

g. $-2(4)^2$ **h.** $-(-2 \cdot 4)^2$

15. Evaluate the algebraic expression for the given values of the variables.

a. $(x + y)(x^2 - xy + y^2)$ for $x = -2$ and $y = 4$

b. $\frac{-b - \sqrt{b^2 - 4ac}}{2a}$ for $a = 2$, $b = -3$, and $c = -2$

16. SAFETY CONES See Illustration 1.

a. Find the area covered by the square rubber base if its sides are 10 inches long.

b. Find the volume of the cone that is centered atop the base. Round to the nearest tenth.

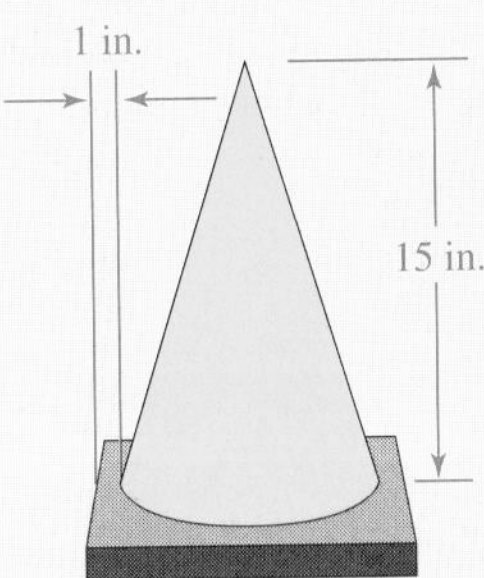

ILLUSTRATION 1

SECTION 1.4 Simplifying Algebraic Expressions

Properties of real numbers:

1. *Associative properties:*
   $(a + b) + c = a + (b + c)$
   $(ab)c = a(bc)$

17. Fill in the blanks to make the statements true by applying the indicated property of the real numbers.

a. $3(x + 7) =$ __________
Distributive property (and simplify)

b. $t \cdot 5 =$ ____
Commutative property of multiplication

2. *Commutative properties:*
$a + b = b + a$
$ab = ba$
3. *Distributive property:*
$a(b + c) = ab + ac$
4. 0 is the *additive identity:*
$a + 0 = 0 + a = a$
5. 1 is the *multiplicative identity:*
$1 \cdot a = a \cdot 1 = a$
6. *Multiplication property of* 0:
$a \cdot 0 = 0 \cdot a = 0$
7. $-a$ is the *negative* (or *additive inverse*) of a:
$a + (-a) = 0$
8. If $a \neq 0$, then $\frac{1}{a}$ is the reciprocal (or multiplicative inverse of a):

$$a \cdot \frac{1}{a} = \frac{1}{a} \cdot a = 1$$

c. $-x + x = $ ___
Additive inverse property

d. $(27 + 1) + 99 = $ ______
Associative property of addition

e. $\frac{1}{8} \cdot 8 = $ ___
Multiplicative inverse property

f. $0 + m = $ ___
Additive identity property

g. ___ $\cdot 9.87 = 9.87$
Multiplicative identity property

h. $5(-9)(0)(2{,}345) = $ ___
Multiplication property of 0

i. $(-3 \cdot 5)2 = $ ______
Associative property of multiplication

j. $(t + z) \cdot t = $ ______
Commutative property of addition

18. Do each division, if possible.

a. $\dfrac{102}{102}$

b. $\dfrac{0}{6}$

c. $\dfrac{-25}{1}$

d. $\dfrac{5.88}{0}$

To *simplify algebraic expressions,* we use properties of real numbers to write them in a less complicated form.

19. Remove the parentheses and simplify.

a. $8(x + 6)$

b. $-6(x - 2)$

c. $-(-4 + 3y)$

d. $(3x - 2y)1.2$

e. $10(2c^2 - c + 1)$

f. $\frac{2}{3}(3t + 9)$

Terms with the same variables and the same exponents are called *like* (similar) *terms.*

To *combine like terms,* add (or subtract) their coefficients and keep the same variables with the same exponents.

20. Simplify each expression.

a. $8(6k)$

b. $(-7x)(-10y)$

c. $-9(-3p)(-7)$

d. $15a + 30a + 7$

e. $3g^2 - 3g^2$

f. $-m + 4(m - 12)$

g. $\frac{7}{8}x + \frac{1}{8}x$

h. $21.45l - 45.99l$

i. $5t^3 + 6t^2 - 2t^2 + t^3$

j. $8(2h + 9) - 5(h - 1)$

SECTION 1.5 Solving Linear Equations and Formulas

The set of numbers that satisfy an equation is called its *solution set.*

21. Determine whether -6 is a solution of each equation.

a. $6 - x = 2x + 24$

b. $\dfrac{5}{3}(x - 3) = -12$

Properties of equality: If $a = b$, then

$$a + c = b + c$$
$$a - c = b - c$$
$$ca = cb \quad (c \neq 0)$$
$$\frac{a}{c} = \frac{b}{c} \quad (c \neq 0)$$

22. Solve each equation and give the solution set.

a. $\dfrac{x}{5} = -45$

b. $t - 3.67 = 4.23$

c. $0.0035 = 0.25g$

d. $0 = x + 4$

To solve a linear equation:

1. Clear the equation of any fractions.
2. Remove all parentheses and combine like terms.
3. Get all variables on one side and all numbers on the other. Combine like terms.
4. Make the coefficient of the variable 1.
5. Check the result.

23. Solve each equation.

a. $5x + 12 = 0$

b. $-3x - 7 + x = 6x + 20 - 5x$

c. $4(y - 1) = 28$

d. $2 - 13(x - 1) = 4 - 6x$

e. $\frac{8(x - 5)}{3} = 2(x - 4)$

f. $\frac{3y}{4} - 14 = -\frac{y}{3} - 1$

g. $-k = -0.06$

h. $\frac{5}{4}p = -10$

i. $\frac{4t + 1}{3} - \frac{t + 5}{6} = \frac{t - 3}{6}$

j. $33.9 - 0.5(75 - 3x) = 0.9$

An *identity* is an equation that is satisfied by every number for which both sides are defined. An *impossible equation* has no solution.

24. Solve each equation. If the equation is an identity or an impossible equation, so state.

a. $2(x - 6) = 10 + 2x$

b. $-5x + 2x - 1 = -(3x + 1)$

To *solve a formula* for a variable, isolate that variable on one side and isolate all other quantities on the other side.

25. Solve each formula for the indicated variable.

a. $V = \pi r^2 h$ for h

b. $Y + 2g = m$ for g

c. $\frac{T}{6} = \frac{1}{6}ab(x + y)$ for x

d. $V = \frac{4}{3}\pi r^3$ for r^3

SECTION 1.6 Using Equations to Solve Problems

Problem-solving strategy:

1. Analyze the problem.
2. Form an equation.
3. Solve the equation.
4. State the conclusion.
5. Check the result.

26. AIRPORTS The world's two busiest airports are Chicago O'Hare International and Hartsfield Atlanta International. Together they served 132 million passengers in 1996, with O'Hare handling 6 million more than Atlanta. How many passengers did each airport serve?

27. TUITION A private school reduces the monthly tuition cost of \$245 by \$5 per child if a family has more than one child attending the school. Write an algebraic expression that gives the monthly tuition cost per child for a family having c children.

Number · value = total value

28. TREASURY BILLS What is the value of five \$1,000 T-bills? What is the value of x \$1,000 T-bills?

29. CABLE TV A 186-foot television cable is to be cut into four pieces. Find the length of each piece if each successive piece is 3 feet longer than the previous one.

30. TOOLING Illustration 2 shows the angle at which a drill is to be held when drilling a hole into a piece of aluminum. Find the measures of both angles labeled in the illustration.

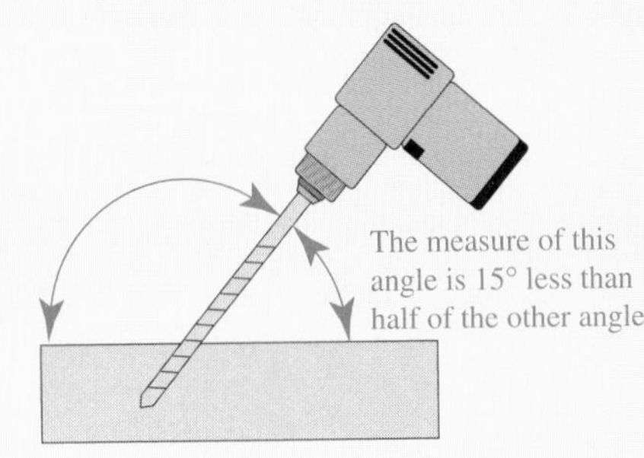
The measure of this angle is 15° less than half of the other angle.

ILLUSTRATION 2

Interest = Principal · rate · time

31. INVESTMENTS Sally has \$25,000 to invest. She invests some money at 10% interest and the rest at 9%. If her total annual income from these two investments is \$2,430, how much does she invest at each rate?

Percent means per one hundred.

Amount = percent · base

32. APPAREL SALES Use the information in the graph in Illustration 3 to determine whether the largest percent increase in apparel sales from 1995 to 1996 occurred in men's, women's, or children's clothing.

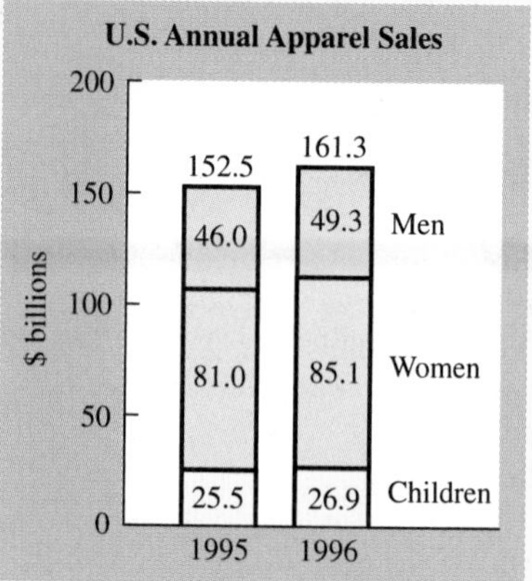

Based on data from *The Wall Street Journal Almanac* (1998)

ILLUSTRATION 3

$Mean = \dfrac{\text{sum of the values}}{\text{number of values}}$.

The *mode* is the value that occurs most often.

The *median* is the middle value.

33. SCHOLASTIC APTITUDE TEST The mean SAT verbal test scores of college-bound seniors for the years 1990–1996 are listed below. Find the mean, median, and mode.

1990	1991	1992	1993	1994	1995	1996
500	499	500	500	499	504	505

Distance = rate · time

34. PAPARAZZI A celebrity leaves a nightclub in his car and travels at 1 mile per minute (60 mph) trying to avoid a tabloid photographer. One minute later, the photographer leaves the nightclub on his motorcycle, traveling at 1.5 miles per minute (90 mph) in pursuit of the celebrity. How long will it take the photographer to catch up with the celebrity?

35. PEST CONTROL How much water must be added to 20 gallons of a 12% pesticide/water solution to dilute it to an 8% solution?

$v = pn$, where v is the value, p is the price per pound, and n is the number of pounds.

36. CANDY SALES Write an algebraic expression that gives the value of a mixture of 3 pounds of Tootsie Rolls with x pounds of Bit O'Honey, if the mix is to sell for \$1.95 per pound.

Chapter 1 Test

In Problems 1 and 2, translate each verbal model into a mathematical model.

1. Each test score T was increased by 10 points to give a new adjusted test score s.

2. The area A of a triangle is the product of one-half the length of the base b and the height h.

In Problems 3–6, consider the set $\left\{-2, \pi, 0, -3\frac{3}{4}, 9.2, \frac{14}{5}, 5, -\sqrt{7}\right\}$.

3. Which numbers are integers?

4. Which numbers are rational numbers?

5. Which numbers are irrational numbers?

6. Which numbers are real numbers?

In Problems 7 and 8, graph each set on the number line.

7. $\left\{\frac{7}{6}, \frac{\pi}{2}, 1.8234503\ldots, \sqrt{3}, 1.\overline{91}\right\}$

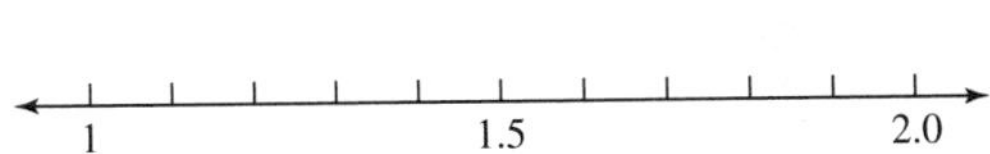

8. The set of prime numbers less than 12

0 1 2 3 4 5 6 7 8 9 10 11 12

In Problems 9 and 10, write each expression without using absolute value symbols.

9. $-|8|$

10. $|-5.5|$

In Problems 11–16, evaluate each expression.

11. $7 - (-5.3)$

12. $-\frac{5}{3}\left(-\frac{4}{25}\right)$

13. $\frac{1}{2} - \left(-\frac{3}{5}\right)$

14. $(-4)^3$

15. $\dfrac{2[-4 - 2(3 - 1)]}{3(\sqrt{9})(2)}$

16. $7 + 2[-1 - 4(5)]$

17. Evaluate the expression for $a = 2$, $b = -3$, and $c = 4$.

$$\frac{-3b + a}{ac - b}$$

18. PEDIATRICS Some doctors use Young's rule in calculating dosage for infants and children.

$$\frac{\text{Age of child}}{\text{Age of child} + 12}\left(\begin{array}{c}\text{average}\\ \text{adult dose}\end{array}\right) = \text{child's dose}$$

The adult dose of Achromycin is 250 mg. What is the dose for an 8-year old child?

In Problems 19 and 20, tell which property of real numbers justifies each statement.

19. $3 + 5 = 5 + 3$

20. $a(b + c) = ab + ac$

In Problems 21–24, simplify each expression.

21. $-y + 3y + 9y$

22. $5(-9q)$

23. $-(4 + t) + t$

24. $12\left(-\frac{4}{3} + \frac{1}{6}a + \frac{3}{2}b\right)$

In Problems 25 and 26, solve each equation.

25. $9(x + 4) + 4 = 4(x - 5)$

26. $\frac{y - 1}{5} + 2 = \frac{2y - 3}{3}$

In Problems 27 and 28, consider the numbers −2, 0, 2, −2, 3, −1, −1, 1, 1, 2.

27. Find the mean.

28. Find the median.

29. Solve $P = L + \frac{s}{f}i$ for i.

30. CALCULATORS The viewing window of a calculator has a perimeter of 26 centimeters and is 5 centimeters longer than it is wide. Find the dimensions of the window.

31. INVESTING An investment club invested part of \$10,000 at 9% annual interest and the rest at 8%. If the annual income from these investments was \$860, how much was invested at 8%?

32. GOLDSMITH How many ounces of a 40% gold alloy must be mixed with 10 ounces of a 10% gold alloy to obtain an alloy that is 25% gold?

33. Explain why the formula is not solved for A.

$$A = \frac{1 + A}{2b - 10}$$

34. What does it mean when we say that $x = 3$ is a solution of the equation $2x - 3 = x$?

Graphs, Equations of Lines, and Functions

2

Campus Connection

The *Earth Science* Department

In a physical geography class, students study the form and structure of the earth. To designate locations on the surface of the earth, geologists use a *coordinate system* consisting of a series of lines representing longitude and latitude. In Chapter 2, we will discuss the rectangular coordinate system. It allows us to make visual representations of mathematical relationships. Learning how to use the rectangular coordinate system will give you the insight you'll need when you encounter coordinate systems in other disciplines such as Earth Science.

MANY RELATIONSHIPS BETWEEN TWO QUANTITIES CAN BE DESCRIBED BY USING A TABLE, A GRAPH, OR AN EQUATION.

2.1 The Rectangular Coordinate System

In this section, you will learn about

- **The rectangular coordinate system**
- **Graphing mathematical relationships**
- **Reading graphs**
- **Step graphs**

INTRODUCTION. It is often said that a picture is worth a thousand words. In this chapter, we will show how numerical relationships can be described by using mathematical pictures called **graphs.** We will also show how graphs are constructed and how we can obtain important information from them.

The rectangular coordinate system

Many cities are laid out on a rectangular grid. For example, on the east side of Rockford, IL, all streets run north and south, and all avenues run east and west. (See Figure 2-1.) If we agree to list the street numbers first, every address can be identified by using an ordered pair of numbers. If Jose lives on the corner of Third Street and Sixth Avenue, his address is given by the ordered pair (3, 6).

This is the street. → (3, 6) ← This is the avenue.

If Lisa has an address of (6, 3), we know that she lives on the corner of Sixth Street and Third Avenue. From the figure, we can see that

- Bob Anderson's address is (4, 1).
- Rosa Vang's address is (7, 5).
- The address of the store is (8, 2).

The idea of associating an ordered pair of numbers with points on a grid is attributed to the 17th-century French mathematician René Descartes. The grid is often called a **rectangular coordinate system,** or **Cartesian coordinate system** after its inventor.

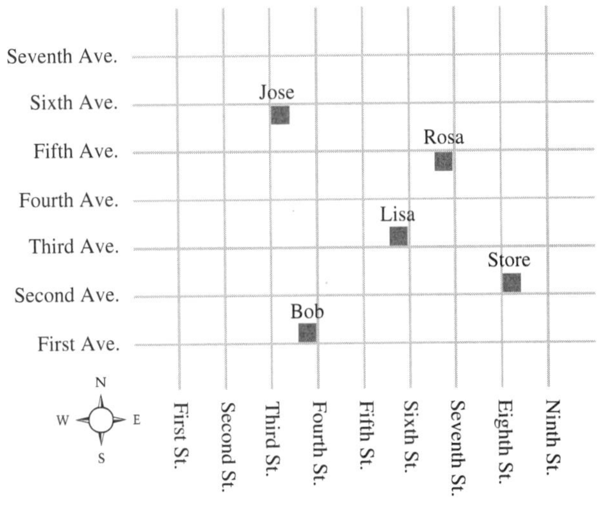

FIGURE 2-1

In general, a rectangular coordinate system is formed by two intersecting perpendicular number lines. (See Figure 2-2.)

- The horizontal number line is usually called the ***x*-axis.**
- The vertical number line is usually called the ***y*-axis.**

The positive direction on the x-axis is to the right, and the positive direction on the y-axis is upward. If no scale is indicated on the axes, we assume that the axes are scaled in units of 1.

The point where the axes cross is called the **origin.** This is the 0 point on each axis. The two axes form a **coordinate plane** and divide it into four regions called **quadrants,** which are numbered as shown in Figure 2-2.

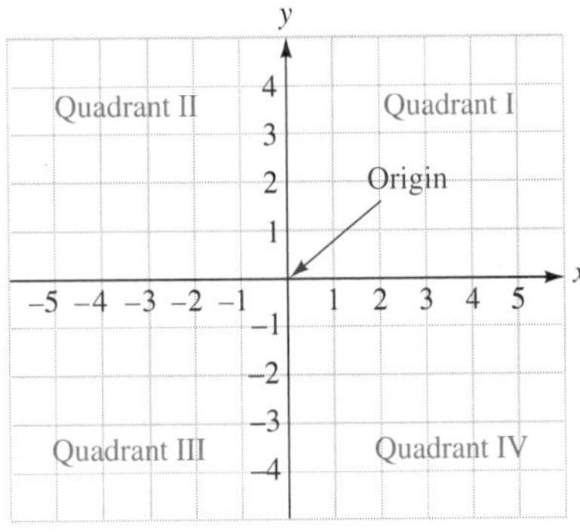

FIGURE 2-2

Every point on a coordinate plane can be identified by a pair of real numbers x and y, written as (x, y). The first number in the pair is the ***x*-coordinate,** and the second number is the ***y*-coordinate.** The numbers are called the **coordinates** of the point. Some examples of ordered pairs are $(-4, 6)$, $(2, 3)$, and $(6, -4)$.

$$(-4, 6)$$

In an ordered pair, the x-coordinate is listed first. — The y-coordinate is listed second.

The process of locating a point in the coordinate plane is called **graphing** or **plotting** the point. In Figure 2-3(a), we show how to graph the point Q with coordinates of

$(-4, 6)$. Since the x-coordinate is negative, we start at the origin and move 4 units to the left along the x-axis. Since the y-coordinate is positive, we then move up 6 units to locate point Q. Point Q is the **graph** of the ordered pair $(-4, 6)$ and lies in quadrant II.

To plot the point $P(2, 3)$, we start at the origin, move 2 units to the right along the x-axis, and then move up 3 units to locate point P. Point P lies in quadrant I. To plot point $R(6, -4)$, we start at the origin and move 6 units to the right and then 4 units down. Point R lies in quadrant IV.

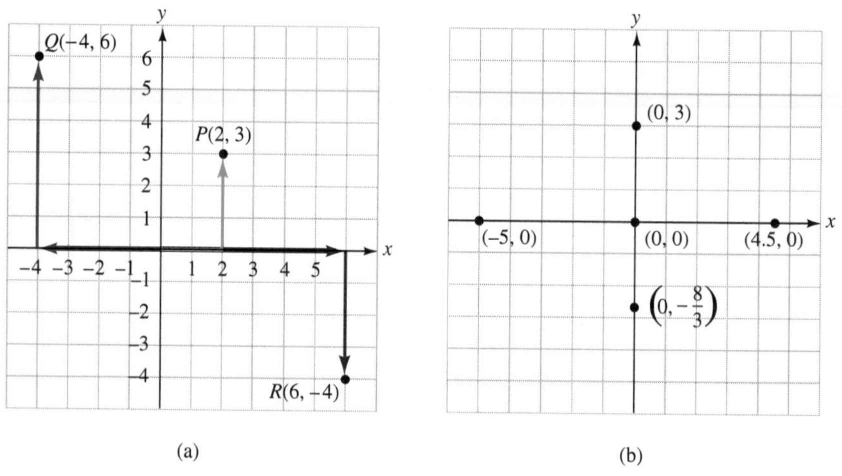

FIGURE 2-3

WARNING! Note that point Q with coordinates of $(-4, 6)$ is not the same as point R with coordinates $(6, -4)$. This illustrates that the order of the coordinates of a point is important. This is why we call the pairs **ordered pairs.**

In Figure 2-3(b), we see that the points $(-5, 0)$, $(0, 0)$, and $(4.5, 0)$ all lie on the x-axis. In fact, every point with a y-coordinate of 0 will lie on the x-axis. Note that the coordinates of the origin are $(0, 0)$.

In Figure 2-3(b), we also see that the points $\left(0, -\frac{8}{3}\right)$, $(0, 0)$, and $(0, 3)$ all lie on the y-axis. In fact, every point with an x-coordinate of 0 will lie on the y-axis.

EXAMPLE 1 ***Halley's comet.*** Halley's comet passes the earth every 76 years as it travels in an elliptical orbit about the sun. Figure 2-4 shows its approximate path. Use the graph to determine the comet's position for the years 1912, 1930, 1948, 1966, 1978, and the most recent time it passed by the earth, 1986.

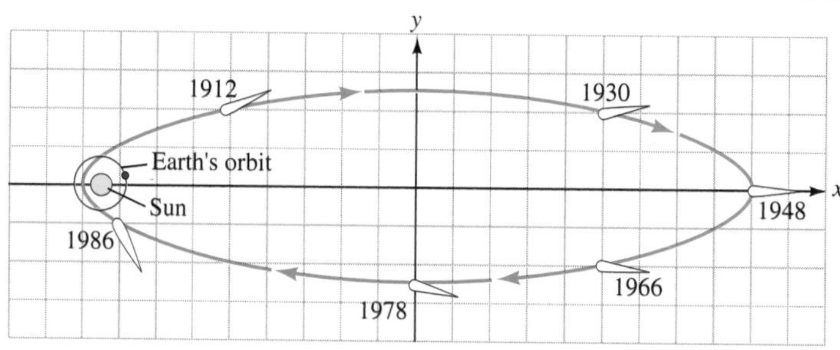

FIGURE 2-4

Solution To find the coordinates of each position, we start at the origin and move left or right along the x-axis to find the x-coordinate and then up or down to find the y-coordinate.

d Length of rental (days)	c Cost (dollars)
2	40
$3\frac{1}{2}$	60
2	40

d. No, the cost per day is not the same. If we look at how the c-coordinates change, we see that the first-day rental fee is \$20. The second day, the cost jumps another \$20. The third day, and all subsequent days, the cost jumps \$10.

Answer: \$50

STUDY SET Section 2.1

VOCABULARY *In Exercises 1–6, fill in the blanks to make the statements true.*

1. The pair of numbers $(6, -2)$ is called an ______________ pair.

2. In the ordered pair $(-2, -9)$, -9 is called the ___ coordinate.

3. The point with coordinates $(0, 0)$ is the ______________.

4. The x- and y-axes divide the coordinate plane into four regions called ______________.

5. Ordered pairs of numbers can be graphed on a ______________ coordinate system.

6. The process of locating a point on a coordinate plane is called ______________ the point.

CONCEPTS *In Exercises 7–8, fill in the blanks to make the statements true.*

7. To plot the point with coordinates $(6, -3.5)$, we start at the ______________ and move 6 units to the ______________ and then 3.5 units ______________.

8. To plot the point with coordinates $\left(-6, \frac{3}{2}\right)$, we start at the ______________ and move 6 units to the ______________ and then $\frac{3}{2}$ units ___.

9. In which quadrant do points with a negative x-coordinate and a positive y-coordinate lie?

10. In which quadrant do points with a positive x-coordinate and a negative y-coordinate lie?

11. Use the graph to complete the table.

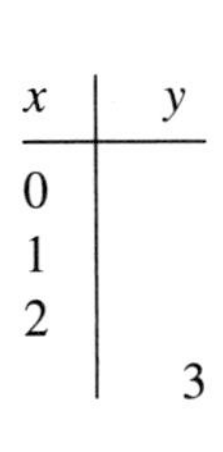

x	y
0	
1	
2	
	3

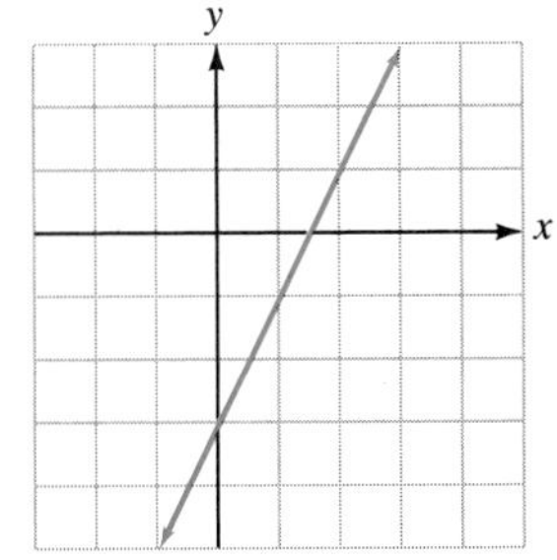

12. Use the graph to complete the table.

x	y
-2	
-1	
	-2
1	
2	

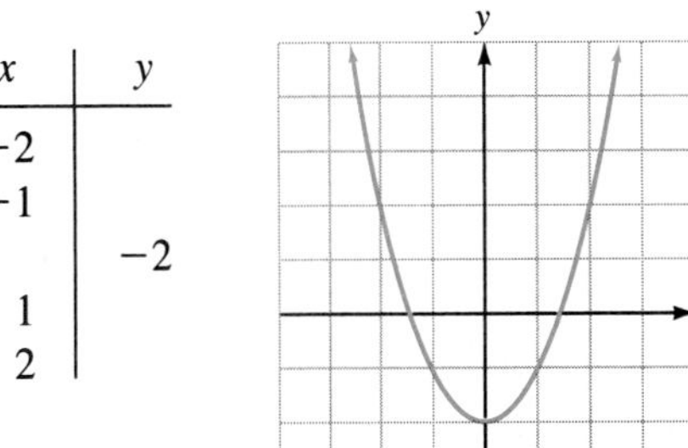

NOTATION

13. For the ordered pair (t, d), which axis is each variable associated with?

14. Do these ordered pairs name the same point?
$\left(5.25, -\frac{3}{2}\right)$, $\left(5\frac{1}{4}, -1.5\right)$, $\left(\frac{21}{4}, -1\frac{1}{2}\right)$

PRACTICE *In Exercises 15–22, plot each point on the rectangular coordinate system shown in Illustration 1.*

15. $A(4, 3)$
16. $B(-2, 1)$
17. $C(3.5, -2)$
18. $D(-2.5, -3)$
19. $E(5, 0)$
20. $F(-4, 0)$
21. $G(\frac{8}{3}, 0)$
22. $H(0, \frac{10}{3})$

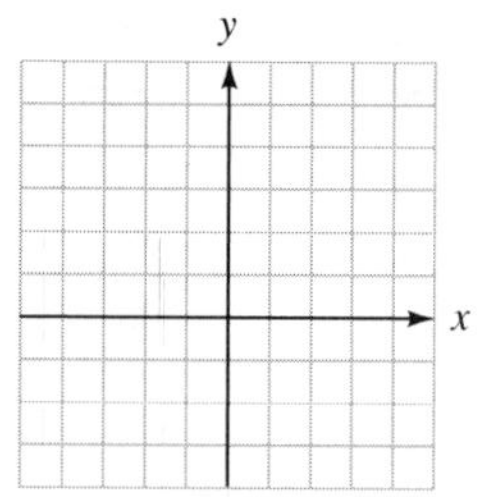

ILLUSTRATION 1

In Exercises 23–30, give the coordinates of each point shown in Illustration 2.

23. A
24. B
25. C
26. D
27. E
28. F
29. G
30. H

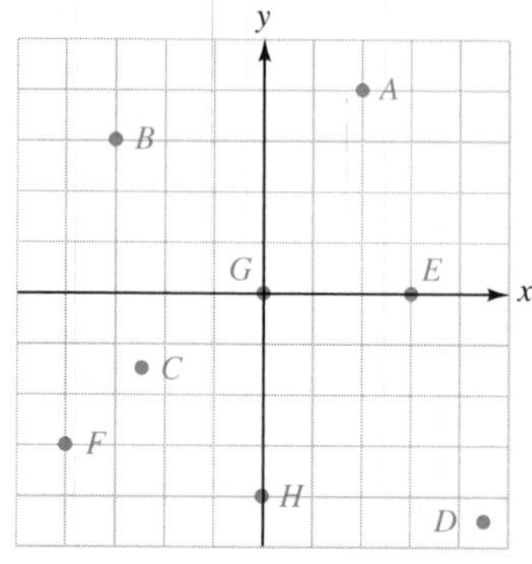

ILLUSTRATION 2

31. Plot the points given in the table and connect them with a line.

x	y
-3	-4
-1	-2
1	0
3	2
4	3

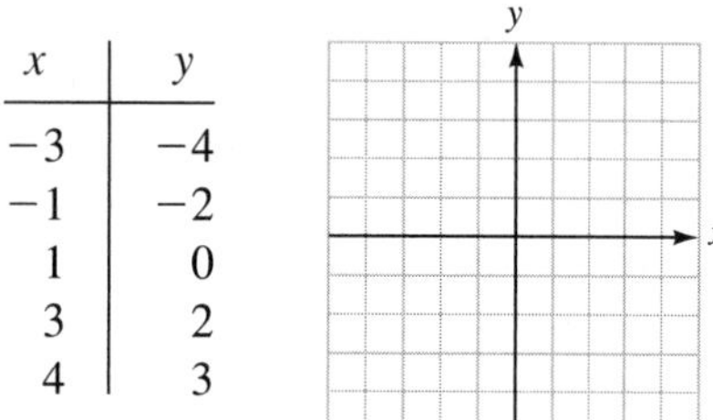

32. Plot the points in the table and connect them with a smooth curve.

x	y
-2	3
-1	1
0	0
1	1
2	3

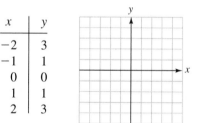

33. The graph in Illustration 3 shows the depths of a submarine at certain times.

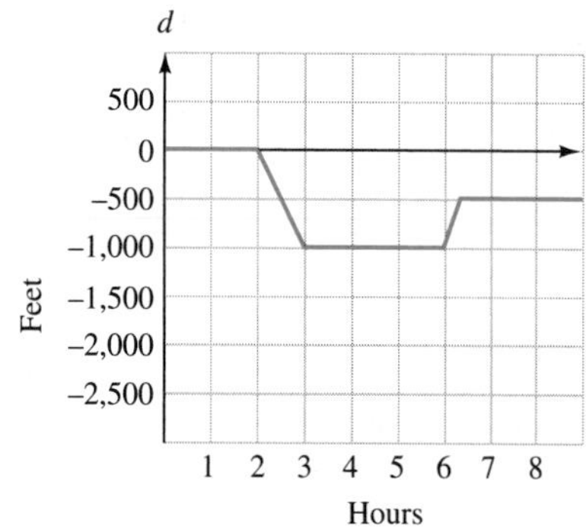

ILLUSTRATION 3

a. Where is the sub when $t = 2$?
b. What is the sub doing as t increases from $t = 2$ to $t = 3$?
c. How deep is the sub when $t = 4$?
d. How large an ascent does the sub begin to make when $t = 6$?

34. The graph in Illustration 4 (on the next page) shows the altitudes of a plane at certain times.
a. Where is the plane when $t = 0$?
b. What is the plane doing as t increases from $t = 1$ to $t = 2$?
c. What is the altitude of the plane when $t = 2$?
d. How much of a descent does the plane begin to make when $t = 4$?

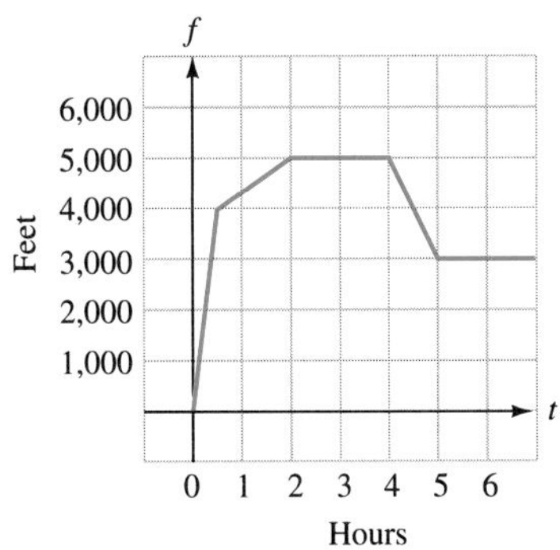

ILLUSTRATION 4

35. Refer to the graph in Illustration 5.

a. When did U.S. petroleum imports decline?

b. When did U.S. petroleum production increase?

c. When did imports surpass production?

d. Estimate the difference in U.S. petroleum imports and production for 1996.

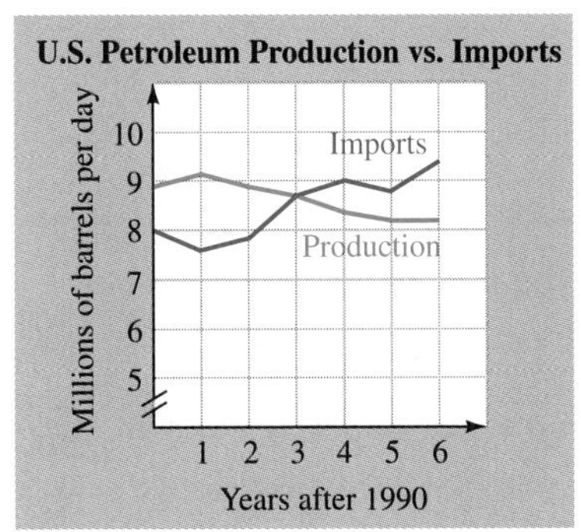

Based on data from the United States Department of Energy

ILLUSTRATION 5

36. Refer to the graph in Illustration 6.

a. Which runner ran faster at the start of the race?

b. Which runner stopped to rest first?

c. Which runner dropped the baton and had to go back and get it?

d. At what times was runner 1 stopped and runner 2 running?

e. Describe what was happening at time D.

f. Which runner won the race?

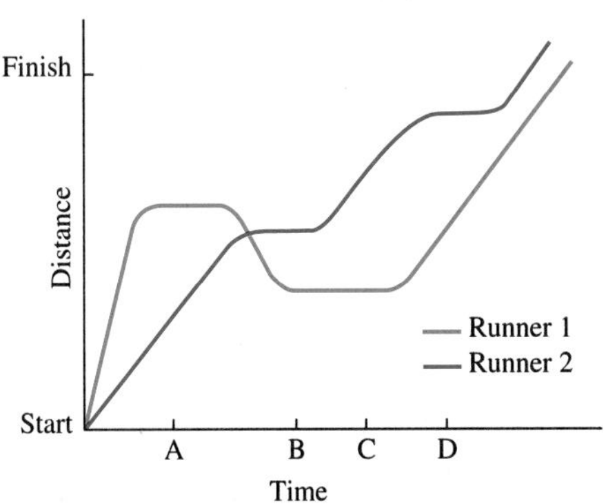

ILLUSTRATION 6

APPLICATIONS

37. ROAD MAPS Road maps have a built-in coordinate system to help locate cities. Use the map in Illustration 7 to find the coordinates of these cities in South Carolina: Jonesville, Easley, Hodges, and Union. Express each answer in the form (number, letter).

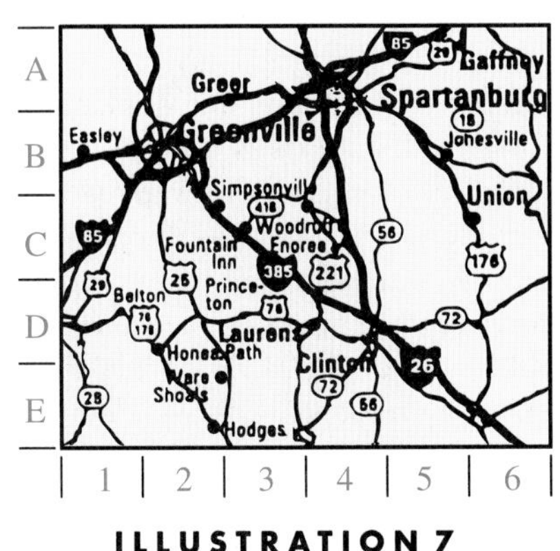

ILLUSTRATION 7

38. TRAIL OF DESTRUCTION A coordinate system that designates the location of places on the surface of the earth uses a series of latitude and longitude lines, as shown in Illustration 8.

a. If we agree to list longitude first, what are the coordinates of New Orleans, expressed as an ordered pair?

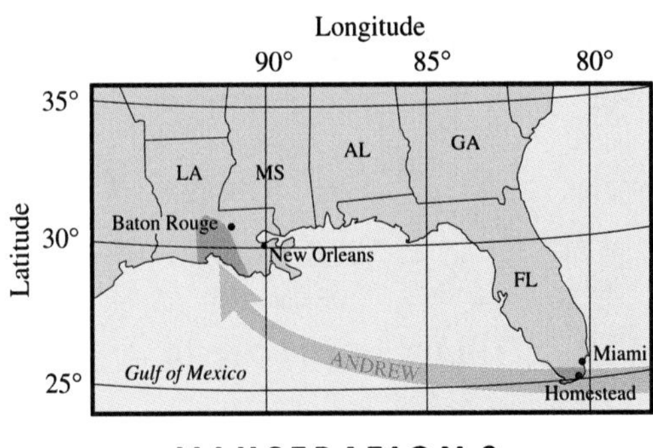

ILLUSTRATION 8

b. In August of 1992, Hurricane Andrew destroyed Homestead, Florida. Estimate the coordinates of Homestead.

c. Estimate the coordinates of where the hurricane hit Louisiana.

39. EARTHQUAKE DAMAGE The map in Illustration 9 shows the area where damage was caused by an earthquake.

a. Find the coordinates of the epicenter (the source of the quake).

b. Was damage done at the point (4, 5)?

c. Was damage done at the point $(-1, -4)$?

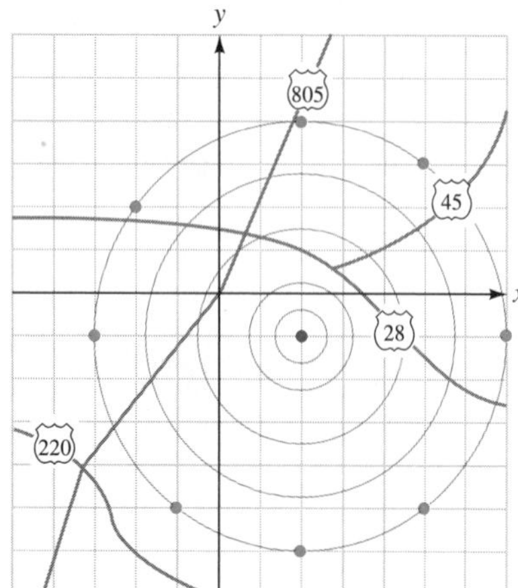

ILLUSTRATION 9

40. GEOGRAPHY Illustration 10 shows a cross-sectional profile of the Sierra Nevada mountain range in California.

a. Estimate the coordinates of Blue Oak, Sagebrush scrub, and Tundra using an ordered pair of the form (distance, elevation).

b. The *treeline* is the highest elevation at which trees grow. Estimate the treeline for this mountain range.

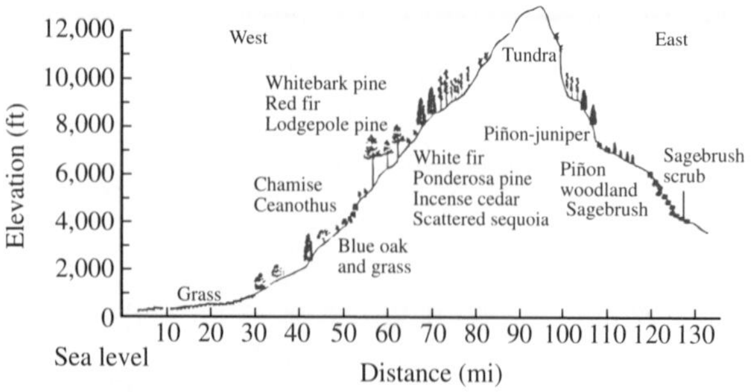

ILLUSTRATION 10

41. VIDEO RENTAL The charges for renting a video are shown in the graph in Illustration 11.

a. Find the charge for a 1-day rental.

b. Find the charge for a 2-day rental.

c. Find the charge if the video is kept for 5 days.

d. Find the charge if the video is kept for a week.

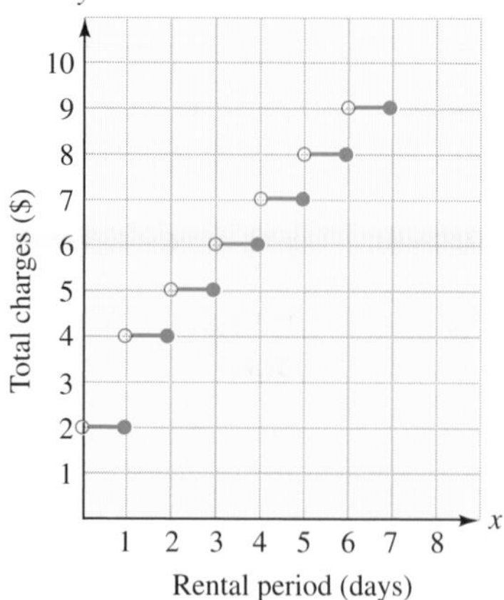

ILLUSTRATION 11

42. POSTAGE RATES The graph shown in Illustration 12 gives the first-class postage rates in 1998 for mailing parcels weighing up to 5 ounces.

a. Find the cost of postage to mail a 3-oz letter.

b. Find the difference in cost for a 2.75-oz letter and a 3.75-oz letter.

c. What is the heaviest letter that can be mailed first class for $1.01?

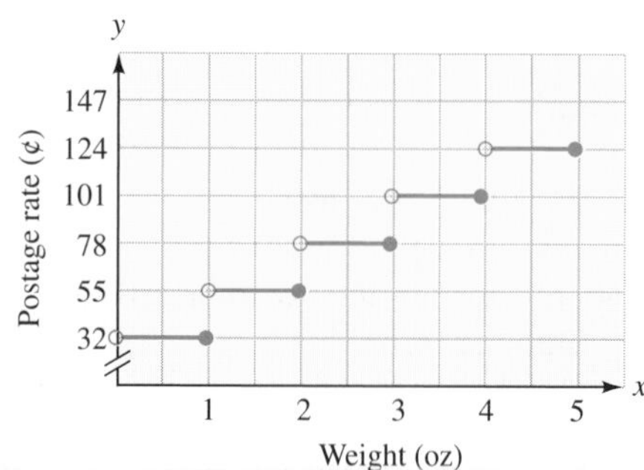

ILLUSTRATION 12

43. BOEING C-17 Engineers use a coordinate system with three axes when designing airplanes. As shown in Illustration 13, the x-axis is used to describe left/right on the airplane, the y-axis forward/backward, and the z-axis up/down. Any point on the airplane can be described by an *ordered triple* of the form (x, y, z). The coordinates of three points on the plane are (0, 181, 56), $(-46, 48, 19)$, and (84, 94, 24). Which highlighted part of the plane corresponds with which ordered triple?

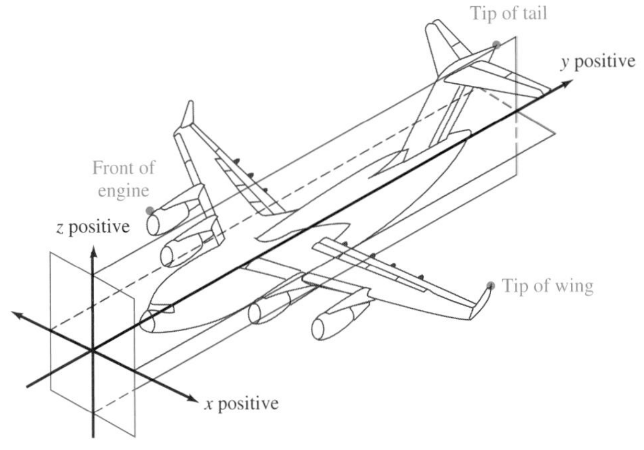

ILLUSTRATION 13

Size	Refrigerator thawing
10 lb to just under 18 lb	3 days
18 lb to just under 22 lb	4 days
22 lb to just under 24 lb	5 days
24 lb to just under 30 lb	6 days

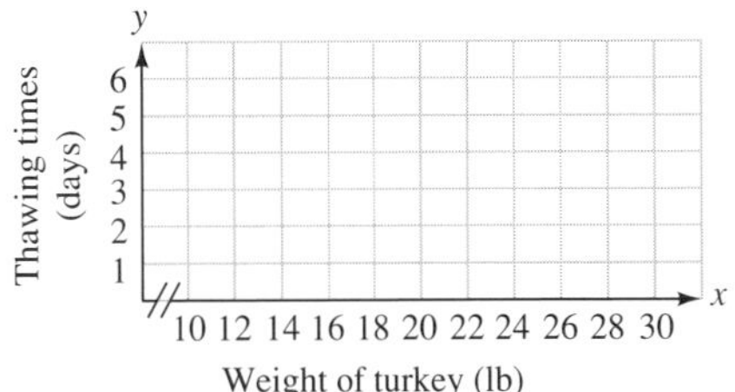

ILLUSTRATION 14

44. ROAST TURKEY The thawing guidelines that appear on the label of a frozen turkey are listed in the table. In Illustration 14, draw a step graph that illustrates these instructions.

WRITING *Write a paragraph using your own words.*

45. Explain how to plot the point with coordinates of $(-2, 5)$.

46. Explain why the coordinates of the origin are $(0, 0)$.

REVIEW *In Exercises 47–50, evaluate each expression.*

47. $-5 - 5(-5)$

48. $(-5)^2 + (-5)$

49. $\dfrac{-3 + 5(2)}{9 + 5}$

50. $|-1 - 9|$

51. Solve $-4x + 0.7 = -2.1$.

52. Solve $P = 2l + 2w$ for w.

2.2 *Graphing Linear Equations*

In this section, you will learn about

- **Graphing linear equations**
- **The intercept method**
- **Horizontal and vertical lines**
- **Linear models**

Introduction. In this section, we will discuss equations that have two variables. Such equations are used to describe relationships between two quantities. To see a picture of the mathematical relationships, we will construct graphs of the equations.

Graphing linear equations

The equation $y = -\frac{1}{2}x + 4$ contains the variables x and y. The solutions of this equation can be written as ordered pairs of real numbers. For example, the ordered pair $(-4, 6)$ is a solution, because the equation is satisfied when $x = -4$ and $y = 6$.

$$y = -\frac{1}{2}x + 4$$

$$6 = -\frac{1}{2}(-4) + 4 \quad \text{Substitute } -4 \text{ for } x \text{ and } 6 \text{ for } y.$$

$$6 = 2 + 4 \quad \text{Do the multiplication: } -\tfrac{1}{2}(-4) = 2.$$

$$6 = 6$$

This pair and others that satisfy the equation are listed in the table of values shown in Figure 2-11.

$$y = -\frac{1}{2}x + 4$$

x	y	(x, y)
−4	6	(−4, 6)
−2	5	(−2, 5)
0	4	(0, 4)
2	3	(2, 3)
4	2	(4, 2)

Note that we choose x-values that are multiples of the denominator 2. This makes the computations easier when multiplying the x value by $-\frac{1}{2}$ to find the corresponding y-value.

FIGURE 2-11

The **graph of the equation** $y = -\frac{1}{2}x + 4$ is the graph of all points (x, y) on the rectangular coordinate system whose coordinates satisfy the equation.

EXAMPLE 1 ***Graphing equations.*** Graph $y = -\frac{1}{2}x + 4$.

Solution

To graph the equation, we plot the five ordered pairs listed in the table shown in Figure 2-11. These points appear to lie on the line shown in Figure 2-12. In fact, if we were to plot many more pairs that satisfied the equation, it would become obvious that the resulting points will all lie on the line.

When we say that the graph of an equation is a line, we imply two things:

1. Every point with coordinates that satisfy the equation will lie on the line.

2. Any point on the line will have coordinates that satisfy the equation.

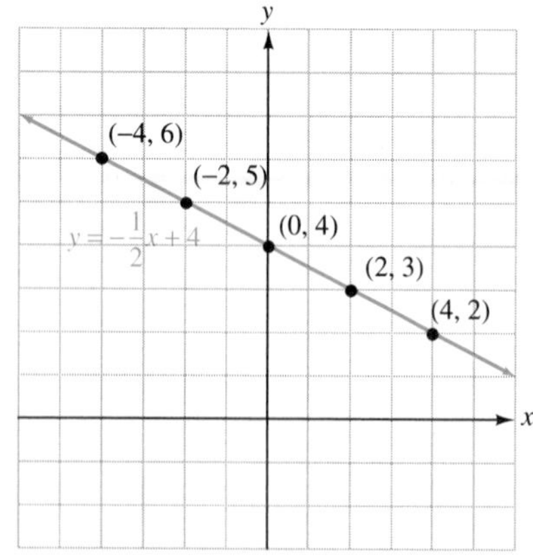

FIGURE 2-12

Self Check

Complete the table of values for $y = 2x - 3$. Then graph the equation.

x	y
−1	
0	
1	
2	
3	

Answer:

x	y
−1	−5
0	−3
1	−1
2	1
3	3

$y = 2x - 3$

When the graph of an equation is a line, we call the equation a **linear equation.** Linear equations are often written in the form $Ax + By = C$, called **general form,** where A, B, and C represent numbers (called **constants**) and x and y are variables.

EXAMPLE 2 ***Graphing linear equations.*** Graph $3x + 2y = 12$.

Self Check

Graph $2x - 3y = -6$

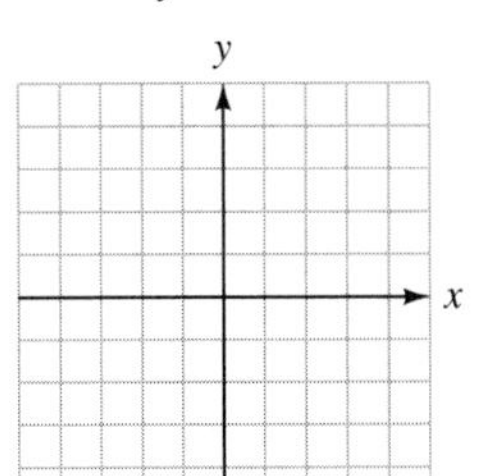

Solution

We can pick values for either x or y, substitute them into the equation, and solve for the other variable. For example, if $x = 2$,

$$\begin{aligned} 3x + 2y &= 12 \\ 3(2) + 2y &= 12 && \text{Substitute 2 for } x. \\ 6 + 2y &= 12 && \text{Do the multiplication: } 3(2) = 6. \\ 2y &= 6 && \text{Subtract 6 from both sides.} \\ y &= 3 && \text{Divide both sides by 2.} \end{aligned}$$

The ordered pair (2, 3) satisfies the equation. If $y = 6$,

$$\begin{aligned} 3x + 2y &= 12 \\ 3x + 2(6) &= 12 && \text{Substitute 6 for } y. \\ 3x + 12 &= 12 && \text{Do the multiplication: } 2(6) = 12. \\ 3x &= 0 && \text{Subtract 12 from both sides.} \\ x &= 0 && \text{Divide both sides by 3.} \end{aligned}$$

A second ordered pair that satisfies the equation is (0, 6).

These pairs and others that satisfy the equation are shown in Figure 2-13. After we plot each pair, we see that they all lie on a line. The graph of the equation is the line shown in the figure.

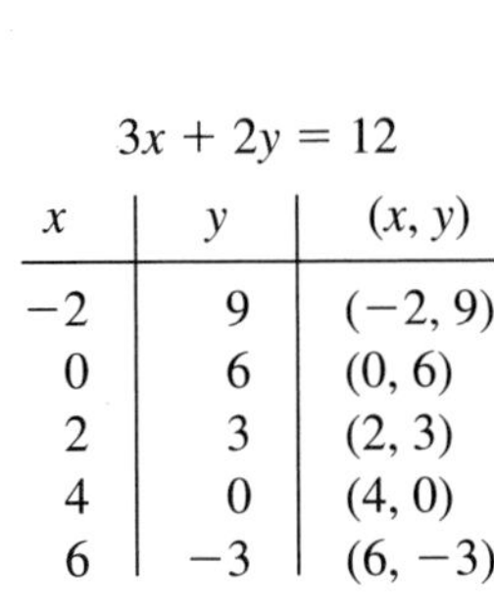

$3x + 2y = 12$

x	y	(x, y)
−2	9	(−2, 9)
0	6	(0, 6)
2	3	(2, 3)
4	0	(4, 0)
6	−3	(6, −3)

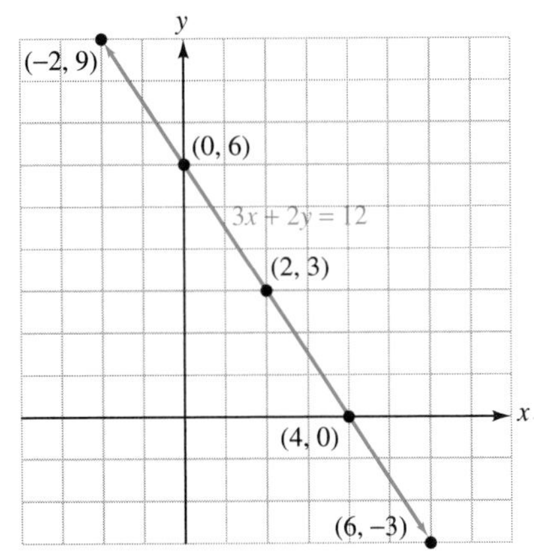

FIGURE 2-13

Answer:

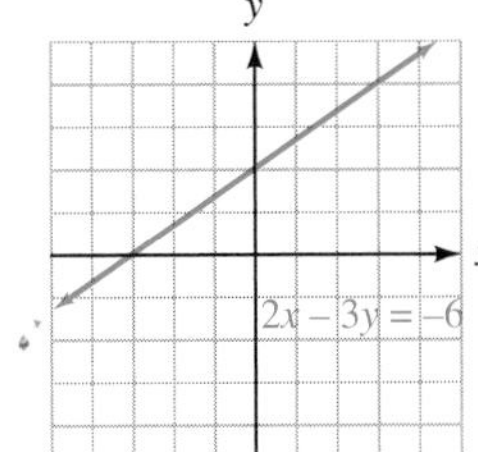

The intercept method

In Example 2, the graph intersected the y-axis at the point with coordinates (0, 6) (called the ***y*-intercept**) and intersected the x-axis at the point with coordinates (4, 0) (called the ***x*-intercept**). In general, we have the following definitions.

Intercepts of a line

The ***y*-intercept** of a line is the point $(0, b)$, where the line intersects the y-axis. To find b, substitute 0 for x in the equation of the line and solve for y.

The ***x*-intercept** of a line is the point $(a, 0)$, where the line intersects the x-axis. To find a, substitute 0 for y in the equation of the line and solve for x.

EXAMPLE 3 ***Graphing using the intercept method.*** Use the x- and y-intercepts to graph $2x + 5y = 10$.

Solution

To find the y-intercept, we substitute 0 for x and solve for y:

$$\begin{aligned} 2x + 5y &= 10 \\ 2(0) + 5y &= 10 && \text{Substitute 0 for } x. \\ 5y &= 10 && \text{Do the multiplication: } 2(0) = 0. \\ y &= 2 && \text{Divide both sides by 5.} \end{aligned}$$

The y-intercept is the point (0, 2). To find the x-intercept, we substitute 0 for y and solve for x:

$$\begin{aligned} 2x + 5y &= 10 \\ 2x + 5(0) &= 10 && \text{Substitute 0 for } y. \\ 2x &= 10 && \text{Do the multiplication: } 5(0) = 0. \\ x &= 5 && \text{Divide both sides by 2.} \end{aligned}$$

The x-intercept is the point (5, 0).

Although two points are enough to draw the line, it is a good idea to find and plot a third point as a check. To find the coordinates of a third point, we can substitute any convenient number (such as -5) for x and solve for y:

$$\begin{aligned} 2x + 5y &= 10 \\ 2(-5) + 5y &= 10 && \text{Substitute } -5 \text{ for } x. \\ -10 + 5y &= 10 && \text{Do the multiplication: } 2(-5) = -10. \\ 5y &= 20 && \text{Add 10 to both sides.} \\ y &= 4 && \text{Divide both sides by 5.} \end{aligned}$$

The line will also pass through the point $(-5, 4)$.

A table of ordered pairs and the graph of $2x + 5y = 10$ are shown in Figure 2-14.

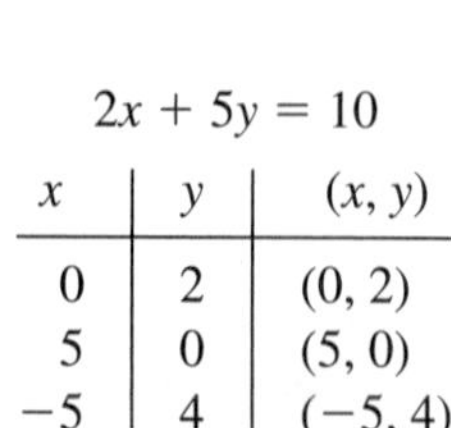

$2x + 5y = 10$

x	y	(x, y)
0	2	(0, 2)
5	0	(5, 0)
-5	4	$(-5, 4)$

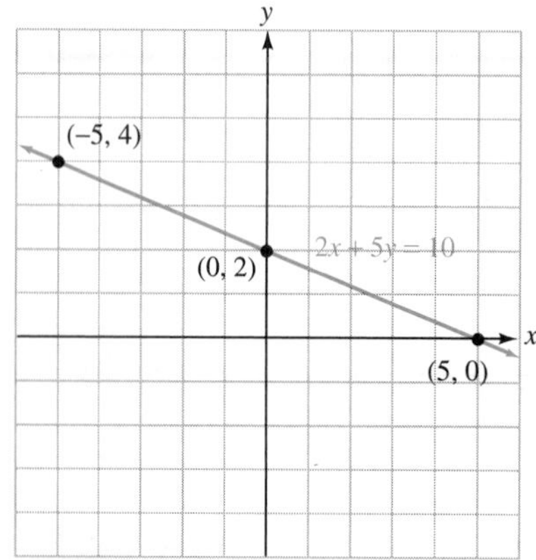

FIGURE 2-14

Self Check

Find the x- and y-intercepts and graph $y = 3x + 4$.

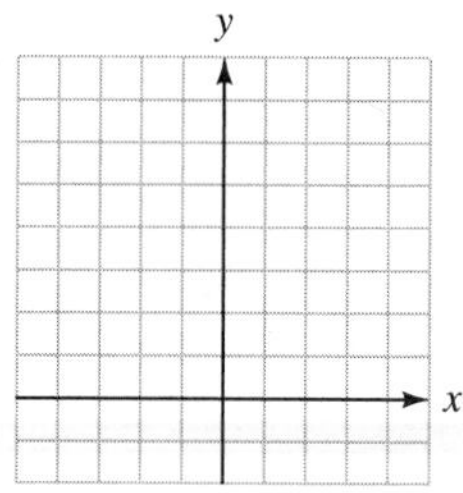

Answer:

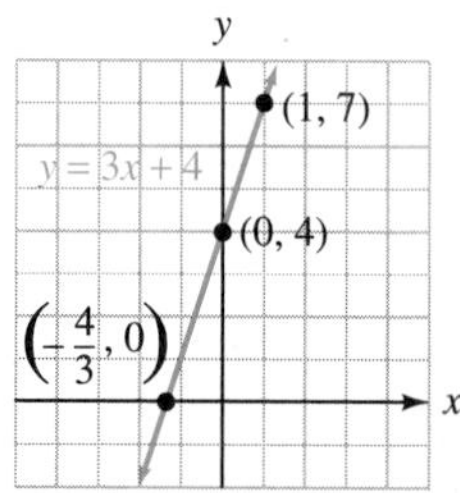

EXAMPLE 4 ***Graphing lines.*** Graph **a.** $y = 3$ and **b.** $x = -2$.

Solution

a. Since the equation $y = 3$ does not contain x, the numbers chosen for x have no effect on y. The value of y is always 3.

After plotting the pairs (x, y) shown in the table in Figure 2-15, we see that the graph is a horizontal line, parallel to the x-axis, with a y-intercept of $(0, 3)$. The line has no x-intercept.

b. Since the equation $x = -2$ does not contain y, the value of y can be any number.

After plotting the pairs (x, y) shown in the table in Figure 2-15, we see that the graph is a vertical line, parallel to the y-axis, with an x-intercept of $(-2, 0)$. The line has no y-intercept.

$y = 3$

x	y	(x, y)
−3	3	(−3, 3)
0	3	(0, 3)
2	3	(2, 3)
4	3	(4, 3)

$x = -2$

x	y	(x, y)
−2	−2	(−2, −2)
−2	0	(−2, 0)
−2	2	(−2, 2)
−2	6	(−2, 6)

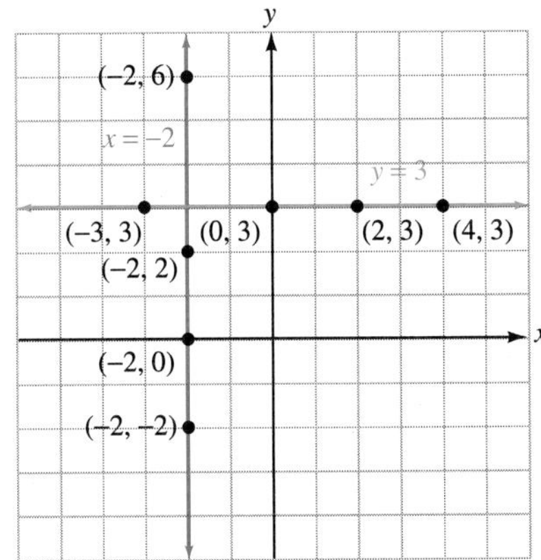

FIGURE 2-15

Self Check

Graph $x = 4$ and $y = -3$ on one set of coordinate axes.

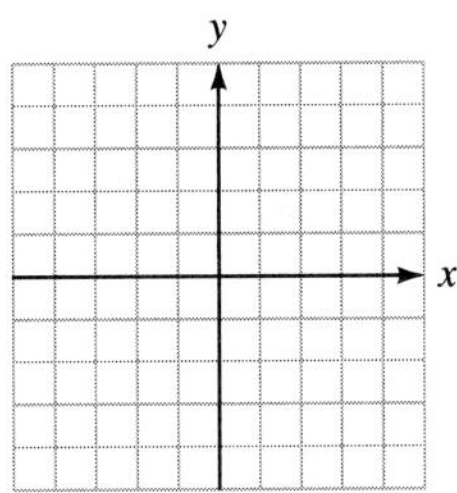

Answer:

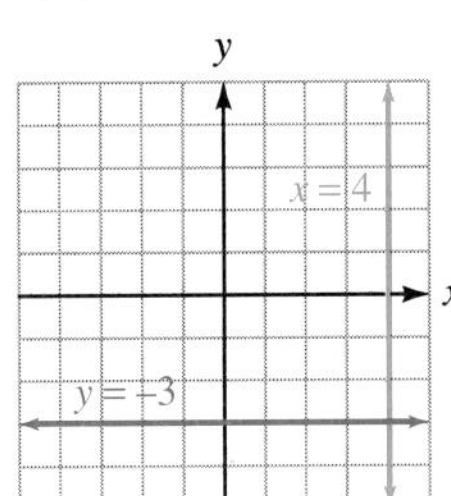

The results of Example 4 suggest the following facts.

Horizontal and vertical lines

If a and b are real numbers, then

The graph of the equation $x = a$ is a vertical line with x-intercept at $(a, 0)$. If $a = 0$, the line is the y-axis.

The graph of the equation $y = b$ is a horizontal line with y-intercept at $(0, b)$. If $b = 0$, the line is the x-axis.

Linear models

In the next two examples, we will see how linear equations can be used to mathematically model real-life situations. In each case, the equations describe a *linear relationship* between two quantities; when they are graphed, the result is a line. We can make observations about what has taken place in the past and what might take place in the future by carefully inspecting the graph.

EXAMPLE 5 ***U.S. workforce demographics.*** The linear equation $p = \frac{3}{5}t + 38$ models the percent of females 16 years or older who were part of the civilian workforce for each of the years 1960–1996. In the equation, t represents the number of years after 1960, and p represents the percent. Graph this equation. What has been the trend over this period?

Solution

The variables t and p are used in the equation. We will associate t with the horizontal axis and p with the vertical axis. Ordered pairs will be of the form (t, p).

To graph the equation, we pick three values for t, substitute them into the equation, and find each corresponding value of p. The results are listed in the table in Figure 2-16.

For $t = 0$

$$p = \frac{3}{5}t + 38$$
$$p = \frac{3}{5}(0) + 38$$
$$p = 38$$

For $t = 10$

$$p = \frac{3}{5}t + 38$$
$$p = \frac{3}{5}(10) + 38$$
$$p = 6 + 38$$
$$p = 44$$

For $t = 20$

$$p = \frac{3}{5}t + 38$$
$$p = \frac{3}{5}(20) + 38$$
$$p = 12 + 38$$
$$p = 50$$

The pairs (0, 38), (10, 44), and (20, 50) satisfy the equation. Next, we plot these points and draw a line through them, as shown in Figure 2-16. From the graph, we see that there has been a steady increase in the percent of the female population 16 years or older that is part of the workforce.

t	p
0	38
10	44
20	50

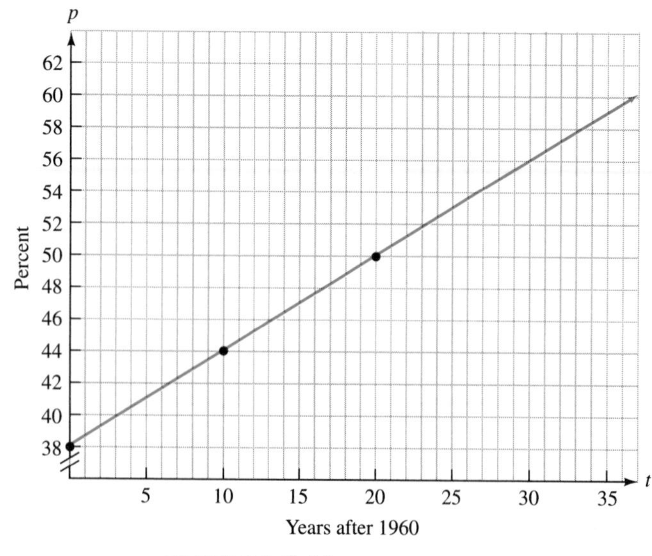

FIGURE 2-16

Self Check

a. Use the equation $p = \frac{3}{5}t + 38$ to determine the percent of females 16 years or older who were part of the workforce in 1975.

b. Use the graph in Figure 2-16 to find the percent of females who were part of the workforce in 1990.

Answers: **a.** 47%, **b.** 56%

EXAMPLE 6 ***Depreciation.*** A copy machine that has been purchased for \$6,750 is expected to depreciate according to the formula $y = -950x + 6{,}750$, where y is the value of the copier after x years. When will the copier have no value?

Solution

The copier will have no value when y is 0. To find x when $y = 0$, we substitute 0 for y and solve for x.

$$y = -950x + 6{,}750$$
$$0 = -950x + 6{,}750$$
$$-6{,}750 = -950x \quad \text{Subtract 6,750 from both sides.}$$
$$7.105263158 = x \quad \text{Divide both sides by } -950.$$

The copier will have no value in about 7.1 years.

The equation $y = -950x + 6{,}750$ is graphed in Figure 2-17. Important information can be obtained from the intercepts of the graph.

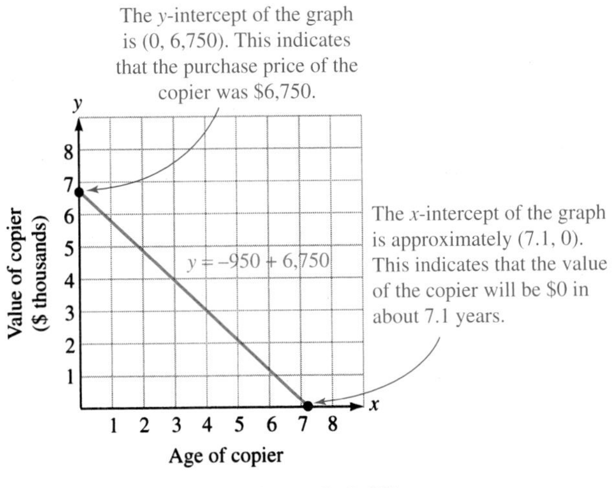

The y-intercept of the graph is (0, 6,750). This indicates that the purchase price of the copier was \$6,750.
y
8
7
6
5
4
3
2
1
Value of copier ($ thousands)
y = −950 + 6,750
The x-intercept of the graph is approximately (7.1, 0). This indicates that the value of the copier will be \$0 in about 7.1 years.
x
1 2 3 4 5 6 7 8
Age of copier

FIGURE 2-17

Self Check

a. Use the equation $y = -950x + 6{,}750$ to determine when the copier will be worth \$3,900.

b. Use the graph in Figure 2-17 to determine when the copier will be worth \$2,000.

Answers: **a.** 3 years, **b.** 5 years

Accent on Technology **Graphing lines**

We have graphed linear equations by finding ordered pairs, plotting points, and drawing lines through those points. Graphing is much easier if we use a graphing calculator.

Graphing calculators have a window to display graphs (see Figure 2-18). To see the proper picture of a graph, we must decide on the minimum and maximum values for the x- and y-coordinates. A window with standard settings of

$$\text{Xmin} = -10 \qquad \text{Xmax} = 10 \qquad \text{Ymin} = -10 \qquad \text{Ymax} = 10$$

will produce a graph where the values of x are in the interval $[-10, 10]$, and y is in the interval $[-10, 10]$.

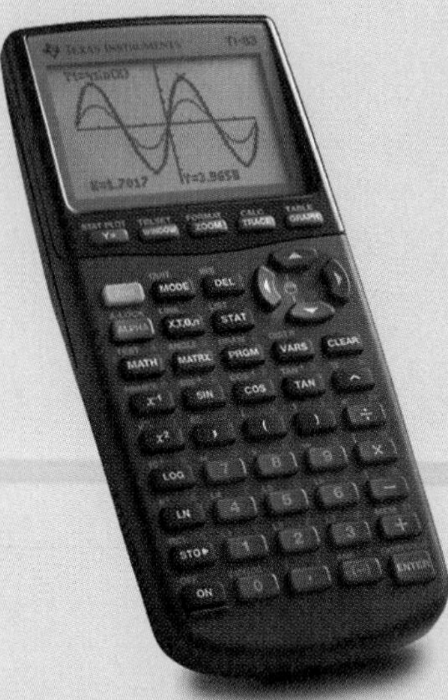

Courtesy of Texas Instruments

FIGURE 2-18

To graph $3x + 2y = 12$, we must first solve the equation for y.

$$3x + 2y = 12$$

$$2y = -3x + 12 \quad \text{Subtract } 3x \text{ from both sides.}$$

$$y = -\frac{3}{2}x + 6 \quad \text{Divide both sides by 2.}$$

To graph the equation, we enter the right-hand side of the equation after a symbol such as $\backslash Y_1 =$ or $f(x) =$. After entering the right-hand side, the display should look like

$$\backslash Y_1 = -(3/2)X + 6 \qquad \text{or} \qquad f(x) = -(3/2)X + 6$$

We then press the GRAPH key to get the graph shown in Figure 2-19(a). To show more detail, we can draw the graph in a different window. A window with settings of $[-1, 5]$ for x and $[-2, 7]$ for y will give the graph shown in Figure 2-19(b).

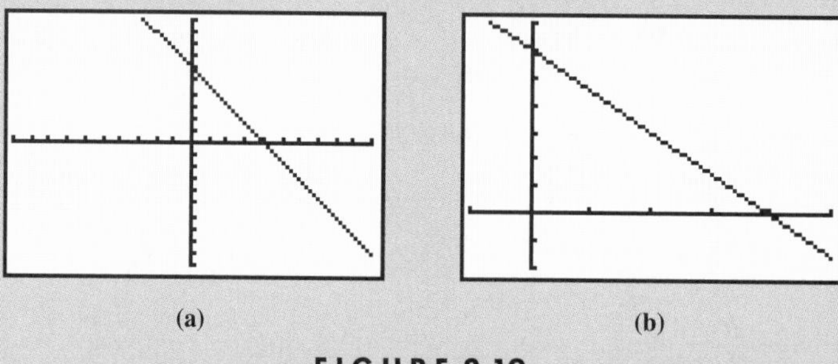

(a) (b)

FIGURE 2-19

We can use the trace command to find the coordinates of any point on a graph. For example, to find the x-intercept of the graph of $2y = -5x - 7$, we first solve for y, then we graph the equation using the standard window settings of $[-10, 10]$ for x and $[-10, 10]$ for y. We then press the TRACE key to get Figure 2-20(a). We can then use the $>$ and $<$ keys to move the cursor along the line toward the x-intercept until we arrive at a point with the coordinates shown in Figure 2-20(b).

To get better results, we can zoom in to get a magnified picture, trace again and move the cursor to the point with coordinates shown in Figure 2-20(c). Since the y-coordinate is nearly 0, this point is nearly the x-intercept. We can achieve better results with more zooms.

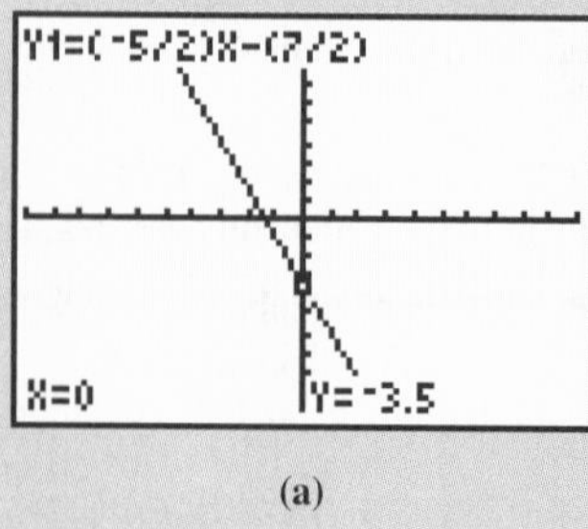

(a)

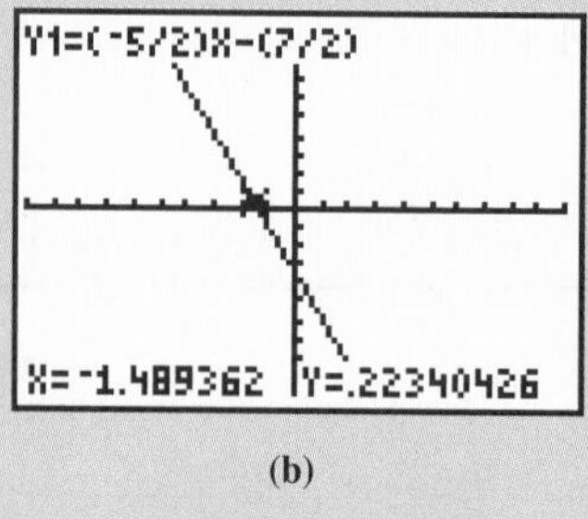

(b)

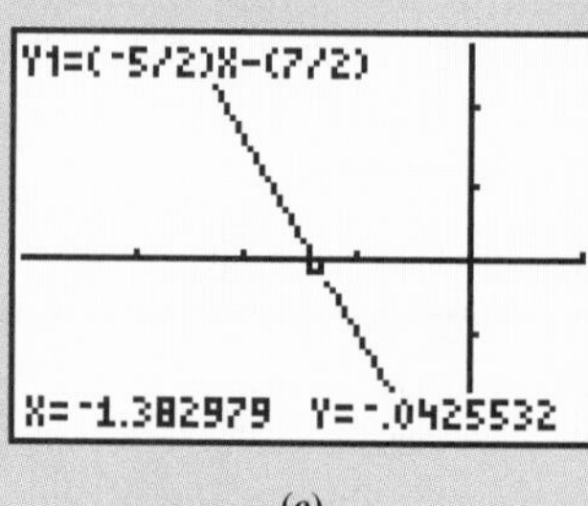

(c)

FIGURE 2-20

STUDY SET Section 2.2

VOCABULARY *In Exercises 1–6, fill in the blanks to make the statements true.*

1. The graph of an equation is the graph of all points (x, y) on the rectangular coordinate system whose coordinates ______________ the equation.

2. Any equation whose graph is a line is called a ______________ equation.

3. The point where a graph intersects the y-axis is called the ______________.

4. The point where a graph intersects the x-axis is called the ______________.

5. The graph of any equation of the form $x = a$ is a ______________ line.

6. The graph of any equation of the form $y = b$ is a ______________ line.

CONCEPTS

7. a. Consider the equation $2x + 4 = 8$, studied in Chapter 1. How many variables does it contain? How many solutions does it have?

b. Consider $2x + 4y = 8$. How many variables does it contain? How many solutions does it have?

8. Consider the linear equation $6x - 4y = -12$.

a. Find the x-intercept of its graph.

b. Find the y-intercept of its graph.

c. Does its graph pass through $(2, 6)$?

9. Linear equations can be written in the form $Ax + By = C$. Write a linear equation with $A = 3$, $B = 5$, and $C = -10$.

10. On which axis does each point lie?

a. $(0, b)$ **b.** $(a, 0)$

11. See Illustration 1.

a. What is the x-intercept and what is the y-intercept of the line?

b. If the coordinates of point M are substituted into the equation of the line that is graphed here, will a true or a false statement result?

12. See Illustration 2. Does the graph indicate a linear relationship between the two quantities? Explain why or why not.

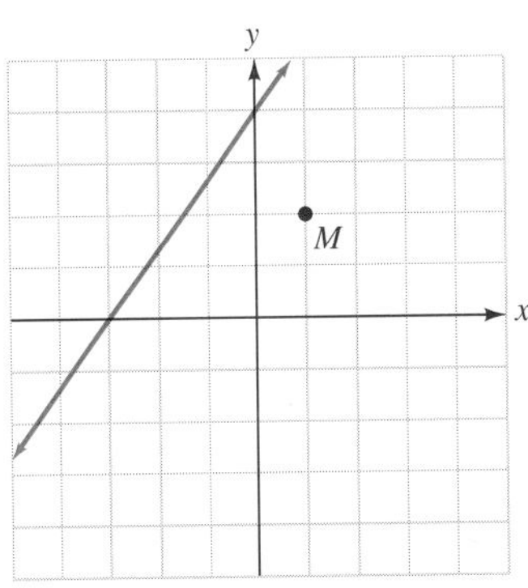

ILLUSTRATION 1

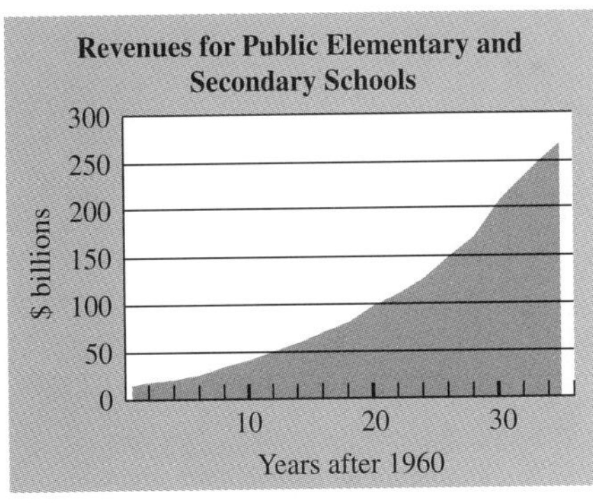

ILLUSTRATION 2

NOTATION *In Exercises 13–14, complete each solution.*

13. Verify that $(-3, -1)$ is a solution of $2x + 2y = -8$.

$2x + 2y = -8$ The original equation.

$2(\square) + 2(-1) \stackrel{?}{=} -8$ Substitute -3 for x and -1 for y.

$-6 + (\square) \stackrel{?}{=} -8$ Do the multiplications.

$-8 = -8$ Simplify the left-hand side.

14. To find the coordinates of a point on the graph of $5x + 2y = 10$, choose $x = 1$ and find y.

$5x + 2y = 10$ The original equation.

$5(\square) + 2y = 10$ Substitute 1 for x.

$5 + \square = 10$ Do the multiplication.

$2y = \square$ Subtract 5 from both sides.

$y = \frac{5}{2}$ Divide both sides by 2.

The point $\square$ is on the graph of $5x + 2y = 10$.

15. The graph of the equation $x = 0$ is which axis?

____

16. Solve the equation $3x + 2y = 9$ for y.

____

PRACTICE *In Exercises 17–20, complete the table of solutions for each equation.*

17. $y = -x + 4$

x	y
-1	
0	
2	

18. $y = x - 2$

x	y
-2	
0	
4	

19. $y = -\frac{1}{3}x - 1$

x	y
-3	
0	
3	

20. $y = -\frac{1}{2}x + \frac{5}{2}$

x	y
-1	
3	
5	

In Exercises 21–24, use the results from Exercises 17–20 to graph each equation.

21. $y = -x + 4$

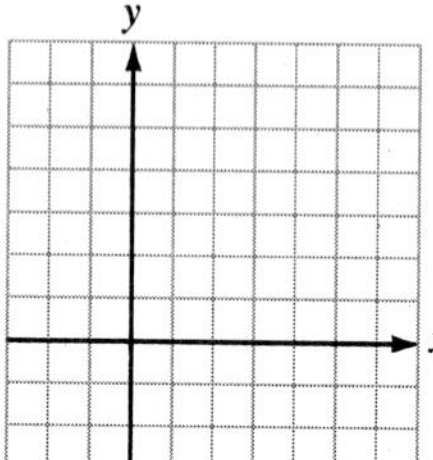

22. $y = x - 2$

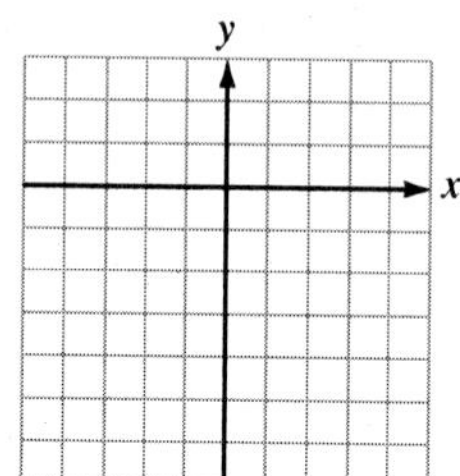

23. $y = -\frac{1}{3}x - 1$

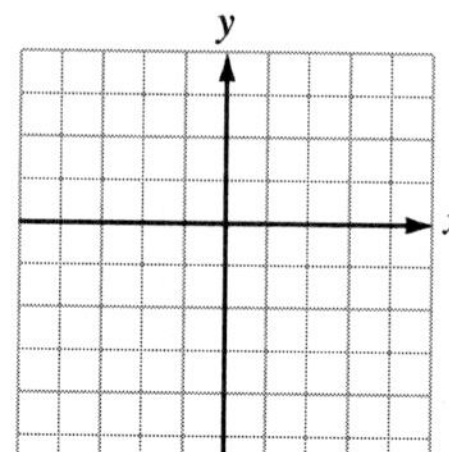

24. $y = -\frac{1}{2}x + \frac{5}{2}$

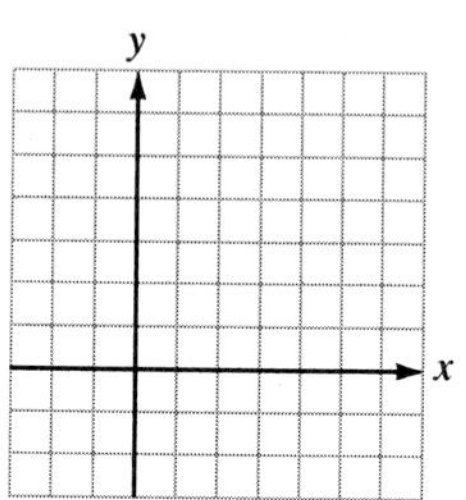

In Exercises 25–32, construct a table of solutions with three entries, and then graph the equation.

25. $y = x$

26. $y = -2x$

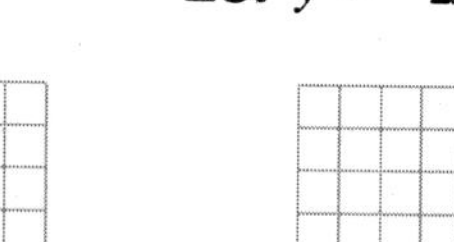

27. $y = -3x + 2$

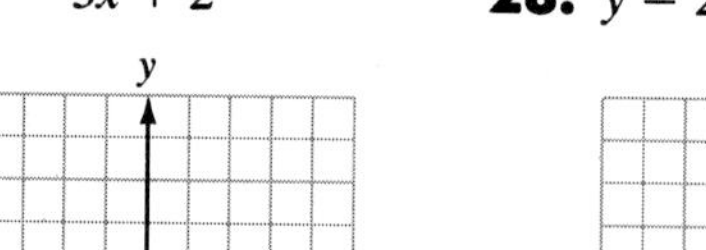

28. $y = 2x - 3$

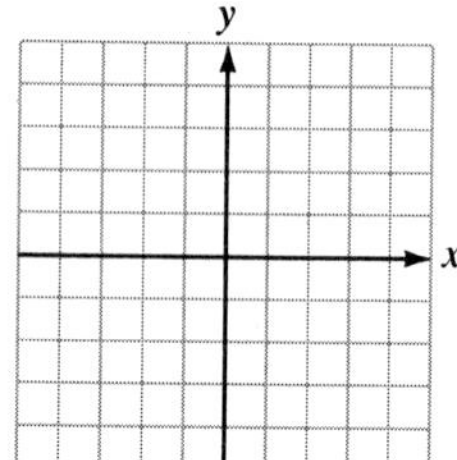

29. $x = 3$

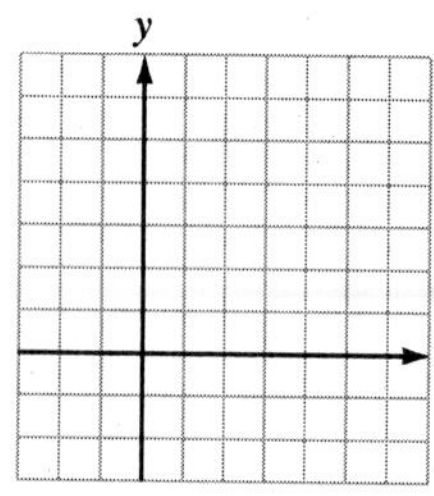

30. $y = -4$

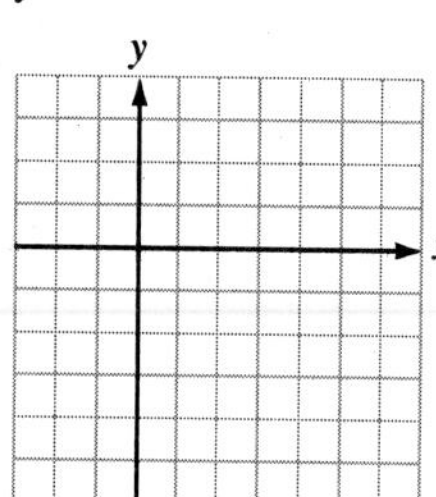

31. $-3y + 2 = 5$

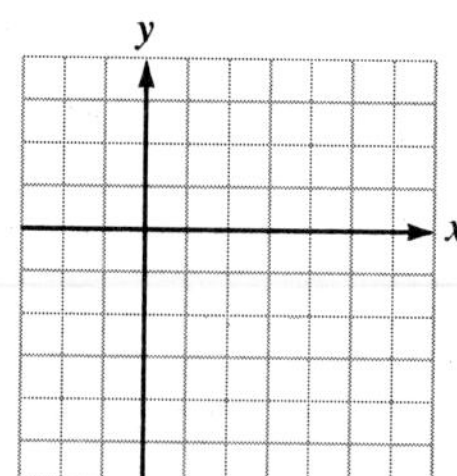

32. $-2x + 3 = 11$

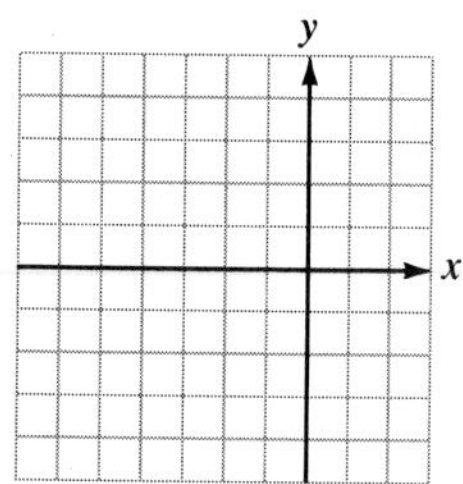

In Exercises 33–40, graph each equation using the intercept method.

33. $3x + 4y = 12$

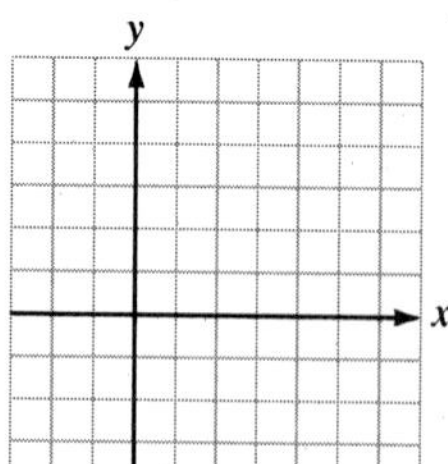

34. $4x - 3y = 12$

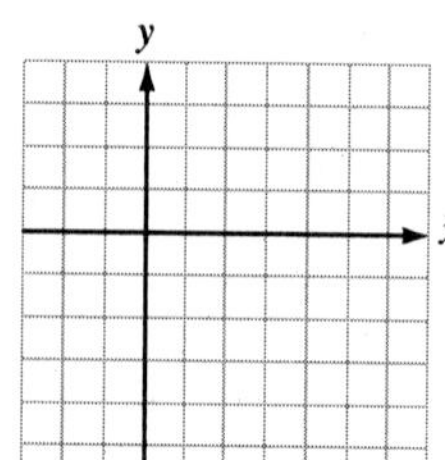

35. $3y = 6x - 9$

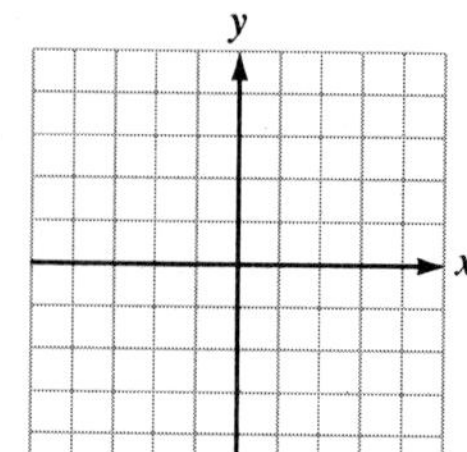

36. $2x = 4y - 10$

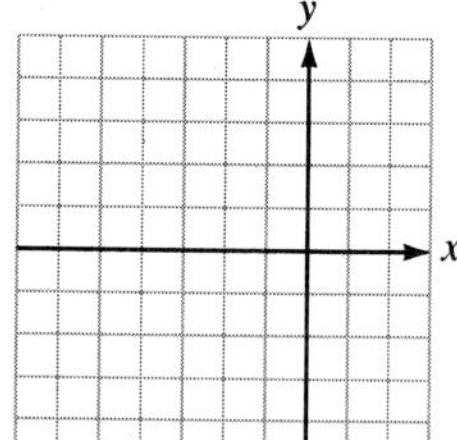

37. $2y + x = -2$

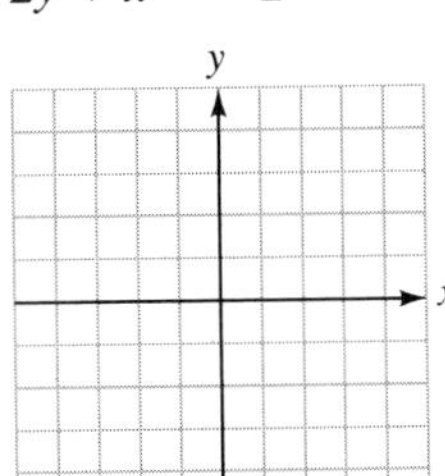

38. $4y + 2x = -8$

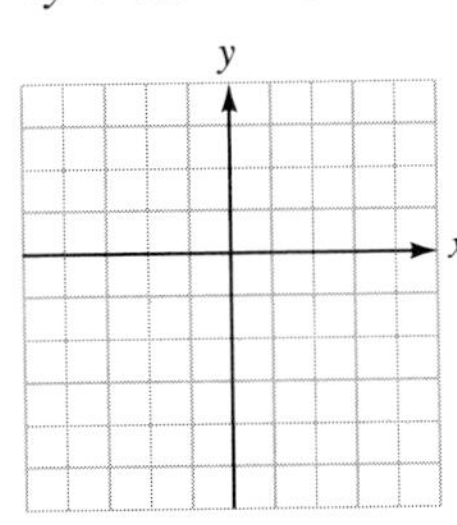

39. $3x + 4y - 8 = 0$

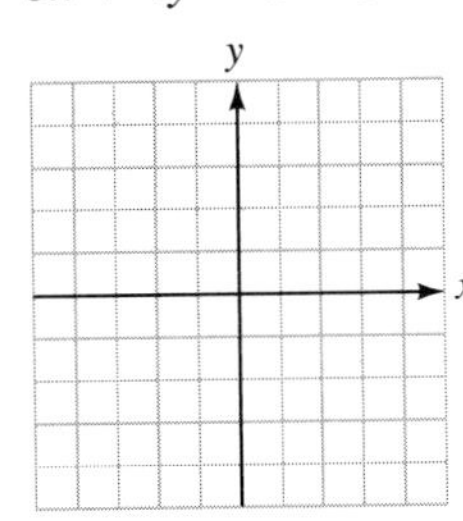

40. $-2y - 3x + 9 = 0$

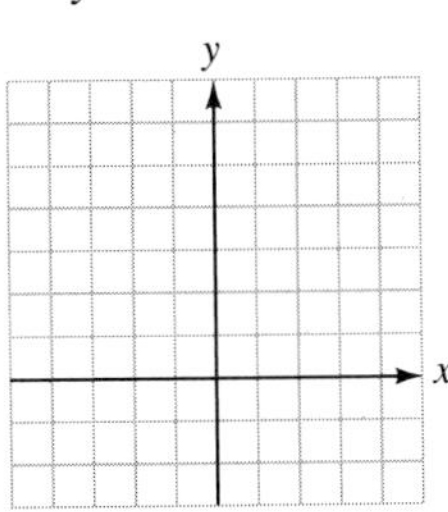

In Exercises 41–44, use a graphing calculator to graph each equation, and then find the x-coordinate of the x-intercept to the nearest hundredth.

41. $y = 3.7x - 4.5$

42. $y = \frac{3}{5}x + \frac{5}{4}$

43. $1.5x - 3y = 7$

44. $0.3x + y = 7.5$

APPLICATIONS

45. HOURLY WAGES The following table gives amount y (in dollars) that a student can earn for working x hours. Plot the ordered pairs and estimate how much the student will earn for working 8 hours.

x	2	4	5	6
y	12	24	30	36

46. VALUE OF A CAR The following table shows the value y (in dollars) of a car that is x years old. Plot the ordered pairs and estimate the value of the car when it is 4 years old.

x	0	1	3
y	15,000	12,000	6,000

47. FARMING The equation $f = -\frac{1}{50}t + 2.4$ models the number of farms in the United States where t is the number of years after 1980, and f is the number of farms in millions. Graph the equation in Illustration 3. Use the graph to estimate the number of farms in 1984 and 1993. Describe the trend.

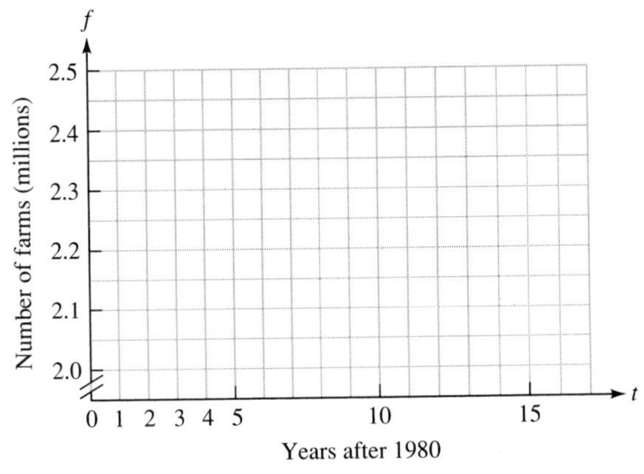

ILLUSTRATION 3

48. TV COVERAGE See Illustration 4. A blimp, hovering over a sports stadium, sends a television signal to the ground along the path described by $y = -\frac{5}{4}x + 5$.

a. Sketch the path of the signal.

b. How far from the entrance of the stadium should a transmitter truck be parked to receive the signal from the blimp?

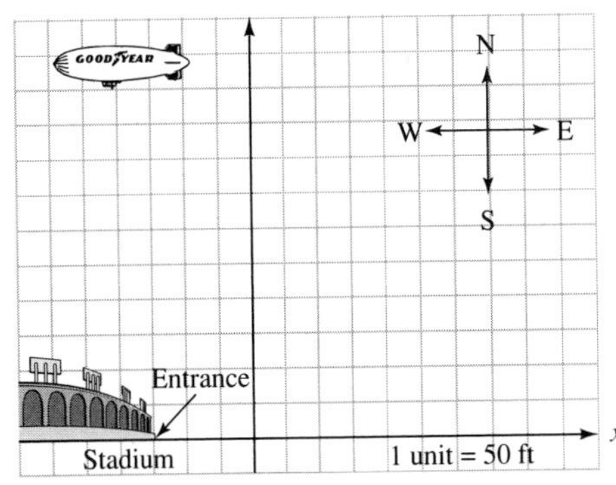

ILLUSTRATION 4

49. HOUSE APPRECIATION A house purchased for \$125,000 is expected to appreciate according to the formula $y = 7{,}500x + 125{,}000$, where y is the value of the house after x years. Find the value of the house 5 years later.

50. CAR DEPRECIATION A car purchased for \$17,000 is expected to depreciate according to the formula $y = -1{,}360x + 17{,}000$. When will the car have no value?

51. DEMAND EQUATION The number of television sets that consumers buy depends on price. The higher the price, the fewer TVs people will buy. The equation that relates price to the number of TVs sold at that price is called a **demand equation.** If the demand equation for a 13-inch TV is $p = -\frac{1}{10}q + 170$, where p is the price and q is the number of TVs sold at that price, how many TVs will be sold at a price of $150?

52. SUPPLY EQUATION The number of television sets that manufacturers produce depends on price. The higher the price, the more TVs manufacturers will produce. The equation that relates price to the number of TVs produced at that price is called a **supply equation.** If the supply equation for a 13-inch TV is $p = \frac{1}{10}q + 130$, where p is the price and q is the number of TVs produced for sale at that price, how many TVs will be produced if the price is $150?

53. BUYING TICKETS Tickets to a circus cost $5 each from Ticketron plus a $2 service fee for each block of tickets.

a. Write a linear equation that gives the cost c for a student buying t tickets.

b. Complete the table in Illustration 5 and graph the equation.

c. Use the graph to estimate the cost of buying 5 tickets.

t	c
1	
2	
3	
4	

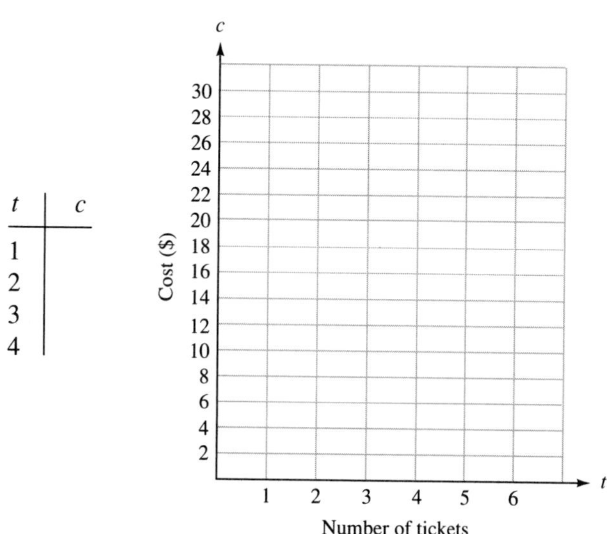

ILLUSTRATION 5

54. TELEPHONE COSTS In a community, the monthly cost of local telephone service is $5 per month, plus 25¢ per call.

a. Write a linear equation that gives the cost c for a person making n calls. Then graph the equation.

b. Complete the table in Illustration 6.

c. Use the graph to estimate the cost of service in a month when 20 calls were made.

n	c
4	
8	
12	
16	

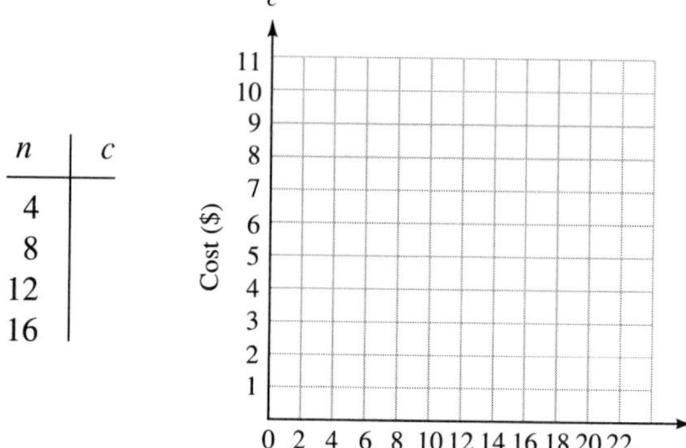

ILLUSTRATION 6

WRITING *Write a paragraph using your own words.*

55. Explain how to graph a line using the intercept method.

56. When graphing a line by plotting points, why is it a good practice to find three solutions instead of two?

REVIEW

57. List the prime numbers between 10 and 30.

58. Write the first ten composite numbers.

59. In what quadrant does the point $(-2, -3)$ lie?

60. What is the formula that gives the area of a circle?

61. Simplify $-4(-20s)$.

62. Approximate π to the nearest thousandth.

63. Remove parentheses: $-(-3x - 8)$.

64. Simplify $\frac{1}{3}b + \frac{1}{3}b + \frac{1}{3}b$.

2.3 Rate of Change and the Slope of a Line

In this section, you will learn about

- **Average rate of change**
- **Slope of a line**
- **Interpretation of slope**
- **Horizontal and vertical lines**
- **Slopes of parallel lines**
- **Slopes of perpendicular lines**

INTRODUCTION. Our world is one of constant change. In this section, we will show how to describe the amount of change in one quantity in relation to the amount of change in another by finding an *average rate of change.*

Average rate of change

The line graphs in Figure 2-21 are models that approximate the number of daily morning newspapers and the number of evening newspapers published in the United States for the years 1990–1996. From the graph, we can see that the number of morning newspapers increased and the number of evening newspapers decreased from 1990 to 1996.

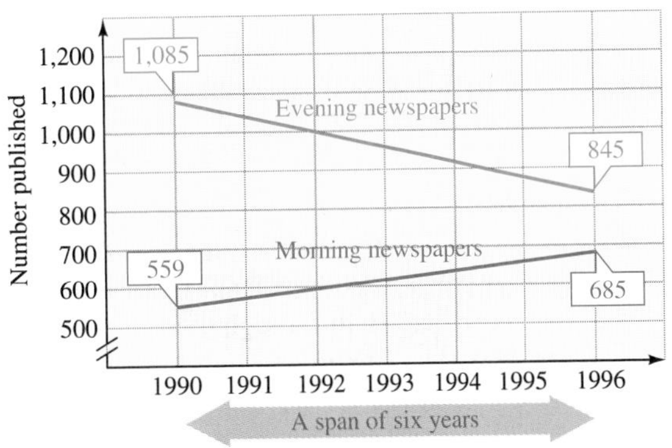

FIGURE 2-21

If we want to know the rate at which the number of morning newspapers increased or the rate at which the number of evening newspapers decreased over this period of time, we can do so by finding an **average rate of change.** To find an average rate of change, we find the **ratio** of the change in the number of newspapers over the length of time in which that change took place.

Ratios and rates

A **ratio** is a comparison of two numbers by their indicated quotient. In symbols, if a and b are two numbers, the ratio of a to b is $\frac{a}{b}$. Ratios that are used to compare quantities with different units are called **rates.**

In Figure 2-21, we see that in 1990, the number of morning newspapers published was 559. In 1996, the number grew to 685. This is a change of $685 - 559$ or 126 over a 6-year time span. So we have

$$\text{Average rate of change} = \frac{\text{change in number of morning newspapers}}{\text{change in time}}$$ The rate of change is a ratio.

$$= \frac{126 \text{ newspapers}}{6 \text{ years}}$$

$$= \frac{21 \cdot \overset{1}{\cancel{6}} \cdot \text{newspapers}}{\underset{1}{\cancel{6}} \text{ years}}$$ Factor 126 as $21 \cdot 6$ and divide out the common factor of 6.

$$= \frac{21 \text{ newspapers}}{1 \text{ year}}$$

The number of morning newspapers published in the United States increased, on average, at a rate of 21 newspapers per year (written 21 newspapers/year) from 1990 through 1996.

In Figure 2-21, we see that in 1990 the number of evening newspapers published was 1,085. In 1996, the number fell to 845. To find the change, we subtract the earlier number from the later number: $845 - 1{,}085 = -240$. The negative result indicates a decline in the number of evening newspapers over the 6-year time span. So we have

$$\text{Average rate of change} = \frac{845 - 1{,}085}{6}$$

$$= \frac{-240}{6}$$

$$= \frac{-40 \cdot \overset{1}{\cancel{6}}}{\underset{1}{\cancel{6}}}$$ Factor -240 as $-40 \cdot 6$.

$$= \frac{-40}{1}$$ Divide out the common factor of 6.

The number of evening newspapers decreased at a rate of -40 newspapers/year. That is, on average, there were 40 fewer evening newspapers per year, every year, from 1990 through 1996.

Slope of a line

In the newspaper example, we measured the steepness of the two lines in the graph in Figure 2-21 to determine the average rates of change. In doing so, we found the **slope** of each line. The slope of a nonvertical line is a number that measures the line's steepness.

To calculate the slope of a line (usually denoted by the letter m), we must first pick two points on the line. To distinguish between the coordinates of two points (say, points P and Q), we use **subscript notation.** Point P can be denoted as $P(x_1, y_1)$, and point Q can be denoted as $Q(x_2, y_2)$. After picking two points on the line, we write the ratio of the vertical change to the corresponding horizontal change as we move from one point to the other.

Slope of a line

The **slope** of a line passing through points $P(x_1, y_1)$ and $Q(x_2, y_2)$ is

$$m = \frac{\text{change in } y}{\text{change in } x} = \frac{y_2 - y_1}{x_2 - x_1} \quad (x_2 \neq x_1)$$

EXAMPLE 1 ***Using the slope formula.*** Find the slope of the line shown in Figure 2-22.

Solution

We can let $P(x_1, y_1) = P(-2, 4)$ and $Q(x_2, y_2) = Q(3, -4)$. Then

$$m = \frac{\text{change in } y}{\text{change in } x}$$

$$= \frac{y_2 - y_1}{x_2 - x_1} \quad \text{The slope formula.}$$

$$= \frac{-4 - 4}{3 - (-2)} \quad \text{Now we substitute } -4 \text{ for } y_2, 4 \text{ for } y_1, 3 \text{ for } x_2, \text{ and } -2 \text{ for } x_1.$$

$$= \frac{-8}{5}$$

$$= -\frac{8}{5}$$

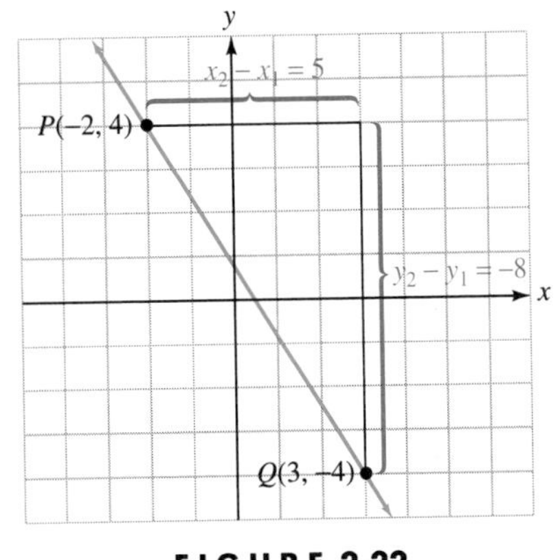

FIGURE 2-22

The slope of the line is $-\frac{8}{5}$. We would obtain the same result if we had let $P(x_1, y_1) = P(3, -4)$ and $Q(x_2, y_2) = Q(-2, 4)$.

Self Check

Find the slope of the line passing through the points $(-3, 6)$ and $(4, -8)$.

Answer: -2

WARNING! When calculating slope, always subtract the y values and the x values in the same order.

$$m = \frac{y_2 - y_1}{x_2 - x_1} \quad \text{or} \quad m = \frac{y_1 - y_2}{x_1 - x_2}$$

However,

$$m \neq \frac{y_2 - y_1}{x_1 - x_2} \quad \text{and} \quad m \neq \frac{y_1 - y_2}{x_2 - x_1}$$

The change in y (often denoted as Δy and read as "delta y") is the **rise** of the line between points P and Q. The change in x (often denoted as Δx and read as "delta x") is the **run.** Using this terminology, we can define slope as the ratio of the rise to the run:

$$m = \frac{\Delta y}{\Delta x} = \frac{\text{rise}}{\text{run}} \quad (\Delta x \neq 0)$$

EXAMPLE 2 ***Finding the slope of a line.*** Find the slope of the line determined by $3x - 4y = 12$.

Solution

We first find the coordinates of two points on the line.

- If $x = 0$, then $y = -3$. The point $(0, -3)$ is on the line.
- If $y = 0$, then $x = 4$. The point $(4, 0)$ is on the line.

Self Check

Find the slope of the line determined by $2x + 5y = 12$.

We then refer to Figure 2-23 and find the slope of the line between $P(0, -3)$ and $Q(4, 0)$ by substituting 0 for y_2, -3 for y_1, 4 for x_2, and 0 for x_1 in the formula for slope.

$$m = \frac{\text{rise}}{\text{run}}$$
$$= \frac{y_2 - y_1}{x_2 - x_1}$$
$$= \frac{0 - (-3)}{4 - (0)}$$
$$= \frac{3}{4}$$

The slope of the line is $\frac{3}{4}$.

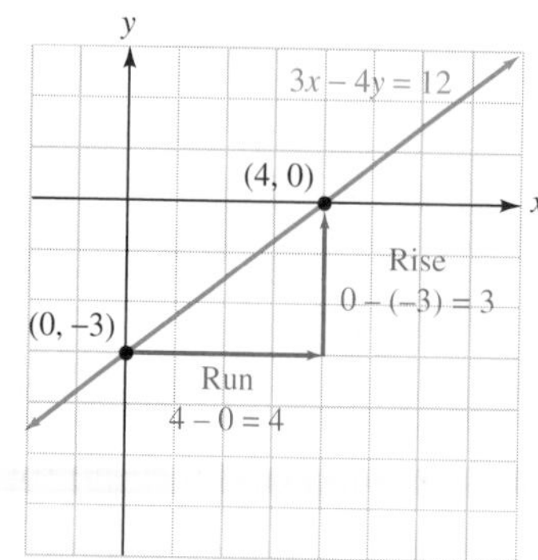

FIGURE 2-23

Answer: $-\frac{2}{5}$

Interpretation of slope

For applied problems, slope can be thought of as the average rate of change in y per unit change in x, where the value of y depends on x.

EXAMPLE 3 ***Carpeting cost.*** A store sells a high-quality carpet for $25 per square yard, plus a $20 delivery charge. The total cost c of n square yards is given by the following formula.

Total cost	=	cost per square yard	·	the number of square yards	+	the delivery charge.
c	=	25	·	n	+	20

Graph this equation and interpret the slope of the line.

Self Check

What information does the c-intercept of the graph give?

Solution

We can graph the equation on a coordinate system with a horizontal n-axis and a vertical c-axis. Figure 2-24 shows a table of ordered pairs and the graph.

$c = 25n + 20$

n	c	(n, c)
10	270	(10, 270)
30	770	(30, 770)
40	1,020	(40, 1,020)

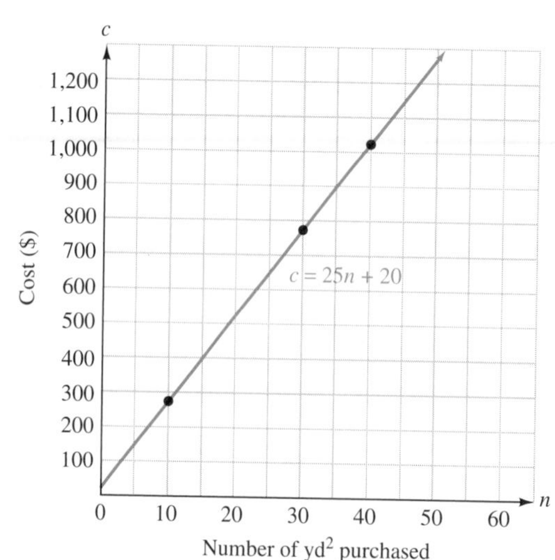

FIGURE 2-24

If we pick the points (30, 770) and (40, 1,020) to find the slope, we have

$$m = \frac{\Delta c}{\Delta n}$$

$$= \frac{c_2 - c_1}{n_2 - n_1}$$ This is the form the slope formula takes when working with ordered pairs of the form (n, c).

$$= \frac{\mathbf{1{,}020} - \mathbf{770}}{\mathbf{40} - \mathbf{30}}$$ Substitute 1,020 for c_2, 770 for c_1, 40 for n_2, and 30 for n_1.

$$= \frac{250}{10}$$

$$= 25$$

The slope of 25 (in dollars/square yard) is the cost per square yard of the carpet.

Answer: The c-coordinate of the c-intercept is the delivery charge, \$20.

EXAMPLE 4 ***Rate of descent.*** It takes a skier 25 minutes to complete the course shown in Figure 2-25. Find his average rate of descent in feet per minute.

Self Check Find the average rate of descent if the skier completes the course in 20 minutes.

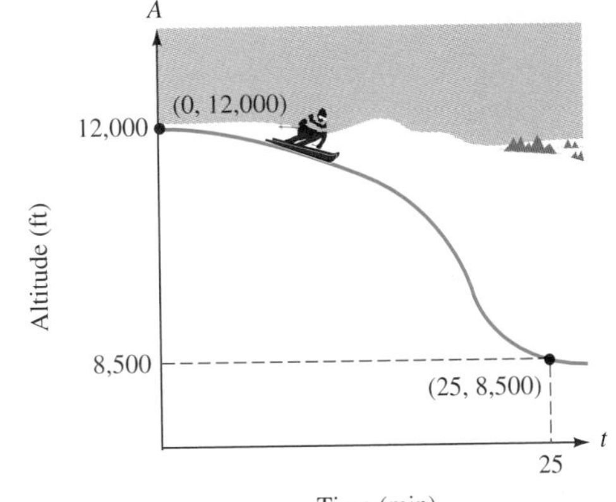

FIGURE 2-25

Solution

To find the average rate of descent, we must find the ratio of the change in altitude (ΔA) to the change in time (Δt). To find this ratio, we calculate the slope of the line passing through the points (0, 12,000) and (25, 8,500).

$$\text{Average rate of descent} = \frac{\Delta A}{\Delta t}$$

$$= \frac{8{,}500 - 12{,}000}{25 - 0}$$ In the numerator, write the change in altitude; in the denominator, the change in time.

$$= \frac{-3{,}500}{25}$$ Do the subtractions.

$$= -140$$

The average rate of descent is 140 feet per minute.

Answer: 175 feet per minute

Horizontal and vertical lines

If $P(x_1, y_1)$ and $Q(x_2, y_2)$ are points on the horizontal line shown in Figure 2-26(a), then $y_1 = y_2$, and the numerator of the fraction

$$\frac{y_2 - y_1}{x_2 - x_1}$$ On a horizontal line, $x_2 \neq x_1$.

is 0. Thus, the value of the fraction is 0, and the slope of the horizontal line is 0.

If $P(x_1, y_1)$ and $Q(x_2, y_2)$ are two points on the vertical line shown in Figure 2-26(b), then $x_1 = x_2$, and the denominator of the fraction

$$\frac{y_2 - y_1}{x_2 - x_1}$$ On a vertical line, $y_2 \neq y_1$.

is 0. Since the denominator of a fraction cannot be 0, a vertical line has no defined slope.

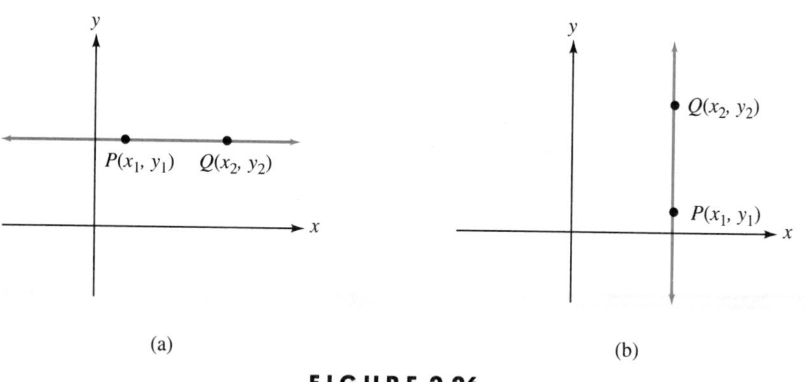

FIGURE 2-26

Slopes of horizontal and vertical lines

All horizontal lines (lines with equations of the form $y = b$) have a slope of 0.

All vertical lines (lines with equations of the form $x = a$) have no defined slope.

If a line rises as we follow it from left to right, as in Figure 2-27(a), its slope is positive. If a line drops as we follow it from left to right, as in Figure 2-27(b), its slope is negative. If a line is horizontal, as in Figure 2-27(c), its slope is 0. If a line is vertical, as in Figure 2-27(d), it has no defined slope.

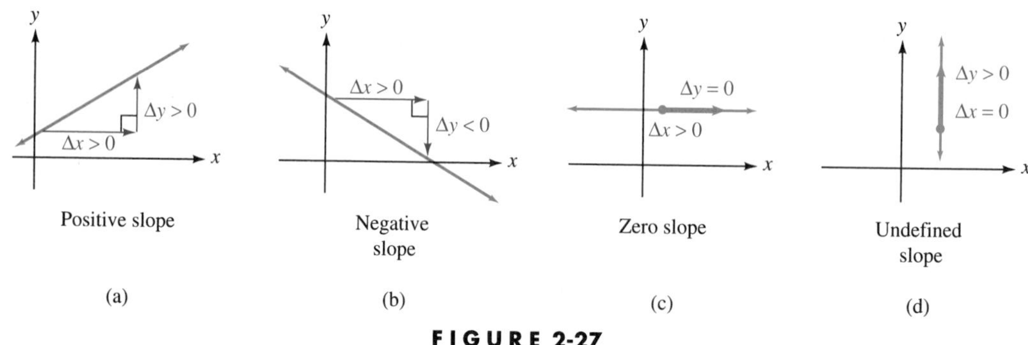

FIGURE 2-27

Slopes of parallel lines

To see a relationship between parallel lines and their slopes, we refer to the parallel lines l_1 and l_2 shown in Figure 2-28, with slopes of m_1 and m_2, respectively. Because right triangles ABC and DEF are similar, it follows that

$$m_1 = \frac{\Delta y \text{ of } l_1}{\Delta x \text{ of } l_1} = \frac{\Delta y \text{ of } l_2}{\Delta x \text{ of } l_2} = m_2$$

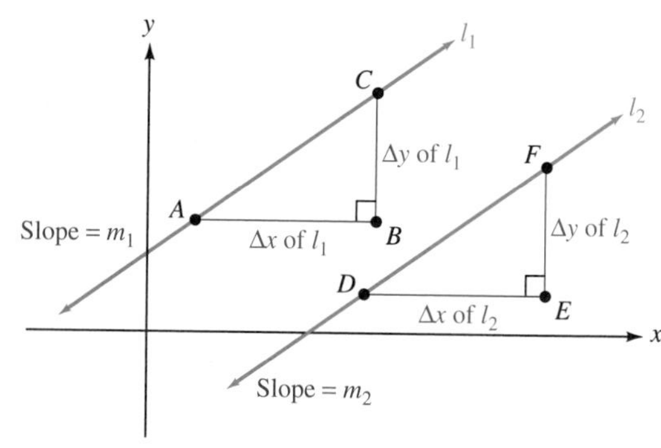

FIGURE 2-28

Thus, if two nonvertical lines are parallel, they have the same slope. It is also true that when two lines have the same slope, they are parallel.

Slopes of parallel lines

Nonvertical parallel lines have the same slope, and different lines having the same slope are parallel. Since vertical lines are parallel, lines with no defined slope are parallel.

EXAMPLE 5 ***Slopes of parallel lines.*** The lines in Figure 2-29 are parallel. Find y.

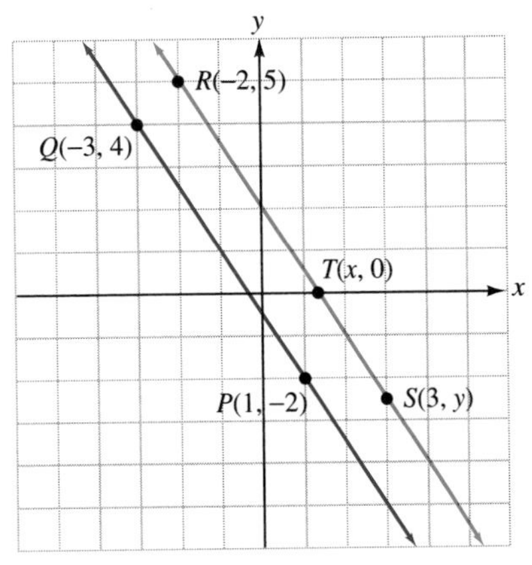

FIGURE 2-29

Solution

Since the lines are parallel, they have equal slopes. To find y, we find the slope of each line, set them equal, and solve the resulting equation.

$$\textit{Slope of PQ} = \textit{Slope of RS}$$

$$\frac{-2-4}{1-(-3)} = \frac{y-5}{3-(-2)}$$

$$\frac{-6}{4} = \frac{y-5}{5} \quad \text{Do the subtractions.}$$

$$-30 = 4(y-5) \quad \text{Multiply both sides by 20.}$$

$$-30 = 4y-20 \quad \text{Use the distributive property.}$$

$$-10 = 4y \quad \text{Add 20 to both sides.}$$

$$-\frac{5}{2} = y \quad \text{Divide both sides by 4 and simplify.}$$

Thus, $y = -\frac{5}{2}$.

Self Check

Find x.

Answer: $\frac{4}{3}$ ■

Slopes of perpendicular lines

Two real numbers a and b are called **negative reciprocals** if $ab = -1$. For example,

$$-\frac{4}{3} \quad \text{and} \quad \frac{3}{4}$$

are negative reciprocals, because $-\frac{4}{3}\left(\frac{3}{4}\right) = -1$.

The following theorem relates perpendicular lines and their slopes.

Slopes of perpendicular lines

If two nonvertical lines are perpendicular, their slopes are negative reciprocals.

If the slopes of two lines are negative reciprocals, the lines are perpendicular.

Because a horizontal line is perpendicular to a vertical line, a line with a slope of 0 is perpendicular to a line with no defined slope.

EXAMPLE 6 ***Slopes of perpendicular lines.*** Are the lines OP and PQ shown in Figure 2-30 perpendicular?

Solution

We find the slopes of the lines and see whether they are negative reciprocals.

$$\text{Slope of } OP = \frac{\Delta y}{\Delta x} = \frac{y_2 - y_1}{x_2 - x_1} = \frac{-4 - 0}{3 - 0} = -\frac{4}{3}$$

and

$$\text{Slope of } PQ = \frac{\Delta y}{\Delta x} = \frac{y_2 - y_1}{x_2 - x_1} = \frac{4 - (-4)}{9 - 3} = \frac{8}{6} = \frac{4}{3}$$

Since their slopes are not negative reciprocals, the lines are not perpendicular.

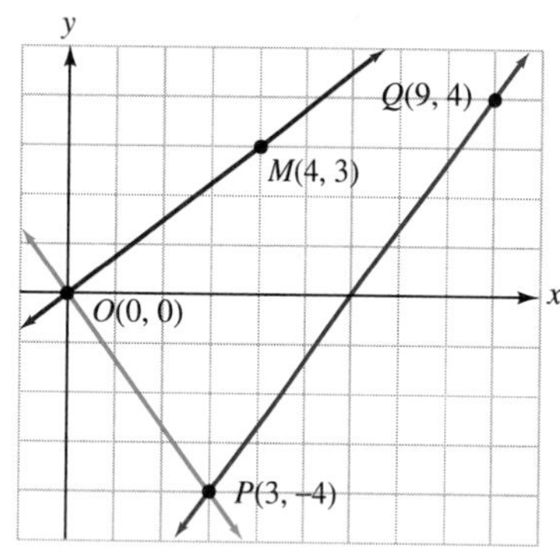

FIGURE 2-30

Self Check

In Figure 2-30, is MO perpendicular to OP?

Answer: yes

STUDY SET Section 2.3

VOCABULARY *In Exercises 1–6, fill in the blanks to make the statements true.*

1. ______________ is defined as the change in y divided by the change in x.
2. A slope is an average ______________ of change.
3. The ______________ in x (denoted as Δx) is the run of the line between points P and Q.
4. The change in y (denoted as Δy) is the ______________ of the line between points P and Q.
5. $\frac{7}{8}$ and $-\frac{8}{7}$ are negative ______________.
6. ______________ lines have the same slope.

CONCEPTS

7. See Illustration 1.
   - **a.** Which line is horizontal? What is its slope?
   - **b.** Which line is vertical? What is its slope?
   - **c.** Which line has a positive slope? What is it?
   - **d.** Which line has a negative slope? What is it?

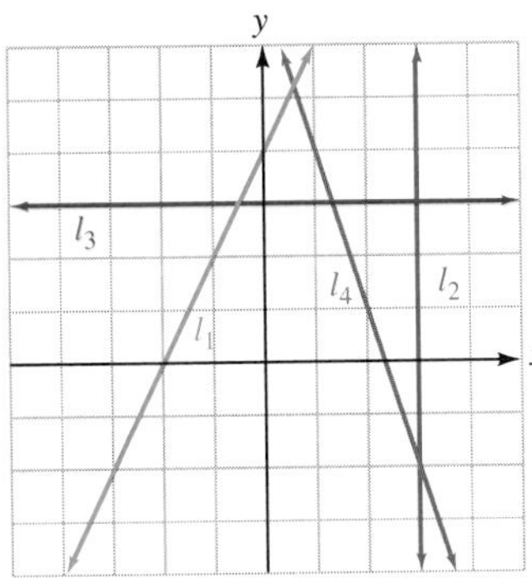

ILLUSTRATION 1

8. See Illustration 2.

a. Find the slopes of lines l_1 and l_2. Are they parallel?

b. Find the slopes of lines l_2 and l_3. Are they perpendicular?

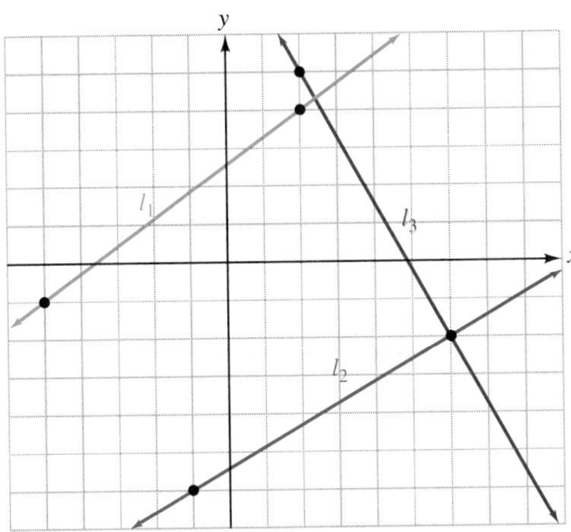

ILLUSTRATION 2

9. EDUCATION The graphs in Illustration 3 are models that approximate the growth in weekly earnings of Americans 25 years and older, by educational attainment, for 1980–1996. Which educational level's weekly earnings are

a. increasing at the fastest rate? Estimate the average rate of change.

b. increasing at the slowest rate? Estimate the average rate of change.

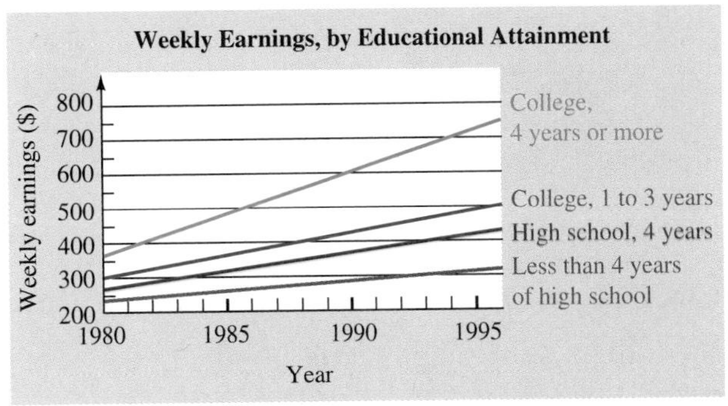

Based on data from *The Wall Street Journal Almanac* (1998)

ILLUSTRATION 3

10. HALLOWEEN A couple kept records of the number of trick-or-treaters who came to their door on Halloween night. (See Illustration 4.) Find the slope of the line. What information does the slope give?

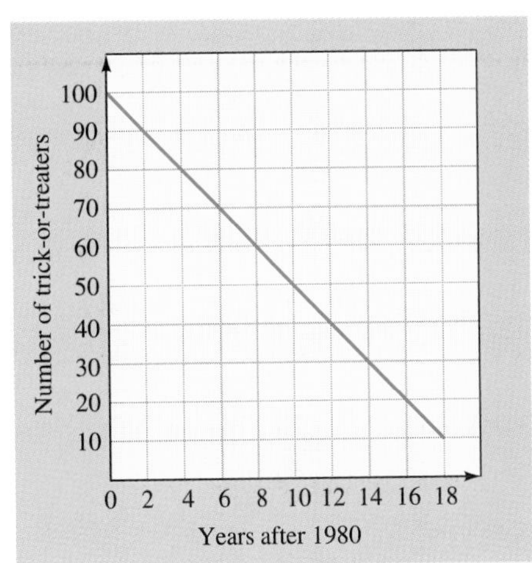

ILLUSTRATION 4

NOTATION

11. What formula is used to find the slope of a line?

12. Tell the difference between x^2 and x_2.

13. See Illustration 5.

a. What is Δy?

b. What is Δx?

c. What is $\frac{\Delta y}{\Delta x}$?

14. See Illustration 6.

a. What is Δy?

b. What is Δx?

c. What is $\frac{\Delta y}{\Delta x}$?

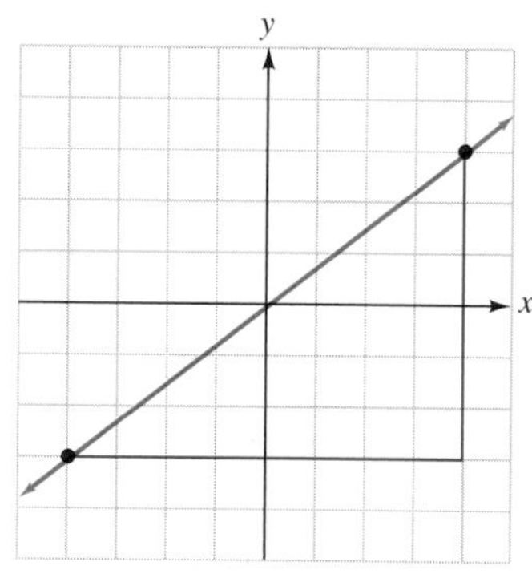

ILLUSTRATION 5

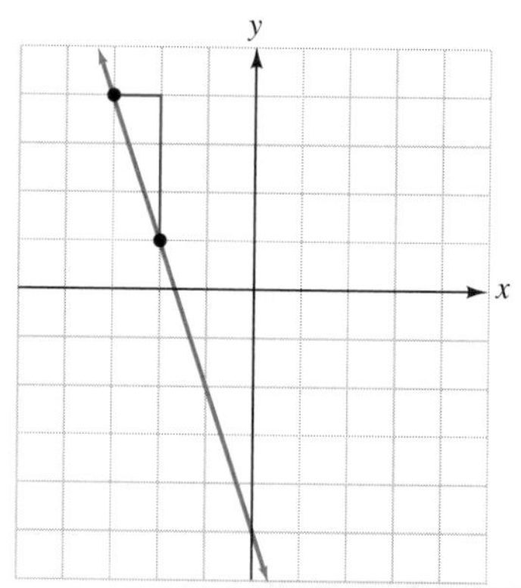

ILLUSTRATION 6

PRACTICE *In Exercises 15–26, find the slope of the line that passes through the given points, if possible.*

15. $(0, 0), (3, 9)$
16. $(9, 6), (0, 0)$
17. $(-1, 8), (6, 1)$
18. $(-5, -8), (3, 8)$
19. $(3, -1), (-6, 2)$
20. $(0, -8), (-5, 0)$
21. $(7, 5), (-9, 5)$
22. $(2, -8), (3, -8)$
23. $(-7, -5), (-7, -2)$
24. $(3, -5), (3, 14)$
25. $(a, b), (b, a)$
26. $(a, b), (-b, -a)$

In Exercises 27–34, find the slope of the line determined by each equation.

27. $3x + 2y = 12$
28. $2x - y = 6$
29. $3x = 4y - 2$
30. $x = y$
31. $y = \frac{x - 4}{2}$
32. $x = \frac{3 - y}{4}$
33. $4y = 3(y + 2)$
34. $x + y = \frac{2 - 3y}{3}$

In Exercises 35–40, tell whether the lines with the given slopes are parallel, perpendicular, or neither.

35. $m_1 = 3, m_2 = -\frac{1}{3}$
36. $m_1 = \frac{1}{4}, m_2 = 4$
37. $m_1 = 4, m_2 = 0.25$
38. $m_1 = -5, m_2 = \frac{1}{-0.2}$
39. $m_1 = \frac{1}{a}, m_2 = a$
40. $m_1 = a, m_2 = -\frac{1}{a}$

In Exercises 41–46, tell whether the line PQ is parallel or perpendicular (or neither) to a line with a slope of −2.

41. $P(3, 4), Q(4, 2)$
42. $P(6, 4), Q(8, 5)$
43. $P(-2, 1), Q(6, 5)$
44. $P(3, 4), Q(-3, -5)$
45. $P(5, 4), Q(6, 6)$
46. $P(-2, 3), Q(4, -9)$

APPLICATIONS

47. LANDING A PLANE A jet descends in a stairstep pattern, as shown in Illustration 7. The required elevations of the plane's path are given. Find the slope of the descent in each of the three parts of its landing that are labeled. Which part is the steepest?

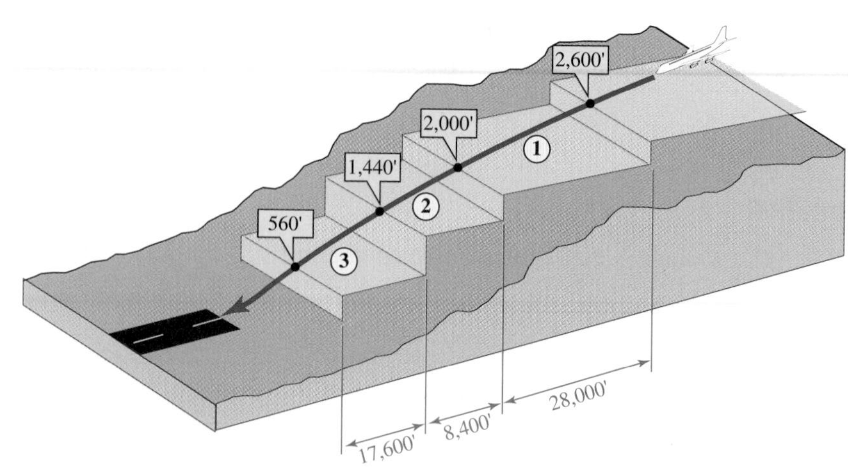

Based on data from *Los Angeles Times* (August 7, 1997), p. A8

ILLUSTRATION 7

48. DROP IN PRICE The price of computers has been dropping for the past ten years. If a desktop PC cost \$5,700 ten years ago, and the same computing power cost \$400 two years ago, find the rate of decrease per year. (Assume a straight-line model.)

49. MAPS Topographic maps have contour lines that connect points of equal elevation on a mountain. The vertical distance between contour lines in Illustration 8 is 50 feet. Find the slope of the west face and the east face of the mountain peak.

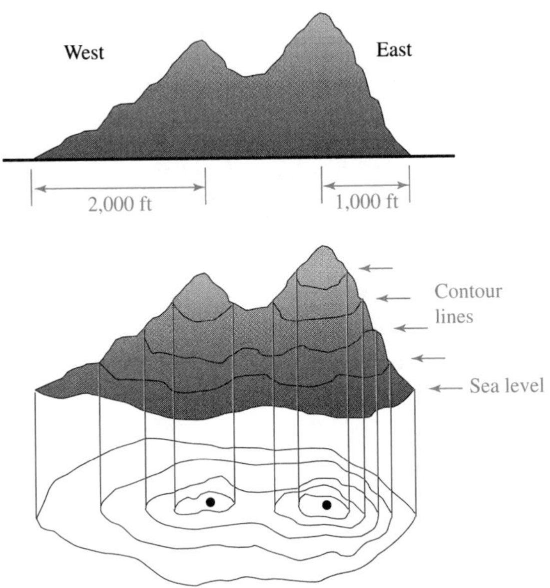

ILLUSTRATION 8

50. SKIING The men's giant slalom course shown in Illustration 9 is longer than the women's course. Does this mean that the men's course is steeper? Use the concept of the slope of a line to explain.

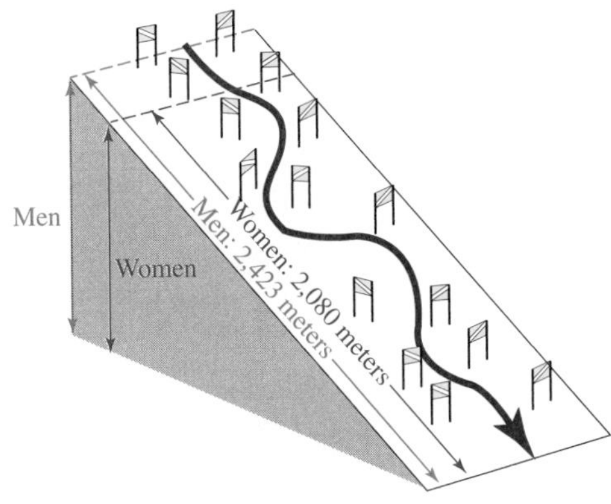

ILLUSTRATION 9

51. ROAD SIGNS Find the slope of the road shown in Illustration 10. Use this information to complete the road warning sign for truckers. (*Hint:* 1 mi = 5,280 ft.)

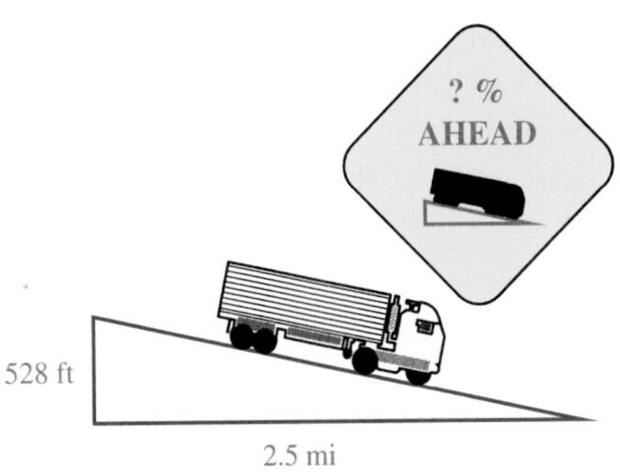

ILLUSTRATION 10

52. GREENHOUSE EFFECT The graphs in Illustration 11 are estimates of future average global temperature rise due to the greenhouse effect. Assume that the models are straight lines. Estimate the average rate of change of each model. Express your answers as fractions.

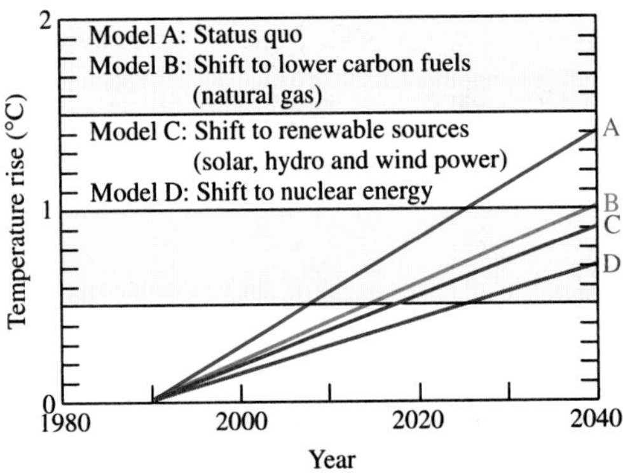

Based on data from *The Blue Planet* (Wiley, 1995)

ILLUSTRATION 11

53. DECK DESIGN See Illustration 12. Find the slopes of the cross-brace and the supports. Is the cross-brace perpendicular to either support?

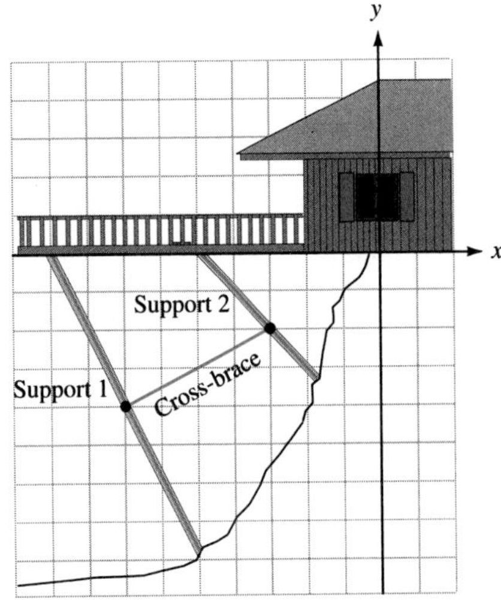

ILLUSTRATION 12

54. AIR PRESSURE Air pressure, measured in units called pascals (Pa), decreases with altitude. Find the rate of change in Pascals for the fastest and the slowest decreasing steps of the graph in Illustration 13.

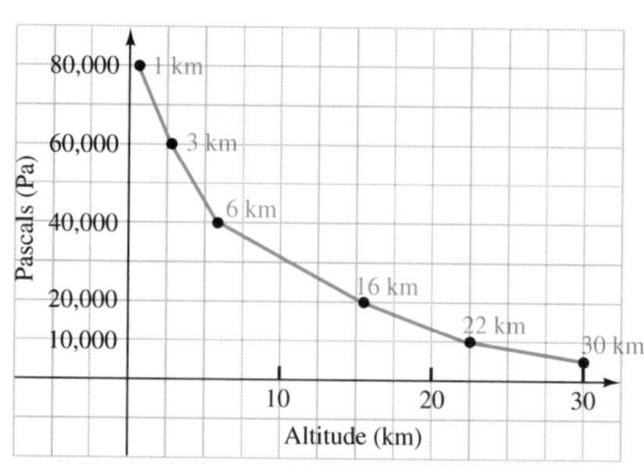

Based on data from *The Blue Planet* (Wiley, 1995)

ILLUSTRATION 13

WRITING *Write a paragraph using your own words.*

55. Explain why a vertical line has no defined slope.

56. Explain how to determine from their slopes whether two lines are parallel, perpendicular, or neither.

REVIEW

57. HALLOWEEN CANDY A candy maker wants to make a 60-pound mixture of two candies to sell for $2 per pound. If black licorice bits sell for $1.90 per pound and orange gumdrops sell for $2.20 per pound, how many pounds of each should be used?

58. MEDICATIONS A doctor prescribes an ointment that is 2% hydrocortisone. A pharmacist has 1% and 5% concentrations in stock. How many ounces of each should the pharmacist use to make a 1-ounce tube?

2.4 *Writing Equations of Lines*

In this section, you will learn about

- **Point–slope form of the equation of a line**
- **Slope–intercept form of the equation of a line**
- **Using slope as an aid in graphing**
- **Parallel and perpendicular lines**
- **Straight-line depreciation**
- **Curve fitting**

INTRODUCTION. We have seen that linear relationships are often presented in graphs. In this section, we begin a discussion of how to write an equation to model a linear relationship.

Point–slope form of the equation of a line

Suppose that line l in Figure 2-31 has a slope of m and passes through $P(x_1, y_1)$. If $Q(x, y)$ is a second point on line l, we have

$$m = \frac{y - y_1}{x - x_1}$$

or if we multiply both sides by $x - x_1$, we have

1. $y - y_1 = m(x - x_1)$

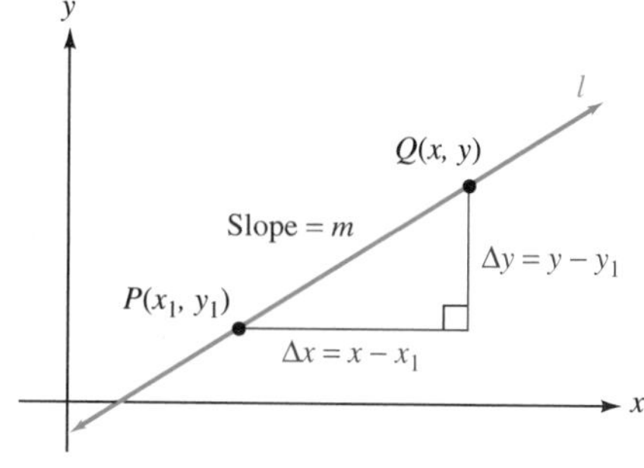

FIGURE 2-31

Because Equation 1 displays the coordinates of the point (x_1, y_1) on the line and the slope m of the line, it is called the **point–slope form** of the equation of a line.

Point–slope form	The equation of the line passing through $P(x_1, y_1)$ and with slope m is $y - y_1 = m(x - x_1)$

EXAMPLE 1 ***Using the point–slope form.*** Write the equation of the line with a slope of $-\frac{2}{3}$ and passing through $P(-4, 5)$.

Solution

We substitute $-\frac{2}{3}$ for m, -4 for x_1, and 5 for y_1 into the point–slope form and simplify.

$$y - y_1 = m(x - x_1) \quad \text{Point–slope form.}$$

$$y - 5 = -\frac{2}{3}[x - (-4)] \quad \text{Substitute } -\tfrac{2}{3} \text{ for } m, -4 \text{ for } x_1, \text{ and 5 for } y_1.$$

$$y - 5 = -\frac{2}{3}(x + 4) \quad \text{Simplify the expression inside the brackets.}$$

$$y - 5 = -\frac{2}{3}x - \frac{8}{3} \quad \text{Use the distributive property to remove parentheses.}$$

$$y = -\frac{2}{3}x + \frac{7}{3} \quad \text{To solve for } y, \text{ add } 5 = \tfrac{15}{3} \text{ to both sides and simplify.}$$

The equation of the line is $y = -\frac{2}{3}x + \frac{7}{3}$.

Self Check

Write the equation of the line with slope of $\frac{5}{4}$ and passing through $Q(0, 5)$.

Answer: $y = \frac{5}{4}x + 5$ ■

EXAMPLE 2 ***Given two points on a line.*** Write the equation of the line passing through $P(-5, 4)$ and $Q(8, -6)$.

Solution

First we find the slope of the line.

$$m = \frac{y_2 - y_1}{x_2 - x_1} \quad \text{The slope formula.}$$

$$= \frac{-6 - 4}{8 - (-5)} \quad \text{Substitute } -6 \text{ for } y_2, 4 \text{ for } y_1, 8 \text{ for } x_2, \text{ and } -5 \text{ for } x_1.$$

$$= -\frac{10}{13}$$

Since the line passes through P and Q, we can choose either point and substitute its coordinates into the point–slope form. If we choose $P(-5, 4)$, we substitute -5 for x_1, 4 for y_1, and $-\frac{10}{13}$ for m and proceed as follows.

$$y - y_1 = m(x - x_1) \quad \text{Point–slope form.}$$

$$y - 4 = -\frac{10}{13}[x - (-5)] \quad \text{Substitute } -\tfrac{10}{13} \text{ for } m, -5 \text{ for } x_1, \text{ and 4 for } y_1.$$

$$y - 4 = -\frac{10}{13}(x + 5) \quad \text{Simplify the expression inside the brackets.}$$

$$y - 4 = -\frac{10}{13}x - \frac{50}{13} \quad \text{Remove parentheses: distribute } -\tfrac{10}{13}.$$

$$y = -\frac{10}{13}x + \frac{2}{13} \quad \text{To solve for } y, \text{ add } 4 = \tfrac{52}{13} \text{ to both sides and simplify.}$$

The equation of the line is $y = -\frac{10}{13}x + \frac{2}{13}$.

Self Check

Write the equation of the line passing through $R(-2, 5)$ and $S(4, -3)$.

Answer: $y = -\frac{4}{3}x + \frac{7}{3}$ ■

Slope–intercept form of the equation of a line

Since the y-intercept of the line shown in Figure 2-32 is the point $(0, b)$, we can write its equation by substituting 0 for x_1 and b for y_1 in the point–slope form and simplifying.

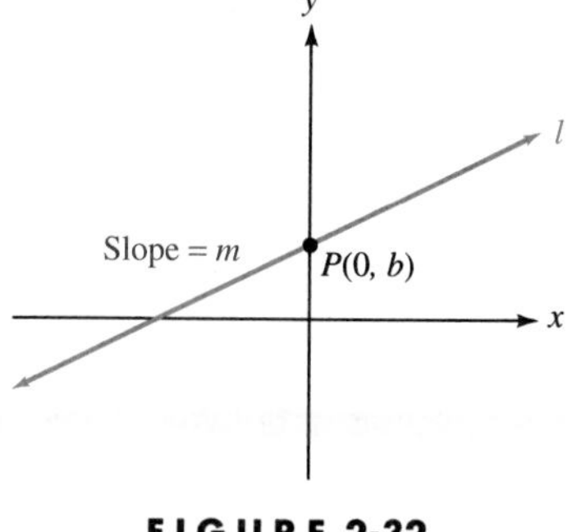

y
l
Slope = m
P(0, b)
x

FIGURE 2-32

$$y - y_1 = m(x - x_1)$$
$$y - b = m(x - 0)$$
$$y - b = mx$$

2. $y = mx + b$ To solve for y, add b to both sides.

Because Equation 2 displays the slope m and the y-coordinate b of the y-intercept, it is called the **slope–intercept form** of the equation of a line.

Slope–intercept form

The equation of the line with slope m and y-intercept $(0, b)$ is

$$y = mx + b$$

EXAMPLE 3 ***Using the slope–intercept form.*** Write the equation of the line with slope 4 that passes through $P(5, 9)$.

Solution

Since we are given that $m = 4$ and that $(5, 9)$ satisfies the equation, we can substitute 5 for x, 9 for y, and 4 for m in the equation $y = mx + b$ and solve for b.

$y = mx + b$ Slope–intercept form.

$9 = 4(5) + b$ Substitute 9 for y, 4 for m, and 5 for x.

$9 = 20 + b$ Do the multiplication.

$-11 = b$ To solve for b, subtract 20 from both sides.

Because $m = 4$ and $b = -11$, the equation is $y = 4x - 11$.

Self Check

Write the equation of the line with slope -2 that passes through the point $Q(-2, 8)$.

Answer: $y = -2x + 4$

When an equation describing a linear relationship between two quantities is written in slope–intercept form, two pieces of information about the relationship are easily seen. As an example, let's consider the equation $L = -0.05t + 7.25$. If we begin with a pencil 7.25 inches long, this linear model gives the new length L in inches of the pencil after it has been inserted into a sharpener and the handle turned t times. (See Figure 2-33.)

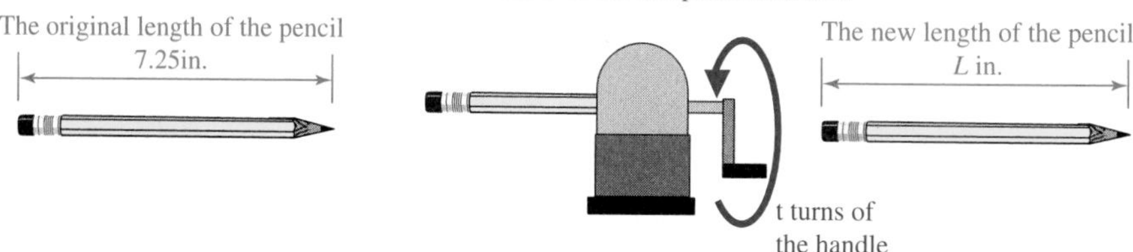

The new length L of the pencil is related to the number of times t the handle of the sharpener is turned.
The original length of the pencil 7.25in.
The new length of the pencil L in.
t turns of the handle

FIGURE 2-33

The value of m (in this case, -0.05) gives the change in the length of the pencil for one turn of the handle. Because the slope is negative, we know that the length of the pencil *decreases* by 0.05 inch for each turn of the handle. The value of b (in this case, 7.25) tells us that before any turns were made (when $t = 0$), the length of the pencil was 7.25 inches.

$$L = -0.05t + 7.25$$

The slope gives the rate of change of the length of the pencil.

The intercept gives the original length of the pencil.

Using slope as an aid in graphing

It is easy to graph a linear equation when it is written in slope–intercept form. For example, to graph $y = \frac{4}{3}x - 2$, we note that $b = -2$ and that the y-intercept is $(0, b) = (0, -2)$. (See Figure 2-34.)

Because the slope of the line is $\frac{\Delta y}{\Delta x} = \frac{4}{3}$, we can locate another point Q on the line by starting at point P and counting 3 units to the right and 4 units up. The change in x from point P to point Q is $\Delta x = 3$, and the corresponding change in y is $\Delta y = 4$. The line joining points P and Q is the graph of the equation.

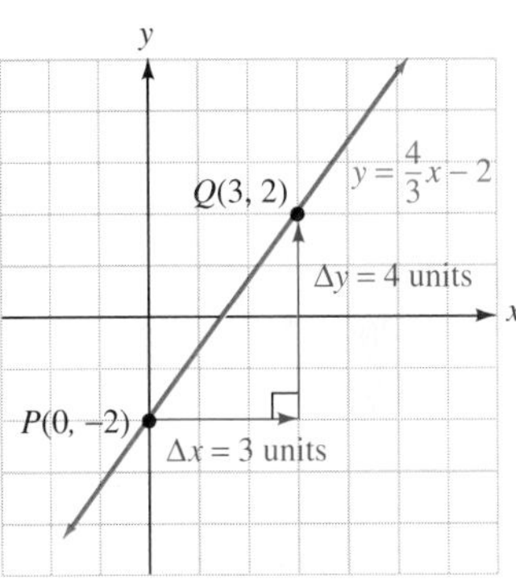

FIGURE 2-34

EXAMPLE 4 Find the slope and the y-intercept of the line with the equation $2x + 3y = -9$. Then graph the line.

Solution

We write the equation in the form $y = mx + b$ to find the slope m and the y-intercept $(0, b)$.

$2x + 3y = -9$	The given equation in general form.
$3y = -2x - 9$	Subtract $2x$ from both sides.
$\frac{3y}{3} = \frac{-2x}{3} - \frac{9}{3}$	Divide both sides by 3.
$y = -\frac{2}{3}x - 3$	Simplify both sides. We see that $m = -\frac{2}{3}$ and $b = -3$.

The slope is $-\frac{2}{3}$, and the y-intercept is $(0, -3)$. To draw the graph, we plot the y-intercept $(0, -3)$ and locate a second point on the line by moving 3 units to the right and 2 units down. We draw a line through the two points to obtain the graph shown in Figure 2-35.

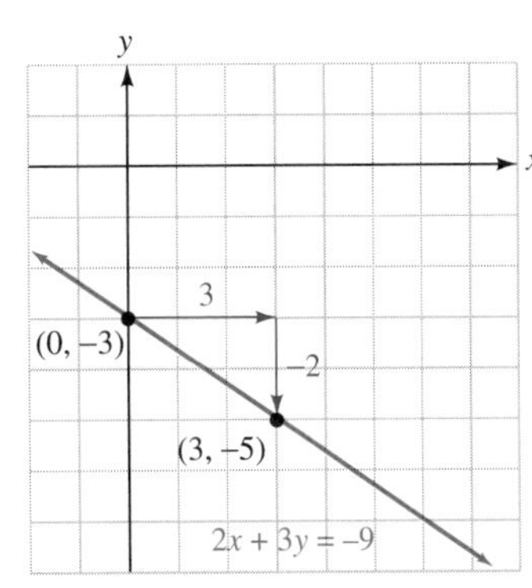

FIGURE 2-35

Self Check

Find the slope and the y-intercept of the line with the equation $3x - 2y = -4$. Then graph the line.

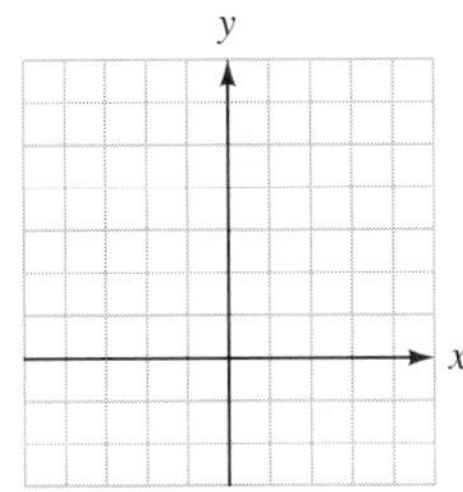

Answer: $m = \frac{3}{2}$, $(0, 2)$

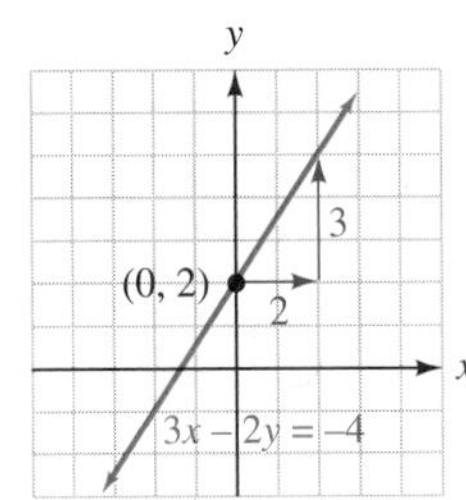

EXAMPLE 5 ***Parallel lines.*** Show that the lines represented by $4x + 8y = 10$ and $2x = 12 - 4y$ are parallel.

Solution

We solve each equation for y to see that the lines are distinct and that their slopes are equal.

$$\begin{aligned} 4x + 8y &= 10 \\ 8y &= -4x + 10 \\ y &= -\frac{1}{2}x + \frac{5}{4} \end{aligned} \qquad \begin{aligned} 2x &= 12 - 4y \\ 4y &= -2x + 12 \\ y &= -\frac{1}{2}x + 3 \end{aligned}$$

Since the values of b in these equations are different ($\frac{5}{4}$ and 3), the lines are distinct. Since the slope of each line is $-\frac{1}{2}$, they are parallel.

Self Check

Are the lines represented by $3x - 2y = 4$ and $6x = 4(y + 1)$ parallel?

Answer: yes

EXAMPLE 6 ***Perpendicular lines.*** Show that the lines represented by $4x + 8y = 10$ and $4x - 2y = 21$ are perpendicular.

Solution

We solve each equation for y to see that the slopes of their straight-line graphs are negative reciprocals.

$$\begin{aligned} 4x + 8y &= 10 \\ 8y &= -4x + 10 \\ y &= -\frac{1}{2}x + \frac{5}{4} \end{aligned} \qquad \begin{aligned} 4x - 2y &= 21 \\ -2y &= -4x + 21 \\ y &= 2x - \frac{21}{2} \end{aligned}$$

Since the slopes are $-\frac{1}{2}$ and 2 (which are negative reciprocals), the lines are perpendicular.

Self Check

Are the lines represented by $3x + 2y = 6$ and $2x - 3y = 6$ perpendicular?

Answer: yes

EXAMPLE 7 ***Parallel lines.*** Write the equation of the line that passes through $P(-2, 5)$ and is parallel to the line $y = 8x - 3$.

Solution

Since the slope of the line given by $y = 8x - 3$ is the coefficient of x, the slope is 8. Since the desired equation is to have a graph that is parallel to the graph of $y = 8x - 3$, its slope must also be 8.

We substitute -2 for x_1, 5 for y_1, and 8 for m in the point–slope form and simplify.

$$\begin{aligned} y - y_1 &= m(x - x_1) && \\ y - 5 &= 8[x - (-2)] && \text{Substitute 5 for } y_1\text{, 8 for } m\text{, and } -2 \text{ for } x_1. \\ y - 5 &= 8(x + 2) && \text{Simplify the expression inside the brackets.} \\ y - 5 &= 8x + 16 && \text{Use the distributive property to remove parentheses.} \\ y &= 8x + 21 && \text{Add 5 to both sides.} \end{aligned}$$

The equation is $y = 8x + 21$.

Self Check

Write the equation of the line that is parallel to the line $y = 8x - 3$ and passes through the origin.

Answer: $y = 8x$

We now summarize the various forms for the equation of a line.

Forms for the equation of a line		
	General form of a linear equation	$Ax + By = C$ A and B cannot both be 0.
	Slope–intercept form of a linear equation	$y = mx + b$ The slope is m, and the y-intercept is $(0, b)$.
	Point–slope form of a linear equation	$y - y_1 = m(x - x_1)$ The slope is m, and the line passes through (x_1, y_1).
	A **horizontal line**	$y = b$ The slope is 0, and the y-intercept is $(0, b)$.
	A **vertical line**	$x = a$ There is no defined slope, and the x-intercept is $(a, 0)$.

Straight-line depreciation

For tax purposes, many businesses use *straight-line depreciation* to find the declining value of aging equipment.

EXAMPLE 8 ***Value of a lathe*** The owner of a machine shop buys a lathe for $1,970 and expects it to last 10 years. The lathe can then be sold as scrap for an estimated *salvage value* of $270. If y represents the value of the lathe after x years of use, and y and x are related by the equation of a line,

a. Find the equation of the line.

b. Find the value of the lathe after $2\frac{1}{2}$ years.

c. Find the economic meaning of the y-intercept of the line.

d. Find the economic meaning of the slope of the line.

Solution **a.** To find the equation of the line, we first find its slope and then use point–slope form to find its equation.

When the lathe is new, its age x is 0, and its value y is $1,970. When the lathe is 10 years old, $x = 10$, and its value is $y = \$270$. Since the line passes through the points (0, 1,970) and (10, 270), as shown in Figure 2-36, the slope of the line is

$$\begin{aligned} m &= \frac{y_2 - y_1}{x_2 - x_1} \\ &= \frac{270 - 1{,}970}{10 - 0} \\ &= \frac{-1{,}700}{10} \\ &= -170 \end{aligned}$$

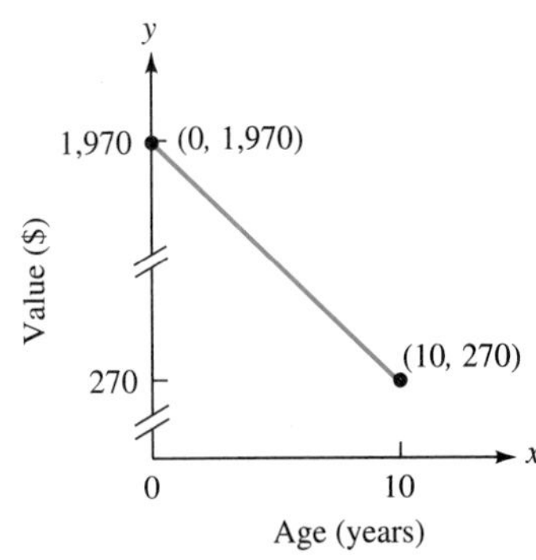

FIGURE 2-36

To find the equation of the line, we substitute -170 for m, 0 for x_1, and 1,970 for y_1 in the point–slope form and simplify.

$$\begin{aligned} y - y_1 &= m(x - x_1) \\ y - 1{,}970 &= -170(x - 0) \\ y &= -170x + 1{,}970 \end{aligned} \quad (3)$$

3.

The current value y of the lathe is related to its age x by the linear model $y = -170x + 1{,}970$.

b. To find the value of the lathe after $2\frac{1}{2}$ years, we substitute 2.5 for x in Equation 3 and solve for y.

$$\begin{aligned} y &= -170x + 1{,}970 \\ &= -170(2.5) + 1{,}970 \\ &= -425 + 1{,}970 \\ &= 1{,}545 \end{aligned}$$

In $2\frac{1}{2}$ years, the lathe will be worth \$1,545.

c. The y-intercept of the graph is $(0, b)$, where b is the value of y when $x = 0$.

$$\begin{aligned} y &= -170x + 1{,}970 \\ y &= -170(0) + 1{,}970 \\ y &= 1{,}970 \end{aligned}$$

Thus, b is the value of a 0-year-old lathe, which is the lathe's original cost, \$1,970.

d. Each year, the value of the lathe decreases by \$170, because the slope of the line is -170. The slope of the depreciation line is the *annual depreciation rate.* ■

Curve fitting

In statistics, the process of using one variable to predict another is called **regression.** For example, if we know a man's height, we can usually make a good prediction about his weight, because taller men usually weigh more than shorter men.

Figure 2-37 shows the results of sampling ten men at random and finding their heights and weights. The graph of the ordered pairs (h, w) is called a **scattergram.**

Man	Height (h) in inches	Weight (w) in pounds
1	66	140
2	68	150
3	68	165
4	70	180
5	70	165
6	71	175
7	72	200
8	74	190
9	75	210
10	75	215

(a)

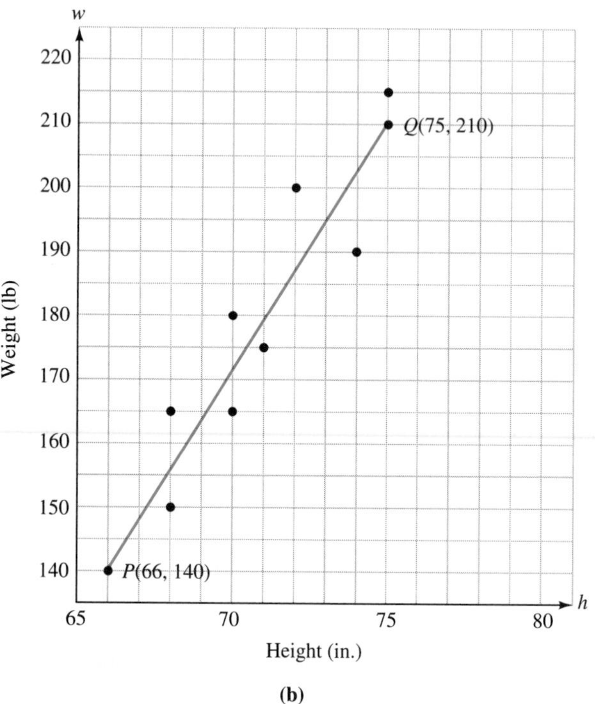

(b)

FIGURE 2-37

To write a *prediction equation* (sometimes called a *regression equation*), we must find the equation of the line that comes closer to all of the points in the scattergram than any other possible line. In statistics, there are exact methods to find this equation. However, here we can only approximate this equation.

To write an approximation of the regression equation, we place a straightedge on the scattergram shown in Figure 2-37 and draw the line joining two points that seems to best fit all of the points. In the figure, line PQ is drawn, where point P has coordinates of (66, 140) and point Q has coordinates of (75, 210).

Our approximation of the regression equation will be the equation of the line passing through points P and Q. To find the equation of this line, we first find its slope.

$$\begin{aligned} m &= \frac{y_2 - y_1}{x_2 - x_1} \\ &= \frac{210 - 140}{75 - 66} \\ &= \frac{70}{9} \end{aligned}$$

We can then use point–slope form to find the equation of the line. Since the line passes through (66, 140) and (75, 210), we can use either point when substituting values for x_1 and y_1.

$$\begin{aligned} y - y_1 &= m(x - x_1) \\ y - 140 &= \frac{70}{9}(x - 66) && \text{Choose (66, 140) for } (x_1, y_1). \\ y &= \frac{70}{9}x - \frac{4{,}620}{9} + 140 && \text{Remove parentheses and add 140 to both sides.} \\ \textbf{4.} \quad y &= \frac{70}{9}x - \frac{1{,}120}{3} && \text{Add: } -\tfrac{4{,}620}{9} + 140 = -\tfrac{4{,}620}{9} + \tfrac{1{,}260}{9} = -\tfrac{3{,}360}{9} = -\tfrac{1{,}120}{3}. \end{aligned}$$

Our approximation of the regression equation is $y = \frac{70}{9}x - \frac{1{,}120}{3}$.

To predict the weight of a man who is 73 inches tall, for example, we substitute 73 for x in Equation 4 and simplify.

$$\begin{aligned} y &= \frac{70}{9}x - \frac{1{,}120}{3} \\ y &= \frac{70}{9}(73) - \frac{1{,}120}{3} \\ y &= \frac{5{,}110}{9} - \frac{1{,}120}{3} \\ y &\approx 194.4444444 \end{aligned}$$

We would predict that a 73-inch tall man chosen at random will weigh about 194 pounds.

STUDY SET Section 2.4

VOCABULARY *In Exercises 1–4, fill in the blanks to make the statements true.*

1. The point–slope form of the equation of a line is ______________.

2. The ______________ form of the equation of a line is $y = mx + b$.

3. Two lines are ______________ when their slopes are negative reciprocals.

4. Two lines are ______________ when they have the same slope.

CONCEPTS

5. If you know the slope of a line, is that enough information about the line to write its equation?

6. If you know a point that a line passes through, is that enough information about the line to write its equation?

7. The line graphed in Illustration 1 passes through the point $(-2, -3)$. Find its slope. Then write its equation. Express your answer in point–slope form.

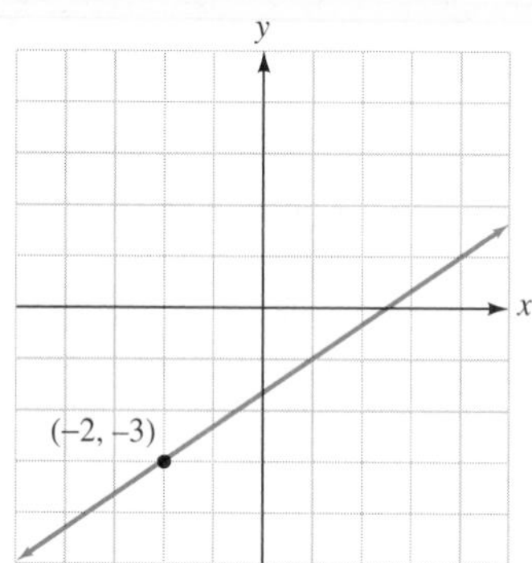

ILLUSTRATION 1

8. For the line graphed in Illustration 2, find the slope and the y-intercept. Then write the equation of the line. Express your answer in slope–intercept form.

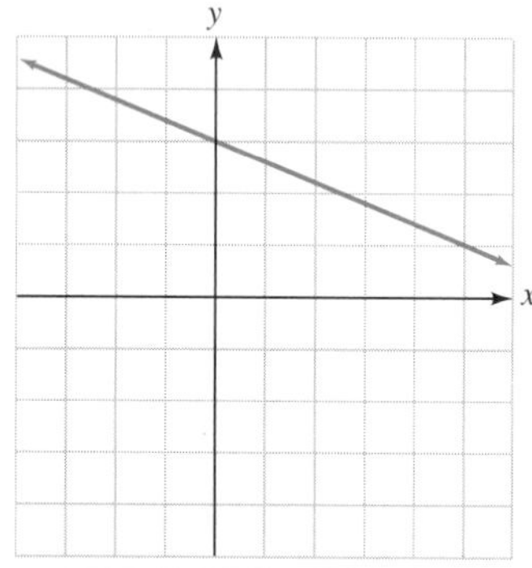

ILLUSTRATION 2

9. When the graph of the line $y = -\frac{2}{3}x + 1$ is drawn, what slope and y-intercept will the line have?

10. When the graph of the line $y - 3 = -\frac{2}{3}(x + 1)$ is drawn, what slope will it have? What point does the equation indicate it will pass through?

11. Is the following statement true? The equations $y - 2 = 3(x - 2)$, $y = 3x - 4$, and $3x - y = 4$ all describe the same line.

12. See the linear model graphed in Illustration 3.

a. What information does the y-intercept give?

b. What information does the slope give?

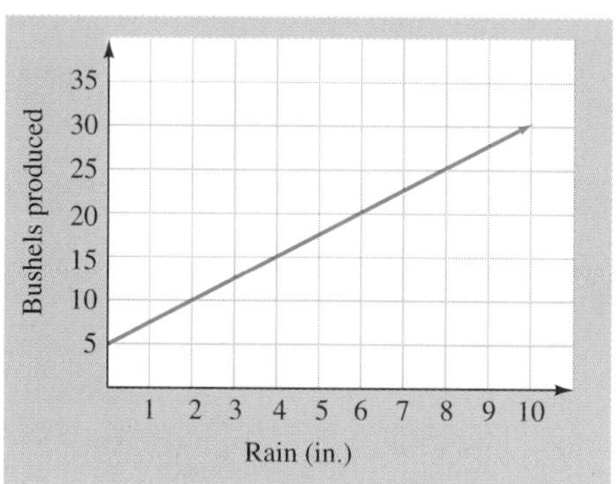

ILLUSTRATION 3

13. When each equation is graphed, what will the y-intercept be?

a. $y = 2x$ **b.** $x = -3$

14. When each equation is graphed, what will be the slope of the line?

a. $y = -x$ **b.** $x = -3$

NOTATION *In Exercises 15–16, complete each solution.*

15. Write $y + 2 = \frac{1}{3}(x + 3)$ in slope–intercept form.

$y + 2 = \frac{1}{3}(x + 3)$	The original equation.
$y + 2 = \square + 1$	Distribute.
$y + 2 - \square = \frac{1}{3}x + 1 - \square$	Subtract 2 from both sides.
$y = \frac{1}{3}x - \square$	Simplify.

$m = \square$, $b = \square$

16. Write the equation of the line with slope -2 that passes through the point $(3, 1)$.

$y - y_1 = m(x - x_1)$	Point–slope form.
$y - \square = -2(x - \square)$	Substitute for x_1, y_1, and m.
$y - 1 = \square + 6$	Distribute.
$y = -2x + 7$	Add 1 to both sides.

PRACTICE *In Exercises 17–20, use point–slope form to write the equation of the line with the given properties. Then write each equation in slope–intercept form.*

17. $m = 5$, passing through $P(0, 7)$

18. $m = -8$, passing through $P(0, -2)$

19. $m = -3$, passing through $P(2, 0)$

20. $m = 4$, passing through $P(-5, 0)$

In Exercises 21–26, use point–slope form to write the equation of the line passing through the two given points. Then write each equation in slope–intercept form.

21. $P(0, 0)$, $Q(4, 4)$

22. $P(-5, 5)$, $Q(0, 0)$

23.

x	y
3	4
0	-3

24.

x	y
4	0
6	-8

25.

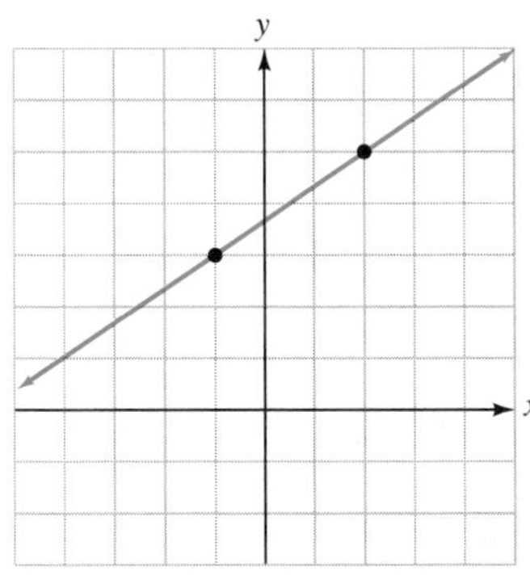

26.

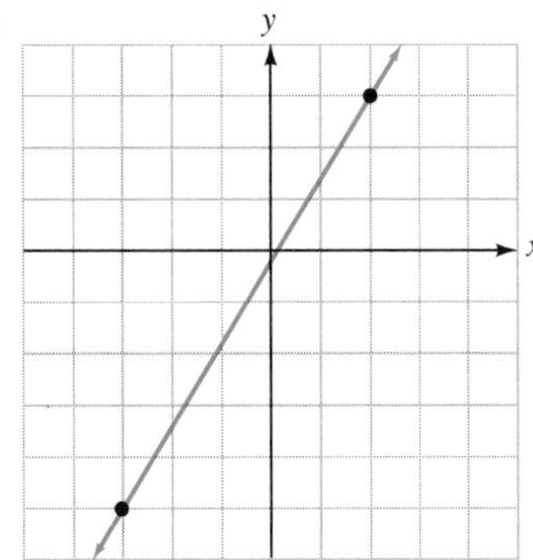

In Exercises 27–34, use the slope–intercept form to write the equation of the line with the given properties.

27. $m = 3$, $b = 17$

28. $m = -2$, $b = 11$

29. $m = -7$, passing through $P(7, 5)$

30. $m = 3$, passing through $P(-2, -5)$

31. $m = 0$, passing through $P(2, -4)$

32. $m = -7$, passing through the origin

33. passing through $P(6, 8)$ and $Q(2, 10)$

34. passing through $P(-4, 5)$ and $Q(2, -6)$

In Exercises 35–38, write each equation in slope–intercept form. Then find the slope and the y-intercept of the line determined by the equation.

35. $3x - 2y = 8$

36. $-2x + 4y = 12$

37. $-2(x + 3y) = 5$

38. $5(2x - 3y) = 4$

In Exercises 39–44, find the slope and y-intercept and use them to draw the line.

39. $y = x - 1$

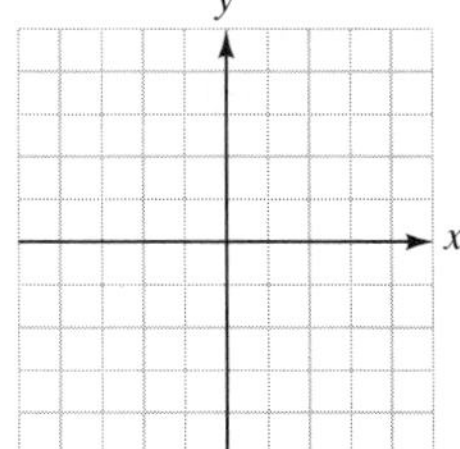

40. $y = -x + 2$

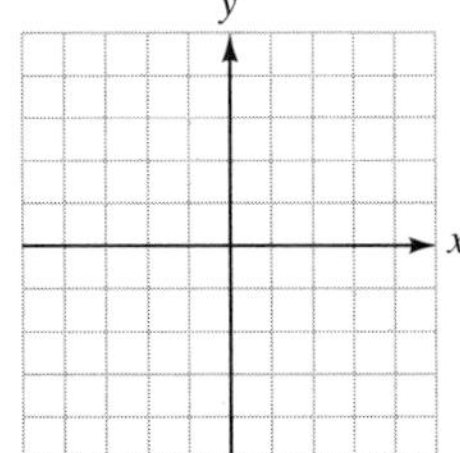

41. $y = \frac{2}{3}x + 2$

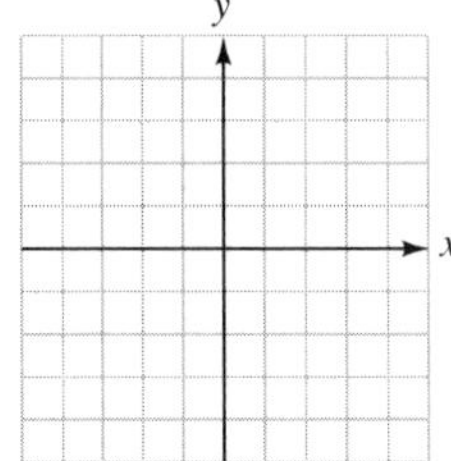

42. $y = -\frac{5}{4}x + \frac{5}{2}$

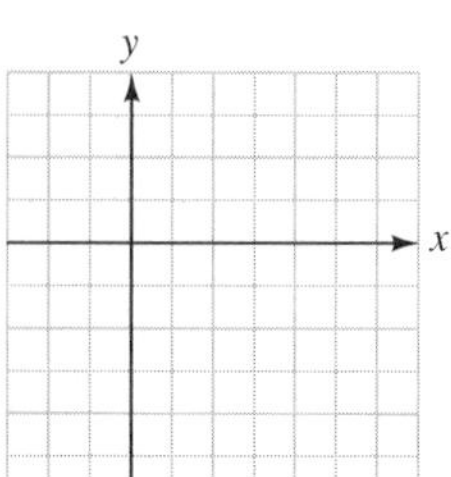

43. $4y - 3 = -3x - 11$

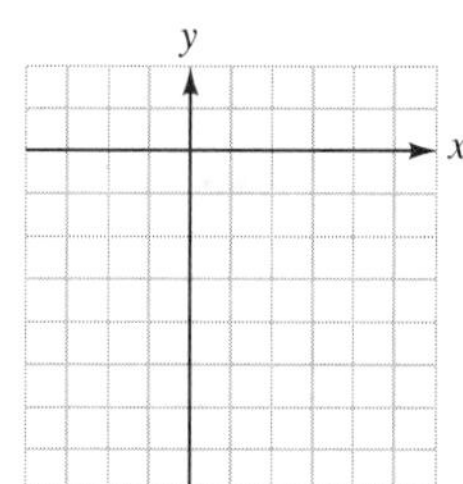

44. $-2x + 4y = 12$

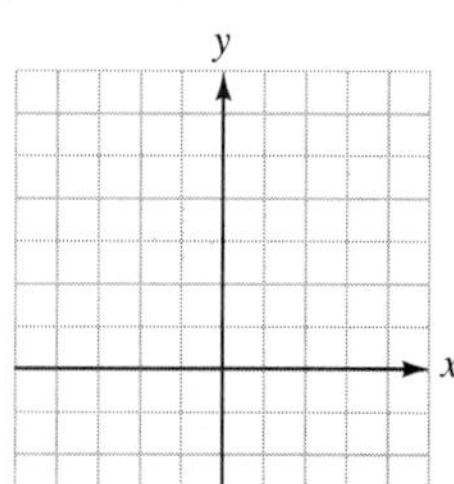

In Exercises 45–52, tell whether the graphs of each pair of equations are parallel, perpendicular, or neither.

45. $y = 3x + 4,\ y = 3x - 7$

46. $y = 4x - 13,\ y = \frac{1}{4}x + 13$

47. $x + y = 2,\ y = x + 5$

48. $x = y + 2,\ y = x + 3$

49. $3x + 6y = 1,\ y = \frac{1}{2}x$

50. $2x + 3y = 9,\ 3x - 2y = 5$

51. $y = 3,\ x = 4$

52. $y = -3,\ y = -7$

In Exercises 53–58, write the equation of the line that passes through the given point and is parallel to the given line. Write the answer in slope–intercept form.

53. $P(0, 0),\ y = 4x - 7$

54. $P(0, 0),\ x = -3y - 12$

55. $P(2, 5),\ 4x - y = 7$

56. $P(-6, 3),\ y + 3x = -12$

57. $P(4, -2),\ x = \frac{5}{4}y - 2$

58. $P(1, -5),\ x = -\frac{3}{4}y + 5$

In Exercises 59–64, write the equation of the line that passes through the given point and is perpendicular to the given line. Write the answer in slope–intercept form.

59. $P(0, 0),\ y = 4x - 7$

60. $P(0, 0),\ x = -3y - 12$

61. $P(2, 5),\ 4x - y = 7$

62. $P(-6, 3),\ y + 3x = -12$

63. $P(4, -2),\ x = \frac{5}{4}y - 2$

64. $P(1, -5),\ x = -\frac{3}{4}y + 5$

APPLICATIONS *In Exercises 65–68, assume straight-line depreciation or straight-line appreciation.*

65. BIG-SCREEN TV Find the linear depreciation equation for the TV in the want ad shown in Illustration 4.

For Sale: 3-year-old 45-inch TV, with matrix surround sound & picture within picture, remote. $1,750 new. Asking $800. Call 875-5555. Ask for Mike.

ILLUSTRATION 4

66. SALVAGE VALUE A truck was purchased for $19,984. Its salvage value at the end of 8 years is expected to be $1,600. Find the depreciation equation.

67. ART In 1987, the painting *Rising Sunflowers* by Vincent van Gogh (Illustration 5) sold for $36,225,000. Suppose that an art appraiser expected the painting to double in value every 20 years. Let x represent the time in years after 1987. Find the appreciation equation.

ILLUSTRATION 5

68. REAL ESTATE LISTING See Illustration 6. Use the information given in the description of the property to write an appreciation equation for the house.

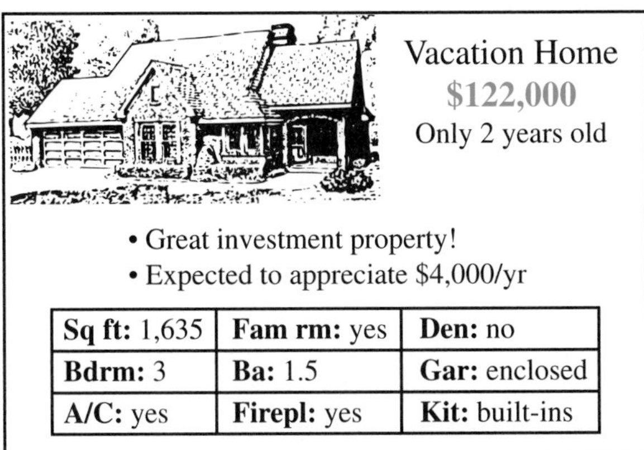

Sq ft: 1,635	**Fam rm:** yes	**Den:** no
Bdrm: 3	**Ba:** 1.5	**Gar:** enclosed
A/C: yes	**Firepl:** yes	**Kit:** built-ins

ILLUSTRATION 6

69. CRIMINOLOGY City growth and the number of burglaries are related by a linear equation. Records show that 575 burglaries were reported in a year when the local population was 77,000 and that the rate of increase in the number of burglaries was 1 for every 100 new residents.

a. Using the variables p for population and b for burglaries, write an equation (in slope–intercept form) that police can use to predict future burglary statistics.

b. How many burglaries can be expected when the population reaches 110,000?

70. CABLE TV Since 1990, when the average monthly basic cable TV rate in the United States was $16.75, the cost has risen by about $1.32 a year.

a. Write an equation in slope–intercept form to predict cable TV costs in the future. Use t to represent time in years after 1990 and C to represent the average basic monthly cost.

b. If the equation in part a were graphed, what would be the meaning of the C-intercept and the slope of the line?

71. PSYCHOLOGY EXPERIMENT The scattergram in Illustration 7 shows the performance of a rat in a maze.

a. Draw a line through (1, 10) and (19, 1). Write its equation using the variables t and E. In psychology, this equation is called the *learning curve* for the rat.

b. What does the slope of the line tell us?

c. What information does the x-intercept of the graph give?

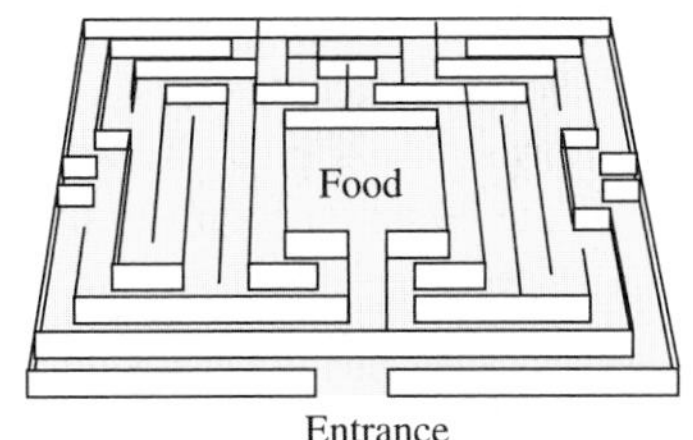

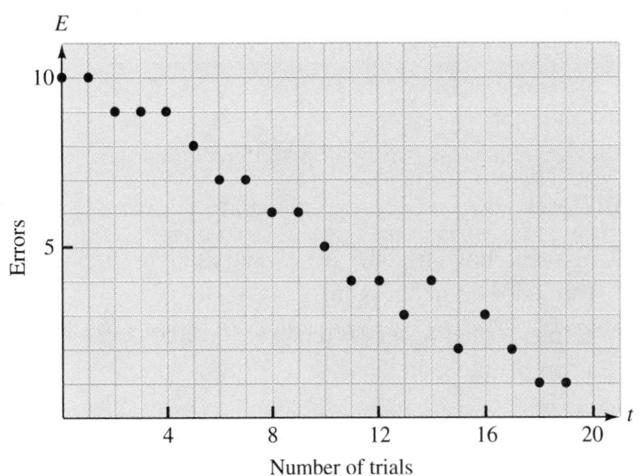

ILLUSTRATION 7

72. UNDERSEA DIVING Illustration 8 shows that the pressure p that divers experience is related to the depth

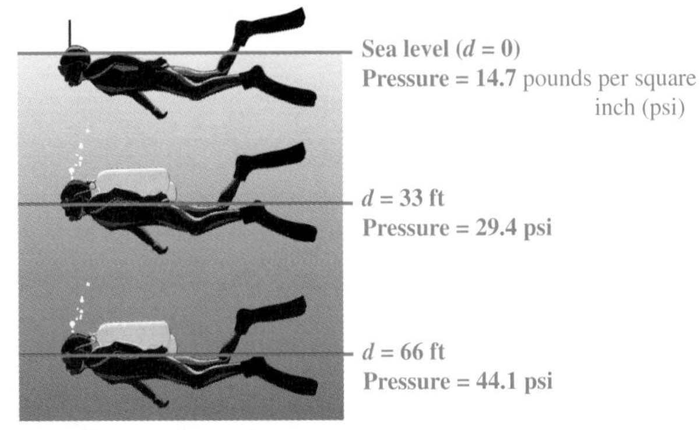

ILLUSTRATION 8

d of the dive. A linear model can be used to describe this relationship.

a. Write the linear model in slope–intercept form.

b. Pearl and sponge divers often reach depths of 100 feet. What pressure do they experience? Round to the nearest tenth.

c. Scuba divers can safely dive to depths of 250 feet. What pressure do they experience? Round to the nearest tenth.

73. WIND-CHILL FACTOR A combination of cold and wind makes a person feel colder than the actual temperature. Illustration 9 shows what temperatures of 35°F and 15°F feel like when a 15-mph wind is blowing. The relationship between the actual temperature and the wind-chill temperature can be modeled with a linear equation.

a. Write the equation that models this relationship. Answer in slope–intercept form.

b. What information is given by the y-intercept of the graph of the equation found in part a?

Actual temperature	Wind-chill temperature
35°F	16°F
15°F	−11°F

ILLUSTRATION 9

74. COMPUTER-AIDED DRAFTING Illustration 10 shows a computer-generated drawing of an airplane part. When the designer clicks the mouse on a line on the drawing, the computer finds the equation of the line. Use a calculator to determine whether the angle where the weld is to be made is a right angle.

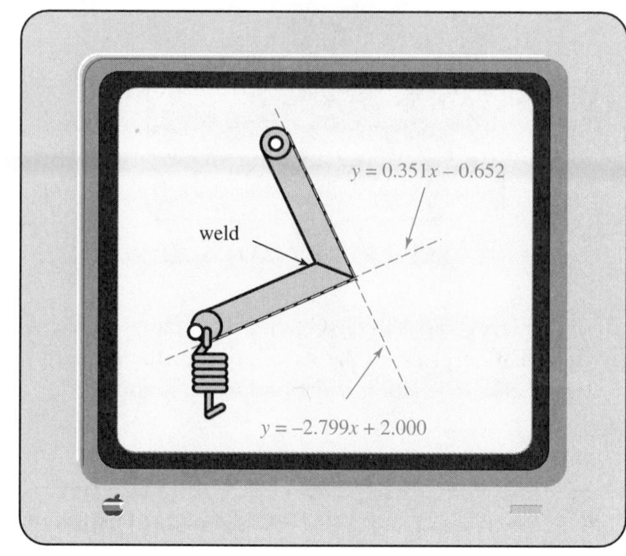

ILLUSTRATION 10

WRITING *Write a paragraph using your own words.*

75. Explain how to find the equation of a line passing through two given points.

76. Explain what m, x_1, and y_1 represent in the point–slope form of the equation of a line.

77. Explain what m and b represent in the slope–intercept form of the equation of a line.

78. Linear relationships between two quantities can be described by an equation or a graph. Which do you think is the more informative? Why?

REVIEW

79. INVESTMENTS Equal amounts are invested at 6%, 7%, and 8% annual interest. The three investments yield a total of $2,037 annual interest. Find the total amount of money invested.

80. MIXING COFFEE To make a mixture of 80 pounds of coffee worth $272, a grocer mixes coffee worth $3.25 a pound with coffee worth $3.85 a pound. How many pounds of cheaper coffee should the grocer use?

2.5 *Introduction to Functions*

In this section, you will learn about

- **Representing functions**
- **Function notation**
- **The graph of a function**
- **Finding the domain and range of a function**
- **The vertical line test**
- **Linear functions**

Introduction. In Chapter 1, we examined the rental hall agreement shown in Figure 2-38. We determined that the cost C of renting the hall depended on the number of hours h the hall was to be rented. We described this dependence in four ways: using words, using an equation, using a table, and using a graph.

Rental Agreement
ROYAL VISTA BANQUET ROOM
Wedding Receptions•Dances•Reunions•Fashion Shows

Rented To____________________ Date____________
Lessee's Address____________________________

Rental Charges
- $100 per hour
- Nonrefundable $200 cleanup fee

Terms and conditions
Lessor leases the undersigned lessee the above described property upon the terms and conditions set forth on this page and on the back of this page. Lessee promises to pay rental cost stated herein.

Words

The cost to rent the hall | is | 100 | times | the number of hours it is rented | plus | 200.

Equation

$C = 100h + 200$

Table

h	C
1	300
2	400
3	500
4	600
5	700
6	800
7	900

Graph

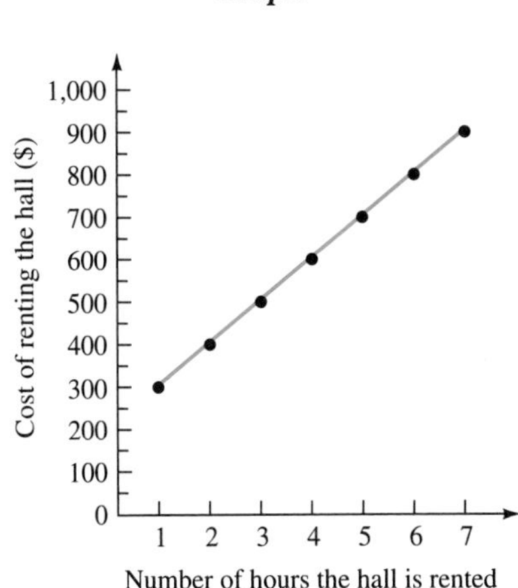

FIGURE 2-38

In this example, a specific rule assigns to each number of hours h the hall is rented a unique number C, the cost to rent the hall. Correspondences of this type are called **functions,** and they are of great importance in mathematics and its applications. In this section, we introduce the concept of function, and we discuss a special notion that is used when working with functions.

Representing functions

If x and y are real numbers, an equation in x and y determines a correspondence between the values of x and y. To see how, we consider the equation $y = \frac{1}{2}x + 3$. To find the value of y (called an **output value**) that corresponds to $x = 4$ (called an **input value**), we multiply 4 by $\frac{1}{2}$ and then add 3.

$$y = \frac{1}{2}x + 3$$

$$y = \frac{1}{2}(4) + 3 \quad \text{Substitute the input value of 4 for } x.$$

$$= 2 + 3$$

$$= 5 \quad \text{The output value is 5.}$$

The ordered pair (4, 5) satisfies the equation and shows that a y value of 5 corresponds to an x value of 4. This ordered pair and others that satisfy the equation appear in the table shown in Figure 2-39. The graph of the equation also appears in the figure.

To see how the table determines the correspondence, we simply find an input in the x column and then read across to find the corresponding output in the y column. For example, if we select 2 as an input value, we get 4 as an output value. Thus, a y value of 4 corresponds to an x value of 2.

To see how the graph determines the correspondence, we draw a vertical and a horizontal line through any point (say, point P) on the graph, as shown in Figure 2-39. Because these lines intersect the x-axis at 4 and the y-axis at 5, the point $P(4, 5)$ associates 5 on the y-axis with 4 on the x-axis. This shows that a y value of 5 is assigned to an x value of 4.

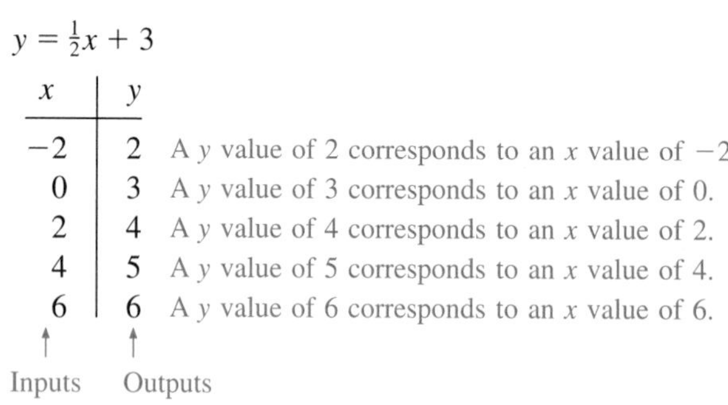

$y = \frac{1}{2}x + 3$

x	y	
−2	2	A y value of 2 corresponds to an x value of −2.
0	3	A y value of 3 corresponds to an x value of 0.
2	4	A y value of 4 corresponds to an x value of 2.
4	5	A y value of 5 corresponds to an x value of 4.
6	6	A y value of 6 corresponds to an x value of 6.
↑ Inputs	↑ Outputs	

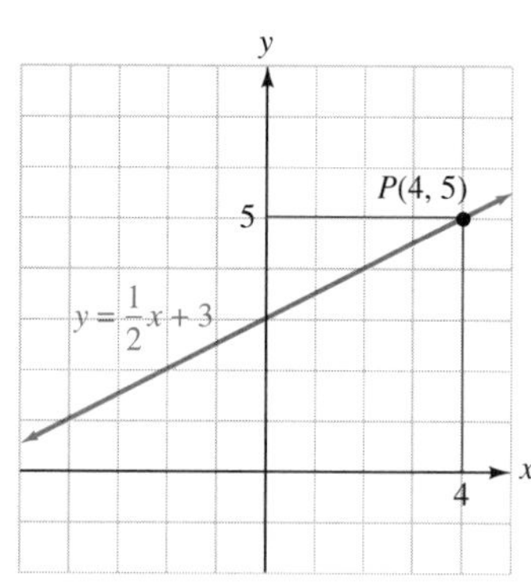

FIGURE 2-39

When a correspondence is described by an equation, by a table, using words, or with a graph, and only one y value corresponds to each x value, we call the correspondence a **function.** The set of input values x is called the **domain** of the function, and the set of output values y is called the **range.** Since the value of y usually depends on the number x, we call y the **dependent variable** and x the **independent variable.**

Functions

A **function** is a correspondence between a set of input values x (called the **domain**) and a set of output values y (called the **range**), where exactly one y value in the range corresponds to each number x in the domain.

EXAMPLE 1 ***Recognizing functions.*** Does $y = 2x - 3$ define y to be a function of x? If so, illustrate the function with a table and a graph.

Self Check

Does $y = -2x + 3$ define y to be a function of x?

Solution

For $y = 2x - 3$ to define a function, every input number x must determine one output value of y. To find y in the equation $y = 2x - 3$, we multiply x by 2 and then subtract 3. Since this arithmetic gives one result, each choice of x determines one value of y. Thus, the equation defines y to be a function of x.

A table of values and the graph appear in Figure 2-40.

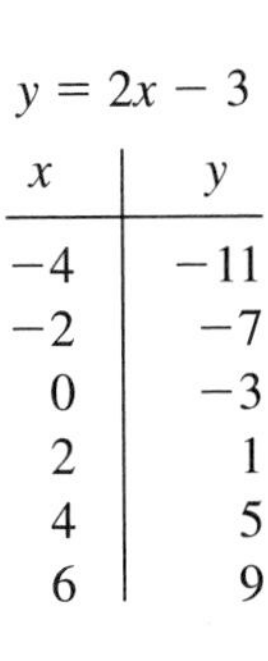

$y = 2x - 3$

x	y
-4	-11
-2	-7
0	-3
2	1
4	5
6	9

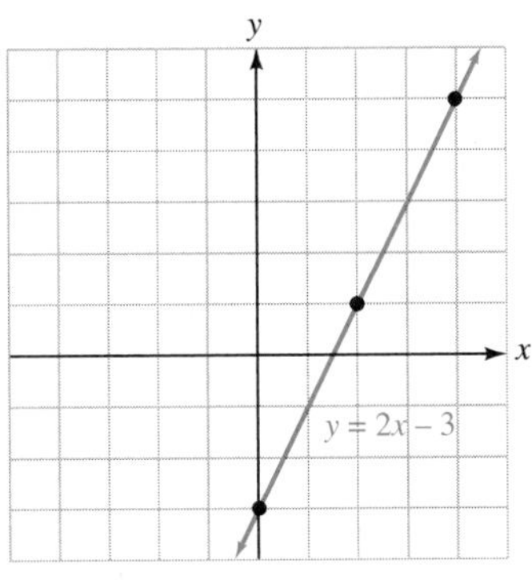

FIGURE 2-40

Answer: yes

As you will see in the next example, not every equation in two variables defines a function.

EXAMPLE 2 ***Recognizing functions.*** Does $y^2 = x$ define y to be a function of x?

Solution

For a function to exist, each input x must determine one output y. If we let $x = 16$, for example, y could be either 4 or -4, because $4^2 = 16$ and $(-4)^2 = 16$. Since more than one value of y is determined when $x = 16$, the equation does not represent a function.

Self Check

Does $|y| = x$ define y to be a function of x?

Answer: no

Function notation

There is a special notation that we will use to denote functions.

Function notation The notation $y = f(x)$ denotes that the variable y is a function of x.

The notation $y = f(x)$ is read as "y equals f of x." Note that y and $f(x)$ are two notations for the same quantity. Thus, the equations $y = 4x + 3$ and $f(x) = 4x + 3$ are equivalent. We read $f(x) = 4x + 3$ as "f of x is equal to $4x + 3$."

This is the variable used to represent input values.

$$f(x) = 4x + 3$$

This is the name of the function.

This expression shows how to obtain an output value from a given input value.

WARNING! The notation $f(x)$ does not mean "f times x."

The notation $y = f(x)$ provides a way of denoting the value of y (the dependent variable) that corresponds to some number x (the independent variable). For example, if $y = f(x)$, the value of y that is determined by $x = 3$ is denoted by $f(3)$.

EXAMPLE 3 ***Evaluating a function.*** Let $f(x) = 4x + 3$. Find **a.** $f(3)$, **b.** $f(-1)$, **c.** $f(0)$, and **d.** $f(r)$.

Solution

In each case, we will substitute the value for x within the parentheses of the function notation and in the algebraic expression $4x + 3$. Then we evaluate the expression.

a. To find $f(3)$, we replace x with 3:

$$\begin{aligned} f(x) &= 4x + 3 \\ f(3) &= 4(3) + 3 \\ &= 12 + 3 \\ &= 15 \end{aligned}$$

b. To find $f(-1)$, we replace x with -1:

$$\begin{aligned} f(x) &= 4x + 3 \\ f(-1) &= 4(-1) + 3 \\ &= -4 + 3 \\ &= -1 \end{aligned}$$

c. To find $f(0)$, we replace x with 0:

$$\begin{aligned} f(x) &= 4x + 3 \\ f(0) &= 4(0) + 3 \\ &= 3 \end{aligned}$$

d. To find $f(r)$, we replace x with r:

$$\begin{aligned} f(x) &= 4x + 3 \\ f(r) &= 4r + 3 \end{aligned}$$

Self Check

If $f(x) = -2x - 1$, find **a.** $f(2)$, **b.** $f(-3)$, and **c.** $f(-t)$

Answers: **a.** -5, **b.** 5, **c.** $2t - 1$

To see why function notation is helpful, we consider the following equivalent sentences:

1. In the equation $y = 4x + 3$, find the value of y when x is 3.
2. In the equation $f(x) = 4x + 3$, find $f(3)$.

Statement 2, which uses $f(x)$ notation, is much more concise.

We can think of a function as a machine that takes some input x and turns it into some output $f(x)$, as shown in Figure 2-41(a). The machine shown in Figure 2-41(b) turns the input number 2 into the output value -3 and turns the input number 6 into the output value -11. The set of numbers that we can put into the machine is the domain of the function, and the set of numbers that comes out is the range.

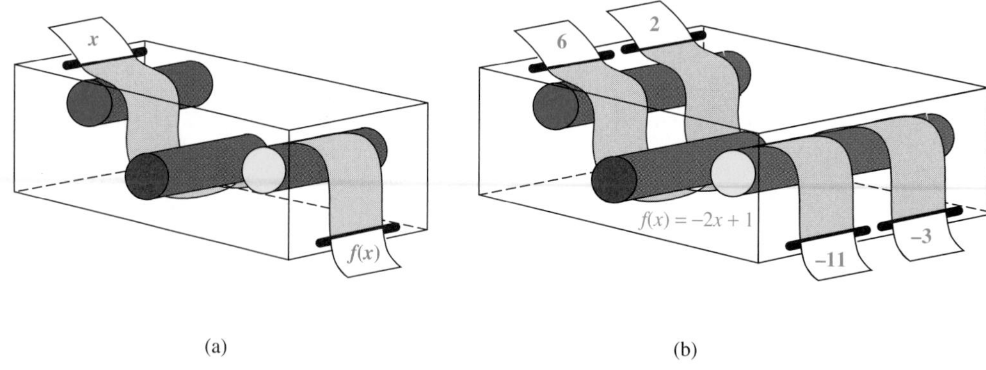

FIGURE 2-41

The letter f used in the notation $y = f(x)$ represents the word *function.* However, other letters can be used to represent functions. For example, the notations $y = g(x)$ and $y = h(x)$ are often used to denote functions involving the independent variable x.

In Example 4, the equation $g(x) = x^2 - 2x$ determines a function, because every possible value of x gives a single value of $g(x)$.

EXAMPLE 4 ***Evaluating a function.*** Let $g(x) = x^2 - 2x$. Find **a.** $g(\frac{2}{5})$, and **b.** $g(-2.4)$.

Solution

a. To find $g(\frac{2}{5})$, we replace x with $\frac{2}{5}$:

$$g(x) = x^2 - 2x$$
$$g\left(\frac{2}{5}\right) = \left(\frac{2}{5}\right)^2 - 2\left(\frac{2}{5}\right)$$
$$= \frac{4}{25} - \frac{4}{5}$$
$$= -\frac{16}{25}$$

b. To find $g(-2.4)$, we replace x with -2.4:

$$g(x) = x^2 - 2x$$
$$g(-2.4) = (-2.4)^2 - 2(-2.4)$$
$$= 5.76 + 4.8$$
$$= 10.56$$

Self Check

Let $h(x) = -\dfrac{x^2 + 2}{2}$. Find **a.** $h(4)$ and **b.** $h(-0.6)$.

Answers: **a.** -9, **b.** -1.18

In the next example, the letter A is chosen to name a function that finds the area of a circle. The letter d is chosen as the independent variable, to help stress the fact that the area of a circle is a function of its diameter.

EXAMPLE 5 ***Archery.*** The area of a circle with a diameter of length d is given by the function $A(d) = \pi(\frac{d}{2})^2$. Find $A(48)$. Round to the nearest tenth. What information does it give about the archery target shown in Figure 2-42?

Solution

Since the diameter of the circular target is 48 inches, $A(48)$ gives the area of the target. To find $A(48)$, we replace d with 48.

$$A(d) = \pi\left(\frac{d}{2}\right)^2$$
$$A(48) = \pi\left(\frac{48}{2}\right)^2 \quad \text{Substitute 48 for } d.$$
$$= \pi(24)^2$$
$$= 576\pi$$
$$\approx 1{,}809.557368 \quad \text{Use a calculator to do the multiplication.}$$

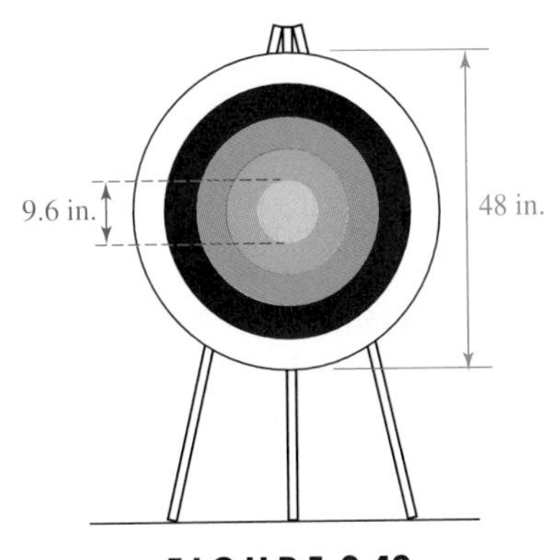

FIGURE 2-42

The area of the entire archery target is approximately 1,809.6 in.^{2}.

Self Check

In Example 5, find $A(9.6)$. Round to the nearest tenth. What information does it give about the archery target in Figure 2-42?

Answers: 72.4; the area of the "bull's eye" is 72.4 in.^{2}

The graph of a function

The *graph of a function* is a "picture" of the ordered pairs $(x, f(x))$ that define the function. From the graph of a function, we can determine function values. To illustrate this, let's consider the graph of function f shown in Figure 2-43 and find $f(4)$.

To find $f(4)$, we need to find the y value that f assigns to an x value of 4. The blue line in the figure illustrates that the graph is 3 units above the point marked 4 on the x-axis. So the point (4, 3) is on the graph of the function, and we can conclude that f assigns a y value of 3 to an x value of 4. Therefore, $f(4) = 3$.

This axis can be labeled $f(x)$ or y.

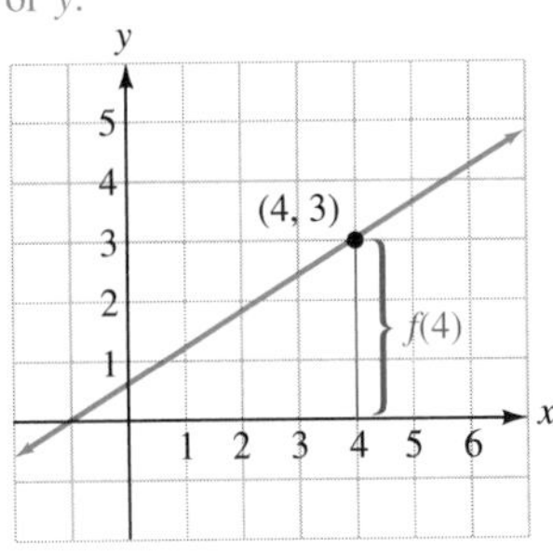

FIGURE 2-43

Finding the domain and range of a function

EXAMPLE 6 ***Domain and range.*** Find the domain and range of each function: **a.** the ordered pairs $(-2, 4)$, $(0, 6)$, and $(2, 8)$, **b.** the equation $y = 3x + 1$, and **c.** the equation $y = \frac{1}{x-2}$.

Solution

a. The ordered pairs set up a correspondence between x and y where a single value of y corresponds to each x. The domain is the set of numbers x: $\{-2, 0, 2\}$. The range is the set of values y: $\{4, 6, 8\}$.

x	y	(x, y)	
-2	4	$(-2, 4)$	4 corresponds to -2.
0	6	$(0, 6)$	6 corresponds to 0.
2	8	$(2, 8)$	8 corresponds to 2.

b. We will be able to evaluate $3x + 1$ for any real-number input x. So the domain of the function is the set of real numbers. Since the output y can be any real number, the range is the set of real numbers.

b. The number 2 cannot be substituted for x, because that would make the denominator equal to zero. Since any real number except 2 can be substituted for x in the equation $y = \frac{1}{x-2}$, the domain is the set of all real numbers except 2.

Since a fraction with a numerator of 1 cannot be 0, the range is the set of all real numbers except 0.

Self Check

Find the domain and range of each function: **a.** the ordered pairs $(-3, 5)$, $(-2, 7)$, and $(1, 11)$ and **b.** the equation $y = \frac{2}{x+3}$.

Answers: **a.** $\{-3, -2, 1\}$, $\{5, 7, 11\}$, **b.** the set of all real numbers except -3, the set of all real numbers except 0.

We can determine the domain and the range of a function from its graph. For the graph in Figure 2-44, the domain is shown on the x-axis, and the range is shown on the y-axis. Note that for any x in the domain, there corresponds a value $y = f(x)$ in the range.

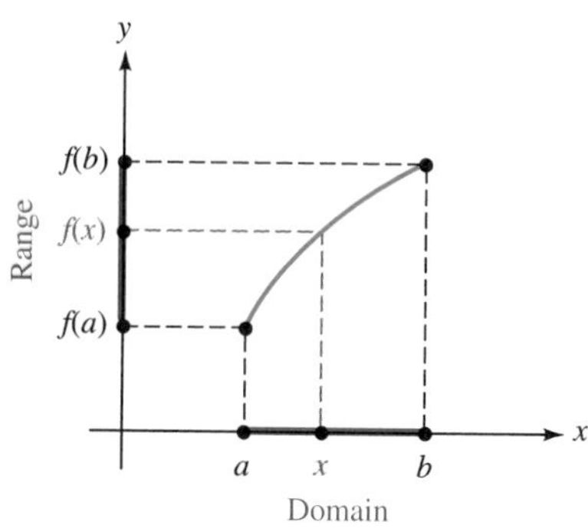

FIGURE 2-44

EXAMPLE 7 ***Finding the domain and range from a graph.*** Find the domain and range of the function graphed in Figure 2-45.

Solution

Since every real number x on the x-axis determines a corresponding value of y, the domain is the set of real numbers. Since the values of y can be any real number on the y-axis, the range is the set of real numbers.

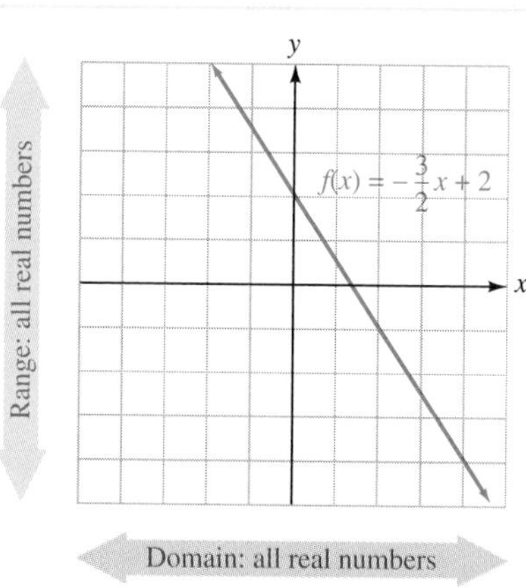

FIGURE 2-45

The vertical line test

The **vertical line test** can be used to determine whether the graph of an equation represents a function. If any vertical line intersects a graph more than once, the graph cannot represent a function, because to one number x there would correspond more than one value of y.

The graph in Figure 2-46(a) represents a function, because every vertical line that intersects the graph does so exactly once. The graph in Figure 2-46(b) does not represent a function, because some vertical lines intersect the graph more than once.

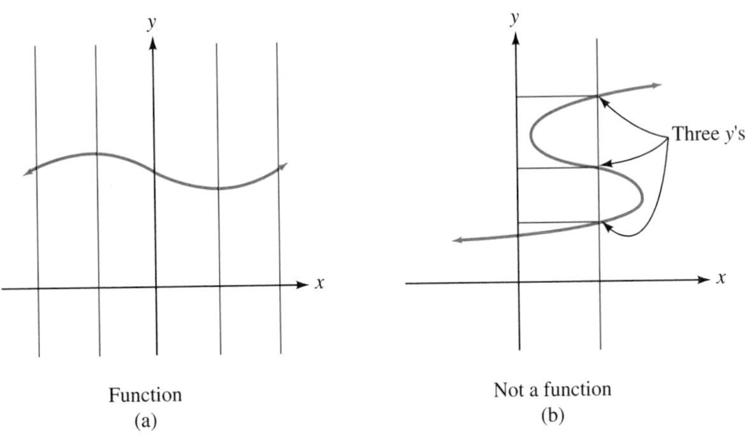

FIGURE 2-46

Linear functions

In Section 2.2, we graphed equations whose graphs were lines. Such equations define a basic type of function called a **linear function.**

Linear functions

A **linear function** is a function defined by an equation that can be written in the form

$$f(x) = mx + b \qquad \text{or} \qquad y = mx + b$$

where m is the slope of the line graph and $(0, b)$ is the y-intercept.

EXAMPLE 8 ***Manicurist.*** A recent graduate of a cosmetology school rents a station from the owner of a beauty salon for \$18 a day. She expects to make \$12 profit from each customer she serves. Write a linear function describing her daily income if she serves c customers per day. Then graph the function.

Solution The manicurist makes a profit of \$12 per customer, so if she serves c customers a day, she will make $\$12c$. To find her income, we must *subtract* the \$18 rental fee she pays from the profit. Therefore the income function is $I(c) = 12c - 18$.

The graph of this linear function, shown in Figure 2-47, is a line with slope 12 and intercept $(0, -18)$. Since the manicurist cannot have a negative number of customers, we do not extend the line into quadrant III.

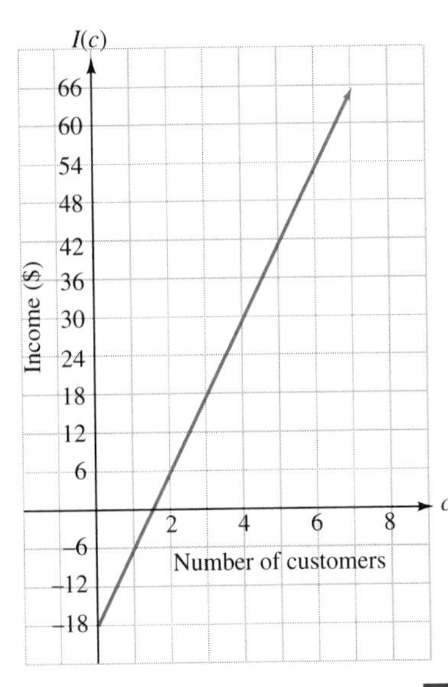

FIGURE 2-47 ■

Accent on Technology **Evaluating Functions**

We can use a graphing calculator to find the value of a function at different numbers x. For example, to find the income earned by the manicurist in Example 8 for different numbers of customers, we first graph the income function $I(c) = 12c - 18$ as $y = 12x - 18$, using window settings of $[0, 10]$ for x and $[0, 100]$ for y to obtain Figure 2-48(a). To find her income when she serves seven customers, we trace and move the cursor until the x-coordinate on the screen is nearly 7, as in Figure 2-48(b). From the screen, we can read that her income is about $66.25.

To find her income when she serves nine customers, we trace and move the cursor until the x-coordinate is nearly 9, as in Figure 2-48(c). From the screen, we can read that her income is about $90.51.

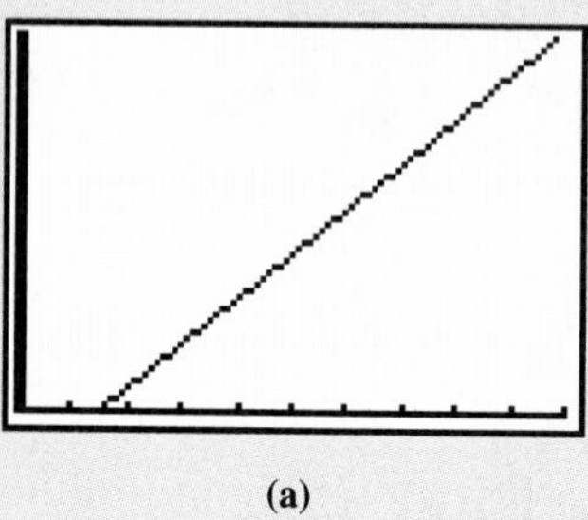

(a)

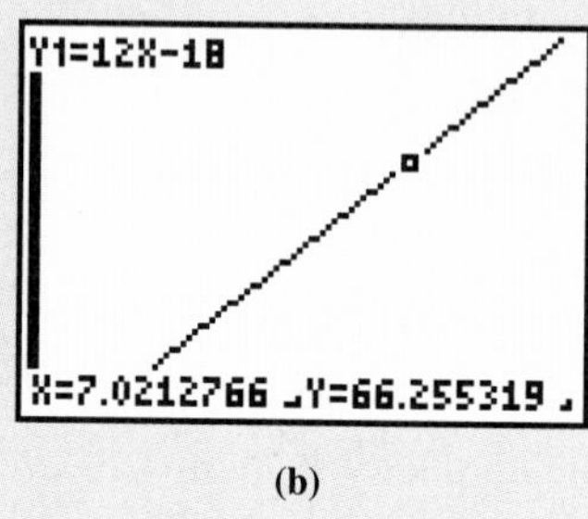

(b)

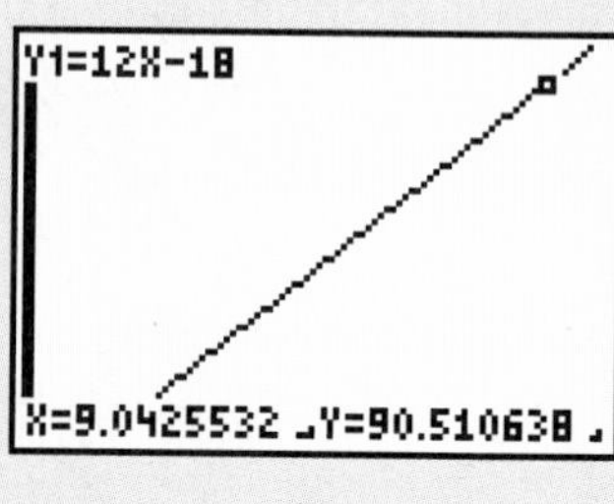

(c)

FIGURE 2-48

STUDY SET **Section 2.5**

VOCABULARY *In Exercises 1–8, fill in the blanks to make the statements true.*

1. A ______________ is a correspondence between a set of input values and a set of output values, where each input value determines ___ output value.

2. When graphing a function, input values are associated with the ______________ axis and output values with the ______________ axis.

3. For a function, the set of all output values is called the ______________ of the function.

4. For a function, the set of all inputs is called the ______________ of the function.

5. For $y = 2x - 9$, the independent variable is ___.

6. For $y = \frac{x}{3} - 9$, if $x = 3$, the corresponding value of ___ is called the output value.

7. For $y = -x + 3$, the dependent variable is ___.

8. Any substitution for x in $f(x) = 5x - 4$ is called an ______________ value.

CONCEPTS

9. Consider the following problems. Fill in the blank so that they ask for the same thing.

1. In the equation $y = -5x + 1$, find the value of y when $x = -1$.
2. In the equation $f(x) = -5x + 1$, find ______________.

10. For the function $f(x) = \frac{1}{x+4}$, why isn't -4 in the domain of f?

11. Consider the graph of the function in Illustration 1.

a. Label each arrow in the illustration with the appropriate term: *domain* or *range*.

b. Give the domain and range.

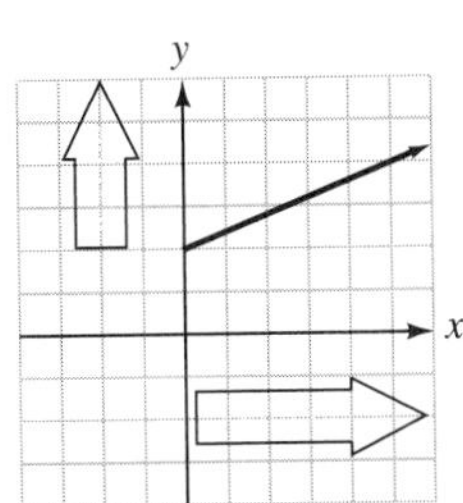

ILLUSTRATION 1

12. Use the graph of function f shown in Illustration 2 to find each of the following.

a. $f(-2)$ **b.** $f(0)$ **c.** $f(1)$

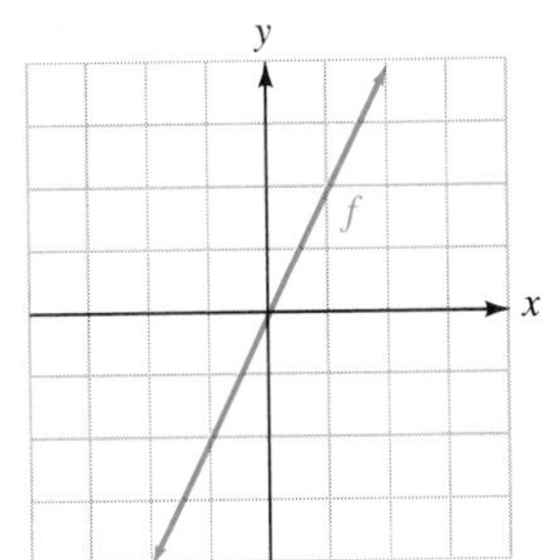

ILLUSTRATION 2

NOTATION *In Exercises 13–14, complete each solution.*

13. If $f(x) = x^2 - 3x$, find $f(-5)$.

$f(x) = x^2 - 3x$

$f(-5) = (-5)^2 - 3(\square)$ Substitute -5 for x.

$= \square + 15$

$= 40$

14. If $g(x) = \dfrac{2 - x}{6}$, find $g(8)$.

$g(x) = \dfrac{2 - x}{6}$

$g(8) = \dfrac{2 - \square}{6}$ Substitute 8 for x.

$= \dfrac{\square}{6}$

$= -1$

15. Complete this sentence: $f(5) = 6$ is read "f ___ 5 is 6."

16. Give two forms in which a linear function can be written.

PRACTICE *In Exercises 17–24, tell whether the equation determines y to be a function of x.*

17. $y = 2x + 3$ **18.** $y = 4x - 1$ **19.** $y = 2x^2$ **20.** $y^2 = x + 1$

21. $y = 3 + 7x^2$ **22.** $y^2 = 3 - 2x$ **23.** $x = |y|$ **24.** $y = |x|$

In Exercises 25–32, find $f(3)$ and $f(-1)$.

25. $f(x) = 3x$ **26.** $f(x) = -4x$ **27.** $f(x) = 2x - 3$ **28.** $f(x) = 3x - 5$

29. $f(x) = 7 + 5x$ **30.** $f(x) = 3 + 3x$ **31.** $f(x) = 9 - 2x$ **32.** $f(x) = 12 + 3x$

In Exercises 33–40, find $g(2)$ and $g(3)$.

33. $g(x) = x^2$ **34.** $g(x) = x^2 - 2$ **35.** $g(x) = x^3 - 1$ **36.** $g(x) = x^3$

37. $g(x) = (x + 1)^2$ **38.** $g(x) = (x - 3)^2$ **39.** $g(x) = 2x^2 - x$ **40.** $g(x) = 5x^2 + 2x$

In Exercises 41–48, find $h(2)$ and $h(-2)$.

41. $h(x) = |x| + 2$ **42.** $h(x) = |x| - 5$ **43.** $h(x) = x^2 - 2$ **44.** $h(x) = x^2 + 3$

45. $h(x) = \dfrac{1}{x + 3}$ **46.** $h(x) = \dfrac{3}{x - 4}$ **47.** $h(x) = \dfrac{x}{x - 3}$ **48.** $h(x) = \dfrac{x}{x^2 + 2}$

In Exercises 49–52, complete each table.

49. $f(x) = |x - 2|$

x	$f(x)$
-1.7	
0.9	
5.4	

50. $f(x) = -2x^2 + 1$

Input	Output
-1.7	
0.9	
5.4	

51. $g(x) = x^3$

Input	Output
$-\frac{3}{4}$	
$\frac{1}{6}$	
$\frac{5}{2}$	

52. $g(x) = 2\left(-x - \frac{1}{4}\right)$

x	$g(x)$
$-\frac{3}{4}$	
$\frac{1}{8}$	
$\frac{5}{2}$	

In Exercises 53–56, find g(w) and g(w + 1).

53. $g(x) = 2x$

54. $g(x) = -3x$

55. $g(x) = 3x - 5$

56. $g(x) = 2x - 7$

In Exercises 57–60, find the domain and range of each function.

57. $\{(-2, 3), (4, 5), (6, 7)\}$

58. $\{(0, 2), (1, 2), (3, 4)\}$

59. $f(x) = \dfrac{1}{x - 4}$

60. $f(x) = \dfrac{5}{x + 1}$

In Exercises 61–64, use the vertical line test to tell whether the given graph represents a function.

61.

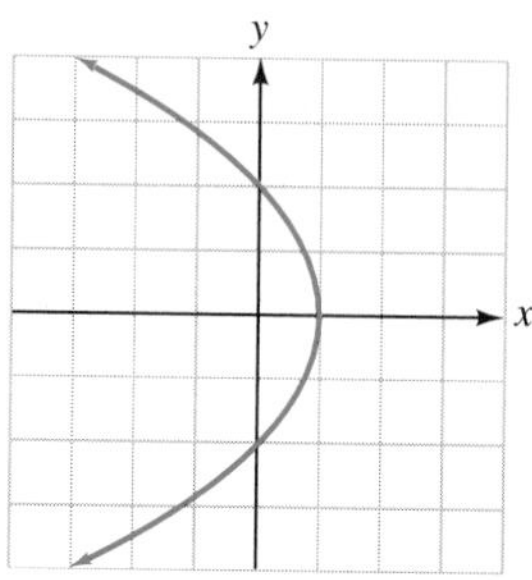

62.

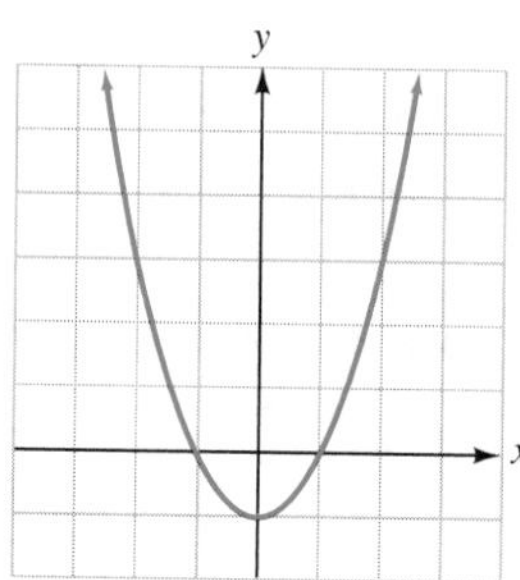

63.

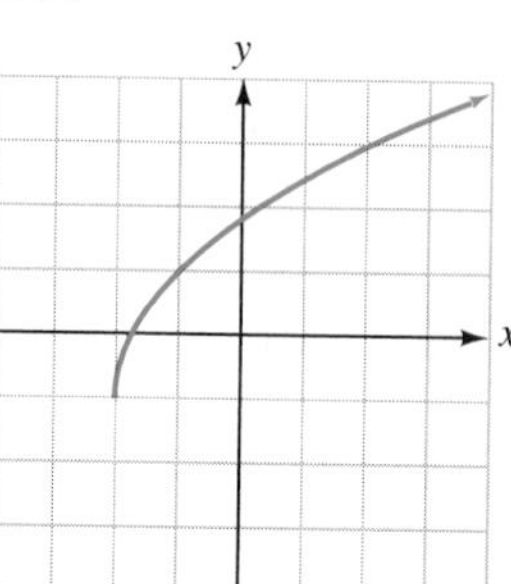

64.

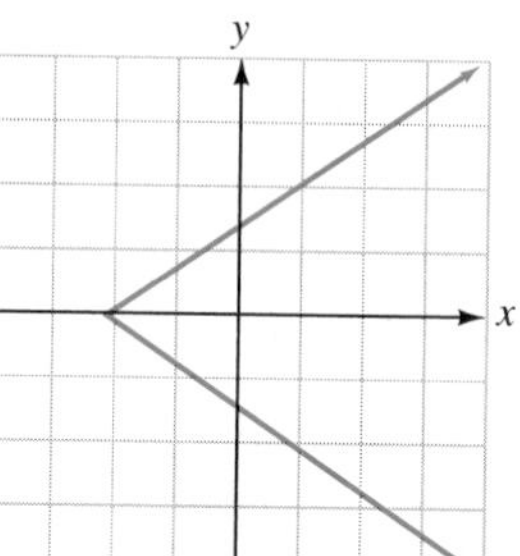

In Exercises 65–68, graph each function. Then find the domain and range.

65. $f(x) = 2x - 1$

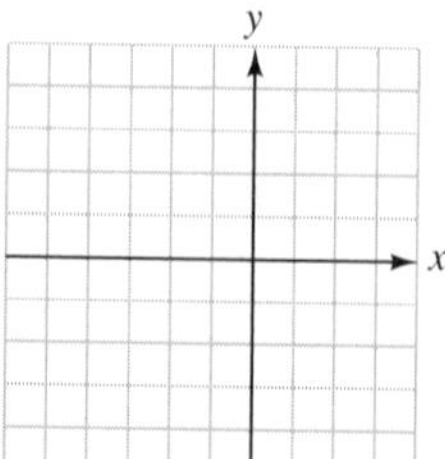

66. $f(x) = -x + 2$

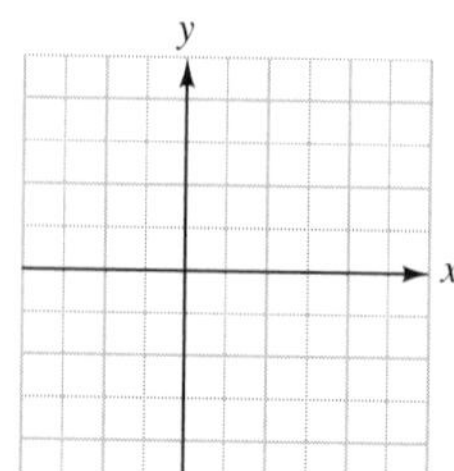

67. $f(x) = \frac{2}{3}x - 2$

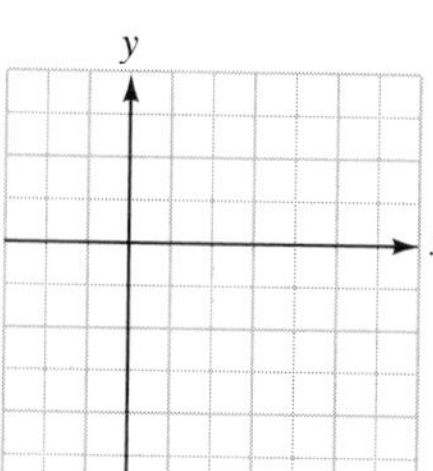

68. $f(x) = -\frac{3}{2}x - 3$

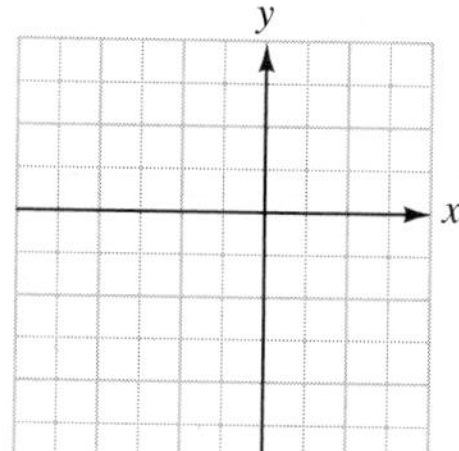

In Exercises 69–72, tell whether each equation defines a linear function.

69. $y = 3x^2 + 2$

70. $y = \frac{x - 3}{2}$

71. $y = x$

72. $y = 3x^3 - 4$

APPLICATIONS

73. DECONGESTANTS The temperature in degrees Celsius that is equivalent to a temperature in degrees Fahrenheit is given by the linear function $C(F) = \frac{5}{9}(F - 32)$. Use this function to find the temperature range, in degrees Celsius, at which a bottle of Dimetapp should be stored. The label directions are shown in Illustration 3.

> **DIRECTIONS:** Adults and children 12 years of age and over: Two teaspoons every 4 hours. DO NOT EXCEED 6 DOSES IN A 24-HOUR PERIOD. Store at a controlled room temperature between 68°F and 77°F.

ILLUSTRATION 3

74. BODY TEMPERATURES The temperature in degrees Fahrenheit that is equivalent to a temperature in degrees Celsius is given by the linear function $F(C) = \frac{9}{5}C + 32$. Convert each of the temperatures in the following excerpt from *The Good Housekeeping Family Health and Medical Guide* to degrees Fahrenheit. (Round to the nearest degree.)

> In disease, the temperature of the human body may vary from about 32.2°C to 43.3°C for a time, but there is grave danger to life should it drop and remain below 35°C or rise and remain at or above 41°C.

75. CONCESSIONAIRE A baseball club pays a peanut vendor \$50 per game for selling bags of peanuts for \$1.75 each.

a. Write a linear function that describes the income the vendor makes for the baseball club during a game if he sells b bags of peanuts.

b. Find the income the baseball club will make if the vendor sells 110 bags of peanuts during a game.

76. NEW HOME CONSTRUCTION In a proposal to some prospective clients, a housing contractor listed the following costs.

Fees, permits, miscellaneous	\$12,000
Construction, per square foot:	\$75

a. Write a linear function that the clients could use to determine the cost of building a home having f square feet.

b. Find the cost to build a home having 1,950 square feet.

77. EARTH'S ATMOSPHERE Illustration 4 shows a graph of the temperatures of the atmosphere at various altitudes above the earth's surface. The temperature is expressed in degrees Kelvin, a scale widely used in scientific work.

a. Estimate the coordinates of three points on the graph that have an x-coordinate of 200.

b. Explain why this is not the graph of a function.

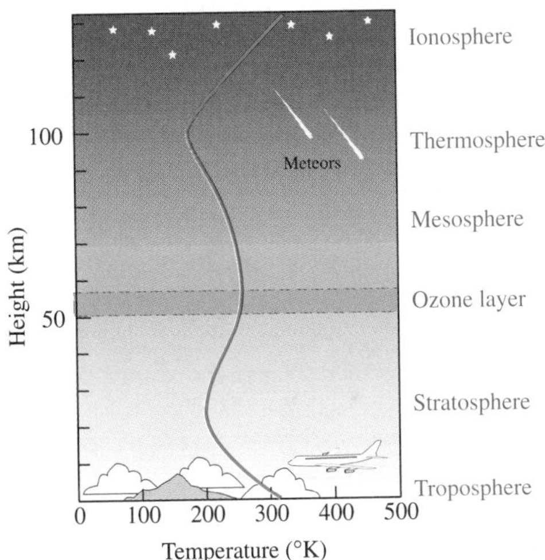

ILLUSTRATION 4

78. CHEMICAL REACTIONS When students in a chemistry laboratory mixed solutions of acetone and chloroform, they found that heat was immediately generated. As time went by, the mixture cooled down. Illustration 5 shows a graph of data points of the form (time, temperature) taken by the students during the experiment.

a. The linear function $T(t) = -\frac{t}{240} + 30$ models the relationship between the elapsed time t since the solutions were combined and the temperature $T(t)$ of the mixture. Graph the function in Illustration 5.

b. Predict the temperature of the mixture immediately after the two solutions are combined.

c. Is $T(180)$ more or less than the temperature recorded by the students for $t = 300$?

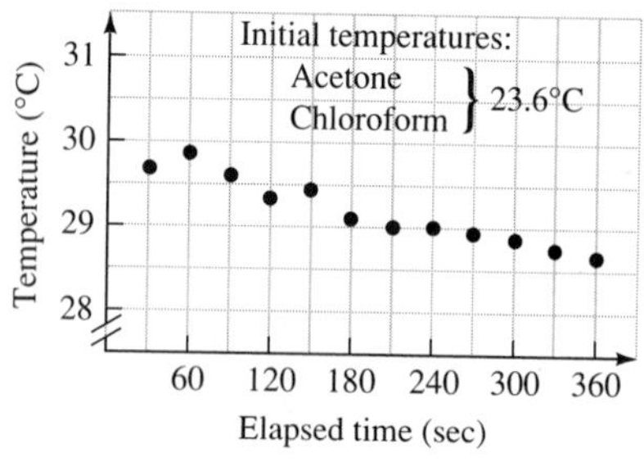

ILLUSTRATION 5

79. INCOME TAX The function

$T(a) = 3{,}697.50 + 0.28(a - 24{,}650)$

(where a is adjusted gross income) is a mathematical model of the instructions given on the second line of the tax rate Schedule X shown in Illustration 6.

a. Find $T(35{,}000)$ and interpret the result.

b. From the directions in the tax rate schedule, what are the domain and the range of the function?

Schedule X–Use if your filing status is Single

If your adjusted gross income is: *Over—*	*But not over—*	Your tax is	*of the amount over—*
$0	$24,650	.... **15%**	**$0**
24,650	59,750	**$3,697.50 + 28%**	**24,650**
59,750	124,650	**13,525.50 + 31%**	**59,750**

ILLUSTRATION 6

c. Write a function that would compute the tax according to the instructions given in the third line of the schedule.

80. COST FUNCTION An electronics firm manufactures tape recorders, receiving $120 for each recorder it makes. If x represents the number of recorders produced, the income received is determined by the *revenue function* $R(x) = 120x$. The manufacturer has fixed costs of $12,000 per month and variable costs of $57.50 for each recorder manufactured. Thus, the *cost function* is $C(x) = 57.50x + 12{,}000$. How many recorders must the company sell for revenue to equal cost? (*Hint:* Set $R(x) = C(x)$.)

81. BALLISTICS The height of a bullet shot from the ground straight upward is given by the function $f(t) = -16t^2 + 256t$.

a. Find the height of the bullet 3 seconds after it is shot.

b. Find $f(16)$. Interpret the result.

82. PLATFORM DIVING The number of feet a diver is above the surface of the water is given by the function $h(t) = -16t^2 + 16t + 32$, where t is the elapsed time in seconds after the diver jumped. Find the height of the diver for the times shown in the table.

t	$h(t)$
0	
0.5	
1.5	
2.0	

WRITING *Write a paragraph using your own words.*

83. Give four ways in which a function can be described. Which do you think is the most informative? Why?

84. Explain why we can think of a function as a machine.

REVIEW *In Exercises 85–88, show that each number is a rational number by expressing it as a ratio of two integers.*

85. $-3\frac{3}{4}$

86. 4.7

87. 0.333. . .

88. $-0.\overline{6}$

2.6 Graphs of Functions

In this section, you will learn about

- **Graphs of nonlinear functions**
- **Translations of graphs**
- **Reflections of graphs**
- **Solving equations graphically**

INTRODUCTION. In the previous section, we saw that many real-world situations can be modeled by linear functions. When these functions are graphed, we can learn information by examining the resulting line. In this section, we will discuss three more types of functions. Like linear functions, they can be used as mathematical models. But unlike linear functions, their graphs are not straight lines; therefore, they are called **nonlinear functions.**

Graphs of nonlinear functions

The first nonlinear function we will discuss is the function $f(x) = x^2$ (or $y = x^2$), called the **squaring function.**

EXAMPLE 1 ***The squaring function.*** Graph the function $f(x) = x^2$ and find the domain and range.

Solution

We substitute values for x in the equation and compute the corresponding values of $f(x)$. For example, if $x = -3$, we have

$$
\begin{aligned}
f(x) &= x^2 \\
f(-3) &= (-3)^2 \quad \text{Substitute } -3 \text{ for } x. \\
&= 9
\end{aligned}
$$

The ordered pair $(-3, 9)$ satisfies the equation and will lie on the graph. We list this pair and others that satisfy the equation in the table shown in Figure 2-49. We plot the points and draw a smooth curve through them to get the graph, called a **parabola.**

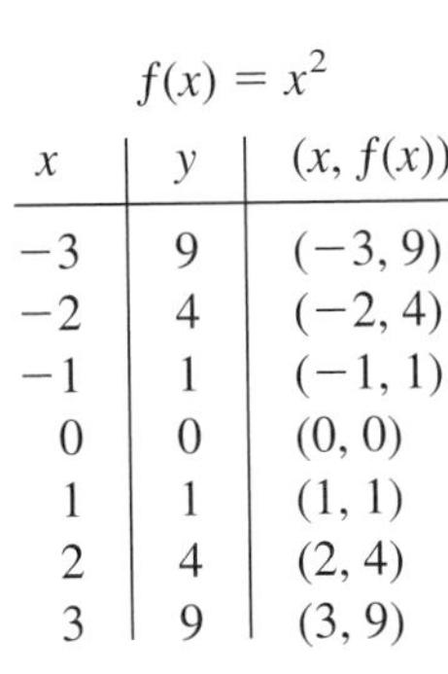

$f(x) = x^2$

x	y	$(x, f(x))$
-3	9	$(-3, 9)$
-2	4	$(-2, 4)$
-1	1	$(-1, 1)$
0	0	$(0, 0)$
1	1	$(1, 1)$
2	4	$(2, 4)$
3	9	$(3, 9)$

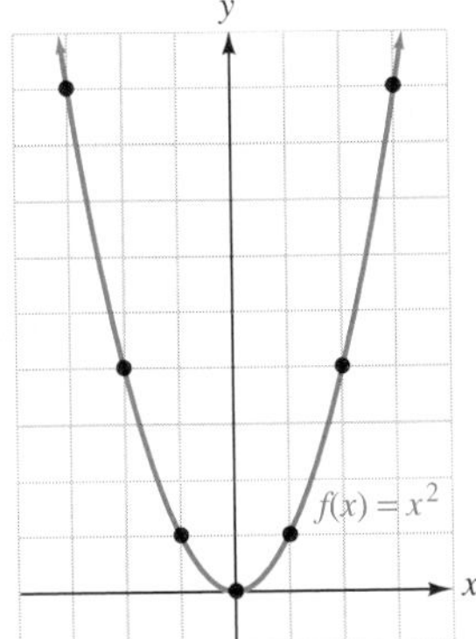

FIGURE 2-49

From the graph, we can see that x can be any real number. This indicates that the domain of the squaring function is the set of real numbers. We can also see that y is always positive or zero. This indicates that the range is the set of nonnegative real numbers.

Self Check

Graph $f(x) = x^2 - 2$. Find the domain and the range. Compare the graph to the graph of $f(x) = x^2$.

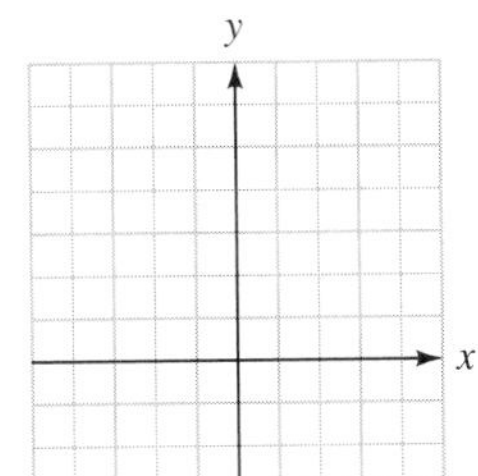

Answers: D = the set of real numbers, R = the set of all real numbers greater than or equal to -2; the graph has the same shape but is 2 units lower

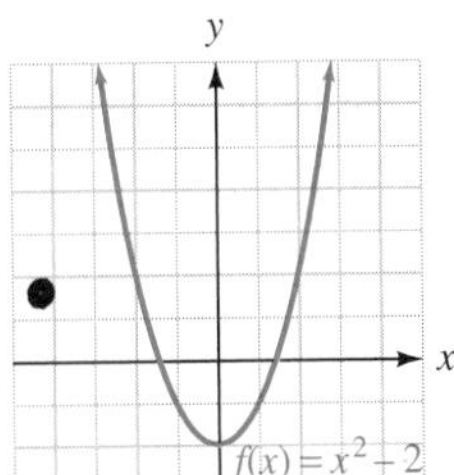

Another important nonlinear function is $f(x) = x^3$ (or $y = x^3$), called the **cubing function.**

EXAMPLE 2 ***The cubing function.*** Graph the function $f(x) = x^3$ and find the domain and range.

Solution

We substitute values for x in the equation and compute the corresponding values of $f(x)$. For example, if $x = -2$, we have

$$f(x) = x^3$$
$$f(-2) = (-2)^3 \quad \text{Substitute } -2 \text{ for } x.$$
$$= -8$$

The ordered pair $(-2, -8)$ satisfies the equation and will lie on the graph. We list this pair and others that satisfy the equation in the table shown in Figure 2-50. We plot the points and draw a smooth curve through them to get the graph.

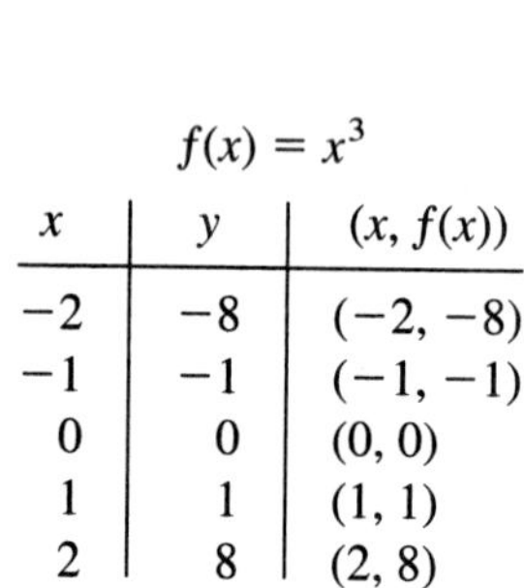

$f(x) = x^3$

x	y	$(x, f(x))$
-2	-8	$(-2, -8)$
-1	-1	$(-1, -1)$
0	0	$(0, 0)$
1	1	$(1, 1)$
2	8	$(2, 8)$

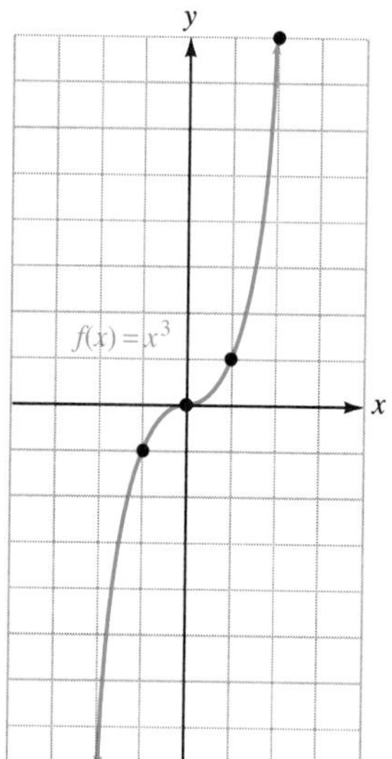

FIGURE 2-50

From the graph, we can see that x can be any real number. This indicates that the domain of the cubing function is the set of real numbers. We can also see that y can be any real number. This indicates that the range is the set of real numbers.

Self Check

Graph $f(x) = x^3 + 1$. Find the domain and the range. Compare the graph to the graph of $f(x) = x^3$.

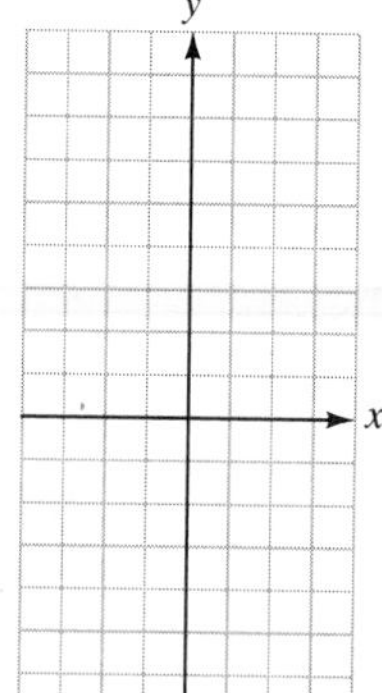

Answers: D = the set of real numbers, R = the set of real numbers; the graph has the same shape but is 1 unit higher

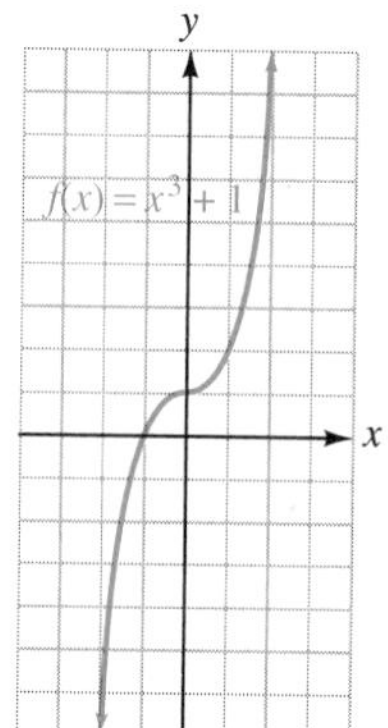

A third nonlinear function is $f(x) = |x|$ (or $y = |x|$), called the **absolute value function.**

EXAMPLE 3 ***The absolute value function.*** Graph the function $f(x) = |x|$ and find the domain and range.

Solution

We substitute values for x in the equation and compute the corresponding values of $f(x)$. For example, if $x = -3$, we have

$$f(x) = |x|$$
$$f(-3) = |-3| \quad \text{Substitute } -3 \text{ for } x.$$
$$= 3$$

The ordered pair $(-3, 3)$ satisfies the equation and will lie on the graph. We list this pair and others that satisfy the equation in the table shown in Figure 2-51. We plot the points and connect them to get the graph.

Self Check

Graph $f(x) = |x - 2|$. Find the domain and the range. Compare the graph to the graph of $f(x) = |x|$.

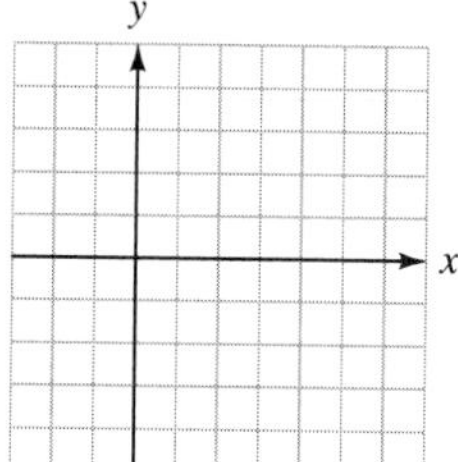

$f(x) = |x|$

x	y	$(x, f(x))$
-3	3	$(-3, 3)$
-2	2	$(-2, 2)$
-1	1	$(-1, 1)$
0	0	$(0, 0)$
1	1	$(1, 1)$
2	2	$(2, 2)$
3	3	$(3, 3)$

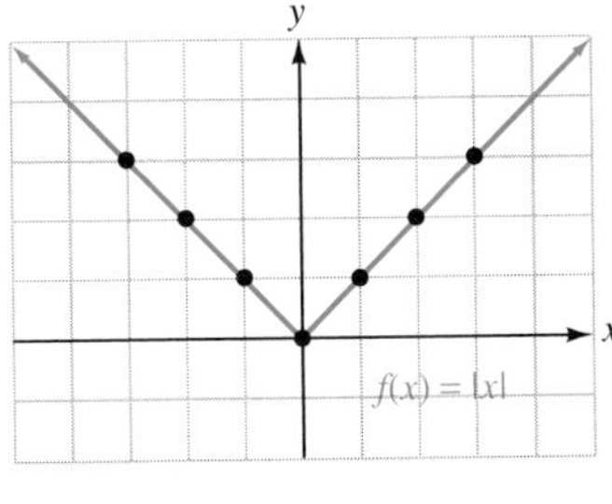

FIGURE 2-51

From the graph, we can see that x can be any real number. So the domain of the absolute value function is the set of real numbers. We can also see that y is always positive or zero. This indicates that the range is the set of nonnegative real numbers.

Answers: D = the set of real numbers, R = the set of nonnegative real numbers; the graph has the same shape but is 2 units to the right

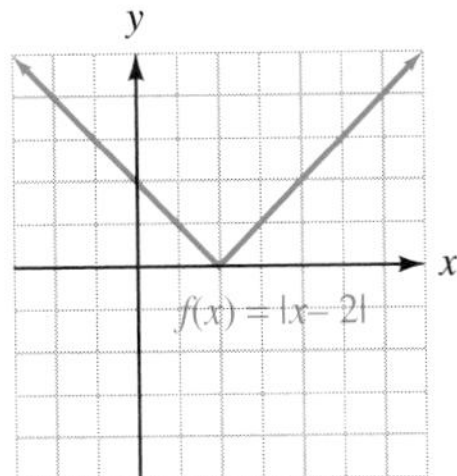

Accent on Technology **Graphing functions**

We can graph nonlinear functions with a graphing calculator. For example, to graph $f(x) = x^2$ in a standard window of $[-10, 10]$ for x and $[-10, 10]$ for y, we enter the function by typing x ^ 2 and then press the GRAPH key. We will obtain the graph shown in Figure 2-52(a).

To graph $f(x) = x^3$, we enter the function by typing x ^ 3 and then press the GRAPH key to obtain the graph in Figure 2-52(b). To graph $f(x) = |x|$, we enter the function by pressing the ABS key (or selecting abs from a menu), typing x, and pressing the GRAPH key to obtain the graph in Figure 2-52(c).

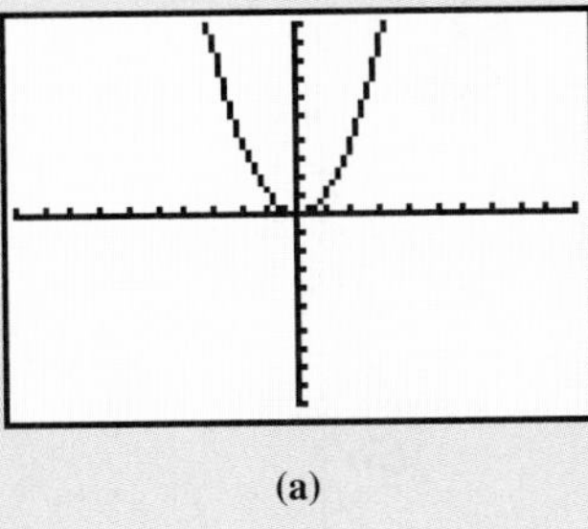

(a)

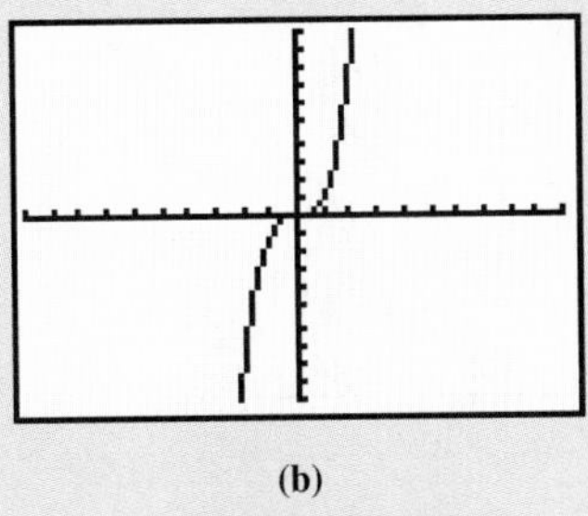

(b)

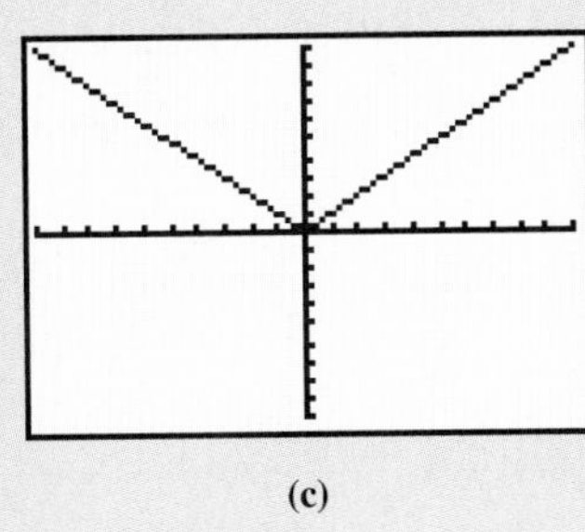

(c)

FIGURE 2-52

When using a graphing calculator, we must be sure that the viewing window does not show a misleading graph. For example, if we graph $f(x) = |x|$ in the window $[0, 10]$ for x and $[0, 10]$ for y, we will obtain a misleading graph that looks like a line. See Figure 2-53. This is not correct. The proper graph is the V-shaped graph shown in Figure 2-52(c). One of the challenges of using graphing calculators is finding an appropriate viewing window.

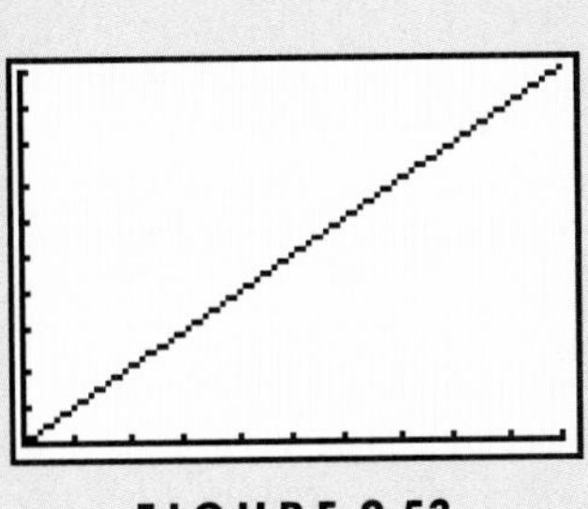

FIGURE 2-53

Translations of graphs

Examples 1, 2, and 3 and their Self Checks suggest that the graphs of different functions may be identical except for their positions in the xy-plane. For example, Figure 2-54 shows the graph of $f(x) = x^2 + k$ for three different values of k. If $k = 0$, we get the graph of $f(x) = x^2$. If $k = 3$, we get the graph of $f(x) = x^2 + 3$, which is identical to the graph of $f(x) = x^2$ except that it is shifted 3 units upward. If $k = -4$, we get the graph of $f(x) = x^2 - 4$, which is identical to the graph of $f(x) = x^2$ except that it is shifted 4 units downward. These shifts are called **vertical translations.**

In general, we can make these observations.

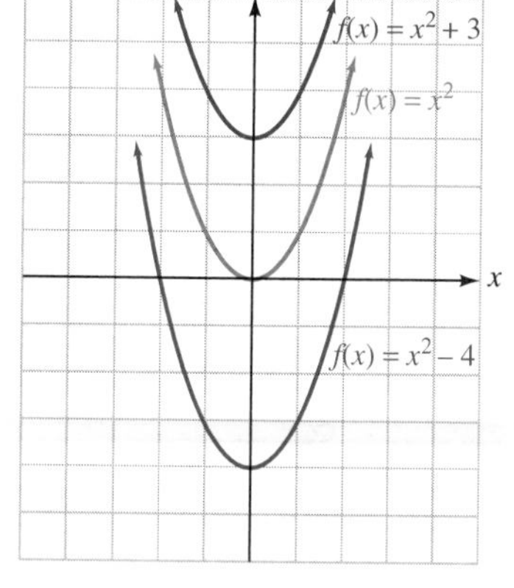

FIGURE 2-54

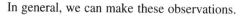

Vertical translations

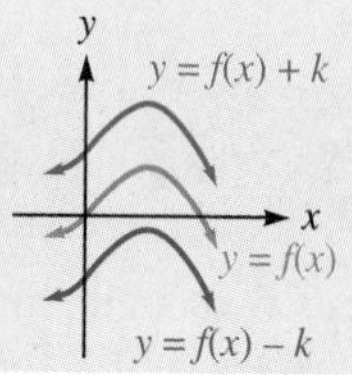

If f is a function and k is a positive number, then

- The graph of $y = f(x) + k$ is identical to the graph of $y = f(x)$ except that it is translated k units upward.
- The graph of $y = f(x) - k$ is identical to the graph of $y = f(x)$ except that it is translated k units downward.

EXAMPLE 4 ***Vertical translations.*** Graph $f(x) = |x| + 2$.

Solution

The graph of $f(x) = |x| + 2$ will be the same V-shaped graph as $f(x) = |x|$, except shifted 2 units up. The graph appears in Figure 2-55.

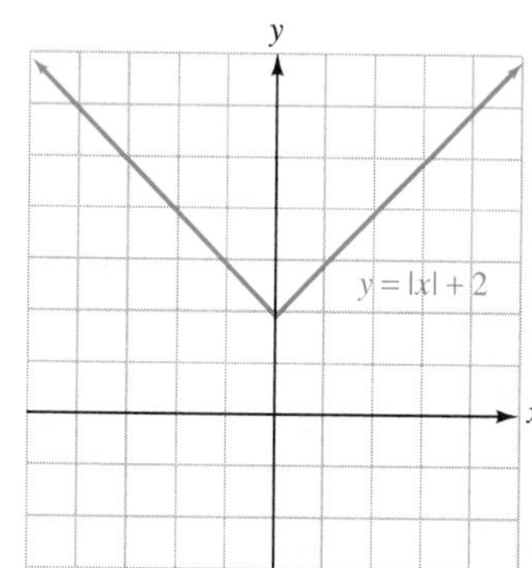

FIGURE 2-55

Self Check

Graph $f(x) = |x| - 3$.

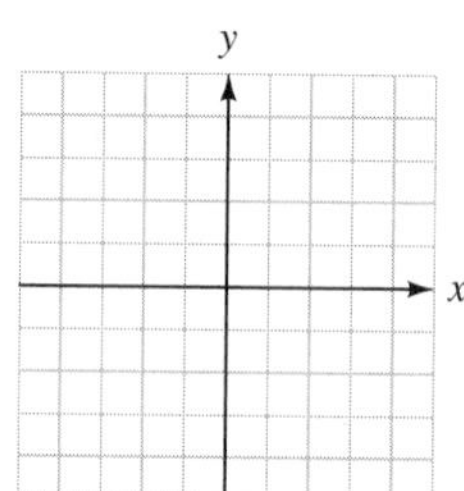

Answer:

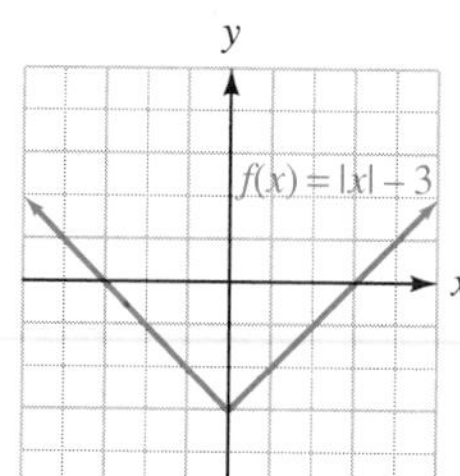

Figure 2-56 shows the graph of $f(x) = (x + h)^2$ for three different values of h. If $h = 0$, we get the graph of $f(x) = x^2$. The graph of $f(x) = (x - 3)^2$ is identical to the graph of $f(x) = x^2$ except that it is shifted 3 units to the right. The graph of $f(x) = (x + 2)^2$ is identical to the graph of $f(x) = x^2$ except that it is shifted 2 units to the left. These shifts are called **horizontal translations.**

In general, we can make these observations.

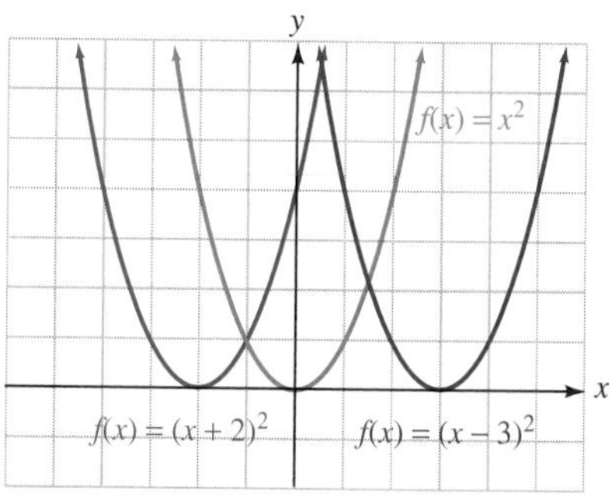

FIGURE 2-56

Horizontal translations

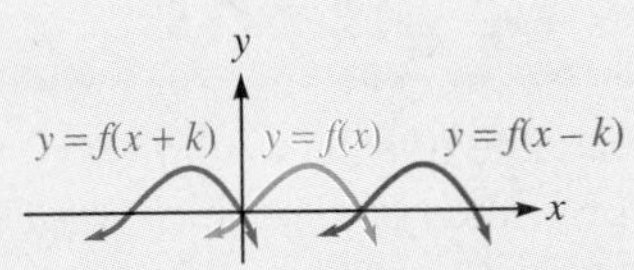

If f is a function and k is a positive number, then

- The graph of $y = f(x - k)$ is identical to the graph of $y = f(x)$ except that it is translated k units to the right.
- The graph of $y = f(x + k)$ is identical to the graph of $y = f(x)$ except that it is translated k units to the left.

EXAMPLE 5 ***Horizontal translations.*** Graph $f(x) = (x + 3)^3$.

Solution

The graph of $f(x) = (x + 3)^3$ will be the same shape as the graph of $f(x) = x^3$ except that it is shifted 3 units to the left. The graph appears in Figure 2-57.

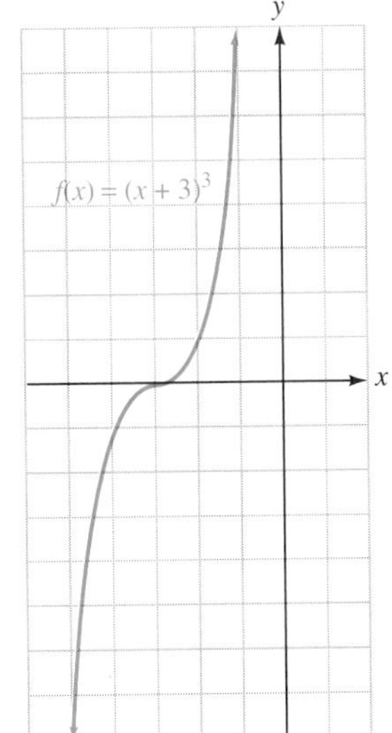

FIGURE 2-57

Self Check

Graph $f(x) = (x - 2)^2$.

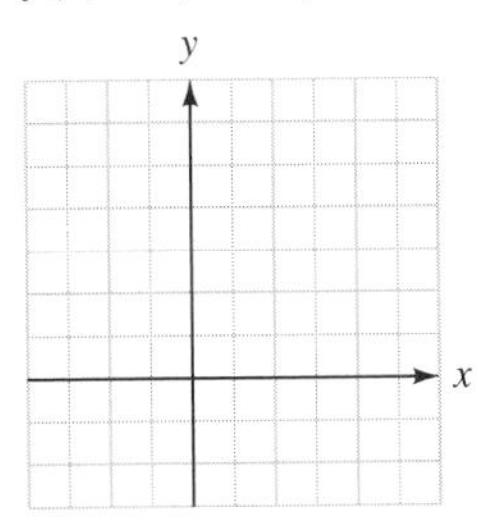

Answer:

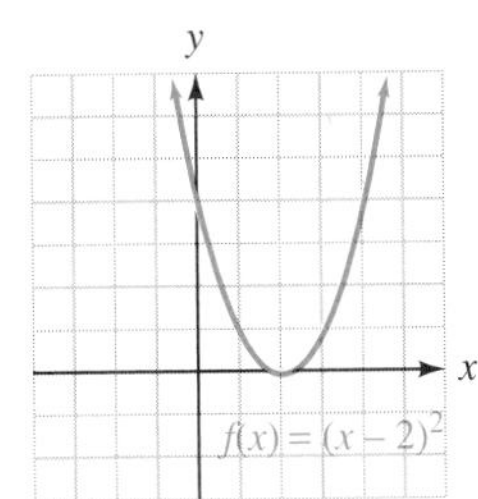

In the next example, two translations are made to the basic graph.

EXAMPLE 6 ***Two translations.*** Graph $f(x) = (x - 3)^2 + 2$.

Solution

We can graph this equation by translating the graph of $f(x) = x^2$ to the right 3 units and then 2 units up, as shown in Figure 2-58.

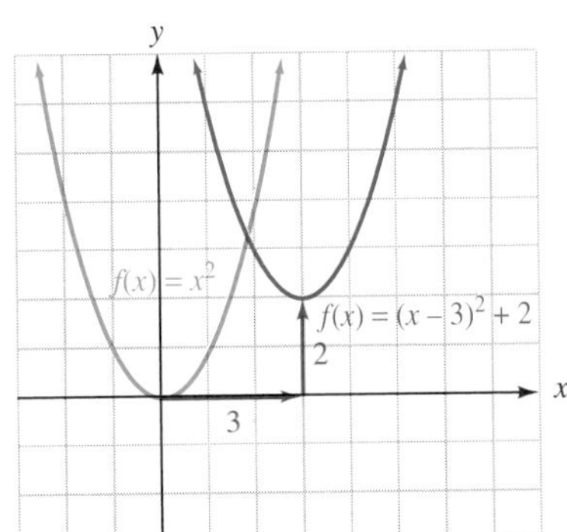

FIGURE 2-58

Self Check

Graph $f(x) = |x + 2| - 3$.

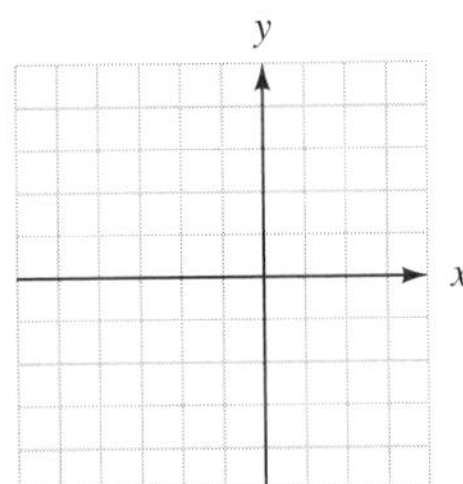

Answer:

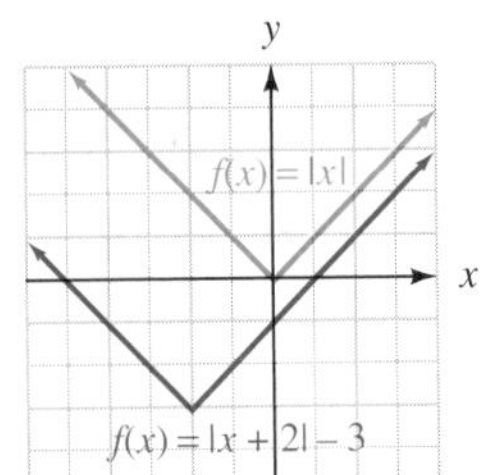

Reflections of graphs

Figure 2-59 shows a table of solutions for $f(x) = x^2$ and for $f(x) = -x^2$. We note that for a given value of x, the corresponding y values in the tables are opposites. When graphed, we see that the $-$ in $f(x) = -x^2$ has the effect of "flipping" the graph of $f(x) = x^2$ over the x-axis so that the parabola opens downward. We say that the graph of $f(x) = -x^2$ is a **reflection** of the graph of $f(x) = x^2$ in the x-axis.

$f(x) = x^2$

x	y	$(x, f(x))$
-2	4	$(-2, 4)$
-1	1	$(-1, 1)$
0	0	$(0, 0)$
1	1	$(1, 1)$
2	4	$(2, 4)$

$f(x) = -x^2$

x	y	$(x, f(x))$
-2	-4	$(-2, -4)$
-1	-1	$(-1, -1)$
0	0	$(0, 0)$
1	-1	$(1, -1)$
2	-4	$(2, -4)$

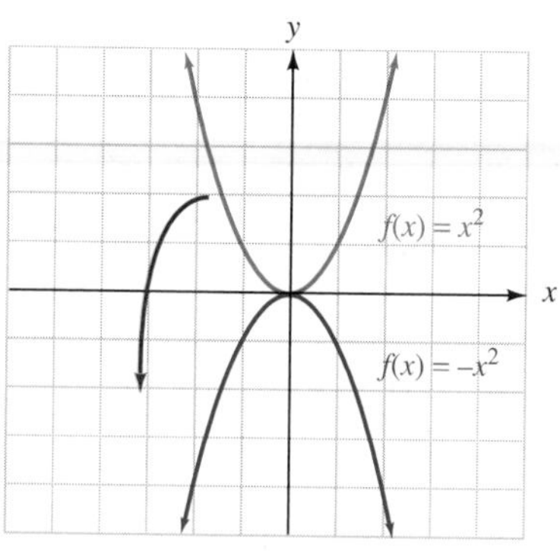

FIGURE 2-59

EXAMPLE 7 ***Reflection of a graph.*** Graph $f(x) = -x^3$.

Solution

To graph $f(x) = -x^3$, we use the graph of $f(x) = x^3$ from Example 2. First, we reflect the portion of the graph of $f(x) = x^3$ in quadrant I to quadrant IV, as shown in Figure 2-60. Then we reflect the portion of the graph of $f(x) = x^3$ in quadrant III to quadrant II.

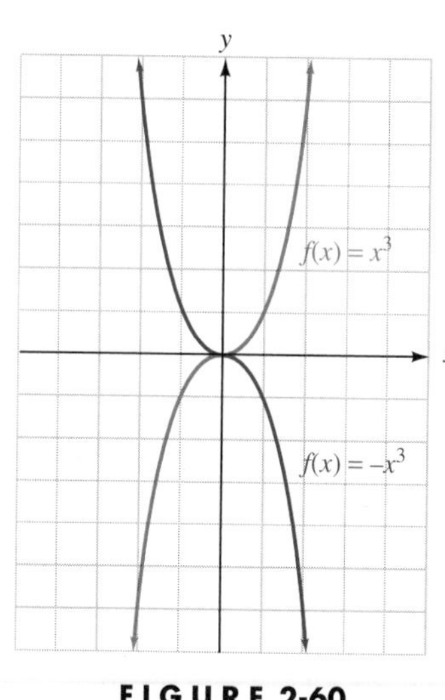

FIGURE 2-60

Self Check

Graph $f(x) = -|x|$.

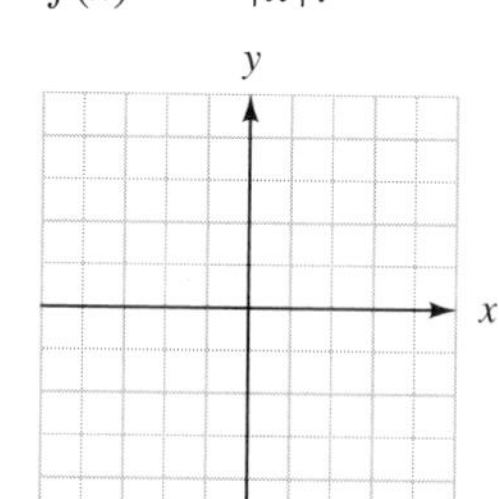

Answer:

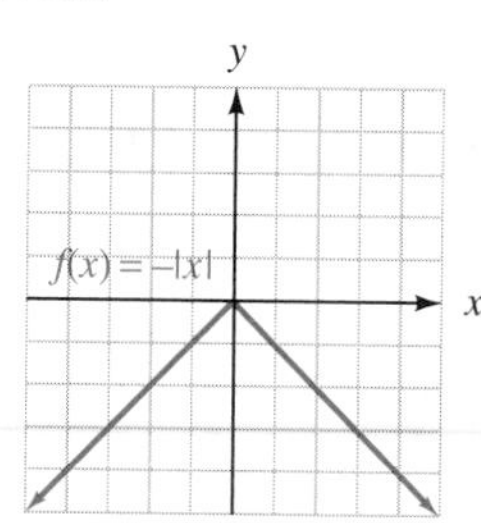

Reflection of a graph	The graph of $y = -f(x)$ is the graph of $y = f(x)$ reflected about the x-axis.

Solving equations graphically

We can use a graphing calculator to solve equations graphically. The next example introduces this process.

EXAMPLE 8 *Solving an equation from a graph.* Solve $2x + 4 = -2$ graphically.

Solution

The graphs of $y = 2x + 4$ and $y = -2$ are shown in Figure 2-61. To solve the equation $2x + 4 = -2$, we need to find the value of x that makes $2x + 4$ equal -2. The point of intersection of the graphs is $(-3, -2)$. This tells us that if $x = -3$, the expression $2x + 4$ equals -2. So the solution of $2x + 4 = -2$ is $x = -3$.

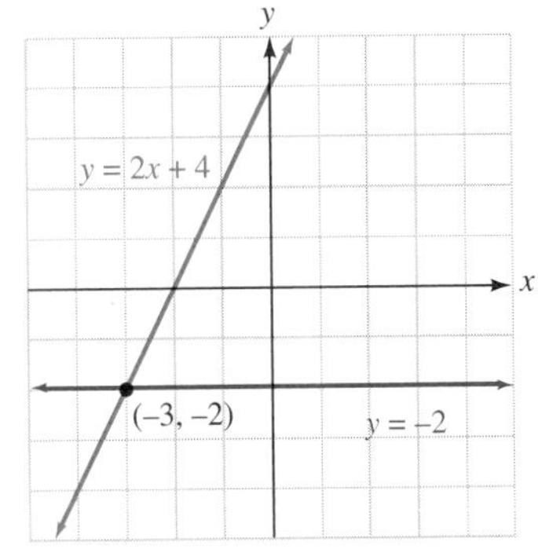

FIGURE 2-61

Self Check

Solve $2x + 4 = 2$ graphically.

Answer: -1

Accent on Technology — Solving equations graphically

To solve the equation $2(x - 3) + 3 = 7$ with a graphing calculator, we graph both the left-hand side and the right-hand side of the equation in the same window, as shown in Figure 2-62(a). We then trace to find the coordinates of the point where the two graphs intersect, as shown in Figure 2-62(b). We can then zoom and trace again to get Figure 2-62(c). Solving the equation algebraically will show that the coordinates of the intersection point are indeed (5, 7). From the figure, we can see that the solution of the equation is $x = 5$.

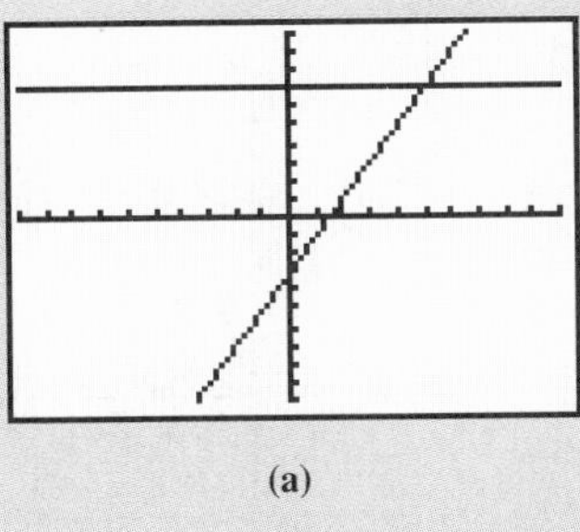

(a)

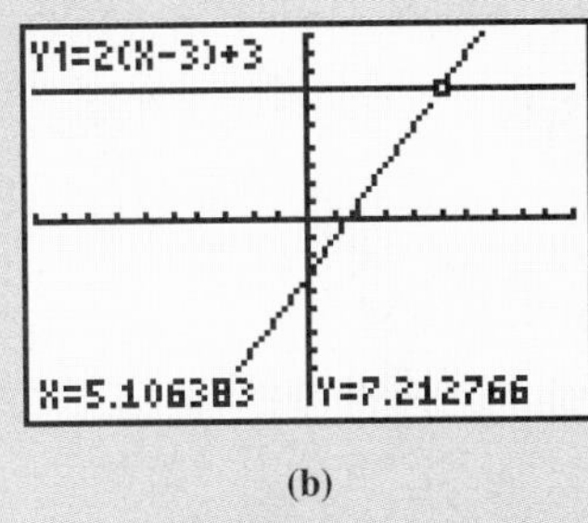

(b)

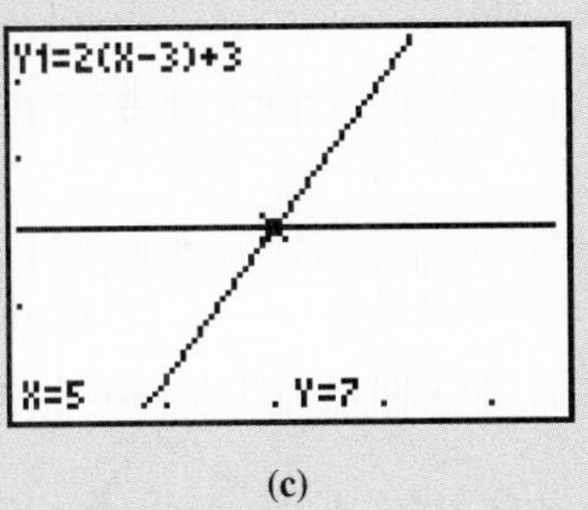

(c)

FIGURE 2-62

STUDY SET Section 2.6

VOCABULARY *In Exercises 1–8, fill in the blanks to make the statements true.*

1. The function $f(x) = x^2$ is called the ____________ function.
2. The function $f(x) = x^3$ is called the ____________ function.
3. The function $f(x) = |x|$ is called the ____________ function.
4. Functions whose graphs are not straight lines are called ____________ functions.
5. Shifting the graph of an equation up or down is called a ____________ translation.
6. Shifting the graph of an equation to the left or to the right is called a horizontal ____________.
7. The graph of $f(x) = -x^2$ is a ____________ of the graph of $f(x) = x^2$ in the x-axis.
8. The set of ____________ real numbers is the set of real numbers greater than or equal to 0.

CONCEPTS *In Exercises 9–12, fill in the blanks to make the statements true.*

9. The graph of $f(x) = (x + 4)^3$ is the same as the graph of $f(x) = x^3$ except that it is shifted ___ units to the ___________.

10. The graph of $f(x) = x^3 - 2$ is the same as the graph of $f(x) = x^3$ except that it is shifted ___ units ___________.

11. The graph of $f(x) = x^2 + 5$ is the same as the graph of $f(x) = x^2$ except that it is shifted ___ units ___.

12. The graph of $f(x) = |x - 5|$ is the same as the graph of $f(x) = |x|$ except that it is shifted ___ units to the ___________.

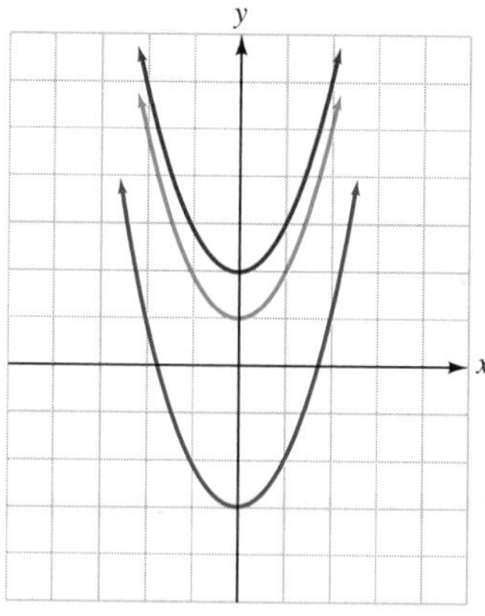

ILLUSTRATION 2

13. Use the graphs in Illustration 1 to solve each equation.

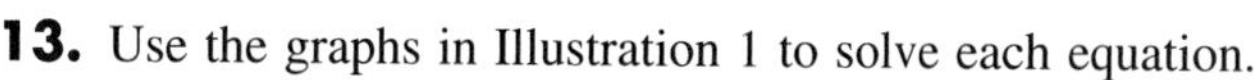

a. $3x - 2 = 4$ **b.** $3x - 2 = -2$

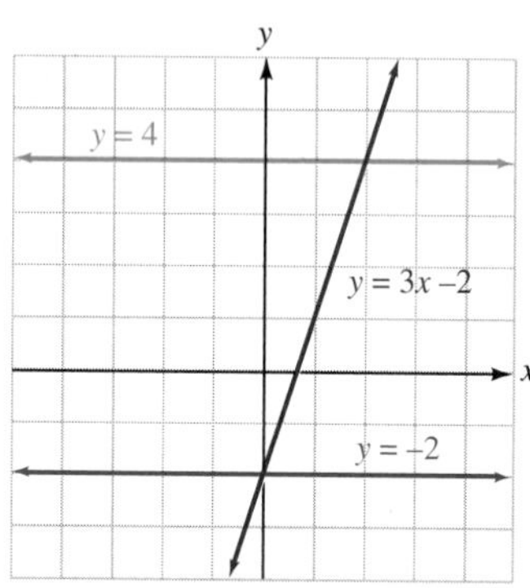

ILLUSTRATION 1

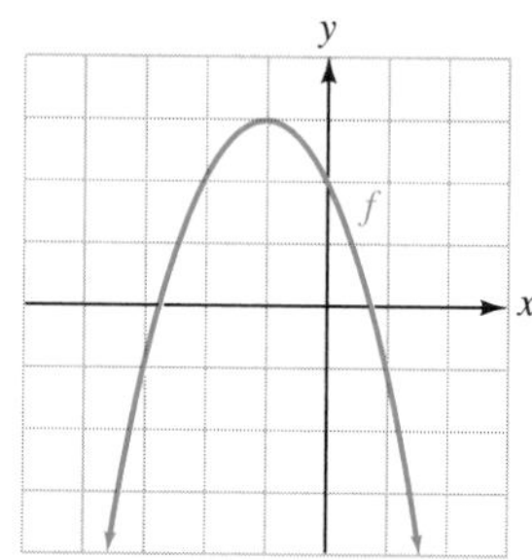

ILLUSTRATION 3

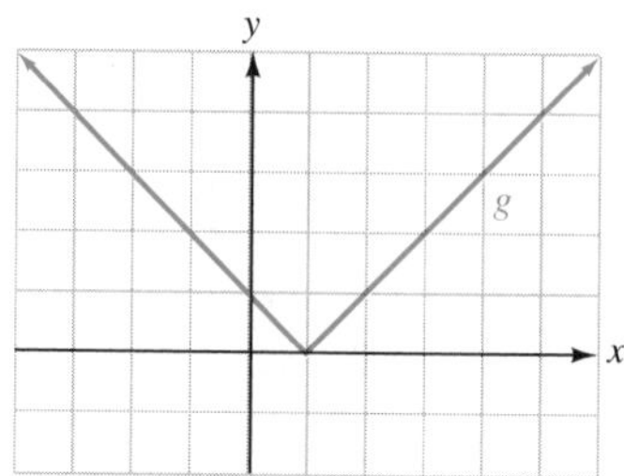

ILLUSTRATION 4

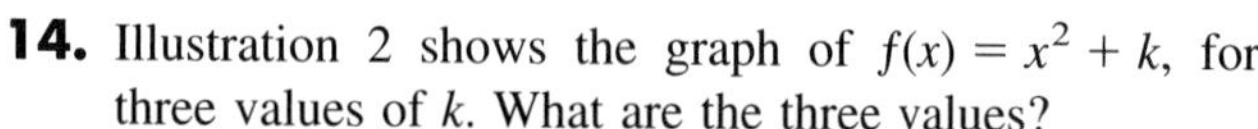

14. Illustration 2 shows the graph of $f(x) = x^2 + k$, for three values of k. What are the three values?

15. Use the graph in Illustration 3 to find each function value.

a. $f(-3)$ **b.** $f(0)$ **c.** $f(1)$

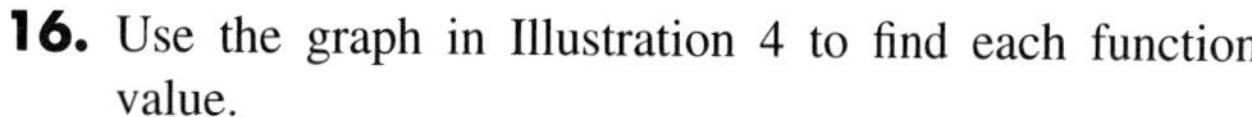

16. Use the graph in Illustration 4 to find each function value.

a. $g(-2)$ **b.** $g(1)$ **c.** $g(2.5)$

PRACTICE *In Exercises 17–24, graph each function by plotting points. Give the domain and range. Check your work with a graphing calculator.*

17. $f(x) = x^2 - 3$

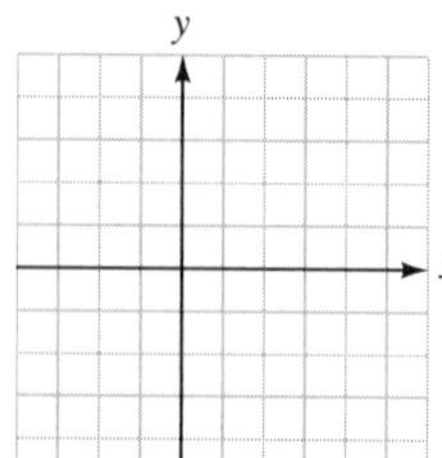

18. $f(x) = x^2 + 2$

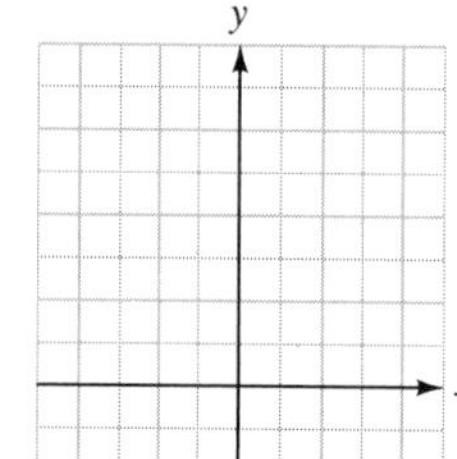

19. $f(x) = (x - 1)^3$

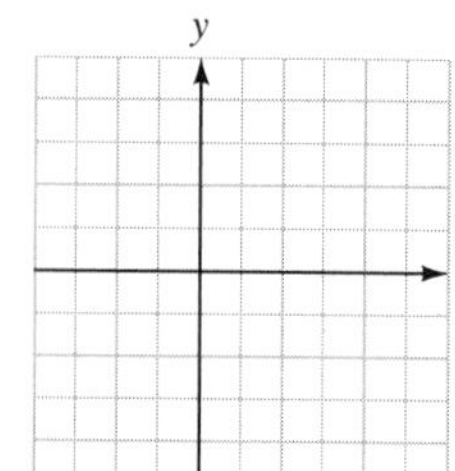

20. $f(x) = (x + 1)^3$

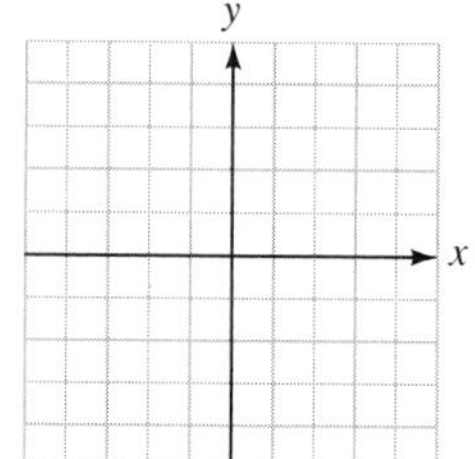

21. $f(x) = |x| - 2$

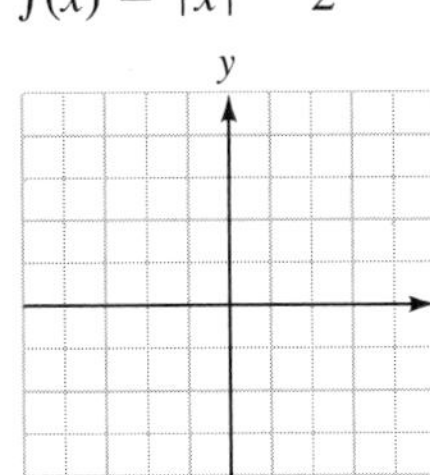

22. $f(x) = |x| + 1$

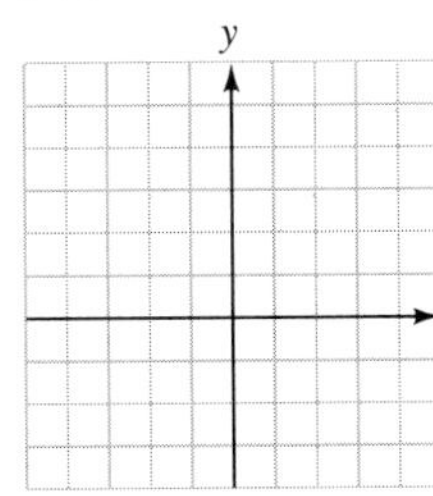

23. $f(x) = |x - 1|$

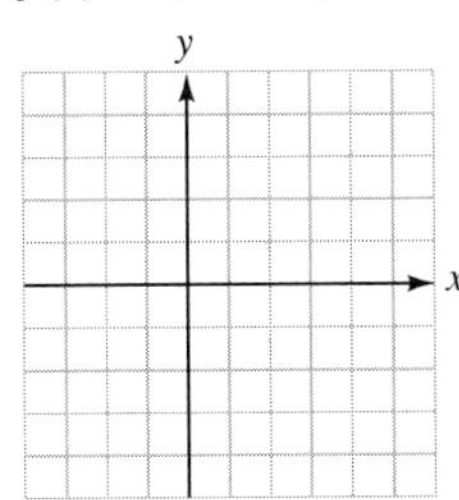

24. $f(x) = |x + 2|$

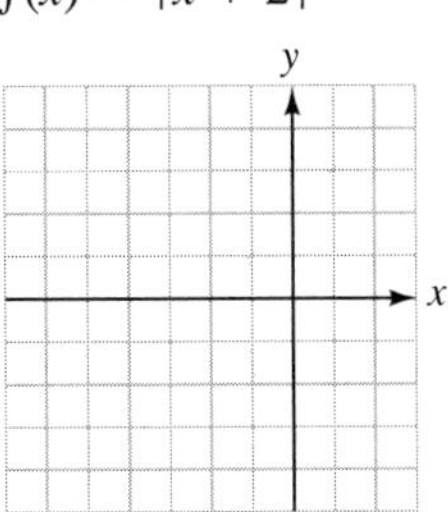

In Exercises 25–32, graph each function using window settings of [−4, 4] for x and [−4, 4] for y. The graph is not what it appears to be. Pick a better viewing window and find the true graph.

25. $f(x) = x^2 + 8$

26. $f(x) = x^3 - 8$

27. $f(x) = |x + 5|$

28. $f(x) = |x - 5|$

29. $f(x) = (x - 6)^2$

30. $f(x) = (x + 9)^2$

31. $f(x) = x^3 + 8$

32. $f(x) = x^3 - 12$

In Exercises 33–44, for each function, first sketch the graph of its associated function, $f(x) = x^2$, $f(x) = x^3$, or $f(x) = |x|$. Then draw each graph using a translation or a reflection.

33. $f(x) = x^2 - 5$

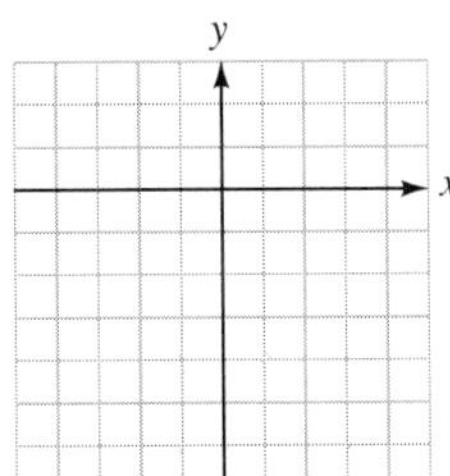

34. $f(x) = x^3 + 4$

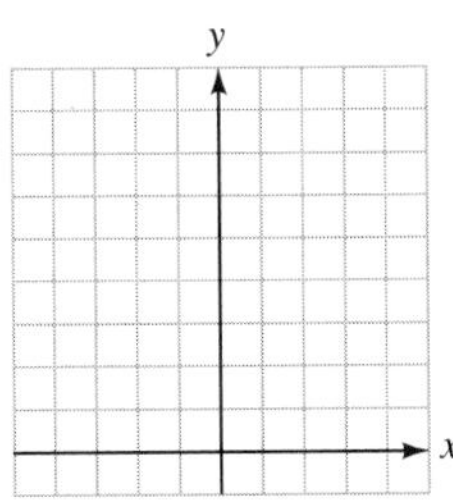

35. $f(x) = (x - 1)^3$

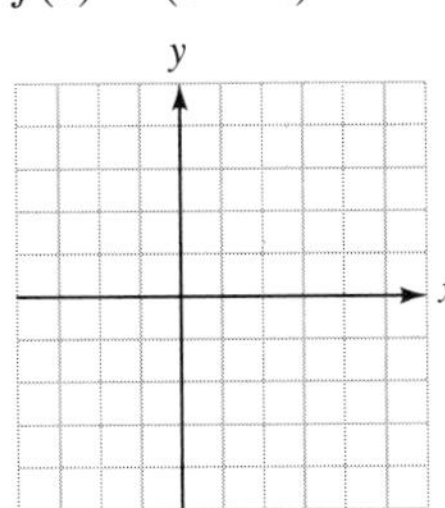

36. $f(x) = (x + 4)^2$

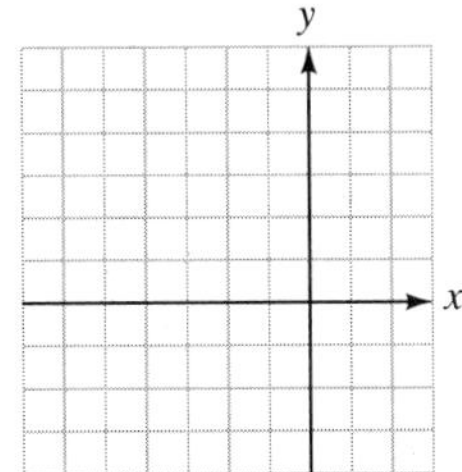

37. $f(x) = |x - 2| - 1$

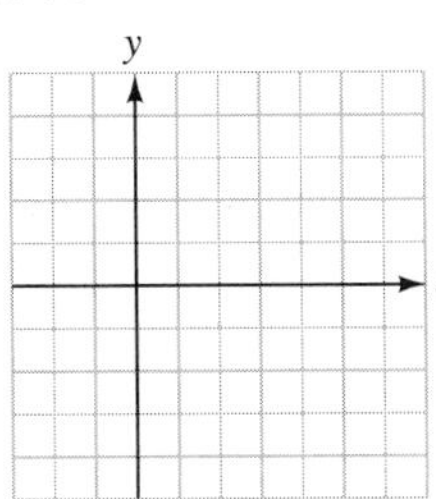

38. $f(x) = (x + 2)^2 - 1$

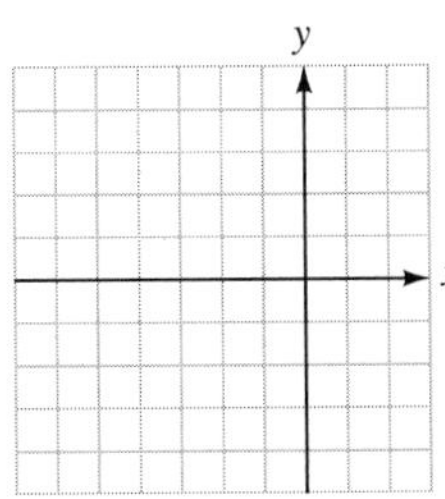

39. $f(x) = (x + 1)^3 - 2$

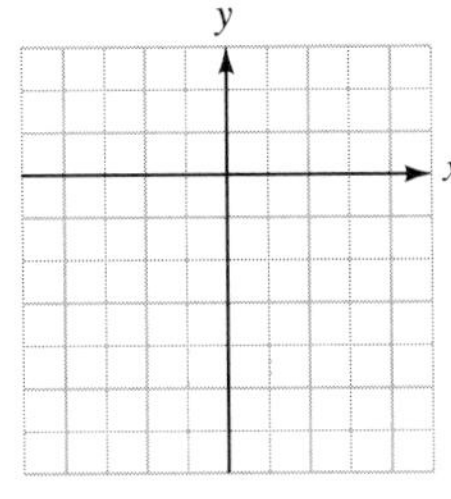

40. $f(x) = |x + 4| + 3$

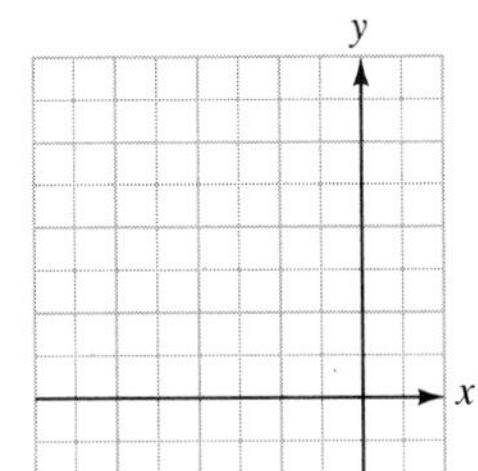

41. $f(x) = -x^3$

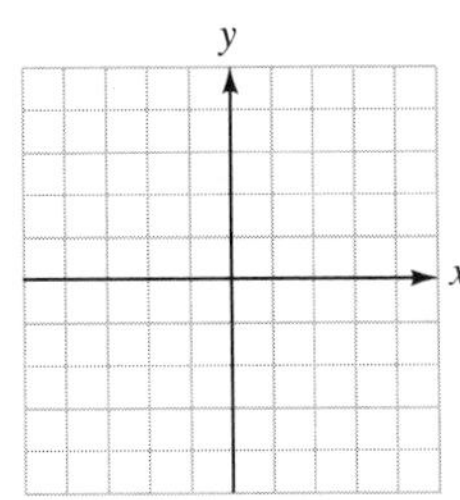

42. $f(x) = -|x|$

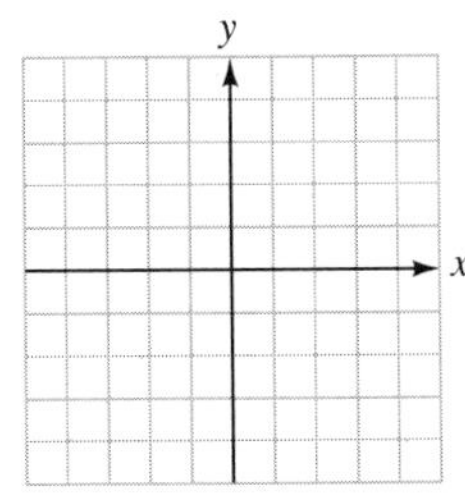

43. $f(x) = -x^2$

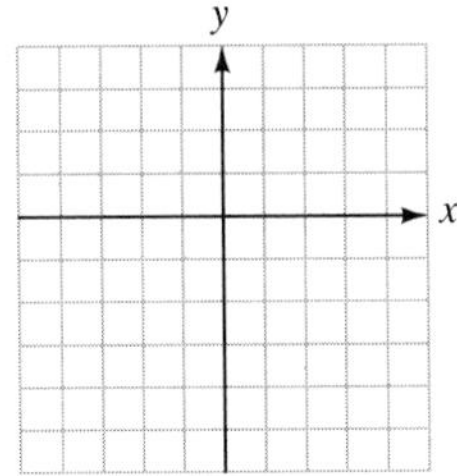

44. $f(x) = -(x + 1)^2$

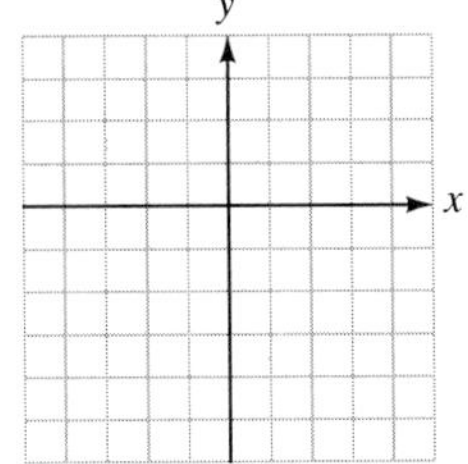

In Exercises 45–48, use a graphing calculator to solve each equation.

45. $4(x - 1) = 3x$

46. $4(x - 3) - x = x - 6$

47. $11x + 6(3 - x) = 3$

48. $2(x + 2) = 2(1 - x) + 10$

APPLICATIONS

49. FREEWAY DESIGN A Grand Avenue exit off the 210 Freeway is to be constructed. (See Illustration 5.) Sketch the off-ramp design on the graph if the sides of the pavement are defined by the functions $f(x) = (x - 1)^3$ and $f(x) = (x - 3)^3$.

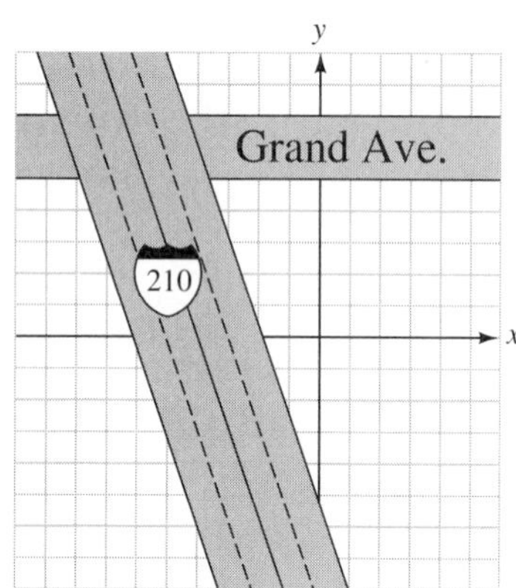

ILLUSTRATION 5

50. OPTICS See Illustration 6. The law of reflection states that the angle of reflection is equal to the angle of incidence. What function studied in this section mathematically models the path of the reflected light beam with an angle of incidence measuring 45°?

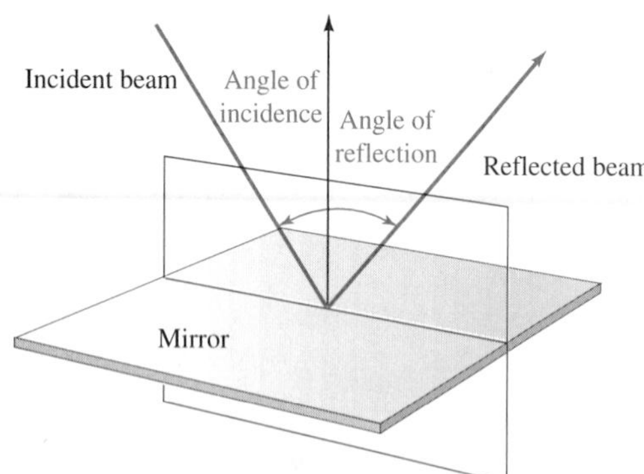

ILLUSTRATION 6

51. CENTER OF GRAVITY See Illustration 7. As a diver performs a $1\frac{1}{2}$-somersault in the tuck position, her center of gravity follows a path that can be described by a graph shape studied in this section. What graph shape is that?

ILLUSTRATION 7

52. FLASHLIGHTS Light beams coming from a flashlight bulb are reflected outward by a parabolic mirror as parallel rays.

a. The cross-sectional view of a parabolic mirror is given by the function $f(x) = x^2$ for the following values of x: $-0.7, -0.6, -0.5, -0.4, -0.3, -0.2, -0.1, 0, 0.1, 0.2, 0.3, 0.4, 0.5, 0.6, 0.7$. Sketch the parabolic mirror using the graph in Illustration 8.

b. From the light bulb filament at $(0, 0.25)$, draw a line segment representing a beam of light that strikes the mirror at $(-0.4, 0.16)$ and then reflects outward, parallel to the y-axis.

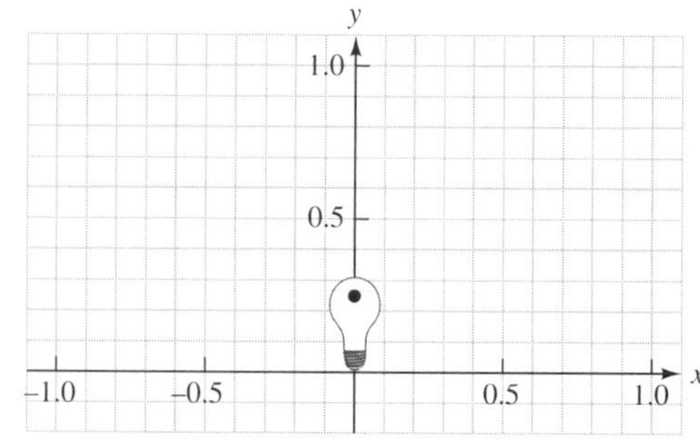

ILLUSTRATION 8

WRITING

53. Explain how to graph an equation by plotting points.

54. Explain why the correct choice of window settings is important when using a graphing calculator.

REVIEW *Solve each formula for the indicated variable.*

55. $T - W = ma$ for W

56. $a + (n - 1)d = l$ for n

57. $s = \frac{1}{2}gt^2 + vt$ for g

58. $e = mc^2$ for m

KEY CONCEPT

Functions

In Chapter 2, we introduced one of the most important concepts in mathematics, that of a **function.** Functions are used to describe relationships where one quantity depends on another.

1. Fill in the blanks to make the statements true.

a. A function is a _____________ between a set of _____________ values x (called the domain) and a set of output values y (called the _____________), where exactly ___ y value in the range corresponds to each number x in the _____________.

b. Since the value of y usually depends on the number x, we call y the _____________ variable and x the _____________ variable.

2. We can think of a function as a machine. Using the words *input, output, domain,* and *range,* explain how the function machine shown below works.

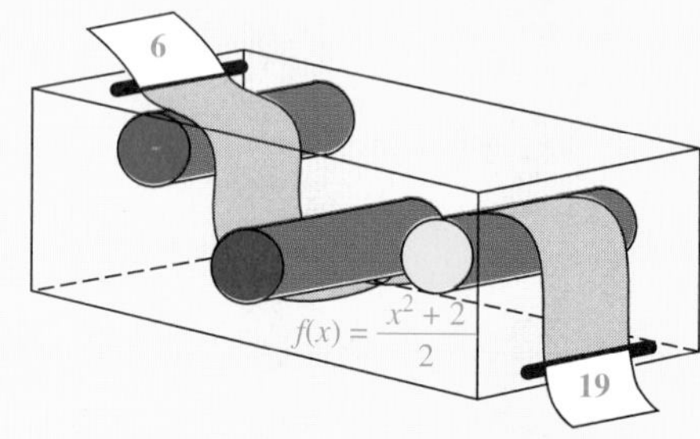

Four Ways to Represent a Function

Functions can be described in words, with an equation, with a table, or with a graph.

3. The equation $y = 2x + 3$ determines a correspondence between the values of x and y. Find the value of y that corresponds to $x = -10$.

4. The area of a circle is the product of π and the radius squared. Use the variables A and r to describe this relationship with an equation.

5. Use the table below to determine the height of a projectile 1.5 seconds after it was shot vertically into the air.

Time t (seconds)	Height h (feet)
0	0
0.5	28
1.0	48
1.5	60
2.0	64

6. Use the graph below to determine what y value the function f assigns to $x = 1$.

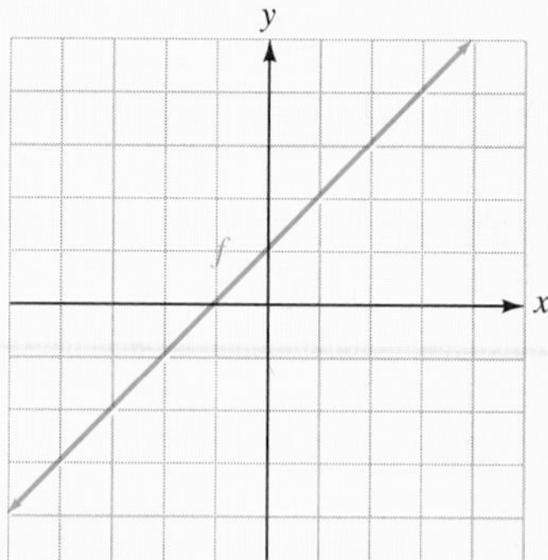

Function Notation

There is a special notation used with functions. The notation $y = f(x)$ denotes that the variable y is a function of x.

7. Write the equation in Problem 3 using function notation. Find $f(0)$.

8. Use function notation to represent the relationship described in Problem 4.

9. If the function $h(t) = -16t^2 + 64t$ gives the height of the projectile described in Problem 5, find $h(4)$ and interpret the result.

10. Use the graph in Problem 6 to determine $f(-3)$ and $f(3)$.

Accent on Teamwork

Section 2.1

CITY MAP Get a map of the city in which you live and use a black marker to draw a rectangular coordinate system on the map. The size of the grid you use will depend on the size of the map. The origin of the coordinate system should be your place of residence. Determine the coordinates of ten important locations in your city, such as the library, fire stations, parks, schools, and city hall.

Section 2.2

GRAPHING LINES Draw rectangular coordinate systems with 1-inch grids on two pieces of white poster board. On the first poster board, graph $y = mx + 2$ for $m = 1, 2, 3, 4$, and 5. Label each line with its respective equation. At the bottom of the poster board, write your observations about how the graph changes for different values of m. On the second poster board, graph $y = 2x + b$ for $b = 1, 2, 3, 4$, and 5. At the bottom of the poster board, write your observations about how the graph changes for different values of b.

Section 2.3

MEASURING SLOPE Use a ruler and a level to find the slopes of five ramps or inclines by measuring $\frac{\text{rise}}{\text{run}}$, as shown in Illustration 1. Record your results in a chart of the form shown below. List the examples in order, from that with the smallest slope to that with the largest slope.

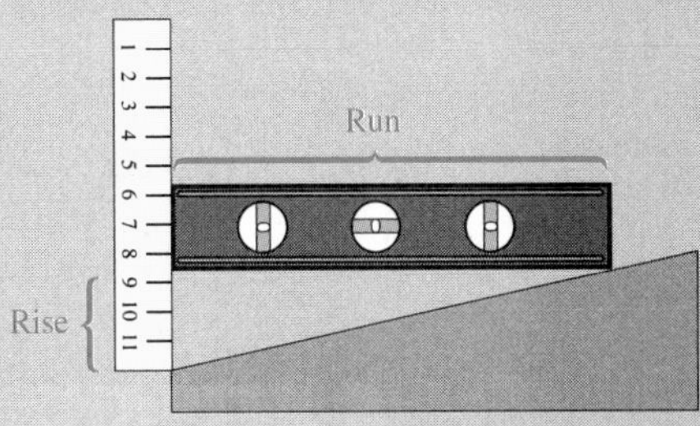

ILLUSTRATION 1

Object/location	Slope
Ramp outside cafeteria	$\frac{\text{Rise}}{\text{Run}} = \frac{4 \text{ in.}}{24 \text{ in.}} = \frac{1}{6}$

Section 2.4

A LINEAR MODEL For this project, you will need a five-gallon pail, a yardstick, a watch that shows seconds, and access to a garden hose.

Tape the yardstick to the pail as shown in Illustration 2. Turn on the water, leaving it running at a constant rate, and begin to fill the pail. Keep track of the time in seconds it takes for the water in the pail to reach heights of 2, 4, 6, 8, 10, and 12 inches.

Draw a coordinate system, with the horizontal axis representing time in seconds and the vertical axis the height of the water in inches. Then plot your data as ordered pairs of the form (time, height). Draw the line that best fits the data and write the equation of the line. Use the equation to predict the height the column of water would be if it were to run for 24 hours, which is 86,400 seconds.

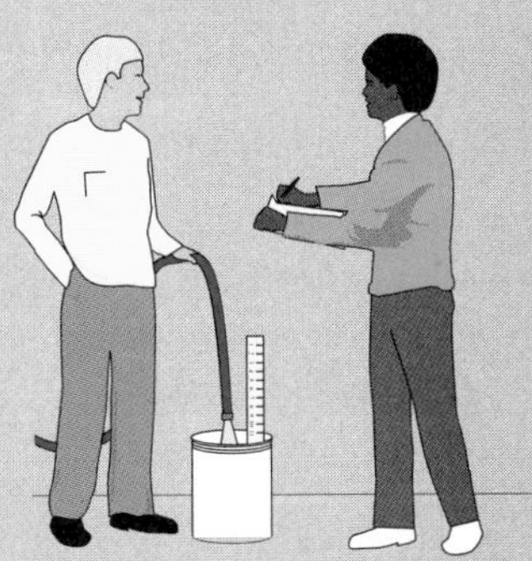

ILLUSTRATION 2

Section 2.5

LINEAR FUNCTIONS Think of a real-life situation like that given in Exercise 75 of Study Set 2.5 and write a linear function that describes it. Graph the function and explain the importance of the slope of the line and its y-intercept.

Section 2.6

GRAPHING FUNCTIONS Draw a rectangular coordinate system using a 1-inch grid on a piece of white poster board. Graph $y = k|x|$ for $k = \frac{1}{2}$, 1, 2, and 3. Label each graph with its respective equation. At the bottom of the poster board, write your observations about how the graph changes for different values of k.

PARABOLAS Borrow some calculus books from your mathematics instructor, or see if you can find several in your school library. Look up the word *parabola* in the index and find some real-life applications of parabolas mentioned in these books. Write a brief description of each use, and include a sketch of the application.

CHAPTER REVIEW

SECTION 2.1 The Rectangular Coordinate System

CONCEPT

A *rectangular coordinate system* is formed by two intersecting perpendicular number lines called the *x-axis* and the *y-axis,* which divide the plane into four *quadrants.*

The process of locating a point in the coordinate plane is called *plotting* or *graphing* that point.

REVIEW EXERCISES

1. Plot each point on the rectangular coordinate system in Illustration 1.

a. (0, 3)
b. (−2, −4)
c. $\left(\frac{5}{2}, -1.75\right)$
d. the origin
e. (2.5, 0)

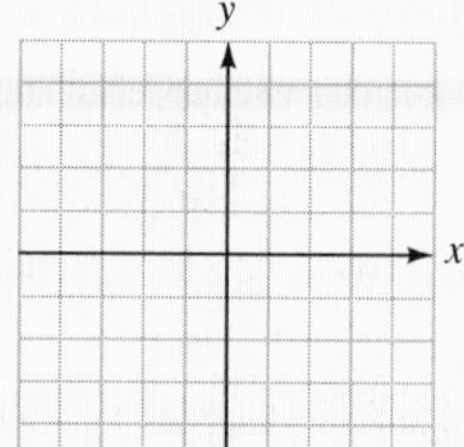

ILLUSTRATION 1

Data in a *table of values* can be expressed as ordered pairs.

2. Use the graph in Illustration 2 to complete the table.

x	y	(x, y)
−3		(−3,)
−2		(−2,)
	3	(, 3)
0		(0,)
1		(1,)
	0	(, 0)
	−5	(, −5)

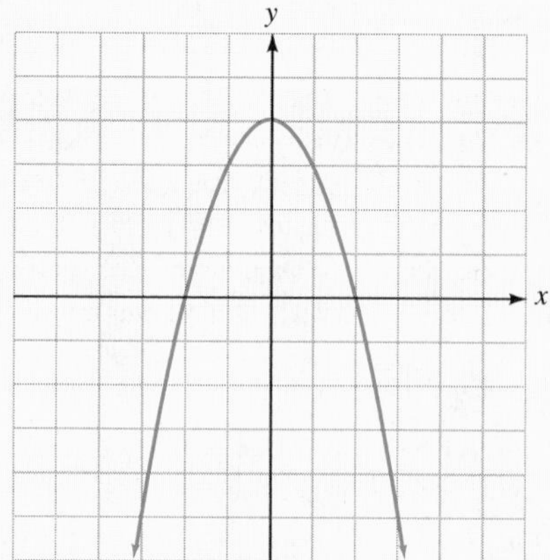

ILLUSTRATION 2

Graphs can be used to visualize relationships between two quantities.

3. The graph in Illustration 3 shows how the height of the water in a flood control channel changed over a seven-day period.

a. Describe the height of the water at the beginning of day 2.

b. By how much did the water level increase or decrease from day 4 to day 5?

c. During what time period did the water level stay the same?

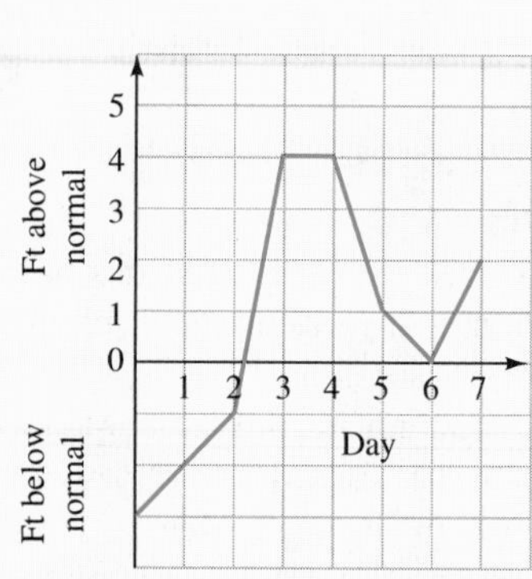

ILLUSTRATION 3

4. AUCTIONS The dollar increments used by an auctioneer during the bidding process depend on what initial price the auctioneer began with for the item. See the step graph in Illustration 4.

a. What increments are used by the auctioneer if the bidding on an item began at \$150?

b. If the first bid on an item being auctioned is \$750, what will be the next price asked for by the auctioneer?

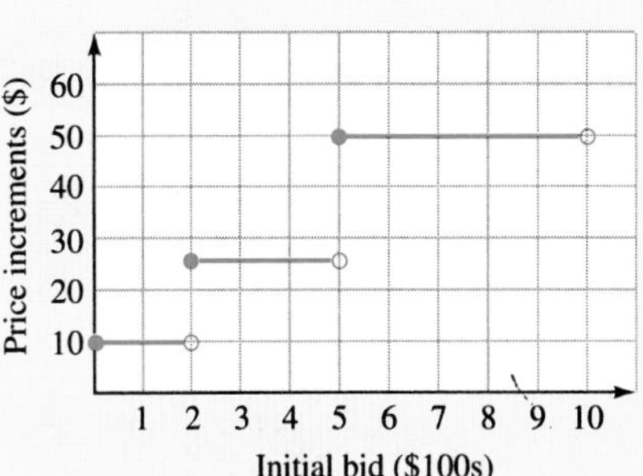

ILLUSTRATION 4

SECTION 2.2 Graphing Linear Equations

The *graph of an equation* is the graph of all points on the rectangular coordinate system whose coordinates satisfy the equation.

5. Graph each equation.

a. $y = 3x + 4$

b. $y = -\frac{1}{3}x - 1$

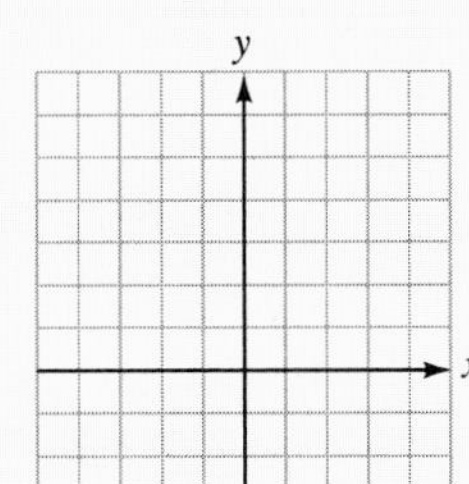

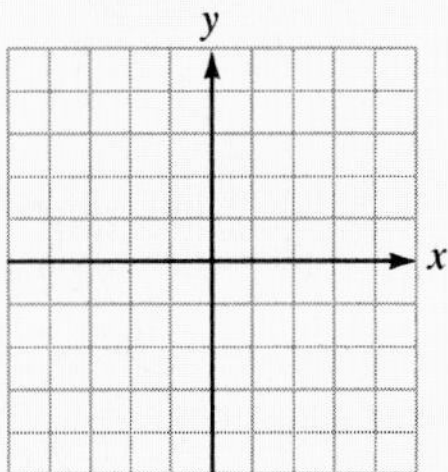

To find the *y-intercept* of a line, substitute 0 for x in the equation and solve for y. To find the x-intercept of a line, substitute 0 for y in the equation and solve for x.

6. Graph each equation using the intercept method.

a. $2x + y = 4$

b. $3x - 4y - 8 = 0$

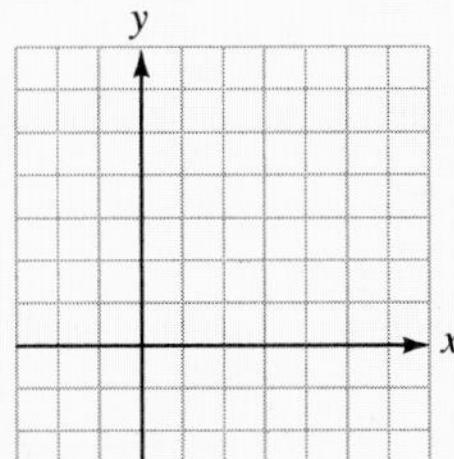

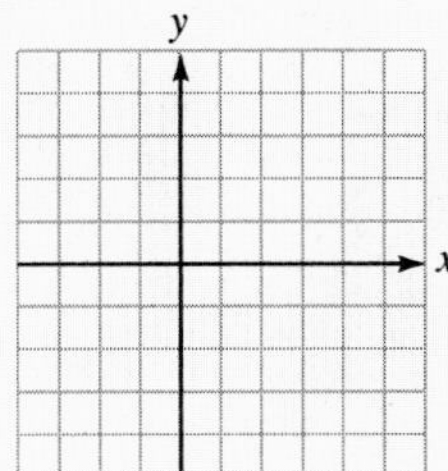

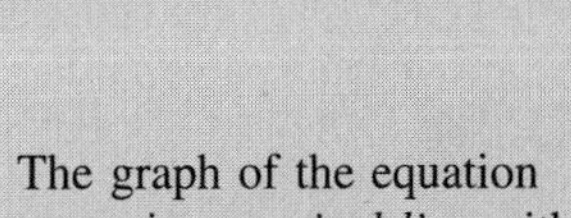

The graph of the equation $x = a$ is a *vertical line* with x-intercept at $(a, 0)$.

The graph of the equation $y = b$ is a *horizontal line* with y-intercept at $(0, b)$.

7. Graph each equation.

a. $y = 4$

b. $x = -2$

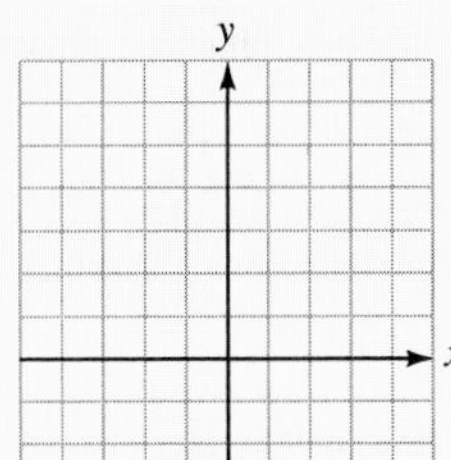

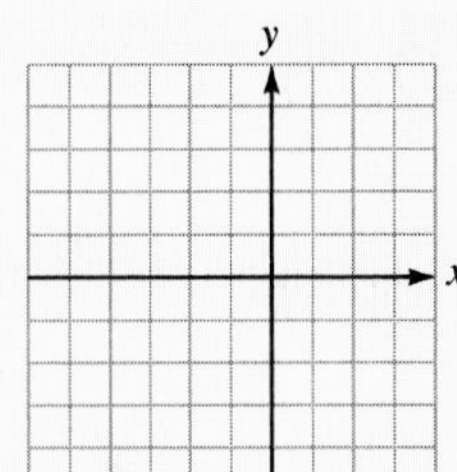

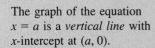

8. Complete the table of solutions for each equation.

a. $y = -3x$

x	y
-3	
0	
3	

b. $y = \frac{1}{2}x - \frac{5}{2}$

x	y
-3	
0	
3	

SECTION 2.3 *Rate of Change and the Slope of a Line*

The *slope* of a nonvertical line is defined to be

$$m = \frac{\text{rise}}{\text{run}} = \frac{\Delta y}{\Delta x}$$

The *slope formula:* If ($x_2 \neq x_1$),

$$m = \frac{y_2 - y_1}{x_2 - x_1}$$

9. Find the slope of lines l_1 and l_2 in Illustration 5.

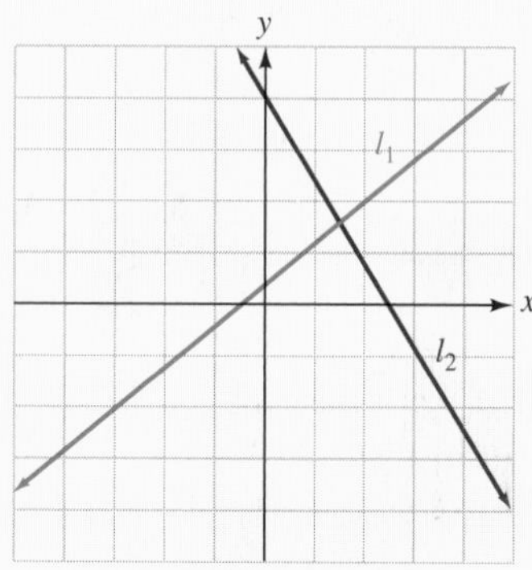

ILLUSTRATION 5

The slope of a line gives the *average rate of change.*

10. U.S. VEHICLE SALES On the graph in Illustration 6, draw a line through the points (0, 21.2) and (16, 43.2). Use this linear model to estimate the rate of increase in the market share of minivans, sport utility vehicles, and light trucks over the years 1980–1996.

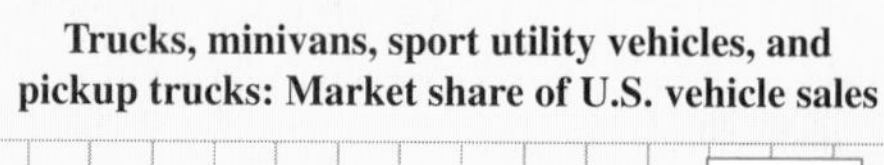

Trucks, minivans, sport utility vehicles, and
pickup trucks: Market share of U.S. vehicle sales

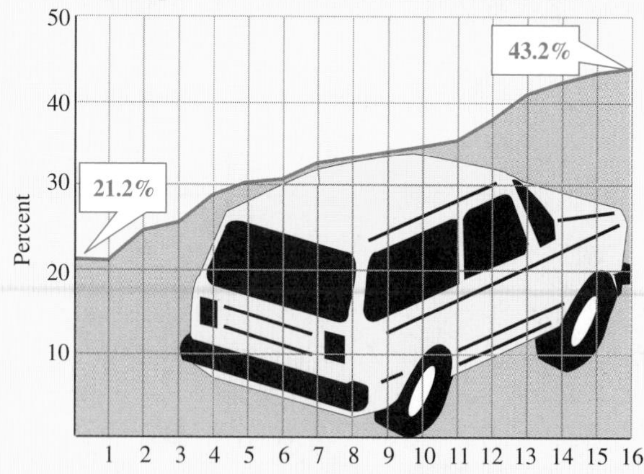

Based on data from the American Automotive Association

ILLUSTRATION 6

Horizontal lines have a slope of 0. Vertical lines have no defined slope.

11. Find the slope of the line passing through points P and Q.

a. $P(2, 5)$ and $Q(5, 8)$
b. $P(3, -2)$ and $Q(-6, 12)$
c. $P(-2, 4)$ and $Q(8, 4)$
d. $P(-5, -4)$ and $Q(-5, 8)$

12. Find the slope of the graph of each equation, if one exists.

a. $y = \frac{2}{3}x + 18$

b. $4x + 2y = 8$

c. $x = 10$

d. $y = 7$

Parallel lines have the same slope. The slopes of two nonvertical *perpendicular lines* are negative reciprocals.

13. Tell whether the lines with the given slopes are parallel, perpendicular, or neither.

a. $m_1 = 4, m_2 = -\frac{1}{4}$

b. $m_1 = 0.5, m_2 = \frac{1}{2}$

SECTION 2.4 Writing Equations of Lines

Equations of a line:

Point–slope form:

$y - y_1 = m(x - x_1)$

Slope–intercept form:

$y = mx + b$

General form:

$Ax + By = C$

14. Write the equation of the line with the given properties. Express the result in slope–intercept form.

a. Slope of 3; passing through $P(-8, 5)$

b. Passing through $(-2, 4)$ and $(6, -9)$

15. Write the equation of the line with the given properties. Write the equation in general form.

a. Passing through $(-3, -5)$; parallel to the graph of $3x - 2y = 7$

b. Passing through $(-3, -5)$; perpendicular to the graph of $3x - 2y = 7$

16. Write $3x + 4y = -12$ in slope–intercept form. Give the slope and y-intercept of the graph of the equation. Then use this information to graph the line.

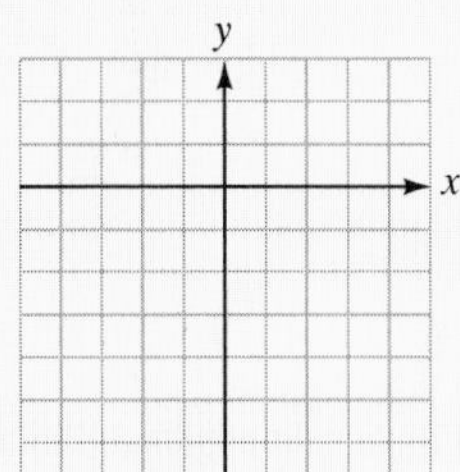

Many real-life situations can be modeled by linear equations.

17. DEPRECIATION A business purchased a copy machine for \$8,700 and will depreciate it on a straight-line basis over the next 5 years. At the end of its useful life, it will be sold as scrap for \$100. Find its depreciation equation.

SECTION 2.5 Introduction to Functions

A *function* is a correspondence between a set of input values x and a set of output values y, where exactly one value of y in the *range* corresponds to each number x in the *domain.*

18. Tell whether each equation determines y to be a function of x.

a. $y = 6x - 4$

b. $y = 4 - x^2$

c. $y^2 = x$

d. $|y| = x$

The notation $y = f(x)$ denotes that the variable y (the *dependent* variable) is a function of x (the *independent* variable).

19. Assume that $f(x) = 3x + 2$ and $g(x) = \frac{x^2 - 4x + 4}{2}$ and find each value.

a. $f(-3)$

b. $g(8)$

c. $g(-2)$

d. $f(t)$

The *domain* of a function is the set of input values. The *range* is the set of output values.

20. Find the domain and range of each function.

a. $f(x) = 4x - 1$

b. $f(x) = x^2 + 1$

c. $f(x) = \dfrac{4}{2 - x}$

d. $y = -|4x|$

The *vertical line test* can be used to determine whether a graph represents a function.

21. Use the vertical line test to determine whether each graph represents a function.

a.

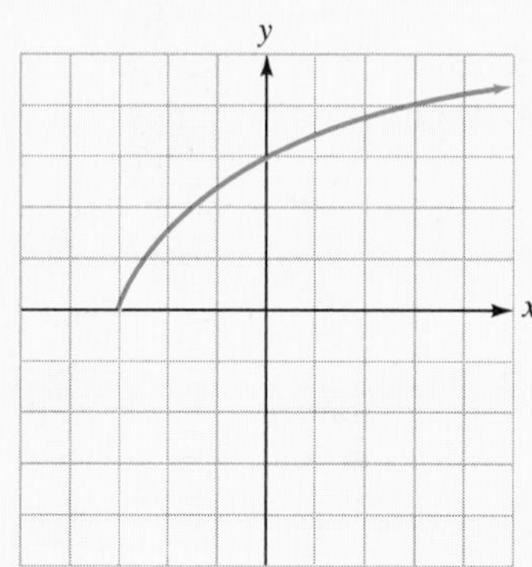

b.

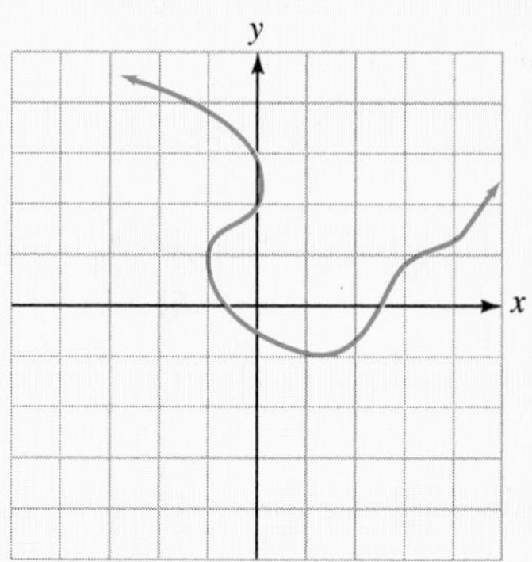

22. MARKET SHARE In Exercise 10 on page 166, a line was drawn on the graph to estimate the rate of increase in the market share of minivans, sport utility vehicles, and light trucks. Use this information to write an equation of the line. Express your result using function notation. Then use the function to predict the market share for the year 2000 if the trend continues.

23. Tell which are linear functions.

a. $f(x) = 3x + 2$

b. $y = x^2 - 25$

SECTION 2.6 Graphs of Functions

Graphs of *nonlinear functions* are not lines.

The *squaring function:*

$f(x) = x^2$

The *cubing function:*

$f(x) = x^3$

The *absolute value function:*

$f(x) = |x|$

24. Graph each function.

a. $f(x) = x^2 - 3$

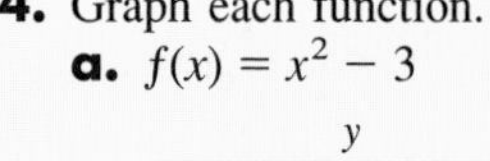

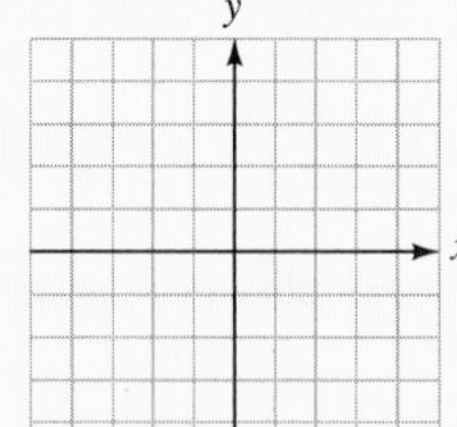

b. $f(x) = (x - 2)^3 + 1$

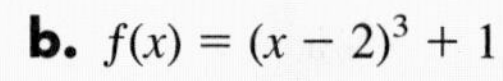

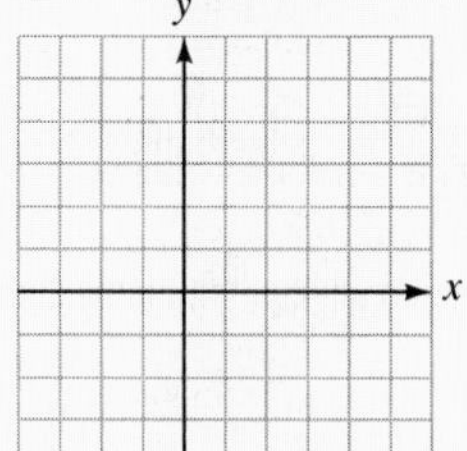

A *horizontal translation* shifts a graph left or right. A *vertical translation* shifts a graph upward or downward. A *reflection* "flips" a graph in the x-axis.

25. Graph $f(x) = |x + 2|$, $g(x) = |x - 2|$, and $h(x) = -|x|$ on the same coordinate system.

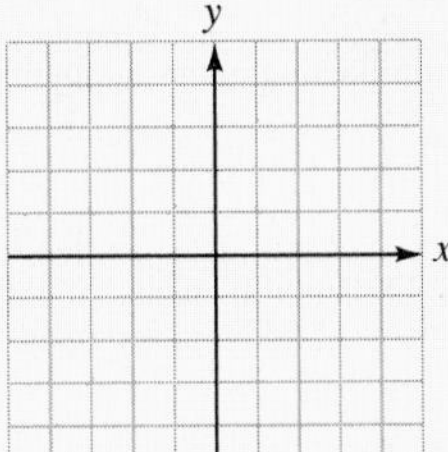

26. Use the graph in Illustration 7 to find each function value.
a. $f(-2)$ **b.** $f(3)$

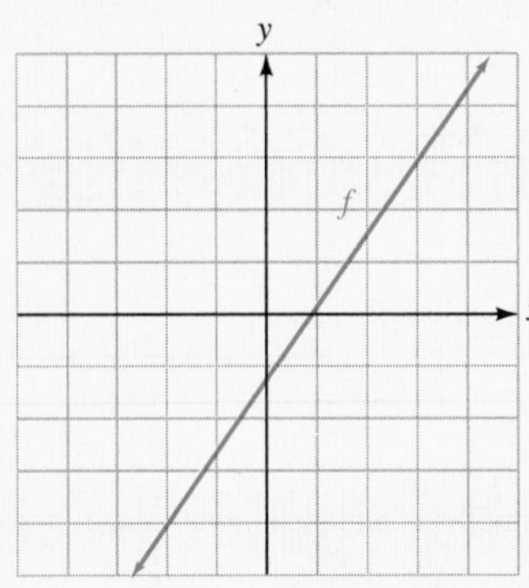

ILLUSTRATION 7

Chapter 2 Test

The graph shown in Illustration 1 shows the height of an object at different times after it was fired straight up into the air.

1. How high was the object 3 seconds into the flight?

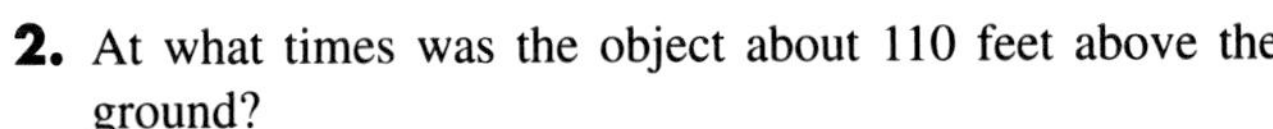

2. At what times was the object about 110 feet above the ground?

3. What was the maximum height reached by the object?

4. How long did the flight take?

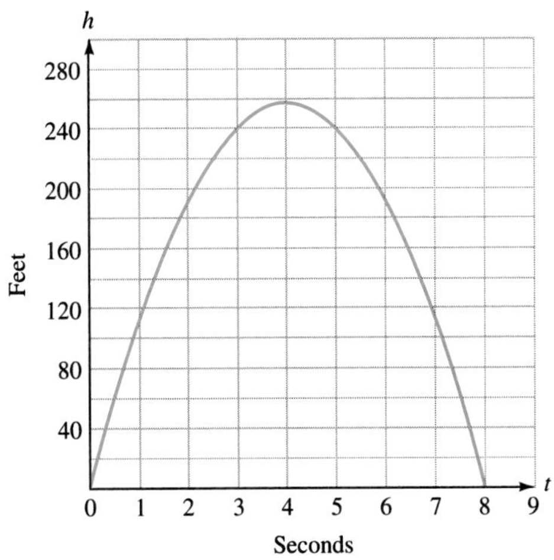

ILLUSTRATION 1

5. Find the x- and y-intercepts of the graph of $2x - 5y = 10$, then graph the equation.

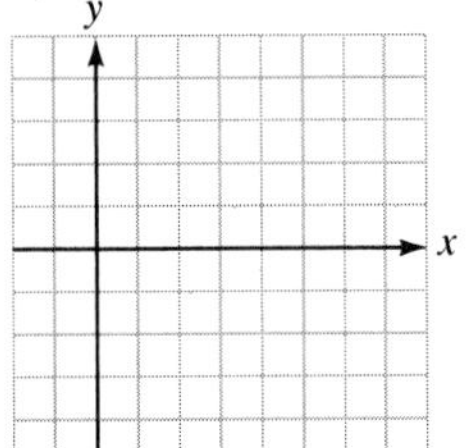

6. Graph the equation $y = -2$.

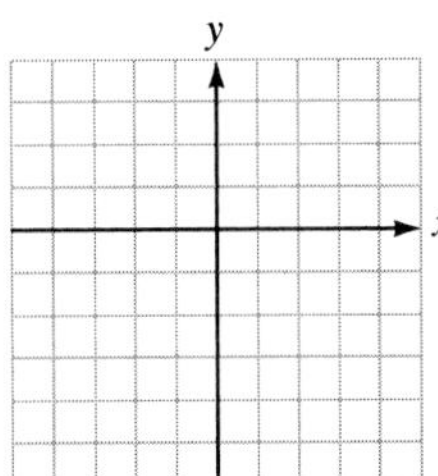

7. Find the slope of the line shown in Illustration 2.

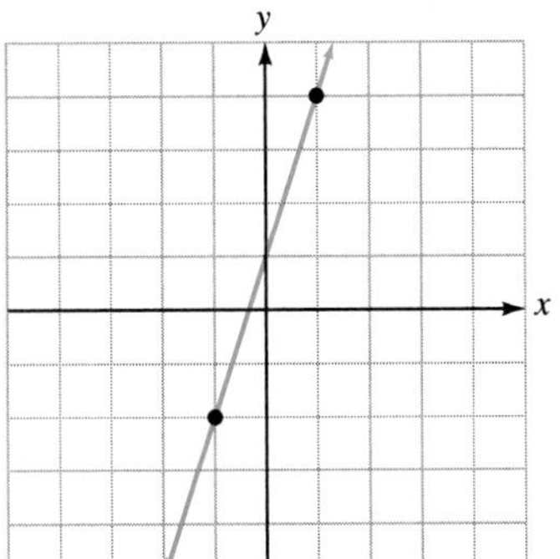

ILLUSTRATION 2

8. Find the rate of change of the temperature for the period of time shown in the graph in Illustration 3.

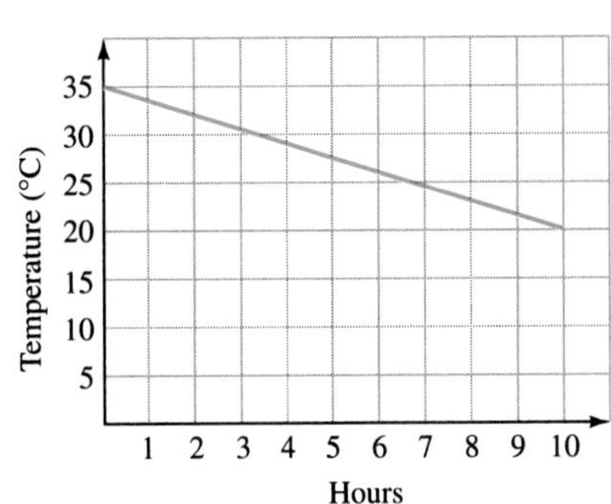

ILLUSTRATION 3

In Problems 9–12, find the slope of each line, if possible.

9. The line through $P(-2, 4)$ and $Q(6, 8)$

10. The graph of $2x - 3y = 8$

11. The graph of $x = 12$

12. The graph of $y = 12$

13. Write the equation of the line that has slope of $\frac{2}{3}$ and passes through $P(4, -5)$. Give the answer in slope–intercept form.

14. Write the equation of the line that passes through $P(-2, 6)$ and $Q(-4, -10)$. Give the answer in general form.

15. Find the slope and the y-intercept of the graph of $-2x + 6 = 6y + 15$.

16. Determine whether the graphs of $4x - y = 12$ and $y = \frac{1}{4}x + 3$ are parallel, perpendicular, or neither.

17. Write the equation of the line that passes through the origin and is parallel to the graph of $y = \frac{3}{2}x - 7$.

18. Does $|y| = x$ define y to be a function of x?

19. Find the domain and range of the function $f(x) = |x|$.

20. Find the domain and range of the function $f(x) = x^3$.

In Problems 21–24, $f(x) = 3x + 1$ and $g(x) = x^2 - 2x - 1$. Find each value.

21. $f(3)$

22. $g(0)$

23. $f\left(\frac{2}{3}\right)$

24. $g(r)$

In Problems 25–26, tell whether each graph represents a function.

25.

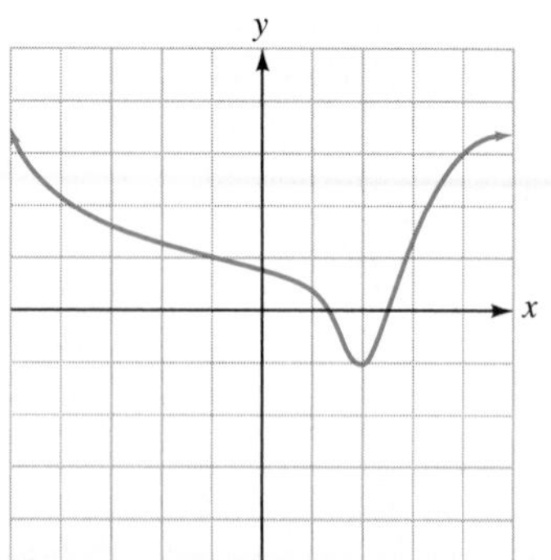

26.

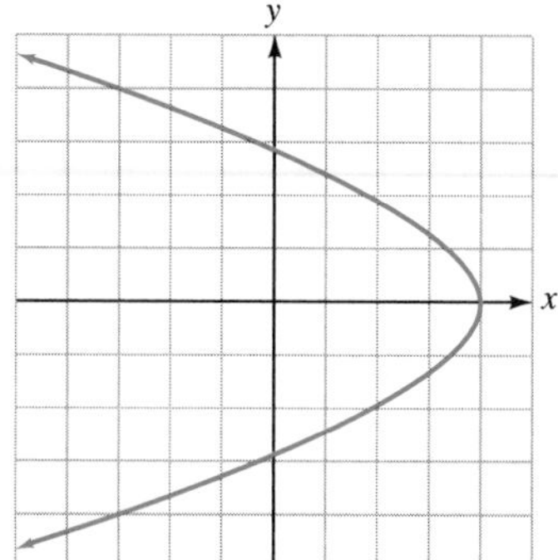

27. Graph $f(x) = x^2 + 3$.

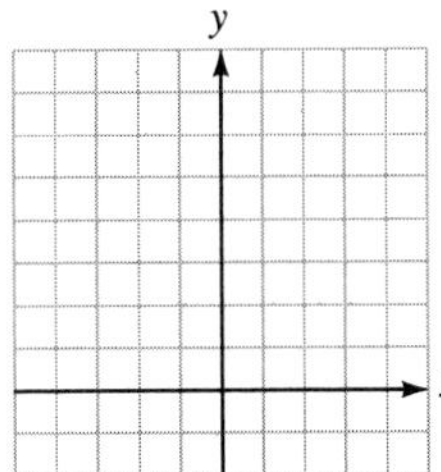

28. Graph $g(x) = -|x + 2|$.

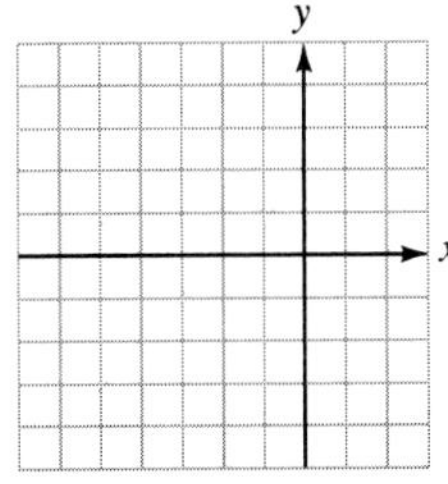

29. Describe four different ways to represent a function.

30. Explain why the graph of a circle does not represent a function.

Chapters 1–2 Cumulative Review Exercises

In Exercises 1–10, tell which numbers in the set $\left\{-2, 0, 1, 2, \frac{13}{12}, 6, 7, \sqrt{5}, \pi\right\}$ *are in each category.*

1. Natural numbers

2. Whole numbers

3. Rational numbers

4. Irrational numbers

5. Negative numbers

6. Real numbers

7. Prime numbers

8. Composite numbers

9. Even numbers

10. Odd numbers

In Exercises 11 and 12, simplify each expression.

11. $-|5| + |-3|$

12. $\dfrac{|-5| + |-3|}{-|4|}$

In Exercises 13–16, evaluate each expression.

13. $2 + 4 \cdot 5$

14. $\dfrac{8 - 4}{2 - 4}$

15. $-\dfrac{16}{5} \div \left(-\dfrac{10}{3}\right)$

16. $\dfrac{(9 - 8)^4 + 21}{3^3 - \left(\sqrt{16}\right)^2}$

In Exercises 17 and 18, evaluate each expression for x = 2 and y = −3.

17. $-x - 2y$

18. $\dfrac{x^2 - y^2}{2x + y}$

In Exercises 19–22, tell which property of real numbers justifies each statement.

19. $(a + b) + c = a + (b + c)$

20. $3(x + y) = 3x + 3y$

21. $(a + b) + c = c + (a + b)$

22. $(ab)c = a(bc)$

In Exercises 23–26, simplify each expression.

23. $12y - 17y$

24. $-7s(-4t)(-1)$

25. $3x^2 + 2x^2 - 5x^2$

26. $-(4 + z) + 2z$

In Exercises 27–32, solve each equation.

27. $2x - 5 = 11$

28. $\dfrac{2x - 6}{3} = x + 7$

29. $4(y - 3) + 4 = -3(y + 5)$

30. $2x - \frac{3(x - 2)}{2} = 7 - \frac{x - 3}{3}$

31. $-3 = -\frac{9}{8}s$

32. $0.04(24) + 0.02x = 0.04(12 + x)$

In Exercises 33–34, solve each formula.

33. $S = \frac{n(a + l)}{2}$ for a

34. $A = \frac{1}{2}h(b_1 + b_2)$ for h

35. INVESTMENTS A woman invested part of $20,000 at 6% and the rest at 7%. If her annual interest is $1,260, how much did she invest at 6%?

36. DRIVING RATES John drove to a distant city in 5 hours. When he returned, there was less traffic, and the trip took only 3 hours. If he drove 26 mph faster on the return trip, how fast did he drive each way?

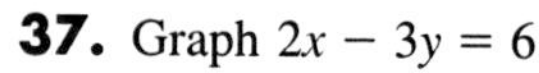

37. Graph $2x - 3y = 6$.

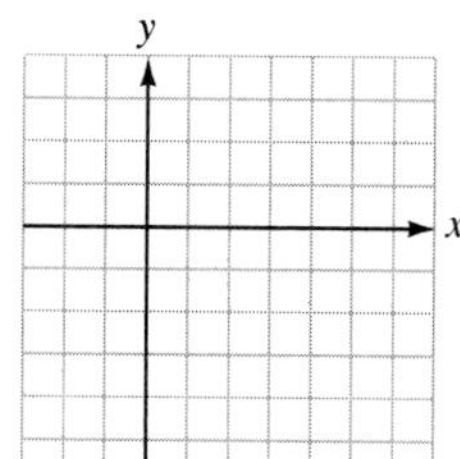

38. Find the slope of the line shown in Illustration 1.

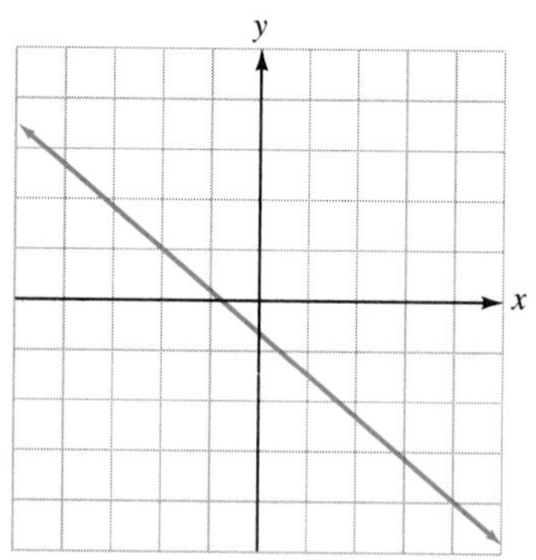

ILLUSTRATION 1

39. Write the equation of the line passing through $P(-2, 5)$ and $Q(8, -9)$.

40. Write the equation of the line passing through $P(-2, 3)$ and parallel to the graph of $3x + y = 8$.

In Exercises 41–44, $f(x) = 3x^2 + 2$ and $g(x) = -2x - 1$. Find each function value.

41. $f(-1)$

42. $g(0)$

43. $g(-2)$

44. $f(-r)$

In Exercises 45–46, graph each equation and tell whether it is a function. If it is a function, give the domain and range.

45. $y = -x^2 + 1$

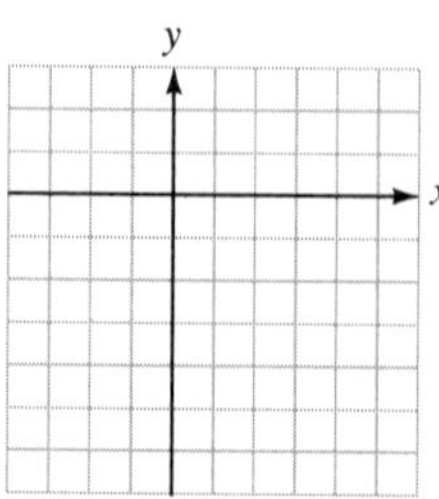

46. $y = |x - 3|$

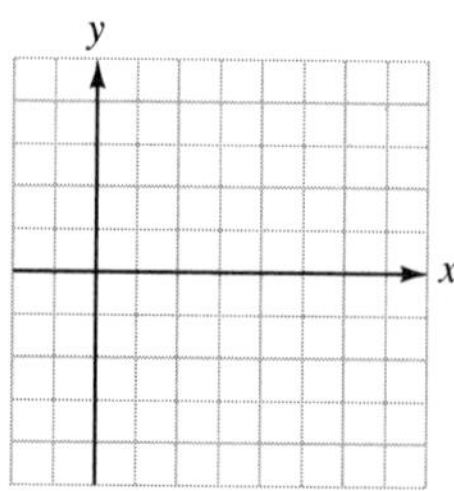

Systems of Equations

3

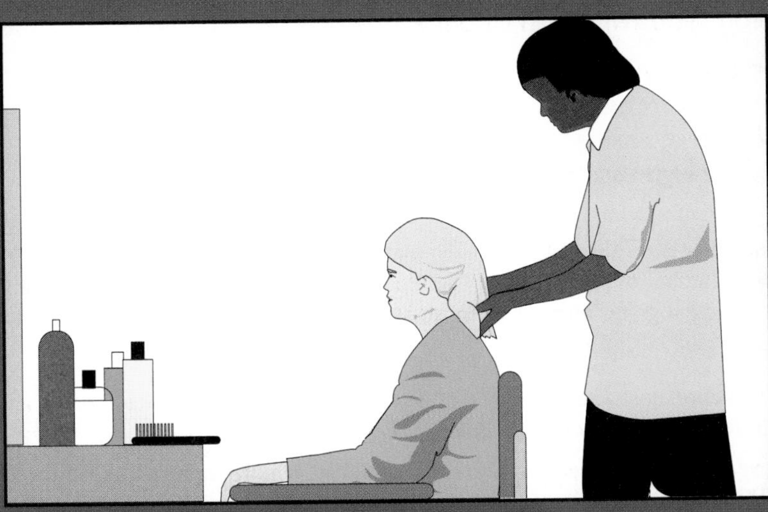

Campus Connection

The *Cosmetology* Department

Cosmetology students learn the techniques of hair cutting, facials, manicuring, and permanents. They also study some of the business practices used in the operation of a beauty shop. In Chapter 3, we will discuss a mathematical procedure that a beauty shop owner can use to determine how much business he or she must have to make a profit. To answer a question such as this, we must write two equations, each containing two variables. We call the pair of equations a *system of linear equations*. After learning how to solve systems of linear equations, you will be able to solve many problems that could not be solved using just one variable.

To solve many problems, we must use two and sometimes three variables. This requires that we solve a system of equations.

3.1 Solving Systems by Graphing

In this section, you will learn about

- **The graphing method**
- **Consistent systems**
- **Inconsistent systems**
- **Dependent equations**

Introduction. The red line graphed in Figure 3-1 shows the cost for a company to produce a given number of skateboards. The blue line shows the revenue the company will earn for selling a given number of those skateboards. The graph offers the company important financial information.

- The production costs exceed the revenue earned if less than 400 skateboards are sold. In this case, the company loses money.
- The revenue earned exceeds the production costs if more than 400 skateboards are sold. In this case, the company makes a profit.
- Production costs equal revenue earned if exactly 400 skateboards are sold. This fact is indicated by the point of intersection of the two lines, (400, 20,000), which is called the **break-even point**.

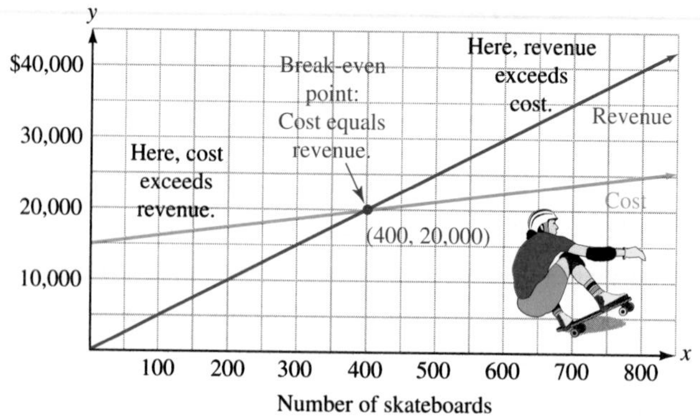

FIGURE 3-1

This example shows that important information can be learned by finding the point of intersection of two lines. In this section, we will discuss how to use the *graphing method* to do this.

The graphing method

In Chapter 2, we discussed linear equations containing the variables x and y. We found that each equation had infinitely many solutions (x, y), and that we could graph each equation on the rectangular coordinate system. In this chapter, we will discuss **systems of linear equations** involving two or three equations.

In the pair of equations

$$\begin{cases} x + 2y = 4 \\ 2x - y = 3 \end{cases} \quad \text{(called a system of two linear equations)}$$

there are infinitely many pairs (x, y) that satisfy the first equation and infinitely many pairs (x, y) that satisfy the second equation. However, there is only one pair (x, y) that satisfies both equations at the same time. The process of finding this pair is called *solving the system.*

To solve a system of two equations in two variables by graphing, we use the following steps.

The graphing method

1. On a single set of coordinate axes, graph each equation.
2. Find the coordinates of the point (or points) where the graphs intersect. These coordinates give the solution of the system.
3. If the graphs have no point in common, the system has no solution.
4. Check the solution in both of the original equations.

Consistent systems

When a system of equations (as in Example 1) has a solution, the system is called a **consistent system.**

EXAMPLE 1 ***The graphing method.*** Solve the system $\begin{cases} x + 2y = 4 \\ 2x - y = 3 \end{cases}$.

Solution

We graph both equations on one set of coordinate axes, as shown in Figure 3-2.

Although infinitely many ordered pairs (x, y) satisfy $x + 2y = 4$, and infinitely many ordered pairs (x, y) satisfy $2x - y = 3$, only the coordinates of the point where the graphs intersect satisfy both equations. Since the intersection point has coordinates of $(2, 1)$, the solution is the ordered pair $(2, 1)$, or $x = 2$ and $y = 1$.

To check the solution, we substitute 2 for x and 1 for y in both equations and verify that $(2, 1)$ satisfies each one.

Self Check

Solve $\begin{cases} 2x + y = 4 \\ x - 3y = -5 \end{cases}$.

y

x

The first equation	The second equation
$x + 2y = 4$	$2x - y = 3$
$2 + 2(1) \stackrel{?}{=} 4$	$2(2) - 1 \stackrel{?}{=} 3$
$2 + 2 \stackrel{?}{=} 4$	$4 - 1 \stackrel{?}{=} 3$
$4 = 4$	$3 = 3$

$x + 2y = 4$

x	y	(x, y)
4	0	(4, 0)
0	2	(0, 2)
−2	3	(−2, 3)

$2x - y = 3$

x	y	(x, y)
$\frac{3}{2}$	0	$\left(\frac{3}{2}, 0\right)$
0	−3	(0, −3)
−1	−5	(−1, −5)

We use the intercept method to graph each line.

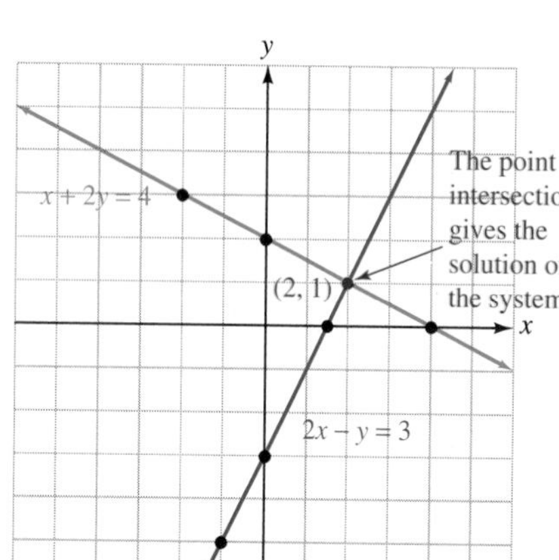

FIGURE 3-2

Answer: (1, 2)

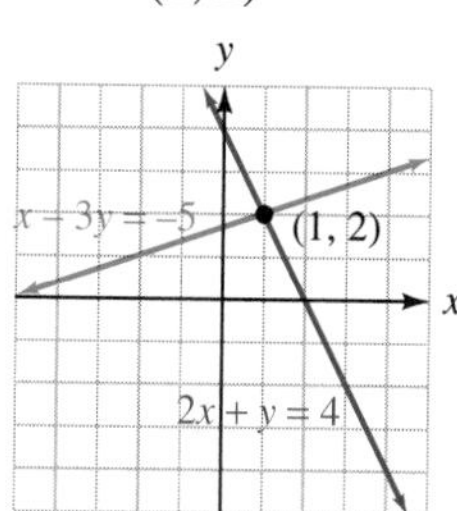

Inconsistent systems

When a system has no solution (as in Example 2), it is called an **inconsistent system.**

EXAMPLE 2 ***The graphing method.*** Solve the system $\begin{cases} 2x + 3y = 6 \\ 4x + 6y = 24 \end{cases}$ if possible.

Solution

We graph both equations on one set of coordinate axes, as shown in Figure 3-3. In this example, the graphs are parallel, because the slopes of the two lines are equal and they have different y-intercepts. We can see that the slope of each line is $-\frac{2}{3}$ by writing each equation in slope–intercept form.

$2x + 3y = 6.$	$4x + 6y = 24$
$3y = -2x + 6$	$6y = -4x + 24$
$y = -\dfrac{2}{3}x + 2$	$y = -\dfrac{2}{3}x + 4$

Since the graphs are parallel lines, the lines do not intersect, and the system does not have a solution. It is an inconsistent system.

Self Check

Solve $\begin{cases} 2x - 3y = 6 \\ y = \frac{2}{3}x + 2 \end{cases}$.

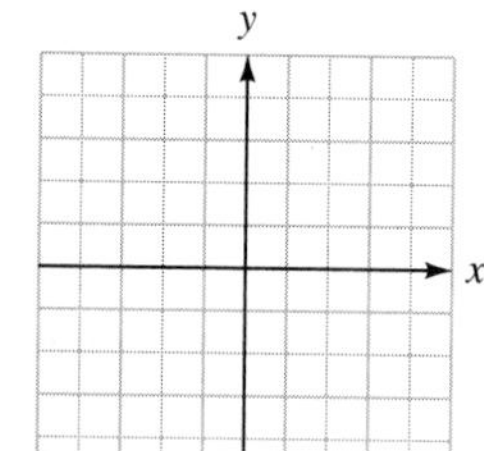

$2x + 3y = 6$		
x	y	(x, y)
3	0	(3, 0)
0	2	(0, 2)
−3	4	(−3, 4)

$4x + 6y = 24$		
x	y	(x, y)
6	0	(6, 0)
0	4	(0, 4)
−3	6	(−3, 6)

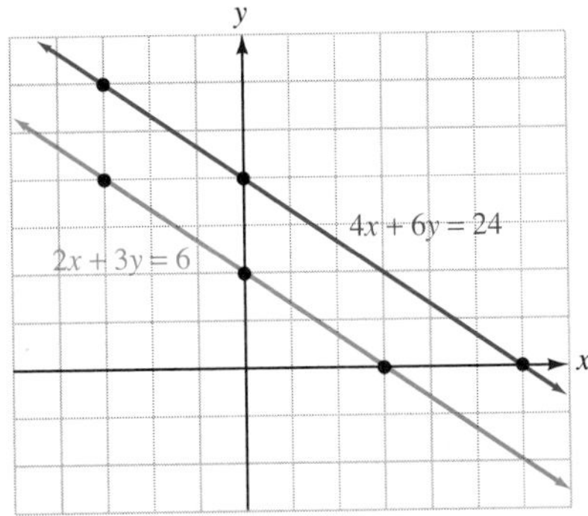

FIGURE 3-3

Answer: no solutions

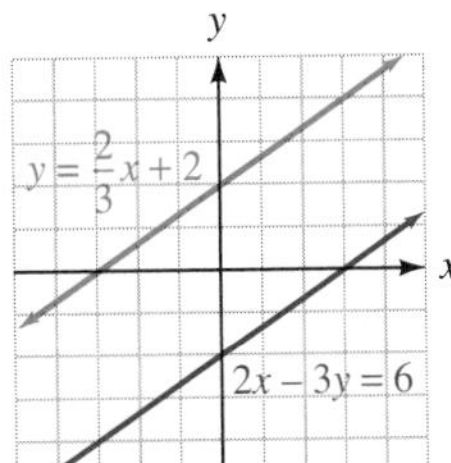

Dependent equations

When the equations of a system have different graphs (as in Examples 1 and 2), the equations are called **independent equations**. Two equations with the same graph are called **dependent equations**.

EXAMPLE 3 ***The graphing method.*** Solve the system $\begin{cases} y = \frac{1}{2}x + 2 \\ 2x + 8 = 4y \end{cases}$.

Solution

We graph each equation on one set of coordinate axes, as shown in Figure 3-4. Since the graphs coincide, the system has infinitely many solutions. Any ordered pair (x, y) that satisfies one equation also satisfies the other.

From the tables of ordered pairs shown in Figure 3-4, we see that (0, 2) and (2, 3) are solutions. We can find infinitely many more solutions by finding additional ordered pairs (x, y) that satisfy either equation.

Because the two equations have the same graph, they are dependent equations.

$y = \frac{1}{2}x + 2$		
x	y	(x, y)
−2	1	(−2, 1)
0	2	(0, 2)
2	3	(2, 3)

$2x + 8 = 4y$		
x	y	(x, y)
−4	0	(−4, 0)
0	2	(0, 2)
2	3	(2, 3)

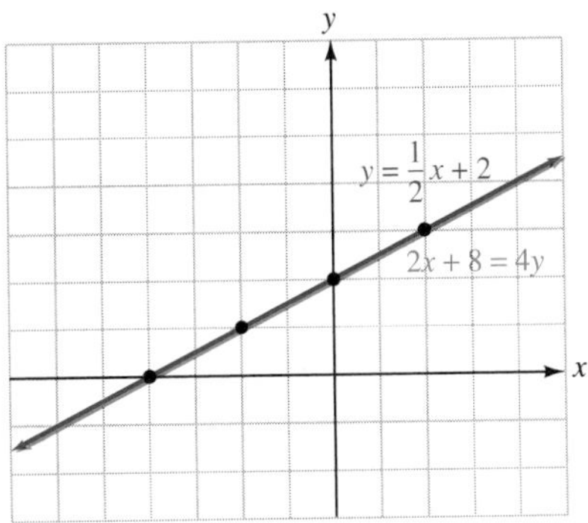

FIGURE 3-4

Self Check

Solve $\begin{cases} 2x - 4 = y \\ x = \frac{1}{2}y + 2 \end{cases}$.

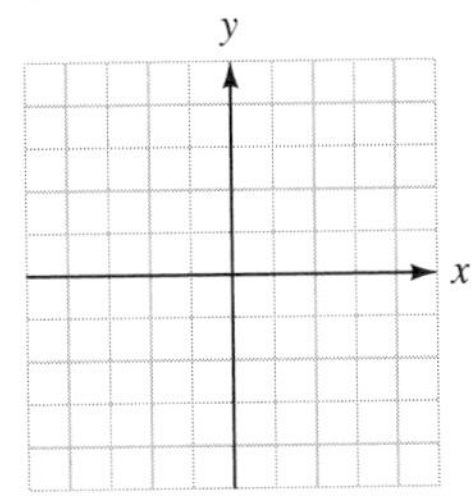

Answer: There are infinitely many solutions; three of them are (0, −4), (2, 0), and (4, 4).

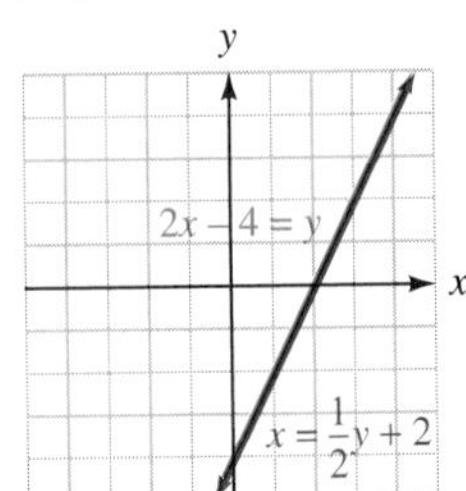

We now summarize the possibilities that can occur when two linear equations, each with two variables, are graphed.

Solving a system of equations by using the graphing method

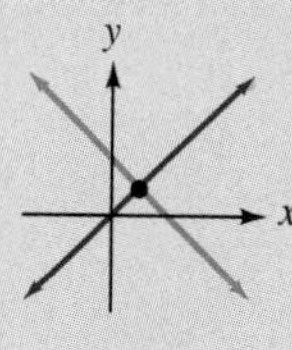

If the lines are different and intersect, the equations are independent, and the system is consistent. **One solution exists.**

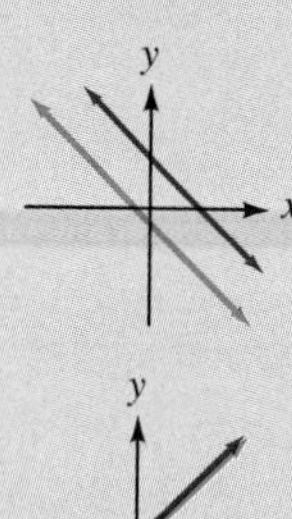

If the lines are different and parallel, the equations are independent, and the system is inconsistent. **No solution exists.**

If the lines coincide, the equations are dependent, and the system is consistent. **Infinitely many solutions exist.**

If the equations in one system are equivalent to the equations in another system, the systems are called **equivalent**.

EXAMPLE 4 ***Equivalent systems.*** Solve the system $\begin{cases} \frac{3}{2}x - y = \frac{5}{2} \\ x + \frac{1}{2}y = 4 \end{cases}$

Solution

We multiply both sides of $\frac{3}{2}x - y = \frac{5}{2}$ by 2 to eliminate the fractions and obtain the equation $3x - 2y = 5$. We multiply both sides of $x + \frac{1}{2}y = 4$ by 2 to eliminate the fractions and obtain the equation $2x + y = 8$.

The new system

$$\begin{cases} 3x - 2y = 5 \\ 2x + y = 8 \end{cases}$$

is equivalent to the original system and is easier to solve, since it has no fractions. If we graph each equation in the new system, as in Figure 3-5, we see that the coordinates of the point where the two lines intersect are (3, 2). Verify that $x = 3$ and $y = 2$ satisfy each equation in the original system.

$3x - 2y = 5$

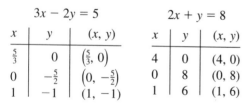

x	y	(x, y)
$\frac{5}{3}$	0	$\left(\frac{5}{3}, 0\right)$
0	$-\frac{5}{2}$	$\left(0, -\frac{5}{2}\right)$
1	-1	$(1, -1)$

$2x + y = 8$

x	y	(x, y)
4	0	(4, 0)
0	8	(0, 8)
1	6	(1, 6)

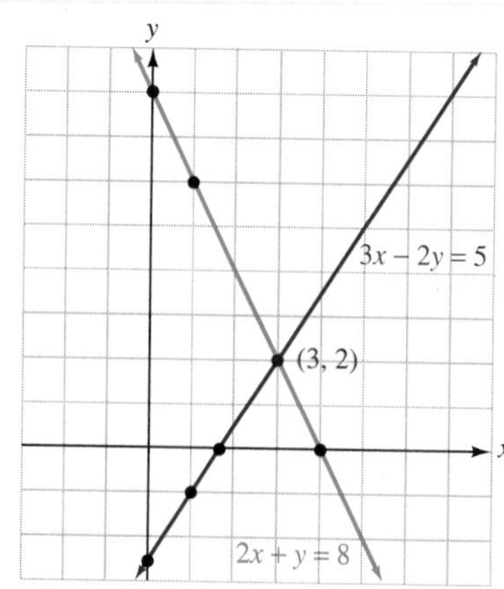

FIGURE 3-5

Self Check

Solve $\begin{cases} \frac{5}{2}x - y = 2 \\ x + \frac{1}{3}y = 3 \end{cases}$ by the graphing method.

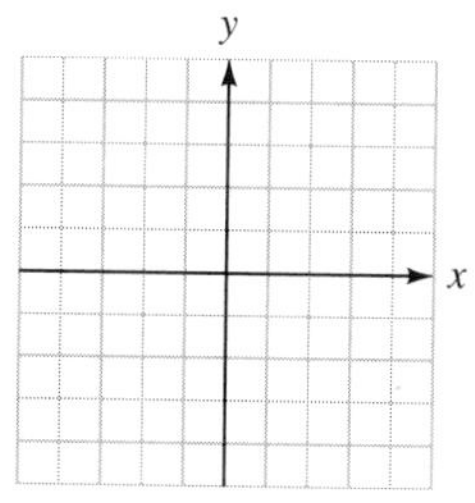

Answer: (2, 3)

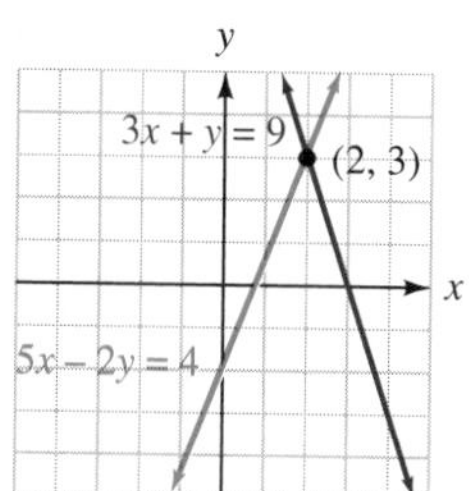

Accent on Technology **Solving Systems by Graphing**

The graphing method has limitations. First, the method is limited to equations with two variables. Systems with three or more variables cannot be solved graphically. Second, it is often difficult to find exact solutions graphically. However, the trace and zoom capabilities of graphing calculators enable us to get very good approximations of such solutions.

To solve the system $\begin{cases} 3x + 2y = 12 \\ 2x - 3y = 12 \end{cases}$

with a graphing calculator, we must first solve each equation for y so that we can enter the equations into the calculator. After solving for y, we obtain the following equivalent system:

$$\begin{cases} y = -\dfrac{3}{2}x + 6 \\ y = \dfrac{2}{3}x - 4 \end{cases}$$

If we use window settings of $[-10, 10]$ for x and $[-10, 10]$ for y, the graphs of the equations will look like those in Figure 3-6(a). If we zoom in on the intersection point of the two lines and trace, we will get an approximate solution like the one shown in Figure 3-6(b). To get better results, we can do more zooms. Verify that the exact solution is $x = \frac{60}{13}$ and $y = -\frac{12}{13}$.

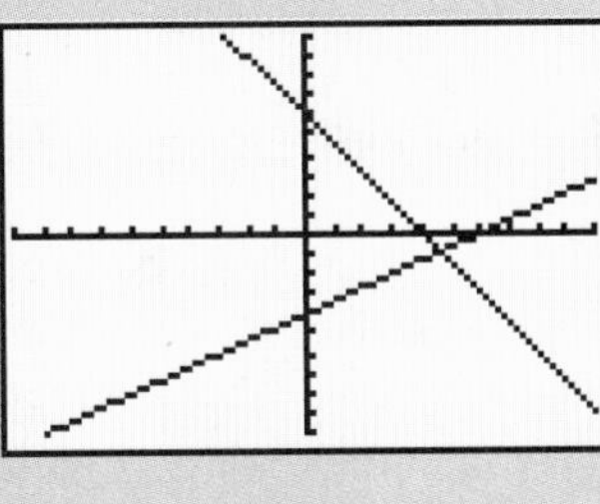

(a)

Y1=-(3/2)X+6

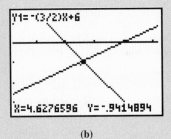

X=4.6276596 Y=-.9414894

(b)

FIGURE 3-6

STUDY SET Section 3.1

VOCABULARY *In Exercises 1–6, fill in the blanks to make the statements true.*

1. $\begin{cases} x + 2y = 4 \\ 2x - y = 3 \end{cases}$ is called a ______________ of linear equations.

2. When a system of equations has one or more solutions, it is called a ______________ system.

3. If a system has no solutions, it is called an ______________ system.

4. If two equations have different graphs, they are called ______________ equations.

5. Two equations with the same graph are called ______________ equations.

6. When solving a system of two linear equations by the graphing method, we look for the point of ______________ of the two lines.

CONCEPTS

7. Refer to Illustration 1. Tell whether a true or a false statement would be obtained when the coordinates of

a. point A are substituted into the equation for line l_1.

b. point B are substituted into the equation for line l_1.

c. point C are substituted into the equation for line l_1.

d. point C are substituted into the equation for line l_2.

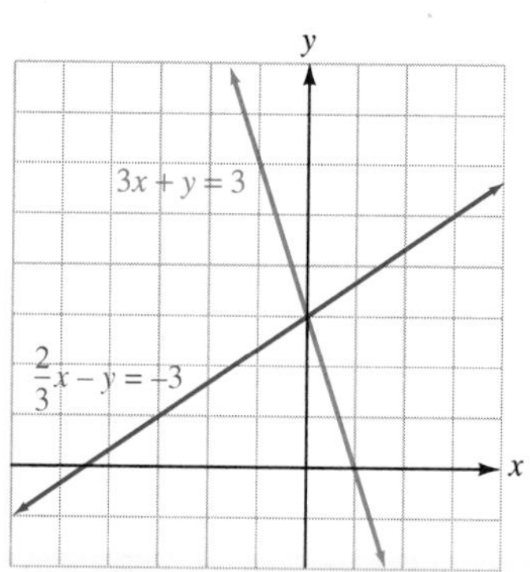

ILLUSTRATION 2

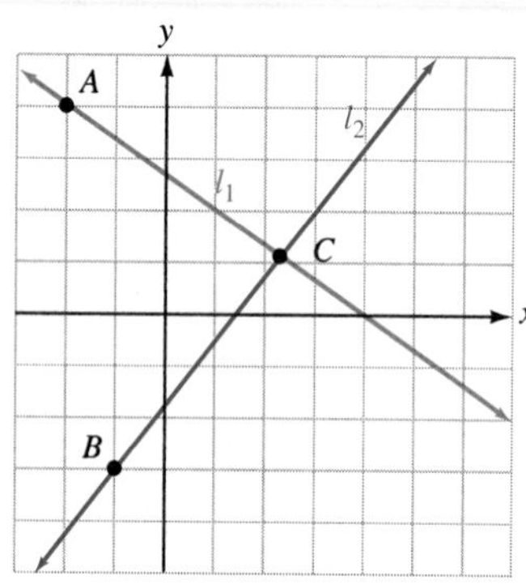

ILLUSTRATION 1

8. Refer to Illustration 2.

a. How many ordered pairs satisfy the equation $3x + y = 3$? Name three.

b. How many ordered pairs satisfy the equation $\frac{2}{3}x - y = -3$? Name three.

c. How many ordered pairs satisfy both equations at the same time? Name it or them.

9. a. The intercept method can be used to graph the equation $2x - 4y = -8$. Complete the following table.

x	y	(x, y)
	0	
0		
2		

b. What is the x-intercept of the graph of $2x - 4y = -8$? What is the y-intercept?

10. a. To graph the equation $y = 3x + 1$, we can pick three numbers for x and find the corresponding values of y. Complete the following table.

x	y	(x, y)
−1		
0		
2		

b. We can also graph the equation $y = 3x + 1$ if we know the slope and the y-intercept of the line. What are they?

PRACTICE *In Exercises 11–14, tell whether the ordered pair is a solution of the system of equations.*

11. $(1, 2)$; $\begin{cases} 2x - y = 0 \\ y = \frac{1}{2}x + \frac{3}{2} \end{cases}$

12. $(-1, 2)$; $\begin{cases} y = 3x + 5 \\ y = x + 4 \end{cases}$

13. $(2, -3)$; $\begin{cases} y + 2 = \frac{1}{2}x \\ 3x + 2y = 0 \end{cases}$

14. $(-4, 3)$; $\begin{cases} 4x - y = -19 \\ 3x + 2y = -6 \end{cases}$

In Exercises 15–34, solve each system by graphing, if possible.

15. $\begin{cases} x + y = 6 \\ x - y = 2 \end{cases}$

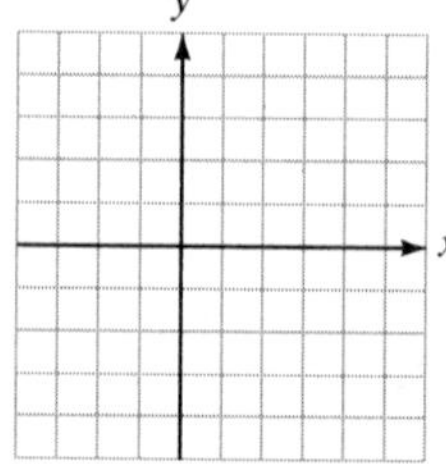

16. $\begin{cases} x - y = 4 \\ 2x + y = 5 \end{cases}$

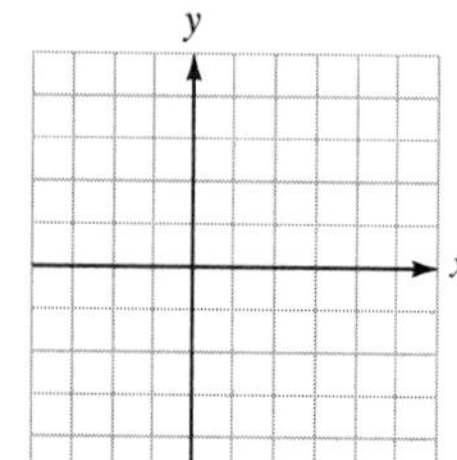

17. $\begin{cases} 2x + y = 1 \\ x - 2y = -7 \end{cases}$

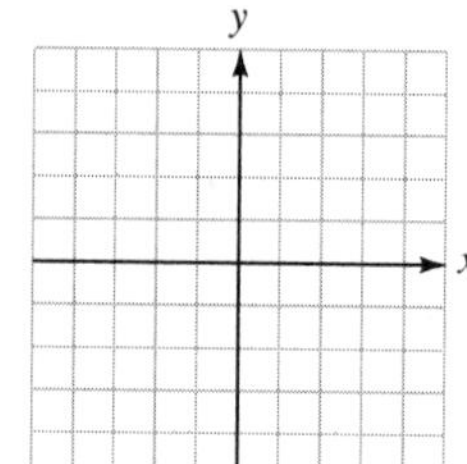

18. $\begin{cases} 3x - y = -3 \\ 2x + y = -7 \end{cases}$

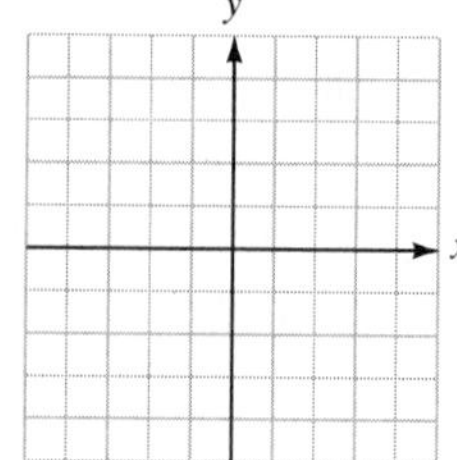

19. $\begin{cases} x = 13 - 4y \\ 3x = 4 + 2y \end{cases}$

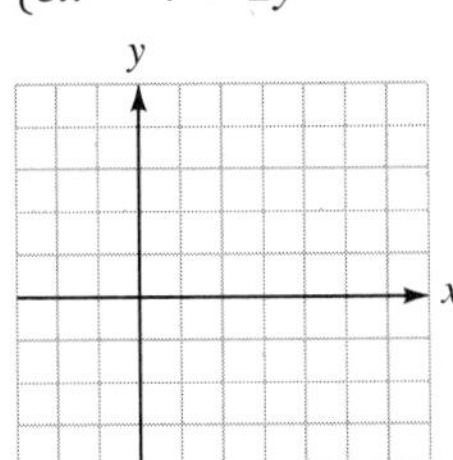

20. $\begin{cases} 3x = 7 - 2y \\ 2x = 2 + 4y \end{cases}$

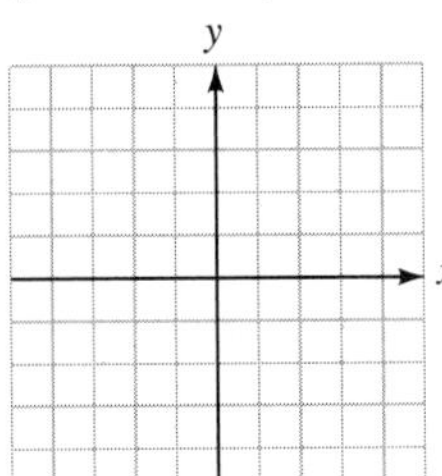

21. $\begin{cases} x = 3 - 2y \\ 2x + 4y = 6 \end{cases}$

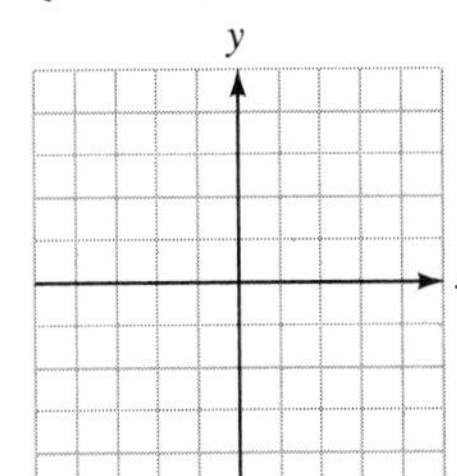

22. $\begin{cases} 3x = 5 - 2y \\ 3x + 2y = 7 \end{cases}$

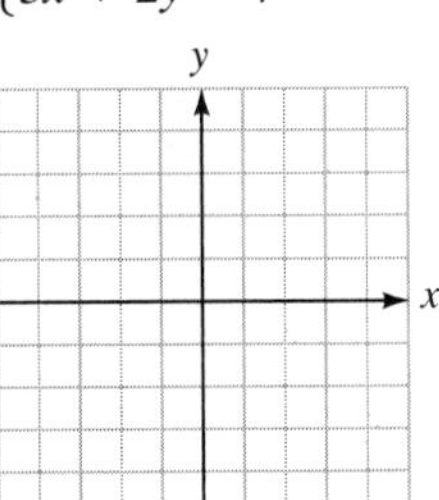

23. $\begin{cases} x = 2 \\ y = -\frac{1}{2}x + 2 \end{cases}$

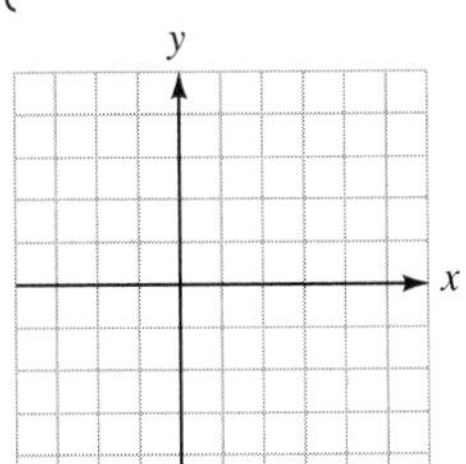

24. $\begin{cases} y = -2 \\ y = \frac{2}{3}x - \frac{4}{3} \end{cases}$

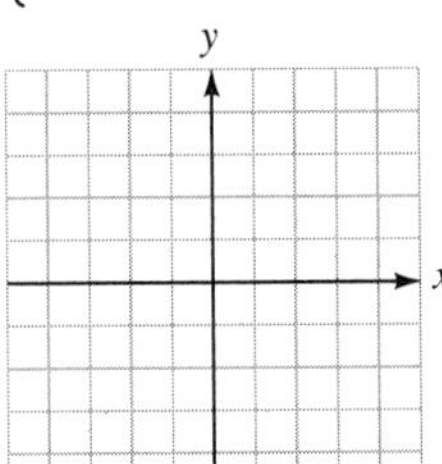

25. $\begin{cases} y = 3 \\ x = 2 \end{cases}$

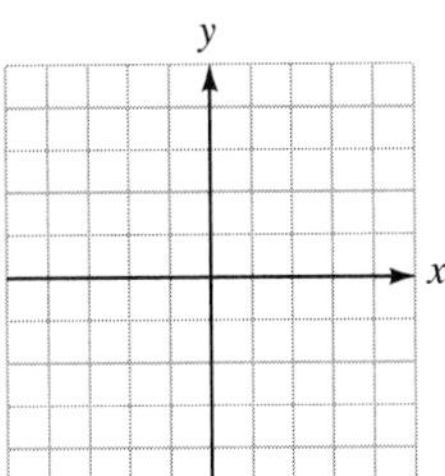

26. $\begin{cases} 2x + 3y = -15 \\ 2x + y = -9 \end{cases}$

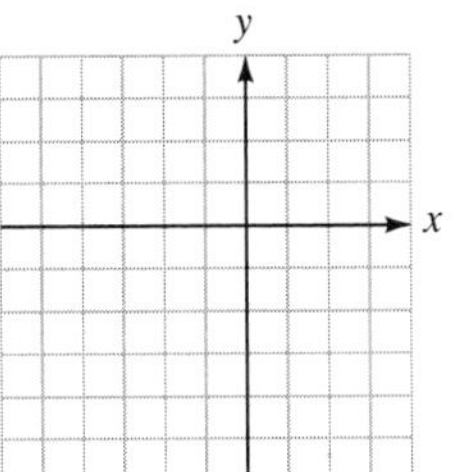

27. $\begin{cases} x = \dfrac{11 - 2y}{3} \\ y = \dfrac{11 - 6x}{4} \end{cases}$

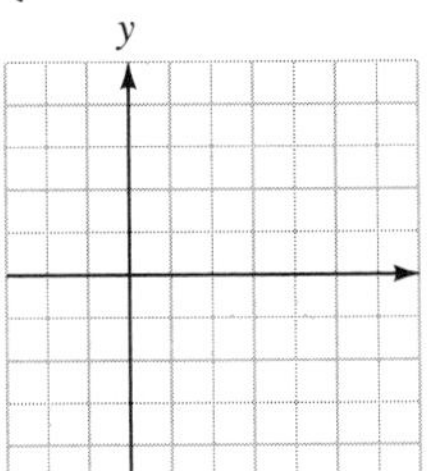

28. $\begin{cases} x = \dfrac{1 - 3y}{4} \\ y = \dfrac{12 + 3x}{2} \end{cases}$

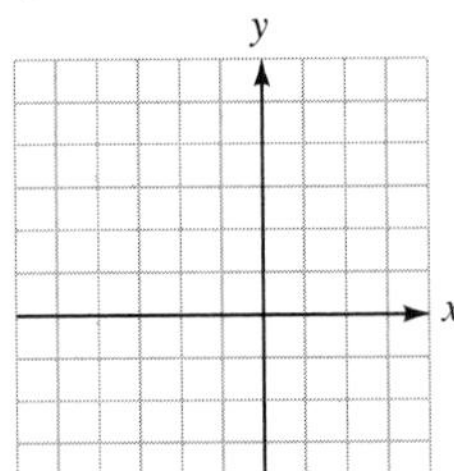

29. $\begin{cases} y = -\frac{5}{2}x + \frac{1}{2} \\ 2x - \frac{3}{2}y = 5 \end{cases}$

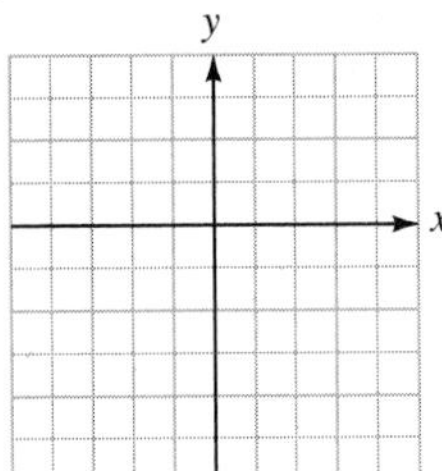

30. $\begin{cases} \frac{5}{2}x + 3y = 6 \\ y = -\frac{5}{6}x + 2 \end{cases}$

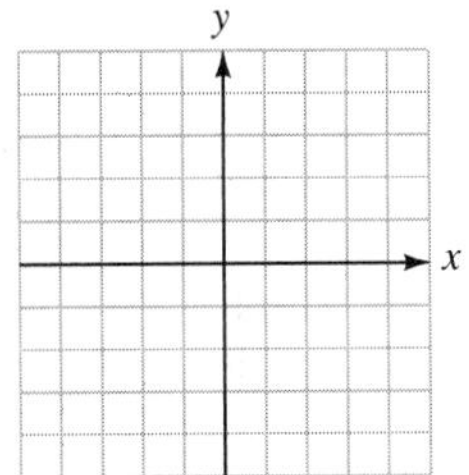

31. $\begin{cases} x = \dfrac{5y - 4}{2} \\ x - \frac{5}{3}y + \frac{1}{3} = 0 \end{cases}$

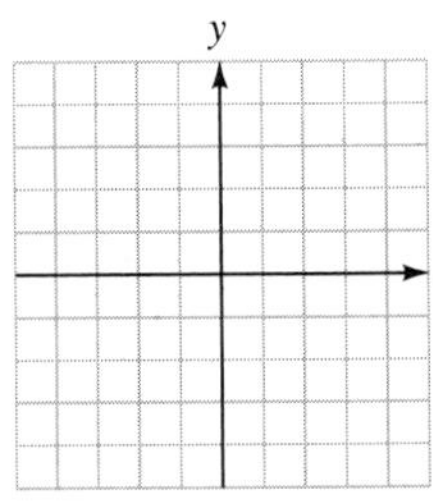

32. $\begin{cases} 2x = 5y - 11 \\ 3x = 2y \end{cases}$

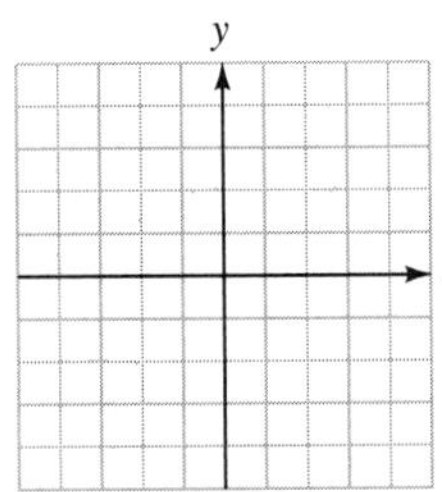

33. $\begin{cases} x = -\frac{3}{2}y \\ x = \frac{3}{2}y - 2 \end{cases}$

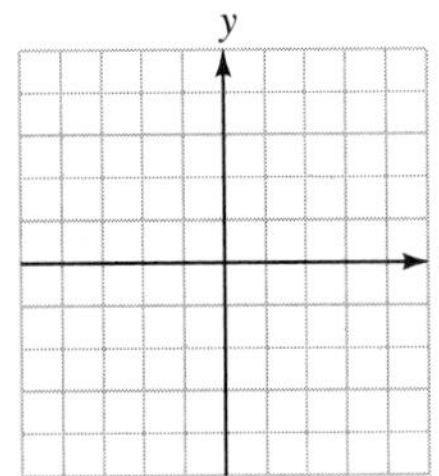

34. $\begin{cases} 4x = 3y - 1 \\ 3y = 4 - 8x \end{cases}$

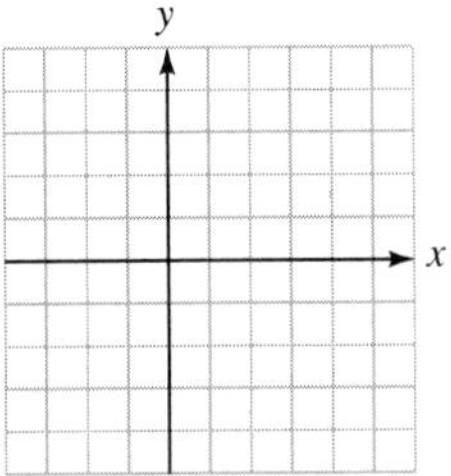

In Exercises 35–38, use a graphing calculator to solve each system. Give answers to the nearest hundredth.

35. $\begin{cases} y = 3.2x - 1.5 \\ y = -2.7x - 3.7 \end{cases}$

36. $\begin{cases} y = -0.45x + 5 \\ y = 5.55x - 13.7 \end{cases}$

37. $\begin{cases} 1.7x + 2.3y = 3.2 \\ y = 0.25x + 8.95 \end{cases}$

38. $\begin{cases} 2.75x = 12.9y - 3.79 \\ 7.1x - y = 35.76 \end{cases}$

APPLICATIONS

39. TAKE-OUT FOOD

a. Estimate the point of intersection of the two graphs shown in Illustration 3. Express your answer in the form (year, number of meals).

b. What information about dining out does the point of intersection give?

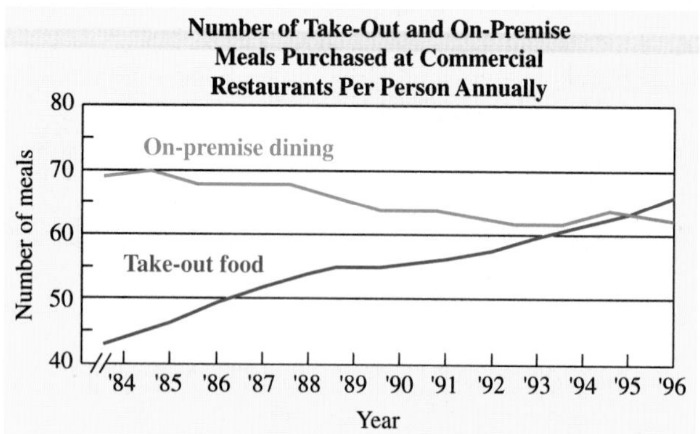

Based on data from *The Wall Street Journal Almanac* (1998)

ILLUSTRATION 3

40. ROAD ATLAS See Illustration 4. Name the cities that lie along Interstate 40. Name the cities that lie along Interstate 25. What city lies on both interstate highways?

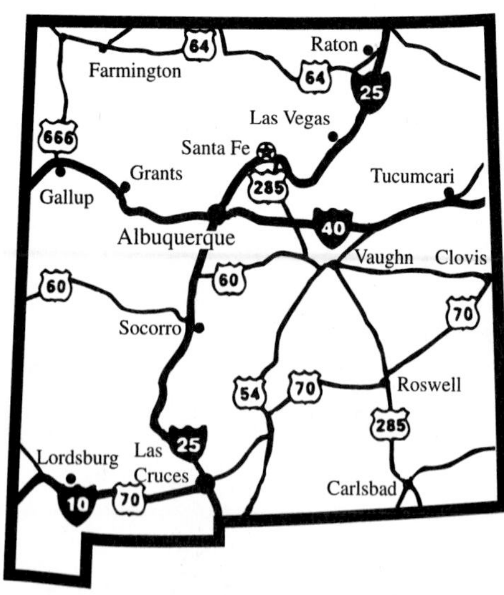

ILLUSTRATION 4

41. LAW OF SUPPLY AND DEMAND The demand function, graphed in Illustration 5, describes the relationship between the price x of a certain camera and the demand for the camera.

a. The supply function, $S(x) = \frac{25}{4}x - 525$, describes the relationship between the price x of the camera and the number of cameras the manufacturer is willing to supply. Graph this function in Illustration 5.

b. For what price will the supply of cameras equal the demand?

c. As the price of the camera is increased, what happens to supply and what happens to demand?

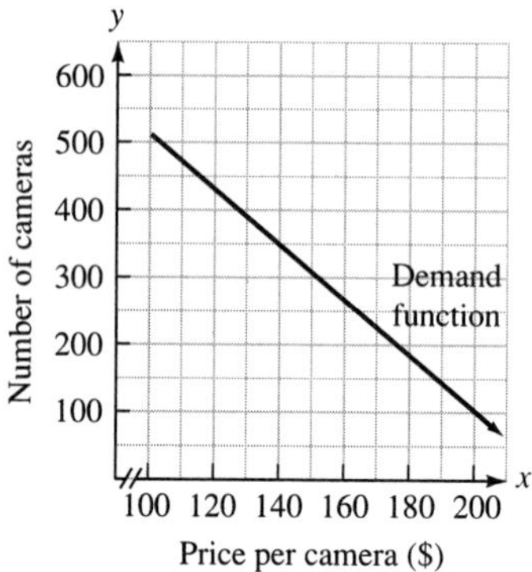

ILLUSTRATION 5

42. COST AND REVENUE The function $C(x) = 200x + 400$ gives the cost for a college to offer x sections of an introductory class in CPR (cardiopulmonary resuscitation). The function $R(x) = 280x$ gives the amount of revenue the college brings in when offering x sections of CPR.

a. Find the *break-even point* (where cost = revenue) by graphing each function on the same coordinate system.

b. How many sections does the college need to offer to make a profit on the CPR training course?

43. NAVIGATION The paths of two ships are tracked on the same coordinate system. One ship is following a path described by the equation $2x + 3y = 6$, and the other is following a path described by the equation $y = \frac{2}{3}x - 3$.

a. Is there a possibility of a collision?

b. What are the coordinates of the danger point?

c. Is a collision a certainty?

44. AIR TRAFFIC CONTROL Two airplanes are tracked using the same coordinate system on a radar screen. One plane is following a path described by the equation $y = \frac{2}{5}x - 2$, and the other is following a path described by the equation $2x = 5y + 7$. Is there a possibility of a collision?

WRITING *Write a paragraph using your own words.*

45. Suppose the solution of a system of two linear equations is $\left(\frac{14}{5}, -\frac{8}{3}\right)$. Knowing this, explain any drawbacks with solving the system by the graphing method.

46. Can a system of two linear equations have exactly two solutions? Why or why not?

REVIEW *In Exercises 47–50, $f(x) = -x^3 + 2x - 2$ and $g(x) = \dfrac{2 - x}{9 + x}$. Find each value.*

47. $f(-1)$

48. $f(10)$

49. $g(2)$

50. $g(-20)$

51. Determine the domain and range of $f(x) = x^2 - 2$.

52. Find the slope of the line passing through the points $(-4, 8)$ and $(3, 8)$.

3.2 *Solving Systems Algebraically*

In this section, you will learn about

- **The substitution method**
- **The addition method**
- **An inconsistent system**
- **A system with infinitely many solutions**
- **Problem solving**

INTRODUCTION. The graphing method provides a way to visualize the process of solving systems of equations. However, this method does have drawbacks. When using the graphing method, it can sometimes be difficult to determine the exact coordinates of the point of intersection. It is also true that we cannot use the graphing method to solve systems of higher order, such as three equations, each with three variables. In this section, we will discuss two other methods, called the *substitution* and the *addition* methods. They can be used to find the exact solutions of systems of equations.

The substitution method

To solve a system of two equations (each containing two variables) by using the **substitution method,** we follow these steps.

The substitution method

1. If necessary, solve one equation for one of its variables—preferably a variable with a coefficient of 1 or -1.
2. Substitute the resulting expression for that variable into the other equation and solve it.
3. Find the value of the other variable by substituting the value of the variable found in step 2 into the equation found in step 1.
4. State the solution.
5. Check the solution in both of the original equations.

EXAMPLE 1 ***The substitution method.*** Solve the system $\begin{cases} 4x + y = 13 \\ -2x + 3y = -17 \end{cases}$.

Solution

Step 1: We solve the first equation for y, because y has a coefficient of 1.

$4x + y = 13$

1. $y = -4x + 13$ To isolate y, subtract $4x$ from both sides.

Step 2: We then substitute $-4x + 13$ for y in the second equation of the system. This step will eliminate the variable y from that equation. The result will be an equation containing only one variable, x.

$-2x + 3y = -17$ The second equation.

$-2x + 3(-4x + 13) = -17$ Substitute $-4x + 13$ for y. The variable y is eliminated from the equation.

$-2x - 12x + 39 = -17$ Use the distributive property to remove parentheses.

$-14x = -56$ To solve for x, first combine like terms and then subtract 39 from both sides.

$x = 4$ Divide both sides by -14.

Step 3: To find y, we substitute 4 for x in Equation 1 and simplify:

$y = -4x + 13$

$= -4(4) + 13$ Substitute 4 for x.

$= -3$

Step 4: The solution is $x = 4$ and $y = -3$, or just $(4, -3)$. The graphs of these two equations would intersect at the point $(4, -3)$.

Step 5: To verify that this solution satisfies both equations, we substitute $x = 4$ and $y = -3$ into each equation in the system and simplify.

The first equation	*The second equation*
$4x + y = 13$	$-2x + 3y = -17$
$4(4) + (-3) \stackrel{?}{=} 13$	$-2(4) + 3(-3) \stackrel{?}{=} -17$
$16 - 3 \stackrel{?}{=} 13$	$-8 - 9 \stackrel{?}{=} -17$
$13 = 13$	$-17 = -17$

Since the ordered pair $(4, -3)$ satisfies both equations of the system, it checks.

Self Check

Solve $\begin{cases} x + 3y = 9 \\ 2x - y = -10 \end{cases}$.

Answer: $(-3, 4)$ ■

EXAMPLE 2 ***The substitution method.*** Solve the system $\begin{cases} \frac{2}{9}x - \frac{2}{9}y = \frac{2}{3} \\ 0.1x = 0.2 - 0.1y \end{cases}$.

Solution

First we find an equivalent system without fractions or decimals. To do this, we multiply both sides of the first equation by 9, which is the lowest common denominator of the fractions in the equation. Then we multiply both sides of the second equation by 10.

2. $2x - 2y = 6$
3. $x = 2 - y$

Since the variable x is isolated in Equation 3, we will substitute $2 - y$ for x in Equation 2. This step will eliminate x from Equation 2, leaving an equation containing only one variable, y. We then proceed to solve for y.

Self Check

Solve $\begin{cases} \frac{x}{8} + \frac{y}{4} = \frac{1}{2} \\ 0.01y = -0.02x + 0.04 \end{cases}$.

$2x - 2y = 6$	Equation 2.
$2(2 - y) - 2y = 6$	Substitute $2 - y$ for x.
$4 - 2y - 2y = 6$	Use the distributive property to remove parentheses.
$-4y = 2$	To solve for y, combine like terms and then subtract 4 from both sides.
$y = -\frac{1}{2}$	Divide both sides by -4 and then simplify the fraction.

We can find x by substituting $-\frac{1}{2}$ for y in Equation 3 and simplifying:

$x = 2 - y$	Equation 3.
$x = 2 - \left(-\frac{1}{2}\right)$	Substitute $-\frac{1}{2}$ for y.
$= 2 + \frac{1}{2}$	
$= \frac{5}{2}$	$2 + \frac{1}{2} = \frac{4}{2} + \frac{1}{2} = \frac{5}{2}$.

The solution is the ordered pair $\left(\frac{5}{2}, -\frac{1}{2}\right)$. Verify that this solution satisfies both equations in the original system.

Answer: $\left(\frac{4}{3}, \frac{4}{3}\right)$ ■

The addition method

Another method for solving a system of linear equations is the **addition method**. In this method, we combine the equations in a way that will eliminate the terms involving one of the variables.

The addition method

1. Write both equations of the system in general form: $(Ax + By = C)$.
2. Multiply the terms of one or both of the equations by constants chosen to make the coefficients of x (or y) differ only in sign.
3. Add the equations and solve the resulting equation, if possible.
4. Substitute the value obtained in step 3 into either of the original equations and solve for the remaining variable.
5. State the solution obtained in step 3 and 4.
6. Check the solution in both of the original equations.

EXAMPLE 3 ***The addition method.*** Solve the system $\begin{cases} 4x + y = 13 \\ -2x + 3y = -17 \end{cases}$.

Self Check

Solve $\begin{cases} 3x + 2y = 0 \\ 2x - y = -7 \end{cases}$.

Solution

Step 1: This is the system discussed in Example 1. In this example, we will solve it by using the addition method. Since both equations are already written in general form, step 1 is unnecessary.

Step 2: We note that the coefficient of x in the first equation is 4. If we multiply both sides of the second equation by 2, the coefficient of x in that equation will be -4. Then the coefficients of x will differ only in sign.

$$\begin{array}{ll} \textbf{4.} & \begin{cases} 4x + y = 13 \\ -4x + 6y = -34 \end{cases} \\ \textbf{5.} & \end{array}$$

Step 3: When these equations are added, the terms involving x drop out (or are eliminated), and we get an equation that contains only the variable y. We then proceed by solving for y.

$$\begin{array}{rrl} & 4x + y = 13 & \text{Add the like terms, column by column.} \\ + & -4x + 6y = -34 & 4x + (-4x) = 0. \\ \hline & 7y = -21 & \\ & y = -3 & \text{To solve for } y\text{, divide both sides by 7.} \end{array}$$

Step 4: To find x, we substitute -3 for y in either of the original equations and solve for x. If we use Equation 4, we have

$$\begin{aligned} 4x + y &= 13 & & \\ 4x + (-3) &= 13 & & \text{Substitute } -3 \text{ for } y. \\ 4x &= 16 & & \text{To solve for } x\text{, add 3 to both sides.} \\ x &= 4 & & \text{Divide both sides by 4.} \end{aligned}$$

Step 5: The solution is $x = 4$ and $y = -3$, or just $(4, -3)$.

Step 6: The check was completed in Example 1.

Answer: $(-2, 3)$

EXAMPLE 4 ***The addition method.*** Solve the system
$\begin{cases} 4x = 3(2 + y) \\ 3(x - 10) = -2y. \end{cases}$

Solution

To use the addition method, we must first write each equation in general form. In each case, the first step is to remove the parentheses.

The first equation	*The second equation*
$4x = 3(2 + y)$	$3(x - 10) = -2y$
$4x = 6 + 3y$	$3x - 30 = -2y$
$4x - 3y = 6$	$3x + 2y = 30$

We now solve the equivalent system

6. $4x - 3y = 6$
7. $3x + 2y = 30$

$$\begin{cases} 4x - 3y = 6 \\ 3x + 2y = 30 \end{cases}$$

To make the y-terms drop out when we add the equations, we multiply both sides of Equation 6 by 2 and both sides of Equation 7 by 3 to get

$$\begin{cases} 8x - 6y = 12 \\ 9x + 6y = 90 \end{cases}$$

When these equations are added, the y-terms drop out, and we get

$$\begin{aligned} 17x &= 102 & & -6y + 6y = 0 \\ x &= 6 & & \text{To solve for } x\text{, divide both sides by 17.} \end{aligned}$$

To find y, we can substitute 6 for x in either of the two original equations, or in Equation 6 or Equation 7. If we substitute 6 for x in Equation 7, we get

$$\begin{aligned} 3x + 2y &= 30 & & \\ 3(6) + 2y &= 30 & & \text{Substitute 6 for } x. \\ 18 + 2y &= 30 & & \text{Do the multiplication.} \\ 2y &= 12 & & \text{Subtract 18 from both sides.} \\ y &= 6 & & \text{Divide both sides by 2.} \end{aligned}$$

The solution is the ordered pair $(6, 6)$.

Self Check

Solve $\begin{cases} 4(2x - y) = 18 \\ 3(x - 3) = 2y - 1 \end{cases}$.

Answer: $\left(1, -\frac{5}{2}\right)$

An inconsistent system

EXAMPLE 5 ***A system with no solution.*** Solve the system $\begin{cases} y = 2x + 4 \\ 8x - 4y = 7 \end{cases}$, if possible.

Solution

Because the first equation is already solved for y, we use the substitution method.

$$\begin{aligned} 8x - 4y &= 7 && \text{The second equation.} \\ 8x - 4(2x + 4) &= 7 && \text{Substitute } 2x + 4 \text{ for } y. \end{aligned}$$

We then solve this equation for x:

$$\begin{aligned} 8x - 8x - 16 &= 7 && \text{Use the distributive property to remove parentheses.} \\ -16 &= 7 && \text{Combine like terms.} \end{aligned}$$

This impossible result shows that the equations in the system are independent and that the system is inconsistent. Since the system has no solution, the graphs of the equations in the system will be parallel.

Self Check

Solve $\begin{cases} x = -2.5y + 8 \\ y = -0.4x + 2 \end{cases}$.

Answer: no solution ■

A system with infinitely many solutions

EXAMPLE 6 ***The addition method.*** Solve the system $\begin{cases} 4x + 6y = 12 \\ -2x - 3y = -6 \end{cases}$.

Solution

Since the equations are written in general form, we use the addition method. We copy the first equation and multiply both sides of the second equation by 2 to get

$$\begin{aligned} 4x + 6y &= 12 \\ -4x - 6y &= -12 \end{aligned}$$

After adding the left-hand sides and the right-hand sides, we get

$$\begin{aligned} 0x + 0y &= 0 \\ 0 &= 0 \end{aligned}$$

Here, both the x- and y-terms drop out. The true statement $0 = 0$ shows that the equations are dependent and that the system is consistent.

Note that the equations of the system are equivalent, because when the second equation is multiplied by -2, it becomes the first equation. The graphs of these equations would coincide. Any ordered pair that satisfies one of the equations also satisfies the other. Some solutions are $(0, 2)$, $(3, 0)$, and $(-3, 4)$.

Self Check

Solve $\begin{cases} x - \frac{5}{2}y = \frac{19}{2} \\ -\frac{2}{5}x + y = -\frac{19}{5} \end{cases}$.

Answer: There are infinitely many solutions; three of them are $(2, -3)$, $(12, 1)$, and $\left(\frac{19}{2}, 0\right)$. ■

Problem solving

To solve problems using two variables, we follow the same problem-solving strategy discussed in Chapters 1 and 2, except that we form two equations instead of one.

EXAMPLE 7 ***Wedding pictures.*** In Figure 3-7, a professional photographer offers two different packages for wedding pictures. Use the information in the figure to determine the cost of an 8×10-in. and a 5×7-in. photograph.

WEDDING PICTURES	
Package 1 includes...	Package 2 includes...
8 - 8 x 10's 12 - 5 x 7's	6 - 8 x 10's 22 - 5 x 7's
Only $133.00	**Only $168.00**

FIGURE 3-7

Analyze the problem From the figure, we see that eight 8×10 and twelve 5×7 pictures cost \$133, and six 8×10 and twenty-two 5×7 pictures cost \$168. We need to find the cost of an 8×10 and a 5×7 photograph.

Form two equations We can let x represent the cost of an 8×10 photograph and let y represent the cost of a 5×7 photograph. For the first package, the cost of eight 8×10 pictures is $8 \cdot \$x = \$8x$, and the cost of twelve 5×7 pictures is $12 \cdot \$y = \$12y$. For the second package, the cost of six 8×10 pictures is $\$6x$, and the cost of twenty-two 5×7 pictures is $\$22y$. To find x and y, we must write and solve two equations.

The cost of eight 8×10 photographs	+	the cost of twelve 5×7 photographs	=	the value of the the first package.
$8x$	+	$12y$	=	133
The cost of six 8×10 photographs	+	the cost of twenty-two 5×7 photographs	=	the value of the second package.
$6x$	+	$22y$	=	168

Solve the system To find the cost of the 8×10 and the 5×7 photographs, we must solve the following system:

8. $\begin{cases} 8x + 12y = 133 \\ 6x + 22y = 168 \end{cases}$
9.

We will use the addition method to solve this system. To make the x-terms drop out, we multiply both sides of Equation 8 by 3. Then we multiply both sides of Equation 9 by -4. We then add the resulting equations and solve for x:

$$\begin{aligned} 24x + 36y &= 399 \\ + \quad -24x - 88y &= -672 \\ \hline -52y &= -273 \end{aligned}$$ Add like terms, column by column. The x-terms drop out.

$$y = 5.25$$ Divide both sides by -52.

To find x, we substitute 5.25 for y in Equation 8 and solve for x:

$$8x + 12y = 133$$
$$8x + 12(\mathbf{5.25}) = 133$$ Substitute 5.25 for y.
$$8x + 63 = 133$$ Do the multiplication.
$$8x = 70$$ Subtract 63 from both sides.
$$x = 8.75$$ Divide both sides by 8.

State the conclusion The cost of an 8×10 photo is \$8.75, and the cost of a 5×7 photo is \$5.25.

Check the result If the first package contains eight 8×10 and twelve 5×7 photographs, the value of the package is $8(\$8.75) + 12(\$5.25) = \$70 + \$63 = \$133$. If the second package contains six 8×10 and twenty-two 5×7 photographs, the value of the package is $6(\$8.75) + 22(\$5.25) = \$52.50 + \$115.50 = \$168$. The answers check. ■

EXAMPLE 8 ***Water treatment.*** A technician determines that 50 fluid ounces of a 15% muriatic acid solution needs to be added to the water in a swimming pool to kill a growth of algae. If the technician has 5% and 20% muriatic solutions on hand, how many ounces of each must be combined to create the 15% solution?

Analyze the problem We need to find the number of ounces of a 5% solution and the number of ounces of a 20% solution that must be combined to obtain 50 ounces of a 15% solution.

Form two equations We can let x represent the number of ounces of the 5% solution and let y represent the number of ounces of the 20% solution that are to be mixed. (See Figure 3-8(a).) Then the amount of muriatic acid in the 5% solution is $0.05x$ ounces, and the amount of muriatic acid in the 20% solution is $0.20y$ ounces. The sum of these amounts is also the amount of muriatic acid in the final mixture, which is 15% of 50 ounces. This information is shown in the table in Figure 3-8(b).

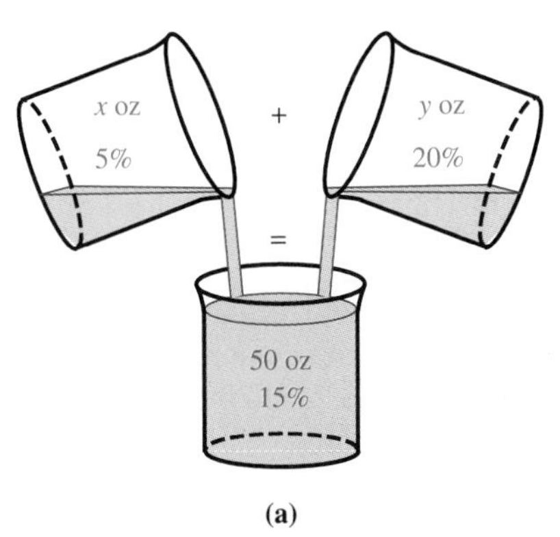

(a)

Solution	% acid	Ounces	Amount of acid
5% solution	0.05	x	$0.05x$
20% solution	0.20	y	$0.20y$
15% mixture	0.15	50	0.15(50)

↑ One equation comes from the information in this column. (Ounces)

↑ Another equation comes from the information in this column. (Amount of acid)

(b)

FIGURE 3-8

The facts of the problem give the following two equations:

The number of ounces of 5% solution	+	the number of ounces of 20% solution	=	the total number of ounces in the 15% mixture.
x	+	y	=	50
The acid in the 5% solution	+	the acid in the 20% solution	=	the acid in the 15% mixture.
$0.05x$	+	$0.20y$	=	0.15(50)

Solve the system To find out how many ounces of each are needed, we solve the following system:

$$\begin{cases} x + y = 50 & (10) \\ 0.05x + 0.20y = 7.5 & (11) \quad 0.15(50) = 7.5. \end{cases}$$

To solve this system by substitution, we can solve the first equation for y:

$$x + y = 50$$

$$y = 50 - x \quad (12) \quad \text{Subtract } x \text{ from both sides.}$$

Then we substitute $50 - x$ for y in Equation 11 and solve for x.

$$
\begin{aligned}
0.05x + 0.20y &= 7.5 && \text{Equation 11.}\\
0.05x + 0.20(\mathbf{50 - x}) &= 7.5 && \text{Substitute } 50 - x \text{ for } y.\\
5x + 20(50 - x) &= 750 && \text{Multiply both sides by 100.}\\
5x + 1{,}000 - 20x &= 750 && \text{Use the distributive property to remove parentheses.}\\
-15x &= -250 && \text{Combine like terms and subtract 1,000 from both sides.}\\
x &= \frac{-250}{-15} && \text{Divide both sides by } -15.\\
x &= \frac{50}{3} && \text{Simplify: } \frac{250}{15} = \frac{\overset{1}{\cancel{5}} \cdot 50}{\underset{1}{\cancel{5}} \cdot 3}.
\end{aligned}
$$

To find y, we can substitute $\frac{50}{3}$ for x in Equation 12:

$$
\begin{aligned}
y &= 50 - x && \text{Equation 12.}\\
&= 50 - \frac{\mathbf{50}}{\mathbf{3}} && \text{Substitute } \tfrac{50}{3} \text{ for } x.\\
&= \frac{100}{3} && 50 = \tfrac{150}{3}.
\end{aligned}
$$

State the conclusion To obtain 50 ounces of a 15% solution, the technician must mix $\frac{50}{3}$ or $16\frac{2}{3}$ ounces of the 5% solution with $\frac{100}{3}$ or $33\frac{1}{3}$ ounces of the 20% solution.

Check the result We note that $16\frac{2}{3}$ ounces of solution plus $33\frac{1}{3}$ ounces of solution equals the required 50 ounces of solution. We also note that 5% of $16\frac{2}{3} \approx 0.83$ and 20% of $33\frac{1}{3} \approx 6.67$, giving a total of 7.5, which is 15% of 50. The answers check. ■

EXAMPLE 9 ***Parallelograms.*** Refer to the parallelogram shown in Figure 3-9 and find the values of x and y.

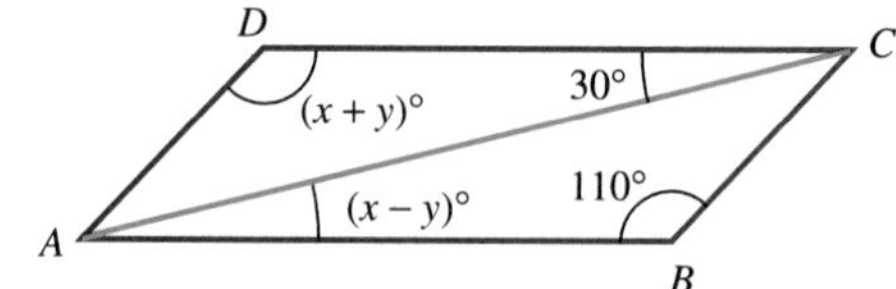

FIGURE 3-9

Solution To solve this problem, we will use two important facts about parallelograms.

- When a diagonal intersects two parallel sides of a parallelogram, pairs of *alternate interior angles* have the same measure. In Figure 3-9, $\angle BAC$ and $\angle DCA$ are alternate interior angles and therefore have the same measure. Thus, $(x - y)° = 30°$.
- *Opposite angles* of a parallelogram have the same measure. Since $\angle B$ and $\angle D$ in Figure 3-9 are opposite angles of the parallelogram, $(x + y)° = 110°$.

We can form the following system of equations and solve it by addition.

$$
\begin{aligned}
&\textbf{13.}\quad x - y = 30\\
&\textbf{14.}\quad \underline{x + y = 110}\\
&\qquad 2x \quad = 140 && \text{Add Equations 13 and 14. The } y\text{-terms drop out.}\\
&\qquad\quad x = 70 && \text{Divide both sides by 2.}
\end{aligned}
$$

We can substitute 70 for x in Equation 14 and solve for y.

$$
\begin{aligned}
x + y &= 110\\
\mathbf{70} + y &= 110 && \text{Substitute 70 for } x.\\
y &= 40 && \text{Subtract 70 from both sides.}
\end{aligned}
$$

Thus, $x = 70$ and $y = 40$. ■

Running a machine involves both *setup costs* and *unit costs*. Setup costs include the cost of preparing a machine to do a certain job. The costs to make one item are unit costs. They depend on the number of items to be manufactured, including costs of raw materials and labor.

EXAMPLE 10 ***Break point.*** The setup cost of a machine that makes wooden coathangers is \$400. After setup, it costs \$1.50 to make each hanger (the unit cost). Management is considering the purchase of a new machine that can manufacture the same type of coathanger at a cost of \$1.25 per hanger. If the setup cost of the new machine is \$500, find the number of coathangers that the company would need to manufacture to make the cost the same using either machine. This is called the **break point**.

Analyze the problem We are to find the number of coat hangers that will cost equal amounts to produce on either machine. The machines have different setup costs and different unit costs.

Form two equations The cost C_1 of manufacturing x coathangers on the machine currently in use is $\$1.50x + \$400$ (the number of coathangers manufactured times \$1.50, plus the setup cost of \$400). The cost C_2 of manufacturing the same number of coathangers on the new machine is $\$1.25x + \$500$. The break point occurs when the costs to make the same number of hangers using either machine are equal ($C_1 = C_2$).

If x represents the number of coat hangers to be manufactured, the cost C_1 using the machine currently in use is

The cost of using the current machine	=	the cost of manufacturing x coathangers	+	the setup cost.
C_1	=	$1.5x$	+	400

The cost C_2 using the new machine to make x coathangers is

The cost of using the new machine	=	the cost of manufacturing x coathangers	+	the setup cost.
C_2	=	$1.25x$	+	500

Solve the system To find the break point, we must solve the system $\begin{cases} C_1 = 1.5x + 400 \\ C_2 = 1.25x + 500 \end{cases}$.

Since the break point occurs when $C_1 = C_2$, we can substitute $1.5x + 400$ for C_2 in the second equation to get

$$\begin{aligned} 1.5x + 400 &= 1.25x + 500 \\ 1.5x &= 1.25x + 100 && \text{Subtract 400 from both sides.} \\ 0.25x &= 100 && \text{Subtract } 1.25x \text{ from both sides.} \\ x &= 400 && \text{Divide both sides by 0.25.} \end{aligned}$$

State the conclusion The break point is 400 coathangers.

Check the result To make 400 coathangers, the cost on the current machine would be $\$400 + \$1.50(400) = \$400 + \$600 = \$1{,}000$. The cost using the new machine would be $\$500 + \$1.25(400) = \$500 + \$500 = \$1{,}000$. Since the costs are equal, the break point is 400. ■

STUDY SET Section 3.2

VOCABULARY *In Exercises 1–4, fill in the blanks to make the statements true.*

1. $Ax + By = C$ is the ____________ form of a linear equation.

2. In the equation $x + 3y = -1$, the x-term has an understood ____________ of 1.

3. When we add the two equations of the system $\begin{cases} x + y = 5 \\ x - y = -3 \end{cases}$, the y-terms are ____________.

4. To solve $\begin{cases} y = 3x \\ x + y = 4 \end{cases}$, we can ____________ $3x$ for y in the second equation.

CONCEPTS

5. If the system $\begin{cases} 4x - 3y = 7 \\ 3x - 2y = 6 \end{cases}$ is to be solved using the addition method, by what constant should each equation be multiplied if

a. the x-terms are to drop out?

b. the y-terms are to drop out?

6. If the system $\begin{cases} 4x - 3y = 7 \\ 3x + y = 6 \end{cases}$ is to be solved using the substitution method, what variable in what equation would it be easier to solve for?

7. Can the system $\begin{cases} 2x + 5y = 7 \\ 4x - 3y = 16 \end{cases}$ be solved more easily by the substitution or the addition method?

8. Given: the equation $3x + y = -4$.

a. Solve for x.

b. Solve for y.

c. Which variable was easier to solve for? Explain why.

9. The substitution method was used to solve three systems of linear equations. The results after y was eliminated and the remaining equation was solved for x are listed below. Match each result with a possible graph of the system from Illustration 1.

a. $-2 = 3$ **b.** $x = 3$ **c.** $3 = 3$

Possible graphs

i ii iii

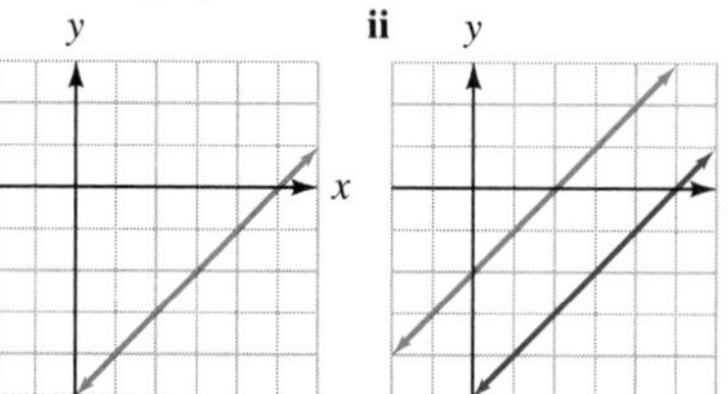

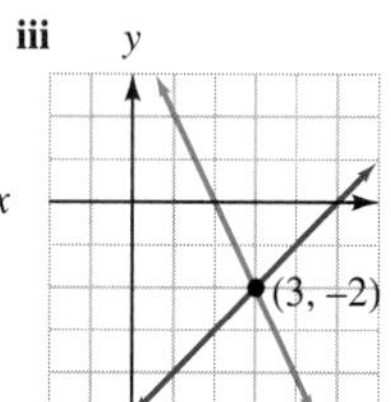

ILLUSTRATION 1

10. Consider the system $\begin{cases} \frac{2}{3}x - \frac{y}{6} = \frac{16}{9} \\ 0.03x + 0.02y = 0.03 \end{cases}$.

a. What algebraic step should be performed to clear the first equation of fractions?

b. What algebraic step should be performed to clear the second equation of decimals?

PRACTICE *In Exercises 11–18, solve each system by substitution, if possible.*

11. $\begin{cases} y = x \\ x + y = 4 \end{cases}$

12. $\begin{cases} y = x + 2 \\ x + 2y = 16 \end{cases}$

13. $\begin{cases} x = 2 + y \\ 2x + y = 13 \end{cases}$

14. $\begin{cases} x = -4 + y \\ 3x - 2y = -5 \end{cases}$

15. $\begin{cases} x + 2y = 6 \\ 3x - y = -10 \end{cases}$

16. $\begin{cases} 2x - y = -21 \\ 4x + 5y = 7 \end{cases}$

17. $\begin{cases} \frac{3}{2}x + 2 = y \\ 0.6x - 0.4y = -0.4 \end{cases}$

18. $\begin{cases} 2x - \frac{5}{2} = y \\ 0.04x - 0.02y = 0.05 \end{cases}$

In Exercises 19–26, solve each system by addition, if possible.

19. $\begin{cases} x - y = 3 \\ x + y = 7 \end{cases}$

20. $\begin{cases} x + y = 1 \\ x - y = 7 \end{cases}$

21. $\begin{cases} 2x + y = -10 \\ 2x - y = -6 \end{cases}$

22. $\begin{cases} x + 2y = -9 \\ x - 2y = -1 \end{cases}$

23. $\begin{cases} 2x + 3y = 8 \\ 3x - 2y = -1 \end{cases}$

24. $\begin{cases} 5x - 2y = 19 \\ 3x + 4y = 1 \end{cases}$

25. $\begin{cases} 4(x - 2) = -9y \\ 2(x - 3y) = -3 \end{cases}$

26. $\begin{cases} 2(2x + 3y) = 5 \\ 8x = 3(1 + 3y) \end{cases}$

In Exercises 27–42, solve each system by any method, if possible.

27. $\begin{cases} 3x - 4y = 9 \\ x + 2y = 8 \end{cases}$

28. $\begin{cases} 3x - 2y = -10 \\ 6x + 5y = 25 \end{cases}$

29. $\begin{cases} 2(x + y) + 1 = 0 \\ 3x + 4y = 0 \end{cases}$

30. $\begin{cases} 5x + 3y = -7 \\ 3(x - y) - 7 = 0 \end{cases}$

31. $\begin{cases} 0.16x - 0.08y = 0.32 \\ 2x - 4 = y \end{cases}$

32. $\begin{cases} 0.6y - 0.9x = -3.9 \\ 3x - 17 = 4y \end{cases}$

33. $\begin{cases} x = \frac{3}{2}y + 5 \\ 2x - 3y = 8 \end{cases}$

34. $\begin{cases} x = \frac{2}{3}y \\ y = 4x + 5 \end{cases}$

35. $\begin{cases} 0.5x + 0.5y = 6 \\ \frac{x}{2} - \frac{y}{2} = -2 \end{cases}$

36. $\begin{cases} \frac{x}{2} - \frac{y}{3} = -4 \\ \frac{x}{2} + \frac{y}{9} = 0 \end{cases}$

37. $\begin{cases} \frac{3}{4}x + \frac{2}{3}y = 7 \\ \frac{3}{5}x - \frac{1}{2}y = 18 \end{cases}$

38. $\begin{cases} \frac{2}{3}x - \frac{1}{4}y = -8 \\ 0.5x - 0.375y = -9 \end{cases}$

39. $\begin{cases} \frac{3x}{2} - \frac{2y}{3} = 0 \\ \frac{3x}{4} + \frac{4y}{3} = \frac{5}{2} \end{cases}$

40. $\begin{cases} \frac{3x}{5} + \frac{5y}{3} = 2 \\ \frac{6x}{5} - \frac{5y}{3} = 1 \end{cases}$

41. $\begin{cases} 12x - 5y - 21 = 0 \\ \frac{3}{4}x - \frac{2}{3}y = \frac{19}{8} \end{cases}$

42. $\begin{cases} 4y + 5x - 7 = 0 \\ \frac{10}{7}x - \frac{4}{9}y = \frac{17}{21} \end{cases}$

In Exercises 43–46, solve each system. To do this, substitute a for $\frac{1}{x}$ and b for $\frac{1}{y}$ and solve for a and b. Then find x and y using the fact that $a = \frac{1}{x}$ and $b = \frac{1}{y}$.

43. $\begin{cases} \frac{1}{x} + \frac{1}{y} = \frac{5}{6} \\ \frac{1}{x} - \frac{1}{y} = \frac{1}{6} \end{cases}$

44. $\begin{cases} \frac{1}{x} + \frac{1}{y} = \frac{9}{20} \\ \frac{1}{x} - \frac{1}{y} = \frac{1}{20} \end{cases}$

45. $\begin{cases} \frac{1}{x} + \frac{2}{y} = -1 \\ \frac{2}{x} - \frac{1}{y} = -7 \end{cases}$

46. $\begin{cases} \frac{3}{x} - \frac{2}{y} = -30 \\ \frac{2}{x} - \frac{3}{y} = -30 \end{cases}$

APPLICATIONS *Use two variables to solve each problem.*

47. ADVERTISING Use the information in the fee schedule shown in Illustration 2 to find the cost of a 15-second and a 30-second radio commercial on radio station KLIZ.

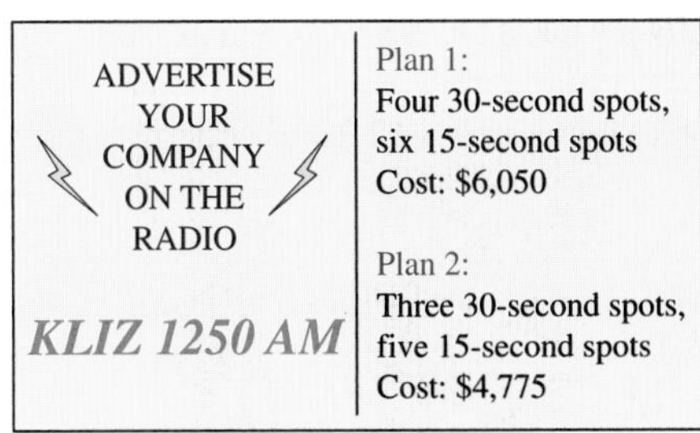

ADVERTISE YOUR COMPANY ON THE RADIO
KLIZ 1250 AM
Plan 1:
Four 30-second spots, six 15-second spots
Cost: $6,050
Plan 2:
Three 30-second spots, five 15-second spots
Cost: $4,775

ILLUSTRATION 2

48. TEMPORARY HELP A law firm had to hire several workers to help finish a large project. From the billing records shown in Illustration 3, determine the daily fee charged by the employment agency for a clerk-typist and for a computer programmer.

TEMPORARY EMPLOYMENT, INC.

We meet your employment needs!

Billed to: Archer Law Offices Attn: B. Kinsell

Day	Posistion/Employee Name	Total cost
Mon. 3/22	*Clerk-typists:* K. Amad, B. Tran, S. Smith *Programmers:* T. Lee, C. Knox	$685
Tues. 3/23	*Clerk-typists:* K. Amad, B. Tran, S. Smith, W. Morada *Programmers:* T. Lee, C. Knox, B. Morales	$975

ILLUSTRATION 3

49. PETS According to the Pet Industry Joint Advisory Council, as of 1998, there were an estimated 124 million dogs and cats in the United States. If there were 8 million more cats than dogs, how many of each type of pet were there in 1998?

50. ELECTRONICS Two resistors in the voltage divider circuit in Illustration 4 have a total resistance of 1,375 ohms. To provide the required voltage, R_1 must be 125 ohms greater than R_2. Find both resistances.

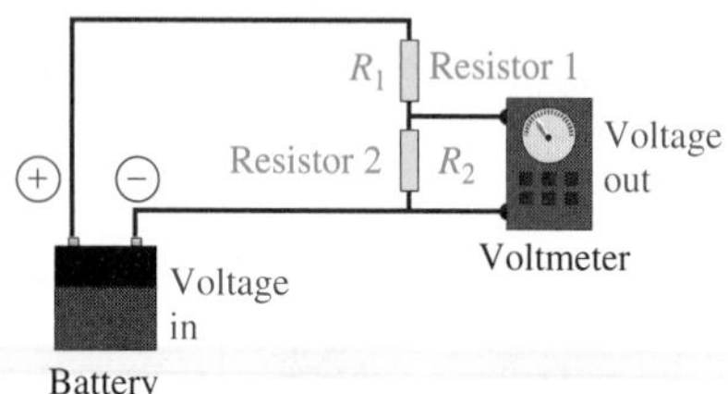

ILLUSTRATION 4

51. FENCING A FIELD The rectangular field in Illustration 5 is surrounded by 72 meters of fencing. If the field is partitioned as shown, a total of 88 meters of fencing is required. Find the dimensions of the field.

ILLUSTRATION 5

52. GEOMETRY In a right triangle, one acute angle is 15° greater than two times the other acute angle. Find the difference between the measures of the angles.

53. BRACING The bracing of a basketball backboard shown in Illustration 6 forms a parallelogram. Find the values of x and y.

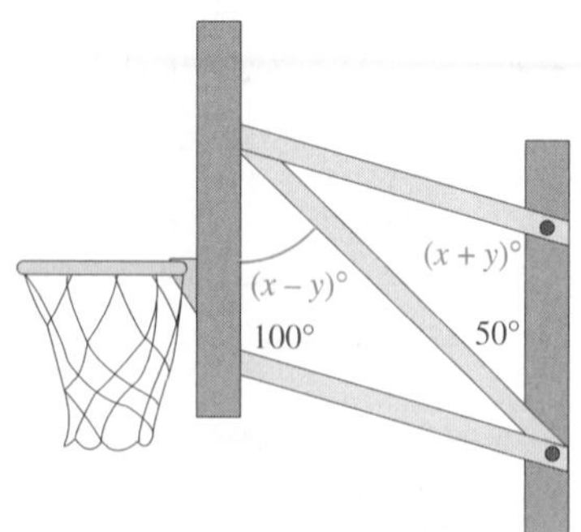

ILLUSTRATION 6

54. TRAFFIC SIGNAL In Illustration 7, brace A and brace B are perpendicular. Find the values of x and y.

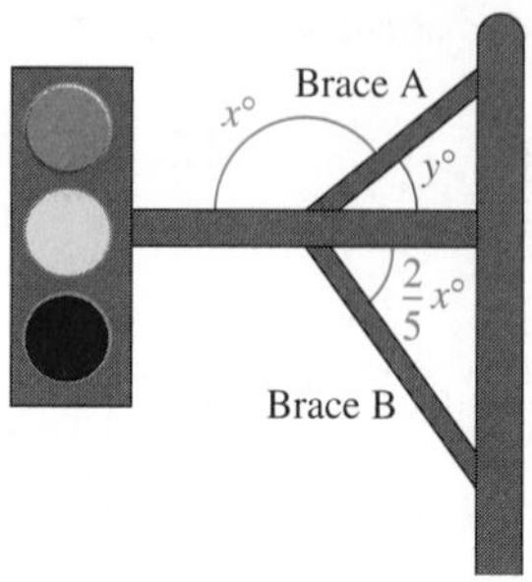

ILLUSTRATION 7

55. INVESTMENT CLUB Part of $8,000 was invested by an investment club at 10% interest and the rest at 12%. If the annual income from these investments is $900, how much was invested at each rate?

56. RETIREMENT INCOME A retired couple invested part of $12,000 at 6% interest and the rest at 7.5%. If their annual income from these investments is $810, how much was invested at each rate?

57. DERMATOLOGY Tests of an antibacterial face-wash cream showed that a mixture containing 0.3% Triclosan (active ingredient) gave the best results. How many grams of cream from each tube shown in Illustration 8 should be used to make an equal-size tube of the 0.3% cream?

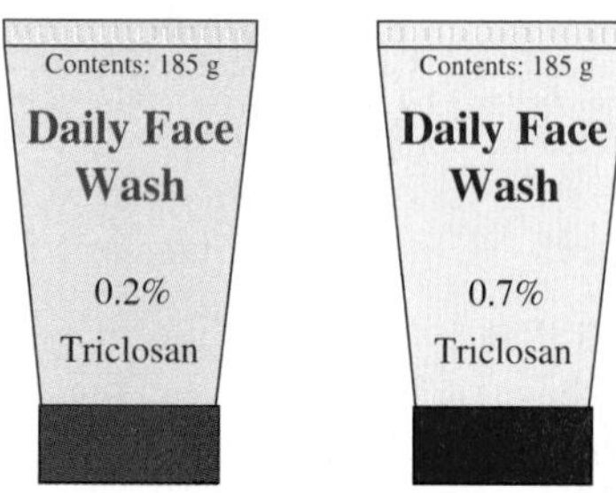

ILLUSTRATION 8

58. MIXING SOLUTIONS How many ounces of the two alcohol solutions in Illustration 9 must be mixed to obtain 100 ounces of a 12.2% solution?

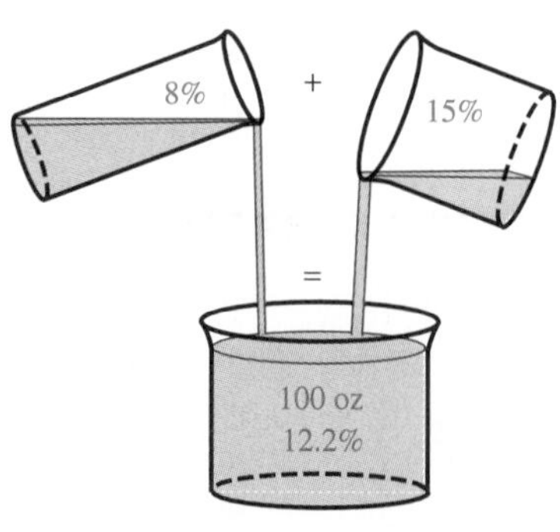

ILLUSTRATION 9

59. TV NEWS A news van and a helicopter left a TV station parking lot at the same time headed in opposite directions to cover breaking news stories that were 145 miles apart. If the helicopter had to travel 55 miles farther than the van, how far did the van have to travel to reach the location of the news story?

60. DELIVERY SERVICE A delivery truck travels 50 miles in the same time that a cargo plane travels 180 miles. The speed of the plane is 143 mph faster than the speed of the truck. Find the speed of the delivery truck.

61. PRODUCTION PLANNING A bicycle manufacturer builds racing bikes and mountain bikes, with the per-unit manufacturing costs shown in Illustration 10. The company has budgeted \$15,900 for labor and \$13,075 for materials. How many bicycles of each type can be built?

Model	Cost of materials	Cost of labor
racing	\$55	\$60
mountain	\$70	\$90

ILLUSTRATION 10

62. FARMING A farmer keeps some animals on a strict diet. Each animal is to recieve 15 grams of protein and 7.5 grams of carbohydrates. The farmer uses two food mixes, with nutrients as shown in Illustration 11. How many grams of each mix should be used to provide the correct nutrients for each animal?

Mix	Protein	Carbohydrates
Mix *A*	12%	9%
Mix *B*	15%	5%

ILLUSTRATION 11

63. RECORDING COMPANY Three people invest a total of \$105,000 to start a record company that will produce reissues of classic jazz. Each release will be a set of 3 CDs that will retail for \$45 per set. If each set can be produced for \$18.95, how many sets must be sold for the investors to make a profit?

64. MACHINE SHOP Two machines can mill a brass plate. One machine has a setup cost of \$300 and a cost per plate of \$2. The other machine has a setup cost of \$500 and a cost per plate of \$1. Find the break point.

65. PUBLISHING A printer has two presses. One has a setup cost of \$210 and can print the pages of a certain book for \$5.98. The other press has a setup cost of \$350 and can print the pages of the same book for \$5.95. Find the break point.

66. MIXING CANDY How many pounds of each candy shown in Illustration 12 must be mixed to obtain 60 pounds of candy that would be worth \$3 per pound?

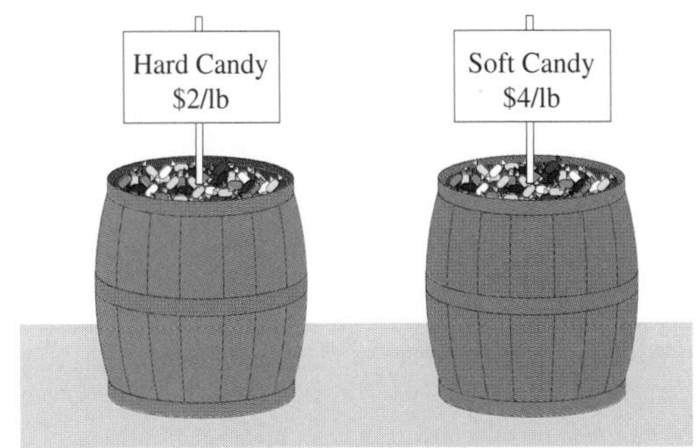

ILLUSTRATION 12

67. COSMETOLOGY A beauty shop specializing in permanents has fixed costs of \$2,101.20 per month. The owner estimates that the cost for each permanent is \$23.60, which covers labor, chemicals, and electricity. If her shop can give as many permanents as she wants at a price of \$44 each, how many must be given each month to break even?

68. PRODUCTION PLANNING A paint manufacturer can choose between two processes for manufacturing house paint, with monthly costs as shown in Illustration 13. Assume that the paint sells for \$18 per gallon.

Process	Fixed costs	Unit cost (per gallon)
A	\$32,500	\$13
B	\$80,600	\$5

ILLUSTRATION 13

a. Find the break point for process A.

b. Find the break point for process B.

c. If expected sales are 7,000 gallons per month, which process should the company use?

69. RETOOLING A manufacturer of automobile water pumps is considering retooling for one of two manufacturing processes, with monthly fixed costs and unit costs as indicated in Illustration 14. Each water pump can be sold for \$50.

Process	Fixed costs	Unit cost
A	\$12,390	\$29
B	\$20,460	\$17

ILLUSTRATION 14

a. Find the break point for process A.

b. Find the break point for process B.

c. If expected sales are 550 per month, which process should be used?

70. SALARY OPTIONS A sales clerk can choose from two salary plans: a straight 7% commision, or \$150 + 2% commission. How much would the clerk have to sell for each plan to produce the same monthly paycheck?

WRITING *Write a paragraph using your own words.*

71. Which method would you use to solve the system $\begin{cases} 4x + 6y = 5 \\ 8x - 3y = 3 \end{cases}$? Explain why.

72. Which method would you use to solve the system $\begin{cases} x - 2y = 2 \\ 2x + 3y = 11 \end{cases}$? Explain why.

73. When solving a problem using two variables, why must we write two equations?

74. Systems of two linear equations can be solved graphically or algebraically. Give an advantage and a drawback of each method.

REVIEW *In Exercises 75–80, find the slope of each line.*

75.

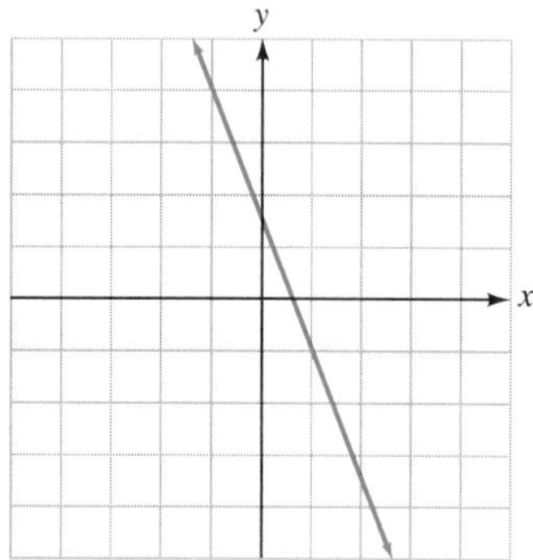

76.

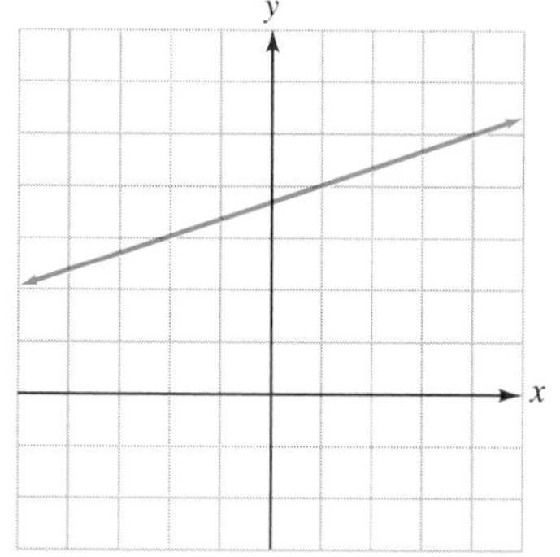

77. The line passing through $(0, -8)$ and $(-5, 0)$

78. The line with equation $y = -3x + 4$

79. The line with equation $4x - 3y = -3$

80. The line with equation $y = 3$

3.3 Systems with Three Variables

In this section, you will learn about

- **Solving three equations with three variables**
- **Consistent systems**
- **An inconsistent system**
- **Systems with dependent equations**
- **Problem solving**
- **Curve fitting**

Introduction. In the two preceding sections, we solved systems of two linear equations with two variables. In this section, we will solve systems of linear equations with three variables by using a combination of the addition method and substitution. We will then use that procedure to solve problems involving three variables.

Solving three equations with three variables

We now extend the definition of a linear equation to include equations of the form $Ax + By + Cz = D$. The solution of a system of three linear equations with three variables is an **ordered triple** of numbers. For example, the solution of the system

$$\begin{cases} 2x + 3y + 4z = 20 \\ 3x + 4y + 2z = 17 \\ 3x + 2y + 3z = 16 \end{cases}$$

is the triple (1, 2, 3), since each equation is satisfied if $x = 1$, $y = 2$, and $z = 3$.

$$\begin{aligned} 2x + 3y + 4z &= 20 \\ 2(1) + 3(2) + 4(3) &= 20 \\ 2 + 6 + 12 &= 20 \\ 20 &= 20 \end{aligned} \qquad \begin{aligned} 3x + 4y + 2z &= 17 \\ 3(1) + 4(2) + 2(3) &= 17 \\ 3 + 8 + 6 &= 17 \\ 17 &= 17 \end{aligned} \qquad \begin{aligned} 3x + 2y + 3z &= 16 \\ 3(1) + 2(2) + 3(3) &= 16 \\ 3 + 4 + 9 &= 16 \\ 16 &= 16 \end{aligned}$$

The graph of an equation of the form $Ax + By + Cz = D$ is a flat surface called a *plane.* A system of three linear equations with three variables is consistent or inconsistent, depending on how the three planes corresponding to the three equations intersect. Figure 3-10 illustrates some of the possibilities.

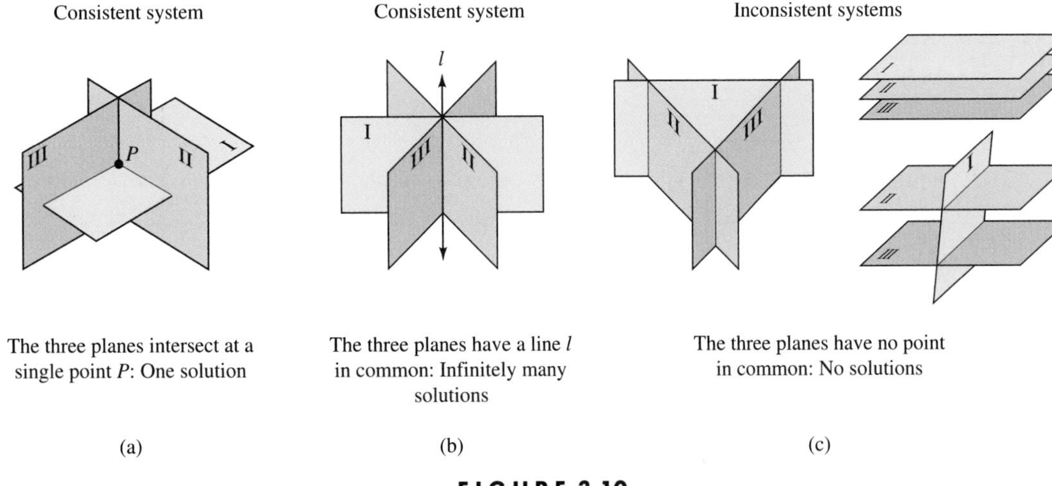

Consistent system
Consistent system
Inconsistent systems
P
I
II
III
l
The three planes intersect at a single point P: One solution
The three planes have a line l in common: Infinitely many solutions
The three planes have no point in common: No solutions
(a)
(b)
(c)

FIGURE 3-10

To solve a system of three linear equations with three variables, we follow these steps.

Solving three equations with three variables

1. Pick any two equations and eliminate a variable.
2. Pick a different pair of equations and eliminate the same variable.
3. Solve the resulting pair of two equations with two variables.
4. To find the value of the third variable, substitute the values of the two variables found in step 3 into any equation containing all three variables and solve the equation.
5. Check the solution in all three of the original equations.

EXAMPLE 1 ***Solving a system of three equations.*** Solve the system $\begin{cases} 2x + y + 4z = 12 \\ x + 2y + 2z = 9 \\ 3x - 3y - 2z = 1 \end{cases}$.

Self Check

Solve $\begin{cases} 2x + y + 4z = 16 \\ x + 2y + 2z = 11 \\ 3x - 3y - 2z = -9 \end{cases}$.

Solution

Step 1: We are given the system

$$\begin{array}{ll} \textbf{1.} & \\ \textbf{2.} & \\ \textbf{3.} & \end{array} \begin{cases} 2x + y + 4z = 12 \\ x + 2y + 2z = 9 \\ 3x - 3y - 2z = 1 \end{cases}$$

If we pick Equations 2 and 3 and add them, the variable z is eliminated.

$$\begin{array}{ll} \textbf{2.} & x + 2y + 2z = 9 \\ \textbf{3.} & \underline{3x - 3y - 2z = 1} \\ \textbf{4.} & 4x - y \qquad = 10 \end{array}$$

Step 2: We now pick a different pair of equations (Equations 1 and 3) and eliminate z again. If each side of Equation 3 is multiplied by 2 and the resulting equation is added to Equation 1, z is eliminated.

$$\begin{array}{lll} \textbf{1.} & 2x + y + 4z = 12 & \\ + & \underline{6x - 6y - 4z = 2} & \text{Multiply both sides of Equation 3 by 2.} \\ \textbf{5.} & 8x - 5y \qquad = 14 & \end{array}$$

Step 3: Equations 4 and 5 form a system of two equations with two variables, x and y.

$$\begin{array}{ll} \textbf{4.} & \\ \textbf{5.} & \end{array} \begin{cases} 4x - y = 10 \\ 8x - 5y = 14 \end{cases}$$

To solve this system, we multiply Equation 4 by -5 and add the resulting equation to Equation 5 to eliminate y:

$$\begin{array}{lll} + & -20x + 5y = -50 & \text{Multiply both sides of Equation 4 by } -5. \\ \textbf{5.} & \underline{8x - 5y = 14} & \\ & -12x \qquad = -36 & \\ & x = 3 & \text{To find } x\text{, divide both sides by } -12. \end{array}$$

To find y, we substitute 3 for x in any equation containing x and y (such as Equation 5) and solve for y:

$$\begin{array}{lll} \textbf{5.} & 8x - 5y = 14 & \\ & 8(3) - 5y = 14 & \text{Substitute 3 for } x. \\ & 24 - 5y = 14 & \text{Simplify.} \\ & -5y = -10 & \text{Subtract 24 from both sides.} \\ & y = 2 & \text{Divide both sides by } -5. \end{array}$$

Step 4: To find z, we substitute 3 for x and 2 for y in an equation containing x, y, and z (such as Equation 1) and solve for z:

1. $2x + y + 4z = 12$

$2(3) + 2 + 4z = 12$ Substitute 3 for x and 2 for y.

$8 + 4z = 12$ Simplify.

$4z = 4$ Subtract 8 from both sides

$z = 1$ Divide both sides by 4.

The solution of the system is $(x, y, z) = (3, 2, 1)$. Because this system has a solution, it is a consistent system.

Step 5: Verify that these values satisfy each equation in the original system.

Answer: (1, 2, 3)

When one or more of the equations of a system is missing a term, the elimination of a variable that is normally performed in step 1 of the solution process can be skipped.

EXAMPLE 2 ***Missing terms.*** Solve the system $\begin{cases} 3x = 6 - 2y + z \\ -y - 2z = -8 - x. \\ x = 1 - 2z \end{cases}$

Self Check

Solve $\begin{cases} x + 2y - z = 1 \\ 2x - y + z = 3. \\ x + z = 3 \end{cases}$

Solution

Step 1: First, we write each equation in $Ax + By + Cz = D$ form.

1. $3x + 2y - z = 6$
2. $x - y - 2z = -8$
3. $x + 2z = 1$

Since Equation 3 does not have a y-term, we can proceed to step 2, where we will find another equation that does not contain a y-term.

Step 2: If each side of Equation 2 is multiplied by 2 and the resulting equation is added to Equation 1, y is eliminated.

1. $3x + 2y - z = 6$
$+\ 2x - 2y - 4z = -16$ Multiply both sides of Equation 2 by 2.
4. $5x - 5z = -10$

Step 3: Equations 3 and 4 form a system of two equations with two variables, x and z:

3. $x + 2z = 1$
4. $5x - 5z = -10$

To solve this system, we multiply Equation 3 by -5 and add the resulting equation to Equation 4 to eliminate x:

$-5x - 10z = -5$ Multiply both sides of Equation 3 by -5.
4. $+\ 5x - 5z = -10$

$-15z = -15$

$z = 1$ To find z, divide both sides by -15.

To find x, we substitute 1 for z in Equation 3.

3. $x + 2z = 1$

$x + 2(1) = 1$ Substitute 1 for z.

$x + 2 = 1$ Multiply.

$x = -1$ Subtract 2 from both sides.

Step 4: To find y, we substitute -1 for x and 1 for z in Equation 1:

1. $3x + 2y - z = 6$

$3(-1) + 2y - 1 = 6$ Substitute -1 for x and 1 for z.

$-3 + 2y - 1 = 6$ Multiply.

$2y = 10$ Add 4 to both sides.

$y = 5$ Divide both sides by 2.

The solution of the system is $(-1, 5, 1)$.

Step 5: Check the solution in all three of the original equations.

Answer: (1, 1, 2)

An inconsistent system

EXAMPLE 3 ***A system with no solution.*** Solve the system

$$\begin{cases} 2a + b - 3c = -3 \\ 3a - 2b + 4c = 2 \\ 4a + 2b - 6c = -7 \end{cases}.$$

Solution

We can multiply the first equation of the system by 2 and add the resulting equation to the second equation to eliminate b:

$$\begin{array}{rl} 4a + 2b - 6c = -6 & \text{Multiply both sides of the first equation by 2.} \\ + \quad 3a - 2b + 4c = 2 & \\ \hline \textbf{1.} \quad 7a \qquad - 2c = -4 & \end{array}$$

Now add the second and third equations of the system to eliminate b again:

$$\begin{array}{r} 3a - 2b + 4c = 2 \\ + \quad 4a + 2b - 6c = -7 \\ \hline \textbf{2.} \quad 7a \qquad - 2c = -5 \end{array}$$

Equations 1 and 2 form the system

$$\begin{array}{l} \textbf{1.} \\ \textbf{2.} \end{array} \begin{cases} 7a - 2c = -4 \\ 7a - 2c = -5 \end{cases}$$

Since $7a - 2c$ cannot equal both -4 and -5, the system is inconsistent and has no solution.

Self Check

Solve $\begin{cases} 2a + b - 3c = 8 \\ 3a - 2b + 4c = 10 \\ 4a + 2b - 6c = -5 \end{cases}$.

Answer: no solution

Systems with dependent equations

When the equations in a system of two equations with two variables are dependent, the system has infinitely many solutions. This is not always true for systems of three equations with three variables. In fact, a system can have dependent equations and still be inconsistent. Figure 3-11 illustrates the different possibilities.

Consistent system

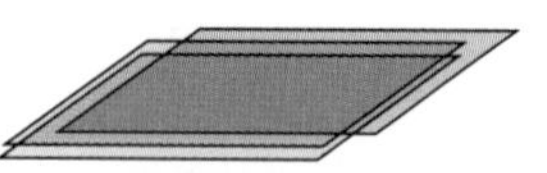

When three planes coincide, the equations are dependent, and there are infinitely many solutions.

(a)

Consistent system

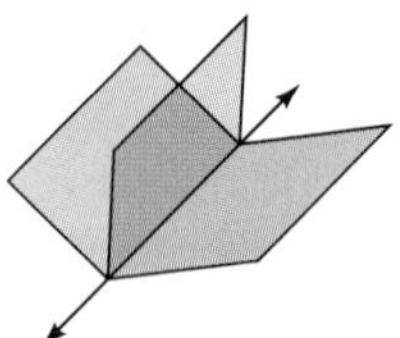

When three planes intersect in a common line, the equations are dependent, and there are infinitely many solutions.

(b)

Inconsistent system

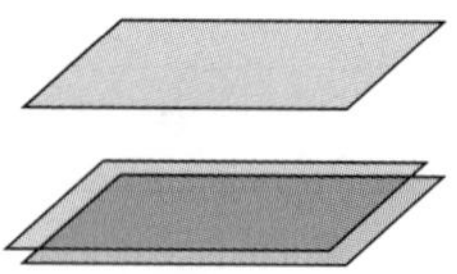

When two planes coincide and are parallel to a third plane, the system is inconsistent, and there are no solutions.

(c)

FIGURE 3-11

EXAMPLE 4 ***A system with infinitely many solutions.*** Solve the system $\begin{cases} 3x - 2y + z = -1 \\ 2x + y - z = 5. \\ 5x - y = 4 \end{cases}$

Solution

We can add the first two equations to get

$$\begin{array}{rl} & 3x - 2y + z = -1 \\ + & 2x + y - z = 5 \\ \hline \textbf{1.} & 5x - y = 4 \end{array}$$

Since Equation 1 is the same as the third equation of the system, the equations of the system are dependent, and there will be infinitely many solutions. From a graphical perspective, the equations represent three planes that intersect in a common line, as shown in Figure 3-11(b).

To write the general solution of this system, we can solve Equation 1 for y to get

$$5x - y = 4$$
$$-y = -5x + 4 \quad \text{Subtract } 5x \text{ from both sides.}$$
$$y = 5x - 4 \quad \text{Multiply both sides by } -1.$$

We can then substitute $5x - 4$ for y in the first equation of the system and solve for z to get

$$3x - 2y + z = -1$$
$$3x - 2(5x - 4) + z = -1 \quad \text{Substitute } 5x - 4 \text{ for } y.$$
$$3x - 10x + 8 + z = -1 \quad \text{Use the distributive property to remove parentheses.}$$
$$-7x + 8 + z = -1 \quad \text{Combine like terms.}$$
$$z = 7x - 9 \quad \text{Add } 7x \text{ and } -8 \text{ to both sides.}$$

Since we have found the values of y and z in terms of x, every solution of the system has the form $(x, 5x - 4, 7x - 9)$, where x can be any real number. For example,

If $x = 1$, a solution is $(1, 1, -2)$. $5(1) - 4 = 1$, and $7(1) - 9 = -2$.
If $x = 2$, a solution is $(2, 6, 5)$. $5(2) - 4 = 6$, and $7(2) - 9 = 5$.
If $x = 3$, a solution is $(3, 11, 12)$. $5(3) - 4 = 11$, and $7(3) - 9 = 12$.

Self Check

Solve $\begin{cases} 3x + 2y + z = -1 \\ 2x - y - z = 5. \\ 5x + y = 4 \end{cases}$

Answer: There are infinitely many solutions. A general solution is $(x, 4 - 5x, -9 + 7x)$. Three solutions are $(1, -1, -2)$, $(2, -6, 5)$, and $(3, -11, 12)$

Problem solving

EXAMPLE 5 ***Tool manufacturing.*** A tool company makes three types of hammers, which are marketed as "good," "better," and "best." The cost of manufacturing each type of hammer is \$4, \$6, and \$7, respectively, and the hammers sell for \$6, \$9, and \$12. Each day, the cost of manufacturing 100 hammers is \$520, and the daily revenue from their sale is \$810. How many hammers of each type are manufactured?

Analyze the problem We need to find how many of each type of hammer are manufactured daily. We must write three equations to find three unknowns.

Form three equations If we let x represent the number of good hammers, y represent the number of better hammers, and z represent the number of best hammers, we know that

The total number of hammers is $x + y + z$.

The cost of manufacturing the good hammers is \$$4x$ (\$4 times x hammers).
The cost of manufacturing the better hammers is \$$6y$ (\$6 times y hammers).
The cost of manufacturing the best hammers is \$$7z$ (\$7 times z hammers).

The revenue received by selling the good hammers is \$6x (\$6 times x hammers).
The revenue received by selling the better hammers is \$9y (\$9 times y hammers).
The revenue received by selling the best hammers is \$12z (\$12 times z hammers).

We can assemble the facts of the problem to write three equations.

The number of good hammers	+	the number of better hammers	+	the number of best hammers	=	the total number of hammers.
x	+	y	+	z	=	100
The cost of good hammers	+	**the cost of better hammers**	+	**the cost of best hammers**	=	**the total cost.**
$4x$	+	$6y$	+	$7z$	=	520
The revenue from good hammers	+	**the revenue from better hammers**	+	**the revenue from best hammers**	=	**the total revenue.**
$6x$	+	$9y$	+	$12z$	=	810

Solve the system We must now solve the system

$$\begin{cases} x + y + z = 100 & \textbf{1.} \\ 4x + 6y + 7z = 520 & \textbf{2.} \\ 6x + 9y + 12z = 810 & \textbf{3.} \end{cases}$$

If we multiply Equation 1 by -7 and add the result to Equation 2, we get

$$\begin{array}{rl} & -7x - 7y - 7z = -700 \\ + & 4x + 6y + 7z = 520 \\ \hline \textbf{4.} & -3x - y = -180 \end{array}$$

If we multiply Equation 1 by -12 and add the result to Equation 3, we get

$$\begin{array}{rl} & -12x - 12y - 12z = -1{,}200 \\ + & 6x + 9y + 12z = 810 \\ \hline \textbf{5.} & -6x - 3y = -390 \end{array}$$

If we multiply Equation 4 by -3 and add it to Equation 5, we get

$$\begin{array}{rl} & 9x + 3y = 540 \\ + & -6x - 3y = -390 \\ \hline & 3x = 150 \end{array}$$

$x = 50$ To find x, divide both sides by 3.

To find y, we substitute 50 for x in Equation 4:

$-3x - y = -180$

$-3(50) - y = -180$ Substitute 50 for x.

$-150 - y = -180$ $-3(50) = -150$.

$-y = -30$ Add 150 to both sides.

$y = 30$ Divide both sides by -1.

To find z, we substitute 50 for x and 30 for y in Equation 1:

$x + y + z = 100$

$50 + 30 + z = 100$

$z = 20$ Subtract 80 from both sides.

State the conclusion The company manufactures 50 good hammers, 30 better hammers, and 20 best hammers each day.

Check the result Check the solution in each equation in the original system. ■

Curve fitting

EXAMPLE 6 ***Finding the equation of a parabola.*** The equation of a parabola opening upward or downward is of the form $y = ax^2 + bx + c$. Find the equation of the parabola shown in Figure 3-12 by determining the values of a, b, and c.

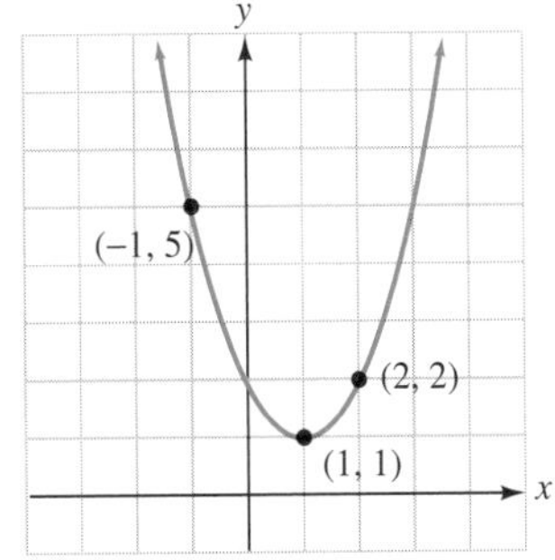

FIGURE 3-12

Solution Since the parabola passes through the points $(-1, 5)$, $(1, 1)$, and $(2, 2)$, each pair of coordinates must satisfy the equation $y = ax^2 + bx + c$. If we substitute the x- and y-coordinates of each point into the equation and simplify, we obtain the following system of three equations with three variables.

$$\begin{cases} a - b + c = 5 & \text{(1) Substitute the coordinates of } (-1, 5) \text{ into } y = ax^2 + bx + c \text{ and simplify.} \\ a + b + c = 1 & \text{(2) Substitute the coordinates of } (1, 1) \text{ into } y = ax^2 + bx + c \text{ and simplify.} \\ 4a + 2b + c = 2 & \text{(3) Substitute the coordinates of } (2, 2) \text{ into } y = ax^2 + bx + c \text{ and simplify.} \end{cases}$$

If we add Equations 1 and 2, we obtain

$$\begin{array}{rl} & a - b + c = 5 \\ + & a + b + c = 1 \\ \hline \textbf{4.} & 2a \qquad + 2c = 6 \end{array}$$

If we multiply Equation 1 by 2 and add the result to Equation 3, we get

$$\begin{array}{rl} & 2a - 2b + 2c = 10 \\ + & 4a + 2b + c = 2 \\ \hline \textbf{5.} & 6a \qquad + 3c = 12 \end{array}$$

We can then divide both sides of Equation 4 by 2 to get Equation 6 and divide both sides of Equation 5 by 3 to get Equation 7. We now have the system

$$\begin{array}{l} \textbf{6.} \\ \textbf{7.} \end{array} \begin{cases} a + c = 3 \\ 2a + c = 4 \end{cases}$$

To eliminate c, we multiply Equation 6 by -1 and add the result to Equation 7. We get

$$\begin{array}{rl} & -a - c = -3 \\ + & 2a + c = 4 \\ \hline & a \qquad = 1 \end{array}$$

To find c, we can substitute 1 for a in Equation 6 and find that $c = 2$. To find b, we can substitute 1 for a and 2 for c in Equation 2 and find that $b = -2$.

After we substitute these values of a, b, and c into the equation $y = ax^2 + bx + c$, we have the equation of the parabola.

$$y = ax^2 + \mathbf{b}x + \mathbf{c}$$
$$y = 1x^2 - \mathbf{2}x + \mathbf{2}$$
$$y = x^2 - 2x + 2$$

■

STUDY SET Section 3.3

VOCABULARY *In Exercises 1–6, fill in the blanks to make the statements true.*

1. $\begin{cases} 2x + y - 3z = 0 \\ 3x - y + 4z = 5 \\ 4x + 2y - 6z = 0 \end{cases}$ is called a ____________ of three linear equations.

2. If the first two equations of the system in Exercise 1 are added, the variable y is ____________.

3. The equation $2x + 3y + 4z = 5$ is a linear equation with ____________ variables.

4. The graph of the equation $2x + 3y + 4z = 5$ is a flat surface called a ____________.

5. When three planes coincide, the equations of the system are ____________, and there are infinitely many solutions.

6. When three planes intersect in a line, the system will have ____________ many solutions.

CONCEPTS

7. For each graph of a system of three equations, tell whether the solution set contains one solution, infinitely many solutions, or no solution.

a.

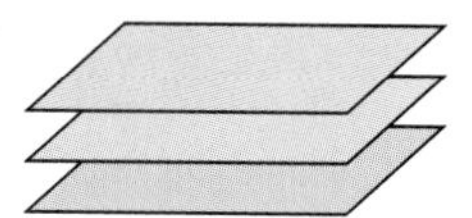

b.

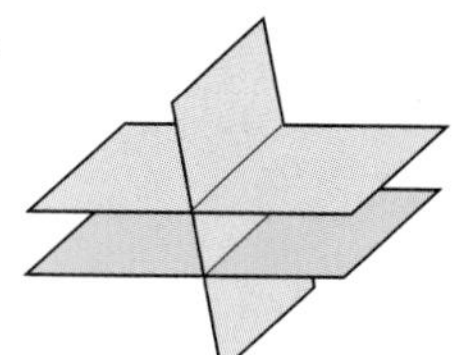

8. Consider the system $\begin{cases} -2x + y + 4z = 3 \\ x - y + 2z = 1 \\ x + y - 3z = 2 \end{cases}$.

a. What is the result if Equation 1 and Equation 2 are added?

b. What is the result if Equation 2 and Equation 3 are added?

c. What variable was eliminated in the steps performed in parts a and b?

NOTATION

9. Write the equation $3z - 2y = x + 6$ in $Ax + By + Cz = D$ form.

10. Fill in the blank to make a true statement: Solutions of a system of three equations in three variables, x, y, and z, are written in the form (x, y, z) and are called ordered ____________.

PRACTICE *In Exercises 11–12, tell whether the given ordered triple is a solution of given system.*

11. $(2, 1, 1)$, $\begin{cases} x - y + z = 2 \\ 2x + y - z = 4 \\ 2x - 3y + z = 2 \end{cases}$

12. $(-3, 2, -1)$, $\begin{cases} 2x + 2y + 3z = -1 \\ 3x + y - z = -6 \\ x + y + 2z = 1 \end{cases}$

In Exercises 13–28, solve each system. If the equations of a system are dependent or if a system is inconsistent, so indicate.

13. $\begin{cases} x + y + z = 4 \\ 2x + y - z = 1 \\ 2x - 3y + z = 1 \end{cases}$

14. $\begin{cases} x + y + z = 4 \\ x - y + z = 2 \\ x - y - z = 0 \end{cases}$

15. $\begin{cases} 2x + 2y + 3z = 10 \\ 3x + y - z = 0 \\ x + y + 2z = 6 \end{cases}$

16. $\begin{cases} x - y + z = 4 \\ x + 2y - z = -1 \\ x + y - 3z = -2 \end{cases}$

17. $\begin{cases} b + 2c = 7 - a \\ a + c = 8 - 2b \\ 2a + b + c = 9 \end{cases}$

18. $\begin{cases} 2a = 2 - 3b - c \\ 4a + 6b + 2c - 5 = 0 \\ a + c = 3 + 2b \end{cases}$

19. $\begin{cases} 2x + y - z = 1 \\ x + 2y + 2z = 2 \\ 4x + 5y + 3z = 3 \end{cases}$

20. $\begin{cases} 4x + 3z = 4 \\ 2y - 6z = -1 \\ 8x + 4y + 3z = 9 \end{cases}$

21. $\begin{cases} a + b + c = 180 \\ \dfrac{a}{4} + \dfrac{b}{2} + \dfrac{c}{3} = 60 \\ 2b + 3c - 330 = 0 \end{cases}$

22. $\begin{cases} 2a + 3b - 2c = 18 \\ 5a - 6b + c = 21 \\ 4b - 2c - 6 = 0 \end{cases}$

23. $\begin{cases} 0.5a + 0.3b = 2.2 \\ 1.2c - 8.5b = -24.4 \\ 3.3c + 1.3a = 29 \end{cases}$

24. $\begin{cases} 4a - 3b = 1 \\ 6a - 8c = 1 \\ 2b - 4c = 0 \end{cases}$

25. $\begin{cases} 2x + 3y + 4z = 6 \\ 2x - 3y - 4z = -4 \\ 4x + 6y + 8z = 12 \end{cases}$

26. $\begin{cases} x - 3y + 4z = 2 \\ 2x + y + 2z = 3 \\ 4x - 5y + 10z = 7 \end{cases}$

27. $\begin{cases} x + \dfrac{1}{3}y + z = 13 \\ \dfrac{1}{2}x - y + \dfrac{1}{3}z = -2 \\ x + \dfrac{1}{2}y - \dfrac{1}{3}z = 2 \end{cases}$

28. $\begin{cases} x - \dfrac{1}{5}y - z = 9 \\ \dfrac{1}{4}x + \dfrac{1}{5}y - \dfrac{1}{2}z = 5 \\ 2x + y + \dfrac{1}{6}z = 12 \end{cases}$

APPLICATIONS

29. MAKING STATUES An artist makes three types of ceramic statues at a monthly cost of \$650 for 180 statues. The manufacturing costs for the three types are \$5, \$4, and \$3. If the statues sell for \$20, \$12, and \$9, respectively, how many of each type should be made to produce \$2,100 in monthly revenue?

30. POTPOURRI The owner of a home decorating shop wants to mix dried rose petals selling for \$6 per pound, dried lavender selling for \$5 per pound, and buckwheat hulls selling for \$4 per pound to get 10 pounds of a mixture that would sell for \$5.50 per pound. She wants to use twice as many pounds of rose petals as lavender. How many pounds of each should she use?

31. DIETITIAN A hospital dietitian is to design a meal that will provide a patient with exactly 14 grams (g) of fat, 9 g of carbohydrates, and 9 g of protein. She is to use a combination of the three foods listed in the table. If one ounce of each of the foods has the nutrient content shown in Illustration 1, how many ounces of each should be used?

Food	Fat	Carbohydrates	Protein
A	2 g	1 g	2 g
B	3 g	2 g	1 g
C	1 g	1 g	2 g

ILLUSTRATION 1

32. NUTRITIONAL PLANNING One ounce of each of three foods has the vitamin and mineral content shown in Illustration 2. How many ounces of each must be used to provide exactly 22 milligrams (mg) of niacin, 12 mg of zinc, and 20 mg of vitamin C?

Food	Niacin	Zinc	Vitamin C
A	1 mg	1 mg	2 mg
B	2 mg	1 mg	1 mg
C	2 mg	1 mg	2 mg

ILLUSTRATION 2

33. CHAINSAW SCULPTING A north woods sculptor carves three types of statues with a chainsaw. The number of hours required for carving, sanding, and painting a totem pole, a bear, and a deer are shown in Illustration 3. How many of each should be produced to use all available labor hours?

	Totem pole	Bear	Deer	Time available
Carving	2 hr	2 hr	1 hr	14 hr
Sanding	1 hr	2 hr	2 hr	15 hr
Painting	3 hr	2 hr	2 hr	21 hr

ILLUSTRATION 3

34. CLOTHING MAKER A clothing manufacturer makes coats, shirts, and slacks. The time required for cutting, sewing, and packaging each item is shown in Illustration 4. How many of each should be made to use all available labor hours?

	Coats	Shirts	Slacks	Time available
Cutting	20 min	15 min	10 min	115 hr
Sewing	60 min	30 min	24 min	280 hr
Packaging	5 min	12 min	6 min	65 hr

ILLUSTRATION 4

35. MAIL The circle graph in Illustration 5 shows the types of mail the average American household receives each week. Use the information in the graph to determine what percent of the week's mail is advertising, is bills and statements, and is personal.

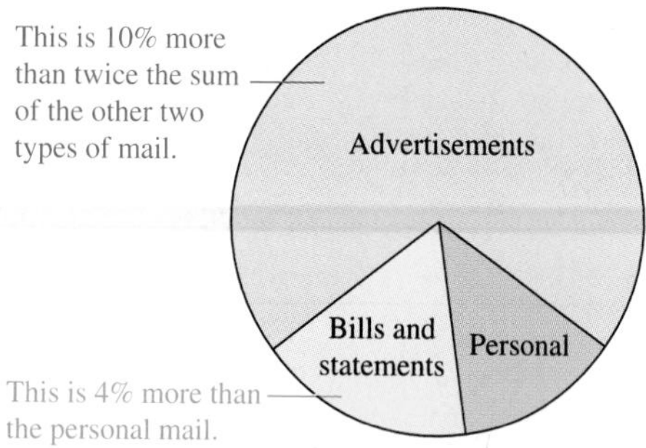

Based on information from *Time* (July 14, 1997).

ILLUSTRATION 5

36. GRAPHS OF SYSTEMS Explain how each picture in Illustration 6 could be thought of as an example of the graph of a system of three equations. Then describe the solution, if there is any.

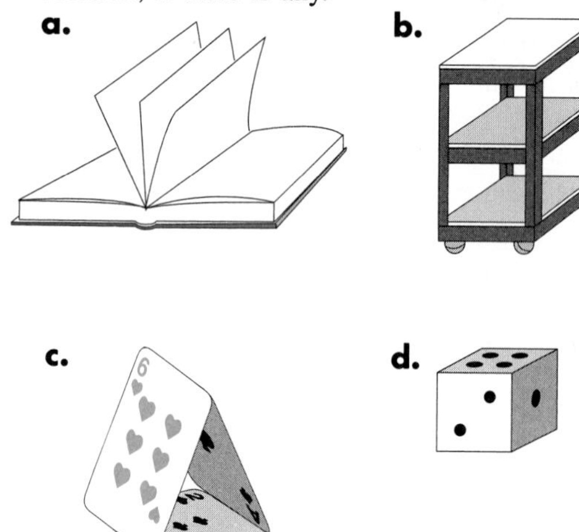

ILLUSTRATION 6

37. ASTRONOMY Comets have elliptical orbits, but the orbits of some comets are so vast that they are indistinguishable from parabolas. Find the equation of the parabola that describes the orbit of the comet shown in Illustration 7.

38. CURVE FITTING Find the equation of the parabola shown in Illustration 8.

39. PARK WALKWAY A circular sidewalk is to be constructed in a city park. The walk is to pass by three particular areas of the park, as shown in the graph in Illustration 9. If an equation of a circle is of the form $x^2 + y^2 + Cx + Dy + E = 0$, find the equation that describes the path of the sidewalk by determining C, D, and E.

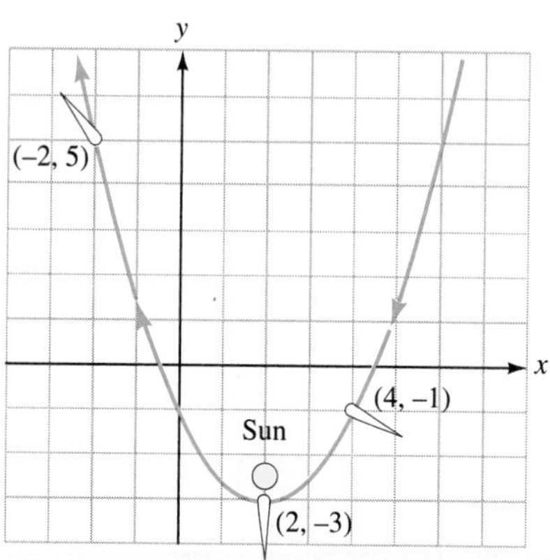

ILLUSTRATION 7

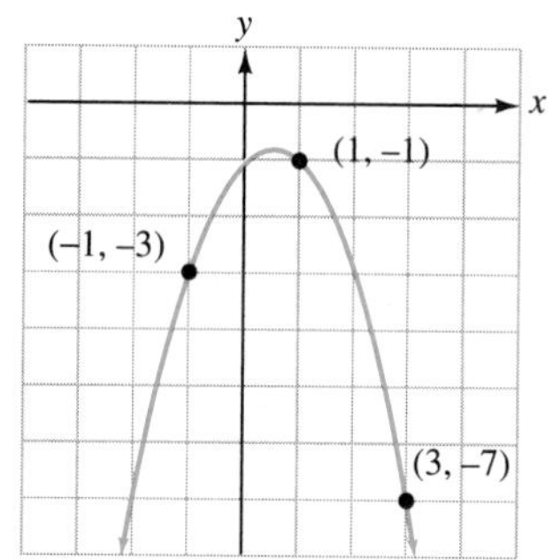

ILLUSTRATION 8

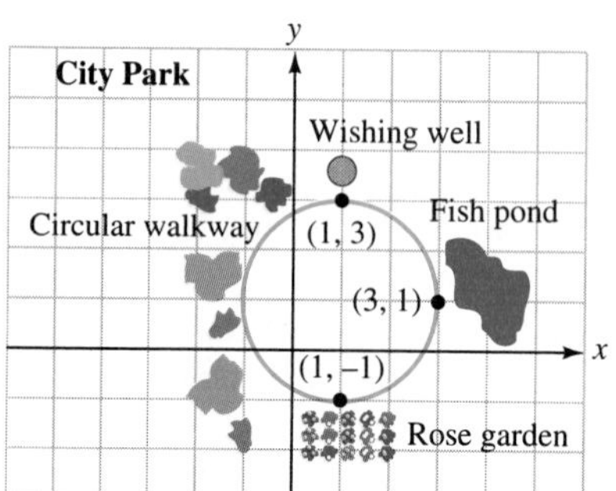

ILLUSTRATION 9

40. CURVE FITTING The equation of a circle is of the form $x^2 + y^2 + Cx + Dy + E = 0$. Find the equation of the circle shown in Illustration 10 by determining C, D, and E.

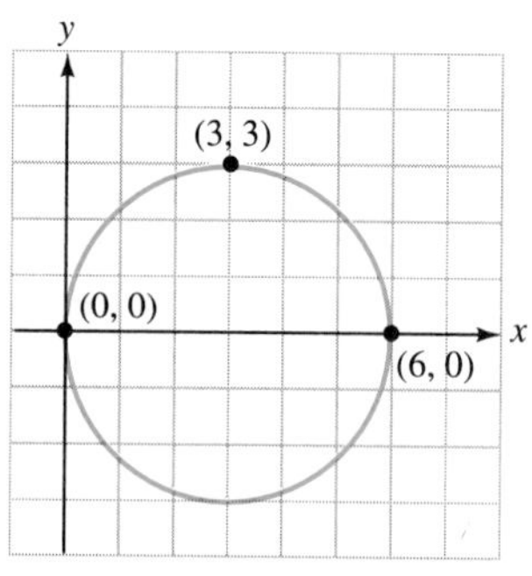

ILLUSTRATION 10

41. TRIANGLES The sum of the angles in any triangle is 180°. In triangle ABC, angle A is 100° less than the sum of angles B and C, and angle C is 40° less than twice angle B. Find each angle.

42. QUADRILATERALS The sum of the angles of any four-sided figure is 360°. In the quadrilateral shown in Illustration 11, the measures of angle A and angle B are the same, angle C is 20° greater than angle A, and angle D measures 40°. Find the angles.

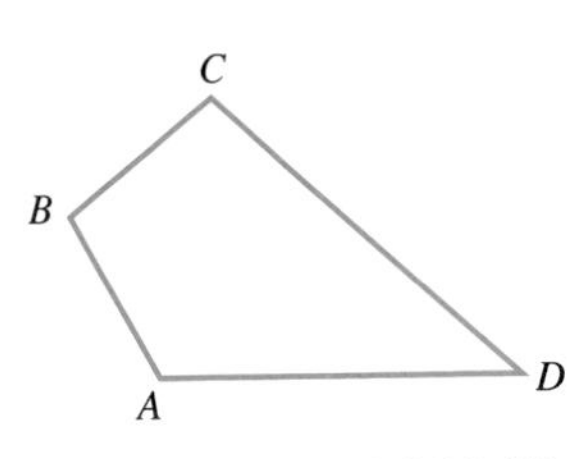

ILLUSTRATION 11

43. INTEGER PROBLEM The sum of three integers is 48. If the first integer is doubled, the sum is 60. If the second integer is doubled, the sum is 63. Find the integers.

44. INTEGER PROBLEM The sum of three integers is 18. The third integer is four times the second, and the second integer is 6 more than the first. Find the integers.

WRITING *Write a paragraph using your own words.*

45. Explain how a system of three equations with three variables can be reduced to a system of two equations with two variables.

46. What makes a system of three equations with three variables inconsistent?

REVIEW *In Exercises 47–50, graph each function.*

47. $f(x) = |x|$

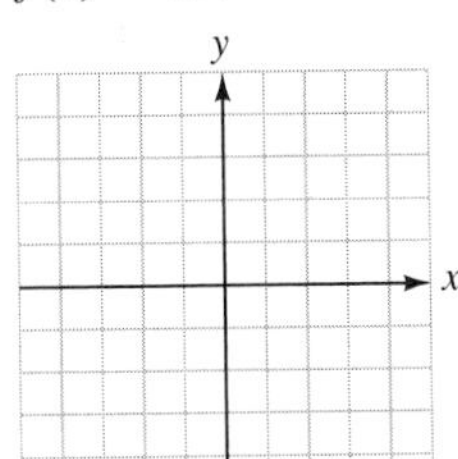

48. $g(x) = x^2$

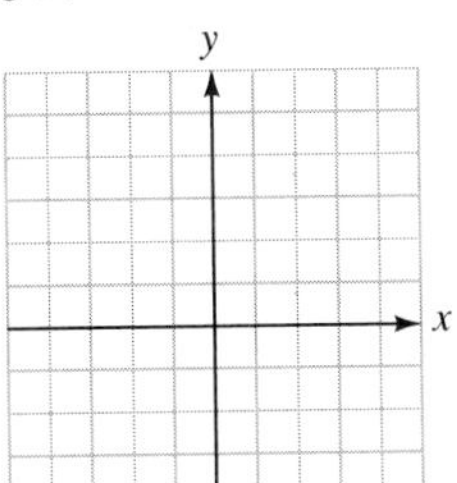

49. $h(x) = x^3$

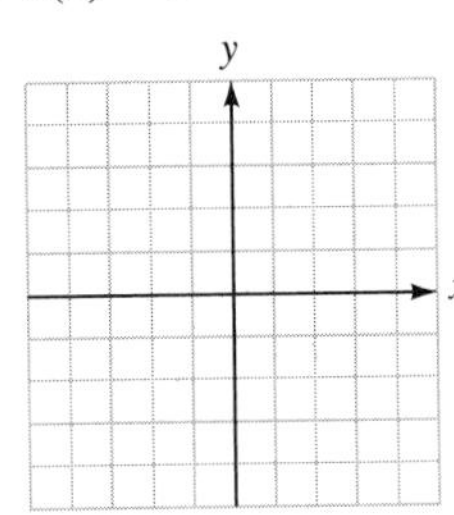

50. $S(x) = x$

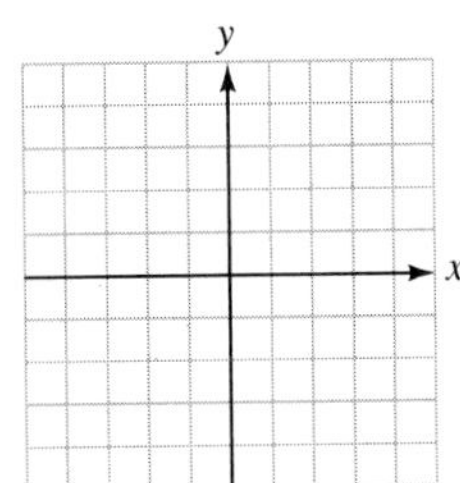

3.4 Solving Systems Using Matrices

In this section, you will learn about

- **Matrices**
- **Solving a system of two equations**
- **Gaussian elimination**
- **Solving a system of three equations**
- **Inconsistent systems and dependent equations**

INTRODUCTION. In this section, we will discuss another method for solving systems of linear equations. This technique uses a mathematical tool called a *matrix* in a series of steps that are based on the addition method.

Matrices

Another method of solving systems of equations involves rectangular arrays of numbers called *matrices.*

Matrix A **matrix** is any rectangular array of numbers arranged in rows and columns, written within brackets.

Some examples of matrices are

$$A = \begin{bmatrix} 1 & -3 & 8 \\ 2 & 5 & -1 \end{bmatrix} \begin{matrix} \leftarrow \text{Row 1} \\ \leftarrow \text{Row 2} \end{matrix} \qquad B = \begin{bmatrix} 1 & 4 & -2 & -4 \\ 6 & -2 & 6 & 1 \\ 3 & 8 & -3 & 12 \end{bmatrix} \begin{matrix} \leftarrow \text{Row 1} \\ \leftarrow \text{Row 2} \\ \leftarrow \text{Row 3} \end{matrix}$$

Column 1, Column 2, Column 3 (matrix A); Column 1, Column 2, Column 3, Column 4 (matrix B)

The numbers in each matrix are called **elements.** Because matrix A has two rows and three columns, it is called a 2×3 matrix (read "2 by 3" matrix). Matrix B is a 3×4 matrix (three rows and four columns).

Solving a system of two equations

To show how to use matrices to solve systems of linear equations, we consider the system

$$\begin{cases} x - y = 4 \\ 2x + y = 5 \end{cases}$$

which can be represented by the following matrix, called an **augmented matrix:**

$$\left[\begin{array}{cc:c} 1 & -1 & 4 \\ 2 & 1 & 5 \end{array}\right]$$

Each row of the augmented matrix represents one equation of the system. The first two columns of the augmented matrix are determined by the coefficients of x and y in the equations of the system. The last column is determined by the constants in the equations.

$$\left[\begin{array}{cc:c} 1 & -1 & 4 \\ 2 & 1 & 5 \end{array}\right] \begin{matrix} \leftarrow \text{This row represents the equation } x - y = 4. \\ \leftarrow \text{This row represents the equation } 2x + y = 5. \end{matrix}$$

Coefficients of x — Coefficients of y — Constants

EXAMPLE 1 ***Augmented matrices.*** Represent each system of equations using an augmented matrix:

a. $\begin{cases} 3x + y = 11 \\ x - 8y = 0 \end{cases}$ and **b.** $\begin{cases} 2a + b - 3c = -3 \\ 9a + 4c = 2 \\ a - b - 6c = -7 \end{cases}$.

Solution

a. $\begin{cases} 3x + y = 11 \\ x - 8y = 0 \end{cases} \begin{matrix} \longleftrightarrow \\ \longleftrightarrow \end{matrix} \left[\begin{array}{cc:c} 3 & 1 & 11 \\ 1 & -8 & 0 \end{array}\right]$

b. $\begin{cases} 2a + b - 3c = -3 \\ 9a + 4c = 2 \\ a - b - 6c = -7 \end{cases} \begin{matrix} \longleftrightarrow \\ \longleftrightarrow \\ \longleftrightarrow \end{matrix} \left[\begin{array}{ccc:c} 2 & 1 & -3 & -3 \\ 9 & 0 & 4 & 2 \\ 1 & -1 & -6 & -7 \end{array}\right]$

Self Check

Represent each system using an augmented matrix:

a. $\begin{cases} 2x - 4y = 9 \\ 5x - y = -2 \end{cases}$ and

b. $\begin{cases} a + b - c = -4 \\ -2b + 7c = 0 \\ 10a + 8b - 4c = 5 \end{cases}$.

Answers:

a. $\left[\begin{array}{cc:c} 2 & -4 & 9 \\ 5 & -1 & -2 \end{array}\right]$,

b. $\left[\begin{array}{ccc:c} 1 & 1 & -1 & -4 \\ 0 & -2 & 7 & 0 \\ 10 & 8 & -4 & 5 \end{array}\right]$ ■

Gaussian elimination

To solve a 2×2 system of equations by **Gaussian elimination,** we transform the augmented matrix into the following matrix, which has 1's down its main diagonal and a 0 below the 1 in the first column.

$$\left[\begin{array}{cc|c} 1 & a & b \\ 0 & 1 & c \end{array}\right] \quad (a, b, \text{ and } c \text{ are real numbers})$$

Main diagonal

To write the augmented matrix in this form, we use three operations called **elementary row operations.**

Elementary row operations

Type 1: Any two rows of a matrix can be interchanged.
Type 2: Any row of a matrix can be multiplied by a nonzero constant.
Type 3: Any row of a matrix can be changed by adding a nonzero constant multiple of another row to it.

- A type 1 row operation corresponds to interchanging two equations of the system.
- A type 2 row operation corresponds to multiplying both sides of an equation by a nonzero constant.
- A type 3 row operation corresponds to adding a nonzero multiple of one equation to another.

None of these row operations will change the solution of the given system of equations.

EXAMPLE 2 ***Row operations.*** Consider the matrices

$$A = \left[\begin{array}{cc|c} 2 & 4 & -3 \\ 1 & -8 & 0 \end{array}\right] \quad B = \left[\begin{array}{cc|c} 1 & -1 & 2 \\ 4 & -8 & 0 \end{array}\right] \quad C = \left[\begin{array}{ccc|c} 2 & 1 & -8 & 4 \\ 0 & 1 & 4 & -2 \\ 0 & 0 & -6 & 24 \end{array}\right]$$

a. Interchange rows 1 and 2 of matrix A.

b. Multiply row 3 of matrix C by $-\frac{1}{6}$.

c. To the numbers in row 2 of matrix B, add the results of multiplying each number in row 1 by -4.

Solution

a. Interchanging the rows of matrix A, we obtain $\left[\begin{array}{cc|c} 1 & -8 & 0 \\ 2 & 4 & -3 \end{array}\right]$.

b. We multiply each number in row 3 by $-\frac{1}{6}$. Rows 1 and 2 remain unchanged.

$$\left[\begin{array}{ccc|c} 2 & 1 & -8 & 4 \\ 0 & 1 & 4 & -2 \\ 0 & 0 & 1 & -4 \end{array}\right]$$

We can represent the instruction to multiply the third row by $-\frac{1}{6}$ with the symbolism $-\frac{1}{6}R_3$.

c. If we multiply each number in row 1 of matrix B by -4, we get

$$-4 \quad 4 \quad -8$$

Self Check

a. Interchange the rows of matrix B.

b. To the numbers in row 1 of matrix A, add the results of multiplying each number in row 2 by -2.

c. Interchange rows 2 and 3 of matrix C.

We then add these numbers to row 2. (Note that row 1 remains unchanged.)

$$\left[\begin{array}{cc|c} 1 & -1 & 2 \\ 4+(-4) & -8+4 & 0+(-8) \end{array}\right]$$

We can abbreviate this procedure using the notation $-4R_1 + R_2$, which means "Multiply row 1 by -4 and add the result to row 2."

After simplifying, we have the matrix

$$\left[\begin{array}{cc|c} 1 & -1 & 2 \\ 0 & -4 & -8 \end{array}\right]$$

Answers:

a. $\left[\begin{array}{cc|c} 4 & -8 & 0 \\ 1 & -1 & 2 \end{array}\right]$,

b. $\left[\begin{array}{cc|c} 0 & 20 & -3 \\ 1 & -8 & 0 \end{array}\right]$,

c. $\left[\begin{array}{ccc|c} 2 & 1 & -8 & 4 \\ 0 & 0 & -6 & 24 \\ 0 & 1 & 4 & -2 \end{array}\right]$ ■

We now solve a system of two linear equations using the **Gaussian elimination** process, which involves a series of elementary row operations.

EXAMPLE 3 ***Solving a system using Gaussian elimination.***

Solve the system $\begin{cases} 2x + y = 5 \\ x - y = 4 \end{cases}$.

Solution

We can represent the system with the following augmented matrix:

$$\left[\begin{array}{cc|c} 2 & 1 & 5 \\ 1 & -1 & 4 \end{array}\right]$$

First, we want to get a 1 in the top row of the first column where the red 2 is. This can be achieved by applying a type 1 row operation: Interchange rows 1 and 2.

$$\left[\begin{array}{cc|c} 1 & -1 & 4 \\ 2 & 1 & 5 \end{array}\right]$$

Interchanging row 1 and row 2 can be abbreviated as $R_1 \leftrightarrow R_2$.

To get a 0 under the 1 in the first column, we use a type 3 row operation. To row 2, add the results of multiplying each number in row 1 by -2.

$$\left[\begin{array}{cc|c} 1 & -1 & 4 \\ 0 & 3 & -3 \end{array}\right]$$

$-2R_1 + R_2$.

To get a 1 in the bottom row of the second column, we use a type 2 row operation: Multiply row 2 by $\frac{1}{3}$.

$$\left[\begin{array}{cc|c} 1 & -1 & 4 \\ 0 & 1 & -1 \end{array}\right]$$

$\frac{1}{3}R_2$.

This augmented matrix represents the equations

$$1x - 1y = 4$$
$$0x + 1y = -1$$

Writing the equations without the coefficients of 1 and -1, we have

1. $x - y = 4$
2. $y = -1$

From Equation 2, we see that $y = -1$. We can **back substitute** -1 for y in Equation 1 to find x.

$x - y = 4$

$x - (-1) = 4$ Substitute -1 for y.

$x + 1 = 4$ $-(-1) = 1$.

$x = 3$ Subtract 1 from both sides.

The solution of the system is $(3, -1)$. Verify that this ordered pair satisfies the original system.

Self Check

Solve $\begin{cases} 3x - 2y = -5 \\ x - y = -4 \end{cases}$.

Answer: $(3, 7)$ ■

In general, if a system of linear equations has a single solution, we can use the following steps to solve the system using matrices.

Solving systems of linear equations using matrices

1. Write an augmented matrix for the system.
2. Use elementary row operations to transform the augmented matrix into a matrix with 1's down its main diagonal and 0's under the 1's.
3. When step 2 is complete, write the resulting system. Then use back substitution to find the solution.

Solving a system of three equations

To show how to use matrices to solve systems of three linear equations containing three variables, we consider the system

$$\begin{cases} x - 2y - z = 6 \\ 2x + 2y - z = 1 \\ -x - y + 2z = 1 \end{cases}$$

which can be represented by the augmented matrix:

$$\left[\begin{array}{ccc|c} 1 & -2 & -1 & 6 \\ 2 & 2 & -1 & 1 \\ -1 & -1 & 2 & 1 \end{array}\right]$$

To solve a 3×3 system of equations by Gaussian elimination, we transform the augmented matrix into a matrix with 1's down its main diagonal and 0's below its main diagonal.

$$\left[\begin{array}{ccc|c} 1 & a & b & c \\ 0 & 1 & d & e \\ 0 & 0 & 1 & f \end{array}\right] \qquad (a, b, c, \ldots, f \text{ are real numbers})$$

Main diagonal

EXAMPLE 4 ***Solving a system using Gaussian elimination.***

Solve the system $\begin{cases} x - 2y - z = 6 \\ 2x + 2y - z = 1 \\ -x - y + 2z = 1 \end{cases}$.

Solution

This system can be represented by the augmented matrix:

$$\left[\begin{array}{ccc|c} 1 & -2 & -1 & 6 \\ 2 & 2 & -1 & 1 \\ -1 & -1 & 2 & 1 \end{array}\right] \qquad \text{We need to get a 0 where the red 2 is.}$$

To get a 0 under the 1 in the first column where the red 2 is, we perform a type 3 row operation: Multiply row 1 by -2 and add the results to row 2.

$$\left[\begin{array}{ccc|c} 1 & -2 & -1 & 6 \\ 0 & 6 & 1 & -11 \\ -1 & -1 & 2 & 1 \end{array}\right] \qquad -2R_1 + R_2.$$

To get a 0 under the 0 in the first column where the red -1 is, we perform another type 3 row operation: To row 3 we add row 1.

$$\left[\begin{array}{ccc|c} 1 & -2 & -1 & 6 \\ 0 & 6 & 1 & -11 \\ 0 & -3 & 1 & 7 \end{array}\right] \qquad R_1 + R_3.$$

Self Check

Solve $\begin{cases} x - 2y - z = 2 \\ 2x + 2y - z = -5. \\ -x - y + 2z = 7 \end{cases}$

To get a 0 under the 6 where the red −3 is, we use another type 3 row operation: Multiply row 2 by $\frac{1}{2}$ and add it to row 3.

$$\left[\begin{array}{ccc|c} 1 & -2 & -1 & 6 \\ 0 & 6 & 1 & -11 \\ 0 & 0 & \frac{3}{2} & \frac{3}{2} \end{array}\right] \quad \tfrac{1}{2}R_2 + R_3.$$

We can use a type 2 row operation to simplify further: Multiply row 3 by $\frac{2}{3}$.

$$\left[\begin{array}{ccc|c} 1 & -2 & -1 & 6 \\ 0 & 6 & 1 & -11 \\ 0 & 0 & 1 & 1 \end{array}\right] \quad \tfrac{2}{3}R_3.$$

The final matrix represents the system

1. $x - 2y - z = 6$
2. $0x + 6y + z = -11$
3. $0x + 0y + z = 1$

From Equation 3, we can read that $z = 1$. To find y, we back substitute 1 for z in Equation 2 and solve for y:

$6y + z = -11$ Equation 2.
$6y + 1 = -11$ Substitute 1 for z.
$6y = -12$ Subtract 1 from both sides.
$y = -2$ Divide both sides by 6.

Thus, $y = -2$. To find x, we back substitute 1 for z and −2 for y in Equation 1 and solve for x:

$x - 2y - z = 6$ Equation 1.
$x - 2(-2) - 1 = 6$ Substitute 1 for z and −2 for y.
$x + 3 = 6$ Simplify.
$x = 3$ Subtract 3 from both sides.

Thus, $x = 3$. The solution to the given system is $(3, -2, 1)$. Verify that this triple satisfies each equation of the original system.

Answer: $(1, -2, 3)$

Inconsistent systems and dependent equations

In the next example, we consider a system with no solution.

EXAMPLE 5 ***An inconsistent system.*** Using matrices, solve the system $\begin{cases} x + y = -1 \\ -3x - 3y = -5 \end{cases}$.

Solution

This system can be represented by the augmented matrix

$$\left[\begin{array}{cc|c} 1 & 1 & -1 \\ -3 & -3 & -5 \end{array}\right]$$

Since the matrix has a 1 in the top row of the first column, we proceed to get a 0 under it by multiplying row 1 by 3 and adding the results to row 2.

$$\left[\begin{array}{cc|c} 1 & 1 & -1 \\ 0 & 0 & -8 \end{array}\right] \quad 3R_1 + R_2.$$

Self Check

Solve $\begin{cases} 4x - 8y = 9 \\ x - 2y = -5 \end{cases}$.

This matrix represents the system

$$\begin{cases} x + y = -1 \\ 0 + 0 = -8 \end{cases}$$

This system has no solution, because the second equation is never true. Therefore, the system is inconsistent. It has no solutions.

Answer: no solution ■

In the next example, we consider a system with infinitely many solutions.

EXAMPLE 6 ***A dependent system.*** Using matrices, solve the system

$$\begin{cases} 2x + 3y - 4z = 6 \\ 4x + 6y - 8z = 12 \\ -6x - 9y + 12z = -18 \end{cases}.$$

Self Check

Solve $\begin{cases} 5x - 10y + 15z = 35 \\ -3x + 6y - 9z = -21. \\ 2x - 4y + 6z = 14 \end{cases}$

Solution

This system can be represented by the augmented matrix

$$\left[\begin{array}{ccc|c} 2 & 3 & -4 & 6 \\ 4 & 6 & -8 & 12 \\ -6 & -9 & 12 & -18 \end{array}\right]$$

To get a 1 in the top row of the first column, we multiply row 1 by $\frac{1}{2}$.

$$\left[\begin{array}{ccc|c} 1 & \frac{3}{2} & -2 & 3 \\ 4 & 6 & -8 & 12 \\ -6 & -9 & 12 & -18 \end{array}\right] \quad \tfrac{1}{2}R_1.$$

Next, we want to get 0's under the 1 in the first column. This can be achieved by multiplying row 1 by -4 and adding the results to row 2, and multiplying row 1 by 6 and adding the results to row 3.

$$\left[\begin{array}{ccc|c} 1 & \frac{3}{2} & -2 & 3 \\ 0 & 0 & 0 & 0 \\ 0 & 0 & 0 & 0 \end{array}\right] \quad \begin{array}{l} \\ -4R_1 + R_2. \\ 6R_1 + R_3. \end{array}$$

The last matrix represents the system

$$\begin{cases} x + \frac{3}{2}y - 2z = 3 \\ 0x + 0y + 0z = 0 \\ 0x + 0y + 0z = 0 \end{cases}$$

If we clear the first equation of fractions, we have the system

$$\begin{cases} 2x + 3y - 4z = 6 \\ 0 = 0 \\ 0 = 0 \end{cases}$$

This system has dependent equations and infinitely many solutions. Solutions of this system would be any triple (x, y, z) that satisfies the equation $2x + 3y - 4z = 6$. Two such solutions would be $(0, 2, 0)$ and $(1, 0, -1)$.

Answer: There are infinitely many solutions—any triple satisfying the equation $x - 2y + 3z = 7$. ■

STUDY SET Section 3.4

VOCABULARY *In Exercises 1–6, fill in the blanks to make the statements true.*

1. A ______________ is a rectangular array of numbers.

2. The numbers in a matrix are called its ______________.

3. A 3×4 matrix has 3 __________ and 4 __________.

4. Elementary __________ operations are used to produce new matrices that lead to the solution of a system.

5. A matrix that represents the equations of a system is called an __________ matrix.

6. The augmented matrix $\left[\begin{array}{cc|c} 1 & 3 & -2 \\ 0 & 1 & 4 \end{array}\right]$ has 1's down its main __________.

CONCEPTS

7. For each matrix, tell the number of rows and the number of columns.

a. $\left[\begin{array}{cc|c} 4 & 6 & -1 \\ \frac{1}{2} & 9 & -3 \end{array}\right]$ **b.** $\left[\begin{array}{ccc|c} 1 & -2 & 3 & 1 \\ 0 & 1 & 6 & 4 \\ 0 & 0 & 1 & \frac{1}{3} \end{array}\right]$

8. For each augmented matrix, give the system of equations it represents.

a. $\left[\begin{array}{cc|c} 1 & 6 & 7 \\ 0 & 1 & 4 \end{array}\right]$ **b.** $\left[\begin{array}{ccc|c} 2 & -2 & 9 & 1 \\ 3 & 1 & 1 & 0 \\ 2 & -6 & 8 & -7 \end{array}\right]$

9. Write the system of equations represented by the augmented matrix. Then use back substitution to find the solution.

$$\left[\begin{array}{cc|c} 1 & -1 & -10 \\ 0 & 1 & 6 \end{array}\right]$$

10. Write the system of equations represented by the augmented matrix. Then use back substitution to find the solution.

$$\left[\begin{array}{ccc|c} 1 & -2 & 1 & -16 \\ 0 & 1 & 2 & 8 \\ 0 & 0 & 1 & 4 \end{array}\right]$$

11. Matrices were used to solve a system of two linear equations. The final matrix is shown here. Explain what the result tells about the system.

$$\left[\begin{array}{cc|c} 1 & 2 & -4 \\ 0 & 0 & 2 \end{array}\right]$$

12. Matrices were used to solve a system of two linear equations. The final matrix is shown here. What does the result tell about the equations?

$$\left[\begin{array}{cc|c} 1 & 2 & -4 \\ 0 & 0 & 0 \end{array}\right]$$

NOTATION

13. Consider the matrix $A = \left[\begin{array}{ccc|c} 3 & 6 & -9 & 0 \\ 1 & 5 & -2 & 1 \\ -2 & 2 & -2 & 5 \end{array}\right]$.

a. Explain what is meant by $\frac{1}{3}R_1$. Then perform the operation on matrix A.

b. Explain what is meant by $-R_1 + R_2$. Then perform the operation on the answer to part a.

14. Consider the matrix $B = \left[\begin{array}{cc|c} -3 & 1 & -6 \\ 1 & -4 & 4 \end{array}\right]$.

a. Explain what is meant by $R_1 \leftrightarrow R_2$. Then perform the operation on matrix B.

b. Explain what is meant by $3R_1 + R_2$. Then perform the operation on the answer to part a.

In Exercises 15 and 16, complete each solution.

15. Solve $\begin{cases} 4x - y = 14 \\ x + y = 6 \end{cases}$.

$$\left[\begin{array}{cc|c} 4 & \square & 14 \\ 1 & 1 & 6 \end{array}\right]$$

$$\left[\begin{array}{cc|c} \square & 1 & 6 \\ 4 & -1 & 14 \end{array}\right] \quad R_1 \leftrightarrow R_2.$$

$$\left[\begin{array}{cc|c} 1 & 1 & 6 \\ 0 & \square & -10 \end{array}\right] \quad -4R_1 + R_2.$$

$$\left[\begin{array}{cc|c} 1 & 1 & 6 \\ 0 & 1 & \square \end{array}\right] \quad -\tfrac{1}{5}R_2.$$

This matrix represents the system

$$\begin{cases} x + y = 6 \\ \square = 2 \end{cases}$$

The solution is $(\square, 2)$.

16. Solve $\begin{cases} 2x + 2y = 18 \\ x - y = 5 \end{cases}$.

$$\left[\begin{array}{cc|c} 2 & 2 & 18 \\ \square & -1 & 5 \end{array}\right]$$

$$\left[\begin{array}{cc|c} 1 & 1 & 9 \\ \square & -1 & 5 \end{array}\right] \quad \tfrac{1}{2}R_1.$$

$$\left[\begin{array}{cc|c} 1 & 1 & 9 \\ 0 & -2 & -4 \end{array}\right] \quad -R_1 + R_2.$$

$$\left[\begin{array}{cc|c} 1 & 1 & 9 \\ 0 & 1 & \square \end{array}\right] \quad -\tfrac{1}{2}R_2.$$

This matrix represents the system

$$\begin{cases} x + y = \square \\ y = 2 \end{cases}$$

The solution is $(\square, 2)$.

PRACTICE *In Exercises 17–32, use matrices to solve each system of equations.*

17. $\begin{cases} x + y = 2 \\ x - y = 0 \end{cases}$

18. $\begin{cases} x + y = 3 \\ x - y = -1 \end{cases}$

19. $\begin{cases} 2x + y = 1 \\ x + 2y = -4 \end{cases}$

20. $\begin{cases} 5x - 4y = 10 \\ x - 7y = 2 \end{cases}$

21. $\begin{cases} 2x - y = -1 \\ x - 2y = 1 \end{cases}$

22. $\begin{cases} 2x - y = 0 \\ x + y = 3 \end{cases}$

23. $\begin{cases} 3x + 4y = -12 \\ 9x - 2y = 6 \end{cases}$

24. $\begin{cases} 2x - 3y = 16 \\ -4x + y = -22 \end{cases}$

25. $\begin{cases} x + y + z = 6 \\ x + 2y + z = 8 \\ x + y + 2z = 9 \end{cases}$

26. $\begin{cases} x - y + z = 2 \\ x + 2y - z = 6 \\ 2x - y - z = 3 \end{cases}$

27. $\begin{cases} 3x + y - 3z = 5 \\ x - 2y + 4z = 10 \\ x + y + z = 13 \end{cases}$

28. $\begin{cases} 2x + y - 3z = -1 \\ 3x - 2y - z = -5 \\ x - 3y - 2z = -12 \end{cases}$

29. $\begin{cases} 3x - 2y + 4z = 4 \\ x + y + z = 3 \\ 6x - 2y - 3z = 10 \end{cases}$

30. $\begin{cases} 2x + 3y - z = -8 \\ x - y - z = -2 \\ -4x + 3y + z = 6 \end{cases}$

31. $\begin{cases} 2a + b + 3c = 3 \\ -2a - b + c = 5 \\ 4a - 2b + 2c = 2 \end{cases}$

32. $\begin{cases} 3a + 2b + c = 8 \\ 6a - b + 2c = 16 \\ -9a + b - c = -20 \end{cases}$

In Exercises 33–44, use matrices to solve each system of equations. If the equations of a system are dependent or if a system is inconsistent, so indicate.

33. $\begin{cases} x - 3y = 9 \\ -2x + 6y = 18 \end{cases}$

34. $\begin{cases} -6x + 12y = 10 \\ 2x - 4y = 8 \end{cases}$

35. $\begin{cases} 4x + 4y = 12 \\ -x - y = -3 \end{cases}$

36. $\begin{cases} 5x - 15y = 10 \\ 2x - 6y = 4 \end{cases}$

37. $\begin{cases} 6x + y - z = -2 \\ x + 2y + z = 5 \\ 5y - z = 2 \end{cases}$

38. $\begin{cases} 2x + 3y - 2z = 18 \\ 5x - 6y + z = 21 \\ 4y - 2z = 6 \end{cases}$

39. $\begin{cases} 2x + y - z = 1 \\ x + 2y + 2z = 2 \\ 4x + 5y + 3z = 3 \end{cases}$

40. $\begin{cases} x - 3y + 4z = 2 \\ 2x + y + 2z = 3 \\ 4x - 5y + 10z = 7 \end{cases}$

41. $\begin{cases} 5x + 3y = 4 \\ 3y - 4z = 4 \\ x + z = 1 \end{cases}$

42. $\begin{cases} y + 2z = -2 \\ x + y = 1 \\ 2x - z = 0 \end{cases}$

43. $\begin{cases} x - y = 1 \\ 2x - z = 0 \\ 2y - z = -2 \end{cases}$

44. $\begin{cases} x + y - 3z = 4 \\ 2x + 2y - 6z = 5 \\ -3x + y - z = 2 \end{cases}$

APPLICATIONS *Write a system of equations to solve each problem. Then use matrices to solve the system.*

45. PHYSICAL THERAPY After an elbow injury, a volleyball player has restricted movement of her arm. (See Illustration 1.) Her range of motion (angle 1) is 28° less than angle 2. Find the measure of each angle.

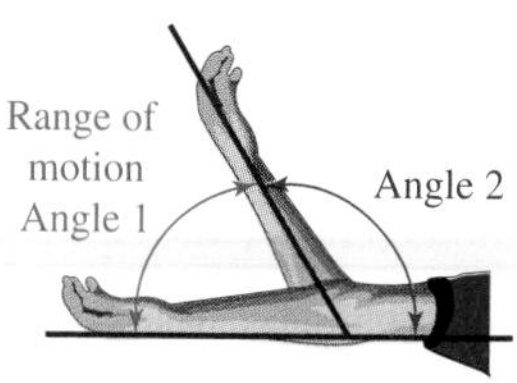

ILLUSTRATION 1

46. TRIANGLES In Illustration 2, $\angle B$ (read "angle B") is 25° more than $\angle A$, and $\angle C$ is 5° less than twice $\angle A$. Find each angle in the triangle. (*Hint:* The sum of the angles in a triangle is 180°.)

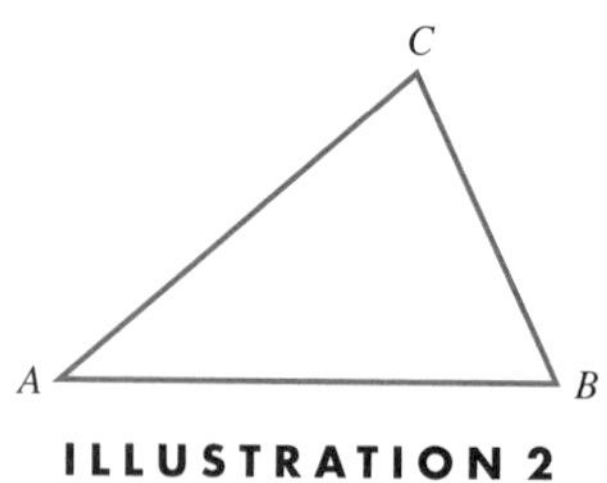

ILLUSTRATION 2

47. THEATER SEATING Illustration 3 shows the cash receipts from two sold-out performances of a play and the ticket prices. Find the number of seats in each of the three sections of the 800-seat theater.

Sunday Ticket Receipts

Matinee	$13,000
Evening	$23,000

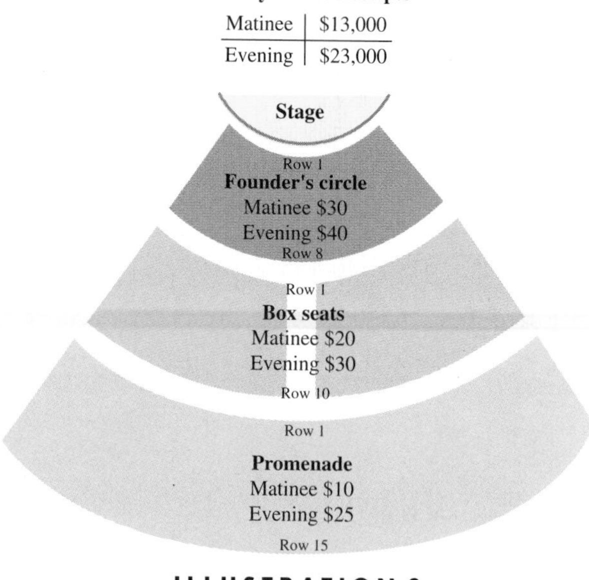

ILLUSTRATION 3

48. ICE SKATING Illustration 4 shows three circles traced out by a figure skater during her performance. If the centers of the circles are the given distances apart, find the radius of each circle.

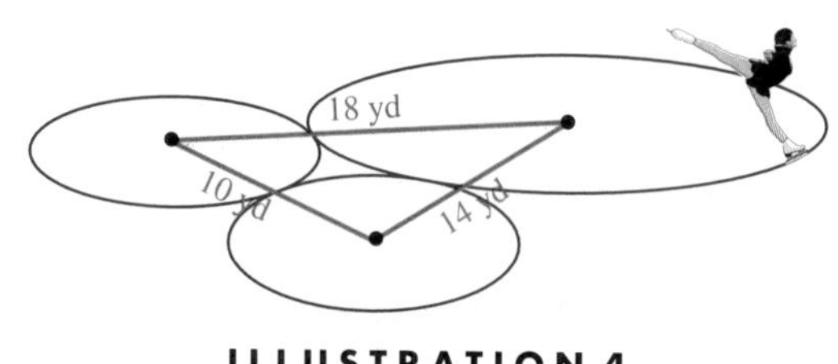

ILLUSTRATION 4

WRITING *Write a paragraph using your own words.*

49. Explain what is meant by the phrase *back substitution.*

50. Explain how a type 3 row operation is similar to the addition method of solving a system of equations.

REVIEW

51. What is the formula used to find the slope of a line, given two points on the line?

52. What is the form of the equation of a horizontal line? Of a vertical line?

53. What is the point–slope form of the equation of a line?

54. What is the slope–intercept form of the equation of a line?

3.5 *Solving Systems Using Determinants*

In this section, you will learn about

- **Determinants**
- **Evaluating a determinant**
- **Using Cramer's rule to solve a system of two equations**
- **Using Cramer's rule to solve a system of three equations**

INTRODUCTION. In this section, we will discuss another method for solving systems of linear equations. With this method, called *Cramer's rule,* we work with combinations of the coefficients and the constants of the equations written as *determinants.*

Determinants

An idea closely related to the concept of matrix is the **determinant.** A determinant is a number that is associated with a **square matrix,** a matrix that has the same number of rows and columns. For any square matrix A, the symbol $|A|$ represents the determinant of A. To write a determinant, we put the elements of a square matrix between two vertical lines.

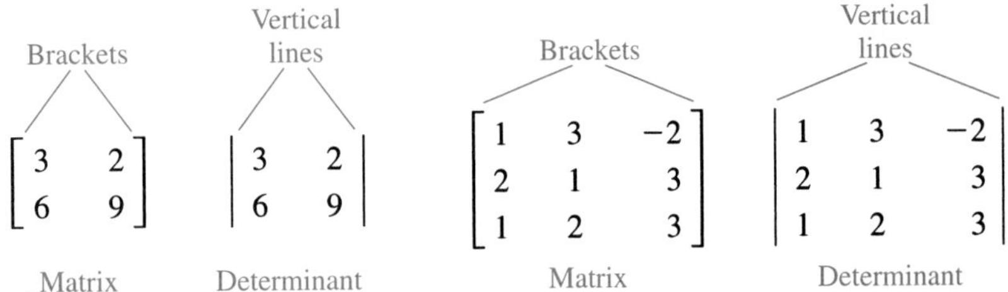

$$\begin{bmatrix} 3 & 2 \\ 6 & 9 \end{bmatrix} \quad \begin{vmatrix} 3 & 2 \\ 6 & 9 \end{vmatrix} \quad \begin{bmatrix} 1 & 3 & -2 \\ 2 & 1 & 3 \\ 1 & 2 & 3 \end{bmatrix} \quad \begin{vmatrix} 1 & 3 & -2 \\ 2 & 1 & 3 \\ 1 & 2 & 3 \end{vmatrix}$$

Like matrices, determinants are classified according to the number of rows and columns they contain. The determinant on the left is a 2×2 determinant. The other is a 3×3 determinant.

Evaluating a determinant

The determinant of a 2×2 matrix is the number that is equal to the product of the numbers on the main diagonal minus the product of the numbers on the other diagonal.

$$\begin{vmatrix} a & b \\ c & d \end{vmatrix} \qquad \begin{vmatrix} a & b \\ c & d \end{vmatrix}$$

Main diagonal Other diagonal

Value of a 2×2 determinant

If a, b, c, and d are numbers, the **determinant** of the matrix $\begin{bmatrix} a & b \\ c & d \end{bmatrix}$ is

$$\begin{vmatrix} a & b \\ c & d \end{vmatrix} = ad - bc$$

EXAMPLE 1 ***Evaluating 2 × 2 determinants.*** Find the value:

a. $\begin{vmatrix} 3 & 2 \\ 6 & 9 \end{vmatrix}$ and **b.** $\begin{vmatrix} -5 & \frac{1}{2} \\ -1 & 0 \end{vmatrix}$.

Solution

From the product of the numbers along the main diagonal, we subtract the product of the numbers along the other diagonal.

a. $\begin{vmatrix} 3 & 2 \\ 6 & 9 \end{vmatrix} = 3(9) - 2(6)$

$= 27 - 12$

$= 15$

b. $\begin{vmatrix} -5 & \frac{1}{2} \\ -1 & 0 \end{vmatrix} = -5(0) - \frac{1}{2}(-1)$

$= 0 + \frac{1}{2}$

$= \frac{1}{2}$

Self Check

Evaluate $\begin{vmatrix} 4 & -3 \\ 2 & 1 \end{vmatrix}$.

Answer: 10

A 3 × 3 determinant is evaluated by **expanding by minors.**

Value of a 3 × 3 determinant

$$\begin{vmatrix} a_1 & b_1 & c_1 \\ a_2 & b_2 & c_2 \\ a_3 & b_3 & c_3 \end{vmatrix} = a_1 \begin{vmatrix} b_2 & c_2 \\ b_3 & c_3 \end{vmatrix} - b_1 \begin{vmatrix} a_2 & c_2 \\ a_3 & c_3 \end{vmatrix} + c_1 \begin{vmatrix} a_2 & b_2 \\ a_3 & b_3 \end{vmatrix}$$

(Minor of a_1, Minor of b_1, Minor of c_1)

To find the minor of a_1, we cross out the elements of the determinant that are in the same row and column as a_1:

$$\begin{vmatrix} a_1 & b_1 & c_1 \\ a_2 & b_2 & c_2 \\ a_3 & b_3 & c_3 \end{vmatrix}$$

The minor of a_1 is $\begin{vmatrix} b_2 & c_2 \\ b_3 & c_3 \end{vmatrix}$.

To find the minor of b_1, we cross out the elements of the determinant that are in the same row and column as b_1:

$$\begin{vmatrix} a_1 & b_1 & c_1 \\ a_2 & b_2 & c_2 \\ a_3 & b_3 & c_3 \end{vmatrix}$$

The minor of b_1 is $\begin{vmatrix} a_2 & c_2 \\ a_3 & c_3 \end{vmatrix}$.

To find the minor of c_1, we cross out the elements of the determinant that are in the same row and column as c_1:

$$\begin{vmatrix} a_1 & b_1 & c_1 \\ a_2 & b_2 & c_2 \\ a_3 & b_3 & c_3 \end{vmatrix}$$

The minor of c_1 is $\begin{vmatrix} a_2 & b_2 \\ a_3 & b_3 \end{vmatrix}$.

EXAMPLE 2 ***Using minors to evaluate a 3 × 3 determinant.***

Find the value of $\begin{vmatrix} 1 & 3 & -2 \\ 2 & 1 & 3 \\ 1 & 2 & 3 \end{vmatrix}$.

Solution

We evaluate this determinant by expanding by minors along the first row of the determinant.

Self Check

Evaluate $\begin{vmatrix} 2 & -1 & 3 \\ 1 & 2 & -2 \\ 3 & 1 & 1 \end{vmatrix}$.

$$\begin{vmatrix} 1 & 3 & -2 \\ 2 & 1 & 3 \\ 1 & 2 & 3 \end{vmatrix} = 1\underbrace{\begin{vmatrix} 1 & 3 \\ 2 & 3 \end{vmatrix}}_{\text{Minor of 1}} - 3\underbrace{\begin{vmatrix} 2 & 3 \\ 1 & 3 \end{vmatrix}}_{\text{Minor of 3}} + (-2)\underbrace{\begin{vmatrix} 2 & 1 \\ 1 & 2 \end{vmatrix}}_{\text{Minor of } -2}$$

$$= 1(3 - 6) - 3(6 - 3) - 2(4 - 1) \quad \text{Evaluate each } 2 \times 2 \text{ determinant.}$$
$$= 1(-3) - 3(3) - 2(3)$$
$$= -3 - 9 - 6$$
$$= -18$$

Answer: 0

We can evaluate a 3×3 determinant by expanding it along any row or column. To determine the signs between the terms of the expansion of a 3×3 determinant, we use the following array of signs.

Array of signs for a 3×3 determinant

$$\begin{matrix} + & - & + \\ - & + & - \\ + & - & + \end{matrix}$$

EXAMPLE 3 ***Expanding along a column.*** Evaluate the determinant $\begin{vmatrix} 1 & 3 & -2 \\ 2 & 1 & 3 \\ 1 & 2 & 3 \end{vmatrix}$ by expanding on the middle column.

Solution

This is the determinant of Example 2. To expand it along the middle column, we use the signs of the middle column of the array of signs:

$$\begin{vmatrix} 1 & 3 & -2 \\ 2 & 1 & 3 \\ 1 & 2 & 3 \end{vmatrix} = -3\underbrace{\begin{vmatrix} 2 & 3 \\ 1 & 3 \end{vmatrix}}_{\text{Minor of 3}} + 1\underbrace{\begin{vmatrix} 1 & -2 \\ 1 & 3 \end{vmatrix}}_{\text{Minor of 1}} - 2\underbrace{\begin{vmatrix} 1 & -2 \\ 2 & 3 \end{vmatrix}}_{\text{Minor of 2}} \quad \text{Use the sign pattern } - + -.$$

$$= -3(6 - 3) + 1[3 - (-2)] - 2[3 - (-4)] \quad \text{Evaluate each } 2 \times 2 \text{ determinant.}$$
$$= -3(3) + 1(5) - 2(7)$$
$$= -9 + 5 - 14$$
$$= -18$$

As expected, we get the same value as in Example 2.

Self Check

Evaluate $\begin{vmatrix} 1 & 3 & -2 \\ 2 & 1 & 3 \\ 1 & 2 & 3 \end{vmatrix}$ by expanding along the last column.

Answer: -18

Accent on Technology ***Evaluating determinants***

It is possible to use a graphing calculator to evaluate determinants. For example, to evaluate the determinant in Example 3, we first enter the matrix by pressing the [MATRIX] key, selecting EDIT, and pressing the [ENTER] key. We then enter the dimensions and the elements of the matrix to get Figure 3-13(a). We then press [2nd] [QUIT] to clear the screen. We then press [MATRIX], select MATH, and press 1 to get Figure 3-13(b). We then press MATRIX, select NAMES, and press 1 to get Figure 3-13(c). To get the value of the determinant, we now press ENTER to get Figure 3-13(d), which shows that the value of the matrix is -18.

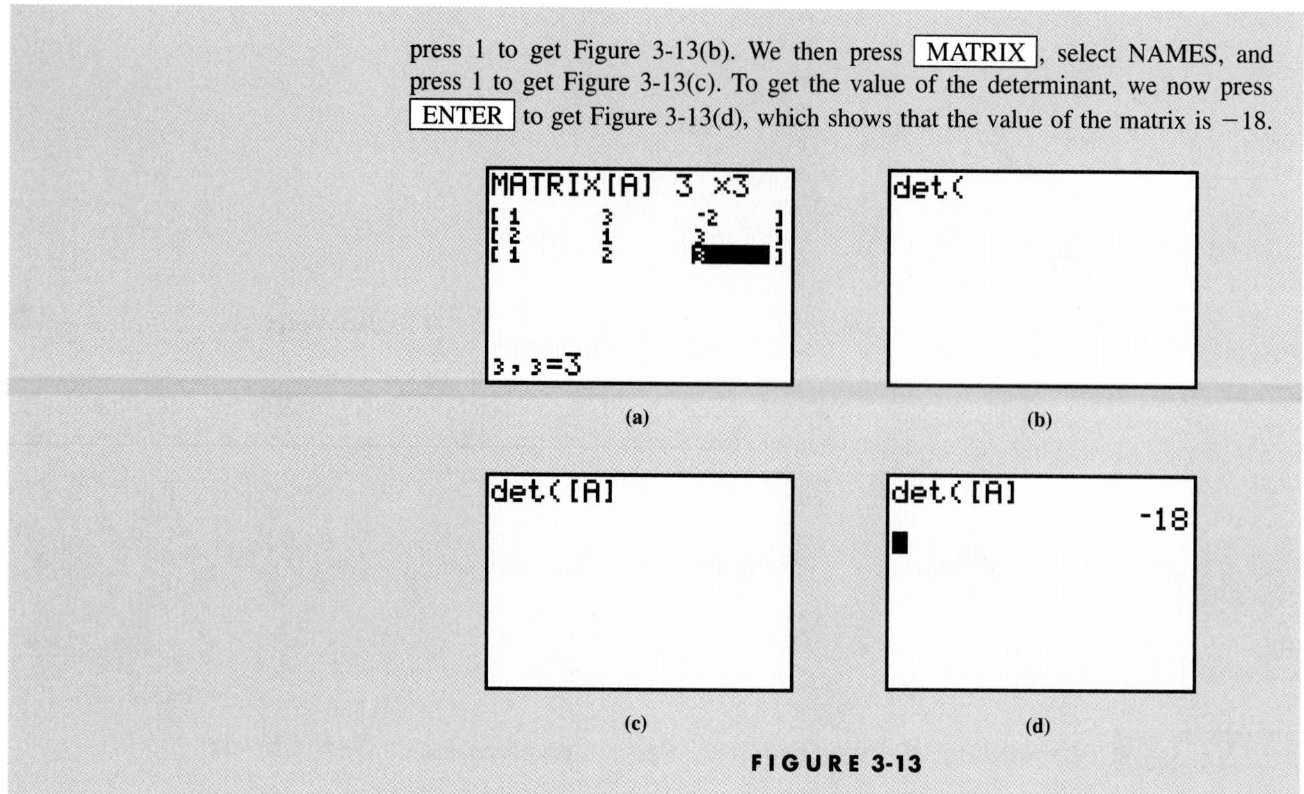

(a) (b) (c) (d)

FIGURE 3-13

Using Cramer's rule to solve a system of two equations

The method of using determinants to solve systems of linear equations is called **Cramer's rule,** named after the 18th-century mathematician Gabriel Cramer. To develop Cramer's rule, we consider the system

$$\begin{cases} ax + by = e \\ cx + dy = f \end{cases}$$

where x and y are variables and a, b, c, d, e, and f are constants.

If we multiply both sides of the first equation by d and multiply both sides of the second equation by $-b$, we can add the equations and eliminate y:

$$\begin{aligned} adx + bdy &= \quad ed \\ -bcx - bdy &= -bf \\ \hline adx - bcx \qquad &= ed - bf \end{aligned}$$

To solve for x, we use the distributive property to write $adx - bcx$ as $(ad - bc)x$ on the left-hand side and divide each side by $ad - bc$:

$$(ad - bc)x = ed - bf$$

$$x = \frac{ed - bf}{ad - bc} \qquad (ad - bc \neq 0)$$

We can find y in a similar manner. After eliminating the variable x, we get

$$y = \frac{af - ec}{ad - bc} \qquad (ad - bc \neq 0)$$

Determinants provide an easy way of remembering these formulas. Note that the denominator for both x and y is

$$\begin{vmatrix} a & b \\ c & d \end{vmatrix} = ad - bc$$

The numerators can be expressed as determinants also:

$$x = \frac{ed - bf}{ad - bc} = \frac{\begin{vmatrix} e & b \\ f & d \end{vmatrix}}{\begin{vmatrix} a & b \\ c & d \end{vmatrix}} \quad \text{and} \quad y = \frac{af - ec}{ad - bc} = \frac{\begin{vmatrix} a & e \\ c & f \end{vmatrix}}{\begin{vmatrix} a & b \\ c & d \end{vmatrix}}$$

If we compare these formulas with the original system

$$\begin{cases} ax + by = e \\ cx + dy = f \end{cases}$$

we note that in the expressions for x and y above, the denominator determinant is formed by using the coefficients a, b, c, and d of the variables in the equations. The numerator determinants are the same as the denominator determinant, except that the column of coefficients of the variable for which we are solving is replaced with the column of constants e and f.

Cramer's rule for two equations in two variables

The solution of the system $\begin{cases} ax + by = e \\ cx + dy = f \end{cases}$ is given by

$$x = \frac{D_x}{D} = \frac{\begin{vmatrix} e & b \\ f & d \end{vmatrix}}{\begin{vmatrix} a & b \\ c & d \end{vmatrix}} \quad \text{and} \quad y = \frac{D_y}{D} = \frac{\begin{vmatrix} a & e \\ c & f \end{vmatrix}}{\begin{vmatrix} a & b \\ c & d \end{vmatrix}}$$

If every determinant is 0, the system is consistent, but the equations are dependent.

If $D = 0$ and D_x or D_y is nonzero, the system is inconsistent. If $D \neq 0$, the system is consistent and the equations are independent.

EXAMPLE 4 ***Cramer's rule.*** Use Cramer's rule to solve $\begin{cases} 4x - 3y = 6 \\ -2x + 5y = 4 \end{cases}$.

Solution

The value of x is the quotient of two determinants, D and D_x. The denominator determinant D is made up of the coefficients of x and y:

$$D = \begin{vmatrix} 4 & -3 \\ -2 & 5 \end{vmatrix}$$

To solve for x, we form the numerator determinant D_x from D by replacing its first column (the coefficients of x) with the column of constants (6 and 4).

To solve for y, we form the numerator determinant D_y from D by replacing the second column (the coefficients of y) with the column of constants (6 and 4).

To find the values of x and y, we evaluate each determinant:

$$x = \frac{D_x}{D} = \frac{\begin{vmatrix} 6 & -3 \\ 4 & 5 \end{vmatrix}}{\begin{vmatrix} 4 & -3 \\ -2 & 5 \end{vmatrix}} = \frac{6(5) - (-3)(4)}{4(5) - (-3)(-2)} = \frac{30 + 12}{20 - 6} = \frac{42}{14} = 3$$

$$y = \frac{D_y}{D} = \frac{\begin{vmatrix} 4 & 6 \\ -2 & 4 \end{vmatrix}}{\begin{vmatrix} 4 & -3 \\ -2 & 5 \end{vmatrix}} = \frac{4(4) - 6(-2)}{14} = \frac{16 + 12}{14} = \frac{28}{14} = 2$$

The solution of this system is (3, 2). Verify that $x = 3$ and $y = 2$ satisfy both equations.

Self Check

Solve $\begin{cases} 2x - 3y = -16 \\ 3x + 5y = 14 \end{cases}$ using Cramer's rule.

Answer: $(-2, 4)$

EXAMPLE 5 ***An inconsistent system.*** Use Cramer's rule to solve $\begin{cases} 7x = 8 - 4y \\ 2y = 3 - \dfrac{7}{2}x \end{cases}$.

Solution

We multiply both sides of the second equation by 2 to eliminate the fraction and write the system in the form

$$\begin{cases} 7x + 4y = 8 \\ 7x + 4y = 6 \end{cases}$$

When we attempt to use Cramer's rule to solve this system for x, we obtain

$$x = \frac{D_x}{D} = \frac{\begin{vmatrix} 8 & 4 \\ 6 & 4 \end{vmatrix}}{\begin{vmatrix} 7 & 4 \\ 7 & 4 \end{vmatrix}} = \frac{8}{0} \quad \text{which is undefined}$$

Since the denominator determinant D is 0 and the numerator determinant D_x is not 0, the system is inconsistent. It has no solutions.

We can see directly from the system that it is inconsistent. For any values of x and y, it is impossible that 7 times x plus 4 times y could be both 8 and 6.

Self Check

Solve $\begin{cases} 3x = 8 - 4y \\ y = \dfrac{5}{2} - \dfrac{3}{4}x \end{cases}$ using Cramer's rule.

Answer: no solutions

Using Cramer's rule to solve a system of three equations

Cramer's rule can be extended to solve systems of three linear equations with three variables.

Cramer's rule for three equations with three variables

The solution of the system $\begin{cases} ax + by + cz = j \\ dx + ey + fz = k \\ gx + hy + iz = l \end{cases}$ is given by

$$x = \frac{D_x}{D}, \quad y = \frac{D_y}{D}, \quad \text{and} \quad z = \frac{D_z}{D}$$

where

$$D = \begin{vmatrix} a & b & c \\ d & e & f \\ g & h & i \end{vmatrix} \qquad D_x = \begin{vmatrix} j & b & c \\ k & e & f \\ l & h & i \end{vmatrix}$$

$$D_y = \begin{vmatrix} a & j & c \\ d & k & f \\ g & l & i \end{vmatrix} \qquad D_z = \begin{vmatrix} a & b & j \\ d & e & k \\ g & h & l \end{vmatrix}$$

If every determinant is 0, the system is consistent, but the equations are dependent.

If $D = 0$ and D_x or D_y or D_z is nonzero, the system is inconsistent. If $D \neq 0$, the system is consistent and the equations are independent.

EXAMPLE 6 ***A system of three equations.*** Use Cramer's rule to solve $\begin{cases} 2x + y + 4z = 12 \\ x + 2y + 2z = 9 \\ 3x - 3y - 2z = 1 \end{cases}$.

Solution

The denominator determinant D is the determinant formed by the coefficients of the variables. The numerator determinants, D_x, D_y, and D_z, are formed by replacing the coefficients of the variable being solved for by the column of constants. We form the quo-

Self Check

Solve $\begin{cases} x + y + 2z = 6 \\ 2x - y + z = 9 \\ x + y - 2z = -6 \end{cases}$ using Cramer's rule.

tients for x, y, and z and evaluate each determinant by expanding by minors about the first row:

$$x = \frac{D_x}{D} = \frac{\begin{vmatrix} 12 & 1 & 4 \\ 9 & 2 & 2 \\ 1 & -3 & -2 \end{vmatrix}}{\begin{vmatrix} 2 & 1 & 4 \\ 1 & 2 & 2 \\ 3 & -3 & -2 \end{vmatrix}}$$

$$= \frac{12\begin{vmatrix} 2 & 2 \\ -3 & -2 \end{vmatrix} - 1\begin{vmatrix} 9 & 2 \\ 1 & -2 \end{vmatrix} + 4\begin{vmatrix} 9 & 2 \\ 1 & -3 \end{vmatrix}}{2\begin{vmatrix} 2 & 2 \\ -3 & -2 \end{vmatrix} - 1\begin{vmatrix} 1 & 2 \\ 3 & -2 \end{vmatrix} + 4\begin{vmatrix} 1 & 2 \\ 3 & -3 \end{vmatrix}}$$

$$= \frac{12(2) - 1(-20) + 4(-29)}{2(2) - 1(-8) + 4(-9)}$$

$$= \frac{-72}{-24}$$

$$= 3$$

$$y = \frac{D_y}{D} = \frac{\begin{vmatrix} 2 & 12 & 4 \\ 1 & 9 & 2 \\ 3 & 1 & -2 \end{vmatrix}}{\begin{vmatrix} 2 & 1 & 4 \\ 1 & 2 & 2 \\ 3 & -3 & -2 \end{vmatrix}}$$

$$= \frac{2\begin{vmatrix} 9 & 2 \\ 1 & -2 \end{vmatrix} - 12\begin{vmatrix} 1 & 2 \\ 3 & -2 \end{vmatrix} + 4\begin{vmatrix} 1 & 9 \\ 3 & 1 \end{vmatrix}}{-24}$$

$$= \frac{2(-20) - 12(-8) + 4(-26)}{-24}$$

$$= \frac{-48}{-24}$$

$$= 2$$

$$z = \frac{D_z}{D} = \frac{\begin{vmatrix} 2 & 1 & 12 \\ 1 & 2 & 9 \\ 3 & -3 & 1 \end{vmatrix}}{\begin{vmatrix} 2 & 1 & 4 \\ 1 & 2 & 2 \\ 3 & -3 & -2 \end{vmatrix}}$$

$$= \frac{2\begin{vmatrix} 2 & 9 \\ -3 & 1 \end{vmatrix} - 1\begin{vmatrix} 1 & 9 \\ 3 & 1 \end{vmatrix} + 12\begin{vmatrix} 1 & 2 \\ 3 & -3 \end{vmatrix}}{-24}$$

$$= \frac{2(29) - 1(-26) + 12(-9)}{-24}$$

$$= \frac{-24}{-24}$$

$$= 1$$

The solution of this system is (3, 2, 1).

Answer: (2, −2, 3)

STUDY SET Section 3.5

VOCABULARY *In Exercises 1–6, fill in the blanks to make the statements true.*

1. $\begin{vmatrix} 2 & 1 \\ -6 & 1 \end{vmatrix}$ is a 2×2 ____________.

2. A ____________ matrix has the same number of rows and columns.

3. The ____________ of b_1 in $\begin{vmatrix} a_1 & b_1 & c_1 \\ a_2 & b_2 & c_2 \\ a_3 & b_3 & c_3 \end{vmatrix}$ is $\begin{vmatrix} a_2 & c_2 \\ a_3 & c_3 \end{vmatrix}$.

4. In $\begin{vmatrix} 7 & -3 \\ 1 & 2 \end{vmatrix}$, 7 and 2 lie along the main ____________.

5. A 3×3 determinant has 3 ____________ and 3 ____________.

6. ____________ rule uses determinants to solve systems of linear equations.

CONCEPTS *In Exercises 7–8, fill in the blanks to make the statements true.*

7. If the denominator determinant D for a system of equations is zero, the equations of the system are ____________ or the system is ____________.

8. To find the minor of 5, we ____________ the elements of the determinant that are in the same row and column as 5.

$$\begin{vmatrix} 3 & 5 & 1 \\ 6 & -2 & 2 \\ 8 & -1 & 4 \end{vmatrix}$$

9. What is the value of $\begin{vmatrix} a & b \\ c & d \end{vmatrix}$?

10. $\begin{vmatrix} 5 & 1 & -1 \\ 8 & 7 & 4 \\ 9 & 7 & 6 \end{vmatrix} = -1\begin{vmatrix} 8 & 7 \\ 9 & 7 \end{vmatrix} - 4\begin{vmatrix} 5 & 1 \\ 9 & 7 \end{vmatrix} + 6\begin{vmatrix} 5 & 1 \\ 8 & 7 \end{vmatrix}$

In evaluating this determinant, about what row or column was it expanded?

11. What is the denominator determinant D for the system $\begin{cases} 3x + 4y = 7 \\ 2x - 3y = 5 \end{cases}$?

12. What is the denominator determinant D for the system $\begin{cases} x + 2y = -8 \\ 3x + y - z = -2 \\ 8x + 4y - z = 6 \end{cases}$?

13. For the system $\begin{cases} 3x + 2y = 1 \\ 4x - y = 3 \end{cases}$, $D_x = -7$, $D_y = 5$, and $D = -11$. What is the solution of the system?

14. For the system $\begin{cases} 2x + 3y - z = -8 \\ x - y - z = -2 \\ -4x + 3y + z = 6 \end{cases}$, $D_x = -28$, $D_y = -14$, $D_z = 14$, and $D = 14$. What is the solution?

NOTATION *In Exercises 15–16, complete the evaluation of each determinant.*

15. $\begin{vmatrix} 5 & -2 \\ -2 & 6 \end{vmatrix}$

$= 5(\square) - (-2)(-2)$

$= \square - 4$

$= 26$

16. $\begin{vmatrix} 2 & 1 & 3 \\ 3 & 4 & 2 \\ 1 & 5 & 3 \end{vmatrix}$

$= 2\begin{vmatrix} 4 & \square \\ 5 & 3 \end{vmatrix} \square 1\begin{vmatrix} 3 & 2 \\ \square & 3 \end{vmatrix} + 3\begin{vmatrix} 3 & 4 \\ 1 & \square \end{vmatrix}$

$= 2(\square - 10) - 1(9 - \square) + 3(15 - \square)$

$= 2(2) - 1(\square) + \square(11)$

$= 4 - 7 + \square$

$= 30$

PRACTICE *In Exercises 17–32, evaluate each determinant.*

17. $\begin{vmatrix} 2 & 3 \\ -2 & 1 \end{vmatrix}$

18. $\begin{vmatrix} 3 & -2 \\ -2 & 4 \end{vmatrix}$

19. $\begin{vmatrix} -1 & 2 \\ 3 & -4 \end{vmatrix}$

20. $\begin{vmatrix} -1 & -2 \\ -3 & -4 \end{vmatrix}$

21. $\begin{vmatrix} 10 & 0 \\ 1 & 20 \end{vmatrix}$

22. $\begin{vmatrix} 1 & 15 \\ 15 & 0 \end{vmatrix}$

23. $\begin{vmatrix} -6 & -2 \\ 15 & 4 \end{vmatrix}$

24. $\begin{vmatrix} 3 & -2 \\ 12 & -8 \end{vmatrix}$

25. $\begin{vmatrix} 1 & 2 & 0 \\ 0 & 1 & 2 \\ 0 & 0 & 1 \end{vmatrix}$

26. $\begin{vmatrix} -1 & 2 & 1 \\ 2 & 1 & -3 \\ 1 & 1 & 1 \end{vmatrix}$

27. $\begin{vmatrix} 1 & -2 & 3 \\ -2 & 1 & 1 \\ -3 & -2 & 1 \end{vmatrix}$

28. $\begin{vmatrix} 1 & 1 & 2 \\ 2 & 1 & -2 \\ 3 & 1 & 3 \end{vmatrix}$

29. $\begin{vmatrix} 1 & 0 & 1 \\ 0 & 1 & 0 \\ 1 & 1 & 1 \end{vmatrix}$

30. $\begin{vmatrix} 3 & 5 & 1 \\ 6 & -2 & 2 \\ 8 & -1 & 4 \end{vmatrix}$

31. $\begin{vmatrix} 1 & 2 & 1 \\ -3 & 7 & 3 \\ -4 & 3 & -5 \end{vmatrix}$

32. $\begin{vmatrix} 1 & 4 & 7 \\ 2 & 5 & 8 \\ 3 & 6 & 9 \end{vmatrix}$

In Exercises 33–54, use Cramer's rule to solve each system of equations, if possible.

33. $\begin{cases} x + y = 6 \\ x - y = 2 \end{cases}$

34. $\begin{cases} x - y = 4 \\ 2x + y = 5 \end{cases}$

35. $\begin{cases} 2x + 3y = 0 \\ 4x - 6y = -4 \end{cases}$

36. $\begin{cases} 4x - 3y = -1 \\ 8x + 3y = 4 \end{cases}$

37. $\begin{cases} 3x + 2y = 11 \\ 6x + 4y = 11 \end{cases}$

38. $\begin{cases} 5x + 6y = 12 \\ 10x + 12y = 24 \end{cases}$

39. $\begin{cases} y = \dfrac{-2x + 1}{3} \\ 3x - 2y = 8 \end{cases}$

40. $\begin{cases} 2x + 3y = -1 \\ x = \dfrac{y - 9}{4} \end{cases}$

41. $\begin{cases} x + y + z = 4 \\ x + y - z = 0 \\ x - y + z = 2 \end{cases}$

42. $\begin{cases} x + y + z = 4 \\ x - y + z = 2 \\ x - y - z = 0 \end{cases}$

43. $\begin{cases} x + y + 2z = 7 \\ x + 2y + z = 8 \\ 2x + y + z = 9 \end{cases}$

44. $\begin{cases} x + 2y + 2z = 10 \\ 2x + y + 2z = 9 \\ 2x + 2y + z = 1 \end{cases}$

45. $\begin{cases} 2x + y + z = 5 \\ x - 2y + 3z = 10 \\ x + y - 4z = -3 \end{cases}$

46. $\begin{cases} 3x + 2y - z = -8 \\ 2x - y + 7z = 10 \\ 2x + 2y - 3z = -10 \end{cases}$

47. $\begin{cases} 4x - 3y = 1 \\ 6x - 8z = 1 \\ 2y - 4z = 0 \end{cases}$

48. $\begin{cases} 4x + 3z = 4 \\ 2y - 6z = -1 \\ 8x + 4y + 3z = 9 \end{cases}$

49. $\begin{cases} 2x + 3y + 4z = 6 \\ 2x - 3y - 4z = -4 \\ 4x + 6y + 8z = 12 \end{cases}$

50. $\begin{cases} x - 3y + 4z - 2 = 0 \\ 2x + y + 2z - 3 = 0 \\ 4x - 5y + 10z - 7 = 0 \end{cases}$

51. $\begin{cases} 2x + y - z - 1 = 0 \\ x + 2y + 2z - 2 = 0 \\ 4x + 5y + 3z - 3 = 0 \end{cases}$

52. $\begin{cases} 2x - y + 4z + 2 = 0 \\ 5x + 8y + 7z = -8 \\ x + 3y + z + 3 = 0 \end{cases}$

53. $\begin{cases} x + y = 1 \\ \dfrac{1}{2}y + z = \dfrac{5}{2} \\ x - z = -3 \end{cases}$

54. $\begin{cases} \dfrac{1}{2}x + y + z + \dfrac{3}{2} = 0 \\ x + \dfrac{1}{2}y + z - \dfrac{1}{2} = 0 \\ x + y + \dfrac{1}{2}z + \dfrac{1}{2} = 0 \end{cases}$

APPLICATIONS *Write a system of equations to solve each problem. Then use Cramer's rule to solve the system.*

55. INVENTORY Illustration 1 on page 230 shows an end-of-the-year inventory report for a warehouse that supplies electronics stores. If the warehouse stocks two models of cordless telephones, one valued at \$67 and the other at \$100, how many of each model of phone did the warehouse have at the time of the inventory?

Item	Number	Merchandise value
Televisions	800	\$1,005,450
Radios	200	\$15,785
Cordless phones	360	\$29,400

ILLUSTRATION 1

56. SIGNALING A system of sending signals uses two flags held in various positions to represent letters of the alphabet. Illustration 2 shows how the letter U is signaled. Find x and y, if y is to be 30° more than x.

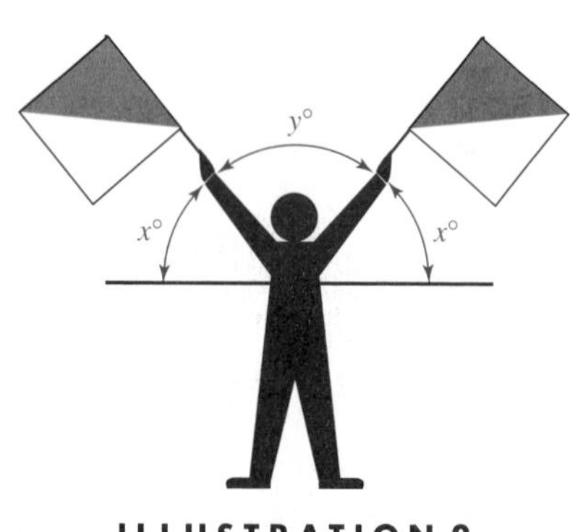

ILLUSTRATION 2

57. INVESTING A student wants to average a 6.6% return by investing \$20,000 in the three stocks listed in Illustration 3. Because HiTech is a high-risk investment, he wants to invest three times as much in SaveTel and OilCo combined as he invests in HiTech. How much should he invest in each stock?

Stock	Rate of return
HiTech	10%
SaveTel	5%
OilCo	6%

ILLUSTRATION 3

58. INVESTING A woman wants to average a $7\frac{1}{3}\%$ return by investing \$30,000 in three certificates of deposit. (See Illustration 4.) She wants to invest five times as much in the 8% CD as in the 6% CD. How much should she invest in each CD?

Type of CD	Rate of return
12 month	6%
24 month	7%
36 month	8%

ILLUSTRATION 4

In Exercises 59–62, use a calculator with matrix capabilities to evaluate each determinant.

59. $\begin{vmatrix} 2 & -3 & 4 \\ -1 & 2 & 4 \\ 3 & -3 & 1 \end{vmatrix}$

60. $\begin{vmatrix} -3 & 2 & -5 \\ 3 & -2 & 6 \\ 1 & -3 & 4 \end{vmatrix}$

61. $\begin{vmatrix} 2 & 1 & -3 \\ -2 & 2 & 4 \\ 1 & -2 & 2 \end{vmatrix}$

62. $\begin{vmatrix} 4 & 2 & -3 \\ 2 & -5 & 6 \\ 2 & 5 & -2 \end{vmatrix}$

WRITING *Write a paragraph using your own words.*

63. Tell how to find the minor of an element of a determinant.

64. Tell how to find x when solving a system of three linear equations by Cramer's rule. Use the words *coefficients* and *constants* in your explanation.

REVIEW

65. Are the lines $y = 2x - 7$ and $x - 2y = 7$ perpendicular?

66. Are the lines $y = 2x - 7$ and $2x - y = 10$ parallel?

67. Are the equations $y = 2x - 7$ and $f(x) = 2x - 7$ the same?

68. How are the graphs of $f(x) = x^2$ and $g(x) = x^2 - 2$ related?

69. For the linear function $y = 2x - 7$, what variable is associated with the domain?

70. Is the graph of a circle the graph of a function?

71. The graph of a line passes through $(0, -3)$. Is this the x-intercept or the y-intercept of the line?

72. What is the name of this function: $f(x) = |x|$?

KEY CONCEPT

Systems of Equations

In Chapter 3, we solved problems involving two and three variables by writing and solving a **system of equations.**

Solutions of a System of Equations

A solution of a system of equations involving two or three variables is an ordered pair or an ordered triple whose coordinates satisfy each equation of the system. In Exercises 1 and 2, decide whether the given ordered pair or ordered triple is a solution of the system.

1. $\begin{cases} 2x - y = 1 \\ 4x + 2y = 0 \end{cases}$ $\left(\frac{1}{4}, -\frac{1}{2}\right)$

2. $\begin{cases} 2x - y + z = 9 \\ 3x + y - 4z = 8 \\ 2x - 7z = -1 \end{cases}$ $(4, 0, 1)$

Methods of Solving Systems of Linear Equations

We have studied several methods for solving systems of two and three linear equations.

3. Solve $\begin{cases} 2x + 5y = 8 \\ y = 3x + 5 \end{cases}$ using the *graphing method.*

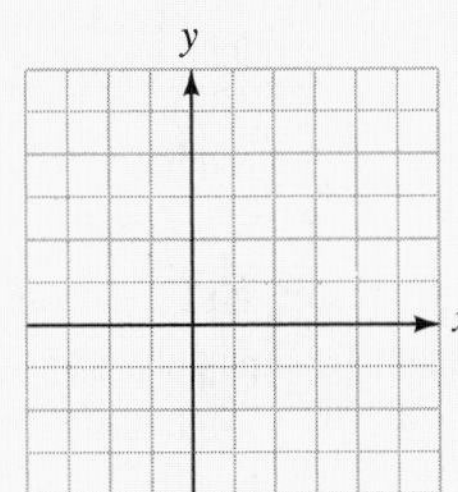

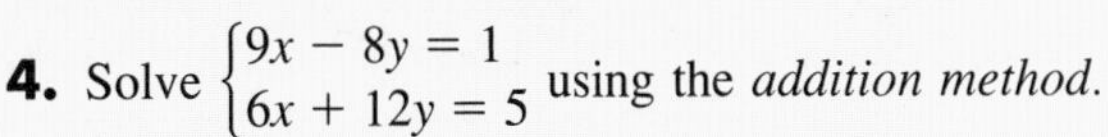

4. Solve $\begin{cases} 9x - 8y = 1 \\ 6x + 12y = 5 \end{cases}$ using the *addition method.*

5. Solve $\begin{cases} 4x - y - 10 = 0 \\ 3x + 5y = 19 \end{cases}$ using the *substitution method.*

6. Solve $\begin{cases} -x + 3y + 2z = 5 \\ 3x + 2y + z = -1 \\ 2x - y + 3z = 4 \end{cases}$ using the *addition method.*

7. Solve $\begin{cases} x - 6y = 3 \\ x + 3y = 21 \end{cases}$ using matrices.

8. Solve $\begin{cases} x + 2z = 7 \\ 2x - y + 3z = 9 \\ y - z = 1 \end{cases}$ using Cramer's rule.

Dependent Equations and Inconsistent Systems

If the equations in a system of two linear equations are dependent, the system has infinitely many solutions. An inconsistent system has no solutions.

9. Suppose you are solving a system of two equations by the addition method, and you obtain the following.

$$\begin{array}{r} 2x - 3y = 4 \\ + \quad -2x + 3y = -4 \\ \hline 0 = 0 \end{array}$$

What can you conclude?

10. Suppose you are solving a system of two equations by the substitution method, and you obtain

$$\begin{aligned} -2(x - 3) + 2x &= 7 \\ -2x + 6 + 2x &= 7 \\ 6 &= 7 \end{aligned}$$

What can you conclude?

Accent on Teamwork

Section 3.1

LINE GRAPHS Find five examples of line graphs where two lines intersect. (See Illustration 1.) Your school library is a good resource. Ask to look through the library's collection of recent magazines and newspapers. For each graph, explain the information given by the point of intersection of the graph.

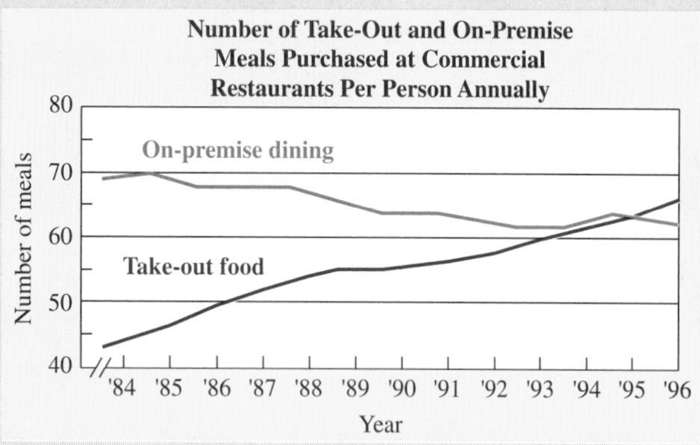

ILLUSTRATION 1

Section 3.2

COMPARING SOLUTIONS Example 8 of Section 3.2, involving water treatment, was solved using two variables. This problem can also be solved using one variable. Solve it using one variable and then comment on which method you like better. What are the advantages and the drawbacks of each method?

BREAK-POINT ANALYSIS Suppose you are a financial analyst for the coathanger company mentioned in Example 10 of Section 3.2. It is your job to decide whether the company should purchase the new machine.

First, graph the equations

$$C = 1.5x + 400$$
$$C = 1.25x + 500$$

on the same coordinate system. Then write a report that could be given to company managers, explaining their options concerning the purchase of the new machine. Under what conditions should they keep the machine now in use? Under what conditions should they buy the new machine?

Section 3.3

GRAPHS OF SYSTEMS OF EQUATIONS From your textbook, find and then graph examples of each of the following types of systems of two linear equations:

- Consistent system
- Inconsistent system
- Dependent equations

Then make cardboard models of each of the graphs of the systems of three linear equations illustrated in Figure 3-10 and Figure 3-11 of Section 3.3. Make a presentation to your class using the graphs and cardboard models as visual aids to help you explain whether each system has a solution, and if so, what form the solution takes.

Section 3.4

GAUSSIAN ELIMINATION Use matrices and elementary row operations to show that the solution to this 4×4 system of linear equations is (1, 1, 0, 1).

$$\begin{cases} x + y + z + w = 3 \\ x - y - z - w = -1 \\ x + y - z - w = 1 \\ x + y - z + w = 3 \end{cases}$$

Section 3.5

METHODS OF SOLUTION Have each person in your group solve the system

$$\begin{cases} x - y = 4 \\ 2x + y = 5 \end{cases}$$

in a different way. The methods to use are graphing, addition, substitution, matrices, and Cramer's rule. Have each person briefly explain his or her method of solution. After everyone has presented a solution, discuss the advantages and drawbacks of each method. Can your group come to consensus? Is there a favorite method?

CHAPTER REVIEW

SECTION 3.1 Solving Systems by Graphing

CONCEPTS

The graph of a linear equation is the graph of all points (x, y) on the rectangular coordinate system whose coordinates satisfy the equation.

To solve a system of two linear equations by the *graphing method*, find the coordinates of the point where the two graphs intersect.

If a system of equations has at least one solution, the system is a *consistent system.* When a system has no solution, it is called an *inconsistent system.*

If the graphs of the equations of a system are distinct, the equations are *independent equations.* Otherwise, the equations are *dependent equations.*

REVIEW EXERCISES

1. See Illustration 1.

a. Give three points that satisfy the equation $2x + y = 5$.

b. Give three points that satisfy the equation $x - y = 4$.

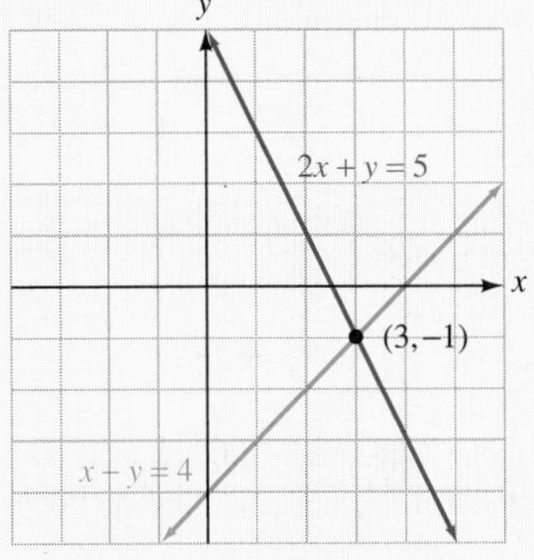

ILLUSTRATION 1

c. What is the solution of $\begin{cases} 2x + y = 5 \\ x - y = 4 \end{cases}$?

2. Solve each system by the graphing method, if possible.

a. $\begin{cases} 2x + y = 11 \\ -x + 2y = 7 \end{cases}$

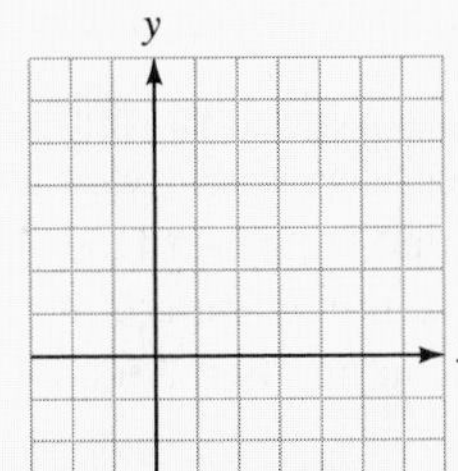

b. $\begin{cases} y = -\frac{3}{2}x \\ 2x - 3y + 13 = 0 \end{cases}$

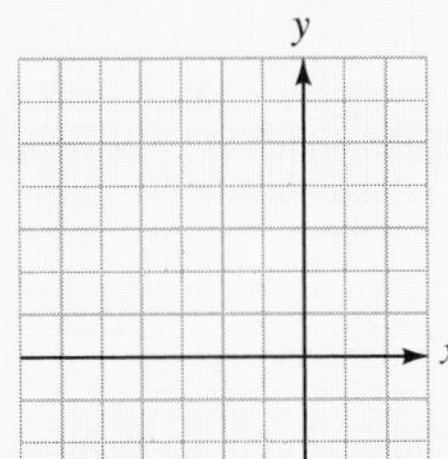

c. $\begin{cases} \frac{1}{2}x + \frac{1}{3}y = 2 \\ y = 6 - \frac{3}{2}x \end{cases}$

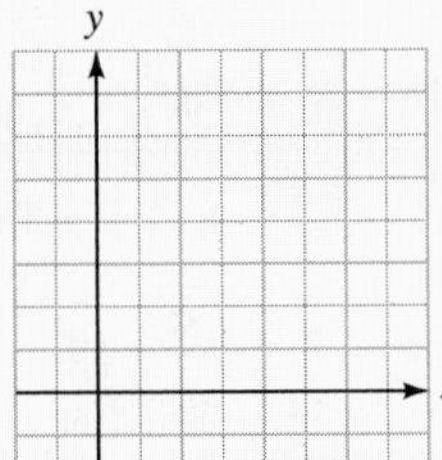

d. $\begin{cases} \frac{x}{3} - \frac{y}{2} = 1 \\ 6x - 9y = 3 \end{cases}$

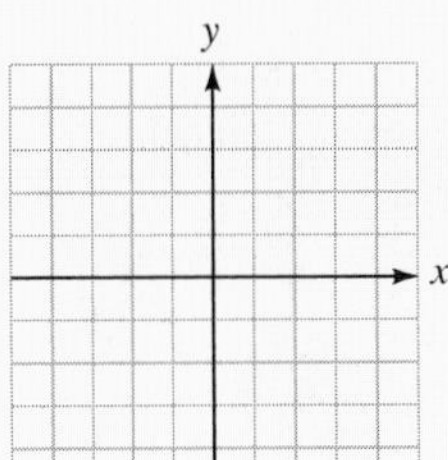

Solving Systems Algebraically

To solve a system by the *substitution method:*

1. Solve one equation for one of its variables.
2. Substitute the resulting expression for that variable into the other equation and solve that equation.
3. Find the value of the other variable by substituting the value of the variable found in step 2 into the equation from step 1.

To solve a system by the *addition method:*

1. Write both equations in general form.
2. Multiply the terms of one or both equations by constants so that the coefficients of one variable differ only in sign.
3. Add the equations from step 2 and solve the resulting equation.
4. Substitute the value obtained in step 3 into either original equation and solve for the remaining variable.

3. Solve each system using the substitution method, if possible.

a. $\begin{cases} x = y - 4 \\ 2x + 3y = 7 \end{cases}$ **b.** $\begin{cases} y = 2x + 5 \\ 3x - 5y = -4 \end{cases}$

c. $\begin{cases} 0.1x + 0.2y = 1.1 \\ 2x - y = 2 \end{cases}$ **d.** $\begin{cases} x = -2 - 3y \\ -2x - 6y = 4 \end{cases}$

4. Solve each system using the addition method, if possible.

a. $\begin{cases} x + y = -2 \\ 2x + 3y = -3 \end{cases}$ **b.** $\begin{cases} 2x - 3y = 5 \\ 2x - 3y = 8 \end{cases}$

c. $\begin{cases} x + \frac{1}{2}y = 7 \\ -2x = 3y - 6 \end{cases}$ **d.** $\begin{cases} y = \dfrac{x - 3}{2} \\ x = \dfrac{2y + 7}{2} \end{cases}$

5. To solve $\begin{cases} 5x - 2y = 19 \\ 3x + 4y = 1 \end{cases}$, which method, addition or substitution, would you use? Explain why.

In Exercises 6–7, use two equations to solve each problem.

6. MILEAGE MAP See Illustration 2. The distance between Austin and Houston is 4 miles less than twice the distance between Austin and San Antonio. The round trip from Houston to Austin to San Antonio and back to Houston is 442 miles. Determine the mileages between Austin and Houston and between Austin and San Antonio.

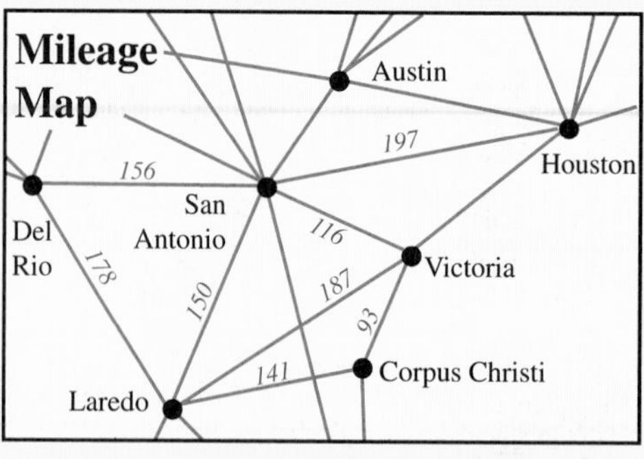

ILLUSTRATION 2

7. RIVERBOAT RIDE A Mississippi riverboat travels 30 miles downstream in three hours and then makes the return trip upstream in five hours. Find the speed of the riverboat in still water and the speed of the current.

SECTION 3.3 Systems with Three Variables

The solution of a system of three linear equations is an *ordered triple.*

8. Tell whether (2, −1, 1) is a solution of the system $\begin{cases} x - y + z = 4 \\ x + 2y - z = -1 \\ x + y - 3z = -1 \end{cases}$.

To solve a system of linear equations with three variables:
1. Pick any two equations and eliminate a variable.
2. Pick a different pair of equations and eliminate the same variable.
3. Solve the resulting pair of equations.
4. Use substitution to find the value of the third variable.

9. Solve each system, if possible.

a. $\begin{cases} x + y + z = 6 \\ x - y - z = -4 \\ -x + y - z = -2 \end{cases}$

b. $\begin{cases} 2x + 3y + z = -5 \\ -x + 2y - z = -6 \\ 3x + y + 2z = 4 \end{cases}$

c. $\begin{cases} x + y - z = -3 \\ x + z = 2 \\ 2x - y + 2z = 3 \end{cases}$

d. $\begin{cases} 3x + 3y + 6z = -6 \\ -x - y - 2z = 2 \\ 2x + 2y + 4z = -4 \end{cases}$

10. MIXING NUTS The owner of a produce store wanted to mix peanuts selling for \$3 per pound, cashews selling for \$9 per pound, and Brazil nuts selling for \$9 per pound to get 50 pounds of a mixture that would sell for \$6 per pound. She used 15 fewer pounds of cashews than peanuts. How many pounds of each did she use?

SECTION 3.4 Solving Systems Using Matrices

A *matrix* is a rectangular array of numbers.

A system of linear equations can be represented by an *augmented matrix.*

11. Represent each system of equations using an augmented matrix.

a. $\begin{cases} 5x + 4y = 3 \\ x - y = -3 \end{cases}$

b. $\begin{cases} x + 2y + 3z = 6 \\ x - 3y - z = 4 \\ 6x + y - 2z = -1 \end{cases}$

Systems of linear equations can be solved using *Gaussian elimination* and *elementary row operations:*
1. Any two rows can be interchanged.
2. Any row can be multiplied by a nonzero constant.
3. Any row can be changed by adding a nonzero constant multiple of another row to it.

12. Solve each system using matrices, if possible.

a. $\begin{cases} x - y = 4 \\ 3x + 7y = -18 \end{cases}$

b. $\begin{cases} x + 2y - 3z = 5 \\ x + y + z = 0 \\ 3x + 4y + 2z = -1 \end{cases}$

c. $\begin{cases} 16x - 8y = 32 \\ -2x + y = -4 \end{cases}$

d. $\begin{cases} x + 2y + 2z = 2 \\ 4x + 5y + 3z = 3 \\ 2x + y - z = 1 \end{cases}$

13. INVESTING One year, a couple invested a total of \$10,000 in two projects. The first investment, a mini-mall, made a 6% profit. The other investment, a skateboard park, made a 12% profit. If their investments made \$960, how much was invested at each rate? To answer this question, write a system of two equations and solve it using matrices.

SECTION 3.5 Solving Systems Using Determinants

A *determinant* of a *square matrix* is a number.

To evaluate a 2×2 determinant:

$$\begin{vmatrix} a & b \\ c & d \end{vmatrix} = ad - bc$$

To evaluate a 3×3 determinant, we expand it by *minors* along any row or column using the *array of signs.*

Cramer's rule can be used to solve systems of linear equations.

14. Evaluate each determinant.

a. $\begin{vmatrix} 2 & 3 \\ -4 & 3 \end{vmatrix}$ **b.** $\begin{vmatrix} -3 & -4 \\ 5 & -6 \end{vmatrix}$

c. $\begin{vmatrix} -1 & 2 & -1 \\ 2 & -1 & 3 \\ 1 & -2 & 2 \end{vmatrix}$ **d.** $\begin{vmatrix} 3 & -2 & 2 \\ 1 & -2 & -2 \\ 2 & 1 & -1 \end{vmatrix}$

15. Use Cramer's rule to solve each system, if possible.

a. $\begin{cases} 3x + 4y = 10 \\ 2x - 3y = 1 \end{cases}$ **b.** $\begin{cases} -6x - 4y = -6 \\ 3x + 2y = 5 \end{cases}$

c. $\begin{cases} x + 2y + z = 0 \\ 2x + y + z = 3 \\ x + y + 2z = 5 \end{cases}$ **d.** $\begin{cases} 2x + 3y + z = 2 \\ x + 3y + 2z = 7 \\ x - y - z = -7 \end{cases}$

16. VETERINARY MEDICINE The daily requirements of a balanced diet for an animal are shown in the nutritional pyramid in Illustration 3. The number of grams per cup of nutrients in three food mixes are shown in the table. How many cups of each mix should be used to meet the daily requirements for protein, carbohydrates, and essential fatty acids in the animal's diet? To answer this problem, write a system of three equations and solve it using Cramer's rule.

	Grams per cup		
	Protein	**Carbohydrates**	**Fatty Acids**
Mix A	5	2	1
Mix B	6	3	2
Mix C	8	3	1

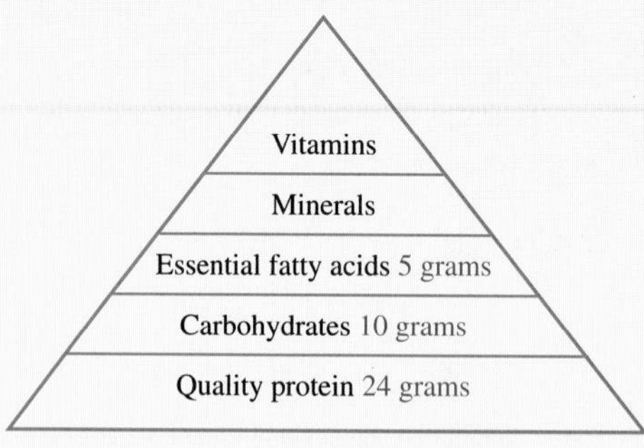

ILLUSTRATION 3

Chapter 3 Test

1. Solve $\begin{cases} 2x + y = 5 \\ y = 2x - 3 \end{cases}$ by graphing.

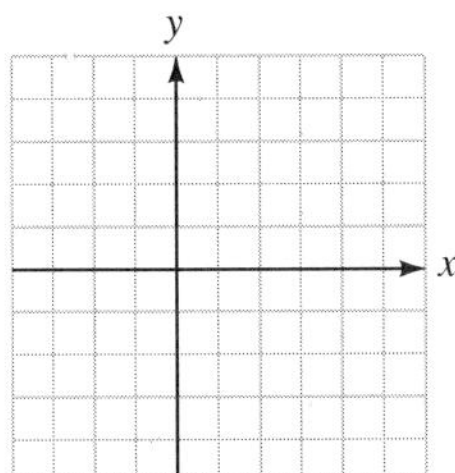

2. Use substitution to solve $\begin{cases} 2x - 4y = 14 \\ x + 2y = 7 \end{cases}$.

3. Use addition to solve $\begin{cases} 2x + 3y = -5 \\ 3x - 2y = 12 \end{cases}$.

4. Are the equations of the system $\begin{cases} 3(x + y) = x - 3 \\ -y = \dfrac{2x + 3}{3} \end{cases}$ dependent or independent?

5. Is $\left(-1, -\frac{1}{2}, 5\right)$ a solution of $\begin{cases} x - 2y + z = 5 \\ 2x + 4y = -4 \\ -6y + 4z = 22 \end{cases}$?

6. Solve the system $\begin{cases} x + y + z = 4 \\ x + y - z = 6 \\ 2x - 3y + z = -1 \end{cases}$ using the addition method.

In Problems 7–8, write a system of equations to solve each problem.

7. In Illustration 1, find x and y, if y is 15 more than x.

ILLUSTRATION 1

8. ANTIFREEZE How much of a 40% antifreeze solution must a mechanic mix with an 80% antifreeze solution if 20 gallons of a 50% antifreeze solution are needed?

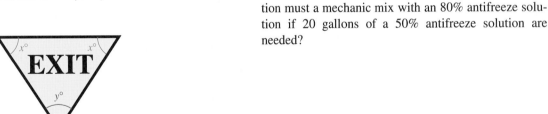

In Problems 9–10, use matrices to solve each system.

9. $\begin{cases} x + y = 4 \\ 2x - y = 2 \end{cases}$

10. $\begin{cases} x + y + 2z = -1 \\ x + 3y - 6z = 7 \\ 2x - y + 2z = 0 \end{cases}$

In Problems 11–12, evaluate each determinant.

11. $\begin{vmatrix} 2 & -3 \\ 4 & 5 \end{vmatrix}$

12. $\begin{vmatrix} 1 & 2 & 0 \\ 2 & 0 & 3 \\ 1 & -2 & 2 \end{vmatrix}$

In Problems 13–16, consider the system $\begin{cases} x - y = -6 \\ 3x + y = -6 \end{cases}$*, which is to be solved with Cramer's rule.*

13. When solving for x, what is the numerator determinant D_x? **(Don't evaluate it.)**

14. When solving for y, what is the denominator determinant D? **(Don't evaluate it.)**

15. Solve the system for x.

16. Solve the system for y.

17. Solve the following system for z only, using Cramer's rule.

$$\begin{cases} x + y + z = 4 \\ x + y - z = 6 \\ 2x - 3y + z = -1 \end{cases}$$

18. MOVIE TICKETS The receipts for one showing of a movie were $410 for an audience of 100 people. The ticket prices are given in the table. If twice as many children's tickets as general admission tickets were purchased, how many of each type of ticket was sold?

Ticket prices	
Children	$3.00
General Admission	$6.00
Seniors	$5.00

19. CANDY SALES Summarize the information that can be learned from the graph in Illustration 2. What do the points of intersection of the graphs tell us?

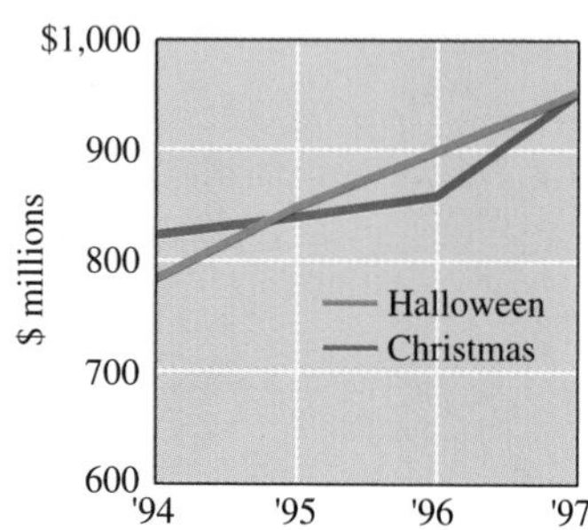

Based on data from *Los Angeles Times* (October 30, 1997)

ILLUSTRATION 2

20. Which method, substitution or addition, would you use to solve the following system?

$$\begin{cases} \frac{x}{2} - \frac{y}{4} = -4 \\ y = -2 - x \end{cases}$$

Explain your reasoning.

Inequalities

Campus Connection

The *Physical Education* Department

In an aerobics class, students perform exercises that improve their cardiovascular fitness. Cardiologists have determined what heart rate range participants need to maintain to get the most out of the training. For example, a 30-year-old woman needs to raise her heart rate to between 137 to 165 beats per minute and then sustain that rate for 10 to 12 minutes. In this chapter, you will see that these suggested ranges can be described using *inequalities.* Learning how to write and solve inequalities will allow you to solve many new types of application problems, including some from physical education.

When working with unequal quantities, we use inequalities instead of equations to describe the situation mathematically.

4.1 Solving Linear Inequalities

In this section, you will learn about

- **Inequalities**
- **Graphs and interval notation**
- **Properties of inequalities**
- **Solving linear inequalities**
- **Problem solving**

Introduction. Traffic signs like the one shown in Figure 4-1 often appear in front of schools. From the sign, a motorist knows that

- A speed *greater than* 25 miles per hour breaks the law and could possibly result in a ticket for speeding.
- A speed *less than or equal to* 25 miles per hour is within the posted speed limit.

Statements such as these can be expressed mathematically using *inequality symbols.*

FIGURE 4-1

Inequalities

Inequalities are statements indicating that two quantities are unequal. Inequalities can be recognized by one or more of the following symbols.

Inequality symbols

$a \neq b$	means	"a is not equal to b."
$a < b$	means	"a is less than b."
$a > b$	means	"a is greater than b."
$a \leq b$	means	"a is less than or equal to b."
$a \geq b$	means	"a is greater than or equal to b."

By definition, $a < b$ means that "a is less than b," but it also means that $b > a$. Furthermore, if a is to the left of b on the number line, then $a < b$. If a is to the right of b on a number line, then $a > b$.

We can use inequality symbols to describe the warning that the traffic sign in Figure 4-1 gives to drivers. If x represents the motorist's speed in miles per hour, he or she is in danger of receiving a speeding ticket if $x > 25$, and he or she is observing the posted speed limit if $x \leq 25$.

Graphs and interval notation

The graphs of sets of real numbers are portions of a number line called **intervals.** The graph shown in Figure 4-2(a) represents all real numbers that are greater than -5. This interval contains the numbers that *satisfy* the inequality $x > -5$, such as -4, -1.8, 0, $2\frac{3}{4}$, π, and 1,050. The parenthesis at -5 indicates that -5 is not included in the interval. We can also express this interval in **interval notation** as $(-5, \infty)$, where ∞ (read as **positive infinity**) indicates that the interval extends indefinitely to the right. The left parenthesis indicates that the endpoint -5 is not included.

Three ways to describe the real numbers greater than -5

1. Graph: (number line, parenthesis at -5)
2. Inequality: $x > -5$
3. Interval notation: $(-5, \infty)$

(a)

Three ways to describe the real numbers less than or equal to 7

1. Graph: (number line, bracket at 7)
2. Inequality: $x \leq 7$
3. Interval notation: $(-\infty, 7]$

(b)

FIGURE 4-2

WARNING! The symbol ∞ does not represent a number. Since it indicates that an interval extends indefinitely to the right, we always use a parenthesis after the symbol ∞.

The interval shown in Figure 4-2(b) is the graph of the real numbers less than or equal to 7. This interval contains the numbers that satisfy the inequality $x \leq 7$. The bracket at 7 indicates that 7 is included in the interval. To express this interval in interval notation, we write $(-\infty, 7]$, where $-\infty$ (read as **negative infinity**) indicates that the interval extends indefinitely to the left. The bracket indicates that 7 is in the interval.

EXAMPLE 1 ***Graphing intervals.*** Write the inequality $x \geq 8$ in interval notation and then graph it.

Solution

The inequality $x \geq 8$ is satisfied by all real numbers that are greater than or equal to 8. This is the interval $[8, \infty)$. The graph is shown in Figure 4-3.

FIGURE 4-3

Self Check

Write the inequality $x < 0$ in interval notation and then graph it.

Answer: $(-\infty, 0)$ (number line, parenthesis at 0)

If an interval extends forever in one direction, as in the previous examples, it is called an **unbounded interval.** The following chart illustrates the various types of unbounded intervals and shows how they are described using an inequality and a graph.

Unbounded intervals		
	The interval (a, ∞) includes all real numbers x such that $x > a$.	(graph: parenthesis at a, shaded right)
	The interval $[a, \infty)$ includes all real numbers x such that $x \geq a$.	(graph: bracket at a, shaded right)
	The interval $(-\infty, a)$ includes all real numbers x such that $x < a$.	(graph: parenthesis at a, shaded left)
	The interval $(-\infty, a]$ includes all real numbers x such that $x \leq a$.	(graph: bracket at a, shaded left)
	The interval $(-\infty, \infty)$ includes all real numbers x. The graph of this interval is the entire number line.	(graph: entire line, 0 marked)

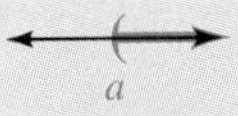

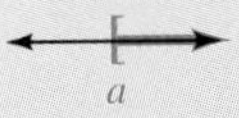

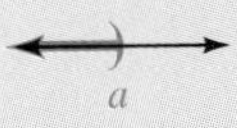

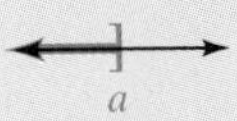

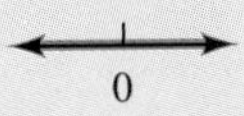

Properties of inequalities

To solve inequalities, we will use three properties of inequalities.

Property 1 of inequalities

Any real number can be added to (or subtracted from) both sides of an inequality to produce another inequality with the same direction.

Property 1 indicates that any number can be added to both sides of a true inequality to get another true inequality with the same direction. To illustrate this property, we consider the inequality $3 < 12$ and add 4 to both sides.

$3 < 12$	A true inequality.
$3 + 4 < 12 + 4$	Add 4 to both sides.
$7 < 16$	The result is a true inequality. Note that the $<$ symbol is unchanged in this process.

Subtracting the same number from both sides of $3 < 12$ doesn't change the direction of the inequality either.

$3 < 12$	A true inequality.
$3 - 4 < 12 - 4$	Subtract 4 from both sides.
$-1 < 8$	The result is a true inequality.

Property 2 of inequalities

If both sides of an inequality are multiplied (or divided) by a positive number, another inequality results, with the same direction as the original one.

Property 2 indicates that both sides of a true inequality can be multiplied by any positive number to get another true inequality with the same direction. To illustrate this property, we consider the inequality $-4 < 6$ and multiply both sides by 2.

$-4 < 6$	A true inequality.
$2(-4) < 2(6)$	Multiply both sides by 2.
$-8 < 12$	The result is a true inequality. The $<$ symbol is unchanged in this process.

Dividing both sides by the same positive number does not change the direction of the inequality either.

$-4 < 6$	A true inequality.
$\frac{-4}{2} < \frac{6}{2}$	Divide both sides by 2.
$-2 < 3$	The result is a true inequality. The $<$ symbol is unchanged in the process.

Property 3 of inequalities

If both sides of an inequality are multiplied (or divided) by a negative number, another inequality results, but with the opposite direction from the original inequality.

Property 3 indicates that if both sides of a true inequality are multiplied by a negative number, another true inequality results, but with the opposite direction. For example, we consider the inequality $-4 < 6$ and multiply both sides by -2.

$-4 < 6$	A true inequality containing a $<$ symbol.
$-2(-4) > -2(6)$	Multiply both sides by -2 and change $<$ to $>$.
$8 > -12$	The result is a true inequality. The inequality would not be true if we did not change the original symbol $<$ to $>$.

Dividing both sides by the same negative number also changes the direction of the inequality.

$$-4 < 6$$ A true inequality.

$$\frac{-4}{-2} > \frac{6}{-2}$$ Divide both sides by -2 and change $<$ to $>$.

$$2 > -3$$ The result is a true inequality.

WARNING! We must remember to change the direction of an inequality symbol every time we multiply or divide both sides of the inequality by a negative number.

Solving linear inequalities

In this section, we will work with **linear inequalities** in one variable.

Linear inequalities

A **linear inequality** in x is any inequality that can be expressed in one of the following forms, where a and c are real numbers and $a \neq 0$.

$$ax + c < 0 \qquad ax + c > 0 \qquad ax + c \leq 0 \qquad \text{or} \qquad ax + c \geq 0$$

To **solve a linear inequality** means to find all the values that, when substituted for the variable, make the inequality true. Most of the inequalities we will solve have infinitely many solutions. The steps we will use to solve an inequality are the same as those used to solve an equation, with one exception. If we multiply or divide both sides of an inequality by a negative number, we must reverse the direction of the inequality symbol.

EXAMPLE 2 ***Solving linear inequalities.*** Solve $3(2x - 9) < 9$. Give the result in interval notation and then graph the interval.

Solution

Our objective is to isolate x on the left-hand side of the inequality. To do that, we use the same strategy as we used to solve equations.

$$3(2x - 9) < 9$$

$$6x - 27 < 9$$ Use the distributive property to remove parentheses.

$$6x < 36$$ To undo the subtraction of 27, add 27 to both sides.

$$x < 6$$ To undo the multiplication by 6, divide both sides by 6.

The solution set is the interval $(-\infty, 6)$, whose graph is shown in Figure 4-4. The parenthesis at 6 indicates that 6 is not included in the solution set.

To check, we pick a number in the graph, such as 4, and see whether it satisfies the inequality.

$$3(2x - 9) < 9$$ The original inequality.

$$3[2(4) - 9] \overset{?}{<} 9$$ Substitute 4 for x.

$$3(8 - 9) \overset{?}{<} 9$$ Do the multiplication: $2(4) = 8$.

$$3(-1) \overset{?}{<} 9$$ Do the subtraction: $8 - 9 = -1$.

$$-3 < 9$$ This is a true statement.

6

FIGURE 4-4

Since $-3 < 9$, 4 satisfies the inequality. The solution appears to be correct.

Self Check

Solve $2(3x + 2) > -44$ and graph the interval.

Answer: $(-8, \infty)$

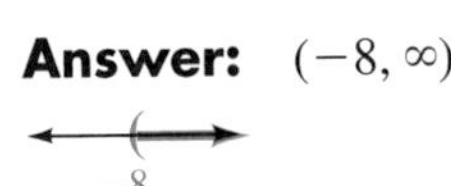

EXAMPLE 3 ***Reversing the inequality symbol.*** Solve $-4(3x + 2) \leq 16$. Give the result in interval notation and then graph the interval.

Solution

To solve this inequality, we need to isolate x.

$$-4(3x + 2) \leq 16$$

$$-12x - 8 \leq 16 \quad \text{Use the distributive property to remove parentheses.}$$

$$-12x \leq 24 \quad \text{To undo the subtraction of 8, add 8 to both sides.}$$

$$x \geq -2 \quad \text{To undo the multiplication by } -12 \text{, divide both sides by } -12.$$

Because we are dividing by a negative number, we reverse the $\leq$ symbol.

The solution set is the interval $[-2, \infty)$, whose graph is shown in Figure 4-5. The bracket at -2 indicates that -2 is included in the solution set.

−2

FIGURE 4-5

Self Check Solve $-3(2x - 2) \leq 0$ and graph it.

Answer: $[1, \infty)$

1

EXAMPLE 4 ***Clearing an inequality of fractions.*** Solve $\frac{2}{3}(x + 2) > \frac{4}{5}(x - 3)$.

Solution

It will be easier to solve the inequality if we clear it of fractions. We do that by multiplying both sides by the LCD of $\frac{2}{3}$ and $\frac{4}{5}$.

$$\frac{2}{3}(x + 2) > \frac{4}{5}(x - 3)$$

$$15 \cdot \frac{2}{3}(x + 2) > 15 \cdot \frac{4}{5}(x - 3) \quad \text{Multiply both sides by the LCD of } \tfrac{2}{3} \text{ and } \tfrac{4}{5} \text{, which is 15.}$$

$$10(x + 2) > 12(x - 3) \quad \text{Simplify: } 15 \cdot \tfrac{2}{3} = 10 \text{ and } 15 \cdot \tfrac{4}{5} = 12.$$

$$10x + 20 > 12x - 36 \quad \text{Distribute the 10 and the 12.}$$

$$-2x + 20 > -36 \quad \text{To eliminate } 12x \text{ on the right-hand side, subtract } 12x \text{ from both sides.}$$

$$-2x > -56 \quad \text{Subtract 20 from both sides.}$$

$$x < 28 \quad \text{Divide both sides by } -2 \text{ and reverse the } > \text{ symbol.}$$

The solution set is the interval $(-\infty, 28)$, whose graph is shown in Figure 4-6.

FIGURE 4-6

Self Check Solve $\frac{3}{2}(x + 2) < \frac{3}{5}(x - 3)$ and graph it.

Answer: $\left(-\infty, -\frac{16}{3}\right)$

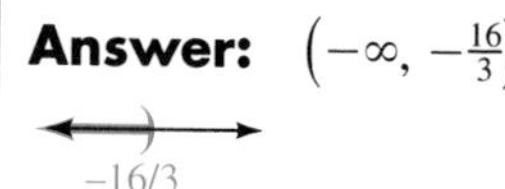

WARNING! When solving an inequality, the variable sometimes ends up on the right-hand side. For instance, suppose we solve an inequality and obtain $-3 < x$. This inequality can be expressed in the equivalent form $x > -3$, which most students find easier to graph and express in interval notation.

Accent on Technology ***Solving linear inequalities***

We can solve linear inequalities with a graphing approach. For example, to solve the inequality $3(2x - 9) < 9$, we can graph $y = 3(2x - 9)$ and $y = 9$ using window settings of $[-10, 10]$ for x and $[-10, 10]$ for y. We get Figure 4-7(a). We can then trace to see that the graph of $y = 3(2x - 9)$ is below the graph of $y = 9$ for x-values in the interval $(-\infty, 6)$. (See Figure 4-7(b).) This interval is the solution, because in this interval, $3(2x - 9) < 9$.

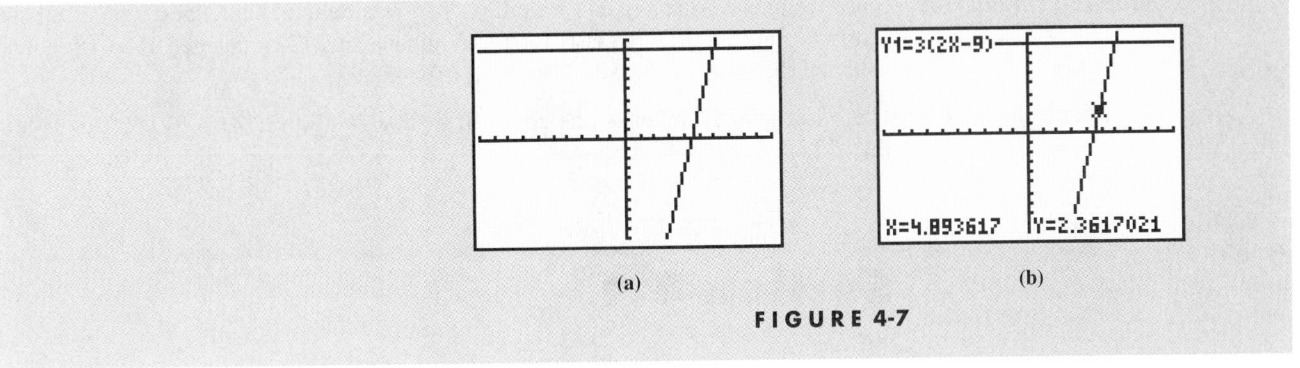

(a) (b)

FIGURE 4-7

Problem solving

In previous chapters, we have used a five-step problem-solving strategy to solve problems. This process involved writing and then solving equations. We will now show how inequalities can be used to solve problems. To decide whether to use an equation or an inequality to solve a problem, you must spot key words and phrases.

EXAMPLE 5 ***Translating from words to symbols*** Translate the sentence to mathematical symbols: *The instructor said that the test would take no more than 50 minutes.*

Solution

Since the test will take no more than 50 minutes, it will take 50 minutes or less to complete. If we let t represent the time it takes to complete the test, then $t \le 50$.

Self Check

Translate the sentence to mathematical symbols: *A PG-13 movie rating means that you must be at least 13 years old to see the movie.*

Answer: $a \ge 13$ ■

EXAMPLE 6 ***Political contributions.*** Some volunteers are making long-distance telephone calls to solicit contributions for their candidate. The calls are billed at the rate of 42¢ for the first three minutes and 11¢ for each additional minute or part thereof. If the campaign chairperson has ordered that the cost of each call is not to exceed \$2.00, for how many minutes can a volunteer talk to a prospective donor on the phone?

Analyze the problem We are given the rate at which a call is billed. Since the cost of a call is not to exceed \$2.00, the cost must be *less than or equal to* \$2.00. This phrase indicates that we should write an *inequality* to find how long a volunteer can talk to a prospective donor.

Form an inequality We will let x represent the total number of minutes that a call can last. Then the cost of a call will be 42¢ for the first three minutes plus 11¢ times the number of additional minutes, where the number of *additional* minutes is $x - 3$ (the total number of minutes minus the first 3 minutes). With this information, we can form an inequality.

The cost of the first three minutes	+	the cost of the additional minutes	is not to exceed	\$2.
0.42	+	$0.11(x - 3)$	$\le$	2

Solve the inequality To simplify the computations, we first clear the inequality of decimals.

$0.42 + 0.11(x - 3) \le 2$

$42 + 11(x - 3) \le 200$ To eliminate the decimals, multiply both sides by 100.

$42 + 11x - 33 \le 200$ Use the distributive property to remove parentheses.

$11x + 9 \le 200$ Combine like terms.

$11x \le 191$ Subtract 9 from both sides.

$x \le 17.\overline{36}$ Divide both sides by 11.

State the conclusion Since the phone company doesn't bill for part of a minute, the longest time a call can last is 17 minutes. If a call lasts for $17.\overline{36}$ minutes, it will be charged as an 18-minute call, and the cost will be $\$0.42 + \$0.11(15) = \$2.07$.

Check the result If the call lasts 17 minutes, the cost will be $\$0.42 + \$0.11(14) = \$1.96$. This is less than \$2.00. The result checks. ■

STUDY SET Section 4.1

VOCABULARY *In Exercises 1–8, fill in the blanks to make the statements true.*

1. $<$, $>$, $\le$, and $\ge$ are ______________ symbols.

2. $(-\infty, 5)$ is an example of an unbounded ______________.

3. The ______________ on the right of the interval notation $(-\infty, 5)$, indicates that 5 is not included in the interval.

4. To ______________ an inequality means to find all values of the variable that make the inequality true.

5. $3x + 2 \ge 7$ is an example of a ______________ inequality.

6. ∞ is a symbol representing positive ______________.

7. The symbol for "______________" is $<$.

8. The symbol for "______________" is $\ge$.

CONCEPTS

9. Describe each set of real numbers using interval notation and then graph it.
a. All real numbers greater than 4
b. All real numbers less than -4
c. All real numbers less than or equal to 4

10. Match each interval with its graph.
a. $(-\infty, -1]$
b. $(-\infty, 1)$
c. $[-1, \infty)$

i
ii
iii

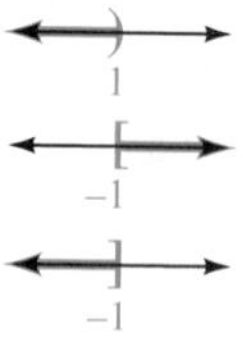

11. Classify each of the following as either an equation, an expression, or an inequality.
a. $-6 - 5x = 8$
b. $5 - 2x$
c. $7x - 5x > -4x$
d. $-(7x - 9)$

12. In each case, tell what is wrong with the interval notation.
a. $(\infty, -3)$
b. $[-\infty, -3)$

13. Perform each step listed below on the inequality $4 > -2$. *Do not reverse the inequality symbol.* Is the resulting statement true?
a. Add 2 to both sides.
b. Subtract 4 from both sides.
c. Multiply both sides by 4.
d. Divide both sides by -2.

14. Consider the linear inequality $3x + 6 \le 6$. Tell whether each value is a solution of the inequality.
a. 0
b. $\frac{2}{3}$
c. -10
d. 1.5

15. What inequality symbol is suggested by each sentence?
a. As many as 16 people were seriously injured.
b. There are no fewer than 10 references to carpools in the speech.

16. Write an equivalent inequality with the variable on the left-hand side.
a. $-10 > x$
b. $\frac{7}{8} < x$
c. $0 \le x$

NOTATION *In Exercises 17–18, complete each solution.*

17. Solve $-5x - 1 \geq -11$.

$-5x - 1 \geq -11$

$-5x \geq \square$ Add 1 to both sides.

$\frac{-5x}{-5} \square \frac{-10}{-5}$ Divide both sides by -5.

$x \leq 2$ Do the divisions.

Using interval notation, the result is $(\square, 2]$.

18. Solve $3 - 6x < 17 + x$.

$3 - 6x < 17 + x$

$3 - \square < 17$ Subtract x from both sides.

$-7x < \square$ Subtract 3 from both sides.

$\frac{-7x}{-7} \square \frac{14}{-7}$ Divide both sides by -7.

$x > -2$ Do the divisions.

Using interval notation, the result is $(-2, \square)$.

PRACTICE *In Exercises 19–42, solve each inequality. Give the result in interval notation and then graph the solution set.*

19. $3x > -9$

20. $4x < -36$

21. $-30y \leq -600$

22. $-6y \geq -600$

23. $0.6x \geq 36$

24. $0.2x < 8$

25. $3 > -\frac{9}{10}x$

26. $-\frac{2}{5} < -\frac{4}{5}x$

27. $x + 4 < 5$

28. $x - 5 > 2$

29. $-5t + 3 \leq 5$

30. $-9t + 6 \geq 16$

31. $7 < \frac{5}{3}a - 3$

32. $5 > \frac{7}{2}a - 9$

33. $0.4x + 0.4 \leq 0.1x + 0.85$

34. $0.05 - 0.5x \leq -0.7 - 0.8x$

35. $3(z - 2) \leq 2(z + 7)$

36. $5(3 + z) > -3(z + 3)$

37. $-11(2 - b) < 4(2b + 2)$

38. $-9(h - 3) + 2h \leq 8(4 - h)$

39. $\frac{1}{2}y + 2 \geq \frac{1}{3}y - 4$

40. $\frac{1}{4}x - \frac{1}{3} \leq x + 2$

41. $\frac{2}{3}x + \frac{3}{2}(x - 5) \leq x$

42. $\frac{5}{9}(x + 3) - \frac{4}{3}(x - 3) \geq x - 1$

APPLICATIONS

43. REAL ESTATE Refer to the graph in Illustration 1. For which regions of the country was the following inequality true in 1997?

Median sales price < U.S. median price

44. MUSIC INDUSTRY Refer to the graph in Illustration 2 on the next page. Determine the first full year in which the following inequality was true.

CDs shipped > cassettes shipped

1997 Median Price of Existing Single-Family Homes

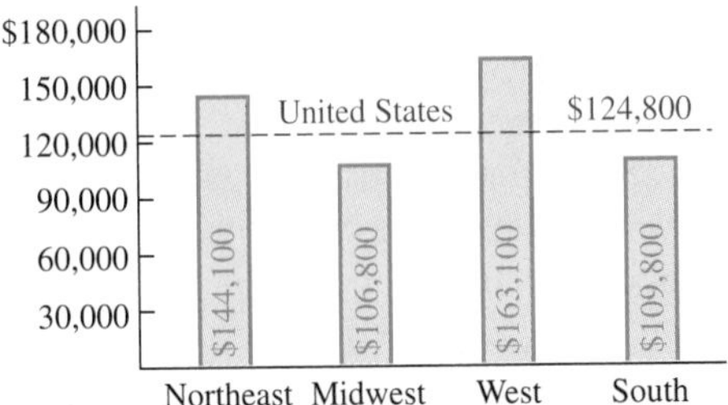

Based on data from the National Association of Realtors

ILLUSTRATION 1

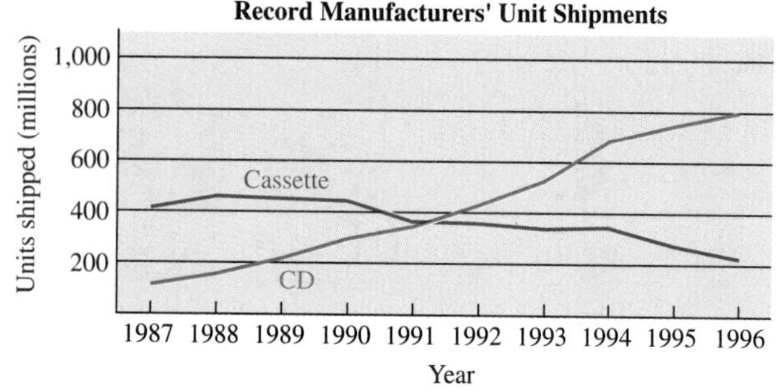

Based on information from *The Wall Street Journal Almanac* (1998), p. 915

ILLUSTRATION 2

45. GEOMETRY The **triangle inequality** states an important relationship between the sides of any triangle:

The sum of the lengths of any two sides of a triangle > the length of the third side

Use the triangle inequality to show that the dimensions of the shuffleboard court shown in Illustration 3 must be mislabeled.

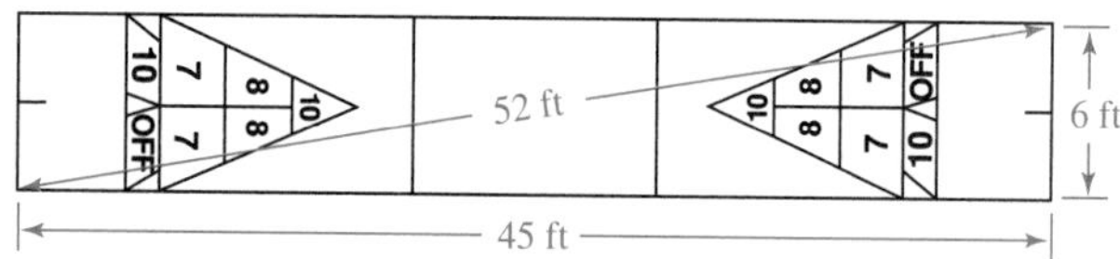

ILLUSTRATION 3

46. COMPUTER PROGRAMMING Flow charts like the one in Illustration 4 are used by programmers to show the step-by-step instructions of a computer program. For each row in the table, work through the steps of the flow chart using the values of a, b, and c, and tell what the computer printout would be in each case.

	a	b	c
Row 1	1	1	1
Row 2	9	−12	4
Row 3	11	−25	−24

47. FUND RAISING A school PTA wants to rent a dunking tank for its annual school fund raising carnival. The cost is \$85.00 for the first three hours and then \$19.50 for each additional hour or part thereof. How long can the tank be rented if up to \$185 is budgeted for this expense?

48. INVESTMENTS If a woman has invested \$10,000 at 8% annual interest, how much more must she invest at 9% so that her annual income will exceed \$1,250?

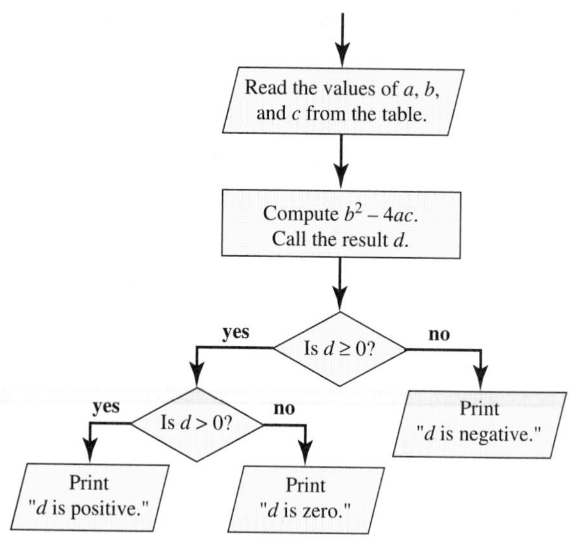

ILLUSTRATION 4

49. BUYING A COMPUTER A student who can afford to spend up to \$2,000 sees the ad shown in Illustration 5. If she decides to buy the computer, find the greatest number of CD-ROMs that she can also purchase. (Disregard sales tax.)

ILLUSTRATION 5

50. AVERAGING GRADES A student has scores of 70, 77, and 85 on three government exams. What score does she need on a fourth exam to give her an average of 80 or better?

51. WORK SCHEDULE A student works two part-time jobs. He earns \$7 an hour for working at the college library and \$12 an hour for construction work. To save time for study, he limits his work to 20 hours a week. If he enjoys the work at the library better, how many hours can he work at the library and still earn at least \$175 a week?

52. SCHEDULING EQUIPMENT An excavating company charges \$300 an hour for the use of a backhoe and \$500 an hour for the use of a bulldozer. (Part of an hour counts as a full hour.) The company employs one operator for 40 hours per week to operate the machinery. If the company wants to bring in at least \$18,500 each week from equipment rental, how many hours per week can it schedule the operator to use a backhoe?

53. MEDICAL PLANS A college provides its employees with a choice of the two medical plans shown in Illustration 6. For what size hospital bills is Plan 2 better for the employee than Plan 1? (*Hint:* The cost to the employee includes both the deductible payment and the employee's coinsurance payment.)

Plan 1	Plan 2
Employee pays \$100 Plan pays 70% of the rest	Employee pays \$200 Plan pays 80% of the rest

ILLUSTRATION 6

54. MEDICAL PLANS To save costs, the college in Exercise 53 raised the employee deductible, as shown in Illustration 7. For what size hospital bills is Plan 2 better for the employee than Plan 1? (*Hint:* The cost to the employee includes both the deductible payment and the employee's coinsurance payment.)

Plan 1	Plan 2
Employee pays \$200 Plan pays 70% of the rest	Employee pays \$400 Plan pays 80% of the rest

ILLUSTRATION 7

In Exercises 55–58, use a graphing calculator to solve each inequality.

55. $2x + 3 < 5$

56. $3x - 2 > 4$

57. $5x + 2 \geq -18$

58. $3x - 4 \leq 20$

WRITING *Write a paragraph using your own words.*

59. The techniques for solving linear equations and linear inequalities are similar, yet different. Explain.

60. Explain how the symbol ∞ is used in this section. Is ∞ a real number?

REVIEW *In Exercises 61–62, use the graph of the function to find $f(-1)$, $f(0)$, and $f(2)$.*

61.

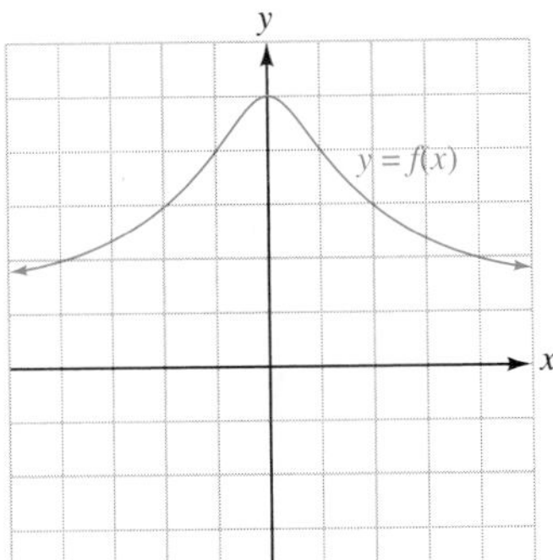

62.

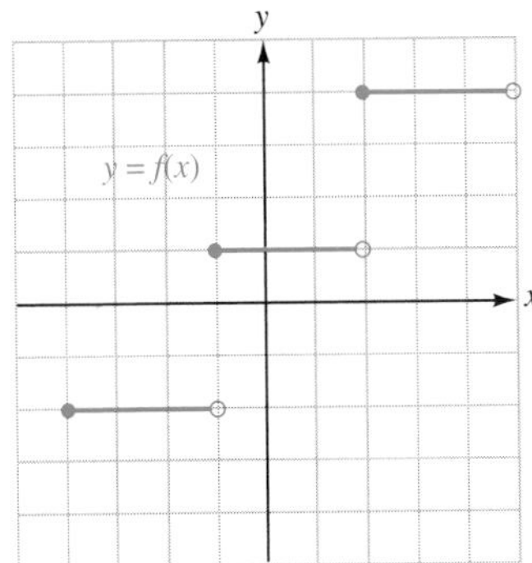

4.2 Solving Compound Inequalities

In this section, you will learn about

- **Solving compound inequalities containing the word *and***
- **Double linear inequalities**
- **Compound inequalities containing the word *or***
- **Solving compound inequalities containing the word *or***

INTRODUCTION. The label on the tube of antibiotic ointment shown in Figure 4-8 advises the user about the temperature at which the medication should be stored. A careful reading of the statement reveals that the storage instruction consists of two parts:

The storage temperature should be at least 59°F

and

The storage temperature should be at most 77°F

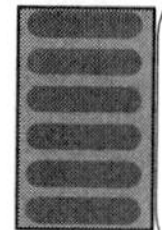

DIRECTIONS: Clean the affected area thoroughly. Apply a small amount of this product (an amount equal to the surface area of the tip of a finger) on the area 1 to 3 times daily. Do not use in eyes. **Store at 59° to 77°F.** Do not use longer than 1 week. Keep this and all drugs out of the reach of children.

FIGURE 4-8

In mathematics, when the words *and* or *or* are used to connect pairs of inequalities, we call these statements *compound inequalities.* In this section, we will discuss the procedures used to solve three types of compound inequalities, as well as the notation used to express their solution sets.

Solving compound inequalities containing the word *and*

When two inequalities are joined with the word *and,* we call the statement a **compound inequality.** Some examples are

$$x \geq -3 \text{ and } x \leq 6$$

$$\frac{x}{2} + 1 > 0 \text{ and } 2x - 3 < 5$$

$$x + 3 \leq 2x - 1 \text{ and } 3x - 2 < 5x - 4$$

The solution set of these inequalities contains all the numbers that make *both* of the inequalities true. For example, we can find the solution set of the compound inequality $x \geq -3$ and $x \leq 6$ by first graphing the solution sets of each inequality on the same number line and then looking for the numbers common to both graphs.

In Figure 4-9(a), the graph of the solution set of $x \geq -3$ is shown in red, and the graph of the solution set of $x \leq 6$ is shown in blue. Figure 4-9(b) shows the graph of the solution of $x \geq -3$ and $x \leq 6$. This portion of the number line is where the red and blue graphs overlap. It represents the numbers common to both graphs.

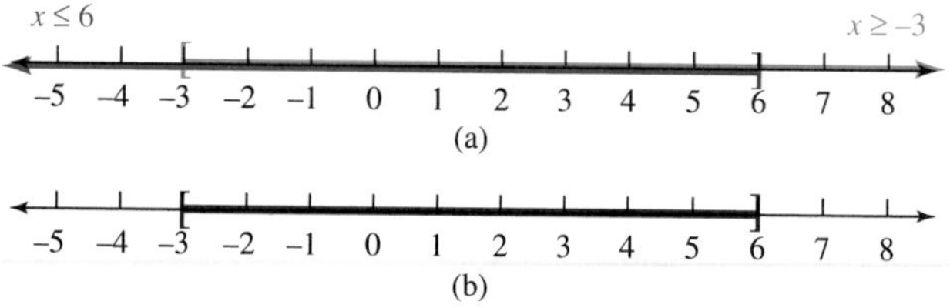

FIGURE 4-9

The solution set of $x \geq -3$ and $x \leq 6$ can be denoted by the **bounded interval** $[-3, 6]$, where the brackets indicate that the endpoints, -3 and 6, are included. It represents all real numbers between -3 and 6, including -3 and 6. Intervals such as this, that contain both endpoints, are called **closed intervals.**

When solving a compound inequality containing *and,* the solution set is the **intersection** of the solution sets of the two inequalities. The intersection of two sets is the set of elements that are common to both sets. We can denote the intersection of two sets using the symbol $\cap$, which is read as "intersection." For the compound inequality $x \geq -3$ and $x \leq 6$, we can write

$$[-3, \infty) \cap (-\infty, 6] = [-3, 6]$$

The solution set of the compound inequality $x \geq -3$ and $x \leq 6$ can be expressed in three other ways:

1. *As a graph:* [−3, 6] on a number line, −3 to 6
2. *In interval notation:* $[-3, 6]$
3. *In words:* all real numbers between -3 and 6, including -3 and 6

EXAMPLE 1 *Solving compound inequalities containing "and."*

Solve $\frac{x}{2} + 1 > 0$ and $2x - 3 < 5$.

Solution

We begin by solving each linear inequality separately.

$$\frac{x}{2} + 1 > 0 \quad \text{and} \quad 2x - 3 < 5$$

$$\frac{x}{2} > -1 \qquad\qquad 2x < 8$$

$$x > -2 \qquad\qquad x < 4$$

Next, we graph the solutions of each inequality on the same number line and determine their intersection. See Figure 4-10.

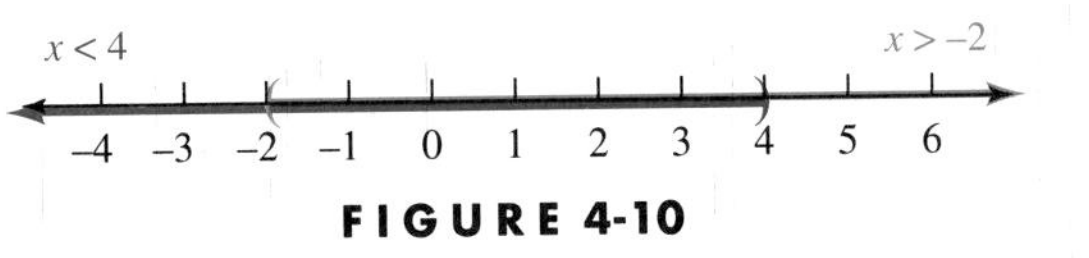

FIGURE 4-10

The intersection of the graphs in Figure 4-10 is the set of all real numbers between -2 and 4. Using interval notation, the solution set is the interval $(-2, 4)$, whose graph is shown in Figure 4-11. This bounded interval, which does not include either endpoint, is called an **open interval.**

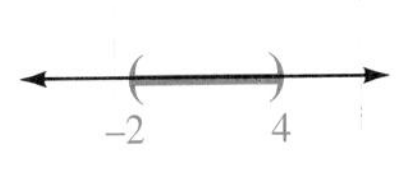

FIGURE 4-11

Self Check

Solve $3x > -18$ and $\frac{x}{5} - 1 \leq 1$ and graph the solution set.

Answer: $(-6, 10]$

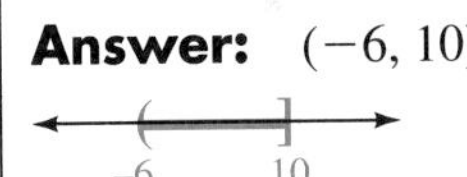

The solution of the compound inequality in the Self Check of Example 1 is the interval $(-6, 10]$. A bounded interval such as this, which includes only one endpoint, is called a **half-open interval.** The following chart shows the various types of bounded intervals, along with the inequalities and interval notation that describes them.

Open intervals	The interval (a, b) includes all real numbers x such that $a < x < b$.	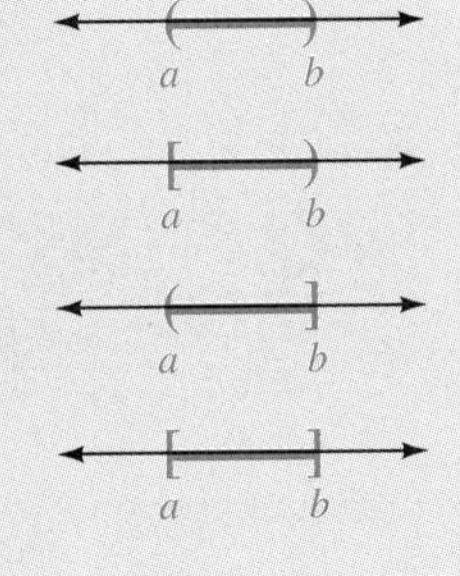
Half-open intervals	The interval $[a, b)$ includes all real numbers x such that $a \leq x < b$.	
	The interval $(a, b]$ includes all real numbers x such that $a < x \leq b$.	
Closed intervals	The interval $[a, b]$ includes all real numbers x such that $a \leq x \leq b$.	

WARNING! Remember that in interval notation, the notation $(-3, 4)$ represents the set of real numbers between -3 and 4, not the coordinates of a point on a graph.

EXAMPLE 2 ***Solving compound inequalities containing "and."*** Solve $x + 3 \le 2x - 1$ and $3x - 2 < 5x - 4$.

Solution

We solve each inequality separately.

$$\begin{array}{lcl} x + 3 \le 2x - 1 & \text{and} & 3x - 2 < 5x - 4 \\ 4 \le x & & 2 < 2x \\ x \ge 4 & & 1 < x \\ & & x > 1 \end{array}$$

Only those x where $x \ge 4$ and $x > 1$ are in the solution set. Since all numbers greater than or equal to 4 are also greater than 1, the solutions are the numbers x where $x \ge 4$. The solution set is the interval $[4, \infty)$, whose graph is shown in Figure 4-12.

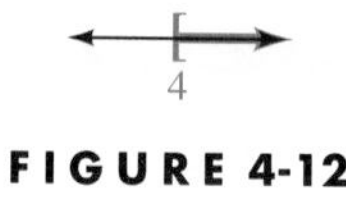

FIGURE 4-12

Self Check Solve $2x + 3 < 4x + 2$ and $3x + 1 < 5x + 3$ and graph the solution set.

Answer: $\left(\frac{1}{2}, \infty\right)$

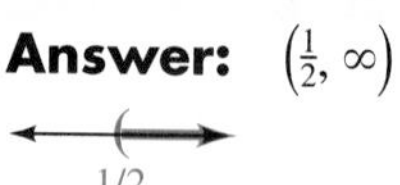

EXAMPLE 3 ***A compound inequality with no solution.*** Solve $x - 1 > -3$ and $2x < -8$.

Solution

We solve each inequality separately.

$$\begin{array}{lcl} x - 1 > -3 & \text{and} & 2x < -8 \\ x > -2 & & x < -4 \end{array}$$

We note that the graphs of the solution sets shown in Figure 4-13 do not intersect.

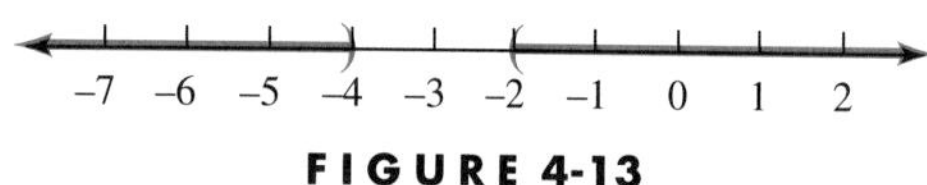

FIGURE 4-13

This means that there are no numbers that make both parts of the original compound inequality true. So the solution set is the empty set, which can be denoted $\varnothing$.

Self Check Solve $2x - 3 < x - 2$ and $0 < x - 3.5$.

Answer: no solution

Double linear inequalities

Inequalities that contain two inequality symbols are called **double inequalities.** An example of a double inequality is

$-3 \le 2x + 5 < 7$ Read as "-3 is less than or equal to $2x + 5$ and $2x + 5$ is less than 7."

Any double linear inequality can be rewritten as a compound inequality containing the word *and.* In general, the following is true.

Double linear inequalities The compound inequality $c < x < d$ is equivalent to $c < x$ and $x < d$.

EXAMPLE 4 ***Solving double linear inequalities.*** Solve $-3 \le 2x + 5 < 7$ and graph the solution set.

Solution

This double inequality $-3 \le 2x + 5 < 7$ means that

$$-3 \le 2x + 5 \text{ and } 2x + 5 < 7$$

Self Check Solve $-5 \le 3x - 8 \le 7$ and graph the solution set.

We could solve each linear inequality separately, but we note that each solution would involve the same steps: subtracting 5 from both sides and dividing both sides by 2. We can solve the double inequality more efficiently by leaving it in its original form and applying these steps to each of its *three parts* to isolate x in the middle.

$$-3 \leq 2x + 5 < 7$$

$$-3 - 5 \leq 2x + 5 - 5 < 7 - 5$$ To undo the addition of 5, subtract 5 from all three parts.

$$-8 \leq 2x < 2$$ Do the subtractions.

$$\frac{-8}{2} \leq \frac{2x}{2} < \frac{2}{2}$$ To undo the multiplication by 2, divide all three parts by 2.

$$-4 \leq x < 1$$ Do the divisions.

The solution set is the half-open interval $[-4, 1)$, whose graph is shown in Figure 4-14.

FIGURE 4-14

Answer: $[1, 5]$

1 5

When multiplying or dividing all three parts of a double equality by a negative number, don't forget to reverse the direction of *both* inequalities. As an example, we will solve $-15 < -5x \leq 25$.

$$-15 < -5x \leq 25$$

$$\frac{-15}{-5} > \frac{-5x}{-5} \geq \frac{25}{-5}$$ Divide all three parts by -5 to isolate x in the middle. Reverse both inequality signs.

$$3 > x \geq -5$$ Do the divisions.

$$-5 \leq x < 3$$ Write an equivalent compound inequality with the smallest number on the left.

Compound inequalities containing the word *or*

A warning on the water temperature gauge of a commercial dishwasher, shown in Figure 4-15, cautions the operator to shut down the unit if

The water temperature goes below 140°

or

The water temperature goes above 160°

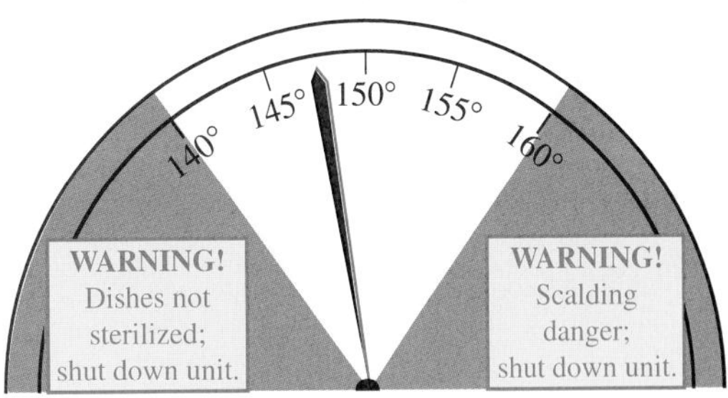

FIGURE 4-15

When two inequalities are joined with the word *or,* we also call the statement a compound inequality. Some examples are

$x < 140$ or $x > 160$

$x \leq -3$ or $x \geq 2$

$\frac{x}{3} > \frac{2}{3}$ or $-(x - 2) > 3$

Solving compound inequalities containing the word *or*

The solution set of a compound inequality containing the word *or* contains all the numbers that make *one or the other or both* inequalities true. For example, we can find the solution set of the compound inequality $x \leq -3$ or $x \geq 2$ by putting the graphs of each inequality on the same number line.

In Figure 4-16(a), the graph of the solution set of $x \leq -3$ is shown in red, and the graph of the solution set of $x \geq 2$ is shown in blue. Figure 4-16(b) shows the graph of the solution set of $x \leq -3$ or $x \geq 2$. This graph is a combination of the two graphs.

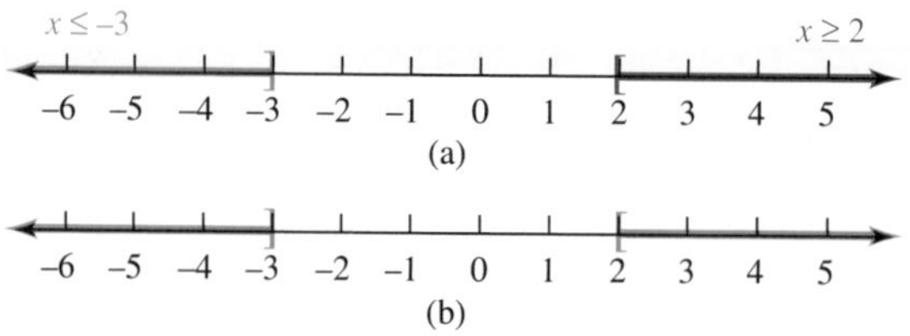

FIGURE 4-16

When solving a compound inequality containing *or*, the solution set is the **union** of the solution sets of the two inequalities. The union of two sets is the set of elements that are in either of the sets or both. We can denote the union of two sets using the symbol $\cup$, which is read as "union." For the compound inequality $x \leq -3$ or $x \geq 2$, we can write the solution set using interval notation as

$$(-\infty, -3] \cup [2, \infty)$$

We can express the solution set of the compound inequality $x \leq -3$ or $x \geq 2$ in three other ways:

1. *As a graph:* 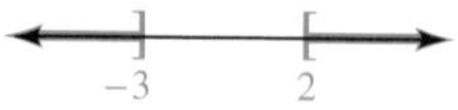

2. *In interval notation:* $(-\infty, -3] \cup [2, \infty)$
3. *In words:* all real numbers less than or equal to -3 or greater than or equal to 2

WARNING! In the statement $x \leq -3$ or $x \geq 2$, it is incorrect to string the inequalities together as $2 \leq x \leq -3$, because that would imply that $2 \leq -3$, which is false.

EXAMPLE 5 ***Solving compound inequalities containing "or."***

Solve $\frac{x}{3} > \frac{2}{3}$ or $-(x - 2) > 3$.

Solution

We solve each inequality separately.

$$\frac{x}{3} > \frac{2}{3} \quad \text{or} \quad -(x - 2) > 3$$

$$x > 2 \qquad\qquad -x + 2 > 3$$

$$-x > 1$$

$$x < -1$$

The graph of the solution set is obtained by graphing the solution sets of each inequality on the same number line. See Figure 4-17.

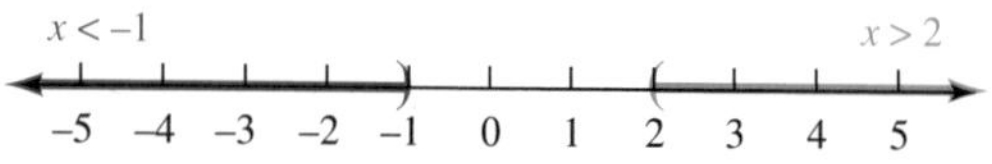

FIGURE 4-17

Self Check

Solve $\frac{x}{2} > 2$ or $-3(x - 2) > 0$

and graph the solution set.

The union of the two solution sets consists of all real numbers less than -1 or greater than 2. Using interval notation, the solution set is the interval $(-\infty, -1) \cup (2, \infty)$. Its graph appears in Figure 4-18.

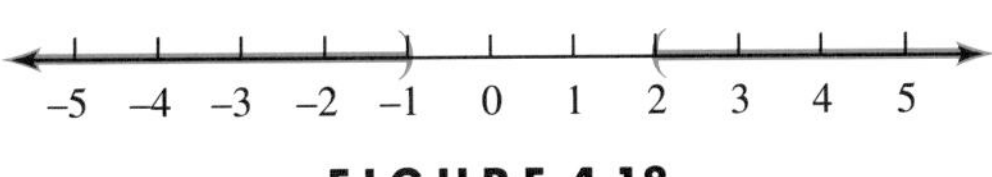

FIGURE 4-18

Answer: $(-\infty, 2) \cup (4, \infty)$

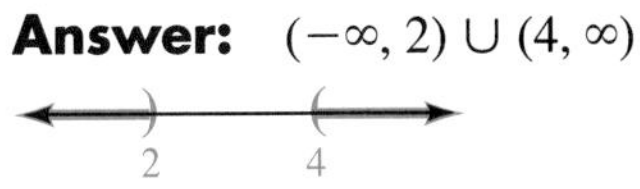

EXAMPLE 6 ***A compound inequality satisfied by all real numbers.*** Solve $x + 3 \geq -3$ or $-x > 0$.

Solution

We solve each inequality separately.

$$x + 3 \geq -3 \quad \text{or} \quad -x > 0$$
$$x \geq -6 \qquad\qquad x < 0$$

We graph the solution set of each inequality on the same number line in Figure 4-19.

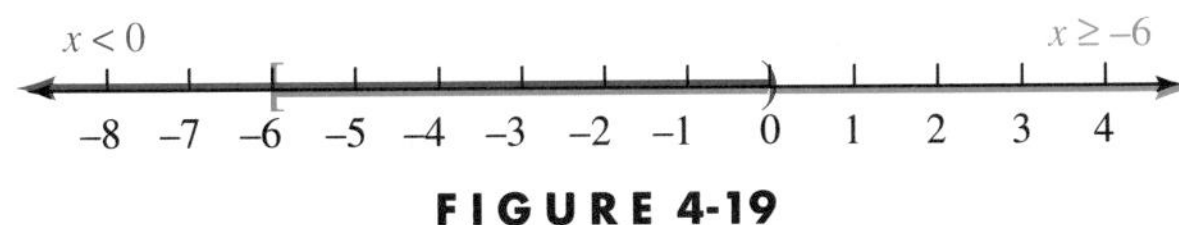

FIGURE 4-19

The entire number line is shaded, which indicates that all real numbers satisfy the original compound inequality. Using interval notation the solution set is denoted $(-\infty, \infty)$. Its graph is shown in Figure 4-20.

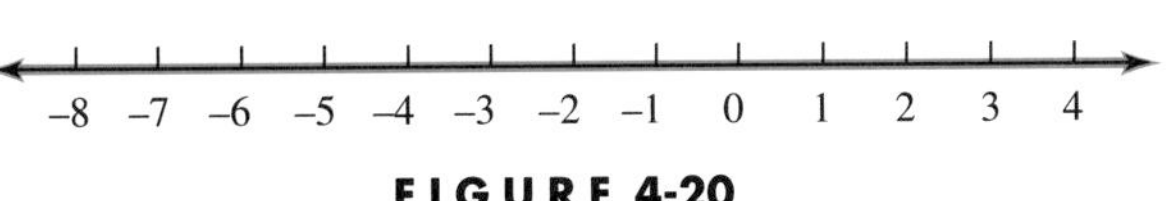

FIGURE 4-20

Self Check

Solve $x - 1 < 5$ or $-2x \leq 10$ and graph the solution set.

Answer: $(-\infty, \infty)$

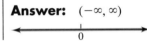

STUDY SET Section 4.2

VOCABULARY *In Exercises 1–4, fill in the blanks to make the statements true.*

1. $x \geq 3$ and $x \leq 4$ is a __________ inequality.
2. $-6 \leq x + 1 < 1$ is a __________ linear inequality.
3. The bounded __________ (2, 8] includes all real numbers x such that $2 < x \leq 8$.
4. $x \leq 3$ or $x > 5$ is a compound __________.

CONCEPTS *In Exercises 5–8, fill in the blanks to make the statements true.*

5. The word *and* between two inequality statements requires that __________ of the inequalities must be true for the entire statement to be true.
6. The word *or* between two inequality statements requires that only __________ of the inequalities must be true for the entire statement to be true.
7. If the three parts of a double inequality are divided by a negative number, the direction of both inequality symbols must be __________.
8. The double inequality $-2 < 3x + 4 < 10$ can be written as $-2 < 3x + 4$ __________ $3x + 4 < 10$.

9. In each case, tell whether $x = -3$ is a solution of the compound inequality.

a. $\frac{x}{3} + 1 \geq 0$ and $2x - 3 < -10$

b. $2x \leq 0$ or $-3x < -5$

10. In each case, tell whether $x = -3$ is a solution of the double linear inequality.

a. $-1 < -3x + 4 < 12$

b. $-1 < -3x + 4 < 14$

11. Give the solution of each inequality in interval notation, if possible.

a. $x < -3$ and $x > 3$

b. $x < 3$ or $x > -3$

12. Give the solution of each inequality in interval notation, if possible.

a. $x < 0$ or $x \geq 0$

b. $x < 0$ and $x > 0$

13. Match each interval with its corresponding graph.

a. $[2, 3)$

b. $(2, 3)$

c. $[2, 3]$

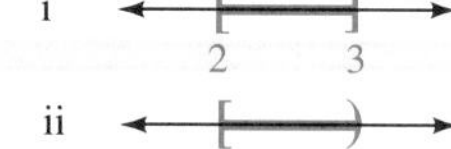

14. Give the interval notation that describes each set. Then graph it.

a. The real numbers between -3 and 3

b. The real numbers less than -3 or greater than 3

c. The real numbers between -3 and 3, including 3

NOTATION

15. a. Graph $(-\infty, 2) \cup [3, \infty)$.

b. Graph $(-\infty, 3) \cap [-2, \infty)$.

16. Classify each interval as open, half-open, or closed.

a. $(-2, 15]$ **b.** $[-2, 15]$

c. $(-2, 15)$ **d.** $[-2, 15)$

17. What is incorrect about the double inequality $3 < -3x + 4 < -3$?

18. What set is denoted by the interval notation $(-\infty, \infty)$? Graph it.

PRACTICE *In Exercises 19–50, solve each compound inequality. Give the result, if one exists, in interval notation and graph the solution set.*

19. $x > -2$ and $x \leq 5$

20. $x \leq -4$ and $x \geq -7$

21. $2.2x < -19.8$ and $-4x < 40$

22. $\frac{1}{2}x \leq 2$ and $0.75x \geq -6$

23. $x + 3 < 3x - 1$ and $4x - 3 \leq 3x$

24. $4x \geq -x + 5$ and $6 \geq 4x - 3$

25. $x + 2 < -\frac{1}{3}x$ and $-6x < 9x$

26. $5(x - 2) \geq 0$ and $-3x < 9$

27. $x - 1 \leq 2(x + 2)$ and $x \leq 2x - 5$

28. $5(x + 1) \leq 4(x + 3)$ and $x + 12 < -3$

29. $4 \leq x + 3 \leq 7$

30. $-5.3 < x - 2.3 < -1.3$

31. $15 > 2x - 7 > 9$

32. $25 > 3x - 2 > 7$

33. $-2 < -b + 3 < 5$

34. $2 < -t - 2 < 9$

35. $-6 < -3(x - 4) \leq 24$

36. $-4 \leq -2(x + 8) < 8$

37. $-4 > \frac{2}{3}x - 2 > -6$

38. $-6 \leq \frac{1}{3}a + 1 < 0$

39. $0 \leq \frac{4 - x}{3} \leq 2$

40. $-2 \leq \frac{5 - 3x}{2} \leq 2$

41. $x \leq -2$ or $x > 6$

42. $x \geq -1$ or $x \leq -3$

43. $x - 3 < -4$ or $x - 2 > 0$

44. $4x < -12$ or $\frac{x}{2} > 4$

45. $3x + 2 < 8$ or $2x - 3 > 11$

46. $3x + 4 < -2$ or $3x + 4 > 10$

47. $-4(x + 2) \geq 12$ or $3x + 8 < 11$

48. $4.5x - 1 < -10$ or $6 - 2x \geq 12$

49. $4.5x - 2 > 2.5$ or $\frac{1}{2}x \leq 1$

50. $0 < x$ or $3x - 5 > 4x - 7$

APPLICATIONS

51. BABY FURNITURE See Illustration 1. A company manufactures various sizes of playpens having perimeters between 128 and 192 inches, inclusive.

a. Complete the double inequality that mathematically describes the range of the perimeters of the playpens.

$$? \leq 4s \leq ?$$

b. Solve the double inequality to find the range of the side lengths of the playpens.

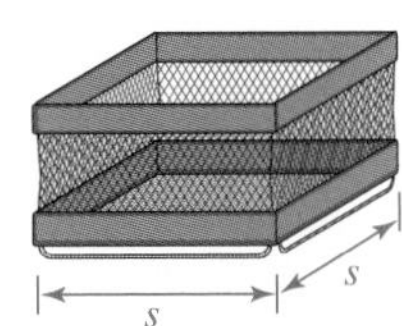

ILLUSTRATION 1

52. TRUCKING The distance that a truck can travel in 8 hours, at a constant rate of r mph, is given by $8r$. A trucker wants to travel at least 350 miles, and company regulations don't allow him to exceed 450 miles in one 8-hour shift.

a. Complete the double inequality that describes the mileage range of the truck.

$$? \leq 8r \leq ?$$

b. Solve the double inequality to find the range of the average rate (speed) of the truck for the 8-hour trip.

53. TREATING A FEVER Use the flow chart shown in Illustration 2 to determine what action should be taken for a 13-month-old child who has had a 99.8° temperature for 3 days and is not suffering any other symptoms. T represents the child's temperature, A the child's age in months, and S the time in hours the child has experienced the symptoms.

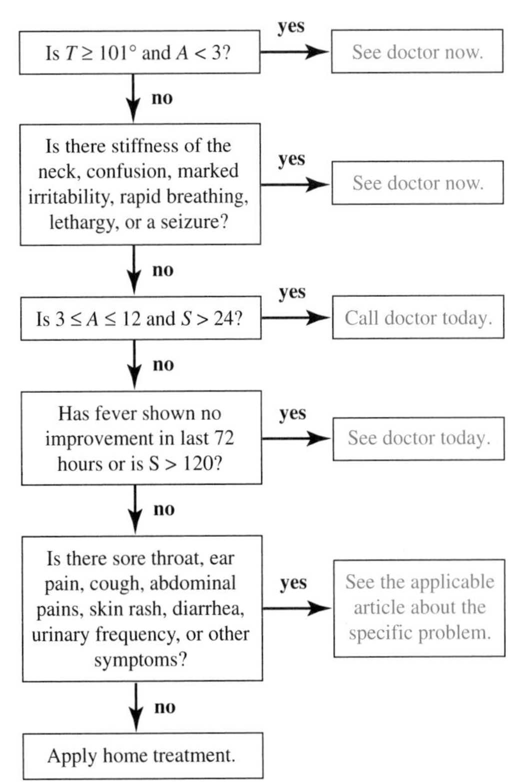

Based on information from *Take Care of Yourself* (Addison-Wesley, 1993)

ILLUSTRATION 2

54. THERMOSTAT The *Temp range* control on the thermostat shown in Illustration 3 directs the heater to come on when the room temperature gets 5 degrees below the *Temp setting;* it directs the air conditioner to come on when the room temperature gets 5 degrees above the *Temp setting.* Use interval notation to describe

a. the temperature range for the room when neither the heater nor the air conditioner will be on.

b. the temperature range for the room when either the heater or air conditioner will be on. (*Note*: The lowest temperature theoretically possible is $-460°$ F, called *absolute zero.*)

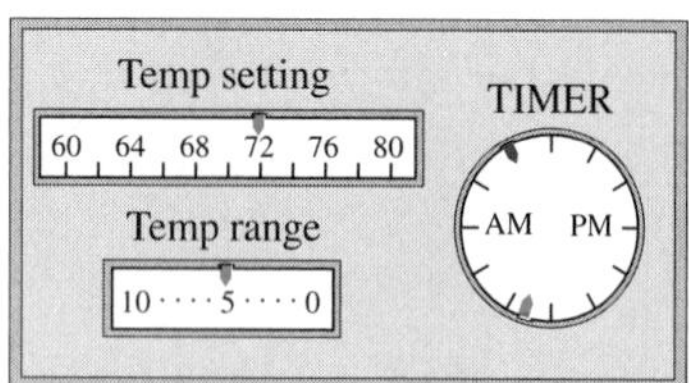

ILLUSTRATION 3

55. HEALTH CARE Refer to the graph in Illustration 4. Let P represent the percent of people covered by private insurance, M the percent covered by Medicare/Medicaid, and N the percent not covered. For what years are the following true?

a. $P \geq 70$ and $N \geq 15$

b. $P \geq 70$ or $N \geq 15$

c. $P \geq 72$ and $M \leq 10$

d. $P \geq 72$ or $M \leq 10$

56. POLLS For each response to the poll question shown in Illustration 5, the *margin of error* is +/− (read as "plus or minus") 3.2%. This means that for the statistical methods used to do the polling, the actual response could be as much as 3.2 points more or 3.2 points less than shown. Use interval notation to describe the possible interval (in percent) for each response.

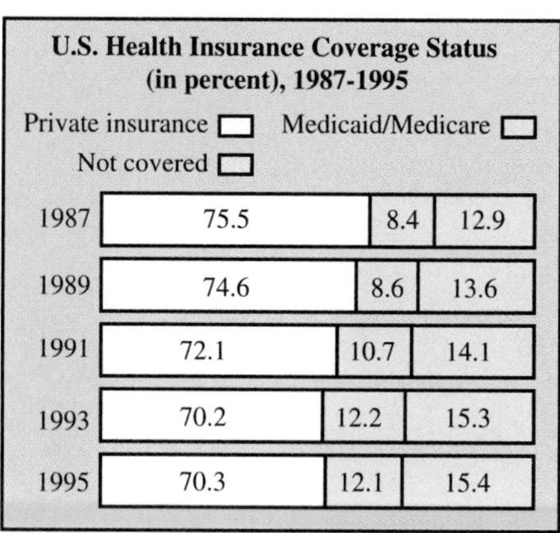

Based on data from U.S. Bureau of the Census

ILLUSTRATION 4

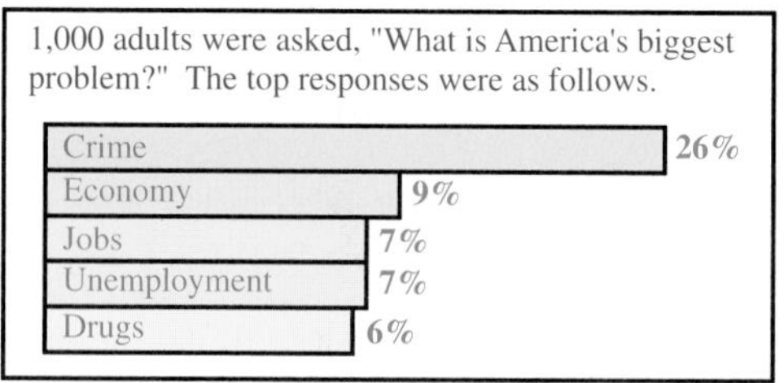

ILLUSTRATION 5

WRITING *Write a paragraph using your own words.*

57. Explain how to find the union and how to find the intersection of $(-\infty, 5)$ and $(-2, \infty)$ graphically.

58. Explain why the double inequality

$2 < x < 8$

can be written in the equivalent form

$2 < x$ and $x < 8$.

REVIEW *In Exercises 59–62, refer to the chart in Illustration 6, which shows the results of each of the games of the eventual champion, the University of Kentucky, in the 1998 NCAA Men's Basketball Tournament. Round to the nearest tenth when necessary.*

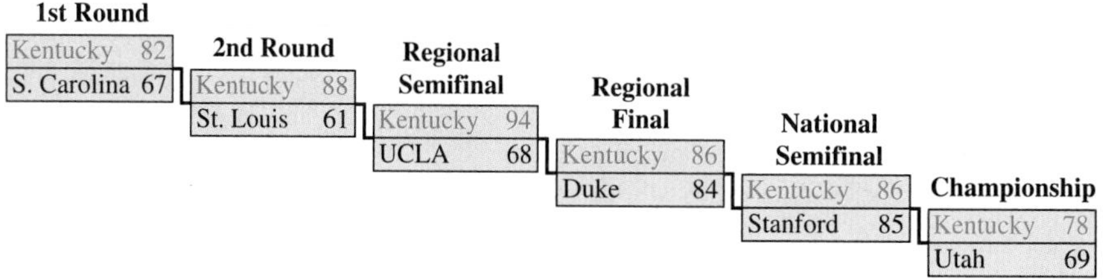

ILLUSTRATION 6

59. What is the mean, median, and mode of the set of Kentucky scores?

60. What is the mean and the median of the set of scores of Kentucky's opponents?

61. Find the margin of victory for Kentucky in each of its games. Then find the average (mean) margin of victory for Kentucky in the tournament.

62. What was the average (mean) combined score for Kentucky and its opponents in the tournament?

4.3 *Solving Absolute Value Equations and Inequalities*

In this section, you will learn about

- **Absolute value**
- **Equations of the form $|x| = k$**
- **Equations with two absolute values**
- **Inequalities of the form $|x| < k$**
- **Inequalities of the form $|x| > k$**

INTRODUCTION. Many quantities studied in mathematics, science, and engineering are expressed as positive numbers. To guarantee that a quantity is positive, we often use the concept of absolute value. In this section, we will work with equations and inequalities containing expressions involving absolute value. Using the definition of absolute value, we will develop procedures to solve absolute value equations and absolute value inequalities.

Absolute value

Recall that the absolute value of any real number is the distance between the number and zero on the number line. For example, the points shown in Figure 4-21 with coordinates of 4 and -4 both lie 4 units from 0. Thus, $|4| = |-4| = 4$.

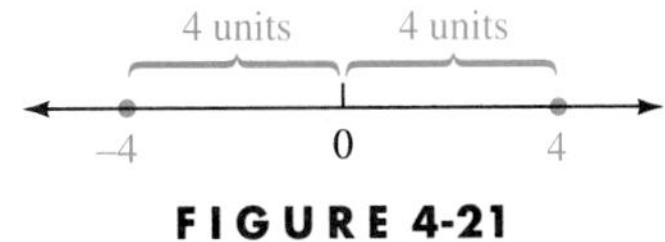

FIGURE 4-21

The absolute value of a real number can be defined more formally.

Absolute value

If $x \geq 0$, then $|x| = x$.
If $x < 0$, then $|x| = -x$.

This definition gives a way for associating a nonnegative real number with any real number.

- If $x \geq 0$, then x (which is positive or 0) is its own absolute value.
- If $x < 0$, then $-x$ (which is positive) is the absolute value.

Either way, $|x|$ is positive or 0. That is, $|x| \geq 0$, for all real numbers x.

EXAMPLE 1 ***Finding absolute values.*** Find **a.** $|9|$, **b.** $|-5.68|$, and **c.** $|0|$.

Solution

a. Since $9 \geq 0$, the number 9 is its own absolute value: $|9| = 9$.

b. Since $-5.68 < 0$, the negative of -5.68 is the absolute value:

$$|-5.68| = -(-5.68) = 5.68$$

c. Since $0 \geq 0$, 0 is its own absolute value: $|0| = 0$.

Self Check

Find **a.** $|-3|$, **b.** $|100.99|$, and **c.** $|-2\pi|$.

Answers: **a.** 3, **b.** 100.99, **c.** 2π

WARNING! The placement of a $-$ sign in an expression containing an absolute value symbol is important. For example, $|-19| = 19$, but $-|19| = -19$.

Equations of the form $|x| = k$

The absolute value of a real number represents the distance on the number line from a point to the origin. To solve the **absolute value equation** $|x| = 5$, we must find the coordinates of all points on the number line that are exactly 5 units from zero. See Figure 4-22. The only two points that satisfy this condition have coordinates 5 and -5. That is, $x = 5$ or $x = -5$.

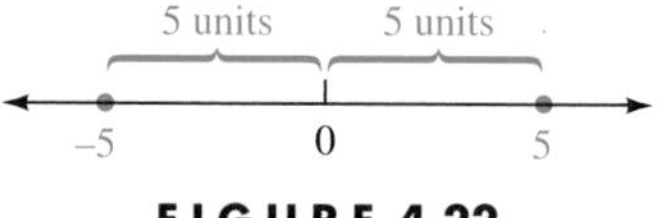

FIGURE 4-22

In general, the solution set of the absolute value equation $|x| = k$, where $k \geq 0$, includes the coordinates of the points on the number line that are k units from the origin. (See Figure 4-23.)

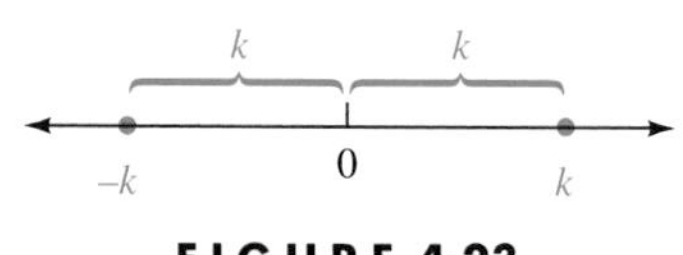

FIGURE 4-23

Absolute value equations	If $k \geq 0$, then $|x| = k$ is equivalent to $x = k$ or $x = -k$

EXAMPLE 2 ***Solving an absolute value equation.*** Solve **a.** $|x| = 8$ and **b.** $|s| = 0.003$.

Solution

a. If $|x| = 8$, then $x = 8$ or $x = -8$.

b. If $|s| = 0.003$, then $s = 0.003$ or $s = -0.003$.

Self Check

Solve **a.** $|y| = 24$ and **b.** $|x| = \frac{1}{2}$.

Answers: **a.** 24, -24 **b.** $\frac{1}{2}, -\frac{1}{2}$

The equation $|x - 3| = 7$ indicates that a point on the number line with a coordinate of $x - 3$ is 7 units from the origin. Thus, $x - 3$ can be either 7 or -7.

$$x - 3 = 7 \quad \text{or} \quad x - 3 = -7$$
$$x = 10 \qquad\qquad x = -4$$

The solutions of the absolute value equation are 10 and -4. We can graph them on a number line, as shown in Figure 4-24. If either of these numbers is substituted for x in $|x - 3| = 7$, the equation is satisfied.

$$|x - 3| = 7 \qquad |x - 3| = 7$$
$$|10 - 3| \stackrel{?}{=} 7 \qquad |-4 - 3| \stackrel{?}{=} 7$$
$$|7| \stackrel{?}{=} 7 \qquad |-7| \stackrel{?}{=} 7$$
$$7 = 7 \qquad 7 = 7$$

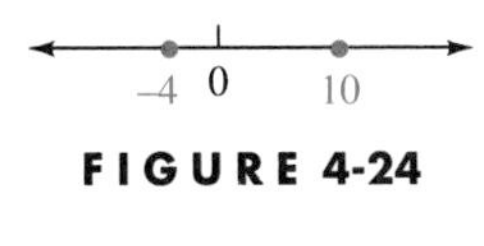

FIGURE 4-24

EXAMPLE 3 ***Solving an absolute value equation.*** Solve $|3x - 2| = 5$.

Solution

We can write $|3x - 2| = 5$ as

$$3x - 2 = 5 \quad \text{or} \quad 3x - 2 = -5$$

and solve each equation for x:

$$3x - 2 = 5 \quad \text{or} \quad 3x - 2 = -5$$
$$3x = 7 \qquad\qquad 3x = -3$$
$$x = \frac{7}{3} \qquad\qquad x = -1$$

Verify that both solutions check.

Self Check

Solve $|2x - 3| = 7$.

Answer: 5, -2 ■

When solving an absolute value equation, we want the absolute value isolated on one side. If this is not the case in a given equation, we use the equation-solving procedures studied earlier to isolate the absolute value first.

EXAMPLE 4 ***Isolating the absolute value.*** Solve $\left|\frac{2}{3}x + 3\right| + 4 = 10$.

Solution

We can isolate $\left|\frac{2}{3}x + 3\right|$ on the left-hand side of the equation by subtracting 4 from both sides.

$$\left|\frac{2}{3}x + 3\right| + 4 = 10$$

1. $$\left|\frac{2}{3}x + 3\right| = 6 \quad \text{Subtract 4 from both sides.}$$

Now that the absolute value is isolated, we can write Equation 1 as

$$\frac{2}{3}x + 3 = 6 \quad \text{or} \quad \frac{2}{3}x + 3 = -6$$

and solve each equation for x:

$$\frac{2}{3}x + 3 = 6 \quad \text{or} \quad \frac{2}{3}x + 3 = -6$$
$$\frac{2}{3}x = 3 \qquad\qquad \frac{2}{3}x = -9$$
$$2x = 9 \qquad\qquad 2x = -27$$
$$x = \frac{9}{2} \qquad\qquad x = -\frac{27}{2}$$

Verify that both solutions check.

Self Check

Solve $|0.4x - 2| - 0.6 = 0.4$.

Answer: 7.5, 2.5 ■

WARNING! Since the absolute value of a quantity cannot be negative, equations such as $|7x + \frac{1}{2}| = -4$ have no solution. Since there are no solutions, their solution sets are empty.

EXAMPLE 5 ***An absolute value equal to 0.***

Solve $3\left|\frac{1}{2}x - 5\right| - 4 = -4$.

Solution

We first isolate $\left|\frac{1}{2}x - 5\right|$ on the left-hand side.

$$3\left|\frac{1}{2}x - 5\right| - 4 = -4$$

$$3\left|\frac{1}{2}x - 5\right| = 0 \quad \text{Add 4 to both sides.}$$

$$\left|\frac{1}{2}x - 5\right| = 0 \quad \text{Divide both sides by 3.}$$

Since 0 is the only number whose absolute value is 0, $\frac{1}{2}x - 5$ must be 0, and we have

$$\frac{1}{2}x - 5 = 0$$

$$\frac{1}{2}x = 5 \quad \text{Add 5 to both sides.}$$

$$x = 10 \quad \text{Multiply both sides by 2.}$$

Verify that 10 satisfies the original equation.

Self Check

Solve $-5\left|\frac{2x}{3} + 4\right| + 1 = 1$.

Answer: -6 ■

Equations with two absolute values

The equation $|a| = |b|$ is true when $a = b$ or when $a = -b$. For example,

$$|3| = |3| \qquad \text{or} \qquad |3| = |-3|$$

The same number. These numbers are opposites.

In general, the following statement is true.

Equations with two absolute values

If a and b represent algebraic expressions, the equation $|a| = |b|$ is equivalent to the pair of equations

$$a = b \quad \text{or} \quad a = -b$$

EXAMPLE 6 ***Solving equations with two absolute values.*** Solve $|5x + 3| = |3x + 25|$.

Solution

This equation is true when $5x + 3 = 3x + 25$, or when $5x + 3 = -(3x + 25)$. We solve each equation for x.

$$5x + 3 = 3x + 25 \quad \text{or} \quad 5x + 3 = -(3x + 25)$$

$$2x = 22 \qquad\qquad 5x + 3 = -3x - 25$$

$$x = 11 \qquad\qquad 8x = -28$$

$$x = -\frac{28}{8}$$

$$x = -\frac{7}{2}$$

Verify that both solutions check.

Self Check

Solve $|2x - 3| = |4x + 9|$.

Answer: $-1, -6$ ■

Inequalities of the form $|x| < k$

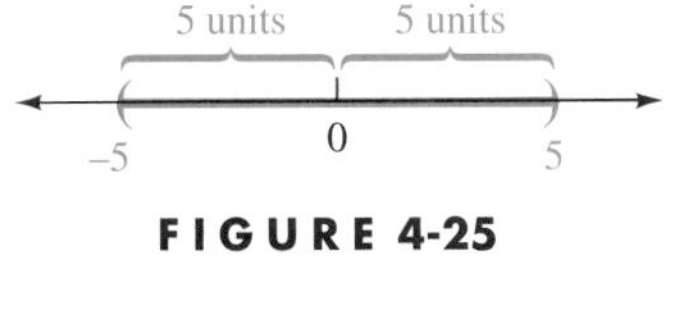

FIGURE 4-25

To solve the **absolute value inequality** $|x| < 5$, we must find the coordinates of all points on a number line that are less than 5 units from the origin. See Figure 4-25. Thus, x is between -5 and 5 and

$$|x| < 5 \quad \text{is equivalent to} \quad -5 < x < 5$$

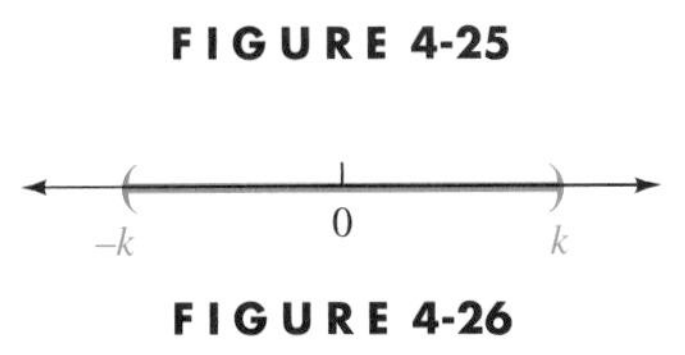

FIGURE 4-26

In general, the solution set of the absolute value inequality $|x| < k$ $(k > 0)$ includes the coordinates of the points on the number line that are less than k units from the origin. See Figure 4-26.

Solving $|x| < k$ and $|x| \le k$	$|x| < k$ is equivalent to $-k < x < k$ $(k > 0)$ $|x| \le k$ is equivalent to $-k \le x \le k$ $(k \ge 0)$

EXAMPLE 7 ***Solving an absolute value inequality.*** Solve $|2x - 3| < 9$ and graph the solution set.

Solution

We write the absolute value inequality as a double inequality and solve for x.

$$|2x - 3| < 9 \quad \text{is equivalent to} \quad -9 < 2x - 3 < 9$$

$$-9 < 2x - 3 < 9$$

$$-6 < 2x < 12 \qquad \text{Add 3 to all three parts.}$$

$$-3 < x < 6 \qquad \text{Divide all parts by 2.}$$

Any number between -3 and 6 is in the solution set. This is the interval $(-3, 6)$, whose graph is shown in Figure 4-27.

FIGURE 4-27

Self Check

Solve $|3x + 2| < 4$ and graph the solution set.

Answer: $\left(-2, \frac{2}{3}\right)$

–2 2/3

EXAMPLE 8 ***Tolerances.*** When manufactured parts are inspected by a quality control engineer, they are classified as acceptable if each dimension falls within a given *tolerance range* of the dimensions listed on the blueprint. For the bracket shown in Figure 4-28, the distance between the two drilled holes is given as 2.900 inches. Because the tolerance is ± 0.015 inch, this distance can be as much as 0.015 inch longer or 0.015 inch shorter, and the part will be considered acceptable. The acceptable distance d between holes can be represented by the absolute value inequality $|d - 2.900| \le 0.015$. Solve the inequality and explain the result.

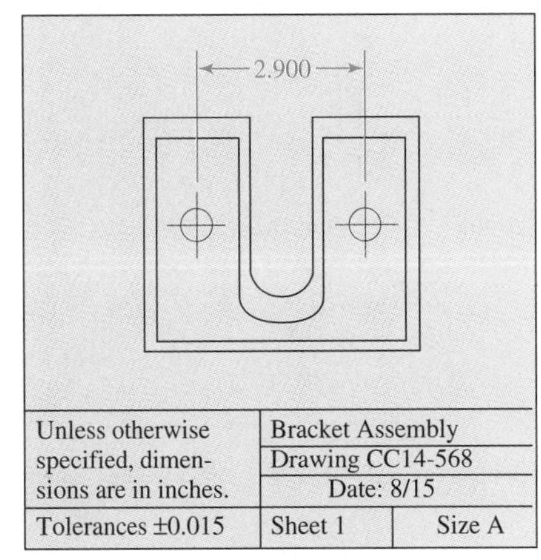

FIGURE 4-28

Solution

We can write the absolute value inequality as a double inequality and solve for d:

$$|d - 2.900| \le 0.015 \quad \text{is equivalent to} \quad -0.015 \le d - 2.900 \le 0.015$$

$$-0.015 \le d - 2.900 \le 0.015$$

$$2.885 \le d \le 2.915 \qquad \text{Add 2.900 to all three parts.}$$

The solution set is the interval [2.885, 2.915]. This means that the distance between the two holes should be between 2.885 and 2.915 inches, inclusive. If the distance is less than 2.885 inches or more than 2.915 inches, the part should be rejected. ■

Inequalities of the form $|x| > k$

To solve the *absolute value inequality* $|x| > 5$, we must find the coordinates of all points on a number line that are more than 5 units from the origin. See Figure 4-29.

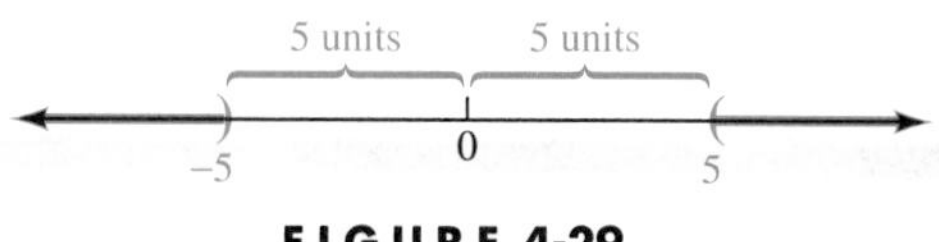

FIGURE 4-29

Thus, $x < -5$ or $x > 5$.

In general, the solution set of $|x| > k$ includes the coordinates of the points on the number line that are more than k units from the origin. See Figure 4-30. Thus,

FIGURE 4-30

$|x| > k$ is equivalent to $x < -k$ or $x > k$

The *or* indicates an either/or situation. It is only necessary that x satisfy one of the two conditions to be in the solution set.

Solving $|x| > k$ and $|x| \geq k$

If $k \geq 0$, then

$|x| > k$ is equivalent to $x < -k$ or $x > k$

$|x| \geq k$ is equivalent to $x \leq -k$ or $x \geq k$

EXAMPLE 9 ***Solving an absolute value inequality.*** Solve $\left|\frac{3-x}{5}\right| \geq 6$ and graph the solution set.

Solution

We write the absolute value inequality as two separate inequalities connected with the word "or."

$$\left|\frac{3-x}{5}\right| \geq 6 \quad \text{is equivalent to} \quad \frac{3-x}{5} \leq -6 \text{ or } \frac{3-x}{5} \geq 6$$

Then we solve each inequality for x:

$$\frac{3-x}{5} \leq -6 \quad \text{or} \quad \frac{3-x}{5} \geq 6$$

$3 - x \leq -30$ $\quad$ $3 - x \geq 30$ $\quad$ Multiply both sides by 5.

$-x \leq -33$ $\quad$ $-x \geq 27$ $\quad$ Subtract 3 from both sides.

$x \geq 33$ $\quad$ $x \leq -27$ $\quad$ Divide both sides by -1 and reverse the direction of the inequality symbol.

The solution set is the interval $(-\infty, -27] \cup [33, \infty)$, whose graph appears in Figure 4-31.

FIGURE 4-31

Self Check

Solve $\left|\frac{2-x}{4}\right| \geq 1$ and graph the solution set.

Answer: $(-\infty, -2] \cup [6, \infty)$

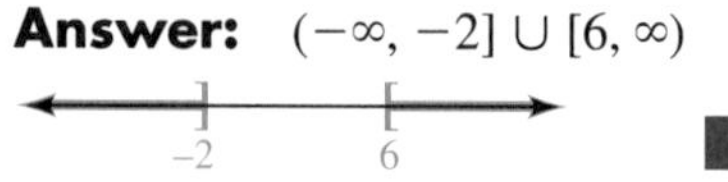

■

EXAMPLE 10 ***Solving an absolute value inequality.*** Solve

$$\left|\frac{2}{3}x - 2\right| - 3 > 6 \text{ and graph the solution set.}$$

Self Check

Solve $\left|\frac{3}{4}x + 2\right| - 1 > 3$ and graph the solution set.

Solution

We begin by adding 3 to both sides to isolate the absolute value on the left-hand side.

$$\left|\frac{2}{3}x - 2\right| - 3 > 6$$

$$\left|\frac{2}{3}x - 2\right| > 9 \quad \text{Add 3 to both sides to isolate the absolute value.}$$

We then proceed as follows:

$$\frac{2}{3}x - 2 < -9 \quad \text{or} \quad \frac{2}{3}x - 2 > 9$$

$$\frac{2}{3}x < -7 \qquad \frac{2}{3}x > 11 \quad \text{Add 2 to both sides.}$$

$$2x < -21 \qquad 2x > 33 \quad \text{Multiply both sides by 3.}$$

$$x < -\frac{21}{2} \qquad x > \frac{33}{2} \quad \text{Divide both sides by 2.}$$

The solution set is $\left(-\infty, -\frac{21}{2}\right) \cup \left(\frac{33}{2}, \infty\right)$, whose graph appears in Figure 4-32.

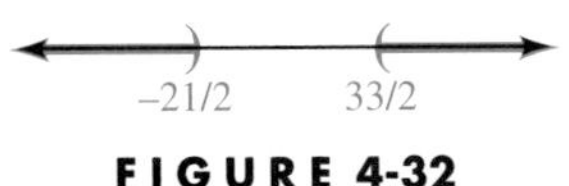

FIGURE 4-32

Answer: $(-\infty, -8) \cup \left(\frac{8}{3}, \infty\right)$

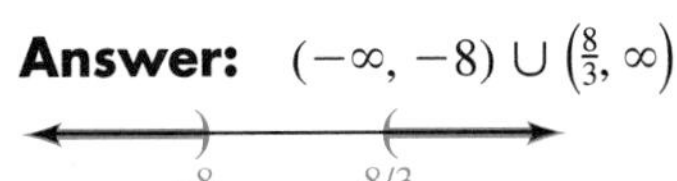

Accent on Technology *Solving absolute value inequalities*

We can also solve absolute value inequalities using a graphing calculator. For example, to solve $|2x - 3| < 9$, we graph the equations $y = |2x - 3|$ and $y = 9$ on the same coordinate system. If we use settings of $[-5, 15]$ for x and $[-5, 15]$ for y, we will get the graph shown in Figure 4-33.

The inequality $|2x - 3| < 9$ will be true for all x-coordinates of points that lie on the graph of $y = |2x - 3|$ and below the graph of $y = 9$. Using the trace feature, we can see that these values of x are in the interval $(-3, 6)$.

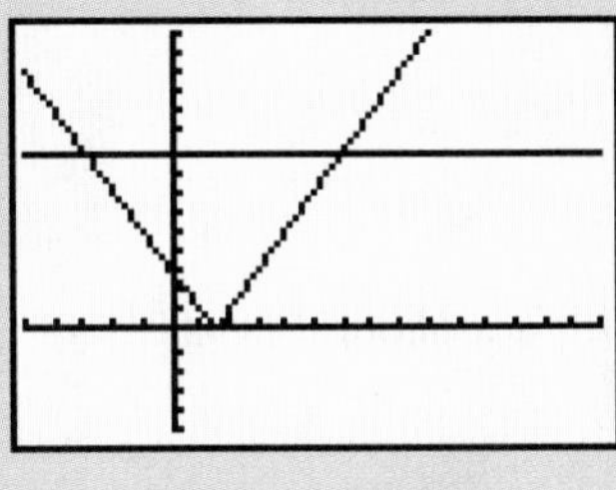

FIGURE 4-33

STUDY SET Section 4.3

VOCABULARY *In Exercises 1–4, fill in the blanks to make the statements true.*

1. $|2x - 1| = 10$ is an absolute value ________.

2. $|2x - 1| > 10$ is an absolute value ________.

3. To ________ the absolute value in $|3 - x| - 4 = 5$, we add 4 to both sides.

4. $|x| = 2$ is ________ to $x = 2$ or $x = -2$.

CONCEPTS *In Exercises 5–10, fill in the blanks to make the statements true.*

5. $|x| \geq$ ___ for all real numbers x.

6. If $x < 0$, $|x| =$ ________.

7. To solve $|x| > 5$, we must find the coordinates of all points on a number line that are ________ 5 units from the origin.

8. To solve $|x| < 5$, we must find the coordinates of all points on a number line that are ________ 5 units from the origin.

9. To solve $|x| = 5$, we must find the coordinates of all points on a number line that are ___ units from the origin.

10. The equation $|a| = |b|$ is true when ________ or when ________.

11. Tell whether $x = -3$ is a solution of the given equation or inequality.

a. $|x - 1| = 4$
b. $|x - 1| > 4$
c. $|x - 1| \leq 4$
d. $|5 - x| = |x + 12|$

12. Write each equation or inequality in its equivalent form.

a. $|x| = 8$
b. $|x| \geq 8$
c. $|x| \leq 8$
d. $|5x - 1| = |x + 3|$

NOTATION

13. Match each equation or inequality with its graph.

a. $|x| = 1$ i 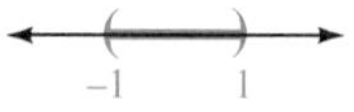

b. $|x| > 1$ ii 

c. $|x| < 1$ iii 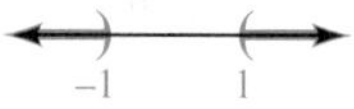

14. Match each graph with its corresponding equation or inequality.

a. 

i $|x| \geq 2$

b. 

ii $|x| \leq 2$

c. 

iii $|x| = 2$

In Exercises 15–18, write each compound inequality as an inequality using absolute values.

15. $-4 < x < 4$

16. $x < -4$ or $x > 4$

17. $x + 3 < -6$ or $x + 3 > 6$

18. $-5 \leq x - 3 \leq 5$

PRACTICE *In Exercises 19–26, find the value of each expression.*

19. $|8|$

20. $|-18|$

21. $-|0.02|$

22. $-|-3.14|$

23. $-\left|-\frac{31}{16}\right|$

24. $-\left|\frac{25}{4}\right|$

25. $|\pi|$

26. $\left|-\frac{\pi}{2}\right|$

In Exercises 27–42, solve each equation, if possible.

27. $|x| = 23$

28. $|x| = 90$

29. $|x - 3.1| = 6$

30. $|x + 4.3| = 8.9$

31. $|3x + 2| = 16$

32. $|5x - 3| = 22$

33. $\left|\frac{7}{2}x + 3\right| = -5$

34. $\left|\frac{2x}{3} + 10\right| = 0$

35. $|3 - 4x| = 5$

36. $|8 - 5x| = 18$

37. $2|3x + 24| = 0$

38. $5|x - 21| = -8$

39. $\left|\frac{3x + 48}{3}\right| = 12$

40. $\left|\frac{4x - 64}{4}\right| = 32$

41. $|x + 3| + 7 = 10$

42. $|2 - x| + 3 = 5$

In Exercises 43–50, solve each equation, if possible.

43. $|2x + 1| = |3x + 3|$

44. $|5x - 7| = |4x + 1|$

45. $|2 - x| = |3x + 2|$

46. $|4x + 3| = |9 - 2x|$

47. $\left|\frac{x}{2} + 2\right| = \left|\frac{x}{2} - 2\right|$

48. $|7x + 12| = |x - 6|$

49. $\left|x + \frac{1}{3}\right| = |x - 3|$

50. $\left|x - \frac{1}{4}\right| = |x + 4|$

In Exercises 51–76, solve each inequality. Write the solution set in interval notation and graph it.

51. $|x| < 4$

52. $|x| < 9$

53. $|x + 9| \le 12$

54. $|x - 8| \le 12$

55. $|3x - 2| < 10$

56. $|4 - 3x| \le 13$

57. $|3x + 2| \le -3$

58. $|5x - 12| < -5$

59. $|x| > 3$

60. $|x| > 7$

61. $|x - 12| > 24$

62. $|x + 5| \ge 7$

63. $|3x + 2| > 14$

64. $|2x - 5| > 25$

65. $|4x + 3| > -5$

66. $|7x + 2| > -8$

67. $|2 - 3x| \ge 8$

68. $|-1 - 2x| > 5$

69. $-|2x - 3| < -7$

70. $-|3x + 1| < -8$

71. $\left|\frac{x - 2}{3}\right| \le 4$

72. $\left|\frac{x - 2}{3}\right| > 4$

73. $|3x + 1| + 2 < 6$

74. $1 + \left|\frac{1}{7}x + 1\right| \le 1$

75. $\left|\frac{1}{3}x + 7\right| + 5 > 6$

76. $-2|3x - 4| < 16$

APPLICATIONS

77. TEMPERATURE RANGES The temperatures on a sunny summer day satisfied the inequality $|t - 78°| \le 8°$, where t is a temperature in degrees Fahrenheit. Solve this inequality and express the range of temperatures as a double inequality.

78. OPERATING TEMPERATURES A car CD player has an operating temperature of $|t - 40°| < 80°$, where t is a temperature in degrees Fahrenheit. Solve the inequality and express this range of temperatures as an interval.

79. AUTO MECHANICS On most cars, the bottoms of the front wheels are closer together than the tops, creating a *camber angle.* This lessens road shock to the steering system. (See Illustration 1.) The specifications for a certain car state that the camber angle c of its wheels should be $0.6° \pm 0.5°$.

a. Express the range with an inequality containing absolute value symbols.

b. Solve the inequality and express this range of camber angles as an interval.

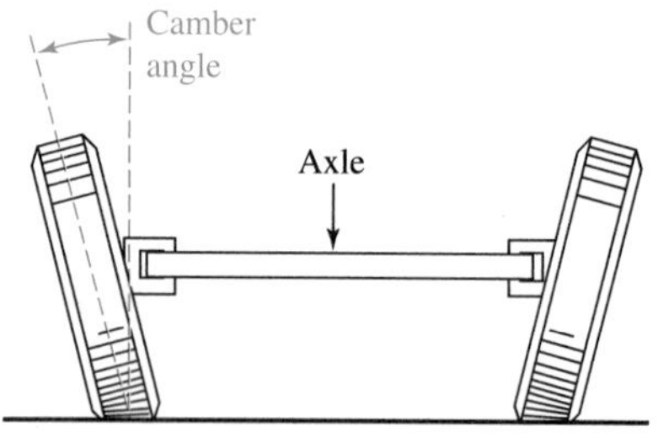

ILLUSTRATION 1

80. STEEL PRODUCTION A sheet of steel is to be 0.250 inch thick with a tolerance of 0.025 inch.

a. Express this specification with an inequality containing absolute value symbols where x is the thickness of a sheet of steel.

b. Solve the inequality and express the range of thickness as an interval.

81. ERROR ANALYSIS In a lab, students measured the percent of copper p in a sample of copper sulfate. The students know that copper sulfate is actually 25.46% copper by mass. They are to compare their results to the actual value and find the amount of *experimental error.*

a. Which measurements shown in Illustration 2 satisfy the absolute value inequality $|p - 25.46| \le 1.00$?

b. What can be said about the amount of error for each of the trials listed in part a?

Lab 4 Section A
Title:
"Percent copper (Cu) in copper sulfate ($CuSO_4 \cdot 5H_2O$)"

Results

	% Copper
Trial #1:	22.91%
Trial #2:	26.45%
Trial #3:	26.49%
Trial #4:	24.76%

ILLUSTRATION 2

82. ERROR ANALYSIS See Exercise 81.

a. Which measurements satisfy the absolute value inequality $|p - 25.46| > 1.00$?

b. What can be said about the amount of error for each of the trials listed in part a?

WRITING *Write a paragraph using your own words.*

83. Explain how to find the absolute value of a given number.

84. Explain why the equation $|x - 4| = -5$ has no solutions.

85. Explain the use of parentheses and brackets when graphing inequalities.

86. Explain the differences between the solution sets of $|x| < 8$ and $|x| > 8$.

REVIEW

87. RAILROAD CROSSING The warning sign in Illustration 3 is to be painted on the street in front of a railroad crossing. If y is 30° more than twice x, find x and y.

88. GEOMETRY Refer to Illustration 3. What is $2x + 2y$?

ILLUSTRATION 3

4.4 Linear Inequalities in Two Variables

In this section, you will learn about

- **Graphing linear inequalities**
- **Horizontal and vertical boundary lines**
- **Problem solving**

Introduction. In the first three sections of this chapter, we have worked with linear inequalities in one variable. Some examples are

$$x \geq -7, \qquad 5 < \frac{7}{2}a - 9, \qquad \text{and} \qquad 5(3 + z) > -3(z + 3)$$

These inequalities have infinitely many solutions. When their solutions are graphed on a real number line, we obtain an interval.

In this section, we will discuss linear inequalities in two variables. Some examples are

$$y > 3x + 2, \qquad 2x - 3y \leq 6, \qquad \text{and} \qquad y < 2x$$

The solutions of these inequalities are ordered pairs. We can graph their solutions on a rectangular coordinate system.

Graphing linear inequalities

The **graph of a linear inequality** in x and y is the graph of all ordered pairs (x, y) that satisfy the inequality.

Linear inequalities

A **linear inequality** in x and y is any inequality that can be written in the form

$$Ax + By < C \quad \text{or} \quad Ax + By > C \quad \text{or} \quad Ax + By \leq C \quad \text{or} \quad Ax + By \geq C$$

where A, B, and C are real numbers and A and B are not both 0.

Because the inequality $y > 3x + 2$ can be written in the form $-3x + y > 2$, it is an example of a linear inequality in x and y.

To graph the linear inequality $y > 3x + 2$, we begin by graphing the linear *equation* $y = 3x + 2$. The graph of $y = 3x + 2$, shown in Figure 4-34(a), is a boundary line that separates the rectangular coordinate plane into two regions called **half-planes.** It is drawn with a broken line to show that it is not part of the graph of $y > 3x + 2$.

To find which half-plane is the graph of $y > 3x + 2$, we can substitute the coordinates of any point in either half-plane. We will choose the origin as the test point because its coordinates, (0, 0), make the computations easy. We substitute 0 for x and 0 for y into the inequality and simplify.

$y > 3x + 2$	The original inequality.
$0 > 3(0) + 2$	Substitute 0 for y and 0 for x.
$0 > 2$	This statement is false.

Since the coordinates of the origin don't satisfy $y > 3x + 2$, the origin is not part of the graph of the inequality. Thus, the half-plane on the other side of the broken line is its graph. We then shade that region, as shown in Figure 4-34(b).

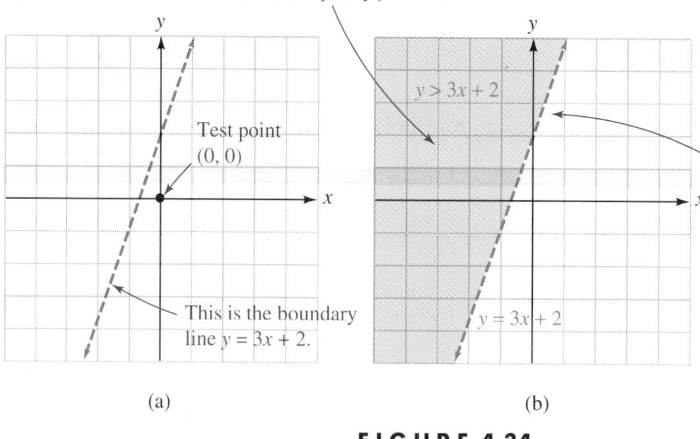

The shaded half-plane represents all the solutions of the inequality y > 3x + 2
Test point (0, 0)
This is the boundary line y = 3x + 2.
y > 3x + 2
y = 3x + 2
The boundary line is often called an edge of the half-plane. In this case, the edge is not included in the graph.
x
y
(a)
(b)

FIGURE 4-34

EXAMPLE 1 ***Graphing a linear inequality.*** Graph $2x - 3y \leq 6$.

Solution

This inequality is the combination of the inequality $2x - 3y < 6$ and the equation $2x - 3y = 6$.

We begin by graphing $2x - 3y = 6$ to find the boundary line that separates the two half-planes. We do so by noting that the line's x-intercept is $(3, 0)$ and its y-intercept is $(0, -2)$. This time, we draw the solid line shown in Figure 4-35(a), because equality is permitted. To decide which half-plane to shade, we check to see whether the coordinates of the origin satisfy the inequality.

$$2x - 3y \leq 6$$

$$2(0) - 3(0) \leq 6 \quad \text{Substitute 0 for } x \text{ and 0 for } y.$$

$$0 \leq 6 \quad \text{This statement is true.}$$

The coordinates of the origin satisfy the inequality. In fact, the coordinates of every point on the same side of the boundary line as the origin satisfy the inequality. We then shade that half-plane to complete the graph of $2x - 3y \leq 6$, shown in Figure 4-35(b).

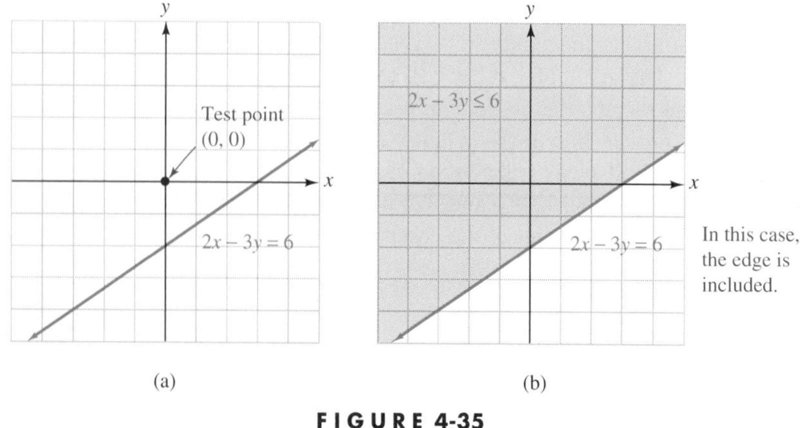

Test point (0, 0)
2x − 3y = 6
2x − 3y ≤ 6
2x − 3y = 6
In this case, the edge is included.
x
y
(a)
(b)

FIGURE 4-35

Self Check

Graph $3x - 2y \geq 6$.

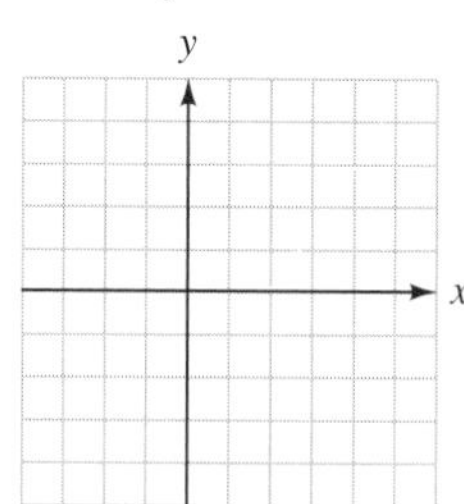

x
y

Answer:

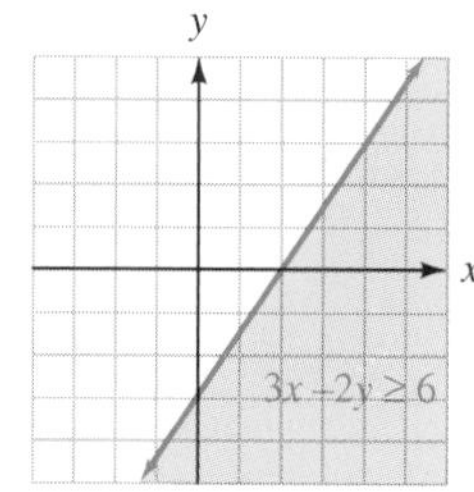

3x − 2y ≥ 6
x
y

EXAMPLE 2 ***Picking a test point other than the origin.*** Graph $y < 2x$.

Solution

To graph $y = 2x$, we use the fact that the equation is in slope–intercept form and that $m = 2 = \frac{2}{1}$ and $b = 0$. Since the symbol $<$ does not include an equals sign, the points on the graph of $y = 2x$ are not on the graph of $y < 2x$. We draw the boundary line as a broken line to show this, as in Figure 4-36(a).

To decide which half-plane is the graph of $y < 2x$, we check to see whether the coordinates of some fixed point satisfy the inequality. We cannot use the origin as a test point, because the boundary line passes through the origin. However, we can choose a different point—say, (3, 1).

$y < 2x$

$\mathbf{1} < 2(\mathbf{3})$ Substitute 1 for y and 3 for x.

$1 < 6$ This is a true statement.

Since $1 < 6$ is a true inequality, the point (3, 1) satisfies the inequality and is in the graph of $y < 2x$. We then shade the half-plane containing (3, 1), as shown in Figure 4-36(b).

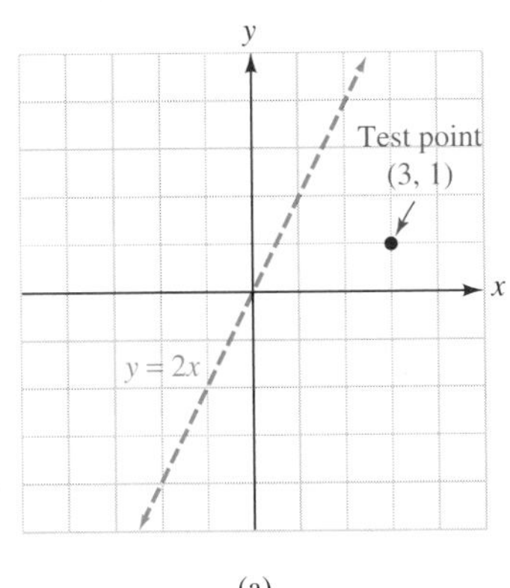

(a)

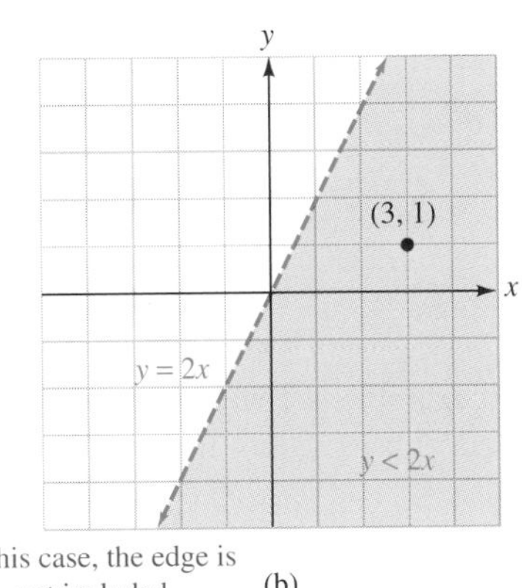

In this case, the edge is not included.

(b)

FIGURE 4-36

Self Check

Graph $y > -x$.

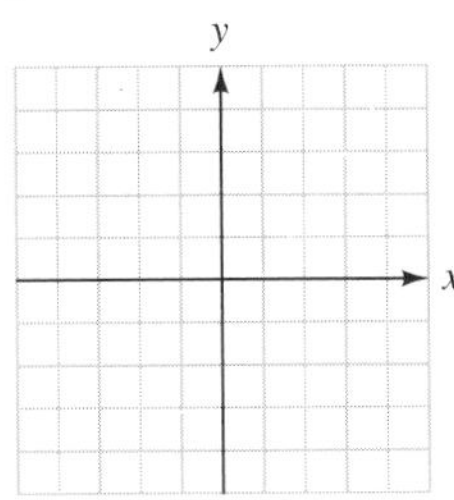

Answer:

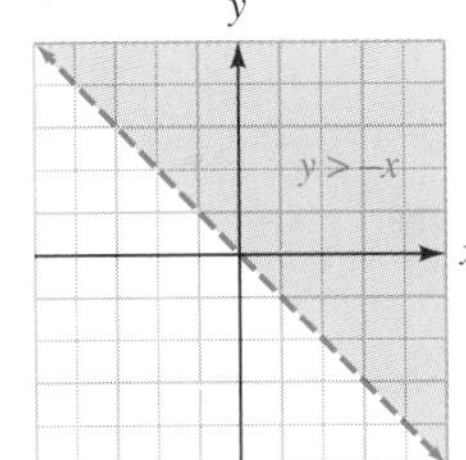

The following is a summary of the procedure for graphing linear inequalities.

Graphing linear inequalities in two variables

1. Graph the boundary line of the region. If the inequality allows the possibility of equality (the symbol is either $\leq$ or $\geq$), draw the boundary line as a solid line. If equality is not allowed ($<$ or $>$), draw the boundary line as a broken line.
2. Pick a test point that is on one side of the boundary line. (Use the origin if possible.) Replace x and y in the original inequality with the coordinates of that point. If the inequality is satisfied, shade the side that contains that point. If the inequality is not satisfied, shade the other side.

Horizontal and vertical boundary lines

Recall from Chapter 2 that the graph of the equation $x = a$ is a vertical line with x-intercept at $(a, 0)$, and the graph of the equation $y = b$ is a horizontal line with y-intercept at $(0, b)$.

EXAMPLE 3 ***A vertical boundary.*** Graph $x \geq -1$.

Solution

The graph of the boundary $x = -1$ is a vertical line passing through $(-1, 0)$. We draw the boundary as a solid line to show that it is part of the solution. See Figure 4-37(a).

In this case, we need not pick a test point. The inequality $x \geq -1$ is satisfied by points with an x-coordinate greater than or equal to -1. Points satisfying this condition lie to the right of the boundary. We shade that half-plane, as shown in Figure 4-37(b), to complete the graph of $x \geq -1$.

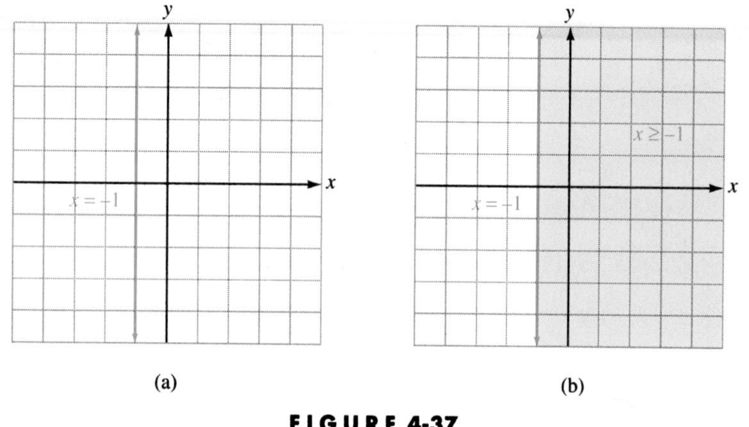

FIGURE 4-37

Self Check

Graph $y < 4$.

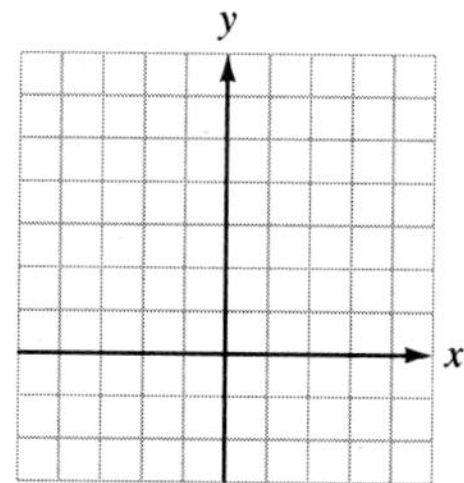

Answer:

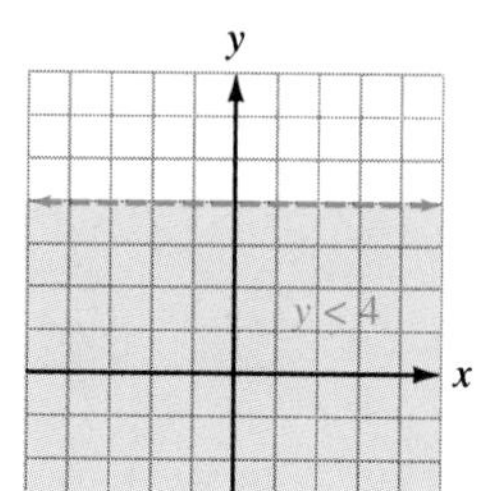

Problem solving

In the next example, we will solve a problem by writing a linear inequality in two variables to model a situation mathematically.

EXAMPLE 4 ***Social Security.*** Retirees, ages 62–65, can earn as much as \$9,120 and still receive their full Social Security benefits. If their annual earnings exceed \$9,120, their benefits are reduced. A 64-year-old retired woman receiving Social Security works two part-time jobs: one at the library, paying \$380 per week and another at a pet store, paying \$285 per week. Write an inequality representing the number of weeks the woman can work at each job during the year without losing any of her Social Security benefits.

Analyze the problem We need to find the various combinations of the number of weeks she can work at the library and at the pet store so that the annual income from these jobs is less than or equal to \$9,120.

Form an inequality If we let x represent the number of weeks she works at the library, she will earn $\$380x$ on the first job annually. If we let y represent the number of weeks she works at the pet store, she will earn $\$285y$ on the second job annually. Combining the income from the two jobs, the total is not to exceed \$9,120.

The weekly rate on the library job	·	the weeks worked on the library job	+	the weekly rate on the pet store job	·	the weeks worked on the pet store job	should not exceed	\$9,120.
\$380	·	x	+	\$285	·	y	$\leq$	\$9,120

Solve the inequality The graph of $380x + 285y \leq 9{,}120$ is shown in Figure 4-38. Any point in the shaded region indicates a way that she can schedule her work weeks and earn \$9,120 or less annually. For example, if she works 8 weeks at the library and 20 weeks at the pet store (represented by the ordered pair (8, 20)), she will earn

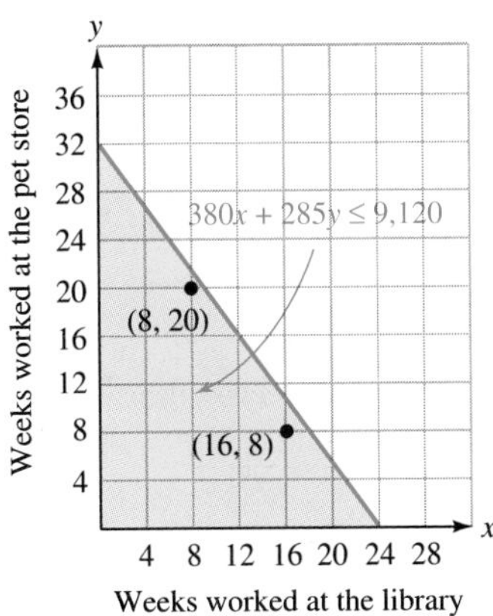

FIGURE 4-38

$$\begin{aligned} \$380(8) + \$285(20) &= \$3{,}040 + \$5{,}700 \\ &= \$8{,}740 \end{aligned}$$

If she works 16 weeks at the library and 8 weeks at the pet store, she will earn

$$\begin{aligned} \$380(16) + \$285(8) &= \$6{,}080 + \$2{,}280 \\ &= \$8{,}360 \end{aligned}$$

Since she cannot work a negative number of weeks, the graph has no meaning when x or y is negative, so only the first quadrant is shown. ■

Accent on Technology **Graphing inequalities**

Some graphing calculators (like the TI-83) have a graph style icon in the $y =$ editor. (See Figure 4-39(a).) Some of the different graph styles are

\	line	A straight line or curved graph is shown.
◥	above	Shading covers the area above a graph.
◣	below	Shading covers the area below a graph.

We can change the icon by placing the cursor on it and pressing the ENTER key.

To graph the inequality of Example 1 using window settings of $x = [-10, 10]$ and $y = [-10, 10]$, we change the graph style icon to above (◥), enter the equation $2x - 3y = 6$ as $y = \frac{2}{3}x - 2$, and press the GRAPH key to get Figure 4-39(b).

To graph the inequality of Example 2 using window settings of $x = [-10, 10]$ and $y = [-10, 10]$, we change the graph style icon to below (◣), enter the equation $y < 2x$, and press the GRAPH key to get Figure 4-39(c).

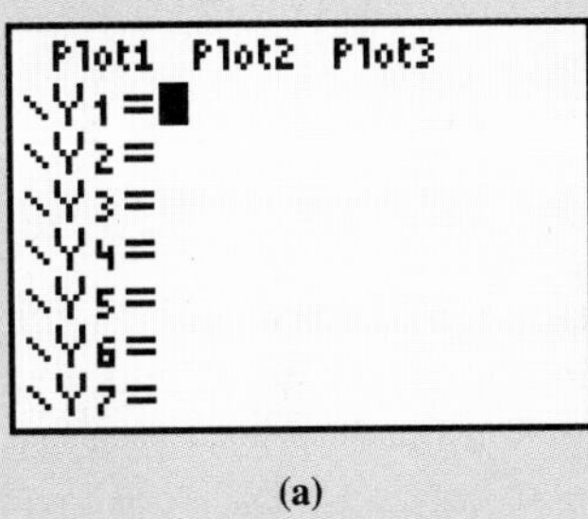

(a)

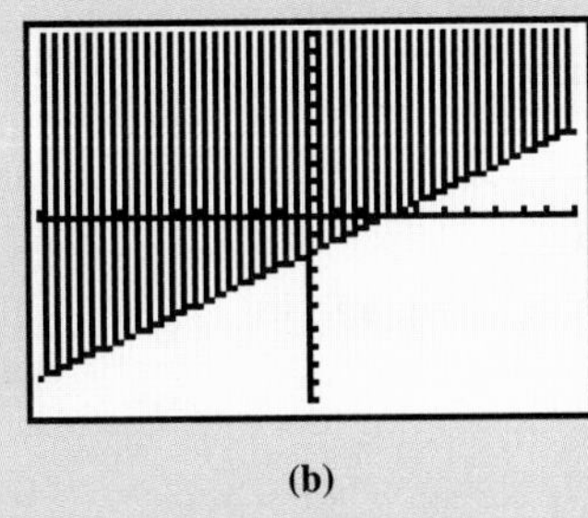

(b)

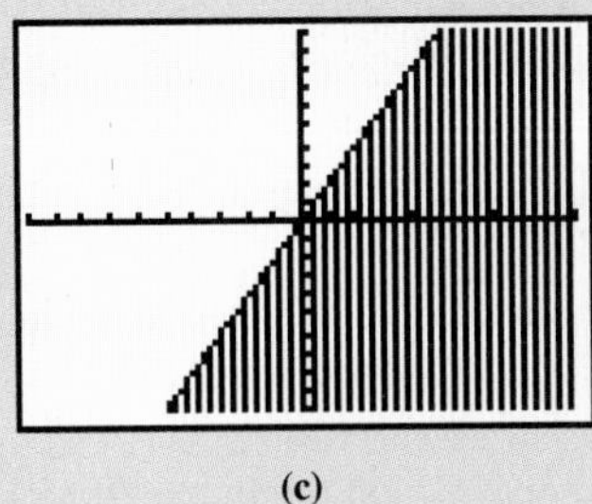

(c)

FIGURE 4-39

If your calculator does not have a graph style icon, you can graph linear inequalities with a shade feature. To do so, consult your owner's manual.

It is important to note that graphing calculators do not distinguish between solid and broken lines to show whether or not the edge of a region is included within the graph.

STUDY SET Section 4.4

VOCABULARY *In Exercises 1–4, fill in the blanks to make the statements true.*

1. $4x - 2y \geq -8$ is an example of a ______________ inequality in ______________ variables.

2. Graphs of linear inequalities are ______________.

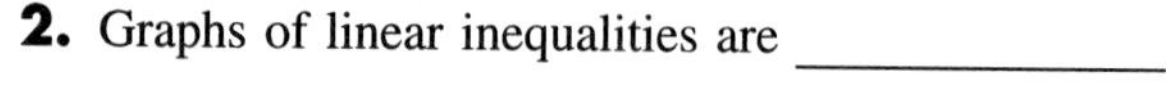

3. The boundary line of a half-plane is called an ______________.

4. The graph of a linear inequality in x and y contains the points (x, y) whose coordinates ______________ the inequality.

CONCEPTS

5. Tell whether each ordered pair is a solution of $3x - 2y \geq 5$.

a. $(3, 1)$ **b.** $(0, 3)$

c. $(-1, -4)$ **d.** $\left(1, \frac{1}{2}\right)$

6. A linear inequality has been graphed in Illustration 1. Tell whether each point satisfies the inequality.

a. $(-1, 4)$

b. $(3, -2)$

c. $(0, 0)$

d. $(-3, -3)$

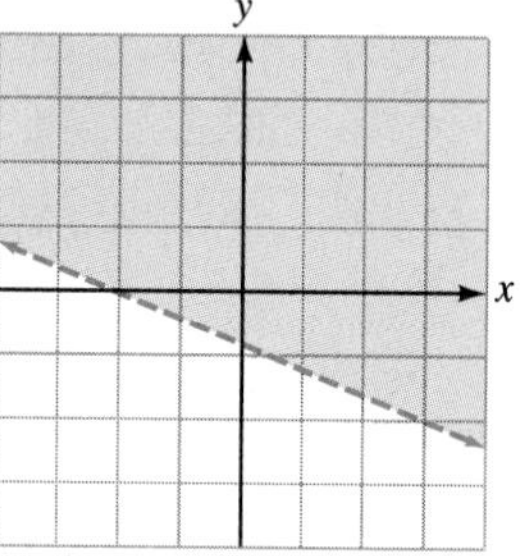

ILLUSTRATION 1

7. To graph the inequality $y > 3x - 1$, we begin by graphing the boundary line $y = 3x - 1$. What is the slope m of the line? What is its y-intercept?

8. To graph the inequality $2x + 3y \leq -6$, we begin by graphing the boundary line $2x + 3y = -6$. What are its x- and y-intercepts?

9. ZOOS To determine the allowable number of juvenile chimpanzees x and adult chimpanzees y that can live in an enclosure, a zookeeper refers to the graph in Illustration 2. Can 7 juvenile and 4 adult chimps be kept in the enclosure?

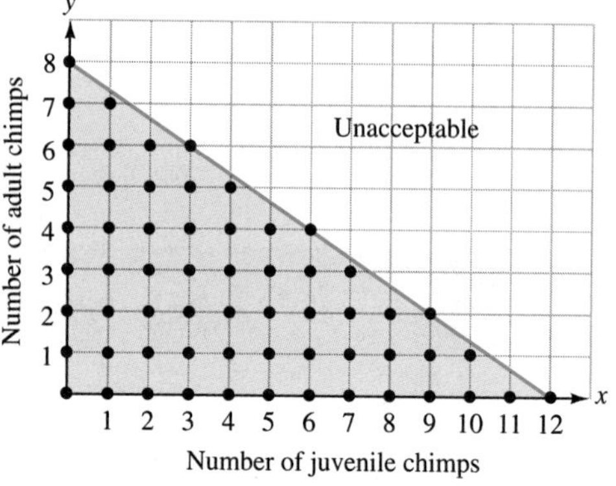

ILLUSTRATION 2

10. The boundary for the graph of a linear inequality is shown in Illustration 3. Why can't the origin be used as a test point to decide which side to shade?

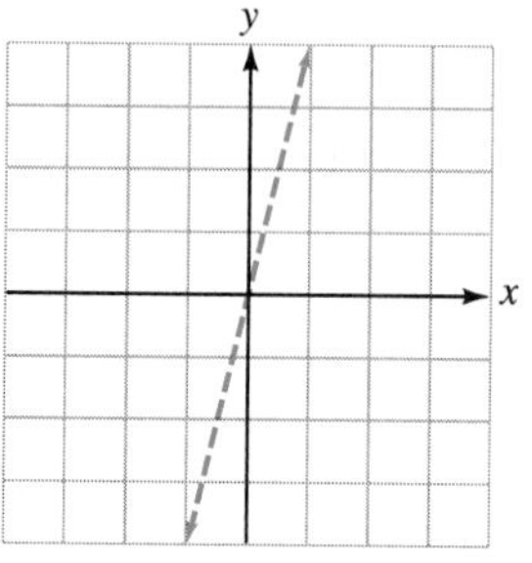

ILLUSTRATION 3

NOTATION

11. a. Solve the inequality in one variable and graph its solution set: $2x + 4 \geq 8$.

b. Graph the inequality in two variables: $2x + 4y \geq 8$.

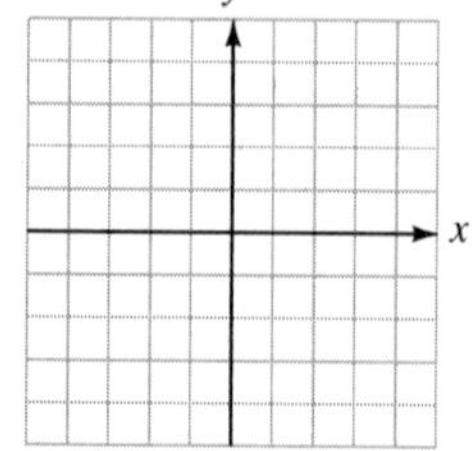

12. Tell whether the graph of each inequality includes the boundary line. In each case, would the boundary be a solid or a broken line?

a. $y < 3x - 1$ **b.** $2x + 3y \geq -6$

c. $y \leq -10$ **d.** $x > 1$

PRACTICE *In Exercises 13–32, graph each inequality.*

13. $y > x + 1$

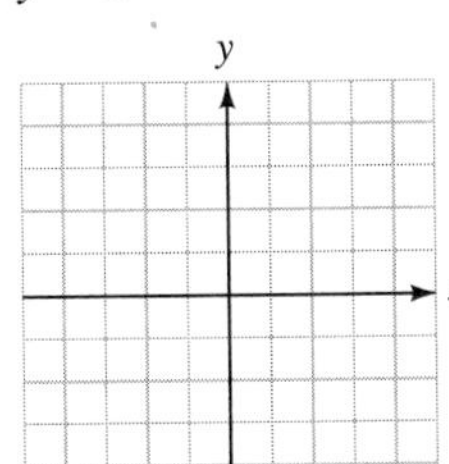

14. $y < 2x - 1$

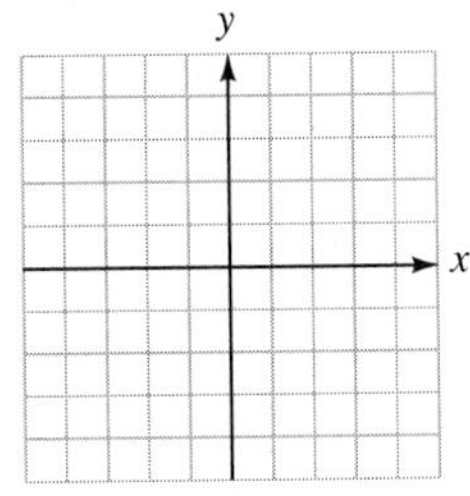

15. $y \geq x$

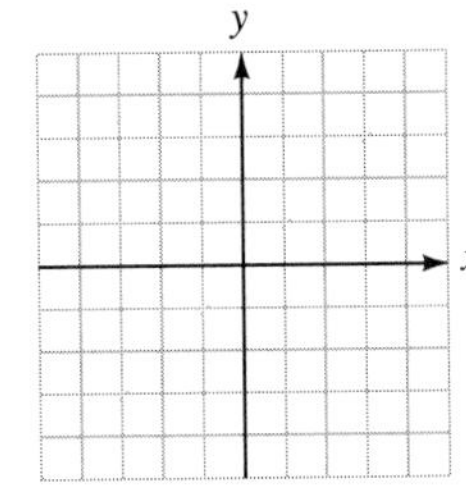

16. $y \leq 2x$

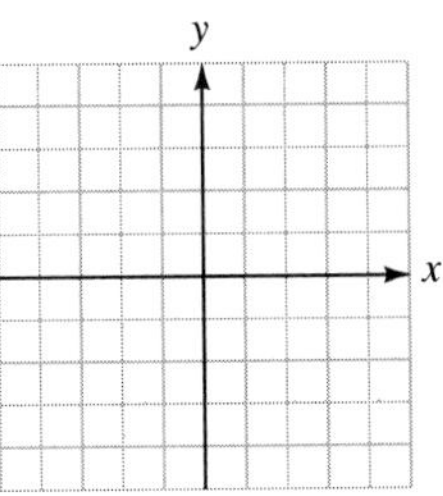

17. $2x + y \leq 6$

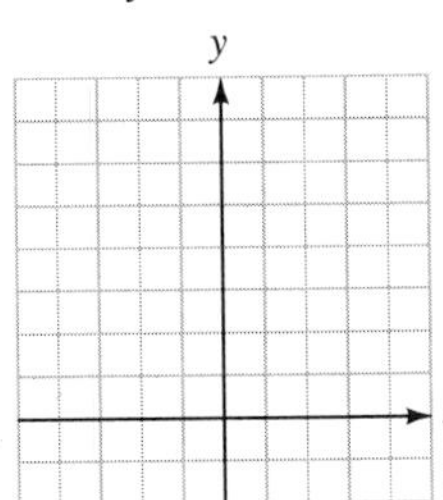

18. $x - 2y \geq 4$

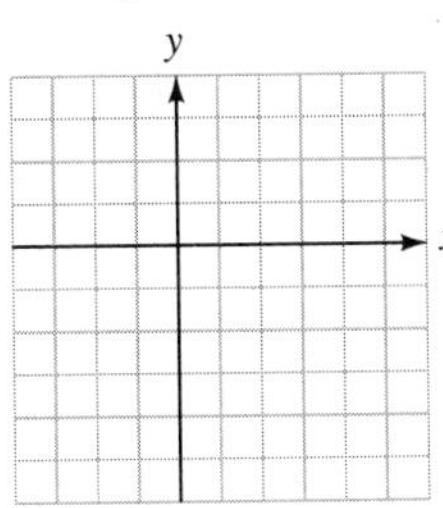

19. $3x \geq -y + 3$

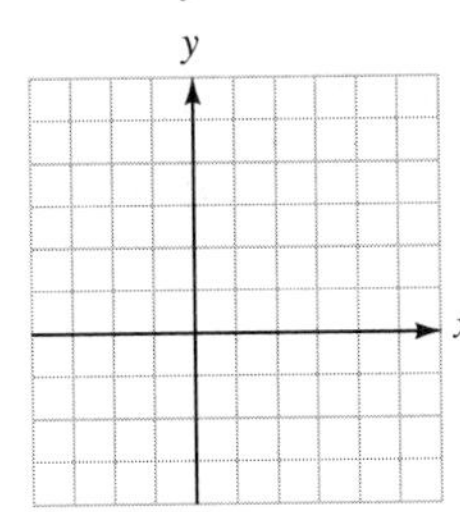

20. $2x \leq -3y - 12$

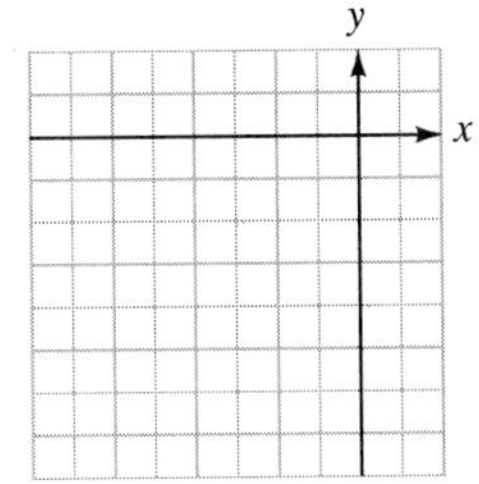

21. $y \geq 1 - \frac{3}{2}x$

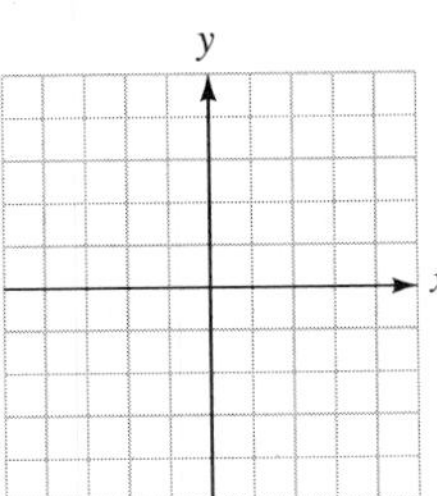

22. $y < \frac{x}{3} - 1$

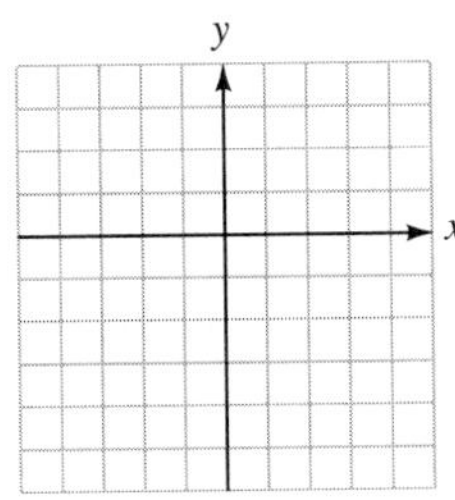

23. $3x + y > 2 + x$

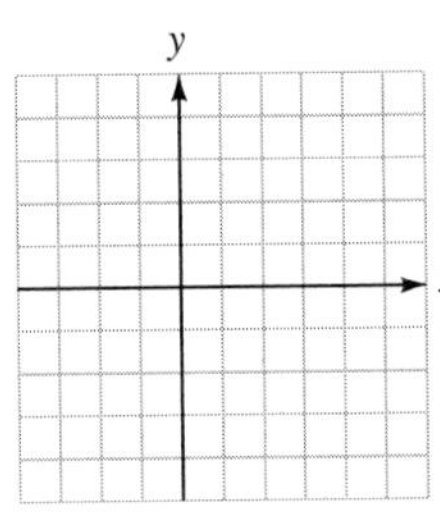

24. $3x - y > 6 + y$

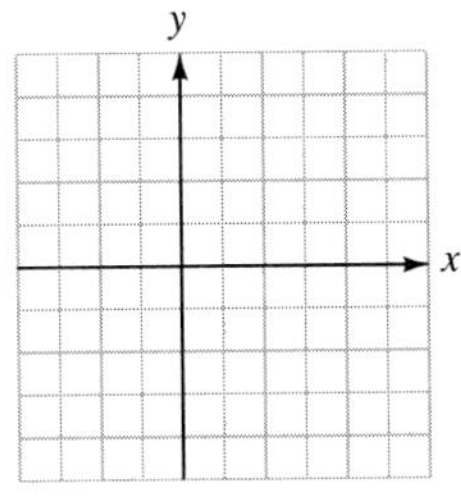

25. $y < -\frac{x}{2}$

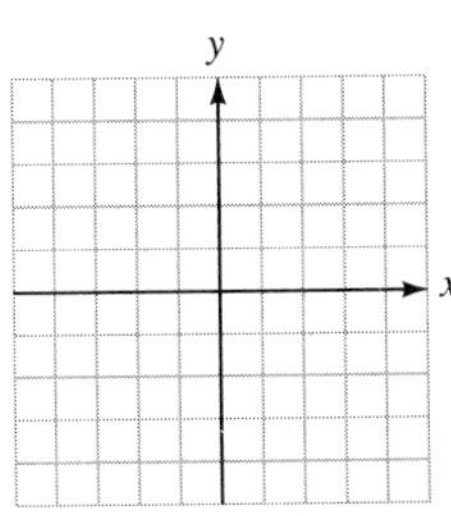

26. $y > \frac{x}{3}$

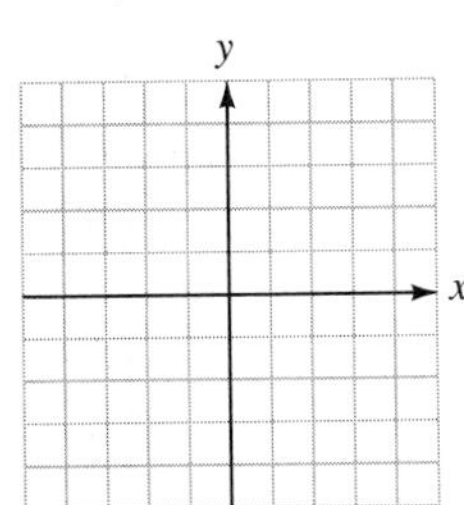

27. $\frac{x}{2} + \frac{y}{2} \leq 2$

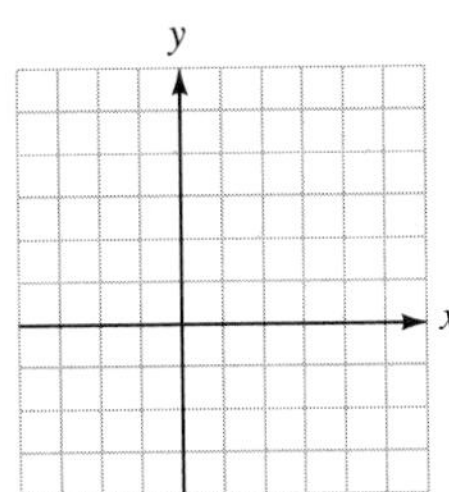

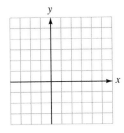

28. $\frac{x}{3} - \frac{y}{2} \geq 1$

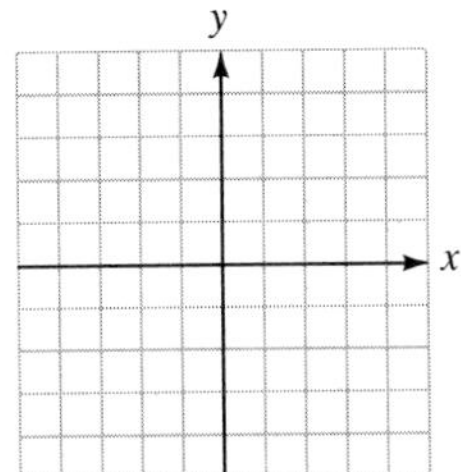

29. $x < 4$

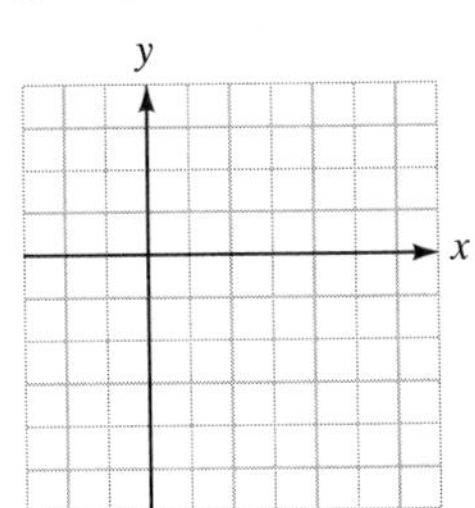

30. $y \geq -2$

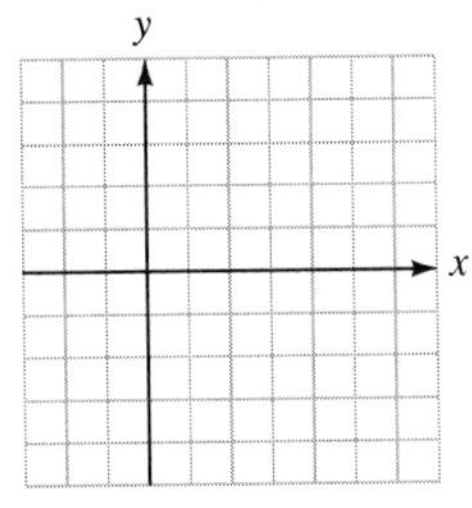

31. $y < 0$

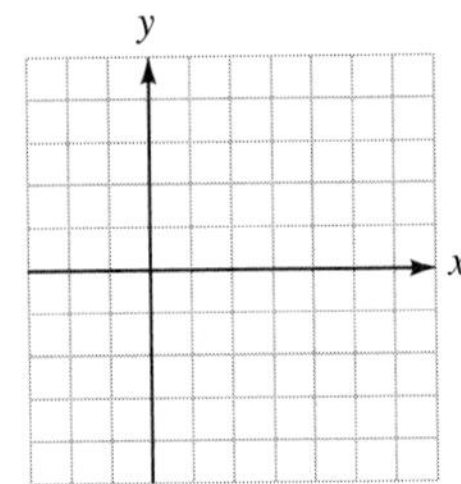

32. $x \geq 0$

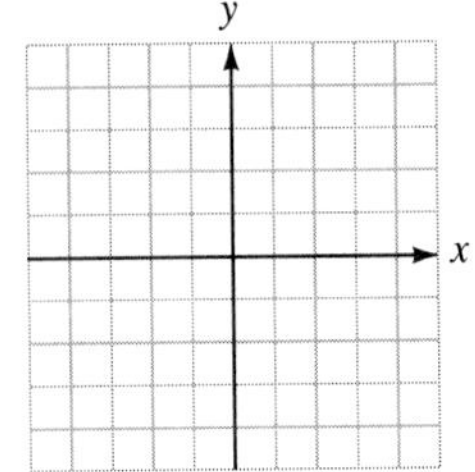

In Exercises 33–36, find the equation of the boundary line. Then give the inequality whose graph is shown.

33.

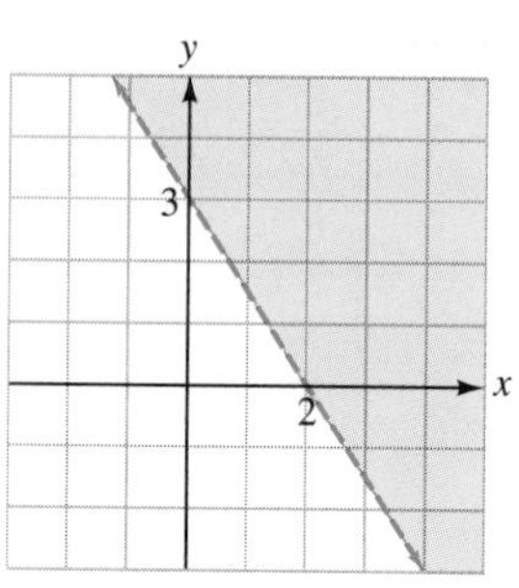

34.

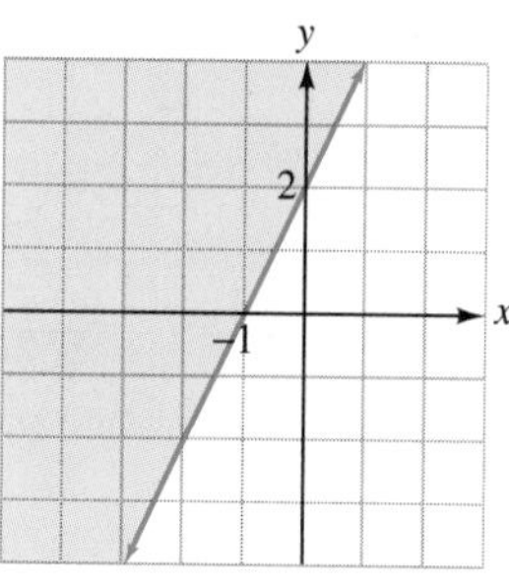

35.

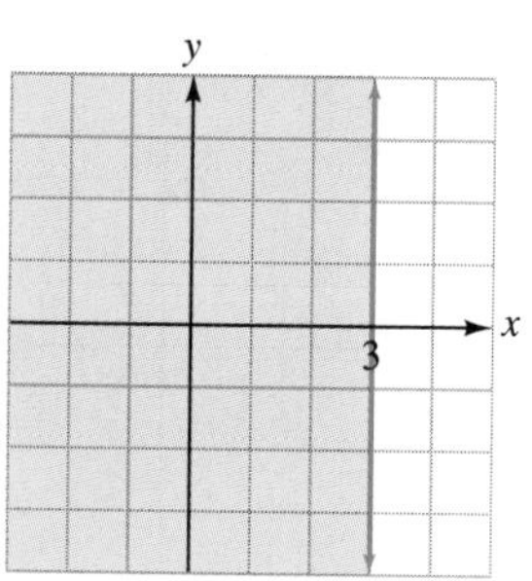

36.

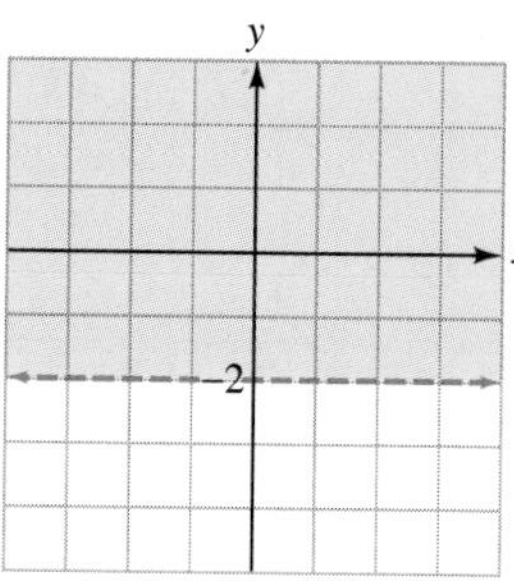

In Exercises 37–40, use a graphing calculator to graph each inequality.

37. $y < 0.27x - 1$

38. $y > -3.5x + 2.7$

39. $y \geq -2.37x + 1.5$

40. $y \leq 3.37x - 1.7$

APPLICATIONS

41. GEOGRAPHY A region of the continental United States is shaded in the map shown in Illustration 4.

a. What is the boundary that separates the shaded and unshaded regions?

b. In words, describe the shaded area with respect to the boundary.

42. KOREAN WAR After World War II, the 38th parallel of north latitude was established as the boundary between North Korea and South Korea. (See Illustration 5.) The Korean War began on June 25, 1950, when the North Korean army crossed this line and invaded South Korea. In the Illustration, shade the region of the Korean Peninsula south of the 38th parallel.

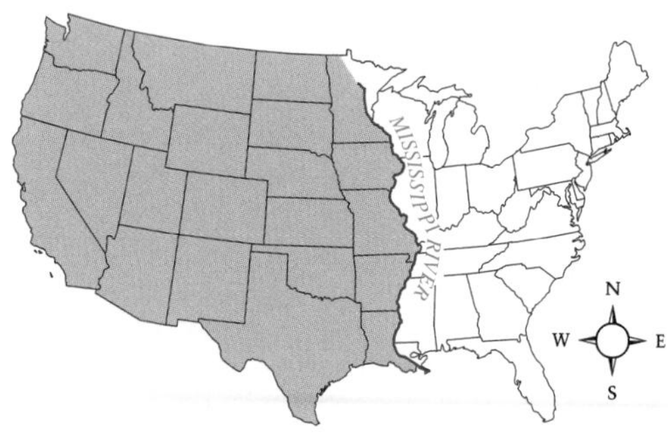

ILLUSTRATION 4

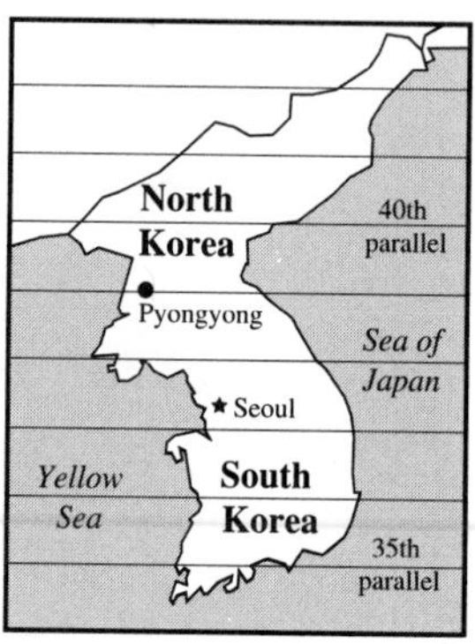

ILLUSTRATION 5

In Exercises 43–46, write a linear inequality that models the situation. Then graph each inequality for nonnegative values of x and y and give three ordered pairs that satisfy the inequality.

43. RESTAURANT SEATING As part of a remodeling project, a restaurant owner will be installing new booths that seat 4 persons, and new tables that seat 6 persons. The overall seating must conform to the sign shown in Illustration 6. Write an inequality that describes the possible combinations of booths (x) and tables (y) that the owner can install.

MAXIMUM OCCUPANCY
NOT TO EXCEED
120
By order of Clake County Fire Marshal

ILLUSTRATION 6

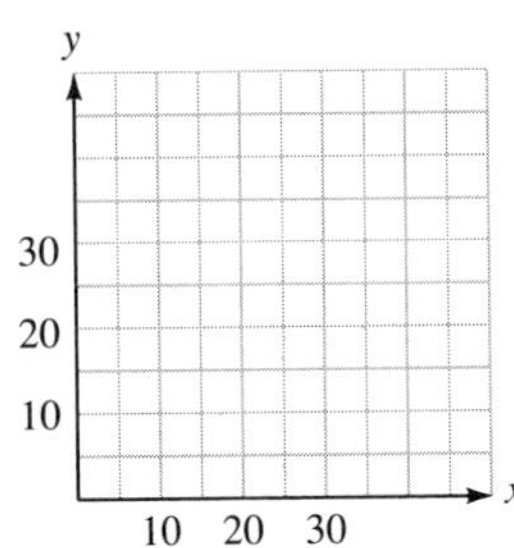

44. GARDENING During an Arbor Day sale, a garden store sold more than \$2,000 worth of trees. If a 6-foot maple costs \$100 and a 5-foot pine costs \$125, write an inequality that shows the possible ways that maple trees (x) and pine trees (y) were sold.

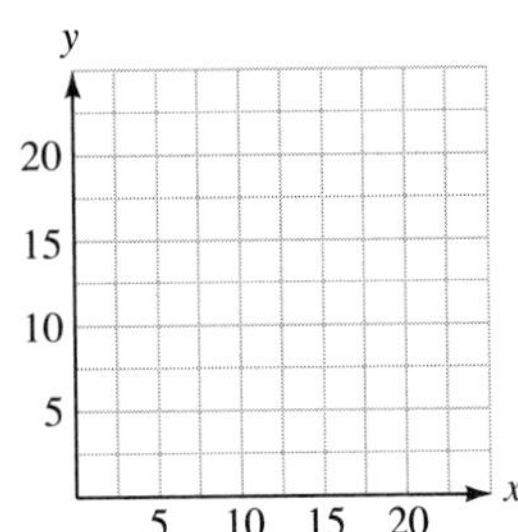

45. SPORTING GOODS A sporting goods manufacturer allocates at least 1,200 units of time per day to make fishing rods and reels. If it takes 10 units of time to make a rod and 15 units of time to make a reel, write an inequality that describes the possible ways to schedule the time to make rods (x) and reels (y).

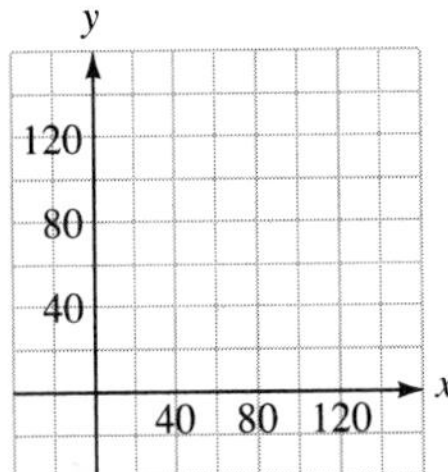

46. HOUSEKEEPING One housekeeper charges \$6 per hour, and another charges \$7 per hour. If Sarah can afford no more than \$42 per week to clean her house, write an inequality that describes the possible ways that she can hire the first housekeeper (x) and the second housekeeper (y).

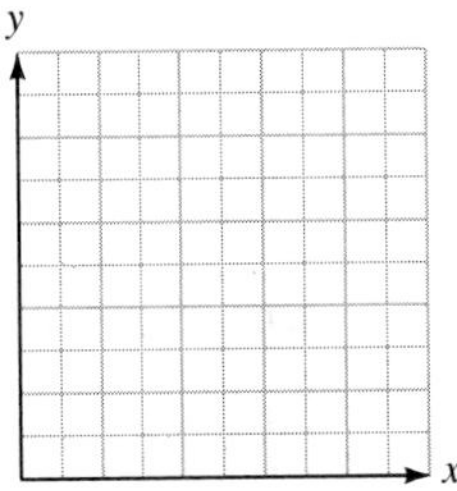

WRITING *Write a paragraph using your own words.*

47. Explain how to decide where to draw the boundary of the graph of a linear inequality, and whether to draw it as a solid or a broken line.

48. Explain how to decide which side of the boundary of the graph of a linear inequality should be shaded.

REVIEW *In Exercises 49–50, tell whether the ordered pair (−4, 3) is a solution of the system of linear equations.*

49. $\begin{cases} 4x - y = -19 \\ 3x + 2y = -6 \end{cases}$

50. $\begin{cases} y = 2x + 11 \\ \dfrac{x}{2} + y = 0 \end{cases}$

In Exercises 51–52, solve each system of equations.

51. $\begin{cases} x + y = 4 \\ x - y = 2 \end{cases}$

52. $\begin{cases} x - \dfrac{y}{2} = -2 \\ 0.01x + 0.02y = 0.03 \end{cases}$

4.5 Systems of Linear Inequalities

In this section, you will learn about

- **Solving systems of linear inequalities**
- **Compound inequalities**
- **Problem solving**

INTRODUCTION. In Chapter 3, we discussed how to solve systems of linear *equations* by the graphing method. For example, to solve

$$\begin{cases} y = -x + 1 \\ 2x - y = 2 \end{cases}$$

by this method, we graph both equations on the same set of coordinate axes and then find the coordinates of the point of intersection of the straight lines.

In this section, we will discuss how to graphically solve systems of linear *inequalities,* such as

$$\begin{cases} y \leq -x + 1 \\ 2x - y > 2 \end{cases}$$

Solving systems of linear inequalities

When the solution of a linear inequality in x and y is graphed, the result is a half-plane. To solve a system of linear inequalities, we graph each of the inequalities on one set of coordinate axes and look for the intersection, or overlap, of the shaded half-planes.

EXAMPLE 1 ***Graphing a system of linear inequalities.*** Graph the solution set of $\begin{cases} y \leq -x + 1 \\ 2x - y > 2 \end{cases}$.

Solution

We graph each inequality on one set of coordinate axes, as shown in Figure 4-40. To make it easy to see the intersection of the two half-planes, we graph one solution set in red and the other in blue.

To graph $y \leq -x + 1$, we first graph the boundary $y = -x + 1$, as shown below. Since the edge is to be included, we draw it as a solid line. To determine which half-plane to shade, we use the origin as a test point. Because the coordinates of the origin satisfy $y \leq -x + 1$, we shade the half-plane containing the origin.

To graph $2x - y > 2$, we first graph the boundary $2x - y = 2$, as shown below. Since the edge is not included, we draw it as a broken line. To determine which half-plane to shade, we again use the origin as a test point. Because the coordinates of the origin don't satisfy $2x - y > 2$, we shade the opposite half-plane.

The area where the half-planes intersect represents the solution of the system of inequalities, because any point in that region has coordinates that will satisfy both inequalities.

$y = -x + 1$

We graph this line using the slope and y-intercept.

$m = -1 = -\frac{1}{1}$

$b = 1$

y-intercept: $(0, 1)$

$2x - y = 2$

We graph this line using the intercept method.

x	y	(x, y)
0	−2	(0, −2)
1	0	(1, 0)

Self Check

Graph $\begin{cases} x + y \geq 1 \\ 2x - y < 2 \end{cases}$.

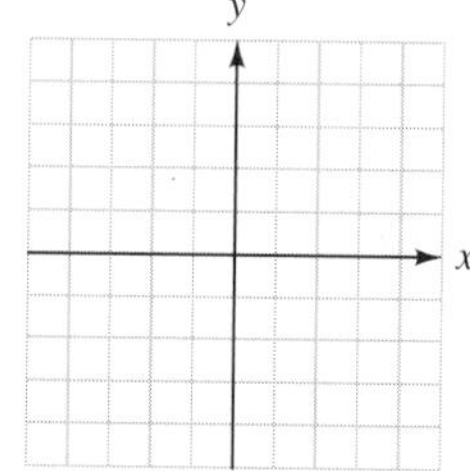

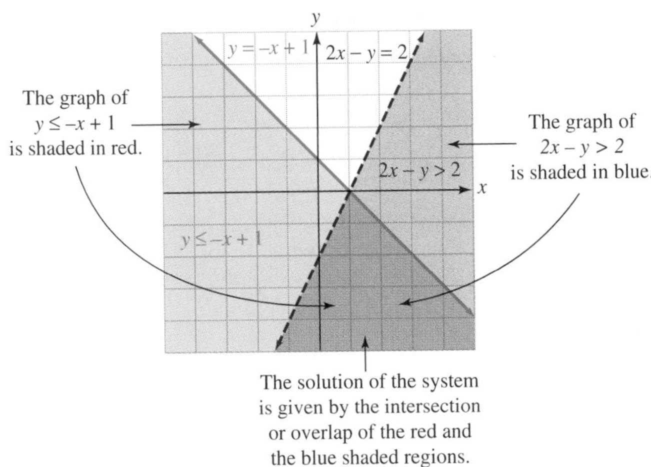

y
$y = -x + 1$
$2x - y = 2$
The graph of
$y \leq -x + 1$
is shaded in red.
The graph of
$2x - y > 2$
is shaded in blue.
$2x - y > 2$
x
$y \leq -x + 1$
The solution of the system
is given by the intersection
or overlap of the red and
the blue shaded regions.

FIGURE 4-40

Answer:

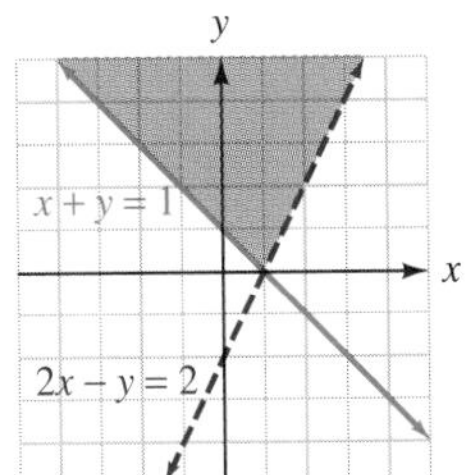

y
$x + y = 1$
x
$2x - y = 2$

EXAMPLE 2 ***A system of three inequalities.*** Graph the solution set of $\begin{cases} x \geq 1 \\ y \geq x \\ 4x + 5y < 20 \end{cases}$.

Solution

We will find the graph of the solution set of the system in stages, using several graphs. Normally, a problem such as this is solved using just one set of axes.

The graph of $x \geq 1$ includes the points that lie on the graph of $x = 1$ and to the right, as shown in red in Figure 4-41(a).

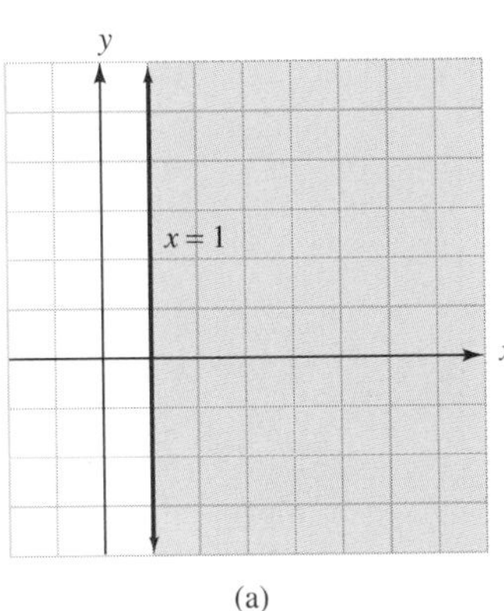

y
$x = 1$
x

(a)

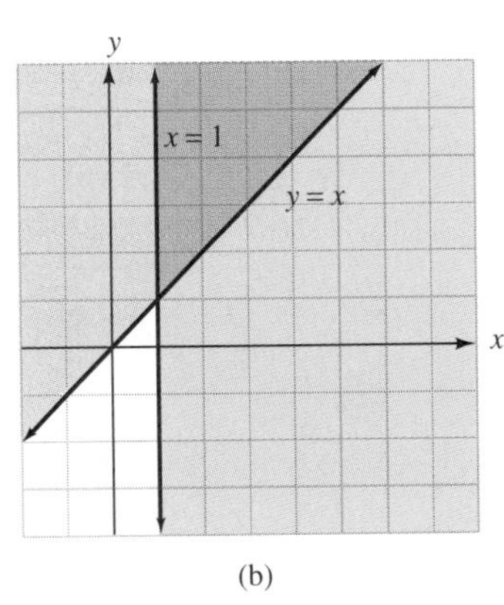

y
$x = 1$
$y = x$
x

(b)

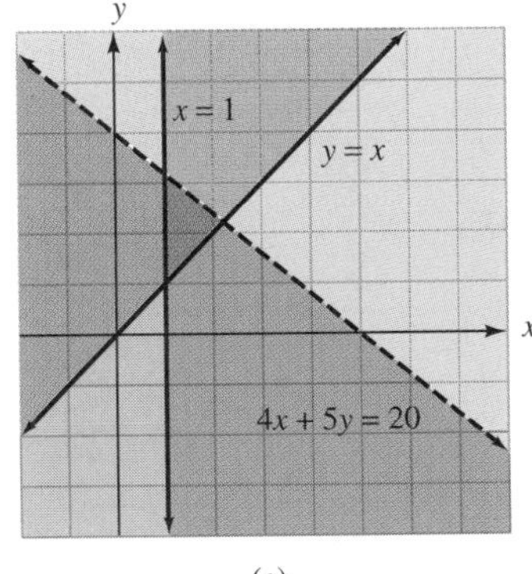

y
$x = 1$
$y = x$
x
$4x + 5y = 20$

(c)

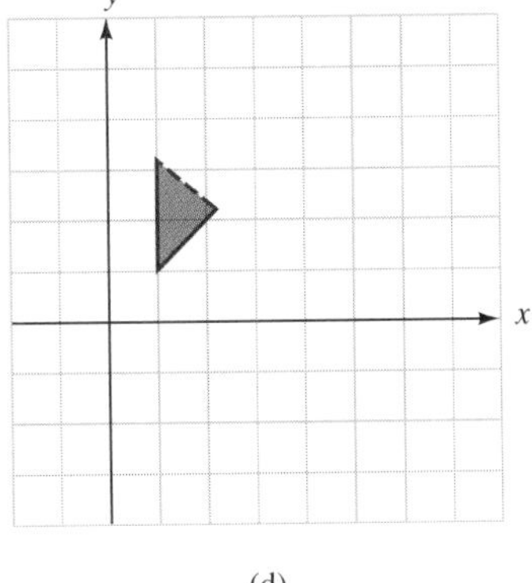

y
x

(d)

FIGURE 4-41

Self Check

Graph $\begin{cases} x \geq 0 \\ y \leq 0 \\ y \geq -2 \end{cases}$.

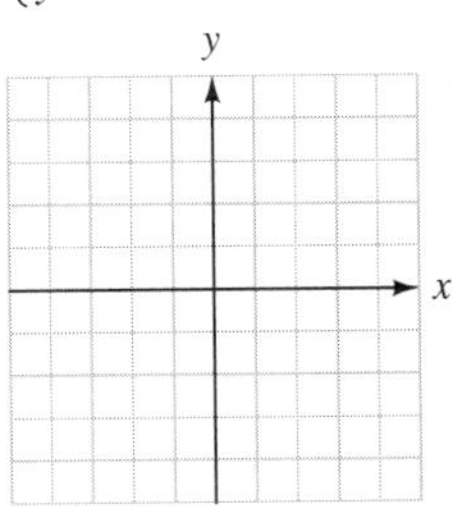

y
x

Figure 4-41(b) shows the graph of $x \geq 1$ and the graph of $y \geq x$. The graph of $y \geq x$, in blue, includes the points that lie on the graph of the boundary $y = x$ and above it.

Figure 4-41(c) shows the graphs of $x \geq 1$, $y \geq x$, and $4x + 5y < 20$. The graph of $4x + 5y < 20$ includes the points that lie below the graph of the boundary $4x + 5y = 20$.

The graph of the solution of the system includes the points that lie within the shaded triangle together with the points on the two sides of the triangle that are drawn with solid line segments, as shown in Figure 4-41(d).

Answer:

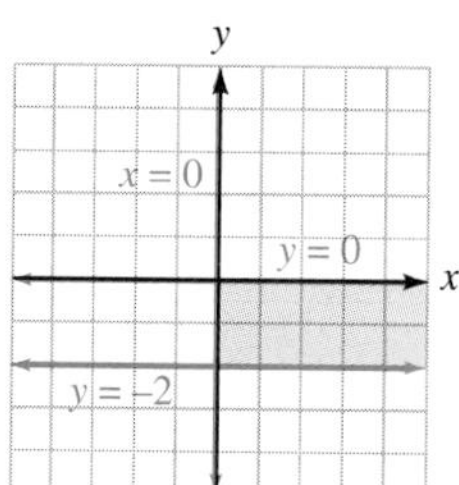

Compound inequalities

We have graphed the solution set of double linear inequalities, such as $2 < x \leq 5$, on a number line. These inequalities contained only one variable. In the next example, we will graph the solution set of $2 < x \leq 5$ in the context of two variables. In this case, we use a rectangular coordinate system.

EXAMPLE 3 ***Graphing a compound inequality.*** Graph $2 < x \leq 5$ in the rectangular coordinate system.

Solution

The compound inequality $2 < x \leq 5$ is equivalent to the following system of two linear inequalities:

$$\begin{cases} 2 < x \\ x \leq 5 \end{cases}$$

The graph of $2 < x$, shown in Figure 4-42 in red, is the half-plane to the right of the vertical line $x = 2$. The graph of $x \leq 5$, shown in the figure in blue, includes the line $x = 5$ and the half-plane to its left. The graph of $2 < x \leq 5$ will contain all points in the plane that satisfy the inequalities $2 < x$ and $x \leq 5$ simultaneously. These points are in the purple-shaded region of the figure.

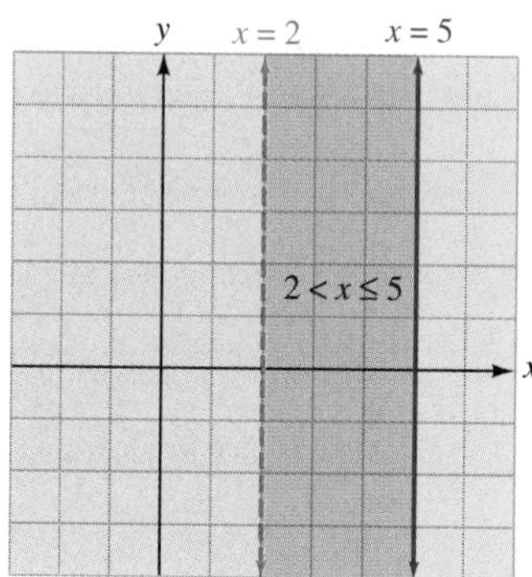

FIGURE 4-42

Self Check

Graph $-2 \leq y < 3$ in the rectangular coordinate system.

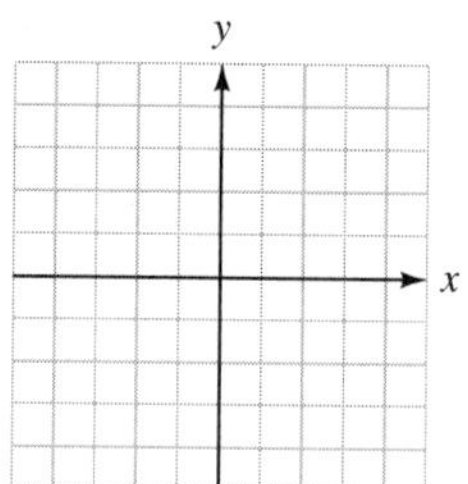

Answer:

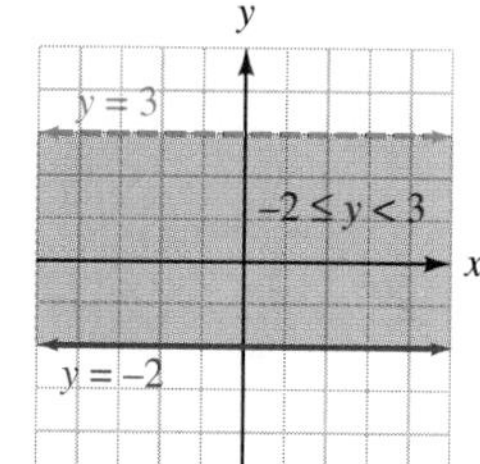

To graph a compound inequality containing the word *or* in the rectangular coordinate system, we sketch the *union* of the solution sets of the inequalities involved. For example, Figure 4-43 shows the graph of the compound inequality

$$x \leq -2 \text{ or } x > 3$$

in the rectangular coordinate system.

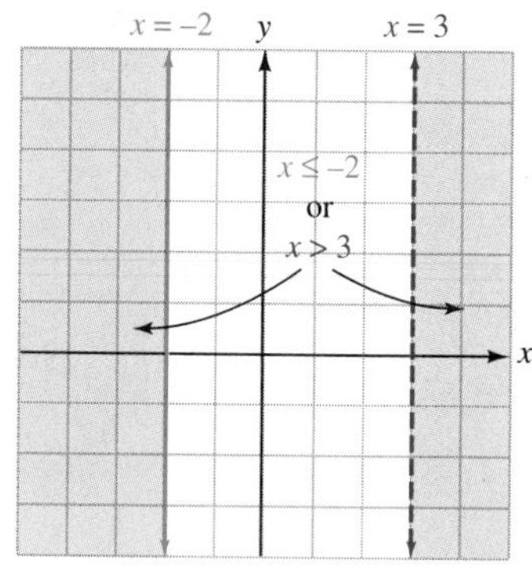

FIGURE 4-43

EXAMPLE 4 ***Landscaping*** A homeowner has a budget of \$300 to \$600 for trees and bushes to landscape his yard. After some comparison shopping, he finds that good trees cost \$150 and mature bushes cost \$75. What combinations of trees and bushes can he afford to buy?

Analyze the problem We need to find the number of trees and the number of bushes that the homeowner can afford. This suggests we should use two variables. We know that he is willing to spend *at least* \$300 and *at most* \$600 for trees and bushes. These phrases suggest that we should write two inequalities that model the situation.

Form two inequalities If x represents the number of trees purchased, then \$150x will be the cost of the trees. If y represents the number of bushes purchased, then \$75y will be the cost of the bushes. We know that the homeowner wants the sum of these costs to be from \$300 to \$600. We can then form the following system of linear inequalities.

The cost of a tree	·	the number of trees purchased	+	the cost of a bush	·	the number of bushes purchased	should be at least	\$300.
\$150	·	x	+	\$75	·	y	$\geq$	\$300
The cost of a tree	·	the number of trees purchased	+	the cost of a bush	·	the number of bushes purchased	should be at most	\$600.
\$150	·	x	+	\$75	·	y	$\leq$	\$600

Solve the system We graph the system

$$\begin{cases} 150x + 75y \geq 300 \\ 150x + 75y \leq 600 \end{cases}$$

as in Figure 4-44. The coordinates of each point shown in the graph give a possible combination of trees (x) and bushes (y) that can be purchased.

State the conclusion The possible combinations of trees and bushes that can be purchased are given by

(0, 4), (0, 5), (0, 6), (0, 7), (0, 8)
(1, 2), (1, 3), (1, 4), (1, 5), (1, 6)
(2, 0), (2, 1), (2, 2), (2, 3), (2, 4)
(3, 0), (3, 1), (3, 2), (4, 0)

The ordered pair (1, 6), for example, indicates that the homeowner can afford 1 tree and 6 bushes.

Only these points can be used, because the homeowner cannot buy a portion of a tree or a bush.

Check the result Check some of the ordered pairs to verify that they satisfy both inequalities.

Because the homeowner cannot buy a negative number of trees or bushes, we graph the system for $x \geq 0$ and $y \geq 0$.

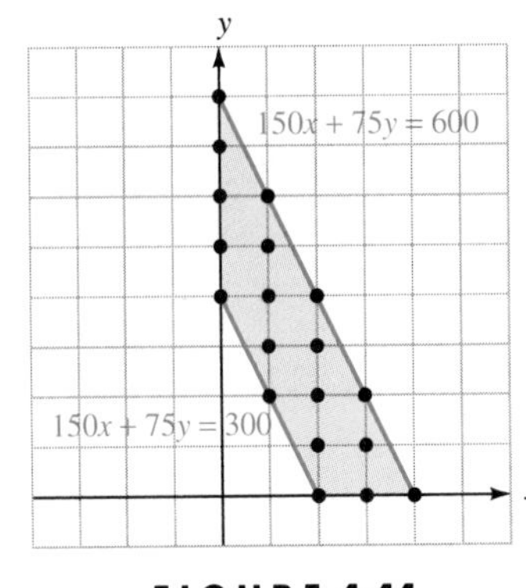

FIGURE 4-44

STUDY SET Section 4.5

VOCABULARY *In Exercises 1–4, fill in the blanks to make the statements true.*

1. $\begin{cases} x + y \le 2 \\ x - 3y > 10 \end{cases}$ is a system of linear ______________.

2. If an edge is included in the graph of an inequality, we draw it as a ______________ line.

3. To solve a system of inequalities by graphing, we graph each inequality. The solution is the region where the graphs overlap or ______________.

4. To determine which half-plane to shade when graphing a linear inequality, we see whether the coordinates of a test ______________ satisfy the inequality.

CONCEPTS

5. Tell whether each point satisfies the system of linear inequalities

$$\begin{cases} x + y \le 2 \\ x - 3y > 10 \end{cases}$$

a. $(2, -3)$ **b.** $(12, -1)$

c. $(0, -3)$ **d.** $(-0.5, -5)$

6. a. Tell whether $(-3, 10)$ satisfies the compound inequality $-5 < x \le 8$ in the rectangular coordinate system.

b. Tell whether $(-3, 3)$ satisfies the compound inequality $y \le 0$ or $y > 4$ in the rectangular coordinate system.

7. In Illustration 1, the solution of one linear inequality is shaded in red, and the solution of a second linear inequality is shaded in blue. Tell whether a true or false statement results if the coordinates of the given point are substituted into the given inequality.

a. A, inequality 1 **b.** A, inequality 2

c. B, inequality 1 **d.** B, inequality 2

e. C, inequality 1 **f.** C, inequality 2

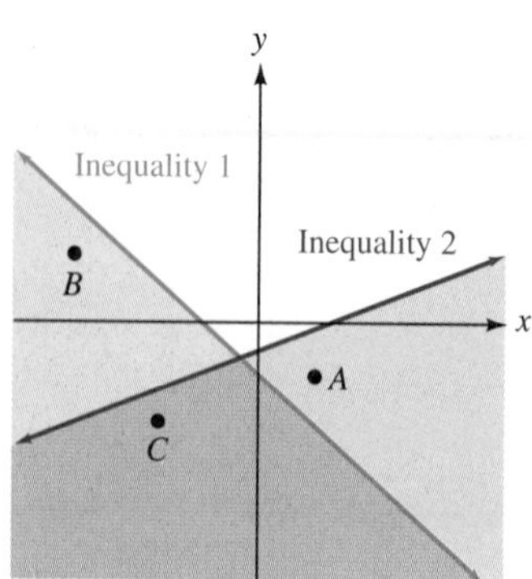

ILLUSTRATION 1

8. Match each equation, inequality, or system with the graph of its solution in Illustration 2.

a. $2x + y = 2$ **b.** $2x + y \ge 2$

c. $\begin{cases} 2x + y = 2 \\ 2x - y = 2 \end{cases}$ **d.** $\begin{cases} 2x + y \ge 2 \\ 2x - y \le 2 \end{cases}$

i
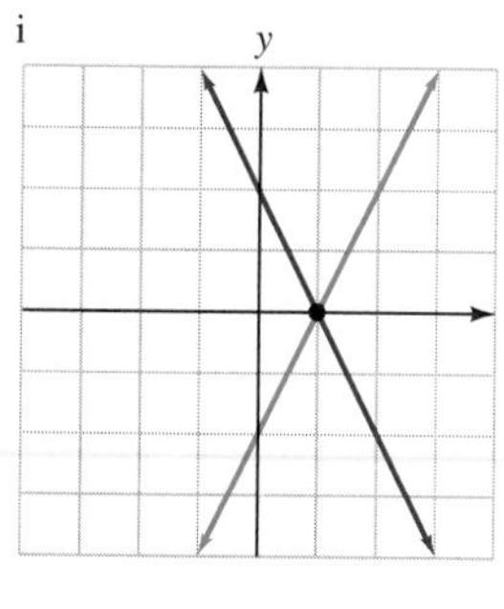

ii
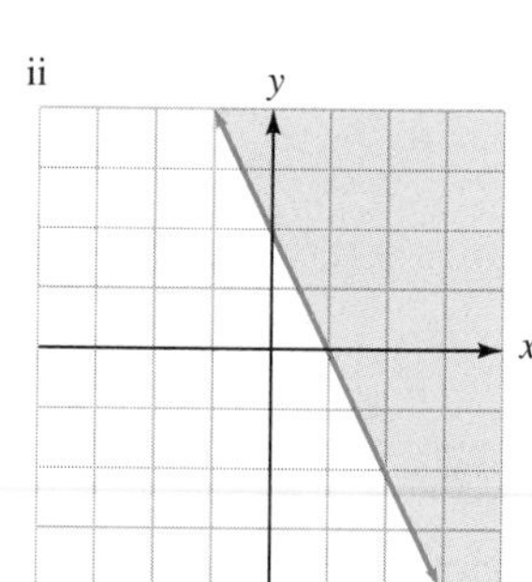

iii
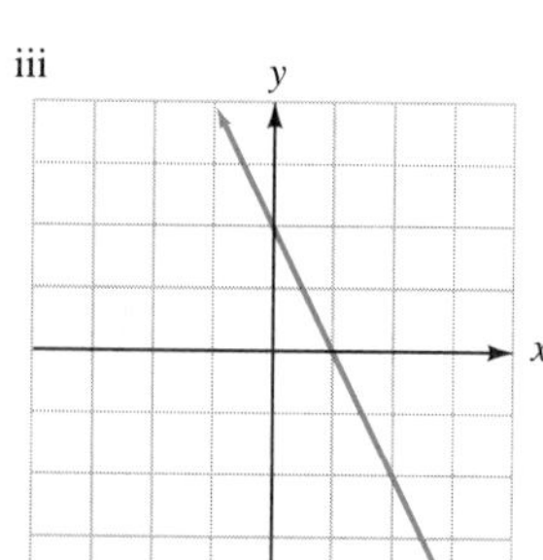

iv
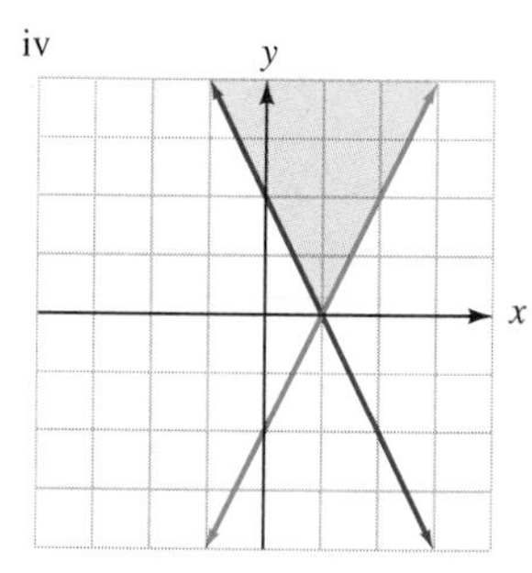

ILLUSTRATION 2

PRACTICE *In Exercises 9–20, graph the solution set of each system of inequalities.*

9. $\begin{cases} y < 3x + 2 \\ y < -2x + 3 \end{cases}$

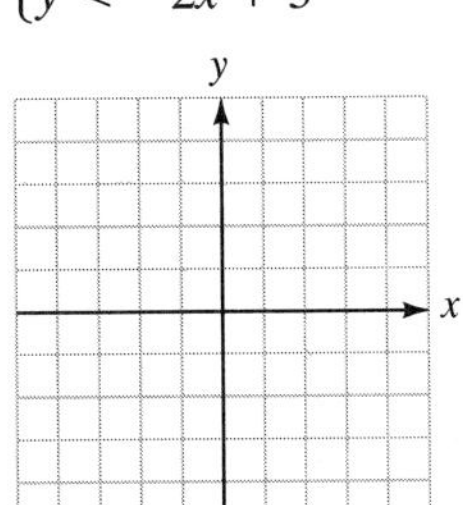

10. $\begin{cases} y \le x - 2 \\ y \ge 2x + 1 \end{cases}$

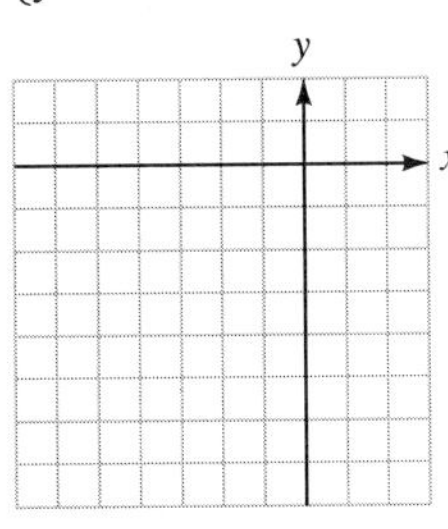

11. $\begin{cases} 3x + 2y > 6 \\ x + 3y \le 2 \end{cases}$

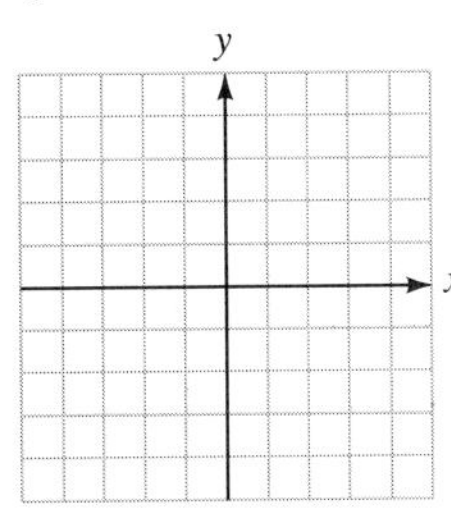

12. $\begin{cases} x + y < 2 \\ x + y \le 1 \end{cases}$

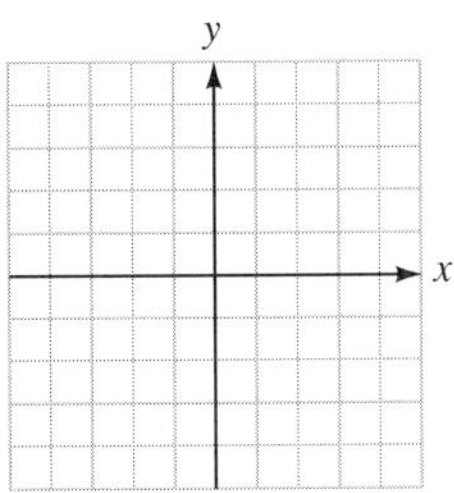

13. $\begin{cases} 3x + y \le 1 \\ -x + 2y \ge 6 \end{cases}$

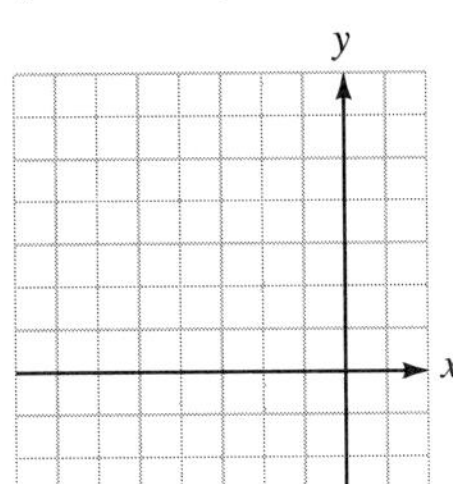

14. $\begin{cases} x + 2y < 3 \\ 2x + 4y < 8 \end{cases}$

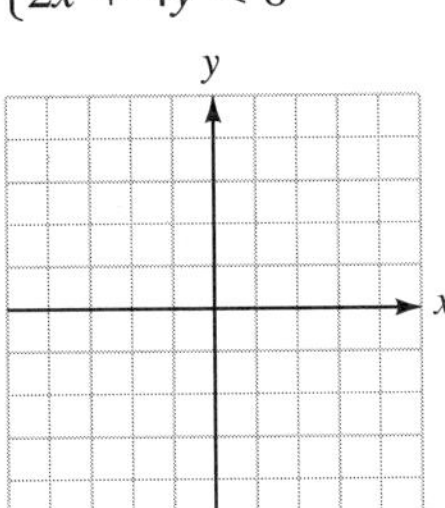

15. $\begin{cases} x > 0 \\ y > 0 \end{cases}$

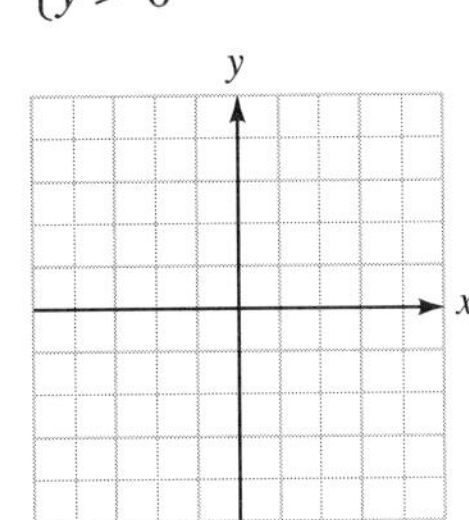

16. $\begin{cases} x \le 0 \\ y < 0 \end{cases}$

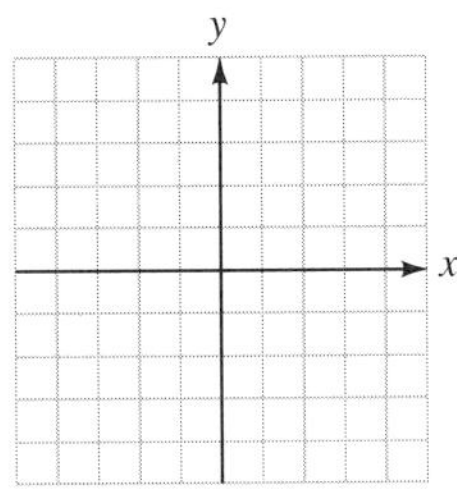

17. $\begin{cases} 2x + 3y \le 6 \\ 3x + y \le 1 \\ x \le 0 \end{cases}$

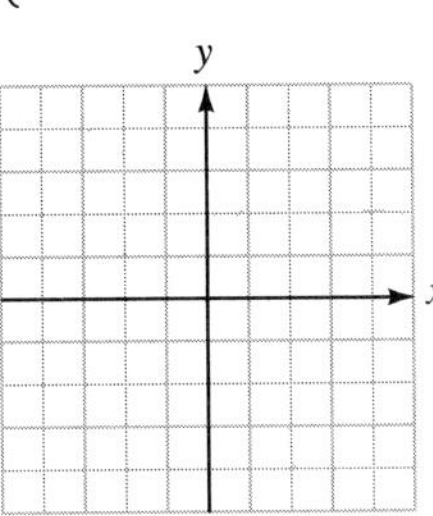

18. $\begin{cases} 2x + y \le 2 \\ y \ge x \\ x \ge 0 \end{cases}$

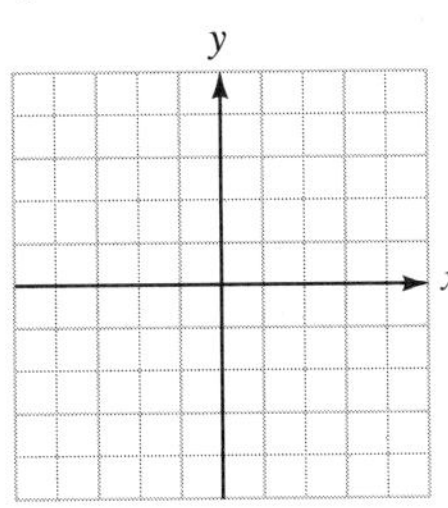

19. $\begin{cases} x - y < 4 \\ y \le 0 \\ x \ge 0 \end{cases}$

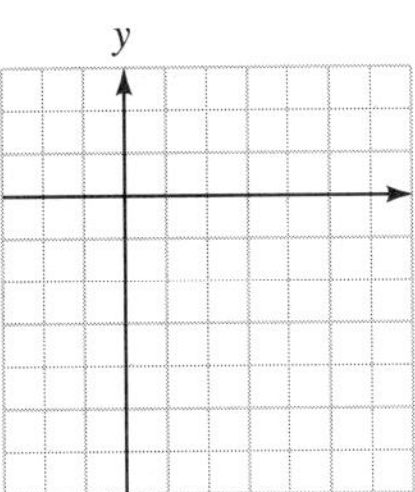

20. $\begin{cases} x \ge 0 \\ y \ge 0 \\ 9x + 3y \le 18 \\ 3x + 6y \le 18 \end{cases}$

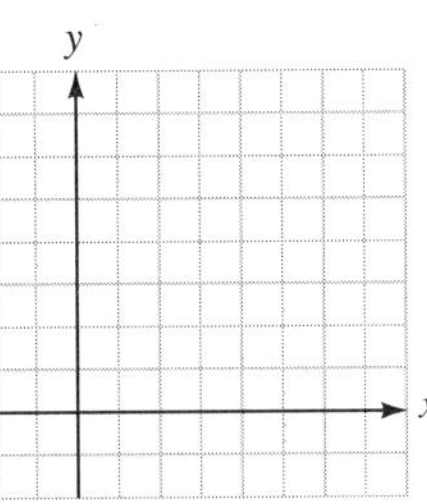

In Exercises 21–24, graph each inequality in the rectangular coordinate system.

21. $-2 \le x < 0$

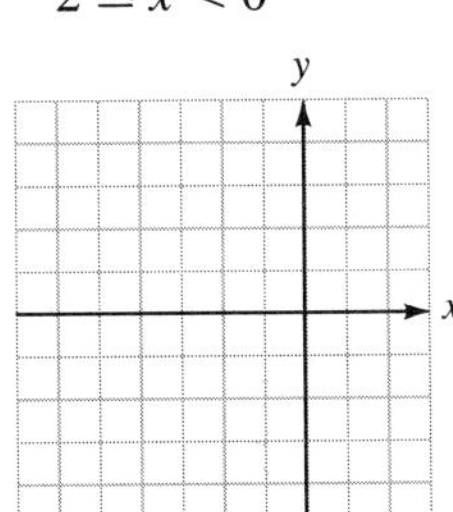

22. $-3 < y \le -1$

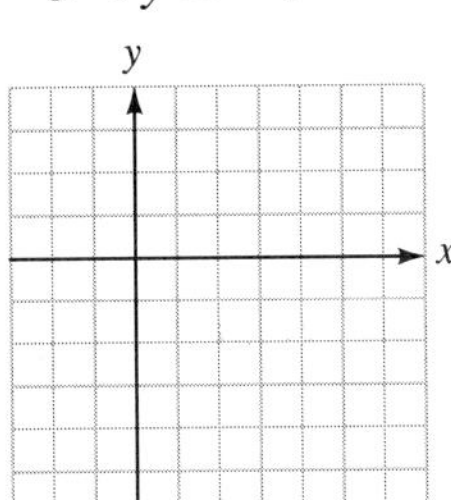

23. $y < -2$ or $y > 3$

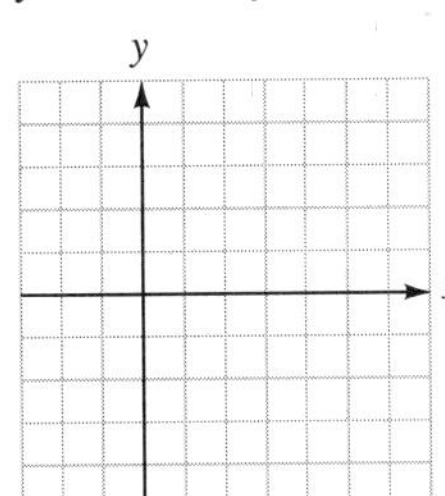

24. $-x \le 1$ or $x \ge 2$

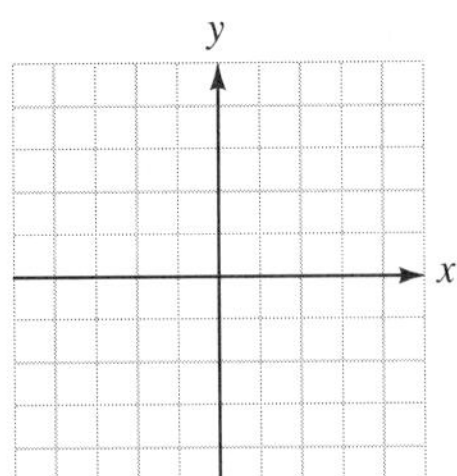

APPLICATIONS

25. PROFESSIONAL FOOTBALL The 1997 Denver Broncos football team scored either a field goal or a touchdown 89.3% of the time when they had possession of the football in their "red zone." If x represents the yard line the ball is on, the red zone is an area on their half of the field that can be described by the system

$$\begin{cases} x > 0 \\ x \le 20 \end{cases}$$

Shade the red zone on the field shown in Illustration 3.

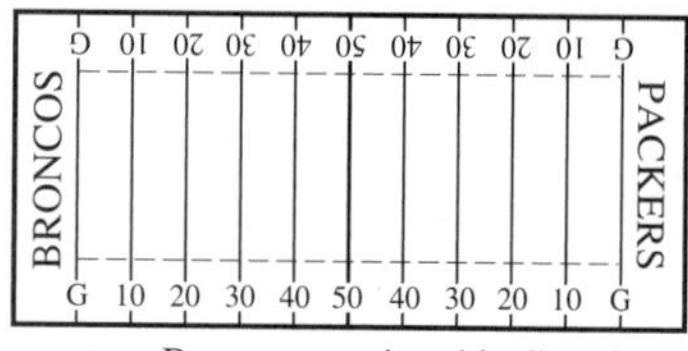

ILLUSTRATION 3

26. TRACK AND FIELD In the shot put, the solid metal ball must land in a marked sector for it to be a fair throw. In Illustration 4, graph the system of inequalities that describes the region in which a shot must land.

$$\begin{cases} y \leq \frac{3}{8}x \\ y \geq -\frac{3}{8}x \\ x \geq 0 \end{cases}$$

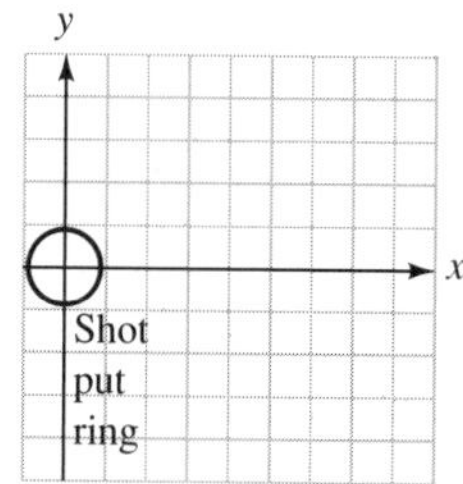

ILLUSTRATION 4

27. NO-FLY ZONES After the Gulf War, U.S. and Allied forces enforced northern and southern "no-fly" zones over Iraq. Iraqi aircraft was prohibited from flying in this air space. If x represents the north latitude parallel measurement, the no-fly zones can be described by

$x \geq 36$ or $x \leq 33$

On the map in Illustration 5, shade the regions of Iraq over which there was a no-fly zone.

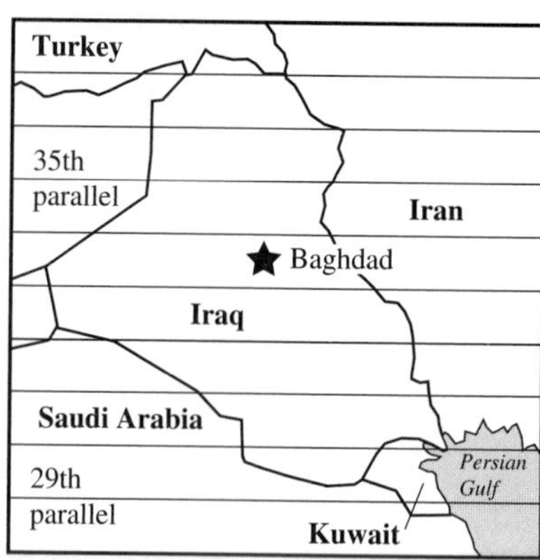

ILLUSTRATION 5

28. CARDIOVASCULAR FITNESS The graph in Illustration 6 shows the range of pulse rates that persons ages 20–90 should maintain during aerobic exercise to get the most benefit from the training. The shaded region "Effective Training Heart Rate Zone" can be described by a system of linear inequalities. Determine what inequality symbol should be inserted in each blank.

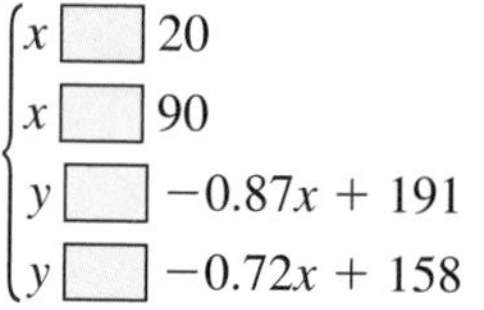

$$\begin{cases} x \;\square\; 20 \\ x \;\square\; 90 \\ y \;\square\; -0.87x + 191 \\ y \;\square\; -0.72x + 158 \end{cases}$$

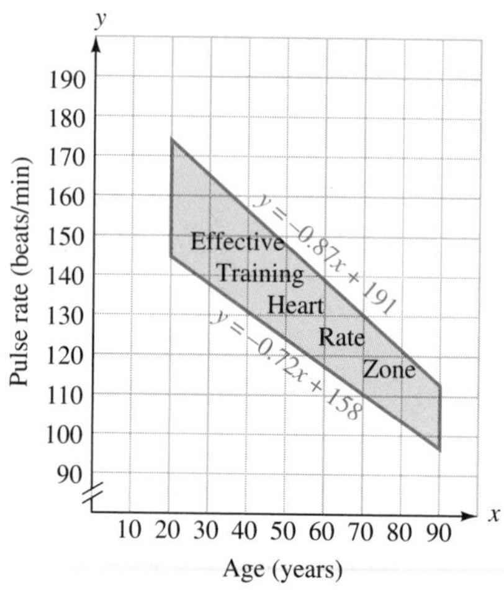

ILLUSTRATION 6

In Exercises 29–32, graph each system of inequalities and give two possible solutions.

29. COMPACT DISCS Melodic Music has compact discs on sale for either \$10 or \$15. If a customer wants to spend at least \$30, but no more than \$60 on CDs, use Illustration 7 to graph a system of inequalities that will show the possible ways a customer can buy \$10 CDs (x) and \$15 CDs (y).

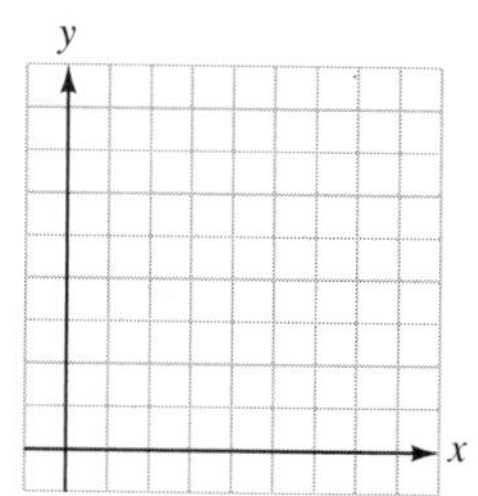

ILLUSTRATION 7

30. BOAT SALES Dry Boat Works wholesales aluminum boats for \$800 and fiberglass boats for \$600. Northland Marina wants to order at least \$2,400 worth, but no more than \$4,800 worth of boats. Use Illustration 8 to graph a system of inequalities that will show the possible combination of aluminum boats (x) and fiberglass boats (y) that can be ordered.

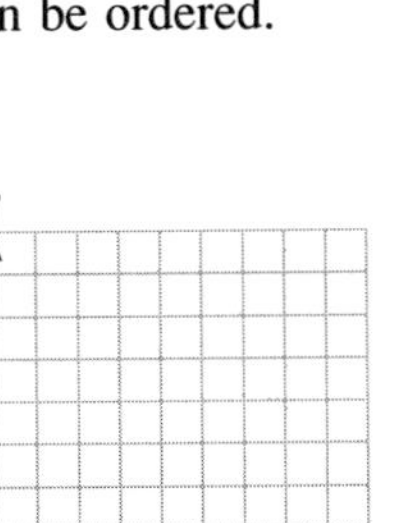

ILLUSTRATION 8

31. FURNITURE SALES A distributor wholesales desk chairs for \$150 and side chairs for \$100. Best Furniture wants to order no more than \$900 worth of chairs, including more side chairs than desk chairs. Use Illustration 9 to graph a system of inequalities that will show the possible combinations of desk chairs (x) and side chairs (y) that can be ordered.

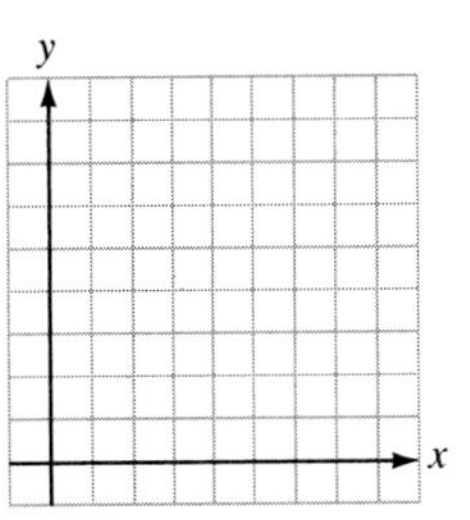

ILLUSTRATION 9

32. FURNACE EQUIPMENT J. Bolden Heating Company wants to order no more than \$2,000 worth of electronic air cleaners and humidifiers from a wholesaler that charges \$500 for air cleaners and \$200 for humidifiers. If Bolden wants more humidifiers than air cleaners, use Illustration 10 to graph a system of inequalities that will show the possible combinations of air cleaners (x) and humidifiers (y) that can be ordered.

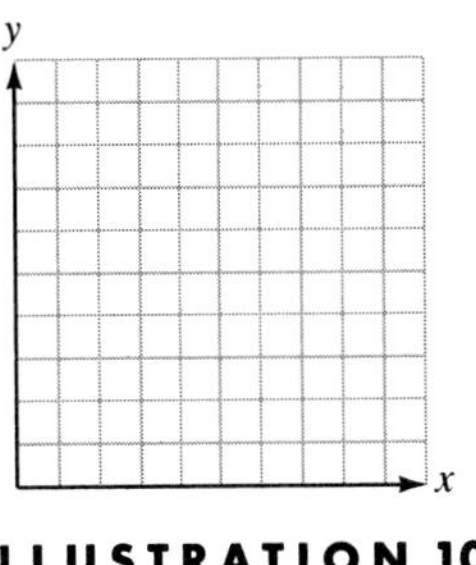

ILLUSTRATION 10

WRITING *Write a paragraph using your own words.*

33. When graphing a system of linear inequalities, explain how to decide which region to shade.

34. Explain how a system of two linear inequalities might have no solution.

REVIEW *In Exercises 35–38, use the given conditions to determine in which quadrant of a rectangular coordinate system each point (x, y) is located.*

35. $x > 0$ and $y < 0$

36. $x < 0$ and $y < 0$

37. $x < 0$ and $y > 0$

38. $x > 0$ and $y > 0$

KEY CONCEPT

Inequalities

Types of Inequalities

An **inequality** is a statement indicating that quantities are unequal. In Chapter 4, we worked with several different types of inequalities and combinations of inequalities.

1. Classify each statement as one of the following: linear inequality in one variable, compound inequality, double linear inequality, absolute value inequality, linear inequality in two variables, system of linear inequalities.

a. $x - 3 < -4$ or $x - 2 > 0$ **b.** $\begin{cases} y < 3x + 2 \\ y < -2x + 3 \end{cases}$ **c.** $|x - 8| \leq 12$

d. $y < \frac{x}{3} - 1$ **e.** $\frac{1}{2}x + 2 \geq \frac{1}{3}x - 4$ **f.** $-6 < -3(x - 4) \leq 24$

g. $5(x - 2) \geq 0$ and $-3x < 9$ **h.** $|-1 - 2x| > 5$ **i.** $y > -x$

Solutions of Inequalities

A solution of a linear inequality in one variable is a value that, when substituted for the variable, makes the inequality true. A solution of a linear inequality in two variables (or a system of linear inequalities), is an ordered pair whose coordinates satisfy the inequality (or inequalities).

2. Tell whether $x = -2$ is a solution of the inequalities in one variable. Tell whether $(-1, 3)$ is a solution of the inequalities (or system of inequalities) in two variables.

a. $x - 3 < -4$ or $x - 2 > 0$ **b.** $\begin{cases} y < 3x + 2 \\ y < -2x + 3 \end{cases}$ **c.** $|x - 8| \leq 12$

d. $y < \frac{x}{3} - 1$ **e.** $\frac{1}{2}x + 2 \geq \frac{1}{3}x - 4$ **f.** $-6 < -3(x - 4) \leq 24$

g. $5(x - 2) \geq 0$ and $-3x < 9$ **h.** $|-1 - 2x| > 5$ **i.** $y > -x$

Graphs of Inequalities

To graph the solution set of an inequality in one variable, we use a number line. To graph the solution set of an inequality in two variables, we use a rectangular coordinate system.

3. Graph the solution set of the linear inequality in one variable: $2x + 1 > 4$.

4. Graph the solution set of the linear inequality in two variables: $2x + y \geq 4$.

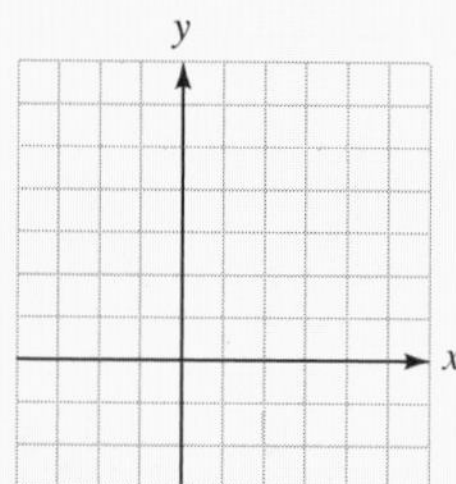

GROUP ACTIVITIES FOR CHAPTER 4

Accent on Teamwork

Section 4.1

INEQUALITY STATEMENTS The warning given on the street sign in Illustration 1 can be described using an inequality. If we let w stand for the weight of a truck, then its weight must be such than $w \leq 3{,}000$ pounds if the truck is to use that street. Find six more real-life situations that can be described using an inequality.

ILLUSTRATION 1

TRIANGLE INEQUALITY The triangle inequality is discussed in Exercise 45 of Study Set 4.1. To demonstrate it, cut two pipe cleaners to lengths of 2 inches and 4 inches, to serve as two sides of a triangle. (See Illustration 2.) Cut a third pipe cleaner to a length of 5 inches. Show that there is no triangle with sides of 2, 4, and 5 inches. Cut another pipe cleaner to a length of 6 inches. Show that there is no triangle with sides of 2, 4, and 6 inches. How long must the third pipe cleaner be for you to be able to form a triangle?

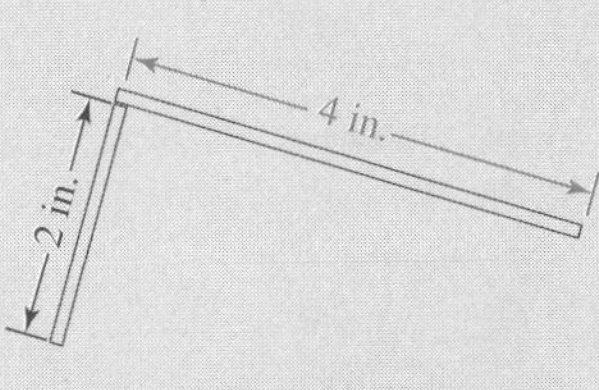

ILLUSTRATION 2

Section 4.2

POLLS In Exercise 56 of Study Set 4.2, the margin of error for a poll of 1,000 adults was discussed. Find a published poll in a magazine or a newspaper in which the margin of error is given. Use interval notation to describe the possible intervals for each response in the poll. Taking the possible error into account, could the rankings of the responses possibly change?

ROOM TEMPERATURE What temperature range do you like to keep inside your home? If t is the temperature, write a double linear inequality that describes the acceptable temperature range. Graph the solution set and express it as an interval. Then write a compound inequality containing the word *or* to describe the unacceptable temper range inside your home. Graph the solution set and express it using interval notation.

VOCABULARY

a. How are the connecting words *and* and *or* used with inequalities? Give examples.

b. Explain the concepts of union and intersection of sets. Give some examples using a number line.

c. How are the symbols ∞ and $-\infty$ used in the context of a number line?

Section 4.3

TOLERANCES Visit a machine shop or automobile repair shop and ask the employees to show you some examples of how they work with tolerances in their profession. Videotape their explanations and then play the video for your class. Show how the tolerances can be described using an absolute value inequality and interval notation. See Example 8 in Section 4.3 for an example.

Section 4.4

INEQUALITIES IN TWO VARIABLES

a. Can an inequality be an identity, one that is satisfied by all (x, y) pairs? Illustrate.

b. Can an inequality have no solutions? Illustrate.

Section 4.5

INTERSECTION Sketch six examples of objects that intersect (overlap). Shade the region where they intersect. For instance, you could draw the intersection of two major streets in your city or the intersection of a chair and the floor.

CHAPTER REVIEW

SECTION 4.1 *Solving Linear Inequalities*

CONCEPTS

To solve an inequality, apply the *properties of inequalities.*

If both sides of an inequality are multiplied (or divided) by a negative number, another inequality results, but with the opposite direction from the original inequality.

REVIEW EXERCISES

1. Solve each inequality. Give each solution set in interval notation and graph it.

a. $5(x - 2) \le 5$

b. $0.3x - 0.4 \ge 1.2 - 0.1x$

c. $-16 < -\frac{4}{5}x$

d. $\frac{7}{4}(x + 3) < \frac{3}{8}(x - 3)$

2. INVESTMENTS A woman has invested \$10,000 at 6% annual interest. How much more must she invest at 7% so that her annual income is at least \$2,000?

SECTION 4.2 *Solving Compound Inequalities*

A solution of a compound inequality containing *and* makes both of the inequalities true.

The solution set of a compound inequality containing *and* is the *intersection* of the two solution sets.

Double linear inequalities:

$c < x < d$

is equivalent to

$c < x$ and $x < d$

A solution of a compound inequality containing the word *or* makes one, or the other, or both inequalities true.

The solution set of a compound inequality containing *or* is the *union* of the two solution sets.

3. Tell whether $x = -4$ is a solution of the compound inequality.

a. $x < 0$ and $x > -5$

b. $x + 3 < -3x - 1$ and $4x - 3 > 3x$

4. Solve each compound inequality. Give the result in interval notation and graph the solution set.

a. $-2x > 8$ and $x + 4 \ge -6$

b. $5(x + 2) \le 4(x + 1)$ and $11 + x < 0$

5. Solve each compound inequality. Give the result in interval notation and graph the solution set.

a. $3 < 3x + 4 < 10$

b. $-2 \le \frac{5 - x}{2} \le 2$

6. Tell whether $x = -4$ is a solution of the compound inequality.

a. $x < 1.6$ or $x > -3.9$

b. $x + 1 < 2x - 1$ or $4x - 3 > 3x$

7. Solve each compound inequality. Give the result in interval notation and graph the solution set.

a. $x + 1 < -4$ or $x - 4 > 0$

b. $\frac{x}{2} + 3 > -2$ or $4 - x > 4$

8. INTERIOR DECORATING
A manufacturer makes a line of decorator rugs that are 4 feet wide and of varying lengths l (in feet). See Illustration 1. The floor area covered by the rugs ranges from 17 ft^2 to 25 ft^2. Write and then solve a double linear inequality to find the range of the lengths of the rugs.

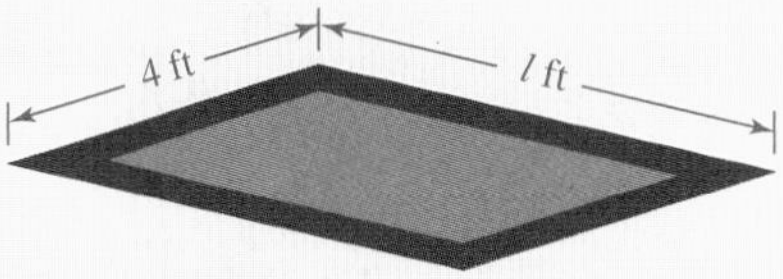

ILLUSTRATION 1

SECTION 4.3 *Solving Absolute Value Equations and Inequalities*

Definition of *absolute value:*

$$\begin{cases} \text{If } x \geq 0, |x| = x. \\ \text{If } x < 0, |x| = -x. \end{cases}$$

9. Find each absolute value.

a. $|-7|$ **b.** $\left|\dfrac{5}{16}\right|$

c. $-|71.05|$ **d.** $-|-12|$

Absolute value equations:

If $k \geq 0$, $|x| = k$ is equivalent to $x = k$ or $x = -k$

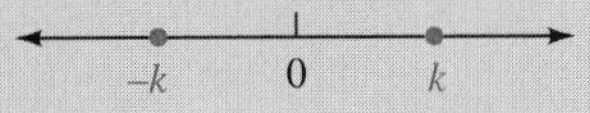

$|a| = |b|$ is equivalent to $a = b$ or $a = -b$

10. Solve each absolute value equation. Check the results.

a. $|4x| = 8$ **b.** $2|3x + 1| = 20$

c. $\left|\dfrac{3}{2}x - 4\right| - 10 = -1$ **d.** $\left|\dfrac{2 - x}{3}\right| = 4$

e. $|3x + 2| = |2x - 3|$ **f.** $\left|\dfrac{3 - 2x}{2}\right| = \left|\dfrac{3x - 2}{3}\right|$

Absolute value inequalities:

If $k > 0$, $|x| < k$ is equivalent to $-k < x < k$

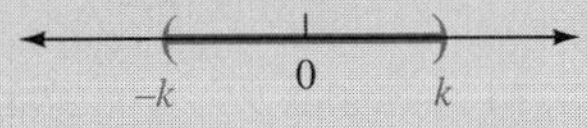

Absolute value inequalities:

$|x| > k$ is equivalent to $x < -k$ or $x > k$

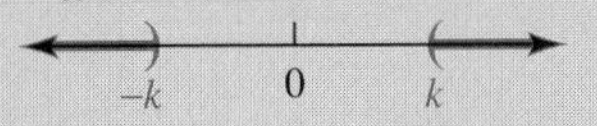

11. Solve each absolute value inequality. Give the solution in interval notation and graph it.

a. $|x| \leq 3$ **b.** $|2x + 7| < 3$

c. $|5 - 3x| \leq 14$ **d.** $\left|\dfrac{2}{3}x + 14\right| < 0$

e. $|x| > 1$ **f.** $\left|\dfrac{1 - 5x}{3}\right| \geq 7$

g. $|3x - 8| > 4$ **h.** $\left|\dfrac{3}{2}x - 14\right| \geq 0$

12. PRODUCE Before packing, freshly picked tomatoes are weighed on the scale shown in Illustration 2. Tomatoes having a weight w (in ounces) that falls within the highlighted range are sold to grocery stores.

a. Express this acceptable weight range using an absolute value inequality.

b. Solve the inequality and express this range as an interval.

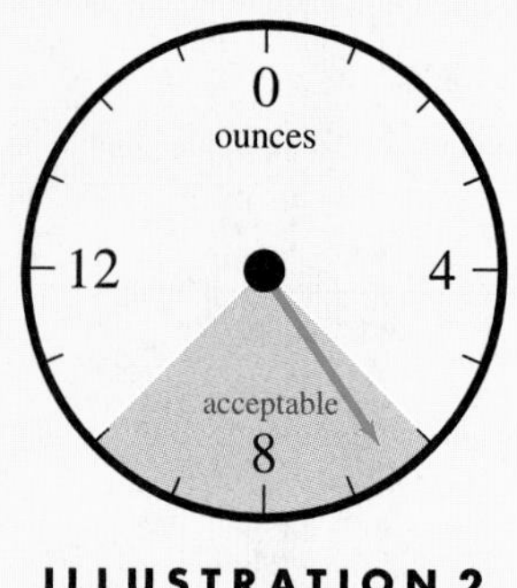

ILLUSTRATION 2

SECTION 4.4 Linear Inequalities in Two Variables

To graph a *linear inequality* in x and y, graph the *boundary line,* and then use a *test point* to decide which side of the boundary should be shaded.

13. Graph each inequality in the rectangular coordinate system.

a. $2x + 3y > 6$

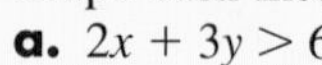

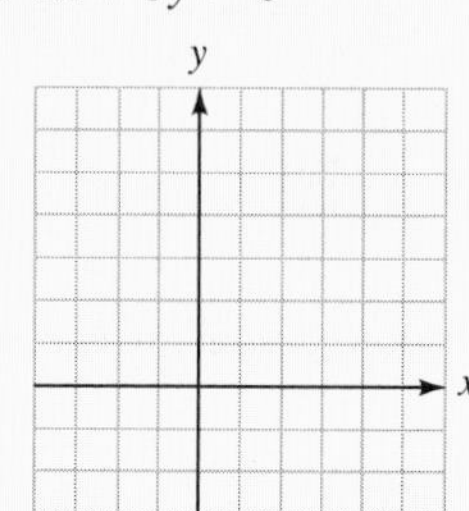

b. $y \leq 4 - x$

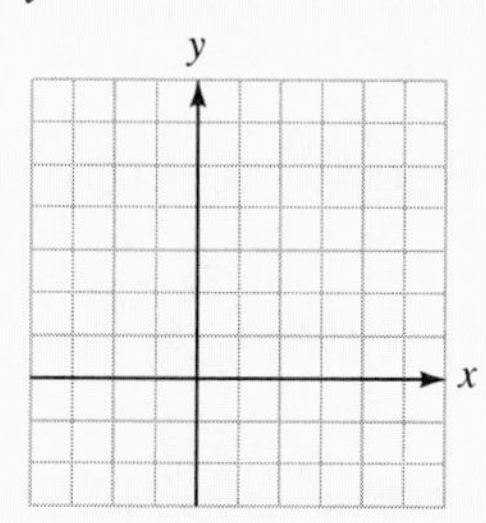

c. $y < \frac{1}{2}x$

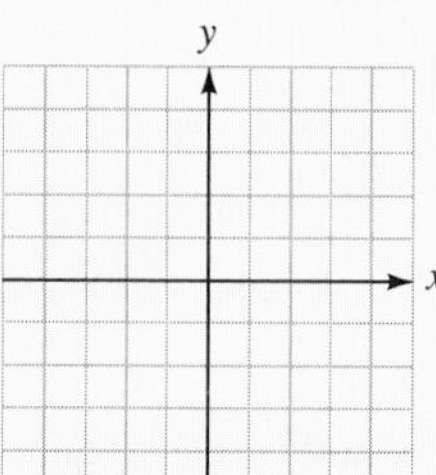

d. $x \geq -\frac{3}{2}$

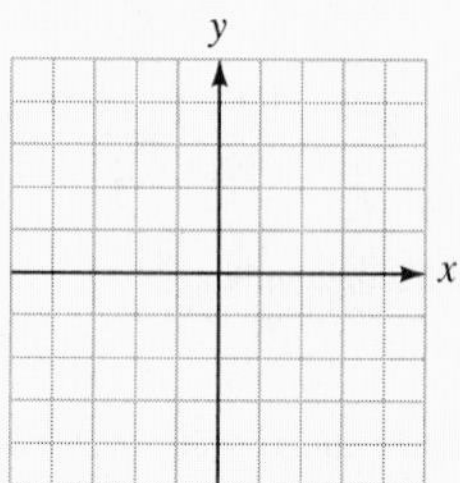

14. CONCERT TICKETS Tickets to a concert cost \$6 for reserved seats and \$4 for general admission. If receipts must be at least \$10,200 to meet expenses, find an inequality that shows the possible ways that the box office can sell reserved seats (x) and general admission tickets (y). Then graph each inequality for nonnegative values of x and y and give three ordered pairs that satisfy the inequality.

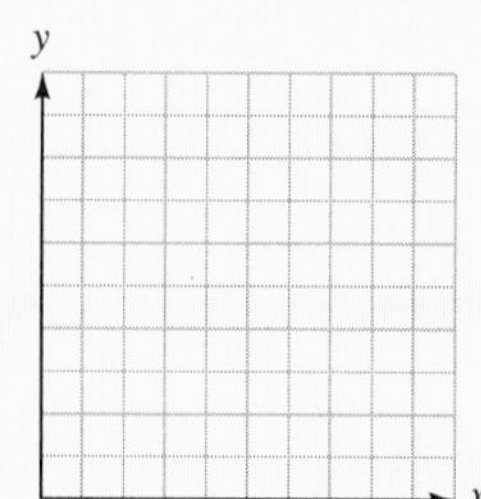

SECTION 4.5 Systems of Linear Inequalities

To solve a *system of linear inequalities,* graph each of the inequalities on the same set of coordinate axes and look for the intersection of the shaded *half-planes.*

15. Graph the solution set of each system of inequalities.

a. $\begin{cases} y \geq x + 1 \\ 3x + 2y < 6 \end{cases}$

b. $\begin{cases} x - y < 3 \\ y \leq 0 \\ x \geq 0 \end{cases}$

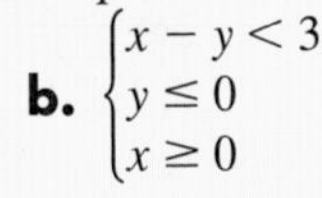

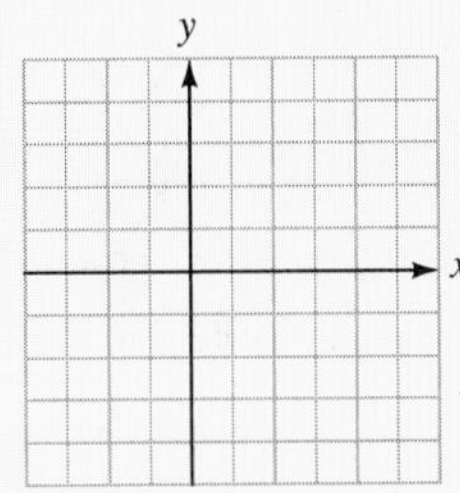

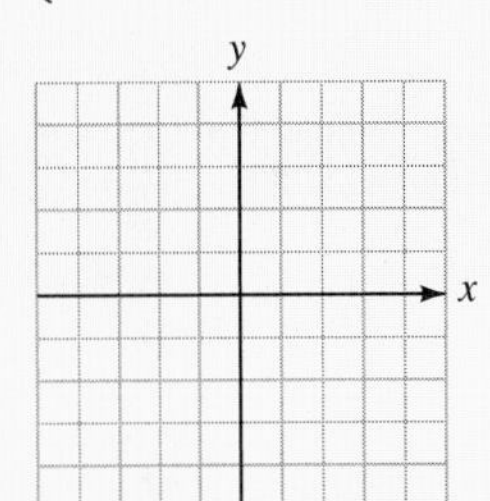

Compound inequalities can be graphed in the rectangular coordinate system.

16. Graph each compound inequality in the rectangular coordinate system.

a. $-2 < x < 4$

b. $y \leq -2$ or $y > 1$

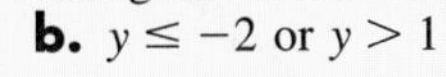

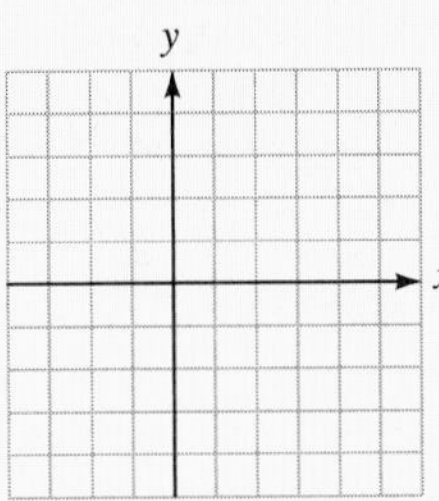

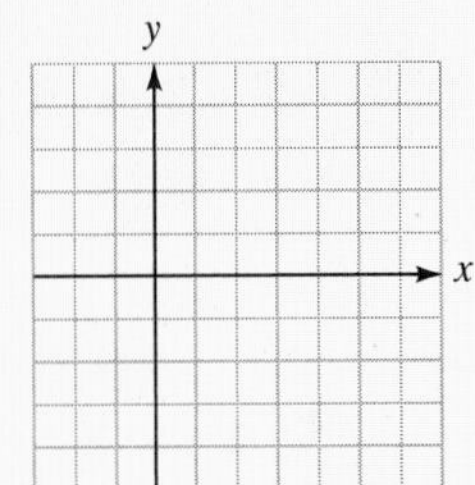

17. PETROLEUM EXPLORATION Organic matter converts to oil and gas within a specific range of temperature and depth called the *petroleum window.* The petroleum window in Illustration 3 can be described by a system of linear inequalities, where x is the temperature in °C of the soil at a depth of y meters. Determine what inequality symbol should be inserted in each blank.

$$\begin{cases} x \;\square\; 35 \\ x \;\square\; 130 \\ y \;\square\; -56x + 280 \\ y \;\square\; -18x + 90 \end{cases}$$

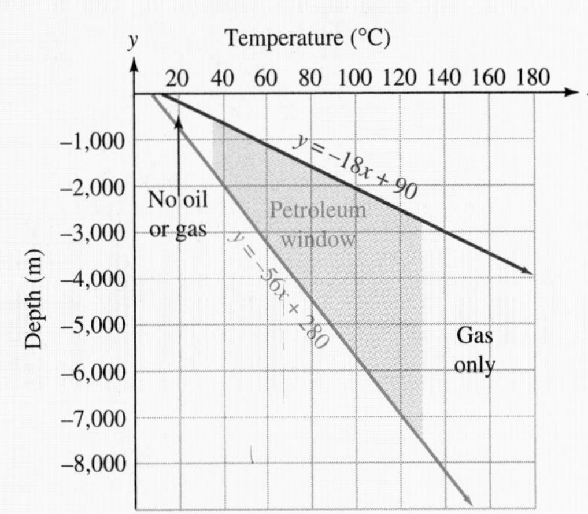

Based on data from *The Blue Planet* (Wiley, 1995)

ILLUSTRATION 3

Chapter 4 Test

1. Tell whether the statement is true or false.
$-5.67 \geq -5$

2. Tell whether $x = -2$ is a solution of the inequality.
$3(x - 2) \leq 2(x + 7)$

In Problems 3–4, graph the solution set of each inequality. Also give the solution in interval notation.

3. $7 < \frac{2}{3}t - 1$

4. $-2(2x + 3) \geq 14$

5. AVERAGING GRADES Use the information from the gradebook in Illustration 1 to determine what score Karen Nelson-Sims needs on the fifth exam to keep her exam average above 80 in the class.

Sociology 101 8:00-10:00 pm MW	Exam 1	Exam 2	Exam 3	Exam 4	Exam 5
Nelson-Sims, Karen	70	79	85	88	

ILLUSTRATION 1

In Problems 6–8, solve each compound inequality. Give the result in interval notation and graph the solution set.

6. $3x \geq -2x + 5$ and $7 \geq 4x - 2$

7. $3x < -9$ or $-\frac{x}{4} < -2$

8. $-2 < \frac{x - 4}{3} < 4$

In Problems 9–10, find the value of each expression.

9. $|8|$

10. $-|-4.75|$

In Problems 11–12, solve each equation.

11. $|4 - 3x| = 19$

12. $|3x + 4| = |x + 12|$

In Problems 13–16, graph the solution set of each inequality. Also give the solution in interval notation.

13. $|x + 3| \leq 4$

14. $|2x - 4| > 22$

15. $|4 - 2x| + 1 > 3$

16. $|2x - 4| \leq 2$

In Problems 17–20, graph the solution set in the rectangular coordinate system.

17. $3x + 2y \geq 6$

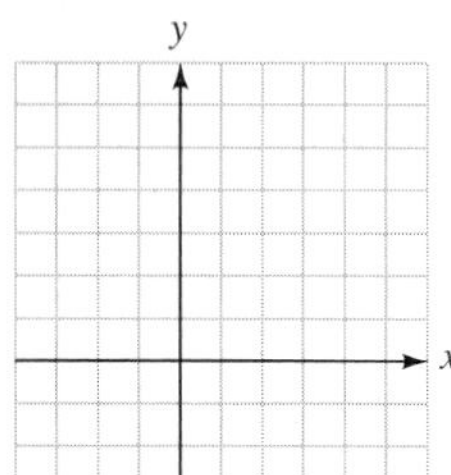

18. $y < x$

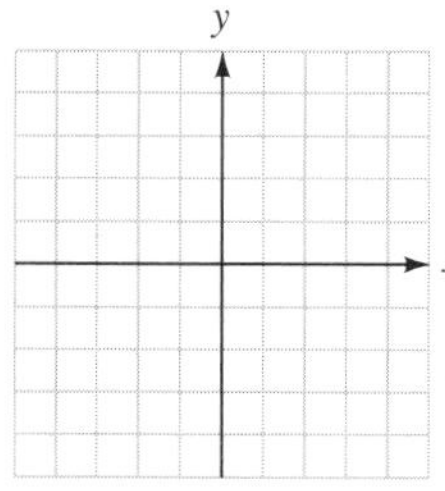

19. $\begin{cases} 2x - 3y \geq 6 \\ y \leq -x + 1 \end{cases}$

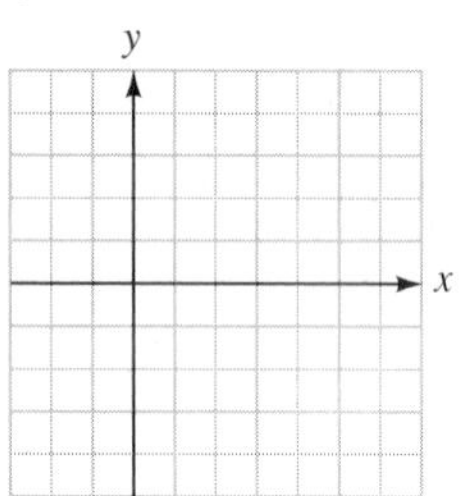

20. $-2 \leq y < 5$

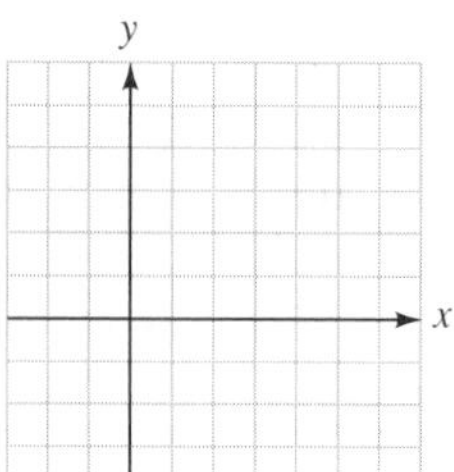

21. ACCOUNTING On average, it takes an accountant 1 hour to complete a simple tax return and 3 hours to complete a complicated return. If the accountant wants to work less than 9 hours per day, find an inequality that shows the number of possible ways that simple returns (x) and complicated returns (y) can be completed each day. Then graph the inequality and give three ordered pairs that satisfy it.

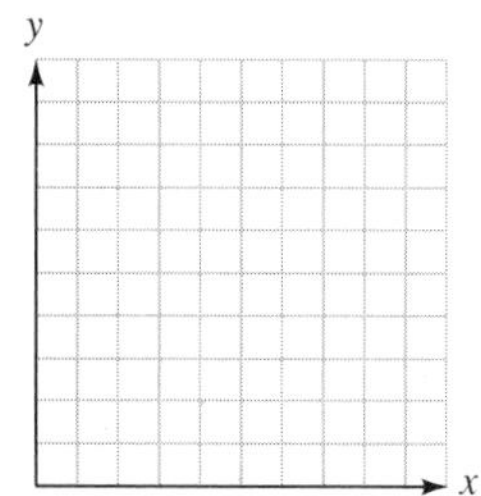

22. Two linear inequalities are graphed on the same coordinate axes in Illustration 2. The solution set of the first inequality is shaded in red, and the solution set of the second in blue. Can we determine from the graph whether the point $(3, -4)$ is a solution of the system of two linear inequalities? Explain your answer.

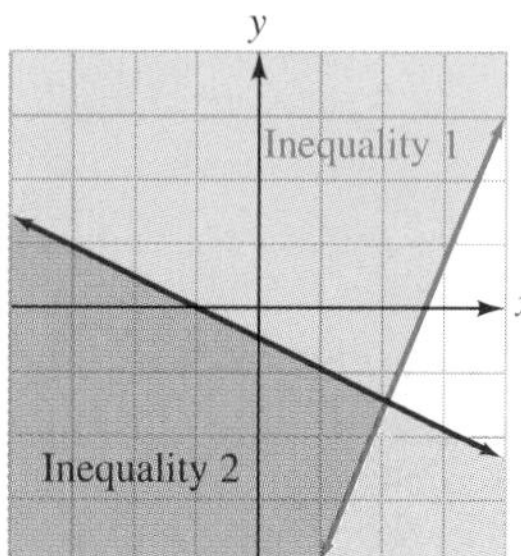

ILLUSTRATION 2

Chapters 1–4 Cumulative Review Exercises

1. The diagram in Illustration 1 shows the sets that compose the set of real numbers. Which of the indicated sets make up the *rational numbers* and the *irrational numbers?*

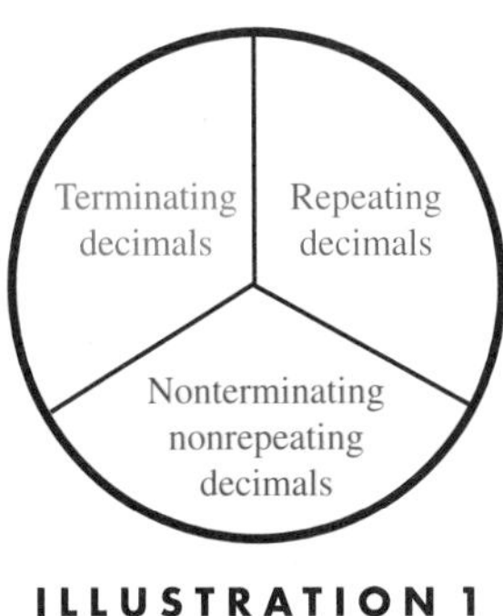

ILLUSTRATION 1

2. SCREWS The thread profile of a screw is determined by the distance between threads. This distance, indicated by the letter p, is known as the *pitch.* If $p = 0.125$, find each of the dimensions labeled in Illustration 2.

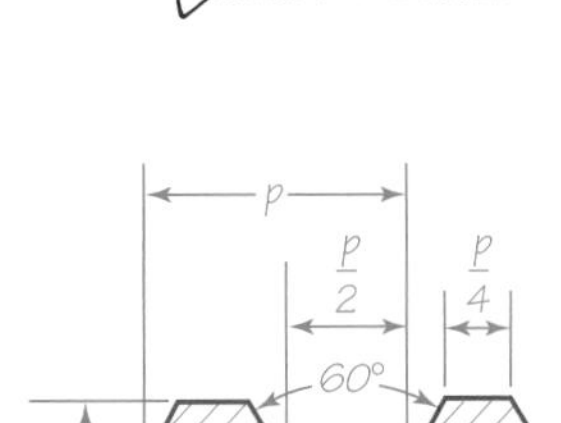

ILLUSTRATION 2

In Exercises 3–4, evaluate each expression when $x = 2$ and $y = -4$.

3. $|x| - xy$

4. $\dfrac{x^2 - y^2}{3x + y}$

In Exercises 5–6, simplify each expression.

5. $3p^2 - 6(5p^2 + p) + p^2$

6. $-(a + 2) - (a - b)$

7. PLASTIC WRAP Estimate the number of *square feet* of plastic wrap on a roll if the dimensions printed on the box describe the roll as 205 feet long by $11\frac{3}{4}$ inches wide.

8. INVESTMENTS Find the amount of money that was invested at $8\frac{7}{8}\%$ if it earned \$1,775 in simple interest in one year.

In Exercises 9–12, solve each equation, if possible.

9. $3x - 6 = 20$

10. $6(x - 1) = 2(x + 3)$

11. $\dfrac{5b}{2} - 10 = \dfrac{b}{3} + 3$

12. $2a - 5 = -2a + 4(a - 2) + 1$

In Exercises 13–14, tell whether the lines represented by the equations are parallel or perpendicular.

13. $3x + 2y = 12,\ 2x - 3y = 5$

14. $3x = y + 4,\ y = 3(x - 4) - 1$

15. Write the equation of the line passing through $P(-2, 3)$ and perpendicular to the graph of $3x + y = 8$. Answer in slope–intercept form.

16. Find the slope of the line that passes through $(0, -8)$, $(-5, 0)$.

17. PRISONS The graph in Illustration 3 shows the growth of the U.S. prison population from 1970 to 1995. Find the rate of change in the prison population from 1970 to 1975.

18. PRISONS Refer to the graph in Illustration 3. During what five-year period was the rate of change in the U.S. prison population the greatest? Find the rate of change.

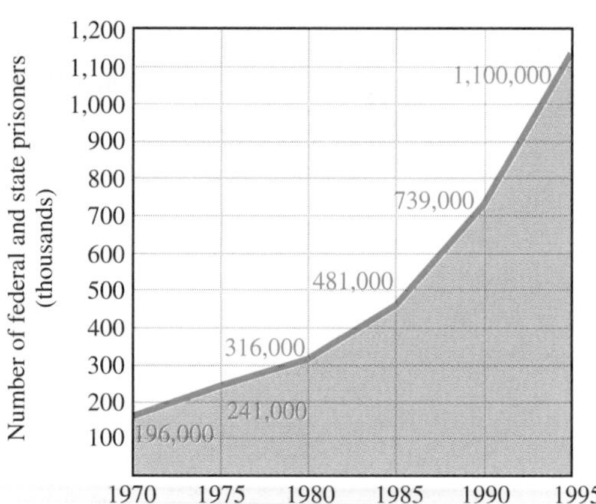

Based on data from *U.S. Statistical Abstract*

ILLUSTRATION 3

In Exercises 19–20, $f(x) = 3x^2 - x$. Find each value.

19. $f(2)$

20. $f(-2)$

21. Using graphing to solve $\begin{cases} 2x + y = 5 \\ x - 2y = 0 \end{cases}$.

22. Use addition to solve $\begin{cases} \dfrac{x}{10} + \dfrac{y}{5} = \dfrac{1}{2} \\ \dfrac{x}{2} - \dfrac{y}{5} = \dfrac{13}{10} \end{cases}$.

23. Use substitution to solve $\begin{cases} y = 4 - 3x \\ 2x - 3y = -1 \end{cases}$.

24. Solve $\begin{cases} x + y + z = 1 \\ 2x - y - z = -4 \\ x - 2y + z = 4 \end{cases}$.

In Exercises 25–26, use Cramer's rule to solve each system.

25. $\begin{cases} 4x - 3y = -1 \\ 3x + 4y = -7 \end{cases}$

26. $\begin{cases} x - 2y - z = -2 \\ 3x + y - z = 6 \\ 2x - y + z = -1 \end{cases}$

27. U.S. WORKERS The graph in Illustration 4 shows how the makeup of the U.S. workforce changed over the years 1900–1990. Estimate the coordinates of the points of intersection in the graph. Explain their significance.

28. WHITE-COLLAR JOBS Refer to the graph in Illustration 4. To model the growth in the percent of white-collar workers, draw a line through (0, 20) and (90, 70) and write an equation for this line. Use the equation to predict the percent of white-collar workers in the year 2010.

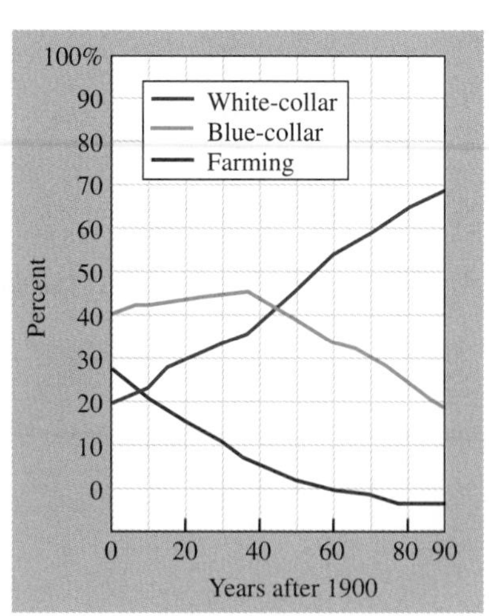

Based on data from *U.S. Statistical Abstract*

ILLUSTRATION 4

29. ENTREPRENEURS A person invests \$18,375 to set up a small business producing a piece of computer software that will sell for \$29.95. If each piece can be produced for \$5.45, how many pieces must be sold to break even?

30. CONCERT TICKETS Tickets for a concert cost \$5, \$3, and \$2. Twice as many \$5 tickets were sold as \$2 tickets. The receipts for 750 tickets were \$2,625. How many tickets were sold at each price?

In Exercises 31–32, solve each equation.

31. $|4x - 3| = 9$

32. $|2x - 1| = |3x + 4|$

In Exercises 33–36, solve each inequality. Give the solution in interval notation and graph it.

33. $-3(x - 4) \geq x - 32$

34. $-8 < -3x + 1 < 10$

35. $|3x - 2| \leq 4$

36. $|2x + 3| - 1 > 4$

In Exercises 37–38, use graphing to solve each inequality or system of inequalities.

37. $2x - 3y \leq 12$

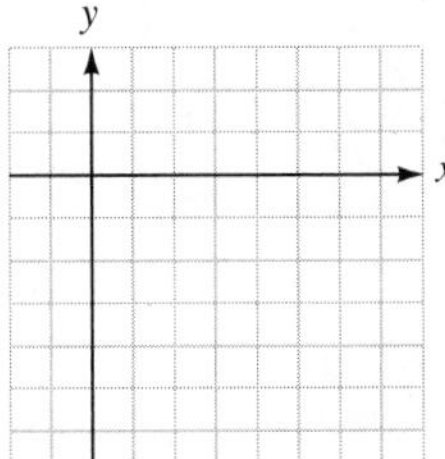

38. $\begin{cases} y < x + 2 \\ 3x + y \leq 6 \end{cases}$

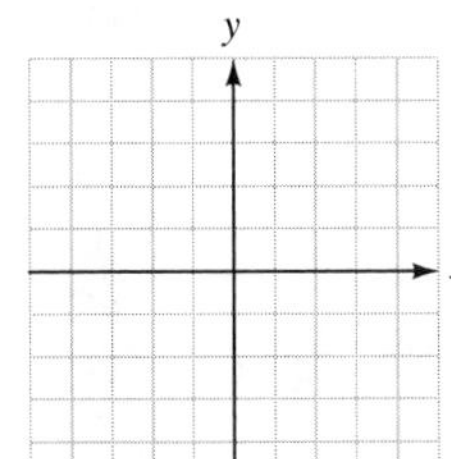

Exponents, Polynomials, and Polynomial Functions

5

Campus Connection

The *Astronomy* Department

In science classes, we often encounter very large and very small numbers. For example, in astronomy, students learn that one of the stars in the constellation known as the Big Dipper is 465,000,000,000,000 miles from earth. Such numbers are awkward to work with when making calculations. In this chapter, we will use *exponents* to write them in a compact form called scientific notation.

POLYNOMIALS ARE ALGEBRAIC EXPRESSIONS THAT CAN BE USED TO MODEL MANY REAL-WORLD SITUATIONS. THEY OFTEN CONTAIN TERMS IN WHICH THE VARIABLES HAVE EXPONENTS.

5.1 Exponents

In this section, you will learn about

- **Exponents**
- **Rules for exponents**
- **Zero exponents**
- **Negative exponents**
- **More rules for exponents**

INTRODUCTION. In Chapter 1, we evaluated exponential expressions having natural-number exponents. In this section, we will extend the definition of exponent to include negative-integer exponents (as in 3^{-2}) and zero exponents (as in 3^0). We will also apply the definition of exponent to develop several rules that can be used to simplify exponential expressions.

Exponents

Exponents provide a way to write products of *repeated factors* in a concise form. For example,

$y \cdot y = y^2$ Read y^2 as "y to the second power" or "y squared."

$z \cdot z \cdot z = z^3$ Read z^3 as "z to the third power" or "z cubed."

$x \cdot x \cdot x \cdot x = x^4$ Read x^4 as "x to the fourth power."

These examples suggest the following definition.

Natural-number exponents

If n is a natural number, then

$$x^n = \overbrace{x \cdot x \cdot x \cdot \cdots \cdot x}^{n \text{ factors of } x}$$

The exponential expression x^n is called a **power of x,** and we read it as "x to the nth power." In this expression, x is called the **base,** and n is called the **exponent.**

Base $\longrightarrow x^n \longleftarrow$ Exponent

A natural-number exponent tells how many times the base of an **exponential expression** is to be used as a factor in a product. When $n = 1$, the exponent is usually omitted. For example, $x^1 = x$.

EXAMPLE 1 ***Identifying the base and the exponent.*** Identify the base and the exponent in each expression: **a.** $(-a)^2$, **b.** $-a^2$, **c.** $5x^3$, and **d.** $(5x)^3$.

Solution

a. For $(-a)^2$, $-a$ is the base and the exponent is 2: $(-a)^2 = (-a)(-a)$.

b. For $-a^2$, the base is a and the exponent is 2: $-a^2 = -(a \cdot a)$.

c. For $5x^3$, x is the base and the exponent is 3: $5x^3 = 5 \cdot x \cdot x \cdot x$.

d. For $(5x)^3$, the base is $5x$ and the exponent is 3: $(5x)^3 = (5x)(5x)(5x)$.

Self Check

Identify the base and the exponent in each expression: **a.** $(-k^2t)^4$, **b.** πr^2, and **c.** $-h^8$.

Answers: **a.** $-k^2t$, 4, **b.** r, 2, **c.** h, 8

Rules for exponents

Several important rules for exponents come directly from the definition of exponent. The first rule we will discuss, called the **product rule for exponents,** gives a quick way to find the result when multiplying exponential expressions that have the same base.

Since x^5 means that x is to be used as a factor five times, and since x^3 means that x is to be used as a factor three times, $x^5 \cdot x^3$ means that x will be used as a factor eight times.

$$x^5x^3 = \overbrace{x \cdot x \cdot x \cdot x \cdot x}^{5 \text{ factors of } x} \cdot \overbrace{x \cdot x \cdot x}^{3 \text{ factors of } x} = \overbrace{x \cdot x \cdot x \cdot x \cdot x \cdot x \cdot x \cdot x}^{8 \text{ factors of } x} = x^8$$

In general,

$$x^mx^n = \overbrace{x \cdot x \cdot x \cdot \cdots \cdot x}^{m \text{ factors of } x} \cdot \overbrace{x \cdot x \cdot \cdots \cdot x}^{n \text{ factors of } x} = \overbrace{x \cdot x \cdot x \cdot x \cdot \cdots \cdot x}^{m+n \text{ factors of } x} = x^{m+n}$$

Thus, *to multiply exponential expressions with the same base, keep the base and add the exponents.*

The product rule for exponents

If m and n are natural numbers, then

$$x^mx^n = x^{m+n}$$

EXAMPLE 2 ***Applying the product rule for exponents.*** Simplify each expression.

a. $x^{11}x^5 = x^{11+5}$
$= x^{16}$

b. $y^5y^4y^3 = (y^5y^4)y^3$
$= y^9y^3$
$= y^{12}$

c. $a^2b^3a^3b^2 = a^2a^3b^3b^2$
$= a^5b^5$

d. $-8x^4\left(\frac{1}{4}x^3\right) = \left(-8 \cdot \frac{1}{4}\right)(x^4x^3)$
$= -2x^7$

Self Check

Simplify each expression:
a. 2^32^5, **b.** $k \cdot k^4$, **c.** $a^2b^3a^3b^4$, and **d.** $-8a^4\left(-\frac{1}{2}a^2b\right)$.

Answers: **a.** $2^8 = 256$, **b.** k^5, **c.** a^5b^7, **d.** $4a^6b$

WARNING! The product rule for exponents applies only to exponential expressions with the same base. The expression x^5y^3, for example, cannot be simplified, because the bases of the exponential expressions are different.

To find another property of exponents, we simplify $(x^4)^3$, which means x^4 cubed or $x^4 \cdot x^4 \cdot x^4$.

$$(x^4)^3 = x^4 \cdot x^4 \cdot x^4 = \overbrace{x \cdot x \cdot x \cdot x}^{x^4} \cdot \overbrace{x \cdot x \cdot x \cdot x}^{x^4} \cdot \overbrace{x \cdot x \cdot x \cdot x}^{x^4} = x^{12}$$

In general, we have

$$(x^m)^n = \overbrace{x^m \cdot x^m \cdot x^m \cdot \cdots \cdot x^m}^{n \text{ factors of } x^m} = \overbrace{x \cdot x \cdot x \cdot x \cdot x \cdot \cdots \cdot x}^{mn \text{ factors of } x} = x^{mn}$$

Thus, *to raise an exponential expression to a power, keep the base and multiply the exponents.*

To find a third property of exponents, we square $3x$ to get

$$(3x)^2 = (3x)(3x) = 3 \cdot 3 \cdot x \cdot x = 3^2x^2 = 9x^2$$

In general, we have

$$(xy)^n = \overbrace{(xy)(xy)(xy) \cdot \cdots \cdot (xy)}^{n \text{ factors of } xy} = \overbrace{xxx \cdot \cdots \cdot x}^{n \text{ factors of } x} \cdot \overbrace{yyy \cdot \cdots \cdot y}^{n \text{ factors of } y} = x^ny^n$$

To find a fourth property of exponents, we cube $\frac{x}{3}$ to get

$$\left(\frac{x}{3}\right)^3 = \frac{x}{3} \cdot \frac{x}{3} \cdot \frac{x}{3} = \frac{x \cdot x \cdot x}{3 \cdot 3 \cdot 3} = \frac{x^3}{3^3} = \frac{x^3}{27}$$

In general, we have

$$\left(\frac{x}{y}\right)^n = \overbrace{\left(\frac{x}{y}\right)\left(\frac{x}{y}\right)\left(\frac{x}{y}\right) \cdot \cdots \cdot \left(\frac{x}{y}\right)}^{n \text{ factors of } \frac{x}{y}} \quad (y \neq 0)$$

$$= \frac{\overbrace{xxx \cdot \cdots \cdot x}^{n \text{ factors of } x}}{\underbrace{yyy \cdot \cdots \cdot y}_{n \text{ factors of } y}}$$ Multiply the numerators and multiply the denominators.

$$= \frac{x^n}{y^n}$$

The previous results are called the **power rules for exponents.**

The power rules for exponents

If m and n are natural numbers, then

$$(x^m)^n = x^{mn} \qquad (xy)^n = x^ny^n \qquad \left(\frac{x}{y}\right)^n = \frac{x^n}{y^n} \quad (y \neq 0)$$

EXAMPLE 3 ***Applying the power rules for exponents.*** Simplify each expression.

a. $(3^2)^3 = 3^{2\cdot 3}$
$= 3^6$
$= 729$

b. $(x^{11})^5 = x^{11\cdot 5}$
$= x^{55}$

c. $(x^2x^3)^6 = (x^5)^6$
$= x^{30}$

d. $(2x^2)^4(x^3)^2 = 2^4x^8x^6$
$= 16x^{14}$

Self Check

Simplify each expression: **a.** $(a^5)^8$, **b.** $(6^3)^5$, **c.** $(a^4a^3)^3$, and **d.** $(3a^3)^3(a^2)^3$.

Answers: **a.** a^{40}, **b.** 6^{15}, **c.** a^{21}, **d.** $27a^{15}$

EXAMPLE 4 ***Applying the power rules for exponents.*** Simplify each expression. Assume that no denominators are zero.

a. $(x^2y)^3 = (x^2)^3y^3$
$= x^6y^3$

b. $(x^3y^4)^4 = (x^3)^4(y^4)^4$
$= x^{12}y^{16}$

c. $\left(\frac{x}{y^2}\right)^4 = \frac{x^4}{(y^2)^4}$
$= \frac{x^4}{y^8}$

d. $\left(\frac{x^3}{y^4}\right)^2 = \frac{(x^3)^2}{(y^4)^2}$
$= \frac{x^6}{y^8}$

Self Check

Simplify each expression: **a.** $(a^4b^5)^2$ and **b.** $\left(\frac{a^5}{b^7}\right)^3$.

Answers: **a.** a^8b^{10}, **b.** $\frac{a^{15}}{b^{21}}$

Zero exponents

Since the rules for exponents hold for exponents of 0, we have

$$x^0x^n = x^{0+n} = x^n = 1x^n$$

Because $x^0x^n = 1x^n$, it follows that $x^0 = 1$ $(x \neq 0)$.

Zero exponents If $x \neq 0$, then $x^0 = 1$.

WARNING! 0^0 is undefined.

Because of the previous definition, any nonzero base raised to the 0th power is 1. For example, if no variables are zero, then

$$3^0 = 1, \quad (-7)^0 = 1, \quad (3ax^3)^0 = 1, \quad \left(\frac{1}{2}x^5y^7z^9\right)^0 = 1$$

Negative exponents

Since the rules for exponents are true for negative-integer exponents, we have

$$x^{-n}x^n = x^{-n+n} = x^0 = 1 \quad (x \neq 0)$$

Because $x^{-n} \cdot x^n = 1$ and $\frac{1}{x^n} \cdot x^n = 1$, we define x^{-n} to be the reciprocal of x^n.

Negative exponents If n is an integer and $x \neq 0$, then

$$x^{-n} = \frac{1}{x^n} \quad \text{and} \quad \frac{1}{x^{-n}} = x^n$$

WARNING! By the definition of negative exponents, a base cannot be 0. Thus, an expression such as 0^{-5} is undefined.

Because of the definition, we can write expressions containing negative exponents as expressions without negative exponents. For example,

$$3^{-2} = \frac{1}{3^2} = \frac{1}{9} \qquad 10^{-3} = \frac{1}{10^3} = \frac{1}{1{,}000} \qquad \frac{1}{4^{-4}} = 4^4 = 256$$

and if b, c, and x are not 0, we have

$$(2c)^{-3} = \frac{1}{(2c)^3} = \frac{1}{8c^3} \qquad 3x^{-1} = 3 \cdot \frac{1}{x} = \frac{3}{x} \qquad \frac{7}{b^{-2}} = 7 \cdot \frac{1}{b^{-2}} = 7b^2$$

EXAMPLE 5 ***Negative exponents.*** Write each expression without negative exponents.

a. $-2m^{-8} = -2 \cdot \frac{1}{m^8} = -\frac{2}{m^8}$

b. $\frac{6a}{y^{-4}} = 6a \cdot \frac{1}{y^{-4}} = 6a \cdot y^4 = 6ay^4$

Self Check

Write each expression without negative exponents:

a. $-3.14t^{-7}$ and **b.** $\frac{32j}{y^{-9}}$.

Answers: **a.** $-\frac{3.14}{t^7}$, **b.** $32jy^9$ ■

EXAMPLE 6 ***Simplifying expressions involving negative exponents.*** Simplify each exponential expression.

a. $x^{-5}x^3 = x^{-5+3}$ Keep the common base and add the exponents.

$= x^{-2}$

$= \frac{1}{x^2}$

b. $(x^{-3})^{-2} = x^{(-3)(-2)}$ Keep the base and multiply the exponents.

$= x^6$

Self Check

Simplify each expression:

a. $a^{-7}a^3$ and **b.** $(a^{-5})^{-3}$.

Answers: **a.** $\frac{1}{a^4}$, **b.** a^{15} ■

More rules for exponents

To develop a rule for dividing exponential expressions, we proceed as follows:

$$\frac{x^m}{x^n} = x^m\left(\frac{1}{x^n}\right) = x^m x^{-n} = x^{m+(-n)} = x^{m-n}$$

Thus, *to divide exponential expressions with the same nonzero base, keep the common base and subtract the exponent in the denominator from the exponent in the numerator.*

The quotient rule for exponents

If m and n are integers, then

$$\frac{x^m}{x^n} = x^{m-n} \quad (x \neq 0)$$

EXAMPLE 7 ***Applying the quotient rule for exponents.*** Simplify each expression. Write each answer without using negative exponents.

a. $\frac{a^5}{a^3} = a^{5-3}$

$= a^2$

b. $\frac{x^{-5}}{x^{11}} = x^{-5-11}$

$= x^{-16}$

$= \frac{1}{x^{16}}$

Self Check

Simplify each expression:

a. $\frac{b^7}{b^5}$ and **b.** $\frac{b^{-3}}{b^3}$.

Answers: **a.** b^2, **b.** $\frac{1}{b^6}$ ■

EXAMPLE 8 *Applying the quotient rule for exponents.* Simplify each expression. Write each answer without using negative exponents.

a. $\frac{x^4x^3}{x^{-5}} = \frac{x^7}{x^{-5}}$
$= x^{7-(-5)}$
$= x^{12}$

b. $\frac{(x^2)^3}{(x^3)^2} = \frac{x^6}{x^6}$
$= x^{6-6}$
$= x^0$
$= 1$

c. $\frac{x^2y^3}{xy^4} = x^{2-1}y^{3-4}$
$= xy^{-1}$
$= x \cdot \frac{1}{y}$
$= \frac{x}{y}$

d. $\left(\frac{a^{-2}b^3}{a^5b^4}\right)^3 = (a^{-2-5}b^{3-4})^3$
$= (a^{-7}b^{-1})^3$
$= \left(\frac{1}{a^7b}\right)^3$
$= \frac{1}{a^{21}b^3}$

Self Check

Simplify each expression:

a. $\frac{(a^{-2})^3}{(a^2)^{-3}}$ and b. $\left(\frac{a^{-2}b^5}{b^8}\right)^{-3}$.

Answers: a. 1, b. a^6b^9 ■

To illustrate another property of exponents, we consider the following simplification of $\left(\frac{2}{3}\right)^{-4}$.

$$\left(\frac{2}{3}\right)^{-4} = \frac{1}{\left(\frac{2}{3}\right)^4} = \frac{1}{\frac{2^4}{3^4}} = 1 \div \frac{2^4}{3^4} = 1 \cdot \frac{3^4}{2^4} = \frac{3^4}{2^4} = \left(\frac{3}{2}\right)^4$$

The example suggests that to raise a fraction to a negative power, we can invert the fraction and then raise it to a positive power.

Fractions to negative powers

If n is an integer, then

$$\left(\frac{x}{y}\right)^{-n} = \left(\frac{y}{x}\right)^n \quad (x \neq 0, y \neq 0)$$

EXAMPLE 9 *Fractions to negative powers.* Write each expression without using parentheses and negative exponents.

a. $\left(\frac{2}{3}\right)^{-4} = \left(\frac{3}{2}\right)^4$
$= \frac{3^4}{2^4}$
$= \frac{81}{16}$

b. $\left(\frac{y^2}{x^3}\right)^{-3} = \left(\frac{x^3}{y^2}\right)^3$
$= \frac{x^9}{y^6}$

c. $\left(\frac{2x^2}{3y^{-3}}\right)^{-4} = \left(\frac{3y^{-3}}{2x^2}\right)^4$
$= \frac{3^4y^{-12}}{2^4x^8}$
$= \frac{81}{16x^8} \cdot y^{-12}$
$= \frac{81}{16x^8} \cdot \frac{1}{y^{12}}$
$= \frac{81}{16x^8y^{12}}$

d. $\left(\frac{a^{-2}b^3}{a^2a^3b^4}\right)^{-3} = \left(\frac{a^2a^3b^4}{a^{-2}b^3}\right)^3$
$= \left(\frac{a^5b^4}{a^{-2}b^3}\right)^3$
$= (a^{5-(-2)}b^{4-3})^3$
$= (a^7b)^3$
$= a^{21}b^3$

Self Check

Write $\left(\frac{3a^3}{2b^{-2}}\right)^{-5}$ without using parentheses.

Answer: $\frac{32}{243a^{15}b^{10}}$ ■

We summarize the rules for exponents as follows.

Rules for exponents	If there are no divisions by 0, then for all integers m and n,
	$x^m x^n = x^{m+n}$ $(x^m)^n = x^{mn}$ $(xy)^n = x^n y^n$ $\left(\frac{x}{y}\right)^n = \frac{x^n}{y^n}$
	$x^0 = 1 \quad (x \neq 0)$ $x^{-n} = \frac{1}{x^n}$ $\frac{x^m}{x^n} = x^{m-n}$ $\left(\frac{x}{y}\right)^{-n} = \left(\frac{y}{x}\right)^n$

STUDY SET Section 5.1

VOCABULARY *In Exercises 1–4, fill in the blanks to make the statements true.*

1. In the exponential expression x^n, x is called the ____________, and n is called the ____________.

2. The expression $x \cdot x \cdot x \cdot x \cdot x$ contains five ____________ of x.

3. $\{1, 2, 3, 4, 5, 6, \ldots\}$ is the set of ____________ numbers.

4. $\{\ldots, -3, -2, -1, 0, 1, 2, 3, \ldots\}$ is the set of ____________.

CONCEPTS *In Exercises 5–12, complete the rules for exponents. Assume that $x \neq 0$ and $y \neq 0$.*

5. $x^m x^n = \square$

6. $(x^m)^n = \square$

7. $(xy)^n = \square$

8. $\left(\frac{x}{y}\right)^n = \square$

9. $x^0 = \square$

10. $x^{-n} = \square$

11. $\frac{x^m}{x^n} = \square$

12. $\left(\frac{x}{y}\right)^{-n} = \square$

13. An expression with a negative exponent can be written as an equivalent expression with a positive exponent. Explain.

14. Explain the difference between the two expressions.
a. $2x$ and x^2 **b.** $-2x$ and x^{-2}

15. A cube is shown in Illustration 1.
a. Find the area of its base.
b. Find its volume.

16. A rectangular prism is shown in Illustration 2.
a. Find the area of its base.
b. Find its volume.

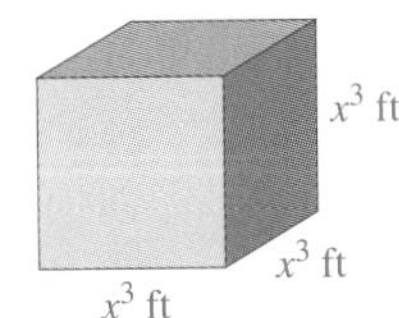

ILLUSTRATION 1

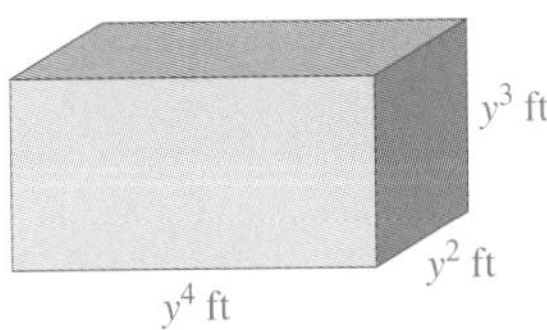

ILLUSTRATION 2

NOTATION *In Exercises 17–18, complete each simplification.*

17. $\frac{x^5x^4}{x^{-2}} = \frac{x^{\square}}{x^{-2}}$ Keep the base and add the exponents.

$= x^{9-\square}$ Keep the base and subtract the exponents.

$= x^{11}$

18. $\left(\frac{a^{-4}}{a^3}\right)^2 = (a^{-4-3})^2$ Keep the base and subtract the exponents.

$= (a^{\square})^2$ Do the subtraction.

$= a^{\square}$ Keep the base and multiply the exponents.

$= \frac{1}{a^{14}}$

PRACTICE *In Exercises 19–26, identify the base and the exponent.*

19. 5^3
20. -7^2
21. $-x^5$
22. $(-t)^4$
23. $2b^6$
24. $(3xy)^5$
25. $\left(\frac{n}{4}\right)^3$
26. $(-pq)^2$

In Exercises 27–90, simplify each expression. Assume that no denominators are zero. Write each answer without using negative exponents.

27. 3^2
28. 3^4
29. -3^2
30. -3^4
31. $(-3)^2$
32. $(-3)^3$
33. 5^{-2}
34. 5^{-4}
35. -5^{-2}
36. -5^{-4}
37. $(-5)^{-2}$
38. $(-5)^{-4}$
39. 8^0
40. -9^0
41. $(-8)^0$
42. $(-9)^0$
43. $(-2x)^5$
44. $(-3a)^3$
45. x^2x^3
46. y^3y^4
47. $x^2x^3x^5$
48. $y^3y^7y^2$
49. k^0k^7
50. x^8x^{11}
51. aba^3b^4
52. $x^2y^3x^3y^2$
53. p^9pp^0
54. z^7z^0z
55. $(-x)^2y^4x^3$
56. $-x^2y^7y^3x^{-2}$
57. $(b^{-8})^9$
58. $(z^{12})^2$
59. $(x^4)^7$
60. $(y^7)^5$
61. $(r^{-3}s)^3$
62. $(m^5n^2)^{-3}$
63. $(a^2a^3)^4$
64. $(bb^2b^3)^4$
65. $(-d^2)^3(d^{-3})^3$
66. $(c^3)^2(c^4)^{-2}$
67. $(3x^3y^4)^3$
68. $\left(\frac{1}{2}a^2b^5\right)^4$
69. $\left(-\frac{1}{3}mn^2\right)^6$
70. $(-3p^2q^3)^5$
71. $\left(\frac{a^3}{b^2}\right)^5$
72. $\left(\frac{a^2}{b^3}\right)^4$
73. $\left(\frac{a^{-3}}{b^{-2}}\right)^{-2}$
74. $\left(\frac{k^{-3}}{k^{-4}}\right)^{-1}$
75. $\frac{a^8}{a^3}$
76. $\frac{c^7}{c^2}$
77. $\frac{c^{12}c^5}{c^{10}}$
78. $\frac{a^{33}}{a^2a^3}$
79. $\left(\frac{2}{3}\right)^{-2}$
80. $\left(\frac{4}{5}\right)^{-3}$
81. $\frac{1}{a^{-4}}$
82. $\frac{3}{b^{-5}}$
83. $\frac{(3x^2)^{-2}}{x^3x^{-4}x^0}$
84. $\frac{y^{-3}y^{-4}y^0}{(2y^{-2})^3}$
85. $\left(\frac{4a^{-2}b}{3ab^{-3}}\right)^3$
86. $\left(\frac{2ab^{-3}}{3a^{-2}b^2}\right)^2$
87. $\left(\frac{3a^{-2}b^2}{17a^2b^3}\right)^0$
88. $\frac{a^0+b^0}{2(a+b)^0}$
89. $\left(\frac{-2a^4b}{a^{-3}b^2}\right)^{-3}$
90. $\left(\frac{-3x^4y^2}{-9x^5y^{-2}}\right)^{-2}$

In Exercises 91–94, use a calculator to find each value.

91. 1.23^6
92. 0.0537^4
93. -6.25^3
94. $(-25.1)^5$

In Exercises 95–100, use a calculator to verify that each statement is true by showing that the values on either side of the equation are equal.

95. $(3.68)^0 = 1$
96. $(2.1)^4(2.1)^3 = (2.1)^7$
97. $(7.2)^2(2.7)^2 = [(7.2)(2.7)]^2$
98. $\left(\frac{5.4}{2.7}\right)^{-4} = \left(\frac{2.7}{5.4}\right)^4$
99. $(3.2)^2(3.2)^{-2} = 1$
100. $(7.23)^{-3} = \frac{1}{(7.23)^3}$

APPLICATIONS

101. MICROSCOPES Illustration 3 shows the relative sizes of some chemical and biological structures, expressed as fractions of a meter (m). Express each fraction shown in the illustration as a power of 10, from the largest to the smallest.

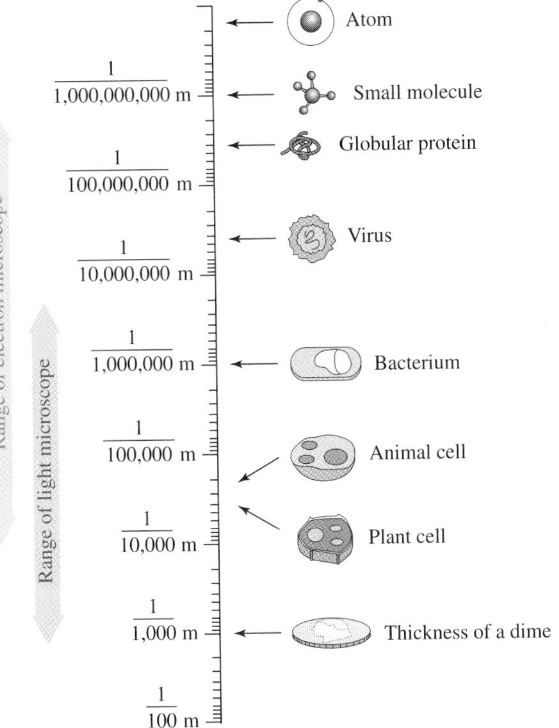

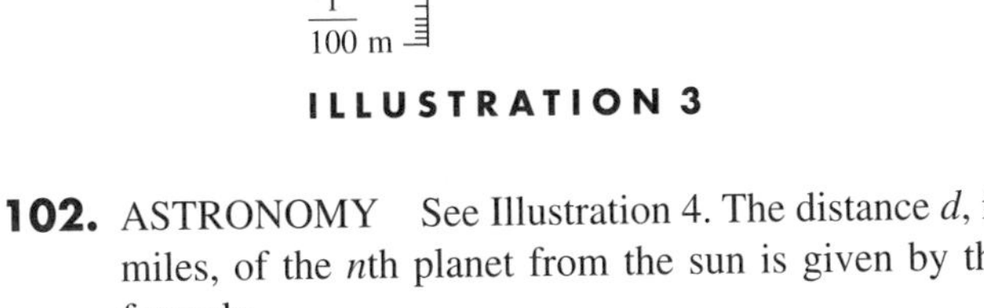

ILLUSTRATION 3

102. ASTRONOMY See Illustration 4. The distance d, in miles, of the nth planet from the sun is given by the formula

$$d = 9{,}275{,}200[3(2^{n-2}) + 4]$$

Find the distance of earth and Mars from the sun.

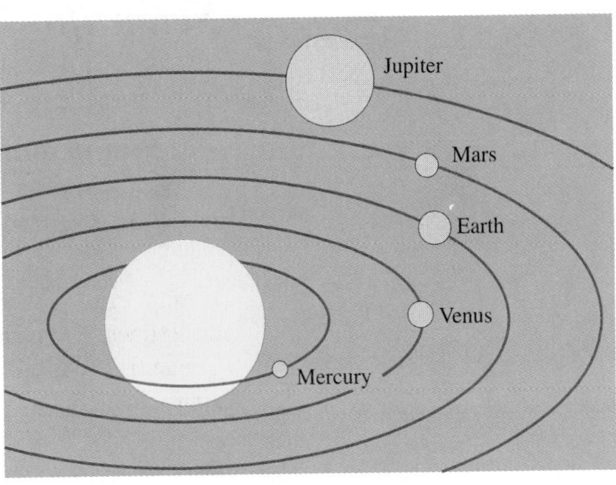

ILLUSTRATION 4

103. LICENSE PLATES The number of different license plates of the form three digits followed by three letters, as in Illustration 5, is $10 \cdot 10 \cdot 10 \cdot 26 \cdot 26 \cdot 26$. Write this expression using exponents. Then evaluate it.

ILLUSTRATION 5

104. PHYSICS Albert Einstein's work in the area of special relativity resulted in the observation that the total energy E of a body is equal to its total mass m times the square of the speed of light c. This relationship is given by the famous equation $E = mc^2$. Identify the base and exponent on the right-hand side.

WRITING *Write a paragraph using your own words.*

105. Explain how an exponential expression with a negative exponent can be expressed as an equivalent expression with a positive exponent. Give an example.

106. In the definition of x^{-n}, x cannot be 0. Why not?

REVIEW *In Exercises 107–110, solve each inequality. Give the result in interval notation and then graph the solution set.*

107. $a + 5 < 6$

108. $-9x + 5 \geq 15$

109. $6(t - 2) \leq 4(t + 7)$

110. $\frac{1}{4}p - \frac{1}{3} \leq p + 2$

5.2 Scientific Notation

In this section, you will learn about

- **Writing numbers in scientific notation**
- **Converting from scientific notation**
- **Using scientific notation to simplify computations**

Introduction. Very large and very small numbers occur often in science and other disciplines. For example, the star nearest to the earth (excluding the sun) is Proxima Centauri, about 24,793,000,000,000 miles away, and the mass of a hydrogen atom is approximately 0.000000000000000000000001673 gram.

Because these numbers contain many zeros, they are difficult to read and cumbersome to work with in computations. In this section, we will discuss a notation that allows us to express such numbers in a more manageable form, using exponents.

Writing numbers in scientific notation

Scientific notation provides a compact way of writing large and small numbers.

Scientific notation

A number is written in **scientific notation** when it is written in the form $N \times 10^n$, where $1 \leq |N| < 10$ and n is an integer.

Each of the following numbers is written in scientific notation.

$$3.67 \times 10^6 \qquad 2.24 \times 10^{-4} \qquad 9.875 \times 10^{22}$$

Every positive number written in scientific notation is the product of a number between 1 (including 1) and 10 and an integer power of 10.

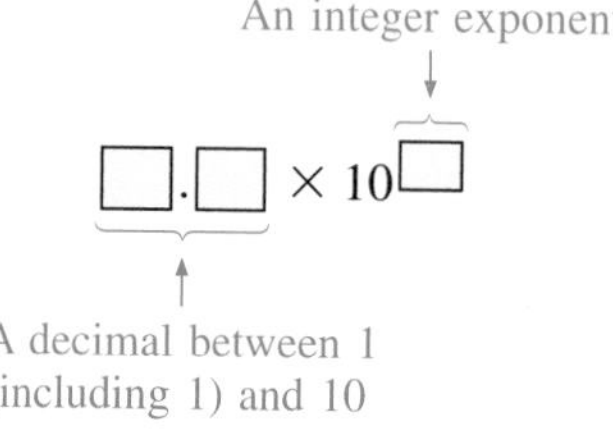

EXAMPLE 1 ***Writing numbers in scientific notation.*** Write each number in scientific notation: **a.** 24,793,000,000,000 and **b.** 0.000000000000000000000001673.

Solution

a. The number 2.4793 is between 1 and 10. To get 24,793,000,000,000, the decimal point in 2.4793 must be moved thirteen places to the *right.* This can be done by multiplying 2.4793 by 10^{13}.

$$24{,}793{,}000{,}000{,}000 = 2.4793 \times 10^{13}$$

b. The number 1.673 is between 1 and 10. To get 0.000000000000000000000001673, the decimal point in 1.673 must be moved twenty-four places to the *left.* This can be done by multiplying 1.673 by 10^{-24}.

$$0.000000000000000000000001673 = 1.673 \times 10^{-24}$$

Self Check

Write each italicized number in scientific notation.

a. In 1996, the country earning the most money in tourism was the United States, *$64,400,000,000.*

b. DNA molecules contain and transmit the information that allows cells to reproduce. They are only *0.000000002* meter wide.

Answers: **a.** 6.44×10^{10}, **b.** 2×10^{-9}

Numbers such as 47.2×10^3 and 0.063×10^{-2} appear to be written in scientific notation, because they are the product of a number and a power of ten. However, they are not. Their first factors (47.2 and 0.063) are not between 1 and 10.

EXAMPLE 2 ***Writing numbers in scientific notation.*** Write **a.** 47.2×10^3 and **b.** 0.063×10^{-2} in scientific notation.

Solution

Since the first factors are not between 1 and 10, neither number is in scientific notation. However, we can change them to scientific notation as follows:

a. $47.2 \times 10^3 = (4.72 \times 10^1) \times 10^3$ — Write 47.2 in scientific notation.

$= 4.72 \times (10^1 \times 10^3)$ — Group the powers of 10 together.

$= 4.72 \times 10^4$ — Apply the product rule for exponents: $10^1 \times 10^3 = 10^{1+3} = 10^4$.

b. $0.063 \times 10^{-2} = (6.3 \times 10^{-2}) \times 10^{-2}$ — Write 0.063 in scientific notation.

$= 6.3 \times (10^{-2} \times 10^{-2})$

$= 6.3 \times 10^{-4}$

Self Check

Write **a.** 27.3×10^2 and **b.** 0.0025×10^{-3} in scientific notation.

Answers: **a.** 2.73×10^3, **b.** 2.5×10^{-6}

Converting from scientific notation

We can change a number written in scientific notation to **standard notation.** For example, to write 9.3×10^7 in standard notation, we multiply 9.3 by 10^7.

$$9.3 \times 10^7 = 9.3 \times 10{,}000{,}000 = 93{,}000{,}000$$ 10^7 is 1 followed by 7 zeros.

EXAMPLE 3 ***Converting to standard notation.*** Change **a.** 8.706×10^5 and **b.** 1.1×10^{-3} to standard notation.

Solution

a. Since multiplication by 10^5 (100,000) moves the decimal point 5 places to the right,

$$8.706 \times 10^5 = 8.70600. = 870{,}600$$

b. Since multiplication by 10^{-3} (0.001) moves the decimal point 3 places to the left,

$$1.1 \times 10^{-3} = 0.001.1 = 0.0011$$

Self Check

Change each number in scientific notation to standard notation.

a. The country with the largest area of forest is Russia, with 1.9×10^9 acres.

b. The average distance between molecules of air in a room is 3.937×10^{-7} inch.

Answers: **a.** 1,900,000,000, **b.** 0.0000003937

Each of the following numbers is written in both scientific and standard notation. In each case, the exponent gives the number of places that the decimal point moves, and the sign of the exponent indicates the direction that it moves:

$5.32 \times 10^4 = 5.3200.$ — 4 places to the right

$6.45 \times 10^7 = 6.4500000.$ — 7 places to the right

$2.37 \times 10^{-4} = 0.0002.37$ — 4 places to the left

$9.234 \times 10^{-2} = 0.09.234$ — 2 places to the left

$4.89 \times 10^0 = 4.89$ — No movement of the decimal point

Using scientific notation to simplify computations

Scientific notation is useful when multiplying and dividing very large or very small numbers.

EXAMPLE 4 ***Astronomy.*** The wheel-shaped galaxy in which we live is called the Milky Way. See Figure 5-1. This system of some 10^{11} stars, one of which is the sun, has a diameter of approximately 100,000 light years. (A light year is the distance light travels in a vacuum in one year: 9.46×10^{15} meters.) What is the diameter of the Milky Way in meters?

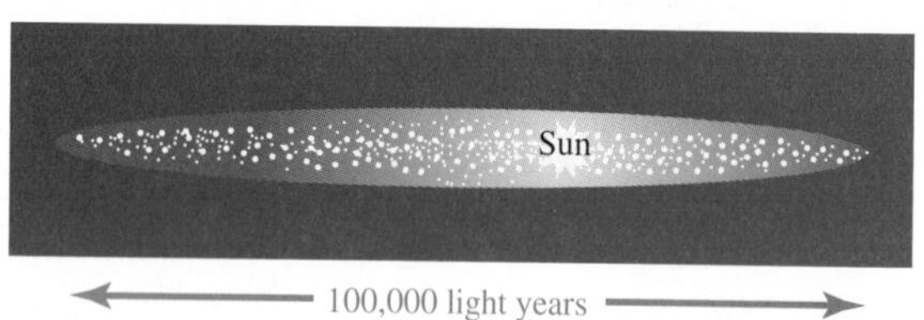

A cross-sectional representation of the Milky Way Galaxy

FIGURE 5-1

Solution

We will multiply the diameter of the Milky Way, expressed in light years, by the number of meters in a light year to find the diameter of the Milky Way in meters. To perform the calculation, we write 100,000 in scientific notation as 1.0×10^5.

$1.0 \times 10^5 \cdot 9.46 \times 10^{15}$

$= (1.0 \cdot 9.46) \times (10^5 \cdot 10^{15})$ Apply the commutative and associative properties of multiplication to group the first factors together and the powers of 10 together.

$= 9.46 \times 10^{5+15}$ Do the multiplication: $1.0 \cdot 9.46 = 9.46$. For the powers of 10, keep the base and add the exponents.

$= 9.46 \times 10^{20}$ Do the addition.

The Milky Way Galaxy is about 9.46×10^{20} meters in diameter.

Self Check

A light year is 5.88×10^{12} miles. What is the diameter of the Milky Way in miles?

Answer: 5.88×10^{17} mi ■

EXAMPLE 5 In an article in *Scientific American,* oil industry experts estimated that there were 1.02×10^{12} barrels of oil reserves in the ground at the start of 1998. At that time, world production was 2.36×10^{10} barrels per year. If annual production remains the same, and if no new oil discoveries are made, when do these experts predict the world's oil supply will run out?

Solution

If we divide the estimated number of barrels of oil in reserve, 1.02×10^{12}, by the number of barrels produced each year, 2.36×10^{10}, we can find the number of years of oil supply left.

$\frac{1.02 \times 10^{12}}{2.36 \times 10^{10}} = \frac{1.02}{2.36} \times \frac{10^{12}}{10^{10}}$ Divide the first factors and the second factors in the numerator and denominator separately.

$\approx 0.43 \times 10^{12-10}$ Do the division: $\frac{1.02}{2.36} \approx 0.43$. For the powers of 10, keep the base and subtract the exponents.

$\approx 0.43 \times 10^2$ Do the subtraction.

≈ 43 Write 0.43×10^2 in standard notation.

According to industry estimates, as of 1998, there were 43 years of oil reserves left. Under these conditions, the world's oil supply will run out in the year 2041. ■

EXAMPLE 6 ***Computing using scientific notation.*** Use scientific notation to evaluate

$$\frac{(0.00000064)(24{,}000{,}000{,}000)}{(400{,}000{,}000)(0.0000000012)}$$

Solution

After writing each number in scientific notation, we can do the arithmetic on the numbers and the exponential expressions separately.

$$\begin{aligned}\frac{(0.00000064)(24{,}000{,}000{,}000)}{(400{,}000{,}000)(0.0000000012)} &= \frac{(6.4 \times 10^{-7})(2.4 \times 10^{10})}{(4 \times 10^{8})(1.2 \times 10^{-9})} \\ &= \frac{(6.4)(2.4)}{(4)(1.2)} \times \frac{10^{-7}10^{10}}{10^{8}10^{-9}} \\ &= \frac{15.36}{4.8} \times 10^{-7+10-8-(-9)} \\ &= 3.2 \times 10^{4}\end{aligned}$$

The result is 3.2×10^4. In standard notation, this is 32,000.

Self Check

Use scientific notation to evaluate

$$\frac{(320)(25{,}000)}{0.00004}$$

Answer:

$2 \times 10^{11} = 200{,}000{,}000{,}000$ ■

Accent on Technology Using scientific notation

Scientific calculators and graphing calculators often give answers in scientific notation. For example, if we use a calculator to find 301.2^8, the display will read

6.77391496 19	On a scientific calculator.
301.2 ∧ 8 6.77391496lE19	On a graphing calculator.

In either case, the answer is given in scientific notation and is to be interpreted as

$6.77391496 \times 10^{19}$

Numbers can also be entered into a calculator in scientific notation. For example, to enter 24,000,000,000 (which is 2.4×10^{10} in scientific notation), we enter these numbers and press these keys:

2.4 [EXP] 10 On most scientific calculators.

2.4 [EE] 10 On a graphing calculator and on some scientific calculators.

To use a scientific calculator to evaluate

$$\frac{(24{,}000{,}000{,}000)(0.00000006495)}{0.00000004824}$$

we must enter each number in scientific notation, because each number has too many digits to be entered directly. In scientific notation, the three numbers are

2.4×10^{10} 6.495×10^{-8} 4.824×10^{-8}

Using a scientific calculator, we enter these numbers and press these keys:

2.4 [EXP] 10 [×] 6.495 [EXP] 8 [+/−] [÷] 4.824 [EXP] 8 [+/−] [=]

The display will read [3.231343284 10]. In standard notation, the answer is 32,313,432,840.

The steps are similar on a graphing calculator.

STUDY SET Section 5.2

VOCABULARY *In Exercises 1–2, fill in the blanks to make the statements true.*

1. 7.4×10^{10} is written in ____________ notation.

2. 10^{-3}, 10^{0}, 10^{1}, and 10^{4} are ____________ of 10.

CONCEPTS *In Exercises 3–6, fill in the blanks to make the statements true.*

3. A number is written in scientific notation when it is written in the form $N \times$ ___, where $1 \le |N| < 10$ and n is an integer.

4. The number 5.3×10^{2} ___ (> or <) the number 5.3×10^{-2}.

5. To change 6.31×10^{-4} to standard notation, we move the decimal point four places to the ____________.

6. To change 9.7×10^{3} to standard notation, we move the decimal point three places to the ____________.

NOTATION

7. Explain why the number 60.22×10^{22} is not written in scientific notation.

8. Explain why the number 0.6022×10^{24} is not written in scientific notation.

PRACTICE *In Exercises 9–24, write each number in scientific notation.*

9. 3,900

10. 1,700

11. 0.0078

12. 0.068

13. 173,000,000,000,000

14. 89,800,000,000

15. 0.0000096

16. 0.000000046

17. 323×10^{5}

18. 689×10^{9}

19. $6{,}000 \times 10^{-7}$

20. 765×10^{-5}

21. 0.0527×10^{5}

22. 0.0298×10^{3}

23. 0.0317×10^{-2}

24. 0.0012×10^{-3}

In Exercises 25–36, write each number in standard notation.

25. 2.7×10^{2}

26. 7.2×10^{3}

27. 3.23×10^{-3}

28. 6.48×10^{-2}

29. 7.96×10^{5}

30. 9.67×10^{6}

31. 3.7×10^{-4}

32. 4.12×10^{-5}

33. 5.23×10^{0}

34. 8.67×10^{0}

35. 23.65×10^{6}

36. 75.6×10^{-5}

In Exercises 37–42, give all answers in scientific notation. Use a calculator to check your results.

37. $(7.9 \times 10^{5})(2.3 \times 10^{6})$

38. $(6.1 \times 10^{8})(3.9 \times 10^{5})$

39. $(9.1 \times 10^{-5})(5.5 \times 10^{12})$

40. $(8.4 \times 10^{-13})(4.8 \times 10^{9})$

41. $\dfrac{4.2 \times 10^{-12}}{8.4 \times 10^{-5}}$

42. $\dfrac{1.21 \times 10^{-15}}{1.1 \times 10^{2}}$

In Exercises 43–50, write each numeral in scientific notation and do the operations. Give all answers in scientific notation and in standard form. Use a calculator to check your results.

43. (89,000,000,000)(4,500,000,000)

44. (0.000000061)(3,500,000,000)

45. $\dfrac{0.00000129}{0.0003}$

46. $\dfrac{4{,}400{,}000{,}000{,}000}{0.0002}$

47. $\dfrac{(220{,}000)(0.000009)}{0.00033}$

48. $\dfrac{(640{,}000)(2{,}700{,}000)}{120{,}000}$

49. $\dfrac{(0.00024)(96{,}000{,}000)}{640{,}000{,}000}$

50. $\dfrac{(0.0000013)(0.00009)}{0.00039}$

APPLICATIONS

51. FIVE-CARD POKER The odds against being dealt the hand shown in Illustration 1 are about 2.6×10^6 to 1. Express the odds using standard notation.

ILLUSTRATION 1

52. ENERGY See Illustration 2. Express each of the following using scientific notation. (1 quadrillion is 10^{15}.)
   a. 1996 U.S. energy consumption
   b. 1996 U.S. energy production
   c. The difference in 1996 consumption and production

1996 U.S. Energy Consumption and Production
(petroleum, natural gas, coal, hydroelectric, nuclear, geothermal, solar, wind)

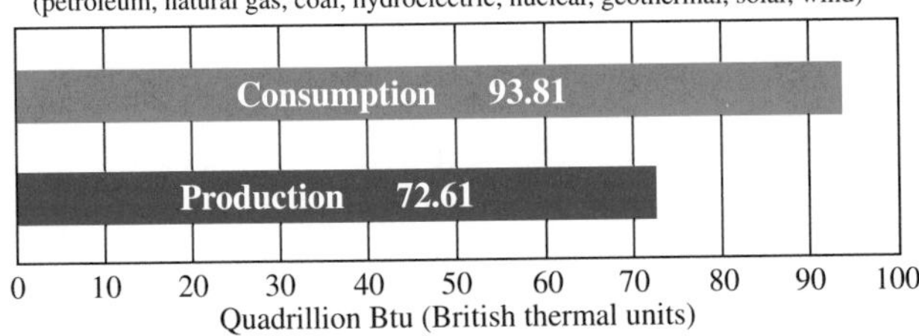

Based on data from the Energy Information Administration, United States Department of Energy

ILLUSTRATION 2

53. THE YEAR 2000 Express each of the dollar amounts appearing in the following excerpt from the *Federal Computer Week* web page (February 16, 1998) in scientific notation.

> President Clinton's fiscal 1999 budget proposal of \$1.7 trillion includes expenditures of about \$3.9 billion to ensure that federal computers can accept dates after Dec. 31, 1999. Clinton has proposed spending \$275 million at the Defense Department and \$312 million at the Treasury Department to fix the year 2000 problem.

54. STAR TREK In the science fiction series *Star Trek*, crew members talk of their spacecraft, the U.S.S. Enterprise, traveling at various "warp speeds." To convert a warp speed, W, to an equivalent velocity in miles per second, v, we can use the equation

$$v = W^3c$$

where c is the speed of light, 1.86×10^5 miles per second. Find the velocity of a spacecraft traveling at warp 2.

55. ATOMS A simple model of a helium atom is shown in Illustration 3. If a proton has a mass of 1.7×10^{-24} grams, and if the mass of an electron is only about $\frac{1}{2{,}000}$ that of a proton, find the mass of an electron.

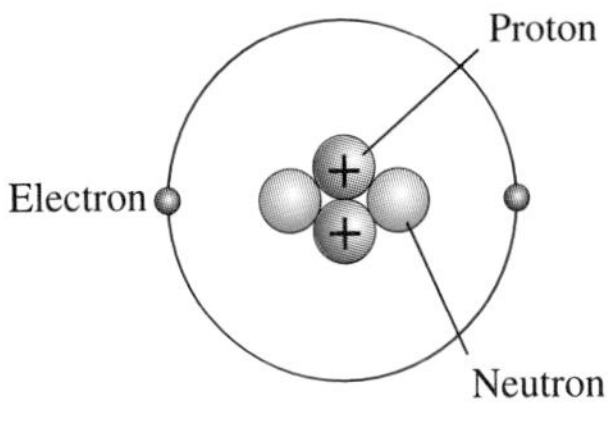

ILLUSTRATION 3

56. OCEANS The mass of the earth's oceans is only about $\frac{1}{4{,}400}$ that of the earth. If the mass of the earth is 6.578×10^{21} tons, find the mass of the oceans.

57. LIGHT YEAR Light travels about 300,000,000 meters per second. A **light year** is the distance that light can travel in one year. Estimate the number of meters in one light year.

58. AQUARIUM Express the volume of the fish tank shown in Illustration 4 in scientific notation.

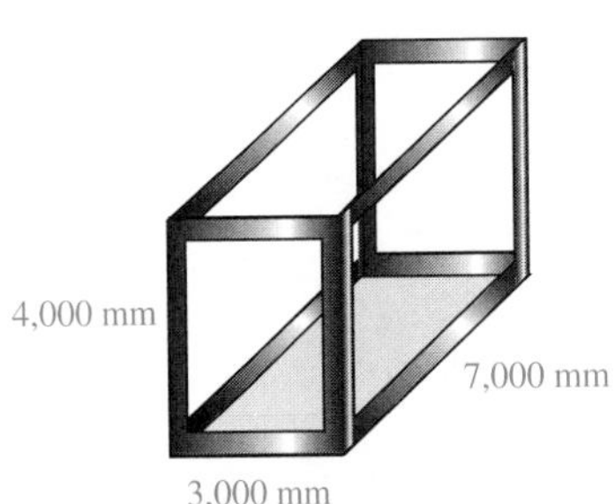

ILLUSTRATION 4

59. THE BIG DIPPER One of the stars in the Big Dipper is named Merak. (See Illustration 5.) It is approximately 4.65×10^{14} miles from the earth.
   a. If light travels about 1.86×10^5 miles/sec, how many seconds does it take light emitted from Merak to reach the earth? (*Hint:* Use the formula $t = \frac{d}{r}$.)
   b. Convert your result from part a to years.

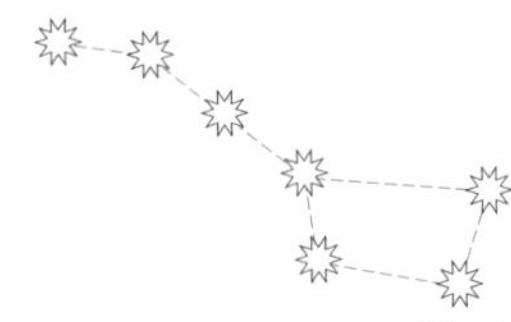

ILLUSTRATION 5

60. BIOLOGY A paramecium is a single-celled organism that propels itself with hair-like projections called *cilia.* Use the scale in Illustration 6 to estimate the length of the paramecium. Express the result in scientific and in standard notation.

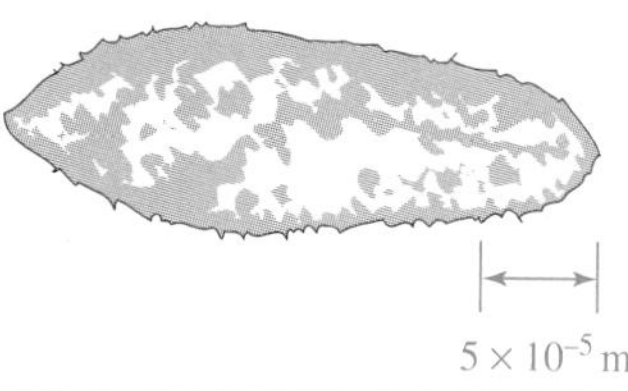

ILLUSTRATION 6

61. COMET HALE-BOPP On March 23, 1997, Comet Hale-Bopp made its closest approach to earth, coming within 1.3 **astronomical units.** One astronomical unit (AU) is the distance from the earth to the sun—about 9.3×10^7 miles. Express this distance in miles, using scientific notation.

62. DIAMONDS The approximate number of atoms of carbon in a $\frac{1}{2}$-carat diamond can be found by computing

$$\frac{6.0 \times 10^{23}}{1.2 \times 10^2}$$

Express the number of carbon atoms in scientific and in standard notation.

WRITING *Write a paragraph using your own words.*

63. Explain how to change a number from standard notation to scientific notation.

64. Explain how to change a number from scientific notation to standard notation.

REVIEW *In Exercises 65–68, solve each compound inequality. Give the result in interval notation and graph the solution set.*

65. $4x \geq -x + 5$ and $6 \geq 4x - 3$

66. $15 > 2x - 7 > 9$

67. $3x + 2 < 8$ or $2x - 3 > 11$

68. $-4(x + 2) \geq 12$ or $3x + 8 < 11$

5.3 *Polynomials and Polynomial Functions*

In this section, you will learn about

- **Polynomials**
- **Degree of a polynomial**
- **Polynomial functions**
- **Evaluating polynomial functions**
- **Graphing polynomial functions**
- **Combining like terms**
- **Adding and subtracting polynomials**

Introduction. In arithmetic, we learned how to add, subtract, multiply, divide, and find powers of numbers. In the next several sections, we will learn how to perform these operations on algebraic expressions called *polynomials.* To begin the discussion of polynomials, we introduce some vocabulary that is used to classify them. Then we will see how *polynomial functions* can be used to model many real-life situations.

Polynomials

Algebraic terms are expressions that contain constants and/or variables. Some examples are

$$17, \qquad 9x, \qquad \frac{15}{16}y^2, \qquad \text{and} \qquad -2.4x^4y^5$$

The **numerical coefficient** of 17 is 17. The numerical coefficients of the remaining terms are the numbers 9, $\frac{15}{16}$, and -2.4, respectively.

Polynomial

A **polynomial** is an algebraic term or the sum of two or more algebraic terms whose variables have whole-number exponents. No variable appears in a denominator.

The following expressions are polynomials in x:

$$-6x, \qquad 3x^2 + 2x, \qquad \frac{3}{2}x^5 - \frac{7}{3}x^4 - \frac{8}{3}x^3, \qquad \text{and} \qquad 19x^{20} + \sqrt{3}x^{14} + 4.5x^{11} - x^2$$

WARNING! The following expressions are not polynomials:

$$\frac{2x}{x^2 + 1}, \qquad x^{1/2} - 8, \qquad \text{and} \qquad x^{-3} + 2x + 24$$

The first expression is a quotient and has a variable in the denominator. The last two have exponents that are not whole numbers.

If any terms of a polynomial contain more than one variable, we say that the polynomial is in more than one variable. Some examples are

$$3xy, \qquad 5x^2y^2 + 2xy - 3y, \qquad \text{and} \qquad u^2v^2w^2 + uv + 1$$

Polynomials can be classified according to the number of terms they have. A polynomial with one term is called a **monomial,** a polynomial with two terms is called a **binomial,** and a polynomial with three terms is called a **trinomial.**

Monomials	*Binomials*	*Trinomials*
$2x^3$	$2x + 5$	$2x^2 + 4x + 3$
a^2b	$-17x^4 - \frac{3}{5}x$	$3mn^3 - m^2n^3 + 7n$
$3x^3y^5z^2$	$32x^{13}y^5z^3 + 47x^3yz$	$-12x^5y^2 + 13x^4y^3 - 7x^3y^3$

Degree of a polynomial

Because the variable x occurs three times as a factor in the monomial $2x^3$, the monomial is called a *third-degree monomial* or a *monomial of degree 3.* The monomial $3x^3y^5z^2$ is called a *monomial of degree 10,* because the variables x, y, and z occur as factors a total of ten times $(3 + 5 + 2)$. These examples illustrate the following definition.

Degree of a monomial

The **degree** of a monomial with one variable is the exponent on the variable. The degree of a monomial in several variables is the sum of the exponents on those variables. If the monomial is a nonzero constant, its degree is 0. The constant 0 has no defined degree.

EXAMPLE 1 ***Determining the degree of a monomial.*** Find the degree of **a.** $3x^4$, **b.** $-4x^2y^3$, and **c.** 3.

Solution

a. $3x^4$ is a monomial of degree 4, because the exponent on the variable is 4.

b. $-4x^2y^3$ is a monomial of degree 5, because the sum of the exponents on the variables is 5.

c. 3 is a monomial of degree 0, because $3 = 3x^0$.

Self Check

Find the degree of **a.** $-12a^2$, **b.** $8a^3b^2$, and **c.** $\frac{1}{2}x^3y^2z^{12}$.

Answers: **a.** 2, **b.** 5, **c.** 17 ■

We determine the degree of a polynomial by considering the degrees of each of its terms.

Degree of a polynomial

The **degree of a polynomial** is the same as the degree of the term in the polynomial with largest degree.

EXAMPLE 2 ***Determining the degree of a polynomial.*** Find the degree of each polynomial: **a.** $3x^5 + 4x^2 + 7$, **b.** $7x^2y^8 - 3x^2y^2$, and **c.** $3x + 2y - xy$.

Solution

a. The terms of $3x^5 + 4x^2 + 7$ have degree 5, 2, and 0, respectively. This trinomial is of degree 5, because the largest degree of the three terms is 5.

b. $7x^2y^8 - 3x^2y^2$ is a binomial of degree 10.

c. $3x + 2y - xy$ is a trinomial of degree 2. (Recall that $xy = x^1y^1$.)

Self Check

Find the degree of
a. $x^2 - x + 1$ and
b. $x^2y^3 - 12x^7y^2 + 3x^9y^3 - 3$.

Answers: **a.** 2, **b.** 12 ■

If the terms of a polynomial in one variable are written so that the exponents decrease as we move from left to right, we say that the terms are written with their exponents in *descending order.* If the terms are written so that the exponents increase as we move from left to right, we say that the terms are written with their exponents in *ascending order.*

$-5x^4 + 2x^3 + 7x^2 + 3x - 1$ This polynomial is written in descending powers of x.

$-1 + 3x + 7x^2 + 2x^3 - 5x^4$ The same polynomial is now written in ascending powers of x.

Polynomial functions

In Chapter 2, we saw that linear functions are defined by equations of the form $f(x) = mx + b$. Some examples of linear functions are

$$f(x) = 3x + 1 \qquad g(x) = -\frac{1}{2}x - 1 \qquad h(x) = 5x$$

In each case, the right-hand side of the equation is a polynomial. For this reason, linear functions are members of a larger class of functions known as **polynomial functions.**

Polynomial functions

A **polynomial function** is a function whose equation is defined by a polynomial in one variable.

Another example of a polynomial function is $f(x) = -x^2 + 6x - 8$. This is a second-degree polynomial function, called a **quadratic function.** Quadratic functions are of the form $f(x) = ax^2 + bx + c$, where $a \neq 0$.

An example of a third-degree polynomial function is $f(x) = x^3 - 3x^2 - 9x + 2$. Third-degree polynomial functions, also called **cubic functions,** are of the form $f(x) = ax^3 + bx^2 + cx + d$, where $a \neq 0$. Polynomial functions of degree 4 or higher, such as $f(x) = x^4 + 2x^3 - 3x + 1$, do not have special names.

Polynomial functions can be used to model many real-life situations. For example, the polynomial function $h(t) = -16t^2 + 128t$ gives the height h of a toy rocket above the ground t seconds after launch. To find the height h for specific times t, we must evaluate the function at those values of t.

Evaluating polynomial functions

To *evaluate a polynomial function* at a specific value, we replace the variable in the defining equation with the value, called the **input.** Then we simplify the resulting expression to find the **output.**

EXAMPLE 3 ***Rocketry.*** If a toy rocket is launched straight up with an initial velocity of 128 feet per second, its height $h(t)$ (in feet) above the ground t seconds after liftoff is given by the function

$$h(t) = -16t^2 + 128t$$

Find the height of the rocket at **a.** 0 second, **b.** 3 seconds, and **c.** 7.9 seconds.

Solution

a. To find the height of the rocket at 0 second, we substitute 0 for t and evaluate the right hand-side.

$$h(t) = -16t^2 + 128t \quad \text{The given function.}$$
$$h(0) = -16(0)^2 + 128(0) \quad \text{The input is 0.}$$
$$= 0 \quad \text{The output is 0.}$$

At 0 second, the rocket's height is 0. It is on the ground waiting to be launched.

b. To find the height at 3 seconds, we substitute 3 for t and simplify.

$$h(t) = -16t^2 + 128t \quad \text{The given function.}$$
$$h(3) = -16(3)^2 + 128(3) \quad \text{The input is 3.}$$
$$= -16(9) + 384$$
$$= -144 + 384$$
$$= 240 \quad \text{The output is 240.}$$

At 3 seconds after liftoff, the height of the rocket is 240 feet.

c. To find the height at 7.9 seconds, we substitute 7.9 for t and simplify.

$$h(t) = -16t^2 + 128t \quad \text{The given function.}$$
$$h(7.9) = -16(7.9)^2 + 128(7.9) \quad \text{The input is 7.9.}$$
$$= -16(62.41) + 1{,}011.2$$
$$= -998.56 + 1{,}011.2$$
$$= 12.64 \quad \text{The output is 12.64.}$$

At 7.9 seconds, the height is 12.64 feet. The rocket has fallen nearly back to earth.

Self Check

Find the height of the rocket 4 seconds after being launched.

Answer: 256 ft

EXAMPLE 4 ***Packaging.*** To make boxes, a manufacturer cuts equal-sized squares from each corner of a 10 in. × 12 in. piece of cardboard, and then folds up the sides. (See Figure 5-2.) The polynomial function $f(x) = 4x^3 - 44x^2 + 120x$ gives the volume (in cubic inches) of the resulting box when a square with sides x inches long is cut from each corner. Find the volume of a box if 3-inch squares are cut out.

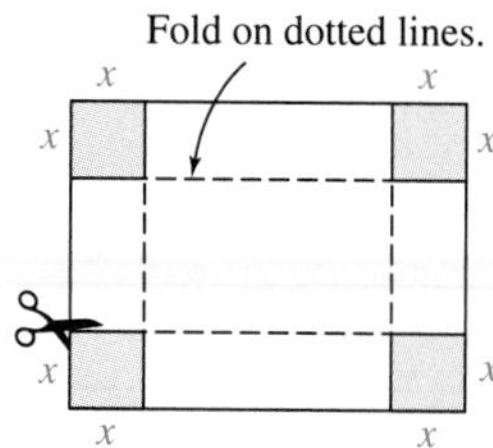

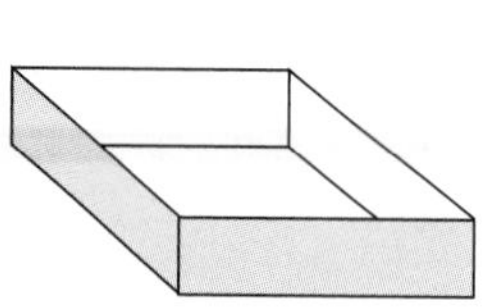

FIGURE 5-2

Solution

To find the volume of the box, we evaluate the function for $x = 3$.

$$
\begin{aligned}
f(x) &= 4x^3 - 44x^2 + 120x \\
f(3) &= 4(3)^3 - 44(3)^2 + 120(3) \quad \text{Substitute 3 for } x. \\
&= 4(27) - 44(9) + 120(3) \\
&= 108 - 396 + 360 \\
&= 72
\end{aligned}
$$

If 3-inch squares are cut out, the box will have a volume of 72 in.^{3}.

Self Check

Find the volume of the resulting box if 2-inch squares are cut from each corner of the cardboard.

Answer: 96 in.^{3}

Graphing polynomial functions

We have previously graphed three basic polynomial functions. The graph of a linear function $f(x) = 3x - 1$, the graph of the squaring function $f(x) = x^2$, and the graph of the cubing function $f(x) = x^3$ are shown in Figure 5-3.

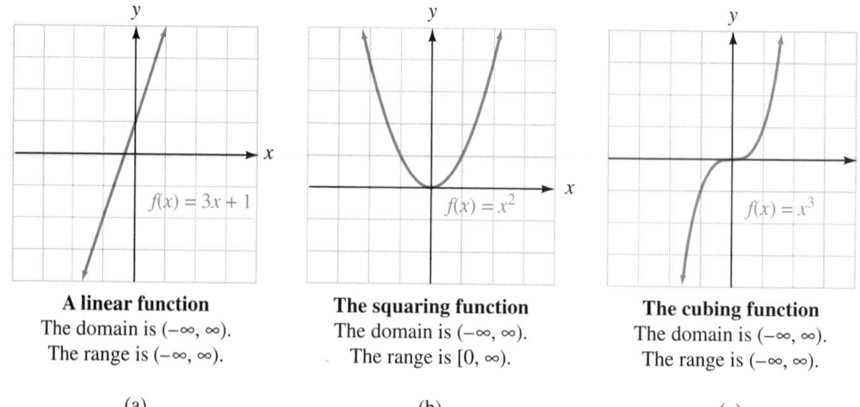

A linear function
The domain is $(-\infty, \infty)$.
The range is $(-\infty, \infty)$.

(a)

The squaring function
The domain is $(-\infty, \infty)$.
The range is $[0, \infty)$.

(b)

The cubing function
The domain is $(-\infty, \infty)$.
The range is $(-\infty, \infty)$.

(c)

FIGURE 5-3

When graphing a linear function, we need to plot only two points, because the graph is a straight line. The graphs of polynomial functions of degree greater than 1 are smooth, continuous curves. To graph them, we must plot many more points.

In Example 3, we saw that the polynomial function $h(t) = -16t^2 + 128t$ gives the height of the rocket t seconds after it has been launched. Since the height of the rocket depends on time, we say that the height is a function of time. To graph this function, we can make a table of values, plot the points, and join them with a smooth curve.

EXAMPLE 5 ***Graphing a polynomial function.*** Graph $h(t) = -16t^2 + 128t$.

Solution

From Example 3, we have seen that

When $t = 0$, $h(t) = 0$

When $t = 3$, $h(t) = 240$

To graph the function, we select other values of t, evaluate the function at those values, and write the ordered pairs in the table shown in Figure 5-4. Then we plot the pairs and join the resulting points to get the parabola shown in the figure. From the graph, we can see that 4 seconds into the flight, the rocket attains a maximum height of 256 feet.

$h(t) = -16t^2 + 128t$

t	$h(t)$	$(t, h(t))$
0	0	(0, 0)
1	112	(1, 112)
2	192	(2, 192)
3	240	(3, 240)
4	256	(4, 256)
5	240	(5, 240)
6	192	(6, 192)
7	112	(7, 112)
8	0	(8, 0)

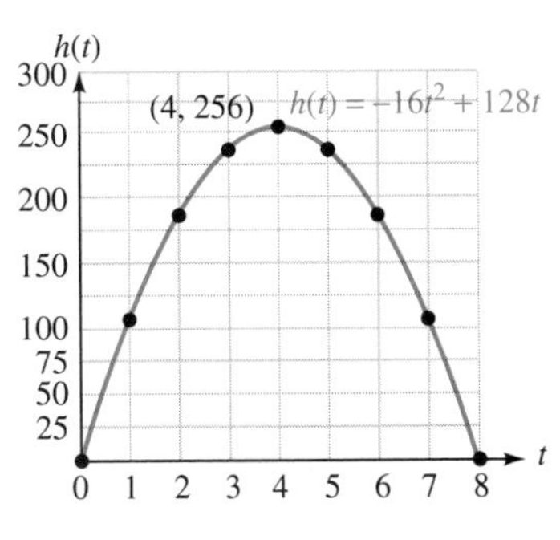

FIGURE 5-4

From the graph, we see the domain of the function is [0, 8] and the range is [0, 256].

WARNING! The parabola shown in the figure describes the height of the rocket in relation to time. It does not show the path of the rocket. The rocket goes straight up and then comes straight down.

Self Check

Use the graph in Figure 5-4 to estimate the height of the rocket at 6.5 seconds.

Answer: 150 ft ■

EXAMPLE 6 ***Graphing a polynomial function.*** Graph $f(x) = x^3 - 3x^2 - 9x + 2$.

Solution

To graph this cubic function, we begin by evaluating it for $x = -3$.

$$
\begin{aligned}
f(x) &= x^3 - 3x^2 - 9x + 2 \\
f(-3) &= (-3)^3 - 3(-3)^2 - 9(-3) + 2 \\
&= -27 - 3(9) - 9(-3) + 2 \\
&= -27 - 27 + 27 + 2 \\
&= -25
\end{aligned}
$$

In the table of values in Figure 5-5, we enter the ordered pair $(-3, -25)$. We continue the evaluating process for $x = -2, -1, 0, 1, 2, 3, 4$, and 5, and list the results in the table. After plotting the ordered pairs, we draw a smooth curve through the points to get the graph of function f.

Self Check

What are the domain and the range of the function graphed in Figure 5-5?

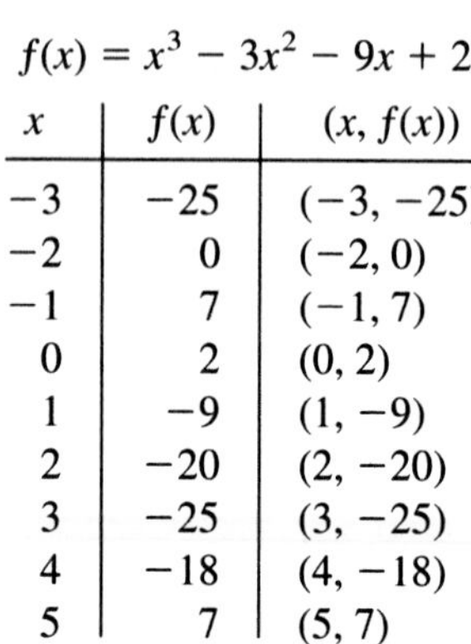

$f(x) = x^3 - 3x^2 - 9x + 2$

x	$f(x)$	$(x, f(x))$
-3	-25	$(-3, -25)$
-2	0	$(-2, 0)$
-1	7	$(-1, 7)$
0	2	$(0, 2)$
1	-9	$(1, -9)$
2	-20	$(2, -20)$
3	-25	$(3, -25)$
4	-18	$(4, -18)$
5	7	$(5, 7)$

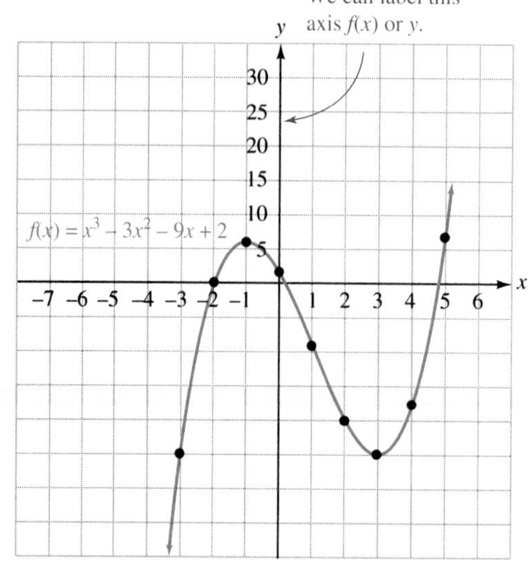

FIGURE 5-5

Answers: domain: $(-\infty, \infty)$, range: $(-\infty, \infty)$

Accent on Technology **Graphing polynomial functions**

We can graph polynomial functions with a graphing calculator. For example, to graph the function from Example 5, $h(t) = -16t^2 + 128t$, we can rewrite it as $f(x) = -16x^2 + 128x$ and use window settings of [0, 8] for x and [0, 260] for y to get the parabola shown in Figure 5-6(a).

We can trace to estimate the height of the rocket for any number of seconds into the flight. Figure 5-6(b) shows that the height of the rocket 1.6 seconds into the flight is approximately 165 feet.

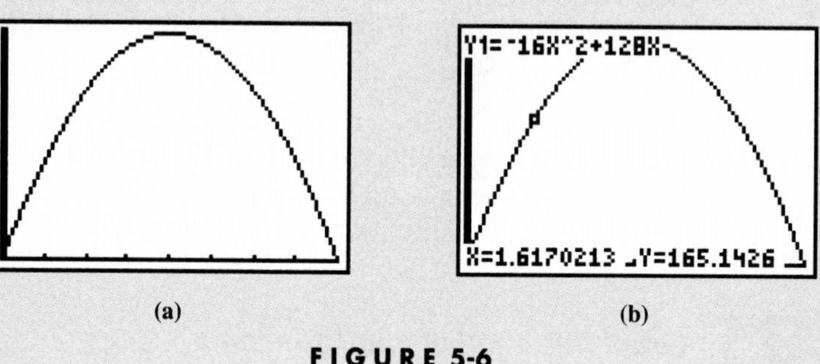

(a) (b)

FIGURE 5-6

Combining like terms

Recall that when terms have the same variables with the same exponents, they are called **like** or **similar terms.**

- $3x^2$ and $7x^2$ are like terms, because they have the same variables with the same exponents.
- $4x^4y^2$ and $98x^7y^9$ are unlike terms. They have the same variables, but with different exponents.
- $3xy$ and $5xz$ are unlike terms. They have different variables.

The distributive property enables us to combine like terms. For example,

$3x^2 + 7x^2 = (3 + 7)x^2$ Use the distributive property.

$= 10x^2$ Do the addition.

$5x^2y^3 + 22x^2y^3 = (5 + 22)x^2y^3$ Use the distributive property.

$= 27x^2y^3$ Do the addition.

The results of the previous examples suggest that to add like terms, *we add their numerical coefficients and keep the same variables with the same exponents.*

WARNING! The terms in the following binomials cannot be combined, because they are not like terms. (Tell why.) In each case, no further simplification is possible.

$$3x^2 - 5y^2, \qquad -2a^2 + 3a^3, \qquad \text{and} \qquad 5y^2 + 17xy$$

Adding and subtracting polynomials

Adding polynomials	To add polynomials, remove parentheses and combine like terms.

EXAMPLE 7 ***Adding polynomials.*** Add $3x^2 - 2x + 4$ and $2x^2 + 4x - 3$.

Solution

$(3x^2 - 2x + 4) + (2x^2 + 4x - 3)$ We are to add two trinomials.

$= 3x^2 - 2x + 4 + 2x^2 + 4x - 3$ Remove the parentheses.

$= 3x^2 + 2x^2 - 2x + 4x + 4 - 3$ Use the commutative and associative properties of addition to rearrange terms so that like terms are together.

$= 5x^2 + 2x + 1$ Combine like terms.

Self Check
Add $2a^2 - 3a + 5$ and $5a^2 + 4a - 2$.

Answer: $7a^2 + a + 3$ ■

EXAMPLE 8 ***Adding polynomials in two variables.*** Add $-5x^3y^2 - 4x^2y^3$ and $2x^3y^2 + 5x^2y^3$.

Solution

$(-5x^3y^2 - 4x^2y^3) + (2x^3y^2 + 5x^2y^3)$

$= -5x^3y^2 + 2x^3y^2 - 4x^2y^3 + 5x^2y^3$ Remove parentheses and rearrange terms.

$= -3x^3y^2 + x^2y^3$ Combine like terms.

Self Check
Add $-6a^2b^3 - 5a^3b^2$ and $3a^2b^3 + 2a^3b^2$.

Answer: $-3a^2b^3 - 3a^3b^2$ ■

The additions in Examples 7 and 8 can be done by aligning the terms vertically.

$$\begin{array}{r} 3x^2 - 2x + 4 \\ + \quad 2x^2 + 4x - 3 \\ \hline 5x^2 + 2x + 1 \end{array} \qquad \begin{array}{r} -5x^3y^2 - 4x^2y^3 \\ + \quad 2x^3y^2 + 5x^2y^3 \\ \hline -3x^3y^2 + x^2y^3 \end{array}$$

To subtract one monomial from another, we add the negative (or opposite) of the monomial that is to be subtracted.

EXAMPLE 9 ***Subtracting monomials.*** Do each subtraction.

a. $8x^2 - 3x^2 = 8x^2 + (-3x^2)$ Add the opposite of $3x^2$, which is $-3x^2$.

$= 5x^2$ Combine like terms.

b. $3x^2y - 9x^2y = 3x^2y + (-9x^2y)$

$= -6x^2y$

c. $-5x^5y^3z^2 - 3x^5y^3z^2 = -5x^5y^3z^2 + (-3x^5y^3z^2)$

$= -8x^5y^3z^2$

Self Check

Subtract

a. $-2a^2b^3 - 5a^2b^3$ and

b. $-2a^2b^3 - (-5a^2b^3)$.

Answers: a. $-7a^2b^3$, b. $3a^2b^3$ ■

Subtracting polynomials To subtract two polynomials, we add the first polynomial and the negative (or opposite) of the second polynomial.

EXAMPLE 10 ***Subtracting two polynomials*** Do each subtraction.

a. $(8x^3y + 2x^2y) - (2x^3y - 3x^2y)$

$= 8x^3y + 2x^2y - 2x^3y + 3x^2y$ Remove the parentheses. Change every sign in the second polynomial.

$= 8x^3y - 2x^3y + 2x^2y + 3x^2y$ Rearrange terms.

$= 6x^3y + 5x^2y$ Combine like terms.

b. $(3rt^2 + 4r^2t^2) - (8rt^2 - 4r^2t^2 + r^3t^2)$

$= 3rt^2 + 4r^2t^2 - 8rt^2 + 4r^2t^2 - r^3t^2$

$= 3rt^2 - 8rt^2 + 4r^2t^2 + 4r^2t^2 - r^3t^2$ Rearrange terms.

$= -5rt^2 + 8r^2t^2 - r^3t^2$ Combine like terms.

Self Check

Subtract $(6a^2b^3 - 2a^2b^2) - (-2a^2b^3 + a^2b^2)$.

Answer: $8a^2b^3 - 3a^2b^2$ ■

To subtract polynomials in vertical form, we add the negative (or opposite) of the polynomial that is being subtracted.

$$\begin{array}{r} 8x^3y + 2x^2y \\ - \quad 2x^3y - 3x^2y \\ \hline \end{array} \Rightarrow \begin{array}{r} 8x^3y + 2x^2y \\ + \quad -2x^3y + 3x^2y \\ \hline 6x^3y + 5x^2y \end{array}$$

STUDY SET Section 5.3

VOCABULARY *In Exercises 1–8, fill in the blanks to make the statements true.*

1. A ______________ is the sum of one or more algebraic terms whose variables have whole-number exponents.
2. A ______________ is a polynomial with one term. A ______________ is a polynomial with two terms. A ______________ is a polynomial with three terms.
3. The ______________ of a monomial with one variable is the exponent on the variable.
4. A second-degree polynomial function is also called a ______________ function.
5. A third-degree polynomial function is also called a ______________ function.
6. The numerical ______________ of the term $-15x^2y^3$ is -15. The ______________ of the term is 5.
7. Terms having the same variables with the same exponents are called ______________ terms.
8. The ______________ of $x^2 + x - 3$ is $-x^2 - x + 3$.

CONCEPTS *In Exercises 9–20, classify each polynomial as a monomial, binomial, trinomial, or none of these. Then determine the degree of the polynomial.*

9. $3x^2$

10. $2y^3 + 4y^2$

11. $3x^2y - 2x + 3y$

12. $a^2 + ab + b^2$

13. $x^2 - y^2$

14. $\frac{17}{2}x^3 + 3x^2 - x - 4$

15. 5

16. $8x^3y^5$

17. $9x^2y^4 - x - y^{10} + 1$

18. x^{17}

19. $4x^9 + 3x^2y^4$

20. -12

21. Write each polynomial with the exponents on x in descending order.
a. $3x - 2x^4 + 7 - 5x^2$
b. $a^2x - ax^3 + 7a^3x^5 - 5a^3x^2$

22. Write each polynomial with the exponents on y in ascending order.
a. $4y^2 - 2y^5 + 7y - 5y^3$
b. $x^3y^2 + x^2y^3 - 2x^3y + x^7y^6 - 3x^6$

In Exercises 23–30, tell whether the terms are like or unlike terms. If they are like terms, combine them.

23. $3x, 7x$

24. $-8x, 3y$

25. $7x, 7y$

26. $3mn, 5mn$

27. $3r^2t^3, -8r^2t^3$

28. $9u^2v, 10u^2v$

29. $9x^2y^3, 3x^2y^2$

30. $27x^6y^4z, 8x^6y^4z^2$

31. Write a polynomial that represents the perimeter of the triangle shown in Illustration 1.

32. Write a polynomial that represents the perimeter of the rectangle shown in Illustration 2.

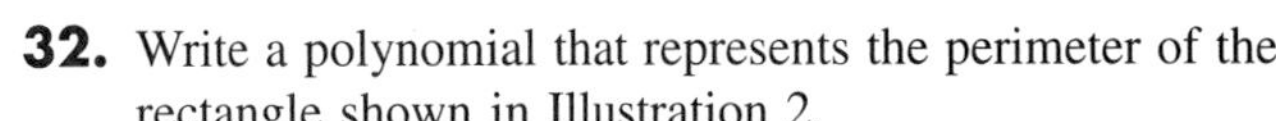

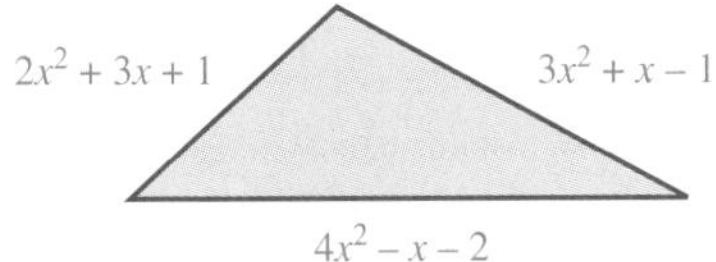

ILLUSTRATION 1

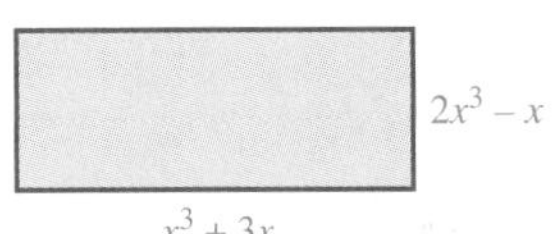

ILLUSTRATION 2

NOTATION *In Exercises 33–34, complete each solution.*

33. If $f(x) = 2x^2 + x + 2$, find $f(-1)$.

$$f(x) = 2x^2 + x + 2$$
$$f(-1) = 2(\square)^2 + (\square) + 2$$
$$= 2(\square) + (-1) + 2$$
$$= 3$$

34. If $h(t) = -t^3 - t^2 + 2t + 1$, find $h(3)$.

$$h(t) = -t^3 - t^2 + 2t + 1$$
$$h(3) = -(\square)^3 - (\square)^2 + 2(3) + 1$$
$$= \square - 9 + 6 + 1$$
$$= -29$$

PRACTICE *In Exercises 35–38, complete each table of values. Then graph each polynomial function. Check your work with a graphing calculator.*

35. $f(x) = 2x^2 - 4x + 2$

x	$f(x)$
-1	
0	
1	
2	
3	

36. $f(x) = -x^2 + 2x + 6$

x	$f(x)$
-2	
-1	
0	
1	
2	
3	
4	

37. $f(x) = 2x^3 - 3x^2 - 11x + 6$

x	$f(x)$
-3	
-2	
-1	
0	
1	
2	
3	
4	

38. $f(x) = -x^3 - x^2 + 6x$

x	$f(x)$
-4	
-3	
-2	
-1	
0	
1	
2	
3	

In Exercises 39–40, use a graphing calculator to graph each polynomial function. Use window settings of $[-4, 6]$ *for x and* $[-5, 5]$ *for y.*

39. $f(x) = 2.75x^2 - 4.7x + 1.5$

40. $f(x) = 0.37x^2 - 1.4x + 1.5$

In Exercises 41–54, do each operation.

41. $(3x^2 + 2x + 1) + (-2x^2 - 7x + 5)$

42. $(-2a^2 - 5a - 7) + (-3a^2 + 7a + 1)$

43. $(-a^2 + 2a + 3) - (4a^2 - 2a - 1)$

44. $(x^2 - 3x + 8) - (3x^2 + x + 3)$

45. $(7y^3 + 4y^2 + y + 3) + (-8y^3 - y + 3)$

46. $(6x^3 + 3x - 2) - (2x^3 + 3x^2 + 5)$

47. $(-2x^2y^3 + 6xy + 5y^2) - (-4x^2y^3 - 7xy + 2y^2)$

48. $(3ax^3 - 2ax^2 + 3a^3) + (4ax^3 + 3ax^2 - 2a^3)$

49.
$$\begin{array}{rr} & 3x^3 - 2x^2 + 4x - 3 \\ + & -2x^3 + 3x^2 + 3x - 2 \\ & 5x^3 - 7x^2 + 7x - 12 \\ \hline \end{array}$$

50.
$$\begin{array}{rr} & 7a^3 \qquad\quad + 3a + 7 \\ + & -2a^3 + 4a^2 \qquad\quad - 13 \\ & 3a^3 - 3a^2 + 4a + 5 \\ \hline \end{array}$$

51.
$$\begin{array}{rr} - & 3x^2 - 4x + 17 \\ & 2x^2 + 4x - 5 \\ \hline \end{array}$$

52.
$$\begin{array}{rr} - & -2y^2 - 4y + 3 \\ & 3y^2 + 10y - 5 \\ \hline \end{array}$$

53.
$$\begin{array}{rr} - & -5y^3 + 4y^2 - 11y + 3 \\ & -2y^3 - 14y^2 + 17y - 32 \\ \hline \end{array}$$

54.
$$\begin{array}{rr} - & 17x^4 - 3x^2 - 65x - 12 \\ & 23x^4 + 14x^2 + 3x - 23 \\ \hline \end{array}$$

APPLICATIONS

55. JUGGLING During a performance, a juggler tosses one ball straight upward, while continuing to juggle three others. (See Illustration 3.) The height $f(t)$, in feet, of the ball is given by the polynomial function $f(t) = -16t^2 + 32t + 4$, where t is the time in seconds since the ball was thrown. Find the height of the ball 1 second after being tossed upward.

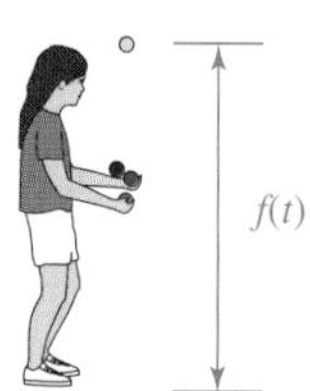

ILLUSTRATION 3

56. STOPPING DISTANCE The number of feet that a car travels before stopping depends on the driver's reaction time and the braking distance. (See Illustration 4.) For one driver, the stopping distance $d(v)$, in feet, is given by the polynomial function $d(v) = 0.04v^2 + 0.9v$, where v is the velocity of the car. Find the stopping distance at 60 mph.

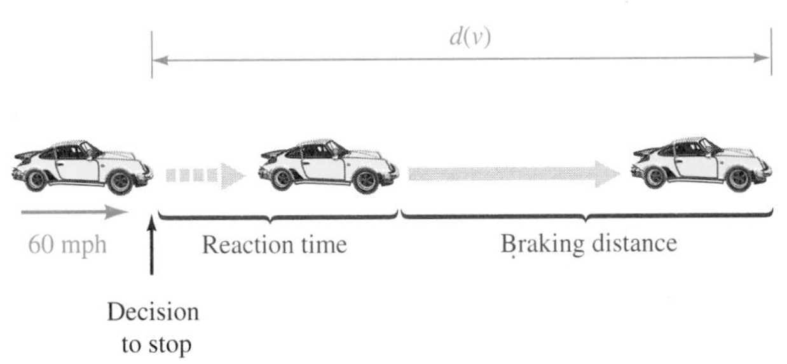

ILLUSTRATION 4

57. STORAGE TANKS See Illustration 5. The volume $V(r)$ of the gasoline storage tank, in cubic feet, is given by the polynomial function $V(r) = 4.2r^3 + 37.7r^2$, where r is the radius in feet of the cylindrical part of the tank. What is the capacity of the tank if its radius is 4 feet?

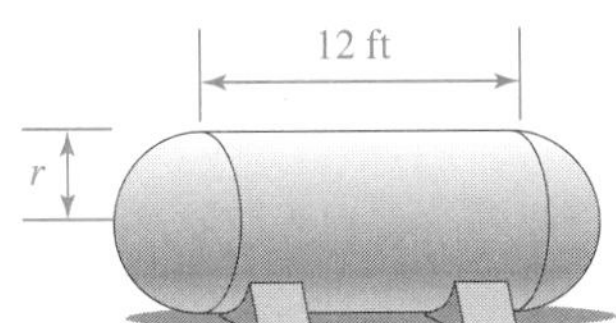

ILLUSTRATION 5

58. ROLLER COASTERS The polynomial function $f(x) = 0.001x^3 - 0.12x^2 + 3.6x + 10$ models the path of a portion of the track of a roller coaster, as shown in Illustration 6. Find the height of the track for $x = 0$, 20, 40, and 60.

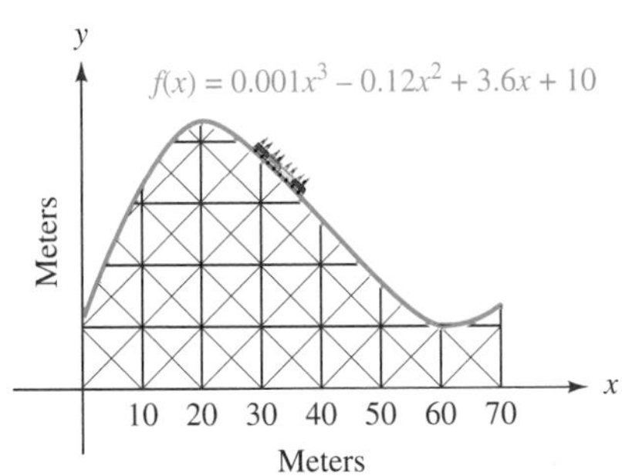

ILLUSTRATION 6

59. RAIN GUTTERS A rectangular sheet of metal will be used to make a rain gutter by bending up its sides, as shown in Illustration 7. If the ends are covered, the capacity $f(x)$ of the gutter is a polynomial function of x: $f(x) = -240x^2 + 1{,}440x$. Find the capacity of the gutter if x is 3 inches.

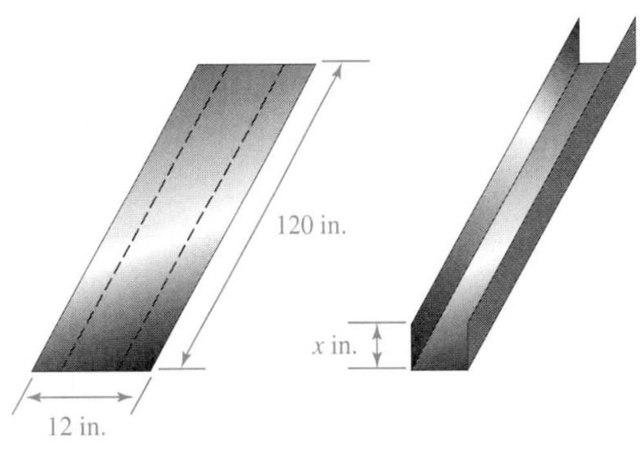

ILLUSTRATION 7

60. CUSTOMER SERVICE A software service hot line has found that on Mondays, the polynomial function $C(t) = -0.0625t^4 + t^3 - 6t^2 + 16t$ approximates the number of callers to the hot line at any one time. Here, t represents the time, in hours, since the hot line opened at 8:00 A.M. How many service technicians should be on duty on Mondays at noon if the company doesn't want any callers to the hot line waiting to be helped by a technician?

61. REAL ESTATE A computer analysis of two properties on the market generated functions to predict the value, in dollars, of each property after x years.

Rental home: $R(x) = 1{,}100x + 125{,}000$

Duplex: $D(x) = 1{,}400x + 150{,}000$

a. Find one polynomial function V that will give the combined value of the two properties after x years.

b. Use your answer to part a to find the combined value of the two properties after 20 years.

62. BUSINESS EXPENSES A company purchased two cars for its sales force to use. The following functions give the respective values of the vehicles after x years.

Toyota Camry LE: $T(x) = -2{,}100x + 16{,}600$

Ford Explorer Sport: $F(x) = -2{,}700x + 19{,}200$

a. Find one polynomial function V that will give the value of both cars after x years.

b. Use your answer in part a to find the combined value of the two cars after 3 years.

WRITING *Write a paragraph using your own words.*

63. Explain why the terms x^2y and xy^2 are not like terms.

64. The family of polynomial functions contains linear, quadratic, and cubic functions. Explain.

REVIEW *In Exercises 65–68, solve each inequality. Give the result in interval notation.*

65. $|x| \leq 5$

66. $|x| > 7$

67. $|x - 4| < 5$

68. $|2x + 1| \geq 7$

5.4 Multiplying Polynomials

In this section, you will learn about

- **Multiplying monomials**
- **Multiplying a polynomial by a monomial**
- **Multiplying a polynomial by a polynomial**
- **The FOIL method**
- **Special products**
- **Applications of multiplying polynomials**

Introduction. In this section, we discuss the procedures used to multiply polynomials. These procedures involve the application of several algebraic concepts introduced in earlier chapters, such as the commutative and associative properties of multiplication, the rules for exponents, and the distributive property.

Multiplying monomials

In Section 1.4, we saw that to multiply one monomial by another, *we multiply the numerical factors and then multiply the variable factors.*

EXAMPLE 1 ***Multiplying monomials.*** Do each multiplication. We can use the commutative and associative properties of multiplication to rearrange the terms and regroup the factors.

a. $(3x^2)(6x^3) = 3 \cdot x^2 \cdot 6 \cdot x^3$

$= (3 \cdot 6)(x^2 \cdot x^3)$

$= 18x^5$ To simplify $x^2 \cdot x^3$, keep the base and add the exponents.

b. $(-8x)(2y)(xy) = -8 \cdot x \cdot 2 \cdot y \cdot x \cdot y$

$= (-8 \cdot 2) \cdot x \cdot x \cdot y \cdot y$

$= -16x^2y^2$

c. $(2a^3b)(-7b^2c)(-12ac^4) = 2 \cdot a^3 \cdot b \cdot (-7) \cdot b^2 \cdot c \cdot (-12) \cdot a \cdot c^4$

$= 2(-7)(-12) \cdot a^3 \cdot a \cdot b \cdot b^2 \cdot c \cdot c^4$

$= 168a^4b^3c^5$

Self Check

Multiply **a.** $(-2a^3)(4a^2)$ and **b.** $(-5b^3)(-3a)(a^2b)$.

Answers: **a.** $-8a^5$, **b.** $15a^3b^4$

Multiplying a polynomial by a monomial

To multiply a polynomial by a monomial, *we multiply each term of the polynomial by the monomial.*

EXAMPLE 2 ***Multiplying by a monomial.*** Do each multiplication. We can use the distributive property to remove parentheses.

a. $3x^2(6xy + 3y^2) = 3x^2 \cdot 6xy + 3x^2 \cdot 3y^2$

$= 18x^3y + 9x^2y^2$ Do the multiplications.

Since $18x^3y$ and $9x^2y^2$ are not like terms, we cannot add them.

b. $5x^3y^2(xy^3 - 2x^2y) = 5x^3y^2 \cdot xy^3 - 5x^3y^2 \cdot 2x^2y$

$= 5x^4y^5 - 10x^5y^3$

c. $-2ab^2(3bz - 2az + 4z^3)$

$= -2ab^2 \cdot 3bz - (-2ab^2) \cdot 2az + (-2ab^2) \cdot 4z^3$

$= -6ab^3z + 4a^2b^2z - 8ab^2z^3$

Self Check

Multiply $-2a^2(a^2 - a + 3)$.

Answer: $-2a^4 + 2a^3 - 6a^2$ ■

Multiplying a polynomial by a polynomial

To multiply a polynomial by a polynomial, we use the distributive property repeatedly.

EXAMPLE 3 ***Multiplying polynomials.*** Do each multiplication. We can use the distributive property to remove parentheses.

a. $(3x + 2)(4x + 9) = (3x + 2) \cdot 4x + (3x + 2) \cdot 9$ Distribute $3x + 2$.

$= 12x^2 + 8x + 27x + 18$ Distribute $4x$ and distribute 9.

$= 12x^2 + 35x + 18$ Combine like terms.

b. $(2a - b)(3a^2 - 4ab + b^2)$

$= (2a - b)3a^2 - (2a - b)4ab + (2a - b)b^2$

$= 6a^3 - 3a^2b - 8a^2b + 4ab^2 + 2ab^2 - b^3$

$= 6a^3 - 11a^2b + 6ab^2 - b^3$

Self Check

Multiply $(2a + b)(3a - 2b)$.

Answer: $6a^2 - ab - 2b^2$ ■

The results of Example 3 suggest that to multiply one polynomial by another, *we multiply each term of one polynomial by each term of the other polynomial.*

In the next example, we organize the work done in Example 3 vertically.

EXAMPLE 4 ***Multiplying vertically.*** Do each multiplication.

a.

$$\begin{array}{rl} 3x + 2 & \\ \underline{4x + 9} & \\ 12x^2 + 8x & \leftarrow \text{This is the result of } 4x(3x + 2). \\ \underline{+ 27x + 18} & \leftarrow \text{This is the result of } 9(3x + 2). \\ 12x^2 + 35x + 18 & \text{Combine like terms, column by column.} \end{array}$$

b.

$$\begin{array}{rl} 3a^2 - 4ab + b^2 & \\ \underline{2a - b} & \\ 6a^3 - 8a^2b + 2ab^2 & \leftarrow 2a(3a^2 - 4ab + b^2) \\ \underline{- 3a^2b + 4ab^2 - b^3} & \leftarrow -b(3a^2 - 4ab + b^2) \\ 6a^3 - 11a^2b + 6ab^2 - b^3 & \end{array}$$

Self Check

Multiply:

$$\begin{array}{r} 3x^2 + 2x - 5 \\ \underline{2x + 1} \end{array}$$

Answer: $6x^3 + 7x^2 - 8x - 5$ ■

The FOIL method

When multiplying two binomials, the distributive property requires that each term of one binomial be multiplied by each term of the other binomial. This fact can be emphasized by drawing arrows to show the indicated products. For example, to multiply $3x + 2$ and $x + 4$, we can write

First terms   Last terms

$$(3x + 2)(x + 4) = 3x \cdot x + 3x \cdot 4 + 2 \cdot x + 2 \cdot 4$$
$$= 3x^2 + 12x + 2x + 8$$
$$= 3x^2 + 14x + 8 \qquad \text{Combine like terms: } 12x + 2x = 14x.$$

Inner terms   Outer terms

We note that

- the product of the **F**irst terms is $3x \cdot x = 3x^2$,
- the product of the **O**uter terms is $3x \cdot 4 = 12x$,
- the product of the **I**nner terms is $2 \cdot x = 2x$, and
- the product of the **L**ast terms is $2 \cdot 4 = 8$.

The procedure is called the **FOIL** method of multiplying two binomials. Foil is an acronym for **F**irst terms, **O**uter terms, **I**nner terms, and **L**ast terms. Of course, the resulting terms of the product must be combined, if possible.

It is easy to multiply binomials by sight using the FOIL method. We find the product of the first terms, then find the products of the outer terms and the inner terms and add them (when possible), and then find the product of the last terms.

EXAMPLE 5 ***The FOIL method.*** Find each product.

a. $(2x - 3)(3x + 2) = 6x^2 - 5x - 6$

The product of the first terms is $2x \cdot 3x = 6x^2$. The middle term in the result comes from combining the outer and inner products of $+4x$ and $-9x$:

$$4x + (-9x) = -5x$$

The product of the last terms is $-3 \cdot 2 = -6$.

b. $(3x + 1)(3x + 4) = 9x^2 + 15x + 4$

The product of the first terms is $3x \cdot 3x = 9x^2$. The middle term in the result comes from combining the products $+12x$ and $+3x$:

$$12x + 3x = 15x$$

The product of the last terms is $1 \cdot 4 = 4$.

c. $(4x - y)(2x + 3y) = 8x^2 + 10xy - 3y^2$

The product of the first terms is $4x \cdot 2x = 8x^2$. The middle term in the result comes from combining the products $+12xy$ and $-2xy$:

$$12xy - 2xy = 10xy$$

The product of the last terms is $-y \cdot 3y = -3y^2$.

Self Check

Multiply $(3a + 4b)(2a - b)$.

Answer: $6a^2 + 5ab - 4b^2$

Special products

It is easy to square a binomial using the FOIL method.

EXAMPLE 6 ***Squaring a binomial.*** Find each square: **a.** $(x + y)^2$ and **b.** $(x - y)^2$.

Solution

We use the FOIL method to multiply each term of one binomial by each term of the other binomial, and then we combine like terms.

a. $(x + y)^2 = (x + y)(x + y)$ — Write $(x + y)$ as a factor twice.

$= x^2 + xy + xy + y^2$ — Use the FOIL method.

$= x^2 + 2xy + y^2$ — Combine like terms: $xy + xy = 2xy$.

We see that the square of the binomial is the square of the first term, plus twice the product of the terms, plus the square of the last term. This product is illustrated graphically in Figure 5-7.

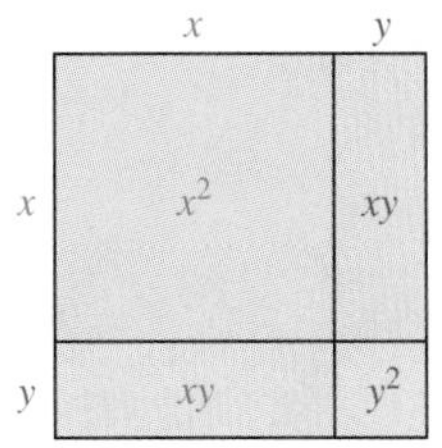

The area of the largest square is the product of its length and width: $(x + y)(x + y) = (x + y)^2$.

The area of the largest square is also the sum of its four pieces: $x^2 + xy + xy + y^2 = x^2 + 2xy + y^2$.

Thus, $(x + y)^2 = x^2 + 2xy + y^2$.

FIGURE 5-7

b. $(x - y)^2 = (x - y)(x - y)$

$= x^2 - xy - xy + y^2$ — Use the FOIL method.

$= x^2 - 2xy + y^2$ — Combine like terms: $-xy - xy = -2xy$.

We see that the square of the binomial is the square of the first term, minus twice the product of the terms, plus the square of the last term.

Self Check

Find the squares:
a. $(a + 2)^2$ and **b.** $(a - 4)^2$.

Answers: **a.** $a^2 + 4a + 4$, **b.** $a^2 - 8a + 16$

EXAMPLE 7 ***Multiplying the sum and difference of two terms.*** Multiply $(x + y)(x - y)$.

Solution

$(x + y)(x - y) = x^2 - xy + xy - y^2$ — Use the FOIL method.

$= x^2 - y^2$ — Combine like terms.

From this example, we see that the product of the sum of two quantities and the difference of two quantities is the square of the first quantity minus the square of the second quantity.

Self Check

Multiply $(a + 3)(a - 3)$.

Answer: $a^2 - 9$

The products discussed in Examples 6 and 7 are called **special products.** Because they occur so often, it is useful to learn their forms.

Special product formulas	
	$(x + y)^2 = (x + y)(x + y) = x^2 + 2xy + y^2$
	$(x - y)^2 = (x - y)(x - y) = x^2 - 2xy + y^2$
	$(x + y)(x - y) = x^2 - y^2$

Because $x^2 + 2xy + y^2 = (x + y)^2$ and $x^2 - 2xy + y^2 = (x - y)^2$, the two trinomials are called **perfect square trinomials.**

WARNING! The squares $(x + y)^2$ and $(x - y)^2$ have trinomials for their products. Don't forget to write the middle terms in these products. Remember that

$$(x + y)^2 \neq x^2 + y^2 \quad \text{and} \quad (x - y)^2 \neq x^2 - y^2$$

Also remember that the product $(x + y)(x - y)$ is the binomial $x^2 - y^2$.

At first, the expression $3[x^2 - 2(x + 3)]$ doesn't look like a polynomial. However, if we remove the parentheses and the brackets and simplify, it takes on the form of a polynomial.

$$\begin{aligned} 3[x^2 - 2(x + 3)] &= 3[x^2 - 2x - 6] \\ &= 3x^2 - 6x - 18 \end{aligned}$$

If an expression has one set of grouping symbols that is enclosed within another set, we always eliminate the inner set first.

EXAMPLE 8 ***Expressions containing grouping symbols.*** Find the product of $-2[y^3 + 3(y^2 - 2)]$ and $5[y^2 - 2(y + 1)]$.

Solution

We change each expression into polynomial form

$$\begin{aligned} &-2[y^3 + 3(y^2 - 2)] \\ &\quad = -2(y^3 + 3y^2 - 6) \\ &\quad = -2y^3 - 6y^2 + 12 \end{aligned} \qquad \begin{aligned} &5[y^2 - 2(y + 1)] \\ &\quad = 5(y^2 - 2y - 2) \\ &\quad = 5y^2 - 10y - 10 \end{aligned}$$

and then do the multiplication vertically:

$$\begin{array}{rrrrrr} & & -2y^3 & -\ 6y^2 & & +\ 12 \\ & & & 5y^2 & -\ 10y & -\ 10 \\ \hline -10y^5 & -\ 30y^4 & & +\ 60y^2 & & \\ & +\ 20y^4 & +\ 60y^3 & & -\ 120y & \\ & & +\ 20y^3 & +\ 60y^2 & & -\ 120 \\ \hline -10y^5 & -\ 10y^4 & +\ 80y^3 & +\ 120y^2 & -\ 120y & -\ 120 \end{array}$$

Self Check

Find the product of $2[a^2 + 3(a - 2)]$ and $3[a^2 + 3(a - 1)]$.

Answer:
$6a^4 + 36a^3 - 162a + 108$

Applications of multiplying polynomials

Profit, revenue, and cost are terms used in the business world. The profit p earned on the sale of one or more items is given by the formula

$$p = r - c$$

where r is the revenue taken in and c is the wholesale cost. If a salesperson has 12 vacuum cleaners and sells them for \$225 each, the revenue will be $r = \$(12 \cdot 225) = \$2,700$. This illustrates the following formula for finding the revenue, r:

$$r = \begin{array}{c}\text{number of} \\ \text{items sold } (x)\end{array} \cdot \begin{array}{c}\text{selling price} \\ \text{of each item } (p)\end{array} = xp = px$$

EXAMPLE 9 ***Selling vacuum cleaners*** Over the years, a saleswoman has found that the number of vacuum cleaners she can sell depends on price. The lower the price, the more she can sell. She has determined that the number of vacuums (x) that she can sell at a price (p) is related by the equation $x = -\frac{2}{25}p + 28$.

a. Find a formula for the revenue r.

b. How much revenue will be taken in if the vacuums are priced at \$250?

Solution **a.** To find a formula for revenue, we substitute $-\frac{2}{25}p + 28$ for x in the formula $r = px$ and solve for r.

$$r = px \quad \text{The formula for revenue.}$$

$$r = p\left(-\frac{2}{25}p + 28\right) \quad \text{Substitute } -\tfrac{2}{25}p + 28 \text{ for } x.$$

$$= -\frac{2}{25}p^2 + 28p \quad \text{Multiply the polynomials.}$$

b. To find how much revenue will be taken in if the vacuums are priced at \$250, we substitute 250 for p in the formula for revenue.

$$r = -\frac{2}{25}p^2 + 28p \quad \text{The formula for revenue.}$$

$$r = -\frac{2}{25}(250)^2 + 28(250) \quad \text{Substitute 250 for } p.$$

$$= -5{,}000 + 7{,}000$$

$$= 2{,}000$$

The revenue will be \$2,000.

STUDY SET Section 5.4

VOCABULARY *In Exercises 1–4, fill in the blanks to make the statements true.*

1. The expression $(x + 4)(x - 5)$ is the ________ of two binomials.

2. The expression $(x + 4)^2$ is the ________ of a binomial.

3. The polynomial $3x^2 - 3x + 2$ contains three ________.

4. To find $-2x(3x^2 - 2)$, we use the ________ property to remove parentheses.

CONCEPTS *In Exercises 5–12, fill in the blanks to make the statements true.*

5. To multiply a monomial by a monomial, we multiply the numerical ________ and then multiply the variable factors.

6. To multiply a polynomial by a monomial, we multiply each ________ of the polynomial by the monomial.

7. To multiply a polynomial by a polynomial, we multiply each ________ of one polynomial by each term of the other polynomial.

8. FOIL is an acronym for ________ terms, ________ terms, ________ terms, and ________ terms.

9. $(x + y)^2 = (x + y)(x + y) =$ ________

10. $(x - y)^2 = (x - y)(x - y) =$ ________

11. $(x + y)(x - y) =$ ________

12. $x^2 + 2xy + y^2$ and $x^2 - 2xy + y^2$ are called ________ trinomials.

13. Write a polynomial that represents the area of the rectangle shown in Illustration 1.

ILLUSTRATION 1

14. Write a polynomial that represents the area of the triangle shown in Illustration 2.

b − 2
b + 5

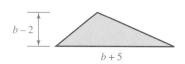

ILLUSTRATION 2

15. Write a polynomial that represents the area of the square shown in Illustration 3.

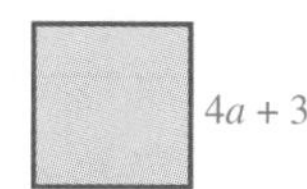

ILLUSTRATION 3

16. Write a polynomial that represents the area of the rectangle shown in Illustration 4.

2a + 3
2a − 3

ILLUSTRATION 4

17. Consider $(2x + 4)(4x - 3)$. Give the

a. First terms
b. Outer terms
c. Inner terms
d. Last terms

18. Find

a. $(4b - 1) + (2b - 1)$
b. $(4b - 1)(2b - 1)$
c. $(4b - 1) - (2b - 1)$

PRACTICE *In Exercises 19–36, find each product.*

19. $(2a^2)(-3ab)$

20. $(-3x^2y)(3xy)$

21. $(-3ab^2c)(5ac^2)$

22. $(-2m^2n)(-4mn^3)$

23. $(4a^2b)(-5a^3b^2)(6a^4)$

24. $(2x^2y^3)(4xy^5)(-5y^6)$

25. $(-5xx^2)(-3xy)^4$

26. $(-2a^2ab^2)^3(-3ab^2b^2)$

27. $3(x + 2)$

28. $-5(a + b)$

29. $3x(x^2 + 3x)$

30. $-2x(3x^2 - 2)$

31. $-2x(3x^2 - 3x + 2)$

32. $3a(4a^2 + 3a - 4)$

33. $7rst(r^2 + s^2 - t^2)$

34. $3x^2yz(x^2 - 2y + 3z^2)$

35. $4m^2n(-3mn)(m + n)$

36. $-3a^2b^3(2b)(3a + b)$

In Exercises 37–52, find each product. If possible, find the product by sight.

37. $(x + 2)(x + 3)$

38. $(y - 3)(y + 4)$

39. $(3t - 2)(2t + 3)$

40. $(p + 3)(3p - 4)$

41. $(3y - z)(2y - z)$

42. $(2m - n)(3m - n)$

43. $(x + 2)^2$

44. $(x - 3)^2$

45. $(a - 4)^2$

46. $(y + 5)^2$

47. $(2a + b)^2$

48. $(a - 2b)^2$

49. $(x + 2)(x - 2)$

50. $(z + 3)(z - 3)$

51. $(2x + 3y)(2x - 3y)$

52. $(3a + 4b)(3a - 4b)$

In Exercises 53–60, find each product.

53. $(x - y)(x^2 + xy + y^2)$

54. $(x + y)(x^2 - xy + y^2)$

55. $(3y + 1)(2y^2 + 3y + 2)$

56. $(a + 2)(3a^2 + 4a - 2)$

57. $(2a - b)(4a^2 + 2ab + b^2)$

58. $(x - 3y)(x^2 + 3xy + 9y^2)$

59. $(a + b)(a - b)(a - 3b)$

60. $(x - y)(x + 2y)(x - 2y)$

In Exercises 61–68, simplify each expression. (Hint: Do any multiplications first, then combine like terms.)

61. $3x(2x + 4) - 3x^2$

62. $2y - 3y(y^2 + 4)$

63. $3pq - p(p - q)$

64. $-4rs(r - 2) + 4rs$

65. $(x + 3)(x - 3) + (2x - 1)(x + 2)$

66. $(2b + 3)(b - 1) - (b + 2)(3b - 1)$

67. $(3x - 4)^2 - (2x + 3)^2$

68. $(3y + 1)^2 + (2y - 4)^2$

In Exercises 69–72, use a calculator to help find each product.

69. $(3.21x - 7.85)(2.87x + 4.59)$

70. $(7.44y + 56.7)(-2.1y - 67.3)$

71. $(-17.3y + 4.35)^2$

72. $(-0.31x + 29.3)(-81x - 0.2)$

APPLICATIONS

73. THE YELLOW PAGES Refer to Illustration 5.

a. Describe the area occupied by the ads for movers by using a product of two binomials.

b. Describe the area occupied by the ad for Budget Moving Co. by using a product. Then do the multiplication.

c. Describe the area occupied by the ad for Snyder Movers by using a product. Then do the multiplication.

d. Explain why your answer to part a is equal to the sum of your answers to parts b and c. What special product does this exercise illustrate?

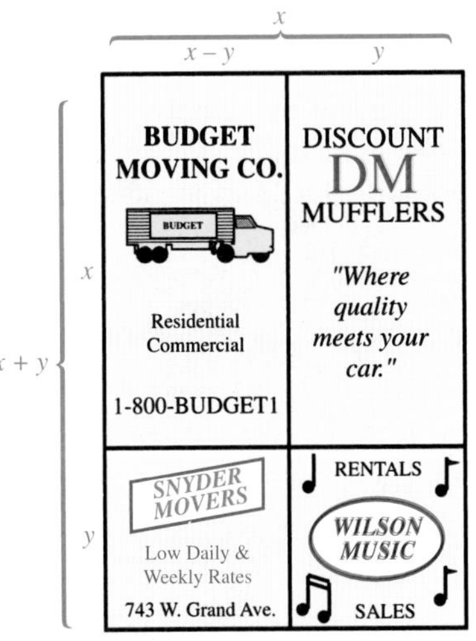

ILLUSTRATION 5

74. HELICOPTER PAD To determine the amount of fluorescent paint needed to paint the circular ring on the landing pad design shown in Illustration 6, painters must find its area. The area of the ring is given by the expression $\pi(R + r)(R - r)$.

a. Find the product $\pi(R + r)(R - r)$.

b. If $R = 25$ feet and $r = 20$ feet, find the area to be painted. Round to the nearest tenth.

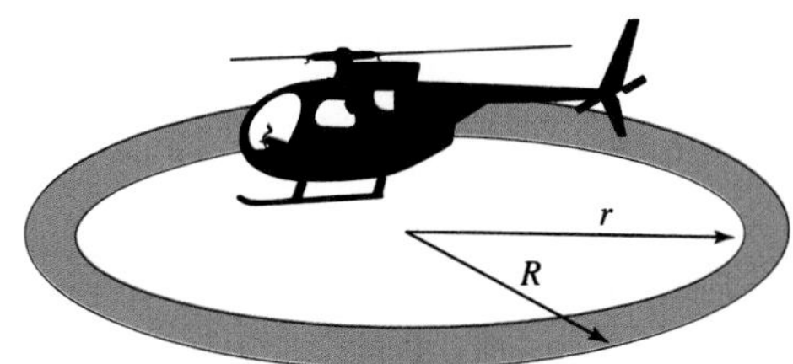

ILLUSTRATION 6

75. GIFT BOXES The corners of a 12-in.-by-12-in. piece of cardboard are creased, folded inward, and glued to make a gift box. (See Illustration 7.) Write a polynomial that gives the volume of the resulting box.

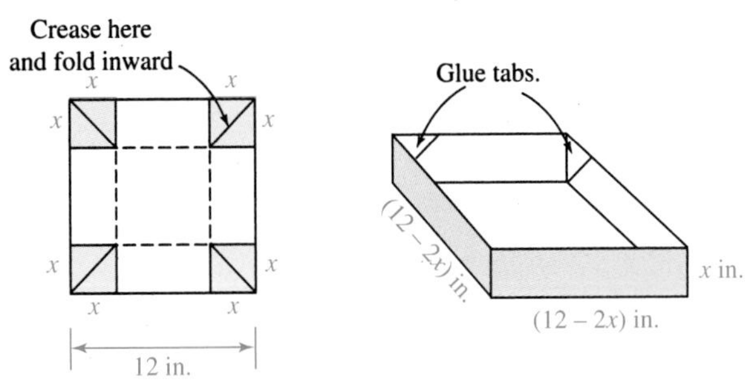

ILLUSTRATION 7

76. CALCULATING REVENUE A salesperson has found that the number (x) of televisions he can sell at a certain price (p) is related by the equation $x = -\frac{1}{5}p + 90$.

a. Find the number of TVs he will sell if the price is \$375.

b. Write a formula for the revenue when x TVs are sold.

c. Find the revenue generated by TV sales if they are priced at \$400 each.

WRITING *Write a paragraph using your own words.*

77. Explain how to use the FOIL method.

78. Explain how you would multiply two trinomials.

79. On a test, when asked to find $(x - y)^2$, a student answered $x^2 - y^2$. What error did the student make?

80. Explain how the distributive property is used to find the product: $2x^3(x^2 - 5x + 1)$.

REVIEW *In Exercises 81–84, graph each inequality or system of inequalities.*

81. $2x + y \le 2$

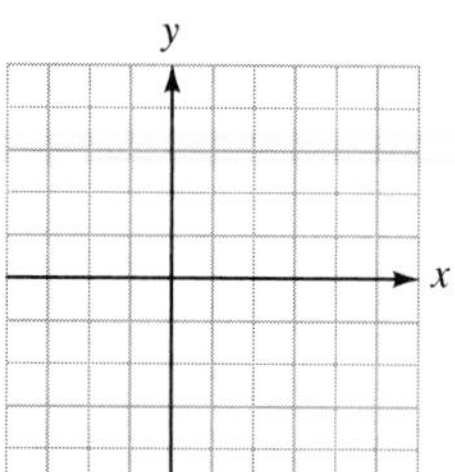

82. $x \ge 2$

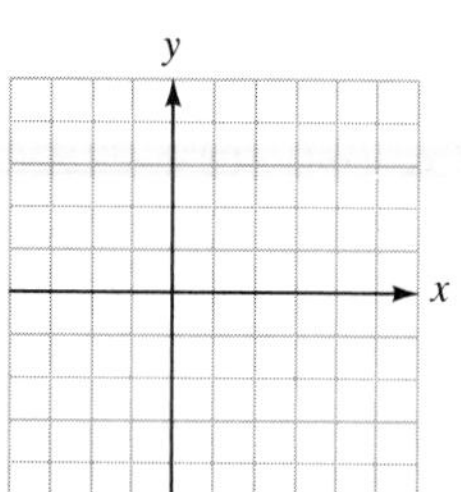

83. $\begin{cases} y - 2 < 3x \\ y + 2x < 3 \end{cases}$

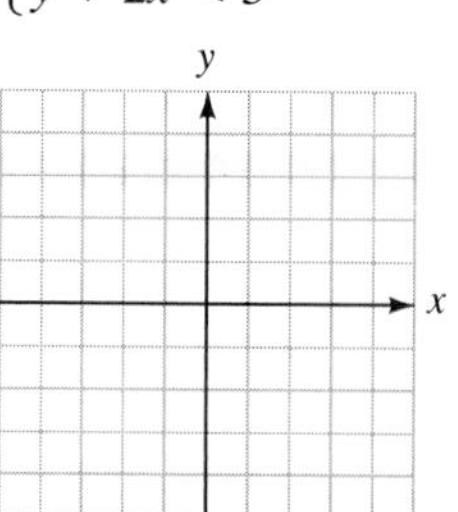

84. $\begin{cases} y < 0 \\ x < 0 \end{cases}$

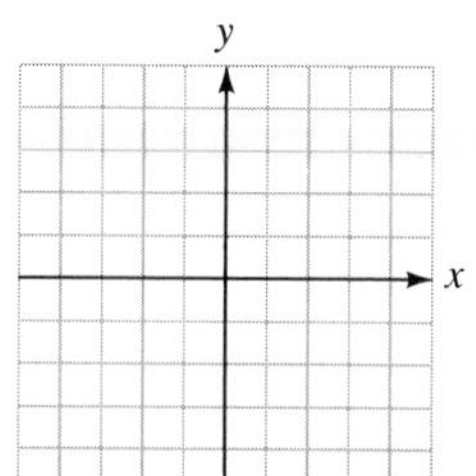

5.5 *The Greatest Common Factor and Factoring by Grouping*

In this section, you will learn about

- **Prime-factored form of a natural number**
- **Factoring out the greatest common factor**
- **Factoring by grouping**
- **Formulas**

INTRODUCTION. In Section 5.4, we discussed how to multiply polynomials. In the next four sections, we will reverse the operation of multiplication and show how to find the factors of a known product. The process of finding the individual factors of a known product is called **factoring.**

Prime-factored form of a natural number

If one number a divides a second number b, then a is called a **factor** of b. For example, because 3 divides 24, it is a factor of 24. Each number in the following list is a factor of 24, because each number divides 24.

1, 2, 3, 4, 6, 8, 12, and 24

To factor a natural number means to write it as a product of other natural numbers. If each factor is a prime number, the natural number is said to be written in **prime-factored form.** Example 1 shows how to find the prime-factored forms of 60, 84, and 180, respectively.

EXAMPLE 1 ***Prime factors.*** Find the prime factorization of each number.

a. $60 = 6 \cdot 10$
$= 2 \cdot 3 \cdot 2 \cdot 5$
$= 2^2 \cdot 3 \cdot 5$

b. $84 = 4 \cdot 21$
$= 2 \cdot 2 \cdot 3 \cdot 7$
$= 2^2 \cdot 3 \cdot 7$

c. $180 = 10 \cdot 18$
$= 2 \cdot 5 \cdot 3 \cdot 6$
$= 2 \cdot 5 \cdot 3 \cdot 3 \cdot 2$
$= 2^2 \cdot 3^2 \cdot 5$

Self Check

Find the prime factorization of 120.

Answer: $2^3 \cdot 3 \cdot 5$ ■

The largest natural number that divides 60, 84, and 180 is called the **greatest common factor (GCF)** of the numbers. Because 60, 84, and 180 all have two factors of 2 and one factor of 3, the GCF of these three numbers is $2^2 \cdot 3 = 12$. We note that

$$\frac{60}{12} = 5, \quad \frac{84}{12} = 7, \quad \text{and} \quad \frac{180}{12} = 15$$

There is no natural number greater than 12 that divides 60, 84, and 180.

Algebraic monomials can also have greatest common factors.

EXAMPLE 2 ***The greatest common factor.*** Find the GCF of $6a^2b^3c$, $9a^3b^2c$, and $18a^4c^3$.

Solution

We begin by factoring each monomial.

$$6a^2b^3c = 3 \cdot 2 \cdot a \cdot a \cdot b \cdot b \cdot b \cdot c$$
$$9a^3b^2c = 3 \cdot 3 \cdot a \cdot a \cdot a \cdot b \cdot b \cdot c$$
$$18a^4c^3 = 2 \cdot 3 \cdot 3 \cdot a \cdot a \cdot a \cdot a \cdot c \cdot c \cdot c$$

Since each monomial has one factor of 3, two factors of a, and one factor of c in common, their GCF is

$$3^1 \cdot a^2 \cdot c^1 = 3a^2c$$

Self Check

Find the GCF of $-24x^2y^3$, $3x^3y$, and $18x^2y^2$.

Answer: $3x^2y$

To find the GCF of several monomials, we follows these steps.

Steps for finding the GCF

1. Find the prime-factored form of each monomial.
2. Identify the prime factors that are common to each monomial.
3. Find the product of the factors found in step 2, with each factor raised to the smallest power that occurs in any one monomial.

Factoring out the greatest common factor

We have seen that the distributive property provides a method for multiplying a polynomial by a monomial. For example,

$$\begin{aligned} 2x^3y^3(3x^2 - 4y^3) &= 2x^3y^3 \cdot 3x^2 - 2x^3y^3 \cdot 4y^3 \\ &= 6x^5y^3 - 8x^3y^6 \end{aligned}$$

If the product of a multiplication is $6x^5y^3 - 8x^3y^6$, we can use the distributive property to find the individual factors.

$$\begin{aligned} 6x^5y^3 - 8x^3y^6 &= 2x^3y^3 \cdot 3x^2 - 2x^3y^3 \cdot 4y^3 \\ &= 2x^3y^3(3x^2 - 4y^3) \end{aligned}$$

Since $2x^3y^3$ is the GCF of the terms of $6x^5y^3 - 8x^3y^6$, this process is called **factoring out the greatest common factor.**

EXAMPLE 3 ***Factoring out the greatest common factor.*** Factor $25a^3b + 15ab^3$.

Solution

We begin by factoring each monomial:

$$25a^3b = 5 \cdot 5 \cdot a \cdot a \cdot a \cdot b$$
$$15ab^3 = 5 \cdot 3 \cdot a \cdot b \cdot b \cdot b$$

Self Check

Factor $9x^4y^2 - 12x^3y^3$.

Since each term has one factor of 5, one factor of a, and one factor of b in common, and there are no other common factors, $5ab$ is the GCF of the two terms. We can use the distributive property to factor it out.

$$\begin{aligned} 25a^3b + 15ab^3 &= 5ab \cdot 5a^2 + 5ab \cdot 3b^2 \\ &= 5ab(5a^2 + 3b^2) \end{aligned}$$

Answer: $3x^3y^2(3x - 4y)$ ■

EXAMPLE 4 ***Factoring out the greatest common factor.*** Factor $3xy^2z^3 + 6xyz^3 - 3xz^2$.

Solution

We begin by factoring each monomial:

$$\begin{aligned} 3xy^2z^3 &= 3 \cdot x \cdot y \cdot y \cdot z \cdot z \cdot z \\ 6xyz^3 &= 3 \cdot 2 \cdot x \cdot y \cdot z \cdot z \cdot z \\ -3xz^2 &= -3 \cdot x \cdot z \cdot z \end{aligned}$$

Since each term has one factor of 3, one factor of x, and two factors of z in common, and because there are no other common factors, $3xz^2$ is the GCF of the three terms. We can use the distributive property to factor it out.

$$\begin{aligned} 3xy^2z^3 + 6xyz^3 - 3xz^2 &= 3xz^2 \cdot y^2z + 3xz^2 \cdot 2yz - 3xz^2 \cdot 1 \\ &= 3xz^2(y^2z + 2yz - 1) \end{aligned}$$

WARNING! The last term $-3xz^2$ of the given trinomial has an understood coefficient of -1. When the $3xz^2$ is factored out, remember to write the -1.

Self Check

Factor $2a^4b^2 + 6a^3b^2 - 4a^2b$.

Answer: $2a^2b(a^2b + 3ab - 2)$ ■

A polynomial that cannot be factored is called a **prime polynomial** or an **irreducible polynomial.**

EXAMPLE 5 ***Factoring out the greatest common factor.*** Factor $3x^2 + 4y + 7$, if possible.

Solution

We factor each monomial:

$$3x^2 = 3 \cdot x \cdot x \qquad 4y = 2 \cdot 2 \cdot y \qquad 7 = 7$$

Since there are no common factors other than 1, this polynomial cannot be factored. It is a prime polynomial.

Self Check

Factor $6a^3 + 7b^2 + 5$.

Answer: a prime polynomial ■

EXAMPLE 6 ***The negative of the greatest common factor.*** Factor the negative of the GCF from $-6u^2v^3 + 8u^3v^2$.

Solution

Because the GCF of the two terms is $2u^2v^2$, the negative of the GCF is $-2u^2v^2$. To factor out $-2u^2v^2$, we proceed as follows:

$$\begin{aligned} -6u^2v^3 + 8u^3v^2 &= -2u^2v^2 \cdot 3v + 2u^2v^2 \cdot 4u \\ &= -2u^2v^2 \cdot 3v - (-2u^2v^2)4u \\ &= -2u^2v^2(3v - 4u) \end{aligned}$$

Self Check

Factor out the negative of the GCF from $-8a^2b^2 - 12ab^3$.

Answer: $-4ab^2(2a + 3b)$ ■

A common factor can have more than one term. For example, in the expression

$$x(a + b) + y(a + b)$$

the binomial $a + b$ is a factor of both terms. We can factor it out to get

$$x(a + b) + y(a + b) = (a + b)x + (a + b)y$$ Use the commutative property of multiplication.

$$= (a + b)(x + y)$$

EXAMPLE 7 ***A common factor having more than one term.*** Factor $a(x - y + z) - b(x - y + z) + 3(x - y + z)$.

Solution

We can factor out the GCF of the three terms, which is $(x - y + z)$.

$$a(x - y + z) - b(x - y + z) + 3(x - y + z)$$
$$= (x - y + z)a - (x - y + z)b + (x - y + z)3$$
$$= (x - y + z)(a - b + 3)$$

Self Check

Factor $x(a + b - c) - y(a + b - c)$.

Answer: $(a + b - c)(x - y)$ ■

Factoring by grouping

Suppose that we wish to factor

$$ac + ad + bc + bd$$

Although there is no factor common to all four terms, there is a common factor of a in the first two terms and a common factor of b in the last two terms. We can factor out these common factors to get

$$ac + ad + bc + bd = a(c + d) + b(c + d)$$

We can now factor out the common factor of $c + d$ on the right-hand side:

$$ac + ad + bc + bd = (c + d)(a + b)$$

The grouping in this type of problem is not always unique. For example, if we write the expression $ac + ad + bc + bd$ in the form

$$ac + bc + ad + bd$$

and factor c from the first two terms and d from the last two terms, we obtain

$$ac + bc + ad + bd = c(a + b) + d(a + b)$$
$$= (a + b)(c + d)$$ This is equivalent to $(c+d)(a+b)$.

The method used in the previous examples is called **factoring by grouping.**

EXAMPLE 8 ***Factoring by grouping.*** Factor $3ax^2 + 3bx^2 + a + 5bx + 5ax + b$.

Solution

Although there is no factor common to all six terms, $3x^2$ can be factored out of the first two terms, and $5x$ can be factored out of the fourth and fifth terms to get

$$3ax^2 + 3bx^2 + a + 5bx + 5ax + b = 3x^2(a + b) + a + 5x(b + a) + b$$

This result can be written in the form

$$3ax^2 + 3bx^2 + a + 5bx + 5ax + b = 3x^2(a + b) + 5x(a + b) + 1(a + b)$$

Since $a + b$ is common to all three terms, it can be factored out to get

$$3ax^2 + 3bx^2 + a + 5bx + 5ax + b = (a + b)(3x^2 + 5x + 1)$$

Self Check

Factor $2x^3 + 3x^2 + x + 2x^2y + 3xy + y$.

Answer: $(x + y)(2x^2 + 3x + 1)$ ■

To factor an expression, it is often necessary to factor more than once, as the following example illustrates.

OO

EXAMPLE 9 ***Factoring completely.*** Factor $3x^3y - 4x^2y^2 - 6x^2y + 8xy^2$.

Solution

We begin by factoring out the common factor of xy.

$$3x^3y - 4x^2y^2 - 6x^2y + 8xy^2 = xy(3x^2 - 4xy - 6x + 8y)$$

We can now factor $3x^2 - 4xy - 6x + 8y$ by grouping:

$$3x^3y - 4x^2y^2 - 6x^2y + 8xy^2$$
$$= xy(3x^2 - 4xy - 6x + 8y)$$
$$= xy[x(3x - 4y) - 2(3x - 4y)]$$ Factor x from $3x^2 - 4xy$ and -2 from $-6x + 8y$.
$$= xy(3x - 4y)(x - 2)$$ Factor out $3x - 4y$.

Because no more factoring can be done, the factorization is complete.

Self Check

Factor $3a^3b + 3a^2b - 2a^2b^2 - 2ab^2$.

Answer: $ab(3a - 2b)(a + 1)$

WARNING! Whenever you factor an expression, always factor it completely. Each factor of a completely factored expression will be prime.

Formulas

Factoring is often required to solve a literal equation for one of its variables.

EXAMPLE 10 ***Electronics*** The formula $r_1r_2 = rr_2 + rr_1$ is used in electronics to relate the combined resistance, r, of two resistors wired in parallel. The variable r_1 represents the resistance of the first resistor, and the variable r_2 represents the resistance of the second. Solve for r_2.

Solution

To isolate r_2 on one side of the equation, we get all terms involving r_2 on the left-hand side and all terms not involving r_2 on the right-hand side. We proceed as follows:

$$r_1r_2 = rr_2 + rr_1$$
$$r_1r_2 - rr_2 = rr_1$$ Subtract rr_2 from both sides.
$$r_2(r_1 - r) = rr_1$$ Factor out r_2 on the left-hand side.
$$r_2 = \frac{rr_1}{r_1 - r}$$ Divide both sides by $r_1 - r$.

Self Check

Solve $A = p + prt$ for p.

Answer: $p = \frac{A}{1 + rt}$

STUDY SET Section 5.5

VOCABULARY *In Exercises 1–6, fill in the blanks to make the statements true.*

1. When we write $2x + 4$ as $2(x + 2)$, we say that we have ______________ $2x + 4$.

2. When we write 100 as $2^2 \cdot 5^2$, we say that we have written 100 in ______________ form.

3. Because 5 divides 20, we say that 5 is a ______________ of 20.

4. The polynomial $2x^2y^3 - 4xy^2 + 6xy$ has three ______________.

5. The abbreviation GCF stands for ______________.

6. If a polynomial cannot be factored, it is called a ______________ polynomial or an irreducible polynomial.

CONCEPTS

7. The prime factorizations of three monomials are shown here. Find their GCF.

$2 \cdot 2 \cdot 3 \cdot x \cdot x \cdot y \cdot y \cdot y$

$2 \cdot 3 \cdot 3 \cdot x \cdot y \cdot y \cdot y \cdot y$

$2 \cdot 3 \cdot 3 \cdot 7 \cdot x \cdot x \cdot x \cdot y \cdot y$

8. a. What property is shown here?

$4a^2b(2ab^3 - 3a^2b^4) = 4a^2b \cdot 2ab^3 - 4a^2b \cdot 3a^2b^4$

b. Explain how we use the distributive property in reverse to factor $8a^3b^4 - 12a^4b^5$.

9. Explain why each factorization of $30t^2 - 20t^3$ is not complete.

a. $5t^2(6 - 4t)$

b. $10t(3t - 2t^2)$

10. a. Factor $-5y^3 - 10y^2 + 15y$ by factoring out the positive GCF.

b. Factor $-5y^3 - 10y^2 + 15y$ by factoring out the negative GCF.

NOTATION *In Exercises 11–14, complete each factorization.*

11. $3a - 12 = 3(a - \square)$

12. $8z^3 + 4z^2 + 2z = 2z(4z^2 + 2z + \square)$

13. $x^3 - x^2 + 2x - 2 = \square(x - 1) + \square(x - 1)$
$= (x - 1)(x^2 + 2)$

14. $-24a^3b^2 + 12ab^2 = -12ab^2(2a^2 \square 1)$

PRACTICE *In Exercises 15–22, find the prime-factored form of each number.*

15. 6
16. 10
17. 135
18. 98
19. 128
20. 357
21. 325
22. 288

In Exercises 23–30, find the GCF of each set of monomials.

23. 36, 48
24. 45, 75
25. 42, 36, 98
26. 16, 40, 60
27. $4a^2b$, $8a^3c$
28. $6x^3y^2z$, $9xyz^2$
29. $18x^4y^3z^2$, $-12xy^2z^3$
30. $6x^2y^3$, $24xy^3$, $40x^2y^2z^3$

In Exercises 31–44, factor each polynomial, if possible.

31. $2x + 8$
32. $3y - 9$
33. $2x^2 - 6x$
34. $3y^3 + 3y^2$
35. $5xy + 12ab^2$
36. $7x^2 + 14x$
37. $15x^2y - 10x^2y^2$
38. $11m^3n^2 - 12x^2y$
39. $14r^2s^3 + 15t^6$
40. $13ab^2c^3 - 26a^3b^2c$
41. $27z^3 + 12z^2 + 3z$
42. $25t^6 - 10t^3 + 5t^2$
43. $45x^{10}y^3 - 63x^7y^7 + 81x^{10}y^{10}$
44. $48u^6v^6 - 16u^4v^4 - 3u^6v^3$

In Exercises 45–54, factor out the negative of the greatest common factor.

45. $-3a - 6$
46. $-6b + 12$
47. $-3x^2 - x$
48. $-4a^3 + a^2$
49. $-6x^2 - 3xy$
50. $-15y^3 + 25y^2$
51. $-18a^2b - 12ab^2$
52. $-21t^5 + 28t^3$
53. $-63u^3v^6z^9 + 28u^2v^7z^2 - 21u^3v^3z^4$
54. $-56x^4y^3z^2 - 72x^3y^4z^5 + 80xy^2z^3$

In Exercises 55–64, factor each expression.

55. $4(x + y) + t(x + y)$

56. $5(a - b) - t(a - b)$

57. $(a - b)r - (a - b)s$

58. $(x + y)u + (x + y)v$

59. $3(m + n + p) + x(m + n + p)$

60. $x(x - y - z) + y(x - y - z)$

61. $(u + v)^2 - (u + v)$

62. $a(x - y) - (x - y)^2$

63. $-a(x + y) + b(x + y)$

64. $-bx(a - b) - cx(a - b)$

In Exercises 65–76, factor by grouping.

65. $ax + bx + ay + by$

66. $ar - br + as - bs$

67. $x^2 + yx + 2x + 2y$

68. $2c + 2d - cd - d^2$

69. $3c - cd + 3d - c^2$

70. $x^2 + 4y - xy - 4x$

71. $a^2 - 4b + ab - 4a$

72. $7u + v^2 - 7v - uv$

73. $ax + bx - a - b$

74. $x^2y - ax - xy + a$

75. $x^2 + xy + xz + xy + y^2 + zy$

76. $ab - b^2 - bc + ac - bc - c^2$

In Exercises 77–82, factor by grouping. Factor out all common monomials first.

77. $mpx + mqx + npx + nqx$

78. $abd - abe + acd - ace$

79. $x^2y + xy^2 + 2xyz + xy^2 + y^3 + 2y^2z$

80. $a^3 - 2a^2b + a^2c - a^2b + 2ab^2 - abc$

81. $2n^4p - 2n^2 - n^3p^2 + np + 2mn^3p - 2mn$

82. $a^2c^3 + ac^2 + a^3c^2 - 2a^2bc^2 - 2bc^2 + c^3$

In Exercises 83–90, solve for the indicated variable.

83. $r_1r_2 = rr_2 + rr_1$ for r_1

84. $r_1r_2 = rr_2 + rr_1$ for r

85. $d_1d_2 = fd_2 + fd_1$ for f

86. $d_1d_2 = fd_2 + fd_1$ for d_1

87. $b^2x^2 + a^2y^2 = a^2b^2$ for a^2

88. $b^2x^2 + a^2y^2 = a^2b^2$ for b^2

89. $S(1 - r) = a - lr$ for r

90. $Sn = (n - 2)180°$ for n.

APPLICATIONS

91. GEOMETRIC FORMULAS

a. Write an expression that gives the area of the portion of the figure in Illustration 1 that is shaded red.

b. Do the same for the portion of the figure that is shaded blue.

c. Add the results from parts a and b and then factor that expression. What important formula from geometry do you obtain?

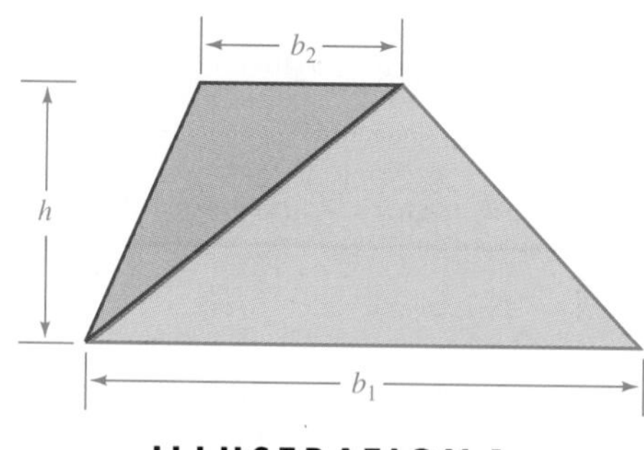

ILLUSTRATION 1

92. PACKAGING The amount of cardboard needed to make the cereal box shown in Illustration 2 can be found by computing the area A, which is given by the formula

$$A = 2wh + 4wl + 2lh$$

where w is the width, h the height, and l the length. Solve the equation for the width.

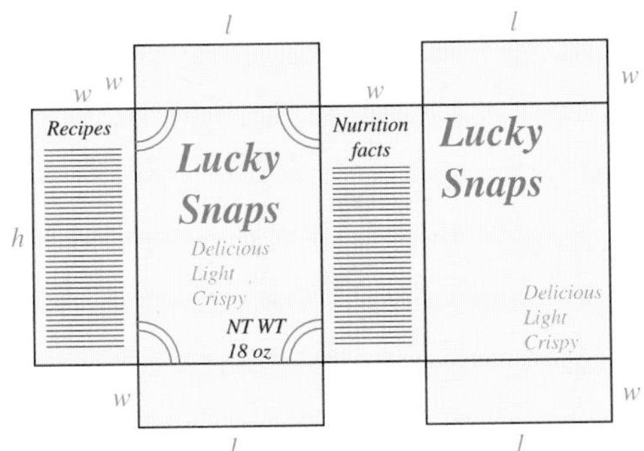

ILLUSTRATION 2

93. LANDSCAPING See Illustration 3. The combined area of the portions of the square lot that the sprinkler doesn't reach is given by $4r^2 - \pi r^2$, where r is the radius of the circular spray. Factor this expression.

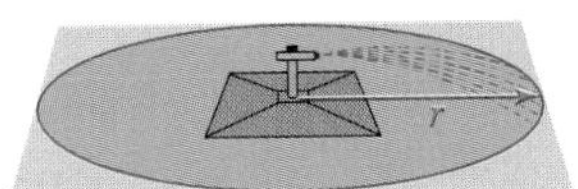

ILLUSTRATION 3

94. CRAYONS The amount of colored wax used to make the crayon shown in Illustration 4 can be found by computing its volume using the formula

$$V = \pi r^2 h_1 + \frac{1}{3}\pi r^2 h_2$$

Factor the expression on the right-hand side of this equation.

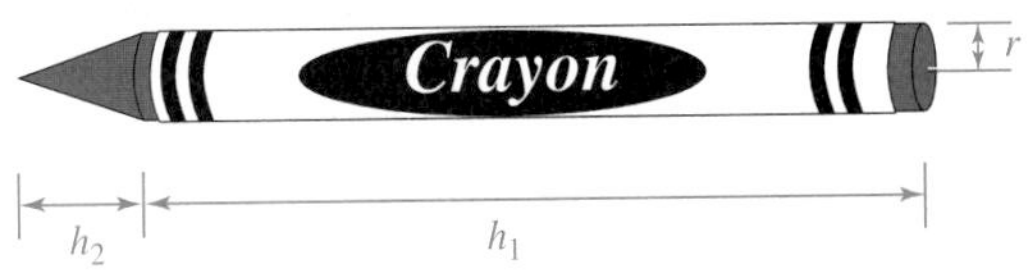

ILLUSTRATION 4

WRITING *Write a paragraph using your own words.*

95. One student commented, "Factoring undoes the distributive property." What do you think she meant?

96. Explain how to find the greatest common factor of two natural numbers.

REVIEW

97. What figure results when the function $f(x) = 3x + 1$ is graphed?

98. What figure results when the function $g(x) = x^2$ is graphed?

99. Solve the inequality $-x > 3$. Express the solution set using interval notation.

100. Evaluate $2|-25 - (-6)(3)|$.

101. If two different lines are parallel, what can be said about their slopes?

102. Are $-3t^2$ and $12t^2$ like terms? If so, combine them.

5.6 *The Difference of Two Squares; the Sum and Difference of Two Cubes*

In this section, you will learn about

- **Perfect squares**
- **The difference of two squares**
- **Perfect cubes**
- **The sum and difference of two cubes**

INTRODUCTION. In this section, we will discuss some special methods of factoring. These methods are applied to polynomials that can be written as the difference of two squares or as the sum or difference of two cubes. To apply these special factoring methods, we must first be able to recognize such polynomials. We begin with a discussion that will help you recognize polynomials with terms that are *perfect squares.*

Perfect squares

To factor the difference of two squares, it is helpful to know the first 20 integers that are **perfect squares.**

1, 4, 9, 16, 25, 36, 49, 64, 81, 100, 121, 144, 169, 196, 225, 256, 289, 324, 361, 400

Expressions such as $x^6y^4z^2$ are also perfect squares, because they can be written as the square of another quantity:

$$x^6y^4z^2 = (x^3y^2z)^2$$

The difference of two squares

In Section 5.4, we developed the special product formula

1. $(x + y)(x - y) = x^2 - y^2$

The binomial $x^2 - y^2$ is called the **difference of two squares,** because x^2 represents the square of x, y^2 represents the square of y, and $x^2 - y^2$ represents the difference of these squares.

Equation 1 can be written in reverse order to give a formula for factoring the difference of two squares.

Factoring the difference of two squares

$$x^2 - y^2 = (x + y)(x - y)$$

If we think of the difference of two squares as the square of a **F**irst quantity minus the square of a **L**ast quantity, we have the formula

$$\text{F}^2 - \text{L}^2 = (\text{F} + \text{L})(\text{F} - \text{L})$$

and we say: *To factor the square of a* ***First*** *quantity minus the square of a* ***Last*** *quantity, we multiply the* ***First*** *plus the* ***Last*** *by the* ***First*** *minus the* ***Last.***

EXAMPLE 1 ***Factoring a difference of two squares.*** Factor $49x^2 - 16$.

Solution

We begin by rewriting the binomial $49x^2 - 16$ as a difference of two squares: $(7x)^2 - (4)^2$. Then we use the formula for factoring the difference of two squares:

$$\begin{array}{ccccccc} \text{F}^2 & - & \text{L}^2 & = (\text{F} & + \text{L})(\text{F} & - & \text{L}) \\ \downarrow & & \downarrow & \downarrow & \downarrow\ \ \downarrow & & \downarrow \\ (7x)^2 & - & 4^2 & = (7x & + 4)(7x & - & 4) \end{array}$$

We can verify this result using the FOIL method to do the multiplication.

$$\begin{aligned}(7x + 4)(7x - 4) &= 49x^2 - 28x + 28x - 16 \\ &= 49x^2 - 16\end{aligned}$$

Self Check

Factor $81p^2 - 25$.

Answer: $(9p + 5)(9p - 5)$ ■

WARNING! Expressions that represent the sum of two squares, such as $(7x)^2 + 4^2$, cannot be factored in the real number system. The binomial $49x^2 + 16$ is a prime binomial.

EXAMPLE 2 ***Factoring a difference of two squares.*** Factor $64a^4 - 25b^2$.

Solution

We can write $64a^4 - 25b^2$ in the form $(8a^2)^2 - (5b)^2$ and use the formula for factoring the difference of two squares.

$$\begin{array}{ccccc} F^2 & - & L^2 & = & (F + L)(F - L) \\ \downarrow & & \downarrow & & \downarrow \quad \downarrow \quad \downarrow \quad \downarrow \\ (8a^2)^2 & - & (5b)^2 & = & (8a^2 + 5b)(8a^2 - 5b) \end{array}$$

Verify by multiplication.

Self Check
Factor $36r^4 - s^2$.

Answer: $(6r^2 + s)(6r^2 - s)$

EXAMPLE 3 ***Using the formula for the difference of two squares twice.*** Factor $x^4 - 1$.

Solution

Because the binomial is the difference of the squares of x^2 and 1, it factors into the sum of x^2 and 1 and the difference of x^2 and 1.

$$\begin{aligned} x^4 - 1 &= (x^2)^2 - (1)^2 \\ &= (x^2 + 1)(x^2 - 1) \end{aligned}$$

The factor $x^2 + 1$ is the sum of two quantities and is prime. However, the factor $x^2 - 1$ is the difference of two squares and can be factored as $(x + 1)(x - 1)$. Thus,

$$\begin{aligned} x^4 - 1 &= (x^2 + 1)(x^2 - 1) \\ &= (x^2 + 1)(x + 1)(x - 1) \end{aligned}$$

Self Check
Factor $a^4 - 81$.

Answer:
$(a^2 + 9)(a + 3)(a - 3)$

WARNING! When asked to factor a polynomial, we must be sure to factor it completely. After factoring a polynomial, always check to see whether any of the factors in the result can be factored further.

EXAMPLE 4 ***Using the formula for the difference of two squares twice.*** Factor $(x + y)^4 - z^4$.

Solution

This expression is the difference of two squares and can be factored:

$$\begin{aligned} (x + y)^4 - z^4 &= [(x + y)^2]^2 - (z^2)^2 \\ &= [(x + y)^2 + z^2][(x + y)^2 - z^2] \end{aligned}$$

The factor $(x + y)^2 + z^2$ is the sum of two squares and is prime. However, the factor $(x + y)^2 - z^2$ is the difference of two squares and can be factored as $(x + y + z)(x + y - z)$. Thus,

$$\begin{aligned} (x + y)^4 - z^4 &= [(x + y)^2 + z^2][(x + y)^2 - z^2] \\ &= [(x + y)^2 + z^2](x + y + z)(x + y - z) \end{aligned}$$

Self Check
Factor $(a - b)^4 - c^4$.

Answer:
$[(a + b)^2 + c^2] \times$
$(a - b + c)(a - b - c)$

When possible, we always factor out a common factor before factoring the difference of two squares. The factoring process is easier when all common factors are factored out first.

EXAMPLE 5 ***Factoring out the GCF first.*** Factor $2x^4y - 32y$.

Solution

$$\begin{aligned} 2x^4y - 32y &= 2y(x^4 - 16) && \text{Factor out the GCF, which is } 2y. \\ &= 2y(x^2 + 4)(x^2 - 4) && \text{Factor } x^4 - 16. \\ &= 2y(x^2 + 4)(x + 2)(x - 2) && \text{Factor } x^2 - 4. \end{aligned}$$

Self Check

Factor $3a^4 - 3$.

Answer: $3(a^2 + 1)(a + 1)(a - 1)$

EXAMPLE 6 ***Factoring by grouping.*** Factor $x^2 - y^2 + x - y$.

Solution

If we group the first two terms and factor the difference of two squares, we have

$$\begin{aligned} x^2 - y^2 + x - y &= (x + y)(x - y) + (x - y) && \text{Factor } x^2 - y^2. \\ &= (x - y)(x + y + 1) && \text{Factor out } x - y. \end{aligned}$$

Self Check

Factor $a^2 - b^2 + a + b$.

Answer: $(a + b)(a - b + 1)$

Perfect cubes

The number 64 is called a perfect cube, because $4^3 = 64$. To factor the sum or difference of two cubes, it is helpful to know the first ten perfect cubes:

1, 8, 27, 64, 125, 216, 343, 512, 729, 1,000

Expressions such as $x^9y^6z^3$ are also perfect cubes, because they can be written as the cube of another quantity:

$$x^9y^6z^3 = (x^3y^2z)^3$$

The sum and difference of two cubes

To find formulas for factoring the sum or difference of two cubes, we use the following product formulas:

2. $(x + y)(x^2 - xy + y^2) = x^3 + y^3$

3. $(x - y)(x^2 + xy + y^2) = x^3 - y^3$

To verify Equation 2, we multiply $x^2 - xy + y^2$ by $x + y$.

$$\begin{aligned} (x + y)(x^2 - xy + y^2) &= (x + y)x^2 - (x + y)xy + (x + y)y^2 \\ &= x \cdot x^2 + y \cdot x^2 - x \cdot xy - y \cdot xy + x \cdot y^2 + y \cdot y^2 \\ &= x^3 + x^2y - x^2y - xy^2 + xy^2 + y^3 \\ &= x^3 + y^3 \end{aligned}$$

Equation 3 can also be verified by multiplication.

If we write Equations 2 and 3 in reverse order, we have the formulas for factoring the sum and difference of two cubes.

Sum and difference of two cubes

$$x^3 + y^3 = (x + y)(x^2 - xy + y^2)$$
$$x^3 - y^3 = (x - y)(x^2 + xy + y^2)$$

If we think of the sum of two cubes as the sum of the cube of a **F**irst quantity plus the cube of a **L**ast quantity, we have the formula

$$\mathrm{F}^3 + \mathrm{L}^3 = (\mathrm{F} + \mathrm{L})(\mathrm{F}^2 - \mathrm{FL} + \mathrm{L}^2)$$

*To factor the cube of a **First** quantity plus the cube of a **Last** quantity, we multiply the sum of the **First** and **Last** by*

- *the **First** squared*
- *minus the **First** times the **Last***
- *plus the **Last** squared.*

The formula for the difference of two cubes is

$$F^3 - L^3 = (F - L)(F^2 + FL + L^2)$$

*To factor the cube of a **First** quantity minus the cube of a **Last** quantity, we multiply the difference of the **First** and **Last** by*

- *The **First** squared*
- *plus the **First** times the **Last***
- *plus the **Last** squared.*

EXAMPLE 7 ***Factoring a sum of two cubes.*** Factor $a^3 + 8$.

Solution

Since $a^3 + 8$ can be written as $a^3 + 2^3$, we have the sum of two cubes, which factors as follows:

$$\begin{aligned} F^3 + L^3 &= (F + L)(F^2 - FL + L^2) \\ a^3 + 2^3 &= (a + 2)(a^2 - a2 + 2^2) \\ &= (a + 2)(a^2 - 2a + 4) \end{aligned}$$

Thus, $a^3 + 8 = (a + 2)(a^2 - 2a + 4)$. Check by multiplication.

Self Check

Factor $p^3 + 27$.

Answer: $(p + 3)(p^2 - 3p + 9)$

EXAMPLE 8 ***Factoring a difference of two cubes.*** Factor $27a^3 - 64b^3$.

Solution

Since $27a^3 - 64b^3$ can be written as $(3a)^3 - (4b)^3$, we have the difference of two cubes, which factors as follows:

$$\begin{aligned} F^3 - L^3 &= (F - L)(F^2 + F\,L + L^2) \\ (3a)^3 - (4b)^3 &= (3a - 4b)[(3a)^2 + (3a)(4b) + (4b)^2] \\ &= (3a - 4b)(9a^2 + 12ab + 16b^2) \end{aligned}$$

Thus, $27a^3 - 64b^3 = (3a - 4b)(9a^2 + 12ab + 16b^2)$. Check by multiplication.

Self Check

Factor $8p^3 - 27q^3$.

Answer: $(2p - 3q)(4p^2 + 6pq + 9q^2)$

EXAMPLE 9 ***Factoring a difference of two cubes.*** Factor $a^3 - (c + d)^3$.

Solution

$$a^3 - (c + d)^3 = [a - (c + d)][a^2 + a(c + d) + (c + d)^2]$$

Now we simplify the expressions inside both sets of brackets.

$$a^3 - (c + d)^3 = (a - c - d)(a^2 + ac + ad + c^2 + 2cd + d^2)$$

Self Check

Factor $(p + q)^3 - r^3$.

Answer: $(p + q - r)(p^2 + 2pq + q^2 + pr + qr + r^2)$

EXAMPLE 10 ***Factoring completely.*** Factor $x^6 - 64$.

Solution

This expression is both the difference of two squares and the difference of two cubes. It is easier to factor it as the difference of two squares first. This expression factors into the product of a sum and a difference.

$$x^6 - 64 = (x^3)^2 - 8^2$$
$$= (x^3 + 8)(x^3 - 8)$$

Each of these factors further, however, for one is the sum of two cubes and the other is the difference of two cubes:

$$x^6 - 64 = (x + 2)(x^2 - 2x + 4)(x - 2)(x^2 + 2x + 4)$$

Self Check

Factor $x^6 - 1$.

Answer: $(x + 1)(x^2 - x + 1)(x - 1)(x^2 + x + 1)$ ■

EXAMPLE 11 ***Factoring out the GCF.*** Factor $2a^5 + 128a^2$.

Solution

We first factor out the common monomial factor of $2a^2$ to obtain

$$2a^5 + 128a^2 = 2a^2(a^3 + 64)$$

Then we factor $a^3 + 64$ as the sum of two cubes to obtain

$$2a^5 + 128a^2 = 2a^2(a + 4)(a^2 - 4a + 16)$$

Self Check

Factor $3x^5 + 24x^2$.

Answer: $3x^2(x + 2)(x^2 - 2x + 4)$ ■

STUDY SET Section 5.6

VOCABULARY *In Exercises 1–2, fill in the blanks to make the statements true.*

1. When the polynomial $4x^2 - 25$ is rewritten as $(2x)^2 - (5)^2$, we see that it is the difference of two ____________.

2. When the polynomial $8x^3 + 125$ is rewritten as $(2x)^3 + (5)^3$, we see that it is the sum of two ____________.

CONCEPTS

3. Write the first ten perfect square natural numbers.

4. Write the first ten perfect cube natural numbers.

5. Use multiplication to verify that the sum of two squares $x^2 + 25$ does not factor as $(x + 5)(x + 5)$.

6. Use multiplication to verify that the difference of two squares $x^2 - 25$ factors as $(x + 5)(x - 5)$.

7. Explain why each factorization is not complete.
   a. $4g^2 - 16 = (2g + 4)(2g - 4)$
   b. $1 - t^8 = (1 + t^4)(1 - t^4)$

8. When asked to factor $81t^2 - 16$, one student answered $(9t - 4)(9t + 4)$, and another answered $(9t + 4)(9t - 4)$. Explain why both students are correct.

9. Factor each polynomial.
   a. $5p^2 + 20$
   b. $5p^2 - 20$

10. Factor each polynomial.
   a. $5p^3 + 20$
   b. $5p^3 + 40$

NOTATION *In Exercises 11–16, complete each factorization.*

11. $p^3 + q^3 = (p + q)\boxed{}$

12. $p^3 - q^3 = (p - q)\boxed{}$

13. $p^2 - q^2 = (p + q)\boxed{}$

14. $p^2q + pq^2 = \boxed{}(p + q)$

15. $36y^2 - 49m^2 = (\boxed{})^2 - (7m)^2$
$= (6y \boxed{} 7m)(6y - \boxed{})$

16. $h^3 - 27k^3 = (h)^3 - (\boxed{})^3$
$= (h \boxed{} 3k)(h^2 + \boxed{} + 9k^2)$

PRACTICE *In Exercises 17–36, factor each polynomial, if possible.*

17. $x^2 - 4$

18. $y^2 - 9$

19. $9y^2 - 64$

20. $16x^4 - 81y^2$

21. $x^2 + 25$

22. $144a^2 - b^4$

23. $625a^2 - 169b^4$

24. $4y^2 + 9z^4$

25. $81a^4 - 49b^2$

26. $64r^6 - 121s^2$

27. $36x^4y^2 - 49z^4$

28. $4a^2b^4c^6 - 9d^8$

29. $(x + y)^2 - z^2$

30. $a^2 - (b - c)^2$

31. $(a - b)^2 - c^2$

32. $(m + n)^2 - p^4$

33. $x^4 - y^4$

34. $16a^4 - 81b^4$

35. $256x^4y^4 - z^8$

36. $225a^4 - 16b^8c^{12}$

In Exercises 37–44, factor each polynomial.

37. $2x^2 - 288$

38. $8x^2 - 72$

39. $2x^3 - 32x$

40. $3x^3 - 243x$

41. $5x^3 - 125x$

42. $6x^4 - 216x^2$

43. $r^2s^2t^2 - t^2x^4y^2$

44. $16a^4b^3c^4 - 64a^2bc^6$

In Exercises 45–50, factor each polynomial by grouping.

45. $a^2 - b^2 + a + b$

46. $x^2 - y^2 - x - y$

47. $a^2 - b^2 + 2a - 2b$

48. $m^2 - n^2 + 3m + 3n$

49. $2x + y + 4x^2 - y^2$

50. $m - 2n + m^2 - 4n^2$

In Exercises 51–60, factor each polynomial.

51. $r^3 + s^3$

52. $t^3 - v^3$

53. $x^3 - 8y^3$

54. $27a^3 + b^3$

55. $64a^3 - 125b^6$

56. $8x^6 + 125y^3$

57. $125x^3y^6 + 216z^9$

58. $1{,}000a^6 - 343b^3c^6$

59. $x^6 + y^6$

60. $x^9 + y^9$

In Exercises 61–68, factor each expression.

61. $5x^3 + 625$

62. $2x^3 - 128$

63. $4x^5 - 256x^2$

64. $2x^6 + 54x^3$

65. $128u^2v^3 - 2t^3u^2$

66. $56rs^2t^3 + 7rs^2v^6$

67. $(a + b)x^3 + 27(a + b)$

68. $(c - d)r^3 - (c - d)s^3$

APPLICATIONS

69. CANDY To find the amount of chocolate used in the outer coating of the malted-milk ball shown in Illustration 1, we can find the volume V of the chocolate shell using the formula

$$V = \frac{4}{3}\pi r_1^{\,3} - \frac{4}{3}\pi r_2^{\,3}$$

Factor the expression on the right-hand side of the formula.

70. MOVIE STUNTS See Illustration 2. The function that gives the distance a stuntwoman is above the ground t seconds after she fell over the side of a 144-foot tall building is

$$h(t) = 144 - 16t^2$$

Factor the right-hand side of the equation.

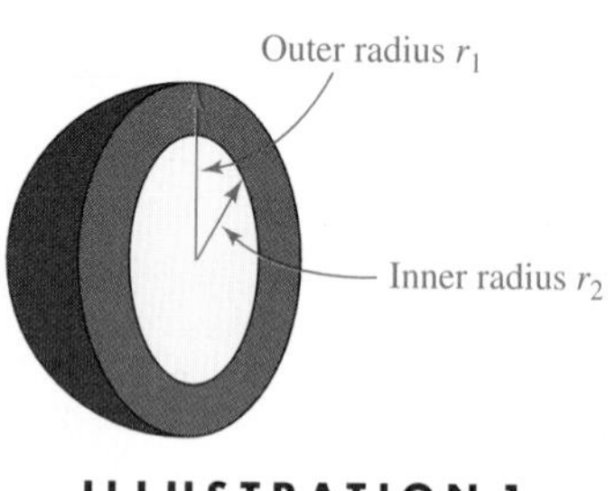

ILLUSTRATION 1

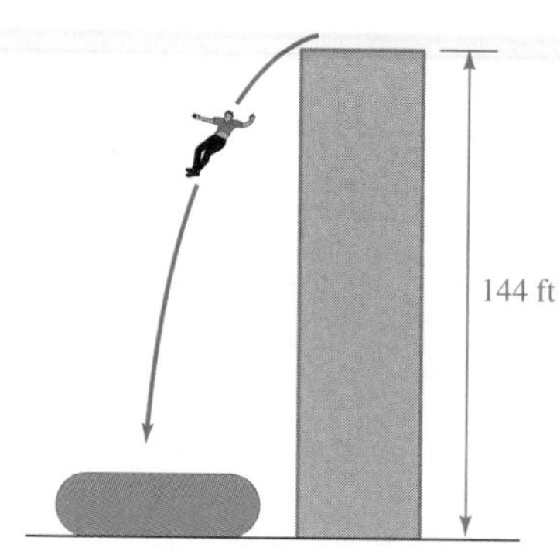

ILLUSTRATION 2

WRITING *Write a paragraph using your own words.*

71. Describe the pattern used to factor the difference of two squares.

72. Describe the patterns used to factor the sum and the difference of two cubes.

REVIEW *In Exercises 73–76, graph the line with the given characteristics.*

73. Passing through $(-2, -1)$; slope $= -\frac{2}{3}$

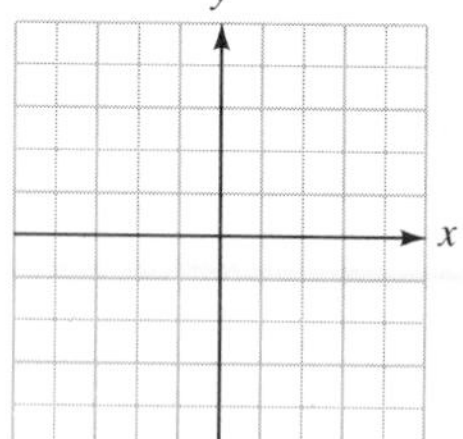

74. y-intercept $(0, -4)$; slope $= 3$

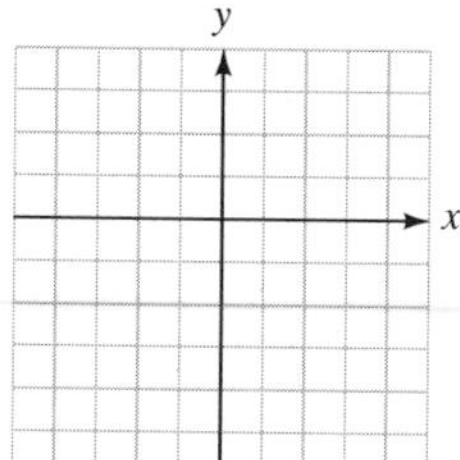

75. Horizontal; y-intercept $(0, -2)$

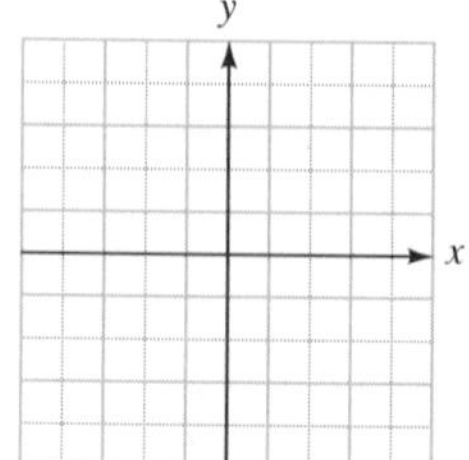

76. Parallel to the y-axis, passing through $(1, 4)$

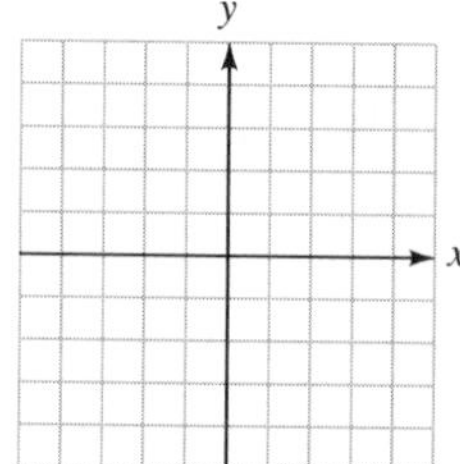

5.7 Factoring Trinomials

In this section, you will learn about

- **Perfect square trinomials**
- **Factoring trinomials with lead coefficients of 1**
- **Factoring trinomials with lead coefficients other than 1**
- **Test for factorability**
- **Using substitution to factor trinomials**
- **Factoring by grouping**
- **Using grouping to factor trinomials**

INTRODUCTION. In this section, we will discuss several techniques for factoring trinomials. These techniques are based on the fact that the product of two binomials is often a trinomial. With that observation in mind, we begin the study of trinomial factoring by considering two special products.

Perfect square trinomials

Many trinomials can be factored by using the following special product formulas.

1. $(x + y)(x + y) = x^2 + 2xy + y^2$
2. $(x - y)(x - y) = x^2 - 2xy + y^2$

To factor $x^2 + 6x + 9$, we note that it can be written in the form $x^2 + 2(3)x + 3^2$. If $y = 3$, this form matches the right-hand side of Equation 1. Thus, $x^2 + 6x + 9$ factors as

$$\begin{aligned} x^2 + 6x + 9 &= x^2 + 2(3)x + 3^2 \\ &= (x + 3)(x + 3) \\ &= (x + 3)^2 \end{aligned}$$

Since $x^2 + 6x + 9$ is the square of $x + 3$, $x^2 + 6x + 9$ is called a **perfect square trinomial.** This result can be verified by multiplication:

$$\begin{aligned} (x + 3)(x + 3) &= x^2 + 3x + 3x + 9 \\ &= x^2 + 6x + 9 \end{aligned}$$

EXAMPLE 1 ***Factoring a perfect square trinomial.*** Factor $x^2 - 4xz + 4z^2$.

Solution

To factor the perfect square trinomial $x^2 - 4xz + 4z^2$, we note that it can be written in the form $x^2 - 2x(2z) + (2z)^2$. If $y = 2z$, this form matches the right-hand side of Equation 2.

$$\begin{aligned} x^2 - 4xz + 4z^2 &= x^2 - 2x(2z) + (2z)^2 \\ &= (x - 2z)(x - 2z) \\ &= (x - 2z)^2 \end{aligned}$$

This result can be verified by multiplication.

Self Check

Factor $b^2 - 10b + 25$.

Answer: $(b - 5)^2$

We begin our discussion of *general trinomials* by considering trinomials with lead coefficients (the coefficient of the squared term) of 1.

Factoring trinomials with lead coefficients of 1

Since the product of two binomials is often a trinomial, we expect that many trinomials will factor as two binomials. For example, to factor $x^2 + 7x + 12$, we must find two binomials $x + a$ and $x + b$ such that

$$x^2 + 7x + 12 = (x + a)(x + b)$$

where $ab = 12$ and $ax + bx = 7x$.

To find the numbers a and b, we list the possible factorizations of 12 and find the one where the sum of the factors is 7.

The one to choose ↓

$$12(1) \quad 6(2) \quad 4(3) \quad -12(-1) \quad -6(-2) \quad -4(-3)$$

Thus, $a = 4$, $b = 3$, and

$$x^2 + 7x + 12 = (x + a)(x + b)$$

3. $x^2 + 7x + 12 = (x + 4)(x + 3)$

This factorization can be verified by multiplying $x + 4$ and $x + 3$ and observing that the product is $x^2 + 7x + 12$.

Because of the commutative property of multiplication, the order of the factors in Equation 3 is not important.

To factor trinomials with lead coefficients of 1, we follow these steps.

Factoring trinomials with lead coefficients of 1

1. Write the trinomial in descending powers of one variable.
2. List the factorizations of the third term of the trinomial.
3. Pick the factorization where the sum of the factors is the coefficient of the middle term.

EXAMPLE 2 ***Factoring a trinomial with lead coefficient of 1.***
Factor $x^2 - 6x + 8$.

Solution

Since the trinomial is written in descending powers of x, we can move to step 2 and list the possible factorizations of the third term, which is 8.

The one to choose ↓

$$8(1) \quad 4(2) \quad -8(-1) \quad -4(-2)$$

In the trinomial, the coefficient of the middle term is -6. The only factorization where the sum of the factors is -6 is $-4(-2)$. Thus, $a = -4$, $b = -2$, and

$$x^2 - 6x + 8 = (x + a)(x + b)$$
$$= (x - 4)(x - 2)$$

We can verify this result by multiplication:

$$(x - 4)(x - 2) = x^2 - 2x - 4x + 8 \quad \text{Use the FOIL method.}$$
$$= x^2 - 6x + 8$$

Self Check

Factor $a^2 - 7a + 12$.

Answer: $(a - 4)(a - 3)$

EXAMPLE 3 ***Factoring a trinomial.*** Factor $-x + x^2 - 12$.

Solution

We begin by writing the trinomial in descending powers of x:

$$-x + x^2 - 12 = x^2 - x - 12$$

The possible factorizations of the third term are

The one to choose ↓

$$12(-1) \quad 6(-2) \quad 4(-3) \quad 1(-12) \quad 2(-6) \quad 3(-4)$$

In the trinomial, the coefficient of the middle term is -1. The only factorization where the sum of the factors is -1 is $3(-4)$. Thus, $a = 3$, $b = -4$, and

$$\begin{aligned} -x + x^2 - 12 &= (x + a)(x + b) \\ &= (x + 3)(x - 4) \end{aligned}$$

Self Check

Factor $-3a + a^2 - 10$.

Answer: $(a + 2)(a - 5)$ ■

EXAMPLE 4 ***Factoring completely.*** Factor $30x - 4xy - 2xy^2$.

Solution

We begin by writing the trinomial in descending powers of y:

$$30x - 4xy - 2xy^2 = -2xy^2 - 4xy + 30x$$

Each term in this trinomial has a common monomial factor of $-2x$, which we will factor out.

$$30x - 4xy - 2xy^2 = -2x(y^2 + 2y - 15)$$

To factor $y^2 + 2y - 15$, we list the factors of -15 and find the pair whose sum is 2.

The one to choose ↓

$$15(-1) \quad 5(-3) \quad 1(-15) \quad 3(-5)$$

The only factorization where the sum of the factors is 2 (the coefficient of the middle term of $y^2 + 2y - 15$) is $5(-3)$. Thus, $a = 5$, $b = -3$, and

$$\begin{aligned} 30x - 4xy - 2xy^2 &= -2x(y^2 + 2y - 15) \\ &= -2x(y + 5)(y - 3) \end{aligned}$$

Self Check

Factor $18a + 3ab - 3ab^2$.

Answer: $-3a(b + 2)(b - 3)$ ■

WARNING! In Example 4, be sure to include all factors in the final result. It is a common error to forget to write the $-2x$.

Factoring trinomials with lead coefficients other than 1

There are more combinations of factors to consider when factoring trinomials with lead coefficients other than 1. To factor $5x^2 + 7x + 2$, for example, we must find two binomials of the form $ax + b$ and $cx + d$ such that

$$5x^2 + 7x + 2 = (ax + b)(cx + d)$$

Since the first term of the trinomial $5x^2 + 7x + 2$ is $5x^2$, the first terms of the binomial factors must be $5x$ and x.

$5x^2$

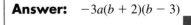

$$5x^2 + 7x + 2 = (5x + b)(x + d)$$

Since the product of the last terms must be 2, and the sum of the products of the outer and inner terms must be $7x$, we must find two numbers whose product is 2 that will give a middle term of $7x$.

$$5x^2 + 7x + 2 = (5x + b)(x + d)$$

2

O + I = $7x$

Since 2(1) and $(-2)(-1)$ give a product of 2, there are four possible combinations to consider:

$$\mathbf{(5x + 2)(x + 1)} \qquad (5x - 2)(x - 1)$$
$$(5x + 1)(x + 2) \qquad (5x - 1)(x - 2)$$

Of these possibilities, only the first one gives the correct middle term of $7x$. Thus,

4. $5x^2 + 7x + 2 = (5x + 2)(x + 1)$

We can verify this result by multiplication:

$$\begin{aligned}(5x + 2)(x + 1) &= 5x^2 + 5x + 2x + 2\\ &= 5x^2 + 7x + 2\end{aligned}$$

Test for factorability

If a trinomial has the form $ax^2 + bx + c$, with integer coefficients and $a \neq 0$, we can test to see whether it is factorable.

- If the value of $b^2 - 4ac$ is a perfect square, the trinomial can be factored using only integers.
- If the value is not a perfect square, the trinomial cannot be factored using only integers.

For example, $5x^2 + 7x + 2$ is a trinomial in the form $ax^2 + bx + c$ with

$$a = 5, \qquad b = 7, \qquad \text{and} \qquad c = 2$$

For this trinomial, the value of $b^2 - 4ac$ is

$$\begin{aligned}b^2 - 4ac &= 7^2 - 4(5)(2)\\ &= 49 - 40\\ &= 9\end{aligned}$$

Since 9 is a perfect square, the trinomial is factorable. Its factorization is shown in Equation 4.

Test for factorability

A trinomial of the form $ax^2 + bx + c$, with integer coefficients and $a \neq 0$, will factor into two binomials with integer coefficients if the value of $b^2 - 4ac$ is a perfect square. If $b^2 - 4ac = 0$, the factors will be the same.

EXAMPLE 5 ***A lead coefficient that is not 1.*** Factor $3p^2 - 4p - 4$.

Solution

In the trinomial, $a = 3$, $b = -4$, and $c = -4$. To see whether it factors, we evaluate $b^2 - 4ac$.

$$\begin{aligned}b^2 - 4ac &= (-4)^2 - 4(3)(-4)\\ &= 16 + 48\\ &= 64\end{aligned}$$

Self Check

Factor $4q^2 - 9q - 9$.

Since 64 is a perfect square, the trinomial is factorable.

To factor the trinomial, we note that the first terms of the binomial factors must be $3p$ and p to give the first term of $3p^2$.

$$3p^2 - 4p - 4 = (3p + ?)(p + ?)$$

The product of the last terms must be -4, and the sum of the products of the outer terms and the inner terms must be $-4p$.

$$3p^2 - 4p - 4 = (3p + ?)(p + ?)$$

$$\text{O} + \text{I} = -4p$$

Because $1(-4)$, $-1(4)$, and $-2(2)$ all give a product of -4, there are six possible combinations to consider:

$(3p + 1)(p - 4)$	$(3p - 4)(p + 1)$
$(3p - 1)(p + 4)$	$(3p + 4)(p - 1)$
$(3p - 2)(p + 2)$	$\mathbf{(3p + 2)(p - 2)}$

Of these possibilities, only the last gives the required middle term of $-4p$. Thus,

$$3p^2 - 4p - 4 = (3p + 2)(p - 2)$$

Answer: $(4q + 3)(q - 3)$

EXAMPLE 6 ***Testing for factorability.*** Factor $4t^2 - 3t - 5$, if possible.

Solution

In the trinomial, $a = 4$, $b = -3$, and $c = -5$. To see whether the trinomial is factorable, we evaluate $b^2 - 4ac$ by substituting the values of a, b, and c.

$$\begin{aligned} b^2 - 4ac &= (-3)^2 - 4(4)(-5) \\ &= 9 + 80 \\ &= 89 \end{aligned}$$

Since 89 is not a perfect square, the trinomial is not factorable using only integer coefficients.

Self Check

Factor $5a^2 - 8a + 2$, if possible.

Answer: a prime polynomial

It is not easy to give specific rules for factoring general trinomials. However, the following hints are helpful.

Factoring a general trinomial

1. Write the trinomial in descending powers of one variable.
2. Factor out any greatest common factor (including -1, if that is necessary to make the coefficient of the first term positive).
3. Test the trinomial for factorability.
4. When the sign of the first term of a trinomial is $+$ and the sign of the third term is $+$, the signs between the terms of each binomial factor are the same as the sign of the middle term of the trinomial.

   When the sign of the first term is $+$ and the sign of the third term is $-$, the signs between the terms of the binomials are opposite.
5. Try various combinations of the factors of the first terms and the last terms until you find the one that works.
6. Check the factorization by multiplication.

EXAMPLE 7 ***A lead coefficient that is not 1.*** Factor $24y + 10xy - 6x^2y$.

Self Check
Factor $9b - 6a^2b - 3ab$.

Solution

We write the trinomial in descending powers of x and factor out the common factor of $-2y$:

$$24y + 10xy - 6x^2y = -6x^2y + 10xy + 24y$$
$$= -2y(3x^2 - 5x - 12)$$

In the trinomial $3x^2 - 5x - 12$, $a = 3$, $b = -5$, and $c = -12$.

$$b^2 - 4ac = (-5)^2 - 4(3)(-12)$$
$$= 25 + 144$$
$$= 169$$

Since 169 is a perfect square, the trinomial will factor.

Since the sign of the first term of $3x^2 - 5x - 12$ is positive and the sign of the third term is negative, the signs between the binomial factors will be opposite. Because the first term is $3x^2$, the first terms of the binomial factors must be $3x$ and x.

$$24y + 10xy - 6x^2y = -2y(3x \quad)(x \quad)$$

(The product of the first terms is $3x^2$.)

The product of the last terms must be -12, and the sum of the product of the outer terms and the product of the inner terms must be $-5x$.

$$24y + 10xy - 6x^2y = -2y(3x \quad ?)(x \quad ?)$$

(The product of the last terms is -12; $\text{O} + \text{I} = -5x$.)

Since $1(-12)$, $2(-6)$, $3(-4)$, $12(-1)$, $6(-2)$, and $4(-3)$ all give a product of -12, there are 12 possible combinations to consider.

	$(3x + 1)(x - 12)$	$(3x - 12)(x + 1)$
	$(3x + 2)(x - 6)$	$(3x - 6)(x + 2)$
	$(3x + 3)(x - 4)$	$(3x - 4)(x + 3)$
	$(3x + 12)(x - 1)$	$(3x - 1)(x + 12)$
	$(3x + 6)(x - 2)$	$(3x - 2)(x + 6)$
The one to choose →	$\mathbf{(3x + 4)(x - 3)}$	$(3x - 3)(x + 4)$

The combinations marked in color cannot work, because one of the factors has a common factor. This implies that $3x^2 - 5x - 12$ would have a common factor, which it doesn't.

After mentally trying the remaining combinations, we find that only $(3x + 4)(x - 3)$ gives the proper middle term of $-5x$.

$$24y + 10xy - 6x^2y = -2y(3x^2 - 5x - 12)$$
$$= -2y(3x + 4)(x - 3)$$

Verify this result by multiplication.

Answer: $-3b(2a + 3)(a - 1)$

EXAMPLE 8 ***A lead coefficient that is not 1.*** Factor $6y + 13x^2y + 6x^4y$.

Solution

We write the trinomial in descending powers of x and factor out the common factor of y to obtain

$$\begin{aligned}6y + 13x^2y + 6x^4y &= 6x^4y + 13x^2y + 6y \\ &= y(6x^4 + 13x^2 + 6)\end{aligned}$$

A test for factorability will show that $6x^4 + 13x^2 + 6$ will factor.

Since the coefficients of the first and last terms of $6x^4 + 13x^2 + 6$ are positive, the signs between the terms in each binomial will be $+$.

Since the first term of the trinomial is $6x^4$, the first terms of the binomial factors must be either $2x^2$ and $3x^2$ or x^2 and $6x^2$.

Since the product of the last terms of the binomial factors must be 6, we must find two numbers whose product is 6 that will lead to a middle term of $7x^2$. After trying some combinations, we find the one that works.

$$\begin{aligned}6y + 13x^2y + 6x^4y &= y(6x^4 + 13x^2 + 6) \\ &= y(2x^2 + 3)(3x^2 + 2)\end{aligned}$$

Verify this result by multiplication.

Self Check

Factor $4b + 11a^2b + 6a^4b$.

Answer: $b(2a^2 + 1)(3a^2 + 4)$ ■

Using substitution to factor trinomials

For more complicated expressions, a substitution sometimes helps to simplify the factoring process.

EXAMPLE 9 ***Using substitution.*** Factor $(x + y)^2 + 7(x + y) + 12$.

Solution

We rewrite the trinomial $(x + y)^2 + 7(x + y) + 12$ as $z^2 + 7z + 12$, where $z = x + y$. The trinomial $z^2 + 7z + 12$ factors as $(z + 4)(z + 3)$.

To find the factorization of $(x + y)^2 + 7(x + y) + 12$, we substitute $x + y$ for z in the expression $(z + 4)(z + 3)$ to obtain

$$\begin{aligned}z^2 + 7z + 12 &= (z + 4)(z + 3) \\ (x + y)^2 + 7(x + y) + 12 &= (x + y + 4)(x + y + 3)\end{aligned}$$

Self Check

Factor $(a + b)^2 - 3(a + b) - 10$.

Answer:
$(a + b + 2)(a + b - 5)$ ■

Factoring by grouping

EXAMPLE 10 ***Factoring by grouping.*** Factor $x^2 + 6x + 9 - z^2$.

Solution

We group the first three terms together and factor the trinomial to get

$$\begin{aligned}x^2 + 6x + 9 - z^2 &= (x + 3)(x + 3) - z^2 \\ &= (x + 3)^2 - z^2\end{aligned}$$

We can now factor the difference of two squares to get

$$x^2 + 6x + 9 - z^2 = (x + 3 + z)(x + 3 - z)$$

Self Check

Factor $a^2 + 4a + 4 - b^2$.

Answer:
$(a + 2 + b)(a + 2 - b)$ ■

Using grouping to factor trinomials

The method of factoring by grouping can be used to help factor trinomials of the form $ax^2 + bx + c$. For example, to factor the trinomial $6x^2 + 7x - 3$, we proceed as follows:

1. First find the product ac: $6(-3) = -18$. This number is called the **key number.**
2. Find two factors of the key number -18 whose sum is $b = 7$:

$$9(-2) = -18 \qquad \text{and} \qquad 9 + (-2) = 7$$

3. Use the factors 9 and -2 as coefficients of two terms to be placed between $6x^2$ and -3:

$$6x^2 + 7x - 3 = 6x^2 + 9x - 2x - 3$$

4. Factor by grouping:

$$\begin{aligned} 6x^2 + 9x - 2x - 3 &= 3x(2x + 3) - 1(2x + 3) \\ &= (2x + 3)(3x - 1) \qquad \text{Factor out } 2x + 3. \end{aligned}$$

We can verify this factorization by multiplication.

EXAMPLE 11 ***The grouping method.*** Factor $10x^2 + 13x - 3$.

Solution

Since $a = 10$ and $c = -3$ in the trinomial, $ac = -30$. We now find two factors of -30 whose sum is $+13$. Two such factors are 15 and -2. We use these factors as coefficients of two terms to be placed between $10x^2$ and -3:

$$10x^2 + 15x - 2x - 3$$

Finally, we factor by grouping.

$$5x(2x + 3) - 1(2x + 3) = (2x + 3)(5x - 1)$$

Self Check

Factor $15a^2 + 17a - 4$.

Answer: $(3a + 4)(5a - 1)$ ■

STUDY SET Section 5.7

VOCABULARY *In Exercises 1–4, fill in the blanks to make the statements true.*

1. A polynomial with three terms, such as $3x^2 - 2x + 4$, is called a ______________.
2. Since $y^2 + 2y + 1$ is the square of $y + 1$, we call $y^2 + 2y + 1$ a ______________ square trinomial.
3. For $a^2 - a - 6$, the ______________ coefficient (the coefficient of the a^2 term) is 1.
4. The trinomial $4a^2 - 5a - 6$ is written in ______________ powers of a.

CONCEPTS

5. Consider $3x^2 - x + 16$. What is the sign of the
   a. First term?
   b. Middle term?
   c. Last term?
6. Explain what is meant when we say that the trinomial $h^2 - 12h + 27$ can be written as the product of two binomials.
7. If $b^2 - 4ac$ is a perfect square, the trinomial $ax^2 + bx + c$ can be factored using what type of coefficients?
8. Use the substitution $x = a + b$ to rewrite the trinomial $6(a + b)^2 - 17(a + b) - 3$.

NOTATION *In Exercises 9–12, find each product.*

9. $(x + y)(x + y) = x^2 + \square$

10. $(x - y)(x - y) = x^2 - \square$

11. $(x + y)(x - y) = \square$

12. $(a + b)(a + b) = \square + b^2$

13. The trinomial $4m^2 - 4m + 1$ is written in $ax^2 + bx + c$ form. Identify a, b, and c.

14. Consider the trinomial $15s^2 + 4s - 4$. Is $b^2 - 4ac$ a perfect square?

PRACTICE *In Exercises 15–20, complete each factorization.*

15. $x^2 + 5x + 6 = (x + 3)\square$

16. $x^2 - 6x + 8 = (x - 4)\square$

17. $x^2 + 2x - 15 = (x + 5)\square$

18. $x^2 - 3x - 18 = (x - 6)\square$

19. $2a^2 + 9a + 4 = \square(a + 4)$

20. $6p^2 - 5p - 4 = \square(2p + 1)$

In Exercises 21–28, use a special product formula to factor each perfect square trinomial.

21. $x^2 + 2x + 1$

22. $y^2 - 2y + 1$

23. $a^2 - 18a + 81$

24. $b^2 + 12b + 36$

25. $4y^2 + 4y + 1$

26. $9x^2 + 6x + 1$

27. $9b^2 - 12b + 4$

28. $4a^2 - 12a + 9$

In Exercises 29–38, test each trinomial for factorability and factor it, if possible.

29. $x^2 - 5x + 6$

30. $y^2 + 7y + 6$

31. $x^2 - 7x + 10$

32. $c^2 - 7c + 12$

33. $b^2 + 8b + 18$

34. $x^2 + 4x - 28$

35. $x^2 - x - 30$

36. $a^2 + 4a - 45$

37. $a^2 + 5a - 50$

38. $b^2 + 9b - 36$

In Exercises 39–46, factor each trinomial. If the lead coefficient is negative, begin by factoring out −1.

39. $3x^2 + 12x - 63$

40. $2y^2 + 4y - 48$

41. $b^2x^2 - 12bx^2 + 35x^2$

42. $c^3x^2 + 11c^3x - 42c^3$

43. $-a^2 + 4a + 32$

44. $-x^2 - 2x + 15$

45. $-3x^2 + 15x - 18$

46. $-2y^2 - 16y + 40$

In Exercises 47–66, factor each trinomial. Factor out all common monomials first (including −1 if the lead coefficient is negative). If a trinomial is prime, so indicate.

47. $6y^2 + 7y + 2$

48. $6x^2 - 11x + 3$

49. $8a^2 + 6a - 9$

50. $15b^2 + 4b - 4$

51. $6x^2 - 5x - 4$

52. $18y^2 - 3y - 10$

53. $5x^2 + 4x + 1$

54. $6z^2 + 17z + 12$

55. $8x^2 - 10x + 3$

56. $4a^2 + 20a + 3$

57. $a^2 - 3ab - 4b^2$

58. $b^2 + 2bc - 80c^2$

59. $3x^3 - 10x^2 + 3x$

60. $3t^3 - 3t^2 + t$

61. $-3a^2 + ab + 2b^2$

62. $-2x^2 + 3xy + 5y^2$

63. $5a^2 + 45b^2 - 30ab$

64. $-4x^2 - 9 + 12x$

65. $21x^4 - 10x^3 - 16x^2$

66. $16x^3 - 50x^2 + 36x$

In Exercises 67–72, factor each trinomial.

67. $x^4 + 8x^2 + 15$

68. $x^4 + 11x^2 + 24$

69. $y^4 - 13y^2 + 30$

70. $y^4 - 13y^2 + 42$

71. $a^4 - 13a^2 + 36$

72. $b^4 - 17b^2 + 16$

In Exercises 73–76, use a substitution to help factor each expression.

73. $(x + a)^2 + 2(x + a) + 1$

74. $(a + b)^2 - 2(a + b) + 1$

75. $(a + b)^2 - 2(a + b) - 24$

76. $(x - y)^2 + 3(x - y) - 10$

In Exercises 77–82, factor each expression by using grouping.

77. $x^2 + 4x + 4 - y^2$

78. $x^2 - 6x + 9 - 4y^2$

79. $x^2 + 2x + 1 - 9z^2$

80. $x^2 + 10x + 25 - 16z^2$

81. $c^2 - 4a^2 + 4ab - b^2$

82. $4c^2 - a^2 - 6ab - 9b^2$

In Exercises 83–88, use grouping to help factor each trinomial.

83. $a^2 - 17a + 16$

84. $b^2 - 4b - 21$

85. $2u^2 + 5u + 3$

86. $6y^2 + 5y - 6$

87. $20r^2 - 7rs - 6s^2$

88. $6s^2 + st - 12t^2$

APPLICATIONS

89. ICE The surface area of the cubical block of ice shown in Illustration 1 is $6x^2 + 36x + 54$. Find the length of an edge of the block.

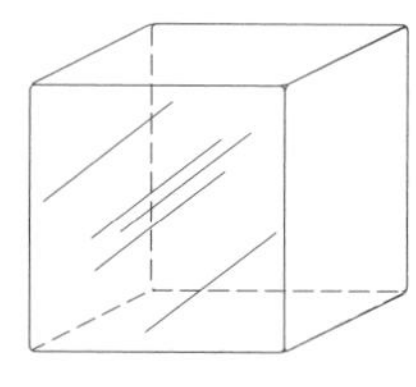

ILLUSTRATION 1

90. CHECKERS The area of the square checkerboard in Illustration 2 is $25x^2 - 40x + 16$. Find the length of a side.

ILLUSTRATION 2

WRITING *Write a paragraph using your own words.*

91. Explain how you would factor -1 from a trinomial.

92. Explain how you would test the polynomial $ax^2 + bx + c$ for factorability.

REVIEW

93. If $f(x) = |2x - 1|$, find $f(-2)$.

94. If $g(x) = 2x^2 - 1$, find $g(-2)$.

95. Solve $-3 = -\frac{9}{8}s$.

96. Solve $2x + 3 = \frac{2}{3}x - 1$.

97. Simplify $3p^2 - 6(5p^2 + p) + p^2$.

98. Solve the system $\begin{cases} 2(2x + 3y) = 5 \\ 8x = 3(1 + 3y) \end{cases}$

5.8 Summary of Factoring Techniques

In this section, you will learn about

- **A general factoring strategy**

INTRODUCTION. Factoring some polynomials involves several steps in which two or more factoring techniques must be used. At times, it can be difficult to decide how to proceed with the factoring process for such polynomials. In this section, we will discuss a general factoring strategy—a step-by-step plan to follow when factoring any polynomial.

A general factoring strategy

In this section we will discuss ways to approach a randomly chosen factoring problem. For example, suppose we wish to factor the trinomial

$$x^2y^2z^3 + 7xy^2z^3 + 6y^2z^3$$

We begin by attempting to identify the problem type. The first possibility to look for is **factoring out a common monomial.** Because the trinomial has a common monomial factor of y^2z^3, we factor it out:

$$x^2y^2z^3 + 7xy^2z^3 + 6y^2z^3 = y^2z^3(x^2 + 7x + 6)$$

We note that $x^2 + 7x + 6$ is a trinomial that can be factored as $(x + 6)(x + 1)$. Thus,

$$\begin{aligned} x^2y^2z^3 + 7xy^2z^3 + 6y^2z^3 &= y^2z^3(x^2 + 7x + 6) \\ &= y^2z^3(x + 6)(x + 1) \end{aligned}$$

To identify the type of factoring problem, we follow these steps.

A general factoring strategy

1. Factor out all common monomial factors.
2. If an expression has two terms, check for the following problem types:
   a. **The difference of two squares:** $(x^2 - y^2) = (x + y)(x - y)$
   b. **The sum of two cubes:** $(x^3 + y^3) = (x + y)(x^2 - xy + y^2)$
   c. **The difference of two cubes:** $(x^3 - y^3) = (x - y)(x^2 + xy + y^2)$
3. If an expression has three terms, attempt to factor the trinomial as a **trinomial.**
4. If an expression has four or more terms, try factoring by **grouping.**
5. Continue until each individual factor is prime.
6. Check the results by multiplying.

EXAMPLE 1 ***Applying the factoring strategy.*** Factor $48a^4c^3 - 3b^4c^3$.

Solution

We begin by factoring out the common monomial factor of $3c^3$:

$$48a^4c^3 - 3b^4c^3 = 3c^3(16a^4 - b^4)$$

Since the expression $16a^4 - b^4$ has two terms, we check to see whether it is the difference of two squares, which it is. As the difference of two squares, it factors as $(4a^2 + b^2)(4a^2 - b^2)$.

$$\begin{aligned} 48a^4c^3 - 3b^4c^3 &= 3c^3(16a^4 - b^4) \\ &= 3c^3(4a^2 + b^2)(4a^2 - b^2) \end{aligned}$$

Self Check

Factor $3p^4r^3 - 3q^4r^3$.

The binomial $4a^2 + b^2$ is the sum of two squares and is prime. However, $4a^2 - b^2$ is the difference of two squares and factors as $(2a + b)(2a - b)$.

$$\begin{aligned} 48a^4c^3 - 3b^4c^3 &= 3c^3(16a^4 - b^4) \\ &= 3c^3(4a^2 + b^2)(4a^2 - b^2) \\ &= 3c^3(4a^2 + b^2)(2a + b)(2a - b) \end{aligned}$$

Since each of the individual factors is prime, the factorization is complete.

Answer: $3r^3(p^2 + q^2)(p + q)(p - q)$ ■

EXAMPLE 2 ***Applying the factoring strategy.*** Factor $x^5y + x^2y^4 - x^3y^3 - y^6$.

Solution

We begin by factoring out the common monomial factor of y:

$$x^5y + x^2y^4 - x^3y^3 - y^6 = y(x^5 + x^2y^3 - x^3y^2 - y^5)$$

Because the expression $x^5 + x^2y^3 - x^3y^2 - y^5$ has four terms, we try factoring by grouping to obtain

$$\begin{aligned} &x^5y + x^2y^4 - x^3y^3 - y^6 \\ &= y(x^5 + x^2y^3 - x^3y^2 - y^5) && \text{Factor out } y. \\ &= y[x^2(x^3 + y^3) - y^2(x^3 + y^3)] && \text{Factor by grouping.} \\ &= y(x^3 + y^3)(x^2 - y^2) && \text{Factor out } x^3 + y^3. \end{aligned}$$

Finally, we factor $x^3 + y^3$ (the sum of two cubes) and $x^2 - y^2$ (the difference of two squares) to obtain

$$x^5y + x^2y^4 - x^3y^3 - y^6 = y(x + y)(x^2 - xy + y^2)(x + y)(x - y)$$

Because each of the individual factors is prime, the factorization is complete.

Self Check

Factor $a^5p - a^3b^2p + a^2b^3p - b^5p$.

Answer: $p(a + b)(a^2 - ab + b^2)(a + b)(a - b)$ ■

EXAMPLE 3 ***Applying the factoring strategy.*** Factor $x^3 + 5x^2 + 6x + x^2y + 5xy + 6y$.

Solution

There are no common monomial factors. Since there are more than three terms, we try factoring by grouping. We can factor x from the first three terms and y from the last three terms.

$$\begin{aligned} &x^3 + 5x^2 + 6x + x^2y + 5xy + 6y \\ &= x(x^2 + 5x + 6) + y(x^2 + 5x + 6) \\ &= (x^2 + 5x + 6)(x + y) && \text{Factor out } x^2 + 5x + 6. \\ &= (x + 3)(x + 2)(x + y) && \text{Factor } x^2 + 5x + 6. \end{aligned}$$

Self Check

Factor $a^3 - 5a^2 + 6a + a^2b - 5ab + 6b$.

Answer: $(a - 2)(a - 3)(a + b)$ ■

EXAMPLE 4 ***Applying the factoring strategy.*** Factor $x^4 + 2x^3 + x^2 + x + 1$.

Solution

There are no common monomial factors. Since there are more than three terms, we try factoring by grouping. We can factor x^2 from the first three terms.

$$\begin{aligned} x^4 + 2x^3 + x^2 + x + 1 &= x^2(x^2 + 2x + 1) + (x + 1) \\ &= x^2(x + 1)(x + 1) + (x + 1) && \text{Factor } x^2 + 2x + 1. \\ &= (x + 1)[x^2(x + 1) + 1] && \text{Factor out } x + 1. \\ &= (x + 1)(x^3 + x^2 + 1) \end{aligned}$$

Self Check

Factor $a^4 - a^3 - 2a^2 + a - 2$.

Answer: $(a - 2)(a^3 + a^2 + 1)$ ■

STUDY SET Section 5.8

VOCABULARY *In Exercises 1–4, fill in the blanks to make the statements true.*

1. The process of finding the individual factors of a known product is called ______________.

2. $x^3 + y^3$ is called a sum of two ______________.

3. $x^3 - y^3$ is called a difference of two ______________.

4. $x^2 - y^2$ is called a ______________ of two squares.

CONCEPTS *In Exercises 5–8, fill in the blanks to make the statements true.*

5. In any factoring problem, always factor out any ______________ factors first.

6. If an expression has two terms, check to see whether the problem type is the ______________ of two squares, the sum of two ______________, or the ______________ of two cubes.

7. If an expression has three terms, try to factor it as a ______________.

8. If an expression has four or more terms, try factoring it by ______________.

9. Explain how to verify that $y^2z^3(x + 6)(x + 1)$ is the factored form of $x^2y^2z^3 + 7xy^2z^3 + 6y^2z^3$.

10. Why is the polynomial $x + 6$ classified as prime?

NOTATION *In Exercises 11–12, complete each factorization.*

11. $18a^3b + 3a^2b^2 - 6ab^3 = \square(6a^2 + ab - 2b^2)$
$= 3ab(3a + \square)(\square - b)$

12. $2x^4 - 1{,}250 = 2(\square)$
$= 2(\square)(x^2 - 25)$
$= 2(x^2 + 25)(x + 5)(\square)$

PRACTICE *In Exercises 13–46, factor each polynomial, if possible.*

13. $x^2 + 16 + 8x$

14. $20 + 11x - 3x^2$

15. $8x^3y^3 - 27$

16. $3x^2y + 6xy^2 - 12xy$

17. $xy - ty + xs - ts$

18. $bc + b + cd + d$

19. $25x^2 - 16y^2$

20. $27x^9 - y^3$

21. $12x^2 + 52x + 35$

22. $12x^2 + 14x - 6$

23. $6x^2 - 14x + 8$

24. $12x^2 - 12$

25. $4x^2y^2 + 4xy^2 + y^2$

26. $100z^2 - 81t^2$

27. $x^3 + (a^2y)^3$

28. $4x^2y^2z^2 - 26x^2y^2z^3$

29. $2x^3 - 54$

30. $4(xy)^3 + 256$

31. $ae + bf + af + be$

32. $a^2x^2 + b^2y^2 + b^2x^2 + a^2y^2$

33. $2(x + y)^2 + (x + y) - 3$

34. $(x - y)^3 + 125$

35. $625x^4 - 256y^4$

36. $2(a - b)^2 + 5(a - b) + 3$

37. $36x^4 - 36$

38. $6x^2 - 63 - 13x$

39. $a^4 - 13a^2 + 36$

40. $x^4 - 17x^2 + 16$

41. $x^2 + 6x + 9 - y^2$

42. $x^2 + 10x + 25 - y^8$

43. $4x^2 + 4x + 1 - 4y^2$

44. $9x^2 - 6x + 1 - 25y^2$

45. $x^2 - y^2 - 2y - 1$

46. $a^2 - b^2 + 4b - 4$

WRITING *Write a paragraph using your own words.*

47. What is your strategy for factoring a polynomial?

48. For the factorization below, explain why the polynomial is not factored completely.

$48a^4c^3 - 3b^4c^3 = 3c^3(16a^4 - b^4)$

REVIEW

49. Tell whether the graphs of $x + y = 2$ and $y = x + 5$ are parallel or perpendicular.

50. When expressed as a decimal, is $\frac{7}{8}$ a terminating or a repeating decimal?

51. Evaluate $\begin{vmatrix} 1 & 15 \\ 15 & 0 \end{vmatrix}$.

52. If a triangle has exactly two sides with equal measures, what type of triangle is it?

5.9 *Solving Equations by Factoring*

In this section, you will learn about

- **Solving quadratic equations**
- **Solving higher-degree polynomial equations**
- **Problem solving**

Introduction. Equations that involve *first-degree* polynomials, such as $9x - 6 = 0$, are called linear equations. Equations such as $9x^2 - 6x = 0$ that involve *second-degree* polynomials are called quadratic equations. The techniques that we have used to solve linear equations cannot be used to solve quadratic equations, because those techniques cannot isolate the variable on one side of the equation. However, we can solve many quadratic equations using factoring.

Solving quadratic equations

An equation such as $3x^2 + 4x - 7 = 0$ or $-5y^2 + 3y + 8 = 0$ is called a **quadratic** or **second-degree** equation.

Quadratic equations

A **quadratic equation** is any equation that can be written in the form

$$ax^2 + bx + c = 0$$

where a, b, and c are real numbers and $a \neq 0$.

Many quadratic equations can be solved by factoring and then by using the **zero-factor property.**

Zero-factor property

If a and b are real numbers, then

If $ab = 0$, then $a = 0$ or $b = 0$.

The zero-factor property states that *if the product of two or more numbers is 0, then at least one of the numbers must be 0.*

To solve the quadratic equation $x^2 + 5x + 6 = 0$, we factor its left-hand side to obtain

$$(x + 3)(x + 2) = 0$$

Since the product of $x + 3$ and $x + 2$ is 0, at least one of the factors must be 0. Thus, we can set each factor equal to 0 and solve each resulting linear equation for x:

$$x + 3 = 0 \quad \text{or} \quad x + 2 = 0$$
$$x = -3 \qquad\qquad x = -2$$

To check these solutions, we substitute -3 and -2 for x in the equation and verify that each number satisfies the equation.

$$\begin{aligned} x^2 + 5x + 6 &= 0 \\ (-3)^2 + 5(-3) + 6 &\stackrel{?}{=} 0 \\ 9 - 15 + 6 &\stackrel{?}{=} 0 \\ 0 &= 0 \end{aligned} \quad \text{or} \quad \begin{aligned} x^2 + 5x + 6 &= 0 \\ (-2)^2 + 5(-2) + 6 &\stackrel{?}{=} 0 \\ 4 - 10 + 6 &\stackrel{?}{=} 0 \\ 0 &= 0 \end{aligned}$$

Both -3 and -2 are solutions, because both satisfy the equation.

EXAMPLE 1 ***Solving quadratic equations by factoring.*** Solve $3x^2 + 6x = 0$.

Solution

To solve the equation, we factor the left-hand side, set each factor equal to 0, and solve each resulting equation for x.

$3x^2 + 6x = 0$

$3x(x + 2) = 0$ — Factor out the common factor of $3x$.

$3x = 0 \quad \text{or} \quad x + 2 = 0$ — By the zero-factor property, at least one of the factors must be equal to zero.

$x = 0 \qquad\quad x = -2$ — Solve each linear equation.

Verify that both solutions check.

Self Check

Solve $4p^2 - 12p = 0$.

Answer: 0, 3

WARNING! In Example 1, do not divide both sides by $3x$, or you will lose the solution $x = 0$.

EXAMPLE 2 ***Solving quadratic equations by factoring.*** Solve $x^2 - 16 = 0$.

Solution

To solve the equation, we factor the difference of two squares on the left-hand side, set each factor equal to 0, and solve each resulting equation.

$$x^2 - 16 = 0$$
$$(x + 4)(x - 4) = 0$$
$$x + 4 = 0 \quad \text{or} \quad x - 4 = 0$$
$$x = -4 \qquad\qquad x = 4$$

Verify that both solutions check.

Self Check

Solve $a^2 - 81 = 0$.

Answer: 9, -9

The following steps can be used to solve a quadratic equation by factoring.

Solving a quadratic equation by using the factoring method

1. Write the equation in $ax^2 + bx + c = 0$ form (called *quadratic* form).
2. Factor the polynomial.
3. Use the zero-factor property to set each factor equal to zero.
4. Solve each resulting equation.
5. Check each solution in the original equation.

Many equations that do not appear to be quadratic can be put into quadratic form and then solved by factoring.

EXAMPLE 3 ***Writing the equation in quadratic form.*** Solve $x = \frac{6}{5} - \frac{6}{5}x^2$.

Solution

We must write the equation in quadratic form. To clear the equation of fractions, we multiply both sides by 5.

$$x = \frac{6}{5} - \frac{6}{5}x^2$$

$$5x = 6 - 6x^2 \quad \text{Multiply both sides by 5.}$$

To use factoring to solve this quadratic equation, one side of the equation must be 0. Since it is easier to factor a second-degree polynomial if the coefficient of the squared term is positive, we add $6x^2$ to both sides and subtract 6 from both sides to obtain

$$6x^2 + 5x - 6 = 0$$

$$(3x - 2)(2x + 3) = 0 \quad \text{Factor the trinomial.}$$

$$3x - 2 = 0 \quad \text{or} \quad 2x + 3 = 0 \quad \text{Set each factor equal to 0 and solve for } x.$$

$$3x = 2 \qquad\qquad 2x = -3$$

$$x = \frac{2}{3} \qquad\qquad x = -\frac{3}{2}$$

Verify that both solutions check.

Self Check

Solve $x = \frac{6}{7}x^2 - \frac{3}{7}$.

Answer: $\frac{3}{2}, -\frac{1}{3}$

WARNING! To solve a quadratic equation by factoring, be sure to set the quadratic polynomial equal to 0 before factoring and applying the zero-factor property. Do not make the following error:

$$6x^2 + 5x = 6$$

$$x(6x + 5) = 6 \quad \text{If the product of two numbers is 6, neither number need be 6. For example, } 2 \cdot 3 = 6.$$

$$x = 6 \quad \text{or} \quad 6x + 5 = 6$$

$$x = \frac{1}{6}$$

Neither solution checks.

Accent on Technology **Solving quadratic equations**

To solve a quadratic equation such as $x^2 + 4x - 5 = 0$ with a graphing calculator, we can use standard window settings of $[-10, 10]$ for x and $[-10, 10]$ for y and graph the quadratic function $y = x^2 + 4x - 5$, as shown in Figure 5-8(a). We can then trace to find the x-coordinates of the x-intercepts of the parabola. See Figures 5-8(b) and 5-8(c). For better results, we can zoom in. Since these are the numbers x that make $y = 0$, they are the solutions of the equation.

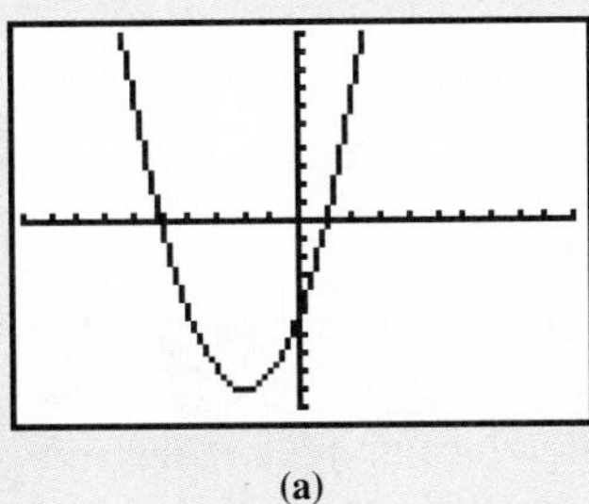

(a)

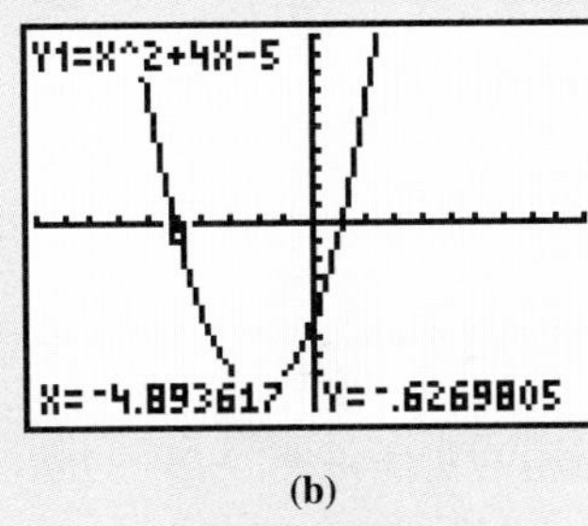

(b)

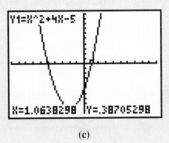

(c)

FIGURE 5-8

Solving higher-degree polynomial equations

We can solve many polynomial equations with degree greater than 2 by factoring and applying an extension of the zero-factor property.

EXAMPLE 4 ***Solving third-degree equations by factoring.*** Solve $6x^3 - x^2 = 2x$.

Solution

First, we subtract $2x$ from both sides so that the right-hand side of the equation is 0.

$$6x^3 - x^2 - 2x = 0$$

Then we factor x from the third-degree polynomial on the left-hand side and proceed as follows:

$$6x^3 - x^2 - 2x = 0$$

$$x(6x^2 - x - 2) = 0 \quad \text{Factor out } x.$$

$$x(3x - 2)(2x + 1) = 0 \quad \text{Factor } 6x^2 - x - 2.$$

$$x = 0 \quad \text{or} \quad 3x - 2 = 0 \quad \text{or} \quad 2x + 1 = 0 \quad \text{Set each of the three factors equal to 0.}$$

$$x = \frac{2}{3} \qquad x = -\frac{1}{2} \quad \text{Solve each equation.}$$

Verify that the three solutions check.

Self Check

Solve $5x^3 + 13x^2 = 6x$.

Answer: $0, \frac{2}{5}, -3$

EXAMPLE 5 ***Solving higher-degree equations by factoring.*** Solve $x^4 + 4 - 5x^2 = 0$.

Solution

First, we write the powers of x in descending order. Then we factor the trinomial on the left-hand side and proceed as follows:

$$x^4 - 5x^2 + 4 = 0$$
$$(x^2 - 1)(x^2 - 4) = 0$$
$$(x + 1)(x - 1)(x + 2)(x - 2) = 0 \quad \text{Factor } x^2 - 1 \text{ and } x^2 - 4.$$
$$x + 1 = 0 \quad \text{or} \quad x - 1 = 0 \quad \text{or} \quad x + 2 = 0 \quad \text{or} \quad x - 2 = 0$$
$$x = -1 \qquad x = 1 \qquad x = -2 \qquad x = 2$$

Verify that each solution checks.

Self Check

Solve $a^4 + 36 - 13a = 0$.

Answer: 2, −2, 3, −3

Accent on Technology *Solving equations*

To solve the equation $x^4 - 5x^2 + 4 = 0$ with a graphing calculator, we can use window settings of $[-6, 6]$ for x and $[-5, 10]$ for y and graph the polynomial function $y = x^4 - 5x^2 + 4$ as shown in Figure 5-9. We can then read the values of x that make $y = 0$. They are $x = -2, -1, 1$, and 2. If the x-coordinates of the x-intercepts were not obvious, we could get their values by using trace and zoom.

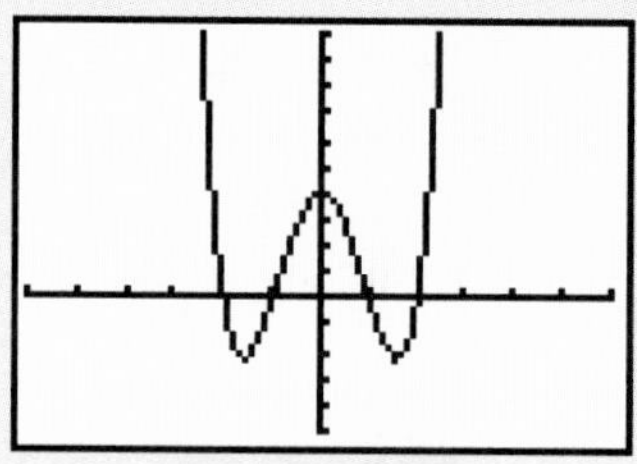

FIGURE 5-9

Problem solving

EXAMPLE 6 ***Stained glass.*** The triangular stained glass window shown in Figure 5-10 is to be installed in a chapel. The length of the base of the window is 3 times its height. The area of the window is 96 square feet. Find its base and height.

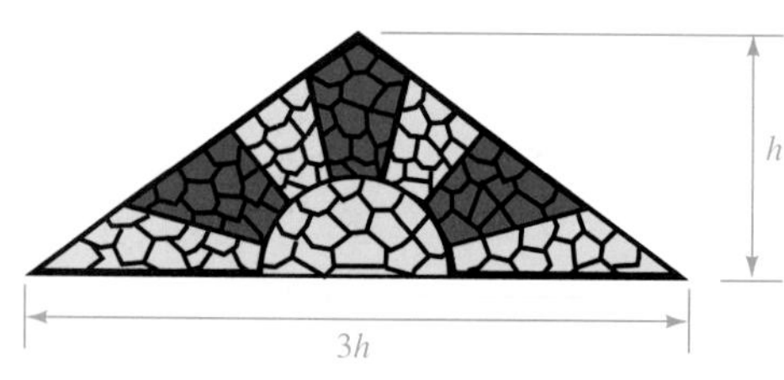

FIGURE 5-10

Analyze the problem We are to find the length of the base and the height of the window. The formula that gives the area of a triangle is $A = \frac{1}{2}bh$, where b is the length of the base and h the height.

Form an equation We can let h be the positive number that represents the height of the window. Then $3h$ represents the length of the base. To form an equation in terms of h, we can substitute $3h$ for b and 96 for A in the formula for the area of a triangle.

$$A = \frac{1}{2}bh$$

$$96 = \frac{1}{2}(3h)h$$

Solve the equation To solve this equation, we must write it in quadratic form.

$$96 = \frac{1}{2}(3h)h$$

$$192 = 3h^2 \quad (3h)h = 3h^2\text{. Multiply both sides by 2.}$$

$$64 = h^2 \quad \text{Divide both sides by 3.}$$

$$0 = h^2 - 64 \quad \text{To obtain 0 on the left-hand side, subtract 64 from both sides.}$$

$$0 = (h + 8)(h - 8) \quad \text{Factor the difference of two squares.}$$

$$h + 8 = 0 \quad \text{or} \quad h - 8 = 0$$

$$h = -8 \qquad\qquad h = 8$$

State the conclusion Since the height of a triangle cannot be negative, we must discard the negative solution. Thus, the height of the window is 8 feet, and the length of its base is 3(8), or 24 feet.

Check the result The area of a triangle with a base of 24 feet and a height of 8 feet is 96 square feet:

$$A = \frac{1}{2}bh = \frac{1}{2}(24)(8) = 12(8) = 96$$

The solution checks. ■

EXAMPLE 7 ***Ballistics.*** If the initial velocity of an object thrown straight up into the air is 176 feet per second, when will the object strike the ground?

Analyze the problem The height of an object thrown straight up into the air with an initial velocity of v feet per second is given by the formula

$$h = vt - 16t^2$$

The height h is in feet, and t represents the number of seconds since the object was released. When the object hits the ground, its height will be 0.

Form an equation In the formula, we set h equal to 0, set v equal to 176, and solve for t.

$$h = vt - 16t^2$$

$$0 = 176t - 16t^2$$

Solve the equation To solve this equation, we will use the factoring method.

$$0 = 176t - 16t^2$$

$$0 = 16t(11 - t) \quad \text{Factor out } 16t.$$

$$16t = 0 \quad \text{or} \quad 11 - t = 0 \quad \text{Set each factor equal to 0.}$$

$$t = 0 \qquad\qquad t = 11$$

State the conclusion When $t = 0$, the object's height above the ground is 0 feet, because it has not been released. When $t = 11$, the height is again 0 feet, and the object has returned to the ground. The solution is 11 seconds.

Check the result Verify that $h = 0$ when $t = 11$. ■

STUDY SET Section 5.9

VOCABULARY *In Exercises 1–2, fill in the blanks to make the statements true.*

1. A ______________ equation is any equation that can be written in the form $ax^2 + bx + c = 0$ $(a \neq 0)$.

2. To ______________ an equation means to find all the values of the variable that make the equation true.

CONCEPTS

3. If the product of two numbers is zero, what must be true about at least one of the numbers?

4. Use a check to determine whether -5 and 4 are solutions of $a^2 - 9a + 20 = 0$.

5. Tell whether each equation is a quadratic equation.

a. $w^2 + 7w + 12 = 0$

b. $6t + 11 = 0$

c. $x(x + 3) = -2$

d. $k^3 - 4k^2 + k - 15 = 0$

6. What is wrong with the work shown here?

Solve for x: $x^2 + 2x = 8$.

$$x(x + 2) = 8$$
$$x = 8 \quad \text{or} \quad x + 2 = 8$$
$$x = 8 \quad \text{or} \quad x = 6$$

7. Use the graph in Illustration 1 to solve the quadratic equation $x^2 - 2x - 3 = 0$.

8. Use the graph in Illustration 2 to solve the polynomial equation $x^3 - 4x^2 + 4x = 0$.

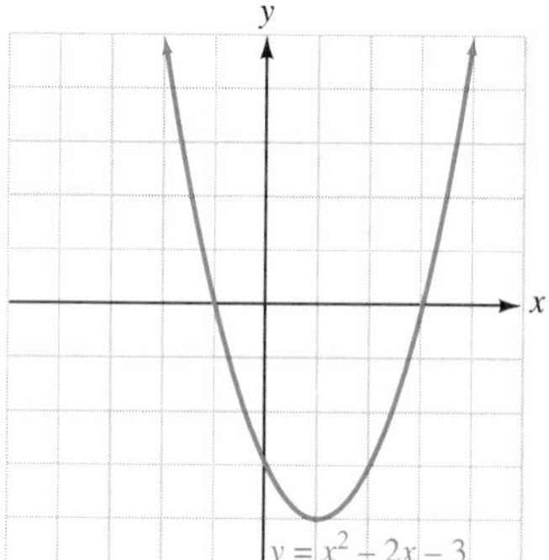

ILLUSTRATION 1

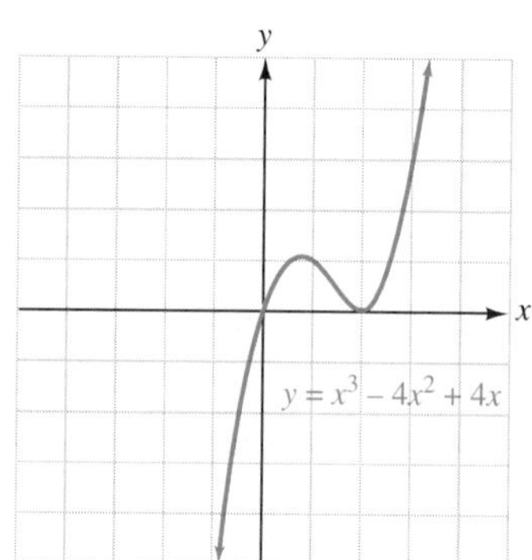

ILLUSTRATION 2

NOTATION *In Exercises 9–10, complete each solution.*

9. Solve $y^2 - 3y - 54 = 0$

$(y - 9)(\square) = 0$

$\square = 0$ or $y + 6 = 0$

$y = 9 \qquad y = \square$

10. Solve $x^2 - x = 12$.

$x^2 - x - 12 = \square$

$(x - 4)(\square) = 0$

$x - 4 = 0$ or $\square = 0$

$x = 4 \qquad x = \square$

PRACTICE *In Exercises 11–28, solve each equation.*

11. $4x^2 + 8x = 0$

12. $x^2 - 9 = 0$

13. $y^2 - 16 = 0$

14. $y^2 - 25 = 0$

15. $x^2 + x = 0$

16. $x^2 - 3x = 0$

17. $5y^2 - 25y = 0$

18. $y^2 - 36 = 0$

19. $z^2 + 8z + 15 = 0$

20. $w^2 + 7w + 12 = 0$

21. $x^2 + 6x + 8 = 0$

22. $x^2 + 9x + 20 = 0$

23. $3m^2 + 10m + 3 = 0$

24. $2r^2 + 5r + 3 = 0$

25. $2y^2 - 5y + 2 = 0$

26. $2x^2 - 3x + 1 = 0$

27. $2x^2 - x - 1 = 0$

28. $2x^2 - 3x - 5 = 0$

In Exercises 29–40, write each equation in quadratic form and then solve it by factoring.

29. $x(x - 6) + 9 = 0$

30. $x^2 + 8(x + 2) = 0$

31. $8a^2 = 3 - 10a$

32. $5z^2 = 6 - 13z$

33. $b(6b - 7) = 10$

34. $2y(4y + 3) = 9$

35. $\frac{3a^2}{2} = \frac{1}{2} - a$

36. $x^2 = \frac{1}{2}(x + 1)$

37. $x^2 + 1 = \frac{5}{2}x$

38. $\frac{3}{5}(x^2 - 4) = -\frac{9}{5}x$

39. $x\left(3x + \frac{22}{5}\right) = 1$

40. $x\left(\frac{x}{11} - \frac{1}{7}\right) = \frac{6}{77}$

In Exercises 41–52, solve each equation by using factoring.

41. $x^3 + x^2 = 0$

42. $2x^4 + 8x^3 = 0$

43. $y^3 - 49y = 0$

44. $2z^3 - 200z = 0$

45. $x^3 - 4x^2 - 21x = 0$

46. $x^3 + 8x^2 - 9x = 0$

47. $z^4 - 13z^2 + 36 = 0$

48. $y^4 - 10y^2 + 9 = 0$

49. $3a(a^2 + 5a) = -18a$

50. $7t^3 = 2t\left(t + \frac{5}{2}\right)$

51. $\frac{x^2(6x + 37)}{35} = x$

52. $x^2 = -\frac{4x^3(3x + 5)}{3}$

53. INTEGER PROBLEM The product of two consecutive even integers is 288. Find the integers. (*Hint:* Let x = the first even integer. Then represent the second even integer in terms of x.)

54. INTEGER PROBLEM The product of two consecutive odd integers is 143. Find the integers. (*Hint:* Let x = the first odd integer. Then represent the second odd integer in terms of x.)

In Exercises 55–58, use a graphing calculator to find the solutions of each equation, if one exists. If an answer is not exact, give the answer to the nearest hundredth.

55. $2x^2 - 7x + 4 = 0$

56. $x^2 - 4x + 7 = 0$

57. $-3x^3 - 2x^2 + 5 = 0$

58. $-2x^3 - 3x - 5 = 0$

APPLICATIONS

59. COOKING The electric griddle shown in Illustration 3 has a cooking surface of 160 square inches. Find the length and the width of the griddle.

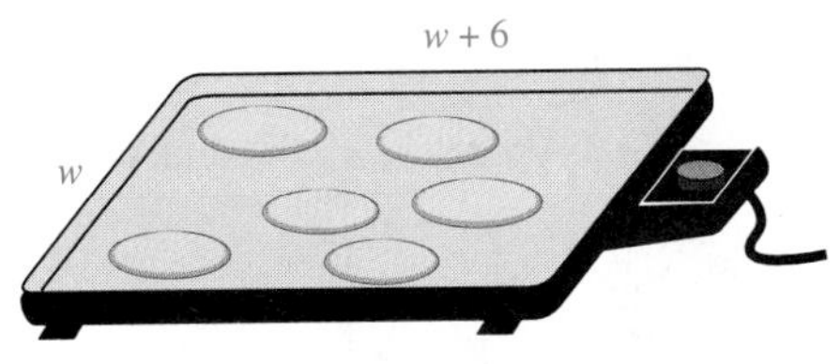

ILLUSTRATION 3

60. STRUCTURAL ENGINEERING The formula for the area of a trapezoid is $A = \frac{h(B + b)}{2}$. The area of the trapezoidal truss in Illustration 4 is 44 square feet. Find the height of the truss if the shorter base is the same as the height.

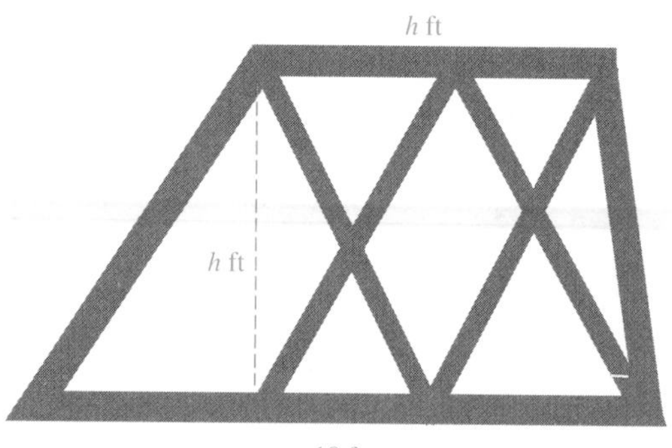

ILLUSTRATION 4

61. SWIMMING POOL DESIGN Building codes require that the rectangular swimming pool in Illustration 5 be surrounded by a uniform-width walkway of at least 516 square feet. The length of the pool is 10 feet less than twice the width. How wide should the border be?

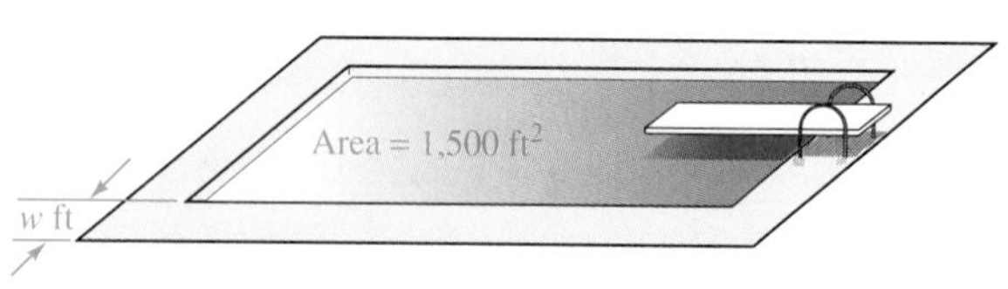

ILLUSTRATION 5

62. FINE ARTS An artist intends to paint a 60-square-foot mural on the large wall shown in Illustration 6. Find the dimensions of the mural if the artist leaves a border of uniform width around it.

ILLUSTRATION 6

63. ARCHITECTURE The rectangular room shown in Illustration 7 is twice as long as it is wide. It is divided into two rectangular parts by a partition, positioned as shown. If the larger part of the room contains 560 square feet, find the dimensions of the entire room.

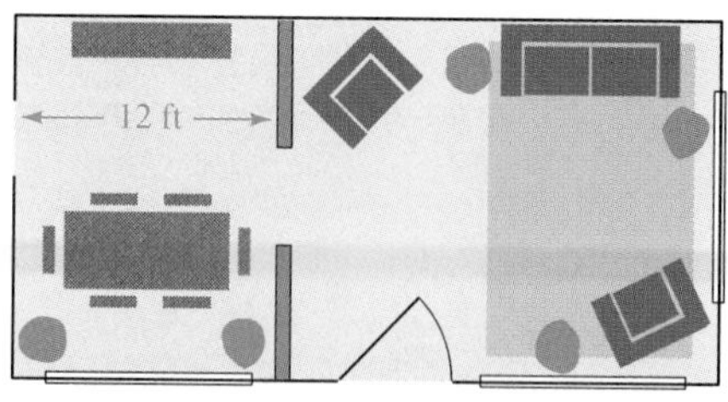

ILLUSTRATION 7

64. WINTER RECREATION The length of the rectangular ice-skating rink in Illustration 8 is 20 meters greater than twice its width. Find the width.

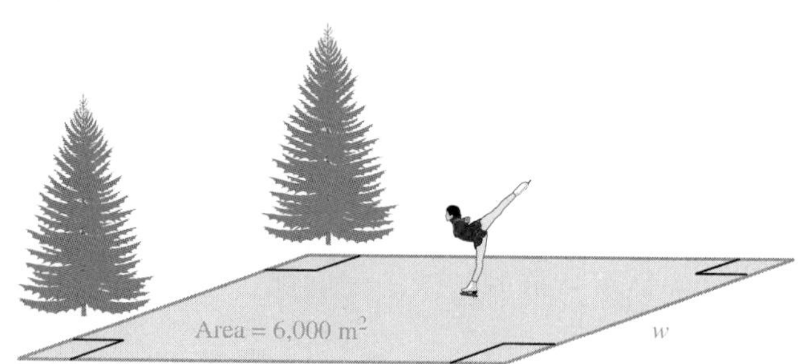

ILLUSTRATION 8

65. BALLISTICS The muzzle velocity of a cannon is 480 feet per second. If a cannonball is fired vertically, at what times will it be at a height of 3,344 feet?

66. SLINGSHOTS A slingshot can provide an initial velocity of 128 feet per second. At what times will a stone, shot vertically upward, be 192 feet above the ground?

67. BUNGEE JUMPING See Illustration 9. The formula $h = -16t^2 + 212$ gives the distance a bungee jumper is from the ground for the free-fall portion of the jump, t seconds after leaping off a bridge. We can find the number of seconds it takes the jumper to reach the point in the fall where the 64-foot bungee chord starts to stretch by substituting 148 for h and solving for t. Find t.

68. MAJOR LEAGUE BASEBALL In 1998, pitcher Roger Clemens, of the Toronto Blue Jays, threw a fastball that was clocked at 97 miles per hour. This is a velocity of approximately 144 feet per second. If he could throw the baseball vertically into the air with this velocity, how long would it take for the ball to fall to the ground?

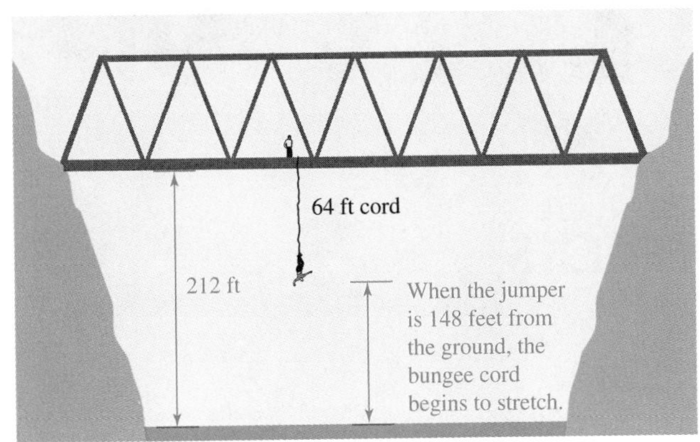

ILLUSTRATION 9

69. FORENSIC MEDICINE The kinetic energy E of a moving object is given by $E = \frac{1}{2}mv^2$, where m is the mass of the object (in kilograms) and v is the object's velocity (in meters per second). Kinetic energy is measured in joules. By measuring the damage done to a victim who has been struck by a 3-kilogram club, a police pathologist finds that the energy at impact was 54 joules. Find the velocity of the club at impact.

70. TRAFFIC ACCIDENTS Investigators at a traffic accident used the function $d(v) = 0.04v^2 + 0.8v$, where v is the velocity of the car (in mph) and $d(v)$ is the stopping distance of the car (in feet), to reconstruct the events leading up to a collision. From physical evidence, it was concluded that it took one car 32 feet to stop. At what velocity was the car traveling prior to the accident?

71. BREAK-EVEN POINT The cost for a guitar maker to hand-craft x guitars is given by the function $C(x) = \frac{1}{8}x^2 - x + 6$. The revenue taken in with the sale of x guitars is given by the function $R(x) = \frac{1}{4}x^2$. Find the number of guitars that must be sold so that the cost equals the revenue.

72. REVENUE Over the years, the manager of a crafts store has found that the number of scented candles x she can sell in a month depends on the price p according to the formula $x = 200 - 10p$. At what price should she sell the candles if she needs to bring in \$750 in revenue a month from their sale? (*Hint:* Revenue = price · number sold = px.)

WRITING *Write a paragraph using your own words.*

73. Explain the zero-factor property.

74. In the work shown below, explain why the student has not solved for x.

Solve: $x^2 + x - 6 = 0$

$$x^2 + x = 6$$
$$x = 6 - x^2$$

REVIEW

75. ALUMINUM FOIL Find the number of square feet of aluminum foil on a roll if it has dimensions of $8\frac{1}{3}$ yards × 12 inches.

76. HOCKEY A hockey puck is a vulcanized rubber disk 2.5 cm (1 in.) thick and 7.6 cm (3 in.) in diameter. Find the volume of a puck in cubic centimeters and cubic inches. Round to the nearest tenth.

KEY CONCEPT

Polynomials

A **polynomial** is an algebraic term or the sum of two or more algebraic terms whose variables have whole-number exponents. No variable appears in a denominator.

Operations with polynomials

In arithmetic, we learned how to add, subtract, multiply, divide, and find powers of numbers. In algebra, we need to be able to perform these operations on polynomials.

In Exercises 1–8, do each operation.

1. $(-2x^2 - 5x - 7) + (-3x^2 + 7x + 1)$

2. $(6s^3 + 3s - 2) - (2s^3 + 3s^2 + 5)$

3. $(3m - 4)(m + 3)$

4. $3r^2st(r^2 - 2s + 3t^2)$

5. $(a - 2d)^2$

6. $(x - 3y)(x^2 + 3xy + 9y^2)$

7. $(3b + 1)(2b^2 + 3b + 2)$

8. $(2y + 3)(y - 1) - (y + 2)(3y - 1)$

Polynomial functions

Polynomial functions can be used to model many real-world situations.

9. WINDOW WASHERS A man on a scaffold, washing the outside windows of a skyscraper, drops a squeegee. As it falls, its distance in feet from the ground $d(t)$, t seconds after being dropped, is given by the polynomial function $d(t) = -16t^2 + 576$. Find $d(6)$ and explain the result.

10. Write a polynomial function $V(x)$ that gives the volume of the ice chest shown in Illustration 1. Then find $V(3)$.

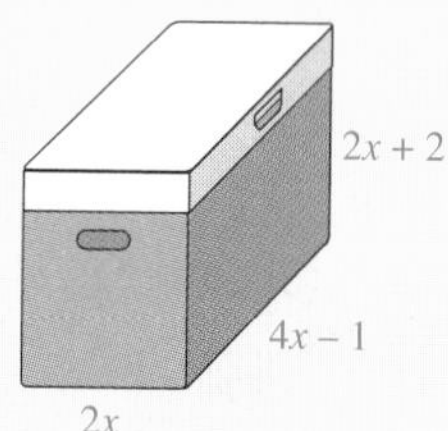

ILLUSTRATION 1

Solving equations by factoring

Quadratic equations can be written in the form $ax^2 + bx + c = 0$. Many quadratic equations can be solved by factoring the polynomial $ax^2 + bx + c$ and applying the zero-factor property. Some higher-degree equations can also be solved by using an extension of this procedure.

In Exercises 11–16, solve each equation by factoring.

11. $x^2 - 81 = 0$

12. $5x^2 - 25x = 0$

13. $z^2 + 8z + 15 = 0$

14. $2r^2 + 5r + 3 = 0$

15. $\frac{3t^2}{2} + t = \frac{1}{2}$

16. $m^3 = 9m - 8m^2$

GROUP ACTIVITIES FOR CHAPTER 5

Accent on Teamwork

Section 5.1

EXPONENTS Have a student in your group write each of the eight rules for exponents listed on page 305 on separate 3×5 cards. On another set of cards, write an explanation of each rule using words. On a third set of cards, write an example of the use of each rule for exponents. Shuffle the cards and work together to match the symbolic description, the word description, and the example for each of the eight rules for exponents.

Section 5.2

SCIENTIFIC NOTATION Go to the library and find five examples of extremely large and five examples of extremely small numbers. Encyclopedias, government statistics books, and science books are good places to look. Write each number in scientific notation on a separate piece of paper. Include a brief explanation of what the number represents. Present the 10 examples in numerical order, beginning with the smallest number first.

Section 5.3

ADDING POLYNOMIALS According to an old adage, "You can't add apples and oranges." Give some examples of how this concept applies when adding two polynomials.

Section 5.4

MULTIPLYING BINOMIALS Use colored construction paper to make a model like that shown in Illustration 1. Use the model as part of a presentation to demonstrate why $(x + y)(x + y) = x^2 + 2xy + y^2$.

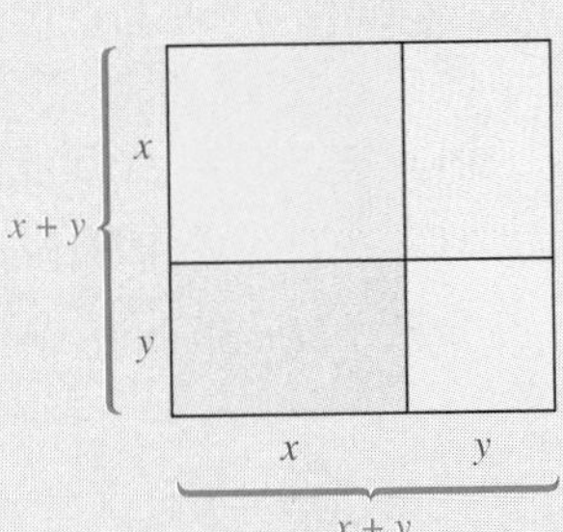

ILLUSTRATION 1

Section 5.5

SOLVING FORMULAS Examine the student's work shown below. Write some comments to the student about what it means to *solve for the indicated variable.* Then write out the correct solution.

Solve for r_1: $r_1 r_2 = rr_2 + rr_1$

$$\frac{r_1 r_2}{r_2} = \frac{rr_2 + rr_1}{r_2}$$

$$r_1 = \frac{rr_2 + rr_1}{r_2}$$

Section 5.6

FACTORING The following expressions contain variables with variable exponents. Use the methods discussed in Section 5.6 to factor each of them.

$x^{2m} - y^{4n}$ $\quad$ $x^{3m} - y^{3n}$ $\quad$ $x^{3m} + y^{3n}$

Section 5.7

AUTHORING A TEXTBOOK Assign each of the 11 examples in Section 5.7 to members of your group. Have them write a new but similar problem for each example, then write a solution complete with an explanation and author notes using the same format used in this book. They should also create an accompanying Self Check problem and include the answer. Compile all 11 examples into a booklet. Make copies of your booklet for the other members of the class.

Section 5.8

FACTORING The following expressions contain variables with variable exponents. Use the factoring methods discussed in Section 5.8 to factor each of them.

$x^{2n} + 2x^n + 1$ $\quad$ $x^{2n} - x^n - 6$ $\quad$ $2x^{6n} - 3x^{3n} - 2$

Section 5.9

QUADRATIC EQUATIONS Suppose you know that the two solutions of a quadratic equation are 4 and -3. Consider the method used to solve quadratic equations studied in Section 5.9. Work backward to find the original equation.

CHAPTER REVIEW

SECTION 5.1 Exponents

CONCEPT

If n is a natural number,

$$x^n = \underbrace{x \cdot x \cdot x \cdot \cdots \cdot x}_{n \text{ factors of } x}$$

where x is the *base* and n the *exponent.*

Rules for exponents:
If there are no divisions by 0, then for all integers m and n,

$x^m x^n = x^{m+n}$ $(x^m)^n = x^{mn}$

$(xy)^n = x^n y^n$ $\left(\frac{x}{y}\right)^n = \frac{x^n}{y^n}$

$x^0 = 1$ $x^{-n} = \frac{1}{x^n}$

$\frac{x^m}{x^n} = x^{m-n}$

$\left(\frac{x}{y}\right)^{-n} = \left(\frac{y}{x}\right)^n$

REVIEW EXERCISES

1. Evaluate each expression.

a. 3^6 **b.** -2^5

c. $(-4)^3$ **d.** 15^1

2. Simplify each expression and write all answers without negative exponents.

a. $x^4 \cdot x^2$ **b.** $a^3b^5a^2b$

c. $(m^6)^3$ **d.** $(-t^2)^2(t^3)^3$

e. $(3x^2y^3)^2$ **f.** $\left(\frac{x^4}{b}\right)^4$

g. $-3x^0$ **h.** $(x^2)^{-5}$

i. -5^{-4} **j.** $\frac{70}{x^{-4}}$

k. $(3x^{-3})^{-2}$ **l.** $2x^{-4}x^3$

m. $-\left(\frac{c^{-3}}{c^{-5}}\right)^5$ **n.** $\left(\frac{4}{5}\right)^{-2}$

o. $\frac{y^{-3}}{y^4y}$ **p.** $\left(\frac{-2a^4b}{a^{-3}b^2}\right)^{-3}$

SECTION 5.2 Scientific Notation

Scientific notation is a compact way of writing large and small numbers. Numbers are written in the form

$N \times 10^n$

where $1 \le |N| < 10$ and n is an integer.

3. Write each number in scientific notation.

a. 19,300,000,000 **b.** 0.00000002735

4. Write each number in standard notation.

a. 7.277×10^7 **b.** 8.3×10^{-9}

In Exercises 5–7, write each number in scientific notation and do the operations. Give answers in scientific notation.

5. THE SPEED OF LIGHT Light travels at about 300,000 kilometers per second. If the average distance from the sun to the planet Mars is approximately 228,000,000 kilometers, how long does it take light from the sun to reach Mars?

6. PROTONS If the mass of 1 proton is 0.00000000000000000000000167248 gram, find the mass of 1 million protons.

7. Evaluate $\dfrac{(616{,}000{,}000)(0.000009)}{0.00066}$.

SECTION 5.3 Polynomials and Polynomial Functions

A *polynomial* is an algebraic term or the sum of two or more algebraic terms whose variables have whole-number exponents. No variable appears in a denominator.

8. Tell whether each expression is a polynomial.

a. $\dfrac{2x^2}{x+1}$

b. $-5x^3 + x^2 - 5x - 4$

c. $2.8y^{15} - y^{10} + y^8 - \dfrac{3}{2}y^6$

d. $x^{-3} + x^{-2} - x^{-1} - 1$

The *degree of a polynomial* is the degree of the term with the highest degree contained within the polynomial.

9. Classify each polynomial as a monomial, binomial, trinomial, or none of these. Then determine the degree of the polynomial.

a. $x^2 - 8$

b. $-15a^3b$

c. $x^4 + x^3 - x^2 + x - 4$

d. $9x^2y + 13x^3y^2 + 8x^4y^4$

To *evaluate a polynomial* function, we replace the variable in the defining equation with its value, called the *input.* Then we simplify to find the *output.*

10. SQUIRT GUN The volume, in cubic inches, of the reservoir on top of the squirt gun shown in Illustration 1 is given by the polynomial function $V(r) = 4.19r^3 + 25.12r^2$, where r is the radius, in inches. Find $V(2)$ to the nearest cubic inch.

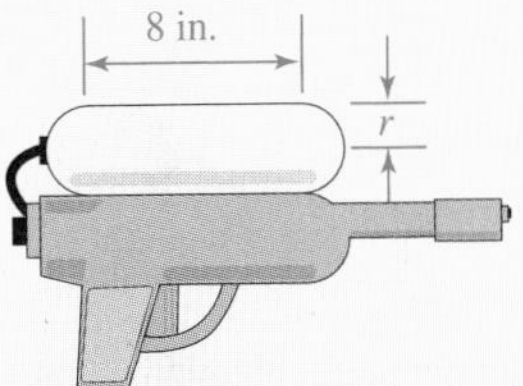

ILLUSTRATION 1

11. Graph each polynomial function.

a. $f(x) = x^2 - 2x$

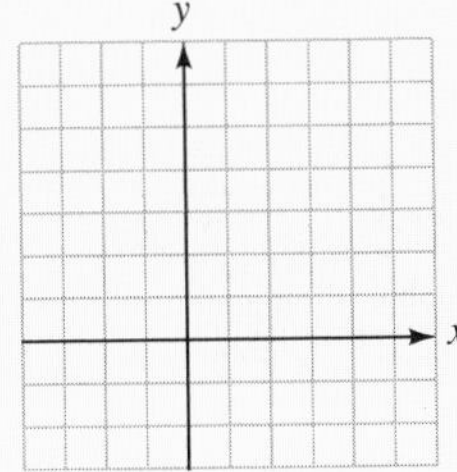

b. $f(x) = x^3 - 3x^2 + 4$

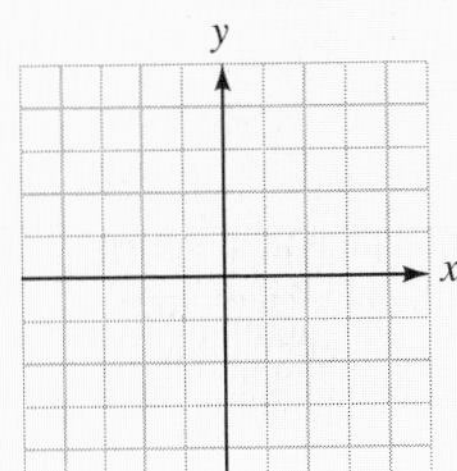

To add polynomials, remove parentheses and *combine like terms* (terms having the same variables with the same exponents).

To subtract polynomials, add the first polynomial and the negative (opposite) of the second polynomial.

12. Do each operation.

a. $(3x^2 + 4x + 9) + (2x^2 - 2x + 7)$

b. $(2x^2y^3 - 5x^2y + 9y) + (x^2y^3 - 3x^2y - y)$

c. $(4x^3 + 4x^2 + 7) - (-2x^3 - x - 2)$

d. $$\begin{array}{r} -10k^4 - 4k^3 + 5k^2 - k + 1 \\ -\quad \underline{-16k^4 + 2k^3 - 4k^2 - k + 3} \end{array}$$

SECTION 5.4 Multiplying Polynomials

To *multiply monomials,* multiply their numerical factors and multiply their variable factors.

13. Find each product.

a. $(8a^2)\left(-\frac{1}{2}a\right)$

b. $(-3xy^2z)(-2xz^3)(xz)$

To *multiply a polynomial by a monomial,* multiply each term of the polynomial by the monomial.

14. Find each product.

a. $2xy^2(x^3y - 4xy^5)$

b. $-a^2b(-a^2 - 2ab + b^2)$

The *FOIL method* is used to multiply two binomials.

To *multiply polynomials,* multiply each term of one polynomial by each term of the other polynomial.

15. Find each product.

a. $(8x - 5)(2x + 3)$

b. $(3x^2 + 2)(2x - 4)$

c. $(5a - 6)^2$

d. $(7c^2 - d)(7c^2 + d)$

e. $(5x^2 - 4x)(3x^2 - 2x + 10)$

f. $(r + s)(r - s)(r - 3s)$

16. SHAVING A razor blade is made from a thin piece of platinum steel. Before its center is punched out, the blade has the shape shown in Illustration 2. Write a polynomial that gives the area.

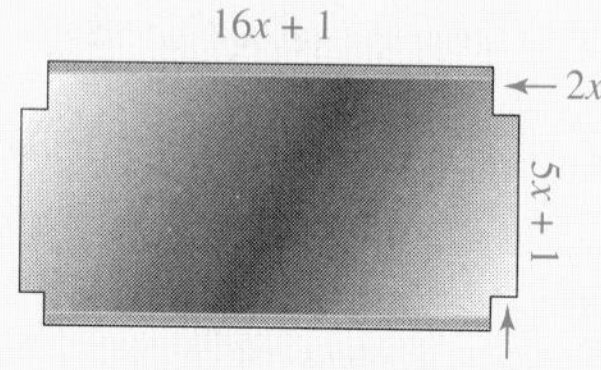

ILLUSTRATION 2

SECTION 5.5 The Greatest Common Factor and Factoring by Grouping

To *prime-factor* a natural number means to write it as a product of prime numbers.

17. Find the prime factorization of 350.

The largest natural number that divides each number in a set of numbers is called their *greatest common factor (GCF).*

18. Find the GCF of each set of monomials.

a. 42, 36, 54

b. $6x^2y^5$, $15xy^3$

The process of finding the individual factors of a known product is called *factoring.*

Always factor out *common monomial factors* as the first step in a factoring problem. Use the distributive property to do this.

19. Factor each polynomial, if possible.

a. $4x + 8$

b. $3x^3 - 6x^2 + 9x$

c. $5x^2y^3 - 11mn^2$

d. $7a^4b^2 + 49a^3b$

e. $5x^2(x + y) - 15x^3(x + y)$

f. $27x^3y^3z^3 + 81x^4y^5z^2 - 90x^2y^3z^7$

A polynomial that cannot be factored is a *prime polynomial.*

20. Factor out the negative of the greatest common factor.

a. $-7b + 14$

b. $-49a^3b^2(a - b)^4 + 63a^2b^4(a - b)^3$

If an expression has four or more terms, try to factor the expression by *grouping*.

21. Factor each polynomial by grouping.
a. $xy + 2y + 4x + 8$
b. $r^2y - ar - ry + a$

22. Solve $m_1m_2 = mm_2 + mm_1$ for m_1.

SECTION 5.6 The Difference of Two Squares; the Sum and Difference of Two Cubes

Factoring the *difference of two squares:*

$$x^2 - y^2 = (x + y)(x - y)$$

23. Factor each expression, if possible.
a. $z^2 - 16$
b. $y^2 - 121$
c. $x^2y^4 - 64z^6$
d. $a^2b^2 + c^2$
e. $c^2 - (a + b)^2$
f. $3x^6 - 300x^2$

Factoring the *sum of two cubes:*

$$x^3 + y^3 = (x + y)(x^2 - xy + y^2)$$

the *difference of two cubes:*

$$x^3 - y^3 = (x - y)(x^2 + xy + y^2)$$

24. Factor each polynomial, if possible.
a. $t^3 + 64$
b. $2x^3y - 54yz^3$

SECTION 5.7 Factoring Trinomials

Perfect square trinomials are the squares of binomials:

$$x^2 + 2xy + y^2 = (x + y)^2$$
$$x^2 - 2xy + y^2 = (x - y)^2$$

25. Factor each trinomial, if possible.
a. $x^2 + 10x + 25$
b. $a^2 - 14a + 49$

Test for factorability: A trinomial of the form $ax^2 + bx + c = 0$ will factor with integer coefficients if $b^2 - 4ac$ is a perfect square.

To factor trinomials with a *lead coefficient of 1,* list the factorizations of the third term.

To factor trinomials with a *lead coefficient other than 1,* use the procedure for factoring a general trinomial.

26. Factor each trinomial, if possible.
a. $y^2 + 21y + 20$
b. $z^2 - 11z + 30$
c. $-x^2 - 3x + 28$
d. $y^2 - 24 - 5y$
e. $4a^2 - 5a + 1$
f. $3b^2 + 2b + 1$
g. $y^3 + y^2 - 2y$
h. $15x^2 - 57xy - 12y^2$
i. $2a^4 + 4a^3 - 6a^2$
j. $v^4 - 13v^2 + 42$

27. Use a substitution to factor $(s + t)^2 - 2(s + t) + 1$.

28. Use grouping to factor $k^2 + 2k + 1 - 9m^2$.

SECTION 5.8 Summary of Factoring Techniques

Use these steps to factor a random expression:

1. Factor out all common monomial factors.
2. If an expression has two terms, check to see if it is
   a. The difference of two squares: $(x^2 - y^2) = (x + y)(x - y)$
   b. The sum of two cubes: $(x^3 + y^3) = (x + y)(x^2 - xy + y^2)$
   c. The difference of two cubes: $(x^3 - y^3) = (x - y)(x^2 + xy + y^2)$
3. If an expression has three terms, attempt to factor it as a *general trinomial.*
4. If an expression has four or more terms, try factoring by *grouping.*
5. Continue until each individual factor is prime.
6. Check the results by multiplying.

29. Factor each expression, if possible.

a. $x^3 + 5x^2 - 6x$

b. $3x^2y - 12xy - 63y$

c. $z^2 - 4 + zx - 2x$

d. $x^2 + 2x + 1 - p^2$

e. $x^2 + 4x + 4 - 4p^4$

f. $y^2 + 3y + 2 + 2x + xy$

g. $4a^3b^3 + 256$

h. $36z^4 - 36$

30. SPANISH ROOF TILE The amount of clay used to make the roof tile shown in Illustration 3 is given by

$$V = \frac{\pi}{2}r_1{}^2h - \frac{\pi}{2}r_2{}^2h$$

Factor the right-hand side of the formula completely.

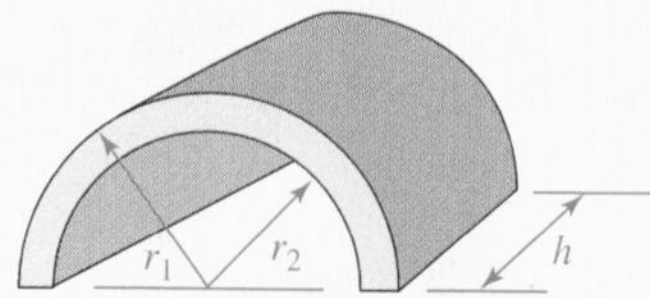

ILLUSTRATION 3

SECTION 5.9 Solving Equations by Factoring

To solve a *quadratic equation* by factoring:

1. Write the equation in the form $ax^2 + bx + c = 0$.
2. Factor the polynomial.
3. Use the zero-factor property to set each factor equal to zero.
4. Solve each resulting equation.
5. Check each solution.

The *zero-factor property:* If a and b are real numbers, then

If $ab = 0$, then $a = 0$ *or* $b = 0$.

31. Solve each equation by factoring.

a. $4x^2 - 3x = 0$

b. $x^2 - 36 = 0$

c. $12x^2 = 5 - 4x$

d. $7y^2 - 37y + 10 = 0$

e. $t^2(15t - 2) = 8t$

f. $u^3 = \frac{u}{3}(19u + 14)$

32. PYRAMIDS The volume of the pyramid in Illustration 4 is given by the formula $V = \frac{Bh}{3}$, where B is the area of its base and h is its height. The volume of the pyramid is 1,020 cubic meters. Find the dimensions of its rectangular base if one edge of the base is 3 meters longer than the other, and the height of the pyramid is 9 meters.

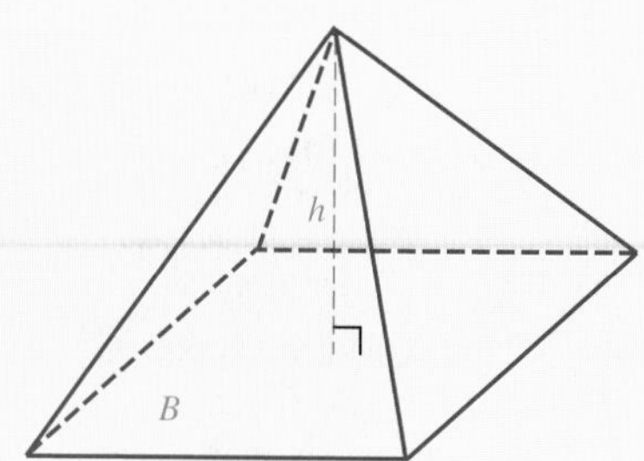

ILLUSTRATION 4

Chapter 5 Test

In Problems 1–4, simplify each expression. Write all answers without using negative exponents. Assume that no denominators are zero.

1. x^3x^5x

2. $(-2x^2y^3)^3$

3. $m^3(m^{-4})^2$

4. $\left(\dfrac{3m^2n^3}{m^4n^{-2}}\right)^{-2}$

5. Write 4,706,000,000,000 in scientific notation.

6. Write 2.45×10^{-4} in standard notation.

7. Evaluate $\dfrac{3.19 \times 10^{15}}{2.2 \times 10^{-4}}$. Express the answer in scientific notation.

8. SPEED OF LIGHT Light travels 1.86×10^5 miles per second. How far does it travel in a minute? Express the answer in scientific notation.

In Problems 9–10, find the degree of each polynomial.

9. $3x^3 - 4x^5 - 3x^2 - 5$

10. $3x^5y^3 - x^8y^2 + 2x^9y^4 - 3x^2y^5 + 4$

11. BOATING The height (in feet) of a warning flare from the surface of the ocean t seconds after being shot into the air is given by the polynomial function $h(t) = -16t^2 + 80t + 10$. What is the height of the flare 2.5 seconds after being fired?

12. STRUCTURAL ENGINEERING Write a polynomial that gives the cross-sectional area of the wooden beam shown in Illustration 1.

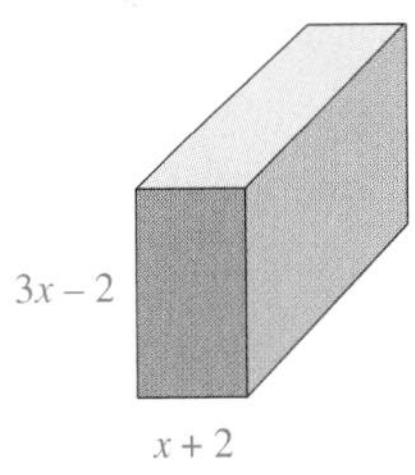

ILLUSTRATION 2

13. Graph the function $f(x) = x^2 + 2x$.

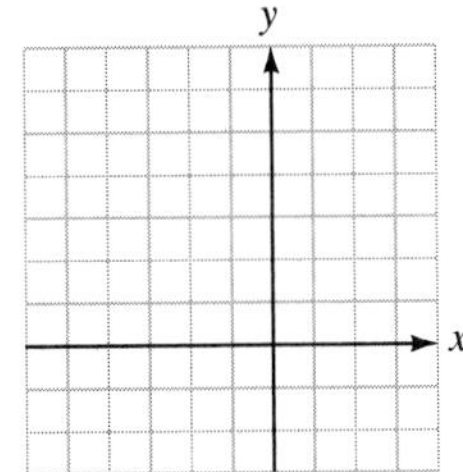

14. Graph the function $f(x) = x^3 + 4x^2 + 4x$.

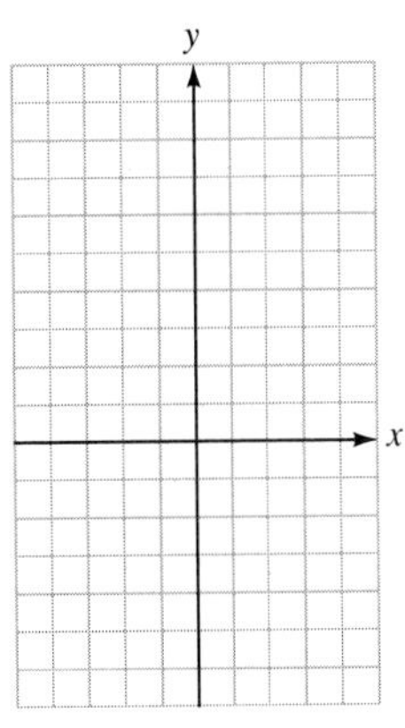

In Problems 15–22, do the operations.

15. $(2y^2 + 4y + 3) + (3y^2 - 3y - 4)$

16. $(-3u^2 + 2u - 7) - (u^2 + 7)$

17. $(3x^3y^2z)(-2xy^{-1}z^3)$

18. $-5a^2b(3ab^3 - 2ab^4)$

19. $(z + 4)(z - 4)$

20. $(3x - 2)(4x - 3)$

21. $(4t - 9)^2$

22. $2s(s - t)(s + t)$

In Problems 23–34, factor each expression, if possible.

23. $3x + 6x^2$

24. $12a^3b^2c - 3a^2b^2c^2 + 6abc^3$

25. $(u - v)r + (u - v)s$

26. $ax - xy + ay - y^2$

27. $x^2 - 49$

28. $4y^4 - 64$

29. $b^2 + 25$

30. $b^3 + 125$

31. $3u^3 - 24$

32. $a^2 - 5a - 6$

33. $6b^2 + b - 2$

34. $x^2 + 6x + 9 - y^2$

35. Solve for x: $x^2 - 5x - 6 = 0$

36. Solve for x: $2x(4x + 3) = 9$

37. Solve for x: $x^2 + 4x = 0$

38. Solve for v: $v_1v_3 - v_3v = v_1v$

39. PREFORMED CONCRETE The slab of concrete in Illustration 2 is twice as long as it is wide. The area in which it is placed includes a 1-foot-wide border of 70 square feet. Find the dimensions of the slab.

2w

w

ILLUSTRATION 2

40. Explain how a factorization of a polynomial can be verified using a check. Give an example.

Rational Expressions

6

Campus Connection

The *Computer Science* Department

In a computer science class, students learn how to access the Internet. Using online services such as CompuServe and America Online, they can browse the World Wide Web to explore vast information resources, including products and services, communication, support forums, and entertainment. In this chapter, we will use a *rational expression* to determine the average cost per hour to use an online service. From a graph, we will see that the average cost decreases as the usage increases. Using rational expressions, we will mathematically describe many other interesting applications from engineering, photography, drafting, electronics, cooking, and transportation.

In this chapter, we extend the concept of fractions to include quotients of two polynomials, which are called rational expressions.

6.1 Rational Functions and Simplifying Rational Expressions

In this section, you will learn about

- **Rational expressions**
- **Rational functions**
- **Graphing rational functions**
- **Finding the domain of a rational function**
- **Simplifying rational expressions**
- **Simplifying rational expressions by factoring out −1**

Introduction. We have seen that linear and polynomial functions can be used to model many real-world situations. In this section, we introduce another family of functions known as *rational functions.* They, too, have applications in a variety of settings. Rational functions get their name from the fact that their defining equation contains a *ratio* (fraction) of two polynomials.

Rational expressions

Rational expressions are fractions that indicate the quotient (or ratio) of two polynomials, such as

$$\frac{-8y^3z^5}{6y^4z^3}, \quad \frac{3x}{x-7}, \quad \frac{5m+n}{8m+16}, \quad \text{and} \quad \frac{6a^2-13a+6}{3a^2+a-2}$$

Since division by 0 is undefined, the value of a polynomial in the denominator of a rational expression cannot be 0. For example, x cannot be 7 in the rational expression $\frac{3x}{x-7}$, because the value of the denominator would be 0. In the rational expression $\frac{5m+n}{8m+16}$, m cannot be -2, because the value of the denominator would be 0.

Rational functions

Rational expressions often define functions. For example, if the cost of subscribing to an online information network is \$6 per month plus \$1.50 per hour of access time, the average (mean) hourly cost of the service is the total monthly cost, divided by the number of hours of access time used that month:

$$\frac{C}{n} = \frac{1.50n + 6}{n}$$ C is the total monthly cost, and n is the number of hours the service is used that month.

The right-hand side of this equation is a rational expression: the quotient of the binomial $1.50n + 6$ and the monomial n.

The rational function that gives the average hourly cost of using the online information network for n hours per month can be written

$$f(n) = \frac{1.50n + 6}{n}$$

We are assuming that at least one access call will be made each month, so the function is defined for $n > 0$.

Rational functions A **rational function** is a function whose equation is defined by a rational expression in one variable, where the value of the polynomial in the denominator is never zero.

EXAMPLE 1 ***Evaluating rational expressions.*** Use the function

$$f(n) = \frac{1.50n + 6}{n}$$

to find the average hourly cost when the network described above is used for **a.** 1 hour and **b.** 9 hours.

Solution

a. To find the average hourly cost for 1 hour of access time, we find $f(1)$:

$$f(1) = \frac{1.50(1) + 6}{1} = 7.5$$ Input 1 for n and simplify.

The average hourly cost for 1 hour of access time is $7.50.

b. To find the average hourly cost for 9 hours of access time, we find $f(9)$:

$$f(9) = \frac{1.50(9) + 6}{9} = 2.166666666\ldots$$ Input 9 for n and simplify.

The average hourly cost for 9 hours of access time is approximately $2.17.

Self Check

Find the average hourly cost when the network is used for **a.** 3 hours and **b.** 100 hours.

Answers: **a.** $3.50, **b.** $1.56

Graphing rational functions

To graph the rational function $f(n) = \frac{1.50n + 6}{n}$, we substitute values for n (the inputs) in the equation, compute the corresponding values of $f(n)$ (the outputs), and express the results as ordered pairs. From the evaluations in Example 1 and its Self Check, we know four ordered pairs that satisfy the equation: (1, 7.50), (3, 3.50), (9, 2.17), and (100, 1.56). Those pairs and others are listed in the table in Figure 6-1 on the next page. We then plot the points and draw a smooth curve through them to get the graph.

From the graph, we can see that the average hourly cost decreases as the number of hours of access time increases. Since the cost of each extra hour of access time is $1.50, the average hourly cost can approach $1.50 but never drop below it. Thus, the graph of the function approaches the line $y = 1.5$ as n increases. When a graph approaches a line, we call the line an **asymptote.** The line $y = 1.5$ is a **horizontal asymptote** of the graph.

As n gets smaller and approaches 0, the graph approaches the y-axis. The y-axis is a **vertical asymptote** of the graph.

$$f(n) = \frac{1.50n + 6}{n}$$

n	$f(n)$
1	7.50
2	4.50
3	3.50
4	3.00
5	2.70
6	2.50
7	2.36
8	2.25
9	2.17
10	2.10
100	1.56

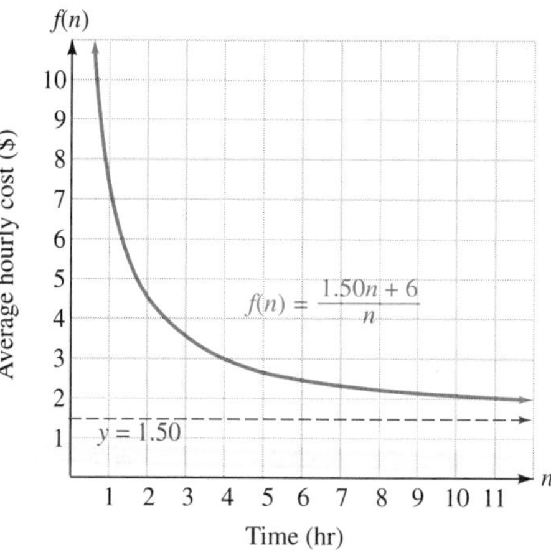

As the access time increases, the graph approaches the line $y = 1.5$, which indicates that the average hourly cost approaches $1.50 as the hours of use increase.

FIGURE 6-1

Finding the domain of a rational function

Since division by 0 is undefined, any values that make the denominator 0 in a rational function must be excluded from the domain of the function.

EXAMPLE 2 ***Finding the domain of a rational function.*** Find the domain of $f(x) = \dfrac{3x + 2}{x^2 + x - 6}$.

Solution

From the set of real numbers, we must exclude any values of x that make the denominator 0. To find these values, we set $x^2 + x - 6$ equal to 0 and solve for x.

$$x^2 + x - 6 = 0$$

$$(x + 3)(x - 2) = 0 \quad \text{Factor the trinomial.}$$

$$x + 3 = 0 \quad \text{or} \quad x - 2 = 0 \quad \text{Set each factor equal to 0.}$$

$$x = -3 \qquad\qquad x = 2 \quad \text{Solve each linear equation.}$$

Thus, the domain of the function is the set of all real numbers except -3 and 2. In interval notation, the domain is $(-\infty, -3) \cup (-3, 2) \cup (2, \infty)$.

Self Check

Find the domain of $f(x) = \dfrac{x^2 + 1}{x - 2}$.

Answer: $(-\infty, 2) \cup (2, \infty)$

Accent on Technology *Finding the domain and range of a rational function*

We can find the domain and range of the function in Example 2 by looking at its graph. If we use window settings of $[-10, 10]$ for x and $[-10, 10]$ for y and graph the function

$$f(x) = \frac{3x + 2}{x^2 + x - 6}$$

we will obtain the graph in Figure 6-2(a).

From the figure, we can see that

- As x approaches -3 from the left, the values of y decrease, and the graph approaches the vertical line $x = -3$.
- As x approaches -3 from the right, the values of y increase, and the graph approaches the vertical line $x = -3$.

From the figure, we can also see that

- As x approaches 2 from the left, the values of y decrease, and the graph approaches the vertical line $x = 2$.
- As x approaches 2 from the right, the values of y increase, and the graph approaches the vertical line $x = 2$.

The lines $x = -3$ and $x = 2$ are vertical asymptotes. Although the vertical lines in the graph appear to be the graphs of $x = -3$ and $x = 2$, they are not. Graphing calculators draw graphs by connecting dots whose x-coordinates are close together. Often when two such points straddle a vertical asymptote and their y-coordinates are far apart, the calculator draws a line between them anyway, producing what appears to be a vertical asymptote. If you set your calculator to dot mode instead of connected mode, the vertical lines will not appear.

From Figure 6-2(a), we can also see that

- As x increases to the right of 2, the values of y decrease and approach the line $y = 0$.
- As x decreases to the left of -3, the values of y increase and approach the line $y = 0$.

The line $y = 0$ (the x-axis) is a horizontal asymptote. Graphing calculators do not draw lines that appear to be horizontal asymptotes.

From the graph, we can see that every real number x, except -3 and 2, gives a value of y. This confirms that the domain of the function is $(-\infty, -3) \cup (-3, 2) \cup (2, \infty)$. We can also see that y can be any value. Thus, the range is $(-\infty, \infty)$.

To find the domain and range of the function $f(x) = \frac{2x+1}{x-1}$, we use a calculator to draw the graph shown in Figure 6-2(b). From this graph, we can see that the line $x = 1$ is a vertical asymptote and that the line $y = 2$ is a horizontal asymptote. Since x can be any real number except 1, the domain is the interval $(-\infty, 1) \cup (1, \infty)$. Since y can be any value except 2, the range is $(-\infty, 2) \cup (2, \infty)$.

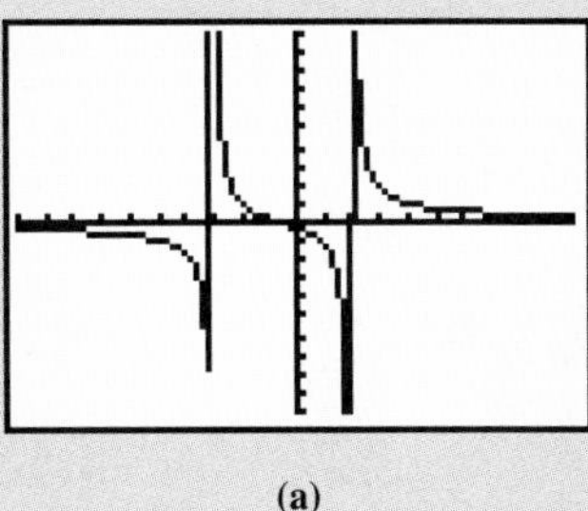

(a)

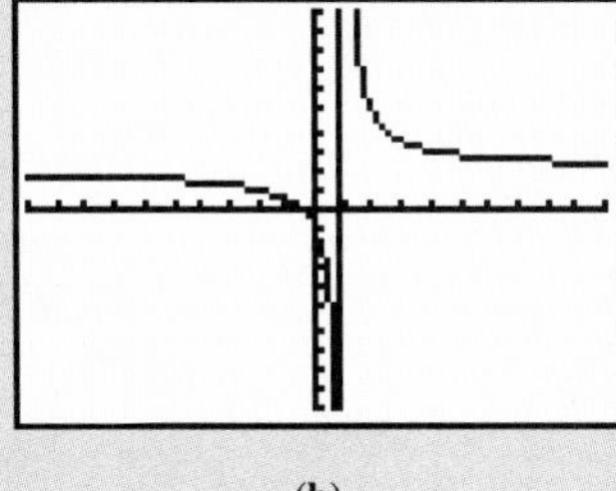

(b)

FIGURE 6-2

Simplifying rational expressions

When we are working with rational expressions, the familiar rules for arithmetic fractions apply.

Properties of fractions

If there are no divisions by 0, then

1. $\frac{a}{b} = \frac{c}{d}$ if and only if $ad = bc$
2. $\frac{a}{1} = a$ and $\frac{a}{a} = 1$
3. $\frac{ak}{bk} = \frac{a}{b} \cdot \frac{k}{k} = \frac{a}{b}$
4. $-\frac{a}{b} = \frac{-a}{b} = \frac{a}{-b}$

Property 3 of fractions is true because

$$\frac{ak}{bk} = \frac{a}{b} \cdot \frac{k}{k} = \frac{a}{b} \cdot 1 = \frac{a}{b} \quad (b \neq 0, k \neq 0)$$ Any number times 1 is the number.

Property 3, which is known as the **fundamental property of fractions,** is used to simplify rational expressions. It enables us to divide out factors that are common to the numerator and the denominator of a fraction.

Simplifying a rational expression

To simplify a rational expression,

1. Completely factor the numerator and the denominator.
2. Divide out the common factors of the numerator and the denominator.

EXAMPLE 3 ***Dividing out common factors*** Simplify **a.** $\frac{10k}{25k^2}$ and **b.** $\frac{-8y^3z^5}{6y^4z^3}$.

Solution

To simplify these rational expressions, we factor each numerator and denominator and divide out all common factors.

a. $$\frac{10k}{25k^2} = \frac{5 \cdot 2 \cdot k}{5 \cdot 5 \cdot k \cdot k}$$

$$= \frac{\overset{1}{\not{5}} \cdot 2 \cdot \overset{1}{\not{k}}}{\underset{1}{\not{5}} \cdot 5 \cdot \underset{1}{\not{k}} \cdot k}$$ Divide out the common factors of 5 and k. Show this using slashes and 1's.

$$= \frac{2}{5k}$$ Do the multiplications in the numerator and the denominator.

b. $$\frac{-8y^3z^5}{6y^4z^3} = \frac{-2 \cdot 4 \cdot y \cdot y \cdot y \cdot z \cdot z \cdot z \cdot z \cdot z}{2 \cdot 3 \cdot y \cdot y \cdot y \cdot y \cdot z \cdot z \cdot z}$$

$$= \frac{-\overset{1}{\not{2}} \cdot 4 \cdot \overset{1}{\not{y}} \cdot \overset{1}{\not{y}} \cdot \overset{1}{\not{y}} \cdot \overset{1}{\not{z}} \cdot \overset{1}{\not{z}} \cdot \overset{1}{\not{z}} \cdot z \cdot z}{\underset{1}{\not{2}} \cdot 3 \cdot \underset{1}{\not{y}} \cdot \underset{1}{\not{y}} \cdot \underset{1}{\not{y}} \cdot y \cdot \underset{1}{\not{z}} \cdot \underset{1}{\not{z}} \cdot \underset{1}{\not{z}}}$$

$$= -\frac{4z^2}{3y}$$

Self Check

Simplify $\frac{-12a^4b^2}{-3ab^4}$.

Answer: $\frac{4a^3}{b^2}$ ■

The fractions in Example 3 can also be simplified using the rules of exponents:

$$\frac{10k}{25k^2} = \frac{5 \cdot 2}{5 \cdot 5}k^{1-2} = \frac{2}{5} \cdot k^{-1} = \frac{2}{5} \cdot \frac{1}{k} = \frac{2}{5k}$$

$$\frac{-8y^3z^5}{6y^4z^3} = \frac{-2 \cdot 4}{2 \cdot 3}y^{3-4}z^{5-3} = \frac{-4}{3} \cdot y^{-1}z^2 = -\frac{4}{3} \cdot \frac{1}{y} \cdot \frac{z^2}{1} = -\frac{4z^2}{3y}$$

EXAMPLE 4 ***Simplifying rational expressions.*** Simplify $\frac{x^2 - 16}{x + 4}$.

Solution

We factor $x^2 - 16$ and use the fact that $\frac{x+4}{x+4} = 1$.

$$\frac{x^2 - 16}{x + 4} = \frac{\overset{1}{\cancel{(x + 4)}}(x - 4)}{\underset{1}{\cancel{(x + 4)}}} \quad \text{Factor the difference of two squares.}$$

$$= \frac{x - 4}{1} \quad \text{Divide out the common factor of } (x + 4).$$

$$= x - 4$$

Self Check

Simplify $\frac{x^2 - 9}{x - 3}$.

Answer: $x + 3$

Accent on Technology *Checking an algebraic simplification*

To show that the simplification in Example 4 is correct, we can graph the functions $f(x) = \frac{x^2 - 16}{x + 4}$ (Figure 6-3(a)) and $g(x) = x - 4$ (Figure 6-3(b)). Except for the point where $x = -4$, the graphs are the same. The point where $x = -4$ is excluded from the graph of $f(x) = \frac{x^2 - 16}{x + 4}$, because -4 is not in the domain of f. However, graphing calculators do not show that this point is excluded. The point where $x = -4$ is included in the graph of $g(x) = x - 4$, because -4 is in the domain of g.

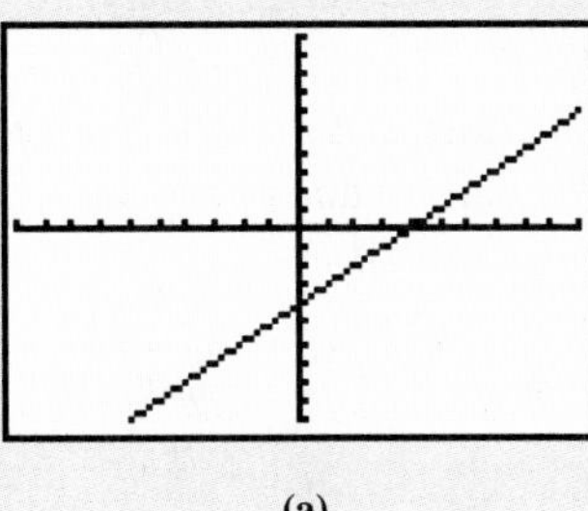

(a)

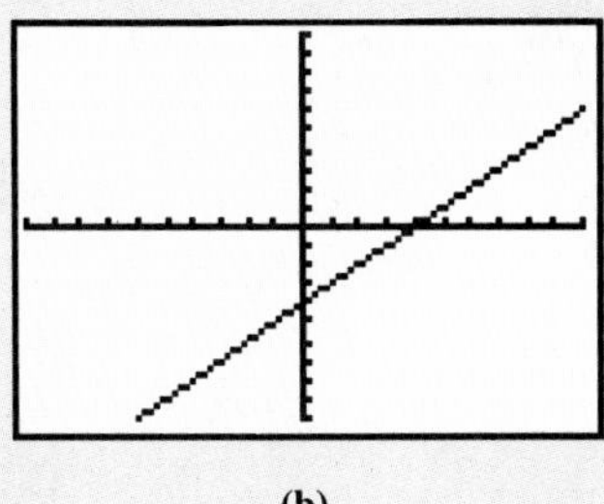

(b)

FIGURE 6-3

EXAMPLE 5 ***Simplifying rational expressions.*** Simplify $\frac{6a^2 - 13a + 6}{3a^2 + a - 2}$.

Solution

We factor the trinomials in the numerator and the denominator and then divide out the common factor.

$$\frac{6a^2 - 13a + 6}{3a^2 + a - 2} = \frac{\overset{1}{\cancel{(3a - 2)}}(2a - 3)}{\underset{1}{\cancel{(3a - 2)}}(a + 1)}$$

$$= \frac{2a - 3}{a + 1}$$

Self Check

Simplify $\frac{2b^2 + 7b - 15}{2b^2 + 13b + 15}$.

WARNING! Do not divide out the a's in $\frac{2a-3}{a+1}$. The a in the numerator is a factor of the first term only, not a factor of the entire numerator. Likewise, the a in the denominator is a factor of the first term only, not a factor of the entire denominator.

Answer: $\dfrac{2b-3}{2b+3}$ ■

We will encounter many fractions that are already in simplified form. For example, to attempt to simplify

$$\frac{x^2 + xa + 2x + 2a}{x^2 + x - 6}$$

we factor the numerator and denominator and divide out any common factors:

$$\frac{x^2 + xa + 2x + 2a}{x^2 + x - 6} = \frac{x(x + a) + 2(x + a)}{(x - 2)(x + 3)} = \frac{(x + a)(x + 2)}{(x - 2)(x + 3)}$$

Since there are no common factors in the numerator and denominator, the fraction is in *lowest terms.* It cannot be simplified.

WARNING! Only factors that are common to the entire numerator and the entire denominator can be divided out. *Terms* common to both the numerator and denominator cannot be divided out. It is incorrect to divide out the common term of 3 in the following simplification, because it gives a wrong answer.

$$\frac{3+7}{3} = \frac{\overset{1}{\cancel{3}} + 7}{\underset{1}{\cancel{3}}} = \frac{1+7}{1} = 8 \quad \text{The correct simplification is } \frac{3+7}{3} = \frac{10}{3}.$$

The 3's in the fraction $\frac{5 + 3(2)}{3(4)}$ cannot be divided out, because the 3 in the numerator is a factor of the second term only. To be divided out, the 3 must be a factor of the entire numerator.

It is not correct to divide out the y in the fraction $\frac{x^2y + 6x}{y}$, because y is not a factor of the entire numerator.

Simplifying rational expressions by factoring out −1

To simplify $\frac{b-a}{a-b}$ $(a \neq b)$, we factor -1 from the numerator and divide out any factors common to both the numerator and the denominator:

$$\frac{b-a}{a-b} = \frac{-a+b}{a-b} \quad \text{Rewrite the numerator.}$$

$$= \frac{-1\overset{1}{\cancel{(a-b)}}}{\underset{1}{\cancel{(a-b)}}} \quad \text{Factor out } -1 \text{ from each term in the numerator. Divide out the common factor.}$$

$$= \frac{-1}{1}$$

$$= -1$$

In general, we have the following principle.

Quotient of a quantity and its opposite

The quotient of any nonzero quantity and its negative (or opposite) is -1.

EXAMPLE 6 ***Factoring out −1.*** Simplify $\frac{3x^2 - 10xy - 8y^2}{4y^2 - xy}$.

Solution

We factor the numerator and denominator. Because $x - 4y$ and $4y - x$ are negatives, their quotient is -1.

$$\frac{3x^2 - 10xy - 8y^2}{4y^2 - xy} = \frac{(3x + 2y)\overset{-1}{\cancel{(x - 4y)}}}{y\underset{1}{\cancel{(4y - x)}}}$$

$$= \frac{-(3x + 2y)}{y}$$

$$= \frac{-3x - 2y}{y}$$

Self Check

Simplify $\frac{2a^2 - 3ab - 9b^2}{3b^2 - ab}$.

Answer:

$-\frac{2a + 3b}{b}$ or $\frac{-2a - 3b}{b}$ ■

STUDY SET Section 6.1

VOCABULARY *In Exercises 1–4, fill in the blanks to make the statements true.*

1. A fraction that is the quotient of two polynomials, such as

$$\frac{x^2 - x - 6}{x^3 - 8}$$

is called a ____________ expression.

2. In the rational expression

$$\frac{(x + 2)(3x - 1)}{(x + 2)(4x + 2)}$$

$(x + 2)$ is a common ____________ of the numerator and the denominator.

3. If a graph approaches a line, the line is called an ____________.

4. The ____________ of a fraction can never be 0.

CONCEPTS *In Exercises 5–8, complete the table of values for each rational function (round to the nearest hundredth when applicable). Then graph it. Each function is defined for $x > 0$. Label the horizontal asymptote.*

5. $f(x) = \frac{6}{x}$

x	$f(x)$
1	
2	
4	
6	
8	
10	
12	

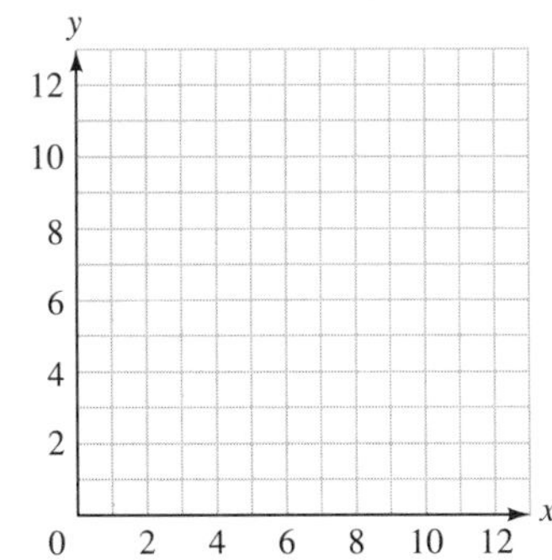

6. $f(x) = \frac{12}{x}$

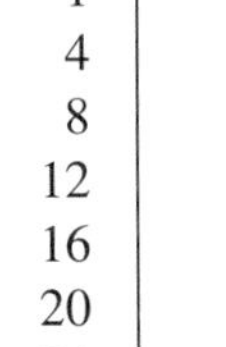

x	$f(x)$
1	
4	
8	
12	
16	
20	
24	

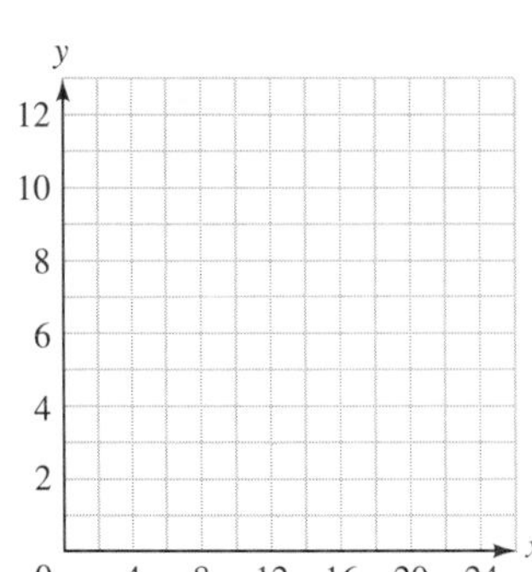

7. $f(x) = \dfrac{x+2}{x}$

x	$f(x)$
1	
2	
4	
6	
8	
10	
12	

8. $f(x) = \dfrac{2x+4}{x}$

x	$f(x)$
1	
4	
8	
12	
16	
20	
24	

9. Simplify each rational expression.

a. $\dfrac{x+8}{x+8}$ **b.** $\dfrac{x+8}{8+x}$

c. $\dfrac{x-8}{x-8}$ **d.** $\dfrac{x-8}{8-x}$

10. Simplify each rational expression, if possible.

a. $\dfrac{x+8}{x}$ **b.** $\dfrac{x+8}{8}$

c. $\dfrac{a^3+8}{2}$ **d.** $\dfrac{x^2+5x+6}{x^2+x-12}$

In Exercises 11–12, refer to the graphs below. Each graph shows the average cost to manufacture a certain item for a given number of units produced.

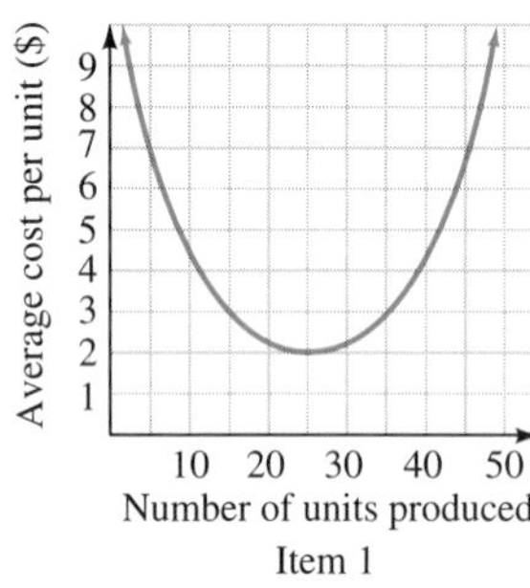

Item 1

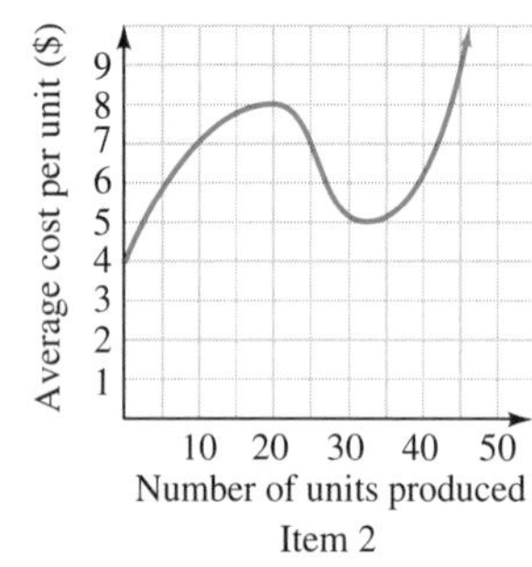

Item 2

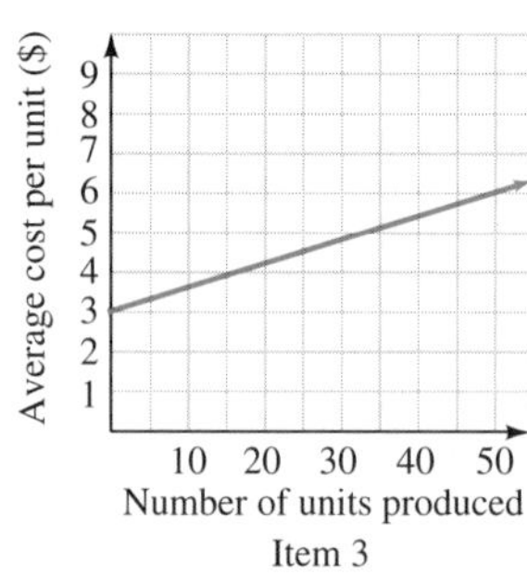

Item 3

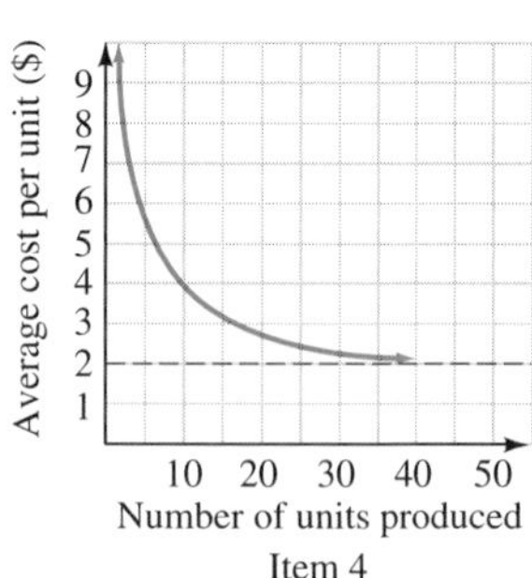

Item 4

11. MANUFACTURING For each graph, briefly describe how the average cost per unit changes as the number of units produced increases.

12. Which graph is best described as the graph of a

a. linear function **b.** quadratic function

c. rational function **d.** polynomial function

NOTATION

13. Tell whether each statement is true or false.

a. $-\dfrac{x-4}{x+4} = \dfrac{4-x}{x+4}$ **b.** $\dfrac{a-3b}{2b-a} = \dfrac{3b-a}{a-2b}$

14. Explain what the slashes and the 1's show.

$$\frac{t^2-4}{t^2+2t} = \frac{\overset{1}{\cancel{(t+2)}}(t-2)}{t\underset{1}{\cancel{(t+2)}}} = \frac{t-2}{t}$$

PRACTICE *In Exercises 15–18, the time t it takes to travel 600 miles is a function of the average rate of speed,* $t = \frac{600}{r}$. *Find t for each value of r.*

15. 30 mph **16.** 40 mph **17.** 50 mph **18.** 60 mph

In Exercises 19–26, find the domain of each rational function. Use interval notation.

19. $f(x) = \frac{2}{x}$

20. $f(x) = \frac{8}{x - 1}$

21. $f(x) = \frac{2}{x + 2}$

22. $f(x) = \frac{2}{x^2 - 2x}$

23. $f(x) = \frac{2}{x - x^2}$

24. $f(x) = \frac{2}{x^2 - 36}$

25. $f(x) = \frac{2}{x^2 - x - 56}$

26. $f(x) = \frac{2}{x^2 + 2x - 24}$

In Exercises 27–70, simplify each rational expression when possible.

27. $\frac{12}{18}$

28. $\frac{25}{55}$

29. $-\frac{112}{36}$

30. $-\frac{49}{21}$

31. $\frac{12x^3}{3x}$

32. $-\frac{15a^2}{25a^3}$

33. $\frac{-24x^3y^4}{18x^4y^3}$

34. $\frac{15a^5b^4}{21b^3c^2}$

35. $-\frac{11x(x - y)}{22(x - y)}$

36. $\frac{x(x - 2)^2}{(x - 2)^3}$

37. $\frac{(a - b)(d - c)}{(c - d)(a - b)}$

38. $\frac{(p + q)(p - r)}{(r - p)(p + q)}$

39. $\frac{y + x}{x^2 - y^2}$

40. $\frac{x - y}{x^2 - y^2}$

41. $\frac{5x - 10}{x^2 - 4x + 4}$

42. $\frac{y - xy}{xy - x}$

43. $\frac{12 - 3x^2}{x^2 - x - 2}$

44. $\frac{x^2 + 2x - 15}{25 - x^2}$

45. $\frac{x^2 + y^2}{x + y}$

46. $\frac{3x + 6y}{2y + x}$

47. $\frac{x^3 + 8}{x^2 - 2x + 4}$

48. $\frac{x^2 + 3x + 9}{x^3 - 27}$

49. $\frac{x^2 + 2x + 1}{x^2 + 4x + 3}$

50. $\frac{6x^2 + x - 2}{8x^2 + 2x - 3}$

51. $\frac{3m - 6n}{3n - 6m}$

52. $\frac{ax + by + ay + bx}{a^2 - b^2}$

53. $\frac{4x^2 + 24x + 32}{16x^2 + 8x - 48}$

54. $\frac{a^2 - 4}{a^3 - 8}$

55. $\frac{3x^2 - 3y^2}{x^2 + 2y + 2x + yx}$

56. $\frac{x^2 + x - 30}{x^2 - x - 20}$

57. $\frac{4x^2 + 8x + 3}{6 + x - 2x^2}$

58. $\frac{6x^2 + 13x + 6}{6 - 5x - 6x^2}$

59. $\frac{a^3 + 27}{4a^2 - 36}$

60. $\frac{a - b}{b^2 - a^2}$

61. $\frac{2x^2 - 3x - 9}{2x^2 + 3x - 9}$

62. $\frac{6x^2 - 7x - 5}{2x^2 + 5x + 2}$

63. $\frac{(m + n)^3}{m^2 + 2mn + n^2}$

64. $\frac{x^3 - 27}{3x^2 - 8x - 3}$

65. $\frac{m^3 - mn^2}{mn^2 + m^2n - 2m^3}$

66. $\frac{p^3 + p^2q - 2pq^2}{pq^2 + p^2q - 2p^3}$

67. $\frac{x^4 - y^4}{(x^2 + 2xy + y^2)(x^2 + y^2)}$

68. $\frac{(x^2 - 1)(x + 1)}{(x^2 - 2x + 1)^2}$

69. $\frac{6xy - 4x - 9y + 6}{6y^2 - 13y + 6}$

70. $\frac{x^2 + 2xy}{x + 2y + x^2 - 4y^2}$

In Exercises 71–74, use a graphing calculator to graph each rational function. From the graph, determine its domain and range.

71. $f(x) = \frac{x}{x - 2}$

72. $f(x) = \frac{x + 2}{x}$

73. $f(x) = \frac{x + 1}{x^2 - 4}$

74. $f(x) = \frac{x - 2}{x^2 - 3x - 4}$

APPLICATIONS

75. ENVIRONMENTAL CLEANUP Suppose the cost (in dollars) of removing $p\%$ of the pollution in a river is given by the rational function

$$f(p) = \frac{50{,}000p}{100 - p} \quad (0 \le p < 100)$$

Find the cost of removing each percent of pollution.
a. 50% **b.** 80%

76. DIRECTORY COSTS The average (mean) cost for a service club to publish a directory of its members is given by the rational function

$$f(x) = \frac{1.25x + 700}{x}$$

where x is the number of directories printed. Find the average cost per directory if
a. 500 directories are printed.
b. 2,000 directories are printed.

77. UTILITY COSTS An electric company charges $7.50 per month plus 9¢ for each kilowatt hour (kwh) of electricity used.
a. Find a linear function that gives the total cost of n kwh of electricity.
b. Find a rational function that gives the average cost per kwh when using n kwh.
c. Find the average cost per kwh when 775 kwh are used.

78. SCHEDULING WORK CREWS The rational function

$$f(t) = \frac{t^2 + 2t}{2t + 2}$$

gives the number of days it would take two construction crews, working together, to frame a house that crew 1 (working alone) could complete in t days and crew 2 (working alone) could complete in $t + 2$ days.
a. If crew 1 could frame a certain house in 15 days, how long would it take both crews working together?
b. If crew 2 could frame a certain house in 20 days, how long would it take both crews working together?

79. FILLING A POOL The rational function

$$f(t) = \frac{t^2 + 3t}{2t + 3}$$

gives the number of hours it would take two pipes, working together, to fill a pool that the larger pipe (working alone) could fill in t hours and the smaller pipe (working alone) could fill in $t + 3$ hours.
a. If the smaller pipe could fill a pool in 7 hours, how long would it take both pipes to fill the pool?
b. If the larger pipe could fill a pool in 8 hours, how long would it take both pipes to fill the pool?

80. RETENTION STUDY After learning a list of words, two subjects were tested over a 28-day period to see what percent of the list they remembered. In both cases, their percent recall could be modeled by rational functions, as shown in Illustration 1.
a. Use the graphs to complete the table.

Days since learning	0	1	2	4	7	14	28
% recall—subject 1							
% recall—subject 2							

b. After 28 days, which subject had the better recall?

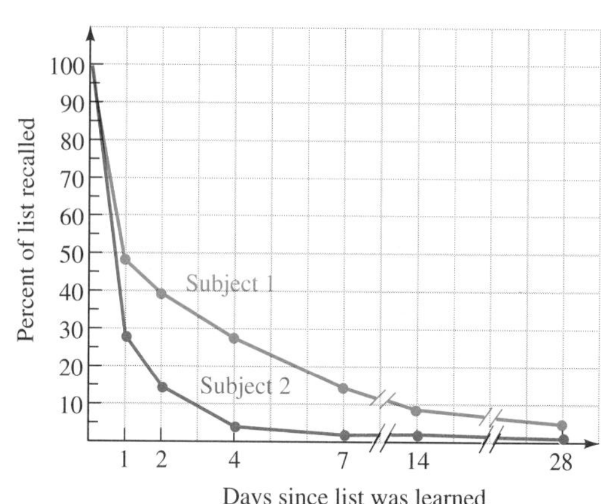

ILLUSTRATION 1

WRITING *Write a paragraph using your own words.*

81. Explain how to simplify a rational expression.

82. Explain how to recognize that a rational expression is in lowest terms.

REVIEW *In Exercises 83–86, factor each expression.*

83. $3x^2 - 9x$

84. $-6t^2 + 5t + 6$

85. $27x^6 + 64y^3$

86. $x^2 + ax + 2x + 2a$

6.2 *Proportion and Variation*

In this section, you will learn about

- **Ratios**
- **Proportions**
- **Solving proportions**
- **Similar triangles**
- **Direct variation**
- **Inverse variation**
- **Joint variation**
- **Combined variation**

INTRODUCTION. In this section, we discuss five mathematical models that have a variety of applications. First, we show how a *ratio-proportion model* can be used to solve shopping problems and to determine the height of a tree given the length of its shadow. Then we introduce four types of *variation models,* each of which expresses a special relationship between two or more quantities. We use these models to solve problems involving travel, lighting, geometry, and highway construction.

Ratios

The quotient of two numbers is often called a **ratio.** For example, the fraction $\frac{2}{3}$ can be read as "the ratio of 2 to 3." Some more examples of ratios are

$$\frac{4x}{7y} \text{ (the ratio of } 4x \text{ to } 7y) \qquad \text{and} \qquad \frac{x-2}{3x} \text{ (the ratio of } x-2 \text{ to } 3x)$$

Ratios are often used to express **unit costs,** such as the cost per pound of ground beef.

$$\begin{matrix} \text{The cost of a package of ground beef} \rightarrow \\ \text{The weight of the package} \rightarrow \end{matrix} \frac{\$7.47}{5 \text{ lb}} \approx \$1.49 \text{ per lb} \leftarrow \text{The cost per pound}$$

Ratios are also used to express **rates,** such as an average rate of speed.

$$\begin{matrix} \text{A distance traveled} \rightarrow \\ \text{in a period of time} \rightarrow \end{matrix} \frac{372 \text{ miles}}{6 \text{ hours}} = 62 \text{ mph} \leftarrow \text{The average rate of speed}$$

Proportions

An equation indicating that two ratios are equal is called a **proportion.** Two examples of proportions are

$$\frac{1}{4} = \frac{2}{8} \qquad \text{and} \qquad \frac{4}{7} = \frac{12}{21}$$

In the proportion $\frac{a}{b} = \frac{c}{d}$, a and d are called the **extremes** of the proportion, and b and c are called the **means.**

To develop a fundamental property of proportions, we suppose that

$$\frac{a}{b} = \frac{c}{d}$$

is a proportion and multiply both sides by bd to obtain

$$bd\left(\frac{a}{b}\right) = bd\left(\frac{c}{d}\right)$$

$$\frac{\not{b}da}{\not{b}} = \frac{b\not{d}c}{\not{d}} \quad \text{Divide out common factors.}$$

$$ad = bc$$

Thus, if $\frac{a}{b} = \frac{c}{d}$, then $ad = bc$. This illustrates the following property.

Fundamental property of proportions In a proportion, the product of the extremes is equal to the product of the means.

Solving proportions

We can solve many problems by writing and then solving a proportion. To solve a proportion, we apply the fundamental property of proportions.

EXAMPLE 1 ***Solving proportions.*** Solve for x: $\frac{x+1}{x} = \frac{x}{x+2}$.

Solution

$$\frac{x+1}{x} = \frac{x}{x+2}$$

$(x+1)(x+2) = x \cdot x$ — In a proportion, the product of the extremes equals the product of the means.

$x^2 + 3x + 2 = x^2$ — Do the multiplications.

$3x + 2 = 0$ — Subtract x^2 from both sides.

$x = -\frac{2}{3}$ — Subtract 2 from both sides and then divide by 3.

Thus, $x = -\frac{2}{3}$.

Self Check

Solve for x: $\frac{x-1}{x} = \frac{x}{x+3}$.

Answer: $\frac{3}{2}$

EXAMPLE 2 ***Solving proportions.*** Solve $\frac{5a+2}{2a} = \frac{18}{a+4}$.

Solution

$$\frac{5a+2}{2a} = \frac{18}{a+4}$$

$(5a+2)(a+4) = 2a(18)$ — In a proportion, the product of the extremes equals the product of the means.

$5a^2 + 22a + 8 = 36a$ — Multiply.

$5a^2 - 14a + 8 = 0$ — Subtract $36a$ from both sides.

$(5a-4)(a-2) = 0$ — Factor to solve the quadratic equation.

$5a - 4 = 0$ or $a - 2 = 0$ — Set each factor equal to 0.

$5a = 4$ $\quad$ $a = 2$ — Solve each linear equation.

$a = \frac{4}{5}$

Thus, $a = \frac{4}{5}$ or $a = 2$.

Self Check

Solve $\frac{3x+1}{12} = \frac{x}{x+2}$.

Answer: $\frac{2}{3}$, 1

EXAMPLE 3 ***Gourmet cooking.*** To make a dessert of Pears Hélène, a chef needs to purchase 14 pears. If they are on sale at 6 for \$2.34, what will 14 cost?

Solution

A proportion can be used to model this situation. First, we let c represent the cost of 14 pears. The price per pear when purchasing 6 pears is $\frac{\$2.34}{6}$, and the price per pear when purchasing 14 pears is $\frac{\$c}{14}$. Since these ratios are equal, we have the following proportion.

$$\frac{2.34}{6} = \frac{c}{14} \quad \text{\$2.34 is to 6 as \$}c\text{ is to 14.}$$

$$14(2.34) = 6c \quad \text{In a proportion, the product of the extremes is equal to the product of the means.}$$

$$32.76 = 6c \quad \text{Multiply.}$$

$$\frac{32.76}{6} = c \quad \text{Divide both sides by 6.}$$

$$c = 5.46 \quad \text{Simplify.}$$

Fourteen pears will cost \$5.46.

Self Check

An HO-scale model railroad engine is 9 inches long. The HO scale is 87 feet to 1 foot. How long is a real engine? (*Hint:* Note the difference in units.)

Answer: $65\frac{1}{4}$ ft

Similar triangles

If two angles of one triangle have the same measure as two angles of a second triangle, the triangles will have the same shape. In this case, we call the triangles **similar triangles.** Here are some facts about similar triangles.

Similar triangles

If two triangles are similar, then

1. the three angles of the first triangle have the same measure, respectively, as the three angles of the second triangle.
2. the lengths of all corresponding sides are in proportion.

The triangles shown in Figure 6-4 are similar triangles.

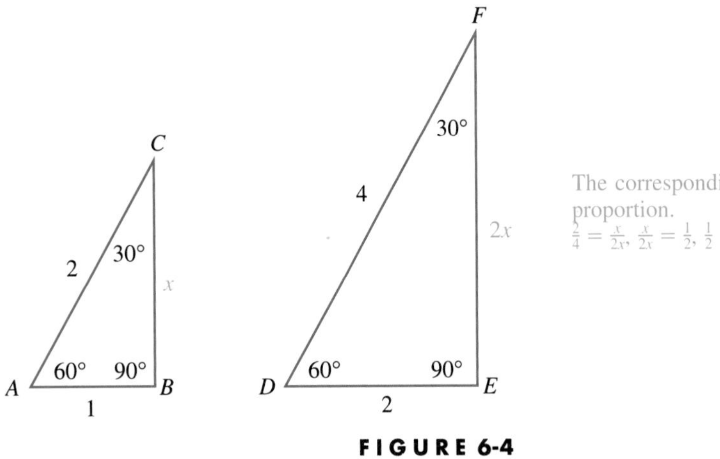

The corresponding sides are in proportion.
$\frac{2}{4} = \frac{x}{2x}$, $\frac{x}{2x} = \frac{1}{2}$, $\frac{1}{2} = \frac{2}{4}$.

FIGURE 6-4

The properties of similar triangles often enable us to determine the lengths of the sides of triangles indirectly. For example, on a sunny day, we can find the height of a tree and stay safely on the ground.

EXAMPLE 4 ***Height of a tree.*** A tree casts a shadow of 29 feet at the same time as a vertical yardstick casts a shadow of 2.5 feet. Find the height of the tree.

Solution Refer to Figure 6-5, which shows the triangles determined by the tree and its shadow and the yardstick and its shadow. Because the triangles have the same shape, they are similar, and the measures of their corresponding sides are in proportion. If we let h represent the height of the tree, we can find h by setting up and solving the following proportion.

$$\frac{h}{3} = \frac{29}{2.5} \quad h \text{ is to 3 as 29 is to 2.5.}$$

$$2.5h = 3(29) \quad \text{In a proportion, the product of the extremes is equal to the product of the means.}$$

$$2.5h = 87 \quad \text{Multiply.}$$

$$h = 34.8 \quad \text{Divide both sides by 2.5.}$$

The tree is about 35 feet tall.

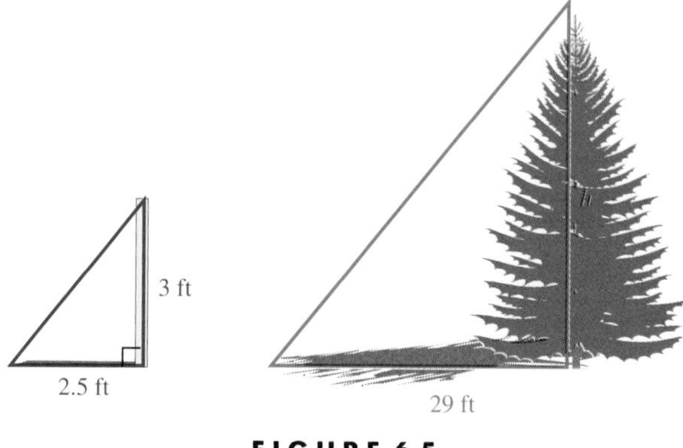

FIGURE 6-5

Direct variation

To introduce direct variation, we consider the formula for the circumference of a circle

$$C = \pi D$$

where C is the circumference, D is the diameter, and $\pi \approx 3.14159$. If we double the diameter of a circle, we determine another circle with a larger circumference C_1 such that

$$C_1 = \pi(2D) = 2\pi D = 2C$$

Thus, doubling the diameter results in doubling the circumference. Likewise, if we triple the diameter, we will triple the circumference.

In this formula, we say that the variables C and D *vary directly,* or that they are *directly proportional.* This is because as one variable gets larger, so does the other, in a predictable way. In this example, the constant π is called the *constant of variation* or the *constant of proportionality.*

Direct variation

The words "y varies directly with x" or "y is directly proportional to x" mean that $y = kx$ for some nonzero constant k. The constant k is called the **constant of variation** or the **constant of proportionality.**

Since the formula for direct variation ($y = kx$) defines a linear function, its graph is always a line with a y-intercept at the origin. The graph of $y = kx$ appears in Figure 6-6 for three positive values of k.

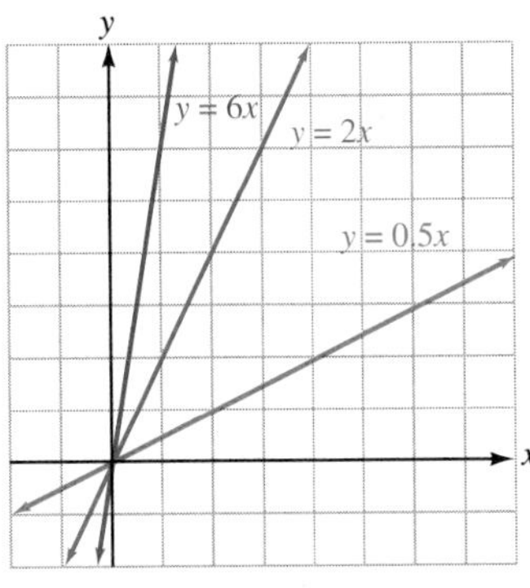

FIGURE 6-6

One example of direct variation is Hooke's law from physics. Hooke's law states that the distance a spring will stretch varies directly with the force that is applied to it.

If d represents a distance and f represents a force, this verbal model of Hooke's law can be expressed mathematically as

$d = kf$ This is a model for direct variation.

where k is the constant of variation. If the spring stretches 10 inches when a weight of 6 pounds is attached, k can be found as follows:

$d = kf$

$10 = k(6)$ Substitute 10 for d and 6 for f.

$\frac{5}{3} = k$

To find the force required to stretch the spring a distance of 35 inches, we can solve the equation $d = kf$ for f, with $d = 35$ and $k = \frac{5}{3}$.

$d = kf$

$35 = \frac{5}{3}f$ Substitute 35 for d and $\frac{5}{3}$ for k.

$105 = 5f$ Multiply both sides by 3.

$21 = f$ Divide both sides by 5.

Thus, the force required to stretch the spring a distance of 35 inches is 21 pounds.

EXAMPLE 5 ***Direct variation*** The distance traveled in a given time is directly proportional to the speed. If a car travels 70 miles at 30 mph, how far will it travel in the same time at 45 mph?

Solution

The verbal model *distance is directly proportional to speed* can be expressed by the equation

1. $d = ks$ This is the direct variation model.

where d is distance, k is the constant of variation, and s is the speed. To find k, we substitute 70 for d and 30 for s and solve for k.

$d = ks$

$70 = k(30)$

$k = \frac{7}{3}$

To find the distance traveled at 45 mph, we substitute $\frac{7}{3}$ for k and 45 for s in Equation 1 and simplify.

$d = ks$

$d = \frac{7}{3}(45)$

$= 105$

In the time it takes to go 70 miles at 30 mph, the car could travel 105 miles at 45 mph.

Self Check

How far will the car travel in the same time at 60 mph?

Answer: 140 mi

Solving variation problems

To solve a variation problem:

1. Translate the verbal model into an equation.
2. Substitute the first set of values into the equation from step 1 to determine the value of k.
3. Substitute the value of k into the equation from step 1.
4. Substitute the remaining set of values into the equation from step 3 and solve for the unknown.

Inverse variation

In the formula $w = \frac{12}{l}$, w gets smaller as l gets larger, and w gets larger as l gets smaller. Since these variables vary in opposite directions in a predictable way, we say that the variables *vary inversely,* or that they are *inversely proportional.* The constant 12 is the constant of variation.

Inverse variation

The words "y varies inversely with x" or "y is inversely proportional to x" mean that $y = \frac{k}{x}$ for some nonzero constant k. The constant k is called the **constant of variation.**

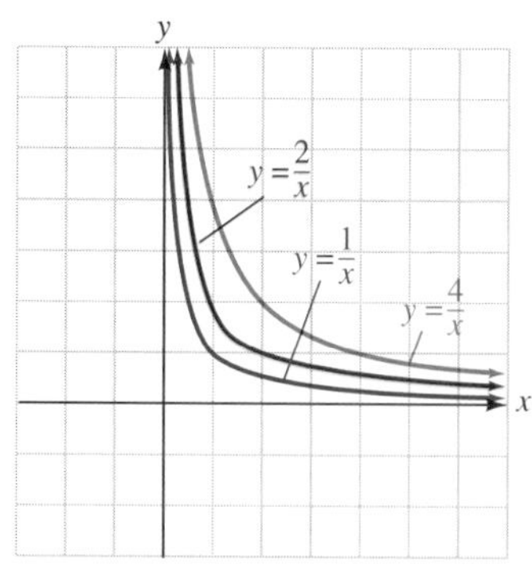

FIGURE 6-7

The formula for inverse variation $\left(y = \frac{k}{x}\right)$ defines a rational function whose graph will have the x- and y-axes as asymptotes. The graph of $y = \frac{k}{x}$ appears in Figure 6-7 for three positive values of k.

Because of gravity, an object in space is attracted to the earth. The force of this attraction varies inversely with the square of the distance from the object to the center of the earth. If f represents the force and d represents the distance, the relationship between f and d can be expressed by the equation

$$f = \frac{k}{d^2} \quad \text{This is an inverse variation model.}$$

If we know that an object 4,000 miles from the center of the earth is attracted to the earth with a force of 90 pounds, we can find k.

$$f = \frac{k}{d^2}$$

$$90 = \frac{k}{\mathbf{4{,}000}^2} \quad \text{Substitute 90 for } f \text{ and 4,000 for } d.$$

$$k = 90(4{,}000)^2 \quad \text{Multiply both sides by } 4{,}000^2 \text{ to solve for } k.$$

$$= 1{,}440{,}000{,}000$$

$$= 1.44 \times 10^9 \quad \text{Write the value of } k \text{ using scientific notation.}$$

To find the force of attraction when the object is 5,000 miles from the center of the earth, we proceed as follows:

$$f = \frac{k}{d^2}$$

$$f = \frac{\mathbf{1.44 \times 10^9}}{\mathbf{5{,}000}^2} \quad \text{Substitute } 1.44 \times 10^9 \text{ for } k \text{ and 5,000 for } d.$$

$$= \frac{1.44 \times 10^9}{2.5 \times 10^7} \quad \text{Write } 5{,}000^2 = 25{,}000{,}000 \text{ using scientific notation.}$$

$$= 57.6$$

The object will be attracted to earth with a force of 57.6 pounds when it is 5,000 miles from the earth's center.

EXAMPLE 6 ***Photography.*** The intensity I of light received from a light source varies inversely with the square of the distance from the light source. If a photographer, 16 feet away from his subject, has a light meter reading of 4 foot-candles of illuminance, what will the meter read if the photographer moves in for a close-up, 4 feet away from the subject?

Solution

The words *intensity varies inversely with the square of the distance d* can be expressed by the equation

$$I = \frac{k}{d^2} \quad \text{This is inverse variation.}$$

To find k, we substitute 4 for I and 16 for d and solve for k.

$$I = \frac{k}{d^2}$$

$$4 = \frac{k}{16^2}$$

$$4 = \frac{k}{256}$$

$$1{,}024 = k$$

To find the intensity when the photographer is 4 feet away from the subject, we substitute 4 for d and 1,024 for k and simplify.

$$I = \frac{k}{d^2}$$

$$I = \frac{1{,}024}{4^2}$$

$$= 64$$

The intensity at 4 feet is 64 foot-candles.

Self Check

Find the intensity when the photographer is 8 feet away from the subject.

Answer: 16 foot-candles ■

Joint variation

There are times when one variable varies with the product of several variables. For example, the area of a triangle varies directly with the product of its base and height:

$$A = \frac{1}{2}bh$$

Such variation is called *joint variation.*

Joint variation If one variable varies directly with the product of two or more variables, the relationship is called **joint variation.** If y varies jointly with x and z, then $y = kxz$. The nonzero constant k is called the **constant of variation.**

EXAMPLE 7 ***Joint variation.*** The volume V of a cone varies jointly with its height h and the area of its base B. If $V = 6 \text{ cm}^3$ when $h = 3$ cm and $B = 6 \text{ cm}^2$, find V when $h = 2$ cm and $B = 8 \text{ cm}^2$.

Solution The words *V varies jointly with h and B* mean that V varies directly as the product of h and B. Thus,

$V = khB$ — The joint variation model can also be read as *V is directly proportional to the product of h and B.*

We can find k by substituting 6 for V, 3 for h and 6 for B.

$$V = khB$$
$$6 = k(3)(6)$$
$$6 = k(18)$$
$$\tfrac{1}{3} = k \quad \text{Divide both sides by 18.}$$

To find V when $h = 2$ and $B = 8$, we substitute these values into the formula $V = 3hB$.

$$V = \tfrac{1}{3}hB$$
$$V = (\tfrac{1}{3})(2)(8)$$
$$= \tfrac{16}{3}$$

The volume will be $\frac{16}{3}$ cm^3 or $5\frac{1}{3}$ cm^3. ■

Combined variation

Many applied problems involve a combination of direct and inverse variation. Such variation is called **combined variation.**

EXAMPLE 8 ***Highway construction.*** The time it takes to build a highway varies directly with the length of the road, but inversely with the number of workers. If it takes 100 workers 4 weeks to build 2 miles of highway, how long will it take 80 workers to build 10 miles of highway?

Solution

We can let t represent the time in weeks, l represent the length in miles, and w represent the number of workers. The relationship between these variables can be expressed by the equation

$$t = \frac{kl}{w} \quad \text{This is a combined variation model.}$$

We substitute 4 for t, 100 for w, and 2 for l to find k:

$$4 = \frac{k(2)}{100}$$
$$400 = 2k \quad \text{Multiply both sides by 100.}$$
$$200 = k \quad \text{Divide both sides by 2.}$$

We now substitute 80 for w, 10 for l, and 200 for k in the equation $t = \frac{kl}{w}$ and simplify:

$$t = \frac{kl}{w}$$
$$t = \frac{200(10)}{80}$$
$$= 25$$

It will take 25 weeks for 80 workers to build 10 miles of highway.

Self Check

How long will it take 60 workers to build 6 miles of highway?

Answer: 20 weeks ■

STUDY SET Section 6.2

VOCABULARY *In Exercises 1–8, fill in the blanks to make the statements true.*

1. ______________ are used to express unit costs and rates.

2. An equation that states that two ratios are equal, such as $\frac{1}{2} = \frac{4}{8}$, is called a ______________.

3. In a proportion, the product of the ______________ is equal to the product of the ______________.

4. If two angles of one triangle have the same measure as two angles of a second triangle, the triangles are ______________.

5. The equation $y = kx$ defines ______________ variation, and the equation $y = \frac{k}{x}$ defines ______________ variation.

6. The equation $y = kxz$ defines ______________ variation, and the equation $y = \frac{kx}{z}$ defines ______________ variation.

7. ______________ variation is represented by a rational function.

8. ______________ variation is represented by a linear function.

CONCEPTS *In Exercises 9–12, decide whether direct or inverse variation applies, and then sketch a reasonable graph for the situation.*

9.

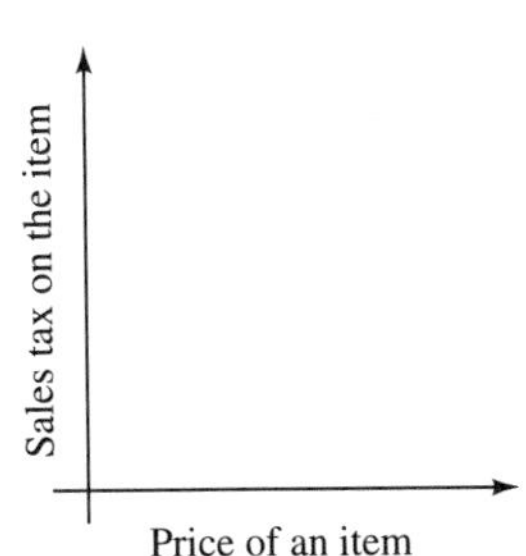

10.

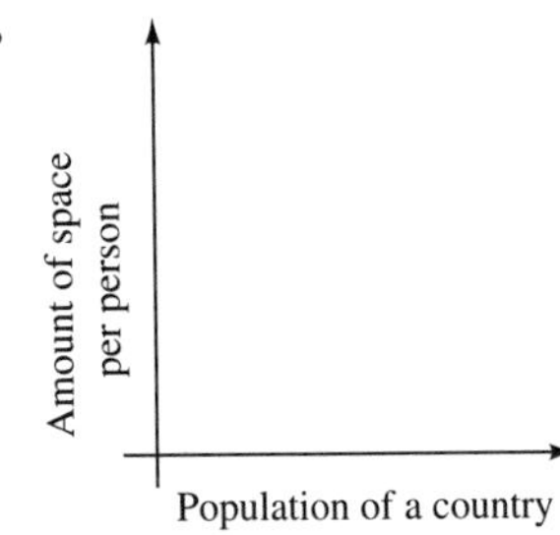

11.

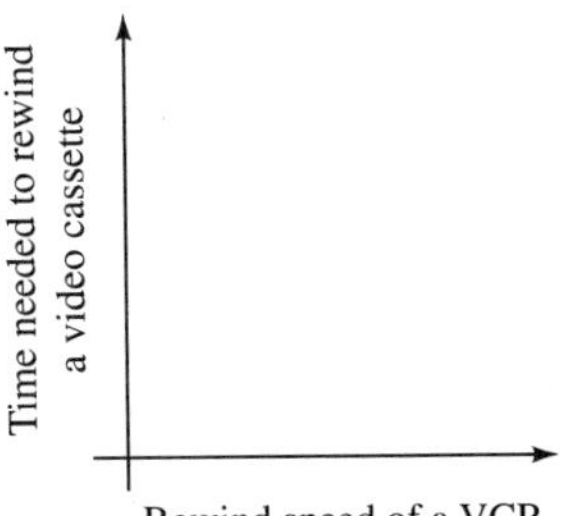

12.

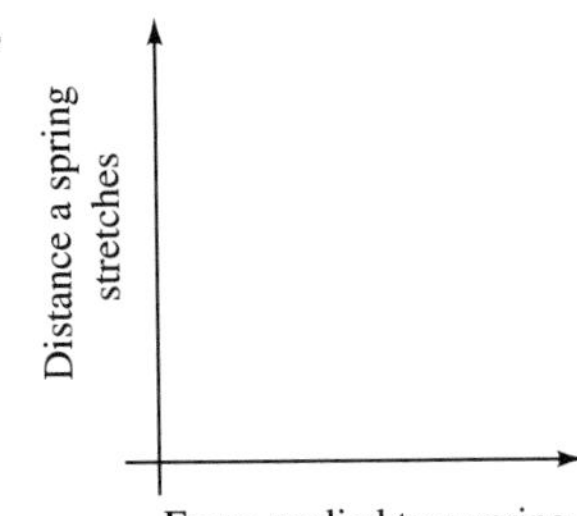

NOTATION *In Exercises 13–14, complete each solution.*

13. Solve $\frac{-7}{6} = \frac{x+3}{12}$.

$$\frac{-7}{6} = \frac{x+3}{12}$$
$$\square(12) = \square(x+3)$$
$$-84 = 6x + \square$$
$$\square = 6x$$
$$-17 = x$$

14. Solve $\frac{18}{2x+1} = \frac{3}{14}$.

$$\frac{18}{2x+1} = \frac{3}{14}$$
$$\square(14) = (\square)3$$
$$252 = \square + 3$$
$$249 = \square$$
$$41.5 = x$$

PRACTICE *In Exercises 15–26, solve each proportion for the variable, if possible.*

15. $\frac{x}{5} = \frac{15}{25}$

16. $\frac{4}{y} = \frac{6}{27}$

17. $\frac{r-2}{3} = \frac{r}{5}$

18. $\frac{x+1}{x-1} = \frac{6}{4}$

19. $\frac{5}{5z+3} = \frac{2z}{2z^2+6}$

20. $\frac{9t+6}{t} = \frac{7}{3}$

21. $\frac{2}{3x} = \frac{6x}{36}$

22. $\frac{y}{4} = \frac{4}{y}$

23. $\frac{2}{c} = \frac{c-3}{2}$

24. $\frac{2}{x+6} = \frac{-2x}{5}$

25. $\frac{1}{x+3} = \frac{-2x}{x+5}$

26. $\frac{x-1}{x+1} = \frac{2}{3x}$

In Exercises 27–32, express each verbal model in symbols.

27. A varies directly with the square of p.

28. z varies inversely with the cube of t.

29. v varies inversely with the square of r.

30. C varies jointly with x, y, and z.

31. P varies directly with the square of a and inversely with the cube of j.

32. M varies inversely with the cube of n and jointly with x and the square of z.

In Exercises 33–36, express each variation model in words. In each equation, k is the constant of variation.

33. $L = kmn$

34. $P = \frac{km}{n}$

35. $R = \frac{kL}{d^2}$

36. $U = krs^2t$

APPLICATIONS *In Exercises 37–44, set up and solve the required proportion.*

37. CAFFEINE Many convenience stores sell super-size 44-ounce soft drinks in refillable cups. For each of the products listed in Illustration 1, find the amount of caffeine contained in one of the large cups. Round to the nearest milligram.

Soft drink, 12 oz	Caffeine (mg)
Mountain Dew	55
Coca-Cola Classic	47
Pepsi	37

Based on data from *Los Angeles Times* (November 11, 1997) p. S4

ILLUSTRATION 1

38. TELEPHONES As of 1997, Sweden had 683 telephone lines per 1,000 people—the highest ratio of any country in the world. If Sweden's population is about 8,901,000, how many telephone lines does the country have?

39. WALL PAPERING The instructions on the label of wallpaper adhesive read as shown in Illustration 2. Estimate the amount of adhesive needed to paper 500 square feet of kitchen walls if a heavy wallpaper will be used.

COVERAGE: One-half gallon will hang approximately 4 single rolls (140 sq ft), depending on the weight of the wall covering and the condition of the wall.

ILLUSTRATION 2

40. RECOMMENDED DOSAGE The recommended child's dose of the sedative hydroxine is 0.006 gram per kilogram of body mass. Find the dosage for a 30-kg child in milligrams.

41. ERGONOMICS The science of ergonomics coordinates the design of working conditions with the requirements of the worker. Illustration 3 gives guidelines for the dimensions (in inches) of a computer work station to be used by a person whose height is 69 inches. Find a set of work station dimensions for a person 5 feet 11 inches tall. Round to the nearest tenth.

Based on information from the Anthropometric Survey of the U.S. Army personnel database

ILLUSTRATION 3

42. SHOPPING A recipe for guacamole dip calls for 5 avocados. If they are advertised at 3 for \$1.98, what will 5 avocados cost?

43. DRAWING See Illustration 4. To make an enlargement of the sailboat, an artist drew a square grid over the smaller picture and transferred the contents of each small square to its corresponding larger square on another sheet of paper. If the smaller picture is 3 in. × 5 in., what are the dimensions of the enlargement?

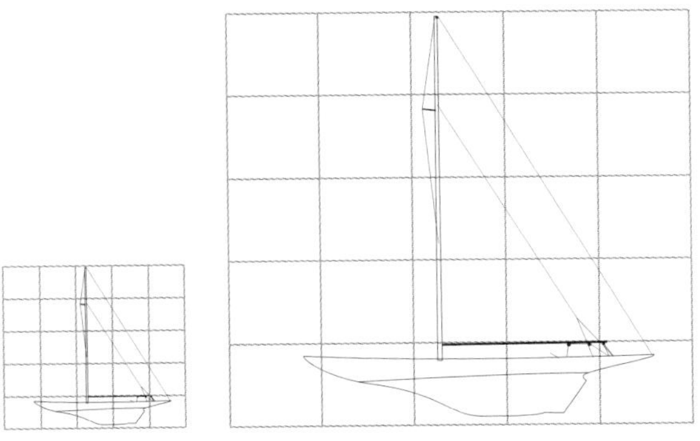

ILLUSTRATION 4

44. DRAFTING In a scale drawing, a 280-foot antenna tower is drawn $7\frac{1}{2}$ inches high. The building next to it is drawn $2\frac{1}{4}$ inches high. How tall is the actual building?

In Exercises 45–50, use similar triangles to help solve each problem.

45. WASHINGTON, DC The Washington Monument casts a shadow of $166\frac{1}{2}$ feet at the same time as a 5-foot-tall tourist casts a shadow of $1\frac{1}{2}$ feet. (See Illustration 5.) Find the height of the monument.

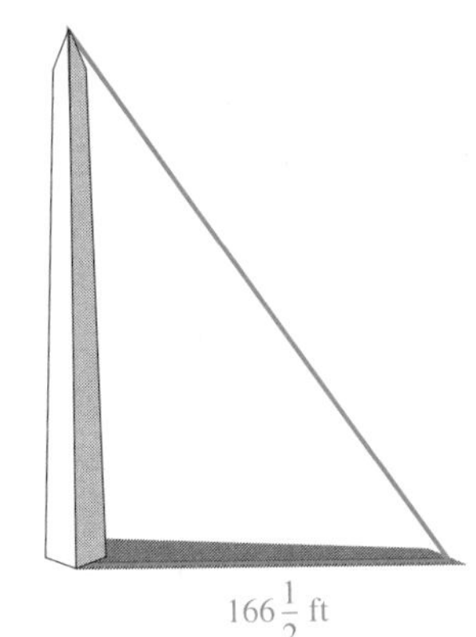

ILLUSTRATION 5

46. HEIGHT OF A FLAGPOLE A man places a mirror on the ground and sees the reflection of the top of a flagpole, as in Illustration 6. The two triangles in the illustration are similar. Find the height h of the flagpole.

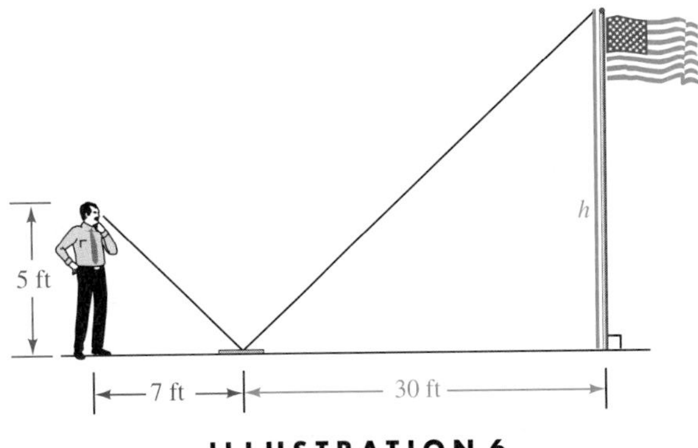

ILLUSTRATION 6

47. WIDTH OF A RIVER Use the dimensions in Illustration 7 to find w, the width of the river. The two triangles in the illustration are similar.

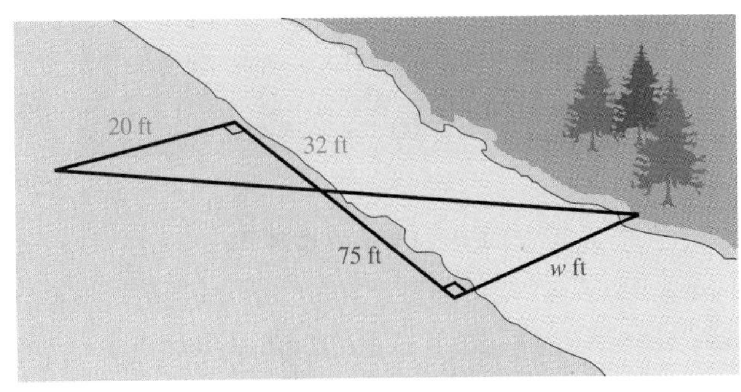

ILLUSTRATION 7

48. FLIGHT PATH An airplane ascends 150 feet as it flies a horizontal distance of 1,000 feet. How much altitude will it gain as it flies a horizontal distance of 1 mile? (See Illustration 8.) (*Hint:* 5,280 feet = 1 mile.)

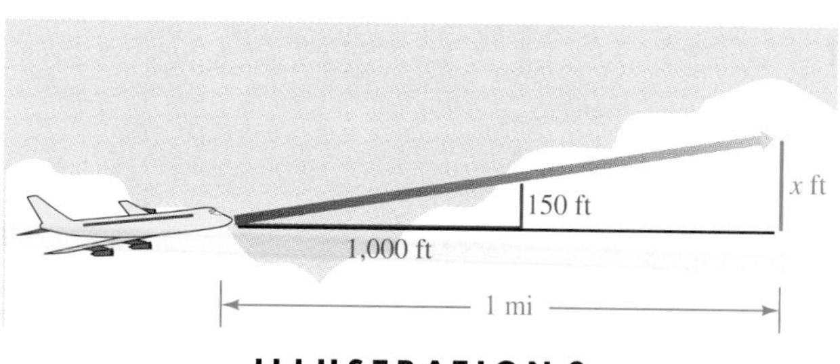

ILLUSTRATION 8

49. SKI RUNS A ski course with $\frac{1}{2}$ mile of horizontal run falls 100 feet in every 300 feet of run. Find the height of the hill.

50. GRAPHIC ARTS The compass shown in Illustration 9 is used to draw circles with different radii (plural for radius). For the setting shown, what radius will the resulting circle have?

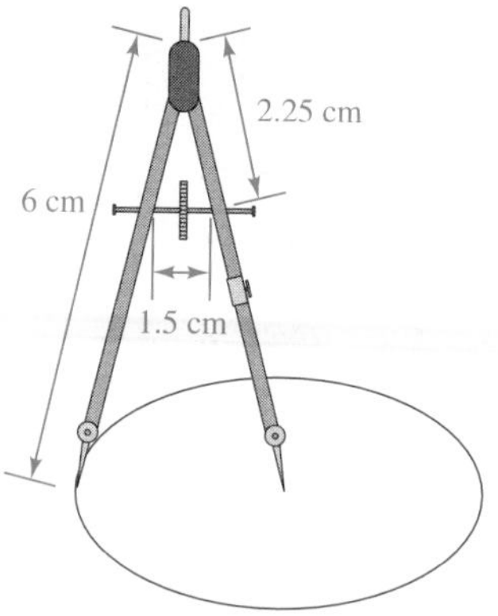

ILLUSTRATION 9

In Exercises 51–64, solve each problem by writing a variation model of the situation.

51. FREE FALL An object in free fall travels a distance s that is directly proportional to the square of the time t. If an object falls 1,024 feet in 8 seconds, how far will it fall in 10 seconds?

52. FINDING DISTANCE The distance that a car can go is directly proportional to the number of gallons of gasoline it consumes. If a car can go 288 miles on 12 gallons of gasoline, how far can it go on a full tank of 18 gallons?

53. FARMING The length of time that a given number of bushels of corn will last when feeding cattle varies inversely with the number of animals. If x bushels will feed 25 cows for 10 days, how long will the feed last for 10 cows?

54. ORGAN PIPES The frequency of vibration of air in an organ pipe is inversely proportional to the length of the pipe. (See Illustration 10.) If a pipe 2 feet long vibrates 256 times per second, how many times per second will a 6-foot pipe vibrate?

l

ILLUSTRATION 10

55. GAS PRESSURE Under constant temperature, the volume occupied by a gas is inversely proportional to the pressure applied. If the gas occupies a volume of 20 cubic inches under a pressure of 6 pounds per square inch, find the volume when the gas is subjected to a pressure of 10 pounds per square inch.

56. REAL ESTATE Illustration 11 shows the listing price for three homes in the same general locality. Write the variation model (direct or inverse) that describes the relationship between the listing price and the number of square feet of a house in this area.

Number of square feet	Listing price
1,720	\$129,000
1,205	\$90,375
1,080	\$81,000

ILLUSTRATION 11

57. TRUCKING COSTS The costs incurred by a trucking company vary jointly with the number of trucks in service and the number of hours they are used. When 4 trucks are used for 6 hours each, the costs are \$1,800. Find the costs of using 10 trucks, each for 12 hours.

58. OIL STORAGE The number of gallons of oil that can be stored in a cylindrical tank varies jointly with the height of the tank and the square of the radius of its base. The constant of proportionality is 23.5. Find the number of gallons that can be stored in the cylindrical tank in Illustration 12.

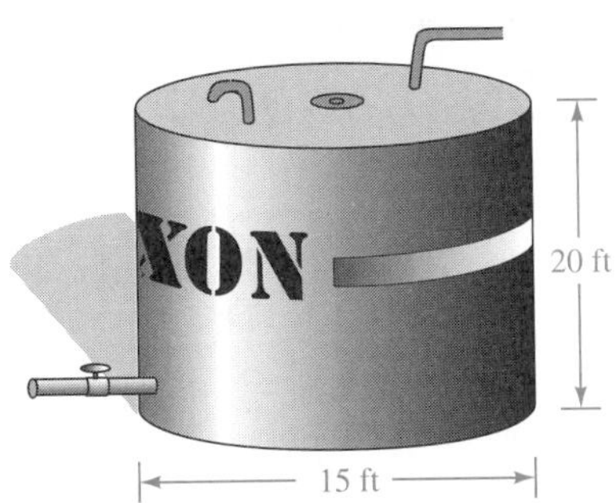

ILLUSTRATION 12

59. ELECTRONICS The voltage (in volts) measured across a resistor is directly proportional to the current (in amperes) flowing through the resistor. The constant of variation is the **resistance** (in ohms). If 6 volts is measured across a resistor carrying a current of 2 amperes, find the resistance.

60. ELECTRONICS The power (in watts) lost in a resistor (in the form of heat) is directly proportional to the square of the current (in amperes) passing through it. The constant of proportionality is the resistance (in ohms). What power is lost in a 5-ohm resistor carrying a 3-ampere current?

61. STRUCTURAL ENGINEERING The deflection of a beam is inversely proportional to its width and the cube of its depth. If the deflection of a 4-inch-by-4-inch beam is 1.1 inches, find the deflection of a 2-inch-by-8-inch beam positioned as in Illustration 13.

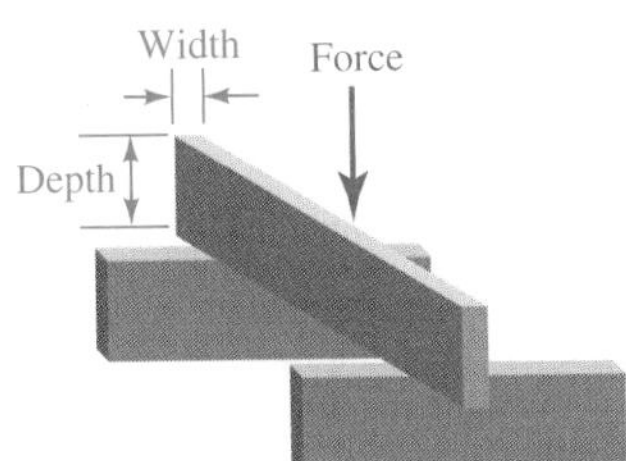

ILLUSTRATION 13

62. STRUCTURAL ENGINEERING Find the deflection of the beam in Exercise 61 when the beam is positioned as in Illustration 14.

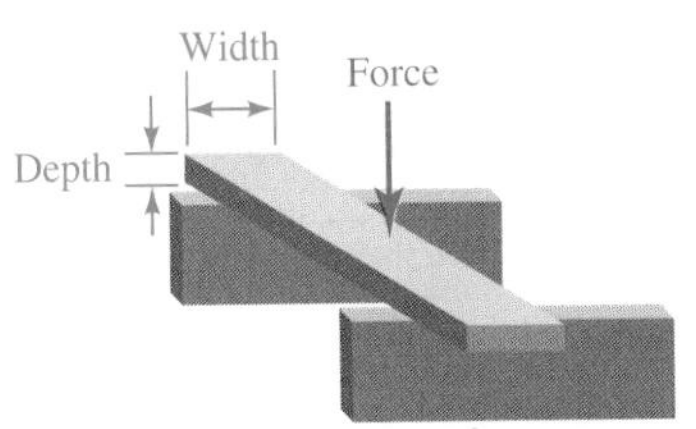

ILLUSTRATION 14

63. TENSION IN A STRING When playing with a Skip It toy, a child swings a weighted ball on the end of a string in a circular motion around one leg while jumping over the revolving string with the other leg. (See Illustration 15.) The tension T in the string is directly proportional to the square of the speed s of the ball and inversely proportional to the radius r of the circle. If the tension in the string is 6 pounds when the speed of the ball is 6 feet per second and the radius is 3 feet, find the tension when the speed is 8 feet per second and the radius is 2.5 feet.

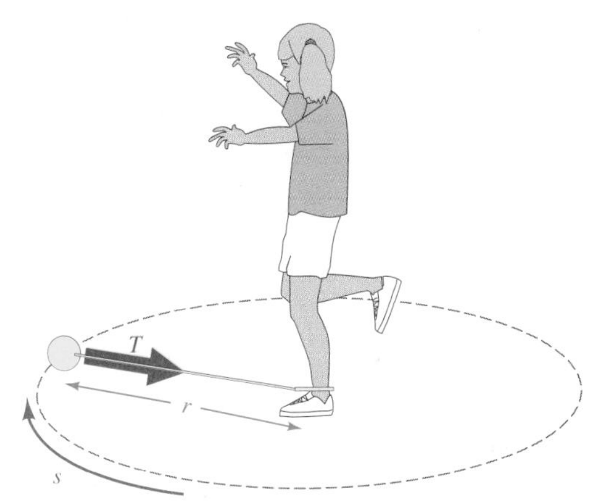

ILLUSTRATION 15

64. GAS PRESSURE The pressure of a certain amount of gas is directly proportional to the temperature (measured in degrees Kelvin) and inversely proportional to the volume. A sample of gas at a pressure of 1 atmosphere occupies a volume of 1 cubic meter at a temperature of 273 Kelvin. When heated, the gas expands to twice its volume, but the pressure remains constant. To what temperature is it heated?

WRITING *Write a paragraph using your own words.*

65. Distinguish between a *ratio* and a *proportion.*

66. From everyday life, give examples of two quantities that vary directly and two quantities that vary inversely.

REVIEW *In Exercises 67–70, simplify each expression.*

67. $(x^2x^3)^2$

68. $\left(\frac{a^3a^5}{a^{-2}}\right)^3$

69. $\frac{b^0 - 2b^0}{b^0}$

70. $\left(\frac{2r^{-2}r^{-3}}{4r^{-5}}\right)^{-3}$

6.3 Multiplying and Dividing Rational Expressions

In this section, you will learn about

- **Multiplying rational expressions**
- **Finding powers of rational expressions**
- **Dividing rational expressions**
- **Mixed operations**

Introduction. In this section, we begin with a review of the rules for multiplying and dividing arithmetic fractions—fractions whose numerators and denominators are integers. Then we use these rules, in combination with the simplification skills learned in Section 6.1, to find products and quotients of rational expressions.

Multiplying rational expressions

In Section 6.1, we introduced four basic properties of fractions. We now present the rule for multiplying fractions.

Multiplying fractions

If no denominators are 0, then

$$\frac{a}{b} \cdot \frac{c}{d} = \frac{a \cdot c}{b \cdot d} = \frac{ac}{bd}$$

To multiply fractions, we multiply the numerators and multiply the denominators.

$$\frac{3}{5} \cdot \frac{2}{7} = \frac{3 \cdot 2}{5 \cdot 7} = \frac{6}{35}$$

$$\frac{4}{7} \cdot \frac{5}{8} = \frac{4 \cdot 5}{7 \cdot 8} = \frac{\overset{1}{\cancel{2}} \cdot \overset{1}{\cancel{2}} \cdot 5}{7 \cdot \underset{1}{\cancel{2}} \cdot \underset{1}{\cancel{2}} \cdot 2} = \frac{5}{14} \qquad \frac{2}{2} = 1.$$

The same rule applies to rational expressions. If $t \neq 0$, then

$$\frac{x^2y}{t} \cdot \frac{xy^3}{t^3} = \frac{x^2y \cdot xy^3}{tt^3} = \frac{x^2x \cdot yy^3}{t^4} = \frac{x^3y^4}{t^4}$$

EXAMPLE 1 ***Multiplying rational expressions.*** Find the product of $\frac{x^2 - 6x + 9}{x}$ and $\frac{x^2}{x - 3}$.

Self Check

Multiply $\frac{a^2 + 6a + 9}{a} \cdot \frac{a^3}{a + 3}$.

Solution

We multiply the numerators and multiply the denominators and then simplify the resulting fraction.

$$\frac{x^2 - 6x + 9}{x} \cdot \frac{x^2}{x - 3} = \frac{(x^2 - 6x + 9)x^2}{x(x - 3)}$$ Multiply the numerators and multiply the denominators.

$$= \frac{(x - 3)(x - 3)xx}{x(x - 3)}$$ Factor in the numerator.

$$= \frac{\overset{1}{\cancel{(x - 3)}}(x - 3)x\overset{1}{\cancel{x}}}{\underset{1}{\cancel{x}}\underset{1}{\cancel{(x - 3)}}}$$ Divide out common factors: $\frac{x-3}{x-3} = 1$ and $\frac{x}{x} = 1$.

$$= x(x - 3)$$

Answer: $a^2(a + 3)$ ■

Accent on Technology **Checking an algebraic simplification**

We can check the simplification in Example 1 by graphing the rational functions $f(x) = \left(\frac{x^2 - 6x + 9}{x}\right)\left(\frac{x^2}{x - 3}\right)$, shown in Figure 6-8(a), and $g(x) = x(x - 3)$, shown in Figure 6-8(b), and observing that the graphs are the same, except that 0 and 3 are not included in the domain of the first function.

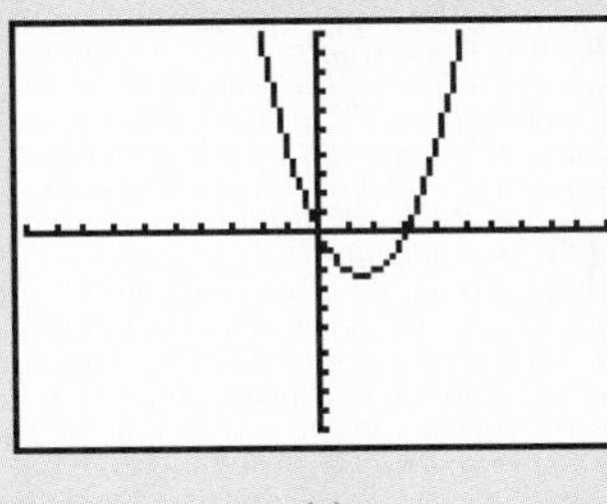
(a)

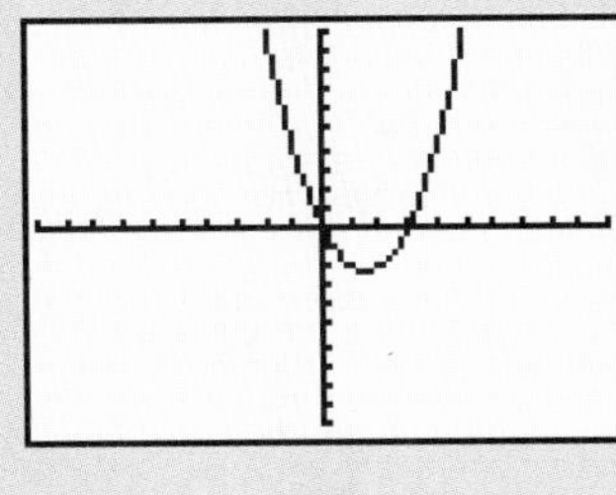
(b)

FIGURE 6-8

EXAMPLE 2 ***Multiplying rational expressions.*** Multiply $\frac{x^2 - x - 6}{x^2 - 4} \cdot \frac{x^2 + x - 6}{x^2 - 9}$.

Solution

$$\frac{x^2 - x - 6}{x^2 - 4} \cdot \frac{x^2 + x - 6}{x^2 - 9}$$

$$= \frac{(x^2 - x - 6)(x^2 + x - 6)}{(x^2 - 4)(x^2 - 9)}$$ Multiply the numerators and multiply the denominators.

$$= \frac{(x - 3)(x + 2)(x + 3)(x - 2)}{(x + 2)(x - 2)(x + 3)(x - 3)}$$ Factor the polynomials.

$$= \frac{\overset{1}{\cancel{(x - 3)}}\overset{1}{\cancel{(x + 2)}}\overset{1}{\cancel{(x + 3)}}\overset{1}{\cancel{(x - 2)}}}{\underset{1}{\cancel{(x + 2)}}\underset{1}{\cancel{(x - 2)}}\underset{1}{\cancel{(x + 3)}}\underset{1}{\cancel{(x - 3)}}}$$ Divide out common factors: $\frac{x-3}{x-3} = 1$, $\frac{x+2}{x+2} = 1$, $\frac{x+3}{x+3} = 1$, and $\frac{x-2}{x-2} = 1$.

$$= 1$$

Self Check

Multiply $\frac{a^2 + a - 56}{a^2 - 49} \cdot \frac{a^2 - a - 56}{a^2 - 64}$.

Answer: 1 ■

WARNING! Note that when all factors divide out, the result is 1 and not 0.

EXAMPLE 3 ***Multiplying rational expressions.*** Multiply

$$\frac{6x^2 + 5x - 4}{2x^2 + 5x + 3} \cdot \frac{8x^2 + 6x - 9}{12x^2 + 7x - 12}.$$

Solution

$$\frac{6x^2 + 5x - 4}{2x^2 + 5x + 3} \cdot \frac{8x^2 + 6x - 9}{12x^2 + 7x - 12}$$

$$= \frac{(6x^2 + 5x - 4)(8x^2 + 6x - 9)}{(2x^2 + 5x + 3)(12x^2 + 7x - 12)}$$ Multiply the numerators and multiply the denominators.

$$= \frac{(3x + 4)(2x - 1)(4x - 3)(2x + 3)}{(2x + 3)(x + 1)(3x + 4)(4x - 3)}$$ Factor the polynomials.

$$= \frac{\overset{1}{\cancel{(3x + 4)}}(2x - 1)\overset{1}{\cancel{(4x - 3)}}\overset{1}{\cancel{(2x + 3)}}}{\underset{1}{\cancel{(2x + 3)}}(x + 1)\underset{1}{\cancel{(3x + 4)}}\underset{1}{\cancel{(4x - 3)}}}$$ Divide out common factors.

$$= \frac{2x - 1}{x + 1}$$

Self Check

Multiply

$$\frac{2a^2 + 5a - 12}{2a^2 + 11a + 12} \cdot \frac{2a^2 - 3a - 9}{2a^2 - a - 3}.$$

Answer: $\frac{a - 3}{a + 1}$ ■

EXAMPLE 4 ***Writing a fraction with a denominator of 1.*** Multiply

$$(2x - x^2) \cdot \frac{x}{x^2 - 5x + 6}.$$

Solution

$$(2x - x^2) \cdot \frac{x}{x^2 - 5x + 6}$$

$$= \frac{2x - x^2}{1} \cdot \frac{x}{x^2 - 5x + 6}$$ Write $2x - x^2$ as $\frac{2x - x^2}{1}$.

$$= \frac{(2x - x^2)x}{1(x^2 - 5x + 6)}$$ Multiply the fractions.

$$= \frac{x\overset{-1}{\cancel{(2 - x)}}x}{1\underset{1}{\cancel{(x - 2)}}(x - 3)}$$ Factor out x in the numerator and factor the trinomial in the denominator. Recall that the quotient of any nonzero quantity and its negative is -1: $\frac{2-x}{x-2} = -1$.

$$= \frac{-x^2}{x - 3}$$

Since $\frac{-a}{b} = -\frac{a}{b}$, the $-$ sign can be written in front of the fraction. Thus, the final result can be written as

$$-\frac{x^2}{x - 3}$$

Self Check

Multiply

$$\frac{x^2 + 5x + 6}{(x^2 + 4x)(x + 2)} \cdot x^3 + 4x^2.$$

Answer: $x(x + 3)$ ■

In Examples 1–4, we would obtain the same answers if we had factored first and divided out the common factors before we multiplied.

Finding powers of rational expressions

EXAMPLE 5 ***Squaring a rational expression.*** Find $\left(\frac{x^2 + x - 1}{2x + 3}\right)^2$.

Solution

To square the rational expression, we write it as a factor twice and do the multiplication.

Self Check

Find $\left(\frac{x + 5}{x^2 - 6x}\right)^2$.

$$\left(\frac{x^2 + x - 1}{2x + 3}\right)^2 = \left(\frac{x^2 + x - 1}{2x + 3}\right)\left(\frac{x^2 + x - 1}{2x + 3}\right)$$
$$= \frac{(x^2 + x - 1)(x^2 + x - 1)}{(2x + 3)(2x + 3)}$$
$$= \frac{x^4 + 2x^3 - x^2 - 2x + 1}{4x^2 + 12x + 9}$$

Answer: $\dfrac{x^2 + 10x + 25}{x^4 - 12x^3 + 36x^2}$ ■

Dividing rational expressions

Here is the rule for dividing fractions.

Dividing fractions

If no denominators are 0, then

$$\frac{a}{b} \div \frac{c}{d} = \frac{a}{b} \cdot \frac{d}{c} = \frac{ad}{bc}$$

We can prove this rule as follows:

$$\frac{a}{b} \div \frac{c}{d} = \frac{\frac{a}{b}}{\frac{c}{d}} = \frac{\frac{a}{b}}{\frac{c}{d}} \cdot 1 = \frac{\frac{a}{b}}{\frac{c}{d}} \cdot \frac{\frac{d}{c}}{\frac{d}{c}} = \frac{\frac{a}{b} \cdot \frac{d}{c}}{\frac{c}{d} \cdot \frac{d}{c}} = \frac{\frac{a}{b} \cdot \frac{d}{c}}{\frac{cd}{cd}} = \frac{\frac{a}{b} \cdot \frac{d}{c}}{1} = \frac{a}{b} \cdot \frac{d}{c}$$

Thus, *to divide two fractions, we can invert the divisor (the second fraction) and multiply.*

$$\frac{3}{5} \div \frac{2}{7} = \frac{3}{5} \cdot \frac{7}{2} = \frac{3 \cdot 7}{5 \cdot 2} = \frac{21}{10}$$

$$\frac{4}{7} \div \frac{2}{21} = \frac{4}{7} \cdot \frac{21}{2} = \frac{4 \cdot 21}{7 \cdot 2} = \frac{\overset{1}{\cancel{2}} \cdot 2 \cdot 3 \cdot \overset{1}{\cancel{7}}}{\underset{1}{\cancel{7}} \cdot \underset{1}{\cancel{2}}} = 6$$

We can state the rule for dividing two fractions in another way: *To divide two fractions, we multiply the first fraction and the reciprocal of the second.* The **reciprocal** of a fraction can be found by interchanging the numerator and denominator. For example, the reciprocal of $\frac{2}{7}$ is $\frac{7}{2}$. Similarly, the reciprocal of the rational expression $\frac{x^2 - 2x + 4}{2x^2 - 2}$ is $\frac{2x^2 - 2}{x^2 - 2x + 4}$.

The rule for division of fractions applies to rational expressions.

$$\frac{x^2}{y^3z^2} \div \frac{x^2}{yz^3} = \frac{x^2}{y^3z^2} \cdot \frac{yz^3}{x^2}$$ Invert the divisor and multiply (or multiply the first fraction by the reciprocal of the second).

$$= \frac{x^2yz^3}{x^2y^3z^2}$$ Multiply the numerators and the denominators.

$$= x^{2-2}y^{1-3}z^{3-2}$$ To divide exponential expressions with the same base, keep the base and subtract the exponents.

$$= x^0y^{-2}z^1$$ Simplify the exponents.

$$= 1 \cdot y^{-2} \cdot z$$ $x^0 = 1$.

$$= \frac{z}{y^2}$$ Write the result without the negative exponent.

EXAMPLE 6 ***Dividing rational expressions.*** Divide $\frac{x^3+8}{x+1} \div \frac{x^2-2x+4}{2x^2-2}$.

Solution

We invert the divisor, which is $\frac{x^2-2x+4}{2x^2-2}$, and multiply.

$$\frac{x^3+8}{x+1} \div \frac{x^2-2x+4}{2x^2-2}$$

$$= \frac{x^3+8}{x+1} \cdot \frac{2x^2-2}{x^2-2x+4}$$

$$= \frac{(x^3+8)(2x^2-2)}{(x+1)(x^2-2x+4)}$$ Multiply the numerators and the denominators.

$$= \frac{(x+2)\overset{1}{\cancel{(x^2-2x+4)}}2\overset{1}{\cancel{(x+1)}}(x-1)}{\underset{1}{\cancel{(x+1)}}\underset{1}{\cancel{(x^2-2x+4)}}}$$ Factor x^3+8 and $2x^2-2$. The polynomial x^2-2x+4 does not factor. Then divide out common factors.

$$= 2(x+2)(x-1)$$

Self Check Divide $\frac{x^3-8}{x-1} \div \frac{x^2+2x+4}{3x^2-3x}$.

Answer: $3x(x-2)$ ■

EXAMPLE 7 ***Division involving two variables.*** Divide $\frac{b^3-4b}{x-1} \div (b-2)$.

Solution

$$\frac{b^3-4b}{x-1} \div (b-2) = \frac{b^3-4b}{x-1} \div \frac{b-2}{1}$$ Write $b-2$ as a fraction with a denominator of 1.

$$= \frac{b^3-4b}{x-1} \cdot \frac{1}{b-2}$$ Invert the divisor and multiply.

$$= \frac{b^3-4b}{(x-1)(b-2)}$$ Multiply the numerators and the denominators.

$$= \frac{b(b+2)\overset{1}{\cancel{(b-2)}}}{(x-1)\underset{1}{\cancel{(b-2)}}}$$ Factor b^3-4b and then divide out common factors.

$$= \frac{b(b+2)}{x-1}$$

Self Check Divide $\frac{m^4-9m^2}{a^2-3a} \div (m^2+3m)$.

Answer: $\frac{m(m-3)}{a(a-3)}$ ■

Mixed operations

EXAMPLE 8 Simplify $\frac{x^2+2x-3}{6x^2+5x+1} \div \frac{2x^2-2}{2x^2-5x-3} \cdot \frac{6x^2+4x-2}{x^2-2x-3}$.

Solution

Since multiplications and divisions are done in order from left to right, we begin by focusing on the division. We introduce grouping symbols to emphasize this. To divide the rational expressions in the parentheses, we invert $\frac{2x^2-2}{2x^2-5x-3}$ and multiply.

$$\left(\frac{x^2+2x-3}{6x^2+5x+1} \div \frac{2x^2-2}{2x^2-5x-3}\right)\frac{6x^2+4x-2}{x^2-2x-3} = \left(\frac{x^2+2x-3}{6x^2+5x+1} \cdot \frac{2x^2-5x-3}{2x^2-2}\right)\frac{6x^2+4x-2}{x^2-2x-3}$$

Next, we multiply the three fractions and simplify the result.

$$= \frac{(x^2+2x-3)(2x^2-5x-3)(6x^2+4x-2)}{(6x^2+5x+1)(2x^2-2)(x^2-2x-3)}$$

$$= \frac{(x+3)\overset{1}{\cancel{(x-1)}}\overset{1}{\cancel{(2x+1)}}\overset{1}{\cancel{(x-3)}}\overset{1}{\cancel{2}}(3x-1)\overset{1}{\cancel{(x+1)}}}{(3x+1)\underset{1}{\cancel{(2x+1)}}\underset{1}{\cancel{2}}(x+1)\underset{1}{\cancel{(x-1)}}\underset{1}{\cancel{(x-3)}}\underset{1}{\cancel{(x+1)}}}$$

$$= \frac{(x+3)(3x-1)}{(3x+1)(x+1)}$$

■

STUDY SET Section 6.3

VOCABULARY *In Exercises 1–4, fill in the blanks to make the statements true.*

1. In the division statement

$$\frac{a^2-9}{a^2-49} \div \frac{a+3}{a+7}$$

the second fraction is called the ______________.

2. The ______________ of $\frac{a+3}{a+7}$ is $\frac{a+7}{a+3}$.

3. In a fraction, the part above the fraction bar is called the ______________.

4. In a fraction, the part below the fraction bar is called the ______________.

CONCEPTS *In Exercises 5–8, fill in the blanks to make the statements true.*

5. To multiply two fractions, we ______________ their numerators and multiply their denominators. In symbols, $\frac{a}{b} \cdot \frac{c}{d} =$ ______________.

6. To divide two fractions, we invert the divisor and ______________. In symbols, $\frac{a}{b} \div \frac{c}{d} =$ ______________

7. The denominator of a fraction cannot be ___.

8. $\frac{a+1}{a+1} =$ ___, provided $a \neq -1$.

NOTATION *In Exercises 9–10, complete each solution.*

9. $\frac{x^2+3x}{5x-25} \cdot \frac{x-5}{x+3} = \frac{(x^2+3x)\square}{\square(x+3)}$

$= \frac{\square(x-5)}{\square(x+3)}$

$= \frac{x}{5}$

10. $\frac{x^2-x-6}{4x^2+16x} \div \frac{x-3}{x+4} = \frac{x^2-x-6}{4x^2+16x} \cdot \square$

$= \frac{\square(x+4)}{(4x^2+16x)\square}$

$= \frac{\square(x+2)(x+4)}{\square(x-3)}$

$= \frac{x+2}{4x}$

11. A student checks her answers with those in the back of her textbook. Tell whether they are equivalent.

Student's answer	Book's answer	Equivalent?
$\dfrac{-x^{10}}{y^2}$	$-\dfrac{x^{10}}{y^2}$	
$\dfrac{x-3}{x+3}$	$\dfrac{3-x}{3+x}$	
$\dfrac{b+a}{(2-x)(d+c)}$	$-\dfrac{a+b}{(x-2)(c+d)}$	

12. a. Write $5x^2 + 35x$ as a fraction.

b. What is the reciprocal of $5x^2 + 35x$?

PRACTICE *In Exercises 13–46, do the operations and simplify.*

13. $\dfrac{3}{4} \cdot \dfrac{5}{3}$

14. $-\dfrac{5}{6} \cdot \dfrac{3}{7}$

15. $-\dfrac{6}{11} \div \dfrac{36}{55}$

16. $\dfrac{17}{12} \div \dfrac{34}{3}$

17. $\dfrac{x^2y^2}{cd} \cdot \dfrac{c^{-2}d^2}{x}$

18. $\dfrac{a^{-2}b^2}{x^{-1}y} \cdot \dfrac{a^4b^4}{x^2y^3}$

19. $\dfrac{-x^2y^{-2}}{x^{-1}y^{-3}} \div \dfrac{x^{-3}y^2}{x^4y^{-1}}$

20. $\dfrac{(a^3)^2}{b^{-1}} \div \dfrac{(a^3)^{-2}}{b^{-1}}$

21. $\dfrac{x^2 + 2x + 1}{x} \cdot \dfrac{x^2 - x}{x^2 - 1}$

22. $\dfrac{2x^2 - x - 3}{x^2 - 1} \cdot \dfrac{x^2 + x - 2}{2x^2 + x - 6}$

23. $\dfrac{x^2 - 16}{x^2 - 25} \div \dfrac{x + 4}{x - 5}$

24. $\dfrac{a^2 - 9}{a^2 - 49} \div \dfrac{a + 3}{a + 7}$

25. $\dfrac{3t^2 - t - 2}{6t^2 - 5t - 6} \cdot \dfrac{4t^2 - 9}{2t^2 + 5t + 3}$

26. $\dfrac{2p^2 - 5p - 3}{p^2 - 9} \cdot \dfrac{2p^2 + 5p - 3}{2p^2 + 5p + 2}$

27. $\dfrac{3n^2 + 5n - 2}{12n^2 - 13n + 3} \div \dfrac{n^2 + 3n + 2}{4n^2 + 5n - 6}$

28. $\dfrac{2p^2 - 5p - 3}{p^2 - 9} \div \dfrac{2p^2 + 5p + 2}{2p^2 + 5p - 3}$

29. $(x + 1) \cdot \dfrac{1}{x^2 + 2x + 1}$

30. $\dfrac{x^2 - 4}{x} \div (x + 2)$

31. $(x^2 - x - 2) \cdot \dfrac{x^2 + 3x + 2}{x^2 - 4}$

32. $(2x^2 - 9x - 5) \cdot \dfrac{x}{2x^2 + x}$

33. $\dfrac{a^2 + 2a - 35}{12x} \div \dfrac{ax - 3x}{a^2 + 4a - 21}$

34. $\dfrac{x^2 - 4}{2b - bx} \div \dfrac{x^2 + 4x + 4}{2b + bx}$

35. $\dfrac{x^3 + y^3}{x^3 - y^3} \div \dfrac{x^2 - xy + y^2}{x^2 + xy + y^2}$

36. $\dfrac{x^2 - 6x + 9}{4 - x^2} \div \dfrac{x^2 - 9}{x^2 - 8x + 12}$

37. $\dfrac{ax + ay + bx + by}{x^3 - 27} \div \dfrac{xc + xd + yc + yd}{x^2 + 3x + 9}$

38. $\dfrac{x^2 + 3x + yx + 3y}{x^2 - 9} \div \dfrac{x + 3}{x - 3}$

39. $\dfrac{x^2 - x - 6}{x^2 - 4} \cdot \dfrac{x^2 - x - 2}{9 - x^2}$

40. $\dfrac{p^3 - q^3}{q^2 - p^2} \cdot \dfrac{q^2 + pq}{p^3 + p^2q + pq^2}$

41. $(4x + 12) \cdot \dfrac{x^2}{2x - 6} \div \dfrac{2}{x - 3}$

42. $(4x^2 - 9) \div \dfrac{2x^2 + 5x + 3}{x + 2} \div (2x - 3)$

43. $(x^2 - x - 6) \div (x - 3) \div (x - 2)$

44. $(x^2 - x - 6) \div [(x - 3) \div (x - 2)]$

45. $\dfrac{2x^2 - 2x - 4}{x^2 + 2x - 8} \cdot \dfrac{3x^2 + 15x}{x + 1} \div \dfrac{4x^2 - 100}{x^2 - x - 20}$

46. $\dfrac{6a^2 - 7a - 3}{a^2 - 1} \div \dfrac{4a^2 - 12a + 9}{a^2 - 1} \cdot \dfrac{2a^2 - a - 3}{3a^2 - 2a - 1}$

In Exercises 47–50, find each power.

47. $\left(\dfrac{x - 3}{x^3 + 4}\right)^2$

48. $\left(\dfrac{2t^2 + t}{t - 1}\right)^2$

49. $\left(\dfrac{2m^2 - m - 3}{x^2 - 1}\right)^2$

50. $\left(\dfrac{-k - 3}{x^2 - x + 1}\right)^2$

APPLICATIONS

51. PHYSICS EXPERIMENT Illustration 1 contains data from a physics experiment. k_1 and k_2 are constants. Complete the table.

Trial	Rate (m/sec)	Time (sec)	Distance (m)
1	$\dfrac{k_1^2 + 3k_1 + 2}{k_1 - 3}$	$\dfrac{k_1^2 - 3k_1}{k_1 + 1}$	
2	$\dfrac{k_2^2 + 6k_2 + 5}{k_2 + 1}$		$k_2^2 + 11k_2 + 30$

ILLUSTRATION 1

52. TRUNK CAPACITY The shape of the storage space in the trunk of the car shown in Illustration 2 is approximately a rectangular solid. Write a simplified rational expression that gives the number of cubic units of storage space in the trunk.

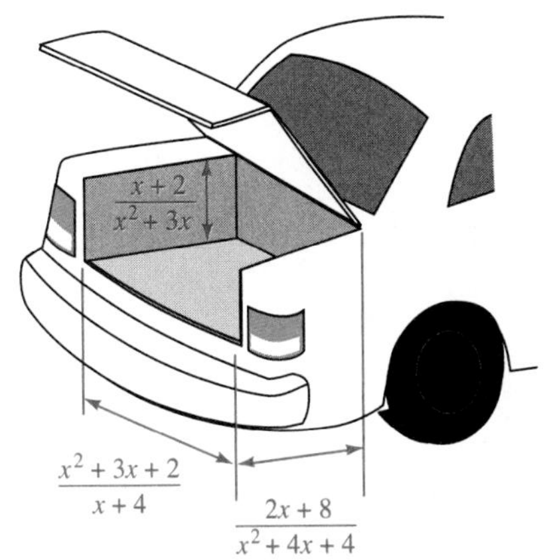

ILLUSTRATION 2

WRITING *Write a paragraph using your own words.*

53. Explain how to multiply two rational expressions.

54. Write some comments to the student who wrote the following solution, explaining the error.

$$\frac{x^2 + x - 2}{x^2 - 4} \cdot \frac{x - 2}{x - 1} = \frac{(x + 2)(x - 1)}{(x + 2)(x - 2)} \cdot \frac{x - 2}{x - 1}$$
$$= \frac{(x+2)(x-1)(x-2)}{(x+2)(x-2)(x-1)}$$
$$= 0$$

REVIEW *In Exercises 55–58, do each operation.*

55. $-2a^2(3a^3 - a^2)$

56. $(2t - 1)^2$

57. $(2g - n)(3g - n)$

58. $(2c - b)(4c^2 + 2cb + b^2)$

6.4 Adding and Subtracting Rational Expressions

In this section, you will learn about

- **Adding and subtracting rational expressions with like denominators**
- **Adding and subtracting rational expressions with unlike denominators**
- **Finding the least common denominator**
- **Mixed operations**

Introduction. The procedures used to add and subtract rational expressions are based on the rules for adding and subtracting arithmetic fractions. In this section, we will add and subtract rational expressions with *like* and *unlike* denominators.

Adding and subtracting rational expressions with like denominators

Fractions with like denominators are added and subtracted according to the following rules.

Adding and subtracting fractions

If there are no divisions by 0, then

$$\frac{a}{b} + \frac{c}{b} = \frac{a+c}{b} \quad \text{and} \quad \frac{a}{b} - \frac{c}{b} = \frac{a-c}{b}$$

In words, *we add (or subtract) fractions with like denominators by adding (or subtracting) the numerators and keeping the common denominator.* Whenever possible, we should simplify the result. These two rules apply to addition and subtraction of rational expressions with like denominators.

EXAMPLE 1 ***Rational expressions with like denominators.***

Perform the operations: **a.** $\frac{4}{3x} + \frac{7}{3x}$, and **b.** $\frac{a^2}{a^2-1} - \frac{a}{a^2-1}$.

Solution

a. $\frac{4}{3x} + \frac{7}{3x} = \frac{4+7}{3x}$ Add the numerators and keep the common denominator.

$= \frac{11}{3x}$ Do the addition in the numerator.

b. $\frac{a^2}{a^2-1} - \frac{a}{a^2-1} = \frac{a^2-a}{a^2-1}$ Subtract the numerators and keep the common denominator.

We note that the polynomials factor in the numerator and the denominator of the result.

Self Check

Perform the operations:

a. $\frac{17}{22} + \frac{13}{22}$, **b.** $\frac{1}{6a} - \frac{7}{6a}$, and

c. $\frac{3a}{a-2} + \frac{2a}{a-2}$.

$$\frac{a^2}{a^2-1} - \frac{a}{a^2-1} = \frac{a\overset{1}{\cancel{(a-1)}}}{(a+1)\underset{1}{\cancel{(a-1)}}}$$ Simplify by dividing out the common factor of $(a - 1)$.

$$= \frac{a}{a+1}$$

Answers: **a.** $\frac{15}{11}$, **b.** $-\frac{1}{a}$, **c.** $\frac{5a}{a-2}$ ■

Accent on Technology **Checking algebra**

We can check the subtraction in part b of Example 1 by graphing the rational functions $f(a) = \frac{a^2}{a^2-1} - \frac{a}{a^2-1}$, shown in Figure 6-9(a), and $g(a) = \frac{a}{a+1}$, shown in Figure 6-9(b), and observing that the graphs are the same. Note that -1 and 1 are not in the domain of the first function and -1 is not in the domain of the second function.

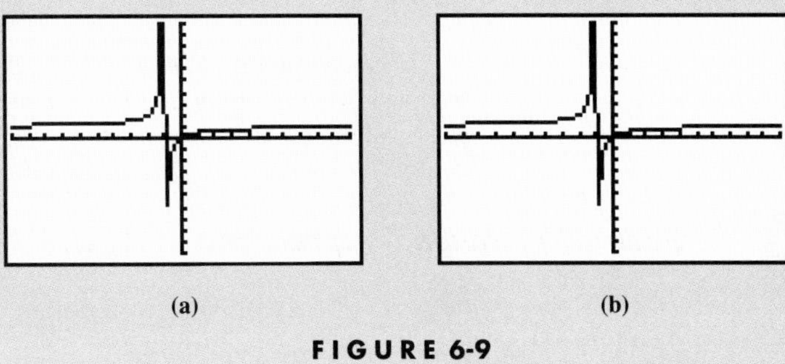

(a) (b)

FIGURE 6-9

Adding and subtracting rational expressions with unlike denominators

To add or subtract fractions with unlike denominators, we change them into fractions with a common denominator. This is done using the **fundamental property of fractions.**

The fundamental property of fractions

If a, b, and k are real numbers, and $b \neq 0$ and $k \neq 0$, then

$$\frac{a}{b} = \frac{a \cdot k}{b \cdot k}$$

In words, *multiplying the numerator and the denominator of a fraction by the same nonzero number does not change the value of the fraction.* This property is true because multiplying the numerator and denominator of a fraction by the same number is equivalent to multiplying it by 1. When a number is multiplied by 1, its value does not change. We use this property to add and subtract rational expressions with unlike denominators.

When adding or subtracting two rational expressions with denominators that are negatives (opposites), we can multiply the numerator and the denominator of one of the expressions by -1 to get a common denominator.

EXAMPLE 2 ***Denominators that are negatives (opposites).*** Add $\frac{x}{x-y} + \frac{y}{y-x}$.

Solution

$$\frac{x}{x-y} + \frac{y}{y-x} = \frac{x}{x-y} + \frac{(-1)y}{(-1)(y-x)}$$ Multiply the numerator and the denominator of the second rational expression by -1.

$$= \frac{x}{x-y} + \frac{-y}{-y+x}$$ Do the multiplication.

$$= \frac{x}{x-y} + \frac{-y}{x-y}$$ Rewrite the second denominator, $-y + x$, as $x - y$. The fractions now have a common denominator.

$$= \frac{x-y}{x-y}$$ Add the numerators and keep the common denominator.

$$= 1$$ Simplify.

Self Check

Add $\frac{2a}{a-b} + \frac{b}{b-a}$.

Answer: $\frac{2a-b}{a-b}$

When adding or subtracting two rational expressions with different denominators, we must often multiply one or both of them by an appropriate expression to get a common denominator. This process is called **building a fraction.**

EXAMPLE 3 ***Subtracting rational expressions.*** Subtract $3 - \frac{7}{x-2}$.

Solution

If we write 3 as $\frac{3}{1}$ and then multiply its numerator and denominator by $x - 2$, the fractions will have a common denominator of $x - 2$.

$$3 - \frac{7}{x-2} = \frac{3}{1} - \frac{7}{x-2}$$ $3 = \frac{3}{1}$.

$$= \frac{3(x-2)}{1(x-2)} - \frac{7}{x-2}$$ Build $\frac{3}{1}$ to a fraction with denominator $x - 2$ by multiplying its numerator and denominator by $x - 2$.

$$= \frac{3x-6}{x-2} - \frac{7}{x-2}$$ Distribute the 3.

$$= \frac{3x-6-7}{x-2}$$ Subtract the numerators and keep the common denominator.

$$= \frac{3x-13}{x-2}$$ Combine like terms in the numerator.

Self Check

Subtract $6 - \frac{5y}{6-y}$.

Answer: $\frac{-11y+36}{6-y}$

EXAMPLE 4 ***Adding rational expressions.*** Add $\frac{3}{x} + \frac{4}{y}$.

Solution

A common denominator for the fractions is xy. We multiply each numerator and denominator by the appropriate factor so that each denominator builds to xy.

$$\frac{3}{x} + \frac{4}{y} = \frac{3 \cdot y}{x \cdot y} + \frac{4 \cdot x}{y \cdot x}$$

$$= \frac{3y}{xy} + \frac{4x}{xy}$$ Multiply in the numerators and in the denominators.

$$= \frac{3y+4x}{xy}$$ Add the numerators and keep the common denominator.

Self Check

Add $\frac{5}{a} + \frac{7}{b}$.

Answer: $\frac{5b+7a}{ab}$

EXAMPLE 5 ***Subtracting rational expressions.*** Subtract

$$\frac{4x}{x+2} - \frac{7x}{x-2}.$$

Solution

By inspection, we see that a common denominator is $(x + 2)(x - 2)$. We multiply the numerator and denominator of each rational expression by the appropriate factor, so that each one has a denominator of $(x + 2)(x - 2)$.

$$\frac{4x}{x+2} - \frac{7x}{x-2}$$

$$= \frac{4x(x-2)}{(x+2)(x-2)} - \frac{(x+2)7x}{(x+2)(x-2)}$$

$$= \frac{4x^2-8x}{(x+2)(x-2)} - \frac{7x^2+14x}{(x+2)(x-2)}$$ Use the distributive property to remove parentheses in each numerator.

$$= \frac{(4x^2-8x)-(7x^2+14x)}{(x+2)(x-2)}$$ Subtract the numerators and keep the common denominator.

$$= \frac{4x^2-8x-7x^2-14x}{(x+2)(x-2)}$$ In the numerator, to subtract the polynomials, add the first and the opposite of the second.

$$= \frac{-3x^2-22x}{(x+2)(x-2)}$$ Combine like terms.

WARNING! Since the $-$ sign between the fractions in step 3 applies to both terms of $7x^2 + 14x$, we must insert parentheses to show this.

Subtract $\frac{3a}{a+3} - \frac{5a}{a-3}$.

Answer: $\frac{-2a^2-24a}{(a+3)(a-3)}$ ■

Finding the least common denominator

When adding or subtracting rational expressions with unlike denominators, it is easiest if we write the rational expressions in terms of the smallest common denominator possible, called the **least** (or lowest) **common denominator (LCD).** To find the least common denominator of several rational expressions, we follow these steps.

Finding the LCD

1. Factor each denominator.
2. List the different factors of each denominator.
3. Write each factor found in step 2 to the highest power that occurs in any one factorization.
4. The LCD is the product of the factors to the highest powers found in step 3.

EXAMPLE 6 ***Finding the LCD of rational expressions.*** Find the LCD of **a.** $\frac{5a}{24b}$ and $\frac{11a}{18b^2}$ and **b.** $\frac{1}{x^2-12x+36}$ and $\frac{3-x}{x^2-6x}$.

Solution

a. We write each denominator as the product of prime numbers and variables.

$$24b = 2 \cdot 2 \cdot 2 \cdot 3 \cdot b = 2^3 \cdot 3 \cdot b$$
$$18b^2 = 2 \cdot 3 \cdot 3 \cdot b \cdot b = 2 \cdot 3^2 \cdot b^2$$

To find the LCD, we form a product using each of these factors the greatest number of times it appears in any one factorization. We use 2 three times, because it appears

Self Check

Find the LCD of **a.** $\frac{3y}{28z^3}$ and $\frac{5x}{21z}$ and **b.** $\frac{a-1}{a^2-25}$ and $\frac{3-a^2}{a^2+7a+10}$.

three times as a factor of 24. We use 3 twice, because it occurs twice as a factor of 18. We use b twice.

$$\begin{aligned} \text{LCD} &= 2 \cdot 2 \cdot 2 \cdot 3 \cdot 3 \cdot b \cdot b \\ &= 72b^2 \end{aligned}$$

b. We factor each denominator completely:

$$x^2 - 12x + 36 = (x - 6)^2$$
$$x^2 - 6x = x(x - 6)$$

To find the LCD, we form a product using the highest power of each of the factors:

$$\text{LCD} = x(x - 6)^2$$

Answers: **a.** $84z^3$, **b.** $(a - 5)(a + 5)(a + 2)$

To add or subtract rational expressions with unlike denominators, we follow these steps.

Adding or subtracting rational expressions with unlike denominators

1. Find the LCD.
2. Express each rational expression with a denominator that is the LCD.
3. Add (or subtract) the resulting rational expressions.
4. Simplify the result if possible.

EXAMPLE 7 ***Adding rational expressions with unlike denominators.*** Add $\frac{5a}{24b} + \frac{11a}{18b^2}$.

Solution

In Example 6, we saw that the LCD of these rational expressions is $72b^2$. We multiply each numerator and denominator by whatever it takes to build the denominator to $72b^2$.

$$\begin{aligned} \frac{5a}{24b} + \frac{11a}{18b^2} &= \frac{5a \cdot 3b}{24b \cdot 3b} + \frac{11a \cdot 4}{18b^2 \cdot 4} \\ &= \frac{15ab}{72b^2} + \frac{44a}{72b^2} && \text{Do the multiplications in the numerators and the denominators.} \\ &= \frac{15ab + 44a}{72b^2} && \text{Add the numerators and keep the common denominator.} \end{aligned}$$

Self Check

Add $\frac{3y}{28z^3} + \frac{5x}{21z}$.

Answer: $\frac{9y + 20xz^2}{84z^3}$

EXAMPLE 8 ***Subtracting rational expressions with unlike denominators.*** Subtract $\frac{x}{x^2 - 2x + 1} - \frac{4}{x^2 - 1}$.

Solution

We factor each denominator to find the LCD:

$$x^2 - 2x + 1 = (x - 1)(x - 1) = (x - 1)^2$$
$$x^2 - 1 = (x + 1)(x - 1)$$

The LCD is $(x - 1)^2(x + 1)$ or $(x - 1)(x - 1)(x + 1)$.

We now write each rational expression with its denominator in factored form. Then we multiply each numerator and denominator by the missing factor, so that each has a denominator of $(x - 1)(x - 1)(x + 1)$.

Self Check

Subtract $\frac{a}{a^2 - 4a + 4} - \frac{2}{a^2 - 4}$.

$$\frac{x}{x^2 - 2x + 1} - \frac{4}{x^2 - 1}$$

$$= \frac{x}{(x - 1)(x - 1)} - \frac{4}{(x + 1)(x - 1)}$$

$$= \frac{x(x + 1)}{(x - 1)(x - 1)(x + 1)} - \frac{4(x - 1)}{(x + 1)(x - 1)(x - 1)}$$

$$= \frac{x^2 + x}{(x - 1)(x - 1)(x + 1)} - \frac{4x - 4}{(x + 1)(x - 1)(x - 1)}$$

$$= \frac{(x^2 + x) - (4x - 4)}{(x - 1)(x - 1)(x + 1)}$$ Subtract the numerators and keep the common denominator.

$$= \frac{x^2 + x - 4x + 4}{(x - 1)(x - 1)(x + 1)}$$ In the numerator, subtract the polynomials.

$$= \frac{x^2 - 3x + 4}{(x - 1)^2(x + 1)}$$ Combine like terms. The result does not simplify.

Answer: $\frac{a^2 + 4}{(a - 2)^2(a + 2)}$ ■

Mixed operations

EXAMPLE 9 Combine: $\frac{2x}{x^2 - 4} - \frac{1}{x^2 - 3x + 2} + \frac{x + 1}{x^2 + x - 2}$.

Solution We factor each denominator to find the LCD:

$$\left.\begin{aligned} x^2 - 4 &= (x - 2)(x + 2) \\ x^2 - 3x + 2 &= (x - 2)(x - 1) \\ x^2 + x - 2 &= (x - 1)(x + 2) \end{aligned}\right\} \text{LCD} = (x - 2)(x + 2)(x - 1)$$

We then write each rational expression as an equivalent rational expression with the LCD as its denominator and do the subtraction and addition.

$$\frac{2x}{x^2 - 4} - \frac{1}{x^2 - 3x + 2} + \frac{x + 1}{x^2 + x - 2}$$

$$= \frac{2x}{(x - 2)(x + 2)} - \frac{1}{(x - 2)(x - 1)} + \frac{x + 1}{(x - 1)(x + 2)}$$

$$= \frac{2x(x - 1)}{(x - 2)(x + 2)(x - 1)} - \frac{1(x + 2)}{(x - 2)(x - 1)(x + 2)} + \frac{(x + 1)(x - 2)}{(x - 1)(x + 2)(x - 2)}$$

$$= \frac{2x(x - 1) - 1(x + 2) + (x + 1)(x - 2)}{(x + 2)(x - 2)(x - 1)}$$

$$= \frac{2x^2 - 2x - x - 2 + x^2 - x - 2}{(x + 2)(x - 2)(x - 1)}$$

$$= \frac{3x^2 - 4x - 4}{(x + 2)(x - 2)(x - 1)}$$ This result can be simplified.

$$= \frac{(3x + 2)\overset{1}{\cancel{(x - 2)}}}{(x + 2)\underset{1}{\cancel{(x - 2)}}(x - 1)}$$ Factor the trinomial and divide out the common factor.

$$= \frac{3x + 2}{(x + 2)(x - 1)}$$ ■

STUDY SET Section 6.4

VOCABULARY *In Exercises 1–2, fill in the blanks to make the statements true.*

1. The least common denominator of a group of rational expressions is ______________ by each of their denominators.

2. The rational expressions $\frac{x}{x-5}$ and $\frac{x+1}{x-5}$ have ______________ denominators. The rational expressions $\frac{2x}{x+5}$ and $\frac{5}{x-5}$ have ______________ denominators.

CONCEPTS *In Exercises 3–8, fill in the blanks to make the statements true.*

3. To subtract fractions with like denominators, we ______________ the numerators and ______________ the common denominator. In symbols,

$$\frac{a}{b} - \frac{c}{b} = \boxed{}$$

4. To add fractions with like denominators, we ______________ the numerators and keep the ______________ denominator. In symbols,

$$\frac{a}{b} + \frac{c}{b} = \boxed{}$$

5. Multiplying the numerator and the denominator of a fraction by the ______________ nonzero number does not change the value of the fraction.

6. To find the LCD of several rational expressions, we ______________ each denominator and use each factor to the ______________ power that it appears in any one factorization.

7. The abbreviation for "least common denominator" is ______________.

8. The denominators of the rational expressions $\frac{a-2}{a-8}$ and $\frac{a}{8-a}$ are ______________.

In Exercises 9–10, consider the two procedures shown here.

i. $\dfrac{x^2 - 2x}{x^2 + 4x - 12} = \dfrac{\overset{1}{\cancel{x(x-2)}}}{(x+6)\underset{1}{\cancel{(x-2)}}} = \dfrac{x}{x+6}$

ii. $\dfrac{x}{x+6} = \dfrac{x(x-2)}{(x+6)(x-2)} = \dfrac{x^2 - 2x}{x^2 + 4x - 12}$

9. a. In which of these procedures are we *building* a rational expression?

b. For what type of problem is this procedure often necessary?

c. What name is used to describe the other procedure?

10. We can think of these two procedures as having the opposite effect. Explain.

NOTATION *In Exercises 11–12, complete each solution.*

11.
$$\frac{6x-1}{3x-1} + \frac{3x-2}{3x-1} = \frac{6x - 1 + \boxed{}}{3x-1}$$
$$= \frac{9x - \boxed{}}{3x-1}$$
$$= \frac{3(\boxed{})}{3x-1}$$
$$= 3$$

12.
$$\frac{8}{3v} - \frac{1}{4v^2} = \frac{8(\boxed{})}{3v(4v)} - \frac{1(3)}{4v^2(\boxed{})}$$
$$= \frac{\boxed{}}{12v^2} - \frac{3}{\boxed{}}$$
$$= \frac{32v - 3}{12v^2}$$

PRACTICE *In Exercises 13–24, do the operations and simplify the result when possible.*

13. $\frac{3}{4} + \frac{7}{4}$

14. $\frac{10}{33} - \frac{21}{33}$

15. $\frac{3}{4y} + \frac{8}{4y}$

16. $\frac{5}{3z^2} - \frac{6}{3z^2}$

17. $\dfrac{3x}{2x+2} + \dfrac{x+4}{2x+2}$

18. $\dfrac{4y}{y-4} - \dfrac{16}{y-4}$

19. $\dfrac{3x}{x-3} - \dfrac{9}{x-3}$

20. $\dfrac{9x}{x-y} - \dfrac{9y}{x-y}$

21. $\dfrac{5x}{x+1} + \dfrac{3}{x+1} - \dfrac{2x}{x+1}$

22. $\dfrac{4}{a+4} - \dfrac{2a}{a+4} + \dfrac{3a}{a+4}$

23. $\dfrac{3(x^2+x)}{x^2-5x+6} + \dfrac{-3(x^2-x)}{x^2-5x+6}$

24. $\dfrac{2x+4}{x^2+13x+12} - \dfrac{x+3}{x^2+13x+12}$

In Exercises 25–32, the denominators of several fractions are given. Find the LCD.

25. $12x,\ 18x^2$

26. $15ab^2,\ 27a^2b$

27. $x^2+3x,\ x^2-9$

28. $3y^2-6y,\ 3y(y-4)$

29. $x^3+27,\ x^2+6x+9$

30. $x^3-8,\ x^2-4x+4$

31. $2x^2+5x+3,\ 4x^2+12x+9,\ x^2+2x+1$

32. $2x^2+5x+3,\ 4x^2+12x+9,\ 4x+6$

In Exercises 33–72, do the operations and simplify the result when possible.

33. $\dfrac{1}{2} + \dfrac{1}{3}$

34. $\dfrac{8}{9} - \dfrac{5}{12}$

35. $\dfrac{3a}{2} - \dfrac{4b}{7}$

36. $\dfrac{a}{2} + \dfrac{2a}{5}$

37. $\dfrac{3}{4x} + \dfrac{2}{3x}$

38. $\dfrac{2}{5a} + \dfrac{3}{2b}$

39. $\dfrac{3a}{2b} - \dfrac{2b}{3a}$

40. $\dfrac{5m}{2n} - \dfrac{3n}{4m}$

41. $\dfrac{3}{ab^2} - \dfrac{5}{a^2b}$

42. $\dfrac{1}{xy^3} - \dfrac{2}{x^2y}$

43. $\dfrac{r}{4b^2} + \dfrac{s}{6b}$

44. $\dfrac{t}{12c^3} + \dfrac{t}{15c^2}$

45. $\dfrac{a+b}{3} + \dfrac{a-b}{7}$

46. $\dfrac{x-y}{2} + \dfrac{x+y}{3}$

47. $\dfrac{3}{x+2} + \dfrac{5}{x-4}$

48. $\dfrac{2}{a+4} - \dfrac{6}{a+3}$

49. $\dfrac{x+2}{x+5} - \dfrac{x-3}{x+7}$

50. $\dfrac{7}{x+3} + \dfrac{4x}{x+6}$

51. $4 + \dfrac{1}{x}$

52. $2 - \dfrac{1}{x+1}$

53. $\dfrac{x+8}{x-3} - \dfrac{x-14}{3-x}$

54. $\dfrac{3-x}{2-x} + \dfrac{x-1}{x-2}$

55. $\dfrac{2a+1}{3a-2} - \dfrac{a-4}{2-3a}$

56. $\dfrac{4}{x-3} + \dfrac{5}{3-x}$

57. $\dfrac{x}{x^2+5x+6} + \dfrac{x}{x^2-4}$

58. $\dfrac{x}{3x^2-2x-1} + \dfrac{4}{3x^2+10x+3}$

59. $\dfrac{4}{x^2-2x-3} - \dfrac{x}{3x^2-7x-6}$

60. $\dfrac{2a}{a^2-2a-8} + \dfrac{3}{a^2-5a+4}$

61. $\dfrac{8}{x^2-9} + \dfrac{2}{x-3} - \dfrac{6}{x}$

62. $\dfrac{x}{x^2-4} - \dfrac{x}{x+2} + \dfrac{2}{x}$

63. $1 + x - \dfrac{x}{x-5}$

64. $2 - x + \dfrac{3}{x-9}$

65. $\dfrac{3x}{2x-1}+\dfrac{x+1}{3x+2}-\dfrac{2x}{6x^3+x^2-2x}$

66. $\dfrac{x+3}{2x^2-5x+2}-\dfrac{3x-1}{x^2-x-2}$

67. $\dfrac{3}{x+1}-\dfrac{2}{x-1}+\dfrac{x+3}{x^2-1}$

68. $\dfrac{2}{x-2}+\dfrac{3}{x+2}-\dfrac{x-1}{x^2-4}$

69. $\dfrac{a}{a-b}+\dfrac{b}{a+b}+\dfrac{a^2+b^2}{b^2-a^2}$

70. $\dfrac{1}{x+y}-\dfrac{1}{x-y}+\dfrac{2y}{y^2-x^2}$

71. $\dfrac{m+1}{m^2+2m+1}+\dfrac{m-1}{m^2-2m+1}+\dfrac{2}{m^2-1}$

(*Hint:* Simplify first.)

72. $\dfrac{a+2}{a^2+3a+2}+\dfrac{a-1}{a^2-1}+\dfrac{3}{a+1}$

(*Hint:* Simplify first.)

APPLICATIONS

73. DRAFTING Among the tools used in drafting are the 45°–45°–90° and the 30°–60°–90° triangles shown in Illustration 1. Find the perimeter of each triangle. Express each result as a single rational expression.

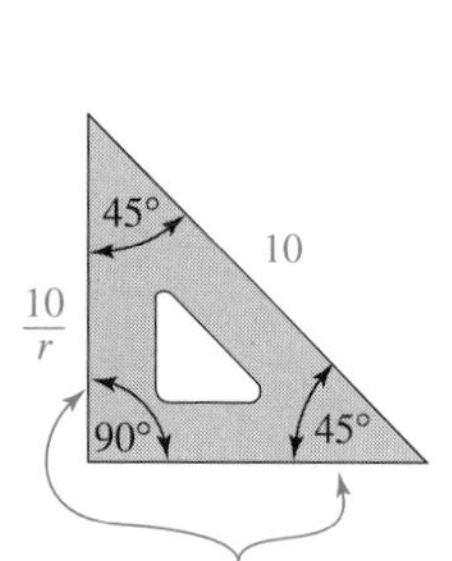

For a 45°-45°-90° triangle, these two sides are the same length.

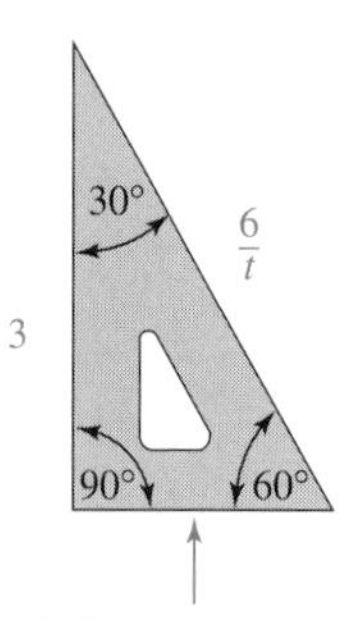

For a 30°-60°-90° triangle, this side is half as long as the hypotenuse.

ILLUSTRATION 1

74. THE AMAZON The Amazon River flows in a general eastern direction to the Atlantic Ocean. In Brazil, when the river is at low stage, the rate of flow is about 5 mph. Suppose that a river guide can canoe in still water at a rate of r mph.

a. Complete the table to find rational expressions that represent the time it would take the guide to canoe 3 miles downriver and to canoe 3 miles upriver on the Amazon.

	Rate (mph)	Time (hr)	Distance (mi)
Downriver			3
Upriver			3

b. Find the difference in the times for the trips downriver and upriver. Express the result as a single rational expression.

WRITING *Write a paragraph using your own words.*

75. Explain how to find the least common denominator of a set of fractions.

76. Explain how to add two rational expressions with unlike denominators.

77. Write some comments to the student who wrote the following solution, explaining his misunderstanding.

$$\frac{1}{x}\cdot\frac{3}{2}=\frac{1\cdot 2}{x\cdot 2}\cdot\frac{3\cdot x}{2\cdot x}$$
$$=\frac{2}{2x}\cdot\frac{3x}{2x}$$
$$=\frac{6x}{2x}$$

78. Write some comments to the student who wrote the following solution, pointing out where she made an error.

$$\frac{1}{x}-\frac{x+1}{x}=\frac{1-x+1}{x}$$
$$=\frac{2-x}{x}$$

REVIEW *In Exercises 79–82, solve each equation.*

79. $a(a-6)=-9$

80. $x^2-\dfrac{1}{2}(x+1)=0$

81. $y^3+y^2=0$

82. $5x^2=6-13x$

6.5 Complex Fractions

In this section, you will learn about

- **Complex fractions**
- **Simplifying complex fractions**

Introduction. We have seen that rational expressions have a polynomial in the numerator and in the denominator. In this section, we will consider *complex rational expressions,* called *complex fractions,* which have numerators and/or denominators that contain rational expressions.

Complex fractions

A **complex fraction** is a fraction that has a fraction (rational expression) in its numerator or its denominator or both. Examples of complex fractions are

$$\frac{\frac{3a}{b}}{\frac{6ac}{b^2}}, \quad \frac{\frac{2}{x}+1}{3+x}, \quad \text{and} \quad \frac{\frac{1}{x}+\frac{1}{y}}{\frac{1}{x}-\frac{1}{y}}$$

Simplifying complex fractions

To *simplify a complex fraction* means to express it as a simple fraction. We can use two methods to simplify the complex fraction

$$\frac{\frac{3a}{b}}{\frac{6ac}{b^2}}$$

With the first method, we eliminate the fractions in the numerator and denominator by writing the complex fraction as a division and using the division rule for fractions:

$$\frac{\frac{3a}{b}}{\frac{6ac}{b^2}} = \frac{3a}{b} \div \frac{6ac}{b^2}$$

$$= \frac{3a}{b} \cdot \frac{b^2}{6ac} \quad \text{Invert the divisor and multiply.}$$

$$= \frac{b}{2c} \quad \text{Multiply the fractions and simplify.}$$

In the other method, we eliminate the fractions in the numerator and denominator by multiplying the fraction by 1, written in the form $\frac{b^2}{b^2}$. We use $\frac{b^2}{b^2}$ because b^2 is the LCD of $\frac{3a}{b}$ and $\frac{6ac}{b^2}$.

$$\frac{\frac{3a}{b}}{\frac{6ac}{b^2}} = \frac{b^2 \cdot \frac{3a}{b}}{b^2 \cdot \frac{6ac}{b^2}}$$

$$= \frac{\frac{3ab^2}{b}}{\frac{6acb^2}{b^2}}$$

$$= \frac{3ab}{6ac}$$ Simplify the fractions in the numerator and denominator.

$$= \frac{b}{2c}$$ Divide out the common factor of $3a$.

With either method, the result is the same.

Methods for simplifying complex fractions

Method 1

Write the numerator and denominator of the complex fraction as single fractions. Then divide the fractions and simplify.

Method 2

Multiply the numerator and denominator of the complex fraction by the LCD of the fractions in its numerator and denominator. Then simplify the results, if possible.

EXAMPLE 1 ***Simplifying a complex fraction.*** Simplify $\frac{\frac{2}{x} + 1}{3 + x}$.

Solution

Method 1

We add the fractions in the numerator and proceed as follows:

$$\frac{\frac{2}{x} + 1}{3 + x} = \frac{\frac{2}{x} + \frac{x}{x}}{\frac{3 + x}{1}}$$ Write 1 as $\frac{x}{x}$ and $3 + x$ as $\frac{3+x}{1}$.

$$= \frac{\frac{2 + x}{x}}{\frac{3 + x}{1}}$$ Add $\frac{2}{x}$ and $\frac{x}{x}$ to get $\frac{2+x}{x}$.

$$= \frac{2 + x}{x} \div \frac{3 + x}{1}$$ Write the complex fraction as a division.

$$= \frac{2 + x}{x} \cdot \frac{1}{3 + x}$$ Invert the divisor and multiply.

$$= \frac{2 + x}{x^2 + 3x}$$ Multiply the numerators and multiply the denominators.

Method 2

To eliminate the denominator of x, we multiply the numerator and the denominator of the complex fraction by x.

$$\frac{\frac{2}{x} + 1}{3 + x} = \frac{x\left(\frac{2}{x} + 1\right)}{x(3 + x)}$$

$$= \frac{2 + x}{x^2 + 3x}$$ Use the distributive property to remove parentheses.

Self Check

Simplify $\frac{\frac{3}{a} + 2}{2 + a}$.

Answer: $\frac{2a + 3}{a^2 + 2a}$

Accent on Technology **Checking algebra**

We can check the simplification in Example 1 by graphing the functions

$$f(x) = \frac{\frac{2}{x} + 1}{3 + x} \quad \text{shown in Figure 6-10(a), and}$$

$$g(x) = \frac{2 + x}{x^2 + 3x}$$

shown in Figure 6-10(b), and observing that the graphs are the same. Each graph has window settings of $[-5, 3]$ for x and $[-10, 10]$ for y.

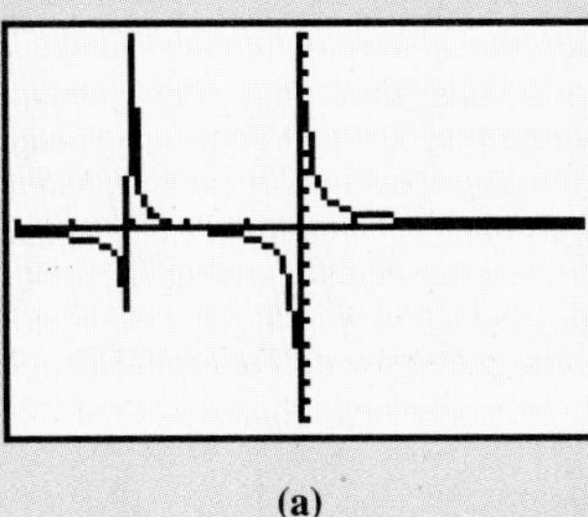

(a)

(b)

FIGURE 6-10

EXAMPLE 2 ***Simplifying a complex fraction.*** Simplify $\dfrac{\frac{1}{x} + \frac{1}{y}}{\frac{1}{x} - \frac{1}{y}}$.

Self Check

Simplify $\dfrac{\frac{1}{a} - \frac{1}{b}}{\frac{1}{a} + \frac{1}{b}}$.

Solution

Method 1

To write the numerator and denominator of the complex fraction as single fractions, we add the fractions in the numerator and subtract the fractions in the denominator.

$$\frac{\frac{1}{x} + \frac{1}{y}}{\frac{1}{x} - \frac{1}{y}} = \frac{\frac{1 \cdot y}{x \cdot y} + \frac{1 \cdot x}{y \cdot x}}{\frac{1 \cdot y}{x \cdot y} - \frac{1 \cdot x}{y \cdot x}}$$

Rewrite the fractions in the numerator in terms of the LCD, xy. Do the same in the denominator.

$$= \frac{\frac{y + x}{xy}}{\frac{y - x}{xy}}$$

Add the fractions in the numerator, and subtract the fractions in the denominator.

$$= \frac{y + x}{xy} \div \frac{y - x}{xy}$$

Write the complex fraction as a division.

$$= \frac{y + x}{xy} \cdot \frac{xy}{y - x}$$

Invert the divisor and multiply.

$$= \frac{(y + x)\overset{1}{\cancel{x}}\overset{1}{\cancel{y}}}{\underset{1}{\cancel{x}}\underset{1}{\cancel{y}}(y - x)}$$

Multiply the fractions and divide out the common factors.

$$= \frac{y + x}{y - x}$$

Method 2

We multiply the numerator and denominator by xy (the LCD of the fractions appearing in the complex fraction) and simplify.

$$\frac{\frac{1}{x}+\frac{1}{y}}{\frac{1}{x}-\frac{1}{y}} = \frac{xy\left(\frac{1}{x}+\frac{1}{y}\right)}{xy\left(\frac{1}{x}-\frac{1}{y}\right)}$$ The LCD of $\frac{1}{x}$ and $\frac{1}{y}$ is xy.

$$= \frac{\frac{xy}{x}+\frac{xy}{y}}{\frac{xy}{x}-\frac{xy}{y}}$$ Use the distributive property to remove parentheses.

$$= \frac{y+x}{y-x}$$ Simplify each fraction.

Answer: $\frac{b-a}{b+a}$

EXAMPLE 3 ***A fraction containing negative exponents.*** Simplify $\frac{x^{-1}+y^{-1}}{x^{-2}-y^{-2}}$.

Self Check

Simplify $\frac{a^{-2}+b^{-2}}{a^{-1}-b^{-1}}$.

Solution

Method 1

We need to write the numerator and denominator of the complex fraction as single fractions.

$$\frac{x^{-1}+y^{-1}}{x^{-2}-y^{-2}} = \frac{\frac{1}{x}+\frac{1}{y}}{\frac{1}{x^2}-\frac{1}{y^2}}$$ Write the fraction without using negative exponents.

$$= \frac{\frac{y}{xy}+\frac{x}{xy}}{\frac{y^2}{x^2y^2}-\frac{x^2}{x^2y^2}}$$ Get a common denominator in the numerator and denominator.

$$= \frac{\frac{y+x}{xy}}{\frac{y^2-x^2}{x^2y^2}}$$ Add the fractions in the numerator and subtract the fractions in the denominator.

$$= \frac{y+x}{xy} \div \frac{y^2-x^2}{x^2y^2}$$ Write the fraction as a division.

$$= \frac{y+x}{xy} \cdot \frac{xxyy}{(y-x)(y+x)}$$ Invert and multiply; factor y^2-x^2 and x^2y^2.

$$= \frac{\overset{1}{\cancel{(y+x)}}\,x\overset{1}{\cancel{x}}y\overset{1}{\cancel{y}}}{\underset{1}{\cancel{x}}\underset{1}{\cancel{y}}(y-x)\underset{1}{\cancel{(y+x)}}}$$ Multiply the numerators and the denominators. Divide out the common factors.

$$= \frac{xy}{y-x}$$

Method 2

We multiply both numerator and denominator by x^2y^2, the LCD of the fractions, and proceed as follows:

$$\frac{x^{-1}+y^{-1}}{x^{-2}-y^{-2}} = \frac{\frac{1}{x}+\frac{1}{y}}{\frac{1}{x^2}-\frac{1}{y^2}}$$ Write the fraction without negative exponents.

$$= \frac{x^2y^2\left(\frac{1}{x}+\frac{1}{y}\right)}{x^2y^2\left(\frac{1}{x^2}-\frac{1}{y^2}\right)}$$ Multiply numerator and denominator by x^2y^2.

$$= \frac{xy^2+yx^2}{y^2-x^2}$$ Use the distributive property to remove parentheses.

$$= \frac{xy(y+x)}{(y+x)(y-x)}$$ Factor the numerator and denominator.

$$= \frac{xy}{y-x}$$ Divide out the common factor $y + x$.

Answer: $\frac{b^2+a^2}{ab(b-a)}$ ■

WARNING! $x^{-1} + y^{-1}$ means $\frac{1}{x} + \frac{1}{y}$, and $(x + y)^{-1}$ means $\frac{1}{x+y}$. Thus,

$$x^{-1} + y^{-1} \neq (x + y)^{-1}$$

EXAMPLE 4 ***Simplifying a complex fraction.*** Simplify

$$\frac{\frac{1}{a^2-3a+2}}{\frac{3}{a-2}-\frac{2}{a-1}}.$$

Solution

We will use Method 2 to do the simplification. To determine the LCD for all the fractions appearing in the complex fraction, we must factor $a^2 - 3a + 2$.

$$\frac{\frac{1}{a^2-3a+2}}{\frac{3}{a-2}-\frac{2}{a-1}} = \frac{\frac{1}{(a-2)(a-1)}}{\frac{3}{a-2}-\frac{2}{a-1}}$$

The LCD of the fractions in the numerator and denominator of the complex fraction is $(a - 2)(a - 1)$. We multiply the numerator and the denominator by the LCD.

$$= \frac{(a-2)(a-1)\left[\frac{1}{(a-2)(a-1)}\right]}{(a-2)(a-1)\left[\frac{3}{a-2}-\frac{2}{a-1}\right]}$$

$$= \frac{\frac{(a-2)(a-1)}{(a-2)(a-1)}}{\frac{3(a-2)(a-1)}{a-2}-\frac{2(a-2)(a-1)}{a-1}}$$ Do the multiplication in the numerator. In the denominator, distribute the LCD.

Self Check

Simplify $\frac{\frac{b}{b+4}+\frac{3}{b+3}}{\frac{b}{b^2+7b+12}}$.

$$= \frac{1}{3(a-1) - 2(a-2)}$$ Simplify each of the three rational expressions by dividing out the common factors.

$$= \frac{1}{3a - 3 - 2a + 4}$$ In the denominator, remove parentheses.

$$= \frac{1}{a+1}$$ Combine like terms.

Answer: $\dfrac{b^2 + 6b + 12}{b}$

STUDY SET Section 6.5

VOCABULARY *In Exercises 1–2, fill in the blanks to make the statements true.*

1. A ______________ fraction is a fraction that has fractions (rational expressions) in its numerator and/or its denominator.

2. To ______________ a complex fraction means to express it as a simple fraction.

CONCEPTS

3. The first step in simplifying a complex fraction using Method 2 is shown below. With this method, the fractions in the numerator and denominator are to be eliminated by multiplying the complex fraction by 1. How is the 1 written in this case?

$$\frac{\frac{4}{t^2}}{\frac{3b}{t}} = \frac{t^2 \cdot \frac{4}{t^2}}{t^2 \cdot \frac{3b}{t}}$$

4. Determine the LCD of the rational expressions appearing in each complex fraction.

a. $\dfrac{1 + \frac{4}{c}}{\frac{2}{c} + c}$

b. $\dfrac{\frac{6}{m^2} + \frac{1}{2m}}{\frac{m^2 - 1}{4}}$

c. $\dfrac{\frac{p}{p+2} + \frac{12}{p+3}}{\frac{p-1}{p^2 + 5p + 6}}$

d. $\dfrac{2 + \frac{3}{x+1}}{\frac{1}{x} + x + x^2}$

NOTATION *In Exercises 5–6, complete each solution.*

5. $\dfrac{\frac{3}{x} - \frac{x}{y}}{6} = \dfrac{\frac{3 \cdot \square}{x \cdot \square} - \frac{x \cdot x}{y \cdot x}}{6}$

$= \dfrac{3y - x^2}{xy} \;\square\; 6$

$= \dfrac{3y - x^2}{xy} \cdot \square$

$= \dfrac{3y - x^2}{6xy}$

6. $\dfrac{\frac{2}{a^2} - \frac{1}{b}}{\frac{2}{a} + \frac{1}{b^2}} = \dfrac{\square\left(\frac{2}{a^2} - \frac{1}{b}\right)}{\square\left(\frac{2}{a} + \frac{1}{b^2}\right)}$

$= \dfrac{\frac{2a^2b^2}{a^2} - \square}{\frac{2a^2b^2}{a} + \square}$

$= \dfrac{2b^2 - a^2b}{2ab^2 + a^2}$

7. The fraction $\dfrac{\frac{a}{b}}{\frac{c}{d}}$ is equivalent to $\dfrac{a}{b}\,\square\,\dfrac{c}{d}$.

8. What is the numerator and what is the denominator of the following complex fraction?

$$\frac{6 - k - \frac{5}{k}}{\frac{6}{k^2} + \frac{4}{k} - 4}$$

PRACTICE *In Exercises 9–50, simplify each complex fraction.*

9. $\dfrac{\frac{1}{2}}{\frac{3}{4}}$

10. $-\dfrac{\frac{3}{4}}{\frac{1}{2}}$

11. $\dfrac{\frac{1}{2} - \frac{2}{3}}{\frac{2}{3} + \frac{1}{2}}$

12. $\dfrac{\frac{1}{4} - \frac{1}{5}}{\frac{1}{3}}$

13. $\dfrac{\frac{4x}{y}}{\frac{6xz}{y^2}}$

14. $\dfrac{\frac{5t^4}{9x}}{\frac{2t}{18x}}$

15. $\dfrac{5ab^2}{\frac{ab}{25}}$

16. $\dfrac{\frac{6a^2b}{4t}}{3a^2b^2}$

17. $\dfrac{\frac{x - y}{xy}}{\frac{y - x}{x}}$

18. $\dfrac{\frac{x^2 + 5x + 6}{3xy}}{\frac{x^2 - 9}{6xy}}$

19. $\dfrac{\frac{1}{x} - \frac{1}{y}}{xy}$

20. $\dfrac{xy}{\frac{1}{x} - \frac{1}{y}}$

21. $\dfrac{\frac{1}{a} + \frac{1}{b}}{\frac{1}{a}}$

22. $\dfrac{\frac{1}{b}}{\frac{1}{a} - \frac{1}{b}}$

23. $\dfrac{1 + \frac{x}{y}}{1 - \frac{x}{y}}$

24. $\dfrac{\frac{x}{y} + 1}{1 - \frac{x}{y}}$

25. $\dfrac{\frac{y}{x} - \frac{x}{y}}{\frac{1}{x} + \frac{1}{y}}$

26. $\dfrac{\frac{y}{x} - \frac{x}{y}}{\frac{1}{y} - \frac{1}{x}}$

27. $\dfrac{\frac{1}{a} - \frac{1}{b}}{\frac{a}{b} - \frac{b}{a}}$

28. $\dfrac{\frac{1}{a} + \frac{1}{b}}{\frac{a}{b} - \frac{b}{a}}$

29. $\dfrac{x + 1 - \frac{6}{x}}{\frac{1}{x}}$

30. $\dfrac{x - 1 - \frac{2}{x}}{\frac{x}{3}}$

31. $\dfrac{5xy}{1 + \frac{1}{xy}}$

32. $\dfrac{3a}{a + \frac{1}{a}}$

33. $\dfrac{a - 4 + \frac{1}{a}}{-\frac{1}{a} - a + 4}$

34. $\dfrac{a + 1 + \frac{1}{a^2}}{\frac{1}{a^2} + a - 1}$

35. $\dfrac{1 + \frac{6}{x} + \frac{8}{x^2}}{1 + \frac{1}{x} - \frac{12}{x^2}}$

36. $\dfrac{1 - x - \frac{2}{x}}{\frac{6}{x^2} + \frac{1}{x} - 1}$

37. $\dfrac{\frac{1}{a+1}+1}{\frac{3}{a-1}+1}$

38. $\dfrac{2+\frac{4}{y-7}}{\frac{4}{y-7}}$

39. $\dfrac{2+\frac{3}{x+1}}{\frac{1}{x}+x}$

40. $\dfrac{\frac{1}{x}-\frac{4}{x-1}}{\frac{3}{x-1}+\frac{2}{x}}$

41. $\dfrac{y}{x^{-1}-y^{-1}}$

42. $\dfrac{x^{-1}+y^{-1}}{(x+y)^{-1}}$

43. $\dfrac{x-y^{-2}}{y-x^{-2}}$

44. $\dfrac{x^{-2}-y^{-2}}{x^{-1}-y^{-1}}$

45. $\dfrac{\frac{t}{x^2-y^2}}{\frac{t}{x+y}}$

46. $\dfrac{\frac{7}{a-b}}{\frac{b}{a^3-b^3}}$

47. $\dfrac{\frac{2}{x+3}-\frac{1}{x-3}}{\frac{3}{x^2-9}}$

48. $\dfrac{2+\frac{1}{x^2-1}}{1+\frac{1}{x-1}}$

49. $\dfrac{\frac{h}{h^2+3h+2}}{\frac{4}{h+2}-\frac{4}{h+1}}$

50. $\dfrac{\frac{1}{r^2+4r+4}}{\frac{r}{r+2}+\frac{r}{r+2}}$

APPLICATIONS

51. ENGINEERING The stiffness k of the shaft shown in Illustration 1 is given by the formula

$$k = \dfrac{1}{\frac{1}{k_1}+\frac{1}{k_2}}$$

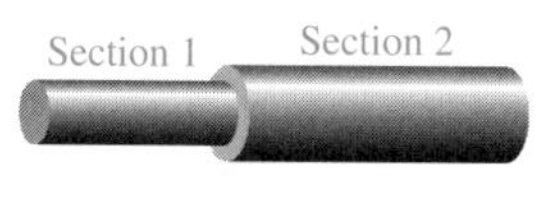

ILLUSTRATION 1

where k_1 and k_2 are the individual stiffnesses of each section. Simplify the complex fraction.

52. TRANSPORTATION If a bus travels a distance d_1 at a speed s_1, and then travels a distance d_2 at a speed s_2, the average (mean) speed $\bar{s}$ is given by the formula

$$\bar{s} = \dfrac{d_1+d_2}{\frac{d_1}{s_1}+\frac{d_2}{s_2}}$$

Simplify the complex fraction.

53. KITCHEN UTENSILS See Illustration 2. What is the ratio of the width of the opening of the ice tongs to the width of the opening of the handles? Express the result in simplest form.

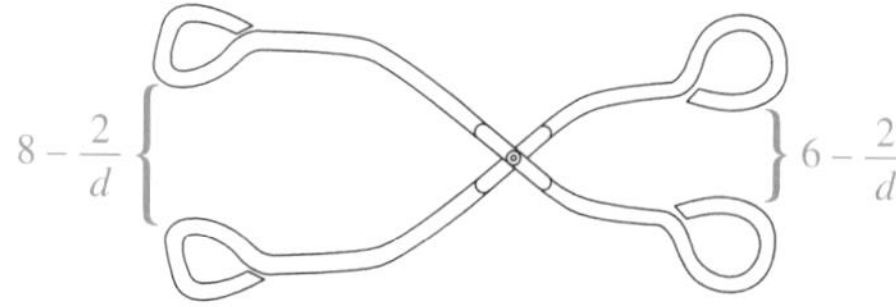

ILLUSTRATION 2

54. DATA ANALYSIS Use the data in the table to find the average measurement for the three-trial experiment. Express the answer as a rational expression.

	Trial 1	Trial 2	Trial 3
Measurement	$\frac{k}{3}$	$\frac{k}{5}$	$\frac{k}{6}$

WRITING *Write a paragraph using your own words.*

55. What is a complex fraction?

56. Two methods can be used to simplify a complex fraction. Which method do you think is simpler? Why?

REVIEW *In Exercises 57–60, solve each equation.*

57. $\dfrac{8(a-5)}{3} = 2(a-4)$

58. $\dfrac{3t^2}{5}+\dfrac{7t}{10}=\dfrac{3t+6}{5}$

59. $a^4-13a^2+36=0$

60. $|2x-1|=9$

6.6 Equations Containing Rational Expressions

In this section, you will learn about

- **Solving rational equations**
- **Extraneous solutions**
- **Solving formulas for a specified variable**
- **Problem solving**

INTRODUCTION. In this section, we will use the five-step problem-solving strategy to solve problems from disciplines such as business, photography, aviation, electronics, and publishing. We will encounter a new type of equation when we write mathematical models of each of these situations. The equations will contain rational expressions and are therefore called *rational equations*. To solve them, we use a technique that clears them of fractions.

Solving rational equations

If an equation contains one or more rational expressions, it is called a **rational equation.** Some examples of rational equations are

$$\frac{3}{5} + \frac{7}{x+2} = 2, \qquad \frac{x+3}{x-3} = \frac{2}{x^2-4}, \qquad \text{and} \qquad \frac{-x^2+10}{x^2-1} + \frac{3x}{x-1} = \frac{2x}{x+1}$$

To solve rational equations, we can multiply both sides of the equation by the LCD of the rational expressions in the equation to clear it of fractions.

EXAMPLE 1 ***Solving rational equations.*** Solve $\frac{3}{5} + \frac{7}{x+2} = 2$.

Solution

We note that x cannot be -2, because this would give a 0 in the denominator of $\frac{7}{x+2}$. If $x \neq -2$, we can multiply both sides of the equation by $5(x + 2)$ to get

$$5(x+2)\left(\frac{3}{5} + \frac{7}{x+2}\right) = 5(x+2)(2)$$ Multiply both sides by the LCD.

$$5(x+2)\left(\frac{3}{5}\right) + 5(x+2)\left(\frac{7}{x+2}\right) = 5(x+2)2$$ On the left-hand side, distribute $5(x + 2)$.

$$\overset{1}{\cancel{5}}(x+2)\left(\frac{3}{\underset{1}{\cancel{5}}}\right) + 5\overset{1}{\cancel{(x+2)}}\left(\frac{7}{\underset{1}{\cancel{(x+2)}}}\right) = 5(x+2)2$$ On the left-hand side, divide out the common factors: $\frac{5}{5} = 1$ and $\frac{x+2}{x+2} = 1$.

$$3(x+2) + 5(7) = 10(x+2)$$ Simplify each side.

The resulting equation does not contain any fractions. We now solve this *linear equation* for x.

$$3x + 6 + 35 = 10x + 20$$ Use the distributive property and simplify.

$$3x + 41 = 10x + 20$$ Combine like terms.

$$-7x = -21$$ Subtract $10x$ and 41 from both sides.

$$x = 3$$ Divide both sides by -7.

Self Check

Solve $\frac{2}{5} + \frac{5}{x-2} = \frac{29}{10}$.

Check: To check, we substitute 3 for x in the original equation and simplify:

$$\frac{3}{5} + \frac{7}{x+2} = 2$$

$$\frac{3}{5} + \frac{7}{3+2} \stackrel{?}{=} 2$$

$$\frac{3}{5} + \frac{7}{5} \stackrel{?}{=} 2$$

$$2 = 2$$

Answer: 4

Accent on Technology **Solving rational equations**

To use a graphing calculator to approximate the solution of $\frac{3}{5} + \frac{7}{x+2} = 2$, we graph the functions $f(x) = \frac{3}{5} + \frac{7}{x+2}$ and $g(x) = 2$. If we use window settings of $[-10, 10]$ for x and $[-10, 10]$ for y, we will obtain the graph shown in Figure 6-11(a).

If we trace and move the cursor closer to the intersection point of the two graphs, we will get the approximate value of x shown in Figure 6-11(b). If we zoom twice and trace again, we get the results shown in Figure 6-11(c). Algebra will show that the exact solution is 3.

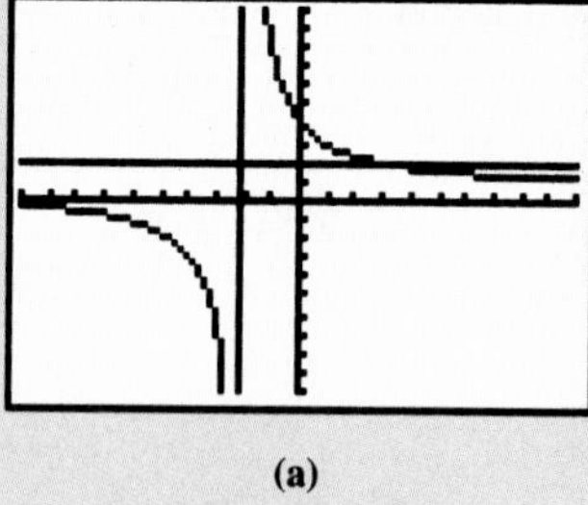

(a)

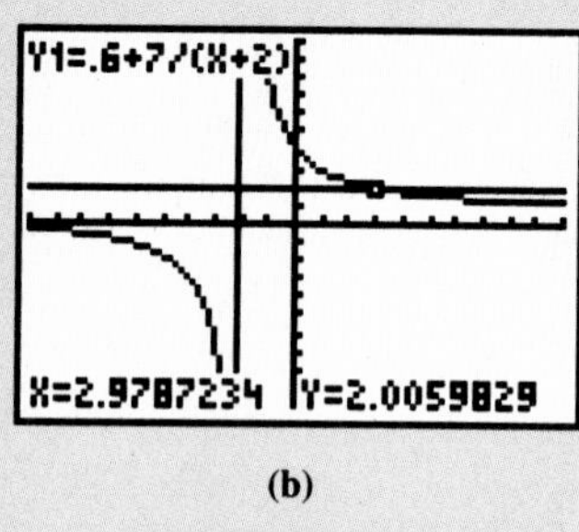

(b)

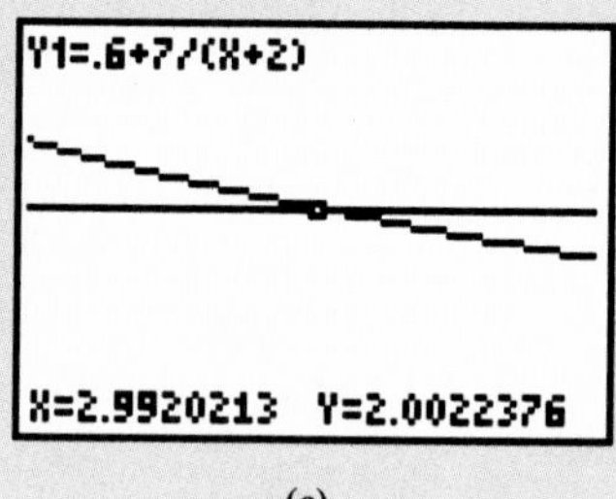

(c)

FIGURE 6-11

EXAMPLE 2 ***Solving rational equations.*** Solve $\frac{-x^2+10}{x^2-1} + \frac{3x}{x-1} = \frac{2x}{x+1}$.

Solution We start by noting that x cannot be 1 or -1, because this would give a 0 in the denominator of a fraction. If $x \neq 1$ and $x \neq -1$, we can clear the equation of fractions by multiplying both sides by the LCD of the three rational expressions and proceeding as follows:

$$\frac{-x^2+10}{x^2-1} + \frac{3x}{x-1} = \frac{2x}{x+1}$$

$$\frac{-x^2+10}{(x+1)(x-1)} + \frac{3x}{x-1} = \frac{2x}{x+1}$$ Factor the denominator $x^2 - 1$. We determine the LCD to be $(x+1)(x-1)$.

$$\frac{(x+1)(x-1)(-x^2+10)}{(x+1)(x-1)} + \frac{3x(x+1)(x-1)}{x-1} = \frac{2x(x+1)(x-1)}{x+1}$$ Multiply both sides by the LCD.

$$\frac{\overset{1}{\cancel{(x+1)}}\overset{1}{\cancel{(x-1)}}(-x^2+10)}{\underset{1}{\cancel{(x+1)}}\underset{1}{\cancel{(x-1)}}} + \frac{3x(x+1)\overset{1}{\cancel{(x-1)}}}{\underset{1}{\cancel{(x-1)}}} = \frac{2x\overset{1}{\cancel{(x+1)}}(x-1)}{\underset{1}{\cancel{(x+1)}}}$$ Divide out the common factors.

$-x^2 + 10 + 3x(x+1) = 2x(x-1)$	Simplify each side. The resulting equation does not contain any fractions.
$-x^2 + 10 + 3x^2 + 3x = 2x^2 - 2x$	Remove parentheses.
$2x^2 + 10 + 3x = 2x^2 - 2x$	Combine like terms.
$10 + 3x = -2x$	Subtract $2x^2$ from both sides.
$10 + 5x = 0$	Add $2x$ to both sides.
$5x = -10$	Subtract 10 from both sides.
$x = -2$	Divide both sides by 5.

Verify that -2 is a solution to the original equation. The solution set is $\{-2\}$. ■

EXAMPLE 3 ***A rational equation that leads to a quadratic equation.*** Solve $\frac{x+1}{5} - 2 = -\frac{4}{x}$.

Solution

We start by noting that x cannot be 0, because this would give a 0 in the denominator of a fraction. If $x \neq 0$, we can clear the equation of fractions by multiplying both sides by $5x$.

$$\frac{x+1}{5} - 2 = -\frac{4}{x}$$

$5x\left(\frac{x+1}{5} - 2\right) = 5x\left(-\frac{4}{x}\right)$	Multiply both sides by the LCD, $5x$.
$5x\left(\frac{x+1}{5}\right) - 5x(2) = 5x\left(-\frac{4}{x}\right)$	Distribute $5x$.
$\overset{1}{\cancel{5}}x\left(\frac{x+1}{\underset{1}{\cancel{5}}}\right) - 5x(2) = 5\overset{1}{\cancel{x}}\left(-\frac{4}{\underset{1}{\cancel{x}}}\right)$	Divide out the common factors.
$x(x+1) - 10x = -20$	Simplify.
$x^2 + x - 10x = -20$	Remove parentheses.

To use factoring to solve this quadratic equation, we must write it in quadratic form ($ax^2 + bx + c = 0$).

$x^2 - 9x + 20 = 0$	Combine like terms and add 20 to both sides.
$(x-5)(x-4) = 0$	Factor $x^2 - 9x + 20$.
$x - 5 = 0$ or $x - 4 = 0$	Set each factor equal to 0.
$x = 5$ $\quad$ $x = 4$	

Since 4 and 5 both satisfy the original equation, the solution set is $\{4, 5\}$.

Self Check

Solve $a + \frac{2}{3} = \frac{2a - 12}{3(a-3)}$.

Answer: 1, 2 ■

We can now summarize the procedure used to solve rational equations.

Solving rational equations

1. Factor all denominators.
2. Multiply both sides of the equation by the LCD of all rational expressions in the equation.
3. Use the distributive property to remove parentheses, divide out any common factors, and write the result in simplified form.
4. Solve the resulting linear or quadratic equation.
5. Check the result(s) in the original equation.

Extraneous solutions

When we multiply both sides of an equation by a quantity that contains a variable, we can get false solutions, called **extraneous solutions.** This happens when we multiply both sides of an equation by 0 and get a solution that gives a 0 in the denominator of a rational expression. Extraneous solutions must be excluded from the solution set of an equation.

EXAMPLE 4 ***Extraneous solutions.*** Solve $\frac{2(x+1)}{x-3} = \frac{x+5}{x-3}$.

Solution

We start by noting that x cannot be 3, because this would give a 0 in the denominator of a fraction. If $x \neq 3$, we can clear the equation of fractions by multiplying both sides by $x - 3$.

$$\frac{2(x+1)}{x-3} = \frac{x+5}{x-3}$$

$$(x-3)\frac{2(x+1)}{x-3} = (x-3)\frac{x+5}{x-3} \quad \text{Multiply both sides by the LCD.}$$

$$2(x+1) = x+5 \quad \text{Divide out the common factor of } (x-3).$$

$$2x+2 = x+5 \quad \text{Remove parentheses.}$$

$$x+2 = 5 \quad \text{Subtract } x \text{ from both sides.}$$

$$x = 3 \quad \text{Subtract 2 from both sides.}$$

Since x cannot be 3, the 3 must be discarded. This equation has no solutions. Its solution set is the empty set, $\varnothing$.

Self Check

Solve $\frac{5}{a} - \frac{3}{2} = 2 + \frac{5}{a}$.

Answer: no solutions; 0 is extraneous ■

Solving formulas for a specified variable

Many formulas must be cleared of fractions before we can solve them for a specific variable.

EXAMPLE 5 ***Physics.*** The *law of gravitation,* formulated by Sir Isaac Newton in 1684, states that if two masses, m_1 and m_2, are separated by a distance of r, the force F exerted by one mass on the other is

$$F = \frac{Gm_1m_2}{r^2}$$

where G is the gravitational constant. (See Figure 6-12.) Solve for m_2.

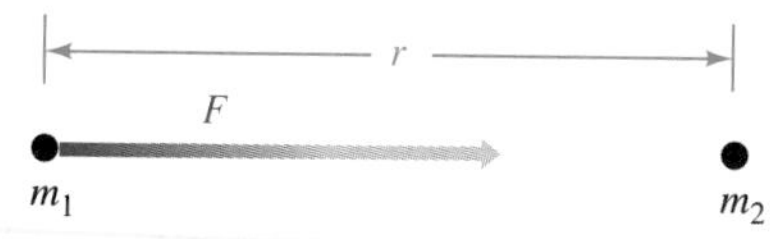

FIGURE 6-12

Solution

$$F = \frac{Gm_1m_2}{r^2}$$

$$r^2(F) = \overset{1}{r^2}\left(\frac{Gm_1m_2}{\underset{1}{r^2}}\right) \quad \text{Multiply both sides by the LCD, } r^2\text{. Divide out the common factors.}$$

$$\frac{r^2F}{Gm_1} = \frac{Gm_1m_2}{Gm_1} \quad \text{To isolate } m_2\text{, divide both sides by } Gm_1.$$

$$\frac{r^2F}{Gm_1} = m_2 \quad \text{Simplify the right-hand side.}$$

$$m_2 = \frac{r^2F}{Gm_1}$$

Self Check

Solve the law of gravitation formula for r^2.

Answer: $r^2 = \frac{Gm_1m_2}{F}$ ■

EXAMPLE 6 ***Electronics.*** In electronic circuits, resistors oppose the flow of an electric current. The total resistance R of a parallel combination of two resistors as shown in Figure 6-13 is given by

$$\frac{1}{R} = \frac{1}{R_1} + \frac{1}{R_2}$$

where R_1 is the resistance of the first resistor and R_2 is the resistance of the second resistor. Solve the formula for R.

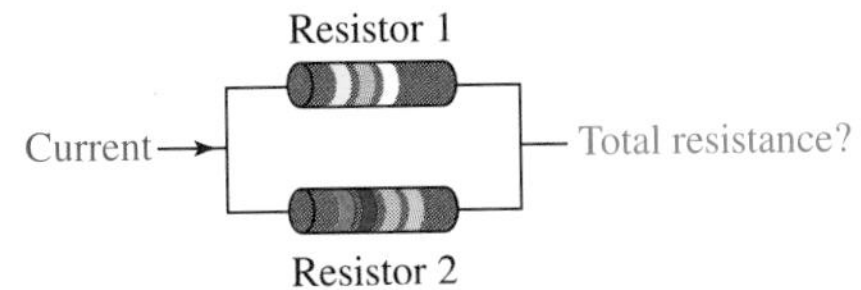

FIGURE 6-13

Solution

We begin by clearing the equation of fractions by multiplying both sides by the LCD, which is RR_1R_2.

$$\frac{1}{R} = \frac{1}{R_1} + \frac{1}{R_2}$$

$$RR_1R_2\left(\frac{1}{R}\right) = RR_1R_2\left(\frac{1}{R_1} + \frac{1}{R_2}\right) \quad \text{Multiply both sides by the LCD.}$$

$$\frac{\overset{1}{\cancel{R}}R_1R_2}{\underset{1}{\cancel{R}}} = \frac{R\overset{1}{\cancel{R_1}}R_2}{\underset{1}{\cancel{R_1}}} + \frac{RR_1\overset{1}{\cancel{R_2}}}{\underset{1}{\cancel{R_2}}} \quad \text{Distribute } RR_1R_2\text{. Then divide out the common factors.}$$

$$R_1R_2 = RR_2 + RR_1 \quad \text{Simplify.}$$

$$R_1R_2 = R(R_2 + R_1) \quad \text{Factor } R \text{ out on the right-hand side.}$$

$$\frac{R_1R_2}{R_2 + R_1} = R \quad \text{To isolate } R\text{, divide both sides by } R_2 + R_1.$$

$$R = \frac{R_1R_2}{R_2 + R_1}$$

Self Check

The *two-intercept form* for the equation of a line is

$$\frac{x}{a} + \frac{y}{b} = 1$$

Solve for b.

Answer: $b = -\frac{ay}{x - a}$ or $b = \frac{ay}{a - x}$

Problem solving

We can use rational equations to model *shared-work* problems. In this case, we assume that the work is being performed at a constant rate by all of those involved.

EXAMPLE 7 ***Drywalling a house.*** A contractor knows that one crew can drywall a house in 4 days and that another crew can drywall the same house in 5 days. One day must be allowed for the plaster coat to dry. If the contractor uses both crews, can the house be ready for painting in 4 days?

Analyze the problem Because 1 day is necessary for drying, the drywallers must complete their work in 3 days. Since the first crew can drywall the house in 4 days, it can do $\frac{1}{4}$ of the job in 1 day. Since the second crew can drywall the house in 5 days, it can do $\frac{1}{5}$ of the job in 1 day. If it takes x days for both crews to finish the house, together they can do $\frac{1}{x}$ of the job in 1 day. The amount of work the first crew can do in 1 day plus the amount of work the second crew can do in 1 day equals the amount of work both crews can do in 1 day working together.

Form an equation If x represents the number of days it takes for both crews, working together, to drywall the house, we can form the equation

What crew 1 can do in one day	+	what crew 2 can do in one day	=	what they can do together in one day.
$\frac{1}{4}$	+	$\frac{1}{5}$	=	$\frac{1}{x}$

Solve the equation We can solve this equation as follows.

$$20x\left(\frac{1}{4} + \frac{1}{5}\right) = 20x\left(\frac{1}{x}\right) \quad \text{Multiply both sides by the LCD, } 20x.$$

$$\frac{20x \cdot 1}{4} + \frac{20x \cdot 1}{5} = \frac{20x \cdot 1}{x} \quad \text{Remove parentheses.}$$

$$\frac{\overset{1}{\cancel{4}} \cdot 5x \cdot 1}{\underset{1}{\cancel{4}}} + \frac{\overset{1}{\cancel{5}} \cdot 4x \cdot 1}{\underset{1}{\cancel{5}}} = \frac{20 \cdot \overset{1}{\cancel{x}} \cdot 1}{\underset{1}{\cancel{x}}} \quad \text{Factor and divide out common factors.}$$

$$5x + 4x = 20 \quad \text{Simplify.}$$

$$9x = 20 \quad \text{Combine like terms.}$$

$$x = \frac{20}{9} \quad \text{Divide both sides by 9.}$$

State the conclusion Since it will take only $\frac{20}{9}$ or $2\frac{2}{9}$ days for both crews to drywall the house and it takes 1 day for drying, it will be ready for painting in $3\frac{2}{9}$ days, which is less than 4 days.

Check the result In $2\frac{2}{9}$ days, the first crew will do $\frac{2\frac{2}{9}}{4}$, or about 56% of the job. The second crew will do $\frac{2\frac{2}{9}}{5}$, or about 44% of the job. When we combine the two efforts, it appears that 100% of the job would be completed. The result seems reasonable. ■

In the next two examples, rational equations are used to model situations involving *uniform motion.*

EXAMPLE 8 ***Driving to a convention.*** A doctor drove 200 miles to attend a national convention. Because of poor weather, her average speed on the return trip was 10 mph less than her average speed going to the convention. If the return trip took 1 hour longer, how fast did she drive in each direction?

Analyze the problem We need to find her rate of speed going to and returning from the convention. They can be represented using a variable. The distance traveled was 200 miles each way. To describe the travel times, we note that

$rt = d$ r is the rate of speed, t is the time, and d is distance.

or

$t = \frac{d}{r}$ Divide both sides by r.

Form an equation Let r represent the average rate of speed going to the meeting. Then $r - 10$ represents the average rate of speed on the return trip. We can organize the facts of the problem in the table shown in Figure 6-14.

Because the return trip took 1 hour longer, we can form the following equation:

The time it took to travel to the convention	+	1	=	the time it took to return.
$\frac{200}{r}$	+	1	=	$\frac{200}{r - 10}$

	Rate	Time	Distance
Going	r	$\frac{200}{r}$	200
Returning	$r - 10$	$\frac{200}{r - 10}$	200

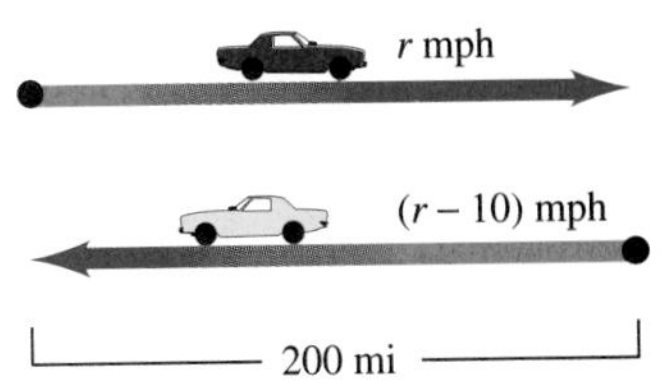

We obtained these two entries by dividing the distance by the rate.

FIGURE 6-14

Solve the equation We can solve the equation as follows:

$$r(r-10)\left(\frac{200}{r}+1\right) = r(r-10)\left(\frac{200}{r-10}\right)$$ Multiply both sides by $r(r - 10)$.

$$\overset{1}{\cancel{r}}(r-10)\frac{200}{\underset{1}{\cancel{r}}} + r(r-10)1 = r\overset{1}{\cancel{(r-10)}}\left(\frac{200}{\underset{1}{\cancel{r-10}}}\right)$$ Distribute $r(r - 10)$. Then divide out the common factors.

$$200(r-10) + r(r-10) = 200r$$ Simplify.

$$200r - 2{,}000 + r^2 - 10r = 200r$$ Remove parentheses. Note that this is a quadratic equation.

$$r^2 - 10r - 2{,}000 = 0$$ Subtract $200r$ from both sides.

$$(r-50)(r+40) = 0$$ Factor $r^2 - 10r - 2{,}000$.

$$r - 50 = 0 \quad \text{or} \quad r + 40 = 0$$ Set each factor equal to 0.

$$r = 50 \qquad\qquad r = -40$$

State the conclusion We must exclude the solution of -40, because a speed cannot be negative. Thus, the doctor averaged 50 mph going to the convention, and she averaged $50 - 10$ or 40 mph returning.

Check the result At 50 mph, the 200-mile trip took 4 hours. At 40 mph, the return trip took 5 hours, which is 1 hour longer. ■

EXAMPLE 9 ***Riverboat cruise.*** The Forest City Queen can make a 9-mile trip down the Rock River and return in a total of 1.6 hours. If the riverboat travels 12 mph in still water, find the speed of the current in the Rock River.

Analyze the problem We can represent the upstream and downstream rates of speed using a variable. In each case, the distance traveled is 9 miles. To write an expression for the time traveled, divide the distance by the rate of speed.

Form an equation We can let c represent the speed of the current. Since the boat travels 12 mph and a current of c mph pushes the boat while it is going downstream, the speed of the boat going downstream is $(12 + c)$ mph. On the return trip, the current pushes against the boat, and its speed is $(12 - c)$ mph. Since $t = \frac{d}{r}$ $\left(\text{time} = \frac{\text{distance}}{\text{rate}}\right)$, the time required for the downstream leg of the trip is $\frac{9}{12 + c}$ hours, and the time required for the upstream leg of the trip is $\frac{9}{12 - c}$ hours. We can organize this information in the table shown in Figure 6-15.

Furthermore, we know that the total time required for the round trip is 1.6 or $\frac{8}{5}$ hours.

The time it takes to travel downstream	+	the time it takes to travel upstream	=	the total time for the round trip.
$\frac{9}{12 + c}$	+	$\frac{9}{12 - c}$	=	$\frac{8}{5}$

	Rate	·	Time	=	Distance
Going downstream	$12 + c$		$\frac{9}{12 + c}$		9
Going upstream	$12 - c$		$\frac{9}{12 - c}$		9

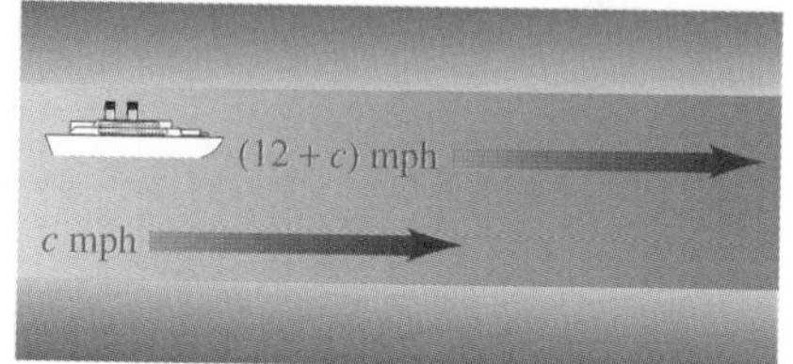

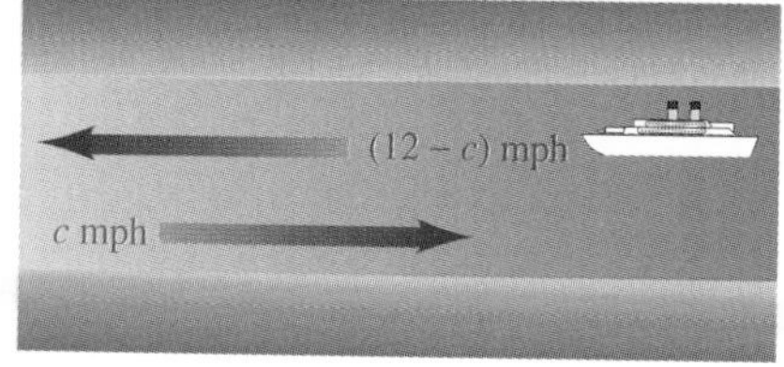

FIGURE 6-15

Solve the equation Multiply both sides of this equation by $5(12 + c)(12 - c)$ to clear it of fractions.

$$5(12 + c)(12 - c)\left(\frac{9}{12 + c} + \frac{9}{12 - c}\right) = 5(12 + c)(12 - c)\left(\frac{8}{5}\right)$$

$$\frac{5\overset{1}{\cancel{(12 + c)}}(12 - c)9}{\underset{1}{\cancel{12 + c}}} + \frac{5(12 + c)\overset{1}{\cancel{(12 - c)}}9}{\underset{1}{\cancel{12 - c}}} = \frac{\overset{1}{\cancel{5}}(12 + c)(12 - c)8}{\underset{1}{\cancel{5}}}$$ Distribute, and then divide out the common factors.

$$45(12 - c) + 45(12 + c) = 8(12 + c)(12 - c)$$ Simplify.

$$540 - 45c + 540 + 45c = 8(144 - c^2)$$ On the left-hand side, distribute. On the right-hand side use the FOIL method.

$$1{,}080 = 1{,}152 - 8c^2$$ Combine like terms and multiply. This is a quadratic equation.

$$8c^2 - 72 = 0$$ Add $8c^2$ and subtract 1,152 from both sides.

$$c^2 - 9 = 0$$ Divide both sides by 8.

$$(c + 3)(c - 3) = 0$$ Factor $c^2 - 9$.

$$c + 3 = 0 \quad \text{or} \quad c - 3 = 0$$ Set each factor each to 0.

$$c = -3 \qquad\qquad c = 3$$

State the conclusion Since the current cannot be negative, the apparent solution -3 must be discarded. The current in the Rock River is 3 mph.

Check the result. ■

STUDY SET Section 6.6

VOCABULARY *In Exercises 1–4, fill in the blanks to make the statements true.*

1. An equation that contains rational expressions, such as $\frac{2}{x} + \frac{3}{2} = \frac{3}{4x}$, is called a __________ equation.

2. A proposed solution to an equation that does not satisfy the equation is called an __________ solution.

3. The lowest __________ of a group of rational expressions is divisible by each of their denominators.

4. To solve an equation means to find all the values of the variable that make the equation a __________ statement.

CONCEPTS

5. Is $x = 2$ a solution of the following equations?

a. $\frac{x+2}{x+3} + \frac{1}{x^2+2x-3} = 1$

b. $\frac{x+2}{x-2} + \frac{1}{x^2-4} = 1$

6. To clear the equation

$$\frac{4}{10} + y = \frac{4y-50}{5y-25}$$

of fractions, by what should both sides be multiplied?

7. Complete the table in Illustration 1.

	r ·	t =	d
Running	x		12
Bicycling	$x + 15$		12

ILLUSTRATION 1

8. Illustration 2 shows the length of time it takes each child in a family to wash their mother's minivan. Complete the table.

	Time to wash the van alone (min)	Amount of van washed in 1 minute
Glenn	25	
Brandon	30	
Kevin	x	

ILLUSTRATION 2

9. Solve $\frac{x}{x+2} = \frac{7}{9}$ by

a. Setting the product of the extremes equal to the product of the means.

b. Multiplying both sides by the LCD.

10. a. What type of equation is $2x + x^2 = 15$?

b. How should it be rewritten if we want to use factoring to solve it?

c. Solve the equation.

NOTATION *In Exercises 11–12, complete each solution.*

11. Solve:

$$\frac{10}{3y} - \frac{7}{30} = \frac{9}{2y}$$

$$\boxed{}\left(\frac{10}{3y} - \frac{7}{30}\right) = 30y\left(\boxed{}\right)$$

$$30y\left(\boxed{}\right) - 30y\left(\frac{7}{30}\right) = 30y\left(\frac{9}{2y}\right)$$

$$100 - \boxed{} = 135$$

$$-7y = \boxed{}$$

$$y = -5$$

12. Solve:

$$\frac{2}{u-1} + \frac{1}{u} = \frac{1}{u^2-u}$$

$$\boxed{} + \frac{1}{u} = \frac{1}{u(u-1)}$$

$$\boxed{}\left(\frac{2}{u-1} + \frac{1}{u}\right) = u(u-1)\left[\frac{1}{u(u-1)}\right]$$

$$u(u-1)\left(\frac{2}{u-1}\right) + u(u-1)\left(\boxed{}\right) = u(u-1)\left[\frac{1}{u(u-1)}\right]$$

$$2u + \left(\boxed{}\right) = \boxed{}$$

$$\boxed{} = 2$$

$$u = \frac{2}{3}$$

PRACTICE *In Exercises 13–40, solve each equation. If a solution is extraneous, so indicate.*

13. $\frac{1}{4} + \frac{9}{x} = 1$

14. $\frac{1}{3} - \frac{10}{x} = -3$

15. $\frac{34}{x} - \frac{3}{2} = -\frac{13}{20}$

16. $\frac{1}{2} + \frac{7}{x} = 2 + \frac{1}{x}$

17. $\frac{3}{y} + \frac{7}{2y} = 13$

18. $\frac{2}{x} + \frac{1}{2} = \frac{7}{2x}$

19. $\frac{x+1}{x} - \frac{x-1}{x} = 0$

20. $\frac{2}{x} + \frac{1}{2} = \frac{9}{4x} - \frac{1}{2x}$

21. $\frac{7}{5x} - \frac{1}{2} = \frac{5}{6x} + \frac{1}{3}$

22. $\frac{x-3}{x-1} - \frac{2x-4}{x-1} = 0$

23. $\frac{3-5y}{2+y} = \frac{3+5y}{2-y}$

24. $\frac{x}{x-2} = 1 + \frac{1}{x-3}$

25. $\frac{a+2}{a+1} - \frac{a-4}{a-3} = 0$

26. $\frac{z+2}{z+8} - \frac{z-3}{z-2} = 0$

27. $\frac{x+2}{x+3} - 1 = \frac{1}{3-2x-x^2}$

28. $\frac{x-3}{x-2} - \frac{1}{x} = \frac{x-3}{x}$

29. $\frac{x}{x+2} = 1 - \frac{3x+2}{x^2+4x+4}$

30. $\frac{3+2a}{a^2+6+5a} + \frac{2-5a}{a^2-4} = \frac{2-3a}{a^2-6+a}$

31. $\frac{2}{x-2} + \frac{1}{x+1} = \frac{1}{x^2-x-2}$

32. $\frac{5}{y-1} + \frac{3}{y-3} = \frac{8}{y-2}$

33. $\frac{a-1}{a+3} - \frac{1-2a}{3-a} = \frac{2-a}{a-3}$

34. $\frac{5}{2z^2+z-3} - \frac{2}{2z+3} = \frac{z+1}{z-1} - 1$

35. $\frac{5}{x+4} + \frac{1}{x+4} = x - 1$

36. $\frac{2}{x-1} + \frac{x-2}{3} = \frac{4}{x-1}$

37. $\frac{3}{x+1} - \frac{x-2}{2} = \frac{x-2}{x+1}$

38. $\frac{x-4}{x-3} + \frac{x-2}{x-3} = x - 3$

39. $\frac{2}{x-3} + \frac{3}{4} = \frac{17}{2x}$

40. $\frac{5}{x+4} - \frac{1}{3} = \frac{x-1}{x}$

In Exercises 41–48, solve each formula for the indicated variable.

41. $I = \frac{E}{R_L + r}$ for r (from physics)

42. $P = \frac{R-C}{n}$ for C (from business)

43. $S = \frac{a - lr}{1 - r}$ for r (from mathematics)

44. $\mu_R = \frac{n_1(n_1 + n_2 + 1)}{2}$ for n_2 (from statistics)

45. $P = \frac{Q_1}{Q_2 - Q_1}$ for Q_1 (from refrigeration/heating)

46. $\frac{P_1V_1}{T_1} = \frac{P_2V_2}{T_2}$ for T_2 (from chemistry)

47. $\frac{1}{R} = \frac{1}{R_1} + \frac{1}{R_2} + \frac{1}{R_3}$ for R (from electronics)

48. $P + \frac{a}{V^2} = \frac{RT}{V-b}$ for b (from physics)

APPLICATIONS

49. PHOTOGRAPHY Illustration 3 shows the relationship between distances when taking a photograph. The design of a camera lens uses the equation

$$\frac{1}{f} = \frac{1}{s_1} + \frac{1}{s_2}$$

which relates the focal length f of a lens to the image distance s_1 and the object distance s_2. Find the focal length of the lens in the illustration. (*Hint:* Convert feet to inches.)

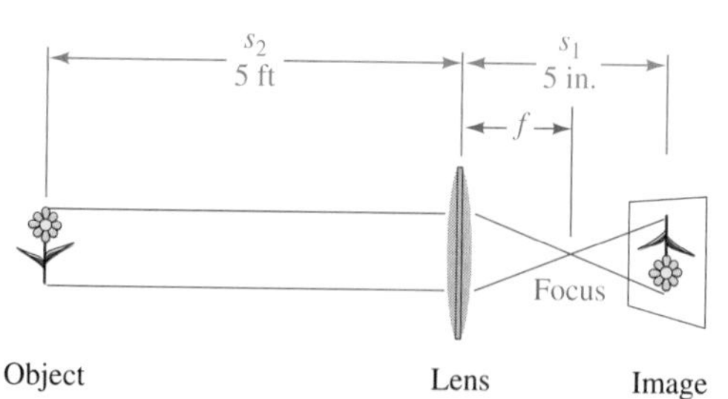

ILLUSTRATION 3

50. OPTICS The focal length, f, of a lens is given by the lensmaker's formula,

$$\frac{1}{f} = 0.6\left(\frac{1}{r_1} + \frac{1}{r_2}\right)$$

where f is the focal length of the lens and r_1 and r_2 are the radii of the two circular surfaces. Find the focal length of the lens in Illustration 4.

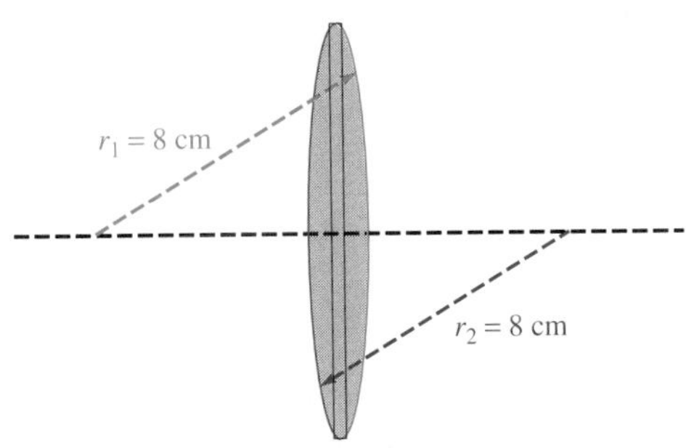

ILLUSTRATION 4

51. TAX ACCOUNTING As a piece of equipment gets older, its value usually lessens. One way to calculate *depreciation* is using the formula

$$V = C - \left(\frac{C - S}{L}\right)N$$

where V denotes the value of the equipment at the end of year N, L is its useful lifetime (in years), C is its cost new, and S is its salvage value at the end of its useful life. Solve for L. Then determine what an accountant considered the useful lifetime of a forklift that cost \$25,000 new, was worth \$13,000 after 4 years, and has a salvage value of \$1,000.

52. MECHANICAL ENGINEERING The equation

$$a = \frac{9.8m_2 - f}{m_2 + m_1}$$

models the system shown in Illustration 5, where a is the acceleration of the suspended block, m_1 and m_2 are the masses of the blocks, and f is the friction force. Solve the equation for m_2.

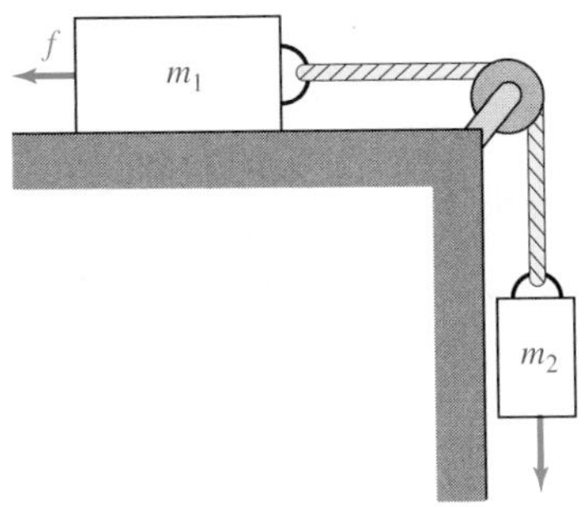

ILLUSTRATION 5

53. HOUSEPAINTING Illustration 6 shows the two bids to paint a house.

a. To get the job done quicker, the homeowner hired both the painters who submitted bids. How long will it take them to paint the house working together?

b. What will the homeowner have to pay each painter?

Santos Painting
Residential
Bid:
3 days
@ \$220 a day
Total: \$660

Mays House Painting
Bid:
\$200 per day
5 days work
Total: \$1,000

ILLUSTRATION 6

54. ROOFING A HOUSE A homeowner estimates that it will take him 7 days to roof his house. A professional roofer estimates that he could roof the house in 4 days. How long will it take if the homeowner helps the roofer?

55. OYSTERS According to the *Guinness Book of World Records,* the record for opening oysters is 100 in 140 seconds by Mike Racz in Invercargill, New Zealand on July 16, 1990. If it would take a novice $8\frac{1}{2}$ minutes to perform the same task, how long would it take them working together to open 100 oysters?

56. FARMING In 10 minutes, a conveyor belt can move 1,000 bushels of corn into the storage bin shown in Illustration 7. A smaller belt can move 1,000 bushels to the storage bin in 14 minutes. If both belts are used, how long it will take to move 1,000 bushels to the storage bin?

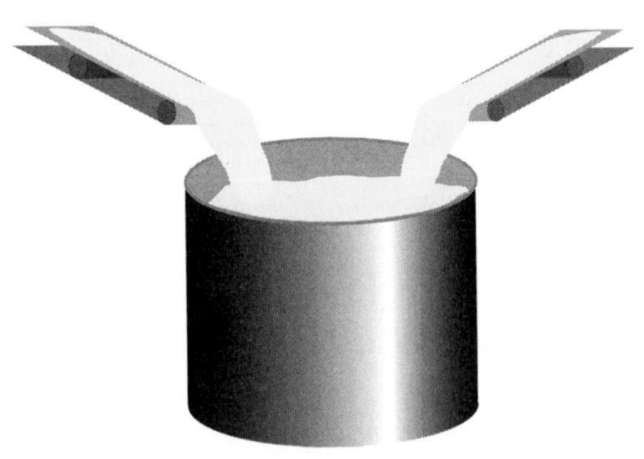

ILLUSTRATION 7

57. FILLING A POND One pipe can fill a pond in 3 weeks, and a second pipe can fill it in 5 weeks. However, evaporation and seepage can empty the pond in 10 weeks. If both pipes are used, how long will it take to fill the pond?

58. HOUSECLEANING Sally can clean the house in 6 hours, and her father can clean the house in 4 hours. Sally's younger brother, Dennis, can completely mess up the house in 8 hours. If Sally and her father clean and Dennis plays, how long will it take to clean the house?

59. BOXING For his morning workout, a boxer bicycles for 8 miles and then jogs back to camp along the same route. If he bicycles 6 mph faster than he jogs, and the entire workout lasts 2 hours, how fast does he jog?

60. DELIVERIES A FedEx delivery van traveled from Rockford to Chicago in 3 hours less time than it took a second FedEx van to travel from Rockford to St. Louis. If the vans traveled at the same average speed, use the information in the mileage chart in Illustration 8 to help determine how long the first driver was on the road.

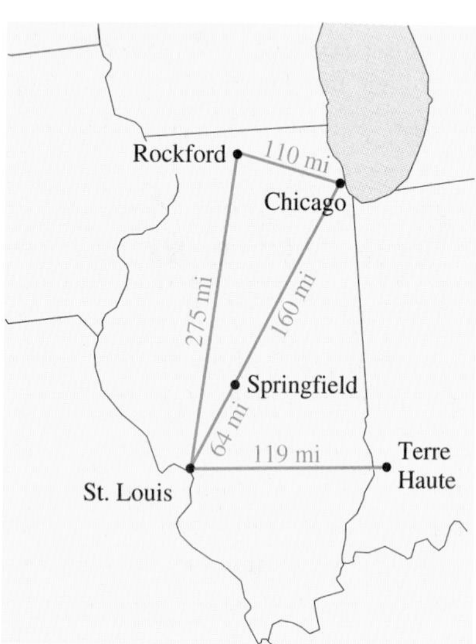

ILLUSTRATION 8

61. FINDING RATES OF SPEED Two trains made the same 315-mile run. Since one train traveled 10 mph faster than the other, it arrived 2 hours earlier. Find the speed of each train.

62. TRAIN TRAVEL A train traveled 120 miles from Freeport to Chicago and returned the same distance in a total time of 5 hours. If the train traveled 20 mph slower on the return trip, how fast did the train travel in each direction?

63. CROP DUSTING A helicopter spraying fertilizer over a field can fly 0.5 mile downwind in the same time as it can fly 0.4 mile upwind. Find the speed of the wind if the helicopter travels 45 mph in still air when dusting crops.

64. BOATING A man can drive a motorboat 45 miles down the Rock River in the same amount of time that he can drive 27 miles upstream. Find the speed of the current if the speed of the boat is 12 mph in still water.

65. UNIT COST One month, an appliance store manager bought several microwave ovens for a total of $1,800. The next month, because the unit cost of the same model of microwave increased by $25, she bought one fewer oven for the same total price. How many ovens did she buy the first month? (*Hint:* Write an expression for the unit cost of a microwave for the second month, then use the formula: Unit cost · number = total cost.)

66. EXTENDED VACATION Use the facts in the E-mail message in Illustration 9 to determine how long the student had originally planned to stay in Europe. (*Hint:* Unit cost · number = total cost.)

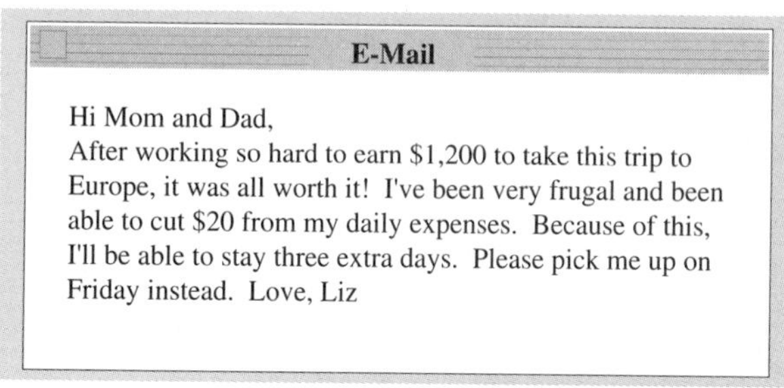

E-Mail

Hi Mom and Dad,
After working so hard to earn $1,200 to take this trip to Europe, it was all worth it! I've been very frugal and been able to cut $20 from my daily expenses. Because of this, I'll be able to stay three extra days. Please pick me up on Friday instead. Love, Liz

ILLUSTRATION 9

WRITING *Write a paragraph using your own words.*

67. Why is it necessary to check the solutions of a rational equation?

68. Explain what it means to *clear* a rational equation of fractions.

REVIEW *In Exercises 69–72, write each italicized number in scientific notation.*

69. OIL The total cost of the Alaskan pipeline, running 800 miles from Prudhoe Bay to Valdez, was *$9,000,000,000*.

70. NATURAL GAS The TransCanada Pipeline transported a record *2,352,000,000,000* cubic feet of gas in 1995.

71. RADIOACTIVITY The least stable radioactive isotope is lithium 5, which decays in *0.00000000000000000000044* second.

72. BALANCES The finest balances in the world are made in Germany. They can weigh objects to an accuracy of 35×10^{-11} ounce.

6.7 Dividing Polynomials

In this section, you will learn about

- **Dividing a monomial by a monomial**
- **Dividing a polynomial by a monomial**
- **Dividing a polynomial by a polynomial**
- **Missing terms**

INTRODUCTION. In Chapter 5, we discussed addition, subtraction, and multiplication of polynomials. We will now introduce the procedures used to divide polynomials. This topic appears in a chapter about rational expressions because rational expressions indicate division of polynomials. For example,

$$\frac{x^2 - 3x + 7}{x + 1} = (x^2 - 3x + 7) \div (x + 1)$$

We begin the discussion of division of polynomials with the simplest case, a monomial divided by a monomial.

Dividing a monomial by a monomial

In Example 1, we review two methods that can be used to divide a monomial by a monomial.

EXAMPLE 1 ***A monomial divided by a monomial.*** Simplify $(3a^2b^3) \div (2a^3b)$.

Self Check

Simplify $\frac{6x^3y^2}{8x^2y^3}$.

Solution

Method 1

We write $(3a^2b^3) \div (2a^3b)$ as a rational expression and divide out all common factors:

$$\frac{3a^2b^3}{2a^3b} = \frac{3\overset{1}{\cancel{a}}\overset{1}{\cancel{a}}\overset{1}{\cancel{b}}bb}{2\underset{1}{\cancel{a}}\underset{1}{\cancel{a}}a\underset{1}{\cancel{b}}}$$

$$= \frac{3b^2}{2a}$$

Method 2

We write $(3a^2b^3) \div (2a^3b)$ as a rational expression and use the rules for exponents:

$$\frac{3a^2b^3}{2a^3b} = \frac{3}{2}a^{2-3}b^{3-1}$$ When dividing like bases, keep the base and subtract exponents.

$$= \frac{3}{2}a^{-1}b^2$$ Subtract.

$$= \frac{3}{2}\left(\frac{1}{a}\right)\frac{b^2}{1}$$ We don't want negative exponents in the result.

$$= \frac{3b^2}{2a}$$ Multiply the numerators. Multiply the denominators.

Answer: $\frac{3x}{4y}$

Dividing monomials	**Method 1** Factor the numerator and denominator completely. Then divide out all common factors. **Method 2** Divide the coefficients. Then use the rules for exponents to simplify the divisions of variable factors with like bases.

Dividing a polynomial by a monomial

In Example 2, we use the fact that $\frac{a}{b} = \frac{1}{b} \cdot a$ $(b \neq 0)$ to divide a polynomial by a monomial.

EXAMPLE 2 ***A polynomial divided by a monomial.*** Divide $4x^3y^2 + 3x^2y^5 - 12xy$ by $3x^2y^3$.

Solution

We rewrite the division as a product and use the distributive property to remove parentheses.

$$\frac{4x^3y^2 + 3x^2y^5 - 12xy}{3x^2y^3} = \frac{1}{3x^2y^3}(4x^3y^2 + 3x^2y^5 - 12xy)$$
$$= \frac{4x^3y^2}{3x^2y^3} + \frac{3x^2y^5}{3x^2y^3} - \frac{12xy}{3x^2y^3}$$

We then simplify each of the rational expressions on the right-hand side of the equals sign to get

$$\frac{4x^3y^2 + 3x^2y^5 - 12xy}{3x^2y^3} = \frac{4x}{3y} + y^2 - \frac{4}{xy^2}$$

Self Check

Divide $\frac{8a^3b^4 - 4a^4b^2 + a^2b^2}{4a^2b^2}$.

Answer: $2ab^2 - a^2 + \frac{1}{4}$

Dividing a polynomial by a monomial	1. Multiply the polynomial by the reciprocal of the monomial. 2. Use the distributive property to remove parentheses. 3. Simplify each of the resulting rational expressions using the rules for simplifying a monomial divided by a monomial.

Dividing a polynomial by a polynomial

There is an **algorithm** (a repeating series of steps) to use when the divisor is not a monomial. To use the division algorithm to divide $x^2 + 7x + 12$ (the **dividend**) by $x + 4$ (the **divisor**), we write the division in long division form and proceed as follows:

$$\begin{array}{r} x \\ x+4\overline{)x^2+7x+12} \end{array}$$

How many times does x divide x^2? $x^2/x = x$. Place the x in the quotient.

$$\begin{array}{r} x \\ x+4\overline{)x^2+7x+12} \\ \underline{x^2+4x} \\ 3x+12 \end{array}$$

Multiply each term in the divisor by x to get $x^2 + 4x$, subtract $x^2 + 4x$ from $x^2 + 7x$, and bring down the 12.

$$\begin{array}{r} x\ +3 \\ x+4\overline{)x^2+7x+12} \\ \underline{x^2+4x} \\ 3x+12 \end{array}$$

How many times does x divide $3x$? $3x/x = +3$. Place the $+3$ in the quotient.

$$\begin{array}{r} x \quad + 3 \\ x + 4\overline{)x^2 + 7x + 12} \\ \underline{x^2 + 4x} \qquad \\ 3x + 12 \\ \underline{3x + 12} \\ 0 \end{array}$$

Multiply each term in the divisor by 3 to get $3x + 12$, and subtract $3x + 12$ from $3x + 12$ to get 0.

The division process stops when the result of the subtraction is a constant or a polynomial with degree less than the degree of the divisor. Here, the quotient is $x + 3$ and the remainder is 0.

We can check the quotient by multiplying the divisor by the quotient. The product should be the dividend.

$$\underbrace{(x + 4)}_{\text{Divisor}} \cdot \underbrace{(x + 3)}_{\text{quotient}} = \underbrace{x^2 + 7x + 12}_{\text{dividend}} \quad \text{The quotient checks.}$$

EXAMPLE 3 Divide $\dfrac{2a^3 + 9a^2 + 5a - 6}{2a + 3}$ using long division

Solution

$$\begin{array}{r} a^2 \qquad\qquad\qquad\quad \\ 2a + 3\overline{)2a^3 + 9a^2 + 5a - 6} \end{array}$$

How many times does $2a$ divide $2a^3$? $2a^3/2a = a^2$. Place a^2 in the quotient.

$$\begin{array}{r} a^2 \qquad\qquad\qquad\quad \\ 2a + 3\overline{)2a^3 + 9a^2 + 5a - 6} \\ \underline{2a^3 + 3a^2} \qquad\qquad \\ 6a^2 + 5a \quad\;\; \end{array}$$

Multiply each term in the divisor by a^2 to get $2a^3 + 3a^2$, subtract $2a^3 + 3a^2$ from $2a^3 + 9a^2$, and bring down the $5a$.

$$\begin{array}{r} a^2 + 3a \qquad\qquad \\ 2a + 3\overline{)2a^3 + 9a^2 + 5a - 6} \\ \underline{2a^3 + 3a^2} \qquad\qquad \\ 6a^2 + 5a \quad\;\; \end{array}$$

How many times does $2a$ divide $6a^2$? $6a^2/2a = 3a$. Place the $+3a$ in the quotient.

$$\begin{array}{r} a^2 + 3a \qquad\qquad \\ 2a + 3\overline{)2a^3 + 9a^2 + 5a - 6} \\ \underline{2a^3 + 3a^2} \qquad\qquad \\ 6a^2 + 5a \quad\;\; \\ \underline{6a^2 + 9a} \quad\;\; \\ -4a - 6 \end{array}$$

Multiply each term in the divisor by $3a$ to get $6a^2 + 9a$, subtract $6a^2 + 9a$ from $6a^2 + 5a$, and bring down -6.

$$\begin{array}{r} a^2 + 3a - 2 \qquad \\ 2a + 3\overline{)2a^3 + 9a^2 + 5a - 6} \\ \underline{2a^3 + 3a^2} \qquad\qquad \\ 6a^2 + 5a \quad\;\; \\ \underline{6a^2 + 9a} \quad\;\; \\ -4a - 6 \end{array}$$

How many times does $2a$ divide $-4a$? $-4a/2a = -2$. Place the -2 in the quotient.

$$\begin{array}{r} a^2 + 3a - 2 \qquad \\ 2a + 3\overline{)2a^3 + 9a^2 + 5a - 6} \\ \underline{2a^3 + 3a^2} \qquad\qquad \\ 6a^2 + 5a \quad\;\; \\ \underline{6a^2 + 9a} \quad\;\; \\ -4a - 6 \\ \underline{-4a - 6} \\ 0 \end{array}$$

Multiply each term in the divisor by -2 to get $-4a - 6$; subtract $-4a - 6$ from $-4a - 6$ to get 0.

Since the remainder is 0, the quotient is $a^2 + 3a - 2$. We can check the quotient by verifying that

$$\underbrace{(2a + 3)}_{\text{Divisor}} \cdot \underbrace{(a^2 + 3a - 2)}_{\text{quotient}} = \underbrace{2a^3 + 9a^2 + 5a - 6}_{\text{dividend}}$$

■

EXAMPLE 4 ***Using long division form.*** Divide $\dfrac{3x^3 + 2x^2 - 3x + 8}{x - 2}$.

Solution

$$
\begin{array}{r}
3x^2 + 8x \quad + 13 \\
x - 2\overline{)3x^3 + 2x^2 - 3x + 8} \\
\underline{3x^3 - 6x^2} \qquad\qquad \\
8x^2 - 3x \qquad \\
\underline{8x^2 - 16x} \qquad \\
13x + 8 \\
\underline{13x - 26} \\
34
\end{array}
$$

This division gives a quotient of $3x^2 + 8x + 13$ and a remainder of 34. It is common to form a fraction with the remainder as the numerator and the divisor as the denominator and to write the result as

$$3x^2 + 8x + 13 + \frac{34}{x - 2}$$

To check, we verify that

$$(x - 2)\left(3x^2 + 8x + 13 + \frac{34}{x - 2}\right) = 3x^3 + 2x^2 - 3x + 8$$

Self Check

Divide $\dfrac{2a^3 + 3a^2 - a + 2}{a - 3}$.

Answer: $2a^2 + 9a + 26 + \dfrac{80}{a - 3}$ ■

EXAMPLE 5 ***Arranging terms in descending powers.*** Divide $(-9x + 8x^3 + 10x^2 - 9) \div (3 + 2x)$.

Solution

The division algorithm works best when the polynomials in the dividend and the divisor are written in descending powers of x. We can use the commutative property of addition to rearrange the terms. Then the division is routine:

$$
\begin{array}{r}
4x^2 - \quad x - 3 \\
2x + 3\overline{)8x^3 + 10x^2 - 9x - 9} \\
\underline{8x^3 + 12x^2} \qquad\qquad \\
-2x^2 - 9x \qquad \\
\underline{-2x^2 - 3x} \qquad \\
-6x - 9 \\
\underline{-6x - 9} \\
0
\end{array}
$$

Thus,

$$\frac{-9x + 8x^3 + 10x^2 - 9}{3 + 2x} = 4x^2 - x - 3$$

Self Check

Divide $2 + 3a\overline{)-4a + 15a^2 + 18a^3 - 4}$.

Answer: $6a^2 + a - 2$ ■

Missing terms

If a power of the variable is missing in the dividend, it is helpful to insert "placeholder" terms, because they aid in the subtraction step of the long division procedure.

EXAMPLE 6 ***Introducing missing terms.*** Divide $8x^3 + 1$ by $2x + 1$.

Solution

When we write the terms in the dividend in descending powers of x, we see that the terms involving x^2 and x are missing. We can introduce the terms $0x^2$ and $0x$ in the dividend or leave spaces for them. Then the division is routine.

Self Check

Divide $27a^3 - 1$ by $3a - 1$.

$$\begin{array}{r}
4x^2 - 2x \ + 1 \\
2x + 1 \overline{)8x^3 + 0x^2 + 0x + 1} \\
\underline{8x^3 + 4x^2} \qquad\qquad \\
-4x^2 + 0x \qquad \\
\underline{-4x^2 - 2x} \qquad \\
2x + 1 \\
\underline{2x + 1} \\
0
\end{array}$$

Thus,

$$\frac{8x^3 + 1}{2x + 1} = 4x^2 - 2x + 1$$

Answer: $9a^2 + 3a + 1$ ■

EXAMPLE 7 ***Arranging the terms in descending powers.*** Divide $-17x^2 + 5x + x^4 + 2$ by $x^2 - 1 + 4x$.

Solution

We write the problem with the divisor and the dividend in descending powers of x. After introducing $0x^3$ for the missing term in the dividend, we proceed as follows:

$$\begin{array}{r}
x^2 - 4x \qquad\qquad\quad \\
x^2 + 4x - 1 \overline{)x^4 + 0x^3 - 17x^2 + 5x + 2} \\
\underline{x^4 + 4x^3 - \quad x^2} \qquad\qquad \\
-4x^3 - 16x^2 + 5x \qquad \\
\underline{-4x^3 - 16x^2 + 4x} \qquad \\
x + 2
\end{array}$$

This division gives a quotient of $x^2 - 4x$ and a remainder of $x + 2$.

$$\frac{-17x^2 + 5x + x^4 + 2}{x^2 - 1 + 4x} = x^2 - 4x + \frac{x + 2}{x^2 + 4x - 1}$$

Self Check

Divide $\dfrac{2a^2 + 3a^3 + a^4 - 7 + a}{a^2 - 2a + 1}$.

Answer:

$a^2 + 5a + 11 + \dfrac{18a - 18}{a^2 - 2a + 1}$ ■

STUDY SET Section 6.7

VOCABULARY *In Exercises 1–4, fill in the blanks to make the statements true.*

1. In the division

$$x - 4 \overline{)x^2 - x - 12} \quad \text{(quotient } x + 3\text{)}$$

$x^2 - x - 12$ is the ____________, $x - 4$ is the ____________, and $x + 3$ is the ____________.

2. For the division shown here, the ____________ is 1.

$$\begin{array}{r}
x + 3 \qquad\quad \\
x + 4 \overline{)x^2 + 7x + 13} \\
\underline{x^2 + 4x} \qquad\quad \\
3x + 13 \\
\underline{3x + 12} \\
1
\end{array}$$

3. The division ____________ is a repeating series of steps used to do a long division.

4. In the polynomial $4x^4 + 2x^3 - x^2 + x + 7$, the powers of x are written in ____________ order.

CONCEPTS *In Exercises 5–6, fill in the blanks to make the statements true.*

5. $\dfrac{a}{b} = \square \cdot a \quad (b \neq 0)$

6. Divisor · ____________ + remainder = dividend

7. Suppose that after dividing $2x^3 + 5x^2 - 11x + 4$ by $2x - 1$, you obtain $x^2 + 3x - 4$. Show how multiplication can be used to check the result.

8. Consider the first step of the division process for

$$2x^2 - 1\overline{)4x^4 + 0x^3 + 0x^2 + 0x - 1}$$

How many times does $2x^2$ divide $4x^4$?

NOTATION *In Exercises 9–10, complete each solution.*

9.

$$\begin{array}{r} 2x \quad + 1 \\ x + 4\overline{)2x^2 + 9x + 4} \\ \underline{\square + 8x} \\ \square + 4 \\ \underline{x + \square} \\ 0 \end{array}$$

10.

$$\begin{array}{r} 2x \quad - 1 \\ 3x + 4\overline{)6x^2 + 5x \quad - 4} \\ \underline{6x^2 + \square} \\ \square - 4 \\ \underline{-3x - \square} \\ 0 \end{array}$$

11. If a polynomial is divided by $3a - 2$ and the quotient is $3a^2 + 5$ with a remainder of 6, how do we write the result?

12. If a polynomial is divided by $3a - 2$ and the quotient is $3a^2 + 5$ with a remainder of -6, how do we write the result?

13. List three ways we can use symbols to write $x^2 - x - 12$ divided by $x - 4$.

14. Tell whether the statement below is true or false. Justify your answer.

$2x^3 - 9 = 2x^3 + 0x^2 + 0x - 9$

PRACTICE *In Exercises 15–26, do each division. Write each answer without using negative exponents.*

15. $\dfrac{4x^2y^3}{8x^5y^2}$

16. $\dfrac{25x^4y^7}{5xy^9}$

17. $-\dfrac{33a^2b^2}{44a^4b^2}$

18. $\dfrac{-63a^4}{81a^6b^3}$

19. $\dfrac{4x + 6}{2}$

20. $\dfrac{11a^3 - 11a^2}{11}$

21. $\dfrac{4x^2 - x^3}{-6x}$

22. $\dfrac{5y^4 + 45y^3}{-15y^2}$

23. $\dfrac{12x^2y^3 + x^3y^2}{6xy}$

24. $\dfrac{54a^3y^2 - 18a^4y^3}{27a^2y^2}$

25. $\dfrac{24x^6y^7 - 12x^5y^{12} + 36xy}{48x^2y^3}$

26. $\dfrac{9x^4y^3 + 18x^2y - 27xy^4}{-9x^3y^3}$

In Exercises 27–54, do each division.

27. $\dfrac{x^2 + 5x + 6}{x + 3}$

28. $\dfrac{x^2 - 5x + 6}{x - 3}$

29. $\dfrac{6x^2 - x - 12}{2x + 3}$

30. $\dfrac{6x^2 - x - 12}{2x - 3}$

31. $\dfrac{3x^3 - 2x^2 + x - 6}{x - 1}$

32. $\dfrac{4a^3 + a^2 - 3a + 7}{a + 1}$

33. $\dfrac{6x^3 - x^2 - 6x - 9}{2x - 3}$

34. $\dfrac{16x^3 + 16x^2 - 9x - 5}{4x + 5}$

35. $(2a + 1 + a^2) \div (a + 1)$

36. $(a - 15 + 6a^2) \div (2a - 3)$

37. $(6y - 4 + 10y^2) \div (5y - 2)$

38. $(-10x + x^2 + 16) \div (x - 2)$

39. $\dfrac{-18x + 12 + 6x^2}{x - 1}$

40. $\dfrac{27x + 23x^2 + 6x^3}{2x + 3}$

41. $\dfrac{13x + 16x^4 + 3x^2 + 3}{4x + 3}$

42. $\dfrac{3x^2 + 9x^3 + 4x + 4}{3x + 2}$

43. $a^3 + 1$ divided by $a - 1$

44. $27a^3 - 8$ divided by $3a - 2$

45. $\dfrac{15a^3 - 29a^2 + 16}{3a - 4}$

46. $\dfrac{4x^3 - 12x^2 + 17x - 12}{2x - 3}$

47. $y - 2\overline{)-24y + 24 + 6y^2}$

48. $3 - a\overline{)21a - a^2 - 54}$

49. $x^2 - 2\overline{)x^6 - x^4 + 2x^2 - 8}$

50. $x^2 + 3\overline{)x^6 + 2x^4 - 6x^2 - 9}$

51. $\dfrac{x^4 + 2x^3 + 4x^2 + 3x + 2}{x^2 + x + 2}$

52. $\dfrac{2x^4 + 3x^3 + 3x^2 - 5x - 3}{2x^2 - x - 1}$

53. $\dfrac{x^3 + 3x + 5x^2 + 6 + x^4}{x^2 + 3}$

54. $\dfrac{x^5 + 3x + 2}{x^3 + 1 + 2x}$

 *In Exercises 55–56, use a calculator to help find each quotient.*

55. $x - 2\overline{)9.8x^2 - 3.2x - 69.3}$

56. $2.5x - 3.7\overline{)-22.25x^2 - 38.9x - 16.65}$

APPLICATIONS

57. ADVERTISING Find the length of one of the longer sides of the billboard shown in Illustration 1 if its area is given by $x^3 - 4x^2 + x + 6$.

ILLUSTRATION 1

58. MASONRY The steel trowel shown in Illustration 2 is in the shape of an isosceles triangle. Find the height if the area is given by $6 + 18t + t^2 + 3t^3$.

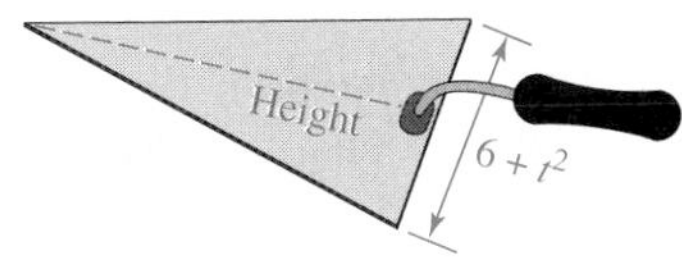

ILLUSTRATION 2

59. WINTER TRAVEL Complete the table in Illustration 3, which lists the rate (mph), time traveled (hr), and distance traveled (mi) by an Alaskan trail guide.

	r ·	t =	d
Dog sled		$4x + 7$	$12x^2 + 13x - 14$
Snowshoes	$3x + 4$		$3x^2 + 19x + 20$

ILLUSTRATION 3

60. PRICING Complete the table in Illustration 4 for two items sold at a produce store.

	Price per lb ·	**Number of lb** =	**Value**
Cashews	$x^2 + 2x + 4$		$x^4 + 4x^2 + 16$
Sunflower seeds		$x^2 + 6$	$x^4 - x^2 - 42$

ILLUSTRATION 4

WRITING *Write a paragraph using your own words.*

61. Explain how to divide a monomial by a monomial.

62. Explain how to check the result of a division problem if there is a nonzero remainder.

REVIEW *In Exercises 63–66, simplify each expression.*

63. $2(x^2 + 4x - 1) + 3(2x^2 - 2x + 2)$

64. $3(2a^2 - 3a + 2) - 4(2a^2 + 4a - 7)$

65. $-2(3y^3 - 2y + 7) - (y^2 + 2y - 4) + 4(y^3 + 2y - 1)$

66. $3(4y^3 + 3y - 2) + 2(3y^2 - y + 3) - 5(2y^3 - y^2 - 2)$

KEY CONCEPT

Expressions and Equations

In this chapter, we have discussed procedures for working with **rational expressions** and procedures for solving **rational equations.**

Rational expressions

The **fundamental property of fractions** is used when simplifying rational expressions and when multiplying and dividing rational expressions: *We can divide out factors that are common to the numerator and the denominator of a fraction.*

1. a. Simplify $\dfrac{6x^2 + x - 2}{8x^2 + 2x - 3}$.

b. What common factor was divided out?

2. a. Multiply $\dfrac{3d^2 - d - 2}{6d^2 - 5d - 6} \cdot \dfrac{4d^2 - 9}{2d^2 + 5d + 3}$.

b. What common factors were divided out?

The fundamental property of fractions also states that *multiplying the numerator and denominator of a fraction by the same nonzero number does not change the value of the fraction.* We use this concept to "build" fractions when adding or subtracting rational expressions with unlike denominators, and when simplifying complex fractions.

3. a. Add $\dfrac{3}{x + 2} + \dfrac{5}{x - 4}$.

b. By what did you multiply the first fraction to rewrite it in terms of the LCD? The second fraction?

4. a. Use Method 2 to simplify $\dfrac{n - 1 - \frac{2}{n}}{\frac{n}{3}}$.

b. By what did you multiply the numerator and denominator to simplify the complex fraction?

Rational equations

The multiplication property of equality states that *if equal quantities are multiplied by the same nonzero number, the results will be equal quantities.* We use this property when solving rational equations. If we multiply both sides of the equation by the LCD of the rational expressions in the equation, we can clear it of fractions.

5. a. Solve $\dfrac{t - 3}{t - 2} - \dfrac{t - 3}{t} = \dfrac{1}{t}$.

b. By what did you multiply both sides to clear the equation of fractions?

6. a. Solve $\dfrac{5}{2x^2 + x - 3} - \dfrac{x + 1}{x - 1} = \dfrac{2}{2x + 3} - 1$.

b. By what did you multiply both sides to clear the equation of fractions?

7. a. Solve $\dfrac{x + 1}{x + 2} = \dfrac{x - 3}{x - 4}$.

b. By what did you multiply both sides to clear the equation of fractions?

8. a. Solve $\dfrac{x^2}{a^2} - \dfrac{y^2}{b^2} = 1$ for a^2.

b. By what did you multiply both sides to clear the equation of fractions?

GROUP ACTIVITIES FOR CHAPTER 6

Accent on Teamwork

Section 6.1

VERTICAL ASYMPTOTES Consider the rational function $f(x) = \frac{1}{x}$. Use your calculator to complete the two tables of values.

Table 1: x approaches 0 from the right

x	3	2	1	0.5	0.25	0.1	0.01	0.001
$f(x)$								

Table 2: x approaches 0 from the left

x	-3	-2	-1	-0.5	-0.25	-0.1	-0.01	-0.001
$f(x)$								

Plot the points from each table on the same graph and draw a smooth curve through them. Explain what happens to the function values $f(x)$ as the values of x approach 0 from the right and approach 0 from the left. Label the vertical asymptote and explain why the function is not defined for $x = 0$.

Section 6.2

COOKING Find a simple recipe for a treat that you can make for your class. Use a proportion to determine the amount of each ingredient needed to make enough of the recipe for the exact number of people in your class. For example, if a recipe makes 2 dozen cookies and there are 30 students in your class, use the ratio $\frac{24}{30}$ in the proportion. Write the old recipe and the new recipe on separate pieces of poster board. Did the recipe serve the correct number of people? Tell the class how you made the calculations and what difficulties you encountered.

Section 6.3

MULTIPLYING RATIONAL EXPRESSIONS

Explain what is wrong with the student's work shown here. Then write a correct solution.

$$\begin{aligned} \frac{3}{4x} + \frac{2}{3x} &= 12x\left(\frac{3}{4x} + \frac{2}{3x}\right) \\ &= 12x \cdot \frac{3}{4x} + 12x \cdot \frac{2}{3x} \\ &= 9 + 8 \\ &= 17 \end{aligned}$$

Section 6.4

ADDING RATIONAL EXPRESSIONS Add the fractions by expressing them in terms of a common denominator $24b^3$. (Note: This is not the LCD.)

$$\frac{r}{4b^2} + \frac{s}{6b}$$

An extra step had to be performed because the lowest common denominator was not used. What was it?

Section 6.5

COMPLEX FRACTIONS Each complex fraction in the list

$$1 + \frac{1}{2},\ 1 + \frac{1}{1 + \frac{1}{2}},\ 1 + \frac{1}{1 + \frac{1}{1 + \frac{1}{2}}},$$

$$1 + \frac{1}{1 + \frac{1}{1 + \frac{1}{1 + \frac{1}{2}}}}, \ldots$$

can be simplified by using the value of the expression preceding it. For example, to simplify the second expression in the list, replace $1 + \frac{1}{2}$ with $\frac{3}{2}$. Show that the expressions in the list simplify to the fractions $\frac{3}{2}, \frac{5}{3}, \frac{8}{5}, \frac{13}{8}, \frac{21}{13}, \frac{34}{21}, \ldots$. Do you see a pattern? Can you predict the next fraction?

Section 6.6

SHARED WORK Use a watch to time (in seconds) how long it takes to fill a pail using a garden hose. From a different faucet, with the water running at a different rate, determine how long it takes a second hose to fill the same pail. Use the concepts studied in Section 6.6 to determine how long it would take to fill the pail with both hoses. Then fill the pail using both hoses, keeping track of the time. How close is the actual time to the predicted time? What are some reasons for any difference between them? Make a video of this project and show it to your class.

Section 6.7

FACTORS Since 6 is a factor of 24, 6 divides 24 exactly with no remainder. Use this same reasoning to determine whether

a. $2x - 3$ is a factor of $10x^2 - x - 21$.
b. $x - 1$ is a factor of $x^5 - 1$.
c. $2x + y$ is a factor of $32x^5 + y^5$.

CHAPTER REVIEW

SECTION 6.1 Rational Functions and Simplifying Rational Expressions

CONCEPTS

Rational expressions are fractions that indicate the quotient of two polynomials.

A *rational function* is defined by a rational expression in one variable where the polynomial in the denominator is never zero.

To *simplify a rational expression:*
1. Completely factor the numerator and the denominator.
2. Divide out the common factors of the numerator and denominator by applying the *fundamental property of fractions:*

$$\frac{ak}{bk} = \frac{a}{b} \quad (b \neq 0 \text{ and } k \neq 0)$$

The quotient of any nonzero quantity and its opposite is -1.

REVIEW EXERCISES

1. Complete the table of values for the rational function $f(x) = \frac{4}{x}$ where $x > 0$. Then graph it. Label the horizontal asymptote.

x	$f(x)$
$\frac{1}{2}$	
1	
2	
3	
4	
5	
6	
7	
8	

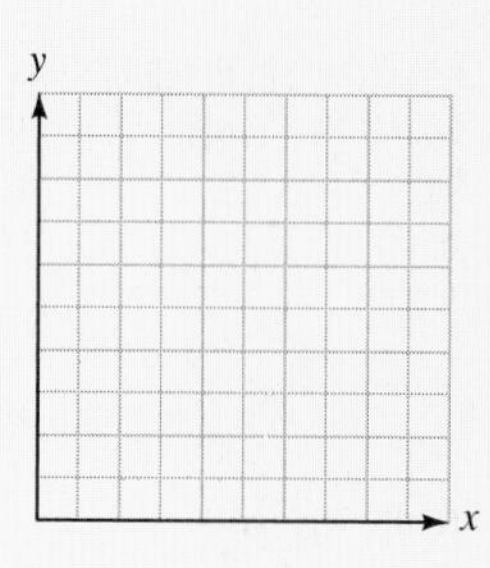

2. Use a graphing calculator to graph the rational function $f(x) = \frac{3x + 2}{x}$. From the graph, determine the equations of the horizontal and vertical asymptotes and the domain and range.

3. Simplify each rational expression.

a. $\frac{248x^2y}{576xy^2}$ **b.** $\frac{x^2 - 49}{x^2 + 14x + 49}$

c. $\frac{x^2 - 2x + 4}{2x^3 + 16}$ **d.** $\frac{x^2 + 6x + 36}{x^3 - 216}$

e. $\frac{ac - ad + bc - bd}{d^2 - c^2}$ **f.** $\frac{m^3 + m^2n - 2mn^2}{mn^2 + m^2n - 2m^3}$

g. $\frac{x - y}{y - x}$ **h.** $\frac{2m - 2n}{n - m}$

SECTION 6.2 Proportion and Variation

In a *proportion,* the product of the *extremes* is equal to the product of the *means.*

4. Solve each proportion.

a. $\frac{x + 1}{8} = \frac{4x - 2}{24}$ **b.** $\frac{1}{x + 6} = \frac{x + 10}{12}$

If two angles of one triangle have the same measure as two angles of a second triangle, the triangles are *similar.* The lengths of corresponding sides of similar triangles are proportional.

5. SIMILAR TRIANGLES Find the height of a tree if it casts a 44-foot shadow when a 4-foot shrub casts a $2\frac{1}{2}$-foot shadow.

Direct variation: As one variable gets larger, the other gets larger as described by the equation $y = kx$, where k is the *constant of proportionality.*

6. PROPERTY TAX The property tax in a certain county varies directly as assessed valuation. If a tax of $1,575 is levied on a single-family home assessed at $90,000, determine the property tax on an apartment complex assessed at $312,000.

Inverse variation: As one variable gets larger, the other gets smaller as described by the equation

$$y = \frac{k}{x} \text{ (}k\text{ is a constant)}$$

7. ELECTRICITY For a fixed voltage, the current in an electrical circuit varies inversely as the resistance in the circuit. If a certain circuit has a current of $2\frac{1}{2}$ amps when the resistance is 150 ohms, find the current in the circuit when the resistance is doubled.

Joint variation: One variable varies with the product of several variables. For example, $y = kxz$ (k is a constant).

8. HURRICANE WINDS The wind force on a vertical surface varies jointly as the area of the surface and the square of the wind's velocity. If a 10-mph wind exerts a force of 1.98 pounds on the sign shown in Illustration 1, find the force on the sign if the wind is blowing at 80 mph.

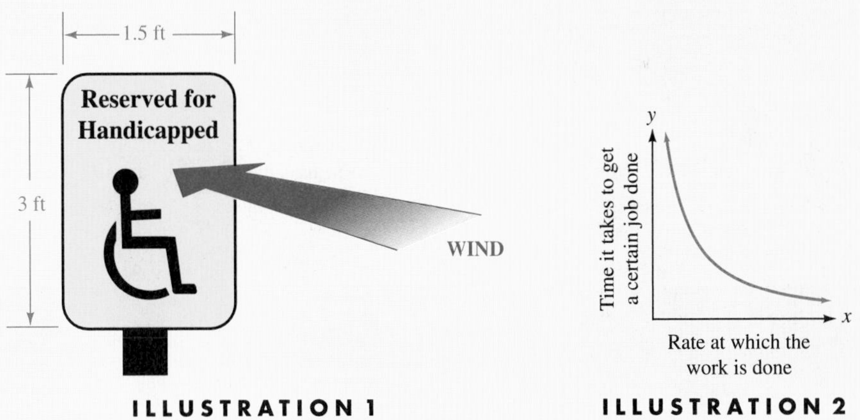

ILLUSTRATION 1

ILLUSTRATION 2

9. Does the graph in Illustration 2 show direct or inverse variation?

Combined variation: a combination of direct and inverse variation. For example,

$$y = \frac{kx}{z} \text{ (}k\text{ is a constant)}$$

10. Assume that x_1 varies directly with the third power of t and inversely with x_2. Find the constant of variation if $x_1 = 1.6$ when $t = 8$ and $x_2 = 64$.

SECTION 6.3 Multiplying and Dividing Rational Expressions

To multiply two fractions, multiply the numerators and multiply the denominators:

$$\frac{a}{b} \cdot \frac{c}{d} = \frac{ac}{bd} \quad (b, d \neq 0)$$

To divide two fractions, invert the divisor and multiply:

$$\frac{a}{b} \div \frac{c}{d} = \frac{a}{b} \cdot \frac{d}{c} \quad (b, d, c \neq 0)$$

(Multiply the first fraction by the *reciprocal* of the second.)

11. Do the operations and simplify:

a. $\dfrac{x^2 + 4x + 4}{x^2 - x - 6} \cdot \dfrac{9 - x^2}{x^2 + 5x + 6}$

b. $\dfrac{2a^2 - 5a - 3}{a^2 - 9} \div \dfrac{2a^2 + 5a + 2}{2a^2 + 5a - 3}$

c. $\left(\dfrac{h - 2}{h^3 + 4}\right)^2$

d. $\dfrac{t^2 - 4}{t} \div (t + 2)$

e. $\dfrac{x^2 + 3x + 2}{x^2 - x - 6} \cdot \dfrac{3x^2 - 3x}{x^2 - 3x - 4} \div \dfrac{x^2 + 3x + 2}{x^2 - 2x - 8}$

SECTION 6.4 Adding and Subtracting Rational Expressions

To add (or subtract) two rational expressions with *like denominators,* add (or subtract) the numerators and keep the common denominator.

12. Do the operations and simplify.

a. $\dfrac{5y}{x - y} - \dfrac{3}{x - y}$

b. $\dfrac{3x - 1}{x^2 + 2} + \dfrac{3(x - 2)}{x^2 + 2}$

c. $\dfrac{4}{t - 3} + \dfrac{6}{3 - t}$

d. $\dfrac{p + 3}{p^2 + 13p + 12} - \dfrac{2p + 4}{p^2 + 13p + 12}$

To find the LCD of several rational expressions, factor each denominator and use each factor the greatest number of times that it appears in any one denominator. The product of these factors is the LCD.

13. The denominators of some rational expressions are given. Find the LCD.

a. $15a^2h$, $20ah^3$

b. $ab^2 - ab$, ab^2, $b^2 - b$

c. $x^2 - 4x - 5$, $x^2 - 25$

d. $m^2 - 4m + 4$, $m^3 - 8$

To add or subtract rational expressions with *unlike denominators,* find the LCD and express each rational expression with a denominator that is the LCD. Add (or subtract) the resulting fractions and simplify the result if possible.

14. Do the operations and simplify.

a. $9 - \dfrac{1}{a + 1}$

b. $\dfrac{x}{4z^2} + \dfrac{y}{6z}$

c. $\dfrac{4x}{x - 4} - \dfrac{3}{x + 3}$

d. $\dfrac{4}{3xy - 6y} - \dfrac{4}{10 - 5x}$

e. $\dfrac{-2(3 + x)}{x^2 + 6x + 9} + \dfrac{3(x + 2)}{x^2 - 6x + 9} - \dfrac{1}{x^2 - 9}$

SECTION 6.5 Complex Fractions

To *simplify a complex fraction:*

Method 1: Write the numerator and denominator as single fractions. Then divide the fractions and simplify.

Method 2: Multiply the numerator and denominator by the LCD of the fractions in the numerator and denominator. Then simplify the results.

15. Simplify each complex fraction.

a. $\dfrac{\dfrac{p^2 - 9}{6pt}}{\dfrac{p^2 + 5p + 6}{3pt}}$

b. $\dfrac{\dfrac{1}{a} + \dfrac{2}{b}}{\dfrac{2}{a} - \dfrac{1}{b}}$

c. $\dfrac{1 - \dfrac{1}{x} - \dfrac{2}{x^2}}{1 + \dfrac{4}{x} + \dfrac{3}{x^2}}$

d. $\dfrac{(x - y)^{-2}}{x^{-2} - y^{-2}}$

SECTION 6.6 Equations Containing Rational Expressions

To solve a *rational equation,* multiply both sides by the LCD of the rational expressions in the equation to clear it of fractions.

Multiplying both sides of an equation by a quantity that contains a variable can lead to *extraneous* (false) solutions. All possible solutions of a rational equation must be checked.

16. Solve each equation, if possible.

a. $\dfrac{4}{x} - \dfrac{1}{10} = \dfrac{7}{2x}$

b. $\dfrac{5}{7 + t} - 1 = \dfrac{-4}{t + 7}$

c. $\dfrac{q - 3}{6} = \dfrac{10}{q + 4}$

d. $\dfrac{2}{x + 5} - \dfrac{1}{6} = \dfrac{1}{x + 4}$

e. $\dfrac{2(x - 5)}{x - 2} = \dfrac{6x + 12}{4 - x^2}$

f. $\dfrac{x + 3}{x - 5} + \dfrac{6 + 2x^2}{x^2 - 7x + 10} = \dfrac{3x}{x - 2}$

17. Solve each formula for the indicated variable.

a. $\dfrac{x^2}{a^2} - \dfrac{y^2}{b^2} = 1$ for y^2

b. $H = \dfrac{2ab}{a + b}$ for b

18. ADVERTISING See Illustration 3. A small plane pulling a banner can fly at a rate of 75 mph in calm air. Flying down the coast, with a tailwind, the plane flew 40 miles in the same time that it took to fly 35 miles up the coast, into a headwind. Find the rate of the wind.

STOP SUNBURN PAIN
WITH SOLARCANE
WIND

ILLUSTRATION 3

19. TRIP LENGTH Traffic reduced a driver's usual speed by 10 mph, which lengthened her 200-mile trip by 1 hour. Find the driver's usual speed.

20. DRAINING A TANK If one outlet pipe can drain a tank in 24 hours and another pipe can drain the tank in 36 hours, how long will it take for both pipes to drain the tank?

21. INSTALLING SIDING Two men have estimated that they can side a house in 8 days. If one of them, who could have sided the house alone in 14 days, gets sick, how long will it take the other man to side the house alone?

22. METALLURGY The stiffness of the flagpole shown in Illustration 4 is given by the formula

$$k = \frac{1}{\frac{1}{k_1} + \frac{1}{k_2}}$$

where k_1 and k_2 are the individual stiffnesses of each section. If the design specifications require that the stiffness k of the entire pole be 1,900,000 in. lb/rad, what must the stiffness of Section 1 be?

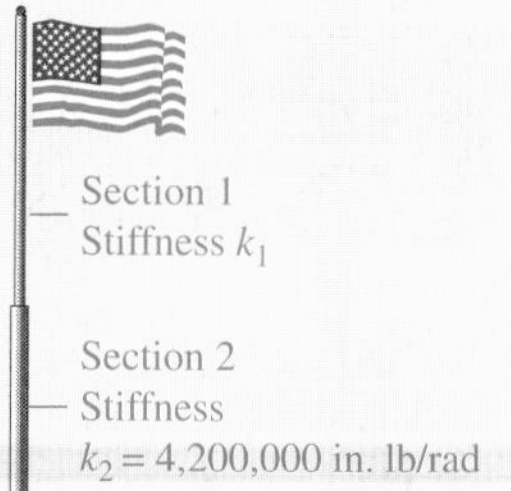

ILLUSTRATION 4

SECTION 6.7 *Dividing Polynomials*

To divide two *monomials:* Factor the numerator and denominator. Then divide out common factors. Or divide the coefficients and apply the rules for exponents.

23. Do each division. Write each answer without negative exponents.

a. $\frac{25h^4k^7}{5hk^9}$ **b.** $(-5x^6y^3) \div (10x^3y^6)$

To divide a *polynomial by a monomial,* multiply the polynomial by the reciprocal of the monomial. Remove parentheses and simplify each resulting rational expression.

24. Find each quotient.

a. $\frac{36a + 32}{6}$ **b.** $\frac{30x^3y^2 - 15x^2y - 10xy^2}{-10xy}$

The *long division* algorithm can be used to divide a *polynomial by a polynomial.* Write the powers of the variable in descending order. If a power of a variable is missing, insert a "placeholder" term with a coefficient of 0.

25. Find each quotient using long division.

a. $b + 5\overline{)b^2 + 9b + 20}$ **b.** $\frac{-33v - 8v^2 + 3v^3 - 10}{1 + 3v}$

c. $x + 2\overline{)x^3 + 8}$ **d.** $(2x^3 + 7x^2 + 3 + 4x) \div (2x + 3)$

Chapter 6 Test

In Problems 1–4, simplify each rational expression.

1. $\dfrac{-12x^2y^3z^2}{18x^3y^4z^2}$

2. $\dfrac{2x + 4}{x^2 - 4}$

3. $\dfrac{3y - 6z}{2z - y}$

4. $\dfrac{2x^2 + 7x + 3}{4x + 12}$

5. HOME RUNS In 1998, Mark McGwire of the St. Louis Cardinals baseball team hit 31 home runs in the first 68 games of the season. If he continued at the same pace, how many home runs would he have hit at the end of a regular season of 162 games? (Round to the nearest home run.)

6. SOUND Sound intensity (loudness) varies inversely as the square of the distance from the source. If a rock band has a sound intensity of 100 decibels 30 feet away from the amplifier, find the sound intensity 60 feet away from the amplifier.

7. Graph the rational function $f(x) = \frac{2}{x}$ for $x > 0$. Label the horizontal asymptote.

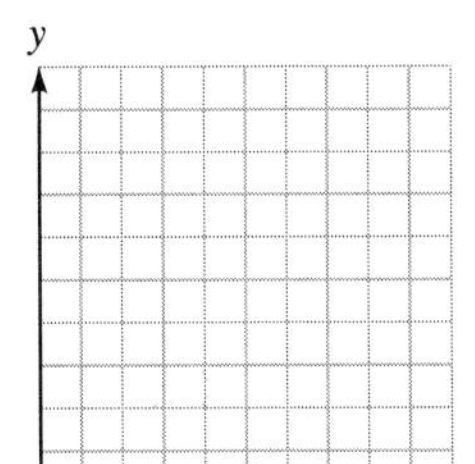

8. Draw a possible graph showing that the weekly salary of a person *varies directly* with the number of hours worked during the week. Label the axes.

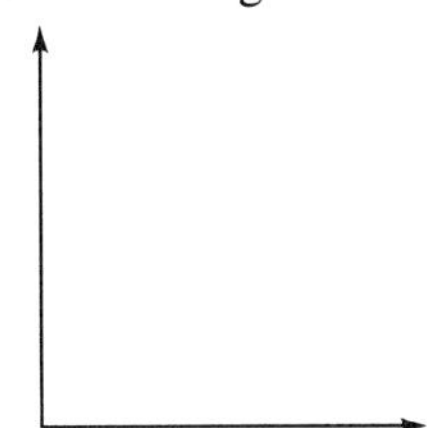

In Problems 9–18, do the operations and simplify, if necessary. Write all answers without negative exponents.

9. $\dfrac{x^2}{x^3z^2y^2} \cdot \dfrac{x^2z^4}{y^2z}$

10. $\dfrac{(x + 1)(x + 2)}{10} \cdot \dfrac{5}{x + 2}$

11. $\dfrac{u^2 + 5u + 6}{u^2 - 4} \cdot \dfrac{u^2 - 5u + 6}{u^2 - 9}$

12. $\dfrac{x^3 + y^3}{4} \div \dfrac{x^2 - xy + y^2}{2x + 2y}$

13. $\dfrac{xu + 2u + 3x + 6}{u^2 - 9} \cdot \dfrac{2u - 6}{x^2 + 3x + 2}$

14. $\dfrac{a^2 + 7a + 12}{a + 3} \div \dfrac{16 - a^2}{a - 4}$

15. $\dfrac{-3t + 4}{t^2 + t - 20} + \dfrac{6 + 5t}{t^2 + t - 20}$

16. $\dfrac{3w}{w - 5} + \dfrac{w + 10}{5 - w}$

17. $8b - 5 + \dfrac{5b + 4}{3b + 1}$

18. $\dfrac{x + 2}{x + 1} - \dfrac{x + 1}{x + 2}$

In Problems 19–20, simplify each complex fraction.

19. $\dfrac{\dfrac{2u^2w^3}{v^2}}{\dfrac{4uw^4}{uv}}$

20. $\dfrac{\dfrac{4}{3k} + \dfrac{k}{k+1}}{\dfrac{k}{k+1} - \dfrac{3}{k}}$

In Problems 21–24, solve each equation.

21. $\dfrac{34}{x} + \dfrac{13}{20} = \dfrac{3}{2}$

22. $\dfrac{u-2}{u-3} + 3 = u + \dfrac{u-4}{3-u}$

23. $\dfrac{3}{x-2} = \dfrac{x+3}{2x}$

24. $\dfrac{4}{m^2-9} + \dfrac{5}{m^2-m-12} = \dfrac{7}{m^2-7m+12}$

In Problems 25–26, solve each formula for the indicated variable.

25. $\dfrac{x^2}{a^2} + \dfrac{y^2}{b^2} = 1$ for a^2

26. $\dfrac{1}{r} = \dfrac{1}{r_1} + \dfrac{1}{r_2}$ for r_2

27. ROOFING One roofing crew can finish a 2,800-square-foot roof in 12 hours, and another crew can do the job in 10 hours. If they work together, can they finish before a predicted rain in 5 hours? If not, how long will they have to work in the rain?

28. TOURING THE COUNTRYSIDE A man bicycles 5 mph faster than he can walk. He bicycles 24 miles and then hikes back along the same route in 11 hours. How fast does he walk?

29. Divide $\dfrac{18x^2y^3 - 12x^3y^2 + 9xy}{-3xy^4}$.

30. Divide $(y^3 - 48) \div (y + 2)$.

31. Explain why $(x + 2)$ can be divided out in the first expression and why it cannot be divided out in the second expression.

1st expression $\dfrac{(x+2)(x-3)}{(x+2)}$

2nd expression $\dfrac{(x+2)+(x-3)}{(x+2)}$

32. Explain what it means for two quantities to vary inversely. Give an example.

Chapters 1–6 Cumulative Review Exercises

1. Solve $-3 = -\frac{9}{8}t$.

2. Solve $\frac{3x - 4}{6} - \frac{x - 2}{2} = \frac{-2x - 3}{3}$.

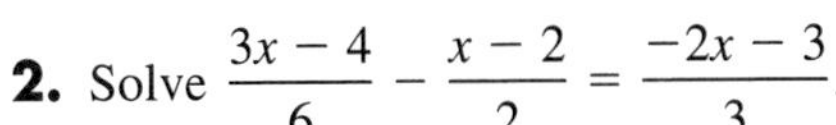

3. AUTO SALES See the graph in Illustration 1. An automobile dealership is going to order 80 new Ford Escorts. According to the survey, exactly how many green Escorts should be purchased to meet the expected customer demand?

4. LIFE EXPECTANCY Determine the predicted rate of change in the life expectancy of females during the years 2000–2050, as shown in the graph in Illustration 2.

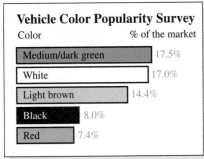

Vehicle Color Popularity Survey

Color	% of the market
Medium/dark green	17.5%
White	17.0%
Light brown	14.4%
Black	8.0%
Red	7.4%

Based on data from DuPont Automotive

ILLUSTRATION 1

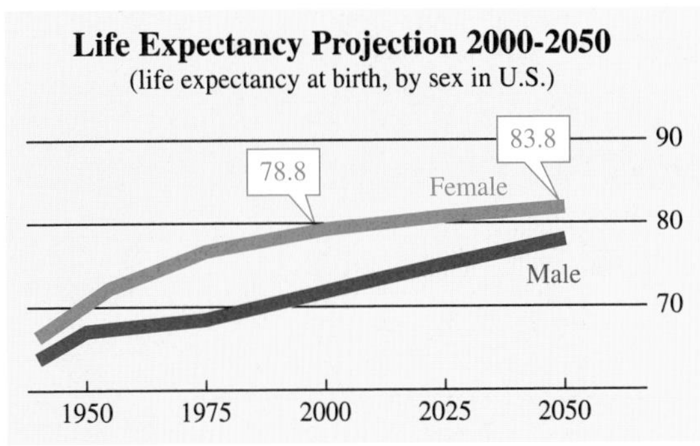

Based on data from the Social Security Administration, Office of Chief Actuary

ILLUSTRATION 2

In Exercises 5–6, write the equation of the line with the given properties. Express your answer in slope–intercept form.

5. Slope of -7, passing through $P(7, 5)$

6. Passing through $P(-4, 5)$ and $Q(2, -6)$

7. Draw the graph of the linear function $f(x) = \frac{2}{3}x - 2$. Then use interval notation to specify the domain and range.

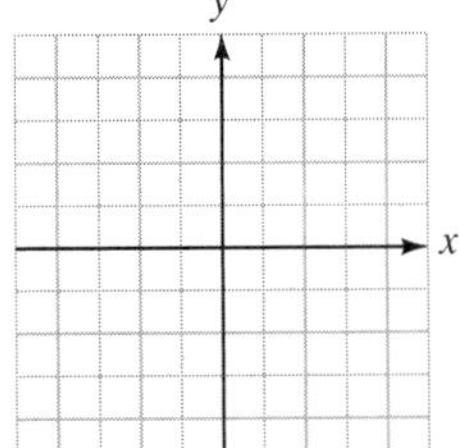

8. MECHANICAL ENGINEERING The tensions T_1 and T_2 (in pounds) in each of the ropes shown in Illustration 3 can be found by solving the system

$$\begin{cases} 0.6T_1 - 0.8T_2 = 0 \\ 0.8T_1 + 0.6T_2 = 100 \end{cases}$$

Find T_1 and T_2.

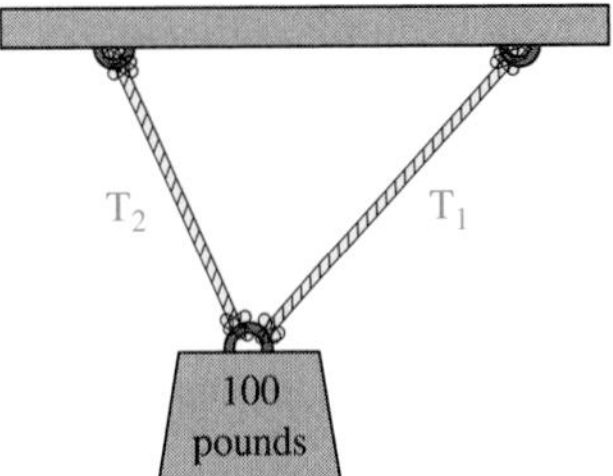

ILLUSTRATION 3

In Exercises 9–12, solve each inequality and show the solution set as a graph on the number line.

9. $-2x < -5$

10. $\left|\frac{3a}{5} - 2\right| + 1 \geq \frac{6}{5}$

11. $5(x + 2) \le 4(x + 1)$ and $11 + x < 0$

12. $x + 1 < -4$ or $x - 4 > 0$

In Exercises 13–16, simplify each expression.

13. $a^3b^2a^5b^2$

14. $\frac{a^3b^6}{a^7b^2}$

15. $\frac{1}{3^{-4}}$

16. $\left(\frac{2x^{-2}y^3}{x^2x^3y^4}\right)^{-3}$

In Exercises 17–18, write each number in standard notation.

17. 4.25×10^4

18. 7.12×10^{-4}

19. Express as a formula: *y varies directly with the product of x and z, and inversely with r.*

20. Evaluate the determinant $\begin{vmatrix} 3 & -2 \\ -2 & 4 \end{vmatrix}$.

21. Graph $y < 4 - x$.

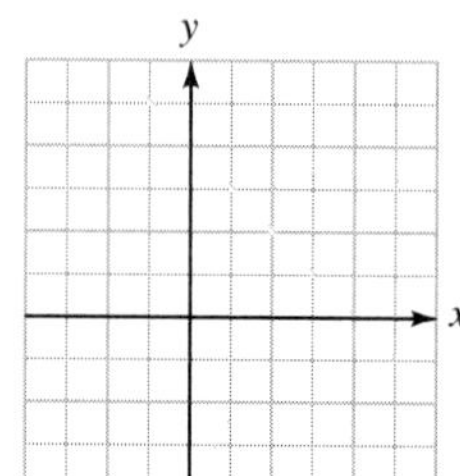

22. Graph $f(x) = 2x^2 - 3$. Then use interval notation to specify the domain and range.

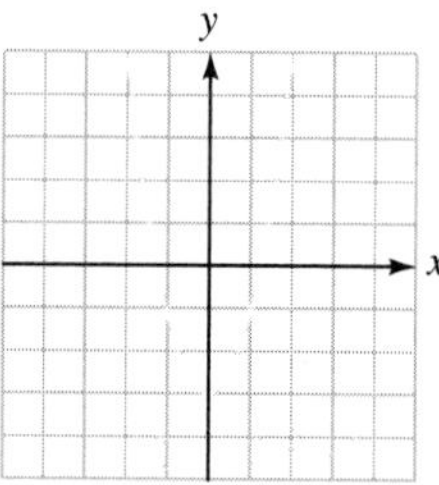

23. If $g(x) = -3x^3 + x - 4$, find $g(-2)$.

24. Find the degree of $3 + x^2y + 17x^3y^4$.

In Exercises 25–28, do the operations and simplify.

25. $(x^3 + 3x^2 - 2x + 7) + (x^3 - 2x^2 + 2x + 5)$

26. $(-5x^2 + 3x + 4) - (-2x^2 + 3x + 7)$

27. $(3x + 4)(2x - 5)$

28. $(2x^3 - 1)^2$

In Exercises 29–32, refer to Illustration 4. The graph shows the correction that must be made to a sundial reading to obtain accurate clock time. The difference is caused by the earth's orbit and tilted axis.

29. Is this the graph of a function?

30. During the year, what is the maximum number of minutes the sundial reading gets ahead of a clock?

31. During the year, what is the maximum number of minutes the sundial reading falls behind a clock?

32. How many times during a year is the sundial reading exactly the same as a clock?

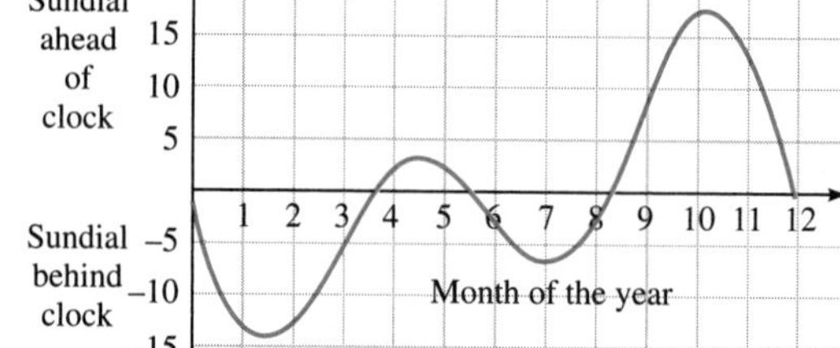

ILLUSTRATION 4

In Exercises 33–40, factor each expression.

33. $3r^2s^3 - 6rs^4$

34. $5(x - y) - a(x - y)$

35. $xu + yv + xv + yu$

36. $81x^4 - 16y^4$

37. $8x^3 - 27y^6$

38. $6x^2 + 5x - 6$

39. $9x^2 - 30x + 25$

40. $15x^2 - x - 6$

41. Solve $6x^2 + 7 = -23x$.

42. Solve $x^3 - 4x = 0$.

43. Solve $b^2x^2 + a^2y^2 = a^2b^2$ for b^2.

44. CAMPING The rectangular-shaped cooking surface of a small camping stove is 108 in.^{2}. If its length is 3 inches longer than its width, what are its dimensions?

In Exercises 45–48, simplify each expression.

45. $\dfrac{2x^2y + xy - 6y}{3x^2y + 5xy - 2y}$

46. $\dfrac{p^3 - q^3}{q^2 - p^2} \cdot \dfrac{q^2 + pq}{p^3 + p^2q + pq^2}$

47. $\dfrac{2}{x + y} + \dfrac{3}{x - y} - \dfrac{x - 3y}{x^2 - y^2}$

48. $\dfrac{\frac{a}{b} + b}{a - \frac{b}{a}}$

49. Solve $\dfrac{5x - 3}{x + 2} = \dfrac{5x + 3}{x - 2}$.

50. Solve $\dfrac{3}{x - 2} + \dfrac{x^2}{(x + 3)(x - 2)} = \dfrac{x + 4}{x + 3}$.

In Exercises 51–52, do the division.

51. $(x^2 + 9x + 20) \div (x + 5)$

52. $(2x^2 + 4x - x^3 + 3) \div (x - 1)$

Radicals and Rational Exponents

7

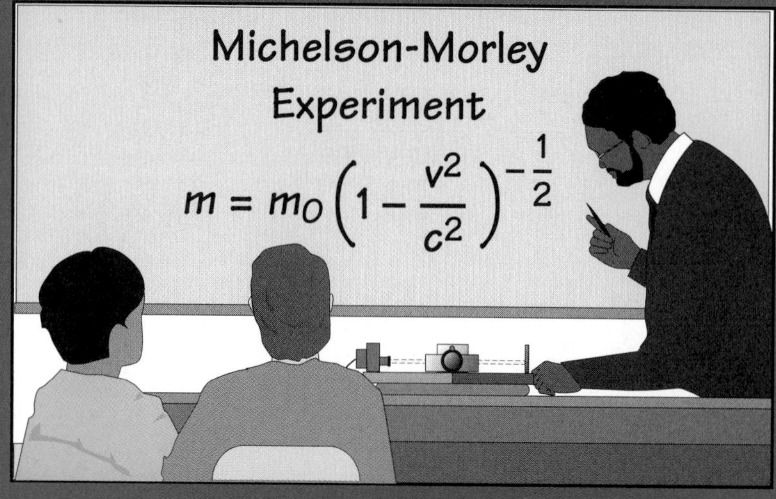

Campus Connection

The *Physics* Department

In a physics class, students study relativity theory. This theory, primarily developed by Albert Einstein in the early 20th century, explains the relationship between matter and energy as well as the properties of light. One equation that mathematically illustrates a major conclusion of relativity theory contains a *rational exponent.* In this chapter, we discuss the meaning of rational (fractional) exponents and consider the relationship between rational exponents and radicals. Rational exponents and radicals occur in formulas and functions used in such diverse disciplines as carpentry, law enforcement, business, theater production, and physics.

RADICAL EXPRESSIONS HAVE THE FORM $\sqrt[n]{a}$. IN THIS CHAPTER, WE WILL SEE HOW THEY ARE USED TO MODEL MANY REAL-WORLD SITUATIONS.

7.1 Radical Expressions and Radical Functions

In this section, you will learn about

- **Square roots**
- **Square roots of expressions containing variables**
- **The square root function**
- **Cube roots**
- ***n*th roots**

INTRODUCTION. In this section, we will reverse the squaring process and learn how to find *square roots* of numbers. Then we will generalize the concept of root and consider cube roots, fourth roots, fifth roots, and so on. We will also discuss a new family of functions, called *radical functions,* and we will see how they have application in many disciplines.

Square roots

When solving problems, we must often find what number must be squared to obtain a second number a. If such a number can be found, it is called a **square root of *a*.** For example,

- 0 is a square root of 0, because $0^2 = 0$.
- 4 is a square root of 16, because $4^2 = 16$.
- -4 is a square root of 16, because $(-4)^2 = 16$.
- $7xy$ is a square root of $49x^2y^2$, because $(7xy)^2 = 49x^2y^2$.
- $-7xy$ is a square root of $49x^2y^2$, because $(-7xy)^2 = 49x^2y^2$.

The preceding examples illustrate the following definition.

Square root of *a*

The number b is a **square root of *a*** if $b^2 = a$.

All positive numbers have two real number square roots, one that is positive and one that is negative.

EXAMPLE 1 ***Square roots.*** Find the two square roots of 121.

Solution

The two square roots of 121 are 11 and −11, because

$11^2 = 121 \quad \text{and} \quad (-11)^2 = 121$

Self Check

Find the square roots of 144.

Answer: 12, −12

In the following definition, the symbol $\sqrt{}$ is called a **radical sign,** and the number x within the radical sign is called a **radicand.**

Principal square root

If $x > 0$, the **principal square root of x** is the positive square root of x, denoted as $\sqrt{x}$.

The principal square root of 0 is 0: $\sqrt{0} = 0$.

By definition, the principal square root of a positive number is always positive. Although 5 and −5 are both square roots of 25, only 5 is the principal square root. The radical expression $\sqrt{25}$ represents 5. The radical expression $-\sqrt{25}$ represents −5. When we write $\sqrt{25} = 5$ or $-\sqrt{25} = -5$, we say that we have *simplified the radical.*

EXAMPLE 2 ***Finding square roots.*** Simplify each radical.

a. $\sqrt{1} = 1$ **b.** $\sqrt{81} = 9$

c. $-\sqrt{81} = -9$ **d.** $-\sqrt{225} = -15$

e. $\sqrt{\frac{1}{4}} = \frac{1}{2}$ **f.** $-\sqrt{\frac{16}{121}} = -\frac{4}{11}$

g. $\sqrt{0.04} = 0.2$ **h.** $-\sqrt{0.0009} = -0.03$

Self Check

Simplify **a.** $-\sqrt{49}$ and **b.** $\sqrt{\frac{25}{49}}$.

Answers: **a.** −7, **b.** $\frac{5}{7}$

Numbers such as 4, 9, 16, 49, and 1,600 are called **integer squares,** because each one is the square of an integer. The square root of every integer square is a rational number.

$\sqrt{4} = 2, \quad \sqrt{9} = 3, \quad \sqrt{16} = 4, \quad \sqrt{49} = 7, \quad \sqrt{1{,}600} = 40$

The square roots of many positive integers are not rational numbers. For example, $\sqrt{11}$ is an *irrational number.* To find an approximate value of $\sqrt{11}$, we enter 11 into a scientific calculator and press the $\boxed{\sqrt{}}$ key.

$\sqrt{11} \approx 3.31662479$

Square roots of negative numbers are not real numbers. For example, $\sqrt{-9}$ is not a real number, because no real number squared equals −9. Square roots of negative numbers come from a set called the **imaginary numbers,** which we will discuss in the next chapter.

Square roots of expressions containing variables

If $x \neq 0$, the positive number x^2 has x and $-x$ for its two square roots. To denote the positive square root of $\sqrt{x^2}$, we must know whether x is positive or negative.

If $x > 0$, we can write

$\sqrt{x^2} = x$ $\sqrt{x^2}$ represents the positive square root of x^2, which is x.

If x is negative, then $-x > 0$, and we can write

$\sqrt{x^2} = -x$ $\quad$ $\sqrt{x^2}$ represents the positive square root of x^2, which is $-x$.

If we don't know whether x is positive or negative, we can use absolute value symbols to guarantee that $\sqrt{x^2}$ is positive.

Definition of $\sqrt{x^2}$

If x can be any real number, then

$$\sqrt{x^2} = |x|$$

EXAMPLE 3 ***Simplifying radical expressions containing variables.*** Simplify **a.** $\sqrt{16x^2}$, **b.** $\sqrt{x^2 + 2x + 1}$, and **c.** $\sqrt{m^4}$.

Solution

If x can be any real number, we have

a. $\sqrt{16x^2} = \sqrt{(4x)^2}$ $\quad$ Write $16x^2$ as $(4x)^2$.

$= |4x|$ $\quad$ Because $(|4x|)^2 = 16x^2$. Since x could be negative, absolute value symbols are needed.

$= 4|x|$ $\quad$ Since 4 is a positive constant in the product $4x$, we can write it outside the absolute value symbols.

b. $\sqrt{x^2 + 2x + 1}$

$= \sqrt{(x + 1)^2}$ $\quad$ Factor $x^2 + 2x + 1$.

$= |x + 1|$ $\quad$ Because $(x + 1)^2 = x^2 + 2x + 1$. Since $x + 1$ can be negative (for example, when $x = -5$), absolute value symbols are needed.

c. $\sqrt{m^4} = m^2$ $\quad$ Because $(m^2)^2 = m^4$. Since $m^2 \geq 0$, no absolute value symbols are needed.

Self Check

Simplify **a.** $\sqrt{25a^2}$ and **b.** $\sqrt{16a^4}$.

Answers: **a.** $5|a|$, **b.** $4a^2$

If we are told that x represents a positive real number in parts a and b of Example 3, then we do not need to use absolute value symbols to guarantee that the answers are positive.

$\sqrt{16x^2} = 4x$ $\quad$ If x is positive, $4x$ is positive.

$\sqrt{x^2 + 2x + 1} = x + 1$ $\quad$ If x is positive, $x + 1$ is positive.

The square root function

Since there is one principal square root for every nonnegative real number x, the equation $f(x) = \sqrt{x}$ determines a function, called a **square root function.** Square root functions belong to a larger family of functions known as **radical functions.**

EXAMPLE 4 ***Graphing a square root function.*** Graph $f(x) = \sqrt{x}$ and find its domain and range.

Solution

To graph this square root function, we will evaluate it for several values of x. We begin with $x = 0$, since 0 is the smallest input for which $\sqrt{x}$ is defined.

$$f(x) = \sqrt{x}$$
$$f(0) = \sqrt{0} = 0$$

We enter the ordered pair (0, 0) in the table of values in Figure 7-1. Then we continue the evaluating process for $x = 1, 4, 9,$ and 16 and list the results in the table. After plotting all the ordered pairs, we draw a smooth curve through the points. This is the

Self Check

Graph $f(x) = \sqrt{x} + 2$. Then give its domain and range and compare the graph of $f(x) = \sqrt{x}$.

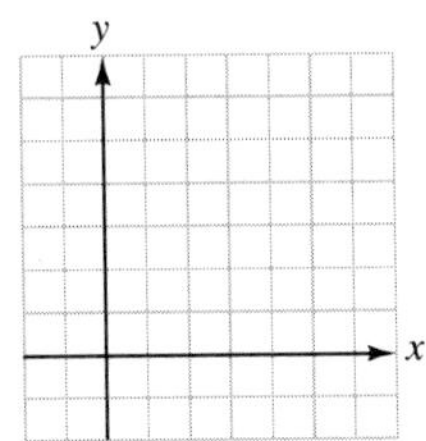

graph of function f (see Figure 7-1(a)). Since the equation defines a function, its graph passes the vertical line test.

We can use a graphing calculator with window settings of $[-1, 9]$ for x and $[-1, 9]$ for y to get the graph shown in Figure 7-1(b). From either graph, we can see that the domain and the range are the set of nonnegative real numbers. Expressed in interval notation, the domain is $[0, \infty)$, and the range is $[0, \infty)$.

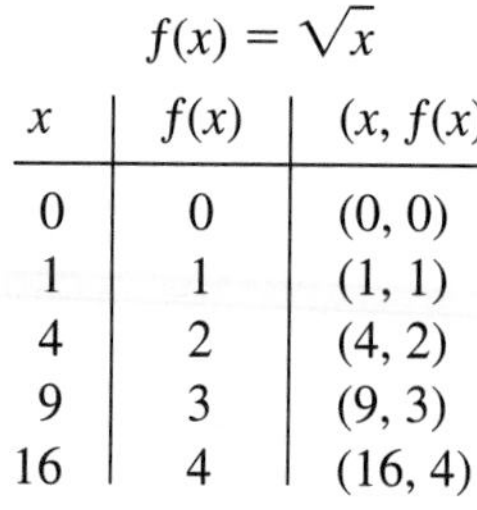

$f(x) = \sqrt{x}$

x	$f(x)$	$(x, f(x))$
0	0	(0, 0)
1	1	(1, 1)
4	2	(4, 2)
9	3	(9, 3)
16	4	(16, 4)

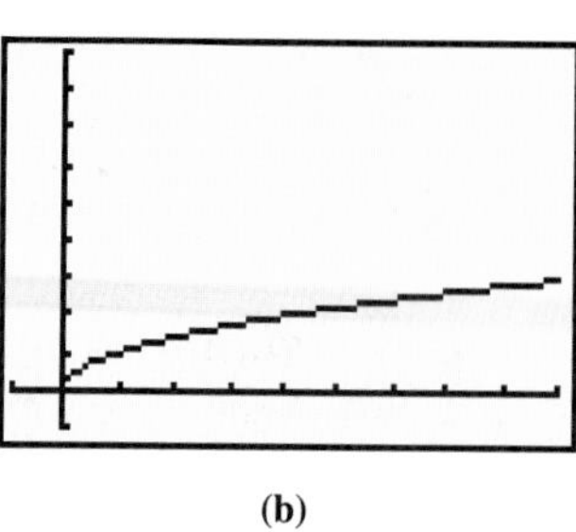

(b)

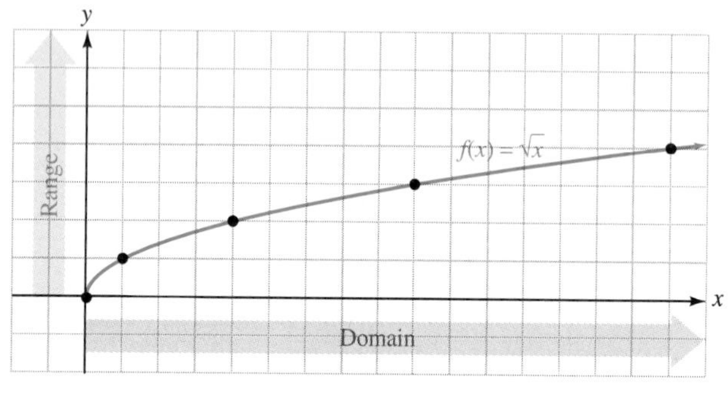

(a)

FIGURE 7-1

Answers:

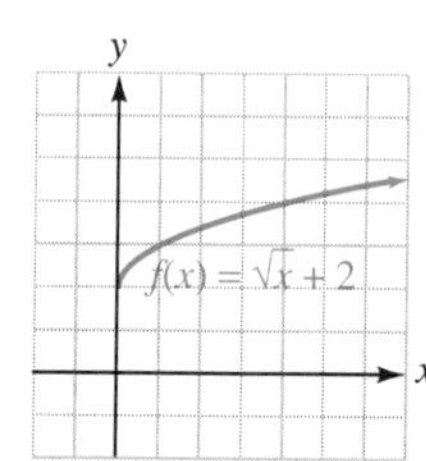

D: $[0, \infty)$, R: $[2, \infty)$; the graph is 2 units higher

The graphs of many radical functions are translations or reflections of the square root function, $f(x) = \sqrt{x}$. For example, if $k > 0$,

- The graph of $f(x) = \sqrt{x} + k$ is the graph of $f(x) = \sqrt{x}$ translated k units up.
- The graph of $f(x) = \sqrt{x} - k$ is the graph of $f(x) = \sqrt{x}$ translated k units down.
- The graph of $f(x) = \sqrt{x + k}$ is the graph of $f(x) = \sqrt{x}$ translated k units to the left.
- The graph of $f(x) = \sqrt{x - k}$ is the graph of $f(x) = \sqrt{x}$ translated k units to the right.
- The graph of $f(x) = -\sqrt{x}$ is the graph of $f(x) = \sqrt{x}$ reflected about the x-axis.

EXAMPLE 5 ***Graphing square root functions.*** Graph $f(x) = -\sqrt{x + 4} - 2$ and find its domain and range.

Solution

This graph will be the reflection of $f(x) = \sqrt{x}$ about the x-axis, translated 4 units to the left and 2 units down. See Figure 7-2(a). We can confirm this graph by using a graphing calculator with window settings of $[-5, 6]$ for x and $[-6, 2]$ for y to get the graph shown in Figure 7-2(b).

From either graph, we can see that the domain is the interval $[-4, \infty)$ and that the range is the interval $(-\infty, -2]$.

Self Check

Graph $f(x) = \sqrt{x - 2} - 4$. Then give the domain and the range.

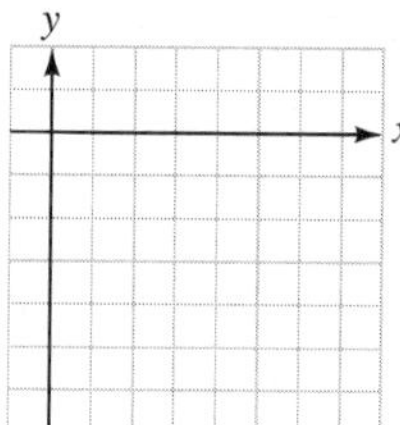

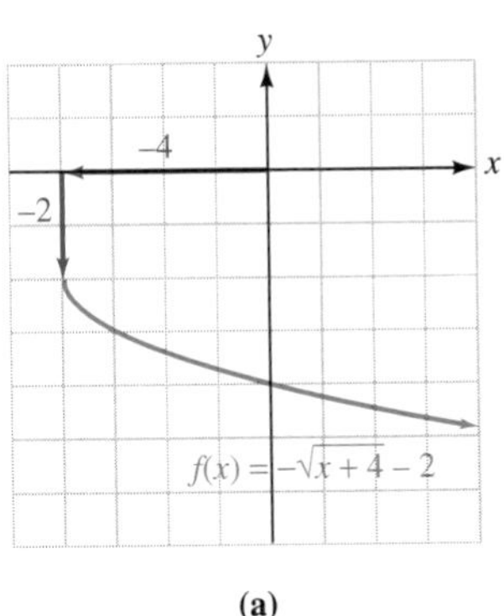

(a)

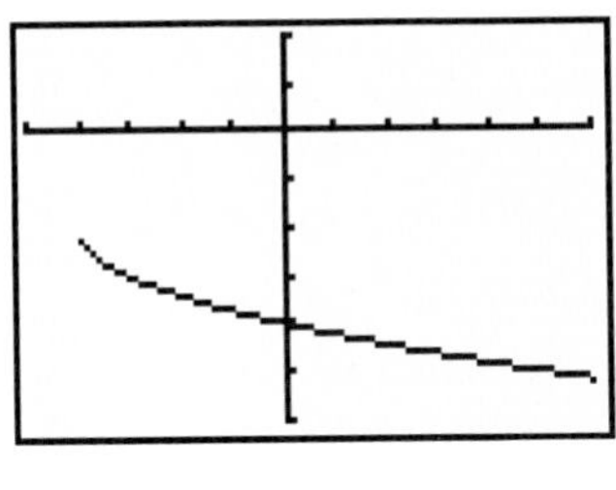
(b)

FIGURE 7-2

Answers:

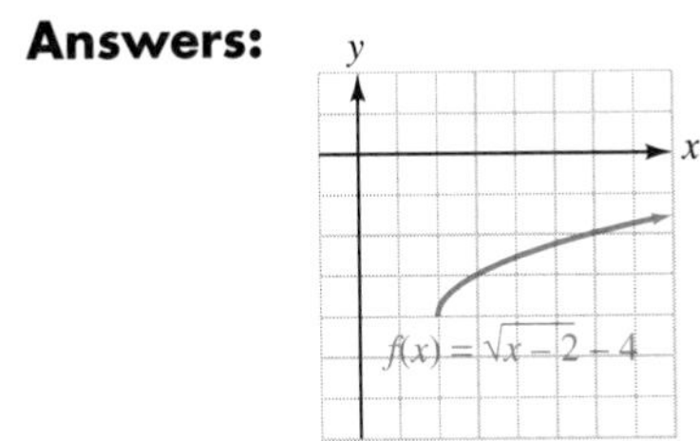

D: $[2, \infty)$, R: $[-4, \infty)$

EXAMPLE 6 ***Square root functions.*** The **period of a pendulum** is the time required for the pendulum to swing back and forth to complete one cycle. The period (in seconds) is a function of the pendulum's length l (in feet) and is given by

$$f(l) = 2\pi\sqrt{\frac{l}{32}}$$

Find the period of the 5-foot-long pendulum of the clock shown in Figure 7-3.

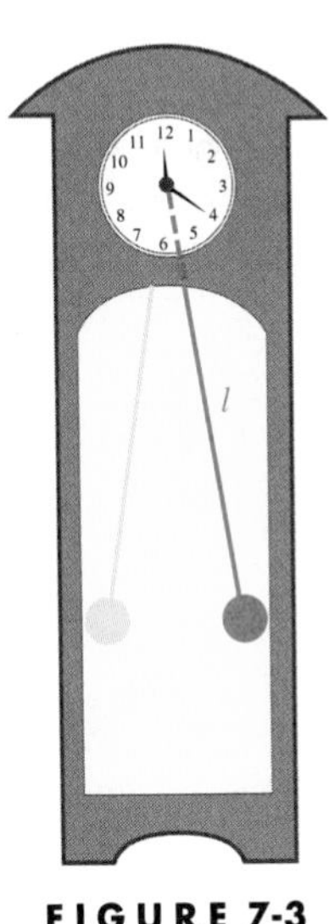

FIGURE 7-3

Solution

To determine the period, we substitute 5 for l.

$$f(l) = 2\pi\sqrt{\frac{l}{32}}$$

$$f(5) = 2\pi\sqrt{\frac{5}{32}}$$

$$\approx 2.483647066$$ Use a calculator to find an approximation.

The period is approximately 2.5 seconds.

Self Check

To the nearest hundredth, find the period of a pendulum that is 3 feet long.

Answer: 1.92 sec

Accent on Technology — Evaluating a square root function using its graph

To solve Example 6 with a graphing calculator with window settings of $[-2, 10]$ for x and $[-2, 10]$ for y, we graph the function $f(x) = 2\pi\sqrt{\frac{x}{32}}$, as in Figure 7-4(a). We then trace and move the cursor toward an x-value of 5 until we see the coordinates shown in Figure 7-4(b). The period is given by the y-value shown on the screen. By zooming in, we can get better results.

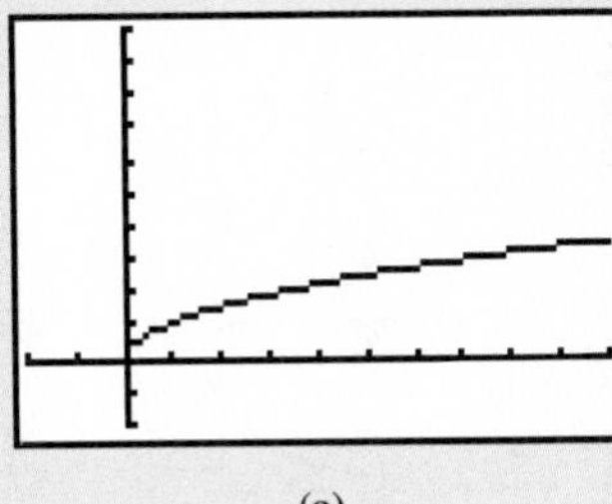
(a)

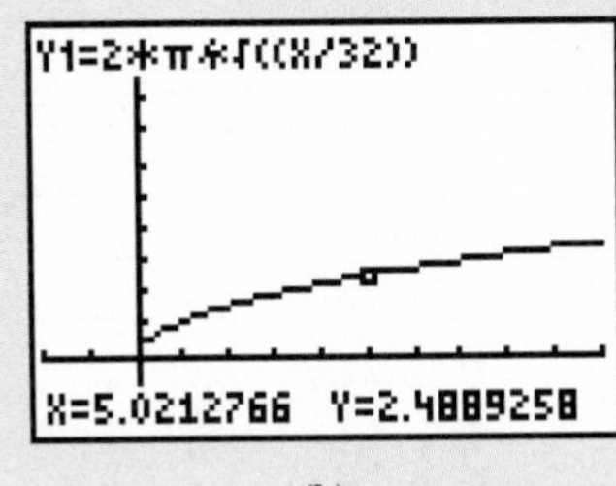

(b)

FIGURE 7-4

Cube roots

The **cube root of x** is any number whose cube is x. For example,

- 4 is a cube root of 64, because $4^3 = 64$.
- $3x^2y$ is a cube root of $27x^6y^3$, because $(3x^2y)^3 = 27x^6y^3$.
- $-2y$ is a cube root of $-8y^3$, because $(-2y)^3 = -8y^3$.

Cube roots

The **cube root of x** is denoted as $\sqrt[3]{x}$ and is defined by

$$\sqrt[3]{x} = y \quad \text{if } y^3 = x$$

We note that 64 has two real-number square roots, 8 and -8. However, 64 has only one real-number cube root 4, because 4 is the only real number whose cube is 64. Since every real number has exactly one real cube root, it is unnecessary to use absolute value symbols when simplifying cube roots.

Definition of $\sqrt[3]{x^3}$

If x is any real number, then

$$\sqrt[3]{x^3} = x$$

EXAMPLE 7 ***Finding cube roots.*** Simplify each radical expression.

a. $\sqrt[3]{125} = 5$ Because $5^3 = 5 \cdot 5 \cdot 5 = 125$.

b. $\sqrt[3]{\dfrac{1}{8}} = \dfrac{1}{2}$ Because $\left(\frac{1}{2}\right)^3 = \frac{1}{2} \cdot \frac{1}{2} \cdot \frac{1}{2} = \frac{1}{8}$.

c. $\sqrt[3]{-27x^3} = -3x$ Because $(-3x)^3 = (-3x)(-3x)(-3x) = -27x^3$.

d. $\sqrt[3]{-\dfrac{8a^3}{27b^3}} = -\dfrac{2a}{3b}$ Because $\left(-\dfrac{2a}{3b}\right)^3 = \left(-\dfrac{2a}{3b}\right)\left(-\dfrac{2a}{3b}\right)\left(-\dfrac{2a}{3b}\right) = -\dfrac{8a^3}{27b^3}$.

e. $\sqrt[3]{0.216x^3y^6} = 0.6xy^2$ Because $(0.6xy^2)^3 = (0.6xy^2)(0.6xy^2)(0.6xy^2) = 0.216x^3y^6$.

Self Check

Simplify **a.** $\sqrt[3]{1{,}000}$, **b.** $\sqrt[3]{\frac{1}{27}}$, and **c.** $\sqrt[3]{125a^3}$.

Answers: **a.** 10, **b.** $\frac{1}{3}$, **c.** $5a$

The equation $f(x) = \sqrt[3]{x}$ defines a **cube root function.** From the graph shown in Figure 7-5(a), we can see that the domain and range of the function $f(x) = \sqrt[3]{x}$ are the set of real numbers. Note that the graph of $f(x) = \sqrt[3]{x}$ passes the vertical line test. Like square root functions, cube root functions are members of the family of radical functions. Figures 7-5(b) and 7-5(c) show several translations of the cube root function.

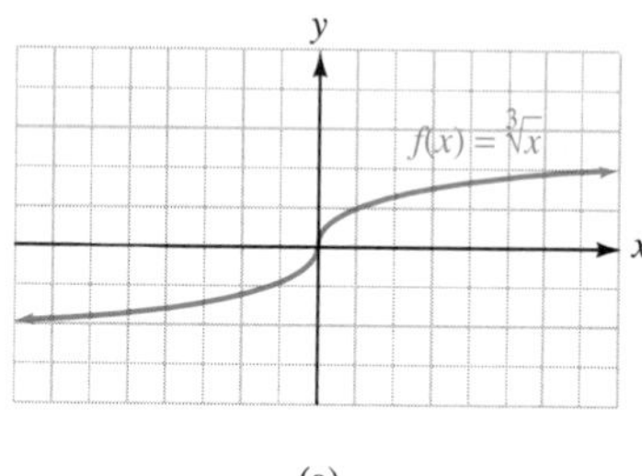

(a)

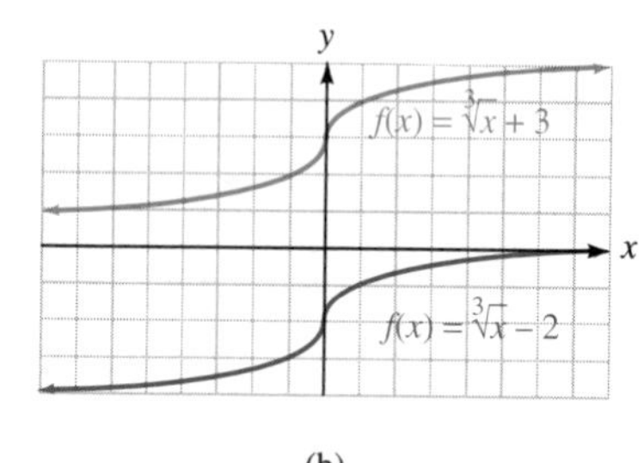

(b)

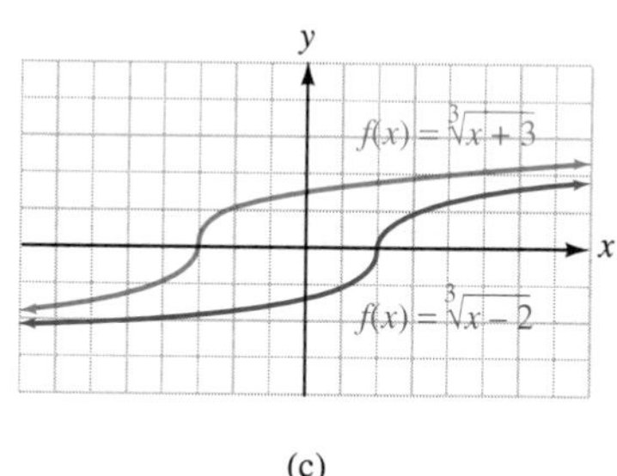

(c)

FIGURE 7-5

nth roots

Just as there are square roots and cube roots, there are fourth roots, fifth roots, sixth roots, and so on.

When n is an odd natural number, the expression $\sqrt[n]{x}$ $(n > 1)$ represents an **odd root.** Since every real number has just one real nth root when n is odd, we don't need to worry about absolute value symbols when finding odd roots. For example,

$\sqrt[5]{243} = \sqrt[5]{3^5} = 3$ Because $3^5 = 243$.

$\sqrt[7]{-128x^7} = \sqrt[7]{(-2x)^7} = -2x$ Because $(-2x)^7 = -128x^7$.

When n is an even natural number, the expression $\sqrt[n]{x}$ $(n > 1, x > 0)$ represents an **even root.** In this case, there will be one positive and one negative real nth root. For example, the real sixth roots of 729 are 3 and -3, because $3^6 = 729$ and $(-3)^6 = 729$. When finding even roots, we can use absolute value symbols to guarantee that the nth root is positive.

$\sqrt[4]{(-3)^4} = |-3| = 3$ We could also simplify this as follows: $\sqrt[4]{(-3)^4} = \sqrt[4]{81} = 3$.

$\sqrt[6]{729x^6} = \sqrt[6]{(3x)^6} = |3x| = 3|x|$ The absolute value symbols guarantee that the sixth root is positive.

In general, we have the following rules.

Rules for $\sqrt[n]{x^n}$

If x is a real number and $n > 1$, then
If n is an odd natural number, $\sqrt[n]{x^n} = x$.
If n is an even natural number, $\sqrt[n]{x^n} = |x|$.

In the radical expression $\sqrt[n]{x}$, n is called the **index** (or **order**) of the radical. When the index is 2, the radical is a square root, and we usually do not write the index.

$$\sqrt{x} = \sqrt[2]{x}$$

WARNING! When n is even $(n > 1)$ and $x < 0$, $\sqrt[n]{x}$ is not a real number. For example, $\sqrt[4]{-81}$ is not a real number, because no real number raised to the fourth power is -81.

EXAMPLE 8 ***Finding even and odd roots.*** Simplify each radical.

a. $\sqrt[4]{625} = 5$, because $5^4 = 625$ Read $\sqrt[4]{625}$ as "the fourth root of 625."

b. $\sqrt[5]{-32} = -2$, because $(-2)^5 = -32$ Read $\sqrt[5]{-32}$ as "the fifth root of -32."

c. $\sqrt[6]{\frac{1}{64}} = \frac{1}{2}$, because $\left(\frac{1}{2}\right)^6 = \frac{1}{64}$ Read $\sqrt[6]{\frac{1}{64}}$ as "the sixth root of $\frac{1}{64}$."

d. $\sqrt[7]{10^7} = 10$, because $10^7 = 10^7$ Read $\sqrt[7]{10^7}$ as "the seventh root of 10^7."

Self Check

Simplify **a.** $\sqrt[4]{\frac{1}{81}}$ and **b.** $\sqrt[5]{10^5}$.

Answers: **a.** $\frac{1}{3}$, **b.** 10

Accent on Technology **Finding roots**

The square root key $\boxed{\sqrt{\ }}$ on a scientific calculator can be used to evaluate square roots. To evaluate roots with an index greater than 2, we can use the root key $\boxed{\sqrt[x]{y}}$. For example, the function

$$r(V) = \sqrt[3]{\frac{3V}{4\pi}}$$

gives the radius of a sphere with volume V. To find the radius of the spherical propane tank shown in Figure 7-6, we substitute 113.1 for V to get

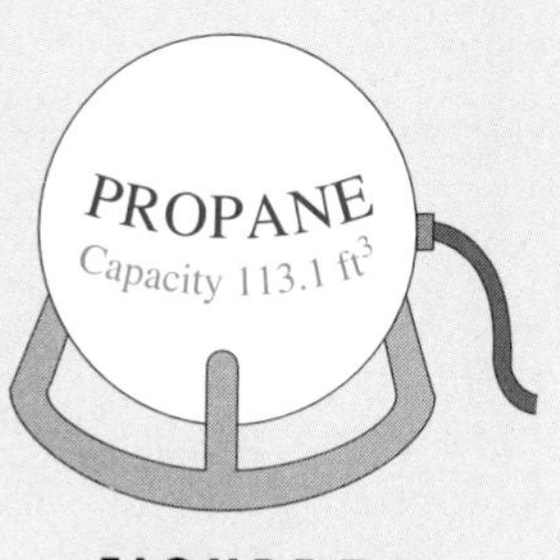

FIGURE 7-6

$$r(V) = \sqrt[3]{\frac{3V}{4\pi}}$$

$$r(113.1) = \sqrt[3]{\frac{3(113.1)}{4\pi}}$$

To evaluate a root, we enter the radicand, then press the root key $\boxed{\sqrt[x]{y}}$ followed by the index of the radical, which in this case is 3.

Keystrokes: 3 [×] 113.1 [÷] [(] 4 [×] [π] [)] [=] [2nd] [$\sqrt[x]{y}$] [3] [=]

[3.000023559]

The radius of the propane tank is about 3 feet.

EXAMPLE 9 ***Simplifying radical expressions containing variables.*** Simplify each radical expression. Assume that x can be any real number.

a. $\sqrt[5]{x^5} = x$ — Since n is odd, absolute value symbols aren't needed.

b. $\sqrt[4]{16x^4} = |2x| = 2|x|$ — Since n is even and x can be negative, absolute value symbols are needed to guarantee that the result is positive.

c. $\sqrt[6]{(x+4)^6} = |x+4|$ — Absolute value symbols are needed to guarantee that the result is positive.

d. $\sqrt[3]{(x+1)^3} = x + 1$ — Since n is odd, absolute value symbols aren't needed.

Self Check

Simplify **a.** $\sqrt[4]{16a^8}$, **b.** $\sqrt[5]{(a+5)^5}$, and **c.** $\sqrt{(x^2+4x+4)^2}$.

Answers: **a.** $2a^2$, **b.** $a + 5$ **c.** $(x + 2)^2$ ■

If we are told that x represents a positive real number in parts b and c of Example 9, we do not need to use absolute value symbols to guarantee that the answers are positive.

$\sqrt[4]{16x^4} = 2x$ — If x is positive, $2x$ is positive.

$\sqrt[6]{(x+4)^6} = x + 4$ — If x is positive, $x + 4$ is positive.

We summarize the definitions concerning $\sqrt[n]{x}$ as follows.

Summary of the definitions of $\sqrt[n]{x}$

If n is a natural number greater than 1 and x is a real number, then

If $x > 0$, then $\sqrt[n]{x}$ is the positive number such that $\left(\sqrt[n]{x}\right)^n = x$.

If $x = 0$, then $\sqrt[n]{x} = 0$.

If $x < 0$ $\begin{cases} \text{and } n \text{ is odd, then } \sqrt[n]{x} \text{ is the real number such that } \left(\sqrt[n]{x}\right)^n = x. \\ \text{and } n \text{ is even, then } \sqrt[n]{x} \text{ is not a real number.} \end{cases}$

STUDY SET Section 7.1

VOCABULARY *In Exercises 1–8, fill in the blanks to make the statements true.*

1. $5x^2$ is the ______________ of $25x^4$, because $(5x^2)^2 = 25x^4$.

2. $f(x) = \sqrt{x}$ and $g(t) = \sqrt[3]{t}$ are ______________ functions.

3. The symbol $\sqrt{}$ is called a ______________ sign.

4. In the expression $\sqrt[3]{27x^6}$, 3 is the ______________ and $27x^6$ is the ______________.

5. When n is an odd number, $\sqrt[n]{x}$ represents an ______________ root.

6. When n is an ______________ number, $\sqrt[n]{x}$ represents an even root.

7. When we write $\sqrt{b^2 + 6b + 9} = |b + 3|$, we say that we have ______________ the radical.

8. 6 is the ______________ of 216 because $6^3 = 216$.

CONCEPTS *In Exercises 9–16, fill in the blanks to make the statements true.*

9. b is a square root of a if ______________.

10. $\sqrt{0} =$ ___ and $\sqrt[3]{0} =$ ___.

11. The number 25 has ______________ square roots. The principal square root of 25 is the ______________ square root of 25.

12. $\sqrt{-4}$ is not a real number, because no real number ______________ equals -4.

13. $\sqrt[3]{x} = y$ if $y^3 =$ ___.

14. $\sqrt{x^2} =$ ___ and $\sqrt[3]{x^3} =$ ___.

15. The graph of $f(x) = \sqrt{x} + 3$ is the graph of $f(x) = \sqrt{x}$ translated ___ units ___.

16. The graph of $f(x) = \sqrt{x + 5}$ is the graph of $f(x) = \sqrt{x}$ translated ___ units to the ______________.

In Exercises 17–18, complete the table of values and then graph the radical function.

17. $f(x) = -\sqrt{x}$

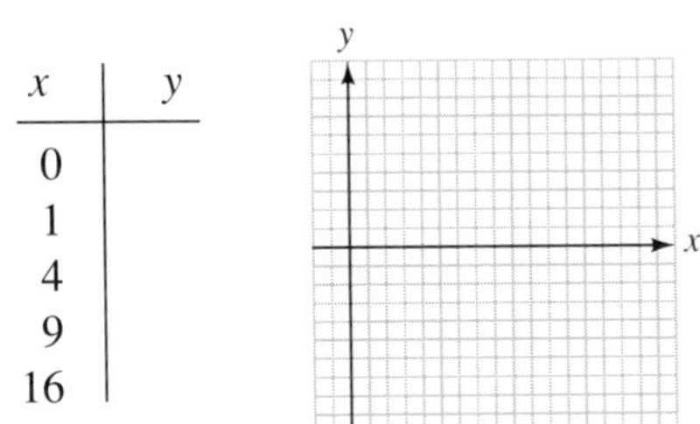

x	y
0	
1	
4	
9	
16	

18. $f(x) = -\sqrt[3]{x}$

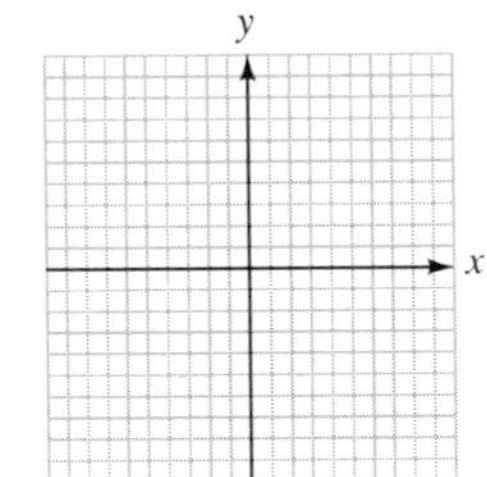

x	y
-8	
-1	
0	
1	
8	

NOTATION *In Exercises 19–22, translate each sentence into mathematical symbols.*

19. The square root of x squared is the absolute value of x.

20. The cube root of x cubed is x.

21. f of x equals the square root of the quantity x minus five.

22. The fifth root of negative thirty-two is negative two.

PRACTICE *In Exercises 23–34, find each square root, if possible.*

23. $\sqrt{121}$

24. $\sqrt{144}$

25. $-\sqrt{64}$

26. $-\sqrt{1}$

27. $\sqrt{\dfrac{1}{9}}$

28. $-\sqrt{\dfrac{4}{25}}$

29. $\sqrt{0.25}$

30. $\sqrt{0.16}$

31. $\sqrt{-25}$

32. $-\sqrt{-49}$

33. $\sqrt{(-4)^2}$

34. $\sqrt{(-9)^2}$

In Exercises 35–38, use a calculator to find each square root. Give the answer to four decimal places.

35. $\sqrt{12}$ **36.** $\sqrt{340}$ **37.** $\sqrt{679.25}$ **38.** $\sqrt{0.0063}$

In Exercises 39–46, find each square root. Assume that all variables are unrestricted, and use absolute value symbols when necessary.

39. $\sqrt{4x^2}$ **40.** $\sqrt{16y^4}$ **41.** $\sqrt{(t+5)^2}$ **42.** $\sqrt{(a+6)^2}$

43. $\sqrt{(-5b)^2}$ **44.** $\sqrt{(-8c)^2}$ **45.** $\sqrt{a^2+6a+9}$ **46.** $\sqrt{x^2+10x+25}$

In Exercises 47–54, find each value given that $f(x) = \sqrt{x-4}$ and $g(x) = \sqrt[3]{x-4}$.

47. $f(4)$ **48.** $f(8)$ **49.** $f(20)$ **50.** $f(29)$

51. $g(12)$ **52.** $g(-4)$ **53.** $g(-996)$ **54.** $g(1{,}004)$

In Exercises 55–58, find each value given that $f(x) = \sqrt{x^2+1}$ and $g(x) = \sqrt[3]{x^2+1}$. Give each answer to four decimal places.

55. $f(4)$ **56.** $f(2.35)$ **57.** $g(6)$ **58.** $g(21.57)$

In Exercises 59–62, graph each radical function and find its domain and range. Check your work with a graphing calculator.

59. $f(x) = \sqrt{x+4}$

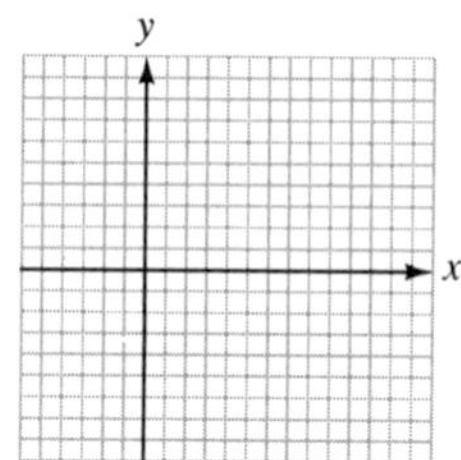

60. $f(x) = -\sqrt{x-1}$

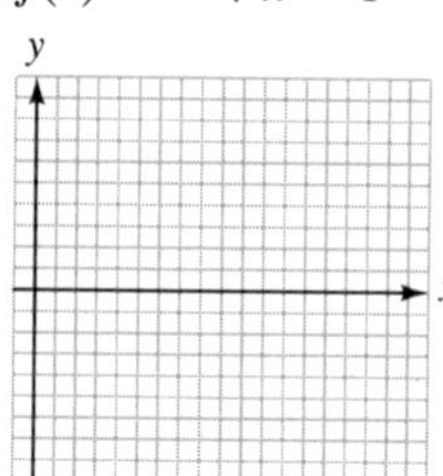

61. $f(x) = -\sqrt[3]{x} - 3$

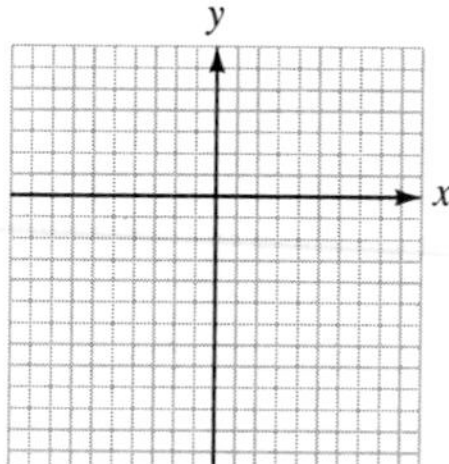

62. $f(x) = -\sqrt[3]{x} - 1$

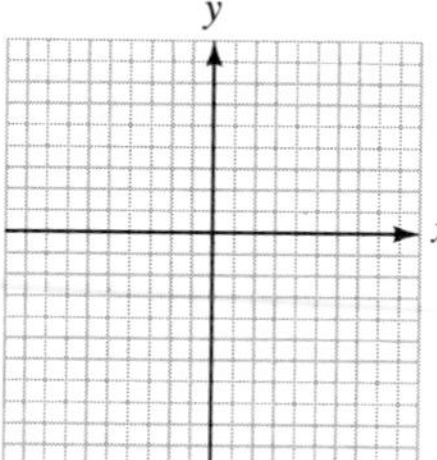

In Exercises 63–74, simplify each cube root.

63. $\sqrt[3]{1}$ **64.** $\sqrt[3]{-125}$ **65.** $\sqrt[3]{-\frac{8}{27}}$ **66.** $\sqrt[3]{\frac{125}{216}}$

67. $\sqrt[3]{0.064}$ **68.** $\sqrt[3]{0.001}$ **69.** $\sqrt[3]{8a^3}$ **70.** $\sqrt[3]{-27x^6}$

71. $\sqrt[3]{-1{,}000p^3q^3}$ **72.** $\sqrt[3]{343a^6b^3}$ **73.** $\sqrt[3]{-0.064s^9t^6}$ **74.** $\sqrt[3]{\frac{27}{1{,}000}a^6b^6}$

In Exercises 75–94, simplify each radical, if possible. Assume that all variables represent positive real numbers.

75. $\sqrt[4]{81}$
76. $\sqrt[6]{64}$
77. $-\sqrt[5]{243}$
78. $-\sqrt[4]{625}$

79. $\sqrt[4]{-256}$
80. $\sqrt[6]{-729}$
81. $\sqrt[4]{\frac{16}{625}}$
82. $\sqrt[5]{-\frac{243}{32}}$

83. $-\sqrt[5]{-\frac{1}{32}}$
84. $-\sqrt[4]{\frac{81}{256}}$
85. $\sqrt[5]{32a^5}$
86. $\sqrt[5]{-32x^5}$

87. $\sqrt[4]{16a^4}$
88. $\sqrt[8]{x^{24}}$
89. $\sqrt[4]{k^{12}}$
90. $\sqrt[6]{64b^6}$

91. $\sqrt[4]{\frac{1}{16}m^4}$
92. $\sqrt[4]{\frac{1}{81}x^8}$
93. $\sqrt[25]{(x+2)^{25}}$
94. $\sqrt[44]{(x+4)^{44}}$

APPLICATIONS

Use a calculator to solve each problem. In each case, round to the nearest tenth.

95. EMBROIDERY The radius r of a circle is given by the formula

$$r = \sqrt{\frac{A}{\pi}}$$

where A is its area. Find the diameter of the embroidery hoop shown in Illustration 1 if there are 38.5 in.^{2} of stretched fabric on which to embroider.

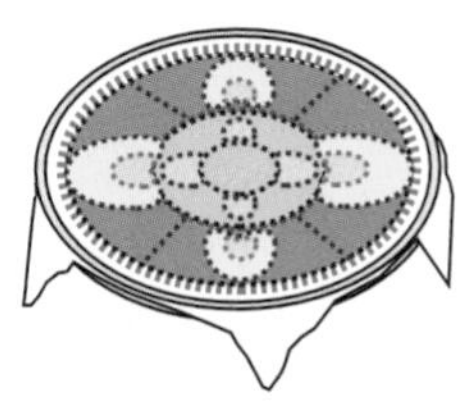

ILLUSTRATION 1

96. SOFTBALL AND BASEBALL The length of a diagonal of a square is given by the function $d(s) = \sqrt{2s^2}$, where s is the length of a side of the square. Find the distance from home plate to second base on a softball diamond and on a baseball diamond. Illustration 2 gives the dimensions of each type of infield.

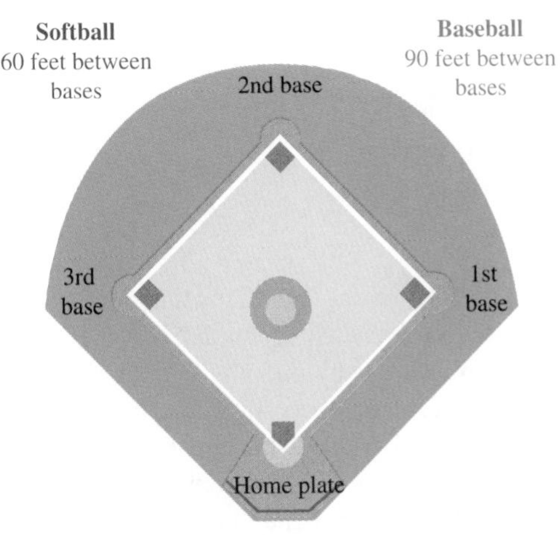

ILLUSTRATION 2

97. PULSE RATE The approximate pulse rate (in beats per minute) of an adult who is t inches tall is given by the function

$$p(t) = \frac{590}{\sqrt{t}}$$

The Guinness Book of World Records 1998 lists Ri Myong-hun of North Korea as the tallest living man, at 7 ft $8\frac{1}{2}$ in. Find his approximate pulse rate as predicted by the function.

98. THE GRAND CANYON The time t (in seconds) that it takes for an object to fall a distance of s feet is given by the formula

$$t = \frac{\sqrt{s}}{4}$$

In some places, the Grand Canyon is one mile (5,280 feet) deep. How long would it take a stone dropped over the edge of the canyon to hit bottom?

99. BIOLOGY Scientists will place five rats inside the controlled environment of a sealed hemisphere to study the rats' behavior. The function

$$d(V) = \sqrt[3]{12\left(\frac{V}{\pi}\right)}$$

gives the diameter of a hemisphere with volume V. Use the function to determine the diameter of the base of the hemisphere, if each rat requires 125 cubic feet of living space.

100. AQUARIUM The function

$$s(g) = \sqrt[3]{\frac{g}{7.5}}$$

determines how long (in feet) an edge of a cube-shaped tank must be if it is to hold g gallons of water. What dimensions should a cube-shaped aquarium have if it is to hold 1,250 gallons of water?

101. COLLECTIBLES The *effective rate of interest r* earned by an investment is given by the formula

$$r = \sqrt[n]{\frac{A}{P}} - 1$$

where P is the initial investment that grows to value A after n years. Determine the effective rate of interest earned by a collector on a Lladró porcelain figurine purchased for \$800 and sold for \$950 five years later.

102. LAW ENFORCEMENT The graphs of the two radical functions shown in Illustration 3 can be used to estimate the speed (in mph) of a car involved in an accident. Suppose a police accident report listed skid marks to be 220 feet long but failed to give the road conditions. Estimate the possible speeds the car was traveling prior to the brakes being applied.

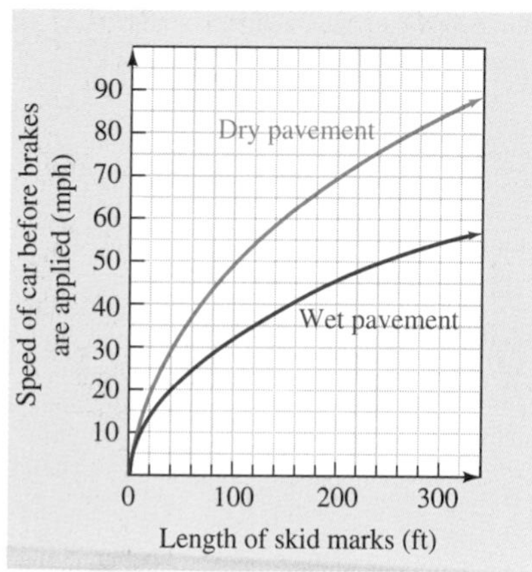

ILLUSTRATION 3

WRITING *Write a paragraph using your own words.*

103. Explain why 36 has two square roots, but $\sqrt{36}$ is just 6 and not -6.

104. If x is any real number, then $\sqrt{x^2} = x$ is not correct. Explain.

REVIEW *In Exercises 105–108, do the operations.*

105. $\dfrac{x^2 - x - 6}{x^2 - 2x - 3} \cdot \dfrac{x^2 - 1}{x^2 + x - 2}$

106. $\dfrac{x^2 - 3x - 4}{x^2 - 5x + 6} \div \dfrac{x^2 - 2x - 3}{x^2 - x - 2}$

107. $\dfrac{3}{m + 1} + \dfrac{3m}{m - 1}$

108. $\dfrac{2x + 3}{3x - 1} - \dfrac{x - 4}{2x + 1}$

7.2 *Radical Equations*

In this section, you will learn about

- **The power rule**
- **Equations containing one radical**
- **Equations containing two radicals**
- **Solving formulas containing radicals**

INTRODUCTION. Many situations can be modeled by equations that contain radicals. In this section, we will develop techniques to solve such equations. For example, to solve the radical equation $\sqrt{x} = 6$, we need to isolate x by undoing the operation performed on it. Recall that $\sqrt{x}$ represents the number that, when squared, gives x. Therefore, if we square $\sqrt{x}$, we will obtain x. From this observation, it is apparent that we can eliminate the radical on the left-hand side of the equation $\sqrt{x} = 6$ by squaring that side. Intuition tells us that we should also square the right-hand side. Squaring both sides of an equation is an application of the *power rule,* which we now consider in more detail.

The power rule

To solve equations containing radicals, we will use the **power rule.**

The power rule

If x, y, and n are real numbers and $x = y$, then

$$x^n = y^n$$

If we raise both sides of an equation to the same power, the resulting equation might not be equivalent to the original equation. For example, if we square both sides of the equation

1. $x = 3$ With a solution set of $\{3\}$

we obtain the equation

2. $x^2 = 9$ With a solution set of $\{3, -3\}$

Equations 1 and 2 are not equivalent, because they have different solution sets, and the solution -3 of Equation 2 does not satisfy Equation 1. Since raising both sides of an equation to the same power can produce an equation with roots that don't satisfy the original equation, we must always check each apparent solution in the original equation and discard any *extraneous solutions.*

Equations containing one radical

Radical equations contain a radical expression with a variable radicand. To solve radical equations, we apply the power rule.

EXAMPLE 1 ***Squaring both sides of an equation.*** Solve $\sqrt{x + 3} = 4$.

Solution

To eliminate the radical, we apply the power rule by squaring both sides of the equation and proceed as follows:

$$\begin{aligned} \sqrt{x+3} &= 4 \\ \left(\sqrt{x+3}\right)^2 &= (4)^2 && \text{Square both sides.} \\ x + 3 &= 16 \\ x &= 13 && \text{Subtract 3 from both sides.} \end{aligned}$$

We must check the apparent solution of 13 to see whether it satisfies the original equation.

Check:

$$\begin{aligned} \sqrt{x+3} &= 4 \\ \sqrt{13+3} &\stackrel{?}{=} 4 && \text{Substitute 13 for } x. \\ \sqrt{16} &\stackrel{?}{=} 4 \\ 4 &= 4 \end{aligned}$$

Since 13 satisfies the original equation, it is a solution.

Self Check

Solve $\sqrt{a - 2} = 3$.

Answer: 11

To solve an equation with radicals, we follow these steps.

Solving an equation containing radicals	1. Isolate one radical expression on one side of the equation. 2. Raise both sides of the equation to the power that is the same as the index of the radical. 3. Solve the resulting equation. If it still contains a radical, go back to step 1. 4. Check the solutions to eliminate extraneous roots.

EXAMPLE 2 ***Amusement park ride.*** The distance d in feet that an object will fall in t seconds is given by the formula

$$t = \sqrt{\frac{d}{16}}$$

If the designers of the amusement park attraction shown in Figure 7-7 want the riders to experience 3 seconds of vertical "free fall," what length of vertical drop is needed?

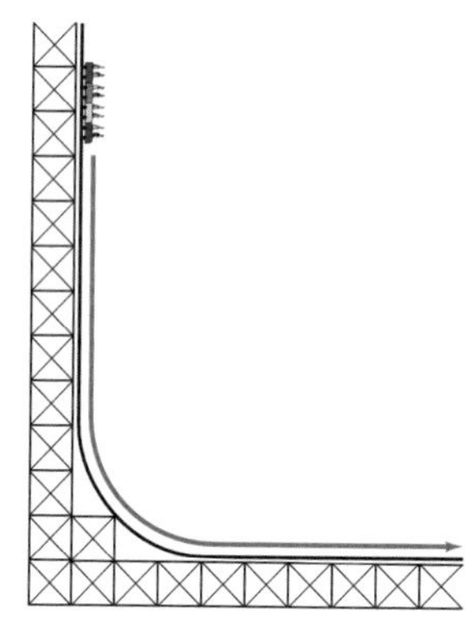

FIGURE 7-7

Solution

We substitute 3 for t in the formula and solve for d.

$t = \sqrt{\frac{d}{16}}$

$3 = \sqrt{\frac{d}{16}}$ Here the radical is isolated on the right-hand side.

$(3)^2 = \left(\sqrt{\frac{d}{16}}\right)^2$ Raise both sides to the second power.

$9 = \frac{d}{16}$ Simplify.

$144 = d$ Solve the resulting equation by multiplying both sides by 16.

The amount of vertical drop needs to be 144 feet.

Self Check

How long a vertical drop is needed if the riders are to "free fall" for 3.5 seconds?

Answer: 196 ft

EXAMPLE 3 ***Isolating the radical.*** Solve $\sqrt{3x + 1} + 1 = x$.

Solution

We first subtract 1 from both sides to isolate the radical. Then, to eliminate the radical, we square both sides of the equation and proceed as follows:

$\sqrt{3x + 1} + 1 = x$

$\sqrt{3x + 1} = x - 1$ Subtract 1 from both sides.

$(\sqrt{3x + 1})^2 = (x - 1)^2$ Square both sides to eliminate the square root.

$3x + 1 = x^2 - 2x + 1$ On the right-hand side, use the FOIL method: $(x - 1)^2 = (x - 1)(x - 1) = x^2 - x - x + 1 = x^2 - 2x + 1$.

$0 = x^2 - 5x$ Subtract $3x$ and 1 from both sides. This is a quadratic equation. Use factoring to solve it.

$0 = x(x - 5)$ Factor $x^2 - 5x$.

$x = 0 \quad \text{or} \quad x - 5 = 0$ Set each factor each to 0.

$x = 0 \qquad\qquad x = 5$

Self Check

Solve $\sqrt{4x + 1} + 1 = x$.

We must check each apparent solution to see whether it satisfies the original equation.

Check:

$$\sqrt{3x+1}+1=x \qquad \sqrt{3x+1}+1=x$$
$$\sqrt{3(0)+1}+1 \stackrel{?}{=} 0 \qquad \sqrt{3(5)+1}+1 \stackrel{?}{=} 5$$
$$\sqrt{1}+1 \stackrel{?}{=} 0 \qquad \sqrt{16}+1 \stackrel{?}{=} 5$$
$$2 \neq 0 \qquad 5 = 5$$

Since 0 does not check, it must be discarded. The only solution of the original equation is 5.

Answer: 6, 0 is extraneous ■

Accent on Technology Solving equations containing radicals

To find approximate solutions for $\sqrt{3x+1}+1=x$ with a graphing calculator, we use window settings of $[-5, 10]$ for x and $[-2, 8]$ for y and graph the functions $f(x)=\sqrt{3x+1}+1$ and $g(x)=x$, as in Figure 7-8(a). We then trace to find the approximate x-coordinate of their intersection point, as in Figure 7-8(b). After repeated zooms, we will see that $x=5$.

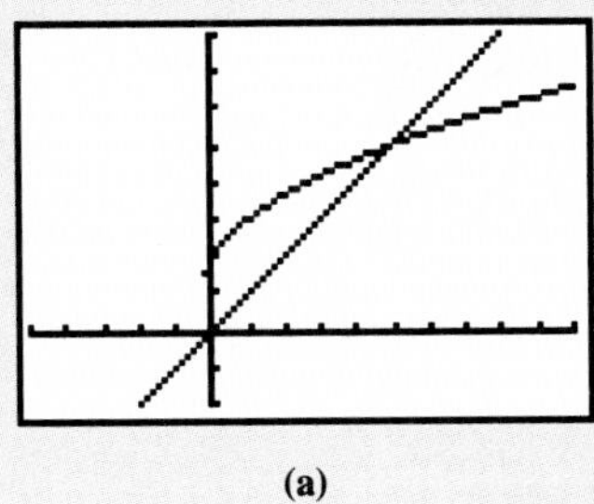

(a)

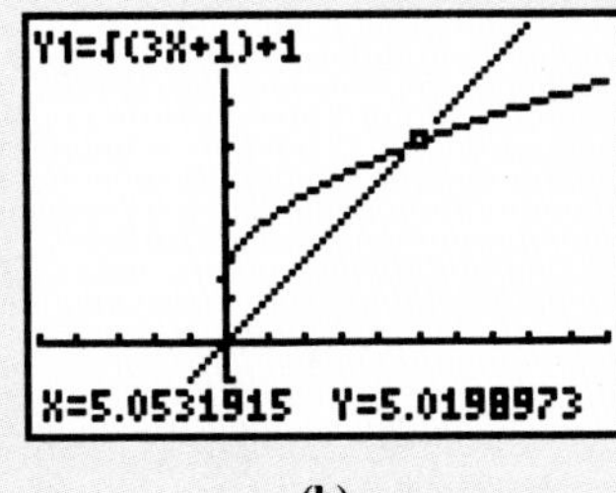

(b)

FIGURE 7-8

EXAMPLE 4 ***Cubing both sides.*** Solve $\sqrt[3]{x^3+7}=x+1$.

Solution

To eliminate the radical, we cube both sides of the equation and proceed as follows:

$$\sqrt[3]{x^3+7}=x+1$$

$$\left(\sqrt[3]{x^3+7}\right)^3=(x+1)^3 \qquad \text{Cube both sides to eliminate the cube root.}$$

$$x^3+7=x^3+3x^2+3x+1 \qquad (x+1)^3=(x+1)(x+1)(x+1).$$

$$0=3x^2+3x-6 \qquad \text{Subtract } x^3 \text{ and 7 from both sides.}$$

$$0=x^2+x-2 \qquad \text{Divide both sides by 3. To solve this quadratic equation, use factoring.}$$

$$0=(x+2)(x-1) \qquad \text{Factor the trinomial.}$$

$$x+2=0 \quad \text{or} \quad x-1=0$$
$$x=-2 \qquad\qquad x=1$$

We check each apparent solution to see whether it satisfies the original equation.

Check:

$$\sqrt[3]{x^3+7}=x+1 \qquad \sqrt[3]{x^3+7}=x+1$$
$$\sqrt[3]{(-2)^3+7} \stackrel{?}{=} -2+1 \qquad \sqrt[3]{1^3+7} \stackrel{?}{=} 1+1$$
$$\sqrt[3]{-8+7} \stackrel{?}{=} -1 \qquad \sqrt[3]{1+7} \stackrel{?}{=} 2$$
$$\sqrt[3]{-1} \stackrel{?}{=} -1 \qquad \sqrt[3]{8} \stackrel{?}{=} 2$$
$$-1=-1 \qquad 2=2$$

Both solutions satisfy the original equation.

Self Check

Solve $\sqrt[3]{x^3+8}=x+2$.

Answer: 0, −2 ■

EXAMPLE 5 ***Solving an equation containing two radicals.*** Solve $\sqrt{5x+9} = 2\sqrt{3x+4}$.

Self Check Solve $\sqrt{x-4} = 2\sqrt{x-16}$.

Solution

Each radical is isolated on one side of the equation, so we square both sides to eliminate them.

$$\sqrt{5x+9} = 2\sqrt{3x+4}$$

$$\left(\sqrt{5x+9}\right)^2 = \left(2\sqrt{3x+4}\right)^2 \quad \text{Square both sides.}$$

$$5x+9 = 4(3x+4) \quad \text{On the right-hand side: } \left(2\sqrt{3x+4}\right)^2 = 2^2\left(\sqrt{3x+4}\right)^2 = 4(3x+4).$$

$$5x+9 = 12x+16 \quad \text{Remove parentheses.}$$

$$-7 = 7x \quad \text{Subtract } 5x \text{ and } 16 \text{ from both sides.}$$

$$-1 = x \quad \text{Divide both sides by 7.}$$

We check the solution by substituting -1 for x in the original equation.

$$\sqrt{5x+9} = 2\sqrt{3x+4}$$

$$\sqrt{5(-1)+9} \stackrel{?}{=} 2\sqrt{3(-1)+4} \quad \text{Substitute } -1 \text{ for } x.$$

$$\sqrt{4} \stackrel{?}{=} 2\sqrt{1}$$

$$2 = 2$$

The solution checks.

Answer: 20 ■

When more than one radical appears in an equation, it is often necessary to apply the power rule more than once.

EXAMPLE 6 ***Solving equations containing two radicals.*** Solve $\sqrt{x} + \sqrt{x+2} = 2$.

Self Check Solve $\sqrt{a} + \sqrt{a+3} = 3$.

Solution

To remove the radicals, we must square both sides of the equation. This is easier to do if one radical is on each side of the equation. So we subtract $\sqrt{x}$ from both sides to isolate $\sqrt{x+2}$ on the left-hand side of the equation.

$$\sqrt{x} + \sqrt{x+2} = 2$$

$$\sqrt{x+2} = 2 - \sqrt{x} \quad \text{Subtract } \sqrt{x} \text{ from both sides.}$$

$$\left(\sqrt{x+2}\right)^2 = \left(2-\sqrt{x}\right)^2 \quad \text{Square both sides to eliminate the square root.}$$

$$x+2 = 4 - 4\sqrt{x} + x \quad \text{Use FOIL: } \left(2-\sqrt{x}\right)^2 = \left(2-\sqrt{x}\right)\left(2-\sqrt{x}\right) = 4 - 4\sqrt{x} + x.$$

$$2 = 4 - 4\sqrt{x} \quad \text{Subtract } x \text{ from both sides.}$$

$$-2 = -4\sqrt{x} \quad \text{Subtract 4 from both sides.}$$

$$\frac{1}{2} = \sqrt{x} \quad \text{Divide both sides by } -4 \text{ and simplify.}$$

$$\frac{1}{4} = x \quad \text{Square both sides.}$$

Check: $\sqrt{x} + \sqrt{x+2} = 2$

$$\sqrt{\frac{1}{4}} + \sqrt{\frac{1}{4} + 2} \stackrel{?}{=} 2$$

$$\frac{1}{2} + \sqrt{\frac{9}{4}} \stackrel{?}{=} 2$$

$$\frac{1}{2} + \frac{3}{2} \stackrel{?}{=} 2$$

$$2 = 2$$

The solution checks.

Answer: 1

Accent on Technology — Solving equations containing radicals

To find approximate solutions for $\sqrt{x} + \sqrt{x+2} = 5$ (an equation similar to that in Example 6) with a graphing calculator, we use window settings of $[-2, 10]$ for x and $[-2, 8]$ for y and graph the functions $f(x) = \sqrt{x} + \sqrt{x+2}$ and $g(x) = 5$, as in Figure 7-9(a). We then trace to find an approximation of the x-coordinate of their intersection point, as in Figure 7-9(b). From the figure, we can see that $x \approx 5.15$. We can zoom to get better results.

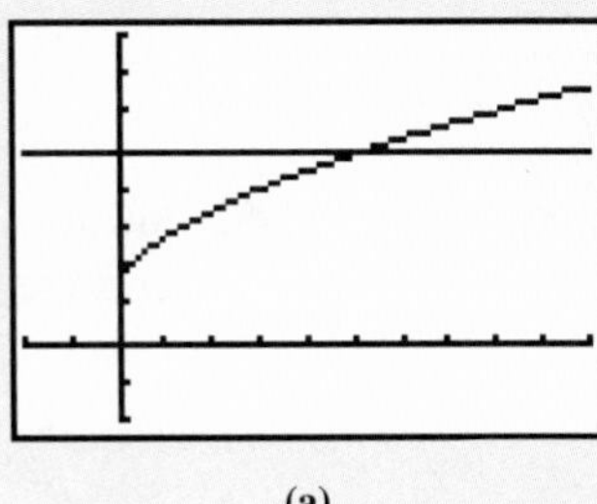

(a)

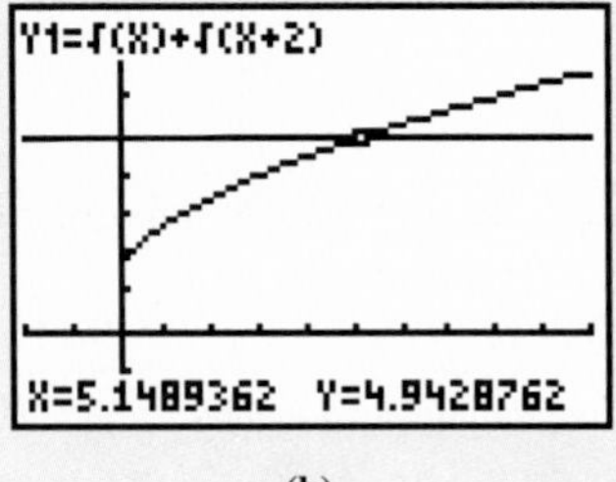

(b)

FIGURE 7-9

Solving formulas containing radicals

To *solve a formula for a variable* means to isolate that variable on one side of the equation, with all other quantities on the other side.

EXAMPLE 7 ***Depreciation rate.*** A piece of office equipment that is now worth V dollars originally cost C dollars 3 years ago. The rate r at which it has depreciated (lost value) is given by

$$r = 1 - \sqrt[3]{\frac{V}{C}}$$

Solve the formula for C.

Solution

We begin by isolating the cube root on the right-hand side of the equation.

$$r = 1 - \sqrt[3]{\frac{V}{C}}$$

$$r - 1 = -\sqrt[3]{\frac{V}{C}} \quad \text{Subtract 1 from both sides.}$$

$$(r-1)^3 = \left[-\sqrt[3]{\frac{V}{C}}\right]^3 \quad \text{Cube both sides.}$$

Self Check

A formula used in statistics to determine the necessary size of a sample to obtain the desired degree of accuracy is

$$e = z_0\sqrt{\frac{pq}{n}}$$

Solve the formula for n.

$$(r-1)^3 = -\frac{V}{C}$$ Simplify the right-hand side.

$$C(r-1)^3 = -V$$ Multiply both sides by C.

$$C = -\frac{V}{(r-1)^3}$$ Divide both sides by $(r-1)^3$.

Answer: $n = \frac{z_0^2 pq}{e^2}$

STUDY SET Section 7.2

VOCABULARY *In Exercises 1–4, fill in the blanks to make each statement true.*

1. Equations such as $\sqrt{x+4} - 4 = 5$ and $\sqrt[3]{x+1} = 12$ are called ______________ equations.

2. When solving equations containing radicals, try to ______________ one radical expression on one side of the equation.

3. Squaring both sides of an equation can introduce ______________ solutions.

4. To ______________ an apparent solution means to substitute it into the original equation and see whether a true statement results.

CONCEPTS

5. What is the first step in solving each equation?

a. $\sqrt{x+4} = 5$

b. $\sqrt[3]{x+4} = 2$

6. Fill in the blank to make a true statement. $\sqrt{x}$ represents the number that, when squared, gives ___.

7. Simplify each expression.

a. $(\sqrt{x})^2$ **b.** $(\sqrt{x-5})^2$

c. $(4\sqrt{2x})^2$ **d.** $(-\sqrt{x+3})^2$

8. Simplify each expression.

a. $(\sqrt[3]{x})^3$ **b.** $(\sqrt[4]{x})^4$

c. $(-\sqrt[3]{2x})^3$ **d.** $(2\sqrt[3]{x+3})^3$

9. What is wrong with the student's work shown below?

Solve $\sqrt{x+1} - 3 = 8$.

$$\sqrt{x+1} = 11$$
$$(\sqrt{x+1})^2 = 11$$
$$x + 1 = 11$$
$$x = 10$$

10. Solve $\sqrt{x-2} + 2 = 4$ graphically, using the graphs in Illustration 1.

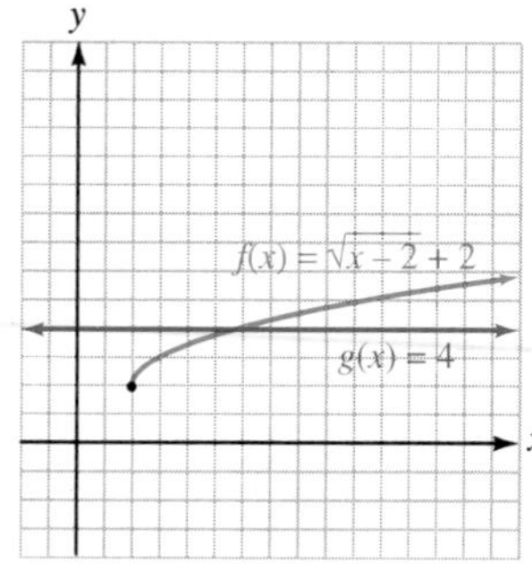

ILLUSTRATION 1

11. Use your own words to restate the power rule: If x, y, and z are real numbers and $x = y$, then $x^n = y^n$.

12. The first step of a student's solution is shown below. What is a better way to begin the solution?

Solve $\sqrt{x} + \sqrt{x+22} = 12$.

$$(\sqrt{x} + \sqrt{x+22})^2 = 12^2$$

NOTATION *In Exercises 13–14, complete each solution.*

13. Solve $2\sqrt{x-2} = 4$.

$$\left(\boxed{}\right)^2 = 4^2$$
$$\boxed{}(x-2) = \boxed{}$$
$$4x - \boxed{} = 16$$
$$4x = \boxed{}$$
$$x = 6$$

14. Solve $\sqrt{1-2x} = \sqrt{x+10}$.

$$\left(\boxed{}\right)^2 = \left(\sqrt{x+10}\right)^2$$
$$\boxed{} = x + 10$$
$$\boxed{} = 9$$
$$x = -3$$

PRACTICE *In Exercises 15–54, solve each equation. Write all apparent solutions. Cross out those that are extraneous.*

15. $\sqrt{5x-6} = 2$

16. $\sqrt{7x-10} = 12$

17. $\sqrt{6x+1} + 2 = 7$

18. $\sqrt{6x+13} - 2 = 5$

19. $2\sqrt{4x+1} = \sqrt{x+4}$

20. $\sqrt{3(x+4)} = \sqrt{5x-12}$

21. $\sqrt[3]{7n-1} = 3$

22. $\sqrt[3]{12m+4} = 4$

23. $\sqrt[4]{10p+1} = \sqrt[4]{11p-7}$

24. $\sqrt[4]{10y+6} = 2\sqrt[4]{y}$

25. $x = \dfrac{\sqrt{12x-5}}{2}$

26. $x = \dfrac{\sqrt{16x-12}}{2}$

27. $\sqrt{x+2} - \sqrt{4-x} = 0$

28. $\sqrt{6-x} - \sqrt{2x+3} = 0$

29. $2\sqrt{x} = \sqrt{5x-16}$

30. $3\sqrt{x} = \sqrt{3x+54}$

31. $r - 9 = \sqrt{2r-3}$

32. $-s - 3 = 2\sqrt{5-s}$

33. $\sqrt{-5x+24} = 6 - x$

34. $\sqrt{-x+2} = x - 2$

35. $\sqrt{y+2} = 4 - y$

36. $\sqrt{22y+86} = y + 9$

37. $\sqrt[3]{x^3-7} = x - 1$

38. $\sqrt[3]{x^3+56} - 2 = x$

39. $\sqrt[4]{x^4+4x^2-4} = -x$

40. $\sqrt[4]{8x-8} + 2 = 0$

41. $\sqrt[4]{12t+4} + 2 = 0$

42. $u = \sqrt[4]{u^4-6u^2+24}$

43. $\sqrt{2y+1} = 1 - 2\sqrt{y}$

44. $\sqrt{u} + 3 = \sqrt{u-3}$

45. $\sqrt{y+7} + 3 = \sqrt{y+4}$

46. $1 + \sqrt{z} = \sqrt{z+3}$

47. $2 + \sqrt{u} = \sqrt{2u+7}$

48. $5r + 4 = \sqrt{5r+20} + 4r$

49. $\sqrt{6t+1} - 3\sqrt{t} = -1$

50. $\sqrt{4s+1} - \sqrt{6s} = -1$

51. $\sqrt{2x+5} + \sqrt{x+2} = 5$

52. $\sqrt{2x+5} + \sqrt{2x+1} + 4 = 0$

53. $\sqrt{x-5} - \sqrt{x+3} = 4$

54. $\sqrt{x+8} - \sqrt{x-4} = -2$

In Exercises 55–62, solve each equation for the indicated variable.

55. $v = \sqrt{2gh}$ for h

56. $d = 1.4\sqrt{h}$ for h

57. $T = 2\pi\sqrt{\dfrac{l}{32}}$ for l

58. $d = \sqrt[3]{\dfrac{12V}{\pi}}$ for V

59. $r = \sqrt[3]{\dfrac{A}{P}} - 1$ for A

60. $r = \sqrt[3]{\dfrac{A}{P}} - 1$ for P

61. $L_A = L_B\sqrt{1 - \dfrac{v^2}{c^2}}$ for v^2

62. $R_1 = \sqrt{\dfrac{A}{\pi} - R_2^{\,2}}$ for A

APPLICATIONS

63. HIGHWAY DESIGN A curved concrete road will accommodate traffic traveling s mph if the radius of the curve is r feet, according to the formula $s = 3\sqrt{r}$. If engineers expect 40-mph traffic, what radius should they specify? Give the result to the nearest foot. (See Illustration 2.)

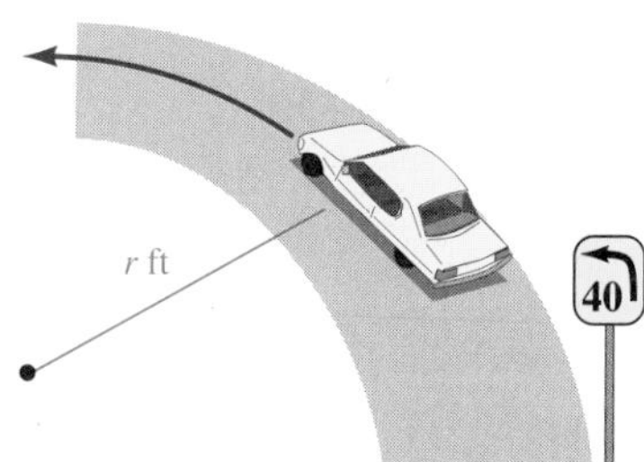

ILLUSTRATION 2

64. FORESTRY The higher a lookout tower is built, the farther an observer can see. See Illustration 3. That distance d (called the *horizon distance,* measured in miles) is related to the height h of the observer (measured in feet) by the formula $d = 1.4\sqrt{h}$. How tall must a lookout tower be to see the edge of the forest, 25 miles away? (Round to the nearest foot.)

ILLUSTRATION 3

65. WIND POWER The power generated by a certain windmill is related to the velocity of the wind by the formula

$$v = \sqrt[3]{\frac{P}{0.02}}$$

where P is the power (in watts) and v is the velocity of the wind (in mph). Find how much power the windmill is generating when the wind is 29 mph.

66. DIAMONDS The *effective rate of interest* r earned by an investment is given by the formula

$$r = \sqrt[n]{\frac{A}{P}} - 1$$

where P is the initial investment that grows to value A after n years. If a diamond buyer got $4,000 for a 1.73-carat diamond that he had purchased 4 years earlier, and earned an annual rate of return of 6.5% on the investment, what did he originally pay for the diamond?

67. THEATER PRODUCTION The ropes, pulleys, and sandbags shown in Illustration 4 are part of a mechanical system used to raise and lower scenery for a stage play. For the scenery to be in the proper position, the following formula must apply:

$$w_2 = \sqrt{w_1^{\,2} + w_3^{\,2}}$$

If $w_2 = 12.5$ lb and $w_3 = 7.5$ lb, find w_1.

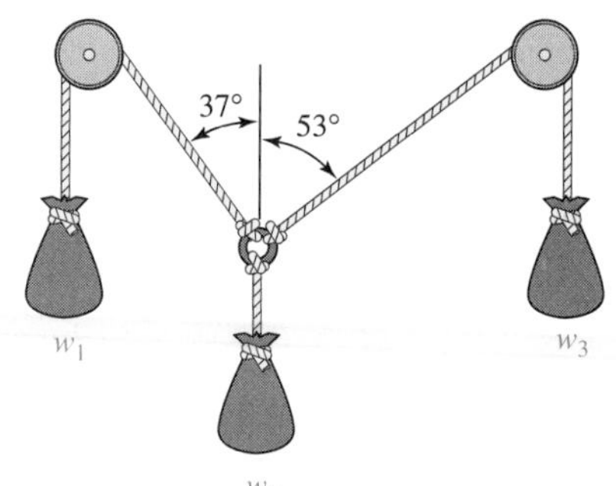

ILLUSTRATION 4

68. CARPENTRY During construction, carpenters often brace walls as shown in Illustration 5, where the length of the brace is given by the formula

$$l = \sqrt{f^2 + h^2}$$

If a carpenter nails a 10-ft brace to the wall 6 feet above the floor, how far from the base of the wall should he nail the brace to the floor?

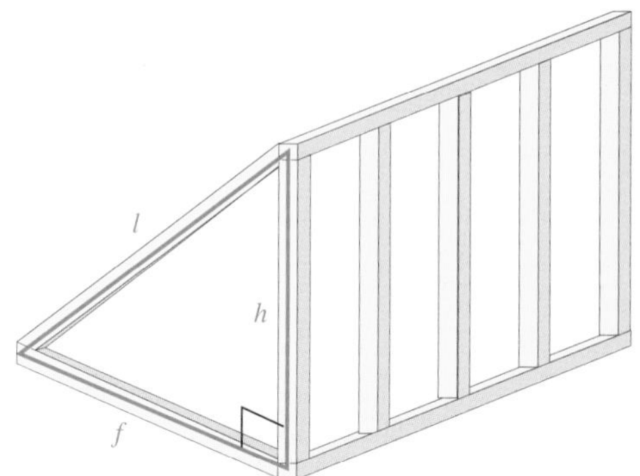

ILLUSTRATION 5

69. SUPPLY AND DEMAND The number of wrenches that will be produced at a given price can be predicted by the formula $s = \sqrt{5x}$, where s is the supply (in thousands) and x is the price (in dollars). The demand d for wrenches can be predicted by the formula $d = \sqrt{100 - 3x^2}$. Find the equilibrium price—that is, find the price at which supply will equal demand.

70. SUPPLY AND DEMAND The number of footballs that will be produced at a given price can be predicted by the formula $s = \sqrt{23x}$, where s is the supply (in thousands) and x is the price (in dollars). The demand d for footballs can be predicted by the formula $d = \sqrt{312 - 2x^2}$. Find the equilibrium price—that is, find the price at which supply will equal demand.

WRITING *Write a paragraph using your own words.*

71. If both sides of an equation are raised to the same power, the resulting equation might not be equivalent to the original equation. Explain.

72. Explain how the radical equation $\sqrt{2x - 1} = x$ can be solved graphically.

REVIEW

73. LIGHTING The intensity of the light reaching you from a light bulb varies inversely as the square of your distance from the bulb. If you are 5 feet away from a light bulb and the intensity is 40 foot-candles, what will the intensity be if you move 20 feet away from the bulb?

74. COMMITTEES What type of variation is shown in Illustration 6? As the number of people on this committee increased, what happened to its effectiveness?

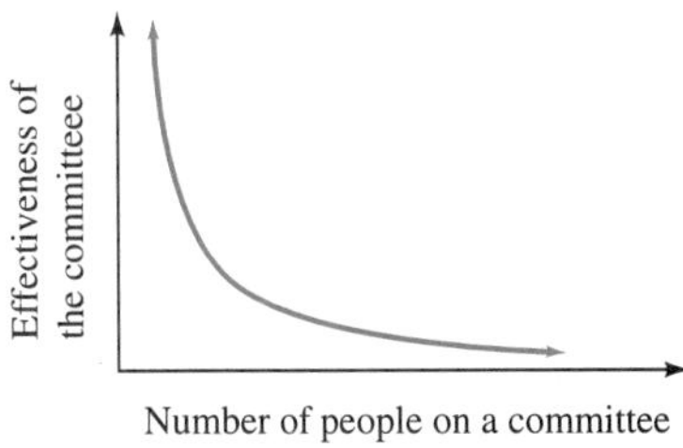

ILLUSTRATION 6

75. TYPESETTING If 12-point type is 0.166044 inch tall, how tall is 30-point type?

76. GUITAR STRINGS The frequency of vibration of a string varies directly as the square root of the tension and inversely as the length of the string. Suppose a string 2.5 feet long, under a tension of 16 pounds, vibrates 25 times per second. Find k, the constant of proportionality.

7.3 *Rational Exponents*

In this section, you will learn about

- **Rational exponents**
- **Exponential expressions with variables in their bases**
- **Rational exponents with numerators other than 1**
- **Negative rational exponents**
- **Applying the rules for exponents**
- **Simplifying radical expressions**

INTRODUCTION. In Chapter 1, we worked with exponential expressions containing natural-number exponents, such as 5^3 and x^2. In Chapter 5, the definition of exponent was extended to include zero and negative integers, which gave meaning to expressions such as 8^{-3} and $(-9xy)^0$. In this section, we will again extend the definition of exponent—this time to include rational (fractional) exponents. We will see how expressions such as $9^{1/2}$, $\left(\frac{1}{16}\right)^{3/4}$, and $(-32x^5)^{-2/5}$ can be simplified by writing them in an equivalent radical form or by using the rules for exponents.

Rational exponents

We have seen that positive-integer exponents indicate the number of times that a base is to be used as a factor in a product. For example, x^5 means that x is to be used as a factor five times.

$$x^5 = \overbrace{x \cdot x \cdot x \cdot x \cdot x}^{\text{5 factors of } x}$$

Furthermore, we recall the following rules for exponents.

Rules for exponents

If there are no divisions by 0, then for all integers m and n,

1. $x^m x^n = x^{m+n}$ **2.** $(x^m)^n = x^{mn}$ **3.** $(xy)^n = x^n y^n$ **4.** $\left(\frac{x}{y}\right)^n = \frac{x^n}{y^n}$

5. $x^0 = 1 \ (x \neq 0)$ **6.** $x^{-n} = \frac{1}{x^n}$ **7.** $\frac{x^m}{x^n} = x^{m-n}$ **8.** $\left(\frac{x}{y}\right)^{-n} = \left(\frac{y}{x}\right)^n$

It is possible to raise many bases to fractional powers. Since we want fractional exponents to obey the same rules as integer exponents, the square of $10^{1/2}$ must be 10, because

$$(10^{1/2})^2 = 10^{(1/2)2} \quad \text{Keep the base and multiply the exponents.}$$
$$= 10^1 \quad \tfrac{1}{2} \cdot 2 = 1.$$
$$= 10 \quad 10^1 = 10.$$

However, we have seen that

$$\left(\sqrt{10}\right)^2 = 10$$

Since $(10^{1/2})^2$ and $\left(\sqrt{10}\right)^2$ both equal 10, we define $10^{1/2}$ to be $\sqrt{10}$. Likewise, we define

$$10^{1/3} \text{ to be } \sqrt[3]{10} \quad \text{and} \quad 10^{1/4} \text{ to be } \sqrt[4]{10}$$

Rational exponents

If n $(n > 1)$ is a natural number and $\sqrt[n]{x}$ is a real number, then

$$x^{1/n} = \sqrt[n]{x}$$

In words, *a rational exponent of* $\frac{1}{n}$ *indicates that the nth root of the base should be found.*

Using this definition, we can simplify the exponential expression $8^{1/3}$. The first step is to write it as an equivalent expression in radical form and proceed as follows:

$$8^{1/3} = \sqrt[3]{8} \quad \text{The base of the exponential expression is the radicand. The denominator of the fractional exponent is the index of the radical.}$$
$$= 2$$

EXAMPLE 1 ***Simplifying expressions containing rational exponents.*** Write each expression in radical form and simplify, if possible.

a. $9^{1/2} = \sqrt{9}$
$= 3$

b. $-\left(\frac{16}{9}\right)^{1/2} = -\sqrt{\frac{16}{9}}$
$= -\frac{4}{3}$

c. $(-64)^{1/3} = \sqrt[3]{-64}$
$= -4$

d. $16^{1/4} = \sqrt[4]{16}$
$= 2$

e. $\left(\frac{1}{32}\right)^{1/5} = \sqrt[5]{\frac{1}{32}}$
$= \frac{1}{2}$

f. $0^{1/8} = \sqrt[8]{0}$
$= 0$

g. $y^{1/4} = \sqrt[4]{y}$

h. $-(2x^2)^{1/5} = -\sqrt[5]{2x^2}$

Self Check

Write each expression in radical form and simplify, if possible:
a. $16^{1/2}$,
b. $\left(-\frac{27}{8}\right)^{1/3}$, and
c. $-(6x^3)^{1/4}$.

Answers: **a.** 4, **b.** $-\frac{3}{2}$, **c.** $-\sqrt[4]{6x^3}$

EXAMPLE 2 ***Writing roots using rational exponents.*** Write $\sqrt{5xyz}$ as an exponential expression with a rational exponent.

Solution

The radicand is $5xyz$, so the base of the exponential expression is $5xyz$. The index of the radical is an understood 2, so the denominator of the fractional exponent is 2.

$$\sqrt{5xyz} = (5xyz)^{1/2}$$

Self Check

Write the radical with a fractional exponent: $\sqrt[6]{7ab}$.

Answer: $(7ab)^{1/6}$ ■

Rational exponents appear in formulas used in many disciplines, such as science and engineering.

EXAMPLE 3 ***Satellites.*** See Figure 7-10. The formula

$$r = \left(\frac{GMP^2}{4\pi^2}\right)^{1/3}$$

gives the orbital radius (in meters) of a satellite circling the earth, where G and M are constants and P is the time in seconds for the satellite to make one complete revolution. Write the formula using a radical.

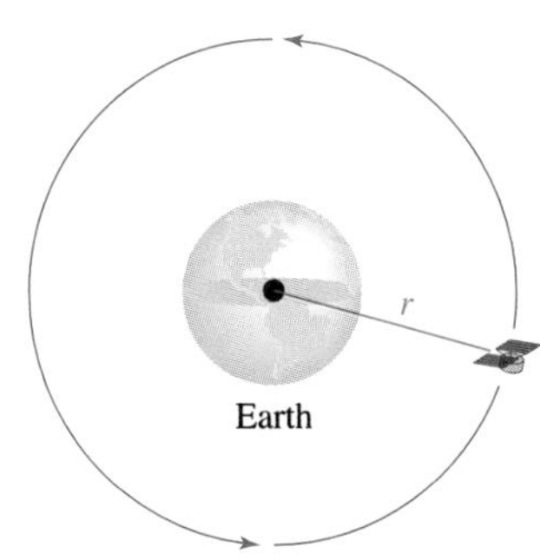

FIGURE 7-10

Solution The fractional exponent $\frac{1}{3}$ has a denominator of 3, which indicates that we are to find the cube root of the base of the exponential expression. So we have

$$r = \sqrt[3]{\frac{GMP^2}{4\pi^2}}$$

■

Exponential expressions with variables in their bases

As with radicals, when n is an *odd natural number* in the expression $x^{1/n}$ $(n > 1)$, there is exactly one real nth root, and we don't have to worry about absolute value symbols.

When n is an *even natural number,* there are two nth roots. Since we want the expression $x^{1/n}$ to represent the positive nth root, we must often use absolute value symbols to guarantee that the simplified result is positive. Thus, if n is even,

$$(x^n)^{1/n} = |x|$$

When n is even and x is negative, the expression $x^{1/n}$ is not a real number.

EXAMPLE 4 ***A variable in the base.*** Simplify each exponential expression. Assume that the variables can be any real number.

a. $(-27x^3)^{1/3} = -3x$ — Because $(-3x)^3 = -27x^3$. Since n is odd, no absolute value symbols are needed.

b. $(256a^8)^{1/8} = 2|a|$ — Because $(2|a|)^8 = 256a^8$. Since n is even and a can be any real number, $2a$ can be negative. Thus, absolute value symbols are needed.

c. $[(y + 4)^2]^{1/2} = |y + 4|$ — Because $|y + 4|^2 = (y + 4)^2$. Since n is even and y can be any real number, $y + 4$ can be negative. Absolute value symbols are needed.

d. $(25b^4)^{1/2} = 5b^2$ — Because $(5b^2)^2 = 25b^4$. Since $b^2 \geq 0$, no absolute value symbols are needed.

e. $(-256x^4)^{1/4}$ is not a real number. — Because no real number raised to the 4th power is $-256x^4$.

Self Check

Simplify each expression:
a. $(625a^4)^{1/4}$ and **b.** $(b^4)^{1/2}$.

Answers: **a.** $5|a|$, **b.** b^2 ■

If we are told that the variables represent positive real numbers in parts b and c of Example 4, the absolute value symbols in the answers are not needed.

$(256a^8)^{1/8} = 2a$ If a represents a positive number, then $2a$ is positive.

$[(y + 4)^2]^{1/2} = y + 4$ If y represents a positive number, then $y + 4$ is positive.

We summarize the cases as follows.

Summary of the definitions of $x^{1/n}$

If n is a natural number greater than 1 and x is a real number,

If $x > 0$, then $x^{1/n}$ is the positive number such that $(x^{1/n})^n = x$.

If $x = 0$, then $x^{1/n} = 0$.

If $x < 0$ $\begin{cases} \text{and } n \text{ is odd, then } x^{1/n} \text{ is the real number such that } (x^{1/n})^n = x. \\ \text{and } n \text{ is even, then } x^{1/n} \text{ is not a real number.} \end{cases}$

Rational exponents with numerators other than 1

We can extend the definition of $x^{1/n}$ to include fractional exponents with numerators other than 1. For example, since $8^{2/3}$ can be written as $(8^{1/3})^2$, we have

$8^{2/3} = (8^{1/3})^2$

$= (\sqrt[3]{8})^2$ Write $8^{1/3}$ in radical form.

$= 2^2$ Find the cube root first: $\sqrt[3]{8} = 2$.

$= 4$ Then find the power.

Thus, we can simplify $8^{2/3}$ by finding the second power of the cube root of 8.

The numerator of the rational exponent is the power.

$$8^{2/3} = (\sqrt[3]{8})^2$$

The base of the exponential expression is the radicand.

The denominator of the exponent is the index of the radical.

We can also simplify $8^{2/3}$ by taking the cube root of 8 squared.

$8^{2/3} = (8^2)^{1/3}$

$= 64^{1/3}$ Find the power first: $8^2 = 64$.

$= \sqrt[3]{64}$ Write $64^{1/3}$ in radical form.

$= 4$ Now find the cube root.

In general, we have the following rule.

Changing from rational exponents to radicals

If m and n are positive integers, $x \geq 0$, and $\frac{m}{n}$ is in simplified form, then

$x^{m/n} = \sqrt[n]{x^m} = (\sqrt[n]{x})^m$

Because of the previous definition, we can interpret $x^{m/n}$ in two ways:

1. $x^{m/n}$ means the nth root of the mth power of x.
2. $x^{m/n}$ means the mth power of the nth root of x.

EXAMPLE 5 *Rational exponents with numerators other than 1.* Simplify each expression in two ways.

a. $9^{3/2} = (\sqrt{9})^3 = 3^3 = 27$ or $9^{3/2} = \sqrt{9^3} = \sqrt{729} = 27$

b. $\left(\frac{1}{16}\right)^{3/4} = \left(\sqrt[4]{\frac{1}{16}}\right)^3 = \left(\frac{1}{2}\right)^3 = \frac{1}{8}$ or $\left(\frac{1}{16}\right)^{3/4} = \left[\left(\frac{1}{16}\right)^3\right]^{1/4} = \left(\frac{1}{4{,}096}\right)^{1/4} = \frac{1}{8}$

c. $(-8x^3)^{4/3} = (\sqrt[3]{-8x^3})^4 = (-2x)^4 = 16x^4$ or $(-8x^3)^{4/3} = \sqrt[3]{(-8x^3)^4} = \sqrt[3]{4{,}096x^{12}} = 16x^4$

Self Check

Simplify **a.** $16^{3/2}$ and **b.** $(-27x^6)^{2/3}$.

Answers: **a.** 64, **b.** $9x^4$

To avoid large numbers, it is usually better to find the root of the base first, as shown with the first solution of each part in Example 5.

Accent on Technology **Rational exponents**

We can evaluate exponential expressions containing rational exponents using the exponential key y^x or x^y on a scientific calculator. For example, to evaluate $10^{2/3}$, we enter these numbers and press these keys:

Keystrokes: 10 y^x (2 ÷ 3) = 4.641588834

To the nearest hundredth, $10^{2/3} \approx 4.64$.

Negative rational exponents

To be consistent with the definition of negative-integer exponents, we define $x^{-m/n}$ as follows.

Definition of $x^{-m/n}$

If m and n are positive integers, $\frac{m}{n}$ is in simplified form, and $x^{1/n}$ is a real number, then

$$x^{-m/n} = \frac{1}{x^{m/n}} \quad \text{and} \quad \frac{1}{x^{-m/n}} = x^{m/n} \quad (x \neq 0)$$

EXAMPLE 6 *Negative rational exponents.* Write each expression without using negative exponents and simplify, if possible.

a. $64^{-1/2} = \frac{1}{64^{1/2}} = \frac{1}{\sqrt{64}} = \frac{1}{8}$

b. $(-16)^{-3/4}$ is not a real number, because $(-16)^{1/4}$ is not a real number.

Self Check

Write without using negative exponents and simplify: **a.** $25^{-3/2}$ and **b.** $(-27a^3)^{-2/3}$.

c. $(-32x^5)^{-2/5} = \frac{1}{(-32x^5)^{2/5}}$

$= \frac{1}{[(-32x^5)^{1/5}]^2}$

$= \frac{1}{\left(\sqrt[5]{-32x^5}\right)^2}$

$= \frac{1}{(-2x)^2}$

$= \frac{1}{4x^2}$

d. $\frac{1}{16^{-3/2}} = 16^{3/2}$

$= (16^{1/2})^3$

$= \left(\sqrt{16}\right)^3$

$= 4^3$

$= 64$

Answers: **a.** $\frac{1}{125}$, **b.** $\frac{1}{9a^2}$

WARNING! By definition, 0^0 is undefined. A base of 0 raised to a negative power is also undefined. For example, 0^{-2} would equal $\frac{1}{0^2}$, which is undefined because we cannot divide by 0.

Applying the rules for exponents

We can use the rules for exponents to simplify many expressions with fractional exponents. If all variables represent positive numbers, no absolute value symbols are necessary.

EXAMPLE 7 ***Simplifying expressions by using the rules for exponents.*** Assume that all variables represent positive numbers. Write all answers without using negative exponents.

a. $5^{2/7}5^{3/7} = 5^{2/7+3/7}$ — Use the rule $x^m x^n = x^{m+n}$.

$= 5^{5/7}$ — Add: $\frac{2}{7} + \frac{3}{7} = \frac{5}{7}$.

b. $(5^{2/7})^3 = 5^{(2/7)(3)}$ — Use the rule $(x^m)^n = x^{mn}$.

$= 5^{6/7}$ — Multiply: $\frac{2}{7}(3) = \frac{6}{7}$.

c. $(a^{2/3}b^{1/2})^6 = (a^{2/3})^6(b^{1/2})^6$ — Use the rule $(xy)^n = x^n y^n$.

$= a^{12/3}b^{6/2}$ — Use the rule $(x^m)^n = x^{mn}$ twice.

$= a^4b^3$ — Simplify the exponent.

d. $\frac{a^{8/3}a^{1/3}}{a^2} = a^{8/3+1/3-2}$ — Use the rules $x^m x^n = x^{m+n}$ and $\frac{x^m}{x^n} = x^{m-n}$.

$= a^{8/3+1/3-6/3}$ — $2 = \frac{6}{3}$.

$= a^{3/3}$ — $\frac{8}{3} + \frac{1}{3} - \frac{6}{3} = \frac{3}{3}$.

$= a$ — $\frac{3}{3} = 1$.

Self Check

Simplify **a.** $(x^{1/3}y^{3/2})^6$ and **b.** $\frac{x^{5/3}x^{2/3}}{x^{1/3}}$.

Answers: **a.** x^2y^9, **b.** x^2

EXAMPLE 8 Assume that all variables represent positive numbers and do the operations. Write all answers without using negative exponents.

a. $a^{4/5}(a^{1/5} + a^{3/5}) = a^{4/5}a^{1/5} + a^{4/5}a^{3/5}$ — Use the distributive property.

$= a^{4/5+1/5} + a^{4/5+3/5}$ — Use the rule $x^m x^n = x^{m+n}$.

$= a^{5/5} + a^{7/5}$ — Simplify the exponents.

$= a + a^{7/5}$ — We cannot add these terms because they are not like terms.

b. $x^{1/2}(x^{-1/2} + x^{1/2}) = x^{1/2}x^{-1/2} + x^{1/2}x^{1/2}$ Use the distributive property.

$= x^{1/2+(-1/2)} + x^{1/2+1/2}$ Use the rule $x^m x^n = x^{m+n}$.

$= x^0 + x^1$ Simplify.

$= 1 + x$ $x^0 = 1$. ■

Simplifying radical expressions

We can simplify many radical expressions by using the following steps.

Using rational exponents to simplify radicals

1. Change the radical expression into an exponential expression.
2. Simplify the rational exponents.
3. Change the exponential expression back into a radical.

EXAMPLE 9 ***Simplifying radical expressions.*** Simplify **a.** $\sqrt[4]{3^2}$, **b.** $\sqrt[8]{x^6}$, and **c.** $\sqrt[9]{27x^6y^3}$.

Self Check

Simplify **a.** $\sqrt[6]{3^3}$ and **b.** $\sqrt[4]{64x^2y^2}$.

Solution

a. $\sqrt[4]{3^2} = (3^2)^{1/4}$ Change the radical to an exponential expression.

$= 3^{2/4}$ Use the rule $(x^m)^n = x^{mn}$.

$= 3^{1/2}$ $\frac{2}{4} = \frac{1}{2}$.

$= \sqrt{3}$ Change back to radical form.

b. $\sqrt[8]{x^6} = (x^6)^{1/8}$ Change the radical to an exponential expression.

$= x^{6/8}$ Use the rule $(x^m)^n = x^{mn}$.

$= x^{3/4}$ $\frac{6}{8} = \frac{3}{4}$.

$= (x^3)^{1/4}$ $\frac{3}{4} = 3\left(\frac{1}{4}\right)$.

$= \sqrt[4]{x^3}$ Change back to radical form.

c. $\sqrt[9]{27x^6y^3} = (3^3x^6y^3)^{1/9}$ Write 27 as 3^3 and change the radical to an exponential expression.

$= 3^{3/9}x^{6/9}y^{3/9}$ Raise each factor to the $\frac{1}{9}$ power by multiplying the fractional exponents.

$= 3^{1/3}x^{2/3}y^{1/3}$ Simplify each fractional exponent.

$= (3x^2y)^{1/3}$ Use the rule $(xy)^n = x^ny^n$ twice.

$= \sqrt[3]{3x^2y}$ Change back to radical form.

Answers: **a.** $\sqrt{3}$, **b.** $\sqrt{8xy}$ ■

STUDY SET Section 7.3

VOCABULARY *In Exercises 1–4, fill in the blanks to make each statement true.*

1. The expressions $4^{1/2}$ and $(-8)^{-2/3}$ have ______________ exponents.

2. In the exponential expression $27^{4/3}$, 27 is the ______________, and 4/3 is the ______________.

3. In the radical expression $\sqrt[3]{4{,}096x^{12}}$, 3 is the ______________, and $4{,}096x^{12}$ is the ______________.

4. $32^{4/5}$ means the fourth ______________ of the fifth ______________ of 32.

CONCEPTS

5. Complete the table by writing the given expression in the alternate form.

Radical form	Exponential form
$\sqrt[5]{25}$	
	$(-27)^{2/3}$
$(\sqrt[4]{16})^{-3}$	
	$81^{3/2}$
$-\sqrt{\frac{9}{64}}$	

6. In your own words, explain the two rules for rational exponents illustrated in the diagrams below.

a. $(-32)^{1/5} = \sqrt[5]{-32}$

b. $125^{4/3} = (\sqrt[3]{125})^4$

7. Graph each number on the number line.

$$\left\{8^{2/3}, (-125)^{1/3}, -16^{-1/4}, 4^{3/2}, -\left(\frac{9}{100}\right)^{-1/2}\right\}$$

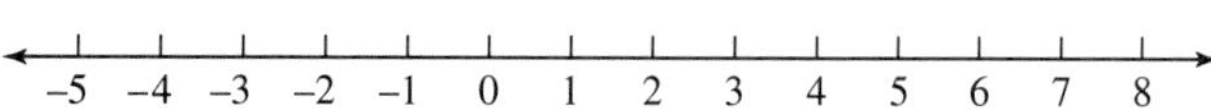

8. Evaluate $25^{3/2}$ in two ways. Which way is easier?

In Exercises 9–14, complete each rule for exponents.

9. $x^m x^n = \square$

10. $(x^m)^n = \square$

11. $\dfrac{x^m}{x^n} = \square$

12. $x^{-n} = \square$

13. $x^{1/n} = \square$

14. $x^{m/n} = \square = \sqrt[n]{x^m}$

In Exercises 15–16, complete each table of values. Then graph the function.

15. $f(x) = x^{1/2}$

x	y
0	
1	
4	
9	
16	

16. $f(x) = x^{1/3}$

x	y
−8	
−1	
0	
1	
8	

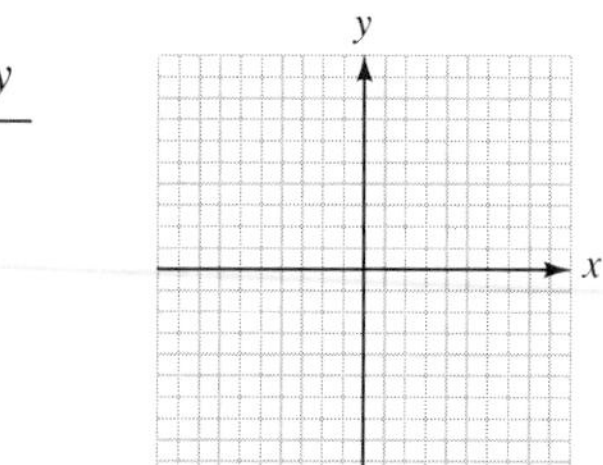

NOTATION *In Exercises 17–18, complete each solution.*

17. Simplify $(100a^4)^{3/2}$.

$$(100a^4)^{3/2} = (\square)^3$$
$$= (\square)^3$$
$$= 1{,}000a^6$$

18. Simplify $(m^{1/3}n^{1/2})^6$.

$$(m^{1/3}n^{1/2})^6 = (\square)^6(n^{1/2})^6$$
$$= m^{\square}n^{6/2}$$
$$= m^2n^3$$

PRACTICE *In Exercises 19–26, write each expression in radical form.*

19. $x^{1/3}$ **20.** $b^{1/2}$ **21.** $(3x)^{1/4}$ **22.** $(4ab)^{1/6}$

23. $\left(\frac{1}{2}x^3y\right)^{1/4}$ **24.** $\left(\frac{3}{4}a^2b^2\right)^{1/5}$ **25.** $(x^2 + y^2)^{1/2}$ **26.** $(x^3 + y^3)^{1/3}$

In Exercises 27–34, change each radical to an exponential expression.

27. $\sqrt{m}$ **28.** $\sqrt[3]{r}$ **29.** $\sqrt[4]{3a}$ **30.** $3\sqrt[5]{a}$

31. $\sqrt[6]{\frac{1}{7}abc}$ **32.** $\sqrt[7]{\frac{3}{8}p^2q}$ **33.** $\sqrt[3]{a^2 - b^2}$ **34.** $\sqrt{x^2 + y^2}$

In Exercises 35–50, simplify each expression, if possible.

35. $4^{1/2}$ **36.** $25^{1/2}$ **37.** $8^{1/3}$ **38.** $125^{1/3}$

39. $16^{1/4}$ **40.** $625^{1/4}$ **41.** $32^{1/5}$ **42.** $0^{1/5}$

43. $\left(\frac{1}{4}\right)^{1/2}$ **44.** $\left(\frac{1}{16}\right)^{1/2}$ **45.** $-16^{1/4}$ **46.** $-125^{1/3}$

47. $(-64)^{1/2}$ **48.** $(-216)^{1/2}$ **49.** $(-27)^{1/3}$ **50.** $(-125)^{1/3}$

In Exercises 51–58, simplify each expression, if possible. Assume that all variables are unrestricted and use absolute value symbols when necessary.

51. $(25y^2)^{1/2}$ **52.** $(-27x^3)^{1/3}$ **53.** $(16x^4)^{1/4}$ **54.** $(-16x^4)^{1/2}$

55. $(243x^5)^{1/5}$ **56.** $[(x + 1)^4]^{1/4}$ **57.** $(-64x^8)^{1/4}$ **58.** $[(x + 5)^3]^{1/3}$

In Exercises 59–70, simplify each expression. Assume that all variables represent positive numbers.

59. $36^{3/2}$ **60.** $27^{2/3}$ **61.** $81^{3/4}$ **62.** $100^{3/2}$

63. $144^{3/2}$ **64.** $1{,}000^{2/3}$ **65.** $\left(\frac{1}{8}\right)^{2/3}$ **66.** $\left(\frac{4}{9}\right)^{3/2}$

67. $(25x^4)^{3/2}$ **68.** $(27a^3b^3)^{2/3}$ **69.** $\left(\frac{8x^3}{27}\right)^{2/3}$ **70.** $\left(\frac{27}{64y^6}\right)^{2/3}$

In Exercises 71–82, write each expression without using negative exponents. Assume that all variables represent positive numbers.

71. $4^{-1/2}$ **72.** $8^{-1/3}$ **73.** $4^{-3/2}$ **74.** $25^{-5/2}$

75. $(16x^2)^{-3/2}$ **76.** $(81c^4)^{-3/2}$ **77.** $(-27y^3)^{-2/3}$ **78.** $(-8z^9)^{-2/3}$

79. $\left(\frac{27}{8}\right)^{-4/3}$ **80.** $\left(\frac{25}{49}\right)^{-3/2}$ **81.** $\left(-\frac{8x^3}{27}\right)^{-1/3}$ **82.** $\left(\frac{16}{81y^4}\right)^{-3/4}$

In Exercises 83–86, use a calculator to evaluate each expression. Round to the nearest hundredth.

83. $\sqrt[3]{15}$ **84.** $\sqrt[4]{50.5}$ **85.** $\sqrt[5]{1.045}$ **86.** $\sqrt[5]{-1{,}000}$

In Exercises 87–106, do the operations. Write the answers without negative exponents. Assume that all variables represent positive numbers.

87. $5^{3/7}5^{2/7}$ **88.** $4^{2/5}4^{2/5}$ **89.** $(4^{1/5})^3$ **90.** $(3^{1/3})^5$

91. $\frac{9^{4/5}}{9^{3/5}}$ **92.** $\frac{7^{2/3}}{7^{1/2}}$ **93.** $6^{-2/3}6^{-4/3}$ **94.** $5^{1/3}5^{-5/3}$

95. $\dfrac{3^{4/3}3^{1/3}}{3^{2/3}}$

96. $\dfrac{2^{5/6}2^{1/3}}{2^{1/2}}$

97. $a^{2/3}a^{1/3}$

98. $b^{3/5}b^{1/5}$

99. $(a^{2/3})^{1/3}$

100. $(t^{4/5})^{10}$

101. $(a^{1/2}b^{1/3})^{3/2}$

102. $(mn^{-2/3})^{-3/5}$

103. $\dfrac{(4x^3y)^{1/2}}{(9xy)^{1/2}}$

104. $\dfrac{(27x^3y)^{1/3}}{(8xy^2)^{2/3}}$

105. $(27x^{-3})^{-1/3}$

106. $(16a^{-2})^{-1/2}$

In Exercises 107–110, do the multiplications. Assume that all variables are positive and write all answers without using negative exponents.

107. $y^{1/3}(y^{2/3} + y^{5/3})$

108. $y^{2/5}(y^{-2/5} + y^{3/5})$

109. $x^{3/5}(x^{7/5} - x^{2/5} + 1)$

110. $x^{4/3}(x^{2/3} + 3x^{5/3} - 4)$

In Exercises 111–114, use rational exponents to simplify each radical. Assume that all variables represent positive numbers.

111. $\sqrt[6]{p^3}$

112. $\sqrt[8]{q^2}$

113. $\sqrt[4]{25b^2}$

114. $\sqrt[9]{-8x^6}$

APPLICATIONS

115. A BALLISTIC PENDULUM See Illustration 1. The formula

$$v = \frac{m + M}{m}(2gh)^{1/2}$$

gives the velocity (in ft/sec) of a bullet with weight m fired into a block with weight M, that raises the height of the block h feet after the collision. The letter g represents the constant, 32. Find the velocity of the bullet to the nearest ft/sec.

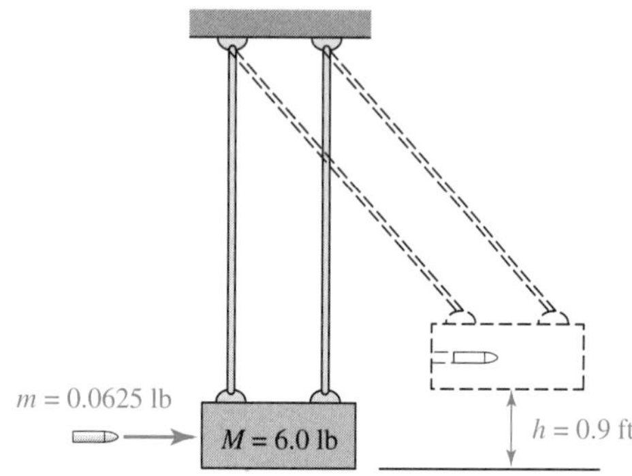

ILLUSTRATION 1

116. GEOGRAPHY The formula

$$A = [s(s - a)(s - b)(s - c)]^{1/2}$$

gives the area of a triangle with sides of length a, b, and c, where s is one-half of the perimeter. Estimate the area of Virginia (to the nearest square mile) using the data given in Illustration 2.

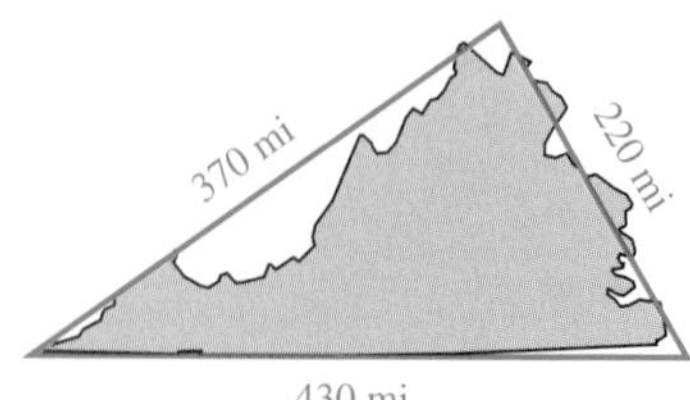

ILLUSTRATION 2

117. RELATIVITY One of the concepts of relativity theory is that an object moving past an observer at a speed near the speed of light appears to have a larger mass because of its motion. If the mass of the object is m_0 when the object is at rest relative to the observer, its mass m will be given by the formula

$$m = m_0(1 - v^2/c^2)^{-1/2}$$

when it is moving with speed v (in miles per second) past the observer. The variable c is the speed of light, 186,000 mi/sec. If a proton with a rest mass of 1 unit is accelerated by a nuclear accelerator to a speed of 160,000 mi/sec, what mass will the technicians observe it to have? Round to the nearest hundredth.

118. LOGGING See Illustration 3. The width w and height h of the strongest rectangular beam that can be cut from a cylindrical log of radius a are given by

$$w = \frac{2a}{3}(3^{1/2}) \qquad h = a\left(\frac{8}{3}\right)^{1/2}$$

Find the width, height, and cross-sectional area of the strongest beam that can be cut from a log with *diameter* 4 feet. Round to the nearest hundredth.

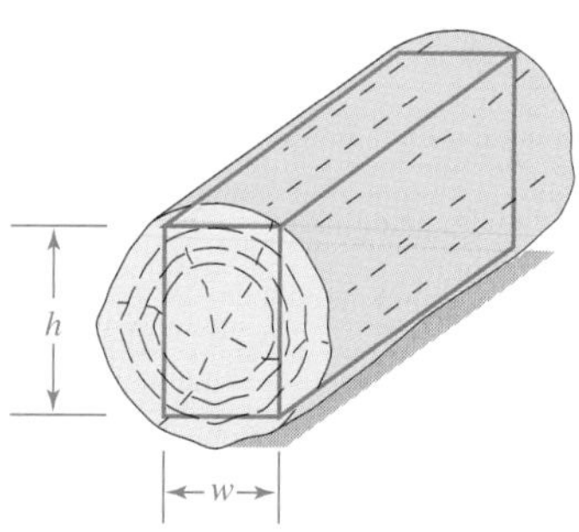

ILLUSTRATION 3

119. CUBICLES The area of the base of a cube is given by the function $A(x) = V^{2/3}$, where V is the volume of the cube. In a preschool room, 18 children's cubicles like that shown in Illustration 4 are placed on the floor around the room. How much floor space is lost to the cubicles? Give your answer in square inches and in square feet.

ILLUSTRATION 4

120. CARPENTRY See Illustration 5. The length L of the longest board that can be carried horizontally around the right-angle corner of two intersecting hallways is given by the formula

$$L = (a^{2/3} + b^{2/3})^{3/2}$$

where a and b represent the widths of the hallways. Find the longest shelf that a carpenter can carry around the corner if $a = 40$ in. and $b = 64$ in. Give your result in inches and in feet. In each case, round to the nearest tenth.

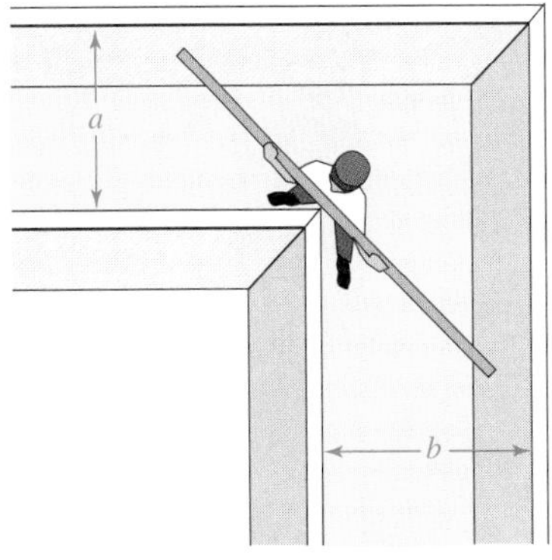

ILLUSTRATION 5

WRITING *Write a paragraph using your own words.*

121. What is a rational exponent? Give some examples.

122. Explain how the root key $\boxed{\sqrt[x]{y}}$ on a scientific calculator can be used in combination with other keys to evaluate the expression $16^{3/4}$.

REVIEW *In Exercises 123–126, solve each inequality. Write each answer as an inequality.*

123. $5x - 4 < 11$

124. $-2(3t - 5) \geq 8$

125. $\frac{4}{5}(r - 3) > \frac{2}{3}(r + 2)$

126. $-4 < 2x - 4 \leq 8$

7.4 Simplifying and Combining Radical Expressions

In this section, you will learn about

- **Properties of radicals**
- **Simplifying radical expressions**
- **Adding and subtracting radical expressions**

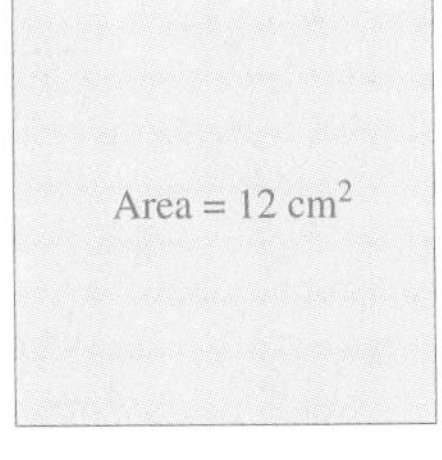

FIGURE 7-11

INTRODUCTION. The square shown in Figure 7-11 has an area of 12 square centimeters. We can determine the length of one side of the square in the following way:

$$A = s^2$$

$$12 = s^2 \quad \text{Substitute 12 for } A.$$

$$\sqrt{12} = \sqrt{s^2} \quad \text{Take the positive square root of both sides.}$$

$$\sqrt{12} = s \quad \text{Each side of the square is } \sqrt{12} \text{ centimeters long.}$$

The form in which we express the length of a side of the square depends upon the situation. If an approximation is acceptable, we can use a calculator to find that $\sqrt{12} \approx 3.464101615$, and then we can round to a specified degree of accuracy. For example, to the nearest hundredth, each side is 3.46 centimeters long.

If the situation calls for the exact length, we must use a radical expression. As you will see in this section, it is common practice to write a radical expression such as $\sqrt{12}$ in *simplified* form. To simplify radicals, we will use the multiplication and division properties of radicals.

Properties of radicals

Many properties of exponents have counterparts in radical notation. For example, because $a^{1/n}b^{1/n} = (ab)^{1/n}$, we have

1. $\sqrt[n]{a}\sqrt[n]{b} = \sqrt[n]{ab}$

For example, if x represents a positive real number,

$$\sqrt{5}\sqrt{5} = \sqrt{5 \cdot 5} = \sqrt{5^2} = 5$$
$$\sqrt[3]{7x}\sqrt[3]{49x^2} = \sqrt[3]{7x \cdot 7^2x^2} = \sqrt[3]{7^3 \cdot x^3} = 7x$$
$$\sqrt[4]{2x^3}\sqrt[4]{8x} = \sqrt[4]{2x^3 \cdot 2^3x} = \sqrt[4]{2^4 \cdot x^4} = 2x$$

These observations suggest the following rule.

Multiplication property of radicals

If $\sqrt[n]{a}$ and $\sqrt[n]{b}$ are real numbers, then
$$\sqrt[n]{ab} = \sqrt[n]{a}\sqrt[n]{b}$$

As long as all radical expressions represent real numbers, *the nth root of the product of two numbers is equal to the product of their nth roots.*

WARNING! The multiplication property of radicals applies to the nth root of the product of two numbers. There is no such property for sums or differences. For example,

$$\sqrt{9+4} \neq \sqrt{9} + \sqrt{4} \qquad \sqrt{9-4} \neq \sqrt{9} - \sqrt{4}$$
$$\sqrt{13} \neq 3 + 2 \qquad \sqrt{5} \neq 3 - 2$$
$$\sqrt{13} \neq 5 \qquad \sqrt{5} \neq 1$$

Thus, $\sqrt{a+b} \neq \sqrt{a} + \sqrt{b}$ and $\sqrt{a-b} \neq \sqrt{a} - \sqrt{b}$.

A second property of radicals involves quotients. Because

$$\frac{a^{1/n}}{b^{1/n}} = \left(\frac{a}{b}\right)^{1/n}$$

it follows that

2. $\dfrac{\sqrt[n]{a}}{\sqrt[n]{b}} = \sqrt[n]{\dfrac{a}{b}} \quad (b \neq 0)$

For example, if x represents a positive real number.

$$\frac{\sqrt{8x^3}}{\sqrt{2x}} = \sqrt{\frac{8x^3}{2x}} = \sqrt{4x^2} = 2x$$
$$\frac{\sqrt[3]{54x^5}}{\sqrt[3]{2x^2}} = \sqrt[3]{\frac{54x^5}{2x^2}} = \sqrt[3]{27x^3} = 3x$$

Equation 2 suggests the following property.

Division property of radicals

If $\sqrt[n]{a}$ and $\sqrt[n]{b}$ are real numbers, then

$$\sqrt[n]{\frac{a}{b}} = \frac{\sqrt[n]{a}}{\sqrt[n]{b}} \quad (b \neq 0)$$

As long as all radical expressions represent real numbers, *the nth root of the quotient of two numbers is equal to the quotient of their nth roots.*

Simplifying radical expressions

A radical expression is said to be in simplest form when each of the following statements is true.

Simplified form of a radical expression

A radical expression is in simplest form when

1. No radicals appear in the denominator of a fraction.
2. The radicand contains no fractions or negative numbers.
3. Each factor in the radicand appears to a power that is less than the index of the radical.

EXAMPLE 1 ***Simplifying radical expressions by using the multiplication property of radicals.*** Simplify **a.** $\sqrt{12}$, **b.** $\sqrt{98}$, **c.** $\sqrt[3]{54}$, and **d.** $-\sqrt[4]{48}$.

Self Check

Simplify **a.** $\sqrt{20}$, **b.** $\sqrt[3]{24}$, and **c.** $\sqrt[5]{-128}$.

Solution

a. Recall that numbers that are squares of integers, such as 1, 4, 9, 16, 25, and 36, are *perfect squares.* To simplify $\sqrt{12}$, we first factor 12 so that one factor is the largest perfect square that divides 12. Since 4 is the largest perfect square factor of 12, we write 12 as $4 \cdot 3$, use the multiplication property of radicals, and simplify.

$\sqrt{12} = \sqrt{4 \cdot 3}$ Write 12 in factored form as $4 \cdot 3$.

$= \sqrt{4}\sqrt{3}$ By the multiplication property of radicals, the square root of a product is equal to the product of the square roots: $\sqrt{4 \cdot 3} = \sqrt{4}\sqrt{3}$.

$= 2\sqrt{3}$ Simplify: $\sqrt{4} = 2$.

b. The largest perfect square factor of 98 is 49. Thus,

$\sqrt{98} = \sqrt{49 \cdot 2}$ Write 98 in factored form: $98 = 49 \cdot 2$.

$= \sqrt{49}\sqrt{2}$ Apply the multiplication property of radicals: $\sqrt{49 \cdot 2} = \sqrt{49}\sqrt{2}$.

$= 7\sqrt{2}$ Simplify: $\sqrt{49} = 7$.

c. Numbers that are cubes of integers, such as 1, 8, 27, 64, 125, and 216, are called *perfect cubes.* Since the largest perfect cube factor of 54 is 27, we have

$\sqrt[3]{54} = \sqrt[3]{27 \cdot 2}$ Write 54 as $27 \cdot 2$.

$= \sqrt[3]{27}\sqrt[3]{2}$ By the multiplication property of radicals, the cube root of a product is equal to the product of the cube roots: $\sqrt[3]{27 \cdot 2} = \sqrt[3]{27}\sqrt[3]{2}$.

$= 3\sqrt[3]{2}$ Simplify: $\sqrt[3]{27} = 3$.

d. The largest perfect fourth-power factor of 48 is 16. Thus,

$$-\sqrt[4]{48} = -\sqrt[4]{16 \cdot 3} \quad \text{Write 48 as } 16 \cdot 3.$$
$$= -\sqrt[4]{16}\sqrt[4]{3} \quad \text{Apply the multiplication property of radicals.}$$
$$= -2\sqrt[4]{3} \quad \text{Simplify: } \sqrt[4]{16} = 2.$$

Answers: **a.** $2\sqrt{5}$, **b.** $2\sqrt[3]{3}$, **c.** $-2\sqrt[5]{4}$ ■

EXAMPLE 2 ***Simplifying radical expressions.*** Simplify **a.** $\sqrt{m^9}$, **b.** $\sqrt{128a^5}$, **c.** $\sqrt[3]{24x^5}$, and **d.** $\sqrt[5]{a^9b^5}$. Assume that all variables represent positive real numbers.

Solution

a. The largest perfect square factor of m^9 is m^8.

$$\sqrt{m^9} = \sqrt{m^8 \cdot m} \quad \text{Write } m^9 \text{ in factored form as } m^8 \cdot m.$$
$$= \sqrt{m^8}\sqrt{m} \quad \text{Apply the multiplication property of radicals.}$$
$$= m^4\sqrt{m} \quad \text{Simplify: } \sqrt{m^8} = m^4.$$

b. Since the largest perfect square factor of 128 is 64 and the largest perfect square factor of a^5 is a^4, the largest perfect square factor of $128a^5$ is $64a^4$. We write $128a^5$ as $64a^4 \cdot 2a$ and proceed as follows:

$$\sqrt{128a^5} = \sqrt{64a^4 \cdot 2a}$$
$$= \sqrt{64a^4}\sqrt{2a} \quad \text{Use the multiplication property of radicals.}$$
$$= 8a^2\sqrt{2a} \quad \text{Simplify: } \sqrt{64a^4} = 8a^2.$$

c. We write $24x^5$ as $8x^3 \cdot 3x^2$ and proceed as follows:

$$\sqrt[3]{24x^5} = \sqrt[3]{8x^3 \cdot 3x^2} \quad 8x^3 \text{ is the largest perfect cube factor of } 24x^5.$$
$$= \sqrt[3]{8x^3}\sqrt[3]{3x^2} \quad \text{Use the multiplication property of radicals.}$$
$$= 2x\sqrt[3]{3x^2} \quad \text{Simplify: } \sqrt[3]{8x^3} = 2x.$$

d. The largest perfect fifth-power factor of a^9 is a^5, and b^5 is a fifth power.

$$\sqrt[5]{a^9b^5} = \sqrt[5]{a^5b^5 \cdot a^4} \quad a^5b^5 \text{ is the largest perfect fifth-power factor of } a^9b^5.$$
$$= \sqrt[5]{a^5b^5}\sqrt[5]{a^4} \quad \text{Use the multiplication property of radicals.}$$
$$= ab\sqrt[5]{a^4} \quad \text{Simplify: } \sqrt[5]{a^5b^5} = ab.$$

Self Check

Simplify **a.** $\sqrt{98b^3}$, **b.** $\sqrt[3]{54y^5}$, and **c.** $\sqrt[4]{t^8u^{15}}$.

Answers: **a.** $7b\sqrt{2b}$, **b.** $3y\sqrt[3]{2y^2}$, **c.** $t^2u^3\sqrt[4]{u^3}$ ■

EXAMPLE 3 ***Simplifying radical expressions using the division property of radicals.*** Simplify **a.** $\sqrt{\frac{7}{64}}$, **b.** $\sqrt{\frac{15}{49x^2}}$, and **c.** $\sqrt[3]{\frac{10x^2}{27y^6}}$. Assume that the variables represent positive real numbers.

Solution

a. We can write the square root of the quotient as the quotient of two square roots.

$$\sqrt{\frac{7}{64}} = \frac{\sqrt{7}}{\sqrt{64}} \quad \text{Apply the division property of radicals.}$$
$$= \frac{\sqrt{7}}{8} \quad \text{Simplify the denominator: } \sqrt{64} = 8.$$

b.
$$\sqrt{\frac{15}{49x^2}} = \frac{\sqrt{15}}{\sqrt{49x^2}} \quad \text{Apply the division property of radicals.}$$
$$= \frac{\sqrt{15}}{7x} \quad \text{Simplify the denominator: } \sqrt{49x^2} = 7x.$$

Self Check

Simplify **a.** $\sqrt{\frac{11}{36a^2}}$ $(a > 0)$

b. $\sqrt[4]{\frac{a^3}{625y^{12}}}$ $(a > 0, y > 0)$.

c. We can write the cube root of the quotient as the quotient of two cube roots. Since $y \neq 0$, we have

$$\sqrt[3]{\frac{10x^2}{27y^6}} = \frac{\sqrt[3]{10x^2}}{\sqrt[3]{27y^6}} \quad \text{Use the division property of radicals.}$$

$$= \frac{\sqrt[3]{10x^2}}{3y^2} \quad \text{Simplify the denominator.}$$

Answers: **a.** $\frac{\sqrt{11}}{6a}$, **b.** $\frac{\sqrt[4]{a^3}}{5y^3}$ ■

EXAMPLE 4 ***Simplifying radical expressions.*** Simplify each expression. Assume that all variables represent positive numbers.

a. $\frac{\sqrt{45xy^2}}{\sqrt{5x}}$ and **b.** $\frac{\sqrt[3]{-432x^5}}{\sqrt[3]{8x}}$.

Solution

a. We can write the quotient of the square roots as the square root of a quotient.

$$\frac{\sqrt{45xy^2}}{\sqrt{5x}} = \sqrt{\frac{45xy^2}{5x}} \quad \text{Use the division property of radicals.}$$

$$= \sqrt{9y^2} \quad \text{Simplify the fraction } \frac{45xy^2}{5x}.$$

$$= 3y \quad \text{Simplify the radical.}$$

b. We can write the quotient of the cube roots as the cube root of a quotient.

$$\frac{\sqrt[3]{-432x^5}}{\sqrt[3]{8x}} = \sqrt[3]{\frac{-432x^5}{8x}} \quad \text{Use the division property of radicals.}$$

$$= \sqrt[3]{-54x^4} \quad \text{Simplify the fraction } \frac{-432x^5}{8x}.$$

$$= \sqrt[3]{-27x^3 \cdot 2x} \quad -27x^3 \text{ is the largest perfect cube that divides } -54x^4.$$

$$= \sqrt[3]{-27x^3}\sqrt[3]{2x} \quad \text{Use the multiplication property of radicals.}$$

$$= -3x\sqrt[3]{2x} \quad \text{Simplify: } \sqrt[3]{-27x^3} = -3x.$$

Self Check

Simplify each expression (assume that all variables represent positive numbers): **a.** $\frac{\sqrt{50ab^2}}{\sqrt{2a}}$ and **b.** $\frac{\sqrt[3]{-2{,}000x^5y^3}}{\sqrt[3]{2x}}$.

Answers: **a.** $5b$, **b.** $-10xy\sqrt[3]{x}$ ■

Adding and subtracting radical expressions

Radical expressions with the same index and the same radicand are called **like** or **similar radicals.** For example, $3\sqrt{2}$ and $2\sqrt{2}$ are like radicals. However,

$3\sqrt{5}$ and $4\sqrt{2}$ are not like radicals, because the radicands are different.

$3\sqrt[4]{5}$ and $2\sqrt[3]{5}$ are not like radicals, because the indexes are different.

For a given expression containing two or more radical terms, we should attempt to combine like radicals, if possible. For example, to simplify the expression $3\sqrt{2} + 2\sqrt{2}$, we use the distributive property to factor out $\sqrt{2}$ and simplify.

$$3\sqrt{2} + 2\sqrt{2} = (3 + 2)\sqrt{2}$$
$$= 5\sqrt{2}$$

Radicals with the same index but different radicands can often be written as like radicals. For example, to simplify the expression $\sqrt{27} - \sqrt{12}$, we simplify both radicals first, and then we combine the like radicals.

$$\sqrt{27} - \sqrt{12} = \sqrt{9 \cdot 3} - \sqrt{4 \cdot 3} \quad \text{Write 27 and 12 in factored form.}$$

$$= \sqrt{9}\sqrt{3} - \sqrt{4}\sqrt{3} \quad \text{Use the multiplication property of radicals.}$$

$$= 3\sqrt{3} - 2\sqrt{3} \quad \text{Simplify } \sqrt{9} \text{ and } \sqrt{4}.$$

$$= (3 - 2)\sqrt{3} \quad \text{Factor out } \sqrt{3}.$$

$$= \sqrt{3} \quad 1\sqrt{3} = \sqrt{3}.$$

As the previous examples suggest, we can use the following rule to add or subtract radicals.

Adding and subtracting radicals	To add or subtract radicals, simplify each radical expression and combine all like radicals. To combine like radicals, add (or subtract) the coefficients and keep the common radical.

EXAMPLE 5 ***Combining like radicals.*** Simplify $2\sqrt{12} - 3\sqrt{48} + 3\sqrt{3}$.

Solution

We simplify $2\sqrt{12}$ and $3\sqrt{48}$ separately and then combine like radicals.

$$\begin{aligned} 2\sqrt{12} - 3\sqrt{48} + 3\sqrt{3} &= 2\sqrt{4 \cdot 3} - 3\sqrt{16 \cdot 3} + 3\sqrt{3} \\ &= 2\sqrt{4}\sqrt{3} - 3\sqrt{16}\sqrt{3} + 3\sqrt{3} \\ &= 2(2)\sqrt{3} - 3(4)\sqrt{3} + 3\sqrt{3} \\ &= 4\sqrt{3} - 12\sqrt{3} + 3\sqrt{3} && \text{All three expressions have the same index and radicand.} \\ &= (4 - 12 + 3)\sqrt{3} && \text{Combine the coefficients of these like radicals and keep } \sqrt{3}. \\ &= -5\sqrt{3} \end{aligned}$$

Self Check

Simplify $3\sqrt{75} - 2\sqrt{12} + 2\sqrt{48}$.

Answer: $19\sqrt{3}$ ■

EXAMPLE 6 ***Combining like radicals.*** Simplify $\sqrt[3]{16} - \sqrt[3]{54} + \sqrt[3]{24}$.

Solution

We begin by simplifying each radical expression separately:

$$\begin{aligned} \sqrt[3]{16} - \sqrt[3]{54} + \sqrt[3]{24} &= \sqrt[3]{8 \cdot 2} - \sqrt[3]{27 \cdot 2} + \sqrt[3]{8 \cdot 3} \\ &= \sqrt[3]{8}\sqrt[3]{2} - \sqrt[3]{27}\sqrt[3]{2} + \sqrt[3]{8}\sqrt[3]{3} \\ &= 2\sqrt[3]{2} - 3\sqrt[3]{2} + 2\sqrt[3]{3} \end{aligned}$$

Now we combine the two radical expressions that have the same index and radicand.

$$\sqrt[3]{16} - \sqrt[3]{54} + \sqrt[3]{24} = -\sqrt[3]{2} + 2\sqrt[3]{3}$$

Self Check

Simplify $\sqrt[3]{24} - \sqrt[3]{16} + \sqrt[3]{54}$.

Answer: $2\sqrt[3]{3} + \sqrt[3]{2}$ ■

WARNING! Even though the radical expressions $-\sqrt[3]{2}$ and $2\sqrt[3]{3}$ in the last line of Example 6 have the same index, we cannot combine them, because their radicands are different. Neither can we combine radical expressions having the same radicand but a different index. For example, the expression $\sqrt[3]{2} + \sqrt[4]{2}$ cannot be simplified.

EXAMPLE 7 ***Combining like radicals.*** Simplify $\sqrt[3]{16x^4} + \sqrt[3]{54x^4} - \sqrt[3]{-128x^4}$.

Solution

We simplify each radical expression separately, factor out $\sqrt[3]{2x}$, and simplify.

$$\begin{aligned} &\sqrt[3]{16x^4} + \sqrt[3]{54x^4} - \sqrt[3]{-128x^4} \\ &= \sqrt[3]{8x^3 \cdot 2x} + \sqrt[3]{27x^3 \cdot 2x} - \sqrt[3]{-64x^3 \cdot 2x} \\ &= \sqrt[3]{8x^3}\sqrt[3]{2x} + \sqrt[3]{27x^3}\sqrt[3]{2x} - \sqrt[3]{-64x^3}\sqrt[3]{2x} \\ &= 2x\sqrt[3]{2x} + 3x\sqrt[3]{2x} + 4x\sqrt[3]{2x} && \text{All three radicals have the same index and radicand.} \\ &= (2x + 3x + 4x)\sqrt[3]{2x} \\ &= 9x\sqrt[3]{2x} && \text{In the parentheses, combine like terms.} \end{aligned}$$

Self Check

Simplify $\sqrt{32x^3} + \sqrt{50x^3} - \sqrt{18x^3}$.

Answer: $6x\sqrt{2x}$ ■

STUDY SET Section 7.4

VOCABULARY *In Exercises 1–4, fill in the blanks to make each statement true.*

1. Radical expressions such as $\sqrt[3]{4}$ and $6\sqrt[3]{4}$ with the same index and the same radicand are called ____________ radicals.

2. Numbers such as 1, 4, 9, 16, 25, and 36 are called perfect ____________. Numbers such as 1, 8, 27, 64, and 125 are called perfect ____________.

3. The largest perfect square ____________ of 27 is 9.

4. "To ____________ $\sqrt{24}$" means to write it as $2\sqrt{6}$.

CONCEPTS *In Exercises 5–6, fill in the blanks to make each statement true.*

5. $\sqrt[n]{ab} = \boxed{}$

In words, the nth root of the ____________ of two numbers is equal to the product of their nth ____________.

6. $\sqrt[n]{\dfrac{a}{b}} = \boxed{}$

In words, the nth root of the ____________ of two numbers is equal to the quotient of their nth ____________.

7. Consider the expressions

$$\sqrt{4 \cdot 5} \text{ and } \sqrt{4} \cdot \sqrt{5}$$

Which side of the equation is

a. the square root of a product?

b. the product of square roots?

c. How are these two expressions related?

8. Consider the expressions

$$\frac{\sqrt[3]{a}}{\sqrt[3]{x^2}} \text{ and } \sqrt[3]{\frac{a}{x^2}}$$

Which side of the equation is

a. the cube root of a quotient?

b. the quotient of cube roots?

c. How are these two expressions related?

9. a. Write two radical expressions that have the same radicand but a different index. Can the expressions be added?

b. Write two radical expressions that have the same index but a different radicand. Can the expressions be added?

10. Explain the mistake in the student's solution shown below.

Simplify $\sqrt[3]{54}$.

$$\begin{aligned}\sqrt[3]{54} &= \sqrt[3]{27 + 27}\\ &= \sqrt[3]{27} + \sqrt[3]{27}\\ &= 3 + 3\\ &= 6\end{aligned}$$

NOTATION *In Exercises 11–12, complete each solution.*

11. Simplify the radical expression.

$$\begin{aligned}\sqrt[3]{32k^4} &= \sqrt[3]{\boxed{} \cdot 4k}\\ &= \sqrt[3]{\boxed{}}\sqrt[3]{4k}\\ &= 2k\sqrt[3]{4k}\end{aligned}$$

12. Simplify the radical expression.

$$\begin{aligned}\frac{\sqrt{80s^2t^4}}{\sqrt{5s^2}} &= \sqrt{\frac{80s^2t^4}{\boxed{}}}\\ &= \sqrt{\boxed{}}\\ &= 4t^2\end{aligned}$$

PRACTICE *In Exercises 13–28, simplify each expression. Assume that all variables represent positive real numbers.*

13. $\sqrt{6}\sqrt{6}$

14. $\sqrt{11}\sqrt{11}$

15. $\sqrt{t}\sqrt{t}$

16. $-\sqrt{z}\sqrt{z}$

17. $\sqrt[3]{5x^2}\sqrt[3]{25x}$ **18.** $\sqrt[4]{25a}\sqrt[4]{25a^3}$ **19.** $\dfrac{\sqrt{500}}{\sqrt{5}}$ **20.** $\dfrac{\sqrt{128}}{\sqrt{2}}$

21. $\dfrac{\sqrt{98x^3}}{\sqrt{2x}}$ **22.** $\dfrac{\sqrt{75y^5}}{\sqrt{3y}}$ **23.** $\dfrac{\sqrt{180ab^4}}{\sqrt{5ab^2}}$ **24.** $\dfrac{\sqrt{112ab^3}}{\sqrt{7ab}}$

25. $\dfrac{\sqrt[3]{48}}{\sqrt[3]{6}}$ **26.** $\dfrac{\sqrt[3]{64}}{\sqrt[3]{8}}$ **27.** $\dfrac{\sqrt[3]{189a^4}}{\sqrt[3]{7a}}$ **28.** $\dfrac{\sqrt[3]{243x^7}}{\sqrt[3]{9x}}$

In Exercises 29–48, simplify each radical expression.

29. $\sqrt{20}$ **30.** $\sqrt{8}$ **31.** $-\sqrt{200}$ **32.** $-\sqrt{250}$

33. $\sqrt[3]{80}$ **34.** $\sqrt[3]{270}$ **35.** $\sqrt[3]{-81}$ **36.** $\sqrt[3]{-72}$

37. $\sqrt[4]{32}$ **38.** $\sqrt[4]{48}$ **39.** $\sqrt[5]{96}$ **40.** $\sqrt[7]{256}$

41. $\sqrt{\dfrac{7}{9}}$ **42.** $\sqrt{\dfrac{3}{4}}$ **43.** $\sqrt[3]{\dfrac{7}{64}}$ **44.** $\sqrt[3]{\dfrac{4}{125}}$

45. $\sqrt[4]{\dfrac{3}{10{,}000}}$ **46.** $\sqrt[5]{\dfrac{4}{243}}$ **47.** $\sqrt[5]{\dfrac{3}{32}}$ **48.** $\sqrt[6]{\dfrac{5}{64}}$

In Exercises 49–68, simplify each radical expression. Assume that all variables represent positive numbers.

49. $\sqrt{50x^2}$ **50.** $\sqrt{75a^2}$ **51.** $\sqrt{32b}$ **52.** $\sqrt{80c}$

53. $-\sqrt{112a^3}$ **54.** $\sqrt{147a^5}$ **55.** $\sqrt{175a^2b^3}$ **56.** $\sqrt{128a^3b^5}$

57. $-\sqrt{300xy}$ **58.** $\sqrt{200x^2y}$ **59.** $\sqrt[3]{-54x^6}$ **60.** $-\sqrt[3]{-81a^3}$

61. $\sqrt[3]{16x^{12}y^3}$ **62.** $\sqrt[3]{40a^3b^6}$ **63.** $\sqrt[4]{32x^{12}y^4}$ **64.** $\sqrt[5]{64x^{10}y^5}$

65. $\sqrt{\dfrac{z^2}{16x^2}}$ **66.** $\sqrt{\dfrac{b^4}{64a^8}}$ **67.** $\sqrt[4]{\dfrac{5x}{16z^4}}$ **68.** $\sqrt[3]{\dfrac{11a^2}{125b^6}}$

In Exercises 69–104, simplify and combine like radicals. All variables represent positive numbers.

69. $4\sqrt{2x} + 6\sqrt{2x}$ **70.** $6\sqrt[3]{5y} + 3\sqrt[3]{5y}$

71. $8\sqrt[5]{7a^2} - 7\sqrt[5]{7a^2}$ **72.** $10\sqrt[6]{12xyz} - \sqrt[6]{12xyz}$

73. $\sqrt{2} - \sqrt{8}$ **74.** $\sqrt{20} - \sqrt{125}$

75. $\sqrt{98} - \sqrt{50}$ **76.** $\sqrt{72} - \sqrt{200}$

77. $3\sqrt{24} + \sqrt{54}$ **78.** $\sqrt{18} + 2\sqrt{50}$

79. $\sqrt[3]{24} + \sqrt[3]{3}$ **80.** $\sqrt[3]{16} + \sqrt[3]{128}$

81. $\sqrt[3]{32} - \sqrt[3]{108}$ **82.** $\sqrt[3]{80} - \sqrt[3]{10{,}000}$

83. $2\sqrt[3]{125} - 5\sqrt[3]{64}$ **84.** $3\sqrt[3]{27} + 12\sqrt[3]{216}$

85. $14\sqrt[4]{32} - 15\sqrt[4]{162}$ **86.** $23\sqrt[4]{768} + \sqrt[4]{48}$

87. $3\sqrt[4]{512} + 2\sqrt[4]{32}$ **88.** $4\sqrt[4]{243} - \sqrt[4]{48}$

89. $\sqrt{98} - \sqrt{50} - \sqrt{72}$ **90.** $\sqrt{20} + \sqrt{125} - \sqrt{80}$

91. $\sqrt{18} + \sqrt{300} - \sqrt{243}$ **92.** $\sqrt{80} - \sqrt{128} + \sqrt{288}$

93. $2\sqrt[3]{16} - \sqrt[3]{54} - 3\sqrt[3]{128}$ **94.** $\sqrt[4]{48} - \sqrt[4]{243} - \sqrt[4]{768}$

95. $\sqrt{25y^2z} - \sqrt{16y^2z}$ **96.** $\sqrt{25yz^2} + \sqrt{9yz^2}$

97. $\sqrt{36xy^2} + \sqrt{49xy^2}$ **98.** $3\sqrt{2x} - \sqrt{8x}$

99. $2\sqrt[3]{64a} + 2\sqrt[3]{8a}$ **100.** $3\sqrt[4]{x^4y} - 2\sqrt[4]{x^4y}$

101. $\sqrt{y^5} - \sqrt{9y^5} - \sqrt{25y^5}$ **102.** $\sqrt{8y^7} + \sqrt{32y^7} - \sqrt{2y^7}$

103. $\sqrt[5]{x^6y^2} + \sqrt[5]{32x^6y^2} + \sqrt[5]{x^6y^2}$ **104.** $\sqrt[3]{xy^4} + \sqrt[3]{8xy^4} - \sqrt[3]{27xy^4}$

APPLICATIONS *In Exercises 105–110, first give the exact answer, expressed as a simplified radical expression. Then give an approximation, rounded to the nearest tenth.*

105. UMBRELLA The surface area of a cone is given by the formula $S = \pi r\sqrt{r^2 + h^2}$, where r is the radius of the base and h is its height. Use this formula to find the number of square feet of waterproof cloth used to make the umbrella shown in Illustration 1.

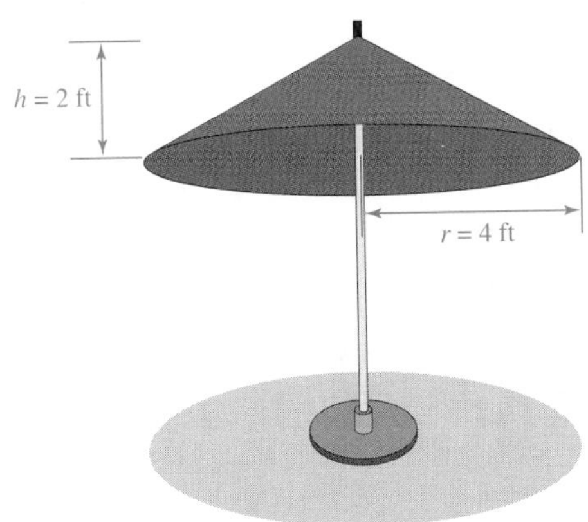

ILLUSTRATION 1

106. STRUCTURAL ENGINEERING Engineers have determined that two additional supports need to be added to strengthen a truss. See Illustration 2. Find the length L of each new support using the formula

$$L = \sqrt{\frac{b^2}{2} + \frac{c^2}{2} - \frac{a^2}{4}}.$$

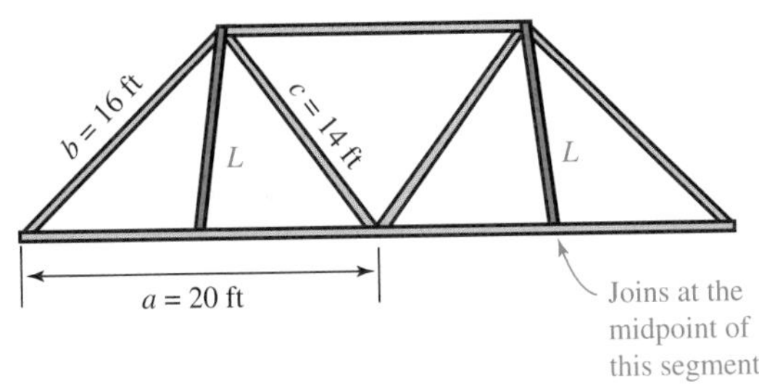

ILLUSTRATION 2

107. BLOW DRYER The current I (in amps), the power P (in watts), and the resistance R (in ohms) are related by the formula $I = \sqrt{\frac{P}{R}}$. What current is needed for a 1,200-watt hair dryer if the resistance is 16 ohms?

108. COMMUNICATIONS SATELLITE Engineers have determined that a spherical communications satellite needs to have a capacity of 565.2 cubic feet to house all of its operating systems. The volume V of a sphere is related to its radius r by the formula $r = \sqrt[3]{\frac{3V}{4\pi}}$. What radius must the satellite have to meet the engineer's specification? Use 3.14 for π.

109. DUCTWORK The pattern shown in Illustration 3 is laid out on a sheet of galvanized tin. Then it is cut out with snips and bent to make an air conditioning duct connection. Find the total length of the cut that must be made with the tin snips. (All measurements are in inches.)

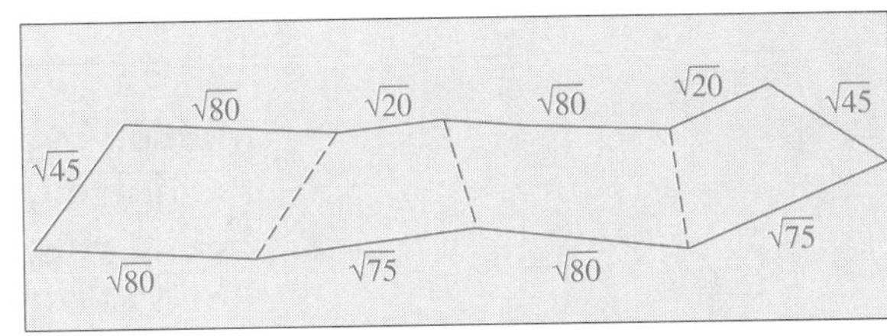

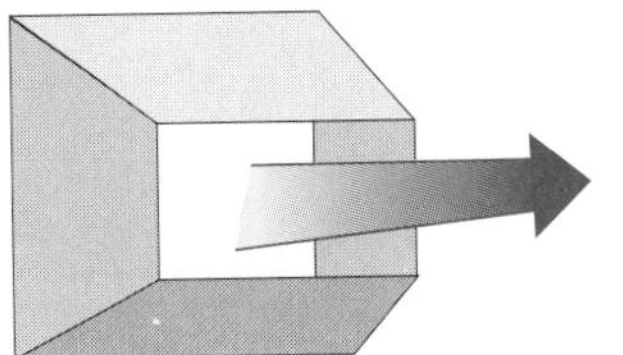

ILLUSTRATION 3

110. OUTDOOR COOKING The diameter of a circle is given by the function $d(A) = 2\sqrt{\frac{A}{\pi}}$, where A is the area of the circle. Find the difference between the diameters of the barbecue grills shown in Illustration 4.

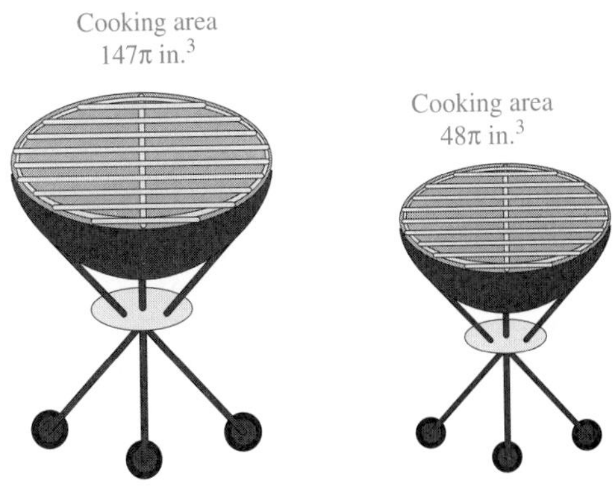

ILLUSTRATION 4

WRITING *Write a paragraph using your own words.*

111. Explain why $\sqrt[3]{9x^4}$ is not in simplified form.

112. How are the procedures used to simplify $3x + 4x$ and $3\sqrt{x} + 4\sqrt{x}$ similar?

REVIEW *In Exercises 113–116, do each operation.*

113. $3x^2y^3(-5x^3y^{-4})$

114. $(2x^2 - 9x - 5) \cdot \dfrac{x}{2x^2 + x}$

115. $2p - 5\overline{)6p^2 - 7p - 25}$

116. $\dfrac{xy}{\frac{1}{x} - \frac{1}{y}}$

7.5 Multiplying and Dividing Radical Expressions

In this section, you will learn about

- **Multiplying a monomial by a monomial**
- **Multiplying a polynomial by a monomial**
- **Multiplying a polynomial by a polynomial**
- **Rationalizing denominators**
- **Rationalizing binomial denominators**

INTRODUCTION. In this section, we will discuss the methods used to multiply and divide radical expressions. Solving these problems often requires the use of procedures and properties studied earlier, such as simplifying radical expressions, combining like radicals, the FOIL method, and the distributive property.

Multiplying a monomial by a monomial

Radical expressions with the same index can be multiplied. For example, to find the product of $\sqrt{5}$ and $\sqrt{10}$, we proceed as follows:

$\sqrt{5}\sqrt{10} = \sqrt{5 \cdot 10}$ By the multiplication property of radicals: $\sqrt[n]{a}\sqrt[n]{b} = \sqrt[n]{ab}$.

$= \sqrt{50}$ Multiply under the radical. Note that $\sqrt{50}$ can be simplified.

$= \sqrt{25 \cdot 2}$ Begin the process of simplifying $\sqrt{50}$ by factoring 50.

$= 5\sqrt{2}$ $\sqrt{25 \cdot 2} = \sqrt{25}\sqrt{2} = 5\sqrt{2}$.

EXAMPLE 1 ***Multiplying radical expressions.*** Multiply $3\sqrt{6}$ by $2\sqrt{3}$.

Solution

We use the commutative and associative properties of multiplication to multiply the coefficients and the radicals separately. Then we simplify any radicals in the product, if possible.

$3\sqrt{6} \cdot 2\sqrt{3} = 3(2)\sqrt{6}\sqrt{3}$ Multiply the coefficients and multiply the radicals.

$= 6\sqrt{18}$ $3(2) = 6$ and $\sqrt{6}\sqrt{3} = \sqrt{18}$.

$= 6\sqrt{9}\sqrt{2}$ Simplify: $\sqrt{18} = \sqrt{9 \cdot 2} = \sqrt{9}\sqrt{2}$.

$= 6(3)\sqrt{2}$ $\sqrt{9} = 3$.

$= 18\sqrt{2}$ Multiply.

Self Check

Multiply $-2\sqrt{7}$ by $5\sqrt{2}$.

Answer: $-10\sqrt{14}$ ■

Multiplying a polynomial by a monomial

To multiply a polynomial by a monomial, we use the distributive property to remove parentheses and then simplify each resulting term, if possible.

EXAMPLE 2 ***Using the distributive property with radical expressions.*** Simplify $3\sqrt{3}(4\sqrt{8} - 5\sqrt{10})$.

Solution

$3\sqrt{3}(4\sqrt{8} - 5\sqrt{10})$

$= 3\sqrt{3} \cdot 4\sqrt{8} - 3\sqrt{3} \cdot 5\sqrt{10}$ Use the distributive property.

$= 12\sqrt{24} - 15\sqrt{30}$ Multiply the coefficients and multiply the radicals.

$= 12\sqrt{4}\sqrt{6} - 15\sqrt{30}$ Simplify: $\sqrt{24} = \sqrt{4 \cdot 6} = \sqrt{4}\sqrt{6}$.

$= 12(2)\sqrt{6} - 15\sqrt{30}$ $\sqrt{4} = 2$.

$= 24\sqrt{6} - 15\sqrt{30}$

Self Check

Simplify $4\sqrt{2}(3\sqrt{5} - 2\sqrt{8})$.

Answer: $12\sqrt{10} - 32$ ■

Multiplying a polynomial by a polynomial

To multiply a binomial by a binomial, we use the FOIL method.

EXAMPLE 3 ***Using the FOIL method with radical expressions.*** Multiply $(\sqrt{7} + \sqrt{2})(\sqrt{7} - 3\sqrt{2})$.

Solution

$(\sqrt{7} + \sqrt{2})(\sqrt{7} - 3\sqrt{2})$

$= \sqrt{7}\sqrt{7} - 3\sqrt{7}\sqrt{2} + \sqrt{2}\sqrt{7} - 3\sqrt{2}\sqrt{2}$ Use the FOIL method.

$= 7 - 3\sqrt{14} + \sqrt{14} - 3(2)$ Do each multiplication: $\sqrt{7}\sqrt{7} = \sqrt{49} = 7$ and $\sqrt{2}\sqrt{2} = \sqrt{4} = 2$.

$= 7 - 2\sqrt{14} - 6$ Combine like radicals: $-3\sqrt{14} + \sqrt{14} = -2\sqrt{14}$.

$= 1 - 2\sqrt{14}$ Combine like terms: $7 - 6 = 1$.

Self Check

Multiply $(\sqrt{5} + 2\sqrt{3})(\sqrt{5} - \sqrt{3})$.

Answer: $-1 + \sqrt{15}$ ■

Technically, the expression $\sqrt{3x} - \sqrt{5}$ is not a polynomial, because the variable does not have a whole-number exponent $(\sqrt{3x} = 3^{1/2}x^{1/2})$. However, we will multiply such expressions as if they were polynomials.

EXAMPLE 4 ***Using the FOIL method with radical expressions.*** Multiply $(\sqrt{3x} - \sqrt{5})(\sqrt{2x} + \sqrt{10})$. Assume that $x > 0$.

Solution

$(\sqrt{3x} - \sqrt{5})(\sqrt{2x} + \sqrt{10})$

$= \sqrt{3x}\sqrt{2x} + \sqrt{3x}\sqrt{10} - \sqrt{5}\sqrt{2x} - \sqrt{5}\sqrt{10}$ Use FOIL.

$= \sqrt{6x^2} + \sqrt{30x} - \sqrt{10x} - \sqrt{50}$ Do each multiplication.

$= \sqrt{6}\sqrt{x^2} + \sqrt{30x} - \sqrt{10x} - \sqrt{25}\sqrt{2}$ Simplify $\sqrt{6x^2}$ and $\sqrt{50}$.

$= \sqrt{6}x + \sqrt{30x} - \sqrt{10x} - 5\sqrt{2}$

Self Check

Multiply $(\sqrt{x} + 1)(\sqrt{x} - 3)$. Assume that $x > 0$.

Answer: $x - 2\sqrt{x} - 3$ ■

WARNING! It is important to draw the radical sign carefully so that it completely covers the radicand, but no more than the radicand. To avoid confusion, we often write an expression such as $\sqrt{6x}$ in the form $x\sqrt{6}$.

Rationalizing denominators

In Section 7.4, we saw that a radical expression is in simplified form when each of the following statements is true.

1. No radicals appear in the denominator of a fraction.
2. The radicand contains no fractions or negative numbers.
3. Each factor in the radicand appears to a power that is less than the index of the radical.

We now consider radical expressions that do not satisfy requirement 1 and radical expressions that do not satisfy requirement 2 of this list. We will introduce an algebraic technique, called *rationalizing the denominator,* that is used to write such expressions in an equivalent simplified form.

To divide radical expressions, we **rationalize the denominator** of a fraction to replace the denominator with a rational number. For example, to divide $\sqrt{70}$ by $\sqrt{3}$, we write the division as the fraction

$$\frac{\sqrt{70}}{\sqrt{3}}$$ The denominator is the irrational number $\sqrt{3}$. This radical expression is not in simplified form, because a radical appears in the denominator.

To eliminate the radical in the denominator, we multiply the numerator and the denominator by a number that will give a perfect square *under the radical in the denominator.* Because $3 \cdot 3 = 9$ and 9 is a perfect square, $\sqrt{3}$ is such a number.

$$\frac{\sqrt{70}}{\sqrt{3}} = \frac{\sqrt{70} \cdot \sqrt{3}}{\sqrt{3} \cdot \sqrt{3}}$$ Multiply numerator and denominator of the fraction by $\sqrt{3}$.

$$= \frac{\sqrt{210}}{\sqrt{9}}$$ Multiply the radicals in the numerator and in the denominator.

$$= \frac{\sqrt{210}}{3}$$ Simplify: $\sqrt{9} = 3$. The denominator is now the rational number 3.

Since there is no radical in the denominator and $\sqrt{210}$ cannot be simplified, the expression $\frac{\sqrt{210}}{3}$ is in simplest form, and the division is complete.

EXAMPLE 5 ***Division of radical expressions.*** Simplify by rationalizing the denominator: **a.** $\sqrt{\frac{20}{7}}$ and **b.** $\frac{4}{\sqrt[3]{2}}$.

Solution

a. This radical expression is not in simplified form, because the radicand contains a fraction. We begin by writing the square root of the quotient as the quotient of two square roots:

$$\sqrt{\frac{20}{7}} = \frac{\sqrt{20}}{\sqrt{7}}$$ Apply the division property of radicals: $\sqrt[n]{\frac{a}{b}} = \frac{\sqrt[n]{a}}{\sqrt[n]{b}}$.

Self Check

Rationalize the denominator:

a. $\sqrt{\frac{24}{5}}$ and **b.** $\frac{5}{\sqrt[4]{3}}$.

To rationalize the denominator, we proceed as follows:

$$\frac{\sqrt{20}}{\sqrt{7}} = \frac{\sqrt{20} \cdot \sqrt{7}}{\sqrt{7} \cdot \sqrt{7}}$$ Multiply the numerator and denominator by $\sqrt{7}$.

$$= \frac{\sqrt{140}}{\sqrt{49}}$$ Multiply the radicals.

$$= \frac{2\sqrt{35}}{7}$$ Simplify: $\sqrt{140} = \sqrt{4 \cdot 35} = \sqrt{4}\sqrt{35} = 2\sqrt{35}$ and $\sqrt{49} = 7$.

b. Here, we must rationalize a denominator that is a cube root. We multiply the numerator and the denominator by a number that will give a perfect cube under the radical sign. Since $2 \cdot 4 = 8$ is a perfect cube, $\sqrt[3]{4}$ is such a number.

$$\frac{4}{\sqrt[3]{2}} = \frac{4 \cdot \sqrt[3]{4}}{\sqrt[3]{2} \cdot \sqrt[3]{4}}$$ Multiply numerator and denominator by $\sqrt[3]{4}$.

$$= \frac{4\sqrt[3]{4}}{\sqrt[3]{8}}$$ Multiply the radicals in the denominator.

$$= \frac{4\sqrt[3]{4}}{2}$$ Simplify: $\sqrt[3]{8} = 2$.

$$= 2\sqrt[3]{4}$$ Simplify: $\frac{4\sqrt[3]{4}}{2} = \frac{\overset{1}{\cancel{2}} \cdot 2\sqrt[3]{4}}{\underset{1}{\cancel{2}}} = 2\sqrt[3]{4}$.

Answers: **a.** $\frac{2\sqrt{30}}{5}$, **b.** $\frac{5\sqrt[4]{27}}{3}$

EXAMPLE 6 ***Photography*** Many camera lenses (see Figure 7-12) have an adjustable opening called the *aperture,* which controls the amount of light passing through the lens. The *f-number* of a lens is its *focal length* divided by the diameter of its circular aperture:

$$f\text{-number} = \frac{f}{d}$$ f is the focal length, and d is the diameter of the aperture.

A lens with a focal length of 12 centimeters and an aperture with a diameter of 6 centimeters has an f-number of $\frac{12}{6}$ and is an $f/2$ lens. If the area of the aperture is reduced to admit half as much light, the f-number of the lens will change. Find the new f-number.

Solution We first find the area of the aperture when its diameter is 6 centimeters.

$$A = \pi r^2$$ The formula for the area of a circle.

$$A = \pi(3)^2$$ Since a radius is half the diameter, substitute 3 for r.

$$A = 9\pi$$

When the size of the aperture is reduced to admit half as much light, the area of the aperture will be $\frac{9\pi}{2}$ square centimeters. To find the diameter d of a circle with this area, we proceed as follows:

FIGURE 7-12

$$A = \pi r^2$$ The formula for the area of a circle.

$$\frac{9\pi}{2} = \pi\left(\frac{d}{2}\right)^2$$ Substitute $\frac{9\pi}{2}$ for A and $\frac{d}{2}$ for r.

$$\frac{9\pi}{2} = \frac{\pi d^2}{4}$$ $\left(\frac{d}{2}\right)^2 = \frac{d^2}{4}$.

$$18 = d^2$$ Multiply both sides by 4 and divide both sides by π.

$$d = 3\sqrt{2}$$ Take the positive square root of both sides. Then simplify: $\sqrt{18} = \sqrt{9}\sqrt{2} = 3\sqrt{2}$.

Since the focal length of the lens is still 12 centimeters and the diameter is now $3\sqrt{2}$ centimeters, the new f-number of the lens is

$$f\text{-number} = \frac{f}{d} = \frac{12}{3\sqrt{2}} \quad \text{Substitute 12 for } f \text{ and } 3\sqrt{2} \text{ for } d.$$

$$= \frac{12\sqrt{2}}{3\sqrt{2}\sqrt{2}} \quad \text{Rationalize the denominator.}$$

$$= \frac{12\sqrt{2}}{3(2)} \quad \text{Simplify the denominator: } \sqrt{2}\sqrt{2} = 2.$$

$$= 2\sqrt{2} \quad \text{Simplify: } \frac{12\sqrt{2}}{3(2)} = \frac{12\sqrt{2}}{6} = \frac{\overset{1}{\cancel{6}} \cdot 2\sqrt{2}}{\underset{1}{\cancel{6}}}.$$

$$\approx 2.828427125 \quad \text{Use a calculator.}$$

The lens is now an $f/2.8$ lens.

EXAMPLE 7 ***Rationalizing a variable denominator.*** Rationalize the denominator of $\frac{\sqrt{5xy^2}}{\sqrt{xy^3}}$ (x and y are positive numbers).

Self Check

Rationalize the denominator of $\frac{\sqrt{4ab^3}}{\sqrt{2a^2b^2}}$.

Solution

In each case, we write the expression as a quotient of radicals and then simplify the radicand.

Method 1

$$\frac{\sqrt{5xy^2}}{\sqrt{xy^3}} = \sqrt{\frac{5xy^2}{xy^3}} = \sqrt{\frac{5}{y}} = \frac{\sqrt{5}}{\sqrt{y}} = \frac{\sqrt{5}\sqrt{y}}{\sqrt{y}\sqrt{y}} = \frac{\sqrt{5y}}{y}$$

Method 2

$$\frac{\sqrt{5xy^2}}{\sqrt{xy^3}} = \sqrt{\frac{5xy^2}{xy^3}} = \sqrt{\frac{5}{y}} = \sqrt{\frac{5 \cdot y}{y \cdot y}} = \frac{\sqrt{5y}}{\sqrt{y^2}} = \frac{\sqrt{5y}}{y}$$

Answer: $\frac{\sqrt{2ab}}{a}$

EXAMPLE 8 ***Simplifying before rationalizing.*** Rationalize the denominator of $\sqrt{\frac{11}{20q^5}}$ ($q > 0$).

Self Check

Rationalize the denominator of $\sqrt[3]{\frac{1}{16h^4}}$.

Solution

We write the expression as a quotient of two radicals. Then we simplify the radical in the denominator before rationalizing.

$$\sqrt{\frac{11}{20q^5}} = \frac{\sqrt{11}}{\sqrt{20q^5}} \quad \text{The square root of a quotient is the quotient of the square roots.}$$

$$= \frac{\sqrt{11}}{\sqrt{4q^4 \cdot 5q}} \quad \text{To simplify } \sqrt{20q^5}\text{, write it as } \sqrt{4q^4 \cdot 5q}.$$

$$= \frac{\sqrt{11}}{2q^2\sqrt{5q}} \quad \sqrt{4q^4 \cdot 5q} = \sqrt{4q^4}\sqrt{5q} = 2q^2\sqrt{5q}.$$

$$= \frac{\sqrt{11}\sqrt{5q}}{2q^2\sqrt{5q}\sqrt{5q}}$$ Rationalize the denominator.

$$= \frac{\sqrt{55q}}{2q^2(5q)}$$ Multiply the radicals: $\sqrt{5q}\sqrt{5q} = 5q$.

$$= \frac{\sqrt{55q}}{10q^3}$$ Multiply in the denominator.

Answer: $\frac{\sqrt[3]{4h^2}}{4h^2}$ ■

EXAMPLE 9 ***Rationalizing a variable denominator.*** Rationalize the denominator of $\frac{\sqrt[3]{5}}{\sqrt[3]{9m}}$.

Solution

We multiply the numerator and the denominator by $\sqrt[3]{3m^2}$, which will produce a perfect cube, $27m^3$, under the radical sign in the denominator.

$$\frac{\sqrt[3]{5}}{\sqrt[3]{9m}} = \frac{\sqrt[3]{5}\sqrt[3]{3m^2}}{\sqrt[3]{9m}\sqrt[3]{3m^2}}$$ Multiply numerator and denominator by $\sqrt[3]{3m^2}$.

$$= \frac{\sqrt[3]{15m^2}}{\sqrt[3]{27m^3}}$$ Multiply the radicals.

$$= \frac{\sqrt[3]{15m^2}}{3m}$$ Simplify: $\sqrt[3]{27m^3} = 3m$.

Self Check

Rationalize the denominator of $\frac{\sqrt{5}}{\sqrt{17b}}$.

Answer: $\frac{\sqrt{85b}}{17b}$ ■

Rationalizing binomial denominators

To rationalize a denominator that contains a binomial expression involving square roots, we multiply its numerator and denominator by the *conjugate* of its denominator. **Conjugate binomials** are binomials with the same terms but with opposite signs between their terms.

Conjugate binomials The **conjugate** of the binomial $a + b$ is $a - b$, and the conjugate of $a - b$ is $a + b$.

When we multiply a binomial expression involving square roots and its conjugate, the product does not contain a radical. For example, if we use the FOIL method to multiply $\sqrt{x} + 2$ and its conjugate, $\sqrt{x} - 2$, we have

$$(\sqrt{x} + 2)(\sqrt{x} - 2) = \sqrt{x}\sqrt{x} - 2\sqrt{x} + 2\sqrt{x} - 4$$
$$= x - 4$$ Combine like terms: $-2\sqrt{x} + 2\sqrt{x} = 0$.

EXAMPLE 10 ***Rationalizing binomial denominators.*** Rationalize the denominator of $\frac{1}{\sqrt{2} + 1}$.

Solution

We multiply the numerator and denominator of the fraction by $\sqrt{2} - 1$, which is the conjugate of the denominator.

Self Check

Rationalize the denominator of $\frac{2}{\sqrt{3} + 1}$.

$$\frac{1}{\sqrt{2}+1} = \frac{1(\sqrt{2}-1)}{(\sqrt{2}+1)(\sqrt{2}-1)}$$

$$= \frac{\sqrt{2}-1}{2-1}$$ In the denominator, multiply the binomials: $(\sqrt{2}+1)(\sqrt{2}-1) = \sqrt{2}\sqrt{2} - \sqrt{2} + \sqrt{2} - 1 = 2 - 1$.

$$= \sqrt{2} - 1$$ Simplify: $\frac{\sqrt{2}-1}{2-1} = \frac{\sqrt{2}-1}{1} = \sqrt{2}-1$.

Answer: $\sqrt{3} - 1$ ■

EXAMPLE 11 ***Rationalizing binomial denominators.*** Rationalize the denominator of $\frac{\sqrt{x}+\sqrt{2}}{\sqrt{x}-\sqrt{2}}$ $(x > 0)$.

Solution

We multiply the numerator and denominator by $\sqrt{x} + \sqrt{2}$, which is the conjugate of $\sqrt{x} - \sqrt{2}$, and simplify.

$$\frac{\sqrt{x}+\sqrt{2}}{\sqrt{x}-\sqrt{2}} = \frac{(\sqrt{x}+\sqrt{2})(\sqrt{x}+\sqrt{2})}{(\sqrt{x}-\sqrt{2})(\sqrt{x}+\sqrt{2})}$$

$$= \frac{x + \sqrt{2x} + \sqrt{2x} + 2}{x - 2}$$ In the numerator and denominator, use the FOIL method.

$$= \frac{x + 2\sqrt{2x} + 2}{x - 2}$$ In the numerator, combine like terms.

Self Check

Rationalize the denominator of $\frac{\sqrt{x}-\sqrt{2}}{\sqrt{x}+\sqrt{2}}$.

Answer: $\frac{x - 2\sqrt{2x} + 2}{x - 2}$ ■

STUDY SET Section 7.5

VOCABULARY *In Exercises 1–6, fill in the blanks to make each statement true.*

1. To multiply $(\sqrt{3}+\sqrt{2})(\sqrt{3}-2\sqrt{2})$, we can use the ______________ method.

2. To multiply $2\sqrt{5}(3\sqrt{8}+\sqrt{3})$, use the ______________ property to remove parentheses.

3. The denominator of the fraction $\frac{4}{\sqrt{5}}$ is an ______________ number.

4. The ______________ of $\sqrt{x} + 1$ is $\sqrt{x} - 1$.

5. To obtain a ______________ cube under the radical in the denominator of $\frac{\sqrt[3]{7}}{\sqrt[3]{5n}}$, we multiply the numerator and denominator by $\sqrt[3]{25n^2}$.

6. To ______________ the denominator of $\frac{4}{\sqrt{5}}$, we multiply the numerator and denominator by $\sqrt{5}$.

CONCEPTS

7. Do each operation, if possible.

a. $4\sqrt{6} + 2\sqrt{6}$ **b.** $4\sqrt{6}(2\sqrt{6})$

c. $3\sqrt{2} - 2\sqrt{3}$ **d.** $3\sqrt{2}(-2\sqrt{3})$

8. Do each operation, if possible.

a. $5 + 6\sqrt[3]{6}$ **b.** $5(6\sqrt[3]{6})$

c. $\frac{30\sqrt[3]{15}}{5}$ **d.** $\frac{\sqrt[3]{15}}{5}$

9. Consider $\frac{\sqrt{3}}{\sqrt{7}} = \frac{\sqrt{3}\sqrt{7}}{\sqrt{7}\sqrt{7}}$. Explain why the expressions on the left-hand side and the right-hand side of the equation are equal.

10. To rationalize the denominator of $\frac{\sqrt[4]{2}}{\sqrt[4]{3}}$, why wouldn't we multiply the numerator and denominator by $\frac{\sqrt[4]{3}}{\sqrt[4]{3}}$?

11. Explain why $\frac{\sqrt[3]{12}}{\sqrt[3]{5}}$ is not in simplified form.

12. Explain why $\sqrt{\frac{3a}{11k}}$ is not in simplified form.

NOTATION *In Exercises 13–14, fill in the blanks to make each statement true.*

13. Multiply $5\sqrt{8} \cdot 7\sqrt{6}$.

$$5\sqrt{8} \cdot 7\sqrt{6} = 5(7)\sqrt{8\square}$$
$$= 35\sqrt{\square}$$
$$= 35\sqrt{\square \cdot 3}$$
$$= 35(\square)\sqrt{3}$$
$$= 140\sqrt{3}$$

14. Rationalize the denominator of $\frac{9}{\sqrt[3]{4a^2}}$.

$$\frac{9}{\sqrt[3]{4a^2}} = \frac{9 \cdot \sqrt[3]{2a}}{\sqrt[3]{4a^2} \cdot \square}$$
$$= \frac{9\sqrt[3]{2a}}{\sqrt[3]{\square}}$$
$$= \frac{9\sqrt[3]{2a}}{2a}$$

PRACTICE *In Exercises 15–42, do each multiplication and simplify, if possible. All variables represent positive real numbers.*

15. $\sqrt{11}\sqrt{11}$
16. $\sqrt{35}\sqrt{35}$
17. $(\sqrt{7})^2$
18. $(\sqrt{23})^2$
19. $\sqrt{2}\sqrt{8}$
20. $\sqrt{3}\sqrt{27}$
21. $\sqrt{5}\sqrt{10}$
22. $\sqrt{7}\sqrt{35}$
23. $2\sqrt{3}\sqrt{6}$
24. $-3\sqrt{11}\sqrt{33}$
25. $\sqrt[3]{5}\sqrt[3]{25}$
26. $-\sqrt[3]{7}\sqrt[3]{49}$
27. $(3\sqrt{2})^2$
28. $(2\sqrt{5})^2$
29. $(-2\sqrt{2})^2$
30. $(-3\sqrt{10})^2$
31. $(3\sqrt[3]{9})(2\sqrt[3]{3})$
32. $(2\sqrt[3]{16})(-\sqrt[3]{4})$
33. $\sqrt[3]{2}\sqrt[3]{12}$
34. $\sqrt[3]{3}\sqrt[3]{18}$
35. $\sqrt{ab^3}\sqrt{ab}$
36. $\sqrt{8x}\sqrt{2x^3y}$
37. $\sqrt{5ab}\sqrt{5a}$
38. $\sqrt{15rs^2}\sqrt{10r}$
39. $-4\sqrt[3]{5r^2s}(5\sqrt[3]{2r})$
40. $-\sqrt[3]{3xy^2}(-\sqrt[3]{9x^3})$
41. $\sqrt{x(x+3)}\sqrt{x^3(x+3)}$
42. $\sqrt{y^2(x+y)}\sqrt{(x+y)^3}$

In Exercises 43–60, do each multiplication and simplify. All variables represent positive real numbers.

43. $3\sqrt{5}(4 - \sqrt{5})$
44. $2\sqrt{7}(3\sqrt{7} - 1)$
45. $3\sqrt{2}(4\sqrt{6} + 2\sqrt{7})$
46. $-\sqrt{3}(\sqrt{7} - \sqrt{15})$
47. $-2\sqrt{5x}(4\sqrt{2x} - 3\sqrt{3})$
48. $3\sqrt{7t}(2\sqrt{7t} + 3\sqrt{3t^2})$
49. $(\sqrt{2} + 1)(\sqrt{2} - 3)$
50. $(2\sqrt{3} + 1)(\sqrt{3} - 1)$
51. $(\sqrt{5z} + \sqrt{3})(\sqrt{5z} + \sqrt{3})$
52. $(\sqrt{3p} - \sqrt{2})(\sqrt{3p} + \sqrt{2})$
53. $(\sqrt{3x} - \sqrt{2y})(\sqrt{3x} + \sqrt{2y})$
54. $(\sqrt{3m} + \sqrt{2n})(\sqrt{3m} + \sqrt{2n})$
55. $(2\sqrt{3a} - \sqrt{b})(\sqrt{3a} + 3\sqrt{b})$
56. $(5\sqrt{p} - \sqrt{3q})(\sqrt{p} + 2\sqrt{3q})$
57. $(3\sqrt{2r} - 2)^2$
58. $(2\sqrt{3t} + 5)^2$
59. $-2(\sqrt{3x} + \sqrt{3})^2$
60. $3(\sqrt{5x} - \sqrt{3})^2$

In Exercises 61–84, simplify each radical expression by rationalizing the denominator. All variables represent positive real numbers.

61. $\sqrt{\frac{1}{7}}$ **62.** $\sqrt{\frac{5}{3}}$ **63.** $\frac{6}{\sqrt{30}}$ **64.** $\frac{8}{\sqrt{10}}$

65. $\frac{\sqrt{5}}{\sqrt{8}}$ **66.** $\frac{\sqrt{3}}{\sqrt{50}}$ **67.** $\frac{\sqrt{8}}{\sqrt{2}}$ **68.** $\frac{\sqrt{27}}{\sqrt{3}}$

69. $\frac{1}{\sqrt[3]{2}}$ **70.** $\frac{2}{\sqrt[3]{6}}$ **71.** $\frac{3}{\sqrt[3]{9}}$ **72.** $\frac{2}{\sqrt[3]{a}}$

73. $\frac{\sqrt[3]{2}}{\sqrt[3]{9}}$ **74.** $\frac{\sqrt[3]{9}}{\sqrt[3]{54}}$ **75.** $\frac{\sqrt{8}}{\sqrt{xy}}$ **76.** $\frac{\sqrt{9xy}}{\sqrt{3x^2y}}$

77. $\frac{\sqrt{10xy^2}}{\sqrt{2xy^3}}$ **78.** $\frac{\sqrt{5ab^2c}}{\sqrt{10abc}}$ **79.** $\frac{\sqrt[3]{4a^2}}{\sqrt[3]{2ab}}$ **80.** $\frac{\sqrt[3]{9x}}{\sqrt[3]{3xy}}$

81. $\frac{1}{\sqrt[4]{4}}$ **82.** $\frac{1}{\sqrt[5]{2}}$ **83.** $\frac{1}{\sqrt[5]{16}}$ **84.** $\frac{4}{\sqrt[4]{32}}$

In Exercises 85–98, rationalize the denominator. All variables represent positive real numbers.

85. $\frac{1}{\sqrt{2}-1}$ **86.** $\frac{3}{\sqrt{3}-1}$ **87.** $\frac{\sqrt{2}}{\sqrt{5}+3}$ **88.** $\frac{\sqrt{3}}{\sqrt{3}-2}$

89. $\frac{\sqrt{3}+1}{\sqrt{3}-1}$ **90.** $\frac{\sqrt{2}-1}{\sqrt{2}+1}$ **91.** $\frac{\sqrt{7}-\sqrt{2}}{\sqrt{2}+\sqrt{7}}$ **92.** $\frac{\sqrt{3}+\sqrt{2}}{\sqrt{3}-\sqrt{2}}$

93. $\frac{2}{\sqrt{x}+1}$ **94.** $\frac{3}{\sqrt{x}-2}$ **95.** $\frac{2z-1}{\sqrt{2z}-1}$ **96.** $\frac{3t-1}{\sqrt{3t}+1}$

97. $\frac{\sqrt{x}-\sqrt{y}}{\sqrt{x}+\sqrt{y}}$ **98.** $\frac{\sqrt{x}+\sqrt{y}}{\sqrt{x}-\sqrt{y}}$

APPLICATIONS

99. STATISTICS An example of a normal distribution curve, or *bell-shaped* curve, is shown in Illustration 1. A fraction that is part of the equation that models this curve is

$$\frac{1}{\sigma\sqrt{2\pi}}$$

where σ is a letter from the Greek alphabet. Rationalize the denominator of the fraction.

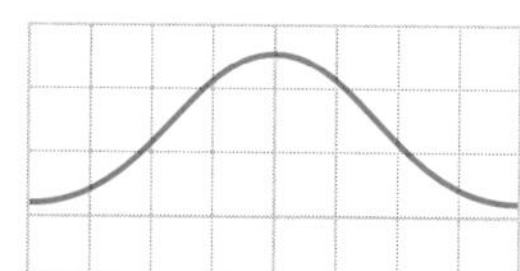

ILLUSTRATION 1

100. ANALYTIC GEOMETRY See Illustration 2. The length of the perpendicular segment drawn from the point $(-2, 2)$ to the line with equation $2x - 4y = 4$ is given by

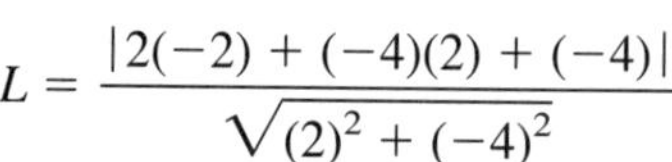

$$L = \frac{|2(-2) + (-4)(2) + (-4)|}{\sqrt{(2)^2 + (-4)^2}}$$

Find L. Express the result in simplified radical form. Then give an approximation to the nearest tenth.

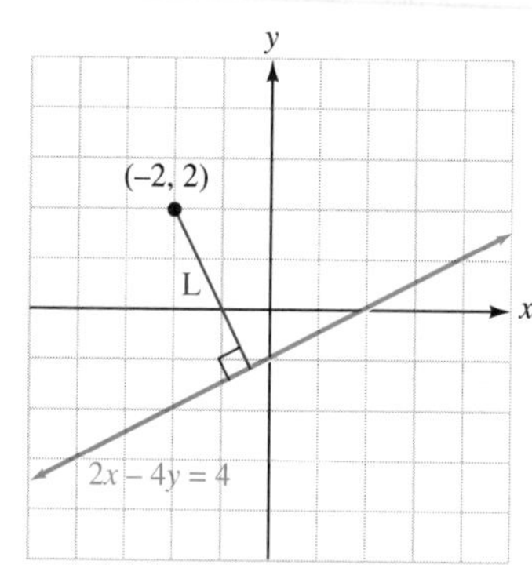

ILLUSTRATION 2

101. TRIGONOMETRY In trigonometry, we must often find the ratio of the lengths of two sides of right triangles. Use the information in Illustration 3 to find the ratio

$$\frac{\text{length of side AC}}{\text{length of side AB}}$$

Write the result in simplified radical form.

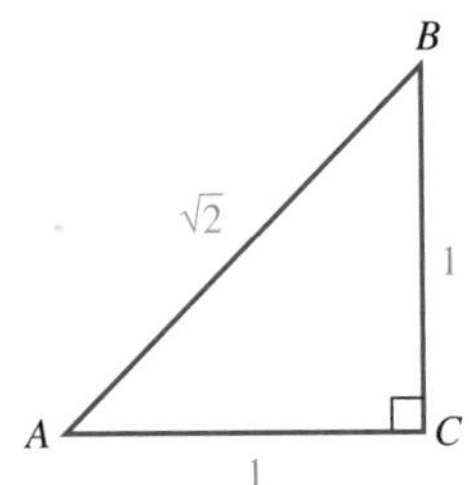

ILLUSTRATION 3

102. MECHANICAL ENGINEERING A measure of how fast the block shown in Illustration 4 will oscillate when the system is set in motion is given by the formula

$$\omega = \sqrt{\frac{k_1 + k_2}{m}}$$

where k_1 and k_2 indicate the stiffness of the springs and m is the mass of the block. Rationalize the right-hand side and restate the formula.

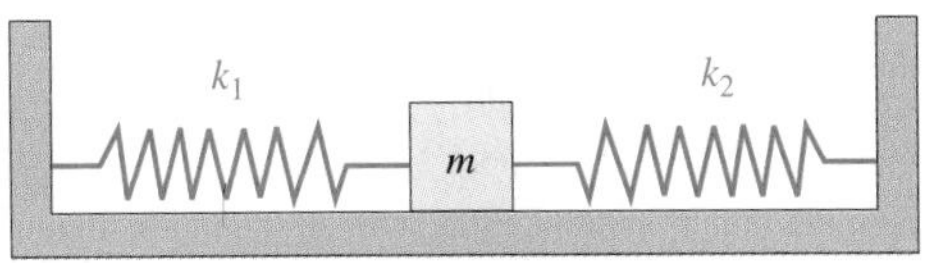

ILLUSTRATION 4

103. PHOTOGRAPHY In Example 6, we saw that a lens with a focal length of 12 centimeters and an aperture $3\sqrt{2}$ centimeters in diameter is an f/2.8 lens. Find the f-number if the area of the aperture is again cut in half.

104. ELECTRONICS A formula that is used when designing AC (alternating current) circuits is

$$f_0 = \frac{1}{2\pi}\sqrt{\frac{1}{LC}}$$

Rationalize the right-hand side and restate the formula.

WRITING *Write a paragraph using your own words.*

105. Explain why $\sqrt{m} \cdot \sqrt{m} = m$ but $\sqrt[3]{m} \cdot \sqrt[3]{m} \neq m$. (Assume that $m > 0$.)

106. Explain why the product of $\sqrt{m} + 3$ and $\sqrt{m} - 3$ does not contain a radical.

REVIEW *In Exercises 107–110, solve the equation.*

107. $\frac{2}{3 - a} = 1$

108. $5(s - 4) = -5(s - 4)$

109. $\frac{8}{b - 2} + \frac{3}{2 - b} = -\frac{1}{b}$

110. $\frac{2}{x - 2} + \frac{1}{x + 1} = \frac{1}{(x + 1)(x - 2)}$

7.6 *Geometric Applications of Radicals*

In this section, you will learn about

- **The Pythagorean theorem**
- **45°–45°–90° triangles**
- **30°–60°–90° triangles**
- **The distance formula**
- **The midpoint formula**

INTRODUCTION. In this section, we will consider several applications of square roots in geometry. Then we will find the distance between two points on a rectangular coordinate system, using a formula that contains a square root. We begin by considering an important theorem (mathematical statement) about right triangles.

The Pythagorean theorem

If we know the lengths of two legs of a right triangle, we can always find the length of the **hypotenuse** (the side opposite the 90° angle) by using the **Pythagorean theorem.**

Pythagorean theorem If a and b are the lengths of two legs of a right triangle and c is the length of the hypotenuse, then

$$a^2 + b^2 = c^2$$

In words, the Pythagorean theorem is expressed as follows:

In any right triangle, the square of the hypotenuse is equal to the sum of the squares of the two legs.

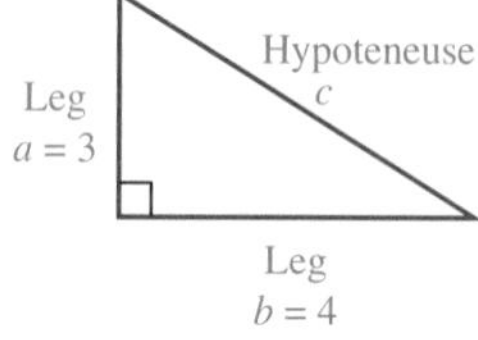

FIGURE 7-13

Suppose the right triangle shown in Figure 7-13 has legs of length 3 and 4 units. To find the length of the hypotenuse, we use the Pythagorean theorem.

$$a^2 + b^2 = c^2$$
$$3^2 + 4^2 = c^2 \quad \text{Substitute 3 for } a \text{ and 4 for } b.$$
$$9 + 16 = c^2$$
$$25 = c^2$$

To solve for c, we need to "undo" the operation performed on it. Since c is squared, we take the positive square root of both sides of the equation.

$$\sqrt{25} = \sqrt{c^2} \quad \text{Since } c \text{ represents the length of a side of a triangle, } c > 0.\ \sqrt{c^2} = c, \text{ because } c \cdot c = c^2.$$
$$5 = c$$

The length of the hypotenuse is 5 units.

EXAMPLE 1 ***Firefighting.*** To fight a forest fire, the forestry department plans to clear a rectangular fire break around the fire, as shown in Figure 7-14. Crews are equipped with mobile communications that have a 3,000-yard range. Can crews at points A and B remain in radio contact?

Solution

Points A, B, and C form a right triangle. To find the distance c from point A to point B, we can use the Pythagorean theorem, substituting 2,400 for a and 1,000 for b and solving for c.

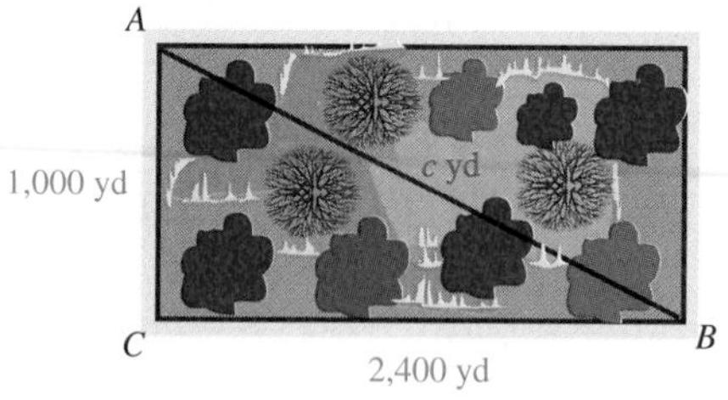

FIGURE 7-14

$$a^2 + b^2 = c^2$$
$$2{,}400^2 + 1{,}000^2 = c^2$$
$$5{,}760{,}000 + 1{,}000{,}000 = c^2$$
$$6{,}760{,}000 = c^2$$
$$\sqrt{6{,}760{,}000} = \sqrt{c^2} \quad \text{Take the positive square root of both sides.}$$
$$2{,}600 = c \quad \text{Use a calculator to find the square root.}$$

The two crews are 2,600 yards apart. Because this distance is less than the range of the radios, they can communicate.

Self Check

Can the crews communicate if $b = 1{,}500$ yards?

Answer: yes

45°–45°–90° triangles

An **isosceles right triangle** is a right triangle with two legs of equal length. Isosceles right triangles have angle measures of 45°, 45°, and 90°. If we know the length of one leg of an isosceles right triangle, we can use the Pythagorean theorem to find the length of the hypotenuse. Since the triangle shown in Figure 7-15 is a right triangle, we have

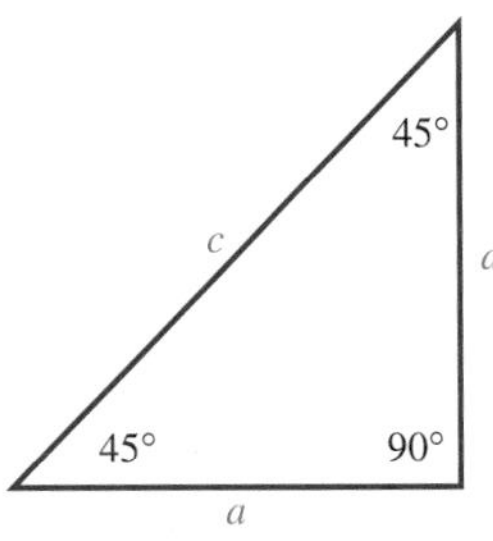

FIGURE 7-15

$$c^2 = a^2 + b^2$$
$$c^2 = a^2 + a^2 \quad \text{Both legs are } a \text{ units long, so replace } b \text{ with } a.$$
$$c^2 = 2a^2 \quad \text{Combine like terms.}$$
$$c = \sqrt{2a^2} \quad \text{Take the positive square root of both sides.}$$
$$c = a\sqrt{2} \quad \text{Simplify the radical: } \sqrt{2a^2} = \sqrt{2}\sqrt{a^2} = \sqrt{2}a = a\sqrt{2}.$$

Thus, *in an isosceles right triangle, the length of the hypotenuse is the length of one leg times* $\sqrt{2}$.

EXAMPLE 2 ***45°–45°–90° triangles.*** If one leg of the isosceles right triangle shown in Figure 7-15 is 10 feet long, find the length of the hypotenuse.

Solution

Since the length of the hypotenuse is the length of a leg times $\sqrt{2}$, we have

$$c = 10\sqrt{2}$$

The length of the hypotenuse is $10\sqrt{2}$ units. To two decimal places, the length is 14.14 units.

Self Check

Find the length of the hypotenuse of an isosceles right triangle if one leg is 12 meters long.

Answer: $12\sqrt{2}$ m ■

If the length of the hypotenuse of an isosceles right triangle is known, we can use the Pythagorean theorem to find the length of each leg.

EXAMPLE 3 Find the exact length of each leg of the isosceles right triangle shown in Figure 7-16.

Solution

We use the Pythagorean theorem.

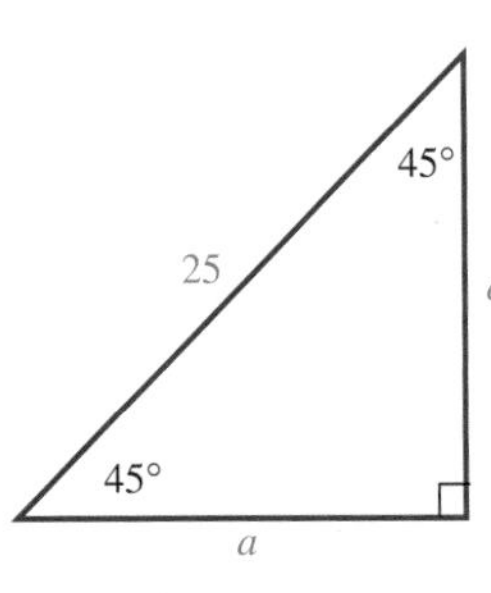

FIGURE 7-16

$$c^2 = a^2 + b^2$$
$$25^2 = a^2 + a^2 \quad \text{Since both legs are } a \text{ units long, substitute } a \text{ for } b. \text{ The hypotenuse is 25 units long. Substitute 25 for } c.$$
$$25^2 = 2a^2 \quad \text{Combine like terms.}$$
$$\frac{625}{2} = a^2 \quad \text{Square 25 and divide both sides by 2.}$$
$$\sqrt{\frac{625}{2}} = a \quad \text{To solve for } a, \text{ take the positive square root of both sides: } \sqrt{a^2} = a.$$
$$\frac{\sqrt{625} \cdot \sqrt{2}}{\sqrt{2} \cdot \sqrt{2}} = a \quad \text{Write } \sqrt{\frac{625}{2}} \text{ as } \frac{\sqrt{625}}{\sqrt{2}}. \text{ Then rationalize the denominator.}$$
$$\frac{25\sqrt{2}}{2} = a \quad \text{In the numerator, simplify the radical: } \sqrt{625} = 25. \text{ In the denominator, do the multiplication: } \sqrt{2} \cdot \sqrt{2} = 2.$$

The exact length of each leg is $\frac{25\sqrt{2}}{2}$ units. To two decimal places, the length is 17.68 units. ■

30°–60°–90° triangles

From geometry, we know that an **equilateral triangle** is a triangle with three sides of equal length and three 60° angles. Each side of the equilateral triangle in Figure 7-17 is $2a$ units long. If an **altitude** is drawn to its base, the altitude bisects the base and divides the equilateral triangle into two 30°–60°–90° triangles. We can see that the shorter leg of each 30°–60°–90° triangle (the side *opposite* the 30° angle) is a units long. Thus,

> *The length of the shorter leg of a 30°–60°–90° right triangle is half as long as the hypotenuse.*

We can discover another important relationship between the legs of a 30°–60°–90° triangle if we find the length of the altitude h in Figure 7-17. We begin by applying the Pythagorean theorem to one of the 30°–60°–90° triangles.

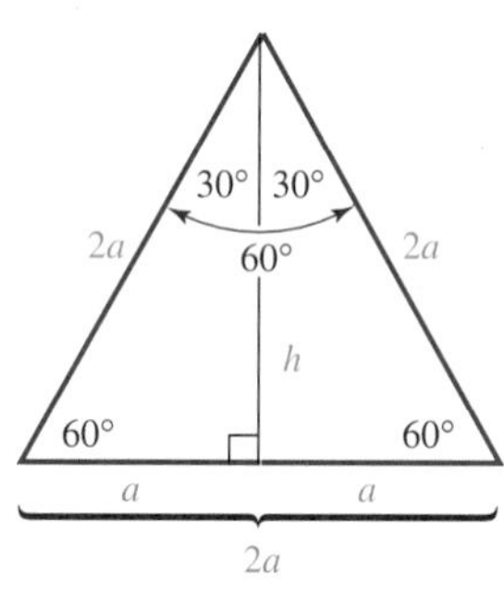

FIGURE 7-17

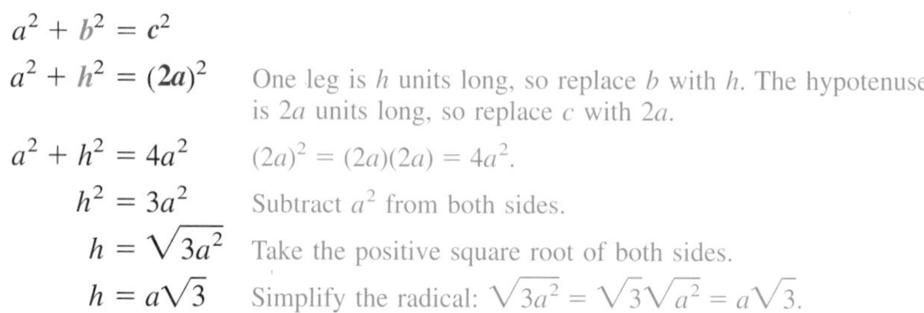

$a^2 + b^2 = c^2$

$a^2 + h^2 = (2a)^2$ One leg is h units long, so replace b with h. The hypotenuse is $2a$ units long, so replace c with $2a$.

$a^2 + h^2 = 4a^2$ $(2a)^2 = (2a)(2a) = 4a^2$.

$h^2 = 3a^2$ Subtract a^2 from both sides.

$h = \sqrt{3a^2}$ Take the positive square root of both sides.

$h = a\sqrt{3}$ Simplify the radical: $\sqrt{3a^2} = \sqrt{3}\sqrt{a^2} = a\sqrt{3}$.

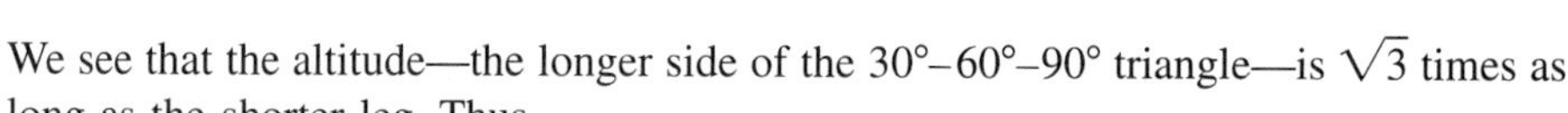

We see that the altitude—the longer side of the 30°–60°–90° triangle—is $\sqrt{3}$ times as long as the shorter leg. Thus,

> *The length of the longer leg of a 30°–60°–90° triangle is the length of the shorter leg times $\sqrt{3}$.*

EXAMPLE 4 ***30°–60°–90° triangles.*** Find the length of the hypotenuse and the longer leg of the right triangle shown in Figure 7-18.

Solution

Since the shorter leg of a 30°–60°–90° triangle is half as long as the hypotenuse, the hypotenuse is 12 centimeters long.

Since the length of the longer leg is the length of the shorter leg times $\sqrt{3}$, the longer leg is $6\sqrt{3}$ (about 10.39) centimeters long.

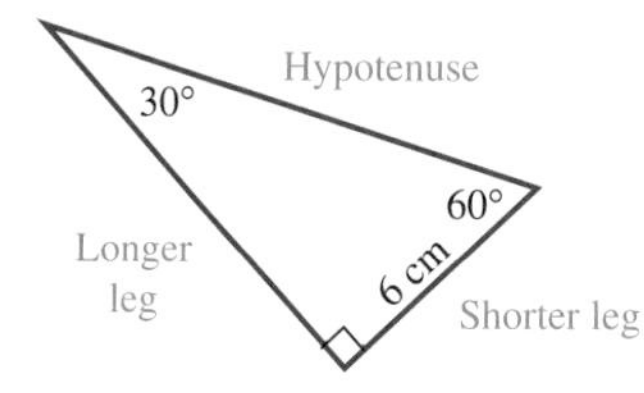

FIGURE 7-18

Self Check

Find the length of the hypotenuse and the longer leg of a 30°–60°–90° triangle if the shorter leg is 8 centimeters long.

Answers: 16 cm, $8\sqrt{3}$ cm

EXAMPLE 5 ***Stretching exercises.*** A doctor prescribed the back-strengthening exercise shown in Figure 7-19(a) for a patient. The patient was instructed to raise his leg to an angle of 60° and hold the position for 10 seconds. If the patient's leg is 36 inches long, how high off the floor will his foot be when his leg is held at the proper angle?

Solution

In Figure 7-19(b), we see that a 30°–60°–90° triangle, which we will call triangle ABC, models the situation. Since the side opposite the 30° angle of a 30°–60°–90° triangle is half as long as the hypotenuse, side AC is 18 inches long.

Since the length of the side opposite the 60° angle is the length of the side opposite the 30° angle times $\sqrt{3}$, side BC is $18\sqrt{3}$, or about 31 inches long. So the patient's foot will be about 31 inches from the floor when his leg is in the proper stretching position.

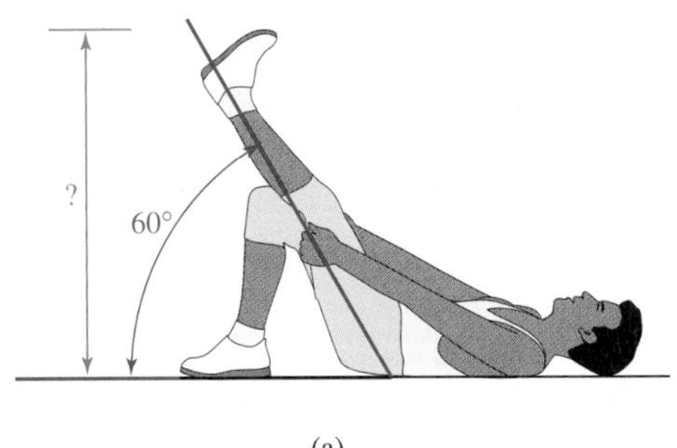

(a)

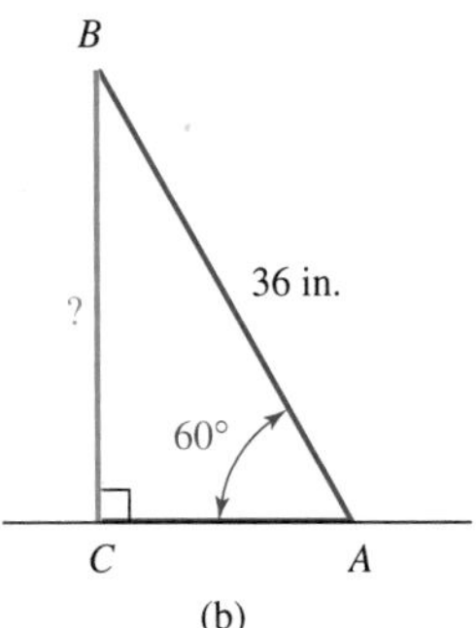

(b)

FIGURE 7-19

The distance formula

With the *distance formula,* we can find the distance between any two points that are graphed on a rectangular coordinate system.

To find the distance d between points $P(x_1, y_1)$ and $Q(x_2, y_2)$ shown in Figure 7-20, we construct the right triangle PRQ. The distance between P and R is $|x_2 - x_1|$, and the distance between R and Q is $|y_2 - y_1|$. We apply the Pythagorean theorem to the right triangle PRQ to get

$$[d(PQ)]^2 = |x_2 - x_1|^2 + |y_2 - y_1|^2$$ Read $d(PQ)$ as "the distance between P and Q."

$$= (x_2 - x_1)^2 + (y_2 - y_1)^2$$ Because $|x_2 - x_1|^2 = (x_2 - x_1)^2$ and $|y_2 - y_1|^2 = (y_2 - y_1)^2$.

Take the positive square root of both sides:

1. $d(PQ) = \sqrt{(x_2 - x_1)^2 + (y_2 - y_1)^2}$

Equation 1 is the called the **distance formula.**

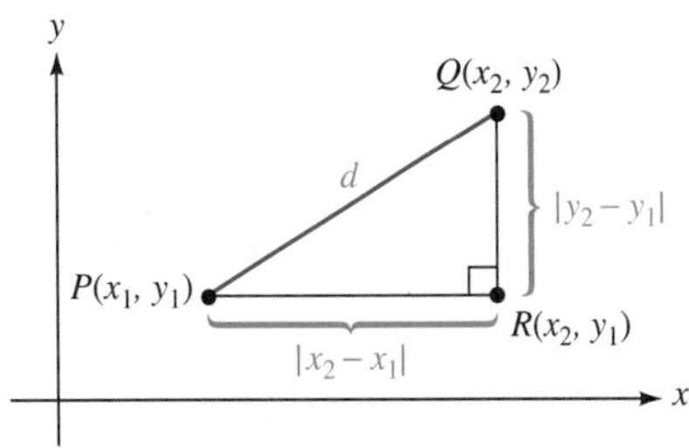

FIGURE 7-20

Distance formula The distance between two points $P(x_1, y_1)$ and $Q(x_2, y_2)$ is given by the formula

$$d(PQ) = \sqrt{(x_2 - x_1)^2 + (y_2 - y_1)^2}$$

EXAMPLE 6 ***Using the distance formula.*** Find the distance between points $P(-2, 3)$ and $Q(4, -5)$.

Solution

To find the distance, we can use the distance formula by substituting 4 for x_2, -2 for x_1, -5 for y_2, and 3 for y_1.

Self Check

Find the distance between $P(-2, -2)$ and $Q(3, 10)$.

$$
\begin{aligned}
d(PQ) &= \sqrt{(x_2 - x_1)^2 + (y_2 - y_1)^2} \\
&= \sqrt{[4 - (-2)]^2 + (-5 - 3)^2} \\
&= \sqrt{(4 + 2)^2 + (-5 - 3)^2} \\
&= \sqrt{6^2 + (-8)^2} \\
&= \sqrt{36 + 64} \\
&= \sqrt{100} \\
&= 10
\end{aligned}
$$

The distance between P and Q is 10 units.

Answer: 13

EXAMPLE 7 ***Robotics.*** Computerized robots are used to weld the parts of an automobile chassis together on an automated production line. To do this, an imaginary coordinate system is superimposed on the side of the vehicle, and the robot is programmed to move to specific positions to make each weld. See Figure 7-21, which is scaled in inches. If the welder unit moves from point to point at an average rate of speed of 48 in./sec, how long will it take it to move from position 1 to position 2?

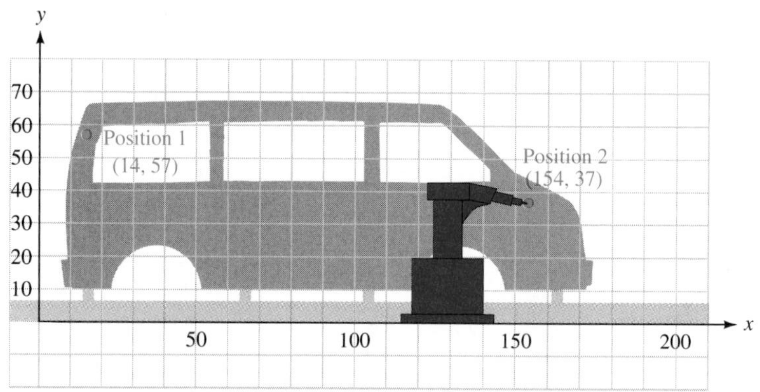

FIGURE 7-21

Solution This is a uniform motion problem. We can use the formula $t = \frac{d}{r}$ to find the time it takes for the welder to move from position 1 (14, 57) to position 2 (154, 37).

We can use the distance formula to find the distance d that the welder unit moves.

$$d = \sqrt{(x_2 - x_1)^2 + (y_2 - y_1)^2}$$

$$d = \sqrt{(154 - 14)^2 + (37 - 57)^2}$$ Substitute 154 for x_2, 14 for x_1, 37 for y_2, and 57 for y_1.

$$= \sqrt{140^2 + (-20)^2}$$

$$= \sqrt{20{,}000}$$ $140^2 + (-20)^2 = 19{,}600 + 400 = 20{,}000$.

$$= 100\sqrt{2}$$ Simplify: $\sqrt{20{,}000} = \sqrt{100 \cdot 100 \cdot 2} = 100\sqrt{2}$.

The welder travels $100\sqrt{2}$ inches as it moves from position 1 to position 2. To find the time this will take, we divide the distance by the average rate of speed, 48 in./sec.

$$t = \frac{d}{r}$$

$$t = \frac{100\sqrt{2}}{48}$$ Substitute $100\sqrt{2}$ for d and 48 for r.

$$t \approx 2.9$$ Use a calculator to find an approximation to the nearest tenth.

It will take the welder about 2.9 seconds to travel from position 1 to position 2.

The midpoint formula

If point M in Figure 7-22 lies midway between points $P(x_1, y_1)$ and $Q(x_2, y_2)$, it is called the **midpoint** of segment PQ. To find the coordinates of point M, we find the mean of the x-coordinates and the mean of the y-coordinates of P and Q.

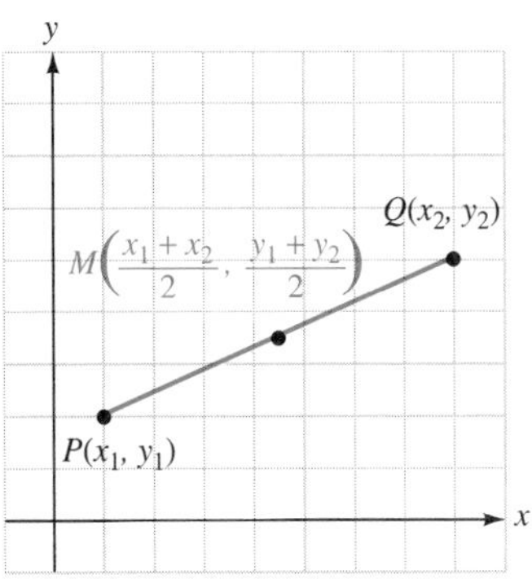

FIGURE 7-22

The midpoint formula The midpoint of the line segment with endpoints at $P(x_1, y_1)$ and $Q(x_2, y_2)$ is the point M with coordinates of

$$\left(\frac{x_1 + x_2}{2}, \frac{y_1 + y_2}{2}\right)$$

EXAMPLE 8 ***Using the midpoint formula.*** Find the midpoint of the line segment joining $P(-2\sqrt{5}, 3)$ and $Q(3\sqrt{5}, -5)$.

Solution

To find the midpoint, we find the mean of the x-coordinates and the mean of the y-coordinates.

$$\frac{x_1 + x_2}{2} = \frac{-2\sqrt{5} + 3\sqrt{5}}{2} = \frac{\sqrt{5}}{2} \quad \text{and} \quad \frac{y_1 + y_2}{2} = \frac{3 + (-5)}{2} = -1$$

The midpoint of segment PQ is the point $M\left(\frac{\sqrt{5}}{2}, -1\right)$.

Self Check

Find the midpoint of the segment joining $P(9, -7\sqrt{2})$ and $Q(-3, 4\sqrt{2})$.

Answer: $\left(3, -\frac{3\sqrt{2}}{2}\right)$ ■

STUDY SET Section 7.6

VOCABULARY *In Exercises 1–4, fill in the blanks to make the statements true.*

1. In a right triangle, the side opposite the 90° angle is called the ______________.

2. An ______________ triangle is a right triangle with two legs of equal length.

3. The ______________ theorem states that in any right triangle, the square of the hypotenuse is equal to the sum of the squares of the lengths of the two legs.

4. An ______________ triangle has three sides of equal length and three 60° angles.

CONCEPTS *In Exercises 5–12, fill in the blanks to make the statements true.*

5. If a and b are the lengths of two legs of a right triangle and c is the length of the hypotenuse, then ______________.

6. In any right triangle, the square of the hypotenuse is equal to the ______________ of the squares of the two ______________.

7. In an isosceles right triangle, the length of the hypotenuse is the length of one leg times ______________.

8. The shorter leg of a 30°–60°–90° triangle is ______________ as long as the hypotenuse.

9. The length of the longer leg of a 30°–60°–90° triangle is the length of the shorter leg times ______________.

10. The formula to find the distance between two points $P(x_1, y_1)$ and $Q(x_2, y_2)$ is ________________.

11. In a right triangle, the shorter leg is opposite the ___ angle, and the longer leg is opposite the ___ angle.

12. The formula to find the coordinates of the midpoint of the line segment with endpoints at $P(x_1, y_1)$ and $Q(x_2, y_2)$ is ______________.

13. **a.** To solve the equation $c^2 = 20$, where c represents the length of the hypotenuse of a right triangle, how do we "undo" the operation performed on c?

b. What is the first step when solving the equation $25 + b^2 = 81$?

14. When the lengths of the sides of the triangle in Illustration 1 are substituted into the equation $a^2 + b^2 = c^2$, the result is a false statement. Explain why.

$$a^2 + b^2 = c^2$$
$$2^2 + 4^2 = 5^2$$
$$4 + 16 = 25$$
$$20 = 25$$

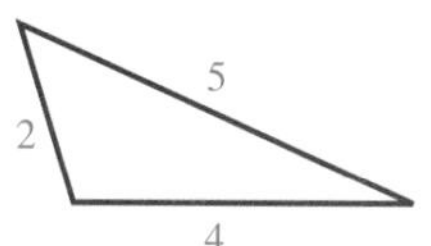

ILLUSTRATION 1

NOTATION *In Exercises 15–16, complete each solution.*

15. Evaluate $\sqrt{(-1 - 3)^2 + [2 - (-4)]^2}$.

$$\begin{aligned}\sqrt{(-1 - 3)^2 + [2 - (-4)]^2} &= \sqrt{(-4)^2 + [\square]^2} \\ &= \sqrt{\square} \\ &= \sqrt{\square \cdot 13} \\ &= \square\sqrt{13} \\ &\approx 7.21\end{aligned}$$

16. Solve $8^2 + 4^2 = c^2$.

$$\begin{aligned}\square + 16 &= c^2 \\ \square &= c^2 \\ \sqrt{\square} &= \sqrt{c^2} \\ \sqrt{\square \cdot 5} &= c \\ \square\sqrt{5} &= c \\ c &\approx 8.94\end{aligned}$$

PRACTICE *In Exercises 17–20, the lengths of two sides of the right triangle ABC shown in Illustration 2 are given. Find the length of the missing side.*

17. $a = 6$ ft and $b = 8$ ft

18. $a = 10$ cm and $c = 26$ cm

19. $b = 18$ m and $c = 82$ m

20. $a = 14$ in. and $c = 50$ in.

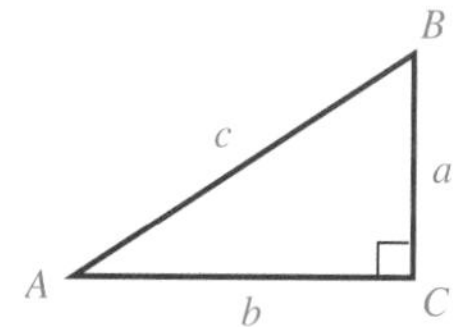

ILLUSTRATION 2

In Exercises 21–24, find the missing lengths in each triangle. Give the exact answer and then an approximation to two decimal places, when applicable.

21.

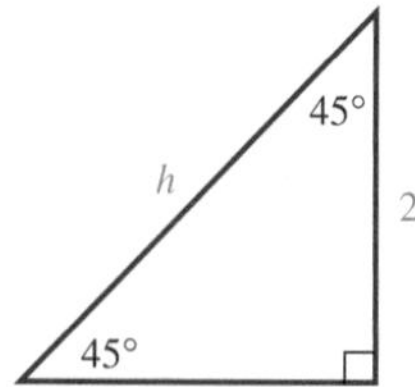

22.

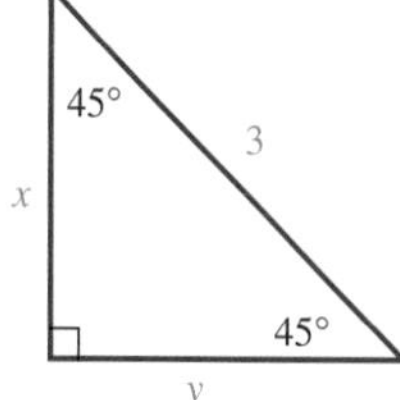

23.

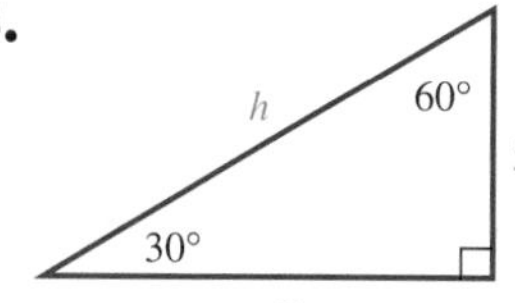

24.

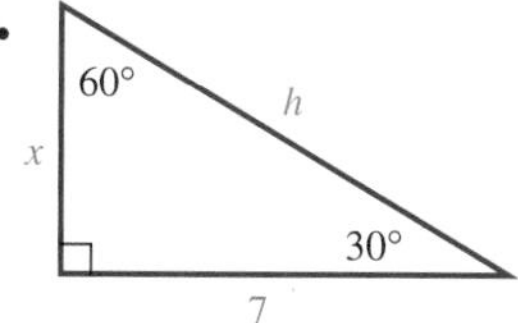

In Exercises 25–28, find the missing lengths in each triangle. Give the answer to two decimal places, when applicable.

25. 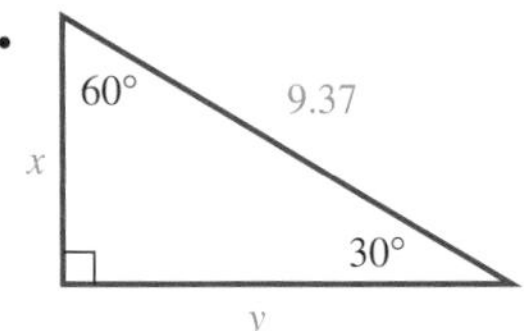

26. 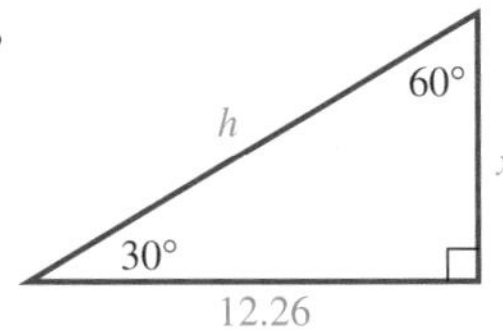

27. 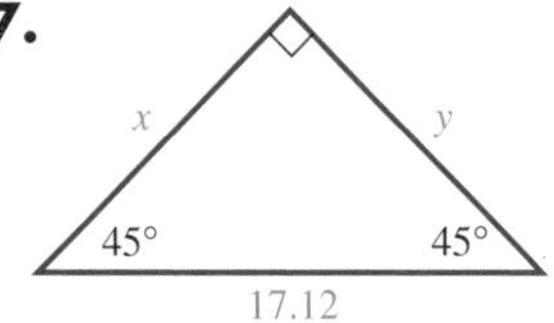

28. 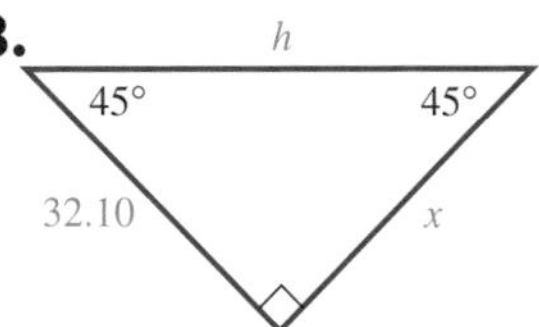

29. GEOMETRY Find the exact length of the diagonal of one of the *faces* of the cube shown in Illustration 3.

30. GEOMETRY Find the exact length of the diagonal of the cube shown in Illustration 3.

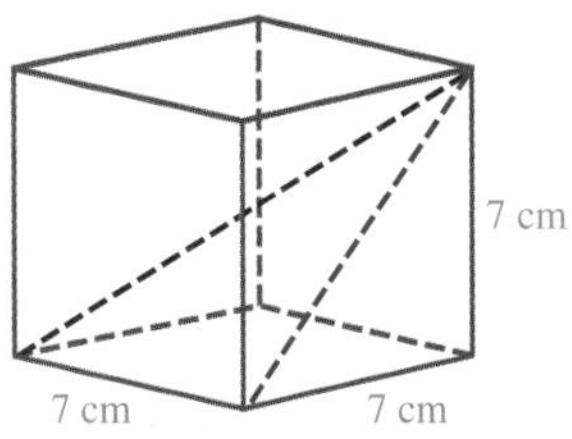

ILLUSTRATION 3

In Exercises 31–38, find the distance between P and Q.

31. $Q(0, 0)$, $P(3, -4)$

32. $Q(0, 0)$, $P(-12, 16)$

33. $P(-2, -8)$, $Q(3, 4)$

34. $P(-5, -2)$, $Q(7, 3)$

35. $P(6, 8)$, $Q(12, 16)$

36. $P(10, 4)$, $Q(2, -2)$

37. $Q(-3, 5)$, $P(-5, -5)$

38. $Q(2, -3)$, $P(4, -8)$

In Exercises 39–46, find the midpoint of the segment PQ.

39. $P(0, 0)$, $Q(6, 8)$

40. $P(10, 12)$, $Q(0, 0)$

41. $P(10, 4)$, $Q(2, -2)$

42. $P(-6, 8)$, $Q(-12, 16)$

43. $P(-11, -8)$, $Q(12, -10)$

44. $P(-10, -3)$, $Q(7, -2)$

45. $Q\left(3\sqrt{2}, -5\right)$, $P\left(-2\sqrt{2}, 3\right)$

46. $Q\left(-3\sqrt{3}, 5\sqrt{3}\right)$, $P\left(3\sqrt{3}, \sqrt{3}\right)$

47. ISOSCELES TRIANGLE Use the distance formula to show that a triangle with vertices at $(-2, 4)$, $(2, 8)$, and $(6, 4)$ is isosceles.

48. RIGHT TRIANGLE Use the distance formula and the Pythagorean theorem to show that a triangle with vertices $(2, 3)$, $(-3, 4)$, and $(1, -2)$ is a right triangle.

APPLICATIONS *In Exercises 49–54, give the exact answer. Then give an approximation to two decimal places.*

49. WASHINGTON, DC The square in Illustration 4 shows the 100-square-mile site selected by George Washington in 1790 to serve as a permanent capital for the United States. In 1847, the part of the district lying on the west bank of the Potomac was returned to Virginia. Find the coordinates of each corner of the original square that outlined the District of Columbia.

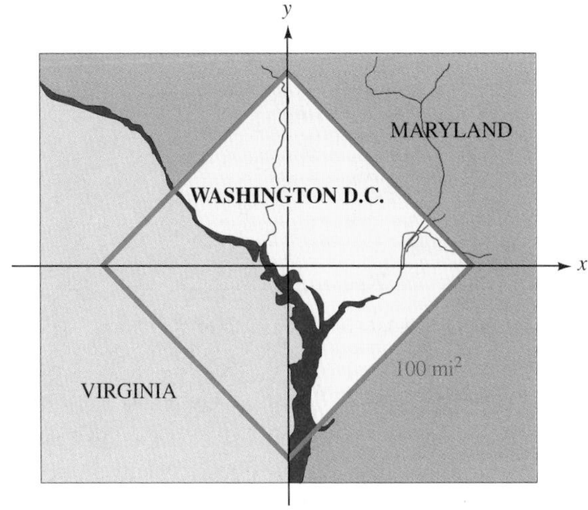

ILLUSTRATION 4

50. PAPER AIRPLANES Illustration 5 gives the directions for making a paper airplane from a square piece of paper with sides 8 inches long. Find the length l of the plane when it is completed.

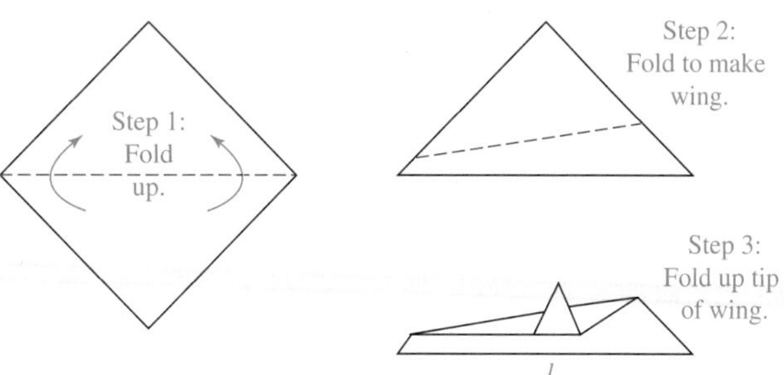

ILLUSTRATION 5

51. HARDWARE The sides of the regular hexagonal nut shown in Illustration 6 are 10 millimeters long. Find the height h of the nut.

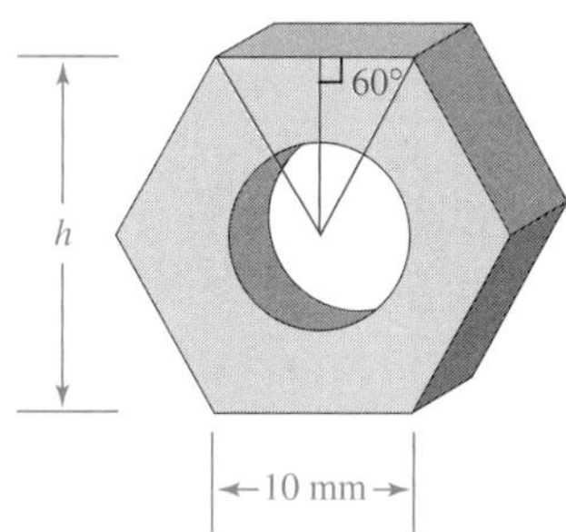

ILLUSTRATION 6

52. IRONING BOARD Find the height h of the ironing board shown in Illustration 7.

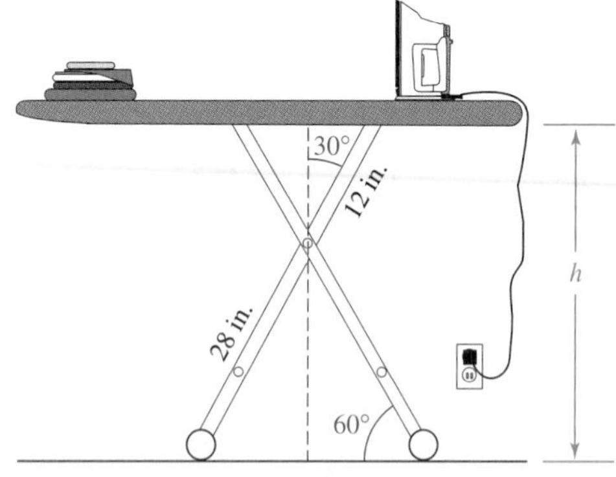

ILLUSTRATION 7

53. BASEBALL The baseball diamond shown in Illustration 8 is a square, 90 feet on a side. If the third baseman fields a ground ball 10 feet directly behind third base, how far must he throw the ball to throw a runner out at first base?

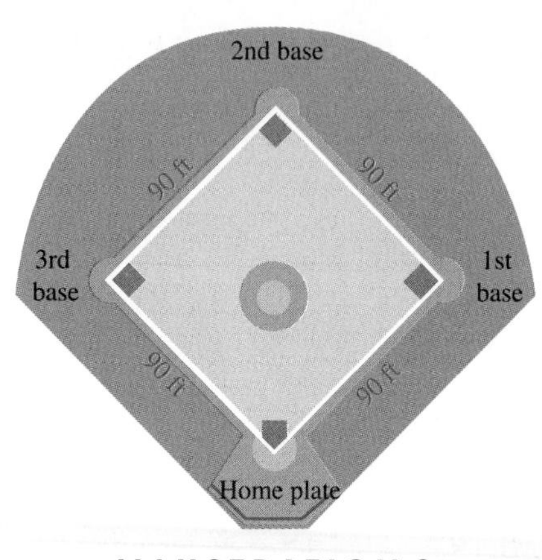

ILLUSTRATION 8

54. BASEBALL A shortstop fields a grounder at a point one-third of the way from second base to third base. (See Illustration 8.) How far will he have to throw the ball to make an out at first base?

55. CLOTHESLINE A pair of damp jeans are hung on a clothesline to dry, as shown in Illustration 9. They pull the center down 1 foot. By how much is the line stretched?

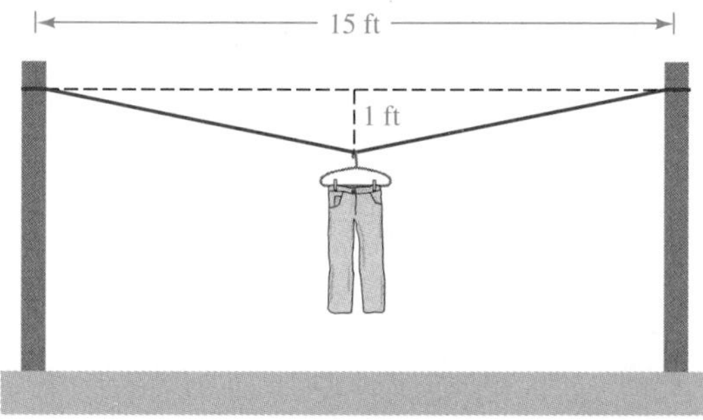

ILLUSTRATION 9

56. FIREFIGHTING The base of the 37-foot ladder in Illustration 10 is 9 feet from the wall. Will the top reach a window ledge that is 35 feet above the ground? Verify your result.

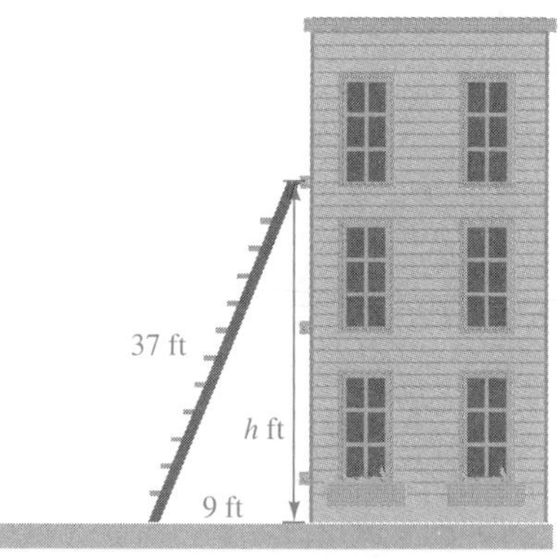

ILLUSTRATION 10

57. ART HISTORY A figure displaying some of the basic characteristics of Egyptian art is shown in Illustration 11. Use the distance formula to find the following dimensions of the drawing. Round your answers to two decimal places.

a. From the foot to the eye

b. From the belt to the hand holding the staff

c. From the shoulder to the symbol held in the hand

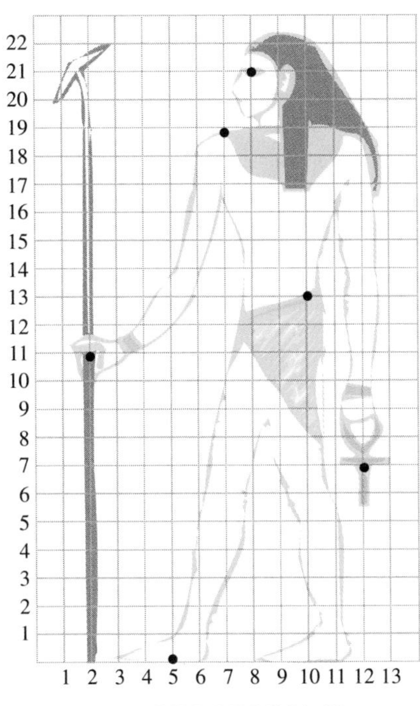

ILLUSTRATION 11

58. PACKAGING The diagonal d of a rectangular box with dimensions $a \times b \times c$ is given by

$$d = \sqrt{a^2 + b^2 + c^2}$$

Will the umbrella fit in the shipping carton in Illustration 12? Verify your result.

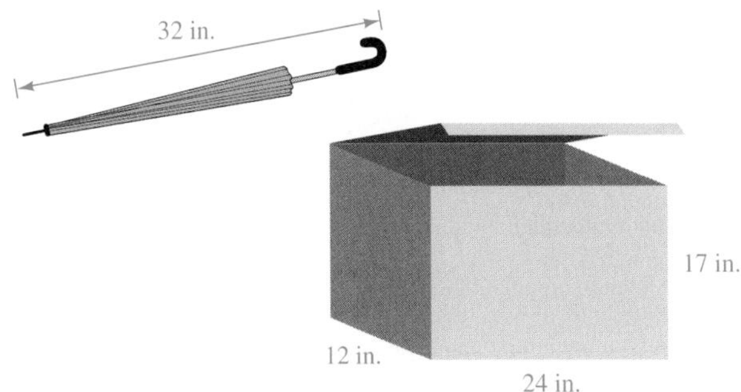

ILLUSTRATION 12

59. PACKAGING An archaeologist wants to ship a 34-inch femur bone. Will it fit in a 4-inch-tall box that has a 24-inch-square base? (See Exercise 58.) Verify your result.

60. TELEPHONE SERVICE The telephone cable in Illustration 13 runs from A to B to C to D. How much cable is required to run from A to D directly?

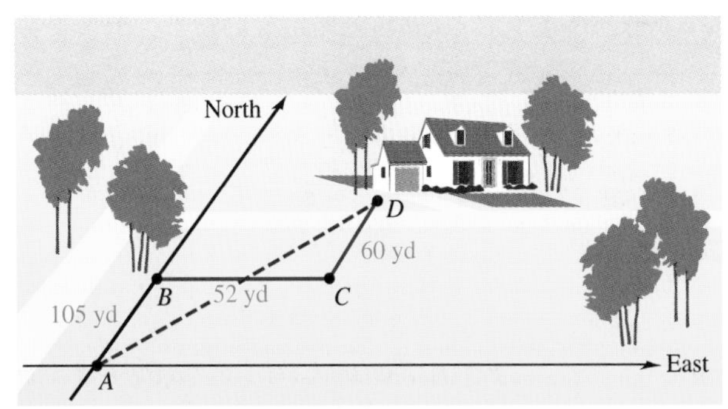

ILLUSTRATION 13

WRITING *Write a paragraph using your own words.*

61. State the Pythagorean theorem.

62. List the facts that you learned about special right triangles in this section.

REVIEW

63. DISCOUNT BUYING A repairman purchased some washing-machine motors for a total of \$224. When the unit cost decreased by \$4, he was able to buy one extra motor for the same total price. How many motors did he buy originally?

64. AVIATION An airplane can fly 650 miles with the wind in the same amount of time as it can fly 475 miles against the wind. If the wind speed is 40 mph, find the speed of the plane in still air.

KEY CONCEPT

Radicals

The expression $\sqrt[n]{a}$ is called a **radical expression.** In this chapter, we have discussed the properties and procedures used when simplifying radical expressions, solving radical equations, and writing radical expressions using rational exponents.

Expressions Containing Radicals

When working with expressions containing radicals, we must often apply the multiplication property and/or the division property of radicals to simplify the expression. Recall that

$$\sqrt[n]{ab} = \sqrt[n]{a}\sqrt[n]{b} \qquad \sqrt[n]{\frac{a}{b}} = \frac{\sqrt[n]{a}}{\sqrt[n]{b}} \quad (b \neq 0)$$

In Exercises 1–8, do each operation and simplify the expression.

1. Simplify: $\sqrt[3]{-54h^6}$.

2. Add: $2\sqrt[3]{64e} + 3\sqrt[3]{8e}$.

3. Subtract: $\sqrt{72} - \sqrt{200}$.

4. Multiply: $-4\sqrt[3]{5r^2s}\left(5\sqrt[3]{2r}\right)$.

5. Multiply: $\left(\sqrt{3s} - \sqrt{2t}\right)\left(\sqrt{3s} + \sqrt{2t}\right)$.

6. Multiply: $-\sqrt{3}\left(\sqrt{7} - \sqrt{5}\right)$.

7. Find the power: $\left(3\sqrt{2n} - 2\right)^2$.

8. Rationalize the denominator: $\dfrac{\sqrt[3]{9j}}{\sqrt[3]{3jk}}$.

Equations Containing Radicals

When solving radical equations, our objective is to rid the equation of the radical. This is achieved by using the *power rule:*

If x, y, and n are real numbers and $x = y$, then $x^n = y^n$.

If we raise both sides of an equation to the same power, the resulting equation might not be equivalent to the original equation. We must always check for extraneous solutions.

In Exercises 9–12, solve each radical equation, if possible.

9. $\sqrt{1 - 2g} = \sqrt{g + 10}$

10. $4 - \sqrt[3]{4 + 12x} = 0$

11. $\sqrt{y + 2} - 4 = -y$

12. $\sqrt[4]{12t + 4} + 2 = 0$

Radicals and Rational Exponents

Radicals can be written using rational (fractional) exponents, and exponential expressions having fractional exponents can be written in radical form. To do this, we use two rules for exponents introduced in this chapter.

$$x^{1/n} = \sqrt[n]{x} \qquad x^{m/n} = \sqrt[n]{x^m} = \left(\sqrt[n]{x}\right)^m$$

13. Express using a rational exponent: $\sqrt[3]{3}$.

14. Express in radical form: $5a^{2/5}$.

GROUP ACTIVITIES FOR CHAPTER 7

Accent on Teamwork

Section 7.1

A SPIRAL OF ROOTS To do this project, you will need a piece of poster board, a protractor, a yardstick, and a pencil. Begin by drawing an isosceles right triangle near the right margin of the poster board. Label the length of each leg as 1 unit. (See Illustration 1.) Use the Pythagorean theorem to determine the length of the hypotenuse. Draw a second right triangle using the hypotenuse of the first triangle as one leg. Draw its second leg with a length of 1 unit. Find the length of the hypotenuse of triangle 2. Continue this process of creating right triangles, using the previous hypotenuse as one leg and drawing a new second leg of length 1 unit each time. Calculate the length of the resulting hypotenuse. What patterns, if any, do you see?

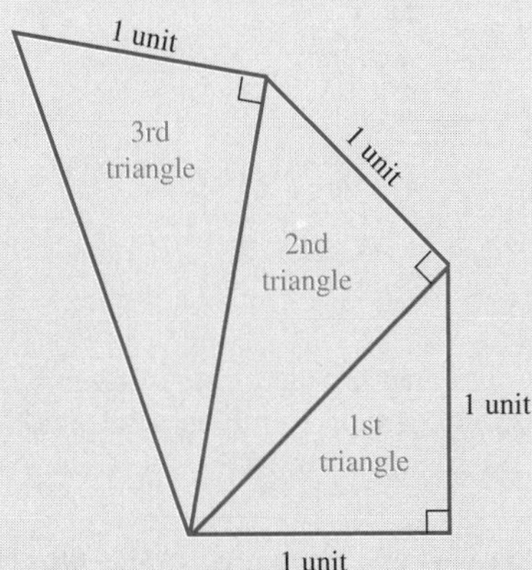

ILLUSTRATION 1

Section 7.2

SOLVING RADICAL EQUATIONS In this chapter, we solved equations that contained two radicals. The radicals in those equations always had the same index. That is not the case for the following two equations:

$$\sqrt[3]{2x} = \sqrt{x} \qquad \sqrt[4]{x} = \sqrt{\frac{x}{4}}$$

Brainstorm in your group to see if you can come up with a procedure that can be used to solve these equations. What are their solutions?

Section 7.3

EXPONENTS Use the $\boxed{y^x}$ key on your calculator to approximate each of the following exponential expressions. Then write them in order from least to greatest. Which of the exponents are not rational exponents? Could you have written the expressions in increasing order without having to approximate them? Explain.

1. $3^{0.999}$ **2.** $3^{3.9}$
3. $3^{1\frac{3}{4}}$ **4.** $3^{\sqrt{2}}$
5. $3^{1.7}$ **6.** 3^{π}
7. $3^{\frac{2}{3}}$ **8.** $3^{\frac{4}{3}}$
9. $3^{2\sqrt{3}}$ **10.** $3^{0.1}$

Section 7.4

COMMON ERRORS In each addition or subtraction problem below, tell what mistake was made. Compare each problem to a similar one involving variables to help clarify your explanation. For example, compare problem 1 to $2a + 3a$ to help explain the correct procedure that should be used to simplify the expression.

1. $2\sqrt{5x} + 3\sqrt{5x} = 5\sqrt{10x}$
2. $30 + 30\sqrt[4]{2} = 60\sqrt[4]{2}$
3. $7\sqrt[3]{y^2} - 5\sqrt[3]{y^2} = 2$
4. $6\sqrt{11ab} - 3\sqrt{5ab} = 3\sqrt{6ab}$

Section 7.5

MULTIPLYING RADICALS In this chapter, when we were asked to find the product of two radical expressions, the radicals always had the same index. Brainstorm in your group to see if you can come up with a procedure that can be used to find

$$\sqrt[3]{3} \cdot \sqrt{3}$$

Keep in mind two things: The indices must be the same to apply the multiplication property of radicals, and radical expressions can be written using rational exponents.

Section 7.6

PYTHAGOREAN THEOREM Demonstrate to your class how the Pythagorean theorem can be used to check whether the sides of a picture frame are perpendicular. Show some other applications where the Pythagorean theorem could be used to check whether two sides of an object are perpendicular.

CHAPTER REVIEW

SECTION 7.1 Radical Expressions and Radical Functions

CONCEPTS

The number b is a *square root of* a if $b^2 = a$.

If $x > 0$, the *principal square root* x is the positive square root of x, denoted $\sqrt{x}$. If x can be any real number, then $\sqrt{x^2} = |x|$.

The *cube root of* x is denoted as $\sqrt[3]{x}$ and is defined by

$\sqrt[3]{x} = y$ if $y^3 = x$

If n is an even natural number, $\sqrt[n]{a^n} = |a|$

If n is an odd natural number, $\sqrt[n]{a^n} = a$

If n is a natural number greater than 1 and x is a real number, then

- If $x > 0$, then $\sqrt[n]{x}$ is the positive number such that $(\sqrt[n]{x})^n = x$.
- If $x = 0$, then $\sqrt[n]{x} = 0$.
- If $x < 0$, and n is odd, $\sqrt[n]{x}$ is the real number such that $(\sqrt[n]{x})^n = x$.
- If $x < 0$, and n is even, $\sqrt[n]{x}$ is not a real number.

REVIEW EXERCISES

1. Simplify each radical expression, if possible. Assume that x can be any real number.

a. $\sqrt{49}$ **b.** $-\sqrt{121}$

c. $\sqrt{\dfrac{225}{49}}$ **d.** $\sqrt{-4}$

e. $\sqrt{0.01}$ **f.** $\sqrt{25x^2}$

g. $\sqrt{x^8}$ **h.** $\sqrt{x^2 + 4x + 4}$

2. Simplify each radical expression.

a. $\sqrt[3]{-27}$ **b.** $-\sqrt[3]{216}$

c. $\sqrt[3]{64a^6b^3}$ **d.** $\sqrt[3]{\dfrac{s^9}{125}}$

3. Simplify each radical expression, if possible. Assume that x and y can be any real number.

a. $\sqrt[4]{625}$ **b.** $\sqrt[5]{-32}$

c. $\sqrt[4]{256x^8y^4}$ **d.** $\sqrt{(-22y)^2}$

e. $-\sqrt[4]{\dfrac{1}{16}}$ **f.** $\sqrt[6]{-1}$

g. $\sqrt{0}$ **h.** $\sqrt[3]{0}$

4. GEOMETRY The side of a square with area A square feet is given by the function $s(A) = \sqrt{A}$. Find the *perimeter* of a square with an area of 144 square feet.

5. VOLUME OF A CUBE The total surface area of a cube is related to its volume V by the function $A(V) = 6\sqrt[3]{V^2}$. Find the surface area of a cube with a volume of 8 cm^3.

6. Graph each radical function. Find the domain and range.

a. $f(x) = \sqrt{x + 2}$ **b.** $f(x) = -\sqrt[3]{x} + 3$

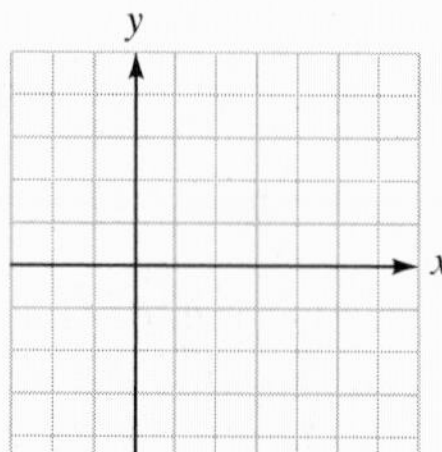

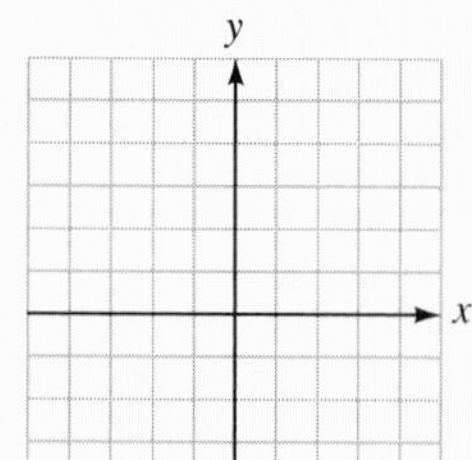

SECTION 7.2 Radical Equations

The power rule:

If $x = y$, then $x^n = y^n$.

Solving equations containing radicals:

1. Isolate one radical expression on one side of the equation.
2. Raise both sides of the equation to the power that is the same as the index.
3. Solve the resulting equation. If it still contains a radical, go back to step 1.
4. Check the solutions to eliminate *extraneous* solutions.

7. Solve each equation. Write all solutions. Cross out those that are extraneous.

a. $\sqrt{7x - 10} - 1 = 11$ **b.** $u = \sqrt{25u - 144}$

c. $2\sqrt{y - 3} = \sqrt{2y + 1}$ **d.** $\sqrt{z + 1} + \sqrt{z} = 2$

e. $\sqrt[3]{x^3 + 56} - 2 = x$ **f.** $\sqrt[4]{8x - 8} + 2 = 0$

8. Solve each equation for the indicated variable.

a. $r = \sqrt{\frac{A}{P}} - 1$ for P **b.** $h = \sqrt[3]{\frac{12I}{b}}$ for I

9. ELECTRONICS The current I (measured in amperes) and the power P (measured in watts) are related by the formula

$$I = \sqrt{\frac{P}{R}}$$

Find the resistance R in a circuit if the current used by an electrical appliance that is rated at 980 watts is 7.4 amperes.

SECTION 7.3 Rational Exponents

If n $(n > 1)$ is a natural number and $\sqrt[n]{x}$ is a real number, then

$$x^{1/n} = \sqrt[n]{x}$$

If n is a natural number greater than 1 and x is a real number,

- If $x > 0$, then $x^{1/n}$ is the positive number such that $(x^{1/n})^n = x$.
- If $x = 0$, then $x^{1/n} = 0$.
- If $x < 0$, and n is odd, then $x^{1/n}$ is the real number such that $(x^{1/n})^n = x$.
- If $x < 0$ and n is even, then $x^{1/n}$ is not a real number.

If m and n are positive integers, $x > 0$, and $\frac{m}{n}$ is in simplest form,

$$x^{m/n} = \sqrt[n]{x^m} = \left(\sqrt[n]{x}\right)^m$$

$$x^{-m/n} = \frac{1}{x^{m/n}}$$

$$\frac{1}{x^{-m/n}} = x^{m/n} \quad (x \neq 0)$$

10. Write each expression in radical form.

a. $t^{1/2}$ **b.** $(5xy^3)^{1/4}$

11. Simplify each expression, if possible. Assume that all variables represent positive real numbers.

a. $25^{1/2}$ **b.** $-36^{1/2}$

c. $(-36)^{1/2}$ **d.** $1^{1/2}$

e. $\left(\frac{9}{x^2}\right)^{1/2}$ **f.** $\left(\frac{1}{27}\right)^{1/3}$

g. $(-8)^{1/3}$ **h.** $625^{1/4}$

i. $(27a^3b)^{1/3}$ **j.** $(81c^4d^4)^{1/4}$

12. Simplify each expression, if possible. Assume that all variables represent positive real numbers.

a. $9^{3/2}$ **b.** $8^{-2/3}$

c. $-49^{5/2}$ **d.** $\frac{1}{100^{-1/2}}$

e. $\left(\frac{4}{9}\right)^{-3/2}$ **f.** $\frac{1}{25^{5/2}}$

g. $(25x^2y^4)^{3/2}$ **h.** $(8u^6v^3)^{-2/3}$

The *rules for exponents* can be used to simplify expressions with fractional exponents.

13. Do the operations. Write answers without negative exponents. Assume that all variables represent positive real numbers.

a. $5^{1/4}5^{1/2}$

b. $a^{3/7}a^{-2/7}$

c. $(k^{4/5})^{10}$

d. $\dfrac{(4g^3h)^{1/2}}{(9gh^{-1})^{1/2}}$

14. Do the multiplications. Assume that all variables represent positive real numbers and write all answers without negative exponents.

a. $u^{1/2}(u^{1/2} - u^{-1/2})$

b. $v^{2/3}(v^{1/3} + v^{4/3})$

15. Simplify $\sqrt[4]{\dfrac{a^2}{25b^2}}$. (The variables represent positive real numbers.)

16. Substitute the x- and y-coordinates of each point labeled in the graph in Illustration 1 into the equation

$x^{2/3} + y^{2/3} = 32$

Show that each one satisfies the equation.

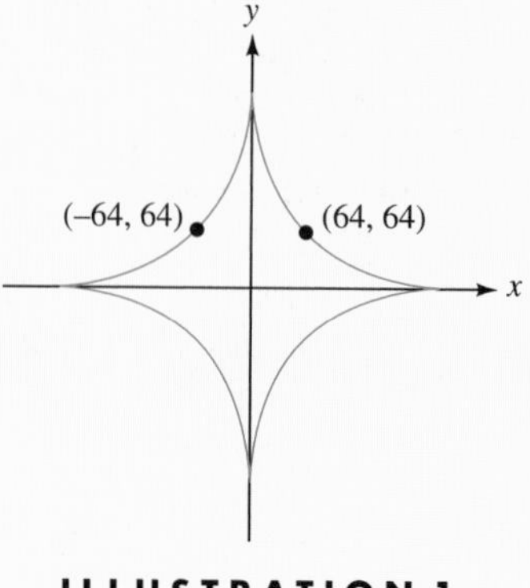

ILLUSTRATION 1

SECTION 7.4 *Simplifying and Combining Radical Expressions*

A radical is in *simplest form* when:

1. No radicals appear in a denominator.
2. The radicand contains no fractions or negative numbers.
3. Each factor in the radicand appears to a power less than the index.

Properties of radicals:

1. Multiplication:

$\sqrt[n]{ab} = \sqrt[n]{a}\sqrt[n]{b}$

2. Division:

$\sqrt[n]{\dfrac{a}{b}} = \dfrac{\sqrt[n]{a}}{\sqrt[n]{b}} \quad (b \neq 0)$

Like radicals can be combined by addition and subtraction.

Radicals that are not similar can often be converted to radicals that are similar and then combined.

17. Simplify each expression. Assume that all variables represent positive real numbers.

a. $\sqrt{240}$

b. $\sqrt[3]{54}$

c. $\sqrt[4]{32}$

d. $-2\sqrt[5]{-96}$

e. $\sqrt{8x^5}$

f. $\sqrt[3]{r^{17}}$

g. $\sqrt[3]{16x^5y^4}$

h. $3\sqrt[3]{27j^7k}$

i. $\dfrac{\sqrt{32x^3}}{\sqrt{2x}}$

j. $\sqrt{\dfrac{17xy}{64a^4}}$

18. Simplify and combine like radicals. Assume that all variables represent positive real numbers.

a. $\sqrt{2} + 2\sqrt{2}$

b. $6\sqrt{20} - \sqrt{5}$

c. $2\sqrt[3]{3} - \sqrt[3]{24}$

d. $-\sqrt[4]{32} - 2\sqrt[4]{162}$

e. $2x\sqrt{8} + 2\sqrt{200x^2} + \sqrt{50x^2}$

f. $\sqrt[3]{54} - 3\sqrt[3]{16} + 4\sqrt[3]{128}$

19. SEWING A corner of fabric is folded over to form a collar and stitched down as shown in Illustration 2. From the dimensions given in the figure, determine the exact number of inches of stitching that must be made. Then give an approximation to one decimal place. (All measurements are in inches.)

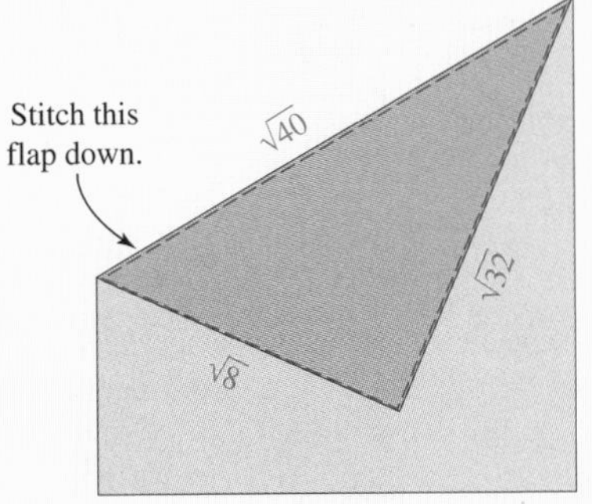

ILLUSTRATION 2

SECTION 7.5 Multiplying and Dividing Radical Expressions

If two radicals have the same index, they can be multiplied:

$$\sqrt[n]{a}\sqrt[n]{b} = \sqrt[n]{ab}$$

20. Simplify each expression. Assume that all variables represent positive real numbers.

a. $\sqrt{7}\sqrt{7}$ **b.** $(2\sqrt{5})(3\sqrt{2})$

c. $(-2\sqrt{8})^2$ **d.** $2\sqrt{6}\sqrt{216}$

e. $\sqrt{9x}\sqrt{x}$ **f.** $(\sqrt{33})^2$

g. $-\sqrt[3]{2x^2}\sqrt[3]{4x}$ **h.** $4\sqrt[3]{9}\sqrt[3]{9}$

i. $\sqrt{2}(\sqrt{8} - 3)$ **j.** $-\sqrt[4]{256x^5y^{11}}\sqrt[4]{625x^9y^3}$

k. $(\sqrt{3b} + \sqrt{3})^2$ **l.** $(2\sqrt{u} + 3)(3\sqrt{u} - 4)$

If a radical appears in a denominator of a fraction, or if a radicand contains a fraction, we can write the radical in simplest form by *rationalizing the denominator.*

To *rationalize the binomial denominator* of a fraction, multiply the numerator and the denominator by the conjugate of the binomial in the denominator.

21. Rationalize each denominator.

a. $\dfrac{10}{\sqrt{3}}$ **b.** $\sqrt{\dfrac{3}{5}}$

c. $\dfrac{x}{\sqrt{xy}}$ **d.** $\dfrac{\sqrt[3]{uv}}{\sqrt[3]{u^5v^7}}$

e. $\dfrac{2}{\sqrt{2} - 1}$ **f.** $\dfrac{\sqrt{a} + 1}{\sqrt{a} - 1}$

22. VOLUME The formula relating the radius r of a sphere and its volume V is $r = \sqrt[3]{\dfrac{3V}{4\pi}}$. Write the radical in simplest form.

SECTION 7.6 Geometric Applications of Radicals

The Pythagorean theorem:

If a and b are the lengths of the *legs* of a right triangle and c is the length of the *hypotenuse,* then $a^2 + b^2 = c^2$.

23. CARPENTRY The gable end of the roof shown in Illustration 3 is divided in half by a vertical brace, 8 feet in height. Find the length of the roof line.

24. SAILING A technique called *tacking* allows a sailboat to make progress into the wind. A sailboat follows the course in Illustration 4. Find d, the distance the boat advances into the wind after tacking.

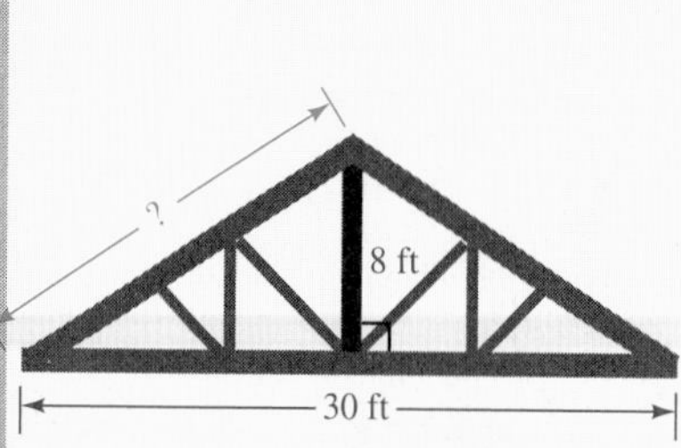

ILLUSTRATION 3

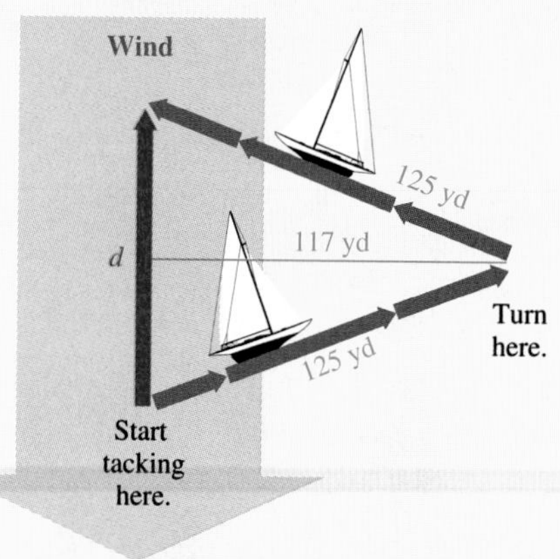

ILLUSTRATION 4

In an *isosceles right triangle,* the length of the hypotenuse is the length of one leg times $\sqrt{2}$.

25. Find the length of the hypotenuse of an isosceles right triangle whose legs measure 7 meters.

The shorter leg of a *30°–60°–90° triangle* (the side opposite the 30° angle) is half as long as the hypotenuse. The longer leg (the side opposite the 60° angle) is the length of the shorter leg times $\sqrt{3}$.

26. The hypotenuse of a 30°–60°–90° triangle measures $12\sqrt{3}$ centimeters. Find the length of each leg.

27. Find x to two decimal places.

a.

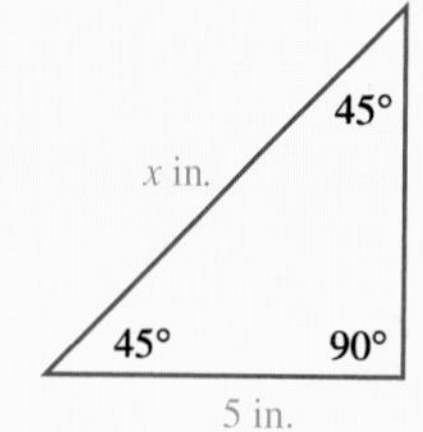

b.

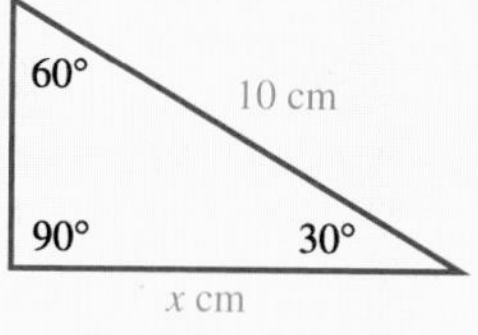

The distance formula:

$$d(PQ) = \sqrt{(x_2 - x_1)^2 + (y_2 - y_1)^2}$$

The *midpoint* of the line segment with endpoints at $P(x_1, y_1)$ and $Q(x_2, y_2)$ has coordinates

$$\left(\frac{x_1 + x_2}{2}, \frac{y_1 + y_2}{2}\right)$$

28. Find the distance between points P and Q. Then find the coordinates of the midpoint M of the line segment with endpoints P and Q.

a. $P(0, 0)$ and $Q(5, -12)$

b. $P(-4, 6)$ and $Q(-2, 8)$

Chapter 7 Test

1. Complete the table of values for $f(x) = \sqrt{x - 1}$. Then graph the function.

x	y
1	
2	
3	
5	
10	
12	
17	

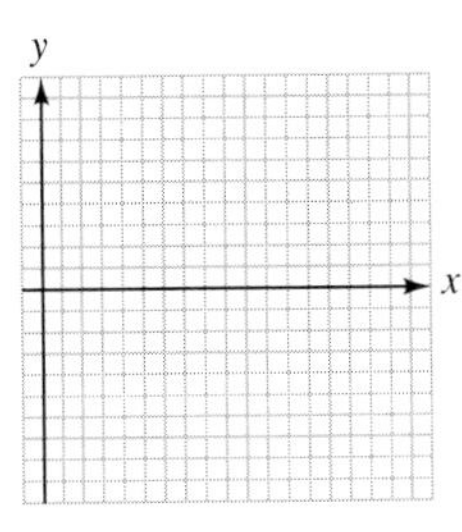

2. SUBMARINE The horizontal distance (measured in miles) that an observer can see is related to the height h (measured in feet) of the observer by the function $d(h) = 1.4\sqrt{h}$. If a submarine's periscope extends 4.7 feet above the surface of the ocean, how far is the horizon? Round to the nearest tenth.

In Problems 3–5, solve and check each equation.

3. $2\sqrt{x} = \sqrt{x + 1}$

4. $\sqrt[3]{6n + 4} - 4 = 0$

5. $1 - \sqrt{u} = \sqrt{u - 3}$

6. Solve $r = \sqrt[3]{\dfrac{GMt^2}{4\pi^2}}$ for G.

In Problems 7–12, simplify each expression. Assume that all variables represent positive real numbers, and write answers without using negative exponents.

7. $16^{1/4}$

8. $27^{2/3}$

9. $36^{-3/2}$

10. $\left(-\dfrac{8}{27}\right)^{-2/3}$

11. $\dfrac{2^{5/3}2^{1/6}}{2^{1/2}}$

12. $\dfrac{(8x^3y)^{1/2}(8xy^5)^{1/2}}{(x^3y^6)^{1/3}}$

In Problems 13–16, simplify each expression. Assume that the variables are unrestricted.

13. $\sqrt{x^2}$

14. $\sqrt{8x^2}$

15. $\sqrt[3]{54x^5}$

16. $\sqrt{18x^4y^8}$

In Problems 17–24, simplify each expression. Assume that all variables represent positive real numbers.

17. $\sqrt[3]{-64x^3y^6}$

18. $\sqrt{\dfrac{4a^2}{9}}$

19. $\sqrt[4]{-16}$

20. $\sqrt[5]{(t + 8)^5}$

21. $\sqrt{48}$

22. $\sqrt{250x^3y^5}$

23. $\dfrac{\sqrt[3]{24x^{15}y^4}}{\sqrt[3]{y}}$

24. $\sqrt{\dfrac{3a^5}{48a^7}}$

In Problems 25–28, simplify and combine like radicals. Assume that all variables represent positive real numbers.

25. $\sqrt{12} - \sqrt{27}$

26. $2\sqrt[3]{40} - \sqrt[3]{5{,}000} + 4\sqrt[3]{625}$

27. $2\sqrt{48y^5} - 3y\sqrt{12y^3}$

28. $\sqrt[4]{768z^5} + z\sqrt[4]{48z}$

In Problems 29–30, do each operation and simplify, if possible. All variables represent positive real numbers.

29. $-2\sqrt{xy}\left(3\sqrt{x} + \sqrt{xy^3}\right)$

30. $\left(3\sqrt{2} + \sqrt{3}\right)\left(2\sqrt{2} - 3\sqrt{3}\right)$

In Problems 31–34, rationalize each denominator.

31. $\dfrac{1}{\sqrt{5}}$

32. $\dfrac{6}{\sqrt[3]{9}}$

33. $\dfrac{-4\sqrt{2}}{\sqrt{5} + 3}$

34. $\dfrac{3t - 1}{\sqrt{3t} - 1}$

In Problems 35–36, find x to two decimal places.

35.

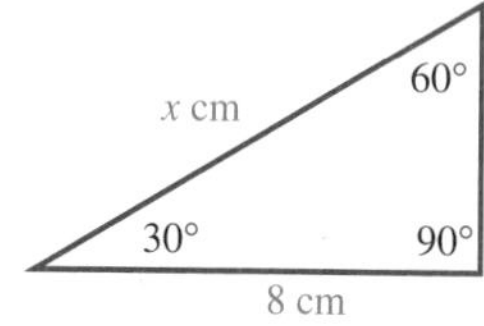

36.

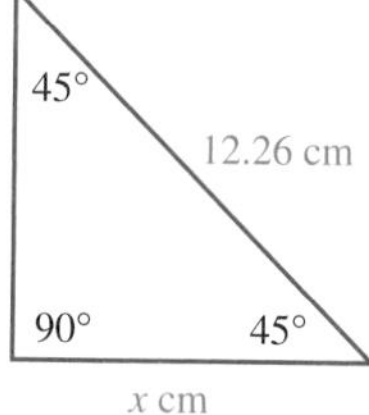

37. Find the distance between $P(-2, 5)$ and $Q(22, 12)$. Then find the coordinates of the midpoint M of the line segment with endpoints P and Q.

38. PENDULUM The time t, in seconds, it takes for a pendulum to swing back and forth to complete one period is given by the formula $t = 2\pi\sqrt{\frac{l}{32}}$, where l is the length of the pendulum in feet. Find the exact period of a 4-foot-long pendulum. Then approximate the period to the nearest tenth.

39. SHIPPING CRATE The diagonal brace on the shipping crate in Illustration 1 is 53 inches. Find the height h of the crate.

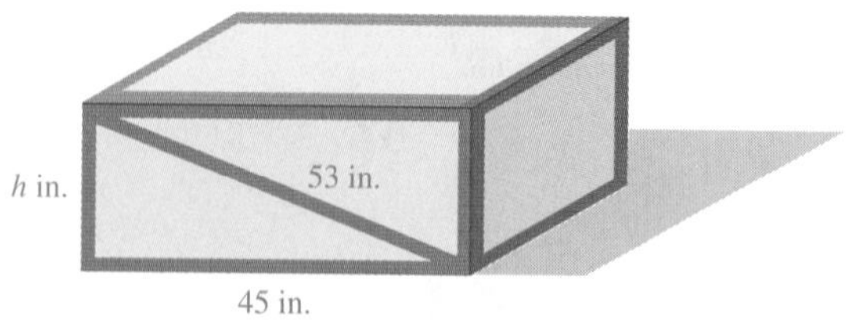

ILLUSTRATION 1

40. Explain why, without having to perform any algebraic steps, it is obvious that the equation $\sqrt{x - 8} = -10$ has no solutions.

Quadratic Equations, Functions, and Inequalities

8

Campus Connection

The *Art* Department

In a watercolor class, students learn that light and shadow, color, and perspective are fundamental components of an attractive painting. They also learn that the appropriate matting and frame can enhance their work. In this section, we will use mathematics to determine the dimensions of a uniform matting that is to have the same area as the picture it frames. To do this, we will write a quadratic equation and then solve it by *completing the square.* The technique of completing the square can be used to derive *the quadratic formula.* This formula is a valuable algebraic tool for solving any quadratic equation.

WE HAVE PREVIOUSLY SEEN HOW TO SOLVE QUADRATIC EQUATIONS BY FACTORING. IN THIS CHAPTER, WE WILL DISCUSS MORE GENERAL METHODS FOR SOLVING QUADRATIC EQUATIONS, AND WE WILL CONSIDER THE GRAPHS OF QUADRATIC FUNCTIONS.

8.1 Completing the Square

In this section, you will learn about

- **A review of solving quadratic equations by factoring**
- **The square root property**
- **Completing the square**
- **Solving equations by completing the square**
- **Problem solving**

INTRODUCTION. We have seen that equations that involve first-degree polynomials, such as $12x - 4 = 0$, are called *linear equations.* We have also seen that equations that involve second-degree polynomials, such as $12x^2 - 4x = 0$, are called *quadratic equations.* In Chapter 5, we learned how to solve quadratic equations by factoring. However, as we shall see, the factoring method has its limitations. In this section, we will introduce a more general method that enables us to solve *any* quadratic equation.

A review of solving quadratic equations by factoring

A *quadratic equation* is an equation of the form $ax^2 + bx + c = 0$ $(a \neq 0)$, where a, b, and c are real numbers. We have discussed how to solve quadratic equations by factoring. For example, to solve $6x^2 - 7x - 3 = 0$, we proceed as follows:

$$6x^2 - 7x - 3 = 0$$

$$(2x - 3)(3x + 1) = 0 \quad \text{Factor.}$$

$$2x - 3 = 0 \quad \text{or} \quad 3x + 1 = 0 \quad \text{Set each factor equal to 0.}$$

$$x = \frac{3}{2} \qquad x = -\frac{1}{3} \quad \text{Solve each linear equation.}$$

Many expressions do not factor as easily as $6x^2 - 7x - 3$. For example, it would be difficult to solve $2x^2 + 4x + 1 = 0$ by factoring, because $2x^2 + 4x + 1$ cannot be factored by using only integers. With this in mind, we will now develop another method of solving quadratic equations. It is based on the **square root property.**

The square root property

To develop general methods for solving all quadratic equations, we first consider the equation $x^2 = c$. If $c > 0$, we can find the real solutions of $x^2 = c$ as follows:

$$x^2 = c$$

$$x^2 - c = 0 \quad \text{Subtract } c \text{ from both sides.}$$

$$x^2 - \left(\sqrt{c}\right)^2 = 0 \quad \text{Replace } c \text{ with } \left(\sqrt{c}\right)^2\text{, since } c = \left(\sqrt{c}\right)^2.$$

$$\left(x + \sqrt{c}\right)\left(x - \sqrt{c}\right) = 0 \quad \text{Factor the difference of two squares.}$$

$$x + \sqrt{c} = 0 \quad \text{or} \quad x - \sqrt{c} = 0 \quad \text{Set each factor equal to 0.}$$

$$x = -\sqrt{c} \qquad x = \sqrt{c} \quad \text{Solve each linear equation.}$$

The two solutions of $x^2 = c$ are $x = \sqrt{c}$ and $x = -\sqrt{c}$.

Square root property If $c > 0$, the equation $x^2 = c$ has two real solutions. They are

$$x = \sqrt{c} \quad \text{and} \quad x = -\sqrt{c}$$

EXAMPLE 1 ***Solving a quadratic equation using the square root property.*** Solve $x^2 - 12 = 0$.

Solution

We can write the equation as $x^2 = 12$ and use the square root property to solve it.

$$x^2 - 12 = 0$$

$$x^2 = 12 \quad \text{Add 12 to both sides.}$$

$$x = \sqrt{12} \quad \text{or} \quad x = -\sqrt{12} \quad \text{Use the square root property.}$$

$$x = 2\sqrt{3} \qquad x = -2\sqrt{3} \quad \text{Simplify: } \sqrt{12} = \sqrt{4}\sqrt{3} = 2\sqrt{3}.$$

Verify that each solution satisfies the equation.

We can write the two solutions of the equation in a more concise form as $x = \pm 2\sqrt{3}$, where $\pm$ is read as "plus or minus."

Self Check

Solve $x^2 - 18 = 0$.

Answer: $\pm 3\sqrt{2}$

EXAMPLE 2 ***Phonograph records.*** Before compact disc (CD) technology, one way of recording music was by engraving grooves on thin vinyl discs called records. (See Figure 8-1.) The vinyl discs used for long-playing records had a surface area of about 111 square inches per side and were played at $33\frac{1}{3}$ revolutions per minute on a turntable. What is the radius of a long-playing record?

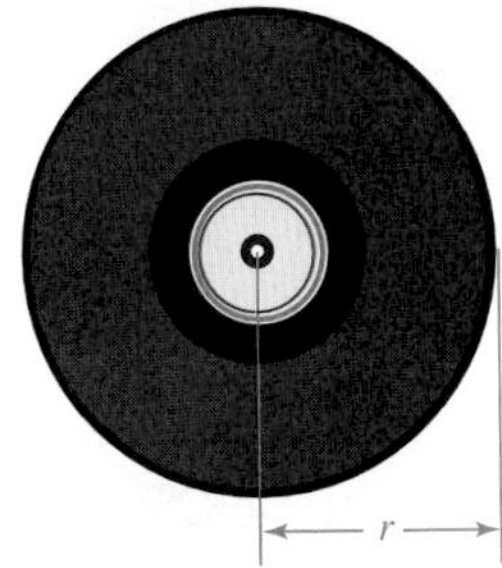

FIGURE 8-1

Solution

The relationship between the area of a circle and its radius is given by the formula $A = \pi r^2$. We can find the radius of a record by substituting 111 for A and solving for r.

$$A = \pi r^2 \quad \text{The formula for the area of a circle.}$$

$$111 = \pi r^2 \quad \text{Substitute 111 for } A.$$

$$\frac{111}{\pi} = r^2 \quad \text{To undo the multiplication by } \pi\text{, divide both sides by } \pi.$$

$$r = \sqrt{\frac{111}{\pi}} \quad \text{or} \quad r = -\sqrt{\frac{111}{\pi}}$$

Use the square root property. Since the radius of the record cannot be negative, discard the second solution.

The radius of a record is $\sqrt{\frac{111}{\pi}}$ inches—to the nearest tenth, 5.9 inches.

EXAMPLE 3 ***Using the square root property to solve an equation.*** Solve $(x - 3)^2 = 16$.

Solution

$$
\begin{aligned}
(x-3)^2 &= 16 \\
x - 3 = \sqrt{16} \quad \text{or} \quad x - 3 &= -\sqrt{16} \qquad \text{Use the square root property.} \\
x - 3 = 4 \qquad x - 3 &= -4 \qquad \text{Simplify: } \sqrt{16} = 4. \\
x = 3 + 4 \qquad x &= 3 - 4 \qquad \text{Add 3 to both sides.} \\
x = 7 \qquad x &= -1 \qquad \text{Simplify.}
\end{aligned}
$$

Verify that each solution satisfies the equation.

Self Check

Solve $(x + 2)^2 = 9$.

Answer: 1, −5 ■

Completing the square

All quadratic equations can be solved by **completing the square.** This method involves the special products

$$x^2 + 2ax + a^2 = (x + a)^2 \qquad \text{and} \qquad x^2 - 2ax + a^2 = (x - a)^2$$

The trinomials $x^2 + \mathbf{2a}x + a^2$ and $x^2 - \mathbf{2a}x + a^2$ are both perfect square trinomials, because both factor as the square of a binomial. In each case, the coefficient of the first term is 1, and if we take one-half of the coefficient of x in the middle term and square it, we obtain the third term.

$$\left[\frac{1}{2}(2a)\right]^2 = a^2 \qquad \left[\frac{1}{2}(-2a)\right]^2 = (-a)^2 = a^2$$

EXAMPLE 4 ***Completing the square*** Add a number to make each binomial a perfect square trinomial: **a.** $x^2 + 10x$, **b.** $x^2 - 6x$, and **c.** $x^2 - 11x$.

Solution

a. To make $x^2 + 10x$ a perfect square trinomial, we find one-half of 10, square it, and add that result to $x^2 + 10x$.

$$
\begin{aligned}
x^2 + \mathbf{10}x + \left[\frac{1}{2}(\mathbf{10})\right]^2 &= x^2 + 10x + (5)^2 \qquad \text{Simplify: } \tfrac{1}{2}(10) = 5. \\
&= x^2 + 10x + 25 \qquad \text{Note that } x^2 + 10x + 25 = (x + 5)^2.
\end{aligned}
$$

b. To make $x^2 - 6x$ a perfect square trinomial, we find one-half of −6, square it, and add that result to $x^2 - 6x$.

$$
\begin{aligned}
x^2 - \mathbf{6}x + \left[\frac{1}{2}(\mathbf{-6})\right]^2 &= x^2 - 6x + (-3)^2 \qquad \text{Simplify: } \tfrac{1}{2}(-6) = -3. \\
&= x^2 - 6x + 9 \qquad \text{Note that } x^2 - 6x + 9 = (x - 3)^2.
\end{aligned}
$$

c. To make $x^2 - 11x$ a perfect square trinomial, we find one-half of −11, square it, and add that result to $x^2 - 11x$.

$$
\begin{aligned}
&x^2 - \mathbf{11}x + \left[\frac{1}{2}(\mathbf{-11})\right]^2 \\
&= x^2 - 11x + \left(-\frac{11}{2}\right)^2 \qquad \text{Simplify: } \tfrac{1}{2}(-11) = -\tfrac{11}{2}. \\
&= x^2 - 11x + \frac{121}{4} \qquad \text{Note that } x^2 - 11x + \tfrac{121}{4} = \left(x - \tfrac{11}{2}\right)^2.
\end{aligned}
$$

Self Check

Add a number to $a^2 - 5a$ to make it a perfect trinomial square.

Answer: $a^2 - 5a + \frac{25}{4}$ ■

Solving equations by completing the square

To solve an equation of the form $ax^2 + bx + c = 0$ by completing the square, we use the following steps.

Completing the square

1. Make sure that the coefficient of x^2 (the **lead coefficient**) is 1. If it is not, make it 1 by dividing both sides of the equation by the coefficient of x^2.
2. If necessary, add a number to both sides of the equation so that the constant term is on the right-hand side of the equals sign.
3. Complete the square:
   a. Find one-half of the coefficient of x and square it.
   b. Add that square to both sides of the equation.
4. Factor the trinomial square.
5. Solve the resulting equation using the square root property.

EXAMPLE 5 ***Completing the square.*** Use completing the square to solve $x^2 + 8x + 7 = 0$.

Solution *Step 1:* In this example, the coefficient of x^2 is an understood 1.

Step 2: We add -7 to both sides so that the constant is on the right-hand side of the equals sign:

$$x^2 + 8x + 7 = 0$$
$$x^2 + 8x = -7$$

Step 3: The coefficient of x is 8, one-half of 8 is 4, and $4^2 = 16$. To complete the square, we add 16 to both sides.

$$x^2 + 8x + 16 = 16 - 7$$

1. $x^2 + 8x + 16 = 9$ Simplify: $16 - 7 = 9$.

Step 4: Since the left-hand side of Equation 1 is a perfect square trinomial, we can factor it to get $(x + 4)^2$.

$$x^2 + 8x + 16 = 9$$

2. $(x + 4)^2 = 9$

Step 5: We then solve Equation 2 by using the square root property.

$$x + 4 = \pm\sqrt{9}$$
$$x + 4 = 3 \quad \text{or} \quad x + 4 = -3$$
$$x = -1 \qquad\qquad x = -7$$

Verify that both solutions satisfy the equation. ■

EXAMPLE 6 ***The lead coefficient is not 1.*** Solve $6x^2 + 5x - 6 = 0$.

Solution *Step 1:* To make the coefficient of x^2 equal to 1, we divide both sides of the equation by 6.

$$6x^2 + 5x - 6 = 0$$
$$\frac{6x^2}{6} + \frac{5}{6}x - \frac{6}{6} = \frac{0}{6} \quad \text{Divide both sides by 6.}$$
$$x^2 + \frac{5}{6}x - 1 = 0 \quad \text{Simplify.}$$

Step 2: We add 1 to both sides so that the constant is on the right-hand side of the equals sign:

$$x^2 + \frac{5}{6}x = 1$$

Step 3: The coefficient of x is $\frac{5}{6}$, one-half of $\frac{5}{6}$ is $\frac{5}{12}$, and $\left(\frac{5}{12}\right)^2 = \frac{25}{144}$. To complete the square, we add $\frac{25}{144}$ to both sides.

$$x^2 + \frac{5}{6}x + \frac{25}{144} = 1 + \frac{25}{144}$$

3. $x^2 + \frac{5}{6}x + \frac{25}{144} = \frac{169}{144}$ Simplify: $1 + \frac{25}{144} = \frac{144}{144} + \frac{25}{144} = \frac{169}{144}$.

Step 4: Since the left-hand side of Equation 3 is a perfect square trinomial, we can factor it to get $\left(x + \frac{5}{12}\right)^2$.

4. $\left(x + \frac{5}{12}\right)^2 = \frac{169}{144}$

Step 5: We can solve Equation 4 by using the square root property.

$$x + \frac{5}{12} = \pm\sqrt{\frac{169}{144}}$$

$x + \frac{5}{12} = \frac{13}{12}$ or $x + \frac{5}{12} = -\frac{13}{12}$ Simplify: $\sqrt{\frac{169}{144}} = \frac{13}{12}$.

$x = -\frac{5}{12} + \frac{13}{12}$ $x = -\frac{5}{12} - \frac{13}{12}$ Subtract $\frac{5}{12}$ from both sides.

$x = \frac{8}{12}$ $x = -\frac{18}{12}$ Simplify.

$x = \frac{2}{3}$ $x = -\frac{3}{2}$ Simplify each fraction.

Verify that both solutions satisfy the original equation. ■

EXAMPLE 7 ***Solving quadratic equations by completing the square.*** Solve $2x^2 + 4x + 1 = 0$.

Solution

$2x^2 + 4x + 1 = 0$

$x^2 + 2x + \frac{1}{2} = 0$ Divide both sides by 2 to make the coefficient of x^2 equal to 1.

$x^2 + 2x = -\frac{1}{2}$ Subtract $\frac{1}{2}$ from both sides.

$x^2 + 2x + \mathbf{1} = \mathbf{1} - \frac{1}{2}$ Square half the coefficient of x and add it to both sides.

$(x + 1)^2 = \frac{1}{2}$ Factor and combine like terms.

$x + 1 = \pm\sqrt{\frac{1}{2}}$ Apply the square root property.

To write $\sqrt{\frac{1}{2}}$ in simplified radical form, we write it as a quotient of square roots and then rationalize the denominator.

Self Check

Solve $3x^2 + 6x + 1 = 0$.

$$x + 1 = \frac{\sqrt{2}}{2} \qquad x + 1 = -\frac{\sqrt{2}}{2} \qquad \sqrt{\frac{1}{2}} = \frac{\sqrt{1}}{\sqrt{2}} = \frac{1 \cdot \sqrt{2}}{\sqrt{2}\sqrt{2}} = \frac{\sqrt{2}}{2}.$$

$$x = -1 + \frac{\sqrt{2}}{2} \qquad x = -1 - \frac{\sqrt{2}}{2} \qquad \text{Subtract 1 from both sides.}$$

We can express each solution in an alternate form if we write -1 as a fraction with a denominator of 2.

$$x = -\frac{2}{2} + \frac{\sqrt{2}}{2} \qquad x = -\frac{2}{2} - \frac{\sqrt{2}}{2} \qquad \text{Write } -1 \text{ as } -\tfrac{2}{2}.$$

$$x = \frac{-2 + \sqrt{2}}{2} \qquad x = \frac{-2 - \sqrt{2}}{2} \qquad \text{Add (subtract) the numerators and keep the common denominator of 2.}$$

The exact solutions are $x = \frac{-2 + \sqrt{2}}{2}$ or $x = \frac{-2 - \sqrt{2}}{2}$, or more concisely, $x = \frac{-2 \pm \sqrt{2}}{2}$. We can use a calculator to approximate them. To the nearest hundredth, $x \approx -0.29$ or $x \approx -1.71$.

Answer: $\frac{-3 \pm \sqrt{6}}{3}$

WARNING! Recall that to simplify a fraction, we divide out common *factors* of the numerator and denominator. In Example 7, since -2 is a *term* of the numerator of $\frac{-2 + \sqrt{2}}{2}$, no further simplification of this expression can be made. In the Self Check, 2 is a common factor and can be divided out: $\frac{-6 \pm 2\sqrt{6}}{6} = \frac{2(-3 \pm \sqrt{6})}{2(3)} = \frac{-3 \pm \sqrt{6}}{3}$.

Problem solving

EXAMPLE 8 ***Graduation announcements.*** In creating the announcement shown in Figure 8-2, the graphic artist wants to follow two design elements.

- A border of uniform width should surround the text.
- Equal areas should be devoted to the text and to the border.

To meet these requirements, how wide should the border be?

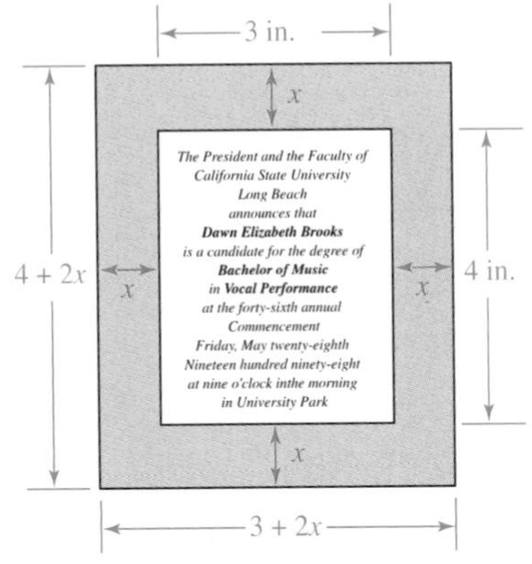

FIGURE 8-2

Analyze the problem The text occupies $4 \cdot 3 = 12$ in.^{2} of space. The border must also have an area of 12 in.^{2}.

Form an equation If we let x represent the width of the border, the length of the announcement is $(4 + 2x)$ inches and the width is $(3 + 2x)$ inches. We can now form the equation.

The area of the announcement	minus	the area of the text	equals	the area of the border.
$(4 + 2x)(3 + 2x)$	$-$	12	$=$	12

Solve the equation

$$(4 + 2x)(3 + 2x) - 12 = 12$$

$$12 + 8x + 6x + 4x^2 - 12 = 12 \qquad \text{On the left-hand side, use the FOIL method.}$$

$$4x^2 + 14x = 12 \qquad \text{Combine like terms.}$$

$$2x^2 + 7x - 6 = 0 \qquad \text{Subtract 12 from both sides. Then divide both sides by 2.}$$

We note that the trinomial on the left-hand side does not factor. We will solve the equation by completing the square.

$$x^2 + \frac{7}{2}x - 3 = 0$$ Divide both sides by 2 so that the coefficient of x^2 is 1.

$$x^2 + \frac{7}{2}x = 3$$ Add 3 to both sides.

$$x^2 + \frac{7}{2}x + \frac{49}{16} = 3 + \frac{49}{16}$$ Half of $\frac{7}{2}$ is $\frac{7}{4}$. Square $\frac{7}{4}$, which is $\frac{49}{16}$, and add it to both sides.

$$\left(x + \frac{7}{4}\right)^2 = \frac{97}{16}$$ On the left-hand side, factor the trinomial. On the right-hand side, $3 = \frac{3 \cdot 16}{1 \cdot 16} = \frac{48}{16}$ and $\frac{48}{16} + \frac{49}{16} = \frac{97}{16}$.

$$x + \frac{7}{4} = \pm\frac{\sqrt{97}}{4}$$ Apply the square root property. On the right-hand side, $\sqrt{\frac{97}{16}} = \frac{\sqrt{97}}{\sqrt{16}} = \frac{\sqrt{97}}{4}$.

$$x = -\frac{7}{4} + \frac{\sqrt{97}}{4} \quad \text{or} \quad x = -\frac{7}{4} - \frac{\sqrt{97}}{4}$$ Subtract $\frac{7}{4}$ from both sides.

$$x = \frac{-7 + \sqrt{97}}{4} \qquad x = \frac{-7 - \sqrt{97}}{4}$$ Write each expression as a single fraction.

State the conclusion The width of the border should be $\frac{-7 + \sqrt{97}}{4} \approx 0.71$ inch. (We discard the solution $\frac{-7 - \sqrt{97}}{4}$, since it is negative.)

Check the result If the border is 0.71 inch wide, the announcement has an area of about $5.42 \cdot 4.42 = 23.9564$ in.^{2}. If we subtract the area of the text from the area of the announcement, we get $23.9564 - 12 = 11.9564$ in.^{2}. This represents the area of the border, which was to be 12 in.^{2}. The answer seems reasonable. ■

STUDY SET Section 8.1

VOCABULARY *In Exercises 1–4, fill in the blanks to make the statements true.*

1. An equation of the form $ax^2 + bx + c = 0$ $(a \neq 0)$ is called a ______ equation.
2. The symbol ± is read as "______."
3. $x^2 + 6x + 9$ is called a ______ square trinomial because it factors as $(x + 3)^2$.
4. The ______ of x^2 in $x^2 - 12x + 36 = 0$ is 1, and the ______ term is 36.

CONCEPTS *In Exercises 5–8, fill in the blanks to make the statements true.*

5. The solutions of $x^2 = c$, where $c > 0$ are ☐ and ☐.
6. To complete the square on x in $x^2 + 6x$, find one-half of ___, square it to get ___, and add ___ to get ______.
7. In the quadratic equation $3x^2 - 2x + 5 = 0$, $a =$ ___, $b =$ ___, and $c =$ ___.
8. In the quadratic equation $ax^2 + bx + c = 0$, a cannot be ___.
9. Check to see if $-2 + \sqrt{2}$ is a solution of $x^2 + 4x + 2 = 0$.
10. Check to see if $-3\sqrt{2}$ is a solution of $x^2 - 18 = 0$.

11. Find one half of the coefficient of x and then square it.

a. $x^2 + 12x$

b. $x^2 - 5x$

c. $x^2 - \dfrac{x}{2}$

12. Add a number to make each binomial a perfect square trinomial. Then factor the result.

a. $x^2 + 8x$

b. $x^2 - 8x$

c. $x^2 - x$

13. What is the first step in solving the equation $x^2 + 12x = 35$

a. by the factoring method?

b. by completing the square?

14. Solve the equation $x^2 = 16$

a. by the factoring method,

b. by the square root method.

15. Explain the error in the work shown below.

$$\frac{4 \pm \sqrt{3}}{8} = \frac{\overset{1}{\cancel{4}} \pm \sqrt{3}}{\underset{1}{\cancel{4}} \cdot 2} = \frac{1 \pm \sqrt{3}}{2}$$

16. Explain the error in the work shown below.

$$\frac{1 \pm \sqrt{5}}{5} = \frac{1 \pm \sqrt{\cancel{5}^{1}}}{\cancel{5}^{1}} = \frac{1 \pm 1}{1}$$

In Exercises 17–18, note that a and b are solutions of the equation $(x - a)(x - b) = 0$.

17. Find a quadratic equation with a solution set of $\{3, 5\}$.

18. Find a quadratic equation with a solution set of $\{-4, 6\}$.

NOTATION

19. In solving a quadratic equation, a student obtains $x = \pm 2\sqrt{5}$.

a. How many solutions are represented by this notation? List them.

b. Approximate the solutions to the nearest hundredth.

20. In solving a quadratic equation, a student obtains $x = \dfrac{-5 \pm \sqrt{7}}{3}$.

a. How many solutions are represented by this notation? List them.

b. Approximate the solutions to the nearest hundredth.

PRACTICE *In Exercises 21–28, use factoring to solve each equation.*

21. $6x^2 + 12x = 0$

22. $5x^2 + 11x = 0$

23. $2y^2 - 50 = 0$

24. $4y^2 - 64 = 0$

25. $r^2 + 6r + 8 = 0$

26. $x^2 + 9x + 20 = 0$

27. $2z^2 = -2 + 5z$

28. $3x^2 = 8 - 10x$

In Exercises 29–40, use the square root property to solve each equation.

29. $x^2 = 36$

30. $x^2 = 144$

31. $z^2 = 5$

32. $u^2 = 24$

33. $3x^2 - 16 = 0$

34. $5x^2 - 49 = 0$

35. $(x + 1)^2 = 1$

36. $(x - 1)^2 = 4$

37. $(s - 7)^2 - 9 = 0$

38. $(t + 4)^2 = 16$

39. $(x + 5)^2 - 3 = 0$

40. $(x + 3)^2 - 7 = 0$

In Exercises 41–44, use the square root property to solve for the indicated variable. Assume that all variables represent positive numbers. Express all radicals in simplified form.

41. $2d^2 = 3h$ for d

42. $2x^2 = d^2$ for d

43. $E = mc^2$ for c

44. $A = \pi r^2$ for r

In Exercises 45–62, use completing the square to solve each equation.

45. $x^2 + 2x - 8 = 0$

46. $x^2 + 6x + 5 = 0$

47. $x + 1 = 2x^2$

48. $-2 = 2x^2 - 5x$

49. $6x^2 + x - 2 = 0$

50. $9 - 6r = 8r^2$

51. $x^2 + 8x + 6 = 0$

52. $x^2 + 6x + 4 = 0$

53. $x^2 - 2x - 17 = 0$

54. $x^2 + 10x - 7 = 0$

55. $3x^2 - 6x = 1$

56. $2x^2 - 6x = -3$

57. $4x^2 - 4x - 7 = 0$

58. $2x^2 - 8x + 5 = 0$

59. $2x^2 + 5x - 2 = 0$

60. $4x^2 - 4x - 1 = 0$

61. $\frac{7x + 1}{5} = -x^2$

62. $\frac{3x^2}{8} = \frac{1}{8} - x$

APPLICATIONS

63. FLAG In 1912, an executive order by President Taft fixed the overall width and length of the U.S. flag in the ratio 1 to 1.9. (See Illustration 1.) If 100 square feet of cloth are to be used to make a U.S. flag, estimate its dimensions to the nearest $\frac{1}{4}$ foot.

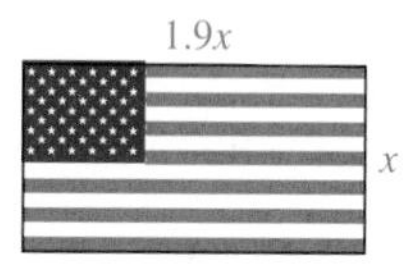

ILLUSTRATION 1

64. MOVIE STUNTS According to the *Guinness Book of World Records, 1998,* stunt man Dan Koko fell a distance of 312 feet into an airbag after jumping from the Vegas World Hotel and Casino. The distance d in feet traveled by a free-falling object in t seconds is given by the formula $d = 16t^2$. To the nearest tenth of a second, how long did the stunt man's free fall last?

65. ACCIDENTS The height h (in feet) of an object that is dropped from a height of s feet is given by the formula $h = s - 16t^2$, where t is the time the object has been falling. A five-foot-tall woman on a sidewalk looks directly overhead and sees a window washer drop a bottle from 4 stories up. How long does she have to get out of the way? Round to the nearest tenth. (A story is 12 feet.)

66. GEOGRAPHY The surface area S of a sphere is given by the formula $S = 4\pi r^2$, where r is the radius of the sphere. An almanac lists the surface area of the earth as 196,938,800 square miles. Assuming the earth to be spherical, what is its radius to the nearest mile?

67. AUTOMOBILE ENGINE As the piston shown in Illustration 2 moves upward, it pushes a "cylinder" of a gasoline/air mixture that is ignited by the spark plug. The formula that gives the volume of a cylinder is $V = \pi r^2 h$, where r is the radius and h the height. Find the radius of the piston (to the nearest hundredth of an inch) if it displaces 47.75 cubic inches of gasoline/air mixture as it moves from its lowest to its highest point.

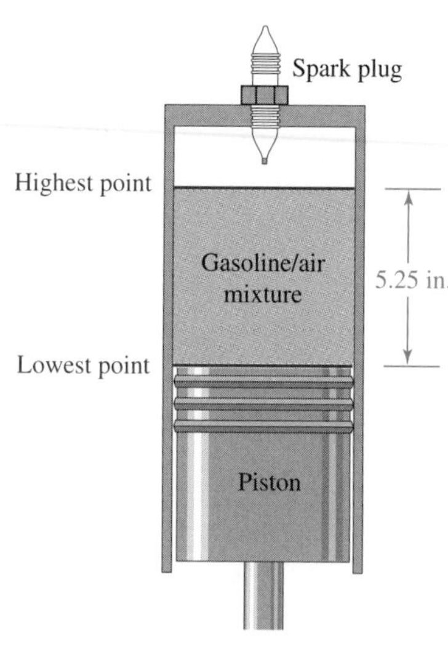

ILLUSTRATION 2

68. INVESTMENTS If P dollars are deposited in an account that pays an annual rate of interest r, then in n years, the amount of money A in the account is given by the formula $A = P(1 + r)^n$. A savings account was opened on January 3, 1996, with a deposit of \$10,000 and closed on January 2, 1998, with an ending balance of \$11,772.25. Find r, the rate of interest.

69. PICTURE FRAMING The matting around the picture in Illustration 3 has a uniform width. How wide is the matting if its area equals the area of the picture? Round to the nearest hundredth of an inch.

ILLUSTRATION 3

70. SWIMMING POOL See the advertisement in Illustration 4. How wide will the "free" concrete decking be if a uniform width is constructed around the perimeter of the pool? Round to the nearest hundredth of a yard. (*Hint:* Note the difference in units.)

ILLUSTRATION 4

71. DIMENSIONS OF A RECTANGLE A rectangle is 4 feet longer than it is wide, and its area is 20 square feet. Find its dimensions to the nearest tenth of a foot.

72. DIMENSIONS OF A TRIANGLE The height of a triangle is 4 meters longer than twice its base. Find the base and height if the area of the triangle is 10 square meters. Round to the nearest hundredth of a meter.

WRITING *Write a paragraph using your own words.*

73. Explain how to complete the square.

74. Tell why a cannot be 0 in the quadratic equation $ax^2 + bx + c = 0$.

REVIEW *In Exercises 75–80, simplify each expression. All variables represent positive real numbers.*

75. $\sqrt[3]{40a^3b^6}$

76. $\sqrt[3]{-27x^6}$

77. $\sqrt[8]{x^{24}}$

78. $\sqrt[4]{\dfrac{16}{625}}$

79. $\sqrt{175a^2b^3}$

80. $\sqrt{\dfrac{z^2}{16x^2}}$

8.2 The Quadratic Formula

In this section, you will learn about

- **The quadratic formula**
- **Solving quadratic equations using the quadratic formula**
- **Problem solving**
- **Quadratic equations as models**

Introduction. We can solve any quadratic equation by the method of completing the square, but the work is often tedious. In this section, we will develop a formula, called the *quadratic formula,* that lets us solve quadratic equations with much less effort.

The quadratic formula

To develop a formula we can use to solve quadratic equations, we solve the general quadratic equation $ax^2 + bx + c = 0$ $(a \neq 0)$.

$$ax^2 + bx + c = 0$$

$$\frac{ax^2}{a} + \frac{bx}{a} + \frac{c}{a} = \frac{0}{a}$$ Since $a \neq 0$, we can divide both sides by a.

$$x^2 + \frac{bx}{a} = -\frac{c}{a}$$ $\frac{0}{a} = 0$; subtract $\frac{c}{a}$ from both sides.

$$x^2 + \frac{b}{a}x + \left(\frac{b}{2a}\right)^2 = \left(\frac{b}{2a}\right)^2 - \frac{c}{a}$$ Complete the square on x. Half of $\frac{b}{a}$ is $\frac{b}{2a}$. Add $\left(\frac{b}{2a}\right)^2$ to both sides.

$$x^2 + \frac{b}{a}x + \frac{b^2}{4a^2} = \frac{b^2}{4a^2} - \frac{4ac}{4aa}$$ Remove parentheses and get a common denominator of $4a^2$ on the right-hand side.

1. $$\left(x + \frac{b}{2a}\right)^2 = \frac{b^2 - 4ac}{4a^2}$$ Factor the left-hand side and add the fractions on the right-hand side.

We can solve Equation 1 using the square root property.

$$x + \frac{b}{2a} = \sqrt{\frac{b^2 - 4ac}{4a^2}} \quad \text{or} \quad x + \frac{b}{2a} = -\sqrt{\frac{b^2 - 4ac}{4a^2}}$$

$$x + \frac{b}{2a} = \frac{\sqrt{b^2 - 4ac}}{\sqrt{4a^2}} \qquad x + \frac{b}{2a} = -\frac{\sqrt{b^2 - 4ac}}{\sqrt{4a^2}}$$

$$x + \frac{b}{2a} = \frac{\sqrt{b^2 - 4ac}}{2a} \qquad x + \frac{b}{2a} = -\frac{\sqrt{b^2 - 4ac}}{2a}$$

$$x = -\frac{b}{2a} + \frac{\sqrt{b^2 - 4ac}}{2a} \qquad x = -\frac{b}{2a} - \frac{\sqrt{b^2 - 4ac}}{2a}$$

$$x = \frac{-b + \sqrt{b^2 - 4ac}}{2a} \qquad x = \frac{-b - \sqrt{b^2 - 4ac}}{2a}$$

These two solutions give the **quadratic formula.**

The quadratic formula

The solutions of $ax^2 + bx + c = 0$ $(a \neq 0)$ are given by the formula

$$x = \frac{-b \pm \sqrt{b^2 - 4ac}}{2a}$$ Read the symbol $\pm$ as "plus or minus."

WARNING! Be sure to draw the fraction bar under both parts of the numerator, and be sure to draw the radical sign exactly over $b^2 - 4ac$. Do not write the quadratic formula as

$$x = -b \pm \frac{\sqrt{b^2 - 4ac}}{2a} \quad \text{or as} \quad x = -b \pm \sqrt{\frac{b^2 - 4ac}{2a}}$$

Solving quadratic equations using the quadratic formula

EXAMPLE 1 ***Solving quadratic equations.*** Solve $2x^2 - 3x - 5 = 0$ using the quadratic formula.

Self Check
Solve $3x^2 - 5x - 2 = 0$.

Solution
In this equation $a = 2$, $b = -3$, and $c = -5$.

$$x = \frac{-b \pm \sqrt{b^2 - 4ac}}{2a}$$

$$= \frac{-(-3) \pm \sqrt{(-3)^2 - 4(2)(-5)}}{2(2)}$$ Substitute 2 for a, -3 for b, and -5 for c.

$$= \frac{3 \pm \sqrt{9 + 40}}{4}$$ Simplify within the radical.

$$= \frac{3 \pm \sqrt{49}}{4}$$ Simplify: $\sqrt{49} = 7$.

$$= \frac{3 \pm 7}{4}$$

This result represents two solutions. To find the first solution, we add in the numerator. To find the second, we subtract in the numerator.

$$x = \frac{3 + 7}{4} \quad \text{or} \quad x = \frac{3 - 7}{4}$$

$$x = \frac{10}{4} \qquad x = \frac{-4}{4}$$

$$x = \frac{5}{2} \qquad x = -1$$

Verify that both solutions satisfy the original equation.

Answer: $2, -\frac{1}{3}$ ■

When solving a quadratic equation using the quadratic formula, the equation should be written in $ax^2 + bx + c = 0$ form so that a, b, and c can be determined.

EXAMPLE 2 ***Writing an equation in quadratic form first.*** Solve $2x^2 = -4x - 1$.

Self Check

Solve $3x^2 - 2x - 3 = 0$.

Solution

We begin by writing the equation in quadratic form:

$2x^2 + 4x + 1 = 0$ Add $4x$ and 1 to both sides.

In this equation, $a = 2$, $b = 4$, and $c = 1$.

$$x = \frac{-b \pm \sqrt{b^2 - 4ac}}{2a}$$

$$= \frac{-4 \pm \sqrt{4^2 - 4(2)(1)}}{2(2)}$$ Substitute 2 for a, 4 for b, and 1 for c.

$$= \frac{-4 \pm \sqrt{16 - 8}}{4}$$ Simplify within the radical.

$$= \frac{-4 \pm \sqrt{8}}{4}$$

$$= \frac{-4 \pm 2\sqrt{2}}{4}$$ Simplify: $\sqrt{8} = \sqrt{4 \cdot 2} = 2\sqrt{2}$.

$$= \frac{-2 \pm \sqrt{2}}{2}$$ $\frac{-4 \pm 2\sqrt{2}}{4} = \frac{2(-2 \pm \sqrt{2})}{4} = \frac{\overset{1}{\cancel{2}}(-2 \pm \sqrt{2})}{\underset{1}{\cancel{2}} \cdot 2} = \frac{-2 \pm \sqrt{2}}{2}$.

The solutions are $x = \frac{-2 + \sqrt{2}}{2}$ or $x = \frac{-2 - \sqrt{2}}{2}$. We can approximate the solutions using a calculator. To two decimal places, $x \approx -0.29$ or $x \approx -1.71$.

Answer: $\frac{1 \pm \sqrt{10}}{3}$ ■

Problem solving

EXAMPLE 3 ***Taking a shortcut.*** Instead of using the existing hallways, students are wearing a path through a planted quad area to walk 195 feet directly from the classrooms to the cafeteria. See Figure 8-3. If the length of the hallway from the office to the cafeteria is 105 feet longer than the hallway from the office to the classrooms, how much walking are the students saving by taking the shortcut?

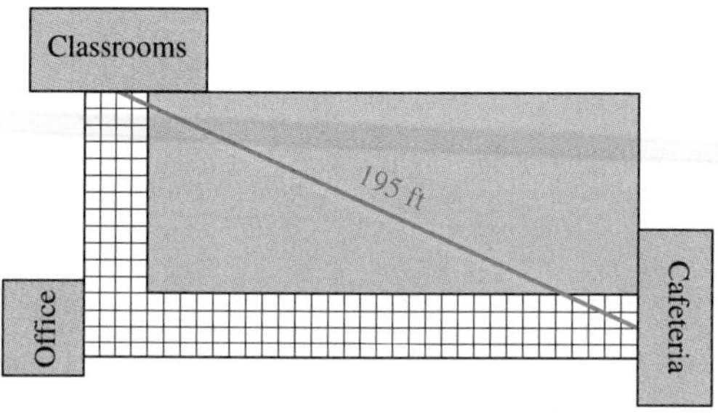

FIGURE 8-3

Analyze the problem The two hallways and the shortcut form a right triangle with a hypotenuse 195 feet long. We will use the Pythagorean theorem to solve this problem.

Form an equation If we let x represent the length (in feet) of the hallway from the classrooms to the office, then the length of the hallway from the office to the cafeteria is $(x + 105)$ feet. Substituting these lengths into the Pythagorean theorem, we have

$$a^2 + b^2 = c^2$$ The Pythagorean theorem.

$$x^2 + (x + 105)^2 = 195^2$$ Substitute x for a, $(x + 105)$ for b, and 195 for c.

$$x^2 + x^2 + 105x + 105x + 11{,}025 = 38{,}025$$ Use the FOIL method to find $(x + 105)^2$.

$$2x^2 + 210x + 11{,}025 = 38{,}025$$ Combine like terms.

$$2x^2 + 210x - 27{,}000 = 0$$ Subtract 38,025 from both sides.

$$x^2 + 105x - 13{,}500 = 0$$ Divide both sides by 2.

Solve the equation To solve $x^2 + 105x - 13{,}500 = 0$, we will use the quadratic formula with $a = 1$, $b = 105$, and $c = -13{,}500$.

$$x = \frac{-b \pm \sqrt{b^2 - 4ac}}{2a}$$

$$x = \frac{-105 \pm \sqrt{(105)^2 - 4(1)(-13{,}500)}}{2(1)}$$

$$x = \frac{-105 \pm \sqrt{65{,}025}}{2}$$ Simplify: $(105)^2 - 4(1)(-13{,}500) = 11{,}025 + 54{,}000 = 65{,}025$.

$$x = \frac{-105 \pm 255}{2}$$ Using a calculator, $\sqrt{65{,}025} = 255$.

$$x = \frac{150}{2} \quad \text{or} \quad x = \frac{-360}{2}$$

$$x = 75 \qquad x = -180$$ Since the length of the hallway can't be negative, discard the solution $x = -180$.

State the conclusion The length of the hallway from the classrooms to the office is 75 feet. The length of the hallway from the office to the cafeteria is $75 + 105 = 180$ feet. Instead of using the hallways, a distance of $75 + 180 = 255$ feet, the students are taking the 195-foot short cut to the cafeteria, a savings of $(255 - 195)$, or 60 feet.

Check the result The length of the 180-foot hallway is 105 feet longer than the length of the 75-foot hallway. The sum of the squares of the lengths of the hallways is $75^2 + 180^2 = 38{,}025$. This equals the square of the length of the 195-foot short cut. The answer checks. ■

EXAMPLE 4 ***Mass transit.*** A bus company has 4,000 passengers daily, each currently paying a 75¢ fare. For each 15¢ fare increase, the company estimates that it will lose 50 passengers. If the company needs to bring in \$6,570 per day to stay in business, what fare must be charged to produce this amount of revenue?

Analyze the problem To understand how a fare increase affects the number of passengers, let's consider what happens if there are two fare increases. We organize the data in a table. The fares are expressed in terms of dollars.

Number of increases	New fare	Number of passengers
One \$0.15 increase	\$0.75 + \$0.15(1) = \$0.90	4,000 − 50(1) = 3,950
Two \$0.15 increases	\$0.75 + \$0.15(2) = \$1.05	4,000 − 50(2) = 3,900

In general, the new fare will be the old fare (\$0.75) plus the number of fare increases times \$0.15. The number of passengers who will pay the new fare is 4,000 minus 50 times the number of \$0.15 fare increases.

Form an equation If we let x represent the number of \$0.15 fare increases necessary to bring in \$6,570 daily, then $\$(0.75 + 0.15x)$ is the fare that must be charged. The number of passengers who will pay this fare is $(4{,}000 - 50x)$. We can now form the equation.

The bus fare	times	the number of passengers who will pay that fare	equals	\$6,570.
$(0.75 + 0.15x)$	$\cdot$	$(4{,}000 - 50x)$	$=$	$6{,}570$

Solve the equation

$$(0.75 + 0.15x)(4{,}000 - 50x) = 6{,}570$$

$$3{,}000 - 37.5x + 600x - 7.5x^2 = 6{,}570$$ Use the FOIL method.

$$-7.5x^2 + 562.5x + 3{,}000 = 6{,}570$$ Combine like terms: $-37.5x + 600x = 562.5x$.

$$-7.5x^2 + 562.5x - 3{,}570 = 0$$ Subtract 6,570 from both sides.

$$7.5x^2 - 562.5x + 3{,}570 = 0$$ Multiply both sides by -1 so that a, 7.5, is positive.

To solve this equation, we will use the quadratic formula.

$$x = \frac{-b \pm \sqrt{b^2 - 4ac}}{2a}$$

$$x = \frac{-(-562.5) \pm \sqrt{(-562.5)^2 - 4(\mathbf{7.5})(\mathbf{3{,}570})}}{2(\mathbf{7.5})}$$ Substitute 7.5 for a, -562.5 for b, and 3,570 for c.

$$x = \frac{562.5 \pm \sqrt{209{,}306.25}}{15}$$ Simplify: $(-562.5)^2 - 4(7.5)(3{,}570) = 316{,}406.25 - 107{,}100 = 209{,}306.25$.

$$x = \frac{562.5 \pm 457.5}{15}$$ Using a calculator, $\sqrt{209{,}306.25} = 457.5$.

$$x = \frac{1{,}020}{15} \quad \text{or} \quad x = \frac{105}{15}$$

$$x = 68 \qquad\qquad x = 7$$

State the conclusion If there are 7 fifteen-cent increases in the fare, the new fare will be $\$0.75 + \$0.15(7) = \$1.80$. If there are 68 fifteen-cent increases in the fare, the new fare will be $\$0.75 + \$0.15(68) = \$10.95$. Although this fare would bring in the necessary revenue, a \$10.95 bus fare is unreasonable, so we discard it.

Check the result A fare of \$1.80 will be paid by $[4{,}000 - 50(7)] = 3{,}650$ bus riders. The amount of revenue brought in would be $\$1.80(3{,}650) = \$6{,}570$. The answer checks. ■

Quadratic equations as models

Many real-life situations can be modeled by quadratic equations.

EXAMPLE 5 ***Lawyers.*** The number of lawyers N in the United States each year from 1980 to 1996 is approximated by the quadratic equation $N = 660x^2 + 11{,}500x + 580{,}000$, where $x = 0$ corresponds to the year 1980, $x = 1$ corresponds to 1981, $x = 2$ corresponds to 1982, and so on. (Thus, $0 \le x \le 16$.) In what year does this model indicate that the United States had three-quarters of a million (750,000) lawyers?

Solution We will substitute 750,000 for N in the equation. Then we can solve for x, which will give the number of years after 1980 that the United States had approximately 750,000 lawyers.

$$N = 660x^2 + 11{,}500x + 580{,}000$$

$$750{,}000 = 660x^2 + 11{,}500x + 580{,}000 \quad \text{Replace } N \text{ with } 750{,}000.$$

$$0 = 660x^2 + 11{,}500x - 170{,}000 \quad \text{Subtract 750,000 from both sides so that the equation is in quadratic form.}$$

We can simplify the computations by dividing both sides of the equation by 20, the common factor of 660, 11,500, and 170,000.

$$0 = 33x^2 + 575x - 8{,}500 \quad \text{Divide both sides by 20.}$$

We solve this equation using the quadratic formula.

$$x = \frac{-b \pm \sqrt{b^2 - 4ac}}{2a}$$

$$x = \frac{-575 \pm \sqrt{(575)^2 - 4(\mathbf{33})(\mathbf{-8{,}500})}}{2(\mathbf{33})} \quad \text{Substitute 33 for } a\text{, 575 for } b\text{, and } -8{,}500 \text{ for } c.$$

$$x = \frac{-575 \pm \sqrt{330{,}625 + 1{,}122{,}000}}{66} \quad \text{Simplify: } (575)^2 - 4(33)(-8{,}500) = 330{,}625 + 1{,}122{,}000.$$

$$x = \frac{-575 \pm \sqrt{1{,}452{,}625}}{66}$$

$$x \approx \frac{630}{66} \quad \text{or} \quad x \approx \frac{-1{,}780}{66} \quad \text{Use a calculator.}$$

$$x \approx 9.5 \qquad x \approx -27.0$$

Since the model is defined only for $0 \le x \le 16$, we discard the second solution.

In 9.5 years after 1980, or midway through 1989, the United States had approximately 750,000 lawyers. ■

STUDY SET Section 8.2

VOCABULARY *In Exercises 1–2, fill in the blanks to make the statements true.*

1. An equation of the form $ax^2 + bx + c = 0$ $(a \neq 0)$ is a ____________ equation.

2. $x = \dfrac{-b \pm \sqrt{b^2 - 4ac}}{2a}$ is called the ____________ formula.

CONCEPTS

3. Consider the quadratic equation $4x^2 - 2x = 0$.
a. Solve it by factoring.
b. What are a, b, and c? Solve it using the quadratic formula.

4. Consider the quadratic equation $x^2 = 8$.
a. Solve it by the square root method.
b. What are a, b, and c? Solve it using the quadratic formula.

5. Tell whether each statement is true or false.
a. Any quadratic equation can be solved by using the quadratic formula.
b. Any quadratic equation can be solved by completing the square.

6. What is wrong with the beginning of the solution shown below?

Solve: $x^2 - 3x = 2$.

a = 1 b = -3 c = 2

7. What form would the quadratic formula take if, in developing it, we began with the general quadratic equation expressed as $rc^2 + sc + t = 0$ $(r \neq 0)$?

8. A student used the quadratic formula to solve a quadratic equation and obtained $x = \dfrac{-2 \pm \sqrt{3}}{2}$.
a. How many solutions does the equation have? What are they exactly?
b. Graph the solutions on a number line.

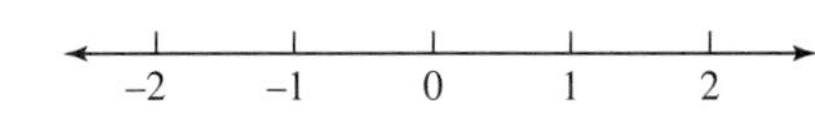

9. Solve $x^2 - 4x + 1 = 0$ using the quadratic formula. Simplify the result.

10. Solve $x^2 + 4x + 2 = 0$ using the quadratic formula. Simplify the result.

NOTATION

11. On a quiz, students were asked to write the quadratic formula. What is wrong with each answer shown below?
a. $x = -b \pm \dfrac{\sqrt{b^2 - 4ac}}{2a}$
b. $x = \dfrac{-b\sqrt{b^2 - 4ac}}{2a}$

12. In reading $\dfrac{-b \pm \sqrt{b^2 - 4ac}}{2a}$, we say, "The ____________ of b, plus or ____________ the square ____________ of b ____________ minus ___ times a times c, all ____________ $2a$."

PRACTICE *In Exercises 13–30, use the quadratic formula to solve each equation.*

13. $x^2 + 3x + 2 = 0$

14. $x^2 - 3x + 2 = 0$

15. $x^2 + 12x = -36$

16. $y^2 - 18y = -81$

17. $5x^2 + 5x + 1 = 0$

18. $4w^2 + 6w + 1 = 0$

19. $8u = -4u^2 - 3$

20. $4t + 3 = 4t^2$

21. $16y^2 + 8y - 3 = 0$

22. $16x^2 + 16x + 3 = 0$

23. $\frac{x^2}{2} + \frac{5}{2}x = -1$

24. $\frac{x^2}{8} - \frac{x}{4} = \frac{1}{2}$

25. $2x^2 - 1 = 3x$

26. $-9x = 2 - 3x^2$

27. $-x^2 + 10x = 18$

28. $-3x = \frac{x^2}{2} + 2$

29. $x^2 - 6x - 391 = 0$

30. $x^2 - 27x - 280 = 0$

In Exercises 31–32, use the quadratic formula and a scientific calculator to solve each equation. Give all answers to the nearest hundredth.

31. $0.7x^2 - 3.5x - 25 = 0$

32. $-4.5x^2 + 0.2x + 3.75 = 0$

APPLICATIONS

33. MOVIE THEATER SCREEN The largest permanent movie screen is in the Panasonic Imax theater at Darling Harbor, Sydney, Australia. The rectangular screen has an area of 11,349 square feet. Find the dimensions of the screen if it is 20 feet longer than it is wide.

34. ROCK CONCERT During a 1997 tour, the rock group U2 used the world's largest LED (Light Emitting Diode) electronic screen as part of the stage backdrop. The screen, with an area of 9,520 square feet, had a length that was 2 feet more than three times its width. Find the dimensions of the LED screen.

35. CENTRAL PARK Central Park is one of New York's best-known landmarks. Rectangular in shape, its length is 5 times its width. When measured in miles, its perimeter numerically exceeds its area by 4.75. Find the dimensions of Central Park if we know that its width is less than 1 mile.

36. ANCIENT HISTORY One of the most important cities of the ancient world was Babylon. Greek historians wrote that the city was square-shaped. Measured in miles, its area numerically exceeded its perimeter by about 124. Find its dimensions. (Round to the nearest tenth.)

37. BADMINTON The person who wrote the instructions for setting up the badminton net shown in Illustration 1 forgot to give the specific dimensions for securing the pole. How long is the support string?

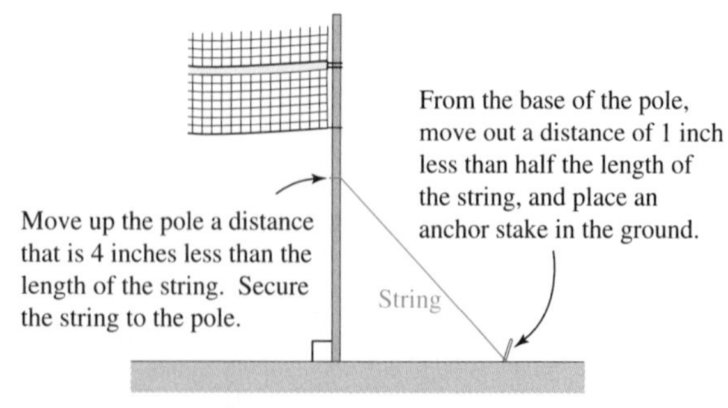
From the base of the pole, move out a distance of 1 inch less than half the length of the string, and place an anchor stake in the ground.
Move up the pole a distance that is 4 inches less than the length of the string. Secure the string to the pole.
String

ILLUSTRATION 1

38. RIGHT TRIANGLE The hypotenuse of a right triangle is 2.5 units long. The longer leg is 1.7 units longer than the shorter leg. Find the lengths of the sides of the triangle.

39. SCHOOL DANCE Tickets to a school dance cost \$4, and the projected attendance is 300 persons. It is further projected that for every 10¢ increase in ticket price, the average attendance will decrease by 5. At what ticket price will the receipts from the dance be \$1,248?

40. TICKET SALES A carnival at a county fair normally sells three thousand 25¢ ride tickets on a Saturday. For each 5¢ increase in price, management estimates that 80 fewer tickets will be sold. What increase in ticket price will produce \$994 of revenue on Saturday?

41. MAGAZINE SALES The *Gazette's* profit is \$20 per year for each of its 3,000 subscribers. Management estimates that the profit per subscriber will increase by 1¢ for each additional subscriber over the current 3,000. How many subscribers will bring a total profit of \$120,000?

42. POLYGONS The five-sided polygon called a *pentagon,* shown in Illustration 2, has 5 diagonals. The number of diagonals d of a polygon of n sides is given by the formula

$$d = \frac{n(n-3)}{2}$$

Find the number of sides of a polygon if it has 275 diagonals.

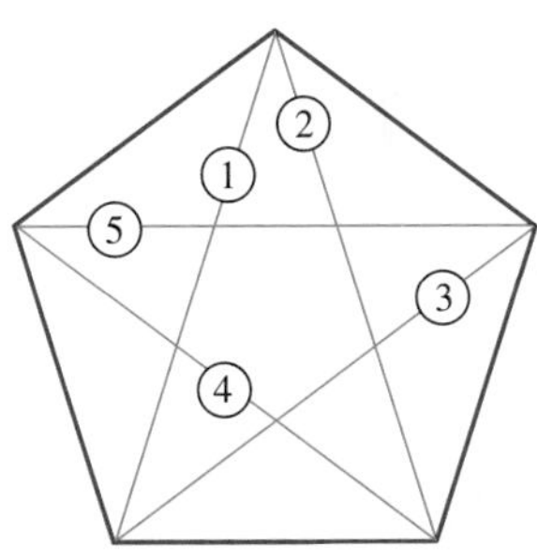

ILLUSTRATION 2

43. INVESTMENT RATES A woman invests \$1,000 in a mutual fund for which interest is compounded annually at a rate r. After one year, she deposits an additional \$2,000. After two years, the balance in the account is

$$\$1{,}000(1 + r)^2 + \$2{,}000(1 + r)$$

If this amount is \$3,368.10, find r.

44. METAL FABRICATION A box with no top is to be made by cutting a 2-inch square from each corner of the square sheet of metal shown in Illustration 3. After bending up the sides, the volume of the box is to be 220 cubic inches. How large should the piece of metal be? Round to the nearest hundredth.

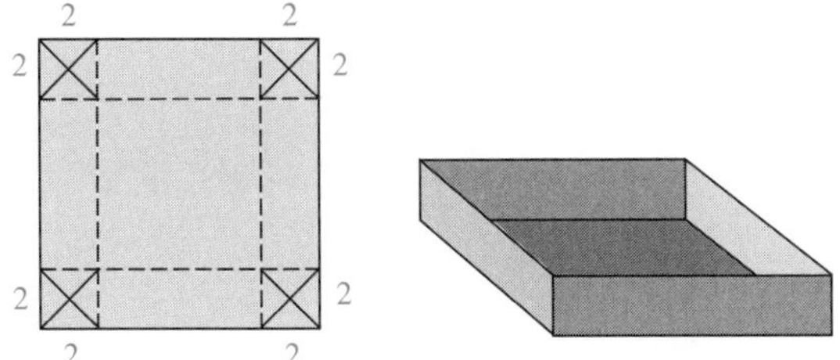

ILLUSTRATION 3

45. RETIREMENT The labor force participation rate P (in percent) for men ages 55–64 from 1970 to 1996 is approximated by the quadratic equation $P = 0.027x^2 - 1.363x + 82.5$, where $x = 0$ corresponds to the year 1970, $x = 1$ corresponds to 1971, $x = 2$ corresponds to 1972, and so on. (Thus, $0 \le x \le 26$.)

a. To the nearest percent, what percent of men ages 55–64 were in the workforce in 1970?

b. When does the model indicate that 75% of the men ages 55–64 were part of the workforce?

46. SPACE PROGRAM The yearly budget B (in billions of dollars) for the National Aeronautics and Space Administration (NASA) is approximated by the quadratic equation $B = -0.1492x^2 + 1.8058x + 9$, where x is the number of years since 1988 and $0 \le x \le 8$.

a. Use the model to find the approximate budget for NASA in 1996. Round to the nearest tenth.

b. In what year does the model indicate that NASA's budget was about \$12 billion?

c. For the years 1988–1996, has NASA's budget ever been \$15 billion? Explain how you know.

WRITING *Write a paragraph using your own words.*

47. Explain why the quadratic formula, in most cases, is less tedious to use in solving a quadratic equation than is the method of completing the square.

48. On an exam, a student was asked to solve the equation $-4w^2 - 6w - 1 = 0$. Her first step was to multiply both sides of the equation by -1. She then used the quadratic formula to solve $4w^2 + 6w + 1 = 0$ instead. Is this a valid approach?

REVIEW *In Exercises 49–52, change each radical to an exponential expression.*

49. $\sqrt{n}$

50. $\sqrt[7]{\frac{3}{8}r^2s}$

51. $\sqrt[4]{3b}$

52. $3\sqrt[3]{c^2 - d^2}$

In Exercises 53–56, write each expression in radical form.

53. $t^{1/3}$

54. $\left(\frac{3}{4}m^2n^2\right)^{1/5}$

55. $(3t)^{1/4}$

56. $(c^2 + d^2)^{1/2}$

8.3 Quadratic Functions and Their Graphs

In this section, you will learn about

- **Quadratic functions**
- **Graphs of $f(x) = ax^2$**
- **Graphs of $f(x) = ax^2 + c$**
- **Graphs of $f(x) = a(x - h)^2$**
- **Graphs of $f(x) = a(x - h)^2 + k$**
- **Graphs of $f(x) = ax^2 + bx + c$**
- **Determining minimum and maximum values**

INTRODUCTION. The graph shown in Figure 8-4 shows the distance a trampolinist is from the ground (in relation to time) as she bounds into the air and then falls back down to the trampoline.

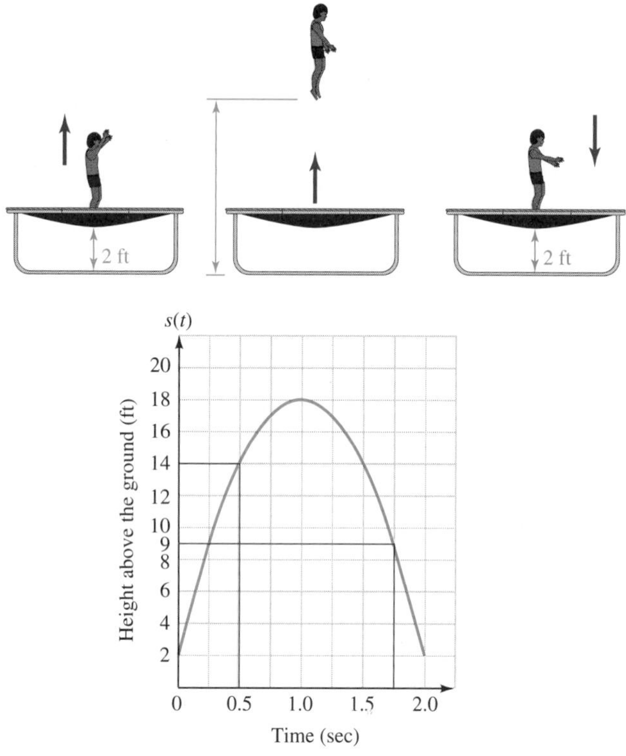

FIGURE 8-4

From the graph, we can see that the trampolinist is 14 feet above the ground 0.5 second after bounding upward and that her height above the ground after 1.75 seconds is 9 feet.

The parabola shown in Figure 8-4 is the graph of the *quadratic function* $s(t) = -16t^2 + 32t + 2$. In this section, we will discuss two forms in which quadratic functions are written and how to graph them.

Quadratic functions

Quadratic functions

A **quadratic function** is a second-degree polynomial function of the form

$$f(x) = ax^2 + bx + c \quad \text{or} \quad y = ax^2 + bx + c \quad (a \neq 0)$$

where a, b, and c are real numbers.

EXAMPLE 1 ***Quadratic functions as models.*** The quadratic function $s(t) = -16t^2 + 32t + 2$ gives the distance (in feet) that the trampolinist shown in Figure 8-4 is from the ground, t seconds after bounding upward. How far is she from the ground after being in the air for $\frac{3}{4}$ second?

Solution

To find her distance from the ground, we find the value of the function for $t = \frac{3}{4} = 0.75$.

$$\begin{aligned} s(t) &= -16t^2 + 32t + 2 \\ s(0.75) &= -16(0.75)^2 + 32(0.75) + 2 \quad \text{Replace } t \text{ with 0.75.} \\ &= -9 + 24 + 2 \\ &= 17 \end{aligned}$$

The trampolinist is 17 feet off the ground $\frac{3}{4}$ second after bounding upward.

Self Check

Find the distance the trampolinist is from the ground after being in the air for $1\frac{1}{2}$ seconds.

Answer: 14 feet

We have seen that important information can be gained from the graph of a quadratic function. We now begin a discussion of how to graph such a function by considering the simplest case, quadratic functions of the form $f(x) = ax^2$.

Graphs of $f(x) = ax^2$

EXAMPLE 2 ***Graphing quadratic functions.*** Graph **a.** $f(x) = x^2$, **b.** $g(x) = 3x^2$, and **c.** $h(x) = \dfrac{1}{3}x^2$.

Solution

We can make a table of ordered pairs that satisfy each equation, plot each point, and join them with a smooth curve, as in Figure 8-5. We note that the graph of $h(x) = \frac{1}{3}x^2$ is wider than the graph of $f(x) = x^2$, and that the graph of $g(x) = 3x^2$ is narrower than the graph of $f(x) = x^2$. In the function $f(x) = ax^2$, the smaller the value of $|a|$, the wider the graph.

$f(x) = x^2$

x	$f(x)$	$(x, f(x))$
-2	4	$(-2, 4)$
-1	1	$(-1, 1)$
0	0	$(0, 0)$
1	1	$(1, 1)$
2	4	$(2, 4)$

$g(x) = 3x^2$

x	$g(x)$	$(x, g(x))$
-2	12	$(-2, 12)$
-1	3	$(-1, 3)$
0	0	$(0, 0)$
1	3	$(1, 3)$
2	12	$(2, 12)$

$h(x) = \frac{1}{3}x^2$

x	$h(x)$	$(x, h(x))$
-2	$\frac{4}{3}$	$\left(-2, \frac{4}{3}\right)$
-1	$\frac{1}{3}$	$\left(-1, \frac{1}{3}\right)$
0	0	$(0, 0)$
1	$\frac{1}{3}$	$\left(1, \frac{1}{3}\right)$
2	$\frac{4}{3}$	$\left(2, \frac{4}{3}\right)$

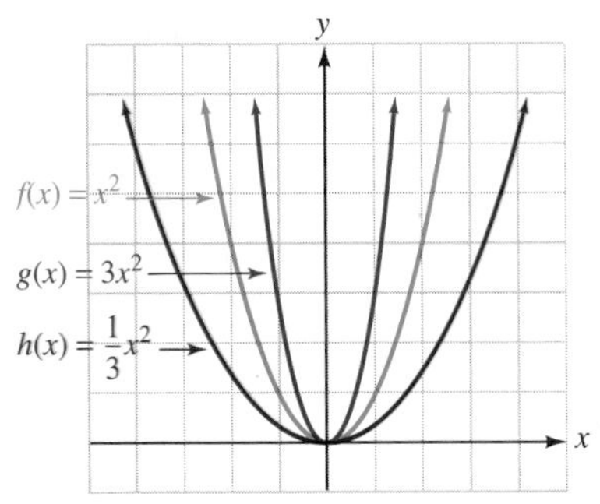

FIGURE 8-5

EXAMPLE 3 ***Graphing quadratic functions.*** Graph $f(x) = -3x^2$.

Solution

We make a table of ordered pairs that satisfy the equation, plot each point, and join them with a smooth curve, as in Figure 8-6. We see that the parabola opens downward and has the same shape as the graph of $f(x) = 3x^2$.

Self Check

Graph $f(x) = -\dfrac{1}{3}x^2$.

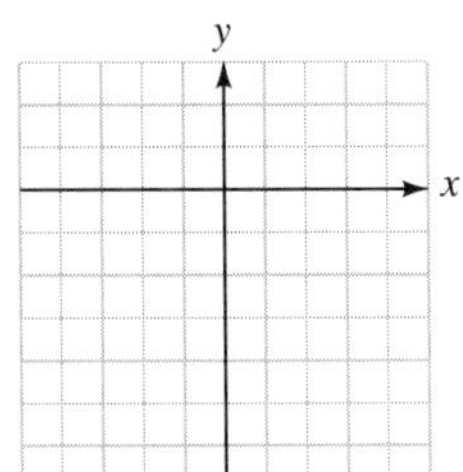

$f(x) = -3x^2$

x	$f(x)$	$(x, f(x))$
-2	-12	$(-2, -12)$
-1	-3	$(-1, -3)$
0	0	$(0, 0)$
1	-3	$(1, -3)$
2	-12	$(2, -12)$

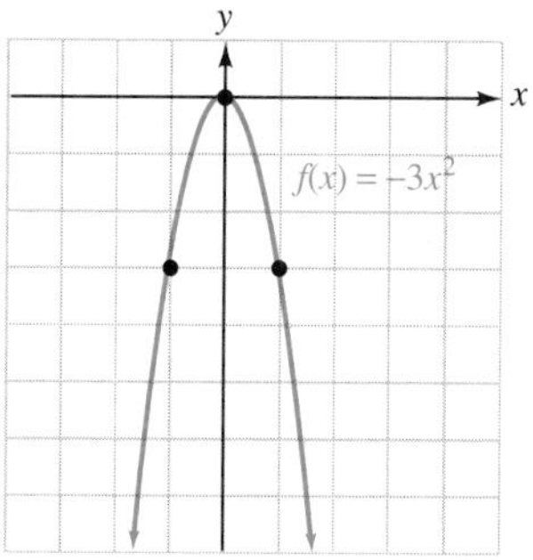

FIGURE 8-6

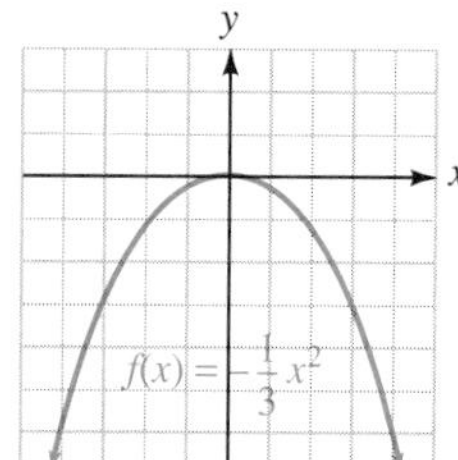

The graphs of quadratic functions of the form $f(x) = ax^2$ are parabolas. The lowest point of a parabola that opens upward, or the highest point of a parabola that opens downward, is called the **vertex** of the parabola. For example, the vertex of the parabola shown in Figure 8-6 is the point (0, 0). The vertical line, called an **axis of symmetry,** that passes through the vertex divides the parabola into two congruent halves. The axis of symmetry of the parabola shown in Figure 8-6 is the line $x = 0$ (the y-axis).

The results of Examples 2 and 3 confirm the following facts.

The graph of $f(x) = ax^2$

The graph of $f(x) = ax^2$ is a parabola opening upward when $a > 0$ and downward when $a < 0$, with vertex at the point (0, 0) and axis of symmetry the line $x = 0$.

Graphs of $f(x) = ax^2 + c$

EXAMPLE 4 ***Graphing quadratic functions.*** Graph **a.** $f(x) = 2x^2$, **b.** $g(x) = 2x^2 + 3$ and **c.** $h(x) = 2x^2 - 3$.

Solution We make a table of ordered pairs that satisfy each equation, plot each point, and join them with a smooth curve, as in Figure 8-7. We note that the graph of $g(x) = 2x^2 + 3$ is identical to the graph of $f(x) = 2x^2$, except that it has been translated 3 units upward. The graph of $h(x) = 2x^2 - 3$ is identical to the graph of $f(x) = 2x^2$, except that it has been translated 3 units downward.

$f(x) = 2x^2$

x	$f(x)$	$(x, f(x))$
-2	8	$(-2, 8)$
-1	2	$(-1, 2)$
0	0	$(0, 0)$
1	2	$(1, 2)$
2	8	$(2, 8)$

$g(x) = 2x^2 + 3$

x	$g(x)$	$(x, g(x))$
-2	11	$(-2, 11)$
-1	5	$(-1, 5)$
0	3	$(0, 3)$
1	5	$(1, 5)$
2	11	$(2, 11)$

$h(x) = 2x^2 - 3$

x	$h(x)$	$(x, h(x))$
-2	5	$(-2, 5)$
-1	-1	$(-1, -1)$
0	-3	$(0, -3)$
1	-1	$(1, -1)$
2	5	$(2, 5)$

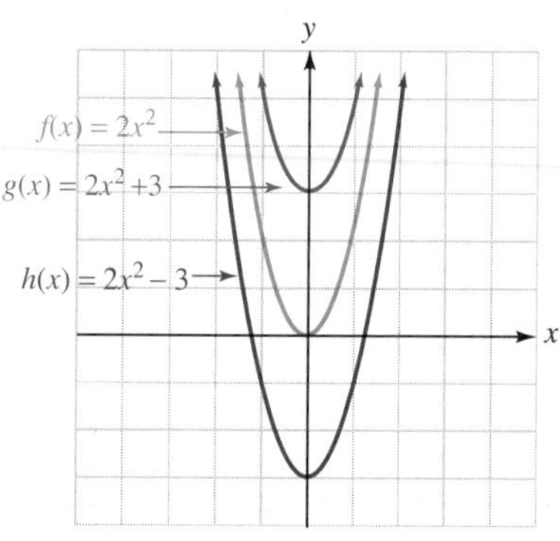

FIGURE 8-7

The results of Example 4 confirm the following facts.

The graph of $f(x) = ax^2 + c$	The graph of $f(x) = ax^2 + c$ is a parabola having the same shape as $f(x) = ax^2$ but translated upward c units if c is positive and downward $\lvert c \rvert$ units if c is negative.

Graphs of $f(x) = a(x - h)^2$

EXAMPLE 5 ***Graphing quadratic functions.*** Graph **a.** $f(x) = 2x^2$, **b.** $g(x) = 2(x - 3)^2$, and **c.** $h(x) = 2(x + 3)^2$.

Solution We make a table of ordered pairs that satisfy each equation, plot each point, and join them with a smooth curve, as in Figure 8-8. We note that the graph of $g(x) = 2(x - 3)^2$ is identical to the graph of $f(x) = 2x^2$, except that it has been translated 3 units to the right. The graph of $h(x) = 2(x + 3)^2$ is identical to the graph of $f(x) = 2x^2$, except that it has been translated 3 units to the left.

$f(x) = 2x^2$

x	$f(x)$	$(x, f(x))$
−2	8	(−2, 8)
−1	2	(−1, 2)
0	0	(0, 0)
1	2	(1, 2)
2	8	(2, 8)

$g(x) = 2(x - 3)^2$

x	$g(x)$	$(x, g(x))$
1	8	(1, 8)
2	2	(2, 2)
3	0	(3, 0)
4	2	(4, 2)
5	8	(5, 8)

$h(x) = 2(x + 3)^2$

x	$h(x)$	$(x, h(x))$
−5	8	(−5, 8)
−4	2	(−4, 2)
−3	0	(−3, 0)
−2	2	(−2, 2)
−1	8	(−1, 8)

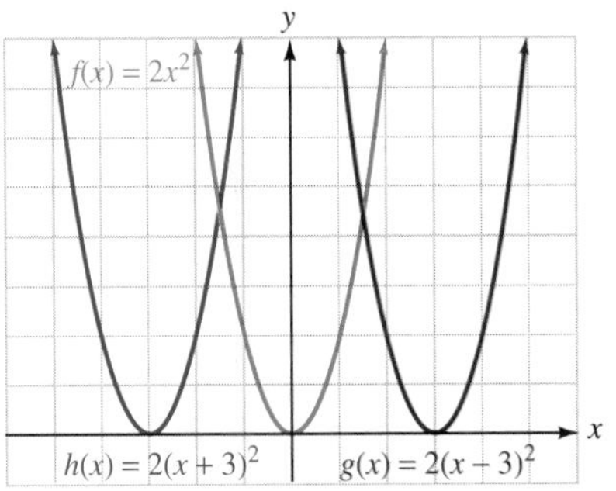

FIGURE 8-8

The results of Example 5 confirm the following facts.

The graph of $f(x) = a(x - h)^2$	The graph of $f(x) = a(x - h)^2$ is a parabola having the same shape as $f(x) = ax^2$ but translated h units to the right if h is positive and $\lvert h \rvert$ units to the left if h is negative.

Graphs of $f(x) = a(x - h)^2 + k$

EXAMPLE 6 ***Graphing quadratic functions.*** Graph $f(x) = 2(x - 3)^2 - 4$.

Solution

The graph of $f(x) = 2(x - 3)^2 - 4$ is identical to the graph of $g(x) = 2(x - 3)^2$, except that it has been translated 4 units downward. The graph of $g(x) = 2(x - 3)^2$ is identical to the graph of $h(x) = 2x^2$, except that it has been translated 3 units to the right. Thus, to graph $f(x) = 2(x - 3)^2 - 4$, we can graph $h(x) = 2x^2$ and shift it 3 units to the right and then 4 units downward, as shown in Figure 8-9.

The vertex of the graph is the point $(3, -4)$, and the axis of symmetry is the line $x = 3$.

Self Check

Graph $f(x) = 2(x + 3)^2 + 1$. Label the vertex and draw the axis of symmetry.

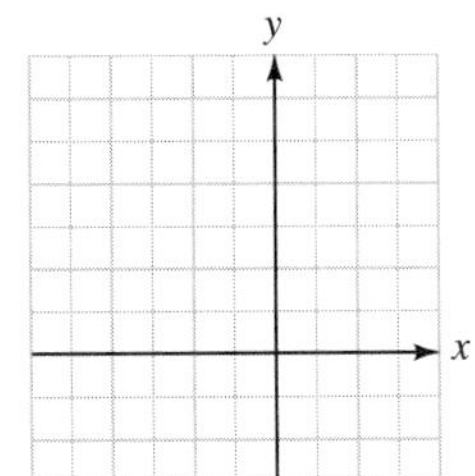

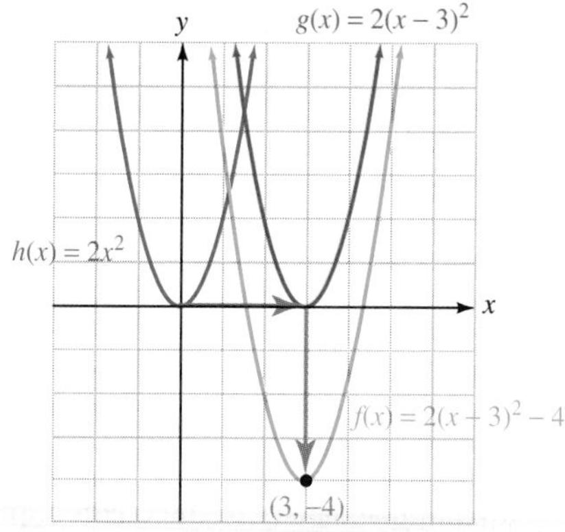

FIGURE 8-9

Answer:

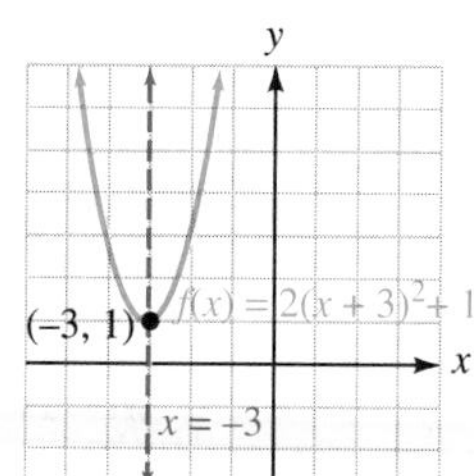

The results of Example 6 confirm the following facts.

The graph of $f(x) = a(x - h)^2 + k$

The graph of the quadratic function

$$f(x) = a(x - h)^2 + k$$

or

$$y = a(x - h)^2 + k$$

(where $a \neq 0$) is a parabola with vertex at (h, k). The parabola opens upward when $a > 0$ and downward when $a < 0$. The axis of symmetry is the line $x = h$.

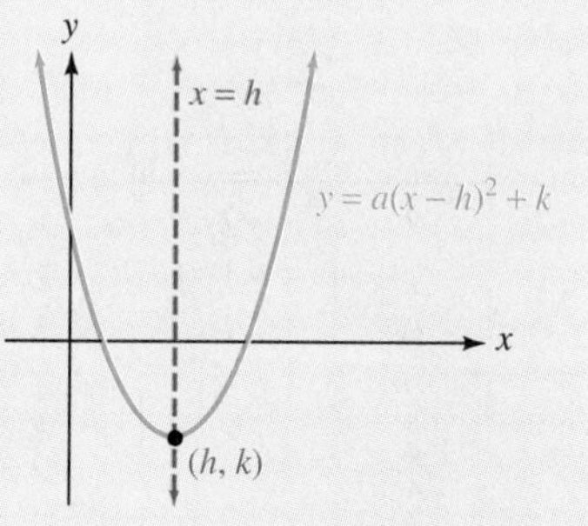

EXAMPLE 7 ***Determining the vertex and the axis of symmetry.*** For the graph of $f(x) = -3(x + 1)^2 - 4$,

a. does the graph open upward or downward?

b. what are the coordinates of the vertex?

c. what is the axis of symmetry?

Solution

Rewriting the given function in $f(x) = a(x - h)^2 + k$ form, we have

We need a minus sign here. We need a plus sign here.

$$f(x) = \underset{a}{-3}\,[x - (\underset{h}{-1})\,]^2 + (\underset{k}{-4})$$

a. Since $a = -3 < 0$, the parabola opens downward.

b. The vertex is $(h, k) = (-1, -4)$.

c. The axis of symmetry is the line $x = h$. In this case, $x = -1$.

Self Check

Answer parts a, b, and c of Example 7 for the graph of $y = 6(x - 5)^2 + 1$.

Answers: **a.** upward, **b.** $(5, 1)$, **c.** $x = 5$

Graphs of $f(x) = ax^2 + bx + c$

To graph functions of the form $f(x) = ax^2 + bx + c$, we complete the square to write the function in the form $f(x) = a(x - h)^2 + k$.

EXAMPLE 8 ***Completing the square to determine the vertex.***
Graph $f(x) = 2x^2 - 4x - 1$.

Solution

Step 1: We complete the square on x to write the given function in the form $f(x) = a(x - h)^2 + k$.

$f(x) = 2x^2 - 4x - 1$

$f(x) = 2(x^2 - 2x) - 1$ Factor 2 from $2x^2 - 4x$.

Now we complete the square on x by adding 1 inside the parentheses. Since this adds 2 to the right-hand side, we also subtract 2 from the right-hand side.

$f(x) = 2(x^2 - 2x + 1) - 1 - 2$ To complete the square, half of $-2 = -1$ and $(-1)^2 = 1$.

1. $f(x) = 2(x - 1)^2 - 3$ Factor $x^2 - 2x + 1$ and combine like terms.

Step 2: From Equation 1, we can see that $h = 1$ and $k = -3$, so the vertex will be at the point $(1, -3)$, and the axis of symmetry is $x = 1$. We plot the vertex and axis of symmetry on the coordinate system in Figure 8-10.

Step 3: Finally, we construct a table of values, plot the points, and draw the graph, which appears in the figure.

$f(x) = 2x^2 - 4x - 1$

x	$f(x)$	$(x, f(x))$
-1	5	$(-1, 5)$
0	-1	$(0, -1)$
2	-1	$(2, -1)$
3	5	$(3, 5)$

↑ The x-coordinate of the vertex is 1. Choose values for x close to 1.

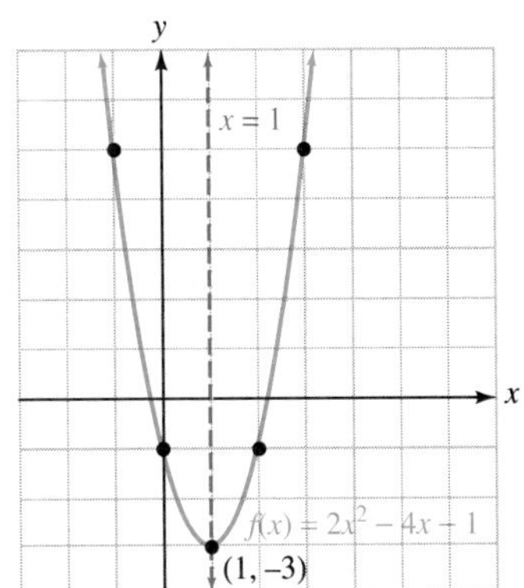

FIGURE 8-10

Self Check

Graph $f(x) = 2x^2 - 4x + 1$.

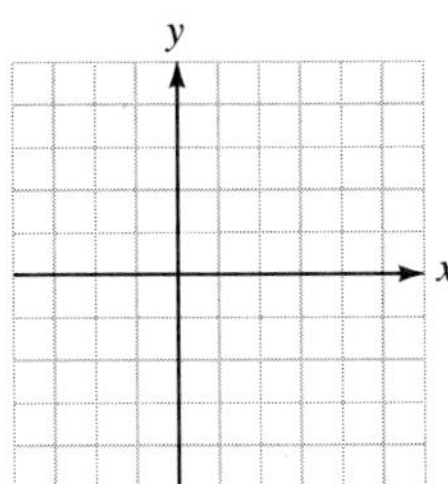

Answer:

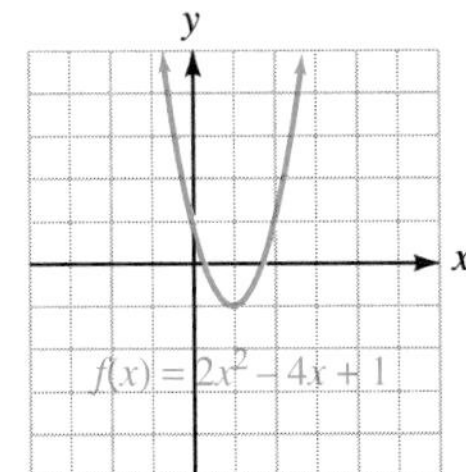

We can derive a formula for the vertex of the graph of $f(x) = ax^2 + bx + c$ by completing the square in the same manner as we did in Example 8. After using similar steps, the result is

$$f(x) = a\left[x - \left(-\frac{b}{2a}\right)\right]^2 + \frac{4ac - b^2}{4a}$$

(with $-\frac{b}{2a}$ marked h and $\frac{4ac - b^2}{4a}$ marked k)

The x-coordinate of the vertex is $-\frac{b}{2a}$. The y-coordinate of the vertex is $\frac{4ac - b^2}{4a}$. However, we can also find the y-coordinate of the vertex by substituting the x-coordinate, $-\frac{b}{2a}$, for x in the quadratic function.

Formula for the vertex of a parabola

The vertex of the graph of the quadratic function $f(x) = ax^2 + bx + c$ is

$$\left(-\frac{b}{2a}, f\left(-\frac{b}{2a}\right)\right)$$

and the axis of symmetry of the parabola is the line $x = -\frac{b}{2a}$.

In Example 8, for the function $f(x) = 2x^2 - 4x - 1$, $a = 2$ and $b = -4$. To find the vertex of its graph, we compute

$$-\frac{b}{2a} = -\frac{-4}{2(2)} = -\frac{-4}{4} = 1 \qquad f\left(-\frac{b}{2a}\right) = f(1) = 2(1)^2 - 4(1) - 1 = -3$$

The vertex is the point $(1, -3)$. This agrees with the result we obtained in Example 8 by completing the square.

Accent on Technology Graphing quadratic functions

To graph the function $f(x) = 2x^2 + 6x - 3$ and find the coordinates of the vertex and the axis of symmetry of the parabola, we can use a graphing calculator with window settings of $[-10, 10]$ for x and $[-10, 10]$ for y. If we enter the function, we will obtain the graph shown in Figure 8-11(a).

We then trace to move the cursor to the lowest point on the graph, as shown in Figure 8-11(b). By zooming in, we can determine that the vertex is the point $\left(-\frac{3}{2}, -\frac{15}{2}\right)$ and that the line $x = -\frac{3}{2}$ is the axis of symmetry.

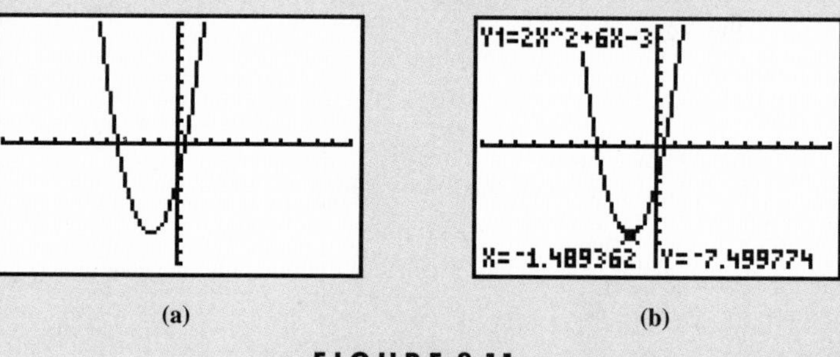

FIGURE 8-11

Because it is easy to graph quadratic functions with a graphing calculator, we can use graphing to find approximate solutions of quadratic equations. For example, the solutions of $0.7x^2 + 2x - 3.5 = \mathbf{0}$ are the numbers x that will make $y = 0$ in the quadratic function $\mathbf{y} = 0.7x^2 + 2x - 3.5$. To approximate these numbers, we graph the quadratic function and read the x-intercepts from the graph.

We can use the standard window settings of $[-10, 10]$ for x and $[-10, 10]$ for y and graph the function, as in Figure 8-12(a). We then trace to move the cursor to each x-intercept, as in Figures 8-12(b) and 8-12(c). From the graph, we can read the approximate value of the x-coordinate of each x-intercept. For better results, we can zoom in.

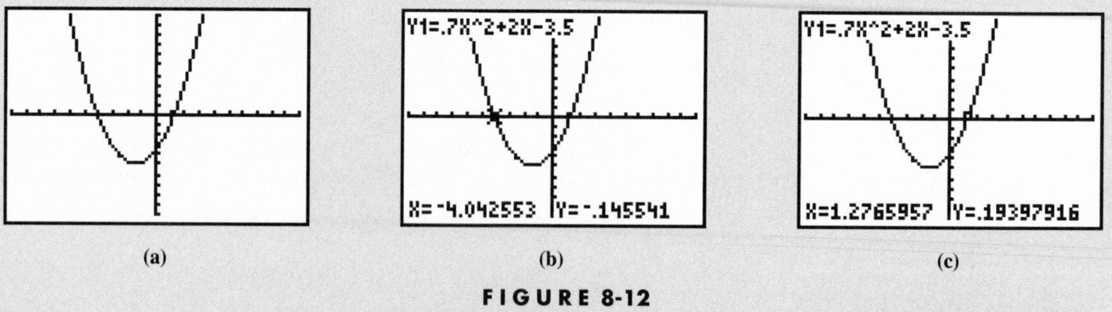

FIGURE 8-12

In the event that the graph of a quadratic function does not intersect the x-axis, we can conclude that the associated quadratic equation has no real-number solutions.

Determining minimum and maximum values

It is often useful to know the smallest or largest possible value a quantity can assume. For example, companies try to minimize their costs and maximize their profits. If the quantity is expressed by a quadratic function, the vertex of the graph of the function gives its minimum or maximum value.

EXAMPLE 9 ***Minimizing costs.*** A glassworks that makes lead crystal vases has daily production costs given by the function $C(x) = 0.2x^2 - 10x + 650$, where x is the number of vases made each day. How many vases should be produced to minimize the per-day costs? What will the costs be?

Solution The graph of $C(x) = 0.2x^2 - 10x + 650$ is a parabola opening upward. The vertex is the lowest point on the graph. To find the vertex, we compute

$$-\frac{b}{2a} = -\frac{-10}{2(0.2)} \quad b = -10 \text{ and } a = 0.2.$$
$$= -\frac{-10}{0.4}$$
$$= 25$$

$$f\left(-\frac{b}{2a}\right) = f(25)$$
$$= 0.2(25)^2 - 10(25) + 650$$
$$= 525$$

The vertex is the point (25, 525), and it indicates that the costs are a minimum of $525 when 25 vases are made daily.

To solve this problem with a graphing calculator with window settings of [0, 50] for x and [0, 1,000] for y, we graph the function $C(x) = 0.2x^2 - 10x + 650$ to get the graph in Figure 8-13. By using trace and zoom, we can determine that the minimum cost is $525 when the number of vases produced is 25.

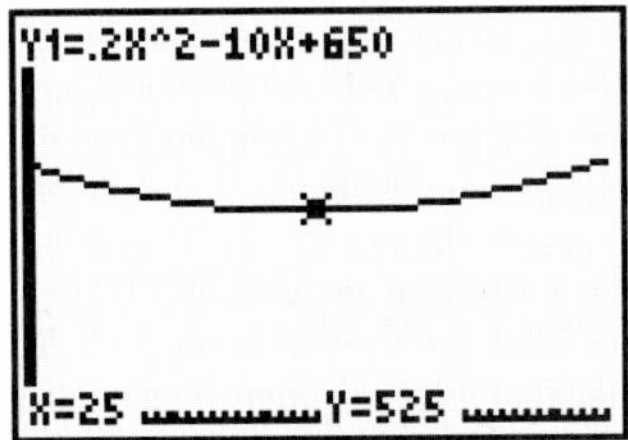

FIGURE 8-13

EXAMPLE 10 ***Maximizing area.*** A kennel owner wants to build the rectangular pen shown in Figure 8-14(a) to house his dog. If he uses one side of his barn, find the maximum area that he can enclose with 80 feet of fencing.

Solution We can let the width of the area be represented by w. Then the length is represented by $80 - 2w$. The function that gives the area enclosed by the pen is

$$A = (80 - 2w)w \quad A = lw.$$

We can find the maximum value of A by determining the vertex of the graph of the function. This time, we will find the vertex by completing the square.

$A = 80w - 2w^2$ — Use the distributive property to remove parentheses.

$= -2(w^2 - 40w)$ — Factor out -2.

$= -2(w^2 - 40w + 400) + 800$ — Complete the square on w. Inside the parentheses, add $\left(\frac{-40}{2}\right)^2 = 400$. Since this subtracts 800 from the right-hand side, add 800 to the right-hand side.

$= -2(w - 20)^2 + 800$ — Factor $w^2 - 40w + 400$.

Thus, the coordinates of the vertex of the graph of the quadratic function are (20, 800), and the maximum area is 800 square feet.

To solve this problem with a graphing calculator with window settings of [0, 50] for x and [0, 1,000] for y, we graph the function $A = -2w^2 + 80w$ to get the graph in Figure 8-14(b). By using trace and zoom, we can determine that the maximum area is 800 square feet when the width is 20 feet.

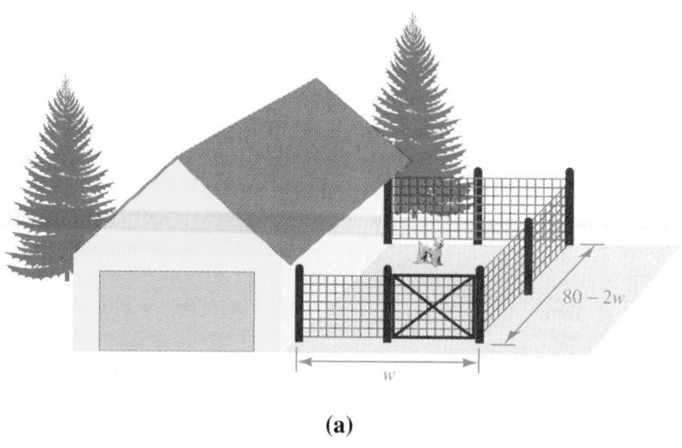

(a)

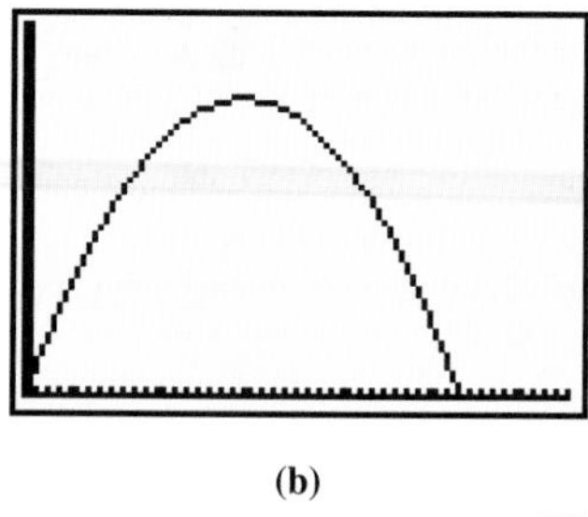

(b)

FIGURE 8-14

STUDY SET Section 8.3

VOCABULARY *In Exercises 1–4, refer to Illustration 1. Fill in the blanks to make the statements true.*

1. The function $f(x) = 2x^2 - 4x + 1$ is graphed in Illustration 1. It is called a ______________ function.

2. The graph in Illustration 1 is called a ______________.

3. The lowest point on the graph is $(1, -1)$. This is called the ______________.

4. The vertical line $x = 1$ divides the parabola into two halves. This line is called the ______________.

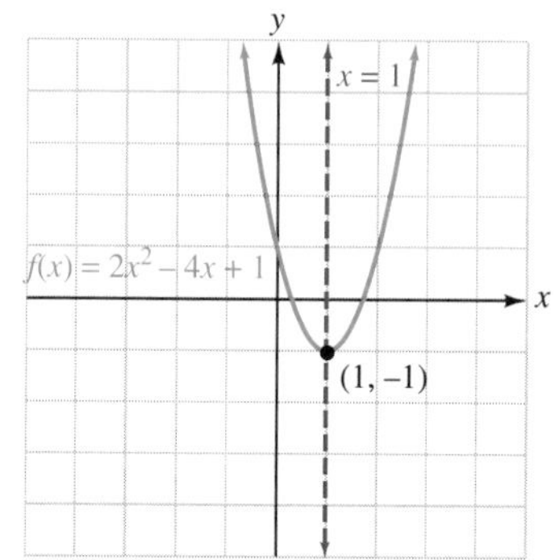

ILLUSTRATION 1

CONCEPTS *In Exercises 5–6, make a table of values for each function. Then graph them on the same coordinate system.*

5. $f(x) = x^2$, $g(x) = 2x^2$, $h(x) = \frac{1}{2}x^2$

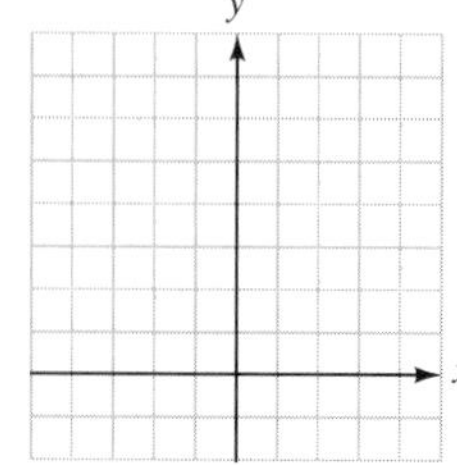

6. $f(x) = -x^2$, $g(x) = -\frac{1}{4}x^2$, $h(x) = -4x^2$

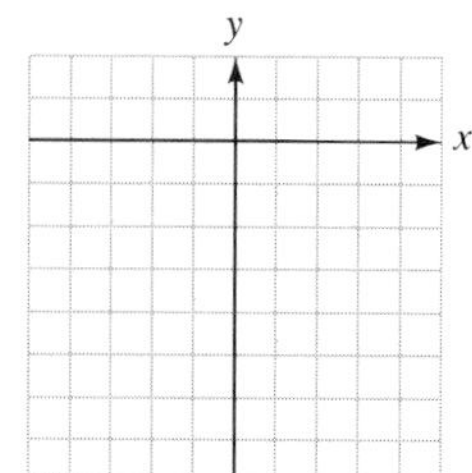

In Exercises 7–8, make a table of values to graph function f. Then use a translation to graph the other two functions on the same coordinate system.

7. $f(x) = 4x^2$, $g(x) = 4x^2 + 3$, $h(x) = 4x^2 - 2$

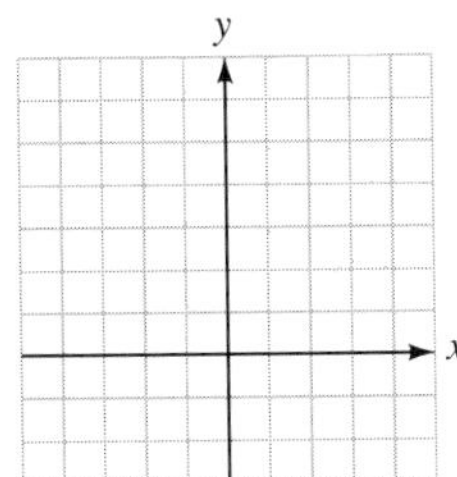

8. $f(x) = 3x^2$, $g(x) = 3(x + 2)^2$, $h(x) = 3(x - 3)^2$

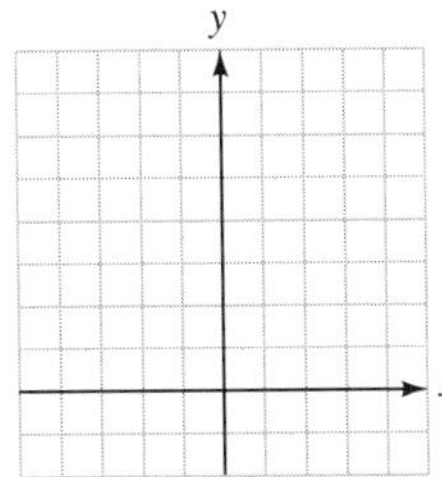

In Exercises 9–10, make a table of values to graph function f. Then use a series of translations to graph function g on the same coordinate system.

9. $f(x) = -3x^2$, $g(x) = -3(x - 2)^2 - 1$

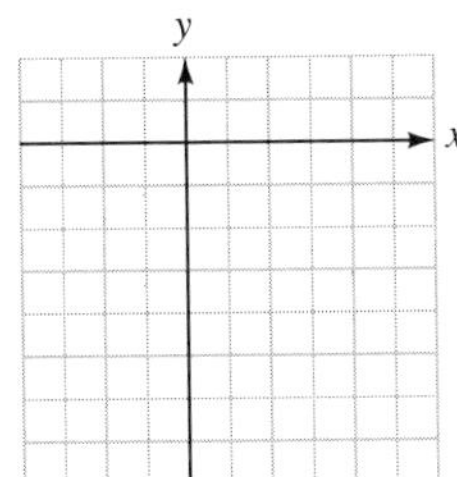

10. $f(x) = -\frac{1}{2}x^2$, $g(x) = -\frac{1}{2}(x + 1)^2 + 2$

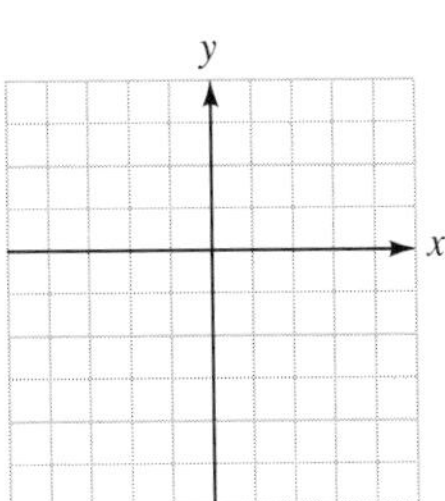

NOTATION

11. The function $f(x) = 2(x + 1)^2 + 6$ is written in the form $f(x) = a(x - h)^2 + k$. Is $h = -1$ or is $h = 1$? Explain.

12. The vertex of a quadratic function $f(x) = ax^2 + bx + c$ is given by the formula $\left(-\frac{b}{2a}, f\left(-\frac{b}{2a}\right)\right)$. Explain what is meant by the notation $f\left(-\frac{b}{2a}\right)$.

PRACTICE *In Exercises 13–22, find the coordinates of the vertex and the axis of symmetry of the graph of each function. If necessary, complete the square on x to write the equation in the form $y = a(x - h)^2 + k$.* ***Do not graph the equation,*** *but tell whether the graph will open upward or downward.*

13. $y = (x - 1)^2 + 2$

14. $y = 2(x - 2)^2 - 1$

15. $f(x) = 2(x + 3)^2 - 4$

16. $f(x) = -3(x + 1)^2 + 3$

17. $y = 2x^2 - 4x$

18. $y = 3x^2 - 3$

19. $y = -4x^2 + 16x + 5$

20. $y = 5x^2 + 20x + 25$

21. $y = 3x^2 + 4x + 2$

22. $y = -6x^2 + 5x - 7$

In Exercises 23–38, first determine the vertex and the axis of symmetry of the graph of the function. Then plot several points and complete the graph. (See Example 8.)

23. $f(x) = (x - 3)^2 + 2$

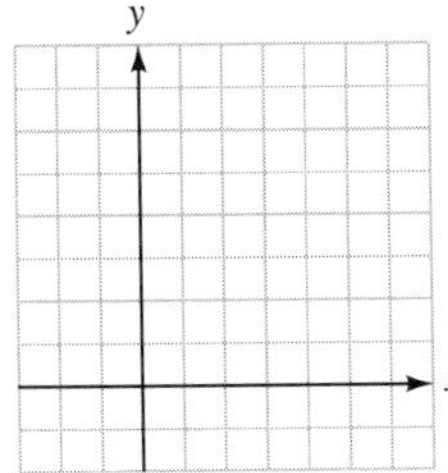

24. $f(x) = (x + 1)^2 - 2$

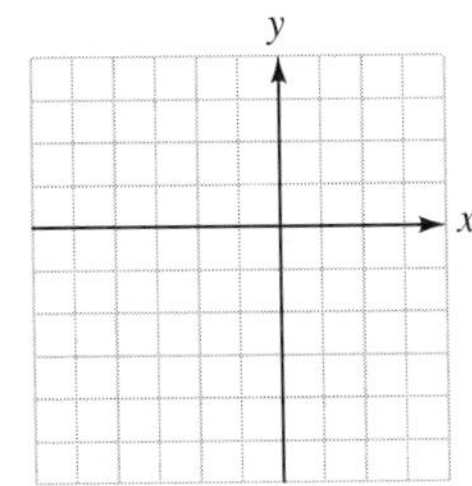

25. $f(x) = -(x - 2)^2$

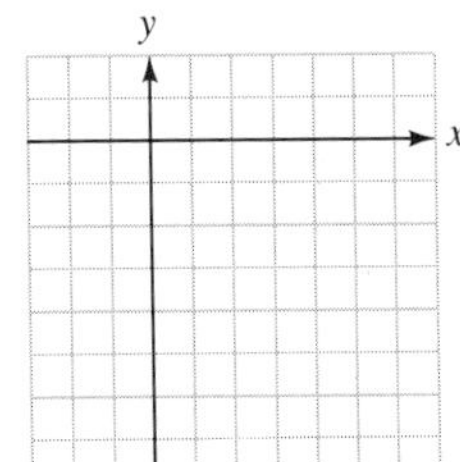

26. $f(x) = -(x + 2)^2$

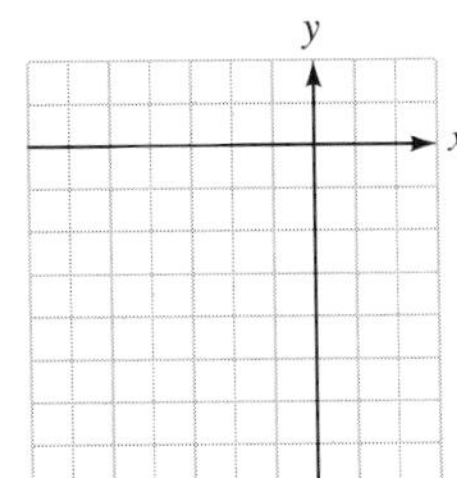

27. $f(x) = -2(x + 3)^2 + 4$

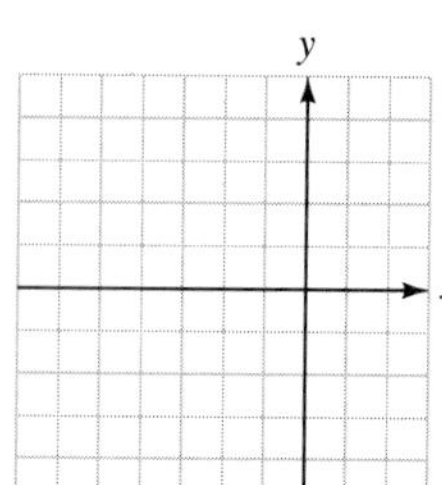

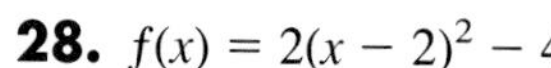

28. $f(x) = 2(x - 2)^2 - 4$

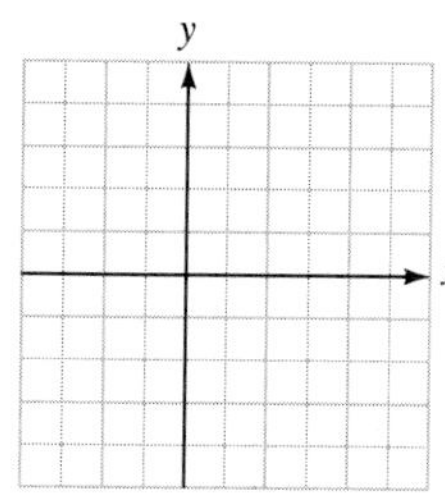

29. $f(x) = (x - 3)^2 + 2$

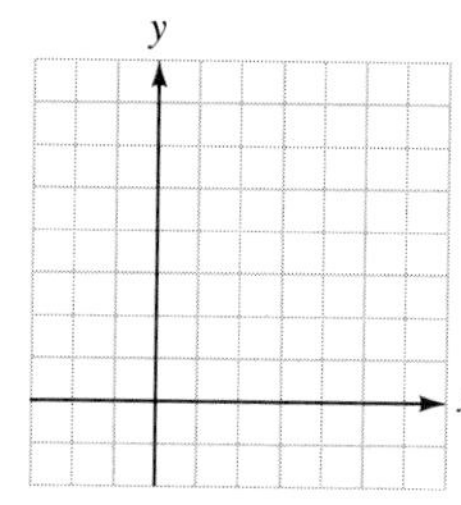

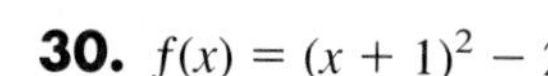

30. $f(x) = (x + 1)^2 - 2$

31. $f(x) = x^2 + x - 6$

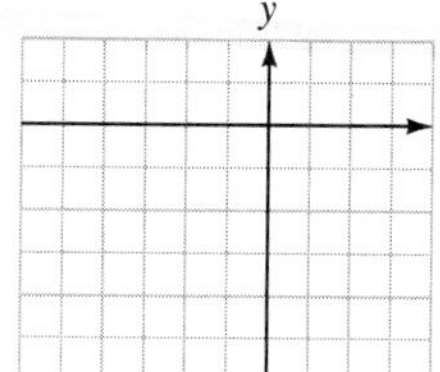

32. $y = x^2 - x - 6$

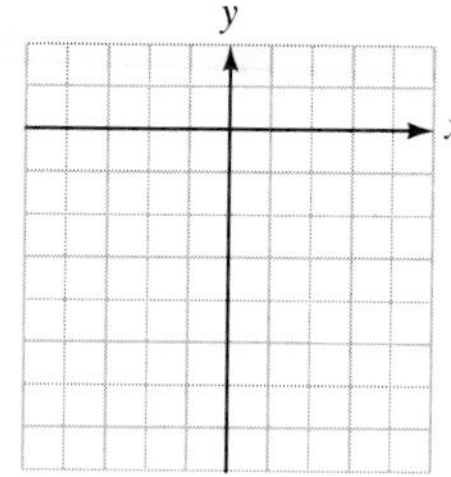

33. $y = -x^2 + 2x + 3$

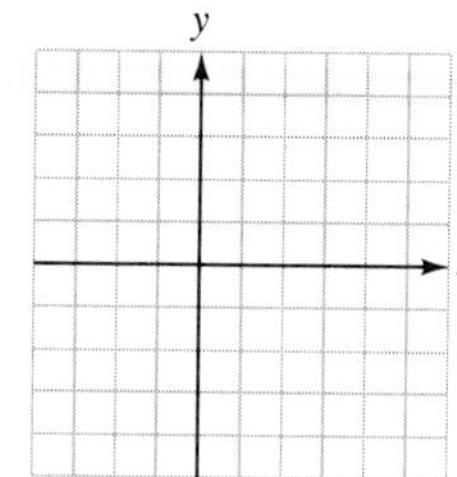

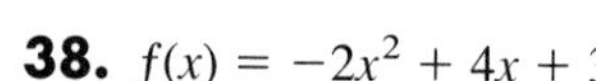

34. $f(x) = -x^2 + x + 2$

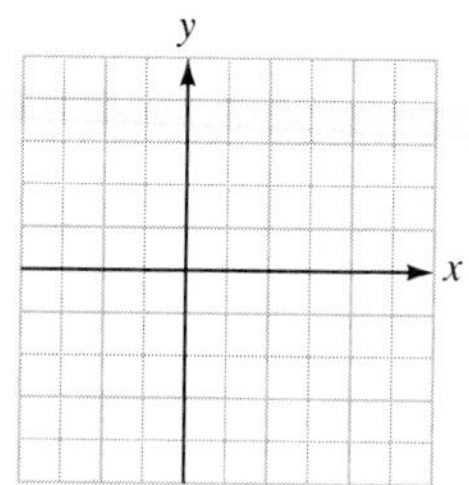

35. $f(x) = -3x^2 + 2x$

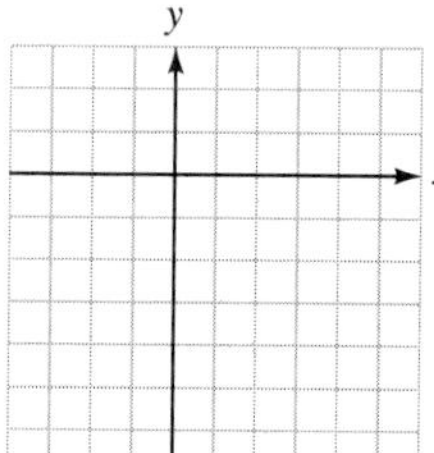

36. $f(x) = 5x + x^2$

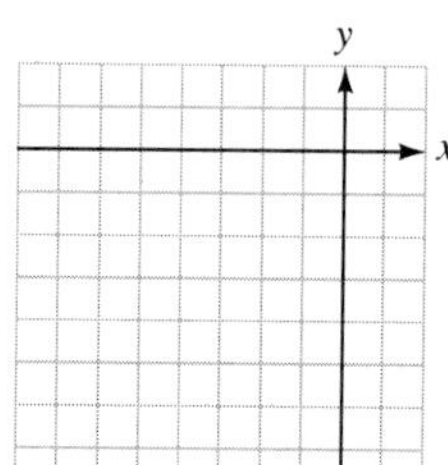

37. $y = -12x^2 - 6x + 6$

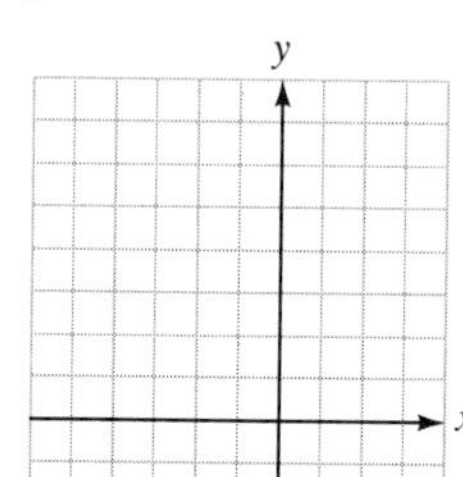

38. $f(x) = -2x^2 + 4x + 3$

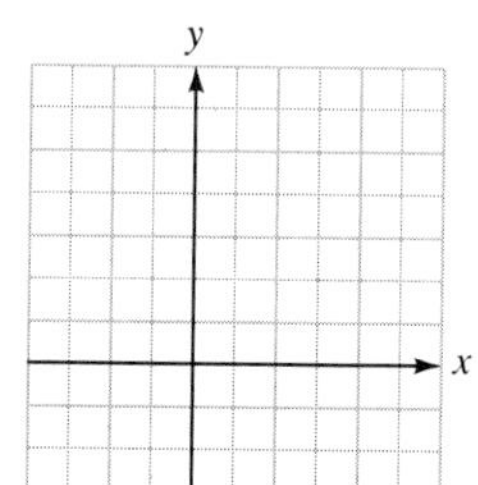

In Exercises 39–42, use a graphing calculator to find the coordinates of the vertex of the graph of each quadratic function. Round to the nearest hundredth.

39. $y = 2x^2 - x + 1$ **40.** $y = x^2 + 5x - 6$ **41.** $y = 7 + x - x^2$ **42.** $y = 2x^2 - 3x + 2$

In Exercises 43–46, use a graphing calculator to solve each equation. If an answer is not exact, round to the nearest hundredth.

43. $x^2 + x - 6 = 0$ **44.** $2x^2 - 5x - 3 = 0$ **45.** $0.5x^2 - 0.7x - 3 = 0$ **46.** $2x^2 - 0.5x - 2 = 0$

APPLICATIONS

47. FIREWORKS A fireworks shell is shot straight up with an initial velocity of 120 feet per second. Its height s after t seconds is given by the equation $s = 120t - 16t^2$. If the shell is designed to explode when it reaches its maximum height, how long after being fired, and at what height, will the fireworks appear in the sky?

48. BALLISTICS From the top of the building in Illustration 2, a ball is thrown straight up with an initial velocity of 32 feet per second. The equation

$$s = -16t^2 + 32t + 48$$

gives the height s of the ball t seconds after it was thrown. Find the maximum height reached by the ball and the time it takes for the ball to hit the ground.

ILLUSTRATION 2

49. FENCING A FIELD See Illustration 3. A farmer wants to fence in three sides of a rectangular field with 1,000 feet of fencing. The other side of the rectangle will be a river. If the enclosed area is to be maximum, find the dimensions of the field.

ILLUSTRATION 3

50. POLICE INVESTIGATION A police officer seals off the scene of a car collision using a roll of yellow police tape that is 300 feet long, as shown in Illustration 4. What dimensions should be used to seal off the maximum rectangular area around the collision? What is the maximum area?

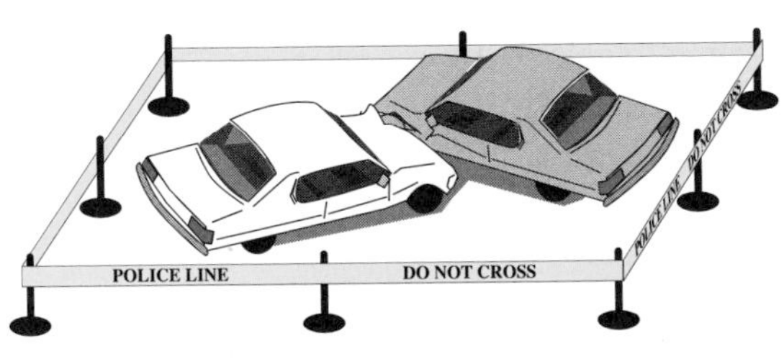

ILLUSTRATION 4

51. OPERATING COSTS The cost C in dollars of operating a certain concrete cutting machine is related to the number of minutes n the machine is run by the function

$$C(n) = 2.2n^2 - 66n + 655$$

For what number of minutes is the cost of running the machine a minimum? What is the minimum cost?

52. WATER USAGE The height (in feet) of the water level in a reservoir over a one-year period is modeled by the function

$$H(t) = 3.3(t - 9)^2 + 14$$

where $t = 1$ represents January, $t = 2$ represents February, and so on. How low did the water level get that year, and when did it reach the low mark?

53. U.S. ARMY The function

$$N(x) = -0.0534x^2 + 0.337x + 0.969$$

gives the number of active-duty military personnel in the United States Army (in millions) for the years 1965–1972, where $x = 0$ corresponds to 1965, $x = 1$ corresponds to 1966, and so on. For this period, when was the army's personnel strength level at its highest, and what was it? Historically, can you explain why?

54. SCHOOL ENROLLMENT The total annual enrollment (in millions) in U.S. elementary and secondary schools for the years 1975–1996 is given by the model

$$E = 0.058x^2 - 1.162x + 50.604$$

where $x = 0$ corresponds to 1975, $x = 1$ corresponds to 1976, and so on.

a. For this period, when was enrollment the lowest? What was it?

b. Use the model to complete the bar graph in Illustration 5.

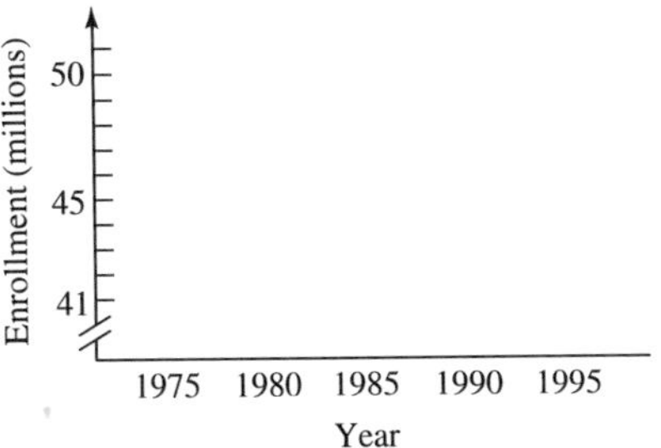

ILLUSTRATION 5

55. MAXIMIZING REVENUE The revenue R received for selling x stereos is given by the formula

$$R = -\frac{x^2}{5} + 80x - 1{,}000$$

How many stereos must be sold to obtain the maximum revenue? Find the maximum revenue.

56. MAXIMIZING REVENUE When priced at \$30 each, a toy has annual sales of 4,000 units. The manufacturer estimates that each \$1 increase in cost will decrease sales by 100 units. Find the unit price that will maximize total revenue. (*Hint:* Total revenue = price · the number of units sold.)

WRITING *Write a paragraph using your own words.*

57. Use the example of a stream of water from a drinking fountain to explain the concepts of the vertex and the axis of symmetry of a parabola.

58. What are some quantities that are good to maximize? What are some quantities that are good to minimize?

REVIEW *In Exercises 59–64, simplify each expression.*

59. $\sqrt{8a}\sqrt{2a^3b}$

60. $(\sqrt{23})^2$

61. $\frac{\sqrt{3}}{\sqrt{50}}$

62. $\frac{3}{\sqrt[3]{9}}$

63. $3(\sqrt{5b} - \sqrt{3})^2$

64. $-2\sqrt{5b}(4\sqrt{2b} - 3\sqrt{3})$

8.4 Complex Numbers

In this section, you will learn about

- **Imaginary numbers**
- **Simplifying imaginary numbers**
- **Complex numbers**
- **Arithmetic of complex numbers**
- **Complex conjugates**
- **Rationalizing the denominator**
- **Powers of *i***

INTRODUCTION. We have seen that negative numbers do not have real-number square roots. For years, people believed that numbers like

$$\sqrt{-1}, \quad \sqrt{-3}, \quad \sqrt{-4}, \quad \text{and} \quad \sqrt{-9}$$

were nonsense. In the 17th century, René Descartes (1596–1650) called them *imaginary numbers.* Today, imaginary numbers have many important uses, such as describing the behavior of alternating current in electronics.

In this section, we will introduce imaginary numbers and discuss a broader set of numbers called *complex numbers.*

Imaginary numbers

So far, all of our work with quadratic equations has involved only real numbers. However, the solutions of many quadratic equations are not real numbers.

EXAMPLE 1 ***A quadratic equation whose solution is not a real number.*** Solve $x^2 + x + 1 = 0$.

Solution

Because $x^2 + x + 1$ is prime and cannot be factored, we will use the quadratic formula, with $a = 1$, $b = 1$, and $c = 1$:

$$x = \frac{-b \pm \sqrt{b^2 - 4ac}}{2a}$$

$$= \frac{-1 \pm \sqrt{1^2 - 4(1)(1)}}{2(1)} \quad \text{Substitute 1 for } a\text{, 1 for } b\text{, and 1 for } c.$$

$$= \frac{-1 \pm \sqrt{1 - 4}}{2} \quad \text{Simplify within the radical.}$$

$$= \frac{-1 \pm \sqrt{-3}}{2}$$

$$x = \frac{-1 + \sqrt{-3}}{2} \quad \text{or} \quad x = \frac{-1 - \sqrt{-3}}{2}$$

Self Check

Solve $a^2 + 3a + 5 = 0$.

Each solution contains the number $\sqrt{-3}$. Since no real number squared is -3, $\sqrt{-3}$ is not a real number, and these two solutions are not real numbers.

To solve equations such as $x^2 + x + 1 = 0$, we must define the square root of a negative number.

Answer: $\dfrac{-3 \pm \sqrt{-11}}{2}$

Square roots of negative numbers are called **imaginary numbers.** The imaginary number $\sqrt{-1}$ is often denoted by the letter i:

$$i = \sqrt{-1}$$

Because i represents the square root of -1, it follows that

$$i^2 = -1$$

Simplifying imaginary numbers

We can use extensions of the multiplication and division properties of radicals discussed in Chapter 6 to simplify imaginary numbers. For example,

$$\sqrt{-25} = \sqrt{25(-1)} = \sqrt{25}\sqrt{-1} = 5i \qquad \text{Replace } \sqrt{-1} \text{ with } i.$$

$$\sqrt{\frac{-100}{49}} = \sqrt{\frac{100}{49}(-1)} = \frac{\sqrt{100}}{\sqrt{49}}\sqrt{-1} = \frac{10}{7}i$$

$$\sqrt{-3} = \sqrt{3(-1)} = \sqrt{3}\sqrt{-1} = \sqrt{3}i$$

The expression $\sqrt{3}i$ is often written as $i\sqrt{3}$ to make it clear that i is not part of the radicand. Do not confuse $\sqrt{3}i$ with $\sqrt{3i}$.

These examples illustrate the following rules.

Properties of radicals

If at least one of a and b is a nonnegative real number, then

$$\sqrt{ab} = \sqrt{a}\sqrt{b} \quad \text{and} \quad \sqrt{\frac{a}{b}} = \frac{\sqrt{a}}{\sqrt{b}} \quad (b \neq 0)$$

EXAMPLE 2 ***Square roots of negative numbers.*** Write each expression in terms of i: **a.** $\sqrt{-81}$, **b.** $-\sqrt{-8}$, and **c.** $3\sqrt{-\frac{98}{16}}$.

Solution

a. $\sqrt{-81} = \sqrt{81(-1)} = \sqrt{81}\sqrt{-1} = 9i$

b. $-\sqrt{-8} = -\sqrt{8(-1)} = -\sqrt{8}\sqrt{-1} = -\sqrt{8}i = -2\sqrt{2}i$

c. $3\sqrt{-\frac{98}{16}} = 3\sqrt{\frac{98}{16}(-1)} = \frac{3\sqrt{49}\sqrt{2}}{\sqrt{16}}\sqrt{-1} = \frac{21\sqrt{2}}{4}i$

Self Check

Write each expression in terms of i: **a.** $-\sqrt{-4}$, **b.** $2\sqrt{-28}$, and **c.** $\sqrt{-\frac{27}{100}}$.

Answers: **a.** $-2i$, **b.** $4\sqrt{7}i$, **c.** $\frac{3\sqrt{3}}{10}i$

EXAMPLE 3 ***Solving quadratic equations.*** Solve $x^2 + 21 = 0$.

Solution

We can write the equation as $x^2 = -21$ and use the square root property to solve it.

$$x^2 + 21 = 0$$

$$x^2 = -21 \qquad \text{Subtract 21 from both sides.}$$

$$x = \pm\sqrt{-21} \qquad \text{Apply the square root property. Note the square root of a negative number.}$$

Self Check

Solve $x^2 + 75 = 0$.

This equation has imaginary solutions, which we can express in terms of i.

$x = \pm\sqrt{21(-1)}$ Use the multiplication property of radicals.

$x = \pm\sqrt{21}i$ $\sqrt{21(-1)} = \sqrt{21}\sqrt{-1} = \sqrt{21}i$.

Verify that $x = \sqrt{21}i$ and $x = -\sqrt{21}i$ satisfy the original equation.

Answer: $\pm 5\sqrt{3}i$ ■

Complex numbers

The imaginary numbers are a subset of a set of numbers called the **complex numbers.**

Complex numbers

A **complex number** is any number that can be written in the form $a + bi$, where a and b are real numbers and $i = \sqrt{-1}$.

In the complex number $a + bi$, a is called the **real part,** and b is called the **imaginary part.**

Some examples of complex numbers are

$3 + 4i$ $a = 3$ and $b = 4$.

$-6 - 5i$ Think of this as $-6 + (-5i)$, where $a = -6$ and $b = -5$. It is common to use $a - bi$ form as a substitute for $a + (-b)i$.

$\frac{1}{2} + 0i$ $a = \frac{1}{2}$ and $b = 0$.

$0 + \sqrt{2}i$ $a = 0$ and $b = \sqrt{2}$.

If $b = 0$, the complex number $a + bi$ is a real number. If $b \neq 0$ and $a = 0$, the complex number $0 + bi$ (or just bi) is an imaginary number. The relationship between the real numbers, the imaginary numbers, and the complex numbers is shown in Figure 8-15.

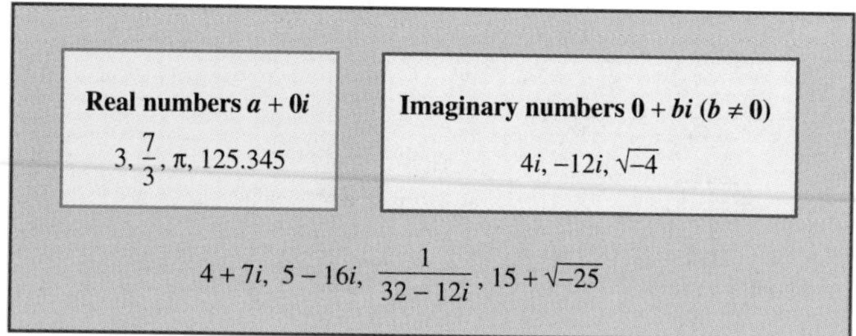

FIGURE 8-15

EXAMPLE 4 ***A quadratic equation with complex-number solutions.*** Solve $4t^2 - 6t + 3 = 0$. Express the solutions in $a + bi$ form.

Self Check

Solve $a^2 + 2a + 3 = 0$.

Solution

We use the quadratic formula to solve the equation.

$$t = \frac{-b \pm \sqrt{b^2 - 4ac}}{2a}$$

$$= \frac{-(-6) \pm \sqrt{(-6)^2 - 4(4)(3)}}{2(4)}$$ Substitute 4 for a, -6 for b, and 3 for c.

$$= \frac{6 \pm \sqrt{36 - 48}}{8}$$ Simplify within the radical.

$$= \frac{6 \pm \sqrt{-12}}{8}$$

$$= \frac{6 \pm 2\sqrt{3}i}{8}$$ $\sqrt{-12} = \sqrt{12(-1)} = \sqrt{12}\sqrt{-1} = 2\sqrt{3}i$.

$$= \frac{3 \pm \sqrt{3}i}{4}$$ $\frac{6 \pm 2\sqrt{3}i}{8} = \frac{\overset{1}{\cancel{2}}(3 \pm \sqrt{3}i)}{\underset{1}{\cancel{2}} \cdot 4} = \frac{3 \pm \sqrt{3}i}{4}$.

Writing each solution as a complex number in $a + bi$ form,

$$t = \frac{3}{4} + \frac{\sqrt{3}}{4}i \quad \text{or} \quad t = \frac{3}{4} - \frac{\sqrt{3}}{4}i$$

Answer: $-1 \pm \sqrt{2}i$

Arithmetic of complex numbers

We now consider the rules used to add, subtract, multiply, and divide complex numbers.

Addition and subtraction of complex numbers

To add (or subtract) two complex numbers, add (or subtract) their real parts and add (or subtract) their imaginary parts.

$$(a + bi) + (c + di) = (a + c) + (b + d)i$$

$$(a + bi) - (c + di) = (a - c) + (b - d)i$$

EXAMPLE 5 ***Adding and subtracting complex numbers.*** Do the operations.

a. $(8 + 4i) + (12 + 8i) = (8 + 12) + (4 + 8)i$ Add the real parts. Add the imaginary parts.

$= 20 + 12i$

b. $\left(7 - \sqrt{-16}\right) + \left(9 + \sqrt{-4}\right)$

$= (7 - 4i) + (9 + 2i)$ Write $\sqrt{-16}$ and $\sqrt{-4}$ in terms of i.

$= (7 + 9) + (-4 + 2)i$ Add the real parts. Add the imaginary parts.

$= 16 - 2i$ Write $16 + (-2i)$ in the form $16 - 2i$.

c. $(-6 + i) - (3 + 2i) = (-6 - 3) + (1 - 2)i$: Subtract the real parts. Subtract the imaginary parts.

$= -9 - i$

Self Check

Do the operations:

a. $(3 - 5i) + (-2 + 7i)$ and

b. $\left(3 - \sqrt{-25}\right) - \left(-2 + \sqrt{-49}\right)$.

Answers: **a.** $1 + 2i$, **b.** $5 - 12i$

To multiply two imaginary numbers, they should first be expressed in terms of i. For example,

$$\sqrt{-2}\sqrt{-20} = \left(\sqrt{2}i\right)\left(2\sqrt{5}i\right)$$ $\sqrt{-20} = \sqrt{20(-1)} = \sqrt{20}i = 2\sqrt{5}i$.

$$= 2\sqrt{10}i^2$$

$$= -2\sqrt{10}$$ $i^2 = -1$.

WARNING! If a and b are both negative, then $\sqrt{ab} \neq \sqrt{a}\sqrt{b}$. For example, if $a = -16$ and $b = -4$,

$$\sqrt{(-16)(-4)} = \sqrt{64} = 8 \quad \text{but} \quad \sqrt{-16}\sqrt{-4} = (4i)(2i) = 8i^2 = 8(-1) = -8$$

To multiply a complex number by a real number, we use the distributive property to remove parentheses and then simplify. For example,

$$\begin{aligned} 6(2 + 9i) &= 6(2) + 6(9i) && \text{Use the distributive property.} \\ &= 12 + 54i && \text{Simplify.} \end{aligned}$$

To multiply a complex number by an imaginary number, we use the distributive property to remove parentheses and then simplify. For example,

$$\begin{aligned} -5i(4 - 8i) &= -5i(4) - (-5i)8i && \text{Use the distributive property.} \\ &= -20i + 40i^2 && \text{Simplify.} \\ &= -40 - 20i && \text{Since } i^2 = -1,\ 40i^2 = 40(-1) = -40. \end{aligned}$$

To multiply two complex numbers, we use the FOIL method to develop the following definition.

Multiplying complex numbers

Complex numbers are multiplied as if they were binomials, with $i^2 = -1$:

$$\begin{aligned} (a + bi)(c + di) &= ac + adi + bci + bdi^2 \\ &= (ac - bd) + (ad + bc)i \end{aligned}$$

EXAMPLE 6 ***Using the FOIL method to multiply complex numbers.*** Multiply the complex numbers.

a. $$\begin{aligned} (2 + 3i)(3 - 2i) &= 6 - 4i + 9i - 6i^2 && \text{Use the FOIL method.} \\ &= 6 + 5i - (-6) && \text{Combine the imaginary terms: } -4i + 9i = 5i. \\ & && \text{Simplify: } i^2 = -1, \text{ so } 6i^2 = -6. \\ &= 6 + 5i + 6 \\ &= 12 + 5i && \text{Combine like terms.} \end{aligned}$$

b. $$\begin{aligned} (-4 + 2i)(2 + i) &= -8 - 4i + 4i + 2i^2 && \text{Use the FOIL method.} \\ &= -8 + 0i - 2 && -4i + 4i = 0i. \text{ Since } i^2 = -1,\ 2i^2 = -2. \\ &= -10 + 0i \end{aligned}$$

Self Check

Multiply $(-2 + 3i)(3 - 2i)$.

Answer: $0 + 13i$ ■

Complex conjugates

Complex conjugates

The complex numbers $a + bi$ and $a - bi$ are called **complex conjugates.**

For example,

$3 + 4i$ and $3 - 4i$ are complex conjugates.

$5 - 7i$ and $5 + 7i$ are complex conjugates.

$8 + 17i$ and $8 - 17i$ are complex conjugates.

EXAMPLE 7 ***Multiplying complex conjugates.*** Find the product of $3 + i$ and its complex conjugate.

Solution

The complex conjugate of $3 + i$ is $3 - i$. We can find the product as follows:

$$(3 + i)(3 - i) = 9 - 3i + 3i - i^2 \quad \text{Use the FOIL method.}$$
$$= 9 - i^2 \quad \text{Combine like terms.}$$
$$= 9 - (-1) \quad i^2 = -1.$$
$$= 10$$

Self Check
Multiply $(2 + 3i)(2 - 3i)$.

Answer: 13 ■

The product of the complex number $a + bi$ and its complex conjugate $a - bi$ is the real number $a^2 + b^2$, as the following work shows:

$$(a + bi)(a - bi) = a^2 - abi + abi - b^2i^2 \quad \text{Use the FOIL method.}$$
$$= a^2 - b^2(-1) \quad \text{Combine like terms. } i^2 = -1.$$
$$= a^2 + b^2$$

Rationalizing the denominator

To divide complex numbers, we often have to rationalize a denominator.

EXAMPLE 8 ***Rationalizing the denominator.*** Divide and write the result in $a + bi$ form: $\frac{1}{3 + i}$.

Solution

The denominator of the fraction $\frac{1}{3+i}$ is not a rational number. We can rationalize the denominator by multiplying both the numerator and the denominator of the fraction by the complex conjugate of the denominator, which is $3 - i$.

$$\frac{1}{3 + i} = \frac{1}{3 + i} \cdot \frac{3 - i}{3 - i} \quad \frac{3-i}{3-i} = 1.$$
$$= \frac{3 - i}{9 - 3i + 3i - i^2} \quad \text{Multiply the numerators and multiply the denominators.}$$
$$= \frac{3 - i}{9 - (-1)} \quad i^2 = -1.$$
$$= \frac{3 - i}{10} \quad \text{Simplify in the denominator.}$$
$$= \frac{3}{10} - \frac{1}{10}i \quad \text{Write the complex number in } a + bi \text{ form.}$$

Self Check
Divide 1 by $5 - i$.

Answer: $\frac{5}{26} + \frac{1}{26}i$ ■

EXAMPLE 9 ***Dividing complex numbers.*** Divide and write in $a + bi$ form: $\frac{3 - i}{2 + i}$.

Solution

We multiply both the numerator and the denominator of the fraction by the complex conjugate of the denominator.

Self Check
Divide $\frac{5 + 4i}{3 + 2i}$.

$$\frac{3-i}{2+i} = \frac{3-i}{2+i} \cdot \frac{2-i}{2-i}$$ $\frac{2-i}{2-i} = 1.$

$$= \frac{6-3i-2i+i^2}{4-2i+2i-i^2}$$ Multiply the numerators and multiply the denominators.

$$= \frac{5-5i}{4-(-1)}$$ $i^2 = -1.$

$$= \frac{5(1-i)}{5}$$ Factor out 5 in the numerator and divide out the common factor of 5.

$$= 1 - i$$ Simplify.

Answer: $\frac{23}{13} + \frac{2}{13}i$ ■

EXAMPLE 10 ***Dividing complex numbers.*** Write $\frac{4+\sqrt{-16}}{2+\sqrt{-4}}$ in $a + bi$ form.

Solution

$$\frac{4+\sqrt{-16}}{2+\sqrt{-4}} = \frac{4+4i}{2+2i}$$ Write the numerator and denominator in $a + bi$ form.

$$= \frac{2\cancel{(2+2i)}^{1}}{\cancel{2+2i}_{1}}$$ Factor out 2 in the numerator and divide out the common factor of $2 + 2i$.

$$= 2 + 0i$$

Self Check

Write: $\frac{3+\sqrt{-9}}{4+\sqrt{-16}}$ in $a + bi$ form.

Answer: $\frac{3}{4} + 0i$ ■

EXAMPLE 11 ***Division involving i.*** Find the quotient: $\frac{7}{2i}$. Express the result in $a + bi$ form.

Solution

The denominator can be expressed as $0 + 2i$. Its conjugate is $0 - 2i$, or just $-2i$.

$$\frac{7}{2i} = \frac{7}{2i} \cdot \frac{-2i}{-2i}$$ $\frac{-2i}{-2i} = 1.$

$$= \frac{-14i}{-4i^2}$$

$$= \frac{-14i}{4}$$ $i^2 = -1.$

$$= \frac{-7i}{2}$$ Simplify.

$$= 0 - \frac{7}{2}i$$ Write in $a + bi$ form.

Self Check

Divide: $\frac{5}{-i}$.

Answer: $0 + 5i$ ■

Powers of *i*

The powers of i produce an interesting pattern:

$$i = \sqrt{-1} = i \qquad i^5 = i^4 i = 1i = i$$
$$i^2 = \left(\sqrt{-1}\right)^2 = -1 \qquad i^6 = i^4 i^2 = 1(-1) = -1$$
$$i^3 = i^2 i = -1i = -i \qquad i^7 = i^4 i^3 = 1(-i) = -i$$
$$i^4 = i^2 i^2 = (-1)(-1) = 1 \qquad i^8 = i^4 i^4 = (1)(1) = 1$$

The pattern continues: $i, -1, -i, 1, \ldots$.

Larger powers of i can be simplified by using the fact that $i^4 = 1$. For example, to simplify i^{29}, we note that 29 divided by 4 gives a quotient of 7 and a remainder of 1. Thus, $29 = 4 \cdot 7 + 1$ and

$$\begin{aligned} i^{29} &= i^{4\cdot 7+1} && 4 \cdot 7 = 28. \\ &= (i^4)^7 \cdot i^1 \\ &= 1^7 \cdot i && i^4 = 1. \\ &= i && 1 \cdot i = i. \end{aligned}$$

The result of this example illustrates the following fact.

Powers of i

If n is a natural number that has a remainder of r when divided by 4, then

$$i^n = i^r$$

EXAMPLE 12 ***Simplifying powers of i.*** Simplify i^{55}.

Solution

We divide 55 by 4 and get a remainder of 3. Therefore,

$$i^{55} = i^3 = -i$$

Self Check

Simplify i^{62}.

Answer: -1

■

STUDY SET Section 8.4

VOCABULARY *In Exercises 1–6, fill in the blanks to make the statements true.*

1. $\sqrt{-1}$, $\sqrt{-3}$ and $\sqrt{-4}$ are examples of ______________ numbers.

2. $3 + 5i$, $2 - 7i$, and $5 - \frac{1}{2}i$ are examples of ______________ numbers.

3. The ______________ part of $5 + 7i$ is 5. The ______________ part is 7.

4. $6 + 3i$ and $6 - 3i$ are called complex ______________.

5. ___________ is a number whose square is -1.

6. i^{25} is called a ______________ of i.

CONCEPTS *In Exercises 7–12, fill in the blanks to make the statements true.*

7. $i =$ ☐

8. $i^2 =$ ☐

9. $i^3 =$ ☐

10. $i^4 =$ ☐

11. We multiply two complex numbers by using the ______________ method.

12. To divide two complex numbers, we ______________ the denominator.

13. Use a check to see if $x = -5\sqrt{2}i$ is a solution of $x^2 + 50 = 0$.

14. Give the complex conjugate of each number.

a. $2 - 3i$ **b.** 2 **c.** $-3i$

15. Complete Illustration 1 to show the relationship between the real numbers, the imaginary numbers, the complex numbers, the rational numbers, and the irrational numbers.

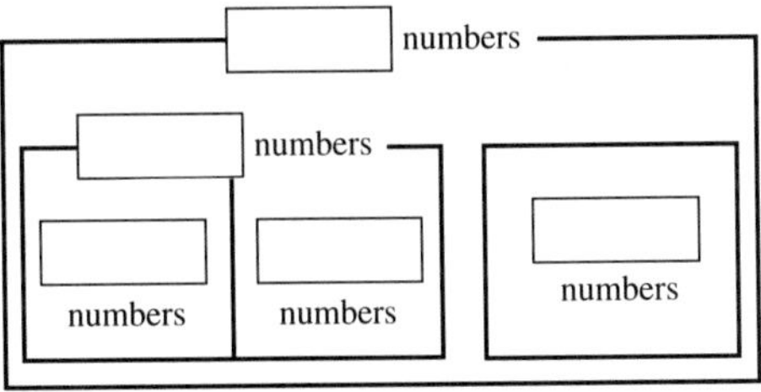

ILLUSTRATION 1

16. Tell whether each statement is true or false.

a. Every complex number is a real number.

b. Every real number is a complex number.

c. Imaginary numbers can always be denoted using i.

d. The square root of a negative number is an imaginary number.

17. Solve each equation.

a. $x^2 - 1 = 0$ **b.** $x^2 + 1 = 0$

18. Is $\sqrt[3]{-64}$ an imaginary number? Explain.

NOTATION *In Exercises 19–20, complete each solution.*

19. $(3 + 2i)(3 - i) = \square - 3i + \square - 2i^2$
$= 9 + 3i + \square$
$= 11 + 3i$

20. $\dfrac{3}{2 - i} = \dfrac{3}{2 - i} \cdot \dfrac{\square}{\square}$
$= \dfrac{6 + \square}{4 + \square - 2i - i^2}$
$= \dfrac{6 + 3i}{\square}$
$= \dfrac{6}{5} + \dfrac{3}{5}i$

21. Tell whether each statement is true or false.

a. $\sqrt{6i} = i\sqrt{6}$ **b.** $\sqrt{8i} = \sqrt{8}i$

c. $\sqrt{-25} = -\sqrt{25}$ **d.** $-i = i$

22. Tell whether each statement is true or false.

a. $\sqrt{-3}\sqrt{-2} = \sqrt{6}$

b. $\sqrt{-36} + \sqrt{-25} = \sqrt{-61}$

PRACTICE *In Exercises 23–34, express each number in terms of i.*

23. $\sqrt{-9}$ **24.** $\sqrt{-4}$ **25.** $\sqrt{-7}$ **26.** $\sqrt{-11}$

27. $\sqrt{-24}$ **28.** $\sqrt{-28}$ **29.** $-\sqrt{-24}$ **30.** $-\sqrt{-72}$

31. $5\sqrt{-81}$ **32.** $6\sqrt{-49}$ **33.** $\sqrt{-\dfrac{25}{9}}$ **34.** $-\sqrt{-\dfrac{121}{144}}$

In Exercises 35–42, simplify each expression.

35. $\sqrt{-1}\sqrt{-36}$ **36.** $\sqrt{-9}\sqrt{-100}$ **37.** $\sqrt{-2}\sqrt{-6}$ **38.** $\sqrt{-3}\sqrt{-6}$

39. $\dfrac{\sqrt{-25}}{\sqrt{-64}}$ **40.** $\dfrac{\sqrt{-4}}{\sqrt{-1}}$ **41.** $-\dfrac{\sqrt{-400}}{\sqrt{-1}}$ **42.** $-\dfrac{\sqrt{-225}}{\sqrt{-16}}$

In Exercises 43–54, solve each equation. Write all solutions in bi or a + bi form.

43. $x^2 + 9 = 0$ **44.** $x^2 + 100 = 0$ **45.** $3x^2 = -16$

46. $2x^2 = -25$ **47.** $x^2 + 2x + 2 = 0$ **48.** $x^2 - 2x + 6 = 0$

49. $2x^2 + x + 1 = 0$ **50.** $3x^2 + 2x + 1 = 0$ **51.** $3x^2 - 4x = -2$

52. $2x^2 + 3x = -3$ **53.** $3x^2 - 2x = -3$ **54.** $5x^2 = 2x - 1$

In Exercises 55–72, do the operations. Write all answers in a + bi form.

55. $(3 + 4i) + (5 - 6i)$ **56.** $(5 + 3i) - (6 - 9i)$ **57.** $(7 - 3i) - (4 + 2i)$

58. $(8 + 3i) + (-7 - 2i)$ **59.** $(8 + \sqrt{-25}) + (7 + \sqrt{-4})$ **60.** $(-7 + \sqrt{-81}) - (-2 - \sqrt{-64})$

61. $3(2 - i)$

62. $-4(3 + 4i)$

63. $-5i(5 - 5i)$

64. $2i(7 + 2i)$

65. $(2 + i)(3 - i)$

66. $(4 - i)(2 + i)$

67. $(2 - \sqrt{-16})(3 + \sqrt{-4})$

68. $(3 - \sqrt{-4})(4 - \sqrt{-9})$

69. $(2 + \sqrt{2}i)(3 - \sqrt{2}i)$

70. $(5 + \sqrt{3}i)(2 - \sqrt{3}i)$

71. $(2 + i)^2$

72. $(3 - 2i)^2$

In Exercises 73–92, write each expression in a + bi form.

73. $\frac{1}{i}$

74. $\frac{1}{i^3}$

75. $\frac{4}{5i^3}$

76. $\frac{3}{2i}$

77. $\frac{3i}{8\sqrt{-9}}$

78. $\frac{5i^3}{2\sqrt{-4}}$

79. $\frac{-3}{5i^5}$

80. $\frac{-4}{6i^7}$

81. $\frac{5}{2 - i}$

82. $\frac{26}{3 - 2i}$

83. $\frac{-12}{7 - \sqrt{-1}}$

84. $\frac{4}{3 + \sqrt{-1}}$

85. $\frac{5i}{6 + 2i}$

86. $\frac{-4i}{2 - 6i}$

87. $\frac{3 - 2i}{3 + 2i}$

88. $\frac{2 + 3i}{2 - 3i}$

89. $\frac{3 + 2i}{3 + i}$

90. $\frac{2 - 5i}{2 + 5i}$

91. $\frac{\sqrt{5} - \sqrt{3}i}{\sqrt{5} + \sqrt{3}i}$

92. $\frac{\sqrt{3} + \sqrt{2}i}{\sqrt{3} - \sqrt{2}i}$

In Exercises 93–100, simplify each expression.

93. i^{21}

94. i^{19}

95. i^{27}

96. i^{22}

97. i^{100}

98. i^{42}

99. i^{97}

100. i^{200}

APPLICATIONS

101. FRACTAL GEOMETRY Complex numbers are fundamental in the creation of the intricate geometric shape shown in Illustration 2, called a *fractal.* Fractal geometry is a rapidly expanding field with applications in science, medicine, and computer graphics. The process of creating this image is based on the following sequence of steps, which begins by picking any complex number, which we will call z.

1. Square z, and then add that result to z.
2. Square the result from step 1, and then add it to z.
3. Square the result from step 2, and then add it to z.

If we begin with the complex number i, what is the result after performing steps 1, 2, and 3?

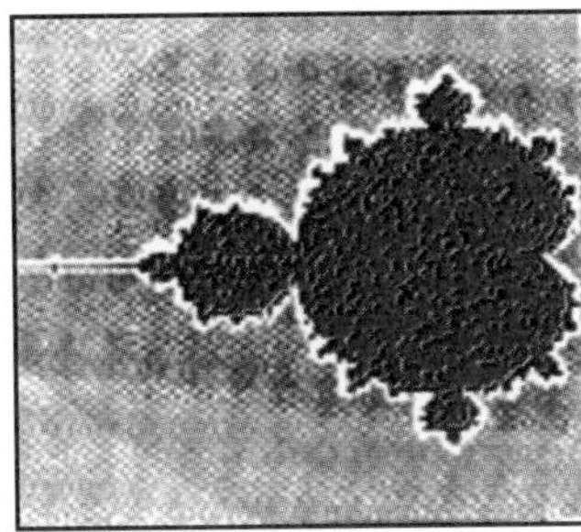

ILLUSTRATION 2

102. ELECTRONICS The impedance Z in an AC (alternating current) circuit is a measure of how much the circuit impedes (hinders) the flow of current through it. The impedance is related to the voltage V and the current I by the formula

$$V = IZ$$

If a circuit has a current of $(0.5 + 2.0i)$ amps and an impedance of $(0.4 - 3.0i)$ ohms, find the voltage.

WRITING *Write a paragraph using your own words.*

103. What is an imaginary number?

104. In Example 1 of this section, what unusual situation illustrated the need to define the square root of a negative number?

REVIEW

105. What are the lengths of the two legs of a 30°–60°–90° triangle if the hypotenuse is 30 units long?

106. What are the lengths of the other two sides of a 45°–45°–90° triangle if one leg is 30 units long?

107. WIND SPEED A plane that can fly 200 mph in still air makes a 330-mile flight with a tail wind and returns, flying into the same wind. Find the speed of the wind if the total flying time is $3\frac{1}{3}$ hours.

108. FINDING RATES A student drove a distance of 135 miles at an average speed of 50 mph. How much faster would he have to drive on the return trip to save 30 minutes of driving time?

8.5 *The Discriminant and Equations That Can Be Written in Quadratic Form*

In this section, you will learn about

- **The discriminant**
- **Equations that can be written in quadratic form**
- **Problem solving**

INTRODUCTION. In this section, we will discuss how to predict what type of solutions a quadratic equation will have without having to solve the equation. We will then solve some special equations that can be written in quadratic form. Finally, we will use the equation-solving methods of this chapter to solve a shared-work problem.

The discriminant

We can predict what type of solutions a particular quadratic equation will have without solving it. To see how, we suppose that the coefficients a, b, and c in the equation $ax^2 + bx + c = 0$ $(a \neq 0)$ are real numbers. Then the solutions of the equation are given by the quadratic formula

$$x = \frac{-b \pm \sqrt{b^2 - 4ac}}{2a} \qquad (a \neq 0)$$

If $b^2 - 4ac \geq 0$, the solutions are real numbers. If $b^2 - 4ac < 0$, the solutions are nonreal complex numbers. Thus, the value of $\boldsymbol{b^2 - 4ac}$, called the **discriminant,** determines the type of solutions for a particular quadratic equation.

The discriminant

If a, b, and c are real numbers and

If $b^2 - 4ac$ *is . . .*	*the solutions are . . .*
positive,	real numbers and unequal
0,	real numbers and equal.
negative,	nonreal complex numbers and complex conjugates.

If a, b, and c are rational numbers and

If $b^2 - 4ac$ *is . . .*	*the solutions are . . .*
a perfect square,	rational numbers and unequal.
positive and not a perfect square,	irrational numbers and unequal.

EXAMPLE 1 ***Using the discriminant.*** Determine the type of solutions for each equation:

a. $x^2 + x + 1 = 0$ **b.** $3x^2 + 5x + 2 = 0$

Solution

a. We calculate the discriminant for $x^2 + x + 1 = 0$:

$$\begin{aligned} b^2 - 4ac &= 1^2 - 4(1)(1) && a = 1,\ b = 1,\text{ and } c = 1. \\ &= -3 && \text{The result is a negative number.} \end{aligned}$$

Since $b^2 - 4ac < 0$, the solutions of $x^2 + x + 1 = 0$ are nonreal complex numbers and complex conjugates.

b. For $3x^2 + 5x + 2 = 0$,

$$\begin{aligned} b^2 - 4ac &= 5^2 - 4(3)(2) && a = 3,\ b = 5,\text{ and } c = 2. \\ &= 25 - 24 \\ &= 1 && \text{The result is a positive number.} \end{aligned}$$

Since $b^2 - 4ac > 0$ and $b^2 - 4ac$ is a perfect square, the solutions of $3x^2 + 5x + 2 = 0$ are rational and unequal.

Self Check

Determine the type of solutions for

a. $x^2 + x - 1 = 0$

b. $4x^2 - 10x + 25 = 0$

Answers: **a.** real numbers that are irrational and unequal, **b.** nonreal numbers that are complex conjugates ■

Equations that can be written in quadratic form

Many equations that are not quadratic can be written in quadratic form ($ax^2 + bx + c = 0$) and then solved using the techniques discussed in previous sections. For example, a careful inspection of the equation $x^4 - 5x^2 + 4 = 0$ leads to the following observations:

$$x^4 - 5x^2 + 4 = 0$$

In the leading term, this power of the variable is the square of . . .

. . . the power of the variable in the middle term.

The last term is a constant.

Equations having these characteristics are said to be *quadratic in form.* One method used to solve them is to make a substitution.

$$\begin{aligned} x^4 - 5x^2 + 4 &= 0 && \text{The given equation.} \\ (x^2)^2 - 5(x^2) + 4 &= 0 && \text{Write } x^4 \text{ as } (x^2)^2. \\ y^2 - 5y + 4 &= 0 && \text{Let } y = x^2. \text{ Replace each } x^2 \text{ with } y. \end{aligned}$$

We can solve this quadratic equation by factoring.

$$\begin{aligned} &(y - 4)(y - 1) = 0 && \text{Factor } y^2 - 5y + 4. \\ &y - 4 = 0 \quad \text{or} \quad y - 1 = 0 && \text{Set each factor equal to 0.} \\ &y = 4 \qquad\qquad\quad y = 1 \end{aligned}$$

These are *not* the solutions for x. To find x, we now "undo" the earlier substitutions by replacing each y with x^2. Then we solve for x.

$$\begin{aligned} &x^2 = 4 \qquad\quad \text{or} \quad x^2 = 1 \\ &x = \pm\sqrt{4} \qquad\quad\ \ x = \pm\sqrt{1} && \text{Use the square root property.} \\ &x = \pm 2 \qquad\qquad\ \ x = \pm 1 \end{aligned}$$

This equation has four solutions: 1, -1, 2, and -2. Verify that each one satisfies the original equation.

EXAMPLE 2 ***Solving equations that are quadratic in form.***
Solve $x - 7\sqrt{x} + 12 = 0$.

Solution

This equation is quadratic in form, because the power of the leading term is the square of the variable factor of the middle term: $x = \left(\sqrt{x}\right)^2$. If we let $y = \sqrt{x}$, then $y^2 = x$. With this substitution, the equation

$$x - 7\sqrt{x} + 12 = 0$$

becomes a quadratic equation that can be solved by factoring.

$$y^2 - 7y + 12 = 0 \qquad \text{Substitute } y^2 \text{ for } x \text{ and } y \text{ for } \sqrt{x}.$$
$$(y - 3)(y - 4) = 0 \qquad \text{Factor } y^2 - 7y + 12 = 0.$$
$$y - 3 = 0 \quad \text{or} \quad y - 4 = 0 \qquad \text{Set each factor equal to 0.}$$
$$y = 3 \qquad\qquad y = 4$$

Replace each y with $\sqrt{x}$ and solve the radical equations by squaring both sides.

$$\sqrt{x} = 3 \quad \text{or} \quad \sqrt{x} = 4$$
$$x = 9 \qquad\qquad x = 16 \qquad \text{Square both sides.}$$

Verify that both solutions satisfy the original equation.

Self Check
Solve $x + x^{1/2} - 6 = 0$.

Answer: 4

EXAMPLE 3 ***Solving equations that are quadratic in form.***
Solve $2m^{2/3} - 2 = 3m^{1/3}$.

Solution

After writing the equation in descending powers of m, we see that

$$2m^{2/3} - 3m^{1/3} - 2 = 0$$

is quadratic in form, because $m^{2/3} = (m^{1/3})^2$. We will use the substitution $y = m^{1/3}$ to write this equation in quadratic form.

$$2m^{2/3} - 3m^{1/3} - 2 = 0$$
$$2(m^{1/3})^2 - 3m^{1/3} - 2 = 0$$
$$2y^2 - 3y - 2 = 0 \qquad \text{Replace } m^{1/3} \text{ with } y.$$
$$(2y + 1)(y - 2) = 0 \qquad \text{Factor } 2y^2 - 3y - 2 = 0.$$
$$2y + 1 = 0 \quad \text{or} \quad y - 2 = 0 \qquad \text{Set each factor equal to 0.}$$
$$y = -\frac{1}{2} \qquad\qquad y = 2$$

Replace each y with $m^{1/3}$ and solve for m.

$$m^{1/3} = -\frac{1}{2} \quad \text{or} \quad m^{1/3} = 2$$
$$(m^{1/3})^3 = \left(-\frac{1}{2}\right)^3 \qquad (m^{1/3})^3 = (2)^3 \qquad \text{Recall that } m^{1/3} = \sqrt[3]{m}. \text{ To solve for } m, \text{ cube both sides.}$$
$$m = -\frac{1}{8} \qquad\qquad m = 8$$

Verify that both solutions satisfy the original equation.

Self Check
Solve $a^{2/3} = -3a^{1/3} + 10$.

Answer: $-125, 8$

EXAMPLE 4 ***Solving equations that are quadratic in form.*** Solve $(4t + 2)^2 - 30(4t + 2) + 224 = 0$.

Solution

If we make the substitution $y = (4t + 2)$, the given equation becomes

$$y^2 - 30y + 224 = 0$$

which can be solved by using the quadratic formula.

$$y = \frac{-b \pm \sqrt{b^2 - 4ac}}{2a}$$

$$= \frac{-(-30) \pm \sqrt{(-30)^2 - 4(1)(224)}}{2(1)}$$ Substitute 1 for a, -30 for b, and 224 for c.

$$= \frac{30 \pm \sqrt{900 - 896}}{2}$$ Simplify within the radical.

$$= \frac{30 \pm 2}{2}$$ $\sqrt{900 - 896} = \sqrt{4} = 2$.

$y = 16$ or $y = 14$

To find t, we replace y with $(4t + 2)$ and solve for t.

$$4t + 2 = 16 \quad \text{or} \quad 4t + 2 = 14$$
$$4t = 14 \qquad\qquad 4t = 12$$
$$t = 3.5 \qquad\qquad t = 3$$

Self Check

Solve $(n + 3)^2 - 6(n + 3) = -8$.

Answer: $-1, 1$

Problem solving

EXAMPLE 5 ***Household appliances.*** A water temperature control on a washing machine is shown in Figure 8-16. When the "warm" setting is selected, both the hot and cold water inlets open to fill the tub in 2 minutes 15 seconds. When the "cold" temperature setting is chosen, the cold water inlet fills the tub 45 seconds faster than when the "hot" setting is used. How long does it take to fill the washing machine with hot water?

Electronic Temperature Control

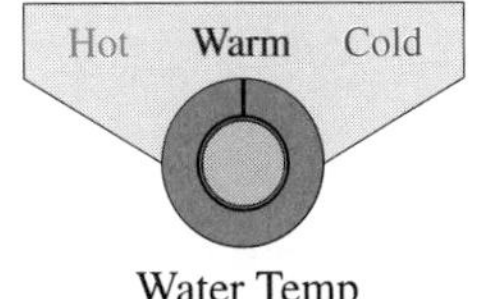

Water Temp

FIGURE 8-16

Analyze the problem The key to solving this problem is to determine how much of the tub is filled by each water temperature setting in 1 second. On the "warm" setting, when the hot and cold inlets are working together, the tub is filled in 2 minutes 15 seconds, or 135 seconds. So in 1 second, they fill $\frac{1}{135}$ of the tub.

Form an equation Let x represent the number of seconds it takes to fill the tub when the "hot" temperature is chosen. In 1 second, the hot water inlet fills $\frac{1}{x}$ of the tub. Since the cold water inlet fills the tub in 45 seconds less time, the washing machine can be filled with cold water in $(x - 45)$ seconds. In 1 second, $\frac{1}{x-45}$ of the tub is filled by the cold water inlet. We can now form an equation.

What the hot water inlet pipe can do in 1 second	plus	what the cold water inlet pipe can do in 1 second	equals	what they can do together in 1 second.
$\frac{1}{x}$	$+$	$\frac{1}{x-45}$	$=$	$\frac{1}{135}$

Solve the equation

$$\frac{1}{x} + \frac{1}{x-45} = \frac{1}{135}$$

$$135x(x-45)\left(\frac{1}{x} + \frac{1}{x-45}\right) = 135x(x-45)\left(\frac{1}{135}\right)$$ Multiply both sides by $135x(x-45)$ to clear the equation of fractions.

$$135(x-45) + 135x = x(x-45)$$ Simplify.

$$135x - 6{,}075 + 135x = x^2 - 45x$$ Use the distributive property to remove parentheses.

$$270x - 6{,}075 = x^2 - 45x$$ Combine like terms.

$$0 = x^2 - 315x + 6{,}075$$ Subtract $270x$ from both sides. Add 6,075 to both sides.

To solve this equation, we will use the quadratic formula, with $a = 1$, $b = -315$, and $c = 6{,}075$.

$$x = \frac{-b \pm \sqrt{b^2 - 4ac}}{2a}$$

$$= \frac{-(-315) \pm \sqrt{(-315)^2 - 4(1)(6{,}075)}}{2(1)}$$ Substitute 1 for a, -315 for b, and 6,075 for c.

$$= \frac{315 \pm \sqrt{99{,}225 - 24{,}300}}{2}$$ Simplify within the radical.

$$\approx \frac{315 \pm \sqrt{74{,}925}}{2}$$ $\sqrt{99{,}225 - 24{,}300} = \sqrt{74{,}925}$.

$$x \approx \frac{589}{2} \quad \text{or} \quad x \approx \frac{41}{2}$$

$$x \approx 294 \qquad x \approx 21$$

State the conclusion We can disregard the solution of 21 seconds, because this would imply that the cold water inlet fills the tub in a negative number of seconds ($21 - 45 = -24$). Therefore, the hot water inlet fills the washing machine tub in about 294 seconds, which is 4 minutes 54 seconds.

Check the result Use estimation to check the result. ■

STUDY SET Section 8.5

VOCABULARY *In Exercises 1–2, fill in the blanks to make the statements true.*

1. For the quadratic equation $ax^2 + bx + c = 0$, the discriminant is ______________.

2. When an equation is written in the form $ax^2 + bx + c = 0$, we say that it is written in ______________ form.

CONCEPTS *In Exercises 3–6, consider the equation $ax^2 + bx + c = 0$, where a, b, and c are rational numbers, and fill in the blanks to make the statement true.*

3. If $b^2 - 4ac < 0$, the solutions of the equation are nonreal complex ______________.

4. If $b^2 - 4ac =$ ____, the solutions of the equation are equal real numbers.

5. If $b^2 - 4ac$ is a perfect square, the solutions are ______________ numbers and ______________.

6. If $b^2 - 4ac$ is positive and not a perfect square, the solutions are ______________ numbers and ______________.

7. Consider $x^4 - 3x^2 + 2 = 0$.
   a. What is the relationship between the powers of x in the first two terms on the left-hand side?
   b. Is this equation quadratic in form?

8. Consider $x^{2/3} + 4x^{1/3} - 5 = 0$.
   a. What is the relationship between the powers of x in the first two terms on the left-hand side?
   b. Is this equation quadratic in form?

NOTATION *In Exercises 9–10, complete each solution.*

9. To find the type of solutions for the equation $x^2 + 5x + 6 = 0$, we compute the discriminant.

$$\begin{aligned} b^2 - 4ac &= \square^2 - 4(1)(\square) \\ &= 25 - \square \\ &= 1 \end{aligned}$$

Since a, b, and c are rational numbers and the value of the discriminant is a perfect square, the solutions are ______________ numbers and unequal.

10. Change $\frac{3}{4} + x = \frac{3x - 50}{4(x - 6)}$ to quadratic form.

$$\begin{aligned} \square\left(\frac{3}{4} + x\right) &= \square\frac{3x - 50}{4(x - 6)} \\ 3(x - 6) + 4x(\square) &= 3x - 50 \\ 3x - \square + 4x^2 - \square &= 3x - 50 \\ 4x^2 - 24x + \square &= 0 \\ \square - 6x + 8 &= 0 \end{aligned}$$

PRACTICE *In Exercises 11–18, use the discriminant to determine what type of solutions exist for each quadratic equation.* ***Do not solve the equation.***

11. $4x^2 - 4x + 1 = 0$
12. $6x^2 - 5x - 6 = 0$
13. $5x^2 + x + 2 = 0$
14. $3x^2 + 10x - 2 = 0$
15. $2x^2 = 4x - 1$
16. $9x^2 = 12x - 4$
17. $x(2x - 3) = 20$
18. $x(x - 3) = -10$

19. Use the discriminant to determine whether the solutions of $1{,}492x^2 + 1{,}776x - 2{,}000 = 0$ are real numbers.

20. Use the discriminant to determine whether the solutions of $1{,}776x^2 - 1{,}492x + 2{,}000 = 0$ are real numbers.

In Exercises 21–50, solve each equation.

21. $x^4 - 17x^2 + 16 = 0$
22. $x^4 - 10x^2 + 9 = 0$
23. $x^4 = 6x^2 - 5$
24. $2x^4 + 24 = 26x^2$
25. $t^4 + 3t^2 = 28$
26. $3h^4 + h^2 - 2 = 0$
27. $2x + \sqrt{x} - 3 = 0$
28. $2x - \sqrt{x} - 1 = 0$
29. $3x + 5\sqrt{x} + 2 = 0$
30. $3x - 4\sqrt{x} + 1 = 0$
31. $x - 6\sqrt{x} = -8$
32. $x - 5x^{1/2} + 4 = 0$
33. $x^{2/3} + 5x^{1/3} + 6 = 0$
34. $x^{2/3} - 7x^{1/3} + 12 = 0$
35. $a^{2/3} - 2a^{1/3} - 3 = 0$
36. $r^{2/3} + 4r^{1/3} - 5 = 0$
37. $2(2x + 1)^2 - 7(2x + 1) + 6 = 0$
38. $3(2 - x)^2 + 10(2 - x) - 8 = 0$
39. $(c + 1)^2 - 4(c + 1) - 8 = 0$
40. $(k - 7)^2 + 6(k - 7) + 10 = 0$
41. $x + 5 + \frac{4}{x} = 0$
42. $x - 4 + \frac{3}{x} = 0$
43. $\frac{1}{x + 2} + \frac{24}{x + 3} = 13$
44. $\frac{3}{x} + \frac{4}{x + 1} = 2$
45. $\frac{2}{x - 1} + \frac{1}{x + 1} = 3$
46. $\frac{3}{x - 2} - \frac{1}{x + 2} = 5$

47. $x^{-4} - 2x^{-2} + 1 = 0$

48. $4x^{-4} + 1 = 5x^{-2}$

49. $x + \dfrac{2}{x - 2} = 0$

50. $x + \dfrac{x + 5}{x - 3} = 0$

APPLICATIONS

51. FLOWER ARRANGING A florist needs to determine the height h of the flowers shown in Illustration 1. The radius r, the width w, and the height h of the circular-shaped arrangement are related by the formula

$$r = \frac{4h^2 + w^2}{8h}$$

If w is to be 34 inches and r is to be 18 inches, find h to the nearest tenth of an inch.

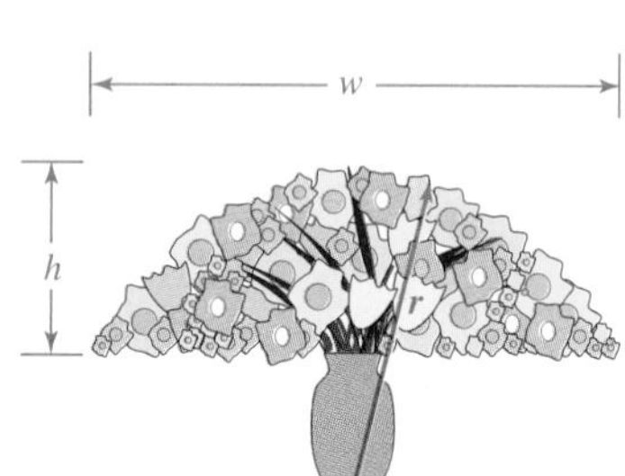

ILLUSTRATION 1

52. ARCHITECTURE A **golden rectangle** is said to be one of the most visually appealing of all geometric forms. The Parthenon, built by the Greeks in the 5th century B.C. and shown in Illustration 2, fits into a golden rectangle once its ruined triangular pediment is drawn in.

In a golden rectangle, the length l and width w must satisfy the equation

$$\frac{l}{w} = \frac{w}{l - w}$$

If a rectangular billboard is to have a width of 20 feet, what should its length be so that it is a golden rectangle? Round to the nearest tenth.

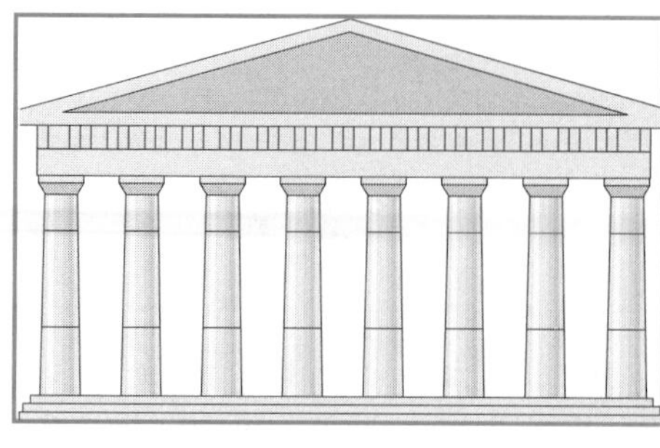

ILLUSTRATION 2

53. SNOWMOBILE A woman drives her snowmobile 150 miles at a rate of r mph. She could have gone the same distance in 2 hours less time if she had increased her speed by 20 mph. Find r.

54. BICYCLING Jeff bicycles 160 miles at the rate of r mph. The same trip would have taken 2 hours longer if he had decreased his speed by 4 mph. Find r.

55. CROWD CONTROL After a sold-out performance at a county fair, security guards have found that the grandstand area can be emptied in 6 minutes if both the east and west exists are opened. If just the east exit is used, it takes 4 minutes longer to clear the grandstand than it does if just the west exit is opened. How long does it take to clear the grandstand if everyone must file through the east exist?

56. PAPER ROUTE When a father, in a car, and his son, on a bicycle, work together to distribute the morning edition, it takes them 35 minutes to complete a paper route. Working alone, it takes the son 25 minutes longer than the father. To the nearest minute, how long does it take the son to cover the paper route on his bicycle?

WRITING *Write a paragraph using your own words.*

57. Describe how to predict what type of solutions the equation $3x^2 - 4x + 5 = 0$ will have.

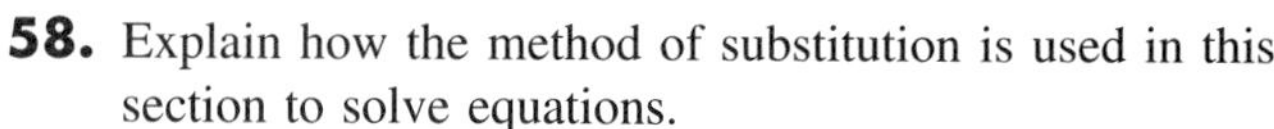

58. Explain how the method of substitution is used in this section to solve equations.

REVIEW *In Exercises 59–60, solve the equation.*

59. $\dfrac{1}{4} + \dfrac{1}{t} = \dfrac{1}{2t}$

60. $\dfrac{p - 3}{3p} + \dfrac{1}{2p} = \dfrac{1}{4}$

61. Find the slope of the line passing through $P(-2, -4)$ and $Q(3, 5)$.

62. Write the equation of the line passing through $P(-2, -4)$ and $Q(3, 5)$ in general form.

8.6 *Quadratic and Other Nonlinear Inequalities*

In this section, you will learn about

- **Solving quadratic inequalities**
- **Solving rational inequalities**
- **Graphs of nonlinear inequalities in two variables**

INTRODUCTION. If $a \neq 0$, inequalities of the form $ax^2 + bx + c < 0$ and $ax^2 + bx + c > 0$ are called *quadratic inequalities.* We will begin this section by showing how to solve these inequalities by making a sign chart. We will then show how to solve other nonlinear inequalities using the same technique. To conclude, we will show how to find graphical solutions of nonlinear inequalities containing two variables.

Solving quadratic inequalities

To solve the inequality $x^2 + x - 6 < 0$, we must find the values of x that make the inequality true. This can be done using a number line. We begin by factoring the trinomial to obtain

$$(x + 3)(x - 2) < 0$$

Since the product of $x + 3$ and $x - 2$ must be less than 0, the values of $x + 3$ and $x - 2$ must be opposite in sign. To find the intervals where this is true, we keep track of their signs by constructing the chart in Figure 8-17. The chart shows that

- $x - 2$ is 0 when $x = 2$, is positive when $x > 2$, and is negative when $x < 2$.
- $x + 3$ is 0 when $x = -3$, is positive when $x > -3$, and is negative when $x < -3$.

The only place where the values of the binomials are opposite in sign is in the interval $(-3, 2)$. Therefore, the solution set of the inequality can be denoted.

$$-3 < x < 2$$

The graph of the solution set is shown on the number line in Figure 8-17.

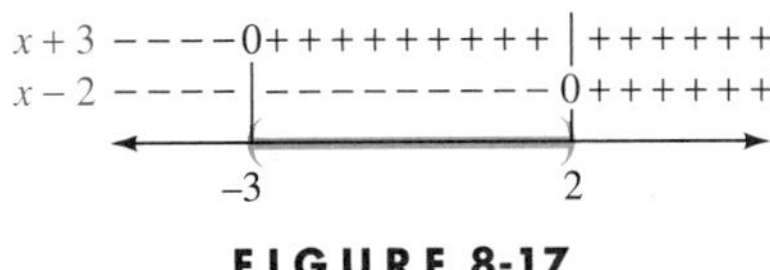

FIGURE 8-17

EXAMPLE 1 ***Constructing a sign chart.*** Solve $x^2 + 2x - 3 \geq 0$.

Solution

We factor the trinomial to get $(x - 1)(x + 3)$ and construct a sign chart, as in Figure 8-18.

- $x - 1$ is 0 when $x = 1$, is positive when $x > 1$, and is negative when $x < 1$.
- $x + 3$ is 0 when $x = -3$, is positive when $x > -3$, and is negative when $x < -3$.

Self Check

Solve $x^2 + 2x - 15 > 0$ and graph the solution set.

The product of $x - 1$ and $x + 3$ will be greater than 0 when the signs of the binomial factors are the same. This occurs in the intervals $(-\infty, -3)$ and $(1, \infty)$. The numbers -3 and 1 are also included, because they make the product equal to 0. Thus, the solution set is

$$(-\infty, -3] \cup [1, \infty) \qquad \text{or} \qquad x \leq -3 \text{ or } x \geq 1$$

The graph of the solution set is shown on the number line in Figure 8-18.

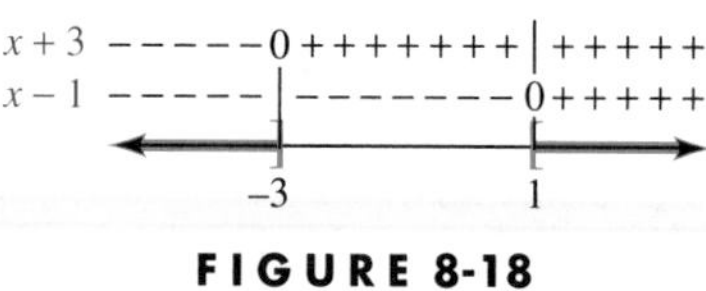

FIGURE 8-18

Answer: $(-\infty, -5) \cup (3, \infty)$

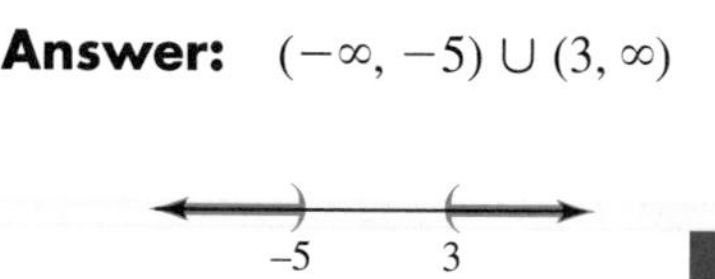

Solving rational inequalities

Making a sign chart is useful for solving many inequalities that are neither linear nor quadratic. In the next three examples, we will use a sign chart to solve *rational inequalities.*

EXAMPLE 2 ***Solving rational inequalities.*** Solve $\frac{1}{x} < 6$.

Self Check

Solve $\frac{3}{x} > 5$.

Solution

We subtract 6 from both sides to make the right-hand side equal to 0. We then find a common denominator and add:

$$\frac{1}{x} < 6$$

$$\frac{1}{x} - 6 < 0 \quad \text{Subtract 6 from both sides.}$$

$$\frac{1}{x} - \frac{6x}{x} < 0 \quad \text{Get a common denominator.}$$

$$\frac{1 - 6x}{x} < 0 \quad \text{Subtract the numerators and keep the common denominator.}$$

We now make a sign chart, as shown in Figure 8-19.

- The denominator x is 0 when $x = 0$, is positive when $x > 0$, and is negative when $x < 0$.
- The numerator $1 - 6x$ is 0 when $x = \frac{1}{6}$, is positive when $x < \frac{1}{6}$, and is negative when $x > \frac{1}{6}$.

The fraction $\frac{1-6x}{x}$ will be less than 0 when the numerator and denominator are opposite in sign. This occurs in the interval

$$(-\infty, 0) \cup \left(\frac{1}{6}, \infty\right) \qquad \text{or} \qquad x < 0 \text{ or } x > \frac{1}{6}$$

The graph of this interval is shown in Figure 8-19.

1 − 6x + + + + + | + + + + + 0 - - - - -
x - - - - - 0 + + + + + | + + + + +
0 1/6

FIGURE 8-19

Answer: $\left(0, \frac{3}{5}\right)$

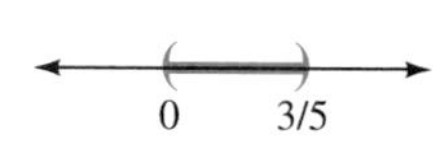

WARNING! Since we don't know whether x is positive, 0, or negative, multiplying both sides of the inequality $\frac{1}{x} < 6$ by x is a three-case situation:

- If $x > 0$, then $1 < 6x$.
- If $x = 0$, then $\frac{1}{x}$ is undefined.
- If $x < 0$, then $1 > 6x$.

If we multiply both sides by x and solve the linear inequality $1 < 6x$, we are considering only one case and will get only part of the answer.

EXAMPLE 3 *Solving rational inequalities.* Solve $\dfrac{x^2 - 3x + 2}{x - 3} \geq 0$.

Solution

We write the fraction with the numerator in factored form.

$$\frac{(x-2)(x-1)}{x-3} \geq 0$$

To keep track of the signs of the three binomials, we construct the sign chart shown in Figure 8-20. The fraction will be positive in the intervals where all factors are positive, or where exactly two factors are negative. The numbers 1 and 2 are included, because they make the numerator (and thus the fraction) equal to 0. The number 3 is not included, because it gives a 0 in the denominator.

The solution is the interval $[1, 2] \cup (3, \infty)$. The graph appears in Figure 8-20.

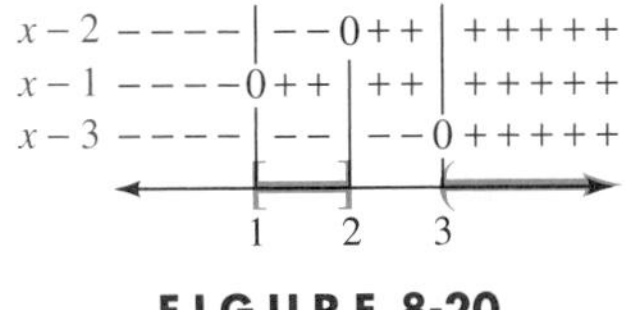

FIGURE 8-20

Self Check

Solve $\dfrac{x+2}{x^2 - 2x - 3} > 0$ and graph the solution set.

Answer: $(-2, -1) \cup (3, \infty)$

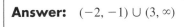

EXAMPLE 4 *Solving rational inequalities.* Solve $\dfrac{3}{x-1} < \dfrac{2}{x}$.

Solution

We subtract $\frac{2}{x}$ from both sides to get 0 on the right-hand side and proceed as follows:

$$\frac{3}{x-1} < \frac{2}{x}$$

$$\frac{3}{x-1} - \frac{2}{x} < 0 \quad \text{Subtract } \tfrac{2}{x} \text{ from both sides.}$$

$$\frac{3x}{(x-1)x} - \frac{2(x-1)}{x(x-1)} < 0 \quad \text{Get a common denominator.}$$

$$\frac{3x - 2x + 2}{x(x-1)} < 0 \quad \text{Keep the denominator and subtract the numerators.}$$

$$\frac{x+2}{x(x-1)} < 0 \quad \text{Combine like terms.}$$

We can keep track of the signs of the three polynomials with the sign chart shown in Figure 8-21. The fraction will be negative in the intervals with either one or three negative factors. The numbers 0 and 1 are not included, because they give a 0 in the denominator, and the number -2 is not included, because it does not satisfy the inequality.

Self Check

Solve $\dfrac{2}{x+1} > \dfrac{1}{x}$ and graph the solution set.

The solution is the interval $(-\infty, -2) \cup (0, 1)$, as shown in Figure 8-21.

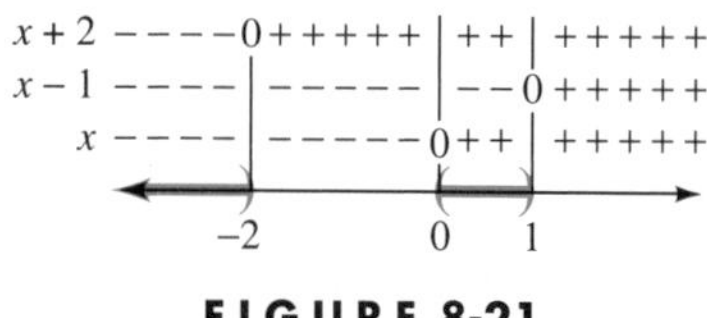

FIGURE 8-21

Answer: $(-1, 0) \cup (1, \infty)$

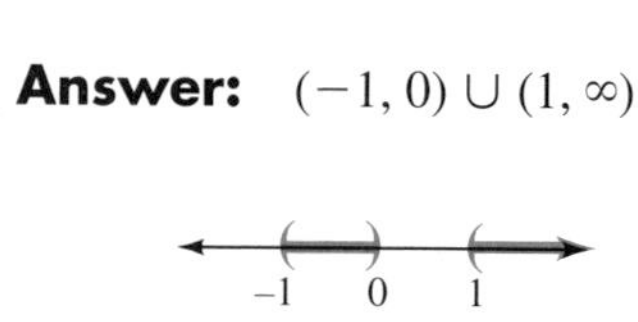

Accent on Technology Solving inequalities graphically

To approximate the solutions of the inequality $x^2 + 2x - 3 \geq 0$ (Example 1) by graphing, we can use the standard window settings of $[-10, 10]$ for x and $[-10, 10]$ for y and graph the quadratic function $y = x^2 + 2x - 3$, as in Figure 8-22. The solution of the inequality will be those numbers x for which the graph of $y = x^2 + 2x - 3$ lies above or on the x-axis. We can trace to find that this interval is $(-\infty, -3] \cup [1, \infty)$.

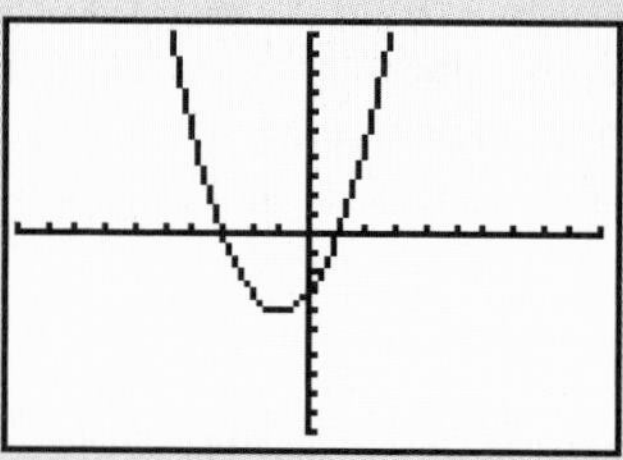

FIGURE 8-22

To approximate the solutions of $\frac{3}{x-1} < \frac{2}{x}$ (Example 4), we first write the inequality in the form

$$\frac{x+2}{x(x-1)} < 0$$

We use window settings of $[-5, 5]$ for x and $[-3, 3]$ for y and graph the function $y = \frac{x+2}{x(x-1)}$, as in Figure 8-23(a). The solution of the inequality will be those numbers x for which the graph lies below the x-axis.

We can trace to see that the graph is below the x-axis when x is less than -2. Since we cannot see the graph in the interval $0 < x < 1$, we redraw the graph using window settings of $[-1, 2]$ for x and $[-25, 10]$ for y. See Figure 8-23(b).

We can now see that the graph is below the x-axis in the interval $(0, 1)$. Thus, the solution to the inequality is the union of two intervals:

$$(-\infty, -2) \cup (0, 1)$$

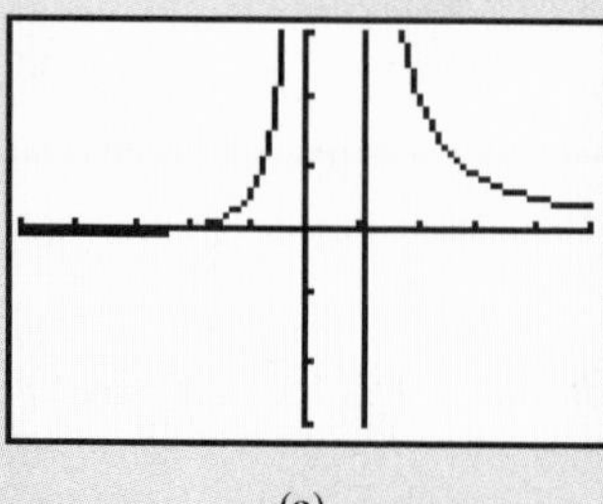

(a)

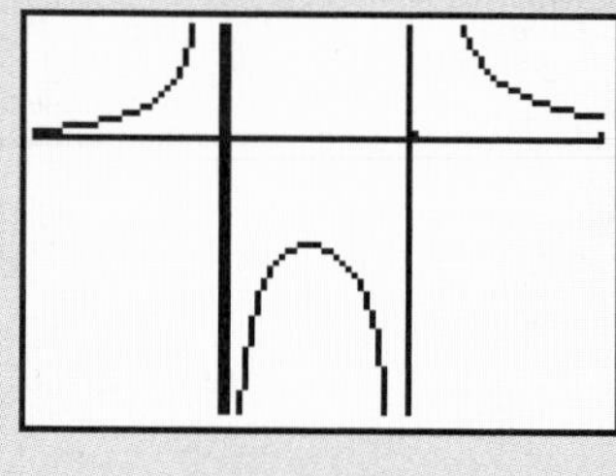

(b)

FIGURE 8-23

Graphs of nonlinear inequalities in two variables

We now consider the graphs of nonlinear inequalities in two variables.

EXAMPLE 5 ***Graphing nonlinear inequalities in two variables.*** Graph $y < -x^2 + 4$.

Solution

The graph of $y = -x^2 + 4$ is the parabolic boundary separating the region representing $y < -x^2 + 4$ and the region representing $y > -x^2 + 4$.

We graph the quadratic function $y = -x^2 + 4$ as a broken parabola, because equality is not permitted. Since the coordinates of the origin satisfy the inequality $y < -x^2 + 4$, the point (0, 0) is in the graph. The complete graph is shown in Figure 8-24.

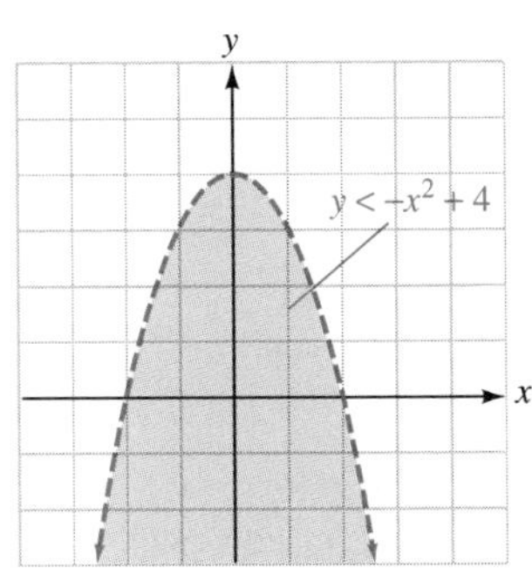

FIGURE 8-24

Self Check

Graph $y \geq -x^2 + 4$.

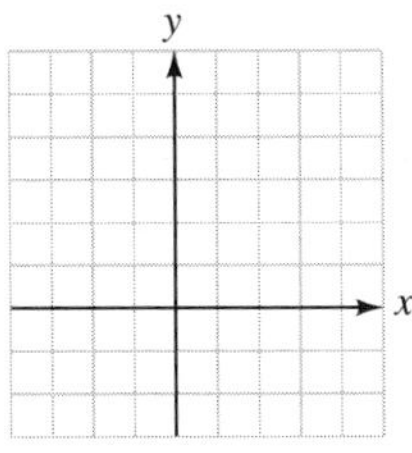

Answer:

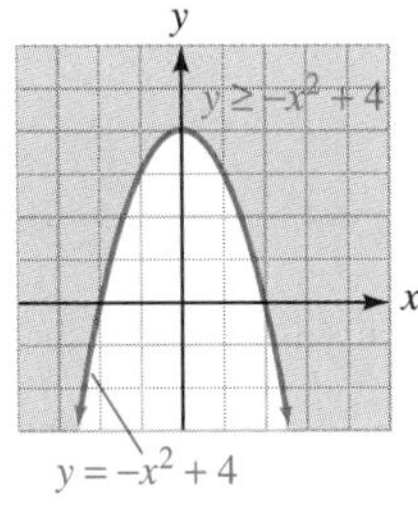

EXAMPLE 6 ***Graphing nonlinear inequalities in two variables.*** Graph $x \leq |y|$.

Solution

We first graph $x = |y|$ as in Figure 8-25(a). We use a solid line, because equality is permitted. Because the origin is on the graph, we cannot use the origin as a test point. However, any another point, such as (1, 0), will do. We substitute 1 for x and 0 for y into the inequality to get

$$x \leq |y|$$
$$1 \leq |0|$$
$$1 \leq 0$$

Since $1 \leq 0$ is a false statement, the point (1, 0) does not satisfy the inequality and is not part of the graph. Thus, the graph of $x \leq |y|$ is to the left of the boundary.

The complete graph is shown in Figure 8-25(b).

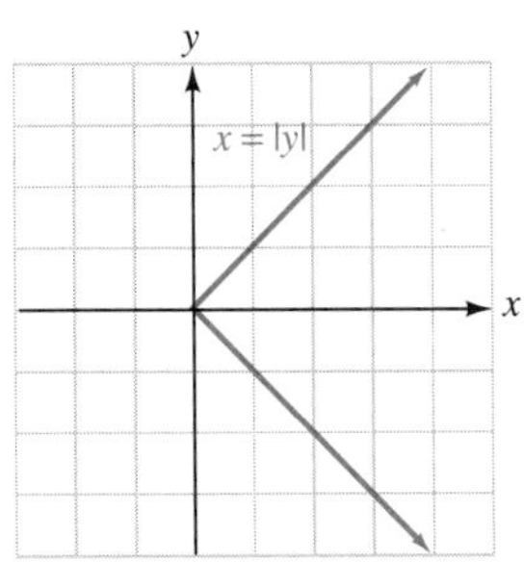

(a)

(b)

FIGURE 8-25

Self Check

Graph $x \geq -|y|$.

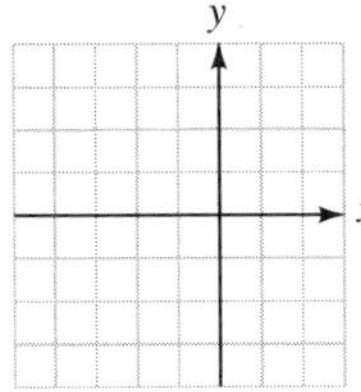

Answer:

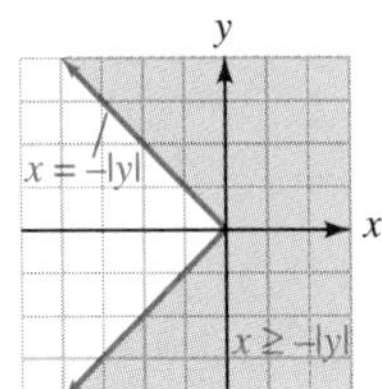

STUDY SET Section 8.6

VOCABULARY *In Exercises 1–4, fill in the blanks to make the statements true.*

1. Any inequality of the form $ax^2 + bx + c > 0$ $(a \neq 0)$ is called a ______________ inequality.

2. The inequality $y < x^2 - 2x + 3$ is a nonlinear inequality in ______________ variables.

3. The ______________ (3, 5) represents the real numbers between 3 and 5.

4. To decide which side of the boundary to shade when solving inequalities in two variables, we pick a ______________ point.

CONCEPTS *In Exercises 5–8, fill in the blanks to make the statements true.*

5. When $x > 3$, the binomial $x - 3$ is ______________ than zero.

6. When $x < 3$, the binomial $x - 3$ is ______________ than zero.

7. If $x = 0$, the fraction $\frac{1}{x}$ is ______________.

8. To keep track of the signs of factors in a product or quotient, we can use a ______________ chart.

NOTATION *In Exercises 9–10, consider the inequality* $(x + 2)(x - 3) > 0$.

9. Since the product of $x + 2$ and $x - 3$ is positive, the values of $x + 2$ and $x - 3$ are both positive or both negative.

a. $x + 2 = 0$ when $x = \square$

b. $x + 2 > 0$ when $x > \square$

c. $x + 2 < 0$ when $x < \square$

d. $x - 3 = 0$ when $x = \square$

e. $x - 3 > 0$ when $x > \square$

f. $x - 3 < 0$ when $x < \square$

10. Use the information in Exercise 9 to make a sign chart of the data and graph the solution set.

PRACTICE *In Exercises 11–38, solve each inequality. Give each result in interval notation and graph the solution set.*

11. $x^2 - 5x + 4 < 0$

12. $x^2 - 3x - 4 > 0$

13. $x^2 - 8x + 15 > 0$

14. $x^2 + 2x - 8 < 0$

15. $x^2 + x - 12 \leq 0$

16. $x^2 - 8x \leq -15$

17. $x^2 + 8x < -16$

18. $x^2 + 6x \geq -9$

19. $x^2 \geq 9$

20. $x^2 \geq 16$

21. $2x^2 - 50 < 0$

22. $3x^2 - 243 < 0$

23. $\frac{1}{x} < 2$

24. $\frac{1}{x} > 3$

25. $-\frac{5}{x} < 3$

26. $\frac{4}{x} \geq 8$

27. $\dfrac{x^2 - x - 12}{x - 1} < 0$

28. $\dfrac{x^2 + x - 6}{x - 4} \geq 0$

29. $\dfrac{6x^2 - 5x + 1}{2x + 1} > 0$

30. $\dfrac{6x^2 + 11x + 3}{3x - 1} < 0$

31. $\dfrac{3}{x - 2} < \dfrac{4}{x}$

32. $\dfrac{-6}{x + 1} \geq \dfrac{1}{x}$

33. $\dfrac{7}{x - 3} \geq \dfrac{2}{x + 4}$

34. $\dfrac{-5}{x - 4} < \dfrac{3}{x + 1}$

35. $\dfrac{x}{x + 4} \leq \dfrac{1}{x + 1}$

36. $\dfrac{x}{x + 9} \geq \dfrac{1}{x + 1}$

37. $(x + 2)^2 > 0$

38. $(x - 3)^2 < 0$

In Exercises 39–42, use a graphing calculator to solve each inequality. Give the answer in interval notation.

39. $x^2 - 2x - 3 < 0$

40. $x^2 + x - 6 > 0$

41. $\dfrac{x + 3}{x - 2} > 0$

42. $\dfrac{3}{x} < 2$

In Exercises 43–50, graph each inequality.

43. $y < x^2 + 1$

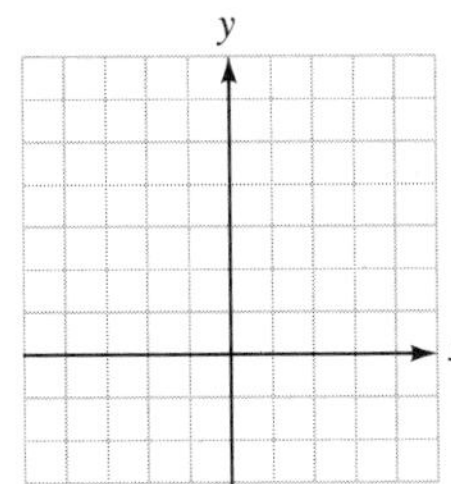

44. 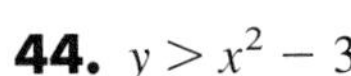 $y > x^2 - 3$

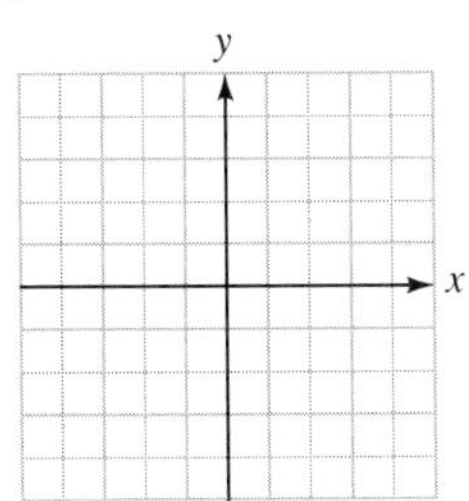

45. $y \leq x^2 + 5x + 6$

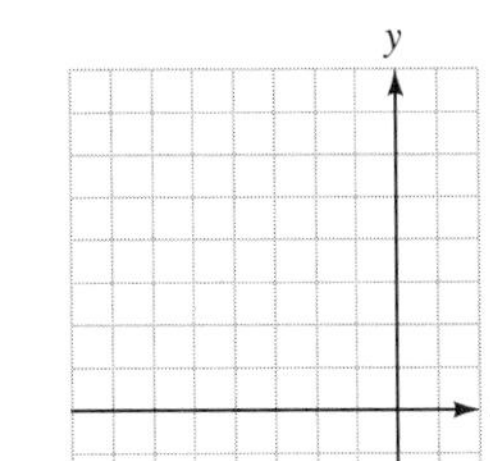

46. $y \geq x^2 + 5x + 4$

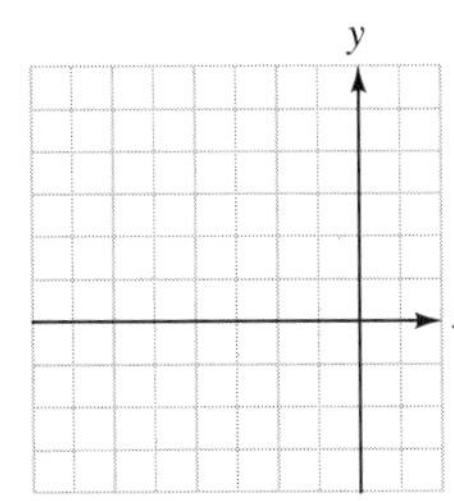

47. $y < |x + 4|$

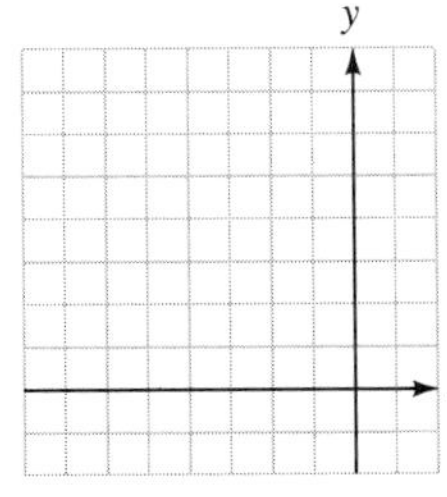

48. 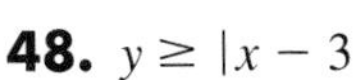 $y \geq |x - 3|$

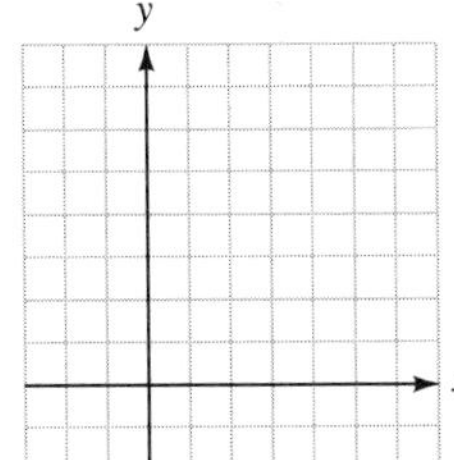

49. $y \leq -|x| + 2$

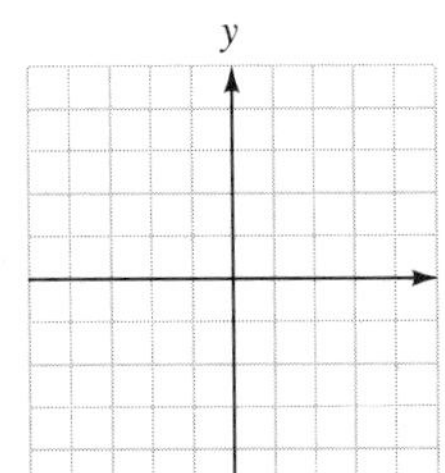

50. $y > |x| - 2$

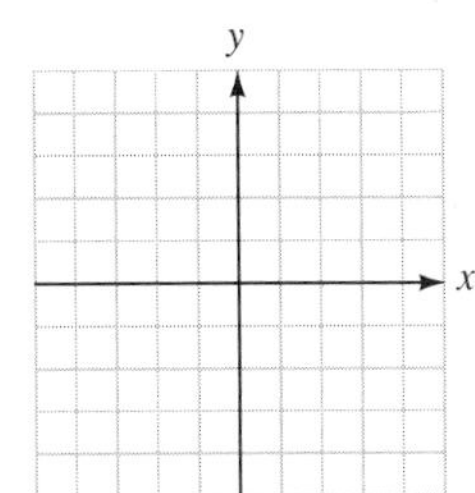

APPLICATIONS

51. SUSPENSION BRIDGE If an x-axis is superimposed over the roadway of the Golden Gate Bridge, with the origin at the center of the bridge as shown in Illustration 1 on the next page, the length L in feet of a vertical support cable can be approximated by the formula

$$L = \frac{1}{9{,}000}x^2 + 5$$

For the Golden Gate Bridge, $-2{,}100 < x < 2{,}100$. For what intervals along the x-axis are the vertical cables more than 95 feet long?

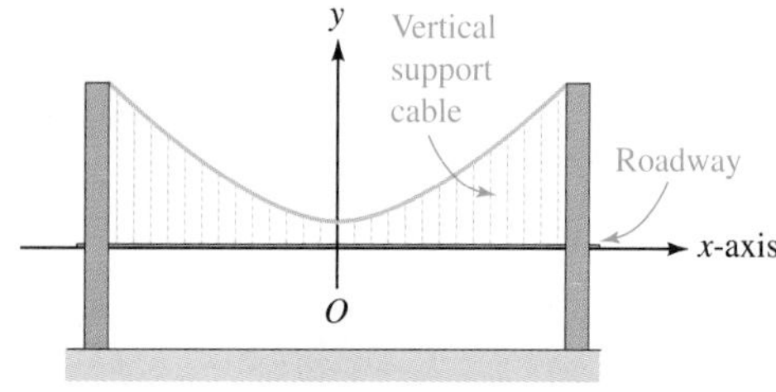

ILLUSTRATION 1

52. MALL The number of people n in a mall is modeled by the formula

$$n = -100x^2 + 1{,}200x$$

where x is the number of hours since the mall opened. If the mall opened at 9 A.M., when were there 2,000 or more people in it?

WRITING *Write a paragraph using your own words.*

53. Explain why $(x - 4)(x + 5)$ will be positive only when the signs of $x - 4$ and $x + 5$ are the same.

54. Explain how to find the graph of $y \geq x^2$.

REVIEW *In Exercises 55–58, write each expression as an equation.*

55. x varies directly with y.

56. y varies inversely with t.

57. t varies jointly with x and y.

58. d varies directly with t and inversely with u^2.

Find the slope of the graph of each linear function.

59. $f(x) = 3x - 4$

60. $f(x) = -x$

KEY CONCEPT

Solving Quadratic Equations

We have discussed five methods for solving **quadratic equations.** Let's review each of them and list an advantage and a drawback of each method.

Factoring

- It can be very quick and simple if the factoring pattern is evident.
- Much of the time, $ax^2 + bx + c$ cannot be factored or is not easily factored.

In Exercises 1–3, solve the equation by factoring.

1. $4k^2 + 8k = 0$ **2.** $z^2 + 8z + 15 = 0$ **3.** $2r^2 + 5r = -3$

The Square Root Method

- If the equation can be written in the form $x^2 = a$ or $(x + d)^2 = a$, where a is a constant, the square root method is a fast method, requiring few computations.
- Most quadratic equations that we must solve are not written in either of these forms.

In Exercises 4–6, solve the equation by the square root method.

4. $u^2 = 24$ **5.** $(s - 7)^2 - 9 = 0$ **6.** $3x^2 - 16 = 0$

Completing the Square

- It can be used to solve any quadratic equation.
- Most often, it involves more steps than the other methods.

In Exercises 7–9, solve the equation by completing the square.

7. $x^2 + 10x - 7 = 0$ **8.** $4x^2 - 4x - 1 = 0$ **9.** $x^2 + 2x + 2 = 0$

The Quadratic Formula

- It simply involves an evaluation of the expression $\dfrac{-b \pm \sqrt{b^2 - 4ac}}{2a}$.
- If applicable, the factoring method and the square root method are usually faster.

In Exercises 10–12, solve the equation by using the quadratic formula.

10. $2x^2 - 1 = 3x$ **11.** $x^2 - 6x - 391 = 0$

12. $3x^2 + 2x + 1 = 0$

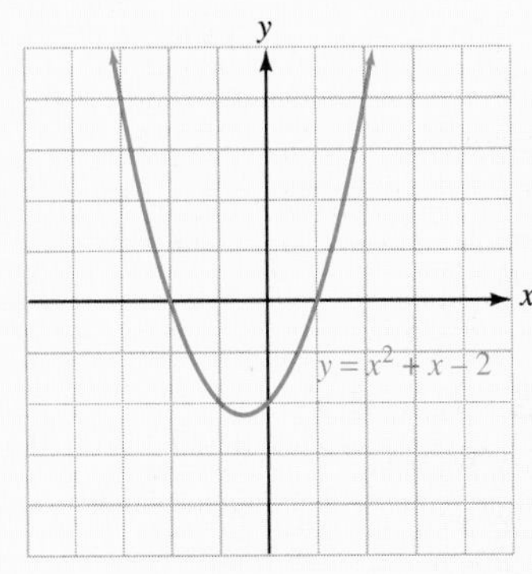

The Graphing Method

- We can solve the equation using a graphing calculator. It doesn't require any computations.
- It usually only gives approximations of the solutions.

13. Use the graph of $y = x^2 + x - 2$ to solve the equation $x^2 + x - 2 = 0$.

Accent on Teamwork

Section 8.1

PICTURE FRAMES Each person in your group is to make an 8 in. × 10 in. collage that illustrates an important aspect of his or her life. Photographs, magazine pictures, drawings, and the like can be used to highlight family, hobbies, jobs, pets, etc. Assign each person the task of making a matting of uniform width to frame his or her collage. Some can make a matting whose area equals that of the picture. Others can make a matting that is half the area, or double the area. When completed, let each person briefly explain his or her collage to the other group members. See Example 8, Section 8.1, for some hints on how to do the mathematics.

Section 8.2

TEAM RELAY Have your group make a presentation to the class showing the derivation of the quadratic formula. (See page 542 of the text.) Begin with the general quadratic equation, $ax^2 + bx + c = 0$, and have each member of your group perform and explain several steps. Make a "baton" from a paper towel roll, like that shown in Illustration 1. As one student finishes with his or her segment of the derivation, the baton is passed to the next student to pick up where he or she left off.

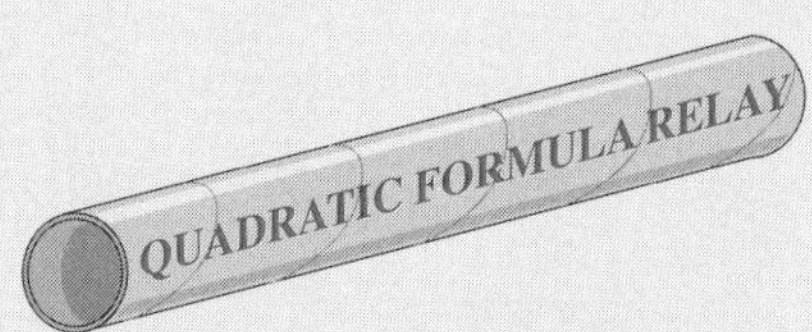

ILLUSTRATION 1

Section 8.3

DETERMINING THE NUMBER OF SOLUTIONS GRAPHICALLY Experiment with various combinations of a's, b's, and c's to write quadratic functions of the form $f(x) = ax^2 + bx + c$ that, when graphed using a calculator, intersect the x-axis **a.** two times, **b.** one time, and **c.** no times. Then give the associated quadratic equation in each case and tell how many real-number solutions the equation has.

Section 8.4

COMPLEX NUMBERS Write a report about complex numbers that answers these questions: When were they first used and by whom? Why were they "invented"? What are some of their applications?

A book about the history of mathematics will be helpful. The school library or perhaps one of your mathematics instructors may have one you can borrow. Some colleges offer a course about complex numbers called Complex Variables. See if you can get a copy of the textbook. Germany once issued a postage stamp honoring Carl Gauss and featuring complex numbers. See if you can find a picture of it. Physics instructors or electronics instructors are two other possible resources for material.

Section 8.5

SOLUTIONS OF A QUADRATIC EQUATION Consider the following property about the solutions of a quadratic equation: If

$$r_1 + r_2 = -\frac{b}{a} \quad \text{and} \quad r_1 r_2 = \frac{c}{a}$$

then r_1 and r_2 are the solutions of a quadratic equation $ax^2 + bx + c = 0$, with $a \neq 0$.

Use this property to show that

a. $\frac{3}{2}$ and $-\frac{1}{3}$ are solutions of $6x^2 - 7x - 3 = 0$.

b. $\sqrt{51}i$ and $-\sqrt{51}i$ are solutions of $x^2 + 51 = 0$.

c. $1 + 3\sqrt{2}$ and $1 - 3\sqrt{2}$ are solutions of $x^2 - 2x - 17 = 0$.

Section 8.6

QUADRATIC INEQUALITIES The fraction shown below has two factors in the numerator and two factors in the denominator.

$$\frac{(x-1)(x+4)}{(x+2)(x+1)}$$

a. With the four factors in mind, under what conditions will the fraction be positive?

b. With the four factors in mind, under what conditions will the fraction be negative?

c. Solve $\frac{(x-1)(x+4)}{(x+2)(x+1)} > 0$.

CHAPTER REVIEW

SECTION 8.1 Completing the Square

CONCEPTS

The square root property:
If $c > 0$, the equation $x^2 = c$ has two real solutions:

$$x = \sqrt{c} \text{ and } x = -\sqrt{c}$$

To *complete the square:*
1. Make sure the coefficient of x^2 is 1.
2. Make sure the constant term is on the right-hand side of the equation.
3. Add the square of one-half of the coefficient of x to both sides.
4. Factor the trinomial.
5. Use the square root property.

REVIEW EXERCISES

1. Solve each equation by factoring or using the square root property.

a. $x^2 + 9x + 20 = 0$ **b.** $6x^2 + 17x + 5 = 0$
c. $x^2 = 28$ **d.** $(t + 2)^2 = 36$
e. $5a^2 + 11a = 0$ **f.** $5x^2 - 49 = 0$

2. What number must be added to $x^2 - x$ to make it a perfect square?

3. Solve each equation by completing the square.

a. $x^2 + 6x + 8 = 0$ **b.** $2x^2 - 6x + 3 = 0$

4. HAPPY NEW YEAR As part of a New Year's Eve celebration, a huge ball is to be dropped from the top of a 605-foot tall building at the proper moment so that it strikes the ground at exactly 12:00 midnight. The distance d in feet traveled by a free-falling object in t seconds is given by the formula $d = 16t^2$. To the nearest second, when should the ball be dropped from the building?

SECTION 8.2 The Quadratic Formula

The quadratic formula:
The solutions of $ax^2 + bx + c = 0$ are given by

$$x = \frac{-b \pm \sqrt{b^2 - 4ac}}{2a} \quad (a \neq 0)$$

5. Solve each equation using the quadratic formula.

a. $-x^2 + 10x - 18 = 0$ **b.** $x^2 - 10x = 0$
c. $2x^2 + 13x = 7$ **d.** $26y - 3y^2 = 2$

6. SPORTS POSTER The design specifications for a poster of tennis star Steffi Graf call for a 615-square-inch photograph to be surrounded by a blue border. (See Illustration 1.) The borders on the sides of the poster are to be half as wide as those at the top and bottom. Find the width of each border.

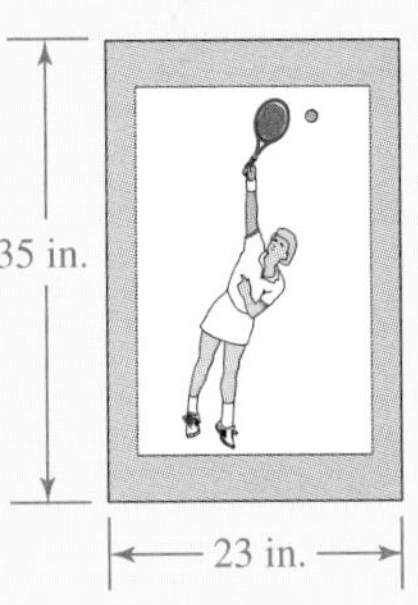

ILLUSTRATION 1

ILLUSTRATION 2

7. ACROBATS To begin his routine on a trapeze, an acrobat is catapulted upward as shown in Illustration 2. His distance d (in feet) from the arena floor during this maneuver is given by the formula $d = -16t^2 + 40t + 5$, where t is the time in seconds since being launched. If the trapeze bar is 25 feet in the air, at what two times will he be able to grab it? Round to the nearest tenth.

SECTION 8.3 Quadratic Functions and Their Graphs

A *quadratic function* is a second-degree polynomial function of the form $f(x) = ax^2 + bx + c$.

The graph of $f(x) = ax^2$ is a *parabola* opening upward when $a > 0$ and downward when $a < 0$, with *vertex* at the point $(0, 0)$ and *axis of symmetry* the line $x = 0$.

Each of the following functions has a graph that is the same shape as $f(x) = ax^2$ but involves a vertical or horizontal translation.

1. $f(x) = ax^2 + c$: translated upward if $c > 0$, downward if $c < 0$.
2. $f(x) = a(x - h)^2$: translated right if $h > 0$, and left if $h < 0$.

If $a \neq 0$, the graph of $f(x) = a(x - h)^2 + k$ is a parabola with vertex at (h, k). It opens upward when $a > 0$ and downward when $a < 0$.

The vertex of the graph of $f(x) = ax^2 + bx + c$ is

$$\left(-\frac{b}{2a}, f\left(-\frac{b}{2a}\right)\right)$$

and the axis of symmetry is the line $x = -\frac{b}{2a}$.

The vertex of the graph of a quadratic function gives the *minimum* or *maximum* value of the function.

8. AEROSPACE INDUSTRY See Illustration 3. The annual sales of the Boeing Company in billions of dollars for the years 1993–1997 can be modeled by the quadratic function

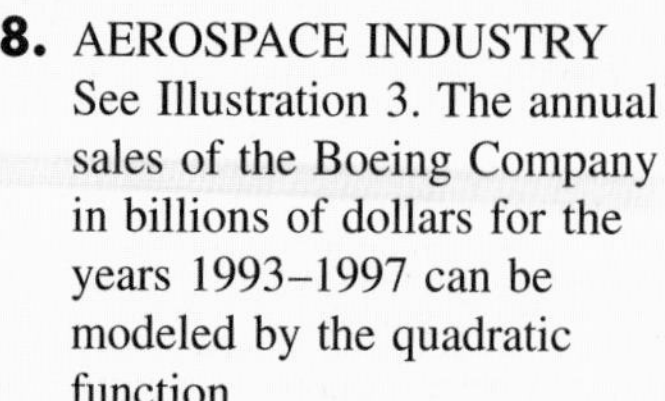

$$S(x) = 2.4375x^2 - 8.225x + 40.7$$

where x is the number of years since 1993. What were the annual sales for 1997?

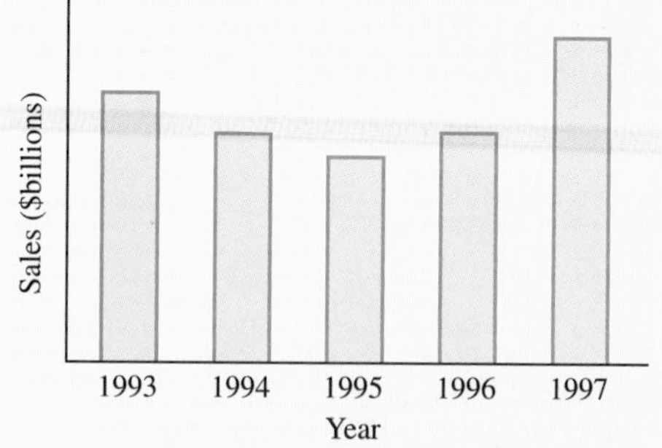

ILLUSTRATION 3

9. Make a table of values to graph function f. Then use a series of translations to graph function g on the same coordinate system.

a. $f(x) = 2x^2$
$g(x) = 2x^2 - 3$

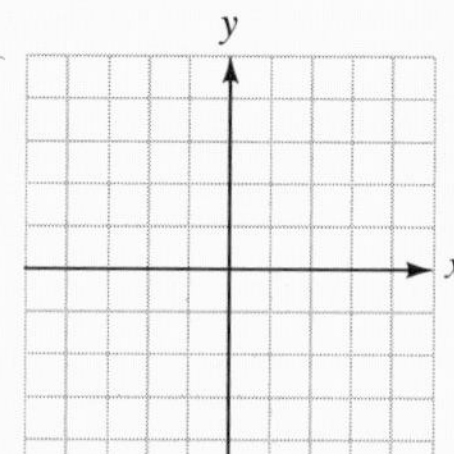

b. $f(x) = -4x^2$
$g(x) = -4(x - 2)^2 + 1$

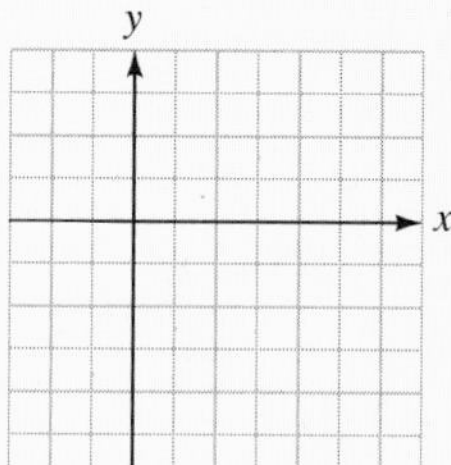

10. First determine the vertex and the axis of symmetry of the graph of each function. Then plot several points and complete the graph.

a. $y = -\left(x + \frac{3}{2}\right)^2 + \frac{5}{2}$

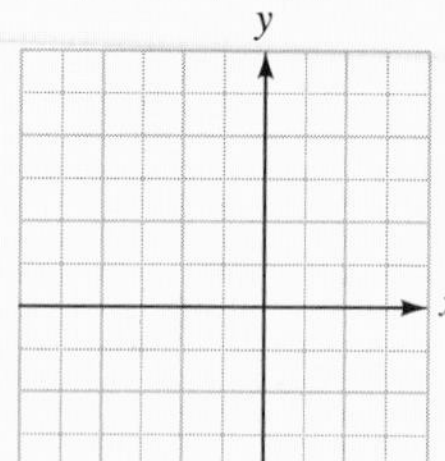

b. $y = 5x^2 + 10x - 1$

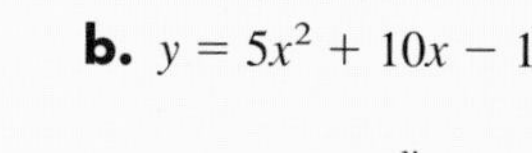

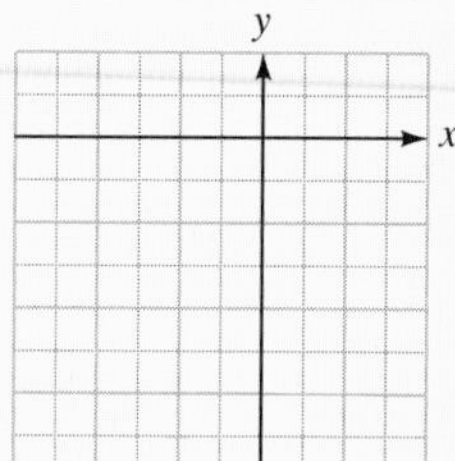

11. FARMING The number of farms in the United States for the years 1870–1970 is modeled by

$$N(x) = -1.46x^2 + 148.82x + 2{,}660$$

where $x = 0$ represents 1870, $x = 1$ represents 1871, and so on. For this period, when was the number of U.S. farms a maximum? How many farms were there?

SECTION 8.4 Complex Numbers

Square roots of negative numbers are called *imaginary numbers.*

$i = \sqrt{-1}$ and $i^2 = -1$

12. Simplify each expression

a. $\sqrt{-4}$ **b.** $\sqrt{-7}$

c. $-3\sqrt{-20}$ **d.** $\sqrt{-\frac{36}{49}}$

e. $\sqrt{-3}\sqrt{-3}$ **f.** $\frac{\sqrt{-64}}{\sqrt{-9}}$

A *complex number* is any number that can be written in the form $a + bi$, where a and b are real numbers and $i^2 = -1$.

13. Solve each equation.

a. $a^2 = -25$ **b.** $x^2 - 2x + 13 = 0$

Adding complex numbers:

$(a + bi) + (c + di) = (a + c) + (b + d)i$

Subtracting complex numbers:

$(a + bi) - (c + di) = (a - c) + (b - d)i$

Multiplying complex numbers:

$(a + bi)(c + di) = (ac - bd) + (ad + bc)i$

14. Do the operations and give all answers in $a + bi$ form.

a. $(5 + 4i) + (7 - 12i)$ **b.** $(-6 - 40i) - (-8 + 28i)$

c. $\left(-8 + \sqrt{-8}\right) + \left(6 - \sqrt{-32}\right)$ **d.** $2i(64 + 9i)$

e. $(2 - 7i)(-3 + 4i)$ **f.** $\left(5 - \sqrt{-27}\right)\left(-6 + \sqrt{-12}\right)$

To divide complex numbers, we often have to *rationalize* a denominator by multiplying the numerator and the denominator by the *complex conjugate* of the denominator.

15. Write each expression in $a + bi$ form.

a. $\frac{6}{2 + i}$ **b.** $\frac{4 + i}{4 - i}$

c. $\frac{\sqrt{3} + \sqrt{-4}}{\sqrt{3} - \sqrt{-4}}$ **d.** $\frac{-2}{5i^3}$

The *powers of i* rotate through a cycle of four numbers: $i = i$, $i^2 = -1$, $i^3 = -i$, $i^4 = 1$

16. Simplify each power of i.

a. i^{65} **b.** i^{48}

SECTION 8.5 The Discriminant and Equations That Can Be Written in Quadratic Form

The *discriminant* predicts the type of solutions of $ax^2 + bx + c = 0$:

1. If $b^2 - 4ac > 0$, the solutions are unequal real numbers.
2. If $b^2 - 4ac = 0$, the solutions are equal real numbers.
3. If $b^2 - 4ac < 0$, the solutions are complex conjugates.

17. Use the discriminant to determine what type of solutions exist for each equation.

a. $3x^2 + 4x - 3 = 0$

b. $4x^2 - 5x + 7 = 0$

c. $9x^2 - 12x + 4 = 0$

Many equations that are not quadratic can be written in quadratic form.

18. Solve each equation.

a. $x - 13\sqrt{x} + 12 = 0$

b. $a^{2/3} + a^{1/3} - 6 = 0$

c. $6x^4 - 19x^2 + 3 = 0$

d. $\dfrac{6}{x+2} + \dfrac{6}{x+1} = 5$

e. $(x-3)^2 - 8(x-3) + 7 = 0$

19. WEEKLY CHORES Working together, two brothers can do the yard work at their house in 45 minutes. When the older boy does it all himself, he can complete the job in 20 minutes less time than it takes the younger boy working alone. How long does it take the older boy to do the yard work?

SECTION 8.6 *Quadratic and Other Nonlinear Inequalities*

To graph a *quadratic inequality in one variable,* get 0 on the right-hand side. Then factor the polynomial on the left-hand side. Use a *sign chart* to determine the solution set.

20. Solve each inequality. Give each result in interval notation and graph the solution set.

a. $x^2 + 2x - 35 > 0$

b. $x^2 - 81 \le 0$

To solve rational inequalities, get 0 on the right-hand side and a single fraction on the left-hand side. Factor the numerator and denominator. Then use a sign chart to determine the solution set.

c. $\dfrac{3}{x} \le 5$

d. $\dfrac{2x^2 - x - 28}{x - 1} > 0$

21. Use a graphing calculator to solve each inequality. Compare the results with Exercise 20.

a. $x^2 + 2x - 35 > 0$

b. $\dfrac{2x^2 - x - 28}{x - 1} > 0$

To graph a *nonlinear inequality in two variables,* first graph the boundary. Then use a test point to determine which half-plane to shade.

22. Graph each inequality.

a. $y < \frac{1}{2}x^2 - 1$

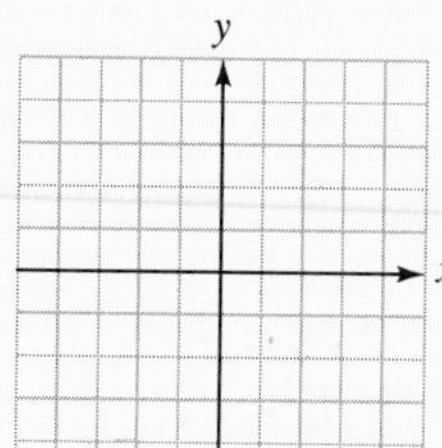

b. $y \ge -|x|$

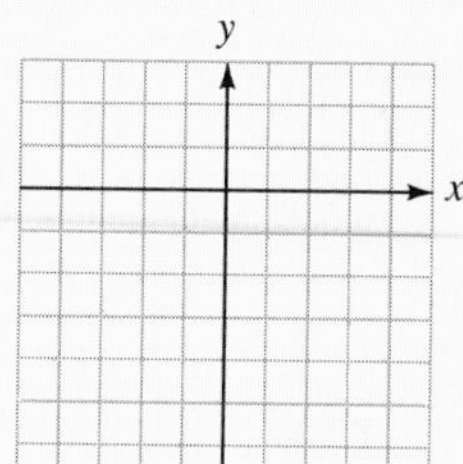

Chapter 8 Test

In Problems 1–2, solve each equation by factoring.

1. $3x^2 + 18x = 0$

2. $x(6x + 19) = -15$

3. Determine what number must be added to $x^2 + 24x$ to make it a perfect square.

4. Solve $x^2 - x - 1 = 0$ by completing the square.

5. Solve the equation $2x^2 - 8x = -5$ using the quadratic formula.

6. TABLECLOTH According to the *Guinness Book of World Records 1998,* the world's longest tablecloth was made in Illinois in 1990 and covered an area of 6,759 square feet. Its length was 3.5 feet more than 333 times its width. Find the dimensions of the tablecloth.

7. Simplify $\sqrt{-48}$.

8. Simplify i^{54}.

In Problems 9–14, do the operations. Give all answers in a + bi form.

9. $(2 + 4i) + (-3 + 7i)$

10. $\left(3 - \sqrt{-9}\right) - \left(-1 + \sqrt{-16}\right)$

11. $2i(3 - 4i)$

12. $(3 + 2i)(-4 - i)$

13. $\dfrac{1}{i^3}$

14. $\dfrac{2 + i}{3 - i}$

15. Determine whether the solutions of $3x^2 + 5x + 17 = 0$ are real or nonreal.

16. Solve $x^2 = -12$.

17. Solve the equation $13 = 4t - t^2$.

18. Solve the equation $2y - 3\sqrt{y} + 1 = 0$.

In Problems 19–20, determine the vertex and the axis of symmetry of the graph of the function. Then graph it.

19. $f(x) = 2x^2 + x - 1$

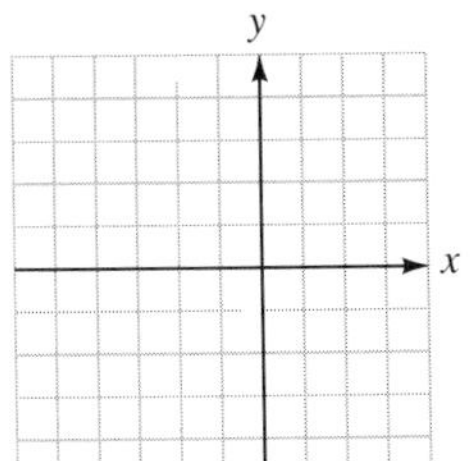

20. $y = -3(x - 1)^2 - 2$

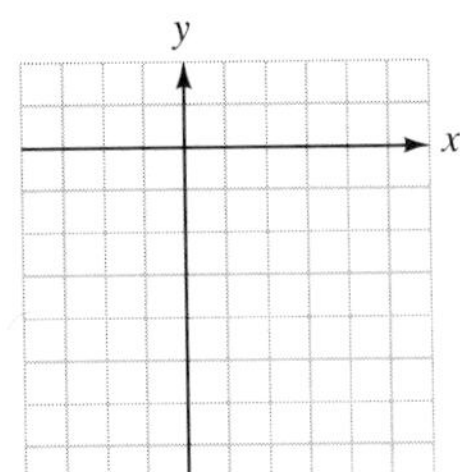

In Problems 21–22, solve the inequality and graph the solution set.

21. $x^2 - 2x - 8 > 0$

22. $\dfrac{x - 2}{x + 3} \leq 0$

23. Graph the inequality $y \leq -x^2 + 3$.

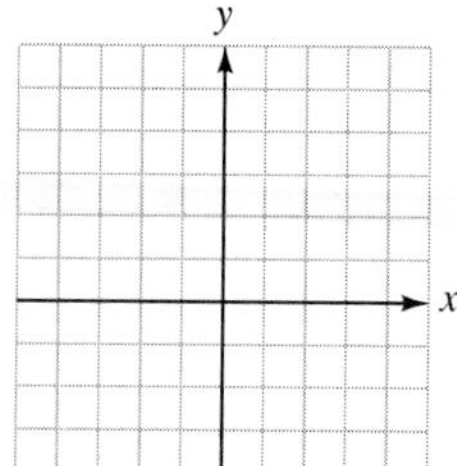

24. DRAWING An artist uses four equal-sized right triangles to block out a perspective drawing of an old hotel. See Illustration 1. For each triangle, one leg is 14 inches longer than the other, and the hypotenuse is 26 inches. On the centerline of the drawing, what is the length of the segment extending from the ground to the top of the building?

ILLUSTRATION 1

25. DISTRESS SIGNAL A flare was fired directly upward into the air from a boat that was experiencing engine problems. The height of the flare (in feet) above the water, t seconds after being fired, is given by the formula $h = -16t^2 + 112t + 15$. If the flare is designed to explode when it reaches its highest point, at what height will this occur?

26. What is an imaginary number? Give some examples.

Chapters 1–8 Cumulative Review Exercises

In Exercises 1–2, write the equation of the line with the given properties.

1. $m = 3$, passing through $(-2, -4)$

2. Parallel to the graph of $2x + 3y = 6$ and passing through $(0, -2)$

3. AIRPORT TRAFFIC From the graph in Illustration 1, determine the projected average rate of change in the number of take-offs and landings at Los Angeles International Airport for the years 2000–2015.

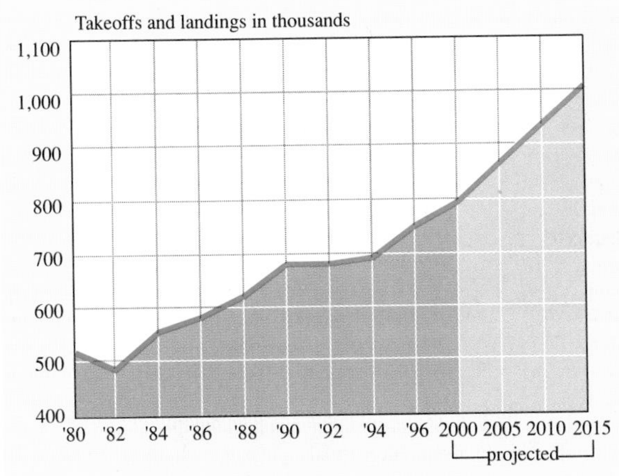

Based on data from *Los Angeles Times* (July 6, 1988) p. B3

ILLUSTRATION 1

4. Solve the system by graphing.

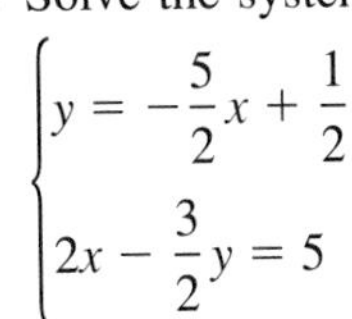

$$\begin{cases} y = -\frac{5}{2}x + \frac{1}{2} \\ 2x - \frac{3}{2}y = 5 \end{cases}$$

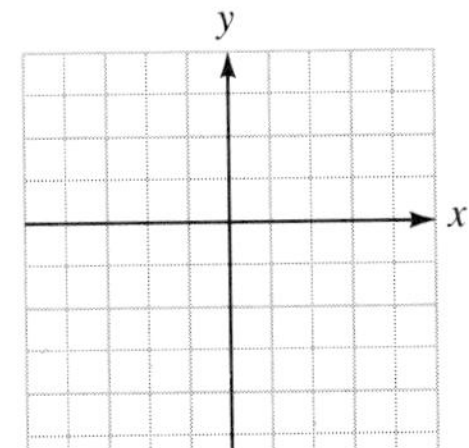

5. Solve the system using Cramer's rule.

$$\begin{cases} x - y + z = 4 \\ x + 2y - z = -1 \\ x + y - 3z = -2 \end{cases}$$

6. Graph the solution set of the system

$$\begin{cases} 3x + 2y > 6 \\ x + 3y \le 2 \end{cases}$$

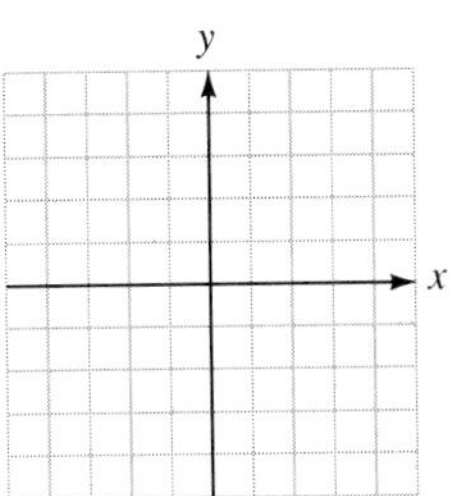

7. Solve the inequality $-9x + 6 > 16$. Give the result in interval notation and graph the solution set.

8. Solve $|2x - 5| \ge 25$. Give the result in interval notation and graph the solution set.

In Exercises 9–10, find the domain and range of each function.

9. $f(x) = 2x^2 - 3$

10. $y = -|x - 4|$

In Exercises 11–12, do each operation.

11. $(2a^2 + 4a - 7) - 2(3a^2 - 4a)$

12. $(3x + 2)(2x - 3)$

In Exercises 13–16, factor each expression.

13. $x^4 - 16y^4$

14. $15x^2 - 2x - 8$

15. $x^2 + 4y - xy - 4x$

16. $8x^6 + 125y^3$

In Exercises 17–20, solve each equation.

17. $x^2 - 5x - 6 = 0$

18. $6a^3 - 2a = a^2$

19. $\dfrac{x-4}{x-3} + \dfrac{x-2}{x-3} = x - 3$

20. $P + \dfrac{a}{V^2} = \dfrac{RT}{V - b}$ solve for b

In Exercises 21–22, simplify each expression.

21. $\dfrac{x^3 + y^3}{x^3 - y^3} \div \dfrac{x^2 - xy + y^2}{x^2 + xy + y^2}$

22. $\dfrac{1}{x+y} - \dfrac{1}{x-y} - \dfrac{2y}{y^2 - x^2}$

In Exercises 23–24, graph each function and give its domain and range.

23. $f(x) = x^3 + x^2 - 6x$

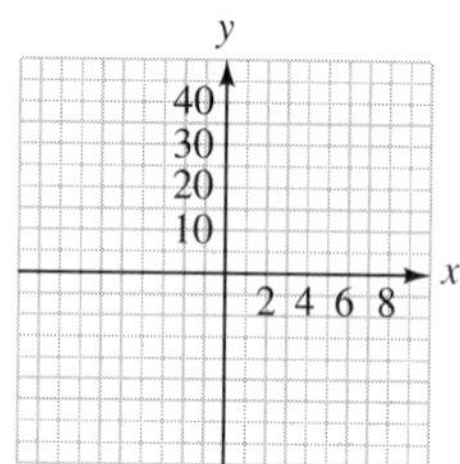

24. $f(x) = \dfrac{4}{x}$ for $x > 0$

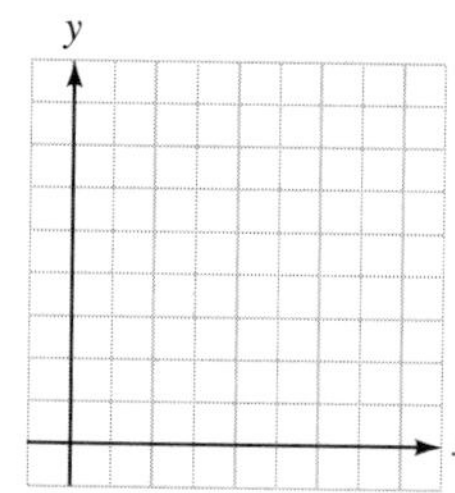

25. LIGHT As light energy radiates away from its source, its intensity varies inversely as the square of the distance from the source. Illustration 2 shows that the light energy passing through an area 1 foot from the source spreads out over 4 units of area two feet from the source. That energy is therefore less intense 2 feet from the source as it was 1 foot from the source. Over how many units of area will the light energy spread out 3 feet from the source?

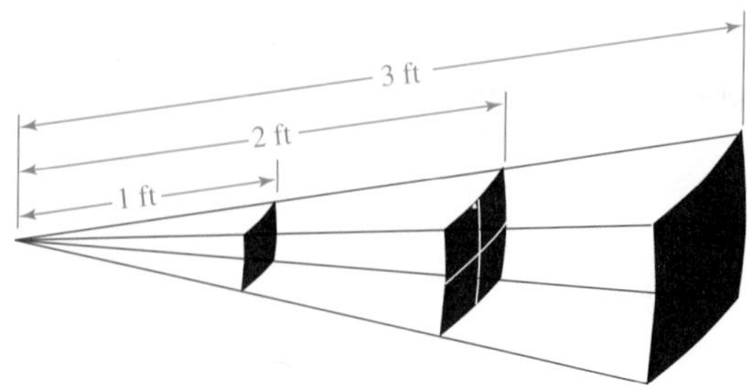

ILLUSTRATION 2

26. Graph the function $f(x) = \sqrt{x - 2}$ and give its domain and range.

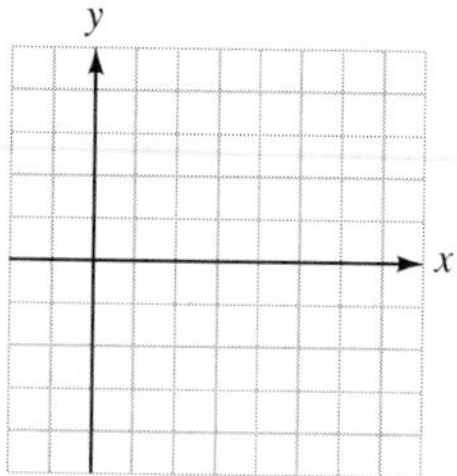

In Exercises 27–30, simplify each expression.

27. $\sqrt[3]{-27x^3}$

28. $\sqrt{48t^3}$

29. $64^{-2/3}$

30. $\dfrac{x^{5/3}x^{1/2}}{x^{3/4}}$

In Exercises 31–34, simplify each expression.

31. $-3\sqrt[4]{32} - 2\sqrt[4]{162} + 5\sqrt[4]{48}$

32. $3\sqrt{2}(2\sqrt{3} - 4\sqrt{12})$

33. $\dfrac{\sqrt{x} + 2}{\sqrt{x} - 1}$

34. $\dfrac{5}{\sqrt[3]{x}}$

In Exercises 35–36, solve each equation.

35. $5\sqrt{x + 2} = x + 8$

36. $\sqrt{x} + \sqrt{x + 2} = 2$

37. Find the length of the hypotenuse of the right triangle shown in Illustration 3.

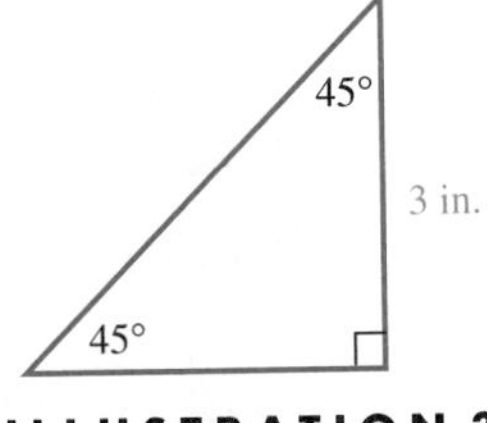

ILLUSTRATION 3

38. Find the length of the hypotenuse of the right triangle shown in Illustration 4.

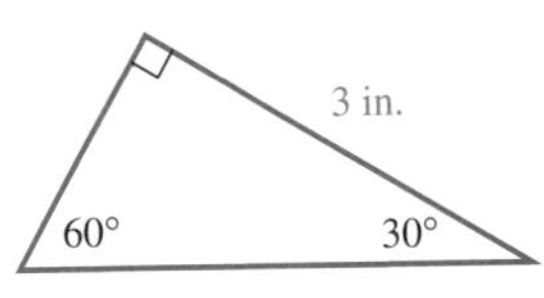

ILLUSTRATION 4

39. Find the distance between $P(-2, 6)$ and $Q(4, 14)$.

40. What number must be added to $x^2 + 6x$ to make a trinomial square?

41. Use the method of completing the square to solve the equation $2x^2 + x - 3 = 0$.

42. Use the quadratic formula to solve the equation $3x^2 + 4x - 1 = 0$.

43. Graph $y = \frac{1}{2}x^2 - x + 1$ and find the coordinates of its vertex.

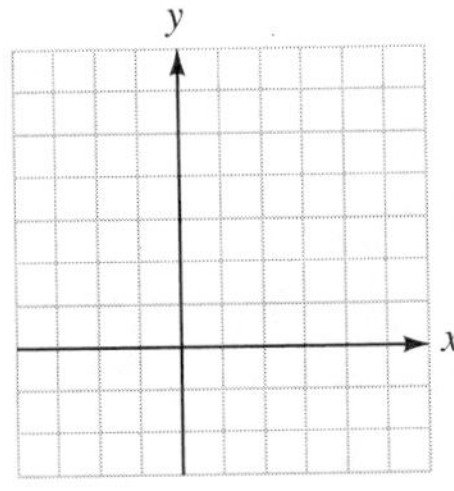

44. Graph $f(x) = -x^2 - 4x$ and find the coordinates of its vertex.

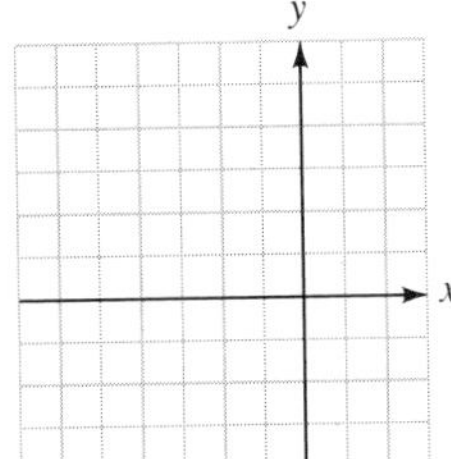

In Exercises 45–46, write each expression in $a + bi$ form.

45. $(3 + 5i) + (4 - 3i)$

46. $\dfrac{5}{3 - i}$

47. Simplify $\sqrt{-64}$.

48. Solve $a - 7a^{1/2} + 12 = 0$

Exponential and Logarithmic Functions

9

Campus Connection

The *Sociology* Department

In a sociology course, students analyze surveys and statistical data to make observations and predictions about human social relations. One area of great interest to sociologists is how population growth affects social structures and institutions, such as the family, the workplace, and the schools. Population growth can often be mathematically modeled by an *exponential function.* In this chapter, we will examine exponential functions and their graphs. We will also introduce an important irrational number that is often used when writing an exponential function to describe growth or decay. Such functions have application in fields as diverse as banking, medicine, and nuclear energy.

In this chapter, we discuss the concept of function in more depth. We also introduce two new families of functions—exponential and logarithmic functions—which have applications in many areas.

9.1 Algebra and Composition of Functions

In this section, you will learn about

- **Algebra of functions**
- **Composition of functions**
- **The identity function**
- **Writing composite functions**

Introduction. So far, we have defined functions with real-number domains and ranges and have graphed them on a rectangular coordinate system. Just as it is possible to perform operations on real numbers, it is also possible to perform operations on functions. We will begin this section by showing how to add, subtract, multiply, and divide functions. Then we will consider another method of combining functions, called *composition of functions*.

Algebra of functions

We now consider how functions can be added, subtracted, multiplied and divided.

Operations on functions

If the domains and ranges of functions f and g are subsets of the real numbers, then
The **sum** of f and g, denoted as $\mathbf{f + g}$, is defined by

$$(f + g)(x) = f(x) + g(x)$$

The **difference** of f and g, denoted as $\mathbf{f - g}$, is defined by

$$(f - g)(x) = f(x) - g(x)$$

The **product** of f and g, denoted as $\mathbf{f \cdot g}$, is defined by

$$(f \cdot g)(x) = f(x)g(x)$$

The **quotient** of f and g, denoted as $\mathbf{f/g}$, is defined by

$$(f/g)(x) = \frac{f(x)}{g(x)} \quad (g(x) \neq 0)$$

The domain of each of these functions is the set of real numbers x that are in the domain of both f and g. In the case of the quotient, there is the further restriction that $g(x) \neq 0$.

EXAMPLE 1 ***Adding and subtracting functions.*** Let $f(x) = 2x^2 + 1$ and $g(x) = 5x - 3$. Find each function and its domain: **a.** $f + g$ and **b.** $f - g$.

Solution

a. $(f + g)(x) = f(x) + g(x)$

$= (2x^2 + 1) + (5x - 3)$

$= 2x^2 + 5x - 2$ Combine like terms.

The domain of $f + g$ is the set of real numbers that are in the domain of both f and g. Since the domain of both f and g is the interval $(-\infty, \infty)$, the domain of $f + g$ is also the interval $(-\infty, \infty)$.

b. $(f - g)(x) = f(x) - g(x)$

$= (2x^2 + 1) - (5x - 3)$

$= 2x^2 + 1 - 5x + 3$ Remove parentheses.

$= 2x^2 - 5x + 4$ Combine like terms.

Since the domain of both f and g is $(-\infty, \infty)$, the domain of $f - g$ is also the interval $(-\infty, \infty)$.

Self Check

Let $f(x) = 3x - 2$ and $g(x) = 2x^2 + 3x$. Find **a.** $f + g$ and **b.** $f - g$.

Answers: **a.** $2x^2 + 6x - 2$, **b.** $-2x^2 - 2$ ■

EXAMPLE 2 ***Multiplying and dividing functions.*** Let $f(x) = 2x^2 + 1$ and $g(x) = 5x - 3$. Find each function and its domain: **a.** $f \cdot g$ and **b.** f/g.

Solution

a. $(f \cdot g)(x) = f(x)g(x)$

$= (2x^2 + 1)(5x - 3)$

$= 10x^3 - 6x^2 + 5x - 3$ Use the FOIL method.

The domain of $f \cdot g$ is the set of real numbers that are in the domain of both f and g. Since the domain of both f and g is the interval $(-\infty, \infty)$, the domain of $f \cdot g$ is also the interval $(-\infty, \infty)$.

b. $(f/g)(x) = \dfrac{f(x)}{g(x)}$

$= \dfrac{2x^2 + 1}{5x - 3}$

Since the denominator of the fraction cannot be 0, $x \neq \frac{3}{5}$. Thus, the domain of f/g is the interval $\left(-\infty, \frac{3}{5}\right) \cup \left(\frac{3}{5}, \infty\right)$.

Self Check

Let $f(x) = 2x^2 - 3$ and $g(x) = x^2 + 1$. Find **a.** $f \cdot g$ and **b.** f/g.

Answers: **a.** $2x^4 - x^2 - 3$, **b.** $\dfrac{2x^2 - 3}{x^2 + 1}$ ■

Composition of functions

We have seen that a function can be represented by a machine: We put in a number from the domain and a number from the range comes out. For example, if we put the number 2 into the machine shown in Figure 9-1(a), the number $f(2) = 8$ comes out. In general, if we put x into the machine shown in Figure 9-1(b), the value $f(x)$ comes out.

Often one quantity is a function of a second quantity that depends, in turn, on a third quantity. For example, the cost of a car trip is a function of the gasoline consumed. The amount of gasoline consumed, in turn, is a function of the number of miles driven. Such chains of dependence can be analyzed mathematically as **compositions of functions.**

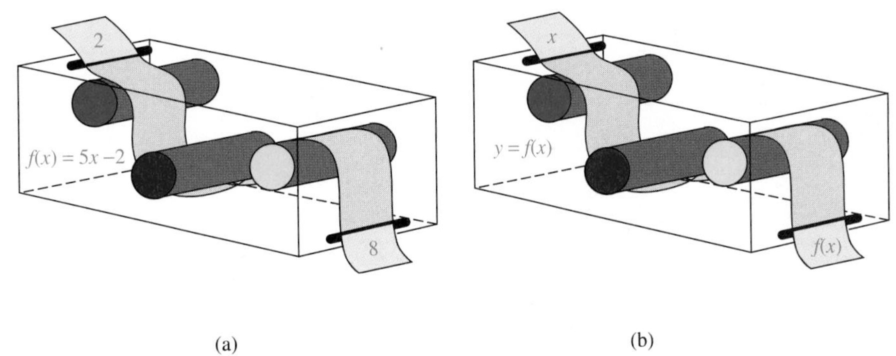

(a) (b)

FIGURE 9-1

Suppose that $y = f(x)$ and $y = g(x)$ define two functions. Any number x in the domain of g will produce the corresponding value $g(x)$ in the range of g. If $g(x)$ is in the domain of function f, then $g(x)$ can be substituted into f, and a corresponding value $f(g(x))$ will be determined. This two-step process defines a new function, called a **composite function,** denoted by $f \circ g$. (This is read as "f composed with g.")

The function machines shown in Figure 9-2 illustrate the composition $f \circ g$. When we put a number x into the function g, $g(x)$ comes out. The value $g(x)$ goes into function f, which transforms $g(x)$ into $f(g(x))$. (This is read as "f of g of x.") If the function machines for g and f were connected to make a single machine, that machine would be named $f \circ g$.

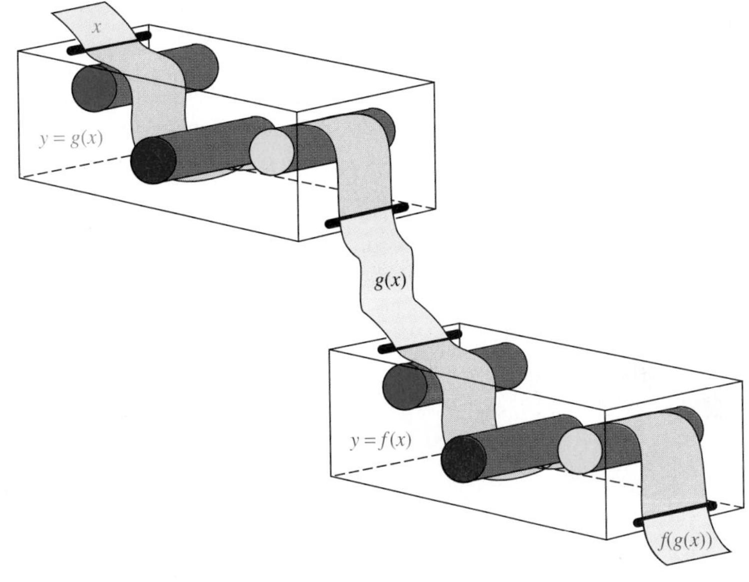

FIGURE 9-2

To be in the domain of the composite function $f \circ g$, a number x has to be in the domain of g. Also, the output of g must be in the domain of f. Thus, the domain of $f \circ g$ consists of those numbers x that are in the domain of g, and for which $g(x)$ is in the domain of f.

Composite functions

The **composite function $f \circ g$** is defined by

$$(f \circ g)(x) = f(g(x))$$

For example, if $f(x) = 4x$ and $g(x) = 3x + 2$, then

$$\begin{aligned}(f \circ g)(x) &= f(g(x)) \\ &= f(3x + 2) \\ &= 4(3x + 2) \\ &= 12x + 8\end{aligned} \qquad \text{or} \qquad \begin{aligned}(g \circ f)(x) &= g(f(x)) \\ &= g(4x) \\ &= 3(4x) + 2 \\ &= 12x + 2\end{aligned}$$

WARNING! Note that in the previous example, $(f \circ g)(x) \neq (g \circ f)(x)$. This shows that the composition of functions is not commutative.

EXAMPLE 3 ***Evaluating composite functions.*** Let $f(x) = 2x + 1$ and $g(x) = x - 4$. Find **a.** $(f \circ g)(9)$, **b.** $(f \circ g)(x)$, and **c.** $(g \circ f)(-2)$.

Solution **a.** $(f \circ g)(9)$ means $f(g(9))$. In Figure 9-3(a), function g receives the number 9, subtracts 4, and releases the number $g(9) = 5$. Then 5 goes into the f function, which doubles 5 and adds 1. The final result, 11, is the output of the composite function $f \circ g$:

$$(f \circ g)(9) = f(g(9)) = f(5) = 2(5) + 1 = 11$$

b. $(f \circ g)(x)$ means $f(g(x))$. In Figure 9-3(a), function g receives the number x, subtracts 4, and releases the number $x - 4$. Then $x - 4$ goes into the f function, which doubles $x - 4$ and adds 1. The final result, $2x - 7$, is the output of the composite function $f \circ g$.

$$(f \circ g)(x) = f(g(x)) = f(x - 4) = 2(x - 4) + 1 = 2x - 7$$

c. $(g \circ f)(-2)$ means $g(f(-2))$. In Figure 9-3(b), function f receives the number -2, doubles it and adds 1, and releases -3 into the g function. Function g subtracts 4 from -3 and outputs a final result of -7. Thus,

$$(g \circ f)(-2) = g(f(-2)) = g(-3) = -3 - 4 = -7$$

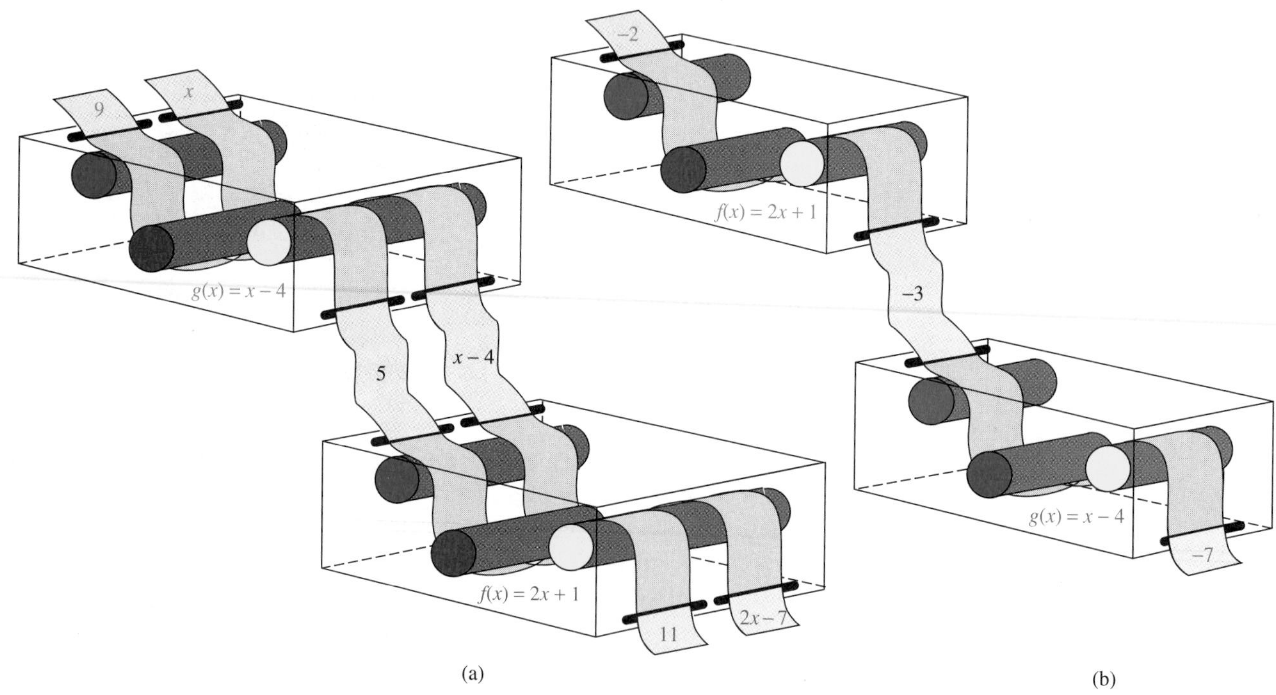

FIGURE 9-3

The identity function

The **identity function** is defined by the equation $I(x) = x$. Under this function, the value that corresponds to any real number x is x itself. For example $I(2) = 2$, $I(-3) = -3$, and $I(7.5) = 7.5$. If f is any function, the composition of f with the identity function is just the function f:

$$(f \circ I)(x) = (I \circ f)(x) = f(x)$$

EXAMPLE 4 ***The identity function.*** Let f be any function and let I be the identity function, $I(x) = x$. Show that **a.** $(f \circ I)(x) = f(x)$ and **b.** $(I \circ f)(x) = f(x)$.

Solution **a.** $(f \circ I)(x)$ means $f(I(x))$. Because $I(x) = x$, we have

$$(f \circ I)(x) = f(I(x)) = f(x)$$

b. $(I \circ f)(x)$ means $I(f(x))$. Because I passes any number through unchanged, we have $I(f(x)) = f(x)$ and

$$(I \circ f)(x) = I(f(x)) = f(x)$$

■

Writing composite functions

EXAMPLE 5 ***Biological research.*** A laboratory specimen is stored in a refrigeration unit at a temperature of 15° Fahrenheit. Biologists remove the specimen and warm it at a controlled rate of 3° F per hour. Express the sample's Celsius temperature as a function of the time t since it was removed from refrigeration.

Solution The temperature of the specimen is 15° F when the time $t = 0$. Because it warms at a rate of 3° F per hour, its initial temperature of 15° increases by $3t$° F in t hours. The Fahrenheit temperature of the specimen is given by the function

$$F(t) = 3t + 15$$

The Celsius temperature C is a function of this Fahrenheit temperature F, given by the function

$$C(F) = \frac{5}{9}(F - 32)$$

To express the specimen's Celsius temperature as a function of *time,* we find the composite function

$$\begin{aligned}
(C \circ F)(t) &= C(F(t)) \\
&= C(3t + 15) && \text{Substitute } 3t + 15 \text{ for } F(t). \\
&= \frac{5}{9}[(3t + 15) - 32] && \text{Substitute } 3t + 15 \text{ for } F \text{ in } \tfrac{5}{9}(F - 32). \\
&= \frac{5}{9}(3t - 17) && \text{Simplify.}
\end{aligned}$$

The composite function, $C(t) = \frac{5}{9}(3t - 17)$, finds the temperature of the specimen in degrees Celsius t hours after it is removed from refrigeration. ■

STUDY SET Section 9.1

VOCABULARY *In Exercises 1–8, fill in the blanks to make the statements true.*

1. The ______________ of f and g, denoted as $f + g$, is defined by $(f + g)(x) = \square$.

2. The ______________ of f and g, denoted as $f - g$, is defined by $(f - g)(x) = \square$.

3. The ______________ of f and g, denoted as $f \cdot g$, is defined by $(f \cdot g)(x) = \square$.

4. The ______________ of f and g, denoted as f/g, is defined by $(f/g)(x) = \square$.

5. In Exercises 1–3, the ______________ of each function is the set of real numbers x that are in the domain of both f and g.

6. The ______________ function $f \circ g$ is defined by $(f \circ g)(x) = f(g(x))$.

7. Under the ______________ function, the value that corresponds to any real number x is x itself.

8. When reading the notation $f(g(x))$, we say "f ___ g ___ x."

CONCEPTS

9. Fill in the blanks to make the statements true.
 a. $(f \circ g)(x) = f(\square)$
 b. To find $f(g(x))$, we must find ______________ and then substitute that value for x in $f(x)$.

10. a. If $f(x) = 3x + 1$ and $g(x) = 1 - 2x$. find $f(g(3))$ and $g(f(3))$.
 b. Is the composition of functions commutative? Explain.

11. Complete the table of values for the identity function, $I(x) = x$. Then graph it.

x	$I(x)$
−3	
−2	
−1	
0	
1	
2	
3	

12. Fill in the blanks in the drawing of the function machines in Illustration 1 that show how to compute $g(f(-2))$.

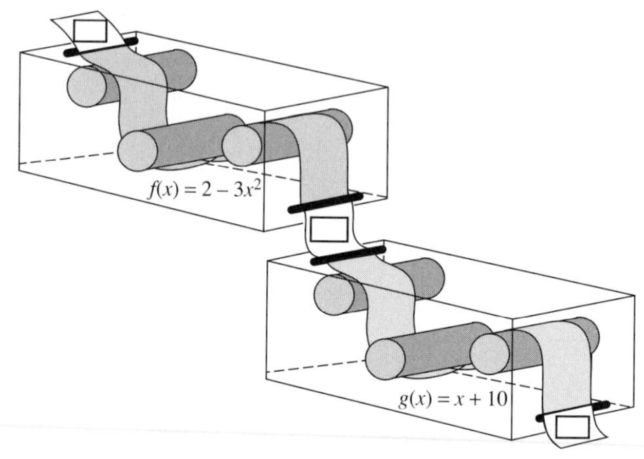

ILLUSTRATION 1

NOTATION *In Exercises 13–14, complete each solution*

13. Let $f(x) = 3x - 1$ and $g(x) = 2x + 3$. Find $f \cdot g$.

$$\begin{aligned}(f \cdot g)(x) &= f(x) \cdot g(x)\\ &= \square(2x + 3)\\ &= 6x^2 + \square - \square - 3\\ &= 6x^2 + 7x - 3\end{aligned}$$

14. Let $f(x) = 3x - 1$ and $g(x) = 2x + 3$. Find $f \circ g$.

$$\begin{aligned}(f \circ g)(x) &= f(g(x))\\ &= f(\square)\\ &= 3(\square) - 1\\ &= \square + \square - 1\\ &= 6x + 8\end{aligned}$$

PRACTICE *In Exercises 15–22, $f(x) = 3x$ and $g(x) = 4x$. Find each function and its domain.*

15. $f + g$
16. $f - g$
17. $f \cdot g$
18. f/g
19. $g - f$
20. $g + f$
21. g/f
22. $g \cdot f$

In Exercises 23–30, $f(x) = 2x + 1$ and $g(x) = x - 3$. Find each function and its domain.

23. $f + g$
24. $f - g$
25. $f \cdot g$
26. f/g
27. $g - f$
28. $g + f$
29. g/f
30. $g \cdot f$

In Exercises 31–34, $f(x) = 3x - 2$ and $g(x) = 2x^2 + 1$. Find each function and its domain.

31. $f - g$
32. $f + g$
33. f/g
34. $f \cdot g$

In Exercises 35–38, $f(x) = x^2 - 1$ and $g(x) = x^2 - 4$. Find each function and its domain.

35. $f - g$
36. $f + g$
37. g/f
38. $g \cdot f$

In Exercises 39–50, $f(x) = 2x + 1$ and $g(x) = x^2 - 1$. Find each value.

39. $(f \circ g)(2)$
40. $(g \circ f)(2)$
41. $(g \circ f)(-3)$
42. $(f \circ g)(-3)$
43. $(f \circ g)(0)$
44. $(g \circ f)(0)$
45. $(f \circ g)\left(\frac{1}{2}\right)$
46. $(g \circ f)\left(\frac{1}{3}\right)$
47. $(f \circ g)(x)$
48. $(g \circ f)(x)$
49. $(g \circ f)(2x)$
50. $(f \circ g)(2x)$

In Exercises 51–58, $f(x) = 3x - 2$ and $g(x) = x^2 + x$. Find each value.

51. $(f \circ g)(4)$
52. $(g \circ f)(4)$
53. $(g \circ f)(-3)$
54. $(f \circ g)(-3)$
55. $(g \circ f)(0)$
56. $(f \circ g)(0)$
57. $(g \circ f)(x)$
58. $(f \circ g)(x)$

59. If $f(x) = x + 1$ and $g(x) = 2x - 5$, show that $(f \circ g)(x) \neq (g \circ f)(x)$.

60. If $f(x) = x^2 + 1$ and $g(x) = 3x^2 - 2$, show that $(f \circ g)(x) \neq (g \circ f)(x)$.

APPLICATIONS

61. METALLURGY A molten alloy must be cooled slowly to control crystallization. When removed from the furnace, its temperature is 2,700° F, and it will be cooled at 200° per hour. Express the Celsius temperature as a function of the number of hours t since cooling began.

62. WEATHER FORECASTING A high pressure area promises increasingly warmer weather for the next 48 hours. The temperature is now 34° Celsius and is expected to rise 1° every 6 hours. Express the Fahrenheit temperature as a function of the number of hours from now. $\left(\textit{Hint: } F = \frac{9}{5}C + 32.\right)$

63. VACATION MILEAGE COSTS

a. Use the graphs in Illustration 2 to determine the cost of the gasoline consumed if a family drove 500 miles on a summer vacation.

b. Write a composition function that expresses the cost of the gasoline consumed on the vacation as a function of the miles driven.

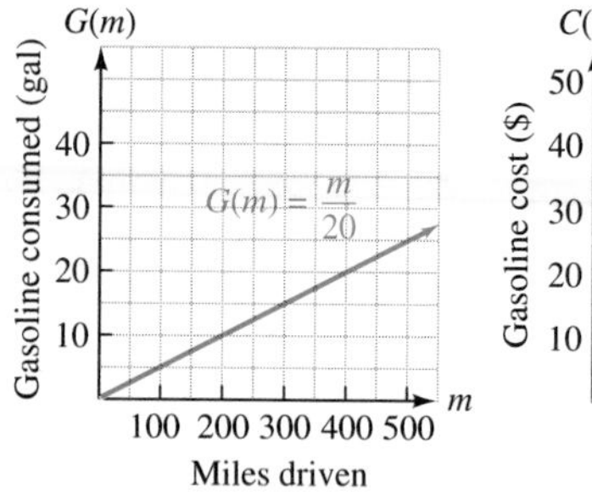

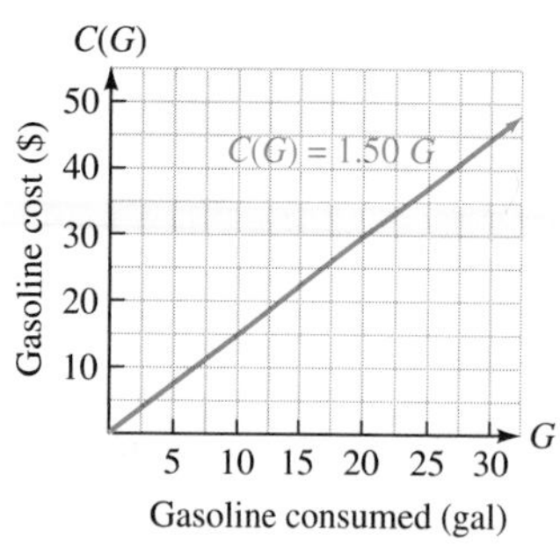

ILLUSTRATION 2

64. HALLOWEEN COSTUMES The tables on the back of the pattern package shown in Illustration 3 can be used to determine the number of yards of material needed to make a rabbit costume for a child.

a. How many yards of material are needed if the child's chest measures 29 inches?

b. In this exercise, one quantity is a function of a second quantity that depends, in turn, on a third quantity. Explain this dependence.

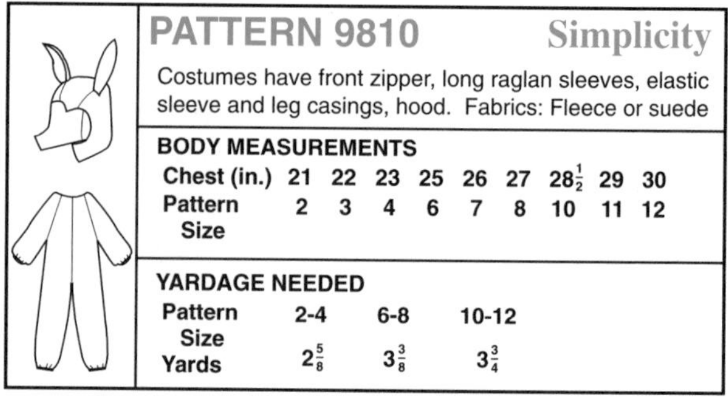

PATTERN 9810 Simplicity

Costumes have front zipper, long raglan sleeves, elastic sleeve and leg casings, hood. Fabrics: Fleece or suede

BODY MEASUREMENTS

Chest (in.)	21	22	23	25	26	27	$28\frac{1}{2}$	29	30
Pattern Size	2	3	4	6	7	8	10	11	12

YARDAGE NEEDED

Pattern Size	2-4	6-8	10-12
Yards	$2\frac{5}{8}$	$3\frac{3}{8}$	$3\frac{3}{4}$

ILLUSTRATION 3

WRITING *Write a paragraph using your own words.*

65. Exercise 63 illustrates a chain of dependence between the cost of the gasoline, the gasoline consumed, and the miles driven. Describe another chain of dependence that could be represented by a composition function.

66. Explain how to add, subtract, multiply, and divide two functions.

REVIEW *In Exercises 67–70, simplify each expression.*

67. $\dfrac{3x^2 + x - 14}{4 - x^2}$

68. $\dfrac{2x^3 + 14x^2}{3 + 2x - x^2} \cdot \dfrac{x^2 - 3x}{x}$

69. $\dfrac{x^2 - 2x - 8}{3x^2 - x - 12} \div \dfrac{3x^2 + 5x - 2}{3x - 1}$

70. $\dfrac{x - 1}{1 + \dfrac{x}{x - 2}}$

9.2 Inverses of Functions

In this section, you will learn about

- **One-to-one functions**
- **The horizontal line test**
- **Finding inverses of functions**

Introduction. The function defined by $C = \frac{5}{9}(F - 32)$ is the formula that we use to convert degrees Fahrenheit to degrees Celsius. If we input a Fahrenheit reading into the formula, the output is a Celsius reading. For example, if we substitute 41° for F, we obtain a Celsius reading of 5°:

$$
\begin{aligned}
C &= \frac{5}{9}(F - 32) \\
&= \frac{5}{9}(41 - 32) \quad \text{Substitute 41 for } F. \\
&= \frac{5}{9}(9) \\
&= 5
\end{aligned}
$$

If we want to find a Fahrenheit reading from a Celsius reading, we need a formula into which we can substitute a Celsius reading and have a Fahrenheit reading come out. Such a formula is $F = \frac{9}{5}C + 32$, which takes the Celsius reading of 5° and turns it back into a Fahrenheit reading of 41°.

$$
\begin{aligned}
F &= \frac{9}{5}C + 32 \\
&= \frac{9}{5}(5) + 32 \quad \text{Substitute 5 for } C. \\
&= 41
\end{aligned}
$$

The functions defined by these two formulas do opposite things. The first turns 41° F into 5° C, and the second turns 5° C back into 41° F. For this reason, we say that the functions are *inverses* of each other.

In this section, we will show how to find inverses of one-to-one functions.

One-to-one functions

Recall that for each input into a function, there is a single output. For some functions, different inputs have the same output, as shown in Figure 9-4(a). For other functions, different inputs have different outputs, as shown in Figure 9-4(b).

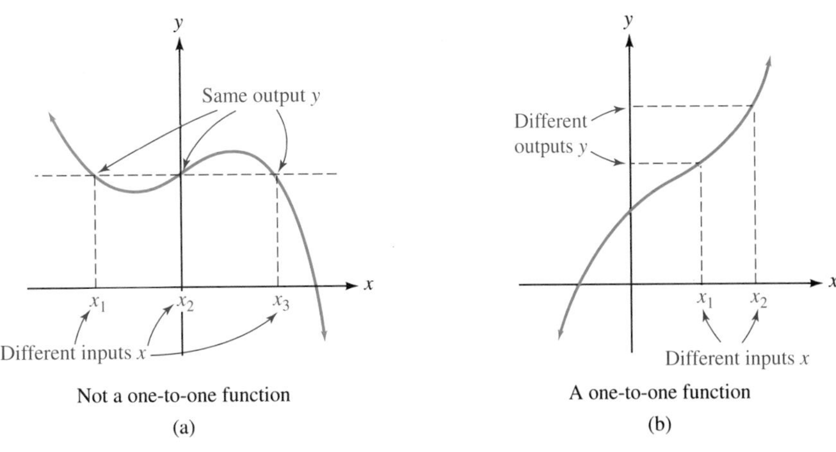

FIGURE 9-4

When every output of a function corresponds to exactly one input, we say that the function is *one-to-one.*

One-to-one functions	A function is called **one-to-one** if each input value of x in the domain determines a different output value of y in the range.

EXAMPLE 1 ***One-to-one functions.*** Determine whether the functions **a.** $f(x) = x^2$ and **b.** $f(x) = x^3$ are one-to-one.

Solution

a. The function $f(x) = x^2$ is not one-to-one, because different input values x can determine the same output value y. For example, inputs of 3 and -3 produce the same output value of 9.

$$f(3) = 3^2 = 9 \quad \text{and} \quad f(-3) = (-3)^2 = 9$$

b. The function $f(x) = x^3$ is one-to-one, because different input values x determine different output values y. This is because different numbers have different cubes.

Self Check

Determine whether $f(x) = 2x + 3$ is one-to-one and explain why or why not.

Answer: yes, because different input values determine different output values

The horizontal line test

The **horizontal line test** can be used to decide whether the graph of a function represents a one-to-one function. If every horizontal line that intersects the graph of a function does so only once, the function is one-to-one. Otherwise, the function is not one-to-one. See Figure 9-5.

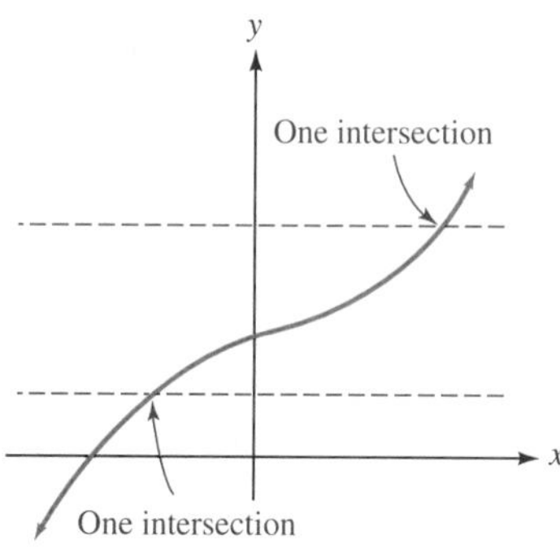

A one-to-one function

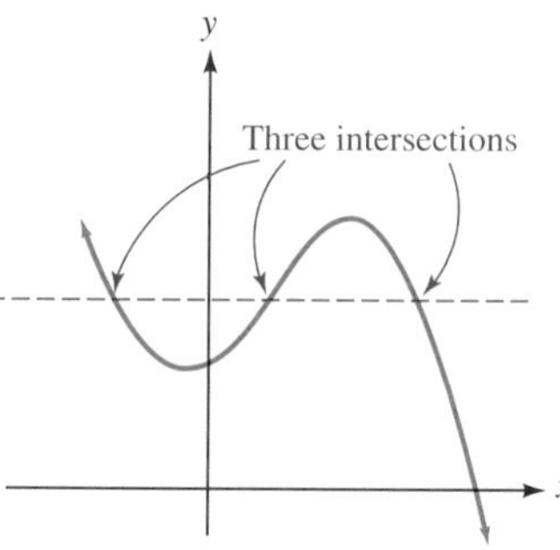

Not a one-to-one function

FIGURE 9-5

EXAMPLE 2 ***Determining one-to-one functions.*** Use the horizontal line test to decide whether the graphs in Figure 9-6 represent one-to-one functions.

Solution

a. Because many horizontal lines intersect the graph shown in Figure 9-6(a) twice, the graph does not represent a one-to-one function.

b. Because each horizontal line that intersects the graph in Figure 9-6(b) does so exactly once, the graph represents a one-to-one function.

Self Check

Determine whether the following graphs represent one-to-one functions.

a.

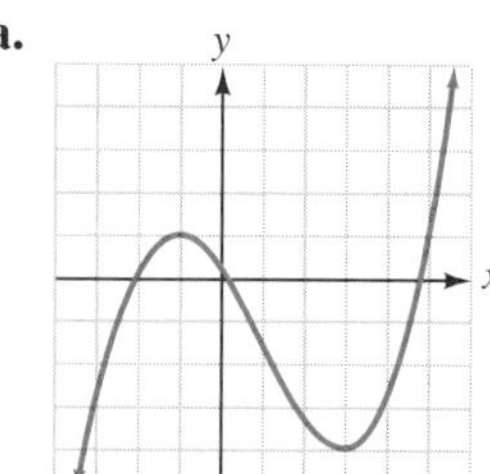

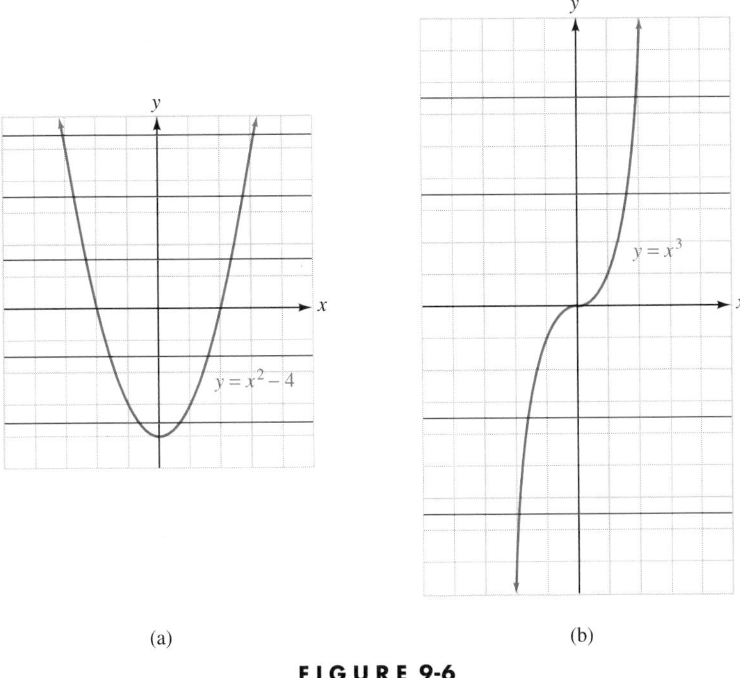

(a) (b)

FIGURE 9-6

b.

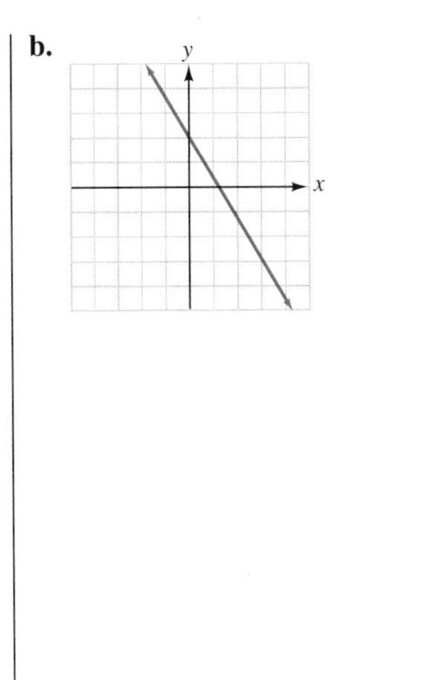

Answers: **a.** no, **b.** yes ■

Finding inverses of functions

If f is the function determined by the ordered pairs in the table shown in Figure 9-7(a), it turns the number 1 into 10, 2 into 20, and 3 into 30. Since the inverse of f must turn 10 back into 1, 20 back into 2, and 30 back into 3, it consists of the ordered pairs shown in Figure 9-7(b).

Function f

x	y
1	10
2	20
3	30
↑ Domain	↑ Range

(a)

Inverse of f

x	y
10	1
20	2
30	3
↑ Domain	↑ Range

(b)

Note that the inverse of f is also a function.

FIGURE 9-7

We note that the domain of f and the range of its inverse is $\{1, 2, 3\}$. The range of f and the domain of its inverse is $\{10, 20, 30\}$.

This example suggests that to form the inverse of a function f, we simply interchange the coordinates of each ordered pair that determines f. When the inverse of a function is also a function, we call it ***f* inverse** and denote it with the symbol f^{-1}.

WARNING! The symbol $f^{-1}(x)$ is read as "the inverse of $f(x)$" or just "f inverse." The -1 in the notation $f^{-1}(x)$ is not an exponent. Remember that $f^{-1}(x) \neq \frac{1}{f(x)}$.

Finding the inverse of a one-to-one function	If a function is one-to-one, we find its inverse as follows: **1.** If the function is written using function notation, replace $f(x)$ with y. **2.** Interchange the variables x and y. **3.** Solve the resulting equation for y. **4.** We can substitute $f^{-1}(x)$ for y.

EXAMPLE 3 ***Finding the inverse of a function.*** If $f(x) = 4x + 2$, find the inverse of f and tell whether it is a function.

Solution

We proceed as follows:

$$f(x) = 4x + 2$$

$$y = 4x + 2 \quad \text{Step 1: Replace } f(x) \text{ with } y.$$

$$x = 4y + 2 \quad \text{Step 2: Interchange the variables } x \text{ and } y.$$

To decide whether the inverse $x = 4y + 2$ is a function, we solve for y (Step 3).

$$x = 4y + 2$$

$$x - 2 = 4y \quad \text{Subtract 2 from both sides.}$$

1. $$y = \frac{x-2}{4} \quad \text{Divide both sides by 4 and write } y \text{ on the left-hand side.}$$

Because each input x that is substituted into Equation 1 gives one output y, the inverse of f is a function, so we can express it in the form

$$f^{-1}(x) = \frac{x-2}{4} \quad \text{Step 4: Substitute } f^{-1}(x) \text{ for } y.$$

Self Check

If $f(x) = -5x - 3$, find the inverse of f and tell whether it is a function.

Answers: $f^{-1}(x) = \dfrac{-x-3}{5}$, yes ■

To emphasize an important relationship between a function and its inverse, we substitute some number x, such as $x = 3$, into the function $f(x) = 4x + 2$ of Example 3. The corresponding value of y produced is

$$f(3) = 4(3) + 2 = 14$$

If we substitute 14 into the inverse function, f^{-1}, the corresponding value of y that is produced is

$$f^{-1}(14) = \frac{14-2}{4} = 3$$

Thus, the function f turns 3 into 14, and the inverse function f^{-1} turns 14 back into 3. In general, the composition of a function and its inverse is the identity function.

To prove that $f(x) = 4x + 2$ and $f^{-1}(x) = \frac{x-2}{4}$ are inverse functions, we must show that their composition (in both directions) is the identity function:

$$\begin{aligned}(f \circ f^{-1})(x) &= f(f^{-1}(x)) \\ &= f\left(\frac{x-2}{4}\right) \\ &= 4\left(\frac{x-2}{4}\right) + 2 \\ &= x - 2 + 2 \\ &= x\end{aligned} \qquad \begin{aligned}(f^{-1} \circ f)(x) &= f^{-1}(f(x)) \\ &= f^{-1}(4x + 2) \\ &= \frac{4x + 2 - 2}{4} \\ &= \frac{4x}{4} \\ &= x\end{aligned}$$

Thus $(f \circ f^{-1})(x) = (f^{-1} \circ f)(x) = x$, which is the identity function $I(x)$.

EXAMPLE 4 ***The graph of an inverse function.*** The set of all pairs (x, y) determined by $3x + 2y = 6$ is a function. Find its inverse function and graph the function and its inverse on one coordinate system.

Solution

To find the inverse function of $3x + 2y = 6$, we interchange x and y to obtain

$$3y + 2x = 6$$

and then solve the equation for y.

$$3y + 2x = 6$$
$$3y = -2x + 6$$
$$y = -\frac{2}{3}x + 2$$

The inverse function of $3x + 2y = 6$ is $y = -\frac{2}{3}x + 2$, which can also be written $f^{-1}(x) = -\frac{2}{3}x + 2$.
The graphs of $3x + 2y = 6$ and its inverse, along with the graph of $y = x$, appear in Figure 9-8.

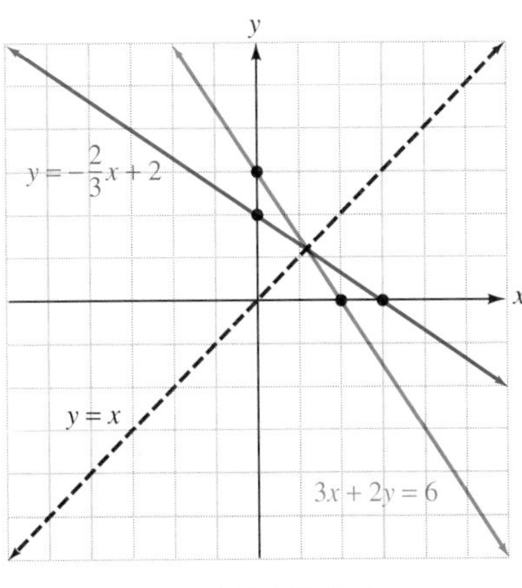

FIGURE 9-8

Self Check

Find the inverse of the function defined by $2x - 3y = 6$. Graph the function and its inverse on one coordinate system. Also graph $y = x$.

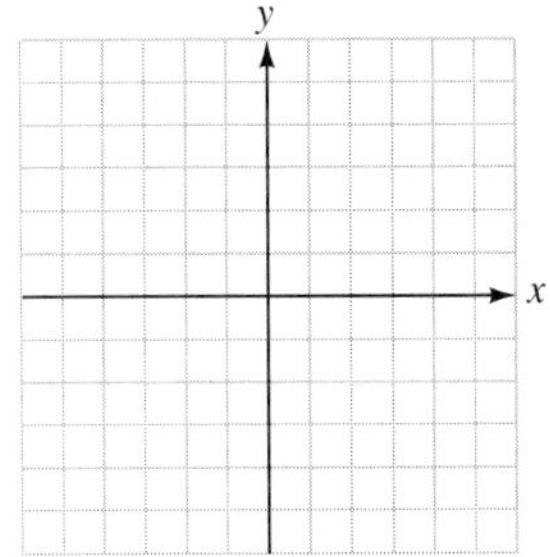

Answer: $f^{-1}(x) = \frac{3}{2}x + 3$

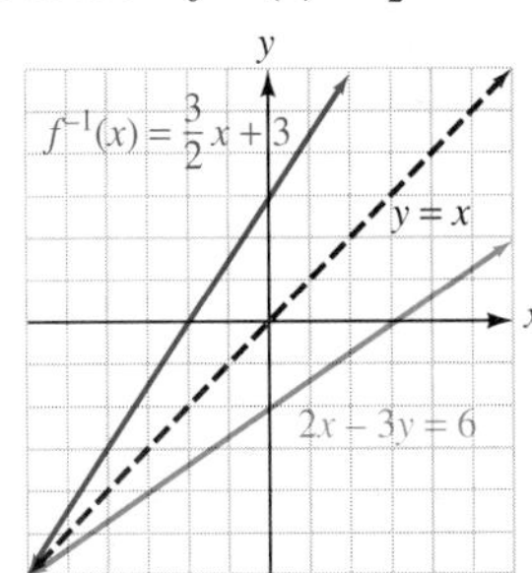

■

In Example 4, the graphs of $3x + 2y = 6$ and $y = -\frac{2}{3}x + 2$, its inverse, are symmetric about the line $y = x$. This is always the case with a function and its inverse, because when the coordinates (a, b) satisfy an equation, the coordinates (b, a) will satisfy its inverse.

In each example so far, the inverse of a function has been another function. This is not always true, as the following example will show.

EXAMPLE 5 ***Finding the inverse of a function.*** Find the inverse of the function determined by $f(x) = x^2$.

Solution

$y = x^2$ Replace $f(x)$ with y.

$x = y^2$ Interchange x and y.

$y = \pm\sqrt{x}$ Use the square root property and write y on the left-hand side.

When the inverse $y = \pm\sqrt{x}$ is graphed, as in Figure 9-9, we see that the graph does not pass the vertical line test. Thus, it is not a function.

The graph of $y = x^2$ is also shown in the figure. Note that it is not one-to-one. The graphs of $y = x^2$ and $y = \pm\sqrt{x}$ are symmetric about the line $y = x$.

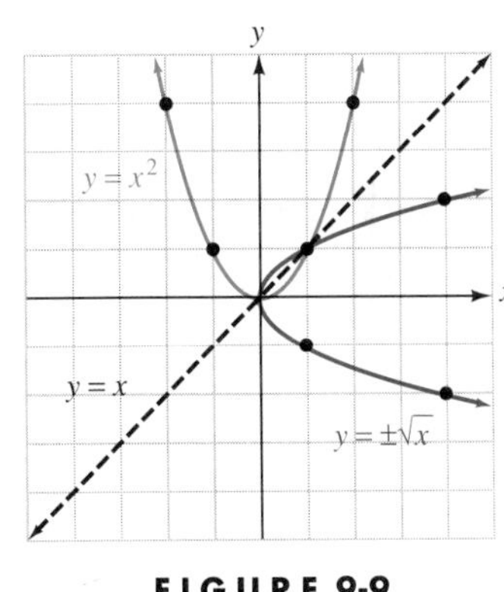

FIGURE 9-9

Self Check

Find the inverse of the function determined by $f(x) = 4x^2$.

Answer: $y = \pm\frac{\sqrt{x}}{2}$ ■

EXAMPLE 6 ***Finding the inverse of a function.*** Find the inverse of $f(x) = x^3$.

Solution

To find the inverse, we proceed as follows:

$y = x^3$ Replace $f(x)$ with y.

$x = y^3$ Interchange the variables x and y.

$\sqrt[3]{x} = y$ Take the cube root of both sides.

We note that to each number x there corresponds one real cube root. Thus, $y = \sqrt[3]{x}$ represents a function. Using $f^{-1}(x)$ notation, the inverse of $f(x) = x^3$ is

$$f^{-1}(x) = \sqrt[3]{x}$$

Self Check

Find the inverse of $f(x) = x^5$.

Answer: $f^{-1}(x) = \sqrt[5]{x}$ ■

If a function is not one-to-one, it is often possible to make it one-to-one by restricting its domain.

EXAMPLE 7 Find the inverse of the function defined by $y = x^2$ with $x \geq 0$. Then tell whether it is a function. Graph the function and its inverse on one set of coordinate axes.

Solution The inverse of the function $y = x^2$ with $x \geq 0$ is

$x = y^2$ with $y \geq 0$ Interchange the variables x and y.

This equation can be written in the form

$$y = \pm\sqrt{x} \quad \text{with } y \geq 0$$

Since $y \geq 0$, each number x gives only one value of y: $y = \sqrt{x}$. Thus, the inverse is a function.

The graphs of the two functions appear in Figure 9-10. The line $y = x$ is included so that we can see that the graphs are symmetric about the line $y = x$.

$y = x^2$ and $x \geq 0$

x	y	(x, y)
0	0	(0, 0)
1	1	(1, 1)
2	4	(2, 4)
3	9	(3, 9)

$x = y^2$ and $y \geq 0$

x	y	(x, y)
0	0	(0, 0)
1	1	(1, 1)
4	2	(4, 2)
9	3	(9, 3)

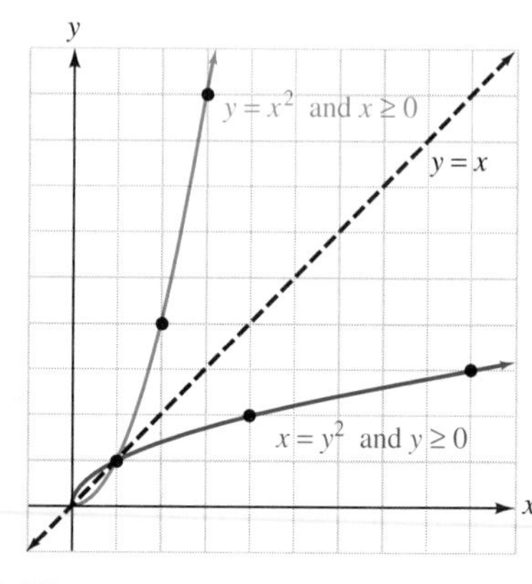

FIGURE 9-10 ■

STUDY SET Section 9.2

VOCABULARY *In Exercises 1–4, fill in the blanks to make the statements true.*

1. A function is called ______________ if each input determines a different output.

2. The ______________ line test can be used to decide whether the graph of a function represents a one-to-one function.

3. The functions f and f^{-1} are ______________.

4. An input value is an element of a function's ______________. An output value is an element of a function's ______________.

CONCEPTS *In Exercises 5–10, fill in the blanks to make the statements true.*

5. If every horizontal line that intersects the graph of a function does so only ______________, the function is one-to-one.

6. If any horizontal line that intersects the graph of a function does so more than once, the function is not ______________.

7. If a function turns an input of 2 into an output of 5, the inverse function will turn an input of 5 into an output of ___.

8. The graphs of a function and its inverse are symmetrical about the line $\square$.

9. $(f \circ f^{-1})(x) = \square$

10. To find the inverse of the function defined by $y = 2x - 3$, we begin by ______________ x and y.

11. Use the table of values of the one-to-one function f to complete a table of values for f^{-1}.

x	$f(x)$
−6	−3
−4	−2
0	0
2	1
8	4

x	$f^{-1}(x)$
−3	
−2	
0	
1	
4	

12. How can we tell that function f is not one-to-one from the table of values?

x	$f(x)$
−2	4
−1	1
0	0
2	4
3	9

13. Is the inverse of a function always a function?

14. Name four points that the line $y = x$ passes through.

15. If f is a one-to-one function, and if $f(2) = 6$, then what is $f^{-1}(6)$?

16. If the point $(2, -4)$ is on the graph of the one-to-one function f, then what point is on the graph of f^{-1}?

NOTATION *In Exercises 17–18, complete each solution.*

17. Find the inverse of $f(x) = 2x - 3$.

$y = \square - 3$ Replace $f(x)$ with y.

$x = \square - 3$ Interchange the variables x and y.

$x + \square = 2y$ Add 3 to both sides.

$\dfrac{x+3}{2} = \square$ Divide both sides by 2.

The inverse of $f(x) = 2x - 3$ is $\square = \dfrac{x+3}{2}$.

18. Find the inverse of $f(x) = \sqrt[3]{x} + 2$.

$\square = \sqrt[3]{x} + 2$ Replace $f(x)$ with y.

$x = \sqrt[3]{\square} + 2$ Interchange the variables x and y.

$x - \square = \sqrt[3]{y}$ Subtract 2 from both sides.

$(x - 2)^3 = \square$ Cube both sides.

The inverse of $f(x) = \sqrt[3]{x} + 2$ is $\square = (x - 2)^3$.

19. The symbol $f^{-1}(x)$ is read as "______________ f" or "f ______________."

20. Explain the difference in the meaning of the -1 in the notation $f^{-1}(x)$ as compared to x^{-1}.

PRACTICE *In Exercises 21–24, determine whether the function is one-to-one.*

21. $f(x) = 2x$

22. $f(x) = |x|$

23. $f(x) = x^4$

24. $f(x) = x^3 + 1$

In Exercises 25–28, each graph represents a function. Use the horizontal line test to decide whether the function is one-to-one.

25.

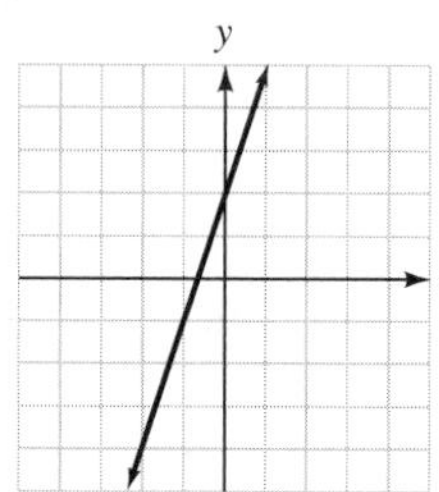

26.

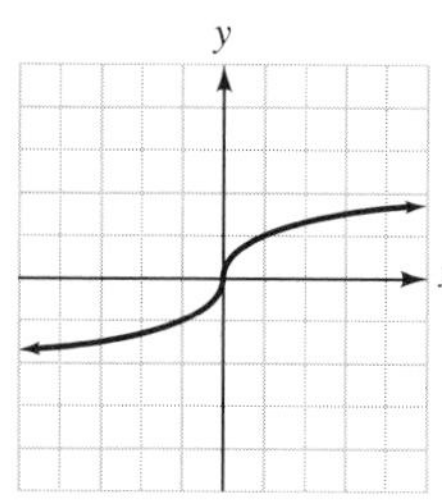

27.

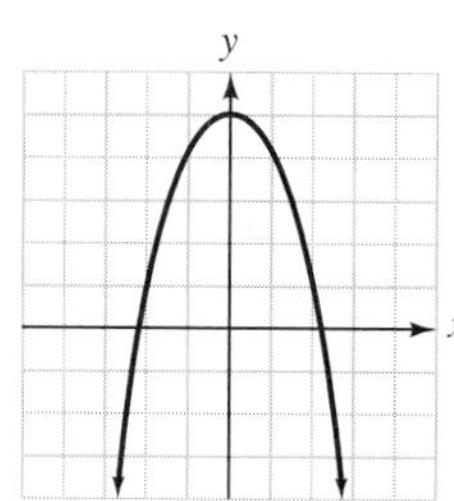

28.

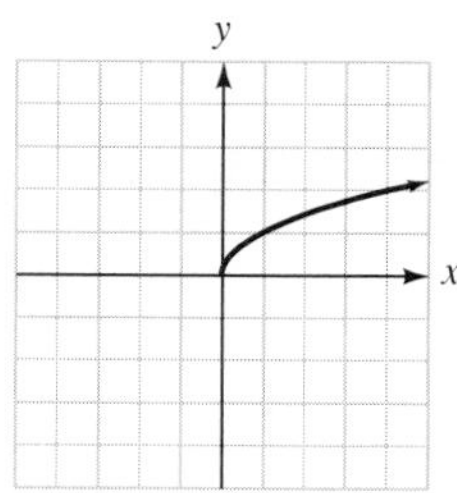

In Exercises 29–32, tell whether the set of ordered pairs (x, y) is a function. Then find the inverse of the set of ordered pairs (x, y) and tell whether the inverse is a function.

29. $\{(3, 2), (2, 1), (1, 0)\}$

30. $\{(4, 1), (5, 1), (6, 1), (7, 1)\}$

31. $\{(1, 1), (2, 1), (3, 1), (4, 1)\}$

32. $\{(1, 2), (2, 3), (1, 3), (1, 5)\}$

In Exercises 33–38, find the inverse of the function and express it using $f^{-1}(x)$ notation.

33. $4x - 5y = 20$

34. $y + 1 = 5x$

35. $x + 4 = 5y$

36. $x = 3y + 1$

37. $f(x) = \dfrac{x - 4}{5}$

38. $f(x) = \dfrac{2x + 6}{3}$

In Exercises 39–42, find the inverse of each function and express it in terms of x and y. Then graph the function and its inverse on one coordinate system. Show the line of symmetry on the graph.

39. $y = 4x + 3$

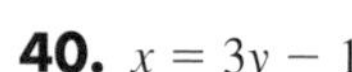

40. $x = 3y - 1$

41. $2x + 3y = 9$

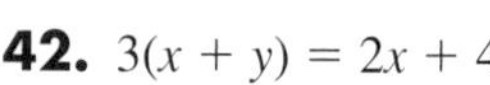

42. $3(x + y) = 2x + 4$

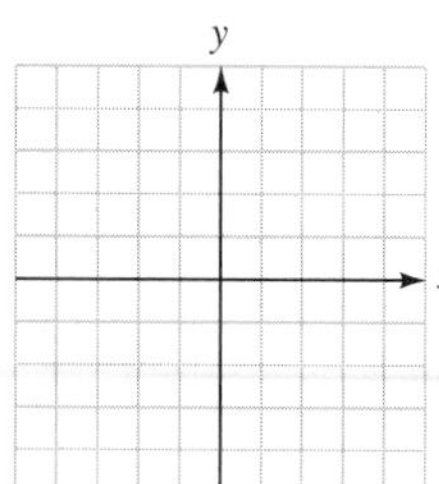

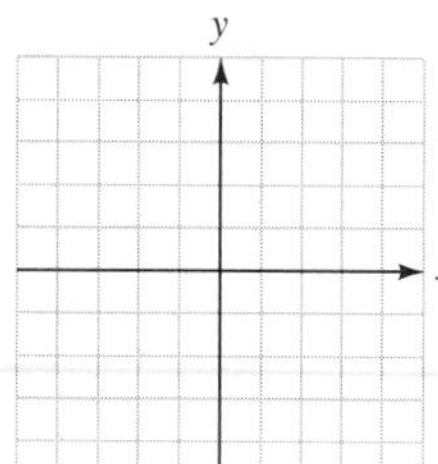

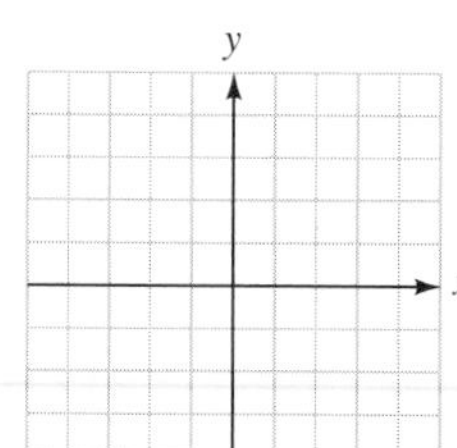

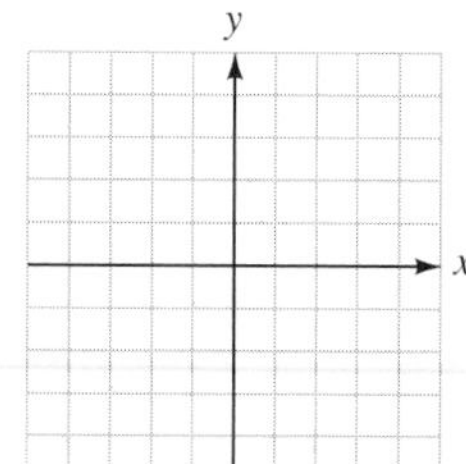

In Exercises 43–50, find the inverse of the function and tell whether it is a function. If it is a function, express it using $f^{-1}(x)$ notation.

43. $y = x^2 + 4$

44. $y = x^2 + 5$

45. $y = x^3$

46. $xy = 4$

47. $y = |x|$

48. $y = \sqrt[3]{x}$

49. $f(x) = 2x^3 - 3$

50. $f(x) = \dfrac{3}{x^3} - 1$

In Exercises 51–54, graph each equation and its inverse on one set of coordinate axes. Show the axis of symmetry.

51. $y = x^2 + 1$

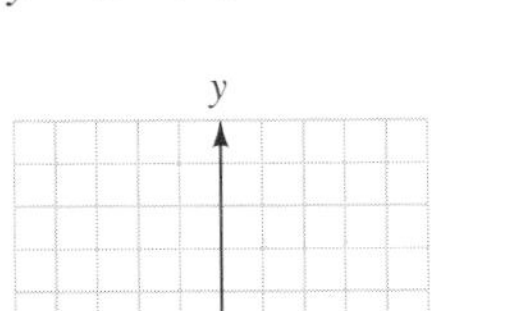

52. $y = \frac{1}{4}x^2 - 3$

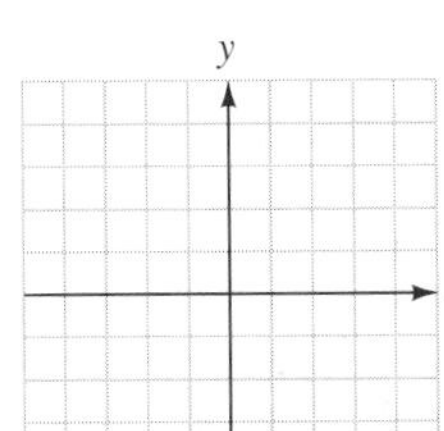

53. $y = \sqrt{x} \quad (x \geq 0)$

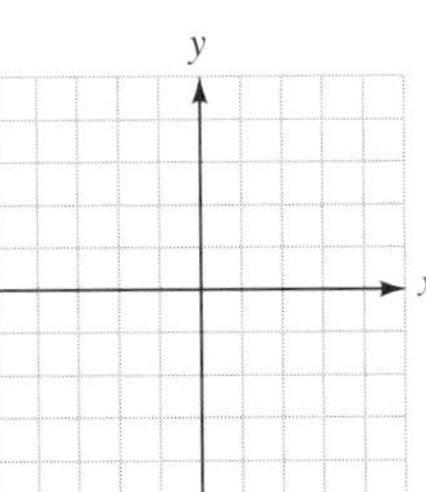

54. $y = |x|$

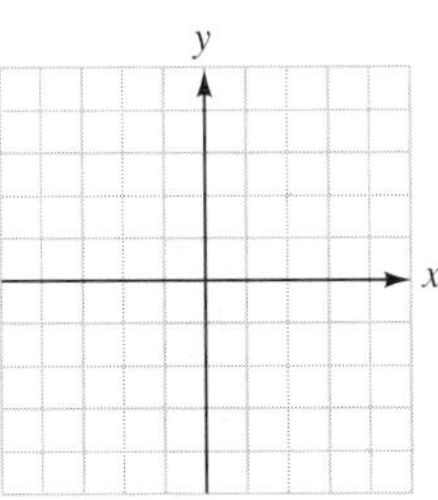

APPLICATIONS

55. INTERPERSONAL RELATIONSHIPS Feelings of anxiety in a relationship can increase or decrease, depending on what is going on in the relationship. The graph in Illustration 1 shows how a person's anxiety might vary as a relationship develops over time.

a. Is this the graph of a function? Is its inverse a function?

b. Does each anxiety level correspond to exactly one point in time? Use the dashed lined labeled *Maximum threshold* to explain.

56. LIGHTING LEVELS The ability of the eye to see detail increases as the level of illumination increases. This relationship can be modeled by a function E, whose graph is shown in Illustration 2.

a. From the graph, determine $E(240)$.

b. Is function E one-to-one? Does E have an inverse?

c. If the effectiveness of seeing in an office is 7, what is the illumination in the office? How can this question be asked using inverse function notation?

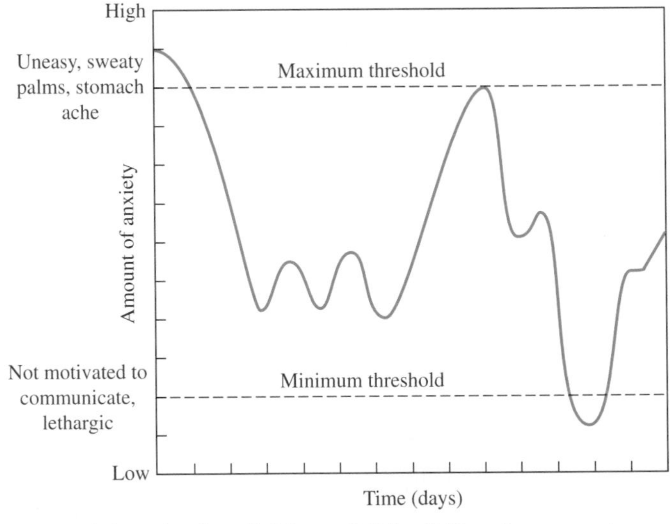

Based on information from Gudykunst, *Building Bridges, Interpersonal Skills for a Changing world* (Houghton Mifflin, 1994)

ILLUSTRATION 1

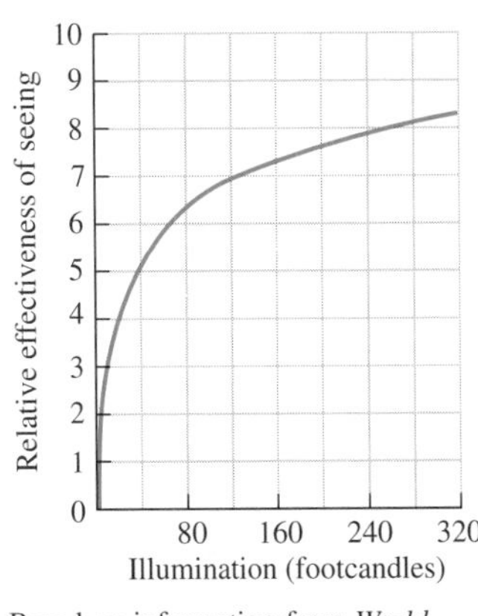

Based on information from *World Book Encyclopedia*

ILLUSTRATION 2

WRITING *Write a paragraph using your own words.*

57. What does it mean when we say that one graph is symmetric to the other about the line $y = x$?

58. Explain the purpose of the horizontal line test.

REVIEW *In Exercises 59–64, simplify each expression and write the result in a + bi form.*

59. $3 - \sqrt{-64}$

60. $(2 - 3i) + (4 + 5i)$

61. $(3 + 4i)(2 - 3i)$

62. $\dfrac{6 + 7i}{3 - 4i}$

63. $(6 - 8i)^2$

64. i^{100}

9.3 Exponential Functions

In this section, you will learn about

- **Irrational exponents**
- **Exponential functions**
- **Graphing exponential functions**
- **Vertical and horizontal translations**
- **Compound interest**
- **Exponential functions as models**

INTRODUCTION. The graph in Figure 9-11 shows the balance in a bank account in which $10,000 was invested in 1998 at 9%, compounded monthly. The graph shows that in the year 2008, the value of the account will be approximately $25,000, and in the year 2028, the value will be approximately $147,000. The curve shown in Figure 9-11 is the graph of a function called an *exponential function.*

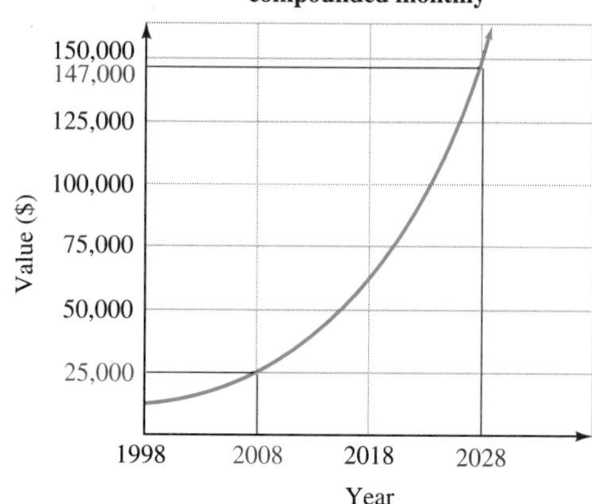

FIGURE 9-11

Exponential functions are also suitable for modeling many other situations, such as population growth, the spread of an epidemic, the temperature of a heated object as it cools, and radioactive decay. Before we can discuss exponential functions in more detail, we must define irrational exponents.

Irrational exponents

We have discussed expressions of the form b^x, where x is a rational number.

$8^{1/2}$ means "the square root of 8."

$5^{1/3}$ means "the cube root of 5."

$3^{-2/5} = \dfrac{1}{3^{2/5}}$ means "the reciprocal of the fifth root of 3^2."

To give meaning to b^x when x is an irrational number, we consider the expression

$$5^{\sqrt{2}} \quad \text{where } \sqrt{2} \text{ is the irrational number } 1.414213562\ldots$$

Each number in the following list is defined, because each exponent is a rational number.

$$5^{1.4},\quad 5^{1.41},\quad 5^{1.414},\quad 5^{1.4142},\quad 5^{1.41421},\quad \ldots$$

Since the exponents are getting closer to $\sqrt{2}$, the numbers in this list are successively better approximations of $5^{\sqrt{2}}$. We can use a calculator to obtain a very good approximation.

Accent on Technology **Evaluating exponential expressions**

To find the value of $5^{\sqrt{2}}$ with a scientific calculator, we enter these numbers and press these keys:

5 [y^x] 2 [$\sqrt{\ }$] [=]

The display will read [9.738517742].

With a graphing calculator, we enter these numbers and press these keys:

5 [^] [$\sqrt{\ }$] 2 [Enter]

The display will read [5^√(2) 9.738517742].

In general, if $b > 0$ and x is a real number, b^x represents a positive number. It can be shown that all of the familiar rules of exponents are also true for irrational exponents.

EXAMPLE 1 ***Exponential expressions having irrational exponents.*** Use the rules of exponents to simplify **a.** $\left(5^{\sqrt{2}}\right)^{\sqrt{2}}$ and **b.** $b^{\sqrt{3}} \cdot b^{\sqrt{12}}$.

Solution

a. $\left(5^{\sqrt{2}}\right)^{\sqrt{2}} = 5^{\sqrt{2}\sqrt{2}}$ Keep the base and multiply the exponents.

$= 5^2$ $\quad \sqrt{2}\sqrt{2} = \sqrt{4} = 2.$

$= 25$

b. $b^{\sqrt{3}} \cdot b^{\sqrt{12}} = b^{\sqrt{3}+\sqrt{12}}$ Keep the base and add the exponents.

$= b^{\sqrt{3}+2\sqrt{3}}$ $\quad \sqrt{12} = \sqrt{4}\sqrt{3} = 2\sqrt{3}.$

$= b^{3\sqrt{3}}$ $\quad \sqrt{3} + 2\sqrt{3} = 3\sqrt{3}.$

Self Check

Simplify: **a.** $\left(3^{\sqrt{2}}\right)^{\sqrt{8}}$ and **b.** $b^{\sqrt{2}} \cdot b^{\sqrt{18}}$.

Answers: **a.** 81, **b.** $b^{4\sqrt{2}}$

Exponential functions

If $b > 0$ and $b \neq 1$, the function $f(x) = b^x$ is an **exponential function.** Since x can be any real number, its domain is the set of real numbers. This is the interval $(-\infty, \infty)$. Since b is positive, the value of $f(x)$ is positive, and the range is the set of positive numbers. This is the interval $(0, \infty)$.

Since $b \neq 1$, an exponential function cannot be the constant function $f(x) = 1^x$, in which $f(x) = 1$ for every real number x.

Exponential functions

An **exponential function with base *b*** is defined by the equation

$$f(x) = b^x \quad \text{or} \quad y = b^x \quad (b > 0,\ b \neq 1, \text{ and } x \text{ is a real number})$$

The **domain of any exponential function** is the interval $(-\infty, \infty)$. The **range** is the interval $(0, \infty)$.

Graphing exponential functions

Since the domain and range of $f(x) = b^x$ are sets of real numbers, we can graph exponential functions on a rectangular coordinate system.

EXAMPLE 2 ***Graphing exponential functions.*** Graph $f(x) = 2^x$.

Solution

To graph $f(x) = 2^x$, we find several points (x, y) whose coordinates satisfy the equation, plot the points, and join them with a smooth curve, as shown in Figure 9-12. For example, if $x = -1$, we have

$$\begin{aligned} f(x) &= 2^x \\ f(-1) &= 2^{-1} \\ &= \frac{1}{2} \end{aligned}$$

The point $\left(-1, \frac{1}{2}\right)$ is on the graph of $f(x) = 2^x$.

$f(x) = 2^x$

x	$f(x)$	$(x, f(x))$
-1	$\frac{1}{2}$	$\left(-1, \frac{1}{2}\right)$
0	1	(0, 1)
1	2	(1, 2)
2	4	(2, 4)
3	8	(3, 8)
4	16	(4, 16)

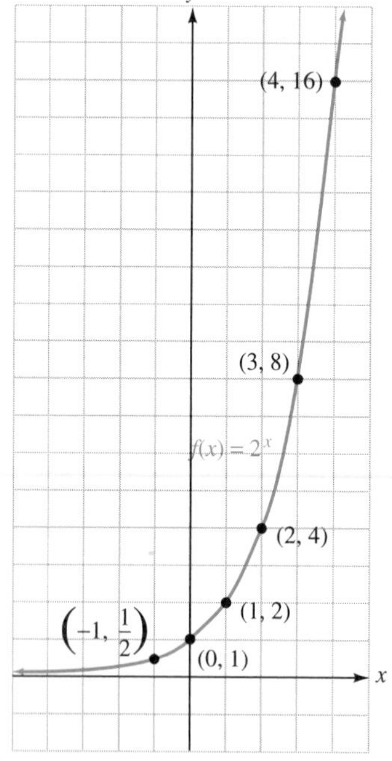

FIGURE 9-12

By looking at the graph, we can verify that the domain is the interval $(-\infty, \infty)$ and that the range is the interval $(0, \infty)$.

Note that as x decreases, the values of $f(x)$ decrease and approach 0. Thus, the x-axis is a horizontal asymptote of the graph.

Also note that the graph of $f(x) = 2^x$ passes through the points (0, 1) and (1, 2).

Self Check

Graph $f(x) = 4^x$.

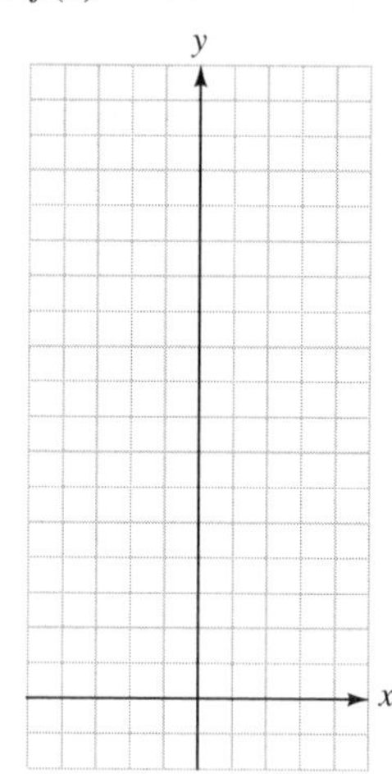

Answer:

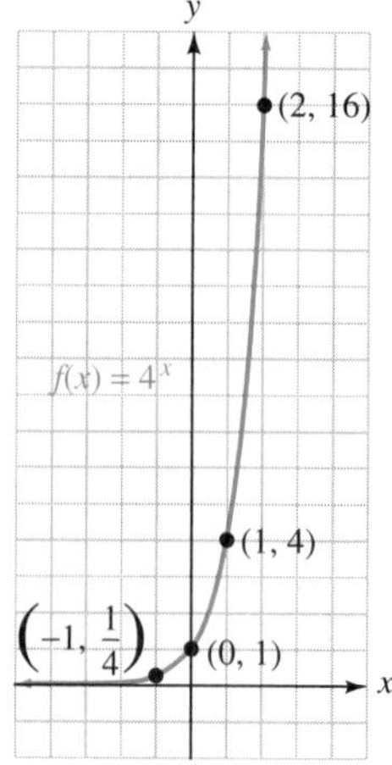

EXAMPLE 3 *Graphing exponential functions.* Graph $f(x) = \left(\frac{1}{2}\right)^x$.

Solution

We find and plot pairs (x, y) that satisfy the equation. The graph of $f(x) = \left(\frac{1}{2}\right)^x$ appears in Figure 9-13. For example, if $x = -4$, we have

$$f(x) = \left(\frac{1}{2}\right)^x$$

$$f(-4) = \left(\frac{1}{2}\right)^{-4}$$

$$= \left(\frac{2}{1}\right)^4$$

$$= 16$$

The point $(-4, 16)$ is on the graph of $f(x) = \left(\frac{1}{2}\right)^x$.

$f(x) = \left(\frac{1}{2}\right)^x$

x	$f(x)$	$(x, f(x))$
-4	16	$(-4, 16)$
-3	8	$(-3, 8)$
-2	4	$(-2, 4)$
-1	2	$(-1, 2)$
0	1	$(0, 1)$
1	$\frac{1}{2}$	$\left(1, \frac{1}{2}\right)$

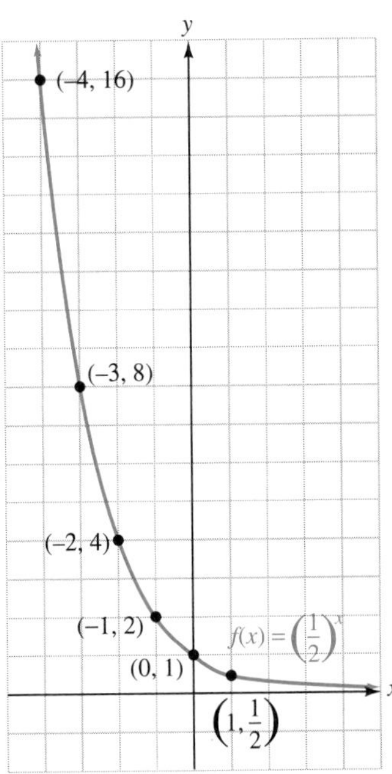

FIGURE 9-13

By looking at the graph, we can verify that the domain is the interval $(-\infty, \infty)$ and that the range is the interval $(0, \infty)$.

In this case, as x increases, the values of $f(x)$ decrease and approach 0. The x-axis is a horizontal asymptote. Note that the graph of $f(x) = \left(\frac{1}{2}\right)^x$ passes through the points $(0, 1)$ and $\left(1, \frac{1}{2}\right)$.

Self Check

Graph $g(x) = \left(\frac{1}{3}\right)^x$.

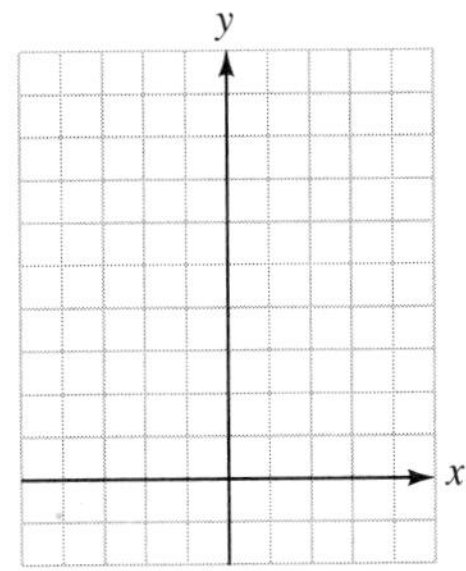

Answer:

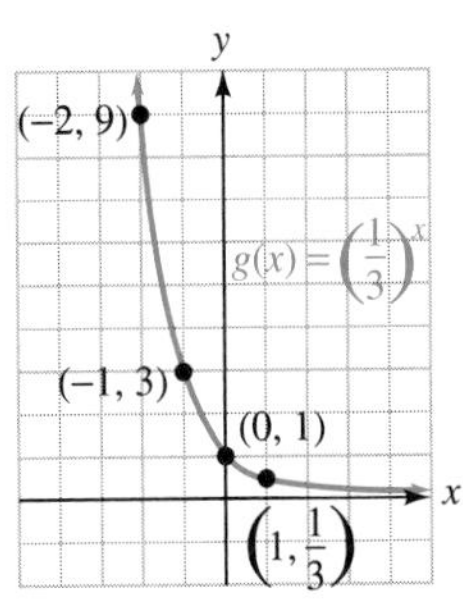

Examples 2 and 3 illustrate the following properties of exponential functions.

Properties of exponential functions

The **domain** of the exponential function $f(x) = b^x$ is the interval $(-\infty, \infty)$.

The **range** is the interval $(0, \infty)$.

The graph has a y-intercept of $(0, 1)$.

The x-axis is an asymptote of the graph.

The graph of $f(x) = b^x$ passes through the point $(1, b)$.

In Example 2 (where $b = 2$), the values of y increase as the values of x increase. Since the graph rises as we move to the right, we call the function an *increasing function.* When $b > 1$, the larger the value of b, the steeper the curve.

In Example 3 $\left(\text{where } b = \frac{1}{2}\right)$, the values of y decrease as the values of x increase. Since the graph drops as we move to the right, we call the function a *decreasing function.* When $0 < b < 1$, the smaller the value of b, the steeper the curve.

In general, the following is true.

Increasing and decreasing functions

If $b > 1$, then $f(x) = b^x$ is an **increasing function.**

If $0 < b < 1$, then $f(x) = b^x$ is a **decreasing function.**

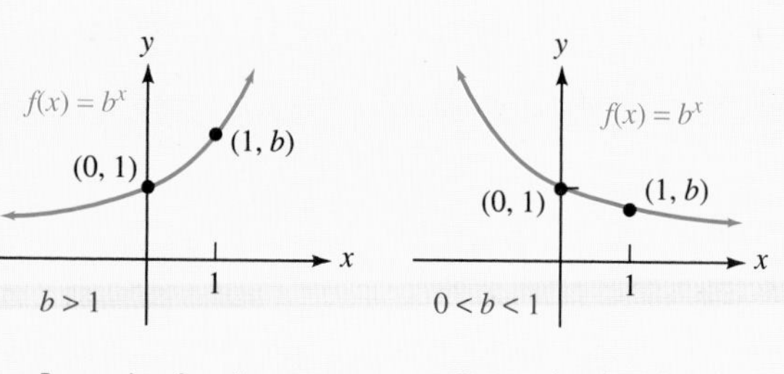

An exponential function with base b is either increasing (for $b > 1$) or decreasing ($0 < b < 1$). Since different real numbers x determine different values of b^x, exponential functions are one-to-one.

Accent on Technology Graphing exponential functions

To use a graphing calculator to graph $f(x) = \left(\frac{2}{3}\right)^x$ and $f(x) = \left(\frac{3}{2}\right)^x$, we enter the right-hand sides of the equations after the symbols $Y_1 =$ and $Y_2 =$. The screen will show the following equations.

$$Y_1 = (2/3) \wedge X$$
$$Y_2 = (3/2) \wedge X$$

If we use window settings of $[-10, 10]$ for x and $[-2, 10]$ for y and press the GRAPH key, we will obtain the graph shown in Figure 9-14.

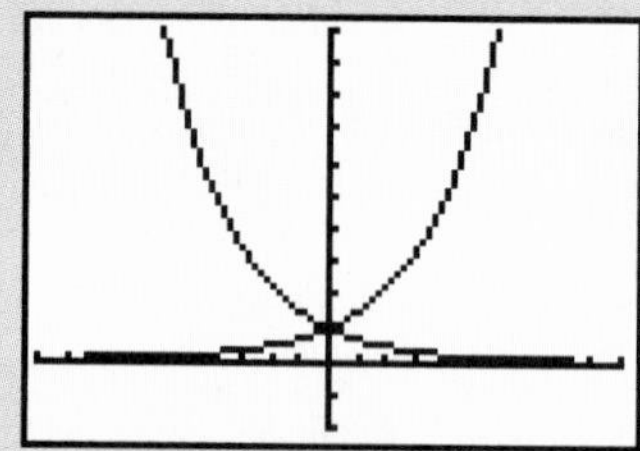

FIGURE 9-14

We note that the graph of $f(x) = \left(\frac{2}{3}\right)^x$ passes through the points $(0, 1)$ and $\left(1, \frac{2}{3}\right)$. Since $\frac{2}{3} < 1$, the function is decreasing.

The graph of $f(x) = \left(\frac{3}{2}\right)^x$ passes through the points $(0, 1)$ and $\left(1, \frac{3}{2}\right)$. Since $\frac{3}{2} > 1$, the function is increasing.

Since both graphs pass the horizontal line test, each function is one-to-one.

Vertical and horizontal translations

EXAMPLE 4 ***Translations.*** On one set of axes, graph $y = 2^x$ and $y = 2^x + 3$, and describe the translation.

Solution

The graph of $y = 2^x + 3$ is identical to the graph of $y = 2^x$, except that it is translated 3 units upward. (See Figure 9-15.)

Self Check

Graph $f(x) = 4^x$ and $g(x) = 4^x - 3$, and describe the translation.

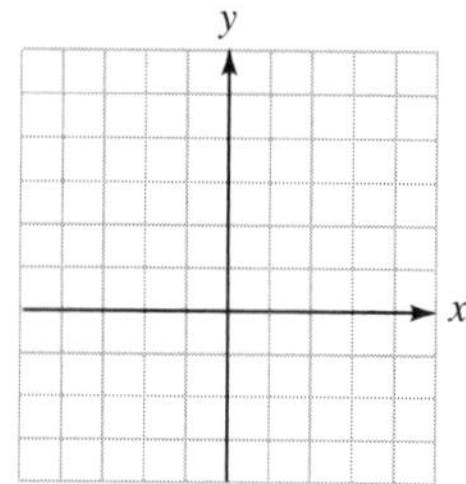

$y = 2^x$

x	y	(x, y)
-4	$\frac{1}{16}$	$\left(-4, \frac{1}{16}\right)$
0	1	(0, 1)
2	4	(2, 4)

$y = 2^x + 3$

x	y	(x, y)
-4	$3\frac{1}{16}$	$\left(-4, 3\frac{1}{16}\right)$
0	4	(0, 4)
2	7	(2, 7)

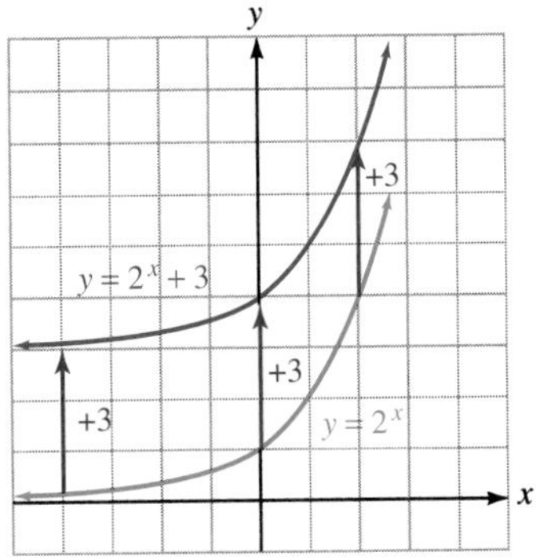

FIGURE 9-15

Answer: The graph of $f(x) = 4^x$ is translated 3 units downward.

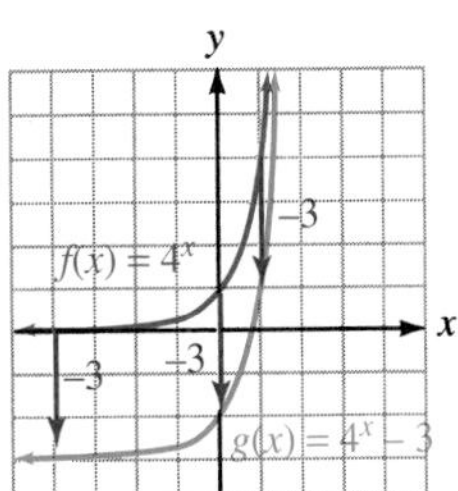

EXAMPLE 5 ***Translations.*** On one set of axes, graph $f(x) = 2^x$ and $g(x) = 2^{x+3}$, and describe the translation.

Solution

The graph of $g(x) = 2^{x+3}$ is identical to the graph of $f(x) = 2^x$, except that it is translated 3 units to the left. (See Figure 9-16.)

$f(x) = 2^x$

x	$f(x)$	$(x, f(x))$
-1	$\frac{1}{2}$	$\left(-1, \frac{1}{2}\right)$
0	1	(0, 1)
1	2	(1, 2)

$g(x) = 2^{x+3}$

x	$g(x)$	$(x, g(x))$
-1	4	$(-1, 4)$
0	8	(0, 8)
1	16	(1, 16)

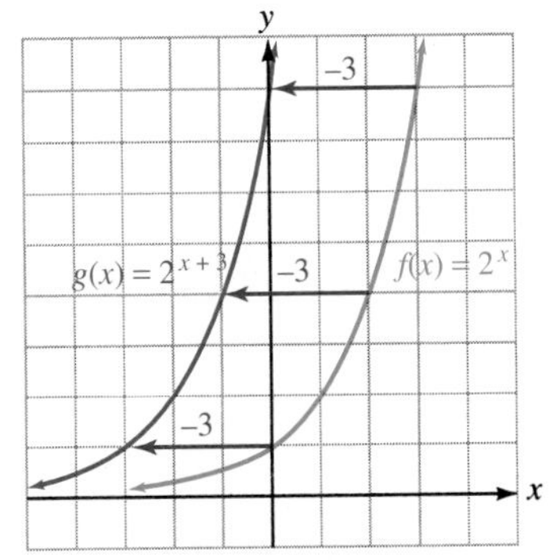

FIGURE 9-16

Self Check

On one set of axes, graph $f(x) = 4^x$ and $g(x) = 4^{x-3}$, and describe the translation.

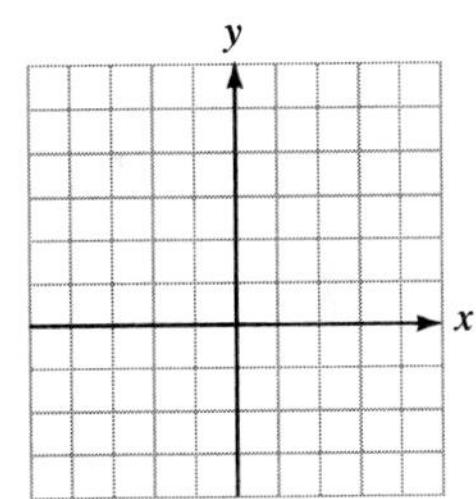

Answer: The graph of $f(x) = 4^x$ is translated 3 units to the right.

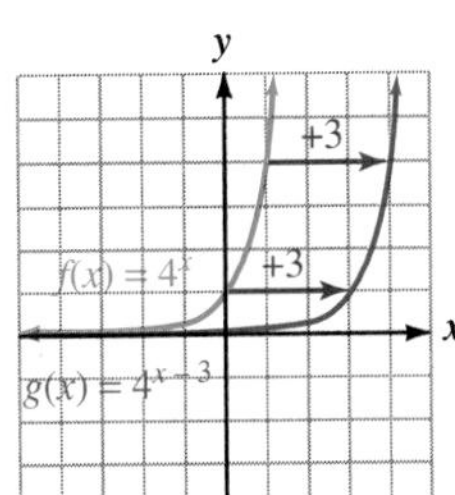

The graphs of $f(x) = kb^x$ and $f(x) = b^{kx}$ are vertical and horizontal stretchings of the graph of $f(x) = b^x$. To graph these functions, we can plot several points and join them with a smooth curve, or use a graphing calculator.

Accent on Technology Graphing exponential functions

To use a graphing calculator to graph the exponential function $f(x) = 3(2^{x/3})$, we enter the right-hand side of the equation after the symbol $Y_1 =$. The display will show the equation

$$Y_1 = 3(2 \wedge (X/3))$$

If we use window settings of $[-10, 10]$ for x and $[-2, 18]$ for y and press the graph key, we will obtain the graph shown in Figure 9-17.

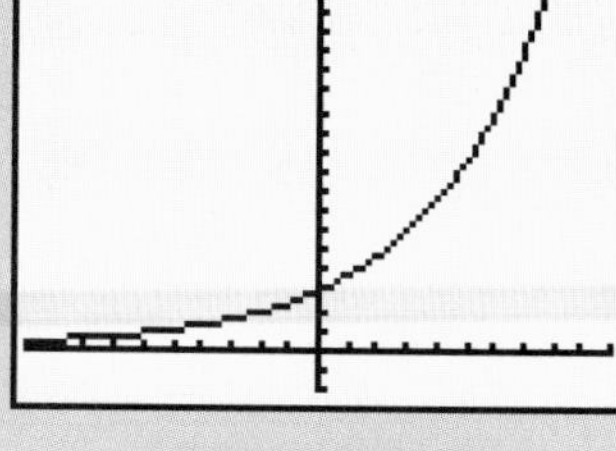

FIGURE 9-17

Compound interest

If we deposit $\$P$ in an account paying an annual interest rate r, we can find the amount A in the account at the end of t years by using the formula $A = P + Prt$, or $A = P(1 + rt)$.

Suppose that we deposit \$500 in such an account that pays interest every six months. Then $P = 500$, and after six months ($\frac{1}{2}$ year), the amount in the account will be

$$A = 500(1 + rt)$$

$$= 500\left(1 + r \cdot \frac{1}{2}\right) \quad \text{Substitute } \tfrac{1}{2} \text{ for } t.$$

$$= 500\left(1 + \frac{r}{2}\right)$$

The account will begin the second six-month period with a value of $\$500\left(1 + \frac{r}{2}\right)$. After the second six-month period, the amount in the account will be

$$A = P(1 + rt)$$

$$A = \left[500\left(1 + \frac{r}{2}\right)\right]\left(1 + r \cdot \frac{1}{2}\right) \quad \text{Substitute } 500\left(1 + \tfrac{r}{2}\right) \text{ for } P \text{ and } \tfrac{1}{2} \text{ for } t.$$

$$= 500\left(1 + \frac{r}{2}\right)\left(1 + \frac{r}{2}\right)$$

$$= 500\left(1 + \frac{r}{2}\right)^2$$

At the end of a third six-month period, the amount in the account will be

$$A = 500\left(1 + \frac{r}{2}\right)^3$$

In this discussion, the earned interest is deposited back in the account and also earns interest. When this is the case, we say that the account is earning **compound interest.** The preceding discussion suggests the following formula for compound interest.

Formula for compound interest

If $\$P$ is deposited in an account and interest is paid k times a year at an annual rate r, the amount A in the account after t years is given by

$$A = P\left(1 + \frac{r}{k}\right)^{kt}$$

EXAMPLE 6 ***Saving for college.*** To save for college, parents invest \$12,000 for their newborn child in a mutual fund that should average a 10% annual return. If the quarterly dividends are reinvested, how much will be available in 18 years?

Solution

We substitute 12,000 for P, 0.10 for r, and 18 for t in the formula for compound interest and find A. Since interest is paid quarterly, $k = 4$.

$$A = P\left(1 + \frac{r}{k}\right)^{kt}$$

$$A = 12{,}000\left(1 + \frac{0.10}{4}\right)^{4(18)}$$

$$= 12{,}000(1 + 0.025)^{72}$$

$$= 12{,}000(1.025)^{72}$$

$$= 71{,}006.74$$

Use a calculator and press these keys: 1.025 [y^x] 72 [=] [×] 12,000 [=].

In 18 years, the account will be worth \$71,006.74.

Self Check

How much would be available if the parents invested \$20,000?

Answer: \$118,344.56

In business applications, the initial amount of money deposited is often called the **present value** (PV). The amount to which the money will grow is called the **future value** (FV). The interest rate used for each compounding period is the **periodic interest rate** (i), and the number of times interest is compounded is the number of **compounding periods** (n). Using these definitions, we have an alternate formula for compound interest.

Formula for compound interest	$FV = PV(1 + i)^n$

This alternate formula appears on business calculators. To use this formula to solve Example 6, we proceed as follows:

$$FV = PV(1 + i)^n$$

$$FV = 12{,}000(1 + 0.025)^{72} \quad i = \tfrac{0.10}{4} = 0.025 \text{ and } n = 4(18) = 72.$$

$$= 71{,}006.74 \quad \text{Use a calculator to evaluate the expression.}$$

Accent on Technology — Solving investment problems

Suppose \$1 is deposited in an account earning 6% annual interest, compounded monthly. To use a graphing calculator to estimate how much will be in the account in 100 years, we can substitute 1 for P, 0.06 for r, and 12 for k in the formula

$$A = P\left(1 + \frac{r}{k}\right)^{kt}$$

$$A = 1\left(1 + \frac{0.06}{12}\right)^{12t}$$

and simplify to get

$$A = (1.005)^{12t}$$

We now graph $A = (1.005)^{12t}$ using window settings of [0, 120] for t and [0, 400] for A to obtain the graph shown in Figure 9-18. We can then trace and zoom to estimate that \$1 grows to be approximately \$397 in 100 years. From the graph, we can see that the money grows slowly in the early years and rapidly in the later years.

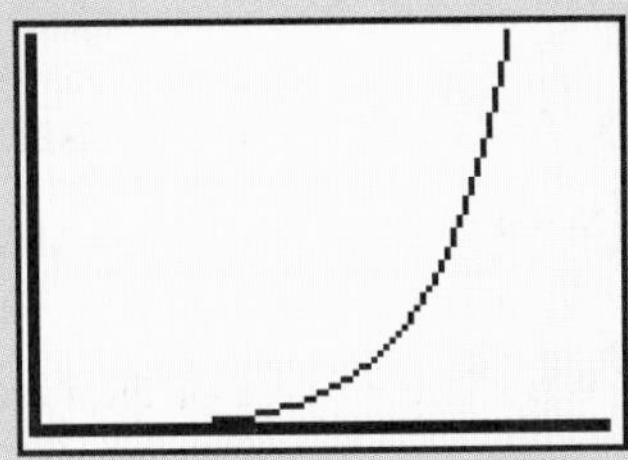

FIGURE 9-18

Exponential functions as models

EXAMPLE 7 ***Cellular telephones.*** For the years 1990–1996, the U.S. cellular telephone industry experienced "exponential growth." (See Figure 9-19.) The exponential function $S(n) = 5.28(1.43)^n$ approximates the number of cellular telephone subscribers in millions, where n is the number of years since 1990 and $0 \leq n \leq 6$.

a. How many subscribers were there in 1990?

b. How many subscribers were there in 1996?

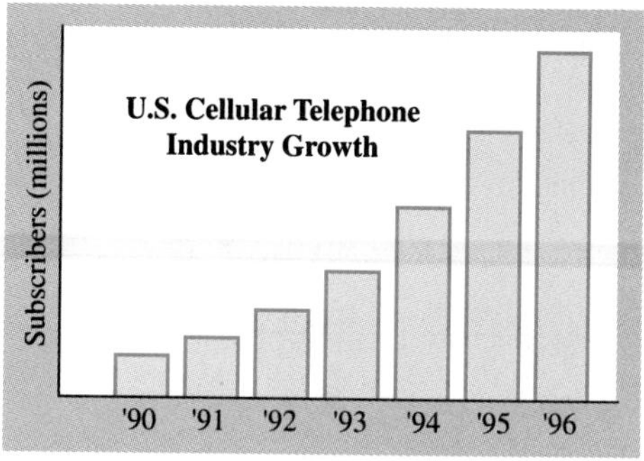

Based on information from *New York Times Almanac,* 1998

FIGURE 9-19

Solution **a.** The year 1990 is 0 years after 1990. To find the number of subscribers in 1990, we substitute 0 for n in the function and find $S(0)$.

$$S(n) = 5.28(1.43)^n$$
$$S(0) = 5.28(1.43)^0$$
$$= 5.28 \cdot 1 \qquad (1.43)^0 = 1.$$
$$= 5.28$$

In 1990, there were approximately 5.28 million cellular telephone subscribers.

b. The year 1996 is 6 years after 1990. We need to find $S(6)$.

$$S(n) = 5.28(1.43)^n$$
$$S(6) = 5.28(1.43)^6 \qquad \text{Substitute 6 for } n.$$
$$\approx 45.14920914 \qquad \text{Use a calculator to find an approximation.}$$

In 1996, there were approximately 45.15 million cellular telephone subscribers. ■

STUDY SET Section 9.3

VOCABULARY *In Exercises 1–8, refer to the graph of $f(x) = 3^x$ in Illustration 1.*

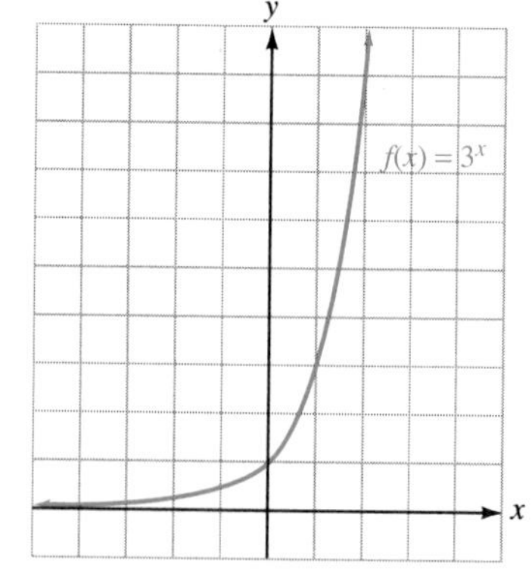

ILLUSTRATION 1

1. What type of function is $f(x) = 3^x$?
2. What is the domain of the function?
3. What is the range of the function?
4. What is the y-intercept of the graph?
5. What is the x-intercept of the graph?
6. What is an asymptote of the graph?
7. Is f an increasing or a decreasing function?
8. The graph passes through the point $(1, y)$. What is y?

CONCEPTS

9. Graph the functions $f(x) = x^2$ and $g(x) = 2^x$ on the same set of coordinate axes.

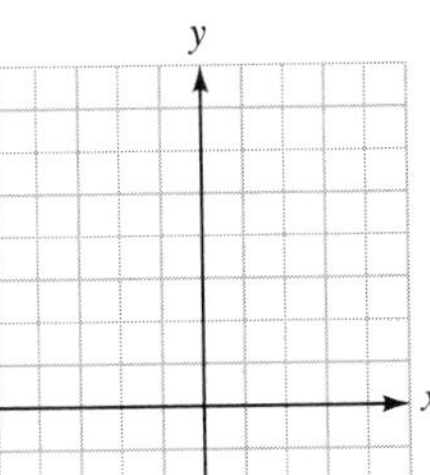

10. Graph the functions $y = x^{\frac{1}{2}}$ and $y = \left(\frac{1}{2}\right)^x$ on the same set of coordinate axes.

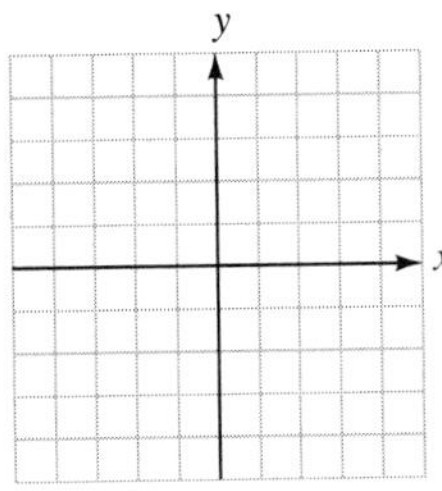

11. What are the two formulas that are used to determine the amount of money in a savings account that is earning compound interest?

12. Explain the order in which the expression $20{,}000(1.036)^{72}$ should be evaluated.

NOTATION

13. In $A = P\left(1 + \frac{r}{k}\right)^{kt}$, what is the base, and what is the exponent?

14. For an exponential function of the form $f(x) = b^x$, what are the restrictions on b?

PRACTICE

In Exercises 15–18, find each value to four decimal places.

15. $2^{\sqrt{2}}$

16. $7^{\sqrt{2}}$

17. $5^{\sqrt{5}}$

18. $6^{\sqrt{3}}$

In Exercises 19–22, simplify each expression.

19. $\left(2^{\sqrt{3}}\right)^{\sqrt{3}}$

20. $3^{\sqrt{2}}3^{\sqrt{18}}$

21. $7^{\sqrt{3}}7^{\sqrt{12}}$

22. $\left(3^{\sqrt{5}}\right)^{\sqrt{5}}$

In Exercises 23–30, graph each exponential function. Check your work with a graphing calculator.

23. $f(x) = 5^x$

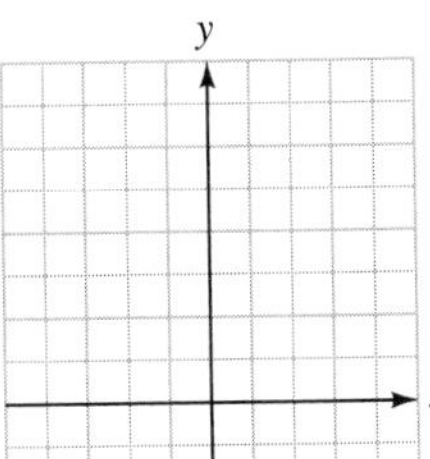

24. $f(x) = 6^x$

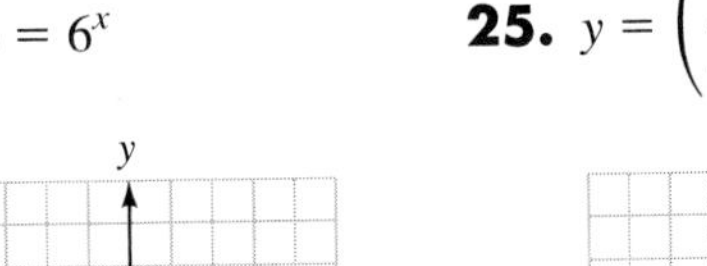

25. $y = \left(\frac{1}{4}\right)^x$

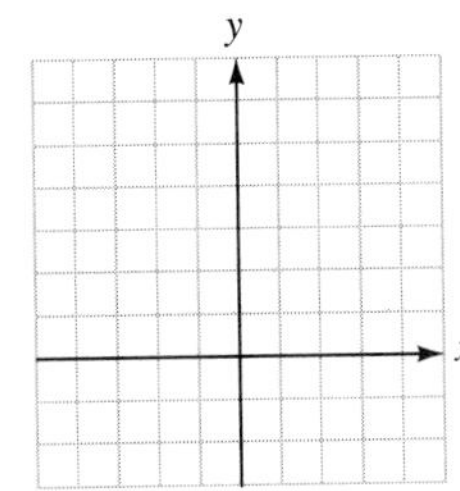

26. $y = \left(\frac{1}{5}\right)^x$

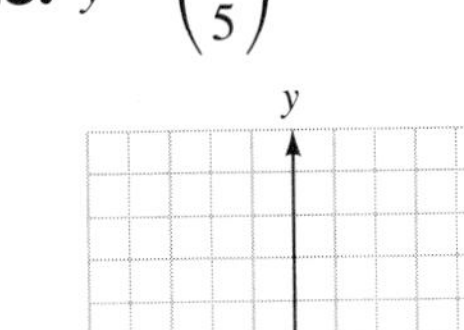

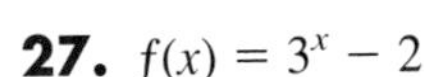

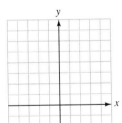

27. $f(x) = 3^x - 2$

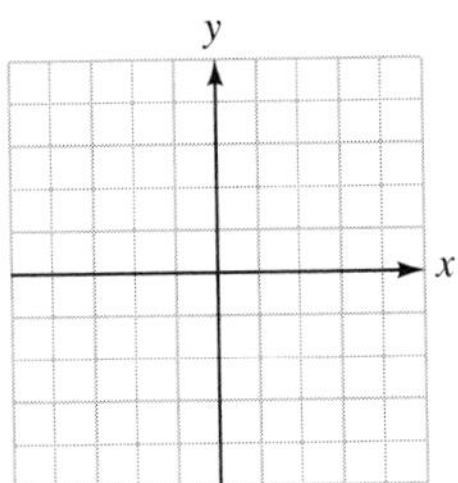

28. $y = 2^x + 1$

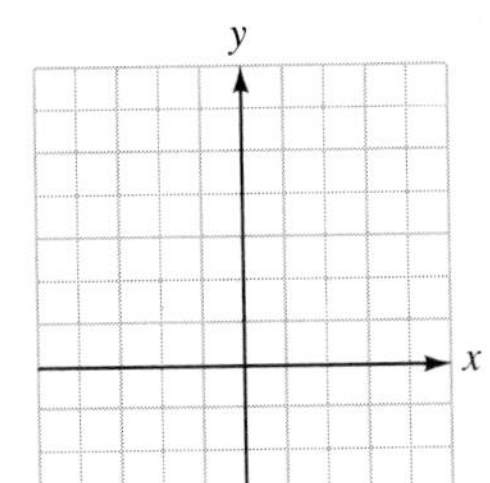

29. $f(x) = 3^{x-1}$

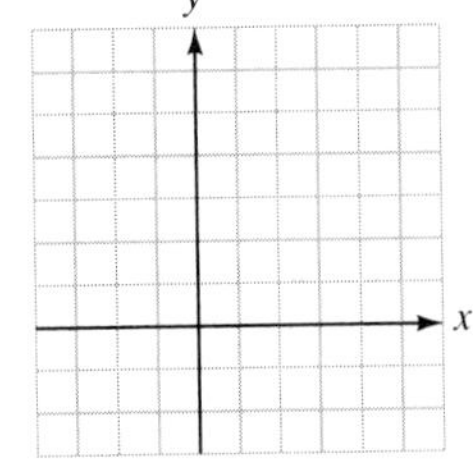

30. $f(x) = 2^{x+1}$

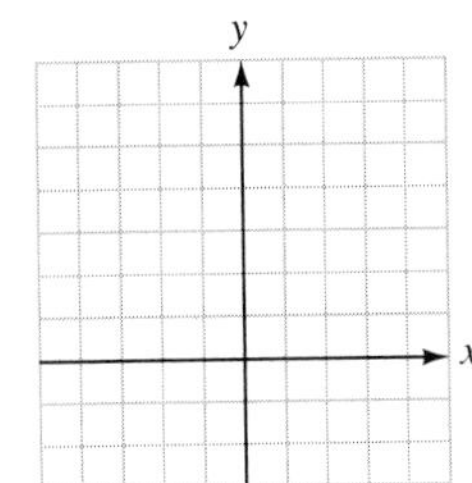

In Exercises 31–34, use a graphing calculator to graph each function. Tell whether the function is an increasing or a decreasing function.

31. $f(x) = \frac{1}{2}(3^{x/2})$

32. $f(x) = -3(2^{x/3})$

33. $y = 2(3^{-x/2})$

34. $y = -\frac{1}{4}(2^{-x/2})$

APPLICATIONS *In Exercises 35–40, assume that there are no deposits or withdrawals.*

35. COMPOUND INTEREST An initial deposit of \$10,000 earns 8% interest, compounded quarterly. How much will be in the account after 10 years?

36. COMPOUND INTEREST An initial deposit of \$10,000 earns 8% interest, compounded monthly. How much will be in the account after 10 years?

37. COMPARING INTEREST RATES How much more interest could \$1,000 earn in 5 years, compounded quarterly, if the annual interest rate were $5\frac{1}{2}\%$ instead of 5%?

38. COMPARING SAVINGS PLANS Which institution in Illustration 2 provides the better investment?

Fidelity Savings & Loan

Earn 5.25%
compounded monthly

Union Trust

Money Market Account
paying 5.35%,
compounded annually

ILLUSTRATION 2

41. WORLD POPULATION See the graph in Illustration 3.

a. Estimate when the world's population reached $\frac{1}{2}$ billion and when it reached 1 billion.

b. Estimate the world's population in the year 2000.

c. What type of function does it appear could be used to model the population growth?

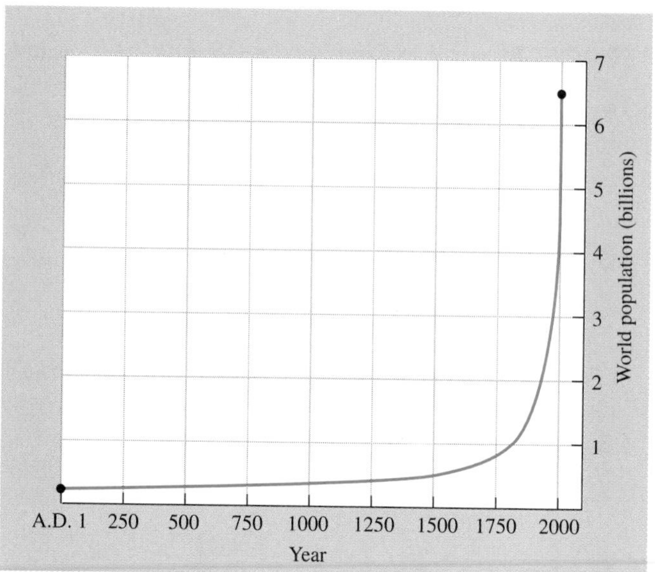

Based on data from *The Blue Planet* (Wiley, 1995)

ILLUSTRATION 3

39. COMPOUND INTEREST If \$1 had been invested on July 4, 1776, at 5% interest, compounded annually, what would it be worth on July 4, 2076?

40. FREQUENCY OF COMPOUNDING \$10,000 is invested in each of two accounts, both paying 6% annual interest. In the first account, interest compounds quarterly, and in the second account, interest compounds daily. Find the difference between the accounts after 20 years.

42. THE STOCK MARKET The Dow Jones Industrial Average is a measure of how well the stock market is doing. Plot the following Dow milestones as ordered pairs of the form (year, average) on the graph in Illustration 4. What type of function does it appear could be used to model the growth of the stock market?

Year	Average	Year	Average
Jan. 1906	100	Feb. 1995	4,000
Mar. 1956	500	Nov. 1995	5,000
Nov. 1972	1,000	Oct. 1996	6,000
Jan. 1987	2,000	Feb. 1997	7,000
Apr. 1991	3,000	July 1997	8,000

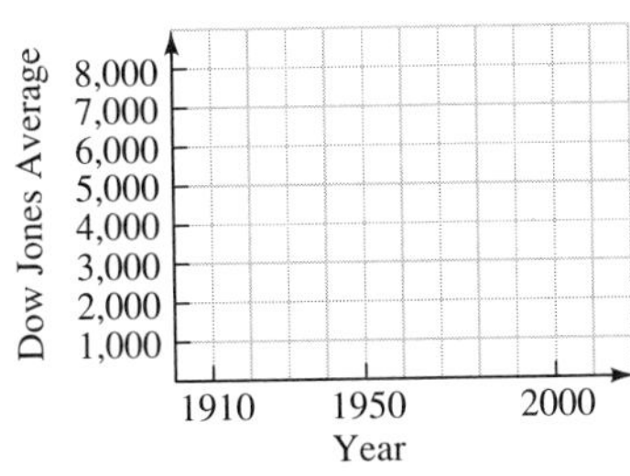

ILLUSTRATION 4

43. MARKET VALUE OF A CAR The graph in Illustration 5 shows how the value of the average car depreciates as a percent of its original value over a 10-year period. It also shows the yearly maintenance costs as a percentage of the car's value.

a. When is the car worth half of its purchase price?

b. When is the car worth a quarter of its purchase price?

c. When do the average yearly maintenance costs surpass the value of the car?

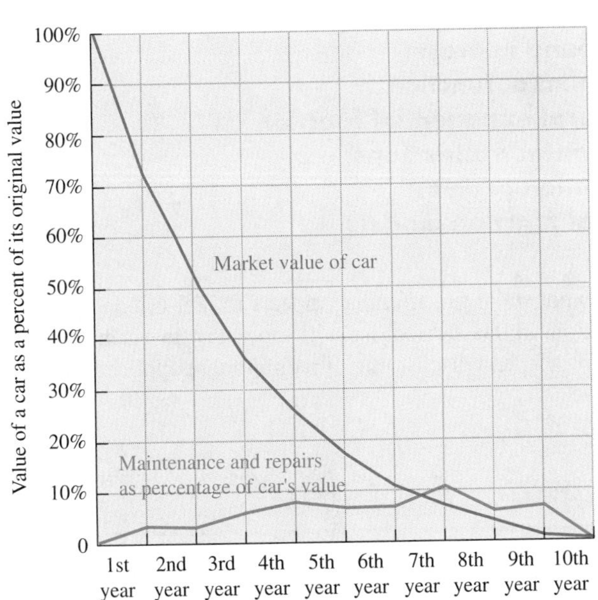

Based on data from the U.S. Department of Transportation

ILLUSTRATION 5

44. DIVING *Bottom time* is the time a scuba diver spends descending plus the actual time spent at a certain depth. On Illustration 6, graph the bottom time limits given in the table.

Bottom time limits			
Depth (ft)	**Bottom time (min)**	**Depth (ft)**	**Bottom time (min)**
30	no limit	80	40
35	310	90	30
40	200	100	25
50	100	110	20
60	60	120	15
70	50	130	10

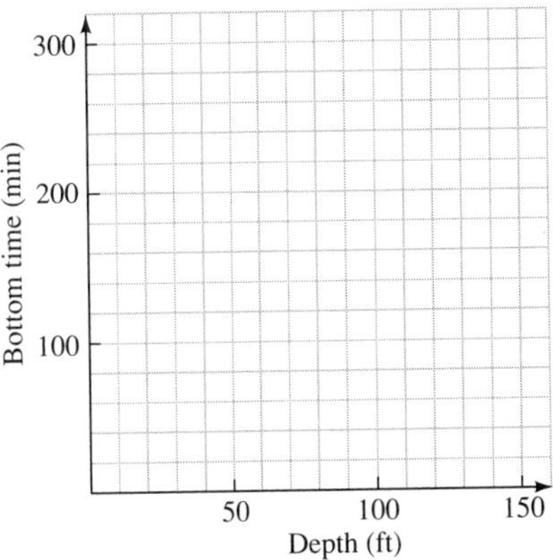

ILLUSTRATION 6

45. BACTERIAL CULTURE A colony of 6 million bacteria is growing in a culture medium. See Illustration 7. The population P after t hours is given by the formula $P = (6 \times 10^6)(2.3)^t$. Find the population after 4 hours.

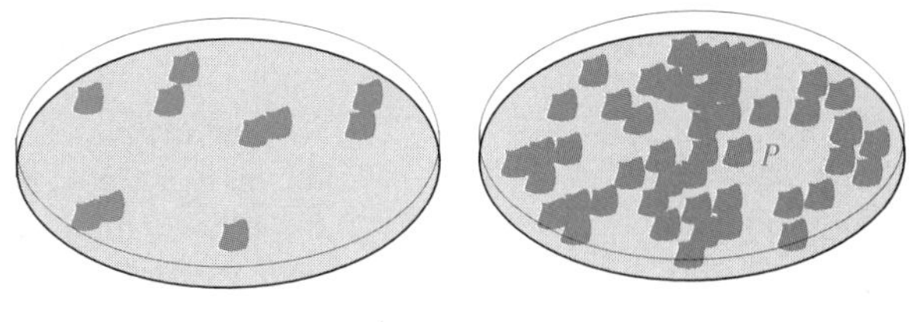

ILLUSTRATION 7

46. RADIOACTIVE DECAY A radioactive material decays according to the formula $A = A_0\left(\frac{2}{3}\right)^t$, where A_0 is the initial amount present and t is measured in years. Find the amount present in 5 years.

47. DISCHARGING A BATTERY The charge remaining in a battery decreases as the battery discharges. The charge C (in coulombs) after t days is given by the formula $C = (3 \times 10^{-4})(0.7)^t$. Find the charge after 5 days.

48. POPULATION GROWTH The population of North Rivers is decreasing exponentially according to the formula $P = 3{,}745(0.93)^t$, where t is measured in years from the present date. Find the population in 6 years, 9 months.

49. SALVAGE VALUE A small business purchased a computer for \$4,700. It is expected that its value each year will be 75% of its value in the preceding year. If the business disposes of the computer after 5 years, find its salvage value (the value after 5 years).

50. LOUISIANA PURCHASE In 1803, the United States negotiated the Louisiana Purchase with France. The country doubled its territory by adding 827,000 square miles of land for \$15 million. If the land appreciated at the rate of 6% each year, what would one square mile of land be worth in 1996?

WRITING *Write a paragraph using your own words.*

51. If world population is increasing exponentially, why is there cause for concern?

52. How do the graphs of $f(x) = 3^x$ and $g(x) = \left(\frac{1}{3}\right)^x$ differ? How are they similar?

REVIEW *In Exercises 53–56, refer to Illustration 8. Lines r and s are parallel.*

53. Find x.

54. Find the measure of $\angle 1$.

55. Find the measure of $\angle 2$.

56. Find the measure of $\angle 3$.

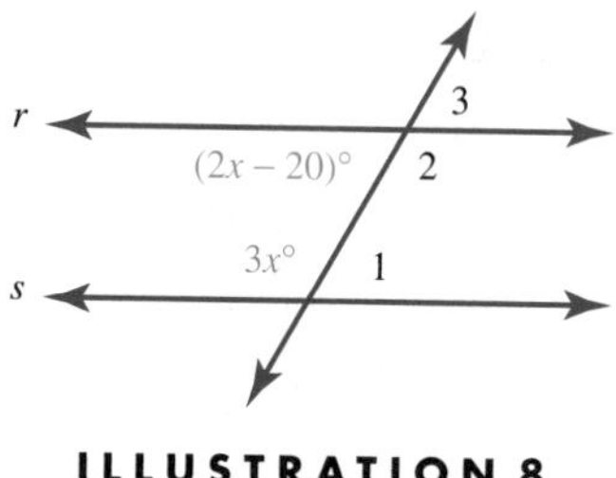

ILLUSTRATION 8

9.4 Base-*e* Exponential Functions

In this section, you will learn about

- **Continuous compound interest**
- **The natural exponential function**
- **Graphing the natural exponential function**
- **Vertical and horizontal translations**
- **Malthusian population growth**
- **Base-e exponential function models**

INTRODUCTION. A special exponential function that appears in real-life applications is the base-e exponential function. In this section, we will show how to evaluate e, graph the exponential function to base e, and discuss one of its important applications in analyzing population growth.

Continuous compound interest

If a bank pays interest twice a year, we say that interest is compounded semiannually. If it pays interest four times a year, we say that interest is compounded quarterly. If it pays interest continuously (infinitely many times in a year), we say that interest is compounded continuously.

To develop the formula for continuous compound interest, we start with the formula

$$A = P\left(1 + \frac{r}{k}\right)^{kt}$$ The formula for compound interest: r is the annual rate and k is the number of times per year interest is paid.

and substitute rn for k. Since r and k are positive numbers, so is n.

$$A = P\left(1 + \frac{r}{rn}\right)^{rnt}$$

We can then simplify the fraction $\frac{r}{rn}$ and use the commutative property of multiplication to change the order of the exponents.

$$A = P\left(1 + \frac{1}{n}\right)^{nrt}$$

Finally, we can use a property of exponents to write the formula as

1. $$A = P\left[\left(1 + \frac{1}{n}\right)^{n}\right]^{rt}$$ Use the property $a^{mn} = (a^m)^n$.

To find the value of $\left(1 + \frac{1}{n}\right)^n$, we evaluate it for several values of n, as shown in Table 9-1.

n	$\left(1 + \frac{1}{n}\right)^n$
1	2
2	2.25
4	2.44140625. . .
12	2.61303529. . .
365	2.71456748. . .
1,000	2.71692393. . .
100,000	2.71826830. . .
1,000,000	2.71828137. . .

TABLE 9-1

The results suggest that as n gets larger, the value of $\left(1 + \frac{1}{n}\right)^n$ approaches the number 2.71828. . . . This number is called e, which has the following value.

$e = 2.718281828459. . .$

In continuous compound interest, k (the number of compoundings) is infinitely large. Since k, r, and n are all positive, r is a fixed rate, and $k = rn$, as k gets very large (approaches infinity), so does n. Therefore, we can replace $\left(1 + \frac{1}{n}\right)^n$ in Equation 1 with e to get

$$A = Pe^{rt}$$

Formula for exponential growth

If a quantity P increases or decreases at an annual rate r, compounded continuously, the amount A after t years is given by

$$A = Pe^{rt}$$

If time is measured in years, then r is called the **annual growth rate.** If r is negative, the "growth" represents a decrease.

The natural exponential function

The number e is often used as the base in applications of exponential functions.

The natural exponential function

The function defined by $f(x) = e^x$ is the **natural exponential function** (or the **base-e exponential function**) where $e = 2.71828. . .$

The $\boxed{e^x}$ key on a calculator is used to evaluate the natural exponential function.

Accent on Technology — The natural exponential function key

To compute the amount to which $12,000 will grow if invested for 18 years at 10% annual interest, compounded continuously, we substitute 12,000 for P, 0.10 for r, and 18 for t in the formula for exponential growth:

$$A = Pe^{rt}$$

$$A = 12{,}000e^{0.10(18)}$$

$$= 12{,}000e^{1.8}$$

To evaluate this expression, we enter these numbers and press these keys on a scientific calculator:

Keystrokes: 1.8 $\boxed{e^x}$ $\boxed{\times}$ 12000 $\boxed{=}$ 72595.76957

After 18 years, the account will contain $72,595.77. This is $1,589.03 more than the result in Example 6 in the previous section, where interest was compounded quarterly.

To evaluate the expression on a graphing calculator, we enter these numbers and press these keys:

Keystrokes: 12000 $\boxed{\times}$ $\boxed{e^x}$ 1.8 $\boxed{\text{ENTER}}$ 72595.76957

EXAMPLE 1 ***Continuous compound interest.*** If $25,000 accumulates interest at an annual rate of 8%, compounded continuously, find the balance in the account in 50 years.

Solution

$$A = Pe^{rt}$$

$$A = 25{,}000e^{(0.08)(50)}$$

$$= 25{,}000e^4$$

$$\approx 1{,}364{,}953.75 \quad \text{Use a calculator.}$$

In 50 years, the balance will be $1,364,953.75—more than a million dollars.

Self Check

Find the balance in 60 years.

Answer: $3,037,760.44 ■

Graphing the natural exponential function

To graph the natural exponential function, we plot several points and join them with a smooth curve, as shown in Figure 9-20.

$f(x) = e^x$

x	$f(x)$	$(x, f(x))$
-2	0.1	$(-2, 0.1)$
-1	0.4	$(-1, 0.4)$
0	1	$(0, 1)$
1	2.7	$(1, 2.7)$
2	7.4	$(2, 7.4)$

↑ The outputs can be found using the e^x key on a calculator.

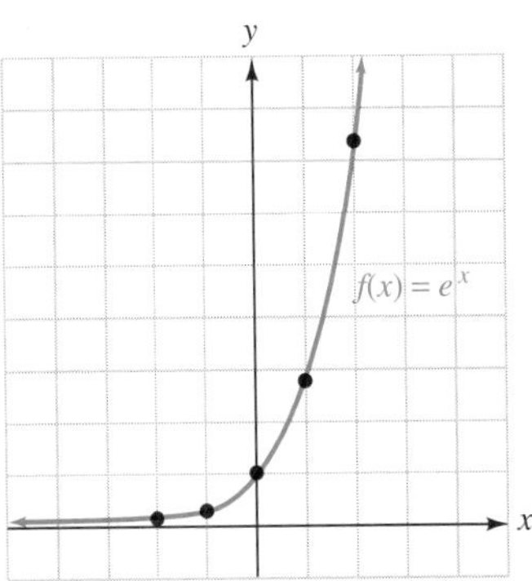

FIGURE 9-20

Vertical and horizontal translations

We can illustrate the effects of vertical and horizontal translations of the natural exponential function by using a graphing calculator.

Accent on Technology — Translations of the natural exponential function

Figure 9-21(a) shows the calculator graphs of $f(x) = e^x$, $g(x) = e^x + 5$, and $h(x) = e^x - 3$. To graph these functions with window settings of $[-3, 6]$ for x and $[-5, 15]$ for y, we enter the right-hand sides of the equations after the symbols Y_1 =, Y_2 =, and Y_3 =. The display will show

$Y_1 = e$ ^ (x)

$Y_2 = e$ ^ $(x) + 5$

$Y_3 = e$ ^ $(x) - 3$

After graphing these functions, we can see that the graph of $g(x) = e^x + 5$ is 5 units above the graph of $f(x) = e^x$, and that the graph of $h(x) = e^x - 3$ is 3 units below the graph of $f(x) = e^x$.

Figure 9-21(b) shows the calculator graphs of $f(x) = e^x$, $g(x) = e^{x+5}$, and $h(x) = e^{x-3}$. To graph these functions with window settings of $[-7, 10]$ for x and $[-5, 15]$ for y, we enter the right-hand sides of the equations after the symbols Y_1 =, Y_2 =, and Y_3 =. The display will show

$Y_1 = e$ ^ (x)

$Y_2 = e$ ^ $(x + 5)$

$Y_3 = e$ ^ $(x - 3)$

After graphing these functions, we can see that the graph of $g(x) = e^{x+5}$ is 5 units to the left of the graph of $f(x) = e^x$, and that the graph of $h(x) = e^{x-3}$ is 3 units to the right of the graph of $f(x) = e^x$.

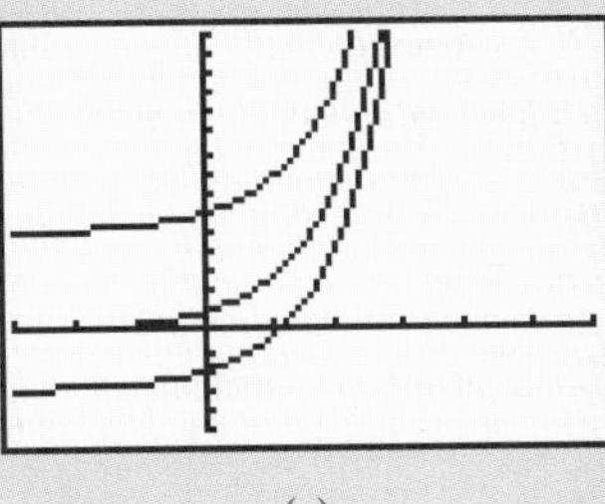

(a)

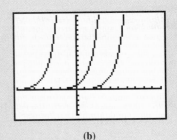

(b)

FIGURE 9-21

Accent on Technology **Graphing exponential functions**

Figure 9-22 shows the calculator graph of $f(x) = 3e^{-x/2}$. To graph this function with window settings of $[-7, 10]$ for x and $[-5, 15]$ for y, we enter the right-hand side of the equation after the symbol $Y_1 =$. The display will show the equation

$$Y_1 = 3(e \wedge (-x/2))$$

The calculator graph appears in Figure 9-22. Explain why the graph has a y-intercept of $(0, 3)$.

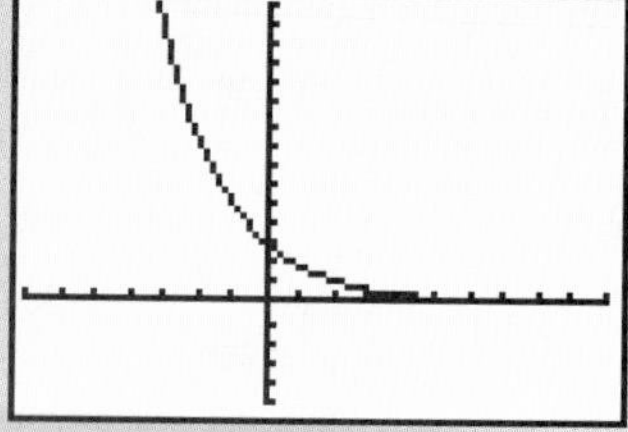

FIGURE 9-22

Malthusian population growth

An equation based on the natural exponential function provides a model for **population growth.** In the **Malthusian model for population growth,** the future population of a colony is related to the present population by the formula $A = Pe^{rt}$.

EXAMPLE 2 ***City planning.*** The population of a city is currently 15,000, but changing economic conditions are causing the population to decrease 2% each year. If this trend continues, find the population in 30 years.

Solution

Since the population is decreasing 2% each year, the annual growth rate is -2%, or -0.02. We can substitute -0.02 for r, 30 for t, and 15,000 for P in the formula for exponential growth and find A.

$$A = Pe^{rt}$$

$$A = 15{,}000e^{-0.02(30)}$$

$$= 15{,}000e^{-0.6}$$

$$\approx 8{,}232.17$$

In 30 years, city planners expect a population of 8,232.

Self Check

Find the population in 50 years.

Answer: 5,518 ■

The English economist Thomas Robert Malthus (1766–1834) pioneered in population study. He believed that poverty and starvation were unavoidable, because the human population tends to grow exponentially, but the food supply tends to grow linearly.

EXAMPLE 3 ***The Malthusian model.*** Suppose that a country with a population of 1,000 people is growing exponentially according to the formula

$$P = 1{,}000e^{0.02t}$$

where t is in years. Furthermore, assume that the food supply F, measured in adequate food per day per person, is growing linearly according to the formula

$$F = 30.625t + 2{,}000 \quad (t \text{ is time in years})$$

In how many years will the population outstrip the food supply?

Self Check

Suppose that the population grows at a 3% rate. Use a graphing calculator to determine for how many years the food supply will be adequate.

Solution

We can use a graphing calculator with window settings of [0, 100] for x and [0, 10,000] for y. After graphing the functions, we obtain Figure 9-23(a). If we trace, as in Figure 9-23(b), we can find the point where the two graphs intersect. From the graph, we can see that the food supply will be adequate for about 71 years. At that time, the population of approximately 4,200 people will begin to have problems.

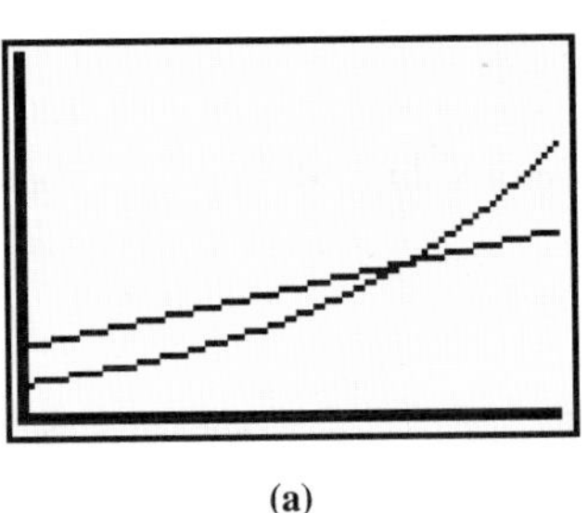
(a)

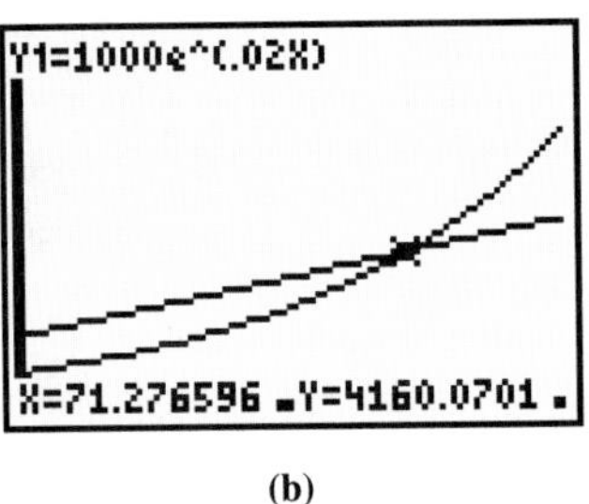

(b)

FIGURE 9-23

Answer: about 38 years ■

Base-e exponential function models

EXAMPLE 4 ***Baking.*** A mother has just taken a cake out of the oven and set it on a rack to cool. The function $T(t) = 68 + 220e^{-0.2t}$ gives the cake's temperature in degrees Fahrenheit after it has cooled for t minutes. If her children will be home from school in 20 minutes, will the cake have cooled enough for the children to eat it?

Solution When the children arrive home, the cake will have cooled for 20 minutes. To find the temperature of the cake at that time, we need to find $T(20)$.

$$\begin{aligned} T(t) &= 68 + 220e^{-0.2t} \\ T(20) &= 68 + 220e^{-0.2(20)} && \text{Substitute 20 for } t. \\ &= 68 + 220e^{-4} \\ &\approx 72.0 && \text{Use a calculator.} \end{aligned}$$

When the children return home, the temperature of the cake will be about 72°, and it can be eaten. ■

STUDY SET Section 9.4

VOCABULARY *In Exercises 1–8, refer to the graph of $f(x) = e^x$ in Illustration 1.*

1. What is the name of the function $f(x) = e^x$?

2. What is the domain of the function?

3. What is the range of the function?

4. What is the y-intercept of the graph?

5. What is the x-intercept of the graph?

6. What is an asymptote of the graph?

7. Is f an increasing or a decreasing function?

8. The graph passes through the point $(1, y)$. What is y?

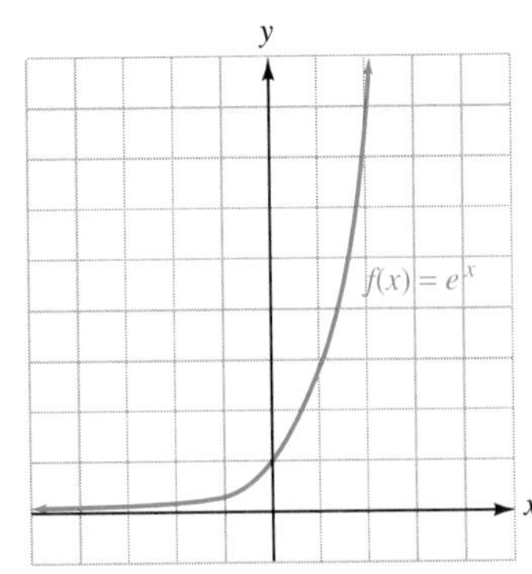

ILLUSTRATION 1

CONCEPTS *In Exercises 9–12, fill in the blanks to make the statements true.*

9. In ______________ compound interest, the number of compoundings is infinitely large.

10. The formula for continuous compound interest is $A =$ ______________.

11. To two decimal places, the value of e is ______________.

12. If n gets larger and larger, the value of $\left(1 + \frac{1}{n}\right)^n$ approaches the value of ______________.

13. Graph each irrational number on the number line. $\left[\pi, e, \sqrt{2}, \frac{\sqrt{3}}{2}\right]$

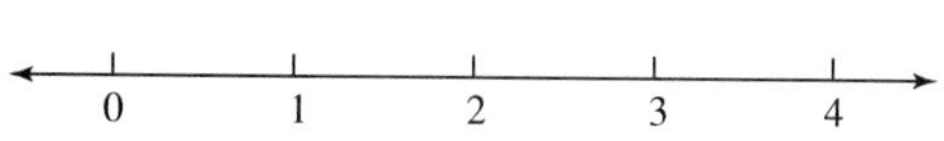

14. Complete the table of values. Round to the nearest hundredth.

x	−2	−1	0	1	2
e^x					

15. POPULATION OF THE UNITED STATES In Illustration 2, graph the U.S. census population figures shown in the table (in millions). What type of function does it appear could be used to model the population?

Year	Population	Year	Population
1790	3.9	1900	76.0
1800	5.3	1910	92.2
1810	7.2	1920	106.0
1820	9.6	1930	123.2
1830	12.9	1940	132.1
1840	17.0	1950	151.3
1850	23.1	1960	179.3
1860	31.4	1970	203.3
1870	38.5	1980	226.5
1880	50.1	1990	248.7
1890	62.9		

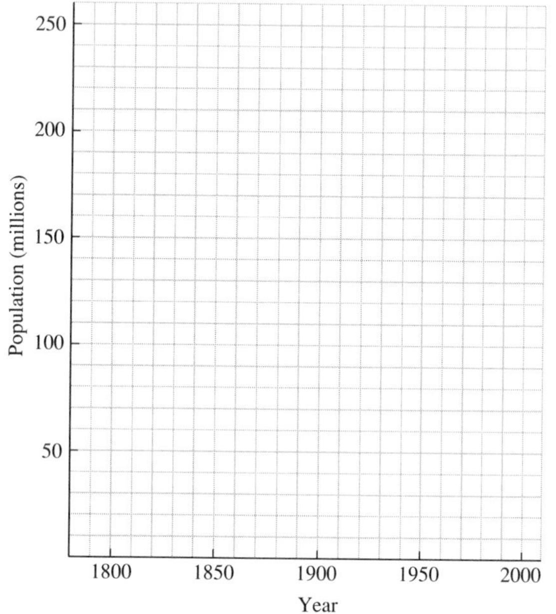

ILLUSTRATION 2

16. What is the Malthusian population growth formula?

NOTATION *In Exercises 17–18, evaluate A in the formula $A = Pe^{rt}$ for the following values of r and t.*

17. $P = 1{,}000$, $r = 0.09$ and $t = 10$

$A = \square e^{(0.09)(\square)}$

$= 1{,}000e^{\square}$

$\approx \square(2.459603111)$

$\approx 2{,}459.603111$

18. $P = 1{,}000$, $r = 0.12$ and $t = 50$

$A = 1{,}000e^{(\square)(50)}$

$= \square e^{6}$

$\approx 1{,}000(\square)$

$\approx 403{,}428.7935$

PRACTICE *In Exercises 19–26, graph each function. Check your work with a graphing calculator. Compare each graph to the graph of $f(x) = e^x$.*

19. $f(x) = e^x + 1$

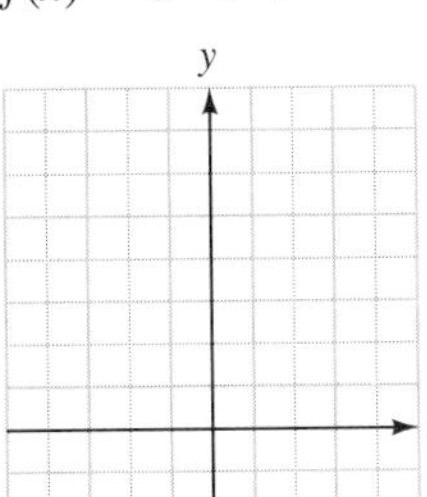

20. $f(x) = e^x - 2$

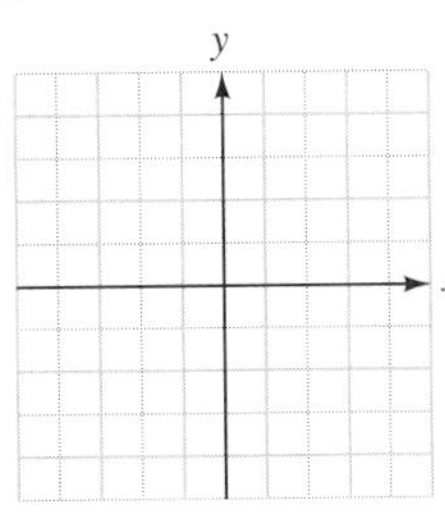

21. $y = e^{x+3}$

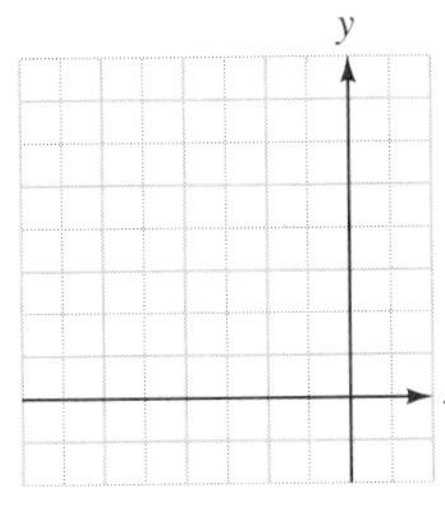

22. $y = e^{x-5}$

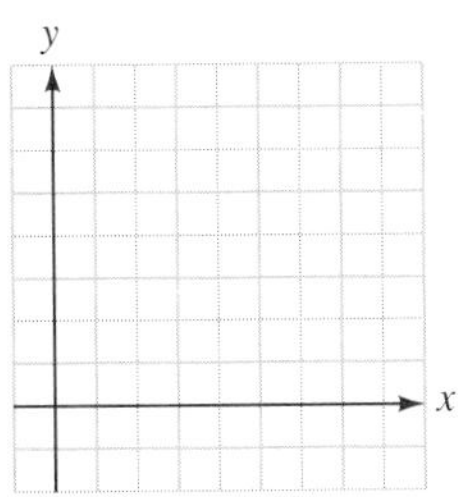

23. $f(x) = -e^x$

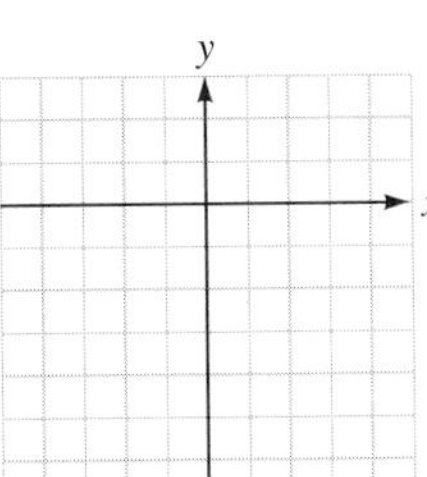

24. $f(x) = -e^x + 1$

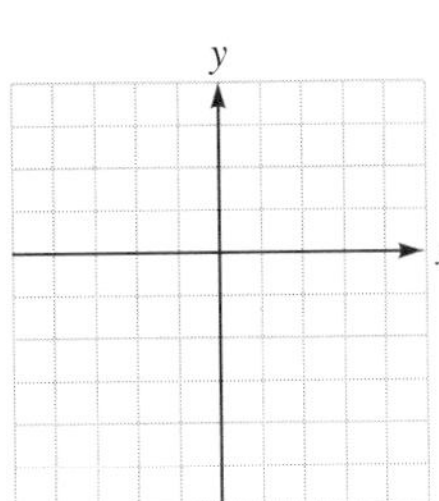

25. $f(x) = 2e^x$

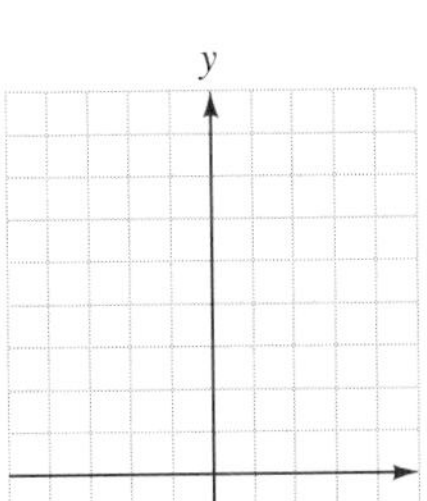

26. $f(x) = \frac{1}{2}e^x$

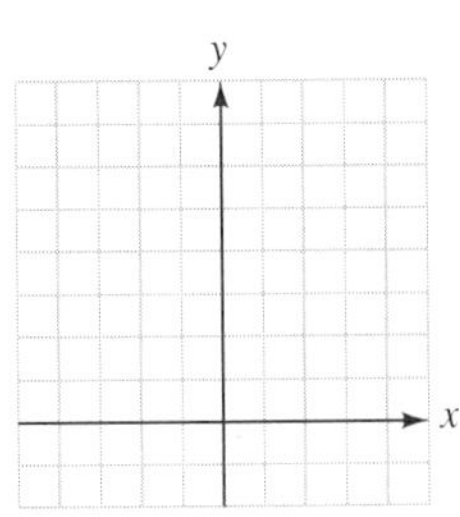

APPLICATIONS *In Exercises 27–32, assume that there are no deposits or withdrawals.*

27. CONTINUOUS COMPOUND INTEREST An initial investment of \$5,000 earns 8.2% interest, compounded continuously. What will the investment be worth in 12 years?

28. CONTINUOUS COMPOUND INTEREST An initial investment of \$2,000 earns 8% interest, compounded continuously. What will the investment be worth in 15 years?

29. COMPARISON OF COMPOUNDING METHODS An initial deposit of \$5,000 grows at an annual rate of 8.5% for 5 years. Compare the final balances resulting from annual compounding and continuous compounding.

30. COMPARISON OF COMPOUNDING METHODS An initial deposit of \$30,000 grows at an annual rate of 8% for 20 years. Compare the final balances resulting from annual compounding and continuous compounding.

31. DETERMINING THE INITIAL DEPOSIT An account now contains \$11,180 and has been accumulating interest at 7% annual interest, compounded continuously, for 7 years. Find the initial deposit.

32. DETERMINING THE PREVIOUS BALANCE An account now contains \$3,610 and has been accumulating interest at 8% annual interest, compounded continuously. How much was in the account 4 years ago?

33. WORLD POPULATION GROWTH The population of the earth is approximately 6 billion people and is growing at an annual rate of 1.9%. Assuming a Malthusian growth model, find the world population in 30 years.

34. WORLD POPULATION GROWTH The population of the earth is approximately 6 billion people and is growing at an annual rate of 1.9%. Assuming a Malthusian growth model, find the world population in 40 years.

35. POPULATION GROWTH The growth of a population is modeled by

$$P = 173e^{0.03t}$$

How large will the population be when $t = 20$?

36. POPULATION DECLINE The decline of a population is modeled by

$$P = (1.2 \times 10^6)e^{-0.008t}$$

How large will the population be when $t = 30$?

37. EPIDEMIC The spread of ungulate fever through a herd of cattle can be modeled by the formula

$$P = P_0e^{0.27t} \quad (t \text{ is in days})$$

If a rancher does not act quickly to treat two cases, how many cattle will have the disease in 12 days?

38. MEDICINE The concentration x of a certain prescription drug in an organ after t minutes is given by

$$x = 0.08(1 - e^{-0.1t})$$

Find the concentration of the drug at 30 minutes.

39. SKYDIVING Before the parachute opens, a skydiver's velocity v in meters per second is given by

$$v = 50(1 - e^{-0.2t})$$

Find the velocity after 20 seconds of free fall.

40. FREE FALL After t seconds a certain falling object has a velocity v given by

$$v = 50(1 - e^{-0.3t})$$

Which is falling faster after 2 seconds—the object or the skydiver in Exercise 39?

In Exercises 41–42, use a graphing calculator to solve each problem.

41. THE MALTHUSIAN MODEL In Example 3, suppose that better farming methods changed the formula for food growth to $y = 31x + 2{,}000$. How long would the food supply be adequate?

42. THE MALTHUSIAN MODEL In Example 3, suppose that a birth-control program changed the formula for population growth to $P = 1{,}000e^{0.01t}$. How long would the food supply be adequate?

WRITING *Write a paragraph using your own words.*

43. Explain why the graph of $y = e^x - 5$ is five units below the graph of $y = e^x$.

44. A feature article in a newspaper stated that the sport of snowboarding was growing *exponentially.* Explain what the author of the article meant by that.

REVIEW *In Exercises 45–48, simplify each expression. Assume that all variables represent positive numbers.*

45. $\sqrt{240x^5}$

46. $\sqrt[3]{-125x^5y^4}$

47. $4\sqrt{48y^3} - 3y\sqrt{12y}$

48. $\sqrt[4]{48z^5} + \sqrt[4]{768z^5}$

9.5 Logarithmic Functions

In this section, you will learn about

- **Logarithms**
- **Graphs of logarithmic functions**
- **Vertical and horizontal translations**
- **Base-10 logarithms**
- **Applications of logarithms**

INTRODUCTION. A function that is closely related to the exponential function is the *logarithmic function.* It can be used to solve many application problems from fields such as electronics, seismology, business, and population growth.

Logarithms

Because an exponential function defined by $y = b^x$ (or $f(x) = b^x$) is one-to-one, it has an inverse function that is defined by the equation $x = b^y$. To express this inverse function in the form $y = f^{-1}(x)$, we must solve the equation $x = b^y$ for y. For this, we need the following definition.

Logarithmic functions	If $b > 0$ and $b \neq 1$, the **logarithmic function with base b** is defined by $y = \log_b x$ if and only if $x = b^y$. The **domain of the logarithmic function** is the interval $(0, \infty)$. The **range** is the interval $(-\infty, \infty)$.

Since the function $y = \log_b x$ is the inverse of the one-to-one exponential function $y = b^x$, the logarithmic function is also one-to-one.

WARNING! Since the domain of the logarithmic function is the set of positive numbers, it is impossible to find the logarithm of 0 or the logarithm of a negative number.

The previous definition guarantees that any pair (x, y) that satisfies the equation $y = \log_b x$ also satisfies the exponential equation $x = b^y$. Because of this relationship, a statement written in logarithmic form can be written in an equivalent exponential form.

Logarithmic form		*Exponential form*
$\log_7 1 = 0$	because	$1 = 7^0$
$\log_5 25 = 2$	because	$25 = 5^2$
$\log_5 \frac{1}{25} = -2$	because	$\frac{1}{25} = 5^{-2}$
$\log_{16} 4 = \frac{1}{2}$	because	$4 = 16^{1/2}$
$\log_2 \frac{1}{8} = -3$	because	$\frac{1}{8} = 2^{-3}$
$\log_b x = y$	because	$x = b^y$

In each of these examples, the logarithm of a number is an exponent. In fact,

$\log_b x$ is the exponent to which b is raised to get x.

Translating this statement into symbols, we have

$$b^{\log_b x} = x$$

EXAMPLE 1 ***Writing an equivalent exponential statement.*** Find y in each equation: **a.** $\log_6 1 = y$, **b.** $\log_3 27 = y$, and **c.** $\log_5 \frac{1}{5} = y$.

Solution

a. We can change the equation $\log_6 1 = y$ into the equivalent exponential equation $1 = 6^y$. Since $1 = 6^0$, it follows that $y = 0$. Thus,

$$\log_6 1 = 0$$

b. $\log_3 27 = y$ is equivalent to $27 = 3^y$. Since $27 = 3^3$, it follows that $3^y = 3^3$, and $y = 3$. Thus,

$$\log_3 27 = 3$$

c. $\log_5 \frac{1}{5} = y$ is equivalent to $\frac{1}{5} = 5^y$. Since $\frac{1}{5} = 5^{-1}$, it follows that $5^y = 5^{-1}$, and $y = -1$. Thus,

$$\log_5 \frac{1}{5} = -1$$

Self Check

Find y in each equation:
a. $\log_3 9 = y$, **b.** $\log_2 16 = y$, and **c.** $\log_5 \frac{1}{125} = y$.

Answers: **a.** 2, **b.** 4, **c.** -3

EXAMPLE 2 ***Writing an equivalent exponential statement.*** Find the value of x in each equation: **a.** $\log_3 81 = x$, **b.** $\log_x 125 = 3$, and **c.** $\log_4 x = 3$.

Solution

a. $\log_3 81 = x$ is equivalent to $3^x = 81$. Because $3^4 = 81$, it follows that $3^x = 3^4$. Thus, $x = 4$.

b. $\log_x 125 = 3$ is equivalent to $x^3 = 125$. Because $5^3 = 125$, it follows that $x^3 = 5^3$. Thus, $x = 5$.

c. $\log_4 x = 3$ is equivalent to $4^3 = x$. Because $4^3 = 64$, it follows that $x = 64$.

Self Check

Find x in each equation:
a. $\log_2 32 = x$, **b.** $\log_x 8 = 3$, and **c.** $\log_5 x = 2$.

Answers: **a.** 5, **b.** 2, **c.** 25

EXAMPLE 3 ***Writing an equivalent exponential statement.*** Find the value of x in each equation: **a.** $\log_{1/3} x = 2$, **b.** $\log_{1/3} x = -2$, and **c.** $\log_{1/3} \frac{1}{27} = x$.

Solution

a. $\log_{1/3} x = 2$ is equivalent to $\left(\frac{1}{3}\right)^2 = x$. Thus, $x = \frac{1}{9}$.

b. $\log_{1/3} x = -2$ is equivalent to $\left(\frac{1}{3}\right)^{-2} = x$. Thus,

$$x = \left(\frac{1}{3}\right)^{-2} = 3^2 = 9$$

c. $\log_{1/3} \frac{1}{27} = x$ is equivalent to $\left(\frac{1}{3}\right)^x = \frac{1}{27}$. Because $\left(\frac{1}{3}\right)^3 = \frac{1}{27}$, it follows that $x = 3$.

Self Check

Find x in each equation.
a. $\log_{1/4} x = 3$ and
b. $\log_{1/4} x = -2$.

Answers: **a.** $\frac{1}{64}$, **b.** 16

Graphs of logarithmic functions

To graph the logarithmic function $y = \log_2 x$, we calculate and plot several points with coordinates (x, y) that satisfy the equivalent equation $x = 2^y$. After joining these points with a smooth curve, we have the graph shown in Figure 9-24(a).

To graph $y = \log_{1/2} x$, we calculate and plot several points with coordinates (x, y) that satisfy the equation $x = \left(\frac{1}{2}\right)^y$. After joining these points with a smooth curve, we have the graph shown in Figure 9-24(b).

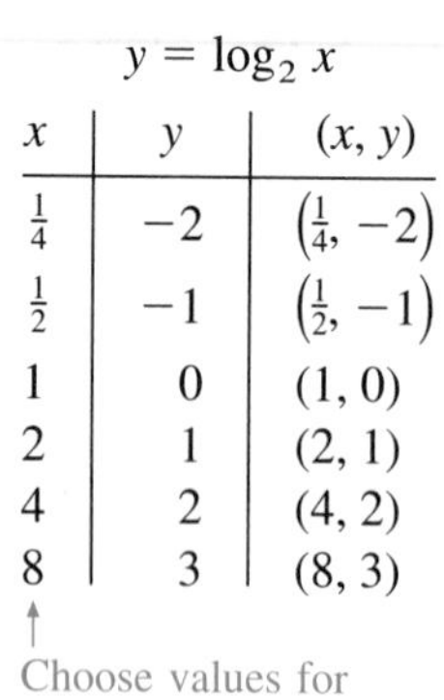

$y = \log_2 x$

x	y	(x, y)
$\frac{1}{4}$	-2	$\left(\frac{1}{4}, -2\right)$
$\frac{1}{2}$	-1	$\left(\frac{1}{2}, -1\right)$
1	0	(1, 0)
2	1	(2, 1)
4	2	(4, 2)
8	3	(8, 3)

↑ Choose values for x that are integer powers of 2.

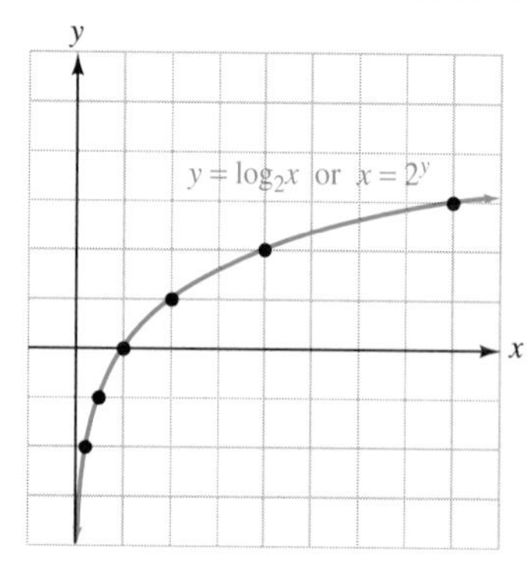

(a)

$y = \log_{1/2} x$

x	y	(x, y)
$\frac{1}{4}$	2	$\left(\frac{1}{4}, 2\right)$
$\frac{1}{2}$	1	$\left(\frac{1}{2}, 1\right)$
1	0	(1, 0)
2	-1	$(2, -1)$
4	-2	$(4, -2)$
8	-3	$(8, -3)$

↑ Choose values for x that are integer powers of $\frac{1}{2}$.

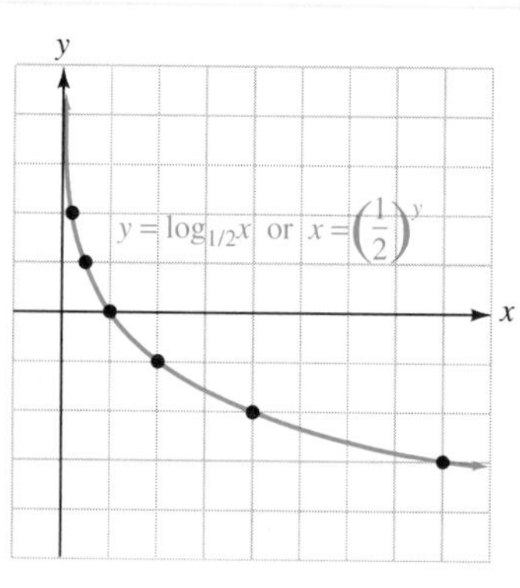

(b)

FIGURE 9-24

The graphs of all logarithmic functions are similar to those in Figure 9-25. If $b > 1$, the logarithmic function is increasing, as in Figure 9-25(a). If $0 < b < 1$, the logarithmic function is decreasing, as in Figure 9-25(b).

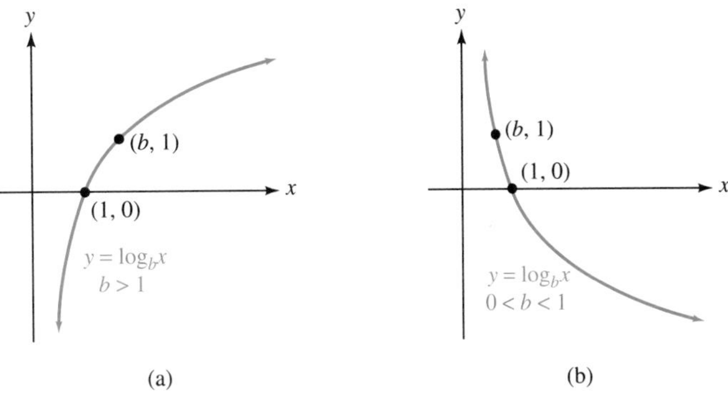

FIGURE 9-25

The graph of $y = \log_b x$ (or $f(x) = \log_b x$) has the following properties.

1. It passes through the point $(1, 0)$.
2. It passes through the point $(b, 1)$.
3. The y-axis (the line $x = 0$) is an asymptote.
4. The domain is $(0, \infty)$ and the range is $(-\infty, \infty)$.

The exponential and logarithmic functions are inverses of each other, so they have symmetry about the line $y = x$. The graphs of $f(x) = \log_b x$ and $g(x) = b^x$ are shown in Figure 9-26(a) when $b > 1$ and in Figure 9-26(b) when $0 < b < 1$.

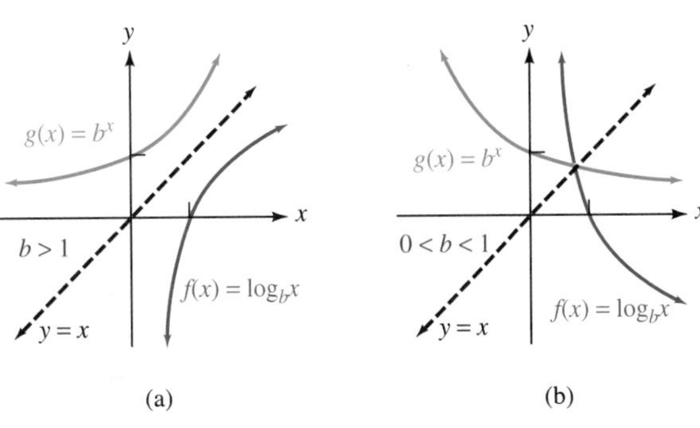

FIGURE 9-26

Vertical and horizontal translations

The graphs of many functions involving logarithms are translations of the basic logarithmic graphs.

EXAMPLE 4 ***Graphing logarithmic functions.*** Graph the function defined by $f(x) = 3 + \log_2 x$ and describe the translation.

Solution

The graph of $f(x) = 3 + \log_2 x$ is identical to the graph of $y = \log_2 x$, except that it is translated 3 units upward. (See Figure 9-27.)

Self Check

Graph $y = \log_3 x - 2$ and describe the translation.

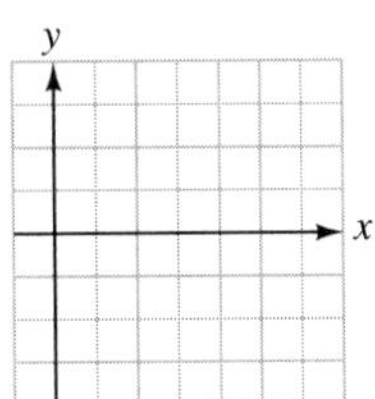

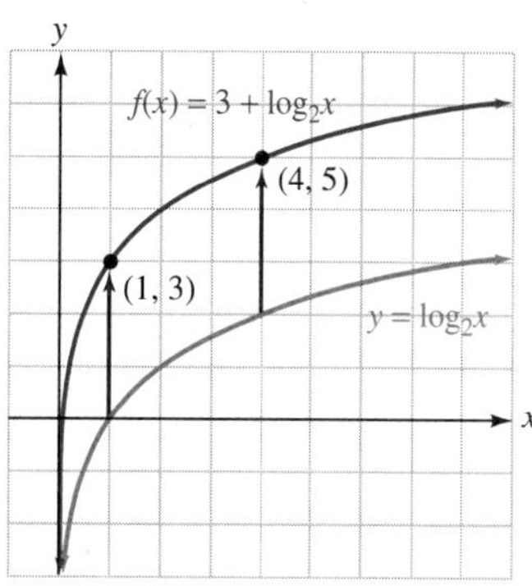

FIGURE 9-27

Answer: The graph of $y = \log_3 x$ is translated 2 units downward.

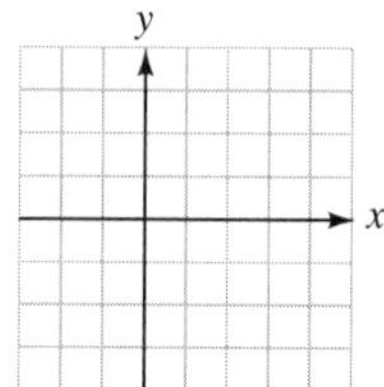

EXAMPLE 5 ***Graphing logarithmic functions.*** Graph $y = \log_{1/2}(x - 1)$ and describe the translation.

Solution

The graph of $y = \log_{1/2}(x - 1)$ is identical to the graph of $y = \log_{1/2} x$, except that it is translated 1 unit to the right. (See Figure 9-28.)

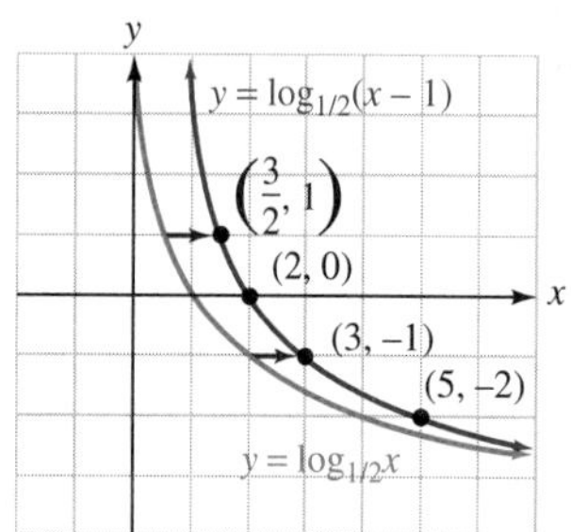

FIGURE 9-28

Self Check

Graph $f(x) = \log_{1/3}(x + 2)$ and describe the translation.

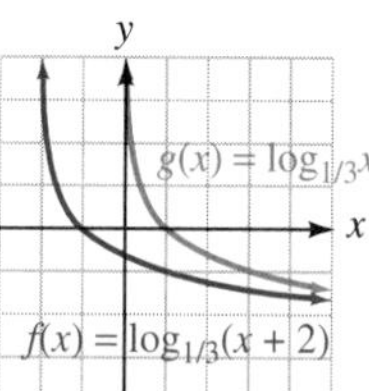

Answer: The graph of $g(x) = \log_{1/3} x$ is translated 2 units to the left.

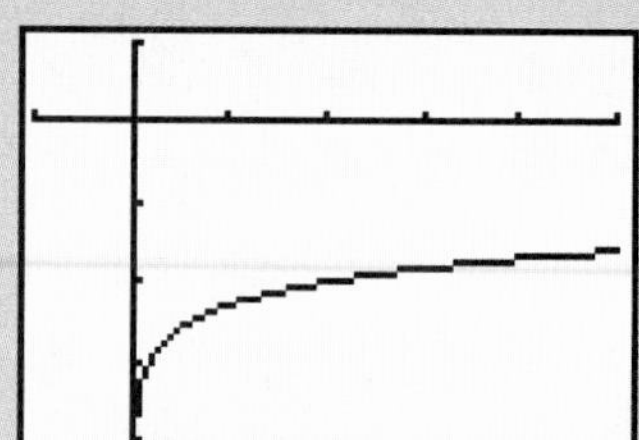

Accent on Technology **Graphing logarithmic functions**

Graphing calculators can draw graphs of logarithmic functions. To use a calculator to graph the logarithmic function $y = -2 + \log_{10}\left(\frac{1}{2}x\right)$, we enter the right-hand side of the equation after the symbol $Y_1 =$. The display will show the equation

$$Y_1 = -2 + \log(1/2x)$$

If we use window settings of $[-1, 5]$ for x and $[-4, 1]$ for y and press the GRAPH key, we will obtain the graph shown in Figure 9-29.

FIGURE 9-29

Base-10 logarithms

For computational purposes and in many applications, we will use base-10 logarithms (also called **common logarithms**). When the base b is not indicated in the notation $\log x$, we assume that $b = 10$:

$$\log x \quad \text{means} \quad \log_{10} x$$

Because base-10 logarithms appear so often, it is a good idea to become familiar with the following base-10 logarithms:

Logarithmic form		*Exponential form*
$\log_{10} \frac{1}{100} = -2$	because	$10^{-2} = \frac{1}{100}$
$\log_{10} \frac{1}{10} = -1$	because	$10^{-1} = \frac{1}{10}$
$\log_{10} 1 = 0$	because	$10^0 = 1$
$\log_{10} 10 = 1$	because	$10^1 = 10$
$\log_{10} 100 = 2$	because	$10^2 = 100$
$\log_{10} 1{,}000 = 3$	because	$10^3 = 1{,}000$

In general, we have

$$\log_{10} 10^x = x$$

Accent on Technology **Finding logarithms**

Before calculators, extensive tables provided logarithms of numbers. Today, logarithms are easy to find with a calculator. For example, to find log 2.34 with a scientific calculator, we enter these numbers and press these keys:

2.34 [LOG]

The display will read [.369215857.] To four decimal places, log 2.34 = 0.3692.

To use a graphing calculator, we enter these numbers and press these keys:

[LOG] 2.34 [ENTER]

The display will read

```
log 2.34
        .3692158574
```

EXAMPLE 6 ***Using a calculator.*** Find x in the equation $\log x = 0.3568$. Round to four decimal places.

Solution

The equation $\log x = 0.3568$ is equivalent to $10^{0.3568} = x$. To find x with a scientific calculator, we enter these numbers and press these keys:

10 [y^x] .3568 [=]

The display will read [2.274049951]. To four decimal places,

$$x = 2.2740$$

If your calculator has a [10^x] key, enter .3568 and press it to get the same result.

Self Check

Solve $\log x = 1.87737$. Round to four decimal places.

Answer: 75.3998 ■

Applications of logarithms

Common logarithms are used in electrical engineering to express the voltage gain (or loss) of an electronic device such as an amplifier. The unit of gain (or loss), called the **decibel,** is defined by a logarithmic relation.

Decibel voltage gain

If E_O is the output voltage of a device and E_I is the input voltage, the decibel voltage gain of the device (db gain) is given by

$$\text{db gain} = 20 \log \frac{E_O}{E_I}$$

EXAMPLE 7 ***Finding db gain.*** If the input to an amplifier is 0.5 volt and the output is 40 volts, find the decibel voltage gain of the amplifier.

Solution We can find the decibel voltage gain by substituting 0.5 for E_I and 40 for E_O into the formula for db gain:

$$\begin{aligned}\text{db gain} &= 20 \log \frac{E_O}{E_I} \\ \text{db gain} &= 20 \log \frac{\mathbf{40}}{\mathbf{0.5}} \\ &= 20 \log 80 \\ &\approx 38 \qquad \text{Use a calculator.}\end{aligned}$$

The amplifier provides a 38-decibel voltage gain. ■

In seismology, common logarithms are used to measure the intensity of earthquakes on the **Richter scale.** The intensity of an earthquake is given by the following logarithmic function.

Richter scale

If R is the intensity of an earthquake, A is the amplitude (measured in micrometers), and P is the period (the time of one oscillation of the earth's surface measured in seconds), then

$$R = \log \frac{A}{P}$$

EXAMPLE 8 ***Measuring earthquakes.*** Find the measure on the Richter scale of an earthquake with an amplitude of 5,000 micrometers (0.5 centimeter) and a period of 0.1 second.

Solution We substitute 5,000 for A and 0.1 for P in the Richter scale formula and simplify:

$$\begin{aligned}R &= \log \frac{A}{P} \\ R &= \log \frac{\mathbf{5{,}000}}{\mathbf{0.1}} \\ &= \log 50{,}000 \\ &\approx 4.698970004 \qquad \text{Use a calculator.}\end{aligned}$$

The earthquake measures about 4.7 on the Richter scale. ■

STUDY SET Section 9.5

VOCABULARY *In Exercises 1–8, refer to the graph of $f(x) = \log_4 x$ in Illustration 1.*

1. What type of function is $f(x) = \log_4 x$?

2. What is the domain of the function?

3. What is the range of the function?

4. What is the y-intercept of the graph?

5. What is the x-intercept of the graph?

6. What is an asymptote of the graph?

7. Is f an increasing or a decreasing function?

8. The graph passes through the point $(4, y)$. What is y?

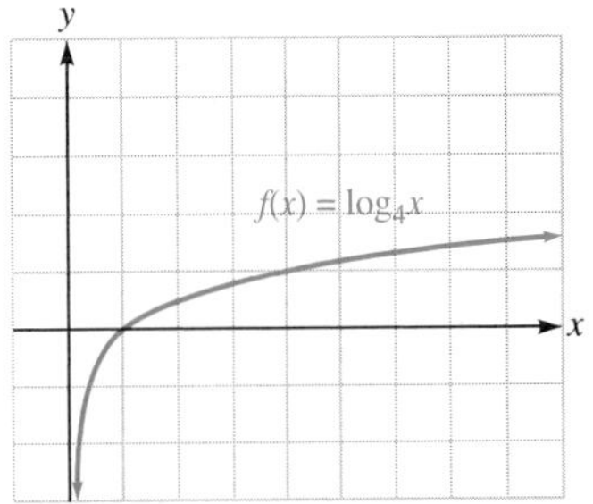

ILLUSTRATION 1

CONCEPTS *In Exercises 9–12, fill in the blanks to make the statements true.*

9. The equation $y = \log_b x$ is equivalent to the exponential equation ______________.

10. $\log_b x$ is the ______________ to which b is raised to get x.

11. The functions $f(x) = \log_b x$ and $f(x) = b^x$ are ______________ functions.

12. The inverse of an exponential function is called a ______________ function.

In Exercises 13–16, complete the table of values. If an evaluation is not possible, write "none."

13. $y = \log x$

x	y
$\frac{1}{100}$	
$\frac{1}{10}$	
1	
10	
100	

14. $f(x) = \log_5 x$

x	$f(x)$
$\frac{1}{25}$	
$\frac{1}{5}$	
1	
5	
25	

15. $f(x) = \log_6 x$

Input	Output
-6	
0	
$\frac{1}{216}$	
$\sqrt{6}$	
6^8	

16. $f(x) = \log_8 x$

x	$f(x)$
-8	
0	
$\frac{1}{8}$	
$\sqrt{8}$	
64	

17. Use a calculator to complete the table of values for $y = \log x$. Round to the nearest hundredth.

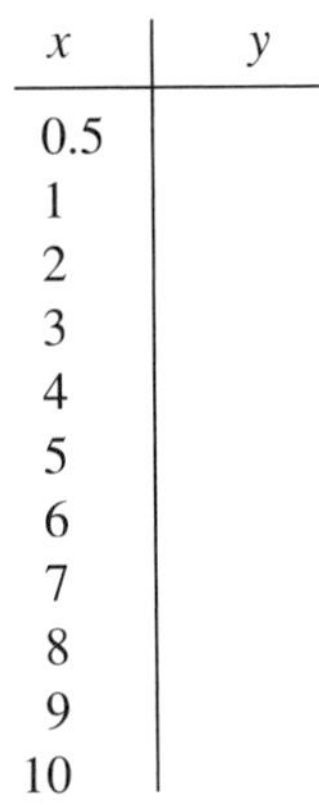

x	y
0.5	
1	
2	
3	
4	
5	
6	
7	
8	
9	
10	

18. Graph $y = \log x$ in Illustration 2. (See Exercise 17.) Note that the units on the x- and y-axes are different.

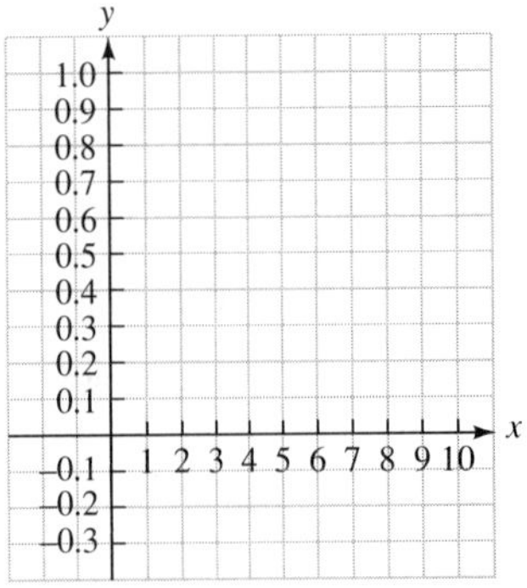

ILLUSTRATION 2

NOTATION *In Exercises 19–20, fill in the blanks to make the statements true.*

19. The notation $\log x$ means $\log_{\square} x$.

20. $\log_{10} 10^x = \square$.

PRACTICE *In Exercises 21–28, write the equation in exponential form.*

21. $\log_3 81 = 4$

22. $\log_7 7 = 1$

23. $\log_{12} 12 = 1$

24. $\log_6 36 = 2$

25. $\log_4 \dfrac{1}{64} = -3$

26. $\log_6 \dfrac{1}{36} = -2$

27. $\log 0.001 = -3$

28. $\log_3 243 = 5$

In Exercises 29–36, write the equation in logarithmic form.

29. $8^2 = 64$

30. $10^3 = 1{,}000$

31. $4^{-2} = \dfrac{1}{16}$

32. $3^{-4} = \dfrac{1}{81}$

33. $\left(\dfrac{1}{2}\right)^{-5} = 32$

34. $\left(\dfrac{1}{3}\right)^{-3} = 27$

35. $x^y = z$

36. $m^n = p$

In Exercises 37–64, find each value of x.

37. $\log_2 8 = x$

38. $\log_3 9 = x$

39. $\log_4 64 = x$

40. $\log_6 216 = x$

41. $\log_{1/2} \dfrac{1}{8} = x$

42. $\log_{1/3} \dfrac{1}{81} = x$

43. $\log_9 3 = x$

44. $\log_{125} 5 = x$

45. $\log_8 x = 2$

46. $\log_7 x = 0$

47. $\log_{25} x = \dfrac{1}{2}$

48. $\log_4 x = \dfrac{1}{2}$

49. $\log_5 x = -2$

50. $\log_3 x = -4$

51. $\log_{36} x = -\dfrac{1}{2}$

52. $\log_{27} x = -\dfrac{1}{3}$

53. $\log_{100} \dfrac{1}{1{,}000} = x$

54. $\log_{5/2} \dfrac{4}{25} = x$

55. $\log_{27} 9 = x$

56. $\log_{12} x = 0$

57. $\log_x 5^3 = 3$

58. $\log_x 5 = 1$

59. $\log_x \dfrac{9}{4} = 2$

60. $\log_x \dfrac{\sqrt{3}}{3} = \dfrac{1}{2}$

61. $\log_x \dfrac{1}{64} = -3$

62. $\log_x \dfrac{1}{100} = -2$

63. $\log_8 x = 0$

64. $\log_4 8 = x$

In Exercises 65–68, use a calculator to find each value. Give answers to four decimal places.

65. $\log 3.25$

66. $\log 0.57$

67. $\log 0.00467$

68. $\log 375.876$

In Exercises 69–72, use a calculator to find each value of y. Give answers to two decimal places.

69. $\log y = 1.4023$

70. $\log y = 0.926$

71. $\log y = -1.71$

72. $\log y = -0.5$

In Exercises 73–76, graph each function. Tell whether each function is an increasing or a decreasing function.

73. $f(x) = \log_3 x$

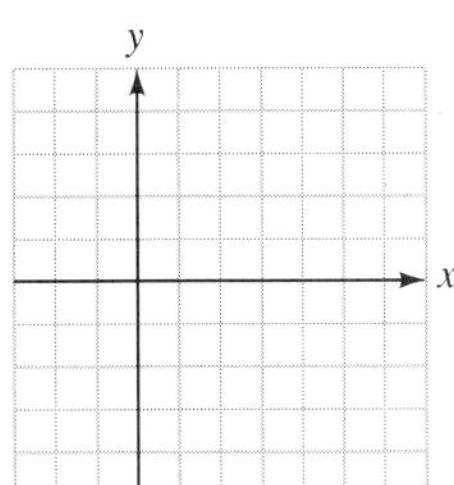

74. $f(x) = \log_{1/3} x$

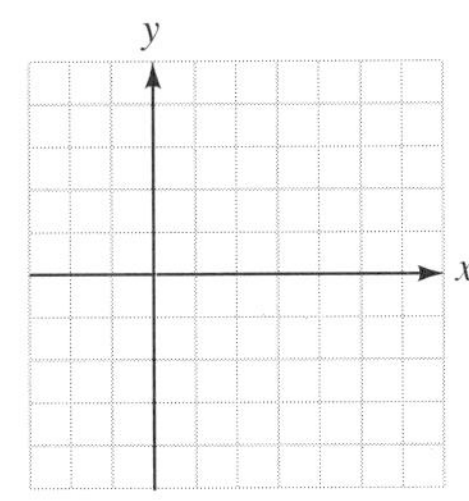

75. $y = \log_{1/2} x$

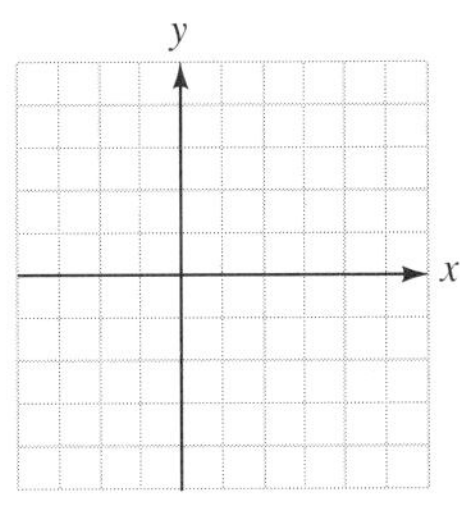

76. $y = \log_4 x$

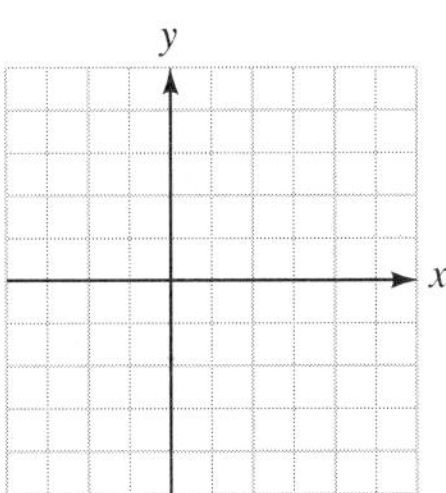

In Exercises 77–80, graph each function.

77. $f(x) = 3 + \log_3 x$

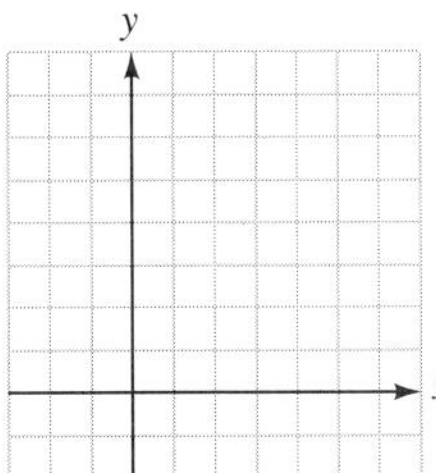

78. $f(x) = \log_{1/3} x - 1$

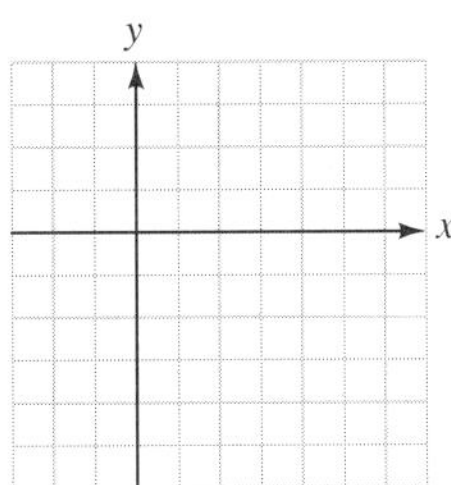

79. $y = \log_{1/2} (x - 2)$

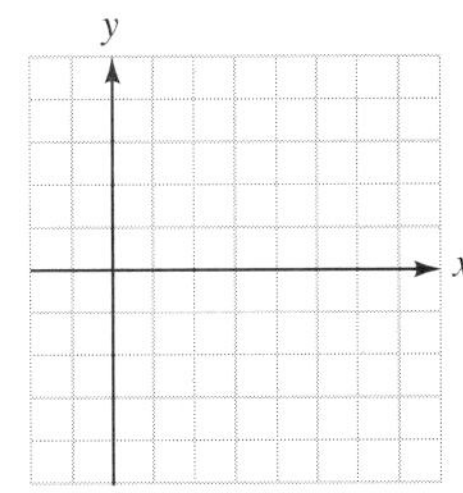

80. $y = \log_4 (x + 2)$

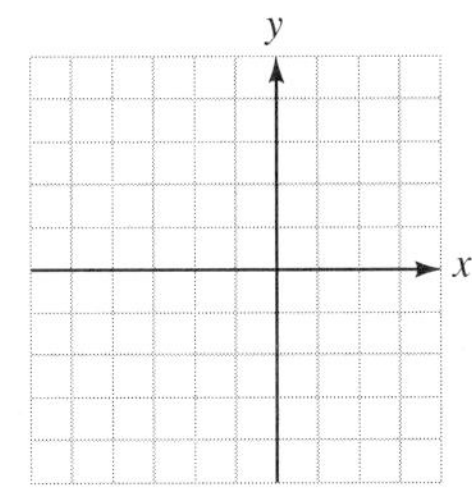

In Exercises 81–84, graph each pair of inverse functions on a single coordinate system. Draw the axis of symmetry.

81. $y = 6^x$
$y = \log_6 x$

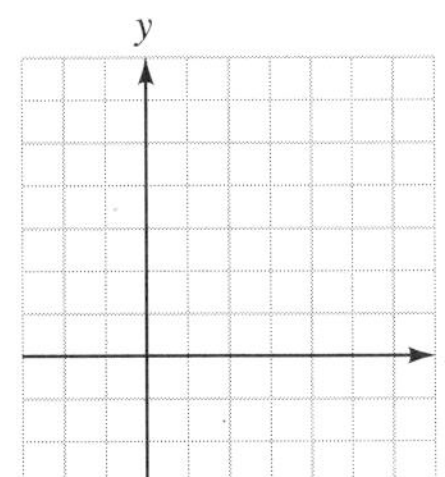

82. $y = 3^x$
$y = \log_3 x$

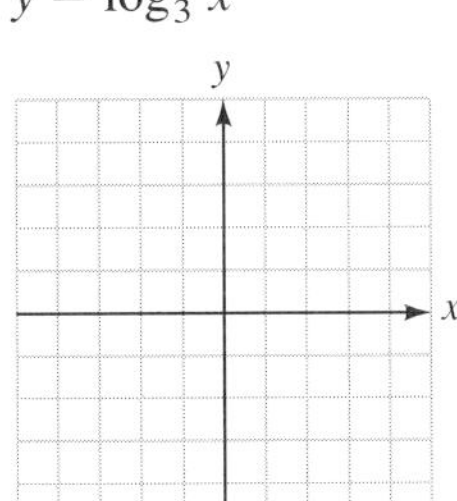

83. $f(x) = 5^x$
$g(x) = \log_5 x$

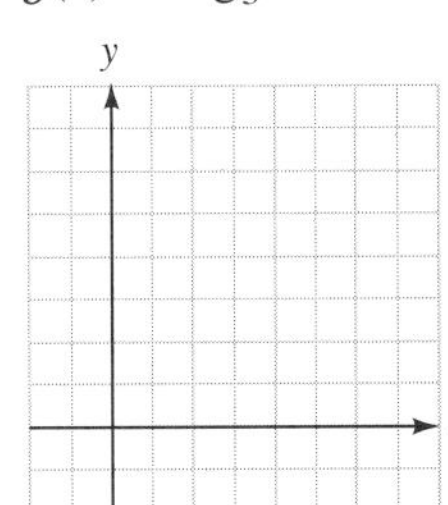

84. $f(x) = 8^x$
$g(x) = \log_8 x$

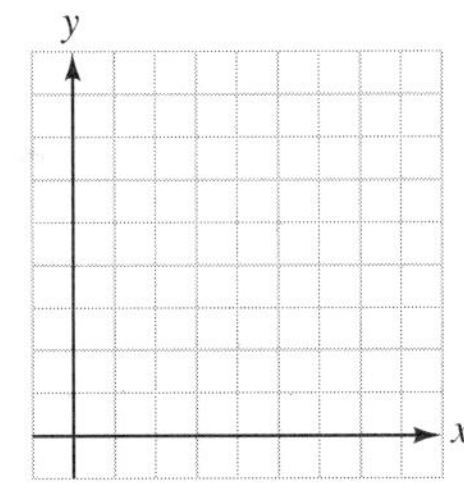

APPLICATIONS

85. FINDING INPUT VOLTAGE Find the db gain of an amplifier if the input voltage is 0.71 volts when the output voltage is 20 volts.

86. FINDING OUTPUT VOLTAGE Find the db gain of an amplifier if the output voltage is 2.8 volts when the input voltage is 0.05 volt.

87. db GAIN OF AN AMPLIFIER Find the db gain of the amplifier shown in Illustration 3.

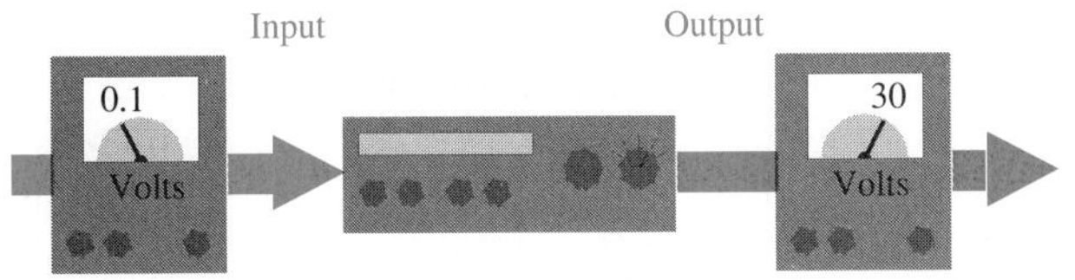

ILLUSTRATION 3

88. db GAIN OF AN AMPLIFIER An amplifier produces an output of 80 volts when driven by an input of 0.12 volts. Find the amplifier's db gain.

89. THE RICHTER SCALE An earthquake has amplitude of 5,000 micrometers and a period of 0.2 second. Find its measure on the Richter scale.

90. EARTHQUAKE Find the period of an earthquake with amplitude of 80,000 micrometers that measures 6 on the Richter scale.

91. EARTHQUAKE An earthquake with a period of $\frac{1}{4}$ second measures 4 on the Richter scale. Find its amplitude.

92. MAJOR EARTHQUAKE In 1985, Mexico City experienced an earthquake of magnitude 8.1 on the Richter scale. In 1989, the San Francisco Bay area was rocked by an earthquake measuring 7.1. By what factor must the amplitude of an earthquake change to increase its severity by 1 point on the Richter scale? (Assume that the period remains constant.)

93. DEPRECIATION In business, equipment is often depreciated using the double declining-balance method. In this method, a piece of equipment with a life expectancy of N years, costing $\$C$, will depreciate to a value of $\$V$ in n years, where n is given by the formula

$$n = \frac{\log V - \log C}{\log\left(1 - \frac{2}{N}\right)}$$

A computer that cost \$37,000 has a life expectancy of 5 years. If it has depreciated to a value of \$8,000, how old is it?

94. DEPRECIATION A typewriter worth \$470 when new had a life expectancy of 12 years. If it is now worth \$189, how old is it? (See Exercise 93.)

95. TIME FOR MONEY TO GROW If $\$P$ is invested at the end of each year in an annuity earning annual interest at a rate r, the amount in the account will be $\$A$ after n years, where

$$n = \frac{\log\left[\frac{Ar}{P} + 1\right]}{\log(1 + r)}$$

If \$1,000 is invested each year in an annuity earning 12% annual interest, how long will it take for the account to be worth \$20,000?

96. TIME FOR MONEY TO GROW If \$5,000 is invested each year in an annuity earning 8% annual interest, how long will it take for the account to be worth \$50,000? (See Exercise 95.)

WRITING *Write a paragraph using your own words.*

97. Explain the mathematical relationship between $y = \log x$ and $y = 10^x$.

98. Explain why it is impossible to find the logarithm of a negative number.

REVIEW *In Exercises 99–102, solve each equation.*

99. $\sqrt[3]{6x + 4} = 4$

100. $\sqrt{3x - 4} = \sqrt{-7x + 2}$

101. $\sqrt{a + 1} - 1 = 3a$

102. $3 - \sqrt{t - 3} = \sqrt{t}$

9.6 Base-e Logarithms

In this section, you will learn about

- **Base-e logarithms**
- **Graphing the natural logarithmic function**
- **An application of base-e logarithms**

INTRODUCTION. A special logarithmic function is the base-e logarithmic function. In this section, we will show how to evaluate base-e logarithms, graph the base-e logarithmic function, and solve some problems that involve the base-e logarithmic function.

Base-e logarithms

We have seen the importance of the number e in mathematical models of events in nature. Base-e logarithms are just as important. They are called **natural logarithms** or **Napierian logarithms** after John Napier (1550–1617). They are usually written as $\ln x$ rather than $\log_e x$:

$$\ln x \quad \text{means} \quad \log_e x$$

Like all logarithmic functions, the domain of $f(x) = \ln x$ is the interval $(0, \infty)$, and the range is the interval $(-\infty, \infty)$.

To find the base-e logarithms of numbers, we can use a calculator.

Accent on Technology **Finding base-e (natural) logarithms**

To use a scientific calculator to find the value of ln 2.34, we enter these numbers and press these keys:

2.34 [LN]

The display will read [.850150929]. To four decimal places, $\ln 2.34 = 0.8502$.

To use a graphing calculator, we enter these numbers and press these keys:

[LN] 2.34 [ENTER]

The display will read [ln 2.34 / .8501509294]

EXAMPLE 1 ***Evaluating base-e logarithms.*** Use a calculator to find each value: **a.** $\ln 17.32$ and **b.** $\ln(-0.05)$.

Solution

a. Enter these numbers and press these keys:

Scientific calculator	*Graphing calculator*
17.32 [LN]	[LN] 17.32 [ENTER]

Either way, the result is 2.851861903.

b. Enter these numbers and press these keys:

Scientific calculator	*Graphing calculator*
0.05 [+/−] [LN]	[LN] [(] [(−)] 0.05 [)] [ENTER]

Either way, we obtain an error, because we cannot take the logarithm of a negative number.

Self Check

Find each value to four decimal places: **a.** $\ln \pi$ and **b.** $\ln 0$.

Answers: **a.** 1.1447, **b.** no value ■

EXAMPLE 2 ***Writing an equivalent exponential statement.*** Solve each equation: **a.** $\ln x = 1.335$ and **b.** $\ln x = -5.5$. Give each result to four decimal places.

Solution

a. Since the base of the natural logarithmic function is e, the equation $\ln x = 1.335$ is equivalent to $e^{1.335} = x$. To use a scientific calculator to find x, press these keys:

1.335 [e^x]

The display will read 3.799995946. To four decimal places,

$$x = 3.8000$$

Self Check

Solve **a.** $\ln x = 1.9344$ and **b.** $-3 = \ln x$. Give each result to four decimal places.

b. The equation $\ln x = -5.5$ is equivalent to $e^{-5.5} = x$. To use a scientific calculator to find x, press these keys:

5.5 [+/−] [e^x]

The display will read 0.004086771. To four decimal places,

$x = 0.0041$

Answers: **a.** 6.9199, **b.** 0.0498

Graphing the natural logarithmic function

The equation $y = \ln x$ is equivalent to the equation $x = e^y$. To get the graph of $\ln x$, we can plot points that satisfy the equation $x = e^y$ and join them with a smooth curve, as shown in Figure 9-30(a). Figure 9-30(b) shows the calculator graph of $y = \ln x$.

$y = \ln x$

x	y	(x, y)
$\frac{1}{e}$	-1	$(\frac{1}{e}, -1)$
1	0	(1, 0)
e	1	$(e, 1)$
e^2	2	$(e^2, 2)$

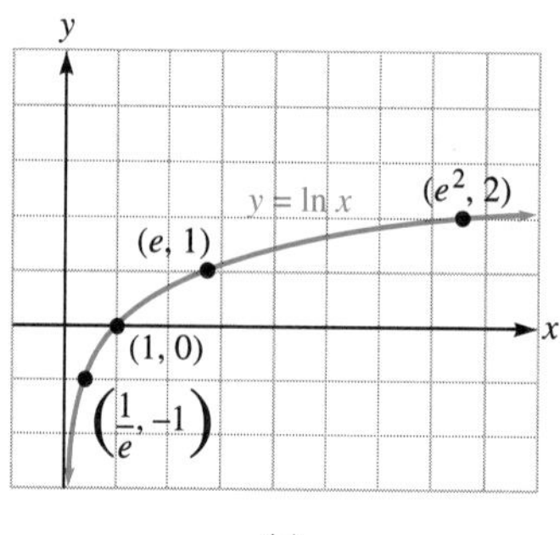

(a)

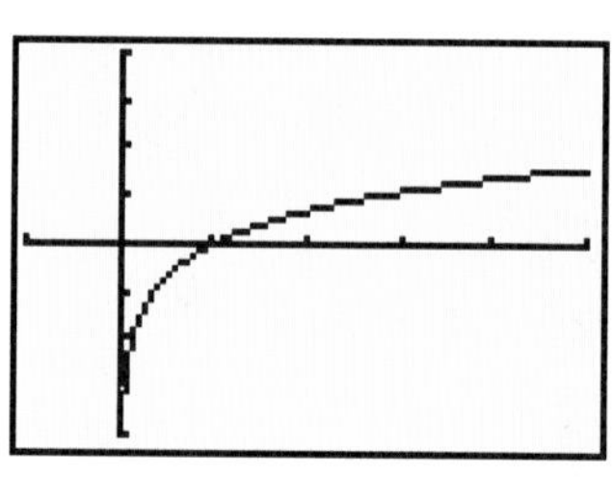
(b)

FIGURE 9-30

The exponential function $f(x) = e^x$ and the natural logarithmic function $f(x) = \ln x$ are inverse functions. Figure 9-31 shows that their graphs are symmetric to the line $y = x$.

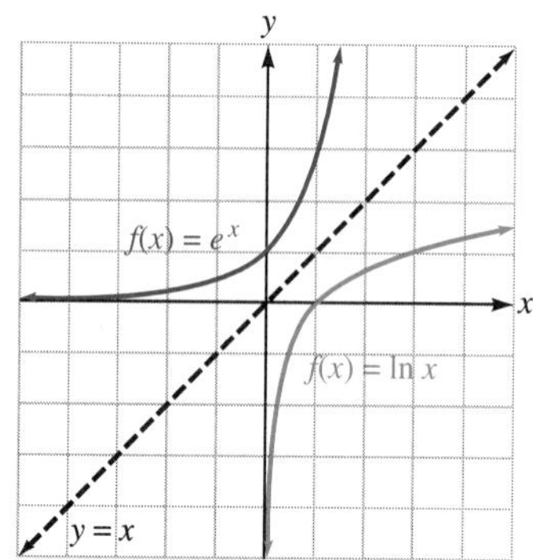

FIGURE 9-31

Accent on Technology Graphing base-e logarithmic functions

Many graphs of logarithmic functions involve translations of the graph of $y = \ln x$. For example, Figure 9-32 shows calculator graphs of the functions $y = \ln x$, $y = \ln x + 2$, and $y = \ln x - 3$.

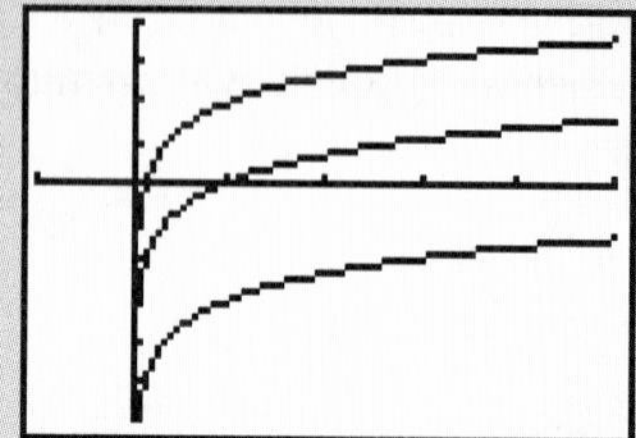

The graph of $y = \ln x + 2$ is 2 units above the graph of $y = \ln x$.

The graph of $y = \ln x - 3$ is 3 units below the graph of $y = \ln x$.

FIGURE 9-32

Figure 9-33 shows the calculator graph of the functions $y = \ln x$, $y = \ln(x - 2)$, and $y = \ln(x + 3)$.

The graph of $y = \ln(x + 3)$ is 3 units to the left of the graph of $y = \ln x$.

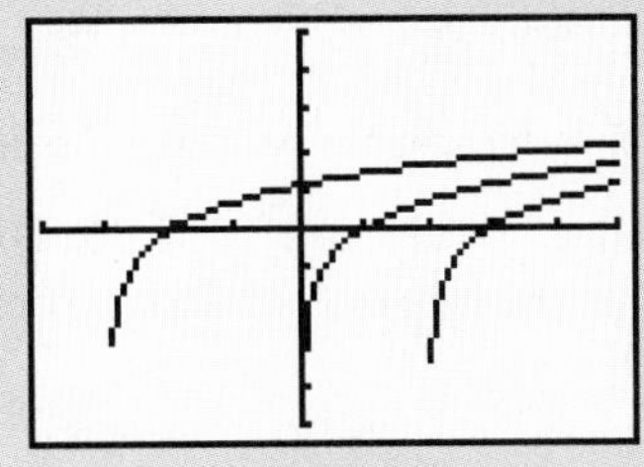

The graph of $y = \ln(x - 2)$ is 2 units to the right of the graph of $y = \ln x$.

FIGURE 9-33

An application of base-e logarithms

Base-e logarithms have many applications. If a population grows exponentially at a certain annual rate, the time required for the population to double is called the **doubling time.** It is given by the following formula.

Formula for doubling time

If r is the annual rate, compounded continuously, and t is time required for a population to double, then

$$t = \frac{\ln 2}{r}$$

EXAMPLE 3 ***Doubling time.*** The population of the earth is growing at the approximate rate of 2% per year. If this rate continues, how long will it take for the population to double?

Solution

Because the population is growing at the rate of 2% per year, we substitute 0.02 for r in the formula for doubling time and simplify.

$$t = \frac{\ln 2}{r}$$

$$t = \frac{\ln 2}{0.02}$$

$$\approx 34.65735903 \quad \text{Use a calculator}$$

The population will double in about 35 years.

Self Check

If the population's annual growth rate could be reduced to 1.5% per year, what would be the doubling time?

Answer: about 46 years ■

EXAMPLE 4 ***Doubling time.*** How long will it take $1,000 to double at an annual rate of 8%, compounded continuously?

Solution

We substitute 0.08 for r and simplify:

$$t = \frac{\ln 2}{r}$$

$$t = \frac{\ln 2}{0.08}$$

$$\approx 8.664339757 \quad \text{Use a calculator.}$$

It will take about $8\frac{2}{3}$ years for the money to double.

Self Check

How long will it take at 9%, compounded continuously?

Answer: about 7.7 years ■

STUDY SET Section 9.6

VOCABULARY *In Exercises 1–2, fill in the blanks to make the statements true.*

1. Base-e logarithms are often called ____________ logarithms.

2. $f(x) = \ln x$ and $f(x) = e^x$ are ____________ functions.

CONCEPTS

3. Use a calculator to complete the table of values for $f(x) = \ln x$. Round to the nearest hundredth.

x	$f(x)$
0.5	
1	
2	
3	
4	
5	
6	
7	
8	
9	
10	

4. Graph $f(x) = \ln x$ in Illustration 1. (See Exercise 3.) Note that the units on the x- and y-axes are different.

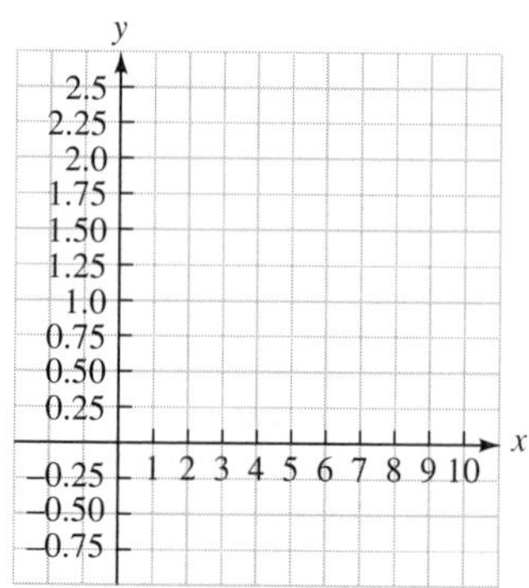

ILLUSTRATION 1

In Exercises 5–10, fill in the blanks to make the statements true.

5. The graph of $f(x) = \ln x$ has the ____________ as an asymptote.

6. The domain of the function $f(x) = \ln x$ is the interval ____________.

7. The range of the function $f(x) = \ln x$ is the interval ____________.

8. The graph of $f(x) = \ln x$ passes through the point $(\square, 0)$.

9. The statement $y = \ln x$ is equivalent to the exponential statement ____________.

10. The logarithm of a negative number is ____________.

NOTATION *In Exercises 11–14, fill in the blanks to make the statements true.*

11. $\ln x$ means $\log_{\square} x$.

12. $\log x$ means $\log_{\square} x$

13. If a population grows exponentially at a rate r, the time it will take the population to double is given by the formula $t =$ ____________.

14. To evaluate a base-10 logarithm with a calculator, use the ____________ key. To evaluate a base-e logarithm, use the ____________ key.

PRACTICE *In Exercises 15–22, use a calculator to find each value, if possible. Express all answers to four decimal places.*

15. $\ln 35.15$
16. $\ln 0.675$
17. $\ln 0.00465$
18. $\ln 378.96$
19. $\ln 1.72$
20. $\ln 2.7$
21. $\ln (-0.1)$
22. $\ln (-10)$

In Exercises 23–30, use a calculator to find x. Express all answers to four decimal places.

23. $\ln x = 1.4023$
24. $\ln x = 2.6490$
25. $\ln x = 4.24$
26. $\ln x = 0.926$
27. $\ln x = -3.71$
28. $\ln x = -0.28$
29. $1.001 = \ln x$
30. $\ln x = -0.001$

In Exercises 31–34, use a graphing calculator to graph each function.

31. $y = \ln\left(\frac{1}{2}x\right)$

32. $y = \ln x^2$

33. $y = \ln(-x)$

34. $y = \ln(3x)$

APPLICATIONS *Use a calculator to solve each problem.*

35. POPULATION GROWTH How long will it take the population of River City to double? See Illustration 2.

River City

A growing community

• 6 parks • 12% annual growth

• 10 churches • Low crime rate

ILLUSTRATION 2

36. DOUBLING MONEY How long will it take \$1,000 to double if it is invested at an annual rate of 5% compounded continuously?

37. POPULATION GROWTH A population growing continuously at an annual rate r will triple in a time t given by the formula

$$t = \frac{\ln 3}{r}$$

How long will it take the population of a town to triple if it is growing at the rate of 12% per year?

38. TRIPLING MONEY Find the length of time for \$25,000 to triple when it is invested at 6% annual interest, compounded continuously. See Exercise 37.

39. FORENSIC MEDICINE To estimate the number of hours t that a murder victim had been dead, a coroner used the formula

$$t = \frac{1}{0.25} \ln \frac{98.6 - T_s}{82 - T_s}$$

where T_s is the temperature of the surroundings where the body was found. If the crime took place in an apartment where the thermostat was set a 70° F, approximately how long ago did the murder occur?

40. MAKING JELL-O After the contents of a package of JELL-O are combined with boiling water, the mixture is placed in a refrigerator whose temperature remains a constant 42° F. Estimate the number of hours t that it will take for the JELL-O to cool to 50° F using the formula

$$t = -\frac{1}{0.9} \ln \frac{50 - T_r}{200 - T_r}$$

where T_r is the temperature of the refrigerator.

WRITING *Write a paragraph using your own words.*

41. Explain the difference between the functions $f(x) = \log x$ and $f(x) = \ln x$.

42. How are the functions $f(x) = \ln x$ and $f(x) = e^x$ related?

REVIEW *In Exercises 43–48, write the equation of the required line.*

43. Parallel to $y = 5x - 8$ and passing through the origin

44. Having a slope of 7 and a y-intercept of 3

45. Passing through the point (3, 2) and perpendicular to the line $y = \frac{2}{3}x - 12$

46. Parallel to the line $3x + 2y = 9$ and passing through the point $(-3, 5)$

47. Vertical line through the point (2, 3)

48. Horizontal line through the point (2, 3)

9.7 Properties of Logarithms

In this section, you will learn about

- **Properties of logarithms**
- **The change-of-base formula**
- **An application from chemistry**

INTRODUCTION. In this section, we will discuss eight properties of logarithms and use them to simplify logarithmic expressions. We will then show how to change a logarithm from one base to another. We conclude the section by solving some problems from the field of chemistry.

Properties of logarithms

Since logarithms are exponents, the properties of exponents have counterparts in the theory of logarithms. We begin with four basic properties.

Properties of logarithms

If b is a positive number and $b \neq 1$, then

1. $\log_b 1 = 0$ **2.** $\log_b b = 1$

3. $\log_b b^x = x$ **4.** $b^{\log_b x} = x \ (x > 0)$

Properties 1 through 4 follow directly from the definition of a logarithm.

1. $\log_b 1 = 0$, because $b^0 = 1$.

2. $\log_b b = 1$, because $b^1 = b$.

3. $\log_b b^x = x$, because $b^x = b^x$.

4. $b^{\log_b x} = x$, because $\log_b x$ is the exponent to which b is raised to get x.

Properties 3 and 4 also indicate that the composition of the exponential and logarithmic functions (in both directions) is the identity function. This is expected, because the exponential and logarithmic functions are inverse functions.

EXAMPLE 1 ***Applying properties of logarithms.*** Simplify each expression: **a.** $\log_5 1$, **b.** $\log_3 3$, **c.** $\log_7 7^3$, and **d.** $b^{\log_b 7}$.

Solution

a. By property 1, $\log_5 1 = 0$, because $5^0 = 1$.

b. By property 2, $\log_3 3 = 1$, because $3^1 = 3$.

c. By property 3, $\log_7 7^3 = 3$, because $7^3 = 7^3$.

d. By property 4, $b^{\log_b 7} = 7$, because $\log_b 7$ is the power to which b is raised to get 7.

Self Check

Simplify **a.** $\log_4 1$, **b.** $\log_4 4$, **c.** $\log_2 2^4$, and **d.** $5^{\log_5 2}$.

Answers: **a.** 0, **b.** 1, **c.** 4, **d.** 2 ■

The next two properties state that

The logarithm of a product is the sum of the logarithms.

The logarithm of a quotient is the difference of the logarithms.

Properties of logarithms

If M, N, and b are positive numbers and $b \neq 1$, then

5. $\log_b MN = \log_b M + \log_b N$ **6.** $\log_b \frac{M}{N} = \log_b M - \log_b N$

Proof To prove property 5, we let $x = \log_b M$ and $y = \log_b N$. We use the definition of logarithm to write each equation in exponential form.

$$M = b^x \quad \text{and} \quad N = b^y$$

Then $MN = b^x b^y$, and a property of exponents gives

$$MN = b^{x+y} \quad \text{Keep the base and multiply the exponents: } b^x b^y = b^{x+y}$$

We write this exponential equation in logarithmic form as

$$\log_b MN = x + y$$

Substituting the values of x and y completes the proof.

$$\log_b MN = \log_b M + \log_b N$$

□

The proof of property 6 is similar.

In the following examples, we will use these properties to write a logarithm of a product as the sum of logarithms and a logarithm of a quotient as the difference of logarithms. Assume that all variables are positive and that $b \neq 1$.

EXAMPLE 2 ***Applying the product rule.*** Use the product rule for logarithms to rewrite each of the following: **a.** $\log_2 (2 \cdot 7)$ and **b.** $\log (100x)$.

Self Check

Rewrite **a.** $\log_3 (4 \cdot 3)$ and **b.** $\log (1{,}000y)$.

Solution

a. $\log_2 (2 \cdot 7) = \log_2 2 + \log_2 7$ The log of a product is the sum of the logs.

$= 1 + \log_2 7$ $\log_2 2 = 1$.

b. $\log (100x) = \log 100 + \log x$ The log of a product is the sum of the logs.

$= 2 + \log x$ $\log 100 = 2$.

Answers: **a.** $\log_3 4 + 1$, **b.** $3 + \log y$ ■

EXAMPLE 3 ***Applying the quotient rule.*** Use the quotient rule for logarithms to rewrite each of the following: **a.** $\ln \frac{10}{7}$ and **b.** $\log_4 \frac{x}{64}$.

Self Check

Rewrite **a.** $\log_6 \frac{6}{5}$ and **b.** $\ln \frac{y}{100}$.

Solution

a. $\ln \dfrac{10}{7} = \ln 10 - \ln 7$ The log of a quotient is the difference of the logs.

b. $\log_4 \dfrac{x}{64} = \log_4 x - \log_4 64$ The log of a quotient is the difference of the logs.

$= \log_4 x - 3$ $\log_4 64 = 3$.

Answers: **a.** $1 - \log_6 5$, **b.** $\ln y - \ln 100$ ■

EXAMPLE 4 ***Applying the product and quotient properties.*** Use logarithm properties to rewrite each expression: **a.** $\log_b xyz$ and **b.** $\log \frac{xy}{z}$.

Self Check

Rewrite $\log_b \dfrac{x}{yz}$.

Solution

a. We observe that $\log_b xyz$ is the logarithm of a product.

$\log_b xyz = \log_b (xy)z$ Group the first two factors together.

$= \log_b (xy) + \log_b z$ The log of a product is the sum of the logs.

$= \log_b x + \log_b y + \log_b z$ The log of a product is the sum of the logs.

b. In the expression $\log \frac{xy}{z}$, we have the logarithm of a quotient.

$\log \frac{xy}{z} = \log (xy) - \log z$	The log of a quotient is the difference of the logs.
$= (\log x + \log y) - \log z$	The log of a product is the sum of the logs.
$= \log x + \log y - \log z$	Remove parentheses.

Answer: $\log_b x - \log_b y - \log_b z$

WARNING! By property 5 of logarithms, the logarithm of a *product* is equal to the *sum* of the logarithms. The logarithm of a sum or a difference usually does not simplify. In general,

$$\log_b (M + N) \neq \log_b M + \log_b N \quad \text{and} \quad \log_b (M - N) \neq \log_b M - \log_b N$$

By property 6, the logarithm of a *quotient* is equal to the *difference* of the logarithms. The logarithm of a quotient is not the quotient of the logarithms:

$$\log_b \frac{M}{N} \neq \frac{\log_b M}{\log_b N}$$

Accent on Technology **Verifying properties of logarithms**

We can use a calculator to illustrate property 5 of logarithms by showing that

$$\ln [(3.7)(15.9)] = \ln 3.7 + \ln 15.9$$

We calculate the left- and right-hand sides of the equation separately and compare the results. To use a scientific calculator to find ln [(3.7)(15.9)], we enter these numbers and press these keys:

3.7 [×] 15.9 [=] [LN]

The display will read [4.074651929].

To find ln 3.7 + ln 15.9, we enter these numbers and press these keys:

3.7 [LN] [+] 15.9 [LN] [=]

The display will read [4.074651929]. Since the left- and right-hand sides are equal, the equation ln [(3.7)(15.9)] = ln 3.7 + ln 15.9 is true.

Two more properties of logarithms state that

The logarithm of a power is the power times the logarithm.

If the logarithms of two numbers are equal, the numbers are equal.

Properties of logarithms

If M, p, and b are positive numbers and $b \neq 1$, then

7. $\log_b M^p = p \log_b M$ **8.** If $\log_b x = \log_b y$, then $x = y$.

Proof To prove property 7, we let $x = \log_b M$, write the expression in exponential form, and raise both sides to the pth power:

$M = b^x$	
$(M)^p = (b^x)^p$	Raise both sides to the pth power.
$M^p = b^{px}$	Keep the base and multiply the exponents.

Using the definition of logarithms gives

$$\log_b M^p = px$$

Substituting the value for x completes the proof.

$$\log_b M^p = p \log_b M$$

□

Property 8 follows from the fact that the logarithmic function is a one-to-one function. Property 8 will be important in the next section, when we solve logarithmic equations.

EXAMPLE 5 ***Applying the power rule.*** Use the power rule for logarithms to rewrite each of the following: **a.** $\log_5 6^2$ and **b.** $\log\sqrt{10}$.

Self Check

Rewrite **a.** $\ln x^4$ and **b.** $\log_2 \sqrt[3]{3}$

Solution

a. $\log_5 6^2 = 2 \log_5 6$ The log of a power is the power times the log.

b. $\log \sqrt{10} = \log (10)^{1/2}$ Write $\sqrt{10}$ using a fractional exponent: $\sqrt{10} = (10)^{1/2}$.

$= \frac{1}{2} \log 10$ The log of a power is the power times the log.

$= \frac{1}{2}$ Simplify: $\log 10 = 1$.

Answers: **a.** $4 \ln x$, **b.** $\frac{1}{3} \log_2 3$ ■

EXAMPLE 6 ***Applying properties of logarithms.*** Use logarithm properties to rewrite each expression: **a.** $\log_b (x^2y^3z)$ and **b.** $\log \frac{y^3\sqrt{x}}{z}$.

Self Check

Rewrite $\log \sqrt[4]{\frac{x^3y}{z}}$.

Solution

a. We begin by recognizing that $\log_b (x^2y^3z)$ is the logarithm of a product.

$\log_b (x^2y^3z) = \log_b x^2 + \log_b y^3 + \log_b z$ The log of a product is the sum of the logs.

$= 2 \log_b x + 3 \log_b y + \log_b z$ The log of a power is the power times the log.

b. The expression $\log \frac{y^3\sqrt{x}}{z}$ is the logarithm of a quotient.

$\log \frac{y^3\sqrt{x}}{z} = \log \left(y^3 \sqrt{x}\right) - \log z$ The log of a quotient is the difference of the logs.

$= \log y^3 + \log \sqrt{x} - \log z$ The log of a product is the sum of the logs.

$= \log y^3 + \log x^{1/2} - \log z$ Write $\sqrt{x}$ as $x^{1/2}$.

$= 3 \log y + \frac{1}{2} \log x - \log z$ The log of a power is the power times the log.

Answer: $\frac{1}{4}(3 \log x + \log y - \log z)$ ■

We can use the properties of logarithms to combine several logarithms into one logarithm.

EXAMPLE 7 ***Condensing logarithmic expressions.*** Write each of the given expressions as one logarithm: **a.** $3 \log_b x + \frac{1}{2} \log_b y$ and **b.** $\frac{1}{2} \log_b (x - 2) - \log_b y + 3 \log_b z$.

Solution

a. We begin by applying the power rule to each term of the expression.

$$3 \log_b x + \frac{1}{2} \log_b y$$

$= \log_b x^3 + \log_b y^{1/2}$ — A power times a log is the log of the power.

$= \log_b (x^3 y^{1/2})$ — The sum of two logs is the log of the product.

b. The first and third terms of this expression can be rewritten using the power rule of logarithms.

$$\frac{1}{2} \log_b (x - 2) - \log_b y + 3 \log_b z$$

$= \log_b (x - 2)^{1/2} - \log_b y + \log_b z^3$ — A power times a log is the log of the power.

$= \log_b \frac{(x - 2)^{1/2}}{y} + \log_b z^3$ — The difference of two logs is the log of the quotient.

$= \log_b \frac{z^3\sqrt{x - 2}}{y}$ — The sum of two logs is the log of the product. Write $(x - 2)^{1/2}$ as $\sqrt{x - 2}$.

Self Check

Write the expression as one logarithm:

$2 \log_b x + \frac{1}{2} \log_b y - 2 \log_b (x - y)$

Answer: $\log_b \frac{x^2\sqrt{y}}{(x - y)^2}$ ■

We summarize the properties of logarithms as follows.

Properties of logarithms

If b, M, and N are positive numbers and $b \neq 1$, then

1. $\log_b 1 = 0$	**2.** $\log_b b = 1$
3. $\log_b b^x = x$	**4.** $b^{\log_b x} = x$
5. $\log_b MN = \log_b M + \log_b N$	**6.** $\log_b \frac{M}{N} = \log_b M - \log_b N$
7. $\log_b M^p = p \log_b M$	**8.** If $\log_b x = \log_b y$, then $x = y$.

EXAMPLE 8 ***Using properties of logarithms.*** Given that $\log 2 \approx 0.3010$ and $\log 3 \approx 0.4771$, find approximations for **a.** $\log 6$ and **b.** $\log 18$.

Solution

a. $\log 6 = \log (2 \cdot 3)$ — Write 6 using the factors 2 and 3.

$= \log 2 + \log 3$ — The log of a product is the sum of the logs.

$\approx 0.3010 + 0.4771$ — Substitute the value of each logarithm.

≈ 0.7781

b. $\log 18 = \log (2 \cdot 3^2)$ — Write 18 using the factors 2 and 3.

$= \log 2 + \log 3^2$ — The log of a product is the sum of the logs.

$= \log 2 + 2 \log 3$ — The log of a power is the power times the log.

$\approx 0.3010 + 2(0.4771)$ — Substitute the value of each logarithm.

≈ 1.2552

Self Check

Find **a.** $\log 1.5$ and **b.** $\log 0.75$.

Answers: **a.** 0.1761, **b.** −0.1249 ■

The change-of-base formula

If we know the base-a logarithm of a number, we can find its logarithm to some other base b with a formula called the **change-of-base formula.**

Change-of-base formula

If a, b, and x are positive, $a \neq 1$, and $b \neq 1$, then

$$\log_b x = \frac{\log_a x}{\log_a b}$$

Proof To prove this formula, we begin with the equation $\log_b x = y$.

$$y = \log_b x$$

$$x = b^y \quad \text{Change the equation from logarithmic to exponential form.}$$

$$\log_a x = \log_a b^y \quad \text{Take the base-}a\text{ logarithm of both sides.}$$

$$\log_a x = y \log_a b \quad \text{The log of a power is the power times the log.}$$

$$y = \frac{\log_a x}{\log_a b} \quad \text{Divide both sides by } \log_a b.$$

$$\log_b x = \frac{\log_a x}{\log_a b} \quad \text{Refer to the first equation and substitute } \log_b x \text{ for } y.$$

□

If we know logarithms to base a (for example, $a = 10$), we can find the logarithm of x to a new base b. We simply divide the base-a logarithm of x by the base-a logarithm of b.

WARNING! $\frac{\log_a x}{\log_a b}$ means that one logarithm is to be divided by the other. They are not to be subtracted.

EXAMPLE 9 *Using the change-of-base formula.* Find $\log_3 5$.

Solution

We can use base-10 logarithms to find a base-3 logarithm. To do this, we substitute 3 for b, 10 for a, and 5 for x in the change-of-base formula:

$$\log_b x = \frac{\log_a x}{\log_a b}$$

$$\log_3 5 = \frac{\log_{10} 5}{\log_{10} 3}$$

$$\approx 1.464973521$$

To four decimal places, $\log_3 5 = 1.4650$.

Self Check

Find $\log_5 3$.

Answer: 0.6826 ■

An application from chemistry

In chemistry, common logarithms are used to express the acidity of solutions. The more acidic a solution, the greater the concentration of hydrogen ions. This concentration is indicated indirectly by the *pH scale,* or *hydrogen ion index.* The pH of a solution is defined by the equation

pH of a solution

If $[H^+]$ is the hydrogen ion concentration in gram-ions per liter, then

$$\text{pH} = -\log [H^+]$$

EXAMPLE 10 ***Finding the pH of a solution.*** Find the pH of pure water, which has a hydrogen ion concentration of 10^{-7} gram-ions per liter.

Solution Since pure water has approximately 10^{-7} gram-ions per liter, its pH is

$$\begin{aligned} pH &= -\log [H^+] \\ pH &= -\log 10^{-7} \\ &= -(-7)\log 10 && \text{The log of a power is the power times the log.} \\ &= -(-7)\cdot 1 && \text{Simplify: } \log 10 = 1. \\ &= 7 \end{aligned}$$

EXAMPLE 11 ***Finding hydrogen ion concentration.*** Find the hydrogen ion concentration of seawater if its pH is 8.5.

Solution To find its hydrogen ion concentration, we solve the following equation for $[H^+]$.

$$\begin{aligned} pH &= -\log [H^+] \\ 8.5 &= -\log [H^+] \\ -8.5 &= \log [H^+] && \text{Multiply both sides by } -1. \\ [H^+] &= 10^{-8.5} && \text{Change the equation to exponential form.} \end{aligned}$$

We can use a calculator to find that

$$[H^+] \approx 3.2 \times 10^{-9} \text{ gram-ions per liter}$$

STUDY SET Section 9.7

VOCABULARY *In Exercises 1–4, fill in the blanks to make the statements true.*

1. The expression $\log_3 (4x)$ is the logarithm of a ____________.

2. The expression $\log_2 \frac{5}{x}$ is the logarithm of a ____________.

3. The expression $\log 4^x$ is the logarithm of a ____________.

4. In the expression $\log_5 4$, the number 5 is the ____________ of the logarithm.

CONCEPTS *In Exercises 5–16, fill in the blanks to make the statements true.*

5. $\log_b 1 = \square$

6. $\log_b b = \square$

7. $\log_b MN = \log_b \square + \log_b \square$

8. $b^{\log_b x} = \square$

9. If $\log_b x = \log_b y$, then $\square = \square$.

10. $\log_b \frac{M}{N} = \log_b M \square \log_b N$

11. $\log_b x^p = p \cdot \log_b \square$

12. $\log_b b^x = \square$

13. $\log_b (A + B) \square \log_b A + \log_b B$

14. $\log_b A + \log_b B \square \log_b AB$

15. $\log_b x = \frac{\log_a x}{\square}$

16. $pH = \square$

In Exercises 17–28, evaluate each expression.

17. $\log_4 1$

18. $\log_4 4$

19. $\log_4 4^7$

20. $\ln e^8$

21. $5^{\log_5 10}$ **22.** $8^{\log_8 10}$ **23.** $\log_5 5^2$ **24.** $\log_4 4^2$

25. $\ln e$ **26.** $\log_7 1$ **27.** $\log_3 3^7$ **28.** $5^{\log_5 8}$

NOTATION *In Exercises 29–30, complete each solution.*

29. $\log_b rst = \log_b (\square)t$
$= \log_b (rs) + \log_b \square$
$= \log_b \square + \log_b \square + \log_b t$

30. $\log \frac{r}{st} = \log r - \log (\square)$
$= \log r - (\log \square + \log t)$
$= \log r - \log s \square \log t$

PRACTICE *In Exercises 31–36, use a calculator to verify each equation.*

31. $\log [(2.5)(3.7)] = \log 2.5 + \log 3.7$

32. $\ln \frac{11.3}{6.1} = \ln 11.3 - \ln 6.1$

33. $\ln (2.25)^4 = 4 \ln 2.25$

34. $\log 45.37 = \frac{\ln 45.37}{\ln 10}$

35. $\log \sqrt{24.3} = \frac{1}{2} \log 24.3$

36. $\ln 8.75 = \frac{\log 8.75}{\log e}$

In Exercises 37–44, use the properties of logarithms to rewrite each expression. Assume that x, y, and z are positive numbers.

37. $\log_2 (4 \cdot 5)$

38. $\log_3 (27 \cdot 5)$

39. $\log_6 \frac{x}{36}$

40. $\log_8 \frac{y}{8}$

41. $\ln y^4$

42. $\ln z^9$

43. $\log \sqrt{5}$

44. $\log \sqrt[3]{7}$

In Exercises 45–56, assume that x, y, z, and b are positive numbers and $b \neq 1$. Use the properties of logarithms to write each expression in terms of the logarithms of x, y, and z.

45. $\log xyz$

46. $\log 4xz$

47. $\log_2 \frac{2x}{y}$

48. $\log_3 \frac{x}{yz}$

49. $\log x^3y^2$

50. $\log xy^2z^3$

51. $\log_b (xy)^{1/2}$

52. $\log_b x^3y^{1/2}$

53. $\ln x\sqrt{z}$

54. $\ln \sqrt{xy}$

55. $\log_b \frac{\sqrt[3]{x}}{\sqrt[4]{yz}}$

56. $\log_b \sqrt[4]{\frac{x^3y^2}{z^4}}$

In Exercises 57–64, assume that x, y, z, and b are positive numbers and $b \neq 1$. Use the properties of logarithms to write each expression as the logarithm of a single quantity.

57. $\log_2 (x + 1) - \log_2 x$

58. $\log_3 x + \log_3 (x + 2) - \log_3 8$

59. $2 \log x + \frac{1}{2} \log y$

60. $-2 \log x - 3 \log y + \log z$

61. $-3 \log_b x - 2 \log_b y + \frac{1}{2} \log_b z$

62. $3 \log_b (x + 1) - 2 \log_b (x + 2) + \log_b x$

63. $\log_b \left(\frac{x}{z} + x\right) - \log_b \left(\frac{y}{z} + y\right)$

64. $\log_b (xy + y^2) - \log_b (xz + yz) + \log_b z$

In Exercises 65–70, tell whether the given statement is true. If a statement is false, explain why.

65. $\log xy = (\log x)(\log y)$

66. $\log ab = \log a + 1$

67. $\log_b (A - B) = \frac{\log_b A}{\log_b B}$

68. $\frac{\log_b A}{\log_b B} = \log_b A - \log_b B$

69. $\log_b \frac{A}{B} = \log_b A - \log_b B$

70. $\log_b 2 = \log_2 b$

In Exercises 71–78, assume that $\log_b 4 = 0.6021$, $\log_b 7 = 0.8451$, and $\log_b 9 = 0.9542$. Use these values and the properties of logarithms to find each value.

71. $\log_b 28$

72. $\log_b \frac{7}{4}$

73. $\log_b \frac{4}{63}$

74. $\log_b 36$

75. $\log_b \frac{63}{4}$

76. $\log_b 2.25$

77. $\log_b 64$

78. $\log_b 49$

In Exercises 79–86, use the change-of-base formula to find each logarithm to four decimal places.

79. $\log_3 7$

80. $\log_7 3$

81. $\log_{1/3} 3$

82. $\log_{1/2} 6$

83. $\log_3 8$

84. $\log_5 10$

85. $\log_{\sqrt{2}} \sqrt{5}$

86. $\log_{\pi} e$

APPLICATIONS

87. pH OF A SOLUTION Find the pH of a solution with a hydrogen ion concentration of 1.7×10^{-5} gram-ions per liter.

88. HYDROGEN ION CONCENTRATION Find the hydrogen ion concentration of a saturated solution of calcium hydroxide whose pH is 13.2.

89. AQUARIUM To test for safe pH levels in a freshwater aquarium, a test strip is compared with the scale shown in Illustration 1. Find the corresponding range in the hydrogen ion concentration.

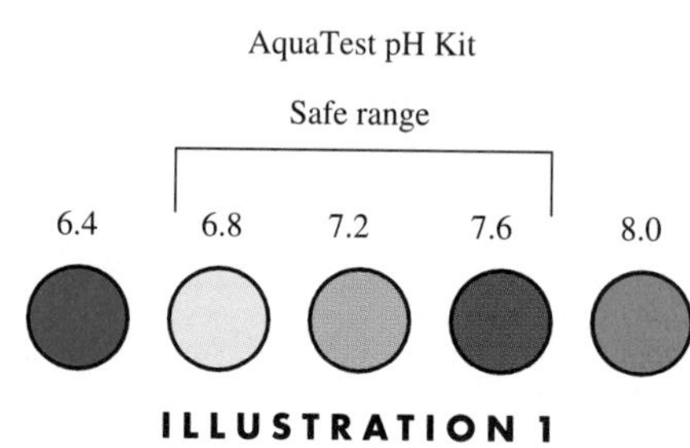

ILLUSTRATION 1

90. pH OF SOUR PICKLES The hydrogen ion concentration of sour pickles is 6.31×10^{-4}. Find the pH.

WRITING

91. Explain the difference between a logarithm of a product and the product of logarithms.

92. How can the LOG key on a calculator be used to find $\log_2 7$?

REVIEW *In Exercises 93–96, consider the line that passes through $P(-2, 3)$ and $Q(4, -4)$.*

93. Find the slope of line PQ.

94. Find the distance PQ.

95. Find the midpoint of segment PQ.

96. Write the equation of line PQ.

9.8 Exponential and Logarithmic Equations

In this section, you will learn about

- **Solving exponential equations**
- **Solving logarithmic equations**
- **Radioactive decay**
- **Population growth**

INTRODUCTION. An **exponential equation** is an equation that contains a variable in one of its exponents. Some examples of exponential equations are

$$3^x = 5, \quad 6^{x-3} = 2^x, \quad \text{and} \quad 2^{x^2+2x} = \frac{1}{2}$$

A **logarithmic equation** is an equation with a logarithmic expression that contains a variable. Some examples of logarithmic equations are

$$\log 2x = 25, \quad \ln x - \ln (x - 12) = 24, \quad \text{and} \quad \log x = \log \frac{1}{x} + 4$$

In this section, we will learn how to solve many of these equations.

Solving exponential equations

EXAMPLE 1 ***Taking the logarithm of both sides of an equation.*** Solve $3^x = 5$.

Solution

Since logarithms of equal numbers are equal, we can take the common logarithm of each side of the equation. The power rule of logarithms then provides a way of moving the variable x from its position as an exponent to a position as a coefficient.

$3^x = 5$

$\log 3^x = \log 5$ Take the common logarithm of each side.

$x \log 3 = \log 5$ The log of a power is the power times the log.

1. $x = \dfrac{\log 5}{\log 3}$ Divide both sides by log 3.

≈ 1.464973521 Use a calculator.

To four decimal places, $x = 1.4650$.

Self Check Solve $5^x = 4$.

Answer: 0.8614

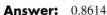

WARNING! A careless reading of Equation 1 leads to a common error. The right-hand side of Equation 1 calls for a division, not a subtraction.

$$\frac{\log 5}{\log 3} \quad \text{means} \quad (\log 5) \div (\log 3)$$

It is the expression $\log \frac{5}{3}$ that means $\log 5 - \log 3$.

EXAMPLE 2 ***Taking the logarithm of both sides.*** Solve $6^{x-3} = 2^x$.

Solution

$$\begin{aligned} 6^{x-3} &= 2^x \\ \log 6^{x-3} &= \log 2^x && \text{Take the common logarithm of each side.} \\ (x-3)\log 6 &= x\log 2 && \text{The log of a power is the power times the log.} \\ x\log 6 - 3\log 6 &= x\log 2 && \text{Use the distributive property.} \\ x\log 6 - x\log 2 &= 3\log 6 && \text{On both sides, add 3 log 6 and subtract } x \text{ log 2.} \\ x(\log 6 - \log 2) &= 3\log 6 && \text{Factor out } x \text{ on the left-hand side.} \\ x &= \frac{3\log 6}{\log 6 - \log 2} && \text{Divide both sides by log 6} - \text{log 2.} \\ x &\approx 4.892789261 && \text{Use a calculator.} \end{aligned}$$

Self Check

Solve $5^{x-2} = 3^x$.

Answer: $\dfrac{2\log 5}{\log 5 - \log 3} \approx 6.3013$ ■

EXAMPLE 3 ***Exponential expressions with the same base.*** Solve $2^{x^2+2x} = \frac{1}{2}$.

Solution

Since $\frac{1}{2} = 2^{-1}$, we can write the equation in the form

$$2^{x^2+2x} = 2^{-1} \quad \text{Each exponential expression has a base of 2.}$$

Since equal quantities with equal bases have equal exponents, we have

$$\begin{aligned} x^2 + 2x &= -1 \\ x^2 + 2x + 1 &= 0 && \text{Add 1 to both sides.} \\ (x+1)(x+1) &= 0 && \text{Factor the trinomial.} \\ x+1 = 0 \quad &\text{or} \quad x+1 = 0 && \text{Set each factor equal to 0.} \\ x = -1 \quad & \qquad\quad x = -1 \end{aligned}$$

Verify that -1 satisfies the equation.

Self Check

Solve $3^{x^2-2x} = \frac{1}{3}$.

Answer: 1, 1 ■

Accent on Technology **Solving exponential equations graphically**

To use a graphing calculator to approximate the solutions of $2^{x^2+2x} = \frac{1}{2}$ (see Example 3), we can subtract $\frac{1}{2}$ from both sides of the equation to get

$$2^{x^2+2x} - \frac{1}{2} = 0$$

and graph the corresponding function

$$y = 2^{x^2+2x} - \frac{1}{2}$$

If we use window settings of $[-4, 4]$ for x and $[-2, 6]$ for y, we obtain the graph shown in Figure 9-34(a).

Since the solutions of the equation are its x-intercepts, we can approximate the solutions by zooming in on the values of the x-intercepts, as in Figure 9-34(b). Since $x = -1$ is the only x-intercept, -1 is the only solution. In this case, we have found an exact solution.

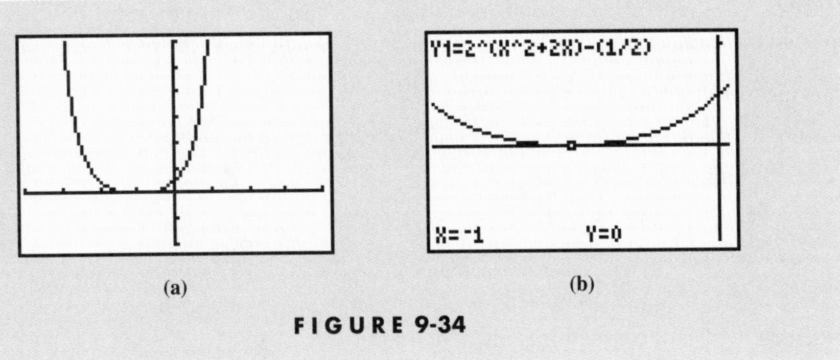

(a)

(b)

FIGURE 9-34

Solving logarithmic equations

EXAMPLE 4 ***Solving logarithmic equations.*** Solve $\log(3x+2) - \log(2x-3) = 0$.

Solution

We isolate each logarithmic expression on one side of the equation.

$$\log(3x+2) - \log(2x-3) = 0$$

$$\log(3x+2) = \log(2x-3) \quad \text{Add } \log(2x-3) \text{ to both sides.}$$

In the previous section, we saw that if the logarithms of two numbers are equal, the numbers are equal. So we have

$$(3x+2) = (2x-3) \quad \text{If } \log r = \log s, \text{ then } r = s.$$

$$x = -5 \quad \text{Subtract } 2x \text{ and } 2 \text{ from both sides.}$$

Check: $\log(3x+2) - \log(2x-3) = 0$

$$\log[3(-5)+2] - \log[2(-5)-3] \stackrel{?}{=} 0$$

$$\log(-13) - \log(-13) \stackrel{?}{=} 0$$

Since the logarithm of a negative number does not exist, the apparent solution of -5 must be discarded. This equation has no solutions.

Self Check

Solve $\log(5x+2) - \log(7x-2) = 0$.

Answer: 2 ■

EXAMPLE 5 ***Using a property of logarithms to solve a logarithmic equation.*** Solve $\log x + \log(x-3) = 1$.

Solution

$$\log x + \log(x-3) = 1$$

$$\log x(x-3) = 1 \quad \text{Use the product rule of logarithms.}$$

$$x(x-3) = 10^1 \quad \text{Use the definition of logarithms to change the equation to exponential form.}$$

$$x^2 - 3x - 10 = 0 \quad \text{Remove parentheses and subtract 10 from both sides.}$$

$$(x+2)(x-5) = 0 \quad \text{Factor the trinomial.}$$

$$x+2 = 0 \quad \text{or} \quad x-5 = 0 \quad \text{Set each factor equal to 0.}$$

$$x = -2 \qquad\qquad x = 5$$

Self Check

Solve $\log x + \log(x+3) = 1$.

Check: The number -2 is not a solution, because it does not satisfy the equation (a negative number does not have a logarithm). We will check the remaining number, 5.

$\log x + \log (x - 3) = 1$	
$\log 5 + \log (5 - 3) \stackrel{?}{=} 1$	Substitute 5 for x.
$\log 5 + \log 2 \stackrel{?}{=} 1$	
$\log 10 \stackrel{?}{=} 1$	Use the product rule of logarithms: $\log 5 + \log 2 = \log (5 \cdot 2) = \log 10$.
$1 = 1$	Simplify: $\log 10 = 1$.

Since 5 satisfies the equation, it is a solution.

Answer: 2

WARNING! Examples 4 and 5 illustrate that we must check the solutions of a logarithmic equation.

Accent on Technology *Solving logarithmic equations graphically*

To use a graphing calculator to approximate the solutions of $\log x + \log (x - 3) = 1$ (see Example 5), we can subtract 1 from both sides of the equation to get

$$\log x + \log (x - 3) - 1 = 0$$

and graph the corresponding function

$$y = \log x + \log (x - 3) - 1$$

If we use window settings of [0, 20] for x and [−2, 2] for y, we obtain the graph shown in Figure 9-35. Since the solution of the equation is the x-intercept, we can find the solution by zooming in on the value of the x-intercept. The solution is $x = 5$.

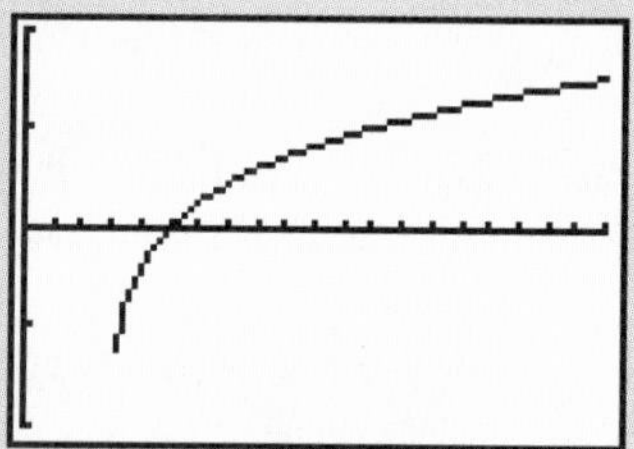

FIGURE 9-35

EXAMPLE 6 ***Solving logarithmic equations.*** Solve $\dfrac{\log_2 (5x - 6)}{\log_2 x} = 2$.

Solution

We can multiply both sides of the equation by $\log_2 x$ to get

$$\log_2 (5x - 6) = 2 \log_2 x$$

and apply the power rule of logarithms to get

$$\log_2 (5x - 6) = \log_2 x^2$$

By property 8 of logarithms, $5x - 6 = x^2$, because they have equal logarithms. Thus,

$$5x - 6 = x^2$$
$$0 = x^2 - 5x + 6$$
$$0 = (x - 3)(x - 2)$$
$$x - 3 = 0 \quad \text{or} \quad x - 2 = 0$$
$$x = 3 \qquad\qquad x = 2$$

Verify that both 2 and 3 satisfy the equation.

Self Check

Solve $\dfrac{\log_3 (5x + 6)}{\log_3 x} = 2$.

Answer: 6

Radioactive decay

Experiments have determined the time it takes for half of a sample of a given radioactive material to decompose. This time is a constant, called the material's **half-life.**

When living organisms die, the oxygen/carbon dioxide cycle common to all living things ceases, and carbon-14, a radioactive isotope with a half-life of 5,700 years, is no longer absorbed. By measuring the amount of carbon-14 present in an ancient object, archaeologists can estimate the object's age by using the radioactive decay formula.

Radioactive decay formula

If A is the amount of radioactive material present at time t, A_0 was the amount present at $t = 0$, and h is the material's half-life, then

$$A = A_0 2^{-t/h}$$

EXAMPLE 7 ***Carbon-14 dating.*** How old is a wooden statue that retains only one-third of its original carbon-14 content?

Solution

To find the time t when $A = \frac{1}{3}A_0$, we substitute $\frac{A_0}{3}$ for A and 5,700 for h in the radioactive decay formula and solve for t:

$$A = A_0 2^{-t/h}$$

$$\frac{A_0}{3} = A_0 2^{-t/5{,}700} \quad \text{The half-life of carbon-14 is 5,700 years.}$$

$$1 = 3(2^{-t/5{,}700}) \quad \text{Divide both sides by } A_0 \text{ and multiply both sides by 3.}$$

$$\log 1 = \log 3(2^{-t/5{,}700}) \quad \text{Take the common logarithm of each side.}$$

$$0 = \log 3 + \log 2^{-t/5{,}700} \quad \log 1 = 0\text{, and use the product rule of logarithms.}$$

$$-\log 3 = -\frac{t}{5{,}700}\log 2 \quad \text{Subtract log 3 from both sides and use the power rule of logarithms.}$$

$$5{,}700\left(\frac{\log 3}{\log 2}\right) = t \quad \text{Multiply both sides by } -\tfrac{5{,}700}{\log 2}.$$

$$t \approx 9{,}034.286254 \quad \text{Use a calculator.}$$

The statue is approximately 9,000 years old.

Self Check

How old is a statue that retains 25% of its original carbon-14 content?

Answer: about 11,400 years

Population growth

Recall that when there is sufficient food and space, populations of living organisms tend to increase exponentially according to the Malthusian growth model.

Malthusian growth model

If P is the population at some time t, P_0 is the initial population at $t = 0$, and k depends on the rate of growth, then

$$P = P_0 e^{kt}$$

EXAMPLE 8 ***Population growth.*** The bacteria in a laboratory culture increased from an initial population of 500 to 1,500 in 3 hours. How long will it take for the population to reach 10,000?

Solution

We substitute 500 for P_0, 1,500 for P, and 3 for t and simplify to find k:

$$P = P_0e^{kt}$$

$$1{,}500 = 500(e^{k3}) \quad \text{Substitute 1,500 for } P\text{, 500 for } P_0\text{, and 3 for } t.$$

$$3 = e^{3k} \quad \text{Divide both sides by 500.}$$

$$3k = \ln 3 \quad \text{Change the equation from exponential to logarithmic form.}$$

$$k = \frac{\ln 3}{3} \quad \text{Divide both sides by 3.}$$

To find when the population will reach 10,000, we substitute 10,000 for P, 500 for P_0, and $\frac{\ln 3}{3}$ for k in the equation $P = P_0e^{kt}$ and solve for t:

$$P = P_0e^{kt}$$

$$10{,}000 = 500e^{[(\ln 3)/3]t}$$

$$20 = e^{[(\ln 3)/3]t} \quad \text{Divide both sides by 500.}$$

$$\left(\frac{\ln 3}{3}\right)t = \ln 20 \quad \text{Change the equation to logarithmic form.}$$

$$t = \frac{3 \ln 20}{\ln 3} \quad \text{Multiply both sides by } \frac{3}{\ln 3}.$$

$$\approx 8.180499084 \quad \text{Use a calculator.}$$

The culture will reach 10,000 bacteria in about 8 hours.

Self Check

How long will it take to reach 20,000?

Answer: about 10 hours

EXAMPLE 9 ***Generation time.*** If a medium is inoculated with a bacterial culture that contains 1,000 cells per milliliter, how many generations will pass by the time the culture has grown to a population of 1 million cells per milliliter?

Solution

During bacterial reproduction, the time required for a population to double is called the *generation time.* If b bacteria are introduced into a medium, then after the generation time of the organism has elapsed, there are $2b$ cells. After another generation, there are $2(2b)$ or $4b$ cells, and so on. After n generations, the number of cells present will be

1. $B = b \cdot 2^n$

To find the number of generations that have passed while the population grows from b bacteria to B bacteria, we solve Equation 1 for n.

$$\log B = \log (b \cdot 2^n) \quad \text{Take the common logarithm of both sides.}$$

$$\log B = \log b + n \log 2 \quad \text{Apply the product and power rules of logarithms.}$$

$$\log B - \log b = n \log 2 \quad \text{Subtract } \log b \text{ from both sides.}$$

$$n = \frac{1}{\log 2}(\log B - \log b) \quad \text{Multiply both sides by } \frac{1}{\log 2}.$$

2. $$n = \frac{1}{\log 2}\left(\log \frac{B}{b}\right) \quad \text{Use the quotient rule of logarithms.}$$

Equation 2 is a formula that gives the number of generations that will pass as the population grows from b bacteria to B bacteria.

To find the number of generations that have passed while a population of 1,000 cells per milliliter has grown to a population of 1 million cells per milliliter, we substitute 1,000 for b and 1,000,000 for B in Equation 2 and solve for n.

$$n = \frac{1}{\log 2} \log \frac{1{,}000{,}000}{1{,}000}$$

$$= \frac{1}{\log 2} \log 1{,}000 \quad \text{Simplify.}$$

$$\approx 3.321928095(3) \quad \frac{1}{\log 2} \approx 3.321928095 \text{ and } \log 1{,}000 = 3.$$

$$\approx 9.965784285$$

Approximately 10 generations will have passed. ■

STUDY SET Section 9.8

VOCABULARY *In Exercises 1–2, fill in the blanks to make the statements true.*

1. An equation with a variable in its exponent, such as $3^{2x} = 8$, is called a(n) ______________ equation.

2. An equation with a logarithmic expression that contains a variable, such as $\log_5 (2x - 3) = \log_5 (x + 4)$, is a(n) ______________ equation.

CONCEPTS *In Exercises 3–4, fill in the blanks to make the statements true.*

3. The formula for radioactive decay is $A =$ ☐.

4. The formula for population growth is $P =$ ☐.

5. Use the graphs in Illustration 1 to estimate the solution of $2^x = 3^{-x+3}$.

6. Use the graphs in Illustration 2 to estimate the solution of $3 \log (x - 1) = 2 \log x$.

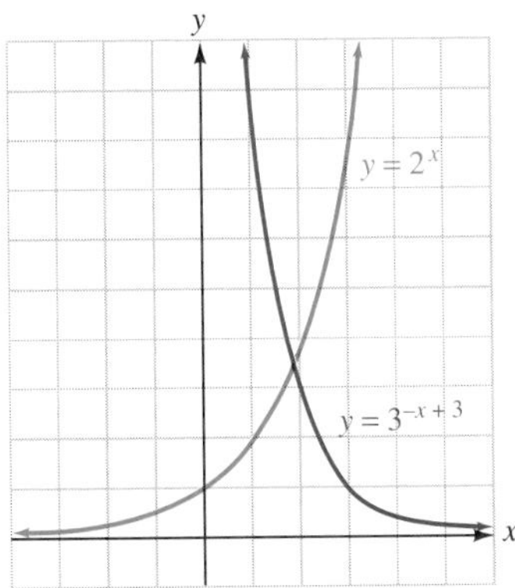

ILLUSTRATION 1

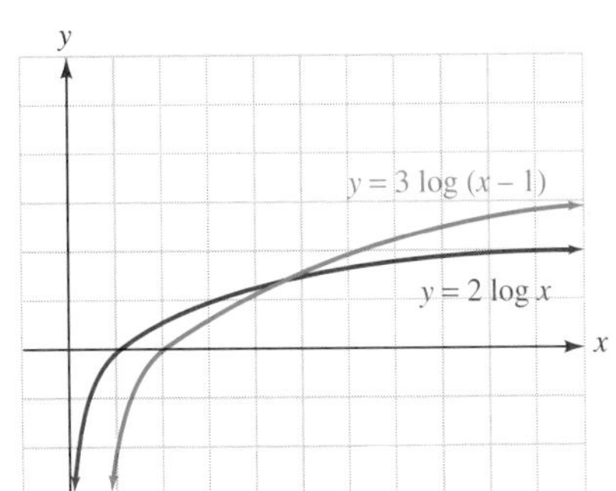

ILLUSTRATION 2

7. Solve each equation. Round to the nearest hundredth.
a. $x^2 = 12$ **b.** $2^x = 12$

8. Solve each equation.
a. $\log (x - 1) = 3$ **b.** $\log (x - 1) = \log 3$

9. Perform a check to see whether $x = -4$ is a solution of $\log_5 (x + 3) = \frac{1}{5}$.

10. Perform a check to see whether $x = -2$ is a solution of $5^{2x+3} = \frac{1}{5}$.

NOTATION *In Exercises 11–12, complete each solution.*

11. Solve $2^x = 7$.

$2^x = 7$

☐ $2^x = \log 7$

x ☐ $= \log 7$

$x = \frac{\log 7}{\log 2}$

12. Solve $\log_2 (2x - 3) = \log_2 (x + 4)$.

$\log_2 (2x - 3) = \log_2 (x + 4)$

☐ $= x + 4$

$x = 7$

PRACTICE *In Exercises 13–32, solve each exponential equation. Give answers to four decimal places when necessary.*

13. $4^x = 5$

14. $7^x = 12$

15. $13^{x-1} = 2$

16. $5^{x+1} = 3$

17. $2^{x+1} = 3^x$

18. $5^{x-3} = 3^{2x}$

19. $2^x = 3^x$

20. $3^{2x} = 4^x$

21. $7^{x^2} = 10$

22. $8^{x^2} = 11$

23. $8^{x^2} = 9^x$

24. $5^{x^2} = 2^{5x}$

25. $2^{x-2} = 64$

26. $3^{-3x+1} = 243$

27. $5^{4x} = \frac{1}{125}$

28. $8^{-x+1} = \frac{1}{64}$

29. $2^{x^2-2x} = 8$

30. $3^{x^2-3x} = 81$

31. $3^{x^2+4x} = \frac{1}{81}$

32. $7^{x^2+3x} = \frac{1}{49}$

In Exercises 33–36, use a graphing calculator to solve each equation. Give all answers to the nearest tenth.

33. $2^{x+1} = 7$

34. $3^{x-1} = 2^x$

35. $4(2^{x^2}) = 8^{3x}$

36. $3^x - 10 = 3^{-x}$

In Exercises 37–64, solve each logarithmic equation.

37. $\log 2x = \log 4$

38. $\log 3x = \log 9$

39. $\log_8 (3x + 1) = \log_8 (x + 7)$

40. $\log_2 (x^2 + 4x) = \log_2(x^2 + 16)$

41. $\log (3 - 2x) - \log (x + 24) = 0$

42. $\log (3x + 5) - \log (2x + 6) = 0$

43. $\log \dfrac{4x + 1}{2x + 9} = 0$

44. $\log \dfrac{2 - 5x}{2(x + 8)} = 0$

45. $\log x^2 = 2$

46. $\log x^3 = 3$

47. $\log x + \log (x - 48) = 2$

48. $\log x + \log (x + 9) = 1$

49. $\log x + \log (x - 15) = 2$

50. $\log x + \log (x + 21) = 2$

51. $\log (x + 90) = 3 - \log x$

52. $\log (x - 90) = 3 - \log x$

53. $\log (x - 6) - \log (x - 2) = \log \dfrac{5}{x}$

54. $\log (3 - 2x) - \log (x + 9) = 0$

55. $\dfrac{\log (3x - 4)}{\log x} = 2$

56. $\dfrac{\log (8x - 7)}{\log x} = 2$

57. $\dfrac{\log (5x + 6)}{2} = \log x$

58. $\dfrac{1}{2} \log (4x + 5) = \log x$

59. $\log_3 x = \log_3 \left(\dfrac{1}{x}\right) + 4$

60. $\log_5 (7 + x) + \log_5 (8 - x) - \log_5 2 = 2$

61. $2 \log_2 x = 3 + \log_2 (x - 2)$

62. $2 \log_3 x - \log_3 (x - 4) = 2 + \log_3 2$

63. $\log (7y + 1) = 2 \log (y + 3) - \log 2$

64. $2 \log (y + 2) = \log (y + 2) - \log 12$

In Exercises 65–68, use a graphing calculator to solve each equation. If an answer is not exact, round to the nearest tenth.

65. $\log x + \log (x - 15) = 2$

66. $\log x + \log (x + 3) = 1$

67. $\ln (2x + 5) - \ln 3 = \ln (x - 1)$

68. $2 \log (x^2 + 4x) = 1$

APPLICATIONS

69. TRITIUM DECAY The half-life of tritium is 12.4 years. How long will it take for 25% of a sample of tritium to decompose?

70. RADIOACTIVE DECAY In two years, 20% of a radioactive element decays. Find its half-life.

71. THORIUM DECAY An isotope of thorium, ^{227}Th, has a half-life of 18.4 days. How long will it take for 80% of the sample to decompose?

72. LEAD DECAY An isotope of lead, ^{201}Pb, has a half-life of 8.4 hours. How many hours ago was there 30% more of the substance?

73. CARBON-14 DATING A bone fragment analyzed by archaeologists contains 60% of the carbon-14 that it is assumed to have had initially. How old is it?

74. CARBON-14 DATING Only 10% of the carbon-14 in a small wooden bowl remains. How old is the bowl?

75. COMPOUND INTEREST If \$500 is deposited in an account paying 8.5% annual interest, compounded semiannually, how long will it take for the account to increase to \$800

76. CONTINUOUS COMPOUND INTEREST In Exercise 75, how long will it take if the interest is compounded continuously?

77. COMPOUND INTEREST If \$1,300 is deposited in a savings account paying 9% interest, compounded quarterly, how long will it take the account to increase to \$2,100?

78. COMPOUND INTEREST A sum of \$5,000 deposited in an account grows to \$7,000 in 5 years. Assuming annual compounding, what interest rate is being paid?

79. RULE OF SEVENTY A rule of thumb for finding how long it takes an investment to double is called the **rule of seventy.** To apply the rule, divide 70 by the interest rate written as a percent. At 5%, doubling requires $\frac{70}{5} = 14$ years to double an investment. At 7%, it takes $\frac{70}{7} = 10$ years. Explain why this formula works.

80. BACTERIAL GROWTH A bacterial culture grows according to the formula

$$P = P_0 a^t$$

If it takes 5 days for the culture to triple in size, how long will it take to double in size?

81. RODENT CONTROL The rodent population in a city is currently estimated at 30,000. If it is expected to double every 5 years, when will the population reach 1 million?

82. POPULATION GROWTH The population of a city is expected to triple every 15 years. When can the city planners expect the present population of 140 persons to double?

83. BACTERIAL CULTURE A bacterial culture doubles in size every 24 hours. By how much will it have increased in 36 hours?

84. OCEANOGRAPHY The intensity I of a light a distance x meters beneath the surface of a lake decreases exponentially. From Illustration 3, find the depth at which the intensity will be 20%.

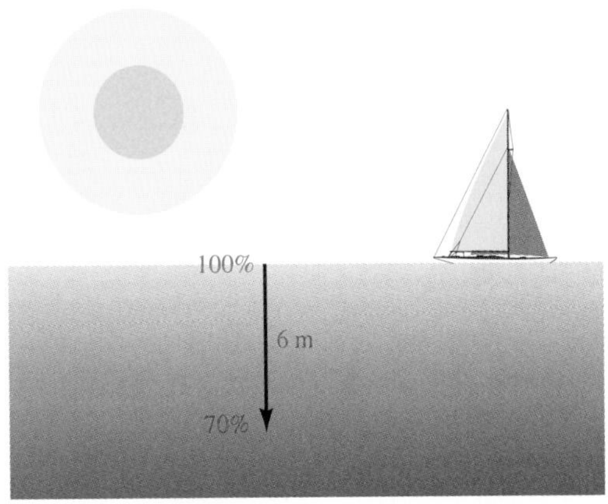

ILLUSTRATION 3

85. MEDICINE If a medium is inoculated with a bacterial culture containing 500 cells per milliliter, how many generations will have passed by the time the culture contains 5×10^6 cells per milliliter?

86. MEDICINE If a medium is inoculated with a bacterial culture containing 800 cells per milliliter, how many generations will have passed by the time the culture contains 6×10^7 cells per milliliter?

87. NEWTON'S LAW OF COOLING Water initially at 100°C is left to cool in a room at temperature 60°C. After 3 minutes, the water temperature is 90°. The water temperature T is a function of time t given by

$$T = 60 + 40e^{kt}$$

Find k.

88. NEWTON'S LAW OF COOLING Refer to Exercise 87 and find the time for the water temperature to reach 70°C.

WRITING

89. Explain how to solve the equation $2^x = 7$.

90. Explain how to solve the equation $2^{x+1} = 32$.

REVIEW *In Exercises 91–94, solve each equation.*

91. $5x^2 - 25x = 0$

92. $4y^2 - 25 = 0$

93. $3p^2 + 10p = 8$

94. $4t^2 + 1 = -6t$

KEY CONCEPT

Inverse Functions

One-to-One Functions

A function is *one-to-one* if each input value x in the domain determines a different output value y in the range. In Exercises 1–3, determine whether the function is one-to-one.

1. $f(x) = x^2$

2. $f(x) = |x|$

3.

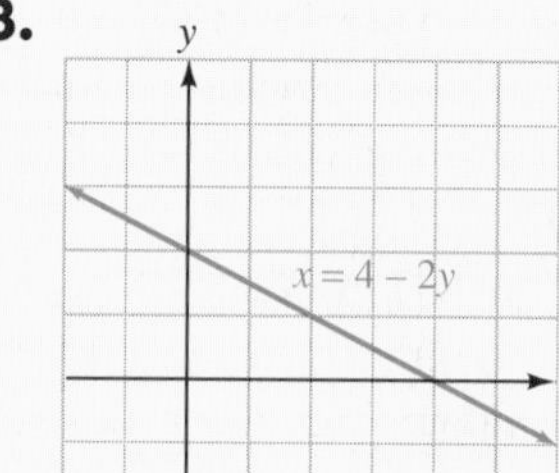

The Inverse of a Function

If a function is one-to-one, its inverse is a function. To find the inverse of a function, interchange x and y and solve for y.

4. Find f^{-1} if $f(x) = -2x - 1$.

5. Given the table of values for a one-to-one function f, complete the table of values for f^{-1}.

x	−2	1	3
$f(x)$	4	−2	−6

x	4	−2	−6
$f^{-1}(x)$			

Exponential and Logarithmic Functions

The exponential function $y = b^x$ and $y = \log_b x$ (where $b > 0$ and $b \neq 1$) are inverse functions.

6. Write the exponential statement $10^3 = 1{,}000$ in logarithmic form.

7. Write the logarithmic statement $\log_2 \frac{1}{8} = -3$ in exponential form.

8. If $\log_4 y = \frac{1}{2}$, what is y?

9. If $\log_y \frac{9}{4} = 2$, what is y?

The Natural Exponential and Natural Logarithmic Functions

A special exponential function that is used in many real-life applications involving growth and decay is the base-e exponential function, $f(x) = e^x$. Its inverse is the natural logarithm function $f(x) = \ln x$.

10. What is an approximate value of e?

11. What is the base of the logarithmic function $f(x) = \ln x$?

12. Use a calculator to find x: $\ln x = -0.28$.

13. Graph $f(x) = e^x$ and $f(x) = \ln x$ on the coordinate system in Illustration 1. Label the axis of symmetry.

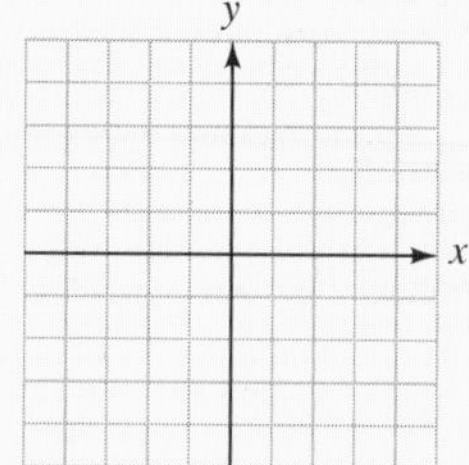

ILLUSTRATION 1

GROUP ACTIVITIES FOR CHAPTER 9

Accent on Teamwork

Section 9.1

COMPOSITION OF FUNCTIONS Consider the functions $f(x) = x^2$, $g(x) = 2x + 1$, and $h(x) = 1 - x$. Determine whether the composition of functions f, g, and h is associative. That is, is the following true?

$$[f \circ (g \circ h)](x) \stackrel{?}{=} [(f \circ g) \circ h](x)$$

Section 9.2

ONE-TO-ONE FUNCTIONS In newspapers, magazines, or books, find line graphs that are graphs of one-to-one functions and some that are not one-to-one. In each case, explain to the other members of your group what relationship the graph illustrates. Then tell whether the graph passes or fails the horizontal line test.

Section 9.3

EXPONENTIAL FUNCTIONS On a piece of poster board, draw a rectangular coordinate system made up of a grid of 1-inch squares. Then graph each of the following exponential functions. How are the graphs similar? How are they different? For positive values of x, as the base increases, what happens to the steepness of the graph?

$$f(x) = 2^x \qquad f(x) = 3^x \qquad f(x) = 4^x$$
$$f(x) = 5^x \qquad f(x) = 6^x \qquad f(x) = 7^x$$

Section 9.4

GROWTH Make two copies of each of the graph shapes shown in Illustration 1.

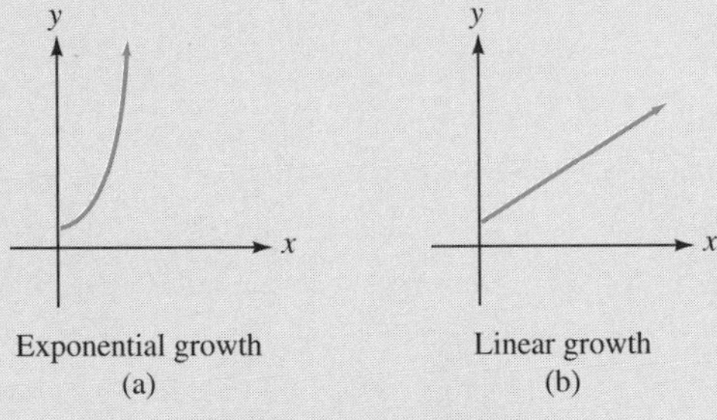

Exponential growth (a) Linear growth (b)

ILLUSTRATION 1

Think of two examples of quantities that, in general, have grown exponentially over the years. Label the horizontal and vertical axes of the graphs with the proper titles. (You do not have to scale the axes in terms of units.) Do the same for two quantities that, in general, have grown linearly over the years. Share your observations with the other members of your group. See if they agree with your models.

Section 9.5

LOGARITHMIC FUNCTIONS On a piece of poster board, draw a rectangular coordinate system made up of a grid of 1-inch squares. Then graph each of the following logarithmic functions. How are the graphs similar? How are they different? For positive values of x, as the base increases, what happens to the steepness of the graph?

$$f(x) = \log_2 x \qquad f(x) = \log_3 x \qquad f(x) = \log_4 x$$
$$f(x) = \log_5 x \qquad f(x) = \log_6 x \qquad f(x) = \log_7 x$$

Section 9.6

THE NUMBER e The value of e can be calculated to any degree of accuracy by adding the terms of the following pattern:

$$1, 1, \frac{1}{2}, \frac{1}{2 \cdot 3}, \frac{1}{2 \cdot 3 \cdot 4}, \frac{1}{2 \cdot 3 \cdot 4 \cdot 5}, \ldots$$

The more terms that are added, the closer the sum will be to e. Write each of the first six terms in simplified form and then add them. To how many decimal places is the sum accurate?

Section 9.7

The properties of logarithms discussed in Section 9.7 hold for logarithms of any base. Use these properties to simplify each of the following natural logarithmic expressions.

a. $\ln x^y + \ln x^z$ **b.** $\ln x^y - \ln x^z$

c. $\ln e^{10}$ **d.** $\ln \frac{x}{y} + \ln \frac{y}{x}$

Section 9.8

To solve exponential equations involving e, it is convenient to take the natural logarithm (instead of the common logarithm) of both sides of the equation. Use this technique to solve each of the following equations. Give all answers to four decimal places.

a. $e^x = 10$ **b.** $e^{2x} = 15$

c. $e^{x+1} = 50.5$ **d.** $3e^{5x-2} = 3$

CHAPTER REVIEW

SECTION 9.1 Algebra and Composition of Functions

CONCEPTS

Functions can be added, subtracted, multiplied, and divided:

$(f + g)(x) = f(x) + g(x)$

$(f - g)(x) = f(x) - g(x)$

$(f \cdot g)(x) = f(x)g(x)$

$(f/g)(x) = \dfrac{f(x)}{g(x)} \quad (g(x) \neq 0)$

Composition of functions:

$(f \circ g)(x) = f(g(x))$

REVIEW EXERCISES

1. Let $f(x) = 2x$ and $g(x) = x + 1$. Find each function or value.

a. $f + g$ **b.** $f - g$

c. $f \cdot g$ **d.** f/g

e. $(f \circ g)(1)$ **f.** $g(f(1))$

g. $(f \circ g)(x)$ **h.** $(g \circ f)(x)$

SECTION 9.2 Inverses of Functions

A function is *one-to-one* if each input value x in the domain determines a different output value y in the range.

2. Determine whether the function is one-to-one.

a. $f(x) = x^2 + 3$ **b.** $f(x) = x + 3$

Horizontal line test: If every horizontal line that intersects the graph of a function does so only once, the function is one-to-one.

3. Use the horizontal line test to decide whether the function is one-to-one.

a.

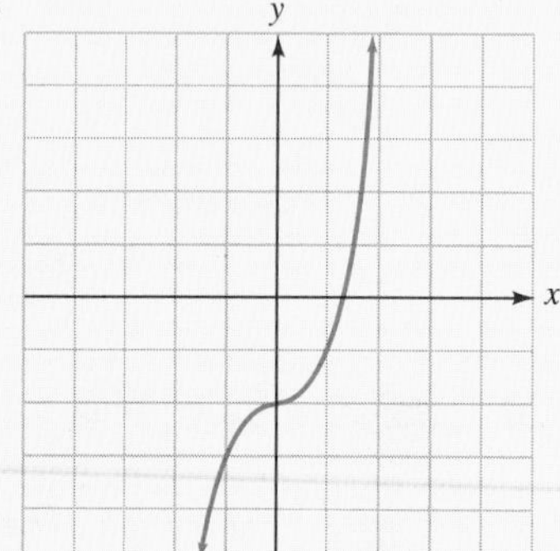

b.

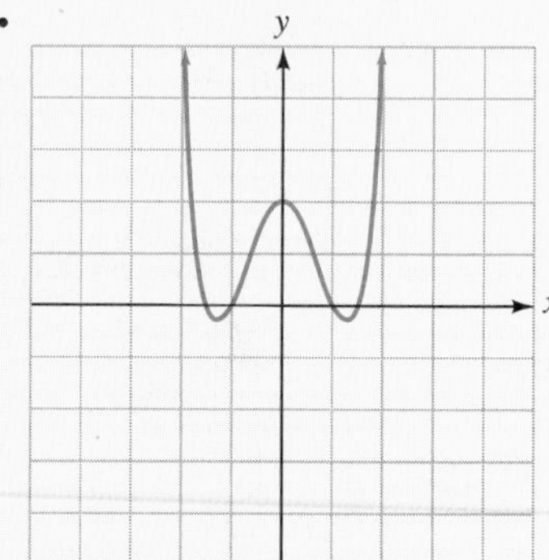

The graph of a function and its inverse are symmetric to the line $y = x$.

4. Given the graph of function f shown in Illustration 1, graph f^{-1} on the same coordinate axes. Label the axis of symmetry.

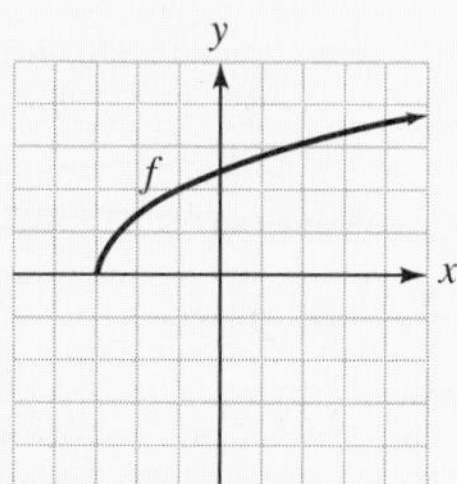

ILLUSTRATION 1

To *find the inverse of a function,* interchange the variables x and y and solve for y.

5. Find the inverse of each function.

a. $f(x) = 6x - 3$

b. $f(x) = 4x + 5$

c. $f(x) = x^3$

d. $y = 2x^2 - 1 \ (x \geq 0)$

SECTION 9.3 *Exponential Functions*

An *exponential function* with base b is defined by the equation

$$f(x) = b^x \quad (b > 0, b \neq 1)$$

If $b > 1$, then $f(x) = b^x$ is an *increasing function.*

If $0 < b < 1$, then $f(x) = b^x$ is a *decreasing function.*

6. Use properties of exponents to simplify each expression.

a. $5^{\sqrt{2}} \cdot 5^{\sqrt{2}}$

b. $\left(2^{\sqrt{5}}\right)^{\sqrt{2}}$

7. Graph each function. Then give the domain and the range.

a. $y = 3^x$

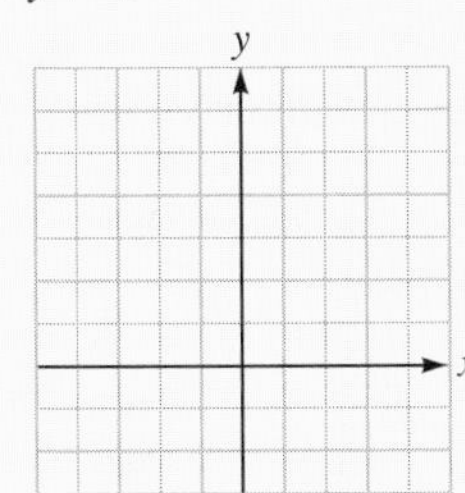

b. $f(x) = \left(\frac{1}{3}\right)^x$

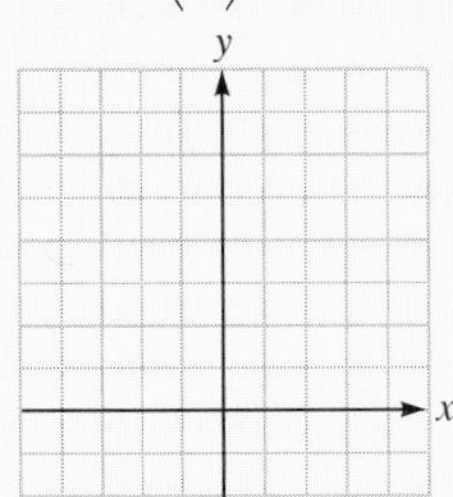

8. Graph each function by using a translation.

a. $f(x) = \left(\frac{1}{2}\right)^x - 2$

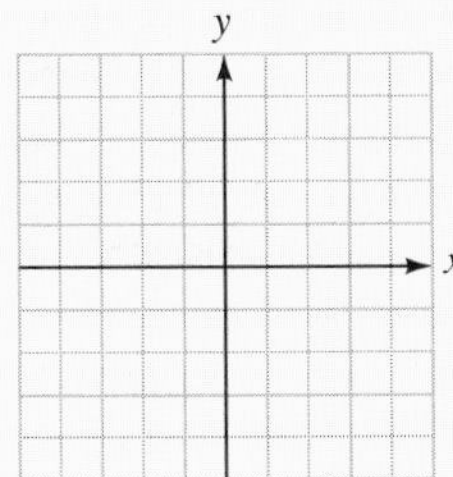

b. $g(x) = \left(\frac{1}{2}\right)^{x+2}$

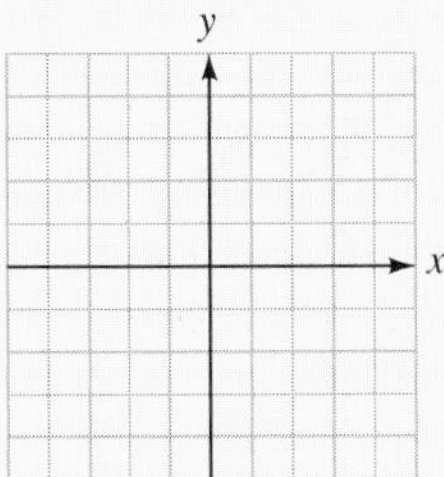

Exponential functions are suitable models for describing many situations involving the *growth* or *decay* of a quantity.

9. COAL PRODUCTION The table gives the number of tons of coal produced in the United States for the years 1800–1920. Graph the data in Illustration 2 on the next page. What type of function does it appear could be used to model coal production over this period?

Year	Tons
1800	108,000
1810	178,000
1820	881,000
1830	1,334,000
1840	2,474,000
1850	8,356,000
1860	20,041,000

Year	Tons
1870	40,429,000
1880	79,407,000
1890	157,771,000
1900	269,684,000
1910	501,596,000
1920	658,265,000

Based on information from *World Book Encyclopedia*

ILLUSTRATION 2

(continues)

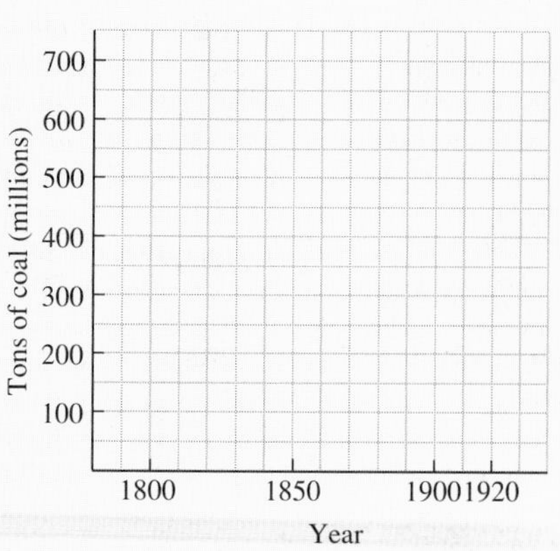

ILLUSTRATION 2 *(continued)*

Compound interest: If $P is the deposit, and interest is paid k times a year at an annual rate r, the amount A in the account after t years is given by

$$A = P\left(1 + \frac{r}{k}\right)^{kt}$$

10. COMPOUND INTEREST How much will $10,500 become if it earns 9% annual interest, compounded quarterly, for 60 years?

SECTION 9.4 Base-e Exponential Functions

The function defined by $f(x) = e^x$ is the *natural exponential function* where $e = 2.718281828459 \ldots$.

11. MORTGAGE RATES The average annual interest rate on a 30-year fixed-rate home mortgage for the years 1980–1996 can be approximated by the function $r(t) = 13.9e^{-0.035t}$, where t is the number of years since 1980. To the nearest hundredth of a percent, what does this model predict was the 30-year fixed rate in 1995?

If a quantity increases or decreases at an annual rate r *compounded continuously,* then the amount A after t years is given by

$$A = Pe^{rt}$$

12. INTEREST COMPOUNDED CONTINUOUSLY If $10,500 accumulates interest at an annual rate of 9%, compounded continuously, how much will be in the account in 60 years?

13. Graph each function. Give the domain and the range.

a. $f(x) = e^x + 1$ **b.** $f(x) = e^{x-3}$

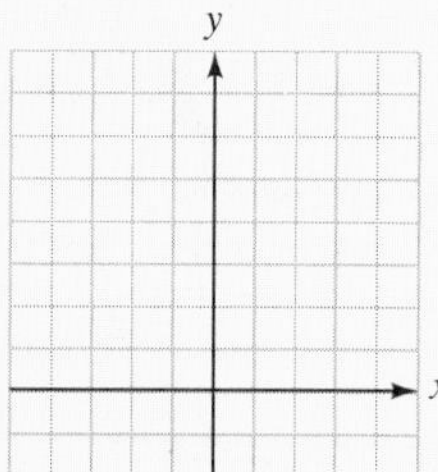

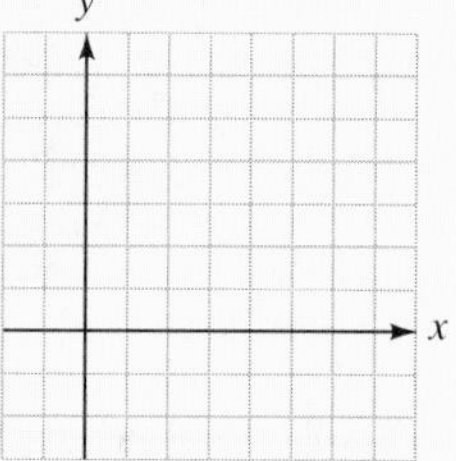

Malthusian population growth is modeled by the formula

$$P = P_0e^{kt}$$

14. POPULATION GROWTH The population of the United States is approximately 275,000,000 people. Find the population in 50 years if $k = 0.015$.

SECTION 9.5 Logarithmic Functions

If $b > 0$ and $b \neq 1$, then the *logarithmic function with base* b is defined by $y = \log_b x$.

$y = \log_b x$ means $x = b^y$.

$\log_b x$ is the exponent to which b is raised to get x.

$b^{\log_b x} = x$

If $b > 1$, then $f(x) = \log_b x$ is an *increasing function.* If $0 < b < 1$, then $f(x) = \log_b x$ is a *decreasing function.*

The exponential function $f(x) = b^x$ and the logarithmic function $f(x) = \log_b x$ are inverses of each other.

Decibel voltage gain:

$$\text{db gain} = 20 \log \frac{E_o}{E_I}$$

The Richter scale:

$$R = \log \frac{A}{P}$$

15. Give the domain and range of the logarithmic function $f(x) = \log x$.

16. **a.** Write the statement $\log_4 64 = 3$ in exponential form.
b. Write the statement $7^{-1} = \frac{1}{7}$ in logarithmic form.

17. Find each value, if possible.

a. $\log_3 9$ **b.** $\log_9 \frac{1}{81}$ **c.** $\log_8 1$

d. $\log_5 (-25)$ **e.** $\log_6 \sqrt{6}$ **f.** $\log 1{,}000$

18. Find the value of x in each equation.

a. $\log_2 x = 5$ **b.** $\log_3 x = -4$ **c.** $\log_x 16 = 2$

d. $\log_x \frac{1}{100} = -2$ **e.** $\log_9 3 = x$ **f.** $\log_{27} x = \frac{2}{3}$

19. Use a calculator to find the value of x to four decimal places.

a. $\log 4.51 = x$ **b.** $\log x = 1.43$

20. Graph each pair of equations on one set of coordinate axes.

a. $y = 4^x$ and $y = \log_4 x$

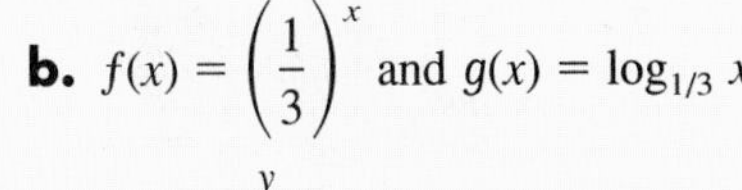

b. $f(x) = \left(\frac{1}{3}\right)^x$ and $g(x) = \log_{1/3} x$

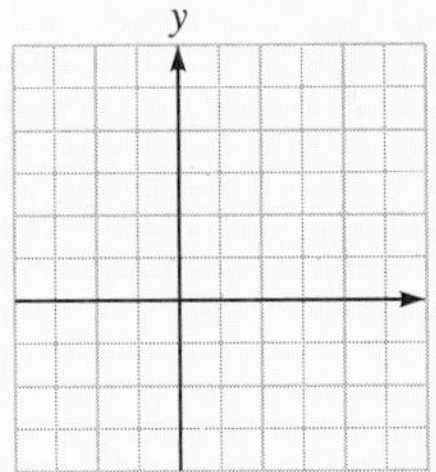

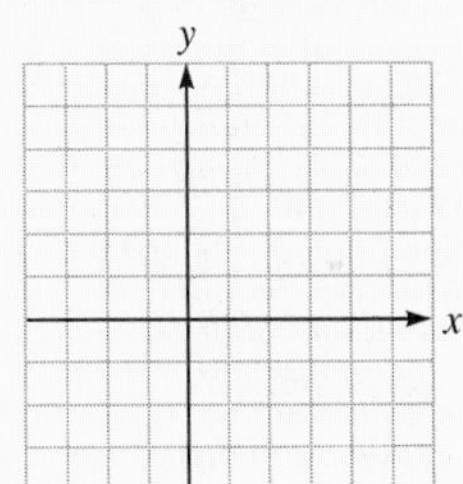

21. Graph each function.

a. $f(x) = \log (x - 2)$ **b.** $f(x) = 3 + \log x$

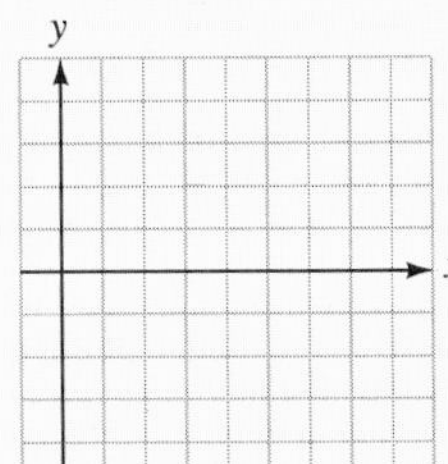

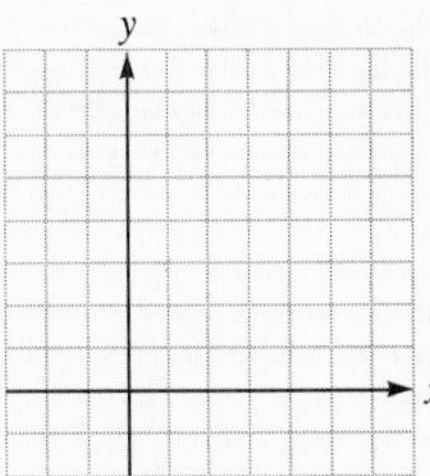

22. ELECTRICAL ENGINEERING An amplifier has an output of 18 volts when the input is 0.04 volt. Find the db gain.

23. EARTHQUAKE An earthquake had a period of 0.3 second and an amplitude of 7,500 micrometers. Find its measure on the Richter scale.

SECTION 9.6 Base-e Logarithms

Natural logarithms:

$\ln x$ means $\log_e x$

24. Use a calculator to find each value to four decimal places.
a. $\ln 452$ **b.** $\ln 0.85$

25. Use a calculator to find the value of x to four decimal places.
a. $\ln x = 2.336$ **b.** $\ln x = -8.8$

26. What function is the inverse of $f(x) = \ln x$?

27. Graph each function.
a. $f(x) = 1 + \ln x$ **b.** $y = \ln(x + 1)$

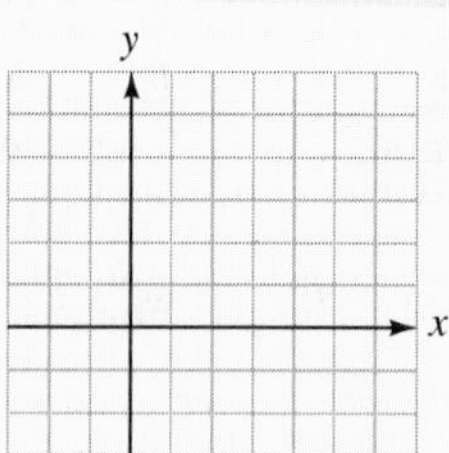

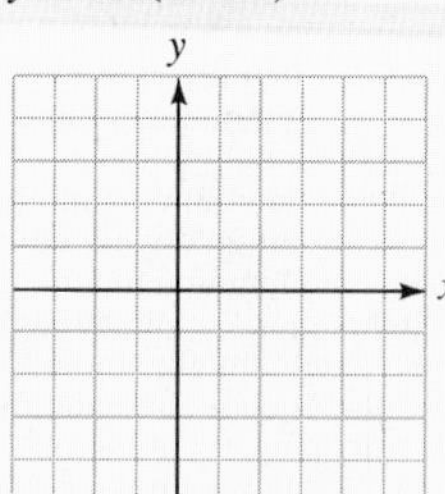

Population doubling time:

$$t = \frac{\ln 2}{r}$$

28. POPULATION GROWTH How long will it take the population of the United States to double if the growth rate is 3% per year?

SECTION 9.7 Properties of Logarithms

Properties of logarithms: If b is a positive number and $b \neq 1$,
1. $\log_b 1 = 0$ 2. $\log_b b = 1$
3. $\log_b b^x = x$ 4. $b^{\log_b x} = x$
5. $\log_b MN = \log_b M + \log_b N$
6. $\log_b \frac{M}{N} = \log_b M - \log_b N$
7. $\log_b M^p = p \log_b M$
8. If $\log_b x = \log_b y$, then $x = y$.

29. Simplify each expression.
a. $\log_7 1$ **b.** $\log_7 7$
c. $\log_7 7^3$ **d.** $7^{\log_7 4}$
e. $\ln e^4$ **f.** $\ln e$

30. Use the properties of logarithms to rewrite each expression.
a. $\log_3 27x$ **b.** $\log \frac{100}{x}$
c. $\log_5 \sqrt{27}$ **d.** $\log 10ab$

31. Write each expression in terms of the logarithms of x, y, and z.
a. $\log_b \frac{x^2y^3}{z}$ **b.** $\log \sqrt{\frac{x}{yz^2}}$

32. Write each expression as the logarithm of one quantity.

a. $3 \log_2 x - 5 \log_2 y + 7 \log_2 z$

b. $-3 \log_b y - 7 \log_b z + \frac{1}{2} \log_b (x + 2)$

33. Assume that $\log_b 5 = 1.1609$ and $\log_b 8 = 1.5000$. Use these values and the properties of logarithms to find each value to four decimal places.
a. $\log_b 40$ **b.** $\log_b 64$

Change-of-base formula:

$$\log_b x = \frac{\log_a x}{\log_a b}$$

34. Find $\log_5 17$ to four decimal places.

pH scale:

$$\text{pH} = -\log [\text{H}^+]$$

35. pH OF GRAPEFRUIT The pH of grapefruit juice is about 3.1 Find its hydrogen ion concentration.

SECTION 9.8 Exponential and Logarithmic Equations

An *exponential equation* is an equation that contains a variable in one of its exponents.

36. Solve each equation for x. Give answers to four decimal places when necessary.

a. $3^x = 7$

b. $5^{x+2} = 625$

c. $2^x = 3^{x-1}$

d. $2^{x^2+4x} = \frac{1}{8}$

A *logarithmic equation* is an equation with a logarithmic expression that contains a variable.

37. Solve each equation for x.

a. $\log x + \log (29 - x) = 2$

b. $\log_2 x + \log_2 (x - 2) = 3$

c. $\dfrac{\log (7x - 12)}{\log x} = 2$

d. $\log_2 (x + 2) + \log_2 (x - 1) = 2$

e. $\log x + \log (x - 5) = \log 6$

f. $\log 3 - \log (x - 1) = -1$

Carbon dating:

$$A = A_0 2^{-t/h}$$

38. CARBON-14 DATING A wooden statue found in Egypt has a carbon-14 content that is two-thirds of that found in living wood. If the half-life of carbon-14 is 5,700 years, how old is the statue?

Chapter 9 Test

In Problems 1–4, $f(x) = 4x$ and $g(x) = x - 1$. Find each function or value.

1. $g + f$

2. $g \cdot f$

3. Find $(g \circ f)(1)$.

4. Find $f(g(x))$.

In Problems 5–6, find the inverse of each function.

5. $3x + 2y = 12$

6. $f(x) = 3x^2 + 4 \quad (x \geq 0)$

In Problems 7–8, graph each function.

7. $f(x) = 2^x + 1$

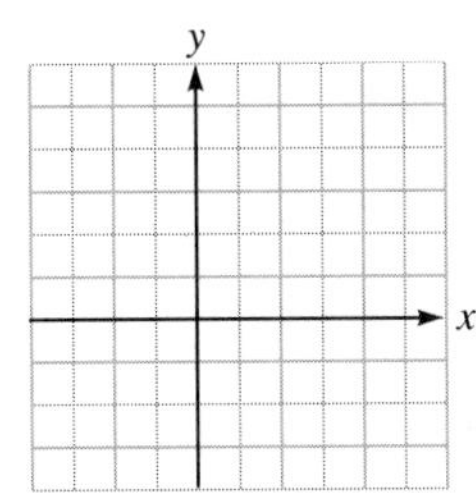

8. $f(x) = 2^{-x}$

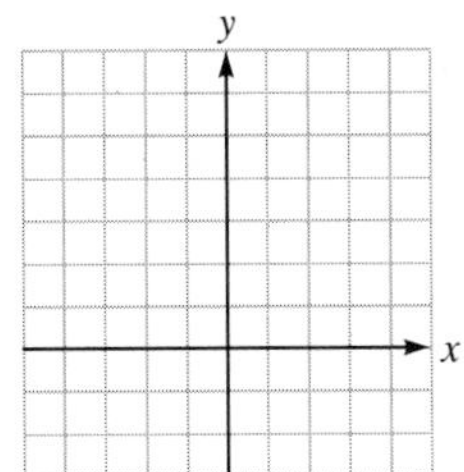

9. RADIOACTIVE DECAY A radioactive material decays according to the formula $A = A_0(2)^{-t}$. How much of a 3-gram sample will be left in 6 years?

10. COMPOUND INTEREST An initial deposit of \$1,000 earns 6% interest, compounded twice a year. How much will be in the account in one year?

11. Graph the function $f(x) = e^x$.

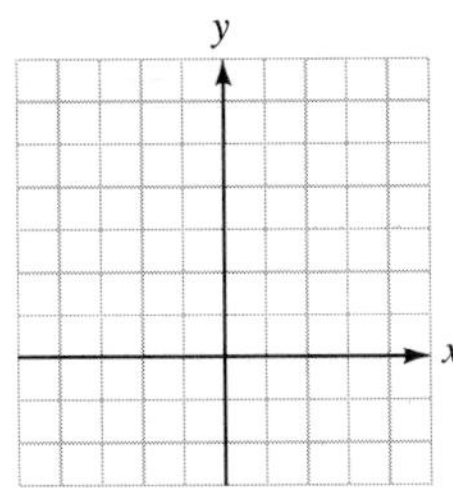

12. CONTINUOUS COMPOUNDING An account contains \$2,000 and has been earning 8% interest, compounded continuously. How much will be in the account in 10 years?

In Problems 13–16, find x.

13. $\log_4 16 = x$

14. $\log_x 81 = 4$

15. $\log_3 x = -3$

16. $\ln x = 1$

17. Write the statement $\log_6 \frac{1}{36} = -2$ in exponential form.

18. Give the domain and range of the function $f(x) = \log x$.

In Problems 19–20, graph each function.

19. $f(x) = -\log_3 x$

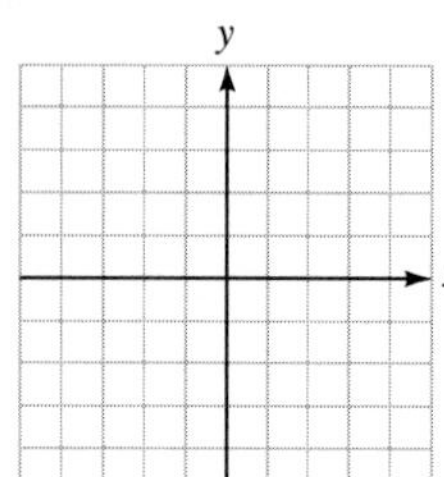

20. $f(x) = \ln x$

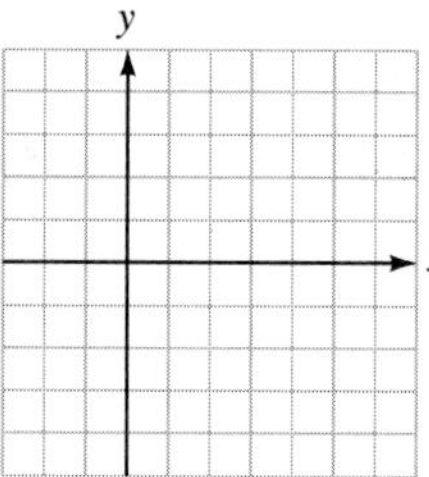

21. Write the expression $\log a^2bc^3$ in terms of the logarithms of a, b, and c.

22. Write the expression $\frac{1}{2}\log(a + 2) + \log b - 3\log c$ as a logarithm of a single quantity.

23. Use the change-of-base formula to find $\log_7 3$ to four decimal places.

24. What function is the inverse of $y = 10^x$?

25. pH Find the pH of a solution with a hydrogen ion concentration of 3.7×10^{-7}. (*Hint:* $\text{pH} = -\log [\text{H}^+]$.)

26. ELECTRONICS Find the db gain of an amplifier when $E_O = 60$ volts and $E_I = 0.3$ volt. (*Hint:* db gain = $20 \log \left(\frac{E_O}{E_I}\right)$).

In Problems 27–30, solve each equation. Round to four decimal places when necessary.

27. $5^x = 3$

28. $3^{x-1} = 27$

29. $\log_2 (5x + 2) = \log_2 (2x + 5)$

30. $\log x + \log (x - 9) = 1$

31. Give an example of a situation studied in this chapter that is modeled by a function with a graph that has the shape shown in Illustration 1. Label the axes. You do not have to scale the axes.

32. Consider the graph shown in Illustration 2.
a. Is it the graph of a function?
b. Is its inverse a function?
c. What is $f^{-1}(260)$? What information does it give?

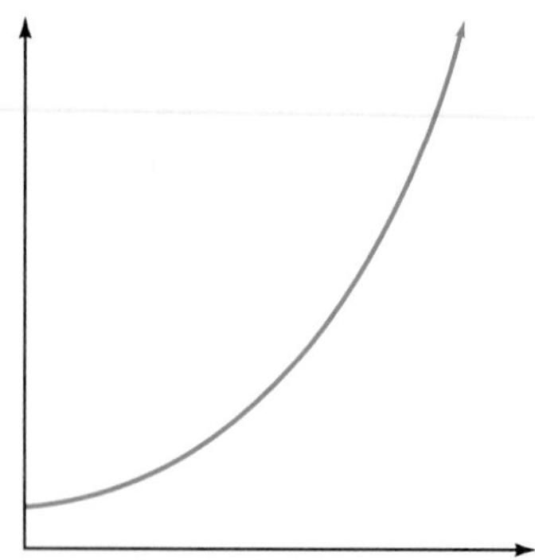

ILLUSTRATION 1

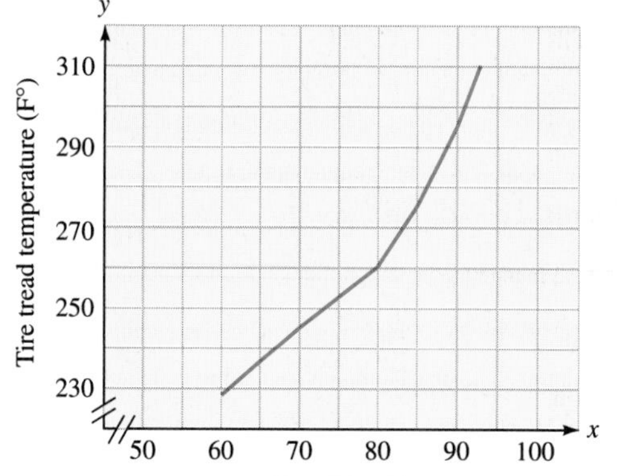

ILLUSTRATION 2

Chapters 1–9 Cumulative Review Exercises

In Exercises 1–4, consider the set $\left\{-\frac{4}{3}, \pi, 5.6, \sqrt{2}, 0, -23, e, 7i\right\}$. *List the elements in the set that are*

1. whole numbers

2. rational numbers

3. irrational numbers

4. real numbers

5. FINANCIAL PLANNING Ana has some money to invest. Her financial planner tells her that if she can come up with \$3,000 more, she will qualify for an 11% simple interest rate. Otherwise, she will have to invest the money at 7.5% annual interest. The financial planner urges her to invest the larger amount, because the 11% investment would yield twice as much annual income as the 7.5% investment. How much does she originally have on hand to invest?

6. BOATING Use the graph in Illustration 1 to determine the average rate of change in the sound level of the engine of a boat in relation to rpm of the engine.

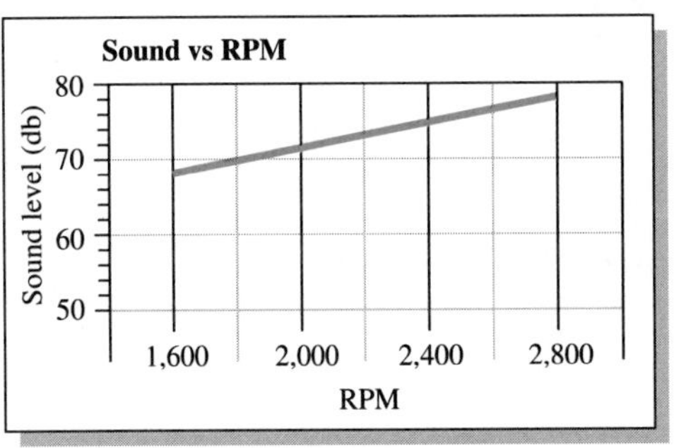

ILLUSTRATION 1

In Exercises 7–8, tell whether the graphs of the linear equations are parallel or perpendicular.

7. $3x - 4y = 12, y = \frac{3}{4}x - 5$

8. $y = 3x + 4, x = -3y + 4$

In Exercises 9–10, write the equation of the line with the given properties.

9. $m = -2$, passing through $(0, 5)$

10. Passing through $P(8, -5)$ and $Q(-5, 4)$

11. Use substitution to solve $\begin{cases} 3x + y = 4 \\ 2x - 3y = -1 \end{cases}$.

12. Use addition to solve $\begin{cases} x + 2y = -2 \\ 2x - y = 6 \end{cases}$.

13. Solve using Cramer's rule: $\begin{cases} 4x - 3y = -1 \\ 3x + 4y = -7 \end{cases}$.

14. Solve $\begin{cases} x + y + z = 1 \\ 2x - y - z = -4 \\ x - 2y + z = 4 \end{cases}$.

15. MARTIAL ARTS Find the measure of each angle of the triangle shown in Illustration 2.

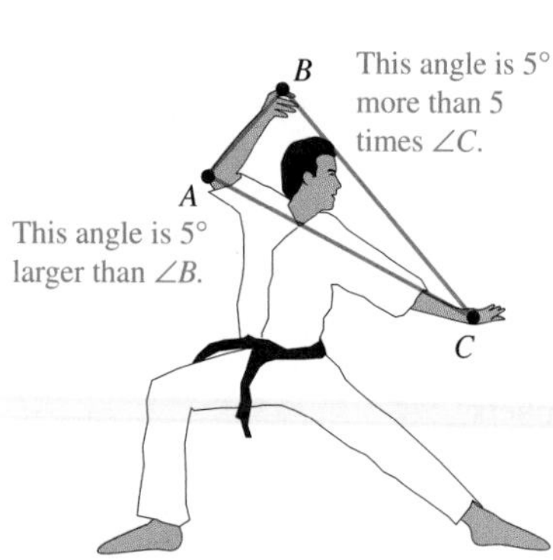

ILLUSTRATION 2

16. Solve 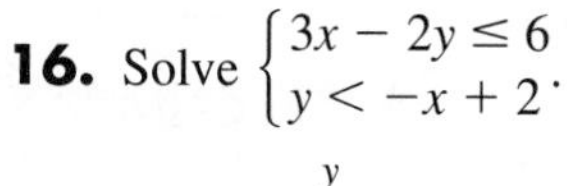 $\begin{cases} 3x - 2y \le 6 \\ y < -x + 2 \end{cases}$.

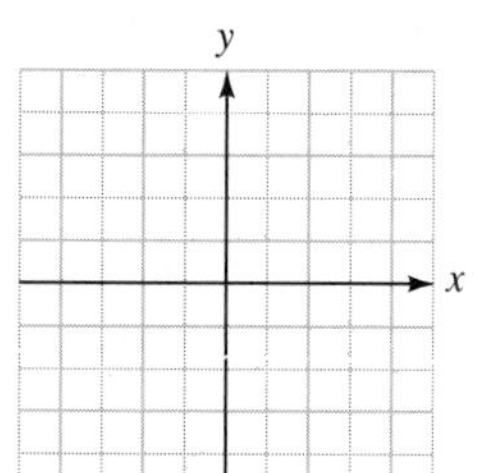

In Exercises 17–18, give the solution in interval notation and graph the solution set.

17. Solve $|5 - 3x| \le 14$.

18. Solve $4.5x - 1 < -10$ or $6 - 2x \ge 12$.

In Exercises 19–22, do the operations.

19. $(4x - 3y)(3x + y)$

20. $(-2x^2y^3 + 6xy + 5y^2) - (-4x^2y^3 - 7xy + 2y^2)$

21. $(a - 2b)^2$

22. $(a + 2)(3a^2 + 4a - 2)$

In Exercises 23–24, factor the expression completely.

23. $3x^3y - 4x^2y^2 - 6x^2y + 8xy^2$

24. $256x^4y^4 - z^8$

In Exercises 25–28, simplify.

25. $\left(\dfrac{4a^{-2}b}{3ab^{-3}}\right)^3$

26. $\dfrac{6x^2 + 13x + 6}{6 - 5x - 6x^2}$

27. $\dfrac{p^3 - q^3}{q^2 - p^2} \cdot \dfrac{q^2 + pq}{p^3 + p^2q + pq^2}$

28. $\dfrac{2}{a - 2} + \dfrac{3}{a + 2} - \dfrac{a - 1}{a^2 - 4}$

29. Solve $\dfrac{x - 4}{x - 3} + \dfrac{x - 2}{x - 3} = x - 3$.

30. Solve $\dfrac{1}{R} = \dfrac{1}{R_1} + \dfrac{1}{R_2} + \dfrac{1}{R_3}$ for R.

31. TIRE WEAR See Illustration 3.

a. What type of function does it appear would model the relationship between the inflation of a tire and the percent of service it gives?

b. At what percent(s) of inflation will a tire offer only 90% of its possible service?

32. CHANGING DIAPERS Illustration 4 shows how to put a diaper on a baby. If the diaper is a square with sides 16 inches long, what is largest waist size that this diaper can wrap around, assuming an overlap of 1 inch to pin the diaper?

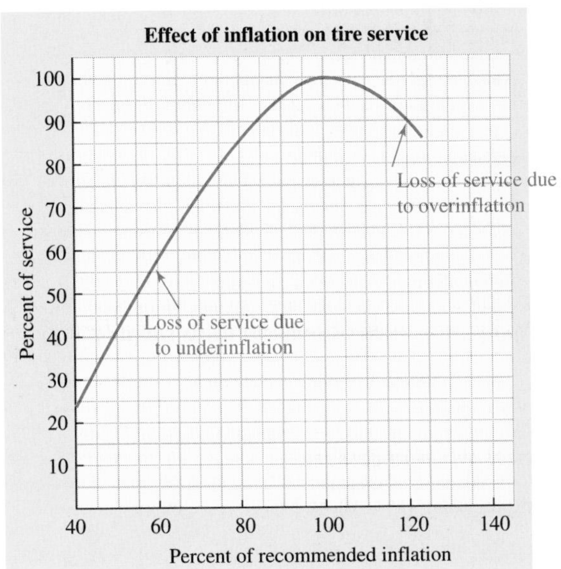

ILLUSTRATION 3

ILLUSTRATION 4

In Exercises 33–34, simplify each expression.

33. $\sqrt{98} + \sqrt{8} - \sqrt{32}$

34. $12\sqrt[3]{648x^4} + 3\sqrt[3]{81x^4}$

35. Evaluate $\left(\frac{25}{49}\right)^{-3/2}$.

36. Rationalize the denominator: $\frac{3t - 1}{\sqrt{3t} + 1}$.

In Exercises 37–38, write the expression in a + bi form.

37. $\left(-7 + \sqrt{-81}\right) - \left(-2 - \sqrt{-64}\right)$

38. $\frac{2 - 5i}{2 + 5i}$

In Exercises 39–44, solve each equation.

39. $\sqrt{3a + 1} = a - 1$

40. $\sqrt{x + 3} - \sqrt{3} = \sqrt{x}$

41. $6a^2 + 5a - 6 = 0$

42. $4w^2 + 6w + 1 = 0$

43. $2(2x + 1)^2 - 7(2x + 1) + 6 = 0$

44. $3x^2 - 4x = -2$

45. If $f(x) = x^2 - 2$ and $g(x) = 2x + 1$, find $(f \circ g)(x)$.

46. Find the inverse function of $f(x) = 2x^3 - 1$.

47. Graph $f(x) = \left(\frac{1}{2}\right)^x$.

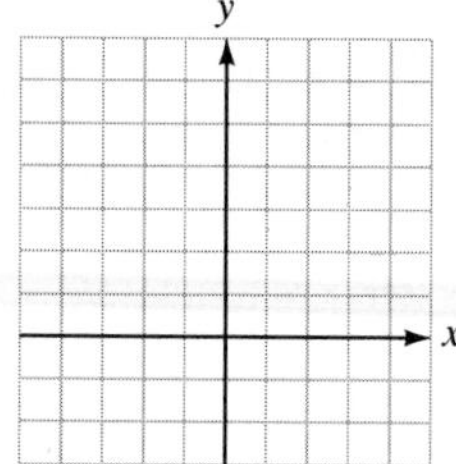

48. Graph $y = e^x$ and its inverse on the same coordinate system. Label the axis of symmetry.

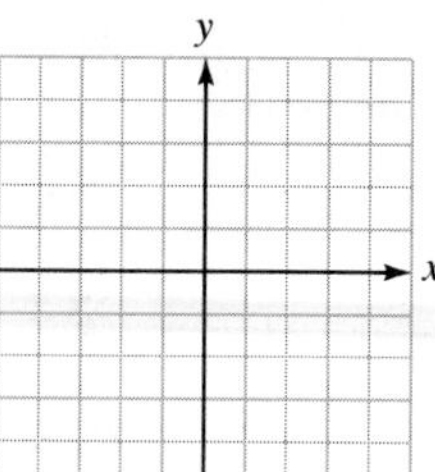

49. Write $y = \log_2 x$ as an exponential equation.

50. Apply properties of logarithms to simplify $\log_6 \frac{x}{36}$.

In Exercises 51–54, solve each equation. Round to four decimal places when necessary.

51. $2^{x+2} = 3^x$

52. $\log x + \log (x + 9) = 1$

53. $5^{4x} = \frac{1}{125}$

54. $\log_3 x = \log_3 \left(\frac{1}{x}\right) + 4$

APPENDIX I
CONIC SECTIONS

I.1 The Circle and the Parabola

In this section, you will learn about:

- **The circle**
- **Problem solving**
- **The parabola**
- **Problem solving**

Introduction. The graphs of second-degree equations in x and y represent figures that have interested people since the time of the ancient Greeks. The equations of these graphs were studied carefully in the 17th century, when René Descartes (1596–1650) and Blaise Pascal (1623–1662) began investigating them.

Descartes discovered that the graphs of second-degree equations fall into one of several categories: a pair of lines, a point, a circle, a parabola, an ellipse, a hyperbola, or no graph at all. Because all of these graphs can be formed by the intersection of a plane and a right-circular cone, they are called **conic sections.** See Figure I-1.

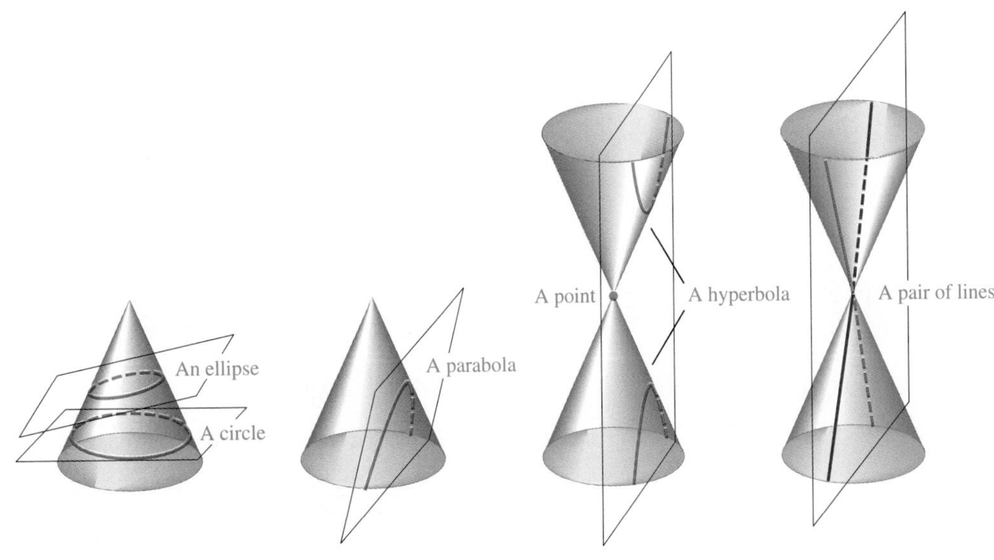

FIGURE I-1

Conic sections have many applications. For example, a parabola can be rotated to generate a dish-shaped surface called a **paraboloid.** Any light or sound placed at a certain point, called the *focus* of the paraboloid, is reflected outward in parallel paths. This property makes parabolic surfaces ideal for flashlight and headlight reflectors.

Using the same property in reverse makes parabolic surfaces good antennas, because signals captured by such an antenna are concentrated at the focus. A parabolic mirror is capable of concentrating the rays of the sun at a single point and thereby generating tremendous heat. This fact is used in the design of certain solar furnaces. Any object that is thrown upward and outward travels in a parabolic path.

In architecture, many arches are parabolic in shape because of their strength, and the cable that supports a suspension bridge hangs in the form of a parabola.

Ellipses have optical and acoustical properties that are useful in architecture and engineering. For example, many arches are portions of an ellipse, because the shape is pleasing to the eye. Gears are often cut into elliptical shapes, to provide nonuniform motion. The planets and some comets have elliptical orbits.

Hyperbolas serve as the basis of a navigational system known as LORAN (LOng RAnge Navigation). They are also used to find the source of a distress signal, are the basis for the design of hypoid gears, and describe the orbits of some comets.

The circle

The circle

A **circle** is the set of all points in a plane that are a fixed distance from a point called its **center.**

The fixed distance is called the **radius** of the circle.

To develop the general equation of a circle, we must write the equation of a circle with a radius of r and with a center at some point $C(h, k)$, as in Figure I-2. This task is equivalent to finding all points $P(x, y)$ such that the length of line segment CP is r. We can use the distance formula to find r.

$$r = \sqrt{(x - h)^2 + (y - k)^2}$$

We then square both sides to obtain

1. $r^2 = (x - h)^2 + (y - k)^2$

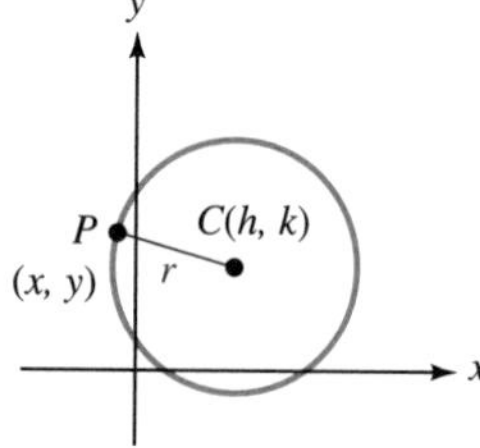

FIGURE I-2

Equation 1 is called the **standard form of the equation of a circle** with a radius of r and center at the point with coordinates (h, k).

Standard equation of a circle with center at (h, k)

Any equation that can be written in the form

$$(x - h)^2 + (y - k)^2 = r^2$$

has a graph that is a circle with radius r and center at point (h, k).

If $r = 0$, the graph reduces to a single point called a **point circle.** If $r^2 < 0$, then a circle does not exist. If both h and k are 0, the center of the circle is the origin.

Standard equation of a circle with center at (0, 0)

Any equation that can be written in the form

$$x^2 + y^2 = r^2$$

has a graph that is a circle with radius r and center at the origin.

EXAMPLE 1 Graph the equation $x^2 + y^2 = 25$.

Solution Because this equation can be written in the form $x^2 + y^2 = r^2$, its graph is a circle with center at the origin. Since $r^2 = 25 = 5^2$, the circle has a radius of 5. The graph appears in Figure I-3.

$x^2 + y^2 = 25$

x	y	(x, y)
-5	0	$(-5, 0)$
-4	± 3	$(-4, \pm 3)$
-3	± 4	$(-3, \pm 4)$
0	± 5	$(0, \pm 5)$
3	± 4	$(3, \pm 4)$
4	± 3	$(4, \pm 3)$
5	0	$(5, 0)$

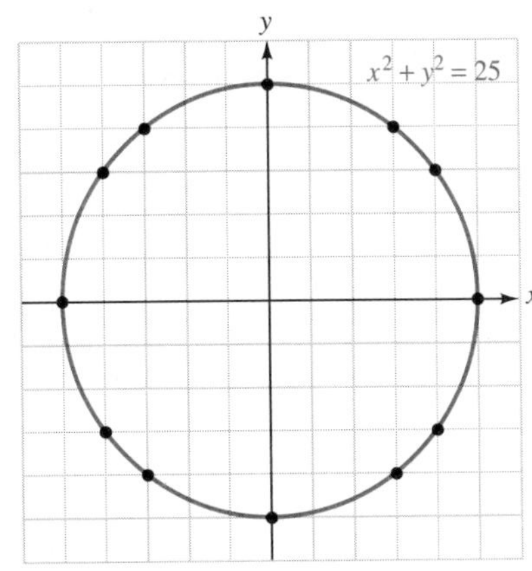

FIGURE I-3

EXAMPLE 2 Find the equation of the circle with radius 5 and center at $C(3, 2)$.

Solution We substitute 5 for r, 3 for h, and 2 for k in standard form and simplify.

$$(x - \mathbf{h})^2 + (y - \mathbf{k})^2 = \mathbf{r}^2$$
$$(x - \mathbf{3})^2 + (y - \mathbf{2})^2 = \mathbf{5}^2$$
$$x^2 - 6x + 9 + y^2 - 4y + 4 = 25$$
$$x^2 + y^2 - 6x - 4y - 12 = 0$$

The equation of the circle is $x^2 + y^2 - 6x - 4y - 12 = 0$.

EXAMPLE 3 Graph the circle $x^2 + y^2 - 4x + 2y = 20$.

Solution Because the equation is not in standard form, the coordinates of the center and the length of the radius are not obvious. To put the equation in standard form, we complete the square on both x and y as follows:

$$x^2 + y^2 - 4x + 2y = 20$$
$$x^2 - 4x + y^2 + 2y = 20$$
$$x^2 - 4x + \mathbf{4} + y^2 + 2y + \mathbf{1} = 20 + \mathbf{4} + \mathbf{1} \quad \text{Add 4 and 1 to both sides to complete the squares.}$$
$$(x - 2)^2 + (y + 1)^2 = 25 \quad \text{Factor } x^2 - 4x + 4 \text{ and } y^2 + 2y + 1.$$
$$(x - 2)^2 + [y - (-1)]^2 = 5^2$$

y
radius 5
x
(2, −1)

$x^2 + y^2 - 4x + 2y = 20$

FIGURE I-4

The radius of the circle is 5, and the coordinates of its center are $h = 2$ and $k = -1$. We plot the center of the circle and draw a circle with a radius of 5 units, as shown in Figure I-4.

Problem solving

EXAMPLE 4 ***Radio translators.*** The effective broadcast area of a television station is bounded by the circle $x^2 + y^2 = 3{,}600$, where x and y are measured in miles. A translator station picks up the signal and retransmits it from the center of a circular area bounded by $(x + 30)^2 + (y - 40)^2 = 1{,}600$. Find the location of the translator and the greatest distance from the main transmitter that the signal can be received.

Solution The coverage of the television station is bounded by $x^2 + y^2 = 60^2$, a circle centered at the origin with a radius of 60 miles, as shown in Figure I-5. Because the translator is at the center of the circle $(x + 30)^2 + (y - 40)^2 = 1{,}600$, it is located at $(-30, 40)$, a point 30 miles west and 40 miles north of the television station. The radius of the translator's coverage is $\sqrt{1{,}600}$, or 40 miles.

As shown in Figure I-5, the greatest distance of reception is the sum of A, the distance of the translator from the television station, and 40 miles, the radius of the translator's coverage.

To find A, we use the distance formula to find the distance between $(x_1, y_1) = (-30, 40)$ and the origin, $(0, 0)$.

$$A = \sqrt{(x_1 - x_2)^2 + (y_1 - y_2)^2}$$
$$A = \sqrt{(-30 - 0)^2 + (40 - 0)^2}$$
$$= \sqrt{(-30)^2 + 40^2}$$
$$= \sqrt{2{,}500}$$
$$= 50$$

Thus, the translator is located 50 miles from the television station, and it broadcasts the signal an additional 40 miles. The greatest reception distance is $50 + 40$, or 90 miles.

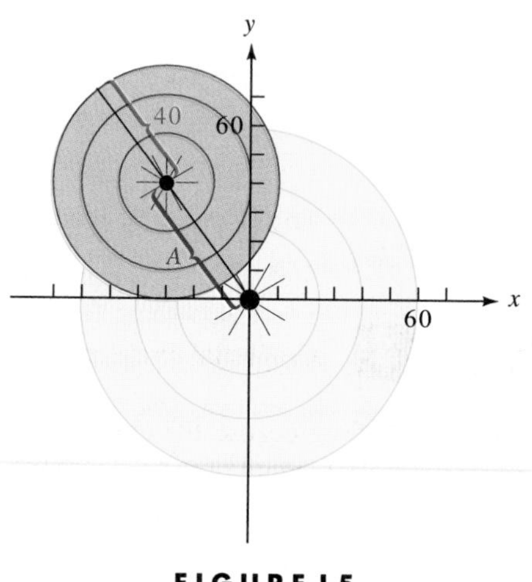

FIGURE I-5 ■

The parabola

We have seen that equations of the form $y = a(x - h)^2 + k$, with $a \neq 0$, represent parabolas with vertex at the point (h, k). They open upward when $a > 0$ and downward when $a < 0$.

Equations of the form $x = a(y - k)^2 + h$ also represent parabolas with vertex at the point (h, k). However, they open to the right when $a > 0$ and to the left when $a < 0$. Parabolas that open to the right or left do not represent functions, because their graphs do not pass the vertical line test.

Several types of parabolas are summarized in the following chart. (In all cases, $a > 0$.)

Equations of parabolas ($a > 0$)

Parabola opening	Vertex at origin	Vertex at (h, k)
Up	$y = ax^2$	$y = a(x - h)^2 + k$
Down	$y = -ax^2$	$y = -a(x - h)^2 + k$
Right	$x = ay^2$	$x = a(y - k)^2 + h$
Left	$x = -ay^2$	$x = -a(y - k)^2 + h$

EXAMPLE 5 Graph the equations **a.** $x = \frac{1}{2}y^2$ and **b.** $x = -2(y - 2)^2 + 3$.

Solution **a.** We make a table of ordered pairs, plot each pair, and draw the parabola, as in Figure I-6(a). Because the equation is of the form $x = ay^2$ with $a > 0$, the parabola opens to the right and has its vertex at the origin.

b. We make a table of ordered pairs, plot each pair, and draw the parabola, as in Figure I-6(b). Because the equation is of the form $x = -a(y - k)^2 + h$, the parabola opens to the left and has its vertex at the point with coordinates (3, 2).

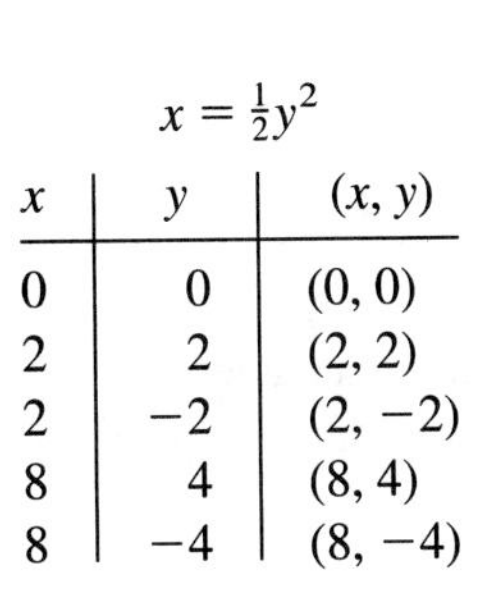

$x = \frac{1}{2}y^2$

x	y	(x, y)
0	0	(0, 0)
2	2	(2, 2)
2	−2	(2, −2)
8	4	(8, 4)
8	−4	(8, −4)

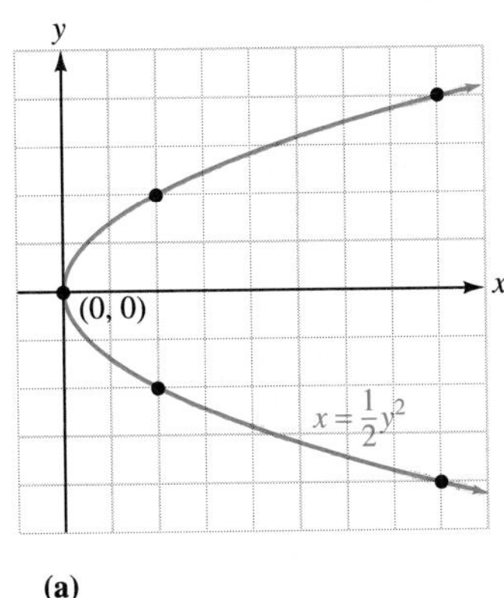

(a)

$x = -2(y - 2)^2 + 3$

x	y	(x, y)
−5	0	(−5, 0)
1	1	(1, 1)
3	2	(3, 2)
1	3	(1, 3)
−5	4	(−5, 4)

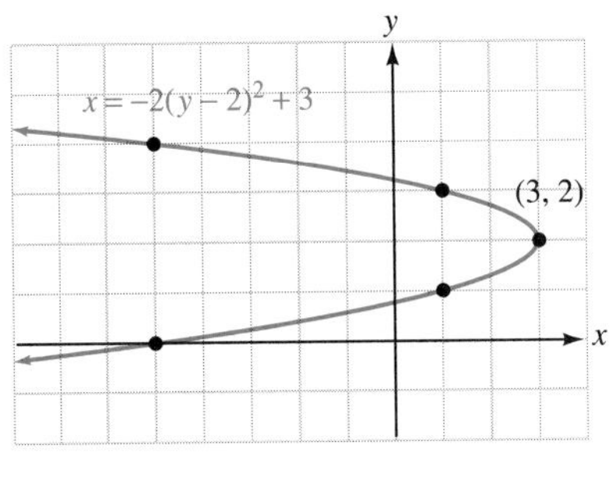

(b)

FIGURE I-6

EXAMPLE 6 Graph the equation $y = -2x^2 + 12x - 15$.

Solution Because the equation is not in standard form, the coordinates of its vertex are not obvious. To put the equation into standard form, we complete the square on x.

$$y = -2x^2 + 12x - 15$$
$$y = -2(x^2 - 6x \quad) - 15 \qquad \text{Factor out } -2 \text{ from } -2x^2 + 12x.$$
$$y = -2(x^2 - 6x + 9 - 9) - 15 \qquad \text{Add and subtract 9.}$$
$$y = -2(x^2 - 6x + 9) + 18 - 15$$
$$y = -2(x - 3)^2 + 3$$

Because the equation is in the form $y = -a(x - h)^2 + k$, we can see that the parabola opens downward and has its vertex at (3, 3). The graph of the equation appears in Figure I-7.

$y = -2x^2 + 12x - 15$

x	y	(x, y)
1	−5	(1, −5)
2	1	(2, 1)
3	3	(3, 3)
4	1	(4, 1)
5	−5	(5, −5)

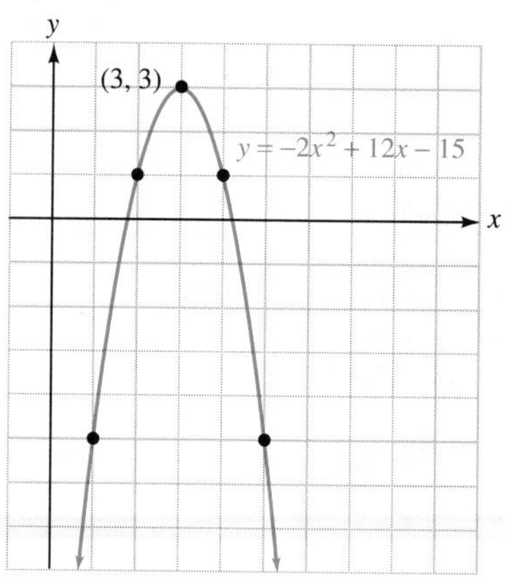

FIGURE I-7

Problem solving

EXAMPLE 7 ***Gateway Arch.*** The shape of the Gateway Arch in St. Louis is approximately a parabola, as shown in Figure I-8(a). How high is the arch 100 feet from its foundation?

Solution We place the parabola in a coordinate system as in Figure I-8(b), with ground level on the x-axis and the vertex of the parabola at the point (0, 630). The equation of this downward-opening parabola has the form

$$y = -a(x - h)^2 + k$$

$$y = -a(x - 0)^2 + 630 \qquad \text{Substitute } h = 0 \text{ and } k = 630.$$

$$y = -ax^2 + 630 \qquad \text{Simplify.}$$

Because the Gateway Arch is 630 feet wide at its base, the parabola passes through the point $\left(\frac{630}{2}, 0\right)$ or (315, 0). To find a in the equation of the parabola, we substitute $x = 315$ and $y = 0$ and proceed as follows:

$$y = -ax^2 + 630$$

$$0 = -a \cdot 315^2 + 630$$

$$\frac{-630}{315^2} = -a \qquad \text{Subtract 630 from both sides and divide both sides by } 315^2.$$

$$\frac{2}{315} = a \qquad \text{Multiply both sides by } -1 \text{ and simplify.}$$

Thus, the equation of the parabola that approximates the shape of the Gateway Arch is

$$y = -\frac{2}{315}x^2 + 630$$

To find the height of the arch at a point 100 feet from its foundation, we substitute 315 − 100, or 215, for x into the equation of the parabola and solve for y.

$$y = -\frac{2}{315}x^2 + 630$$

$$y = -\frac{2}{315}(215)^2 + 630$$

$$\approx 336.5 \qquad \text{Use a calculator.}$$

At a point 100 feet from the foundation, the height of the arch is 336.5 feet.

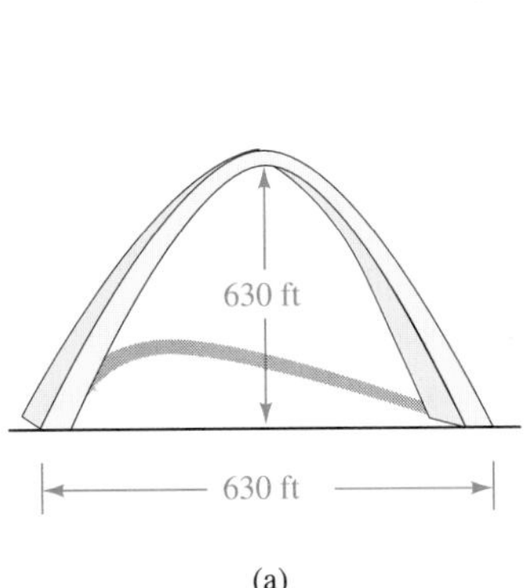

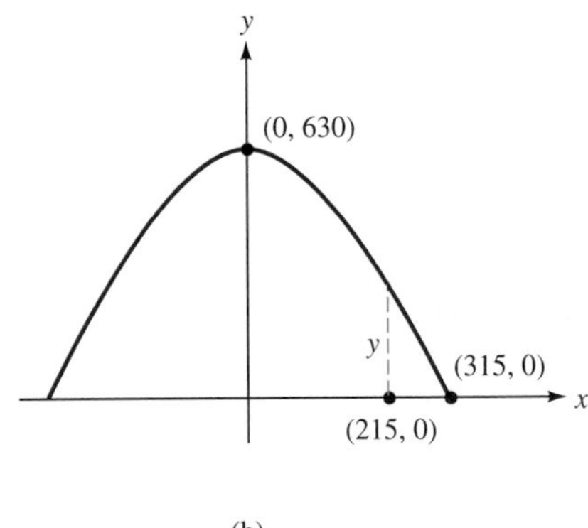

FIGURE I-8

STUDY SET Section I.1

VOCABULARY *In Exercises 1–4, fill in the blanks to make the statements true.*

1. A ______________ is the set of all points in a plane that are a fixed distance from a point called its ______________.

2. The fixed distance in Exercise 1 is called a ______________ of the circle.

3. The graph of the equation $y = a(x - h)^2 + k$, with $a \neq 0$, is called a ______________.

4. Parabolas that open upward or downward represent ______________.

CONCEPTS *In Exercises 5–10, fill in the blanks to make the statements true.*

5. The standard equation of a circle with center at (0, 0) is ______________.

6. The standard equation of a circle with center at (h, k) is ______________.

7. If $a > 0$, the graph of $y = a(x - h)^2 + k$ opens ______________.

8. The vertex of $y = 3(y - 2)^2 + 4$ is the point ______________.

9. The graph of $y = -ax^2$ $(a > 0)$ is a ______________ opening ______________.

10. The graph of $x = a(y - k)^2 + h$ $(a > 0)$ is a ______________ opening to the ______________.

PRACTICE *In Exercises 11–20, graph each equation.*

11. $x^2 + y^2 = 9$

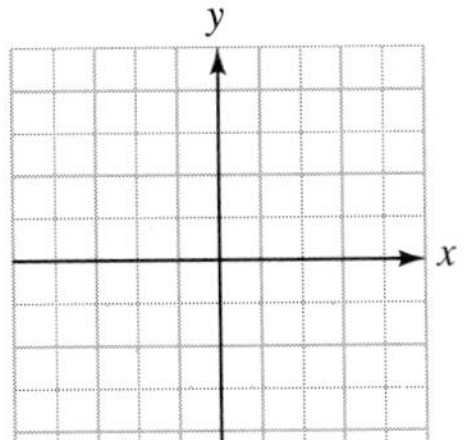

12. $x^2 + y^2 = 16$

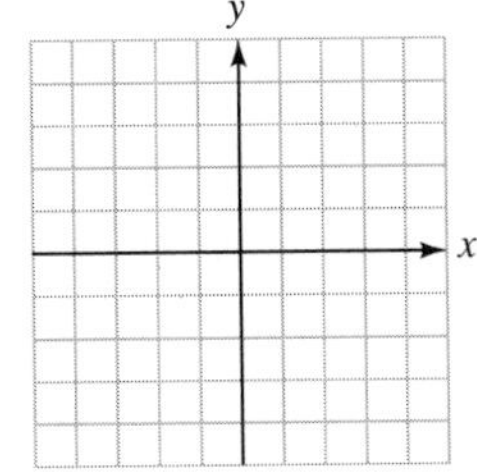

13. $(x - 2)^2 + y^2 = 9$

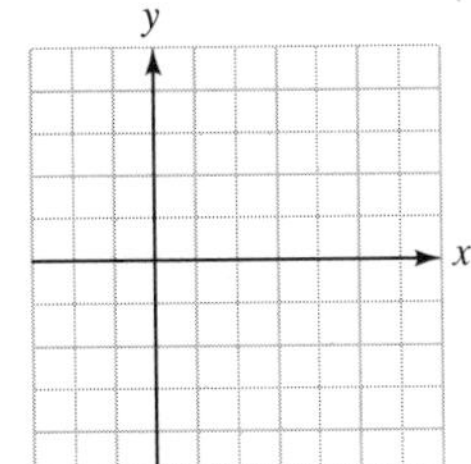

14. $x^2 + (y - 3)^2 = 4$

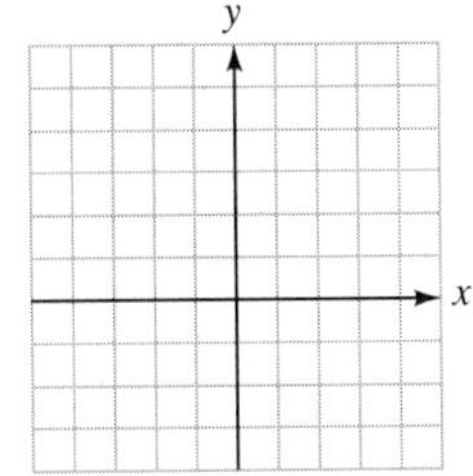

15. $(x-2)^2+(y-4)^2=4$

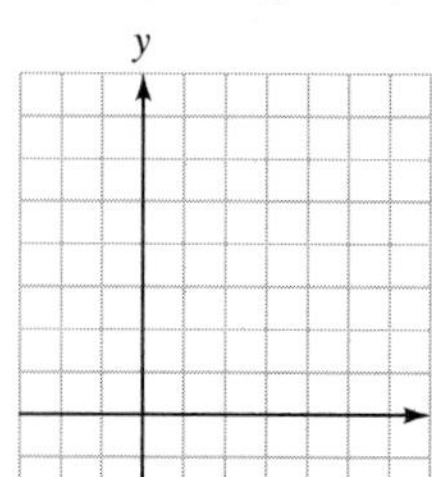

16. $(x-3)^2+(y-2)^2=4$

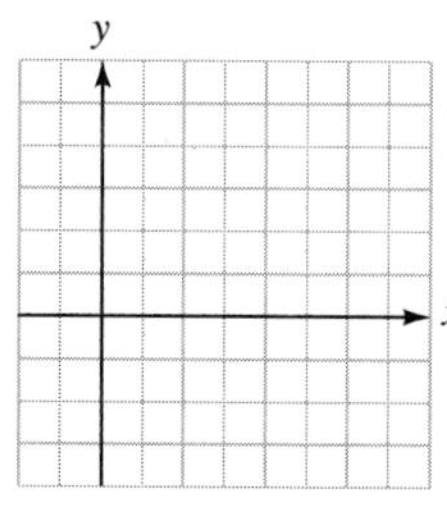

17. $(x+3)^2+(y-1)^2=16$

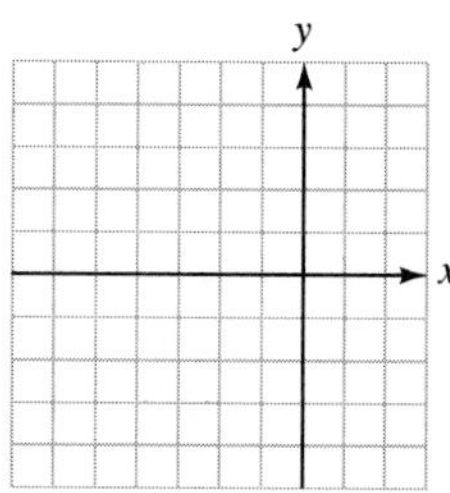

18. $(x-1)^2+(y+4)^2=9$

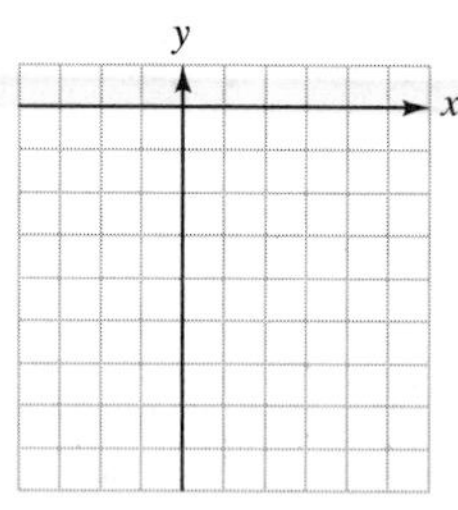

19. $x^2+(y+3)^2=1$

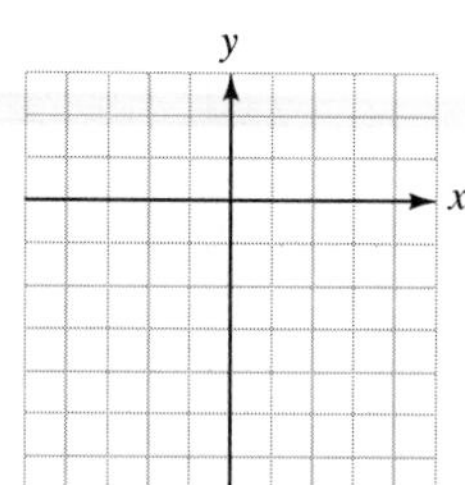

20. $(x+4)^2+y^2=1$

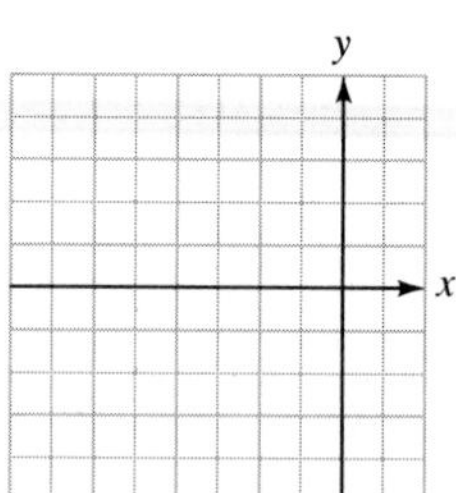

In Exercises 21–28, write the equation of the circle with the following properties.

21. Center at the origin; radius of 1

22. Center at the origin; radius of 4

23. Center at (6, 8); radius of 5

24. Center at (5, 3); radius of 2

25. Center at (−2, 6); radius of 12

26. Center at (5, −4); radius of 6

27. Center at the origin; diameter of $2\sqrt{2}$

28. Center at the origin; diameter of $4\sqrt{3}$

In Exercises 29–36, graph each circle. Give the coordinates of the center.

29. $x^2+y^2+2x-8=0$

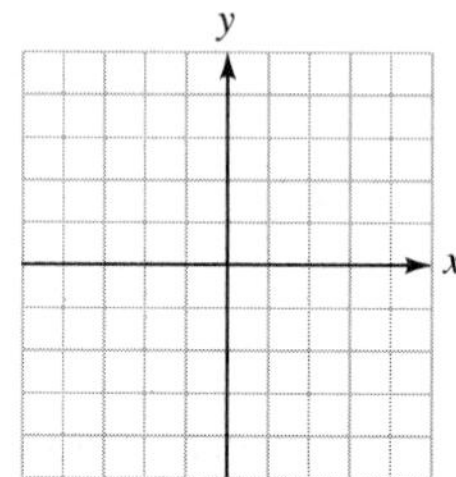

30. $x^2+y^2-4y=12$

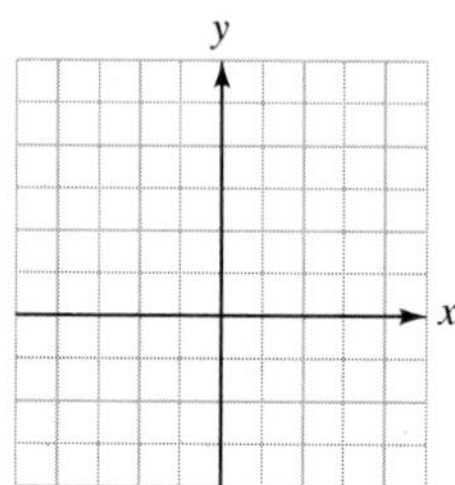

31. $9x^2+9y^2-12y=5$

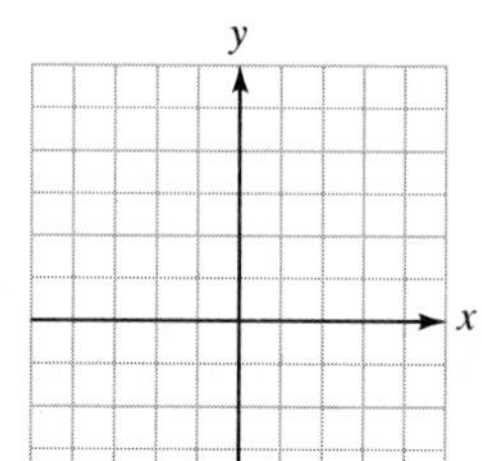

32. $4x^2+4y^2+4y=15$

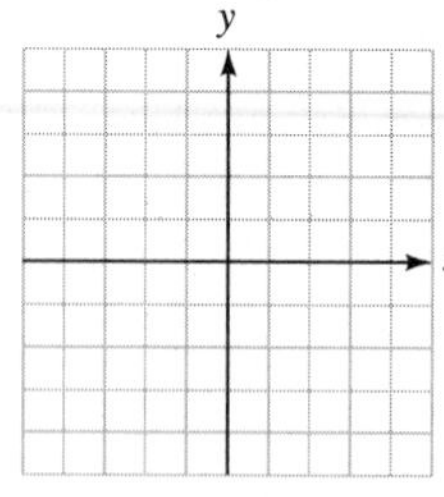

33. $x^2+y^2-2x+4y=-1$

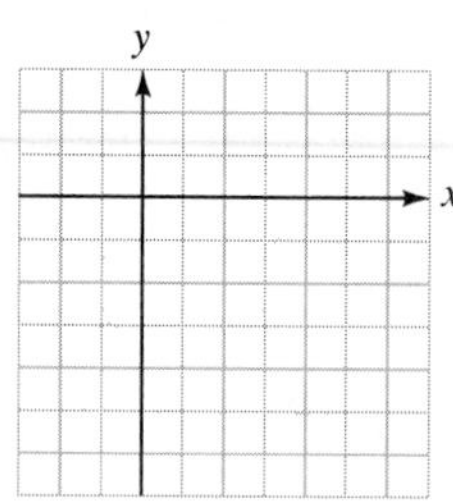

34. $x^2+y^2+4x+2y=4$

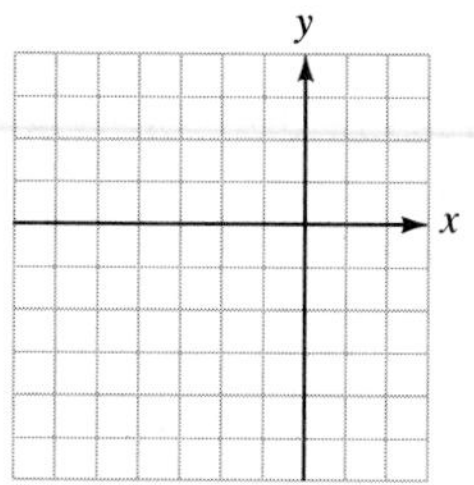

35. $x^2+y^2+6x-4y=-12$

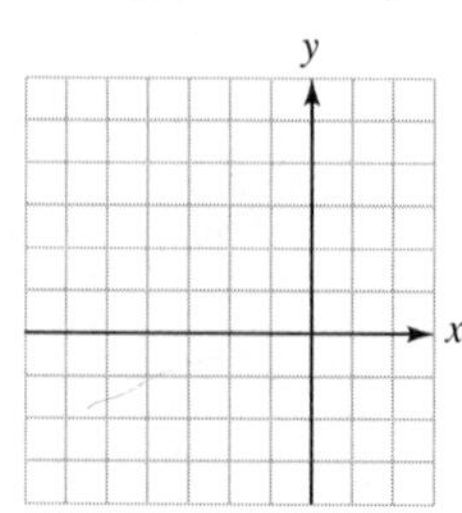

36. $x^2+y^2+8x+2y=-13$

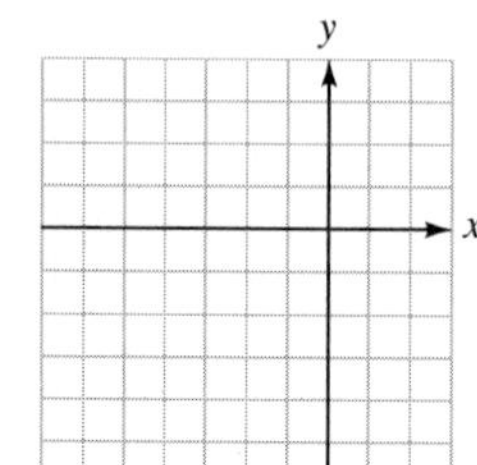

In Exercises 37–50, find the vertex of each parabola. Then graph the parabola.

37. $x = y^2$

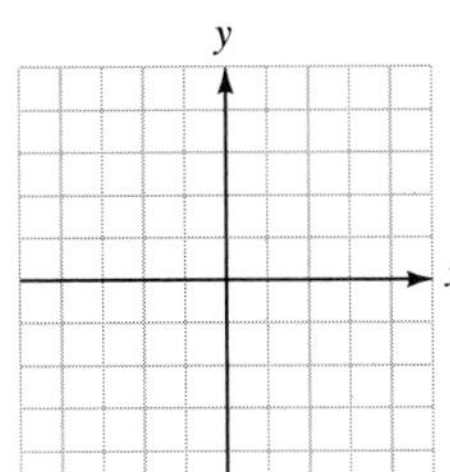

38. $x = -y^2 + 1$

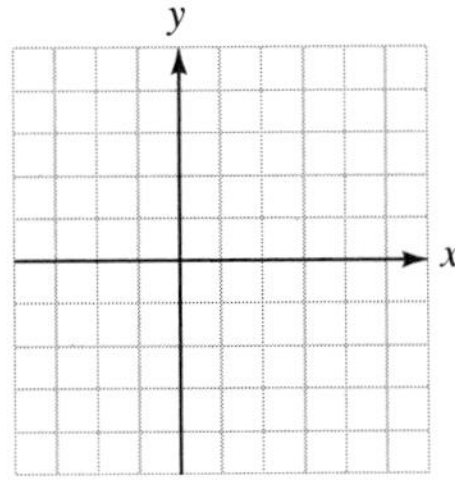

39. $x = -\frac{1}{4}y^2$

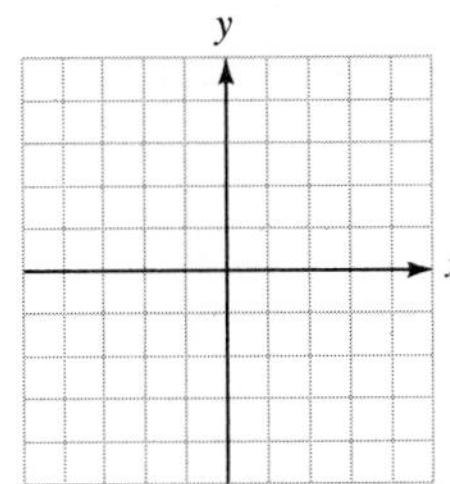

40. $x = 4y^2$

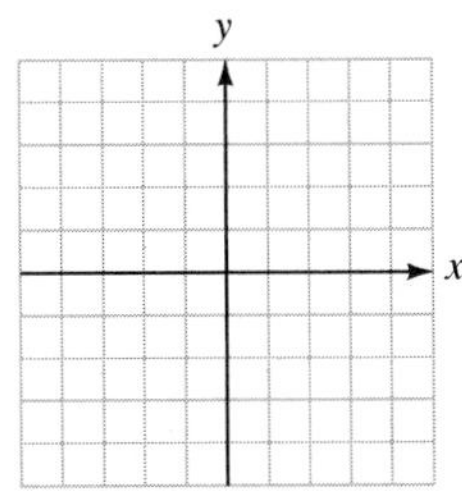

41. $y = x^2 + 4x + 5$

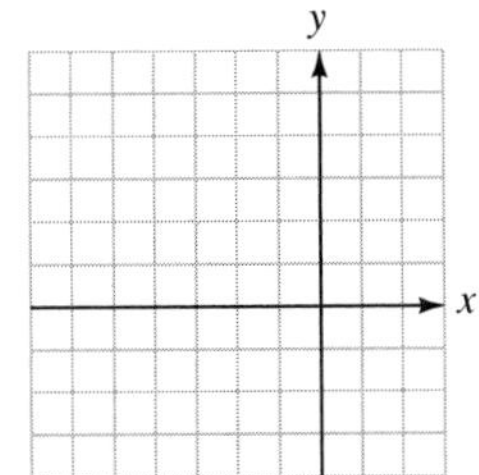

42. $y = -x^2 - 2x + 3$

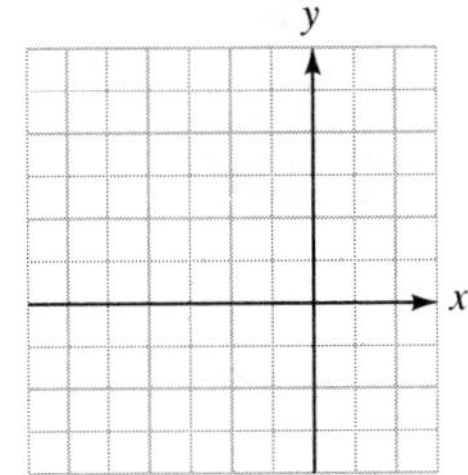

43. $y = -x^2 - x + 1$

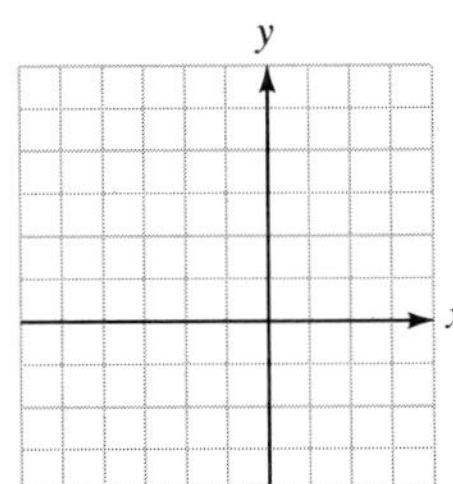

44. $x = \frac{1}{2}y^2 + 2y$

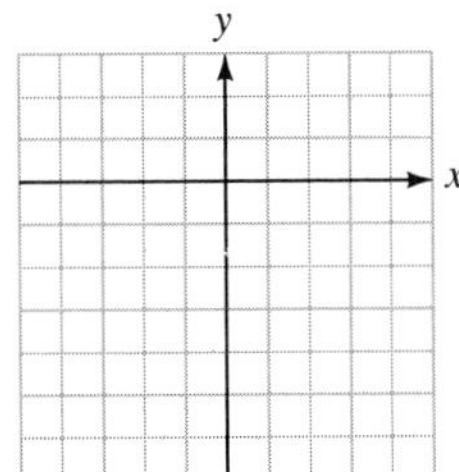

45. $y^2 + 4x - 6y = -1$

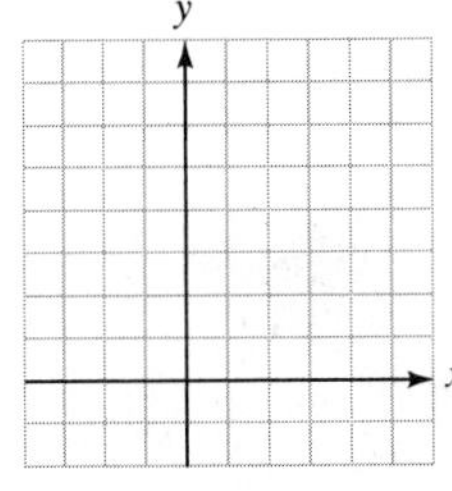

46. $x^2 - 2y - 2x = -7$

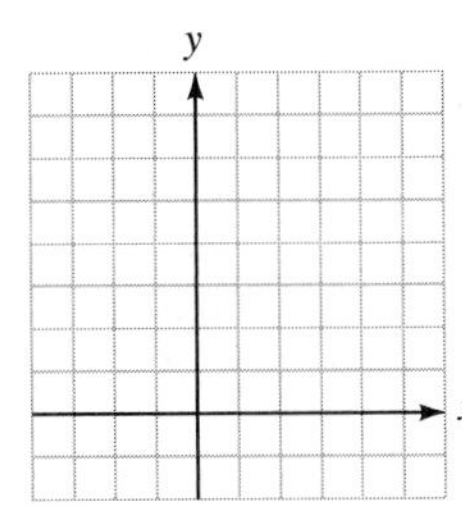

47. $y = 2(x - 1)^2 + 3$

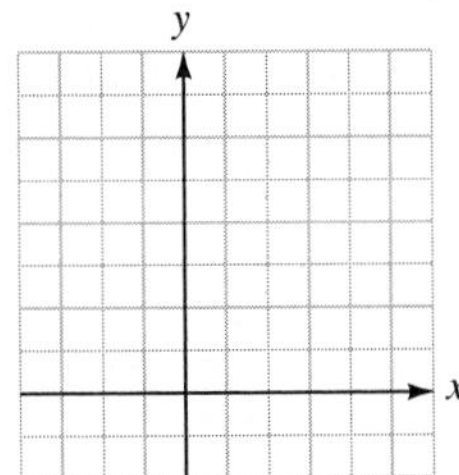

48. $y = -2(x + 1)^2 + 2$

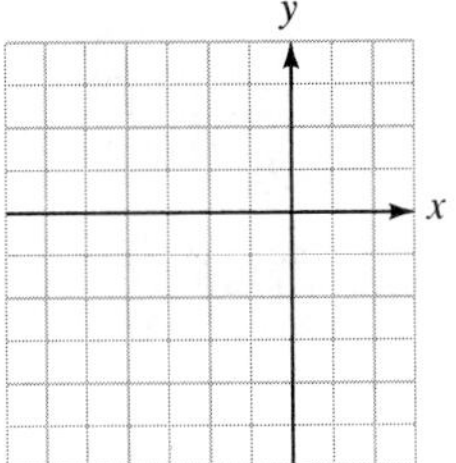

49. $x = -3(y + 2)^2 - 2$

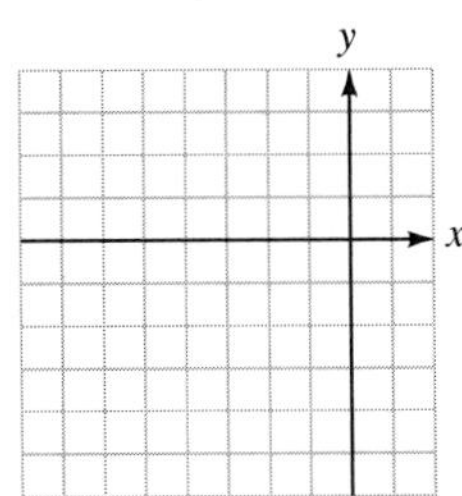

50. $x = 2(y - 3)^2 - 4$

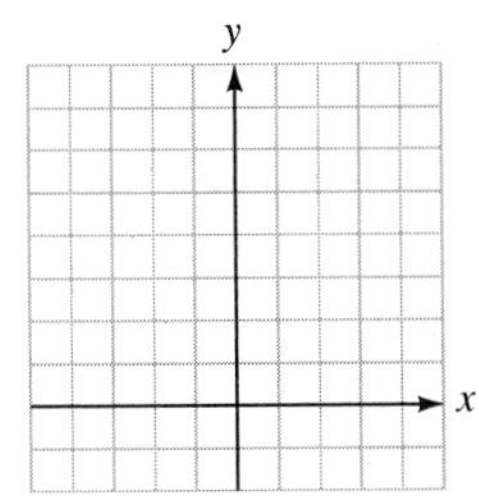

APPLICATIONS

51. MESHING GEARS For design purposes, the large gear in Illustration 1 is the circle $x^2 + y^2 = 16$. The smaller gear is a circle centered at (7, 0) and tangent to the larger circle. Find the equation of the smaller gear.

ILLUSTRATION 1

52. WIDTH OF A WALKWAY The walkway in Illustration 2 is bounded by the two circles $x^2 + y^2 = 2{,}500$ and $(x - 10)^2 + y^2 = 900$, measured in feet. Find the largest and the smallest width of the walkway.

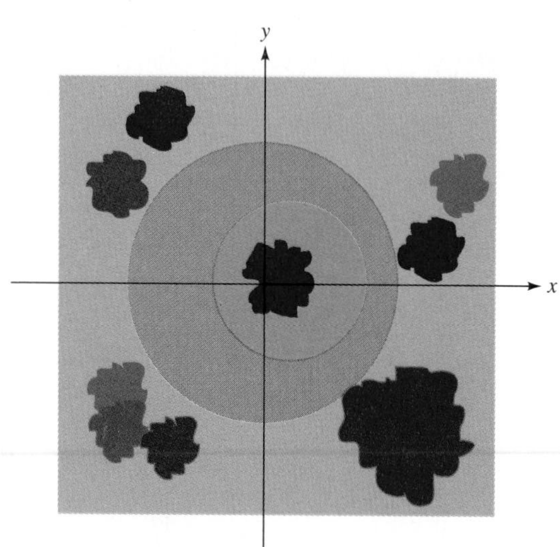

ILLUSTRATION 2

53. BROADCAST RANGES Radio stations applying for licensing may not use the same frequency if their broadcast areas overlap. One station's coverage is bounded by $x^2 + y^2 - 8x - 20y + 16 = 0$, and the other's by $x^2 + y^2 + 2x + 4y - 11 = 0$. May they be licensed for the same frequency?

54. HIGHWAY CURVES Highway design engineers want to join two sections of highway with a curve that is one-quarter of a circle, as in Illustration 3. The equation of the circle is $x^2 + y^2 - 16x - 20y + 155 = 0$, where distances are measured in kilometers. Find the locations (relative to the center of town at the origin) of the intersections of the highway with State and with Main.

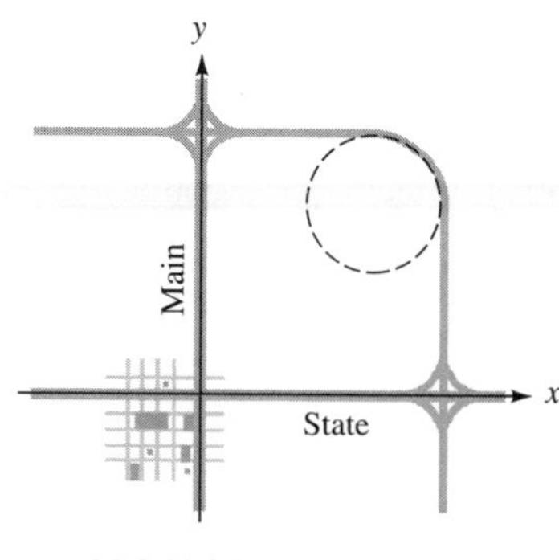

ILLUSTRATION 3

55. FLIGHT OF A PROJECTILE The cannonball in Illustration 4 follows the parabolic trajectory $y = 30x - x^2$. How far away does it land?

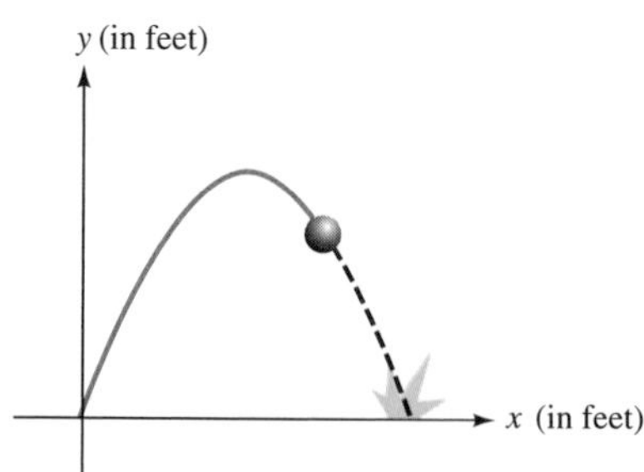

ILLUSTRATION 4

56. FLIGHT OF A PROJECTILE In Exercise 55, how high does the cannonball get?

57. ORBIT OF A COMET If the orbit of the comet shown in Illustration 5 is given by the equation $2y^2 - 9x = 18$, how far is it from the center of the sun at the vertex of its orbit? Distances are in astronomical units (AU).

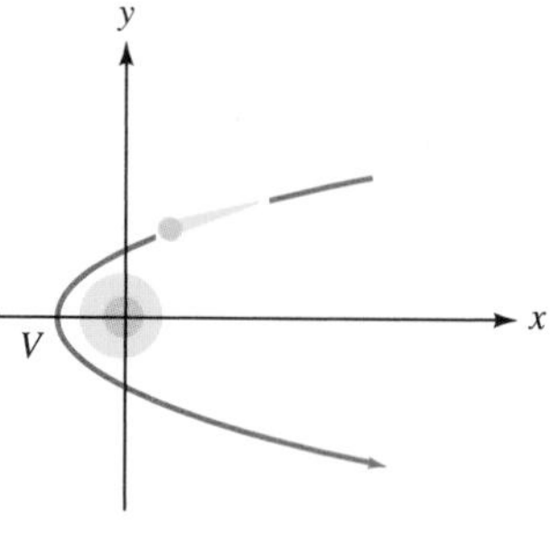

ILLUSTRATION 5

58. SATELLITE ANTENNA The cross section of the satellite antenna in Illustration 6 is a parabola given by the equation $y = \frac{1}{16}x^2$, with distances measured in feet. If the dish is 8 feet wide, how deep is it?

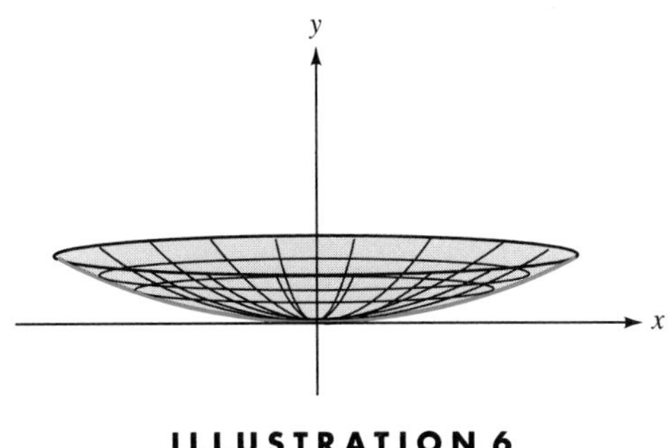

ILLUSTRATION 6

WRITING *Write a paragraph using your own words.*

59. Explain how to decide from its equation whether the graph of a parabola opens up, down, right, or left.

60. From the equation of a circle, explain how to determine the radius and the coordinates of the center.

REVIEW *Solve each equation.*

61. $|3x - 4| = 11$

62. $\left|\frac{4 - 3x}{5}\right| = 12$

63. $|3x + 4| = |5x - 2|$

64. $|6 - 4x| = |x + 2|$

I.2 The Ellipse and the Hyperbola

In this section, you will learn about:

- **The ellipse**
- **Problem solving**
- **The hyperbola**
- **Problem solving**

The ellipse

The ellipse

An **ellipse** is the set of all points P in the plane the sum of whose distances from two fixed points is a constant. See Figure I-9, in which $d_1 + d_2$ is a constant.

Each of the two points is called a **focus.** Midway between the foci is the **center** of the ellipse.

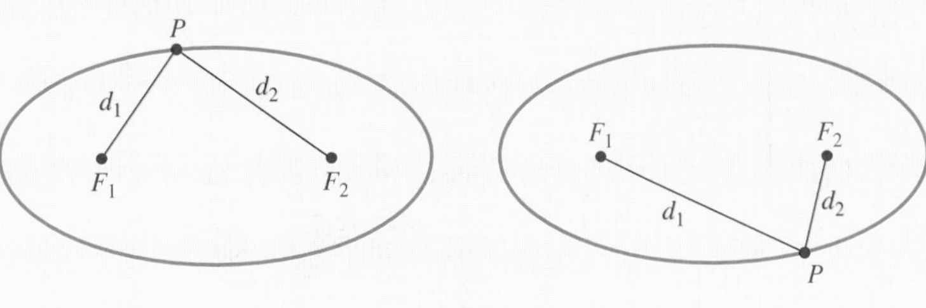

FIGURE I-9

Because of the previous definition, we can construct an ellipse by placing two thumbtacks fairly close together, as in Figure I-10. We then tie each end of a piece of string to a thumbtack, catch the loop with the point of a pencil, and (while keeping the string taut) draw the ellipse.

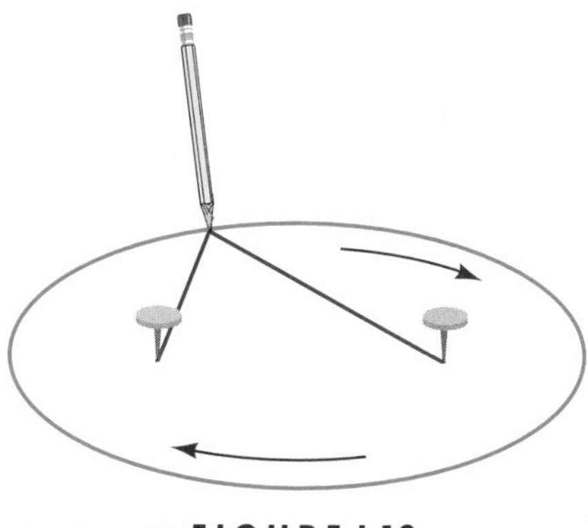

FIGURE I-10

The graph of the equation

$$\frac{x^2}{36} + \frac{y^2}{9} = 1$$

is an ellipse. To graph the ellipse, we make a table of ordered pairs, plot each pair, and join them with a smooth curve, as shown in Figure I-11.

The center of the ellipse is at the origin; it intersects the x-axis at the points (6, 0) and (−6, 0) and the y-axis at the points (0, 3) and (0, −3).

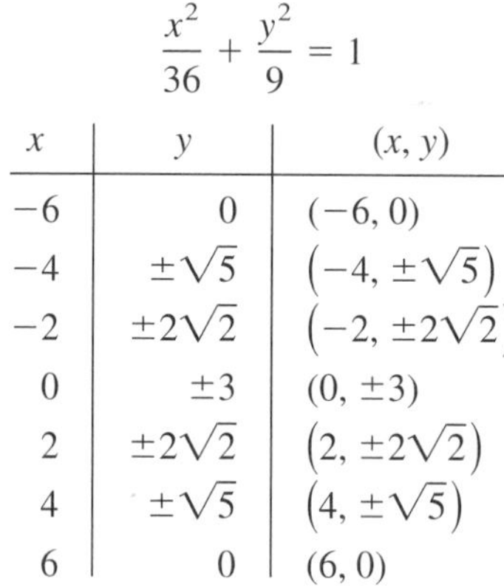

$$\frac{x^2}{36} + \frac{y^2}{9} = 1$$

x	y	(x, y)
-6	0	$(-6, 0)$
-4	$\pm\sqrt{5}$	$(-4, \pm\sqrt{5})$
-2	$\pm 2\sqrt{2}$	$(-2, \pm 2\sqrt{2})$
0	± 3	$(0, \pm 3)$
2	$\pm 2\sqrt{2}$	$(2, \pm 2\sqrt{2})$
4	$\pm\sqrt{5}$	$(4, \pm\sqrt{5})$
6	0	$(6, 0)$

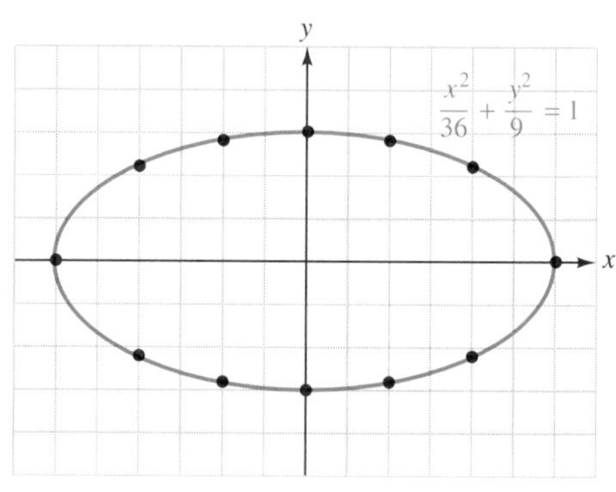

FIGURE I-11

The preceding discussion illustrates this general theorem.

Equations of an ellipse centered at the origin

Any equation that can be written in the form

$$\frac{x^2}{a^2} + \frac{y^2}{b^2} = 1 \quad (a > b)$$

has a graph that is an **ellipse** centered at the origin, as in Figure I-12(a). The x-intercepts are the **vertices** $V(a, 0)$ and $V'(-a, 0)$. (Read V' as "V prime.") The y-intercepts are the points $(0, b)$ and $(0, -b)$.

Any equation that can be written in the form

$$\frac{x^2}{b^2} + \frac{y^2}{a^2} = 1 \quad (a > b)$$

has a graph that is also an ellipse centered at the origin, as in Figure I-12(b). The y-intercepts are the vertices $V(0, a)$ and $V'(0, -a)$. The x-intercepts are the points $(b, 0)$ and $(-b, 0)$.

The point midway between the vertices is the **center** of the ellipse.

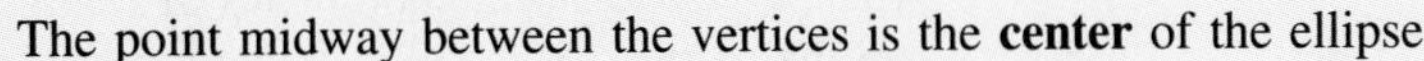

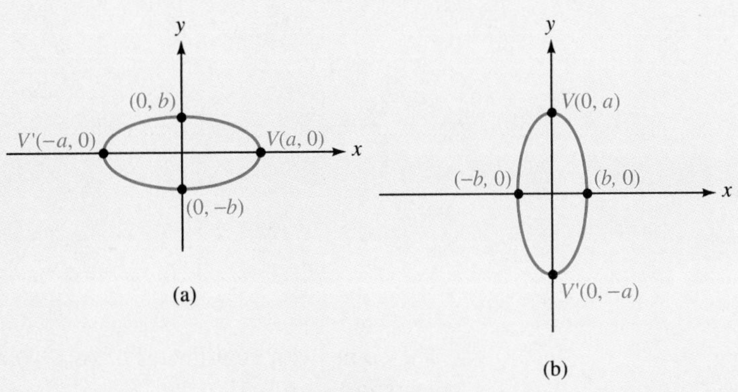

FIGURE I-12

The following theorem gives the equation of an ellipse with center at a point with coordinates (h, k).

Equations of ellipses centered at (h, k)

Any equation that can be written in the form

$$\frac{(x-h)^2}{a^2} + \frac{(y-k)^2}{b^2} = 1 \quad \text{or} \quad \frac{(x-h)^2}{b^2} + \frac{(y-k)^2}{a^2} = 1$$

(with $a > b$) is an ellipse with center at the point (h, k).

EXAMPLE 1 Graph the ellipse $\frac{(x-2)^2}{16} + \frac{(y+3)^2}{25} = 1$.

Solution We write the equation in the form

$$\frac{(x-2)^2}{16} + \frac{[y-(-3)]^2}{25} = 1$$

to see that the center of the ellipse is at the point $(2, -3)$. To find some points on the ellipse, we let $x = 2$ and solve for y.

$$\begin{aligned}
\frac{(x-2)^2}{16} + \frac{(y+3)^2}{25} &= 1 \\
\frac{(2-2)^2}{16} + \frac{(y+3)^2}{25} &= 1 \\
0 + \frac{(y+3)^2}{25} &= 1 \\
(y+3)^2 &= 25 \\
y+3 = 5 \quad \text{or} \quad y+3 &= -5 \\
y = 2 \qquad\qquad y &= -8
\end{aligned}$$

Thus, the points with coordinates of (2, 2) and (2, −8) lie on the graph. These points are the vertices of the ellipse. Now we let $y = -3$ and solve for x.

$$\frac{(x-2)^2}{16} + \frac{(y+3)^2}{25} = 1$$
$$\frac{(x-2)^2}{16} + \frac{(-3+3)^2}{25} = 1$$
$$(x-2)^2 = 16$$
$$x - 2 = 4 \quad \text{or} \quad x - 2 = -4$$
$$x = 6 \qquad\qquad x = -2$$

The points with coordinates of (6, −3) and (−2, −3) lie on the graph. The graph appears in Figure I-13.

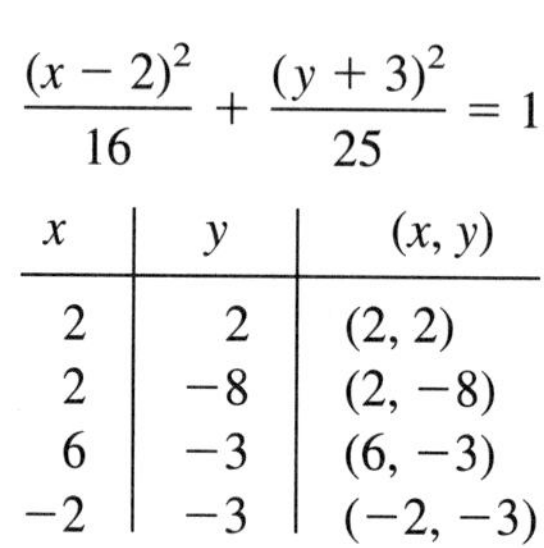

$$\frac{(x-2)^2}{16} + \frac{(y+3)^2}{25} = 1$$

x	y	(x, y)
2	2	(2, 2)
2	−8	(2, −8)
6	−3	(6, −3)
−2	−3	(−2, −3)

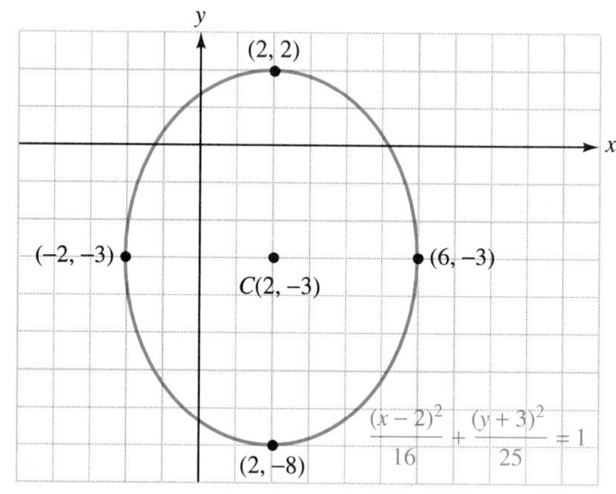

FIGURE I-13

Problem solving

EXAMPLE 2 ***Landscape design.*** A landscape architect is designing an elliptical pool that will fit in the center of a 20-by-30-foot rectangular garden, leaving at least 5 feet of space on all sides. Find the equation of the ellipse.

Solution We place the rectangular garden in a coordinate system, as in Figure I-14. To maintain 5 feet of clearance at the ends of the ellipse, the vertices must be the points $V(10, 0)$ and $V'(-10, 0)$. Similarly, the y-intercepts are the points (0, 5) and (0, −5).

The equation of the ellipse has the form

$$\frac{x^2}{a^2} + \frac{y^2}{b^2} = 1$$

with $a = 10$ and $b = 5$. Thus, the equation of the boundary of the pool is

$$\frac{x^2}{100} + \frac{y^2}{25} = 1$$

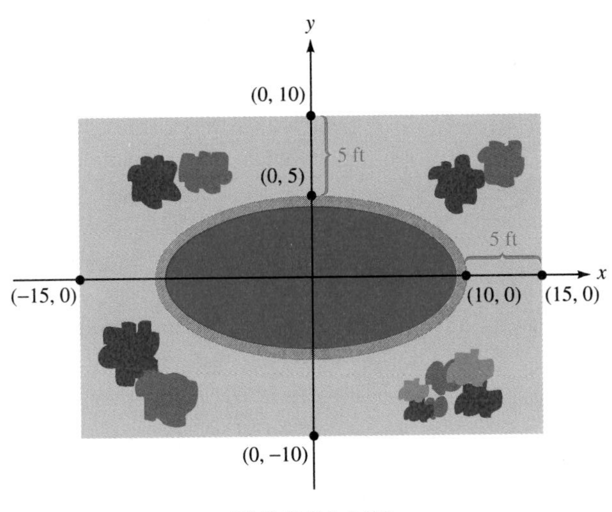

FIGURE I-14

The hyperbola

The hyperbola

A **hyperbola** is the set of all points P in the plane for which the difference of the distances of each point on the hyperbola from two fixed points is a constant. See Figure I-15, in which $d_1 - d_2$ is a constant.

Each of the two points is called a **focus.** Midway between the foci is the **center** of the hyperbola.

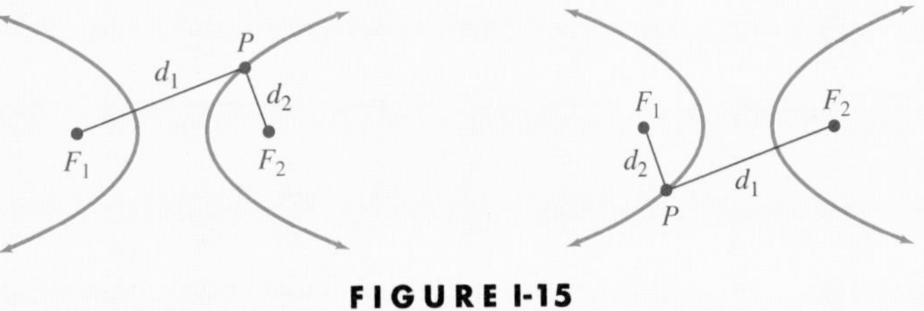

FIGURE I-15

The graph of the equation

$$\frac{x^2}{25} - \frac{y^2}{9} = 1$$

is a hyperbola. To graph this hyperbola, we make a table of ordered pairs, plot each pair and join the points with a smooth curve, as in Figure I-16.

$$\frac{x^2}{25} - \frac{y^2}{9} = 1$$

x	y	(x, y)
−7	±2.9	(−7, ±2.9)
−6	±2.0	(−6, ±2.0)
−5	0	(−5, 0)
5	0	(5, 0)
6	±2.0	(6, ±2.0)
7	±2.9	(7, ±2.9)

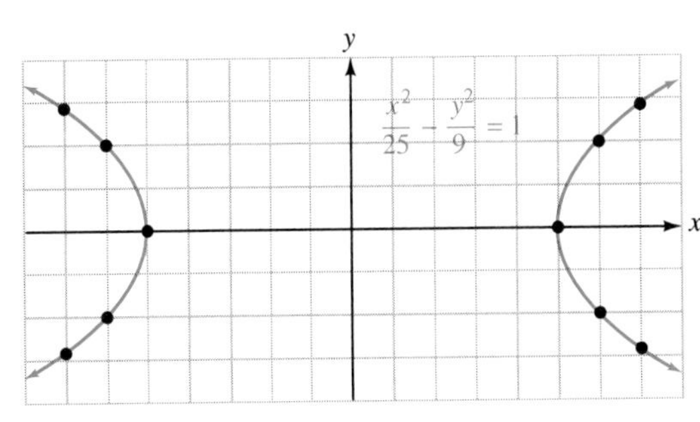

FIGURE I-16

This graph is centered at the origin and intersects the x-axis at $(5, 0)$ and $(-5, 0)$. We also note that the graph does not intersect the y-axis.

Although it is possible to draw any hyperbola by plotting many points and joining them with a smooth curve, there is an easier way. For example, if we want to graph the hyperbola with an equation of

$$\frac{x^2}{a^2} - \frac{y^2}{b^2} = 1$$

we first look at the x- and y-intercepts. To find the x-intercepts, we let $y = 0$ and solve for x:

$$\frac{x^2}{a^2} - \frac{0^2}{b^2} = 1$$
$$x^2 = a^2$$
$$x = \pm a$$

Thus, this hyperbola crosses the x-axis at the points $V(a, 0)$ and $V'(-a, 0)$, called the **vertices** of the hyperbola. (See Figure I-17.)

To attempt to find the y-intercepts, we let $x = 0$ and solve for y:

$$\frac{0^2}{a^2} - \frac{y^2}{b^2} = 1$$
$$y^2 = -b^2$$
$$y = \pm\sqrt{-b^2}$$

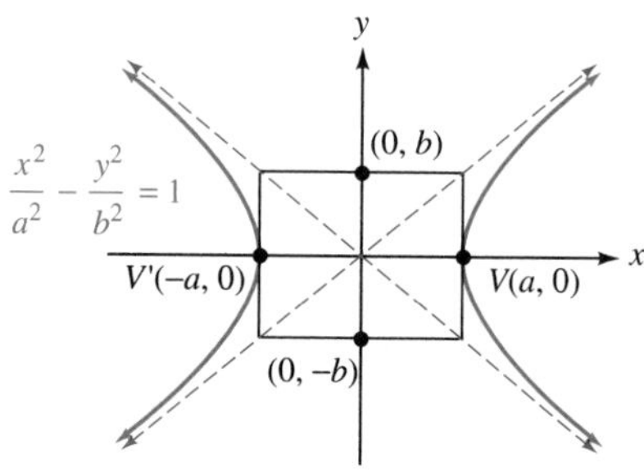

FIGURE I-17

Since $b^2 > 0$, $\sqrt{-b^2}$ is an imaginary number. This means that the hyperbola does not cross the y-axis.

If we construct a rectangle whose sides pass horizontally through $\pm b$ on the y-axis and vertically through $\pm a$ on the x-axis, the extended diagonals of the rectangle will be a useful aid in drawing the graph. As points on the branches of the hyperbola move further away from the origin, they get closer to these extended diagonals. The extended diagonals of the rectangle, which is called the **fundamental rectangle,** are called **asymptotes.**

The preceding discussion illustrates the following theorem.

Equation of a hyperbola centered at the origin

Any equation that can be written in the form

$$\frac{x^2}{a^2} - \frac{y^2}{b^2} = 1$$

has a graph that is a **hyperbola** centered at the origin, as in Figure I-18. The x-intercepts are the **vertices** $V(a, 0)$ and $V'(-a, 0)$. There are no y-intercepts.

The **asymptotes** of the hyperbola are the extended diagonals of the rectangle in the figure.

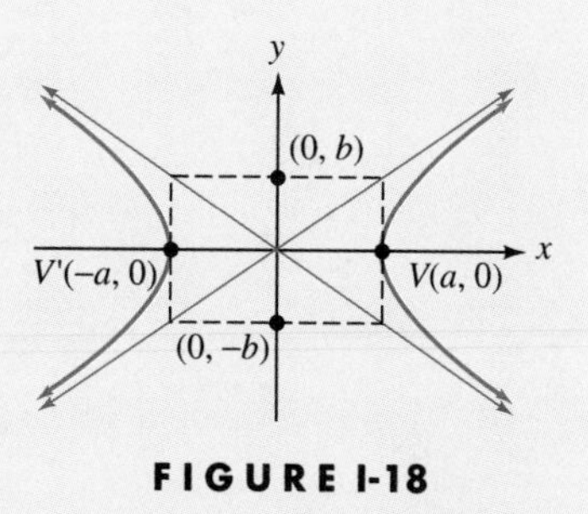

FIGURE I-18

The branches of the hyperbola in previous discussions open to the left and to the right. It is possible for hyperbolas to have different orientations with respect to the x-

and y-axes. For example, the branches of a hyperbola can open upward and downward. In that case, the following theorem applies.

Equation of a hyperbola centered at the origin

Any equation that can be written in the form

$$\frac{y^2}{a^2} - \frac{x^2}{b^2} = 1$$

has a graph that is a **hyperbola** centered at the origin, as in Figure I-19. The y-intercepts are the **vertices** $V(0, a)$ and $V'(0, -a)$. There are no x-intercepts.

The **asymptotes** of the hyperbola are the extended diagonals of the rectangle in the figure.

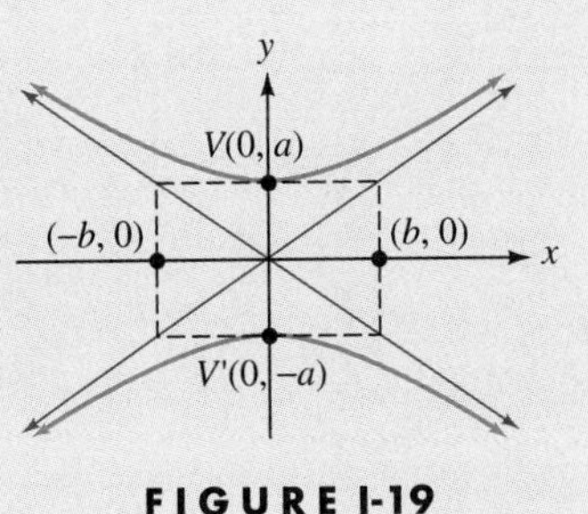

FIGURE I-19

EXAMPLE 3 Graph the equation $9y^2 - 4x^2 = 36$.

Solution To write the equation in standard form, we divide both sides by 36 to obtain

$$\frac{9y^2}{36} - \frac{4x^2}{36} = 1$$

$$\frac{y^2}{4} - \frac{x^2}{9} = 1 \quad \text{Simplify each fraction.}$$

We then find the y-intercepts by letting $x = 0$ and solving for y:

$$\frac{y^2}{4} - \frac{0^2}{9} = 1$$

$$y^2 = 4$$

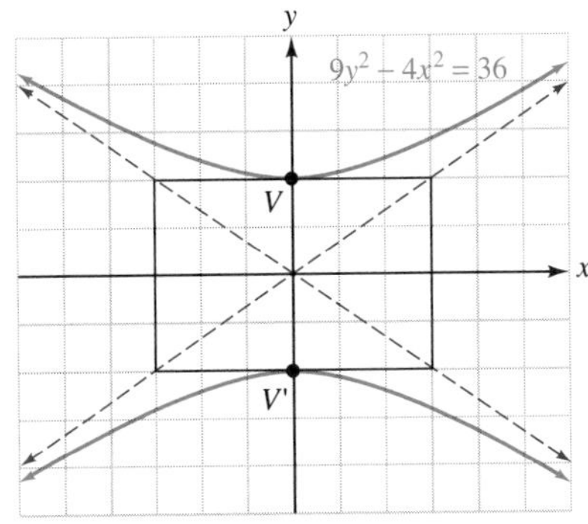

FIGURE I-20

Thus, $y = \pm 2$, and the vertices of the hyperbola are $V(0, 2)$ and $V'(0, -2)$. See Figure I-20.

Since $b = \pm\sqrt{9} = \pm 3$, we use the points $(3, 0)$ and $(-3, 0)$ on the x-axis to help draw the fundamental rectangle. We then draw its extended diagonals and sketch the hyperbola. ■

If a hyperbola is centered at a point with coordinates (h, k), the following theorem applies.

Equations of hyperbolas centered at (h, k)

Any equation that can be written in the form

$$\frac{(x-h)^2}{a^2} - \frac{(y-k)^2}{b^2} = 1$$

is a hyperbola with center at (h, k) that opens left and right.

Any equation of the form

$$\frac{(y-k)^2}{a^2} - \frac{(x-h)^2}{b^2} = 1$$

is a hyperbola with center at (h, k) that opens up and down.

EXAMPLE 4 Graph the hyperbola $\frac{(x-3)^2}{16} - \frac{(y+1)^2}{4} = 1$.

Solution We write the equation in the form

$$\frac{(x-3)^2}{16} - \frac{[y-(-1)]^2}{4} = 1$$

to see that the hyperbola is of the form that opens left and right and is centered at the point $(h, k) = (3, -1)$. Because this hyperbola has its center at $(3, -1)$, its vertices are located at $a = 4$ units to the right and left of the center, at $(7, -1)$ and $(-1, -1)$. Since $b = 2$, we can locate the points 2 units above and 2 units below the center. With these points, we can draw the fundamental rectangle, along with its extended diagonals. We can then sketch the hyperbola, as shown in Figure I-21.

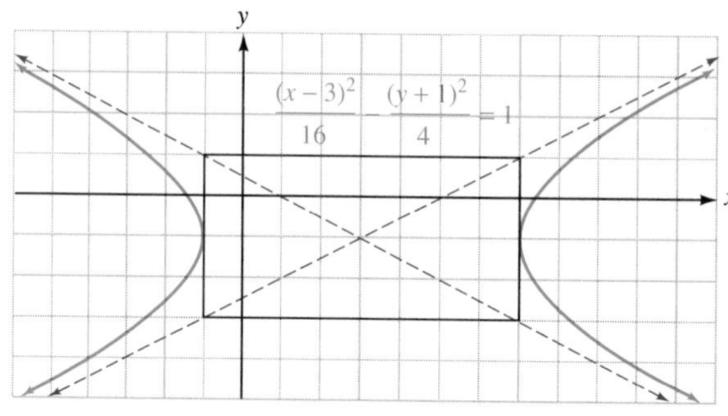

FIGURE I-21

There is a special type of hyperbola (also centered at the origin) that does not intersect either the x- or the y-axis. These hyperbolas have equations of the form $xy = k$, where $k \neq 0$.

EXAMPLE 5 Graph the equation $xy = -8$.

Solution We make a table of ordered pairs, plot each pair, and join the points with a smooth curve to obtain the hyperbola in Figure I-22.

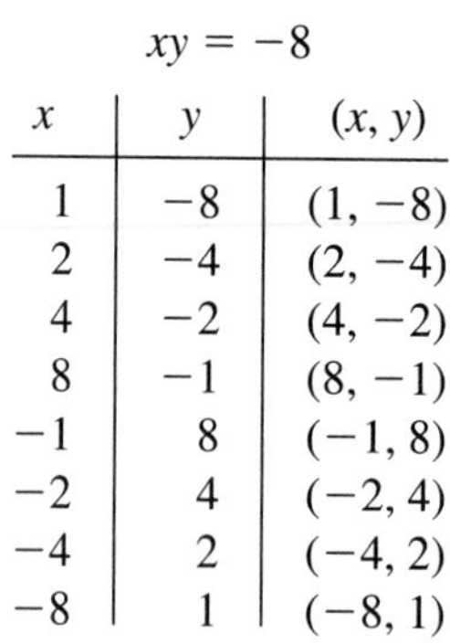

$xy = -8$

x	y	(x, y)
1	−8	(1, −8)
2	−4	(2, −4)
4	−2	(4, −2)
8	−1	(8, −1)
−1	8	(−1, 8)
−2	4	(−2, 4)
−4	2	(−4, 2)
−8	1	(−8, 1)

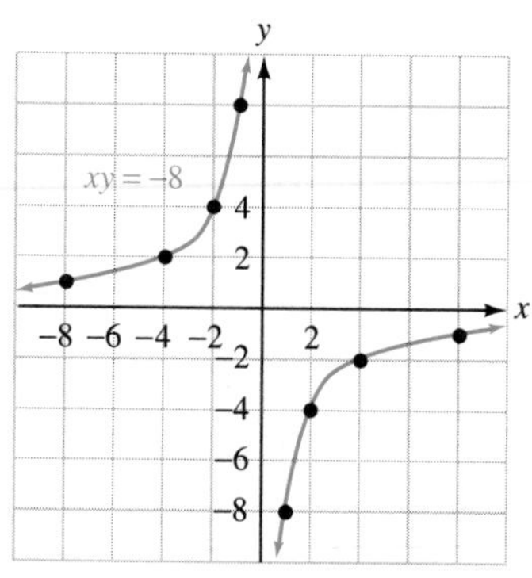

FIGURE I-22

The result in Example 5 suggests the following theorem.

Theorem Any equation of the form $xy = k$, where $k \neq 0$, has a graph that is a **hyperbola**, which does not intersect either the x- or the y-axis.

Problem solving

EXAMPLE 6 ***Atomic structure.*** In an experiment that led to the discovery of the atomic structure of matter, Lord Rutherford (1871–1937) shot high-energy alpha particles toward a thin sheet of gold. Because many were reflected, Rutherford showed the existence of the nucleus of a gold atom. The alpha particle in Figure I-23 is repelled by the nucleus at the origin, and it travels along the hyperbolic path given by $4x^2 - y^2 = 16$. How close does the particle come to the nucleus?

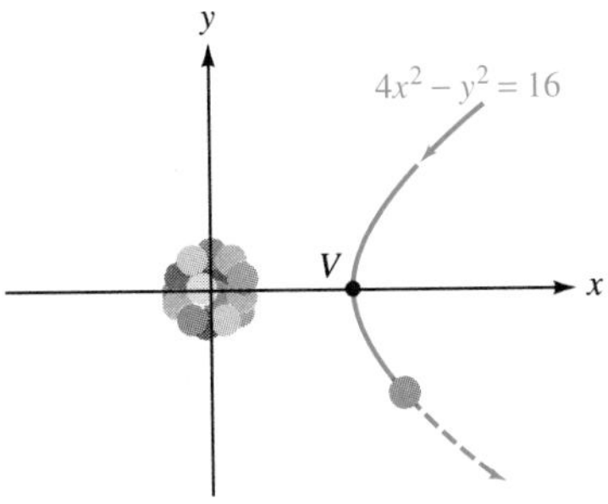

Solution To find the distance from the nucleus at the origin, we must find the coordinates of the vertex V. To do so, we write the equation of the particle's path in standard form:

$$4x^2 - y^2 = 16$$

$$\frac{4x^2}{16} - \frac{y^2}{16} = \frac{16}{16} \quad \text{Divide both sides by 16.}$$

$$\frac{x^2}{4} - \frac{y^2}{16} = 1 \quad \text{Simplify.}$$

$$\frac{x^2}{2^2} - \frac{y^2}{4^2} = 1 \quad \text{Write 4 as } 2^2 \text{ and 16 as } 4^2.$$

This equation is in the form $\frac{x^2}{a^2} - \frac{y^2}{b^2} = 1$, with $a = 2$. Thus, the vertex of the path is (2, 0). The particle is never closer than 2 units from the nucleus. ■

y

$4x^2 - y^2 = 16$

V

x

FIGURE I-23

STUDY SET Section I.2

VOCABULARY *In Exercises 1–6, fill in the blanks to make the statements true.*

1. An ellipse is the set of all points in a plane the ___ of whose distances from two fixed points is a _____________.

2. Each of the fixed points in Exercise 1 is called a _____________ of the ellipse. Midway between the _____________ is the _____________ of the ellipse.

3. A hyperbola is the set of all points in a plane the _____________ of whose distances from two fixed points is a _____________.

4. Each of the fixed points in Exercise 3 is called a _____________ of the hyperbola.

5. The vertices of the hyperbola

$$\frac{x^2}{a^2} - \frac{y^2}{b^2} = 1$$

are the points _____________ and _____________.

6. The extended diagonals of the fundamental rectangle are _____________ of the graph of a hyperbola.

CONCEPTS

7. The standard equation of an ellipse centered at the origin with vertices at $(a, 0)$ and $(-a, 0)$ is _____________.

8. The center of the ellipse

$$\frac{(x-2)^2}{9} + \frac{(y+3)^2}{4} = 1$$

is the point _____________.

9. The graph of

$$\frac{y^2}{a^2} - \frac{x^2}{b^2} = 1$$

is a _____________ with vertices at _____________ and _____________.

10. The graph of

$$\frac{x^2}{9} - \frac{y^2}{16} = 1$$

is a _____________ with vertices at _____________ and _____________.

PRACTICE *In Exercises 11–20, graph each equation.*

11. $\frac{x^2}{4} + \frac{y^2}{9} = 1$

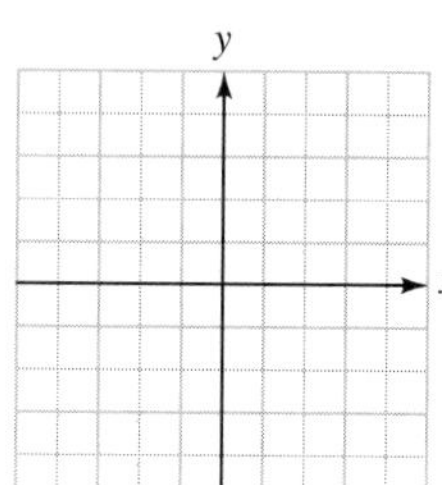

12. $x^2 + \frac{y^2}{9} = 1$

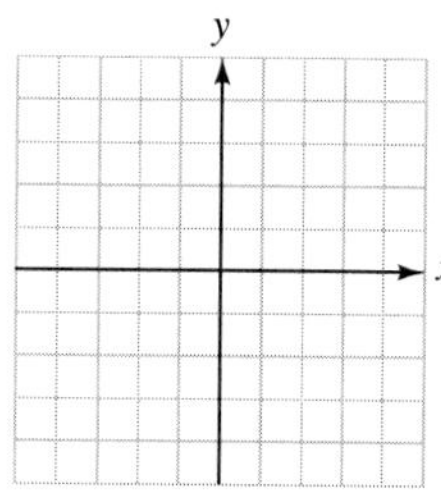

13. $x^2 + 9y^2 = 9$

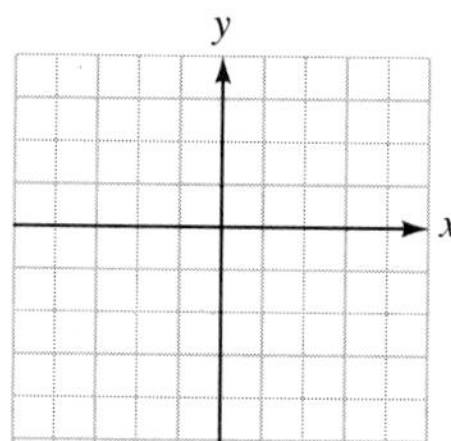

14. $25x^2 + 9y^2 = 225$

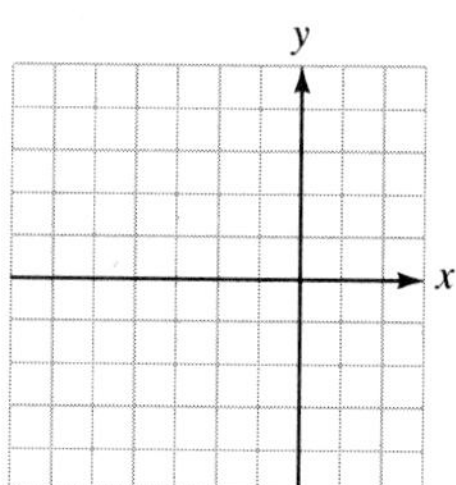

15. $16x^2 + 4y^2 = 64$

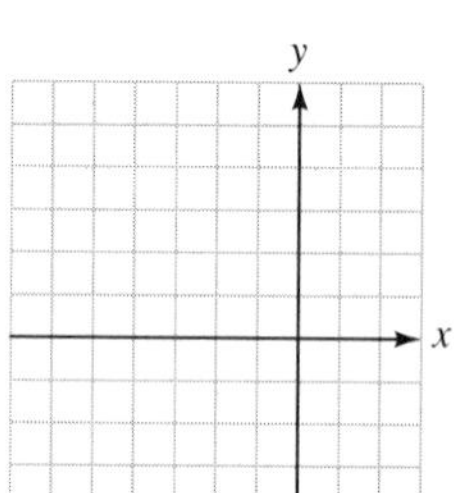

16. $4x^2 + 9y^2 = 36$

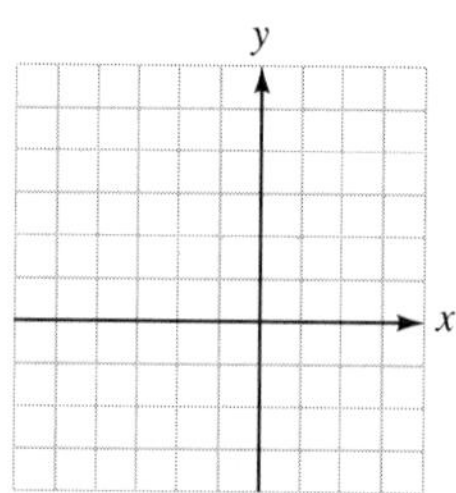

17. $\frac{(x-2)^2}{9} + \frac{(y-1)^2}{4} = 1$

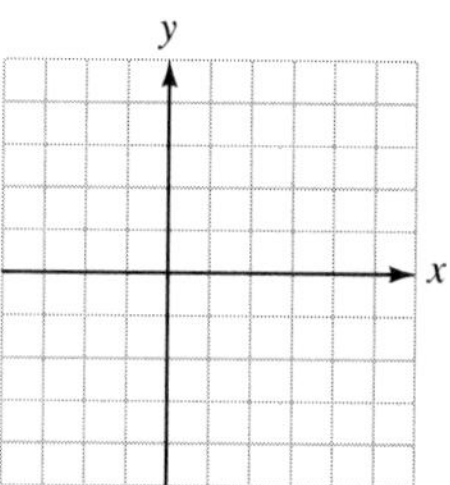

18. $\frac{(x-1)^2}{9} + \frac{(y-3)^2}{4} = 1$

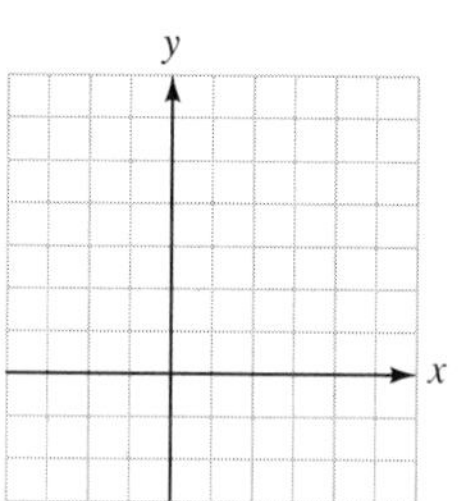

19. $(x+1)^2 + 4(y+2)^2 = 4$

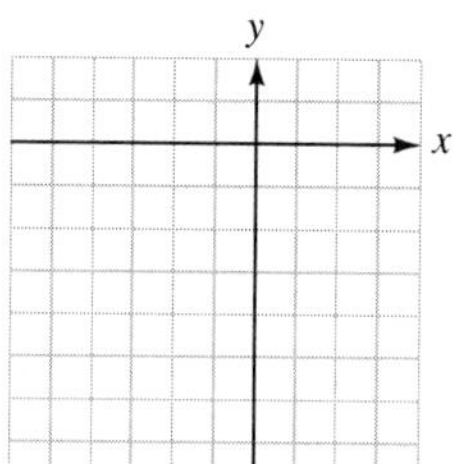

20. $25(x+1)^2 + 9y^2 = 225$

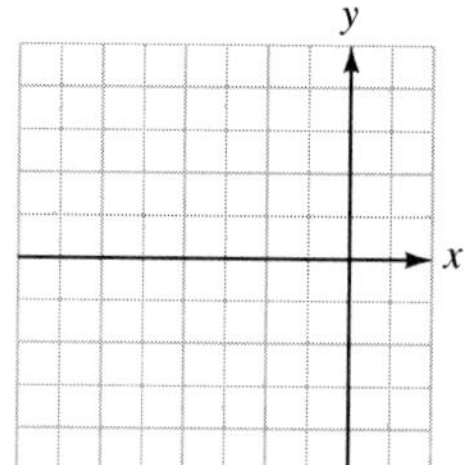

In Exercises 21–32, graph each hyperbola.

21. $\frac{x^2}{9} - \frac{y^2}{4} = 1$

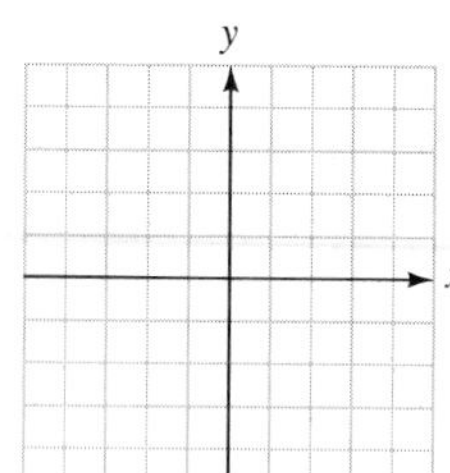

22. $\frac{x^2}{4} - \frac{y^2}{4} = 1$

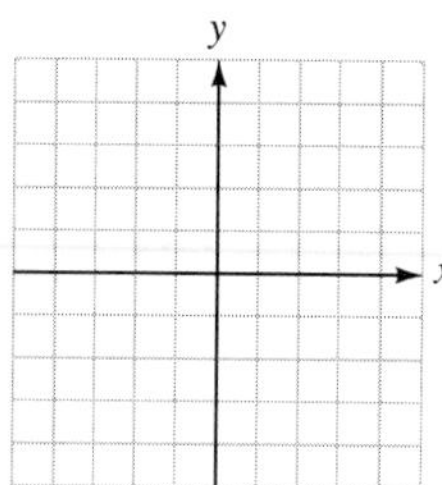

23. $\frac{y^2}{4} - \frac{x^2}{9} = 1$

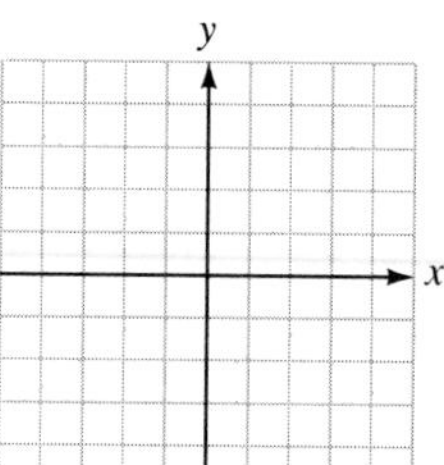

24. $\frac{y^2}{4} - \frac{x^2}{64} = 1$

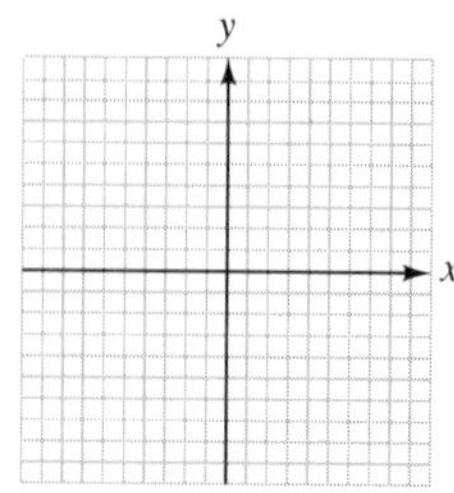

25. $25x^2 - y^2 = 25$

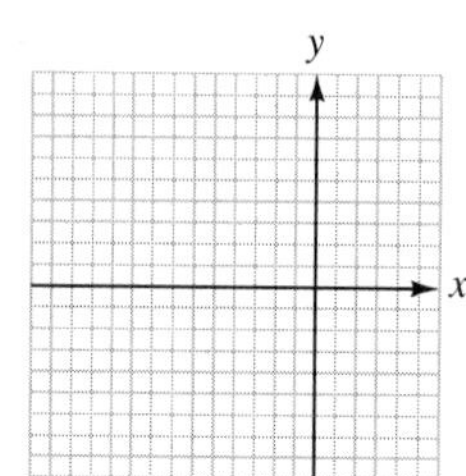

26. $9x^2 - 4y^2 = 36$

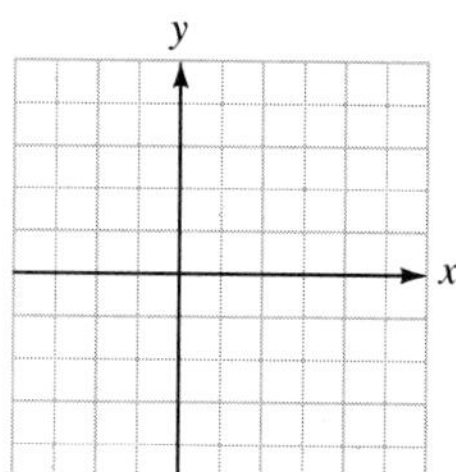

27. $\dfrac{(x-2)^2}{9} - \dfrac{y^2}{16} = 1$

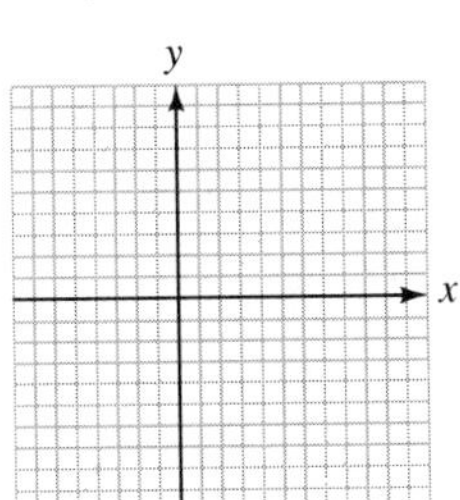

28. $\dfrac{(x+2)^2}{16} - \dfrac{(y-3)^2}{25} = 1$

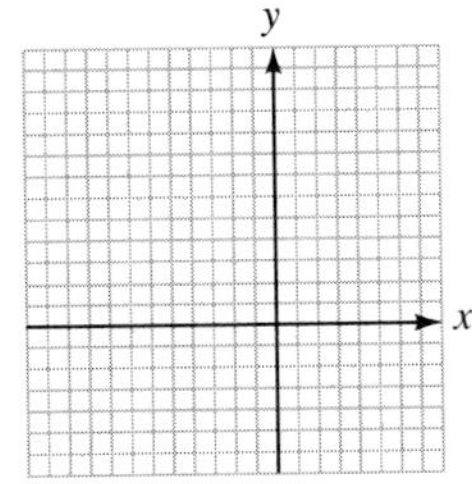

29. $4(x+3)^2 - (y-1)^2 = 4$

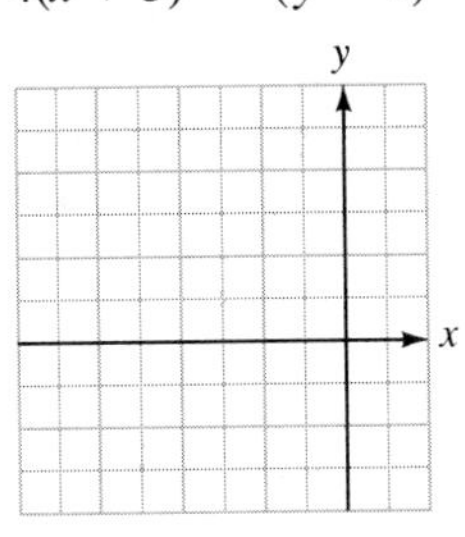

30. $(x+5)^2 - 16y^2 = 16$

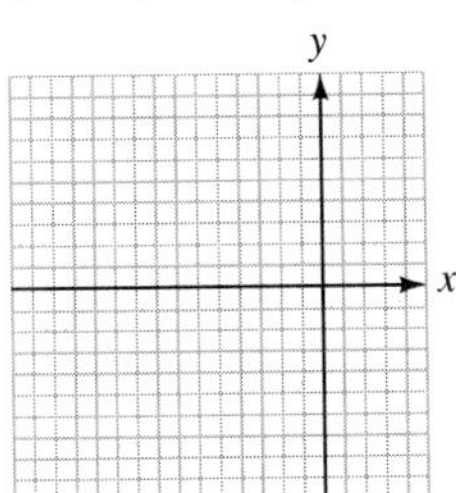

31. $xy = 8$

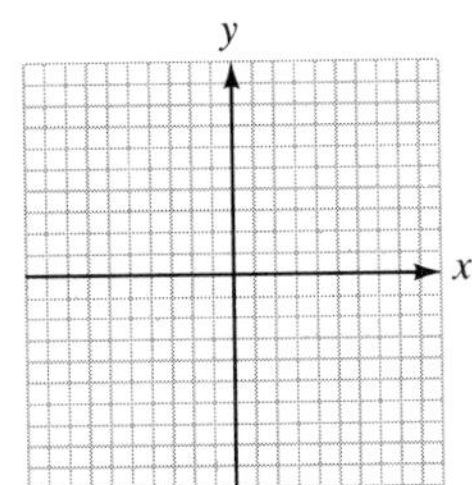

32. $xy = -10$

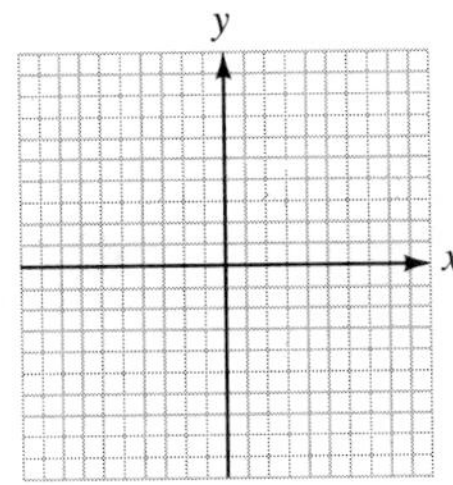

APPLICATIONS

33. DESIGNING AN UNDERPASS The arch of the underpass in Illustration 1 is a part of an ellipse. Find the equation of the ellipse.

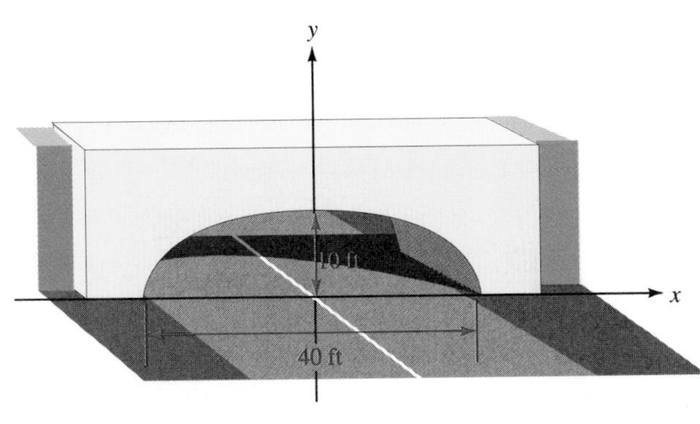

ILLUSTRATION 1

34. CALCULATING CLEARANCE Find the height of the elliptical arch in Exercise 33 at a point 10 feet from the center of the roadway.

35. AREA OF AN ELLIPSE The area A of the ellipse

$$\frac{x^2}{a^2} + \frac{y^2}{b^2} = 1$$

is given by $A = \pi ab$. Find the area of the ellipse $9x^2 + 16y^2 = 144$.

36. AREA OF A TRACK The elliptical track in Illustration 2 is bounded by the ellipses

$$4x^2 + 9y^2 = 576 \text{ and } 9x^2 + 25y^2 = 900$$

Find the area of the track. (See Exercise 35.)

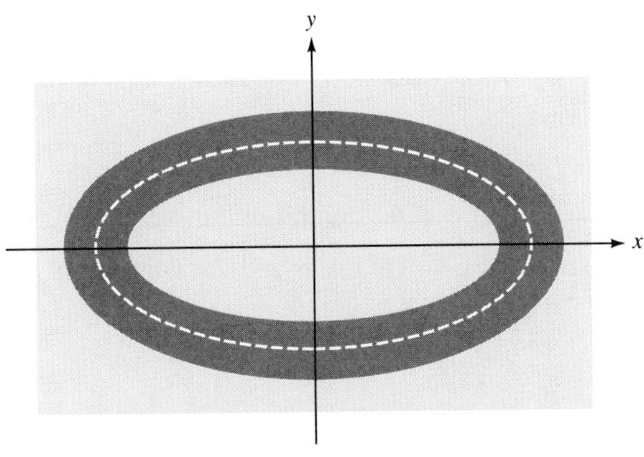

ILLUSTRATION 2

37. ALPHA PARTICLES The particle in Illustration 3 approaches the nucleus at the origin along the path $9y^2 - x^2 = 81$. How close does the particle come to the nucleus?

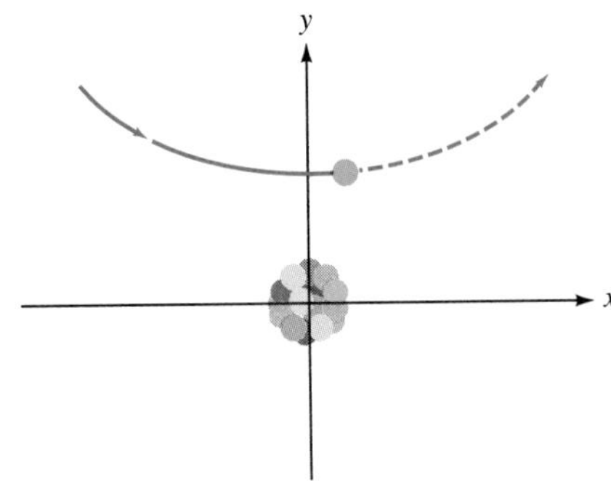

ILLUSTRATION 3

38. LORAN By determining the difference of the distances between the ship in Illustration 4 and two radio transmitters, the LORAN system places the ship on the hyperbola $x^2 - 4y^2 = 576$. If the ship is also 5 miles out to sea, find its coordinates.

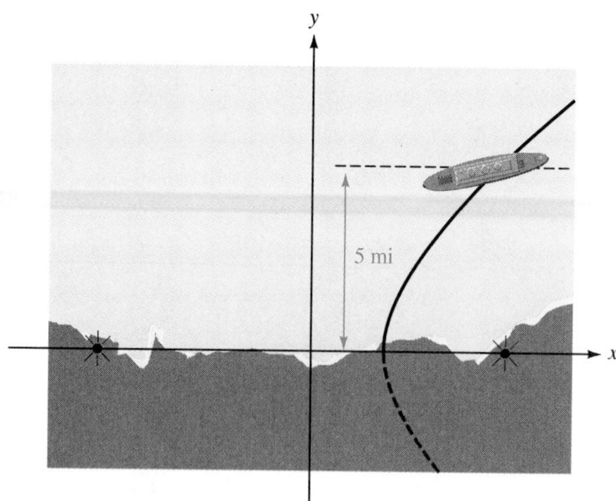

ILLUSTRATION 4

39. SONIC BOOM The position of the sonic boom caused by the faster-than-sound aircraft in Illustration 5 is the hyperbola $y^2 - x^2 = 25$ in the coordinate system shown. How wide is the hyperbola 5 units from its vertex?

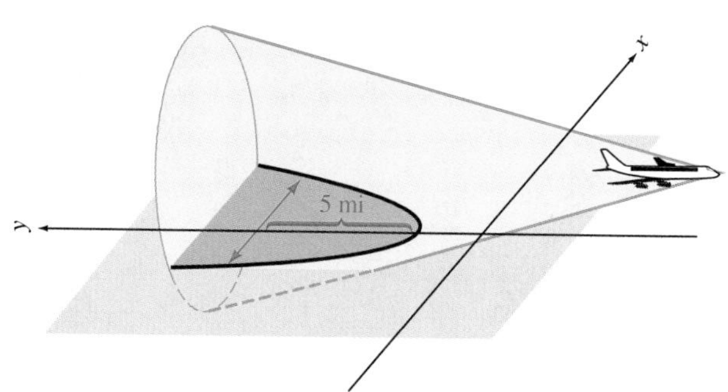

ILLUSTRATION 5

40. ELECTROSTATIC REPULSION Two similarly charged particles are shot together for an almost head-on collision, as in Illustration 6. They repel each other and travel the two branches of the hyperbola given by $x^2 - 4y^2 = 4$. How close do they get?

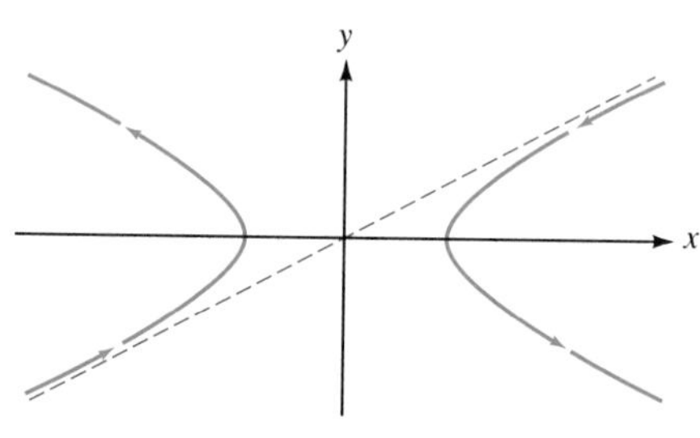

ILLUSTRATION 6

WRITING *Write a paragraph using your own words.*

41. Explain how to find the x- and the y-intercepts of the graph of the ellipse

$$\frac{x^2}{a^2} + \frac{y^2}{b^2} = 1$$

42. Explain why the graph of the hyperbola

$$\frac{x^2}{a^2} - \frac{y^2}{b^2} = 1$$

has no y-intercept.

REVIEW *In Exercises 43–44, find each product.*

43. $3x^{-2}y^2(4x^2 + 3y^{-2})$

44. $(2a^{-2} - b^{-2})(2a^{-2} + b^{-2})$

In Exercises 45–46, write each expression without using negative exponents.

45. $\dfrac{x^{-2} + y^{-2}}{x^{-2} - y^{-2}}$

46. $\dfrac{2x^{-3} - 2y^{-3}}{4x^{-3} + 4y^{-3}}$

APPENDIX II THE BINOMIAL THEOREM

II.1 The Binomial Theorem

In this section, you will learn about:

- **Pascal's triangle**
- **Factorial notation**
- **The binomial theorem**
- **The *n*th term of a binomial expansion**

We have discussed how to raise binomials to positive integral powers. For example, we know that

$$(a + b)^2 = a^2 + 2ab + b^2$$

and that

$$\begin{aligned}(a + b)^3 &= (a + b)(a + b)^2\\ &= (a + b)(a^2 + 2ab + b^2)\\ &= a^3 + 2a^2b + ab^2 + ba^2 + 2ab^2 + b^3\\ &= a^3 + 3a^2b + 3ab^2 + b^3\end{aligned}$$

To show how to raise binomials to positive integral powers without doing the actual multiplications, we consider the following binomial expansions:

$$\begin{aligned}(a + b)^0 &= 1\\ (a + b)^1 &= a + b\\ (a + b)^2 &= a^2 + 2ab + b^2\\ (a + b)^3 &= a^3 + 3a^2b + 3ab^2 + b^3\\ (a + b)^4 &= a^4 + 4a^3b + 6a^2b^2 + 4ab^3 + b^4\\ (a + b)^5 &= a^5 + 5a^4b + 10a^3b^2 + 10a^2b^3 + 5ab^4 + b^5\\ (a + b)^6 &= a^6 + 6a^5b + 15a^4b^2 + 20a^3b^3 + 15a^2b^4 + 6ab^5 + b^6\end{aligned}$$

Several patterns appear in these expansions:

1. Each expansion has one more term than the power of the binomial.
2. The degree of each term in each expansion is equal to the exponent of the binomial that is being expanded.
3. The first term in each expansion is a, raised to the power of the binomial.
4. The exponents of a decrease by 1 in each successive term. The exponents of b, beginning with $b^0 = 1$ in the first term, increase by 1 in each successive term. Thus, the variables have the pattern

$$a^n, a^{n-1}b, a^{n-2}b^2, \ldots, ab^{n-1}, b^n$$

Pascal's triangle

To see another pattern, we write the coefficients of each binomial expansion in the following triangular array:

$$
\begin{array}{ccccccccccccc}
 & & & & & & 1 & & & & & & \\
 & & & & & 1 & & 1 & & & & & \\
 & & & & 1 & & 2 & & 1 & & & & \\
 & & & 1 & & 3 & & 3 & & 1 & & & \\
 & & 1 & & 4 & & 6 & & 4 & & 1 & & \\
 & 1 & & 5 & & 10 & & 10 & & 5 & & 1 & \\
1 & & 6 & & 15 & & 20 & & 15 & & 6 & & 1
\end{array}
$$

In this array, called **Pascal's triangle,** each entry between the 1's is the sum of the closest pair of numbers in the line immediately above it. For example, the first 15 in the bottom row is the sum of the 5 and 10 immediately above it. Pascal's triangle continues with the same pattern forever. The next two lines are

$$
\begin{array}{ccccccccccccccccc}
 & 1 & & 7 & & 21 & & 35 & & 35 & & 21 & & 7 & & 1 & \\
1 & & 8 & & 28 & & 56 & & 70 & & 56 & & 28 & & 8 & & 1
\end{array}
$$

EXAMPLE 1 Expand $(x + y)^5$.

Solution The first term in the expansion is x^5, and the exponents of x decrease by 1 in each successive term. A y first appears in the second term, and the exponents of y increase by 1 in each successive term, concluding when the term y^5 is reached. Thus, the variables in the expansion are

$$x^5, \quad x^4y, \quad x^3y^2, \quad x^2y^3, \quad xy^4, \quad y^5$$

The coefficients of these variables are given in Pascal's triangle in the row whose second entry is 5:

$$1 \quad 5 \quad 10 \quad 10 \quad 5 \quad 1$$

Putting these two pieces of information together gives the required expansion:

$$(x + y)^5 = x^5 + 5x^4y + 10x^3y^2 + 10x^2y^3 + 5xy^4 + y^5$$

■

EXAMPLE 2 Expand $(u - v)^4$.

Solution We note that the expression $(u - v)^4$ can be written in the form $[u + (-v)]^4$. The variables in this expansion are

$$u^4, \quad u^3(-v), \quad u^2(-v)^2, \quad u(-v)^3, \quad (-v)^4$$

and the coefficients are given in Pascal's triangle in the row whose second entry is 4:

$$1 \quad 4 \quad 6 \quad 4 \quad 1$$

Thus, the required expansion is

$$
\begin{aligned}
(u - v)^4 &= u^4 + 4u^3(-v) + 6u^2(-v)^2 + 4u(-v)^3 + (-v)^4 \\
&= u^4 - 4u^3v + 6u^2v^2 - 4uv^3 + v^4
\end{aligned}
$$

■

Factorial notation

Although Pascal's triangle gives the coefficients of the terms in a binomial expansion, it is not the best way to expand a binomial. To develop another way to expand a binomial, we introduce **factorial notation.**

Factorial notation

If n is a natural number, the symbol $n!$ (read as ***n* factorial** or as **factorial *n***) is defined as

$$n! = n(n-1)(n-2)(n-3) \cdot \cdot \cdot (3)(2)(1)$$

EXAMPLE 3 Find **a.** $2!$, **b.** $5!$, **c.** $-9!$, and **d.** $(n-2)!$.

Solution

a. $2! = 2 \cdot 1 = 2$

b. $5! = 5 \cdot 4 \cdot 3 \cdot 2 \cdot 1 = 120$

c. $-9! = -9 \cdot 8 \cdot 7 \cdot 6 \cdot 5 \cdot 4 \cdot 3 \cdot 2 \cdot 1 = -362{,}880$

d. $(n-2)! = (n-2)(n-3)(n-4) \cdot \cdot \cdot \cdot \cdot 3 \cdot 2 \cdot 1$

WARNING! According to the previous definition, part d is meaningful only if $n - 2$ is a natural number. ■

We define zero factorial as follows.

Zero factorial

$$0! = 1$$

We note that

$$5 \cdot 4! = 5 \cdot 4 \cdot 3 \cdot 2 \cdot 1 = 5!$$
$$7 \cdot 6! = 7 \cdot 6 \cdot 5 \cdot 4 \cdot 3 \cdot 2 \cdot 1 = 7!$$
$$10 \cdot 9! = 10 \cdot 9 \cdot 8 \cdot 7 \cdot 6 \cdot 5 \cdot 4 \cdot 3 \cdot 2 \cdot 1 = 10!$$

These examples suggest the following theorem.

Theorem

If n is a positive integer, then $n(n-1)! = n!$.

The binomial theorem

We now state the binomial theorem.

The binomial theorem

If n is any positive integer, then

$$(a+b)^n = a^n + \frac{n!}{1!(n-1)!}a^{n-1}b + \frac{n!}{2!(n-2)!}a^{n-2}b^2 + \frac{n!}{3!(n-3)!}a^{n-3}b^3 + \cdot \cdot \cdot + \frac{n!}{r!(n-r)!}a^{n-r}b^r + \cdot \cdot \cdot + b^n$$

In the binomial theorem, the exponents of the variables follow the familiar pattern:

- The sum of the exponents of a and b in each term is n.
- The exponents of a decrease.
- The exponents of b increase.

Only the method of finding the coefficients is different. Except for the first and last terms, the numerator of each coefficient is $n!$. If the exponent of b in a particular term is r, the denominator of the coefficient of that term is $r!(n-r)!$.

EXAMPLE 4 Use the binomial theorem to expand $(a+b)^3$.

Solution We can substitute directly into the binomial theorem and simplify:

$$\begin{aligned}(a+b)^3 &= a^3 + \frac{3!}{1!(3-1)!}a^2b + \frac{3!}{2!(3-2)!}ab^2 + b^3\\ &= a^3 + \frac{3!}{1!2!}a^2b + \frac{3!}{2!1!}ab^2 + b^3\\ &= a^3 + \frac{3\cdot2\cdot1}{1\cdot2\cdot1}a^2b + \frac{3\cdot2\cdot1}{2\cdot1\cdot1}ab^2 + b^3\\ &= a^3 + 3a^2b + 3ab^2 + b^3\end{aligned}$$

■

EXAMPLE 5 Use the binomial theorem to expand $(x-y)^4$.

Solution We can write $(x-y)^4$ in the form $[x+(-y)]^4$, substitute directly into the binomial theorem, and simplify:

$$\begin{aligned}(x-y)^4 &= [x+(-y)]^4\\ &= x^4 + \frac{4!}{1!(4-1)!}x^3(-y) + \frac{4!}{2!(4-2)!}x^2(-y)^2 + \frac{4!}{3!(4-3)!}x(-y)^3 + (-y)^4\\ &= x^4 - \frac{4\cdot3!}{1!3!}x^3y + \frac{4\cdot3\cdot2!}{2!2!}x^2y^2 - \frac{4\cdot3!}{3!1!}xy^3 + y^4\\ &= x^4 - 4x^3y + 6x^2y^2 - 4xy^3 + y^4\end{aligned}$$

■

EXAMPLE 6 Use the binomial theorem to expand $(3u-2v)^4$.

Solution We write $(3u-2v)^4$ in the form $[3u+(-2v)]^4$ and let $a=3u$ and $b=-2v$. Then we can use the binomial theorem to expand $(a+b)^4$.

$$\begin{aligned}(a+b)^4 &= a^4 + \frac{4!}{1!(4-1)!}a^3b + \frac{4!}{2!(4-2)!}a^2b^2 + \frac{4!}{3!(4-3)!}ab^3 + b^4\\ &= a^4 + 4a^3b + 6a^2b^2 + 4ab^3 + b^4\end{aligned}$$

Now we can substitute $3u$ for a and $-2v$ for b and simplify:

$$\begin{aligned}(3u-2v)^4 &= (3u)^4 + 4(3u)^3(-2v) + 6(3u)^2(-2v)^2 + 4(3u)(-2v)^3 + (-2v)^4\\ &= 81u^4 - 216u^3v + 216u^2v^2 - 96uv^3 + 16v^4\end{aligned}$$

■

The *n*th term of a binomial expansion

To find the fourth term of the expansion of $(a+b)^9$, we could raise the binomial $a+b$ to the ninth power and look at the fourth term. However, this task would be very tedious. By using the binomial theorem, we can construct the fourth term without finding the complete expansion of $(a+b)^9$.

EXAMPLE 7 Find the fourth term in the expansion of $(a + b)^9$.

Solution Since b^1 appears in the second term, b^2 appears in the third term, and so on, the exponent of b in the fourth term is 3. Since the exponent of b added to the exponent of a must equal 9, the exponent of a must be 6. Thus, the variables of the fourth term are

a^6b^3 The sum of the exponents must be 9.

Because of the binomial theorem, the coefficient of the variables must be

$$\frac{n!}{r!(n-r)!} = \frac{9!}{3!(9-3)!}$$

Thus, the complete fourth term is

$$\begin{aligned}\frac{9!}{3!(9-3)!}a^6b^3 &= \frac{9 \cdot 8 \cdot 7 \cdot 6!}{3 \cdot 2 \cdot 1 \cdot 6!}a^6b^3 \\ &= 84a^6b^3\end{aligned}$$

■

EXAMPLE 8 Find the sixth term in the expansion of $(x - y)^7$.

Solution We find the sixth term of $[x + (-y)]^7$. In the sixth term, the exponent of $(-y)$ is 5. Thus, the variables in the sixth term are

$x^2(-y)^5$ The sum of the exponents must be 7.

The coefficient of these variables is

$$\frac{n!}{r!(n-r)!} = \frac{7!}{5!(7-5)!}$$

The complete sixth term is

$$\begin{aligned}\frac{7!}{5!(7-5)!}x^2(-y)^5 &= -\frac{7 \cdot 6 \cdot 5!}{5! \cdot 2 \cdot 1}x^2y^5 \\ &= -21x^2y^5\end{aligned}$$

■

EXAMPLE 9 Find the fourth term of the expansion of $(2x - 3y)^6$.

Solution We can let $a = 2x$ and $b = -3y$ and find the fourth term of the expansion of $(a + b)^6$:

$$\begin{aligned}\frac{6!}{3!(6-3)!}a^3b^3 &= \frac{6 \cdot 5 \cdot 4 \cdot 3!}{3 \cdot 2 \cdot 1 \cdot 3!}a^3b^3 \\ &= 20a^3b^3\end{aligned}$$

We can now substitute $2x$ for a and $-3y$ for b and simplify:

$$\begin{aligned}20a^3b^3 &= 20(2x)^3(-3y)^3 \\ &= -4{,}320x^3y^3\end{aligned}$$

The fourth term in the expansion of $(2x - 3y)^6$ is $-4{,}320x^3y^3$. ■

STUDY SET Section II.1

VOCABULARY *In Exercises 1–2, fill in the blanks to make the statements true.*

1. We can use ____________ triangle to find the coefficients of a binomial expansion.

2. 4! is read as "4 ____________" or "____________ 4."

CONCEPTS

3. $5! = 5 \cdot 4 \cdot$ ____________

4. $0! =$ ___

5. $7! =$ ___ $\cdot\, 6!$

6. The exponent on b in the fourth term of the expansion of $(a + b)^5$ is ___.

PRACTICE *In Exercises 7–26, evaluate each expression.*

7. $3!$
8. $7!$
9. $-5!$
10. $-6!$
11. $3! + 4!$
12. $2!(3!)$
13. $3!(4!)$
14. $4! + 4!$
15. $8(7!)$
16. $4!(5)$
17. $\dfrac{9!}{11!}$
18. $\dfrac{13!}{10!}$
19. $\dfrac{49!}{47!}$
20. $\dfrac{101!}{100!}$
21. $\dfrac{5!}{3!(5-3)!}$
22. $\dfrac{6!}{4!(6-4)!}$
23. $\dfrac{7!}{5!(7-5)!}$
24. $\dfrac{8!}{6!(8-6)!}$
25. $\dfrac{5!(8-5)!}{4!7!}$
26. $\dfrac{6!7!}{(8-3)!(7-4)!}$

In Exercises 27–42, use the binomial theorem to expand each expression.

27. $(x + y)^2$
28. $(x + y)^4$
29. $(x - y)^4$
30. $(x - y)^3$
31. $(2x + y)^3$
32. $(x + 2y)^3$
33. $(x - 2y)^3$
34. $(2x - y)^3$
35. $(2x + 3y)^3$
36. $(3x - 2y)^3$
37. $\left(\dfrac{x}{2} - \dfrac{y}{3}\right)^3$
38. $\left(\dfrac{x}{3} + \dfrac{y}{2}\right)^3$
39. $(3 + 2y)^4$
40. $(2x + 3)^4$
41. $\left(\dfrac{x}{3} - \dfrac{y}{2}\right)^4$
42. $\left(\dfrac{x}{2} + \dfrac{y}{3}\right)^4$

In Exercises 43–60, use the binomial theorem to find the required term of each expression.

43. $(a + b)^3$; second term
44. $(a + b)^3$; third term
45. $(x - y)^4$; fourth term
46. $(x - y)^5$; second term
47. $(x + y)^6$; fifth term
48. $(x + y)^7$; fifth term
49. $(x - y)^8$; third term
50. $(x - y)^9$; seventh term
51. $(x + 3)^5$; third term
52. $(x - 2)^4$; second term
53. $(4x + y)^5$; third term
54. $(x + 4y)^5$; fourth term
55. $(x - 3y)^4$; second term
56. $(3x - y)^5$; third term
57. $(2x - 5)^7$; fourth term
58. $(2x + 3)^6$; sixth term
59. $(2x - 3y)^5$; fifth term
60. $(3x - 2y)^4$; second term

WRITING *Write a paragraph using your own words.*

61. Tell how to construct Pascal's triangle.

62. Tell how to find the variables in the expansion of $(r + s)^4$.

63. Tell how to find the coefficients in the expansion of $(x + y)^5$.

64. Explain why the signs alternate in the expansion of $(x - y)^9$.

REVIEW *In Exercises 65–66, solve each system of equations.*

65. $\begin{cases} 3x + 2y = 12 \\ 2x - y = 1 \end{cases}$

66. $\begin{cases} a + b + c = 6 \\ 2a + b + 3c = 11 \\ 3a - b - c = 6 \end{cases}$

In Exercises 67–68, evaluate each determinant.

67. $\begin{vmatrix} 2 & -3 \\ 4 & -2 \end{vmatrix}$

68. $\begin{vmatrix} 1 & 2 & 3 \\ 4 & 5 & 0 \\ -1 & -2 & 1 \end{vmatrix}$

APPENDIX III
TABLES

TABLE A Powers and Roots

n	n^2	$\sqrt{n}$	n^3	$\sqrt[3]{n}$	n	n^2	$\sqrt{n}$	n^3	$\sqrt[3]{n}$
1	1	1.000	1	1.000	51	2,601	7.141	132,651	3.708
2	4	1.414	8	1.260	52	2,704	7.211	140,608	3.733
3	9	1.732	27	1.442	53	2,809	7.280	148,877	3.756
4	16	2.000	64	1.587	54	2,916	7.348	157,464	3.780
5	25	2.236	125	1.710	55	3,025	7.416	166,375	3.803
6	36	2.449	216	1.817	56	3,136	7.483	175,616	3.826
7	49	2.646	343	1.913	57	3,249	7.550	185,193	3.849
8	64	2.828	512	2.000	58	3,364	7.616	195,112	3.871
9	81	3.000	729	2.080	59	3,481	7.681	205,379	3.893
10	100	3.162	1,000	2.154	60	3,600	7.746	216,000	3.915
11	121	3.317	1,331	2.224	61	3,721	7.810	226,981	3.936
12	144	3.464	1,728	2.289	62	3,844	7.874	238,328	3.958
13	169	3.606	2,197	2.351	63	3,969	7.937	250,047	3.979
14	196	3.742	2,744	2.410	64	4,096	8.000	262,144	4.000
15	225	3.873	3,375	2.466	65	4,225	8.062	274,625	4.021
16	256	4.000	4,096	2.520	66	4,356	8.124	287,496	4.041
17	289	4.123	4,913	2.571	67	4,489	8.185	300,763	4.062
18	324	4.243	5,832	2.621	68	4,624	8.246	314,432	4.082
19	361	4.359	6,859	2.668	69	4,761	8.307	328,509	4.102
20	400	4.472	8,000	2.714	70	4,900	8.367	343,000	4.121
21	441	4.583	9,261	2.759	71	5,041	8.426	357,911	4.141
22	484	4.690	10,648	2.802	72	5,184	8.485	373,248	4.160
23	529	4.796	12,167	2.844	73	5,329	8.544	389,017	4.179
24	576	4.899	13,824	2.884	74	5,476	8.602	405,224	4.198
25	625	5.000	15,625	2.924	75	5,625	8.660	421,875	4.217
26	676	5.099	17,576	2.962	76	5,776	8.718	438,976	4.236
27	729	5.196	19,683	3.000	77	5,929	8.775	456,533	4.254
28	784	5.292	21,952	3.037	78	6,084	8.832	474,552	4.273
29	841	5.385	24,389	3.072	79	6,241	8.888	493,039	4.291
30	900	5.477	27,000	3.107	80	6,400	8.944	512,000	4.309
31	961	5.568	29,791	3.141	81	6,561	9.000	531,441	4.327
32	1,024	5.657	32,768	3.175	82	6,724	9.055	551,368	4.344
33	1,089	5.745	35,937	3.208	83	6,889	9.110	571,787	4.362
34	1,156	5.831	39,304	3.240	84	7,056	9.165	592,704	4.380
35	1,225	5.916	42,875	3.271	85	7,225	9.220	614,125	4.397
36	1,296	6.000	46,656	3.302	86	7,396	9.274	636,056	4.414
37	1,369	6.083	50,653	3.332	87	7,569	9.327	658,503	4.431
38	1,444	6.164	54,872	3.362	88	7,744	9.381	681,472	4.448
39	1,521	6.245	59,319	3.391	89	7,921	9.434	704,969	4.465
40	1,600	6.325	64,000	3.420	90	8,100	9.487	729,000	4.481
41	1,681	6.403	68,921	3.448	91	8,281	9.539	753,571	4.498
42	1,764	6.481	74,088	3.476	92	8,464	9.592	778,688	4.514
43	1,849	6.557	79,507	3.503	93	8,649	9.644	804,357	4.531
44	1,936	6.633	85,184	3.530	94	8,836	9.695	830,584	4.547
45	2,025	6.708	91,125	3.557	95	9,025	9.747	857,375	4.563
46	2,116	6.782	97,336	3.583	96	9,216	9.798	884,736	4.579
47	2,209	6.856	103,823	3.609	97	9,409	9.849	912,673	4.595
48	2,304	6.928	110,592	3.634	98	9,604	9.899	941,192	4.610
49	2,401	7.000	117,649	3.659	99	9,801	9.950	970,299	4.626
50	2,500	7.071	125,000	3.684	100	10,000	10.000	1,000,000	4.642

TABLE B Base-10 Logarithms

N	0	1	2	3	4	5	6	7	8	9
1.0	.0000	.0043	.0086	.0128	.0170	.0212	.0253	.0294	.0334	.0374
1.1	.0414	.0453	.0492	.0531	.0569	.0607	.0645	.0682	.0719	.0755
1.2	.0792	.0828	.0864	.0899	.0934	.0969	.1004	.1038	.1072	.1106
1.3	.1139	.1173	.1206	.1239	.1271	.1303	.1335	.1367	.1399	.1430
1.4	.1461	.1492	.1523	.1553	.1584	.1614	.1644	.1673	.1703	.1732
1.5	.1761	.1790	.1818	.1847	.1875	.1903	.1931	.1959	.1987	.2014
1.6	.2041	.2068	.2095	.2122	.2148	.2175	.2201	.2227	.2253	.2279
1.7	.2304	.2330	.2355	.2380	.2405	.2430	.2455	.2480	.2504	.2529
1.8	.2553	.2577	.2601	.2625	.2648	.2672	.2695	.2718	.2742	.2765
1.9	.2788	.2810	.2833	.2856	.2878	.2900	.2923	.2945	.2967	.2989
2.0	.3010	.3032	.3054	.3075	.3096	.3118	.3139	.3160	.3181	.3201
2.1	.3222	.3243	.3263	.3284	.3304	.3324	.3345	.3365	.3385	.3404
2.2	.3424	.3444	.3464	.3483	.3502	.3522	.3541	.3560	.3579	.3598
2.3	.3617	.3636	.3655	.3674	.3692	.3711	.3729	.3747	.3766	.3784
2.4	.3802	.3820	.3838	.3856	.3874	.3892	.3909	.3927	.3945	.3962
2.5	.3979	.3997	.4014	.4031	.4048	.4065	.4082	.4099	.4116	.4133
2.6	.4150	.4166	.4183	.4200	.4216	.4232	.4249	.4265	.4281	.4298
2.7	.4314	.4330	.4346	.4362	.4378	.4393	.4409	.4425	.4440	.4456
2.8	.4472	.4487	.4502	.4518	.4533	.4548	.4564	.4579	.4594	.4609
2.9	.4624	.4639	.4654	.4669	.4683	.4698	.4713	.4728	.4742	.4757
3.0	.4771	.4786	.4800	.4814	.4829	.4843	.4857	.4871	.4886	.4900
3.1	.4914	.4928	.4942	.4955	.4969	.4983	.4997	.5011	.5024	.5038
3.2	.5051	.5065	.5079	.5092	.5105	.5119	.5132	.5145	.5159	.5172
3.3	.5185	.5198	.5211	.5224	.5237	.5250	.5263	.5276	.5289	.5302
3.4	.5315	.5328	.5340	.5353	.5366	.5378	.5391	.5403	.5416	.5428
3.5	.5441	.5453	.5465	.5478	.5490	.5502	.5514	.5527	.5539	.5551
3.6	.5563	.5575	.5587	.5599	.5611	.5623	.5635	.5647	.5658	.5670
3.7	.5682	.5694	.5705	.5717	.5729	.5740	.5752	.5763	.5775	.5786
3.8	.5798	.5809	.5821	.5832	.5843	.5855	.5866	.5877	.5888	.5899
3.9	.5911	.5922	.5933	.5944	.5955	.5966	.5977	.5988	.5999	.6010
4.0	.6021	.6031	.6042	.6053	.6064	.6075	.6085	.6096	.6107	.6117
4.1	.6128	.6138	.6149	.6160	.6170	.6180	.6191	.6201	.6212	.6222
4.2	.6232	.6243	.6253	.6263	.6274	.6284	.6294	.6304	.6314	.6325
4.3	.6335	.6345	.6355	.6365	.6375	.6385	.6395	.6405	.6415	.6425
4.4	.6435	.6444	.6454	.6464	.6474	.6484	.6493	.6503	.6513	.6522
4.5	.6532	.6542	.6551	.6561	.6571	.6580	.6590	.6599	.6609	.6618
4.6	.6628	.6637	.6646	.6656	.6665	.6675	.6684	.6693	.6702	.6712
4.7	.6721	.6730	.6739	.6749	.6758	.6767	.6776	.6785	.6794	.6803
4.8	.6812	.6821	.6830	.6839	.6848	.6857	.6866	.6875	.6884	.6893
4.9	.6902	.6911	.6920	.6928	.6937	.6946	.6955	.6964	.6972	.6981
5.0	.6990	.6998	.7007	.7016	.7024	.7033	.7042	.7050	.7059	.7067
5.1	.7076	.7084	.7093	.7101	.7110	.7118	.7126	.7135	.7143	.7152
5.2	.7160	.7168	.7177	.7185	.7193	.7202	.7210	.7218	.7226	.7235
5.3	.7243	.7251	.7259	.7267	.7275	.7284	.7292	.7300	.7308	.7316
5.4	.7324	.7332	.7340	.7348	.7356	.7364	.7372	.7380	.7388	.7396

TABLE B *(continued)*

N	0	1	2	3	4	5	6	7	8	9
5.5	.7404	.7412	.7419	.7427	.7435	.7443	.7451	.7459	.7466	.7474
5.6	.7482	.7490	.7497	.7505	.7513	.7520	.7528	.7536	.7543	.7551
5.7	.7559	.7566	.7574	.7582	.7589	.7597	.7604	.7612	.7619	.7627
5.8	.7634	.7642	.7649	.7657	.7664	.7672	.7679	.7686	.7694	.7701
5.9	.7709	.7716	.7723	.7731	.7738	.7745	.7752	.7760	.7767	.7774
6.0	.7782	.7789	.7796	.7803	.7810	.7818	.7825	.7832	.7839	.7846
6.1	.7853	.7860	.7868	.7875	.7882	.7889	.7896	.7903	.7910	.7917
6.2	.7924	.7931	.7938	.7945	.7952	.7959	.7966	.7973	.7980	.7987
6.3	.7993	.8000	.8007	.8014	.8021	.8028	.8035	.8041	.8048	.8055
6.4	.8062	.8069	.8075	.8082	.8089	.8096	.8102	.8109	.8116	.8122
6.5	.8129	.8136	.8142	.8149	.8156	.8162	.8169	.8176	.8182	.8189
6.6	.8195	.8202	.8209	.8215	.8222	.8228	.8235	.8241	.8248	.8254
6.7	.8261	.8267	.8274	.8280	.8287	.8293	.8299	.8306	.8312	.8319
6.8	.8325	.8331	.8338	.8344	.8351	.8357	.8363	.8370	.8376	.8382
6.9	.8388	.8395	.8401	.8407	.8414	.8420	.8426	.8432	.8439	.8445
7.0	.8451	.8457	.8463	.8470	.8476	.8482	.8488	.8494	.8500	.8506
7.1	.8513	.8519	.8525	.8531	.8537	.8543	.8549	.8555	.8561	.8567
7.2	.8573	.8579	.8585	.8591	.8597	.8603	.8609	.8615	.8621	.8627
7.3	.8633	.8639	.8645	.8651	.8657	.8663	.8669	.8675	.8681	.8686
7.4	.8692	.8698	.8704	.8710	.8716	.8722	.8727	.8733	.8739	.8745
7.5	.8751	.8756	.8762	.8768	.8774	.8779	.8785	.8791	.8797	.8802
7.6	.8808	.8814	.8820	.8825	.8831	.8837	.8842	.8848	.8854	.8859
7.7	.8865	.8871	.8876	.8882	.8887	.8893	.8899	.8904	.8910	.8915
7.8	.8921	.8927	.8932	.8938	.8943	.8949	.8954	.8960	.8965	.8971
7.9	.8976	.8982	.8987	.8993	.8998	.9004	.9009	.9015	.9020	.9025
8.0	.9031	.9036	.9042	.9047	.9053	.9058	.9063	.9069	.9074	.9079
8.1	.9085	.9090	.9096	.9101	.9106	.9112	.9117	.9122	.9128	.9133
8.2	.9138	.9143	.9149	.9154	.9159	.9165	.9170	.9175	.9180	.9186
8.3	.9191	.9196	.9201	.9206	.9212	.9217	.9222	.9227	.9232	.9238
8.4	.9243	.9248	.9253	.9258	.9263	.9269	.9274	.9279	.9284	.9289
8.5	.9294	.9299	.9304	.9309	.9315	.9320	.9325	.9330	.9335	.9340
8.6	.9345	.9350	.9355	.9360	.9365	.9370	.9375	.9380	.9385	.9390
8.7	.9395	.9400	.9405	.9410	.9415	.9420	.9425	.9430	.9435	.9440
8.8	.9445	.9450	.9455	.9460	.9465	.9469	.9474	.9479	.9484	.9489
8.9	.9494	.9499	.9504	.9509	.9513	.9518	.9523	.9528	.9533	.9538
9.0	.9542	.9547	.9552	.9557	.9562	.9566	.9571	.9576	.9581	.9586
9.1	.9590	.9595	.9600	.9605	.9609	.9614	.9619	.9624	.9628	.9633
9.2	.9638	.9643	.9647	.9652	.9657	.9661	.9666	.9671	.9675	.9680
9.3	.9685	.9689	.9694	.9699	.9703	.9708	.9713	.9717	.9722	.9727
9.4	.9731	.9736	.9741	.9745	.9750	.9754	.9759	.9763	.9768	.9773
9.5	.9777	.9782	.9786	.9791	.9795	.9800	.9805	.9809	.9814	.9818
9.6	.9823	.9827	.9832	.9836	.9841	.9845	.9850	.9854	.9859	.9863
9.7	.9868	.9872	.9877	.9881	.9886	.9890	.9894	.9899	.9903	.9908
9.8	.9912	.9917	.9921	.9926	.9930	.9934	.9939	.9943	.9948	.9952
9.9	.9956	.9961	.9965	.9969	.9974	.9978	.9983	.9987	.9991	.9996

TABLE C Base-*e* Logarithms

N	0	1	2	3	4	5	6	7	8	9
1.0	.0000	.0100	.0198	.0296	.0392	.0488	.0583	.0677	.0770	.0862
1.1	.0953	.1044	.1133	.1222	.1310	.1398	.1484	.1570	.1655	.1740
1.2	.1823	.1906	.1989	.2070	.2151	.2231	.2311	.2390	.2469	.2546
1.3	.2624	.2700	.2776	.2852	.2927	.3001	.3075	.3148	.3221	.3293
1.4	.3365	.3436	.3507	.3577	.3646	.3716	.3784	.3853	.3920	.3988
1.5	.4055	.4121	.4187	.4253	.4318	.4383	.4447	.4511	.4574	.4637
1.6	.4700	.4762	.4824	.4886	.4947	.5008	.5068	.5128	.5188	.5247
1.7	.5306	.5365	.5423	.5481	.5539	.5596	.5653	.5710	.5766	.5822
1.8	.5878	.5933	.5988	.6043	.6098	.6152	.6206	.6259	.6313	.6366
1.9	.6419	.6471	.6523	.6575	.6627	.6678	.6729	.6780	.6831	.6881
2.0	.6931	.6981	.7031	.7080	.7129	.7178	.7227	.7275	.7324	.7372
2.1	.7419	.7467	.7514	.7561	.7608	.7655	.7701	.7747	.7793	.7839
2.2	.7885	.7930	.7975	.8020	.8065	.8109	.8154	.8198	.8242	.8286
2.3	.8329	.8372	.8416	.8459	.8502	.8544	.8587	.8629	.8671	.8713
2.4	.8755	.8796	.8838	.8879	.8920	.8961	.9002	.9042	.9083	.9123
2.5	.9163	.9203	.9243	.9282	.9322	.9361	.9400	.9439	.9478	.9517
2.6	.9555	.9594	.9632	.9670	.9708	.9746	.9783	.9821	.9858	.9895
2.7	.9933	.9969	1.0006	.0043	.0080	.0116	.0152	.0188	.0225	.0260
2.8	1.0296	.0332	.0367	.0403	.0438	.0473	.0508	.0543	.0578	.0613
2.9	.0647	.0682	.0716	.0750	.0784	.0818	.0852	.0886	.0919	.0953
3.0	1.0986	.1019	.1053	.1086	.1119	.1151	.1184	.1217	.1249	.1282
3.1	.1314	.1346	.1378	.1410	.1442	.1474	.1506	.1537	.1569	.1600
3.2	.1632	.1663	.1694	.1725	.1756	.1787	.1817	.1848	.1878	.1909
3.3	.1939	.1969	.2000	.2030	.2060	.2090	.2119	.2149	.2179	.2208
3.4	.2238	.2267	.2296	.2326	.2355	.2384	.2413	.2442	.2470	.2499
3.5	1.2528	.2556	.2585	.2613	.2641	.2669	.2698	.2726	.2754	.2782
3.6	.2809	.2837	.2865	.2892	.2920	.2947	.2975	.3002	.3029	.3056
3.7	.3083	.3110	.3137	.3164	.3191	.3218	.3244	.3271	.3297	.3324
3.8	.3350	.3376	.3403	.3429	.3455	.3481	.3507	.3533	.3558	.3584
3.9	.3610	.3635	.3661	.3686	.3712	.3737	.3762	.3788	.3813	.3838
4.0	1.3863	.3888	.3913	.3938	.3962	.3987	.4012	.4036	.4061	.4085
4.1	.4110	.4134	.4159	.4183	.4207	.4231	.4255	.4279	.4303	.4327
4.2	.4351	.4375	.4398	.4422	.4446	.4469	.4493	.4516	.4540	.4563
4.3	.4586	.4609	.4633	.4656	.4679	.4702	.4725	.4748	.4770	.4793
4.4	.4816	.4839	.4861	.4884	.4907	.4929	.4951	.4974	.4996	.5019
4.5	1.5041	.5063	.5085	.5107	.5129	.5151	.5173	.5195	.5217	.5239
4.6	.5261	.5282	.5304	.5326	.5347	.5369	.5390	.5412	.5433	.5454
4.7	.5476	.5497	.5518	.5539	.5560	.5581	.5602	.5623	.5644	.5665
4.8	.5686	.5707	.5728	.5748	.5769	.5790	.5810	.5831	.5851	.5872
4.9	.5892	.5913	.5933	.5953	.5974	.5994	.6014	.6034	.6054	.6074
5.0	1.6094	.6114	.6134	.6154	.6174	.6194	.6214	.6233	.6253	.6273
5.1	.6292	.6312	.6332	.6351	.6371	.6390	.6409	.6429	.6448	.6467
5.2	.6487	.6506	.6525	.6544	.6563	.6582	.6601	.6620	.6639	.6658
5.3	.6677	.6696	.6715	.6734	.6752	.6771	.6790	.6808	.6827	.6845
5.4	.6864	.6882	.6901	.6919	.6938	.6956	.6974	.6993	.7011	.7029

Use the properties of logarithms and ln 10 ≈ 2.3026 to find logarithms of numbers less than 1 or greater than 10.

TABLE C (*continued*)

N	0	1	2	3	4	5	6	7	8	9
5.5	1.7047	.7066	.7084	.7102	.7120	.7138	.7156	.7174	.7192	.7210
5.6	.7228	.7246	.7263	.7281	.7299	.7317	.7334	.7352	.7370	.7387
5.7	.7405	.7422	.7440	.7457	.7475	.7492	.7509	.7527	.7544	.7561
5.8	.7579	.7596	.7613	.7630	.7647	.7664	.7681	.7699	.7716	.7733
5.9	.7750	.7766	.7783	.7800	.7817	.7834	.7851	.7867	.7884	.7901
6.0	1.7918	.7934	.7951	.7967	.7984	.8001	.8017	.8034	.8050	.8066
6.1	.8083	.8099	.8116	.8132	.8148	.8165	.8181	.8197	.8213	.8229
6.2	.8245	.8262	.8278	.8294	.8310	.8326	.8342	.8358	.8374	.8390
6.3	.8405	.8421	.8437	.8453	.8469	.8485	.8500	.8516	.8532	.8547
6.4	.8563	.8579	.8594	.8610	.8625	.8641	.8656	.8672	.8687	.8703
6.5	1.8718	.8733	.8749	.8764	.8779	.8795	.8810	.8825	.8840	.8856
6.6	.8871	.8886	.8901	.8916	.8931	.8946	.8961	.8976	.8991	.9006
6.7	.9021	.9036	.9051	.9066	.9081	.9095	.9110	.9125	.9140	.9155
6.8	.9169	.9184	.9199	.9213	.9228	.9242	.9257	.9272	.9286	.9301
6.9	.9315	.9330	.9344	.9359	.9373	.9387	.9402	.9416	.9430	.9445
7.0	1.9459	.9473	.9488	.9502	.9516	.9530	.9544	.9559	.9573	.9587
7.1	.9601	.9615	.9629	.9643	.9657	.9671	.9685	.9699	.9713	.9727
7.2	.9741	.9755	.9769	.9782	.9796	.9810	.9824	.9838	.9851	.9865
7.3	.9879	.9892	.9906	.9920	.9933	.9947	.9961	.9974	.9988	2.0001
7.4	2.0015	.0028	.0042	.0055	.0069	.0082	.0096	.0109	.0122	.0136
7.5	2.0149	.0162	.0176	.0189	.0202	.0215	.0229	.0242	.0255	.0268
7.6	.0281	.0295	.0308	.0321	.0334	.0347	.0360	.0373	.0386	.0399
7.7	.0412	.0425	.0438	.0451	.0464	.0477	.0490	.0503	.0516	.0528
7.8	.0541	.0554	.0567	.0580	.0592	.0605	.0618	.0631	.0643	.0656
7.9	.0669	.0681	.0694	.0707	.0719	.0732	.0744	.0757	.0769	.0782
8.0	2.0794	.0807	.0819	.0832	.0844	.0857	.0869	.0882	.0894	.0906
8.1	.0919	.0931	.0943	.0956	.0968	.0980	.0992	.1005	.1017	.1029
8.2	.1041	.1054	.1066	.1078	.1090	.1102	.1114	.1126	.1138	.1150
8.3	.1163	.1175	.1187	.1199	.1211	.1223	.1235	.1247	.1258	.1270
8.4	.1282	.1294	.1306	.1318	.1330	.1342	.1353	.1365	.1377	.1389
8.5	2.1401	.1412	.1424	.1436	.1448	.1459	.1471	.1483	.1494	.1506
8.6	.1518	.1529	.1541	.1552	.1564	.1576	.1587	.1599	.1610	.1622
8.7	.1633	.1645	.1656	.1668	.1679	.1691	.1702	.1713	.1725	.1736
8.8	.1748	.1759	.1770	.1782	.1793	.1804	.1815	.1827	.1838	.1849
8.9	.1861	.1872	.1883	.1894	.1905	.1917	.1928	.1939	.1950	.1961
9.0	2.1972	.1983	.1994	.2006	.2017	.2028	.2039	.2050	.2061	.2072
9.1	.2083	.2094	.2105	.2116	.2127	.2138	.2148	.2159	.2170	.2181
9.2	.2192	.2203	.2214	.2225	.2235	.2246	.2257	.2268	.2279	.2289
9.3	.2300	.2311	.2322	.2332	.2343	.2354	.2364	.2375	.2386	.2396
9.4	.2407	.2418	.2428	.2439	.2450	.2460	.2471	.2481	.2492	.2502
9.5	2.2513	.2523	.2534	.2544	.2555	.2565	.2576	.2586	.2597	.2607
9.6	.2618	.2628	.2638	.2649	.2659	.2670	.2680	.2690	.2701	.2711
9.7	.2721	.2732	.2742	.2752	.2762	.2773	.2783	.2793	.2803	.2814
9.8	.2824	.2834	.2844	.2854	.2865	.2875	.2885	.2895	.2905	.2915
9.9	.2925	.2935	.2946	.2956	.2966	.2976	.2986	.2996	.3006	.3016

APPENDIX IV
ANSWERS TO SELECTED EXERCISES

Study Set Section 1.1 (page 8)

1. equation **3.** algebraic expressions **5.** formula **7.** perimeter **9.** algebraic expression **11.** equation **13.** equation **15.** algebraic expression **17. a.** a line graph **b.** 1 hour, 2 inches **c.** 7 in.; 0 in. **19.** $c = 13u + 24$ **21.** $w = \frac{c}{75}$ **23.** $A = t + 15$ **25.** $c = 12b$ **27.** $b = t - 10$; the height of the base is 10 ft less than the height of the tower **29. a.** D **b.** 2, 4 **c.** multiplication, addition **31.** 2, 6, 15 **33.** 22.44, 21.43, 0 **35.** 37 in. **37.** 8 yd **39. a.** The rental cost is the product of 10 and the number of hours it is rented, increased by 20. **b.** $C = 10h + 20$ **c.** 30, 40, 50, 60, 100

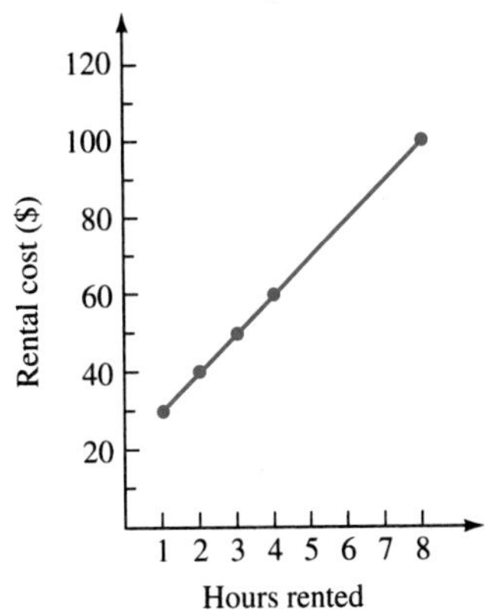

41. a. The measure of the angle on the scrap piece of molding is the difference of 180° and the measure of the angle on the finish piece of molding. **b.** $s = 180° - f$ **c.** 150, 135, 90, 45, 30

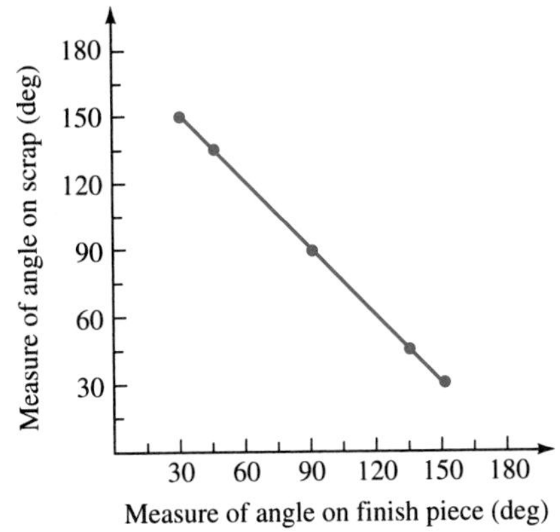

Study Set Section 1.2 (page 18)

1. rational **3.** absolute value **5.** Irrational **7.** composite **9.** 1, 2, 9 **11.** −3, 0, 1, 2, 9 **13.** $\sqrt{3}$, π **15.** 2 **17.** 2 **19.** 9 **21.** nonrepeating, irrational **23.** repeating, rational **25.** $7 = \frac{7}{1}$, $-7\frac{3}{5} = \frac{-38}{5}$, $0.007 = \frac{7}{1{,}000}$, $700.1 = \frac{7{,}001}{10}$ **27.** 3.5 or −3.5 **29.** **31.** is less than **33.** braces **35.** 0.875, terminating **37.** $-0.7\overline{3}$, repeating **39.** **41.** **43.** **45.** **47.**

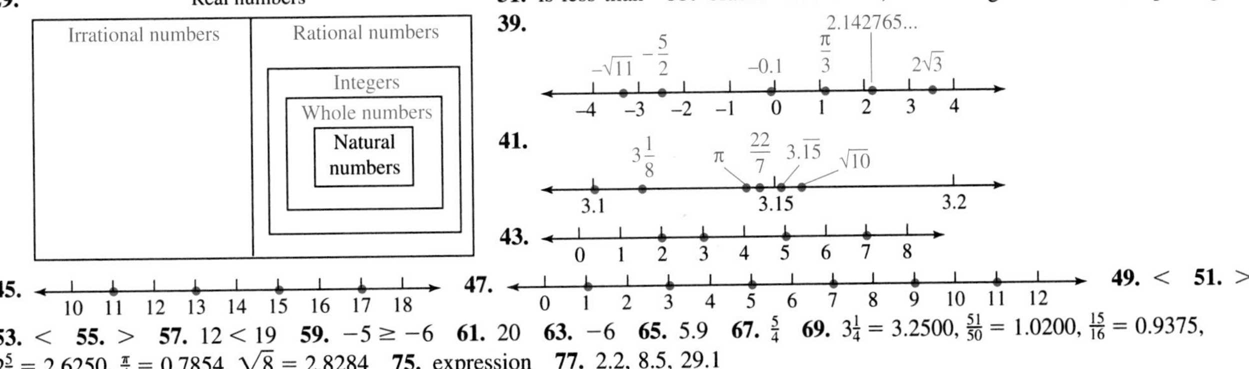

49. < **51.** > **53.** < **55.** > **57.** 12 < 19 **59.** −5 ≥ −6 **61.** 20 **63.** −6 **65.** 5.9 **67.** $\frac{5}{4}$ **69.** $3\frac{1}{4} = 3.2500$, $\frac{51}{50} = 1.0200$, $\frac{15}{16} = 0.9375$, $2\frac{5}{8} = 2.6250$, $\frac{\pi}{4} = 0.7854$, $\sqrt{8} = 2.8284$ **75.** expression **77.** 2.2, 8.5, 29.1

Study Set Section 1.3 (page 31)

1. sum, difference **3.** evaluate **5.** squared, cubed **7.** base, exponent **9.** opposite **11. a.** addition first and multiplication first **b.** 12; multiplication is to be done before addition **13. a.** area **b.** volume **15.** 8, 160 **17.** radical sign

19. 1, 8, 27, 64

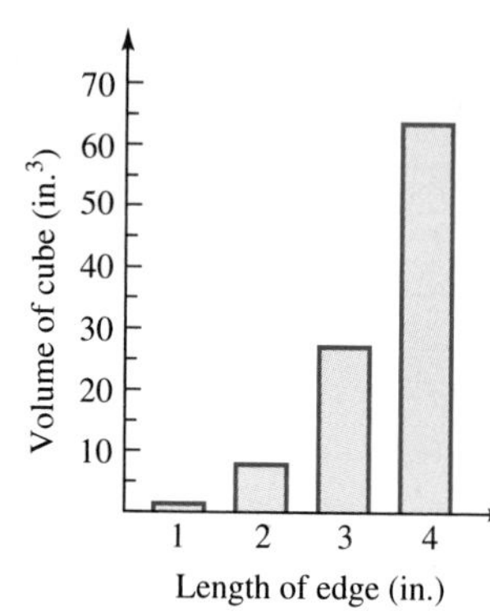

21. -8 **23.** -4.3 **25.** -7 **27.** 0 **29.** -12 **31.** -2 **33.** $\frac{1}{6}$ **35.** $\frac{11}{10}$ **37.** $-\frac{6}{7}$ **39.** $\frac{24}{25}$ **41.** 144 **43.** -25 **45.** 32 **47.** 1.69 **49.** 8 **51.** $-\frac{3}{4}$ **53.** -17 **55.** 64 **57.** 13 **59.** -2 **61.** -8 **63.** 10,000 **65.** -39 **67.** -32 **69.** 5 **71.** $-\frac{1}{2}$ **73.** 2 **75.** 91,985 **77.** -2 **79.** 61 **81.** 1 **83.** 10 **85.** 25 ft^2 **87.** 1st term: area of bottom flap; 2nd term: area of left and right flaps; 3rd term: area of top flap; 4th term: area of face; 42.5 in.^{2} **89.** \$(967) **91.** yes **93.** 9

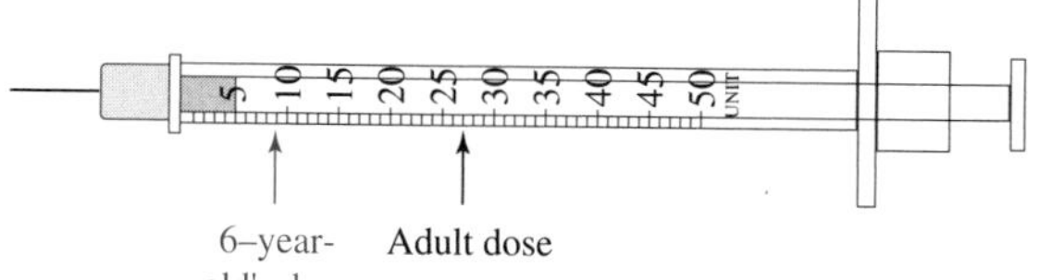

97. -7 and 3 **99.** $\{\ldots, -2, -1, 0, 1, 2, \ldots\}$ **101.** true

Study Set Section 1.4 (page 42)

1. like **3.** term **5.** simplify **7.** undefined **9.** $(x + y) + z = x + (y + z)$ **11.** $r(s + t) = rs + rt$ **13. a.** 5 **b.** -5 **15. a.** $\frac{16}{15}$ **b.** $-\frac{1}{20}$ **c.** 2 **d.** $\frac{1}{x}$ **17. a.** $2x^2, -x, 6$ **b.** 2, -1, 6 **19.** yes, $8x$ **21.** no **23.** yes, $-2x^2$ **25.** no **27.** multiplication by -1 **29.** $7 + 3$ **31.** $3 \cdot 2 + 3d$ **33.** c **35.** 1 **37.** $(8 + 7) + a$ **39.** $2(x + y)$ **41.** assoc. prop. of add. **43.** distrib. prop. **45.** $-4t + 12$ **47.** $-t + 3$ **49.** $2s - 6$ **51.** $0.7s + 1.4$ **53.** $4x - 5y + 1$ **55.** $72m$ **57.** $-45q$ **59.** $30bp$ **61.** $80ry$ **63.** $18x$ **65.** $13x^2$ **67.** 0 **69.** 0 **71.** $6x$ **73.** 0 **75.** $3.1h$ **77.** $\frac{4}{5}t$ **79.** $-4y + 36$ **81.** $7z - 20$ **83.** $14c + 62$ **85.** $-2x^2 + 15x$ **87.** $-2a + b - 2$ **89. a.** $20(x + 6)$ m^2 **b.** $(20x + 120)$ m^2 **c.** $20(x + 6) = 20x + 120$; distrib. prop. **95.** 0 **97.** $-\frac{7}{8}$ **99.** 1,000 **101.** -3

Study Set Section 1.5 (page 53)

1. equation **3.** satisfies **5.** values **7.** c, c **9. a.** $2y + 2$ **b.** 3 **c.** 18 **11.** No; a is not isolated on one side. **13.** It clears the equation of fractions. **15.** $-2x$, 14, 14, 34, -2, -2 **17. a.** -1 **b.** $\frac{2}{3}$ **19.** yes **21.** no **23.** 28 **25.** -20 **27.** -166 **29.** 2.52 **31.** $\frac{2}{3}$ **33.** -8 **35.** 1.395 **37.** 3 **39.** 13 **41.** -11 **43.** -8 **45.** -2 **47.** 24 **49.** 0 **51.** 6 **53.** 3 **55.** 24 **57.** identity **59.** impossible **61.** identity **63.** $B = \frac{3V}{h}$ **65.** $t = \frac{I}{Pr}$ **67.** $w = \frac{P - 2l}{2}$ **69.** $B = \frac{2A}{h} - b$ or $B = \frac{2A - bh}{h}$ **71.** $x = \frac{y - b}{m}$ **73.** $v_0 = 2\bar{v} - v$ **75.** $l = \frac{a - S + Sr}{r}$ **77.** $l = \frac{2S - na}{n}$ or $l = \frac{2S}{n} - a$ **79.** $C = \frac{5(F - 32)}{9}$; 432, -179, 58, -89, 17, -66 **81.** $d = \dfrac{360A}{\pi(r_1{}^2 - r_2{}^2)}$; 140, 160 **83.** $n = \frac{PV}{R(T + 273)}$; 0.008, 0.090 **85.** $n = \frac{C - 6.50}{0.07}$; 621, 1,000, about 1,692.9 kwh **87.** $h = \dfrac{A - 2\pi r^2}{2\pi r}$ **93.** $t - 4$ **95.** $-4b + 32$ **97.** $2.9b$ **99.** t

Study Set Section 1.6 (page 63)

1. acute **3.** complementary **5.** right **7.** $d + 15$, $2d - 10$, $2d + 20$, $\frac{d}{2} - 10$, $2d$ **9. a.** $\frac{2}{3}x$ **b.** $2x$ **c.** $x + 2x + \frac{2}{3}x$ **11.** $26.5 - x$ **13.** Frosted Flakes: \$295,800,000; Cheerios: \$286,400,000 **15.** 20 **17.** 305 mi **19.** 300 shares of BB, 200 shares of SS **21.** 35 \$18 calculators, 50 \$87 calculators **23.** 7 ft, 15 ft **25.** 30°, 150° **27.** 10° **29.** 50° **31.** 60° **33.** ii **35.** 156 ft by 312 ft **37.** 10 ft **39.** 6 in. **43.** repeating **45.** $\{\ldots, -4, -3, -2, -1, 0, 1, 2, 3, 4, \ldots\}$ **47.** $\frac{5}{4}$

Study Set Section 1.7 (page 76)

1. principal **3.** median **5.** amount, base **7.** \$90, \$0.0565x, \$0.07$(850 - x)$ **9. a.** \$25,000 **b.** \$$(30{,}000 - x)$ **11. a.** 80% water **b.** 4 pt, $0.1x$ pt **c.** $(20 + x)$ pt **d.** $(4 + 0.1x)$ pt **13.** $x = 0.05 \cdot 10.56$ **15.** $32.5 = 0.74x$ **17.** 100, 100, $9x$, 40,000, 16,000 **19.** $I = Prt$ **21.** $v = pn$ **23.** 356.83 quadrillion **25.** 20% **27.** \$50 **29.** 9% **31.** 4.5%, 11.9% **33.** city: median 29.5, mode 29; hwy: median 38, mode 38 **35.** 84 **37.** \$2,000 at 8%, \$10,000 at 9% **39.** \$45,000 **41.** $\frac{1}{4}$ hr = 15 min **43.** 3:30 P.M. **45.** $1\frac{1}{2}$ hr **47.** 20 lb of 95¢ candy; 10 lb of \$1.10 candy **49.** 10 oz **51.** 2 gal **55.** 1 **57.** 8

Key Concept (page 81)

1. given: 48 states, 4 more lie east of the Miss. River than west; find: how many states lie west **2.** given: one angle is 5° more than twice another, one angle is 90°, the sum of the angles' measures is 180°; find: the measure of the smallest angle **3.** Let x = the length of the shortest piece. **4.** Let x = the number of miles he can drive. **5. a.** subtraction **b.** multiplication **c.** addition **d.** division **6. a.** $d = rt$ **b.** $I = Prt$ **c.** $v = pn$ **d.** $P = 2l + 2w$ **7.** $0.15(15) + 0.50x = 0.40(15 + x)$; $0.15(15)$, $0.50x$, $0.40(15 + x)$ **8.** $450x + 500x = 2{,}850$

Chapter Review (page 83)

1. a. $C = 2t + 15$ **b.** $l = \frac{25}{w}$ **c.** $P = u - 3$ **2.** 180, 195, 210, 225, 240

3. a. Cooking time (min) 270 240 210 180 150 6.0 7.0 8.0 Weight (lb) **b.**

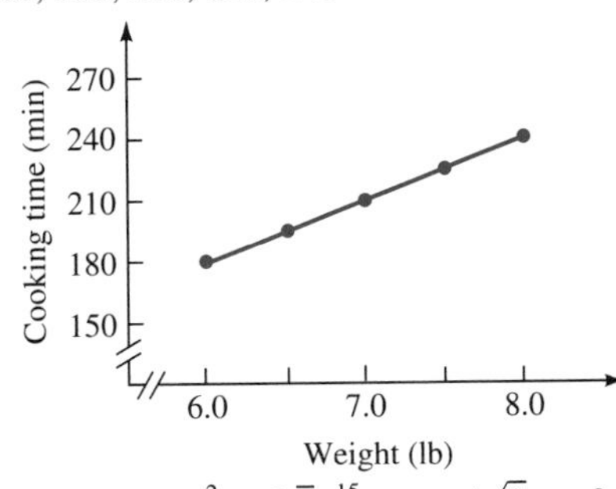

4. 17.5 in. **5. a.** 7 **b.** 0, 7 **c.** −5, 0, 7 **d.** $-5, 0, 2.4, 7, -\frac{2}{3}, -3.\overline{6}, \frac{15}{4}$ **e.** $-\sqrt{3}, \pi, 0.13242368\ldots$ **f.** all **g.** $-5, -\sqrt{3}, -\frac{2}{3}, -3.\overline{6}$ **h.** $2.4, 7, \pi, \frac{15}{4}, 0.13242368\ldots$ **i.** 7 **j.** none **k.** 0 **l.** −5, 7 **6. a.** > **b.** < **7. a.** false **b.** true

8. 20 21 22 23 24 25 26 27 28 29 30 **9.** $2.\overline{3}$ $\frac{3\pi}{4}$ $\sqrt{7}$ $\frac{8}{3}$ 2.75 2 2.5 3 **10. a.** 18 **b.** −6.26

11. a. −7 **b.** 10.1 **c.** $-\frac{3}{4}$ **d.** 2 **e.** 12.6 **f.** $-\frac{1}{32}$ **g.** 0.2 **h.** $-\frac{3}{56}$ **i.** −33 **j.** −5.7 **k.** −120 **l.** 1 **12. a.** −243 **b.** $\frac{4}{9}$ **c.** 0.027 **d.** −25 **13. a.** 2 **b.** −10 **c.** $\frac{3}{5}$ **d.** 0.8 **14. a.** 44 **b.** 1 **c.** −12 **d.** 58 **e.** 8 **f.** 3 **g.** −32 **h.** −64 **15. a.** 56 **b.** $-\frac{1}{2}$ **16. a.** 100 in.^{2} **b.** 251.3 in.^{3} **17. a.** $3x + 21$ **b.** $5t$ **c.** 0 **d.** $27 + (1 + 99)$ **e.** 1 **f.** m **g.** 1 **h.** 0 **i.** $-3(5 \cdot 2)$ **j.** $(z + t) \cdot t$ **18. a.** 1 **b.** 0 **c.** −25 **d.** undefined **19. a.** $8x + 48$ **b.** $-6x + 12$ **c.** $4 - 3y$ **d.** $3.6x - 2.4y$ **e.** $20c^2 - 10c + 10$ **f.** $2t + 6$ **20. a.** $48k$ **b.** $70xy$ **c.** $-189p$ **d.** $45a + 7$ **e.** 0 **f.** $3m - 48$ **g.** x **h.** $-24.54l$ **i.** $6t^3 + 4t^2$ **j.** $11h + 77$ **21. a.** yes **b.** no **22. a.** {−225} **b.** {7.9} **c.** {0.014} **d.** {−4} **23. a.** $-\frac{12}{5}$ **b.** −9 **c.** 8 **d.** $\frac{11}{7}$ **e.** 8 **f.** 12 **g.** 0.06 **h.** −8 **i.** 0 **j.** 3 **24. a.** impossible equation **b.** identity **25. a.** $h = \frac{V}{\pi r^2}$ **b.** $g = \frac{m - Y}{2}$ **c.** $x = \frac{T}{ab} - y$ or $x = \frac{T - aby}{ab}$ **d.** $r^3 = \frac{3V}{4\pi}$ **26.** O'Hare: 69 million; Atlanta: 63 million **27.** $245 - 5c$ **28.** \$5,000; \$1,000x **29.** 42 ft, 45 ft, 48 ft, 51 ft **30.** 50°, 130° **31.** \$18,000 at 10%, \$7,000 at 9% **32.** men's: about 7.2% **33.** 501, 500, 500 **34.** 3 min **35.** 10 gal **36.** $1.95(x + 3)$

Chapter 1 Test (page 89)

1. $s = T + 10$ **2.** $A = \frac{1}{2}bh$ **3.** −2, 0, 5 **4.** $-2, 0, -3\frac{3}{4}, 9.2, \frac{14}{5}, 5$ **5.** $\pi, -\sqrt{7}$ **6.** all

7. $\frac{7}{6}$ $\frac{\pi}{2}$ $\sqrt{3}$ 1.8234503... 1.91 1 1.5 2.0 **8.** 0 1 2 3 4 5 6 7 8 9 10 11 12 **9.** −8

10. 5.5 **11.** 12.3 **12.** $\frac{4}{15}$ **13.** $\frac{11}{10}$ **14.** −64 **15.** $-\frac{8}{9}$ **16.** −35 **17.** 1 **18.** 100 mg **19.** commut. prop. of add. **20.** distrib. prop. **21.** $11y$ **22.** $-45q$ **23.** −4 **24.** $-16 + 2a + 18b$ **25.** −12 **26.** 6 **27.** 0.3 **28.** 0.5 **29.** $i = \frac{f(P - L)}{s}$ **30.** 4 cm by 9 cm **31.** \$4,000 **32.** 10 oz

Study Set Section 2.1 (page 99)

1. ordered **3.** origin **5.** rectangular **7.** origin, right, down **9.** II **11.** −3, −1, 1, 3 **13.** t: horizontal; d: vertical

15–22.

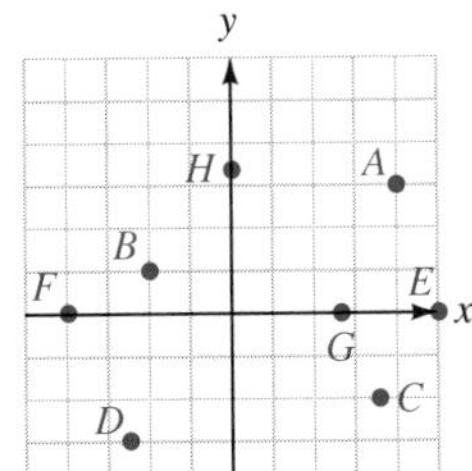

23. (2, 4) **25.** (−2.5, −1.5) **27.** (3, 0) **29.** (0, 0) **31.**

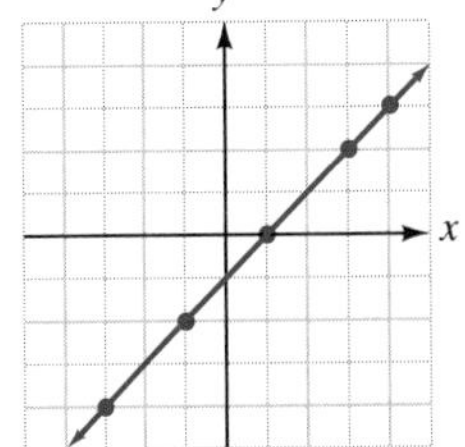

33. a. on the surface **b.** diving **c.** 1,000 ft **d.** 500 ft **35. a.** 1990–1991, 1994–1995 **b.** 1990–1991 **c.** 1993–1996 **d.** Imports exceeded production by about 1.2 million barrels per day. **37.** Jonesville (5, B), Easley (1, B), Hodges (2, E), Union (6, C) **39. a.** (2, −1) **b.** no **c.** yes **41. a.** \$2 **b.** \$4 **c.** \$7 **d.** \$9 **43.** tip of tail, front of engine, tip of wing **47.** 20 **49.** $\frac{1}{2}$ **51.** 0.7

Study Set Section 2.2 (page 111)

1. satisfy **3.** y-intercept **5.** vertical **7. a.** 1, 1 **b.** 2, infinitely many **9.** $3x + 5y = -10$ **11. a.** $(-3, 0)$; $(0, 4)$ **b.** false **13.** -3, -2 **15.** the y-axis **17.** 5, 4, 2 **19.** 0, -1, -2 **21.**

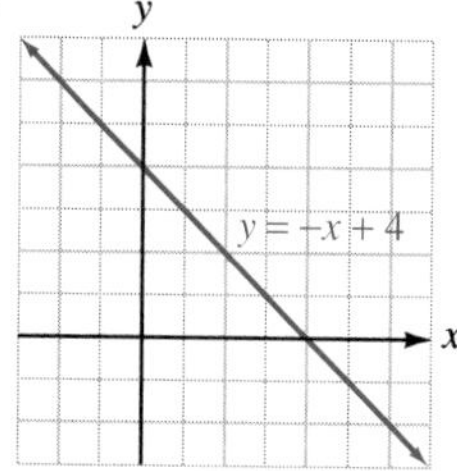

23.

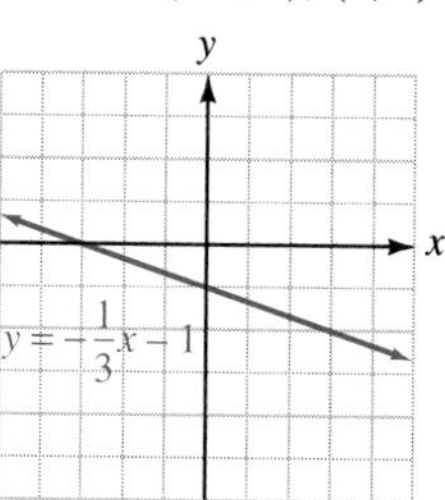

25.

27.

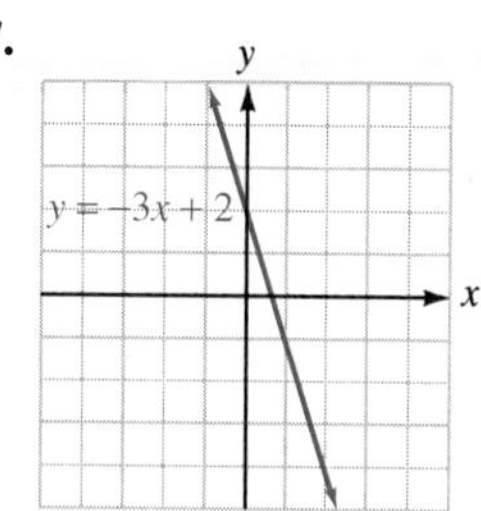

29.

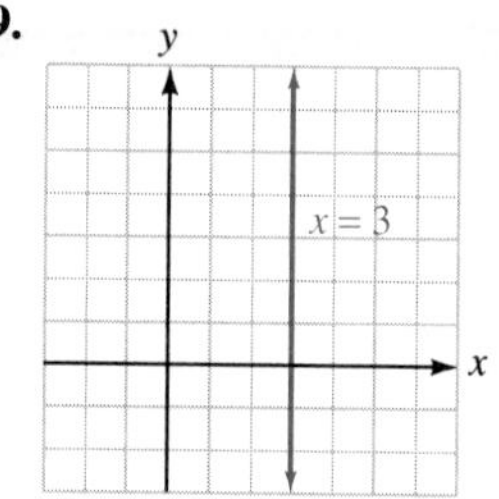

31.

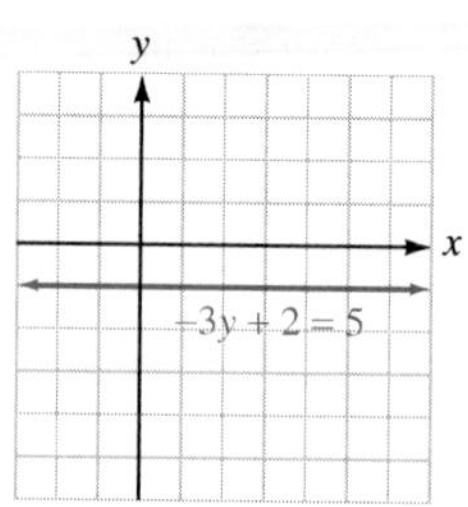

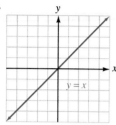

33.

35.

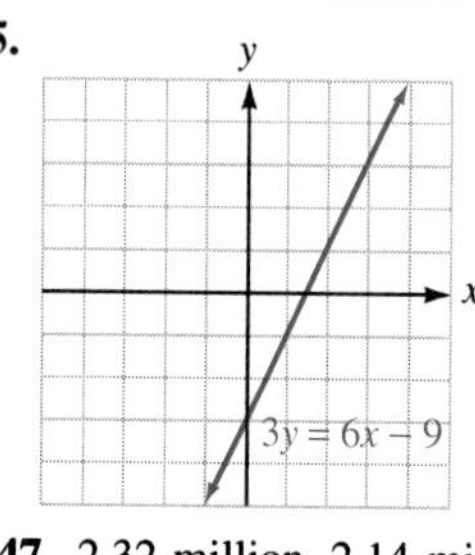

37.

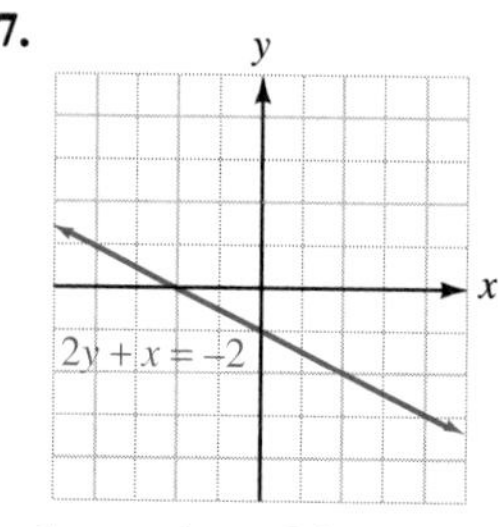

39.

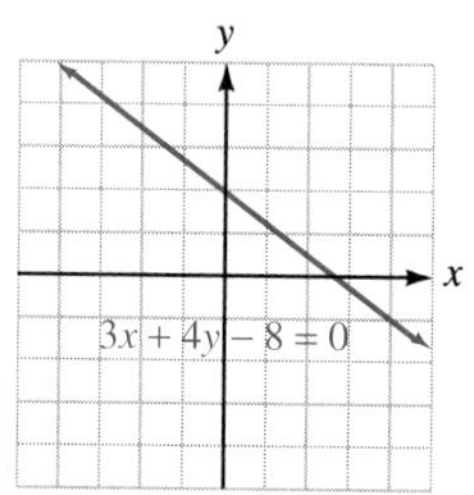

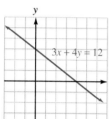

41. 1.22 **43.** 4.67 **45.** \$48 **47.** 2.32 million, 2.14 million; the number of farms is steadily decreasing **49.** \$162,500 **51.** 200 **53. a.** $c = 5t + 2$ **b.** 7, 12, 17, 22 **c.** \$27

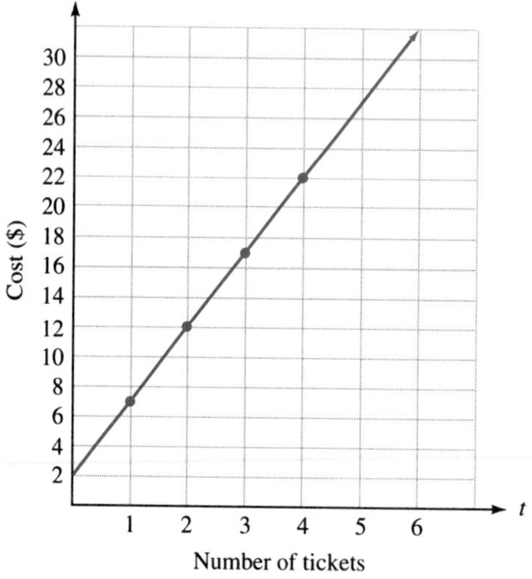

57. 11, 13, 17, 19, 23, 29 **59.** III **61.** $80s$ **63.** $3x + 8$

Study Set Section 2.3 (page 122)

1. slope **3.** change **5.** reciprocals **7. a.** l_3; 0 **b.** l_2; undefined **c.** l_1; 2 **d.** l_4; -3 **9. a.** college, 4 yr or more; about \$25 a week/yr **b.** less than 4 yr h.s.; about \$5 a week/yr **11.** $m = \frac{y_2 - y_1}{x_2 - x_1}$ **13. a.** 6 **b.** 8 **c.** $\frac{3}{4}$ **15.** 3 **17.** -1 **19.** $-\frac{1}{3}$ **21.** 0 **23.** undefined **25.** -1 **27.** $-\frac{3}{2}$ **29.** $\frac{3}{4}$ **31.** $\frac{1}{2}$ **33.** 0 **35.** perpendicular **37.** neither **39.** neither **41.** parallel **43.** perpendicular **45.** neither **47.** $\frac{3}{140}$, $\frac{1}{15}$, $\frac{1}{20}$; part 2 **49.** $\frac{1}{10}$; $\frac{1}{4}$ **51.** $\frac{1}{25}$; 4% **53.** brace: $\frac{1}{2}$; support #1: -2; support #2: -1; yes, to support 1 **57.** 40 lb licorice; 20 lb gumdrops

Study Set Section 2.4 (page 133)

1. $y - y_1 = m(x - x_1)$ **3.** perpendicular **5.** no **7.** $m = \frac{2}{3}$; $y + 3 = \frac{2}{3}(x + 2)$ **9.** $m = -\frac{2}{3}$; $(0, 1)$ **11.** yes **13. a.** $(0, 0)$ **b.** none **15.** $\frac{1}{3}x$, 2, 2, 1, $\frac{1}{3}$, -1 **17.** $y = 5x + 7$ **19.** $y = -3x + 6$ **21.** $y = x$ **23.** $y = \frac{7}{3}x - 3$ **25.** $y = \frac{2}{3}x + \frac{11}{3}$ **27.** $y = 3x + 17$

29. $y = -7x + 54$ **31.** $y = -4$ **33.** $y = -\frac{1}{2}x + 11$ **35.** $\frac{3}{2}$, $(0, -4)$ **37.** $-\frac{1}{3}$, $\left(0, -\frac{5}{6}\right)$

39. 1, $(0, -1)$

41. $\frac{2}{3}$, $(0, 2)$

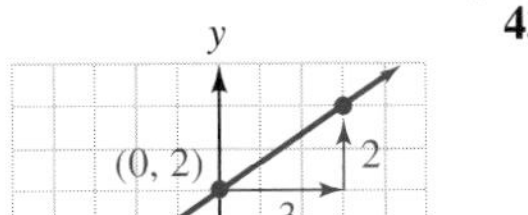

43. $-\frac{3}{4}$, $(0, -2)$

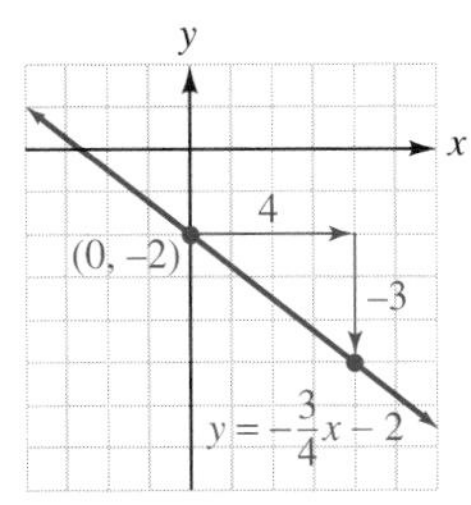

45. parallel **47.** perpendicular **49.** neither **51.** perpendicular **53.** $y = 4x$ **55.** $y = 4x - 3$ **57.** $y = \frac{4}{5}x - \frac{26}{5}$ **59.** $y = -\frac{1}{4}x$ **61.** $y = -\frac{1}{4}x + \frac{11}{2}$ **63.** $y = -\frac{5}{4}x + 3$ **65.** $y = -\frac{950}{3}x + 1{,}750$ **67.** $y = 1{,}811{,}250x + 36{,}225{,}000$ **69. a.** $b = \frac{1}{100}p - 195$ **b.** 905 **71. a.** $E = -\frac{1}{2}t + \frac{21}{2}$ **b.** The number of errors is reduced by 1 for every 2 trials. **c.** On the 21st trial, the rat should make no errors. **73. a.** $y = 1.35x - 31.25$ **b.** When the actual temperature is 0°F, the wind-chill temperature is about −31°F. **79.** \$29,100

Study Set Section 2.5 (page 146)

1. function, one **3.** range **5.** x **7.** y **9.** $f(-1)$ **11. a.**

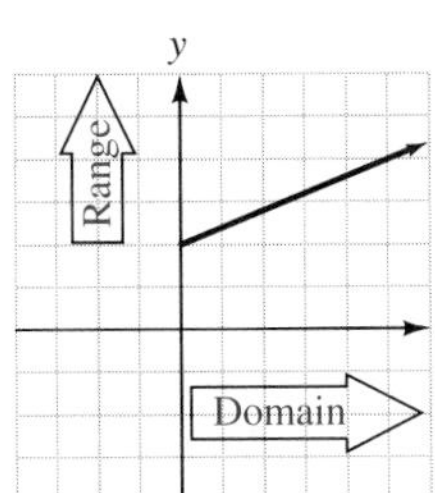

b. D = all real numbers greater than or equal to 0; R = all real numbers greater than or equal to 2 **13.** −5, 25 **15.** of **17.** yes **19.** yes **21.** yes **23.** no **25.** 9, −3 **27.** 3, −5 **29.** 22, 2 **31.** 3, 11 **33.** 4, 9 **35.** 7, 26 **37.** 9, 16 **39.** 6, 15 **41.** 4, 4 **43.** 2, 2 **45.** $\frac{1}{5}$, 1 **47.** $-2, \frac{2}{5}$ **49.** 3.7, 1.1, 3.4 **51.** $-\frac{27}{64}, \frac{1}{216}, \frac{125}{8}$ **53.** $2w$, $2w + 2$ **55.** $3w - 5$, $3w - 2$ **57.** D = {−2, 4, 6}, R = {3, 5, 7} **59.** D = the set of all real numbers except 4, R = the set of all real numbers except 0 **61.** not a function **63.** a function

65.

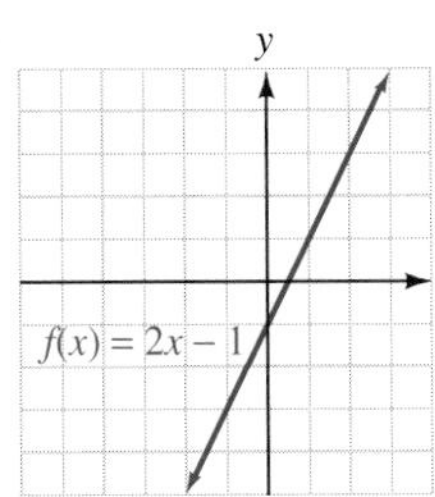

D = the set of all real numbers; R = the set of all real numbers

67.

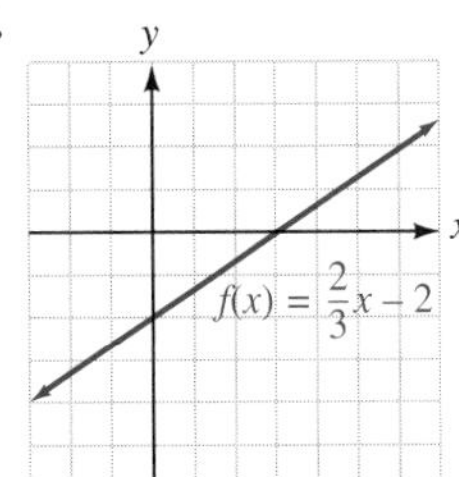

D = the set of all real numbers; R = the set of all real numbers

69. no **71.** yes **73.** between 20°C and 25°C **75. a.** $I(b) = 1.75b - 50$ **b.** \$142.50 **77. a.** (200, 25), (200, 90), (200, 105) **b.** It doesn't pass the vertical line test. **79. a.** 6,595.50; the tax on an income of \$35,000 is \$6,595.50 **b.** D = greater than \$24,650 and less than or equal to \$59,750; R = greater than or equal to \$3,697.50 and less than or equal to \$13,525.50 **c.** $T(a) = 13{,}525.50 + 0.31(a - 59{,}750)$ **81. a.** 624 ft **b.** 0; the bullet strikes the ground 16 seconds after being shot **85.** $\frac{-15}{4}$ **87.** $\frac{1}{3}$

Study Set Section 2.6 (page 157)

1. squaring **3.** absolute value **5.** vertical **7.** reflection **9.** 4, left **11.** 5, up **13. a.** 2 **b.** 0 **15. a.** −1 **b.** 2 **c.** −1

17. D = the set of real numbers, R = the set of all real numbers greater than or equal to −3

19. D = the set of real numbers, R = the set of real numbers

21. D = the set of real numbers, R = the set of all real numbers greater than or equal to −2

23.

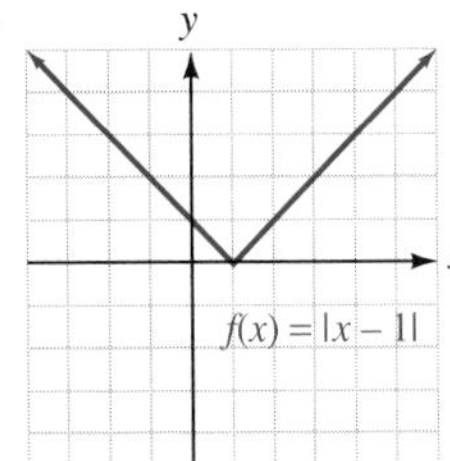

D = the set of real numbers, R = the set of real numbers greater than or equal to 0

25. **27.** 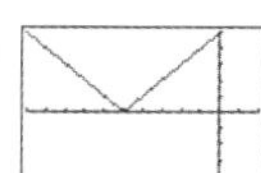 **29.** 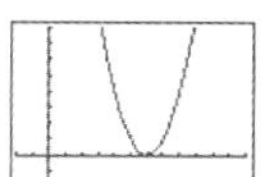 **31.** 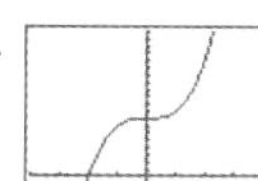

33. 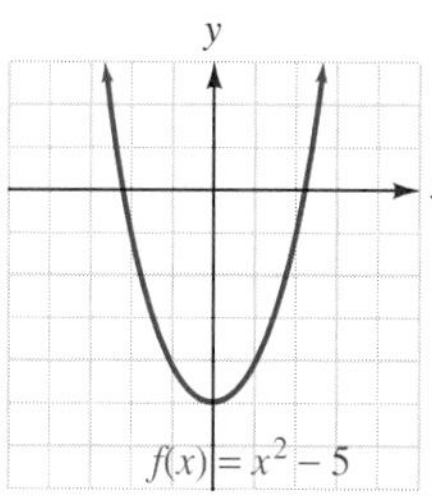

35. 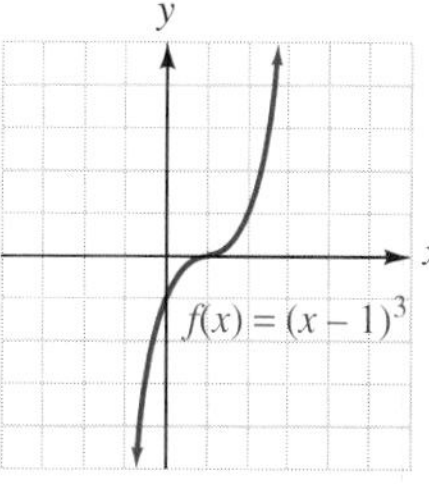

37. 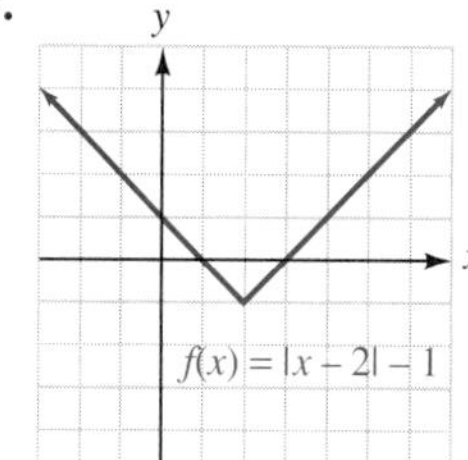

39. 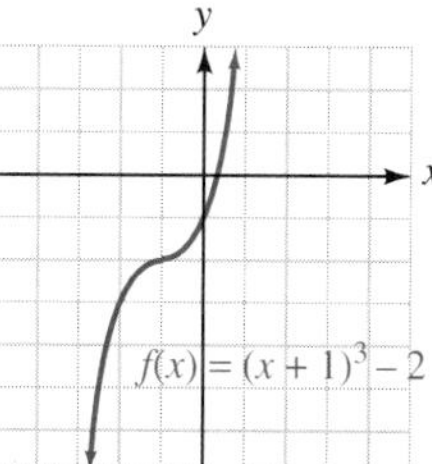

41. 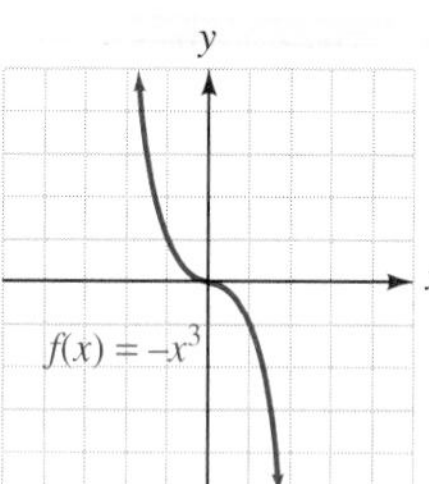

43. 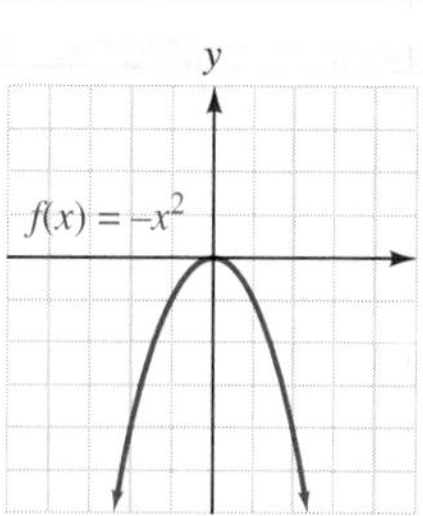

45. 4 **47.** -3 **49.**

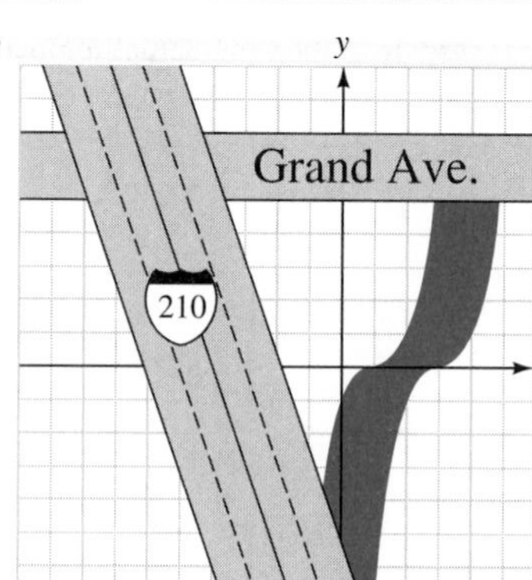

51. a parabola

55. $W = T - ma$ **57.** $g = \dfrac{2(s - vt)}{t^2}$

Key Concept (page 162)

1. a. correspondence, input, range, one, domain **b.** dependent, independent **3.** -17 **4.** $A = \pi r^2$ **5.** 60 ft **6.** 2 **7.** $f(x) = 2x + 3$, 3 **8.** $A(r) = \pi r^2$ **9.** 0; the projectile will strike the ground 4 seconds after being shot into the air **10.** -2, 4

Chapter Review (page 164)

1. 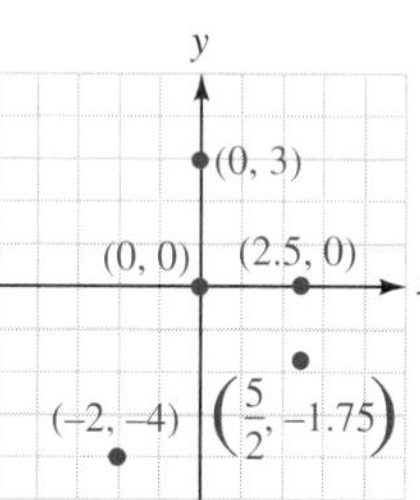

2. $-5, -5$; 0, 0; $-1, -1$; 4, 4; 3, 3; 2, 2; 3, 3 **3. a.** 1 ft below its normal level **b.** decreased by 3 ft **c.** from day 3 to the beginning of day 4 **4. a.** \$10 increments **b.** \$800

5. a. 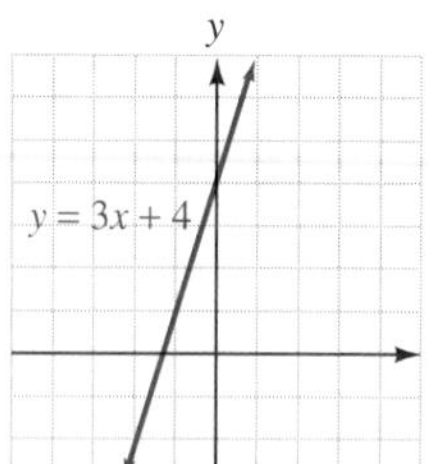

b. 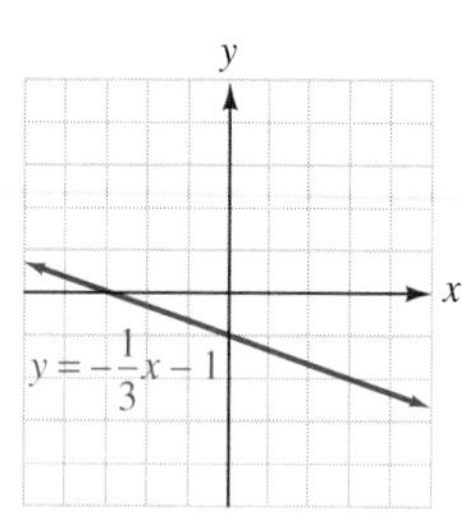

6. a. 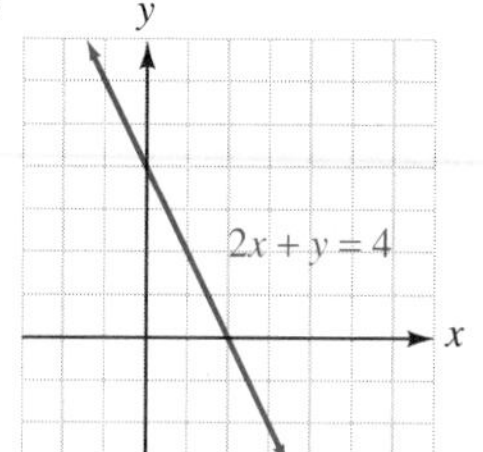

b. 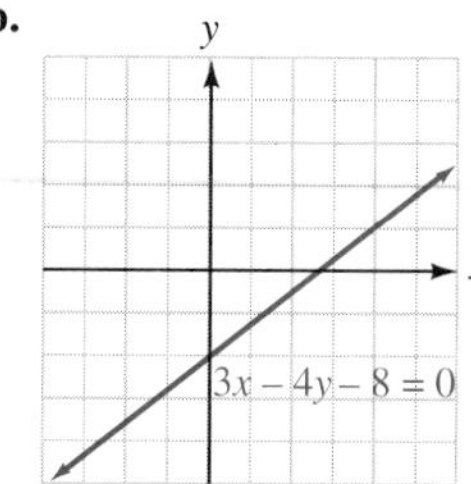

7. a. 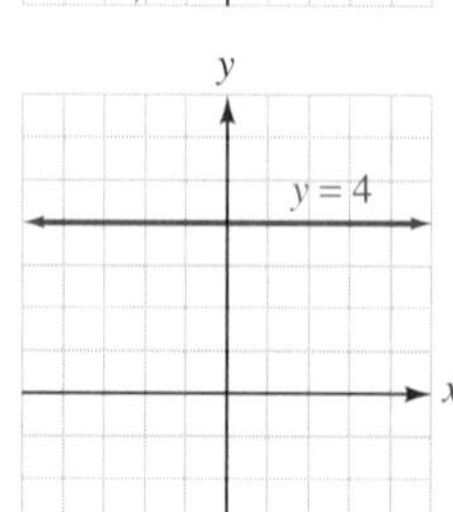

b. 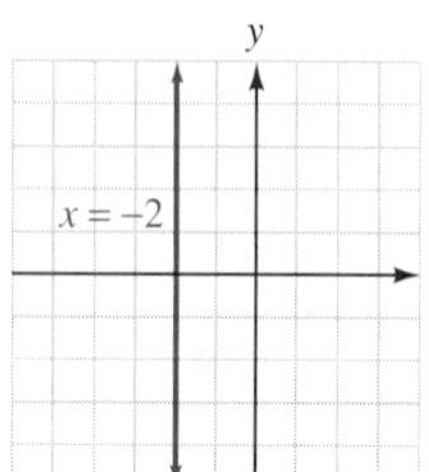 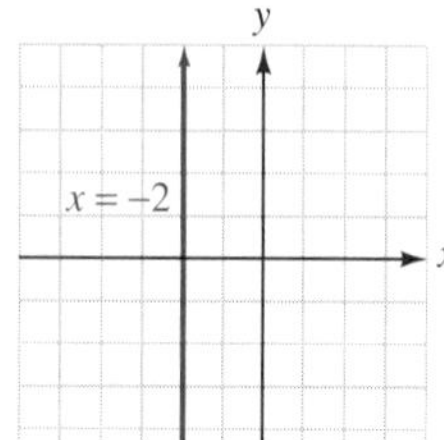

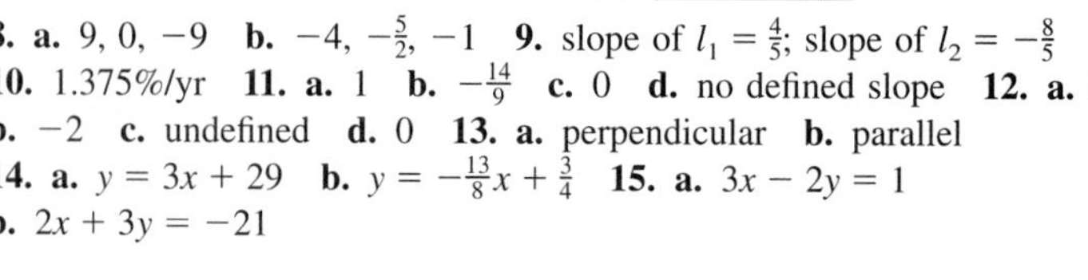

8. a. 9, 0, -9 **b.** -4, $-\frac{5}{2}$, -1 **9.** slope of $l_1 = \frac{4}{5}$; slope of $l_2 = -\frac{8}{5}$ **10.** 1.375%/yr **11. a.** 1 **b.** $-\frac{14}{9}$ **c.** 0 **d.** no defined slope **12. a.** $\frac{2}{3}$ **b.** -2 **c.** undefined **d.** 0 **13. a.** perpendicular **b.** parallel **14. a.** $y = 3x + 29$ **b.** $y = -\frac{13}{8}x + \frac{3}{4}$ **15. a.** $3x - 2y = 1$ **b.** $2x + 3y = -21$

16. $y = -\frac{3}{4}x - 3$; $m = -\frac{3}{4}$, $(0, -3)$

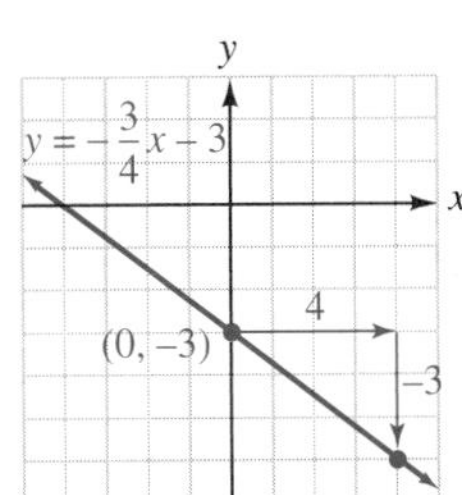

17. $y = -1{,}720x + 8{,}700$ **18. a.** yes **b.** yes **c.** no **d.** no **19. a.** -7 **b.** 18 **c.** 8 **d.** $3t + 2$ **20. a.** D = the set of real numbers, R = the set of real numbers **b.** D = the set of real numbers, R = the set of all real numbers greater than or equal to 1 **c.** D = the set of all real numbers except 2, R = the set of all real numbers except 0 **d.** D = the set of real numbers, R = the set of nonpositive real numbers **21. a.** function **b.** not a function **22.** $f(t) = 1.375t + 21.2$; 48.7% **23. a.** yes **b.** no

24. a.

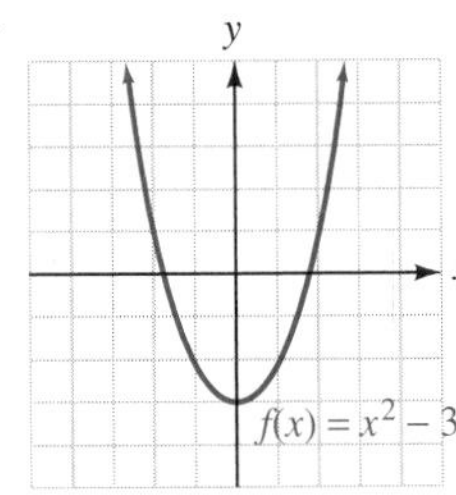

b.

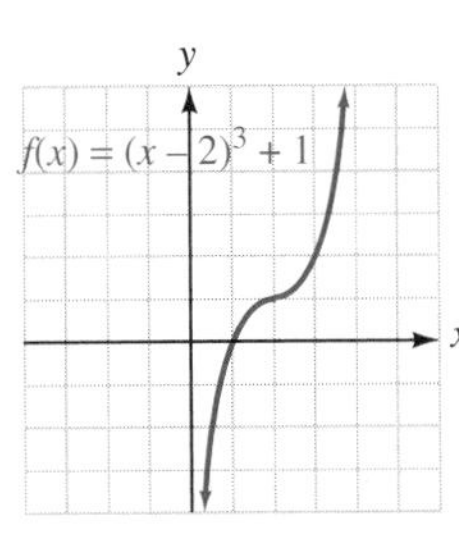

25.

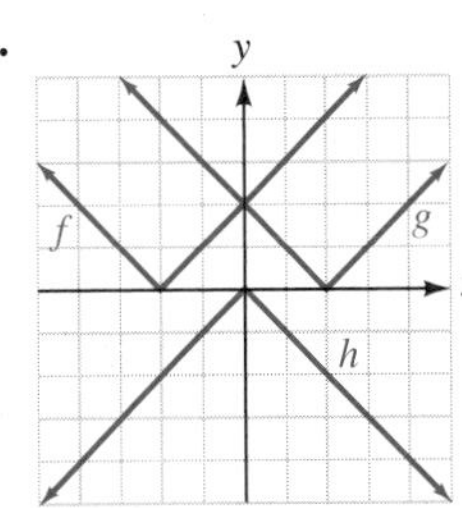

26. a. -4 **b.** 3

Chapter 2 Test (page 171)

1. 240 ft **2.** 1 sec and 7 sec **3.** about 256 ft **4.** 8 sec **5.** $(5, 0)$, $(0, -2)$

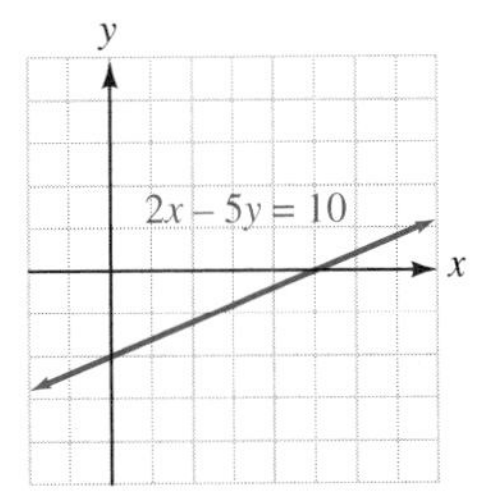

6.

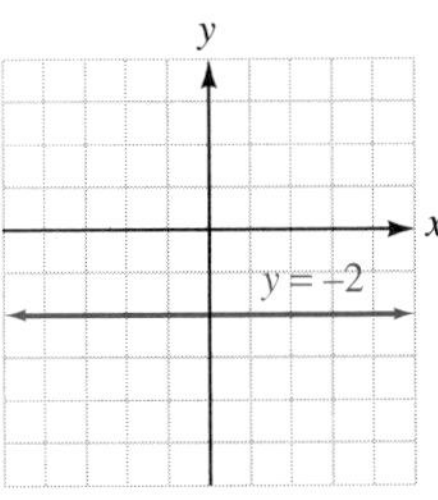

7. 3 **8.** -1.5 degree/hr **9.** $\frac{1}{2}$ **10.** $\frac{2}{3}$ **11.** undefined **12.** 0 **13.** $y = \frac{2}{3}x - \frac{23}{3}$ **14.** $8x - y = -22$ **15.** $m = -\frac{1}{3}$, $\left(0, -\frac{3}{2}\right)$ **16.** neither **17.** $y = \frac{3}{2}x$ **18.** no **19.** D = the set of real numbers, R = the set of nonnegative real numbers **20.** D = the set of real numbers, R = the set of real numbers **21.** 10 **22.** -1 **23.** 3 **24.** $r^2 - 2r - 1$ **25.** function **26.** not a function

27.

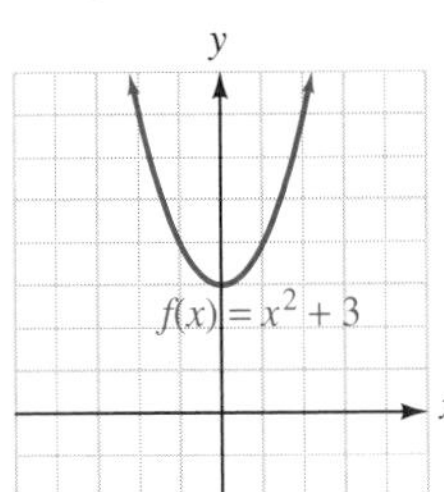

28.

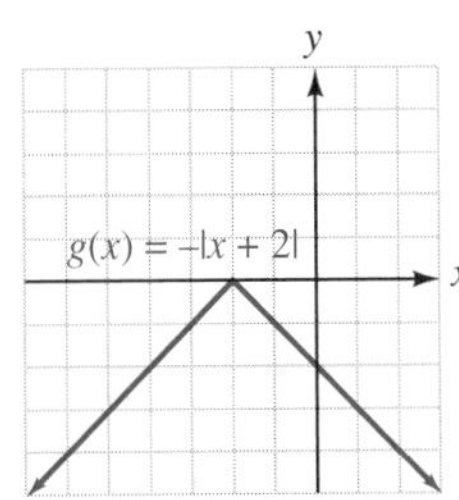

Cumulative Review Exercises (page 175)

1. 1, 2, 6, 7 **2.** 0, 1, 2, 6, 7 **3.** $-2, 0, 1, 2, \frac{13}{12}, 6, 7$ **4.** $\sqrt{5}, \pi$ **5.** -2 **6.** $-2, 0, 1, 2, \frac{13}{12}, 6, 7, \sqrt{5}, \pi$ **7.** 2, 7 **8.** 6 **9.** $-2, 0, 2, 6$ **10.** 1, 7 **11.** -2 **12.** -2 **13.** 22 **14.** -2 **15.** $\frac{24}{25}$ **16.** 2 **17.** 4 **18.** -5 **19.** assoc. prop. of add. **20.** distrib. prop. **21.** commut. prop. of add. **22.** assoc. prop. of mult. **23.** $-5y$ **24.** $-28st$ **25.** 0 **26.** $z - 4$ **27.** 8 **28.** -27 **29.** -1 **30.** 6 **31.** $\frac{8}{3}$ **32.** 24 **33.** $a = \frac{2S}{n} - l$ or $a = \frac{2S - ln}{n}$ **34.** $h = \frac{2A}{b_1 + b_2}$ **35.** \$14,000 **36.** 39 mph going, 65 mph returning

37.

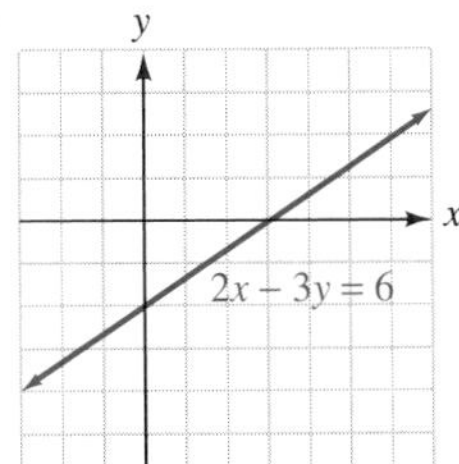

38. $-\frac{5}{6}$ **39.** $y = -\frac{7}{5}x + \frac{11}{5}$ **40.** $y = -3x - 3$ **41.** 5 **42.** -1 **43.** 3 **44.** $3r^2 + 2$

45. a function; D = the set of real numbers, R = the set of all real numbers less than or equal to 1

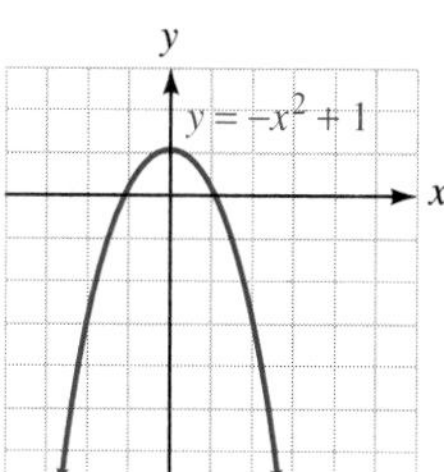

46. a function; D = the set of real numbers, R = the set of nonnegative real numbers

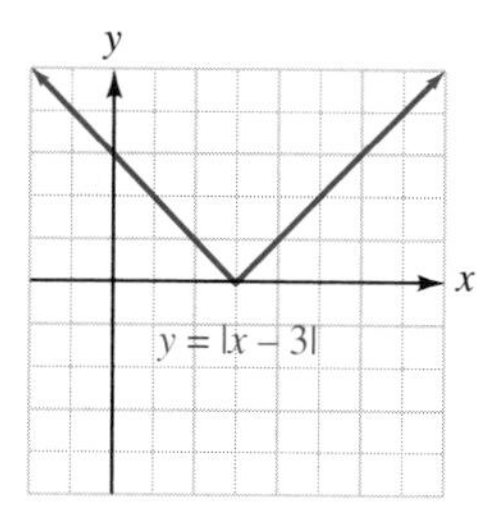

Study Set Section 3.1 (page 183)

1. system **3.** inconsistent **5.** dependent **7. a.** true **b.** false **c.** true **d.** true **9. a.** −4, (−4, 0), 2, (0, 2), 3, (2, 3) **b.** (−4, 0); (0, 2) **11.** yes **13.** no **15.**

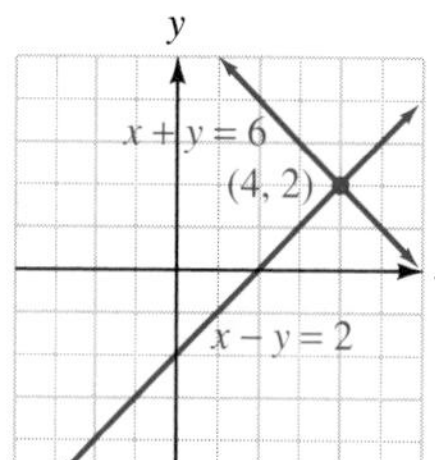

17.

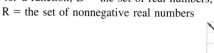

y
(−1, 3)
$x - 2y = -7$
x
$2x + y = 1$

19.

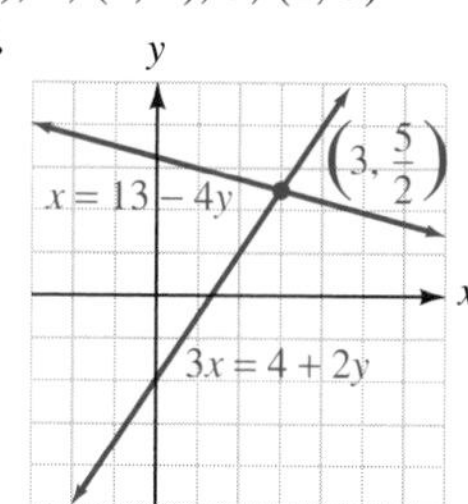

21.

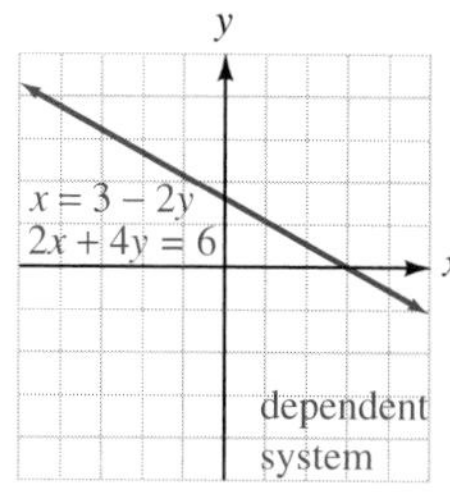

23.

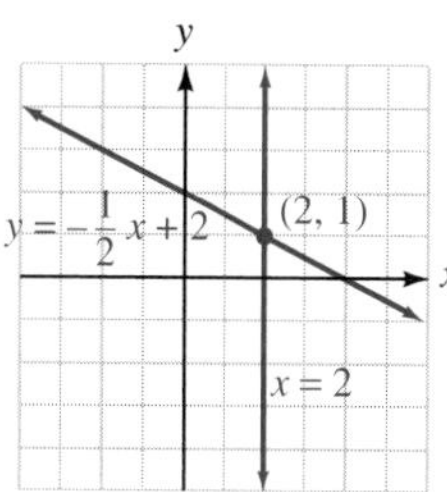

25.

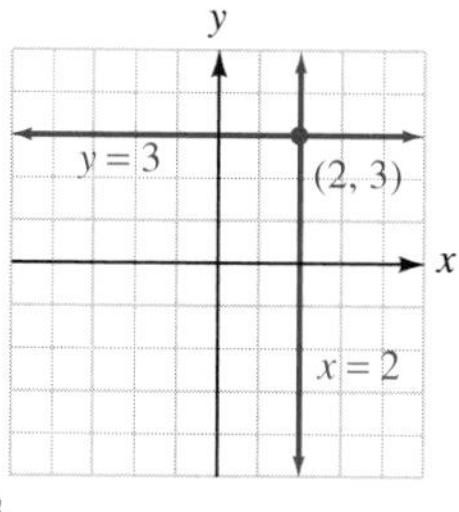

27.

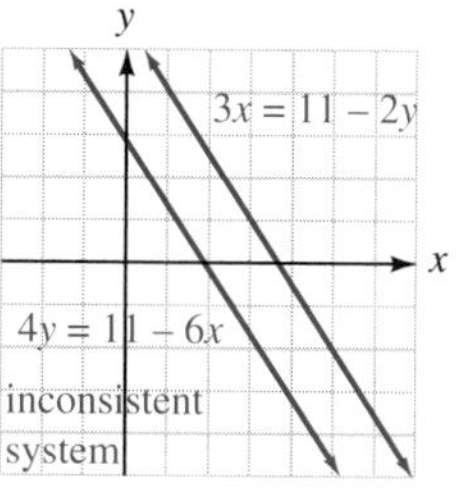

29.

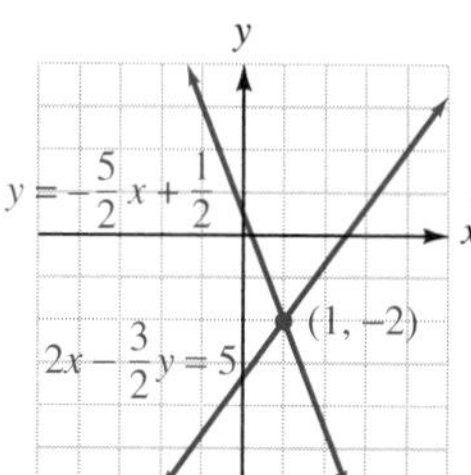

31.

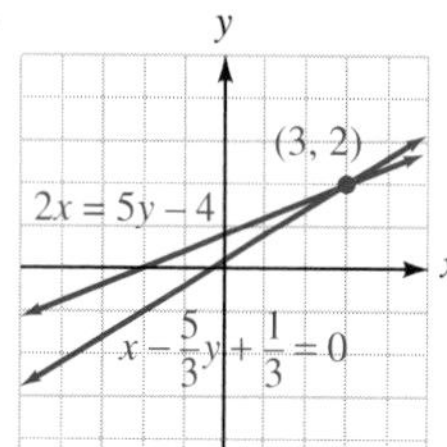

33.

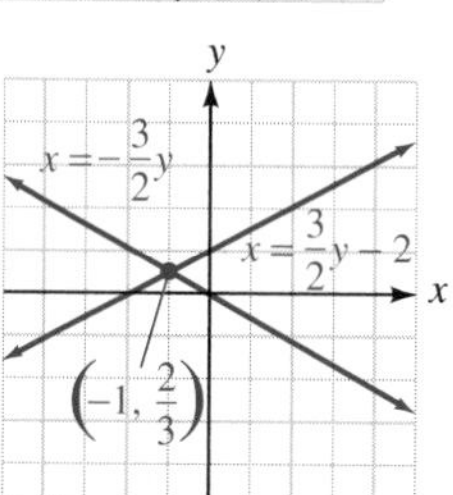

35. (−0.37, −2.69) **37.** (−7.64, 7.04)

39. a. (95, 64) **b.** In 1995, the number of take-out meals and on-premise meals were the same: 64

41. a.

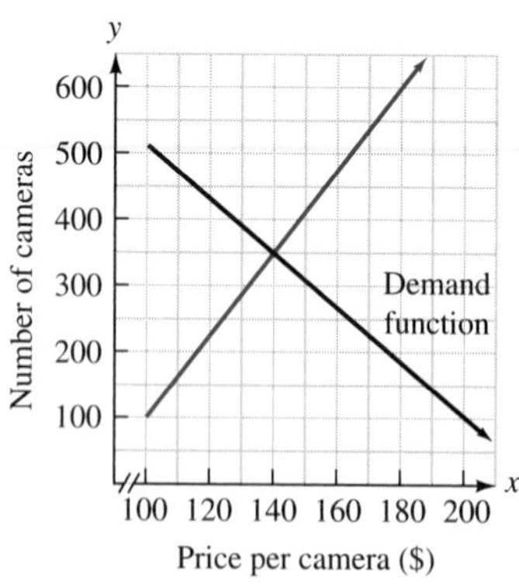

b. \$140 **c.** Supply increases, and demand decreases **43. a.** yes **b.** (3.75, −0.5) **c.** no **47.** −3 **49.** 0 **51.** D = the set of real numbers, R = the set of all real numbers greater than or equal to −2

Study Set Section 3.2 (page 196)

1. general **3.** eliminated **5. a.** 3; −4 (answers may vary) **b.** 2; −3 (answers may vary) **7.** addition method **9. a.** ii **b.** iii **c.** i **11.** (2, 2) **13.** (5, 3) **15.** (−2, 4) **17.** no solution **19.** (5, 2) **21.** (−4, −2) **23.** (1, 2) **25.** $\left(\frac{1}{2}, \frac{2}{3}\right)$ **27.** $\left(5, \frac{3}{2}\right)$ **29.** $\left(-2, \frac{3}{2}\right)$ **31.** dependent equations **33.** no solution **35.** (4, 8) **37.** (20, −12) **39.** $\left(\frac{2}{3}, \frac{3}{2}\right)$ **41.** $\left(\frac{1}{2}, -3\right)$ **43.** (2, 3) **45.** $\left(-\frac{1}{3}, 1\right)$ **47.** \$475, \$800 **49.** dogs: 58 million; cats: 66 million **51.** 16 m by 20 m **53.** 75°, 25° **55.** \$3,000 at 10%, \$5,000 at 12%

57. 148 g of the 0.2%, 37 g of the 0.7% **59.** 45 mi **61.** 85 racing bikes, 120 mountain bikes **63.** 4,031 **65.** $4{,}666\frac{2}{3}$ books **67.** 103 **69. a.** 590 units per month **b.** 620 units per month **c.** A (smaller loss) **75.** $-\frac{5}{2}$ **77.** $-\frac{8}{5}$ **79.** $\frac{4}{3}$

Study Set Section 3.3 (page 208)

1. system **3.** three **5.** dependent **7. a.** no solution **b.** no solution **9.** $x + 2y - 3z = -6$ **11.** yes **13.** (1, 1, 2) **15.** (0, 2, 2) **17.** (3, 2, 1) **19.** inconsistent system **21.** (60, 30, 90) **23.** (2, 4, 8) **25.** dependent equations **27.** (2, 6, 9) **29.** 30 expensive, 50 middle-priced, 100 inexpensive **31.** 2, 3, 1 **33.** 3 poles, 2 bears, 4 deer **35.** 70%, 17%, 13% **37.** $y = \frac{1}{2}x^2 - 2x - 1$ **39.** $x^2 + y^2 - 2x - 2y - 2 = 0$ **41.** $A = 40°$, $B = 60°$, $C = 80°$ **43.** 12, 15, 21

47.

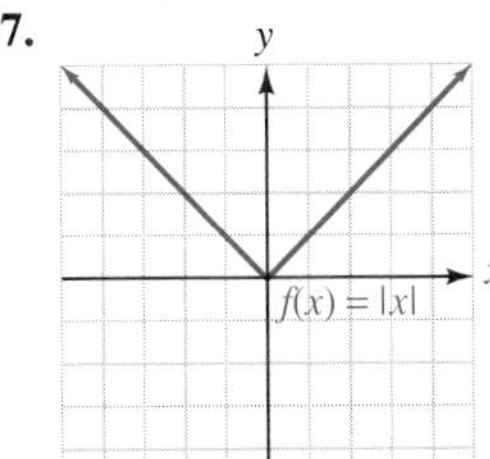

49.

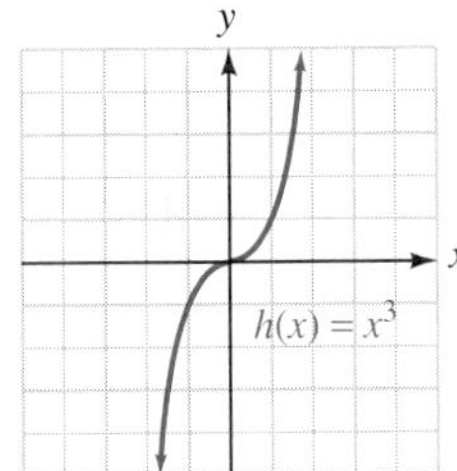

Study Set Section 3.4 (page 217)

1. matrix **3.** rows, columns **5.** augmented **7. a.** 2×3 **b.** 3×4 **9.** $\begin{cases} x - y = -10 \\ y = 6 \end{cases}$; (−4, 6) **11.** It is inconsistent.

13. a. multiply row 1 by $\frac{1}{3}$; $\left[\begin{array}{ccc|c} 1 & 2 & -3 & 0 \\ 1 & 5 & -2 & 1 \\ -2 & 2 & -2 & 5 \end{array}\right]$ **b.** to row 2, add −1 times row 1; $\left[\begin{array}{ccc|c} 1 & 2 & -3 & 0 \\ 0 & 3 & 1 & 1 \\ -2 & 2 & -2 & 5 \end{array}\right]$

15. −1, 1, −5, 2, *y*, 4 **17.** (1, 1) **19.** (2, −3) **21.** (−1, −1) **23.** (0, −3) **25.** (1, 2, 3) **27.** (4, 5, 4) **29.** (2, 1, 0) **31.** (−1, −1, 2) **33.** inconsistent **35.** dependent **37.** (0, 1, 3) **39.** inconsistent **41.** (−4, 8, 5) **43.** dependent equations **45.** 76°, 104° **47.** founder's circle: 100; box seats: 300; promenade: 400 **51.** $m = \frac{y_2 - y_1}{x_2 - x_1}$ $(x_2 \neq x_1)$ **53.** $y - y_1 = m(x - x_1)$

Study Set Section 3.5 (page 228)

1. determinant **3.** minor **5.** rows, columns **7.** dependent, inconsistent **9.** $ad - bc$ **11.** $\begin{vmatrix} 3 & 4 \\ 2 & -3 \end{vmatrix}$ **13.** $\left(\frac{7}{11}, -\frac{5}{11}\right)$ **15.** 6, 30 **17.** 8 **19.** −2 **21.** 200 **23.** 6 **25.** 1 **27.** 26 **29.** 0 **31.** −79 **33.** (4, 2) **35.** $\left(-\frac{1}{2}, \frac{1}{3}\right)$ **37.** no solution **39.** (2, −1) **41.** (1, 1, 2) **43.** (3, 2, 1) **45.** (3, −2, 1) **47.** $\left(-\frac{1}{2}, -1, -\frac{1}{2}\right)$ **49.** dependent equations **51.** no solution **53.** (−2, 3, 1) **55.** 200 of the \$67 phones, 160 of the \$100 phones **57.** \$5,000 in HiTech, \$8,000 in SaveTel, \$7,000 in OilCo **59.** −23 **61.** 26 **65.** no **67.** yes **69.** *x* **71.** *y*-intercept

Key Concept (page 231)

1. yes **2.** no **3.**

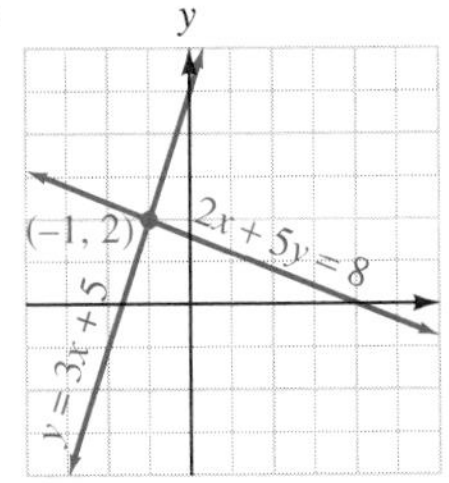

4. $\left(\frac{1}{3}, \frac{1}{4}\right)$ **5.** (3, 2) **6.** (−1, 0, 2) **7.** (15, 2) **8.** (3, 3, 2) **9.** The equations of the system are dependent. **10.** The system is inconsistent.

Chapter Review (page 233)

1. a. (1, 3), (2, 1), (4, −3) **b.** (0, −4), (2, −2), (4, 0) **c.** (3, −1)

2. a.

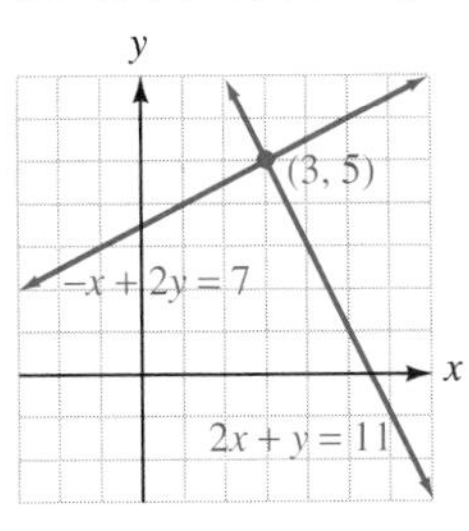

b.

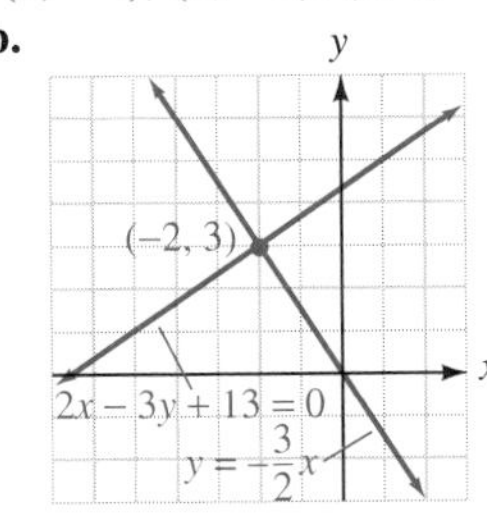

c.

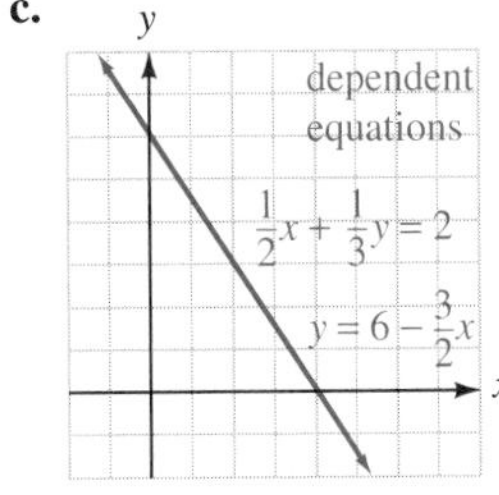

d.

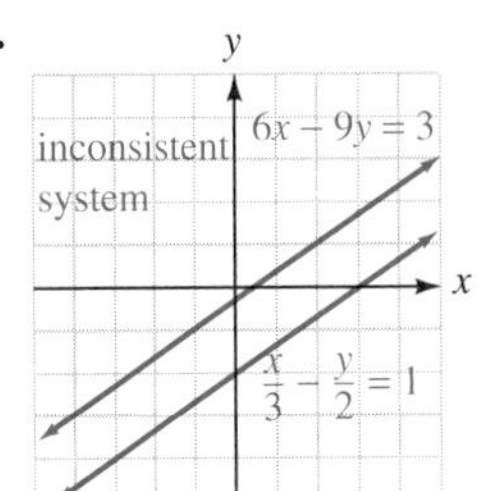

3. a. $(-1, 3)$ **b.** $(-3, -1)$ **c.** $(3, 4)$ **d.** dependent equations **4. a.** $(-3, 1)$ **b.** no solution **c.** $(9, -4)$ **d.** $\left(4, \frac{1}{2}\right)$ **5.** Using the addition method, the computations are easier. **6.** A-H: 162 mi, A-SA: 83 mi **7.** 8 mph, 2 mph **8.** no **9. a.** $(1, 2, 3)$ **b.** inconsistent system **c.** $(-1, 1, 3)$ **d.** dependent equations **10.** 25 lb peanuts, 10 lb cashews, 15 lb Brazil nuts

11. a. $\left[\begin{array}{cc|c} 5 & 4 & 3 \\ 1 & -1 & -3 \end{array}\right]$ **b.** $\left[\begin{array}{ccc|c} 1 & 2 & 3 & 6 \\ 1 & -3 & -1 & 4 \\ 6 & 1 & -2 & -1 \end{array}\right]$ **12. a.** $(1, -3)$ **b.** $(5, -3, -2)$ **c.** dependent system **d.** inconsistent system **13.** \$4,000 at 6%, \$6,000 at 12% **14. a.** 18 **b.** 38 **c.** -3 **d.** 28 **15. a.** $(2, 1)$ **b.** no solution **c.** $(1, -2, 3)$ **d.** $(-3, 2, 2)$ **16.** 2 cups mix A, 1 cup mix B, 1 cup mix C

Chapter 3 Test (page 237)

1.

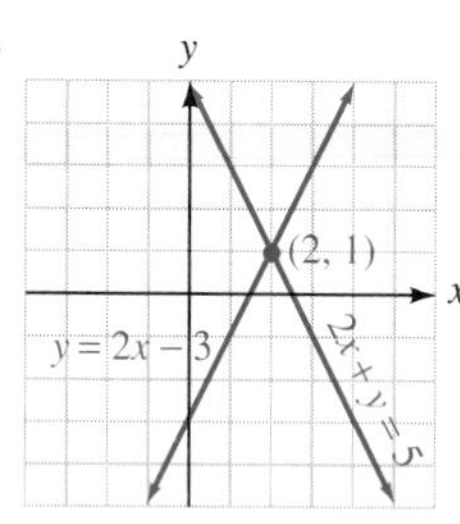

2. $(7, 0)$ **3.** $(2, -3)$ **4.** dependent **5.** no **6.** $(3, 2, -1)$ **7.** 55, 70 **8.** 15 gal 40%, 5 gal 80% **9.** $(2, 2)$ **10.** $(1, 0, -1)$ **11.** 22 **12.** 4 **13.** $\begin{vmatrix} -6 & -1 \\ -6 & 1 \end{vmatrix}$ **14.** $\begin{vmatrix} 1 & -1 \\ 3 & 1 \end{vmatrix}$ **15.** -3 **16.** 3 **17.** -1 **18.** C: 60, GA: 30, S: 10

Study Set Section 4.1 (page 246)

1. inequality **3.** parenthesis **5.** linear **7.** is less than **9. a.** $(4, \infty)$ 4 **b.** $(-\infty, -4)$ −4 **c.** $(-\infty, 4]$ 4 **11. a.** equation **b.** expression **c.** inequality **d.** expression **13. a.** yes **b.** yes **c.** yes **d.** no **15. a.** the number of seriously injured ≤ 16 **b.** the number of references to carpools ≥ 10 **17.** $-10, \leq, -\infty$ **19.** $(-3, \infty)$ −3 **21.** $[20, \infty)$ 20 **23.** $[60, \infty)$ 60 **25.** $\left(-\frac{10}{3}, \infty\right)$ −10/3 **27.** $(-\infty, 1)$ 1 **29.** $\left[-\frac{2}{5}, \infty\right)$ −2/5 **31.** $(6, \infty)$ 6 **33.** $(-\infty, 1.5]$ 1.5 **35.** $(-\infty, 20]$ 20 **37.** $(-\infty, 10)$ 10 **39.** $[-36, \infty)$ −36 **41.** $\left(-\infty, \frac{45}{7}\right]$ 45/7 **43.** Midwest, South **45.** $6 + 45 \not> 52$ **47.** 8 hr **49.** 15 **51.** 13 hr **53.** anything over \$900 **55.** $x < 1$ **57.** $x \geq -4$ **61.** 4, 5, 3

Study Set Section 4.2 (page 255)

1. compound **3.** interval **5.** both **7.** reversed **9. a.** no **b.** yes **11. a.** no solution **b.** $(-\infty, \infty)$ **13. a.** ii **b.** iii **c.** i **15. a.** 2 3 **b.** −2 3 **17.** $3 \not< -3$ **19.** $(-2, 5]$ −2 5 **21.** $(-10, -9)$ −10 −9 **23.** $(2, 3]$ 2 3 **25.** no solution **27.** $[5, \infty)$ 5 **29.** $[1, 4]$ 1 4 **31.** $(8, 11)$ 8 11 **33.** $(-2, 5)$ −2 5 **35.** $[-4, 6)$ −4 6 **37.** $(-6, -3)$ −6 −3 **39.** $[-2, 4]$ −2 4 **41.** $(-\infty, -2] \cup (6, \infty)$ −2 6 **43.** $(-\infty, -1) \cup (2, \infty)$ −1 2 **45.** $(-\infty, 2) \cup (7, \infty)$ 2 7 **47.** $(-\infty, 1)$ 1 **49.** $(-\infty, \infty)$ 0 **51. a.** 128, 192 **b.** $32 \leq s \leq 48$ **53.** See doctor today. **55. a.** 1993, 1995 **b.** 1987, 1989, 1991, 1993, 1995 **c.** 1987, 1989 **d.** 1987, 1989, 1991 **57.** crime: [22.8, 29.2], economy: [5.8, 12.2], jobs: [3.8, 10.2], unemployment: [3.8, 10.2], drugs: [2.8, 9.2] **59.** 85.7, 86, 86 **61.** 13.3 pts/game

Study Set Section 4.3 (page 266)

1. equation **3.** isolate **5.** 0 **7.** more than **9.** 5 **11. a.** yes **b.** no **c.** yes **d.** no **13. a.** ii **b.** iii **c.** i **15.** $|x| < 4$ **17.** $|x + 3| > 6$ **19.** 8 **21.** -0.02 **23.** $-\frac{31}{16}$ **25.** π **27.** 23, -23 **29.** 9.1, -2.9 **31.** $\frac{14}{3}$, -6 **33.** no solution **35.** 2, $-\frac{1}{2}$ **37.** -8 **39.** -4, -28 **41.** 0, -6 **43.** -2, $-\frac{4}{5}$ **45.** 0, -2 **47.** 0 **49.** $\frac{4}{3}$ **51.** $(-4, 4)$ −4 4 **53.** $[-21, 3]$ −21 3 **55.** $\left(-\frac{8}{3}, 4\right)$ −8/3 4 **57.** no solution **59.** $(-\infty, -3) \cup (3, \infty)$ −3 3 **61.** $(-\infty, -12) \cup (36, \infty)$ −12 36 **63.** $\left(-\infty, -\frac{16}{3}\right) \cup (4, \infty)$ −16/3 4 **65.** $(-\infty, \infty)$ 0 **67.** $(-\infty, -2] \cup \left[\frac{10}{3}, \infty\right)$ −2 10/3 **69.** $(-\infty, -2) \cup (5, \infty)$ −2 5 **71.** $[-10, 14]$ −10 14

73. $\left(-\frac{5}{3}, 1\right)$ −5/3 1 **75.** $(-\infty, -24) \cup (-18, \infty)$ −24 −18 **77.** $70° \le t \le 86°$ **79. a.** $|c - 0.6°| \le 0.5°$ **b.** $[0.1°, 1.1°]$ **81. a.** 26.45%, 24.76% **b.** It is less than or equal to 1%. **87.** 50°, 130°

Study Set Section 4.4 (page 274)

1. linear, two **3.** edge **5. a.** yes **b.** no **c.** yes **d.** no **7.** $m = 3$; $(0, -1)$ **9.** no **11. a.** $x \ge 2$ 2

b.

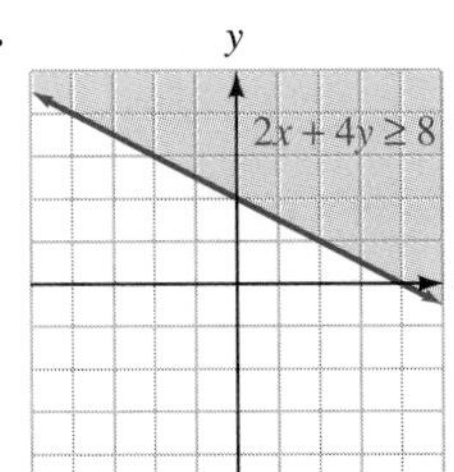

13.

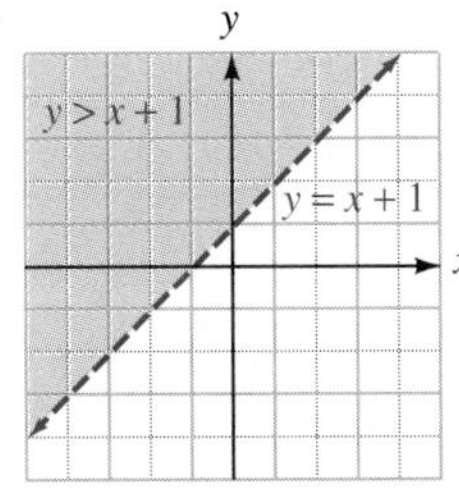

15.

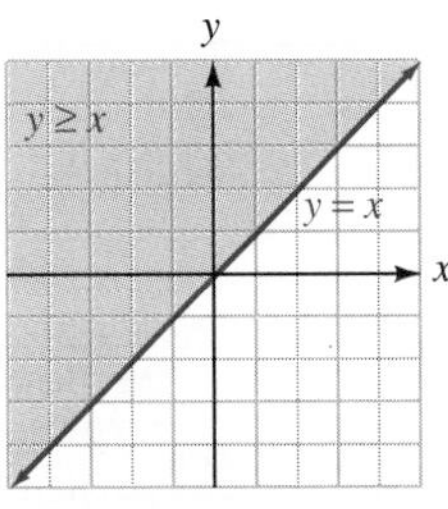

17.

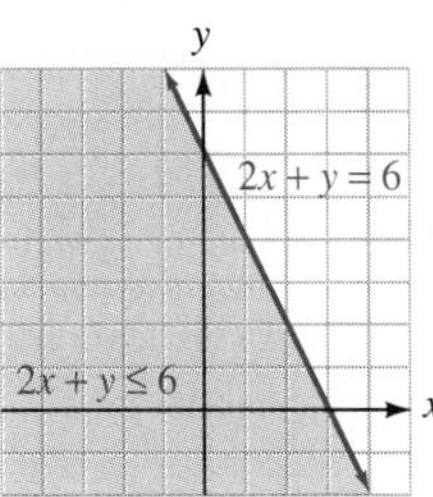

19.

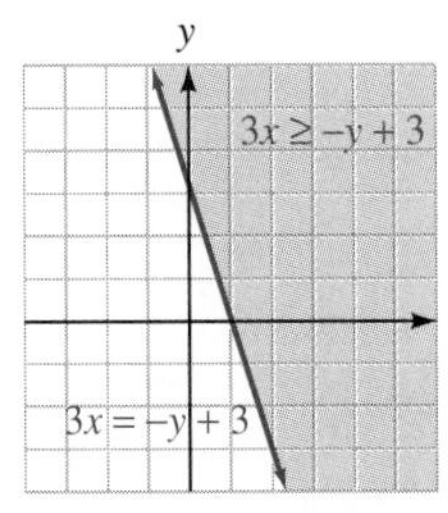

21.

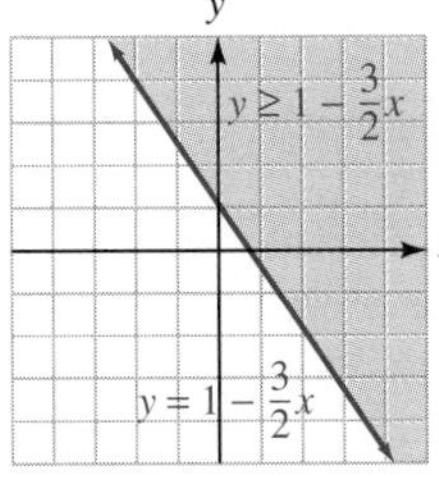

23.

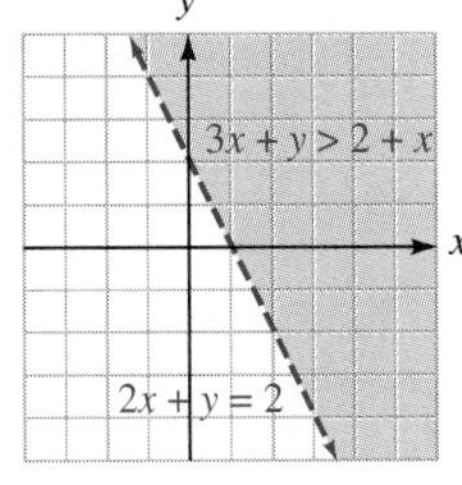

25.

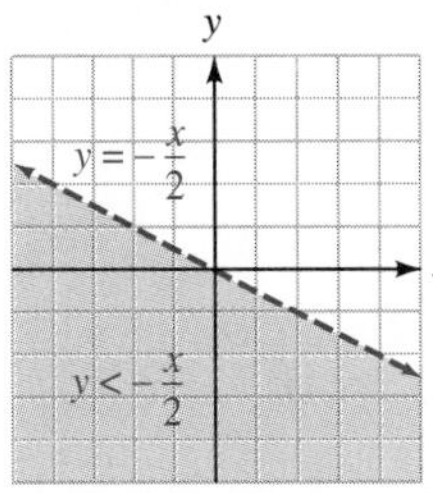

27.

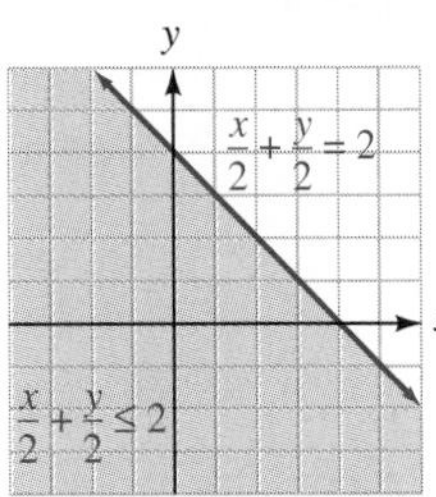

29.

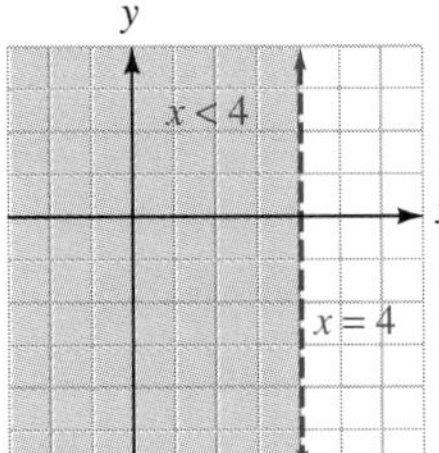

31.

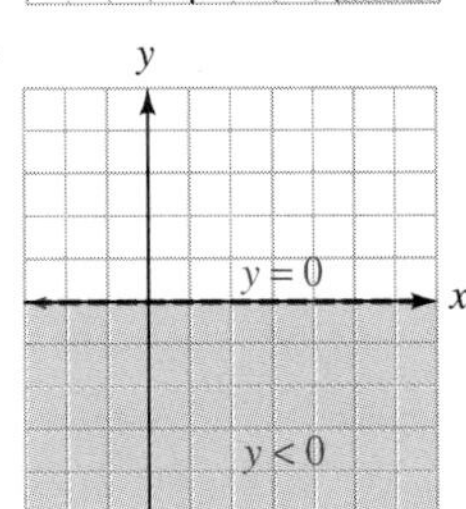

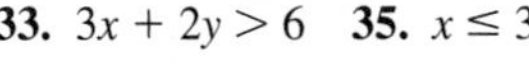

33. $3x + 2y > 6$ **35.** $x \le 3$

37.

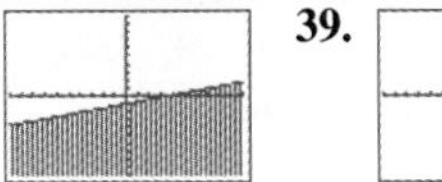

39.

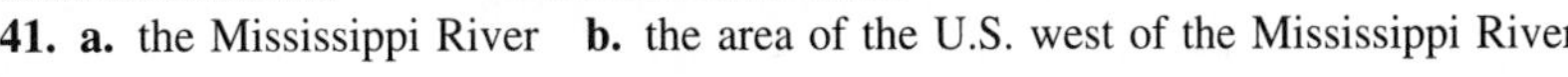

41. a. the Mississippi River **b.** the area of the U.S. west of the Mississippi River

43. (5, 15), (15, 10), (20, 5)

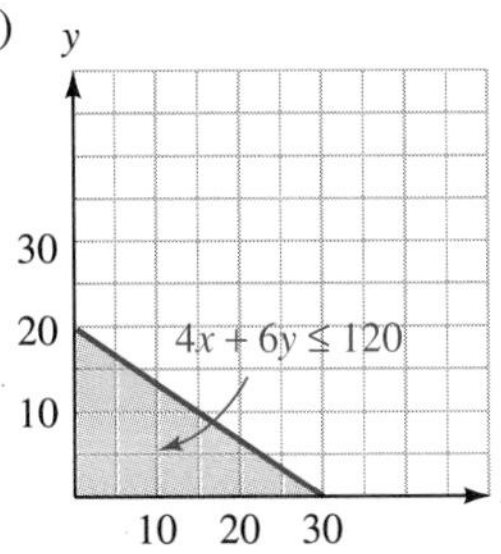

45. (40, 80), (80, 80), (120, 40)

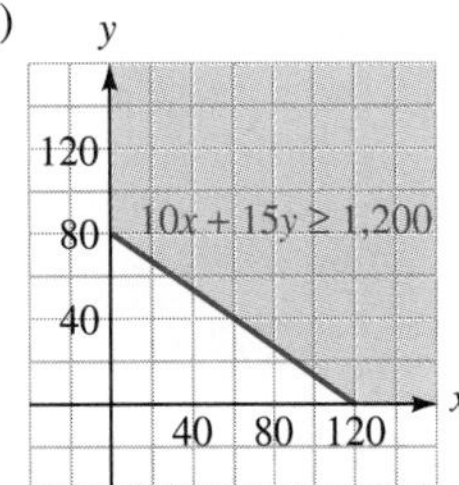

49. yes **51.** (3, 1)

Study Set Section 4.5 (page 282)

1. inequalities **3.** intersect **5. a.** yes **b.** no **c.** no **d.** yes **7. a.** false **b.** true **c.** true **d.** false **e.** true **f.** true

9.

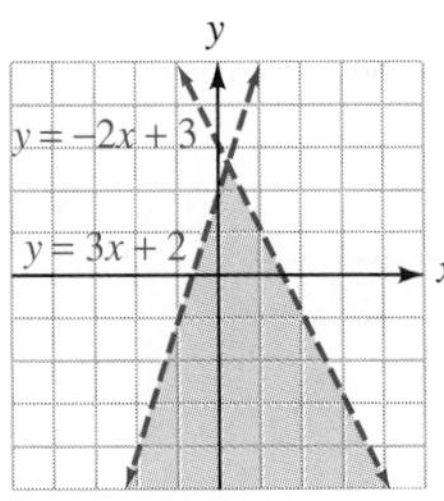

11.

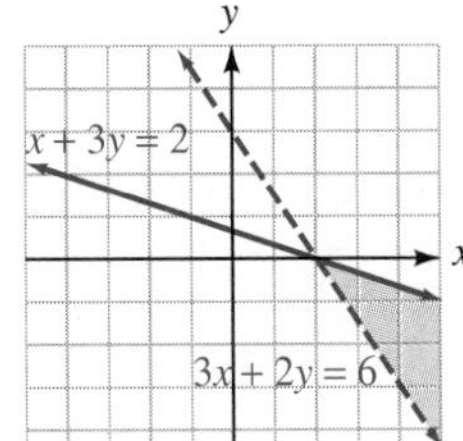

13.

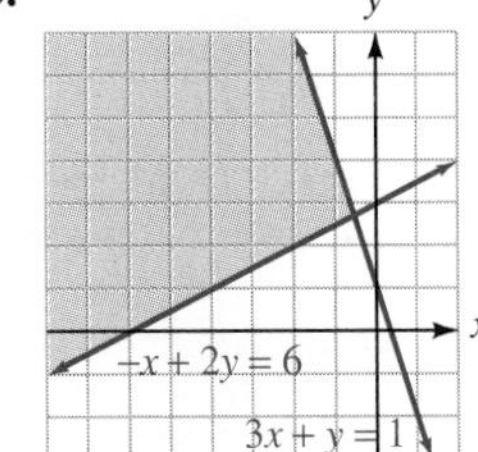

15.

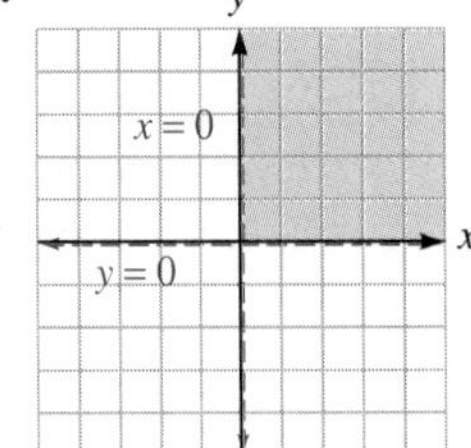

17.

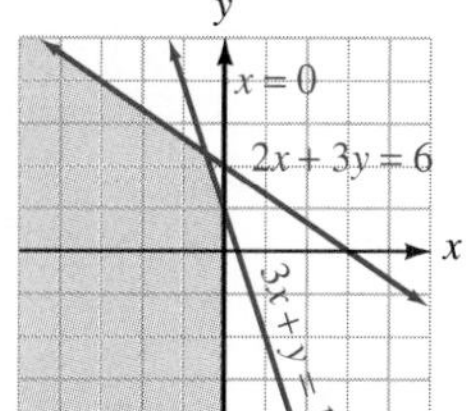

19.

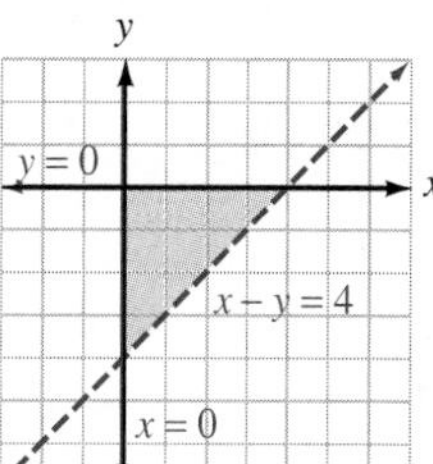

21.

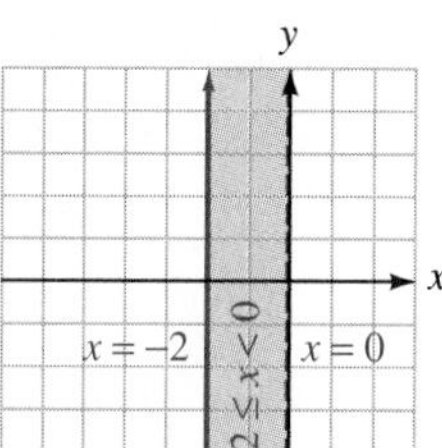

23.

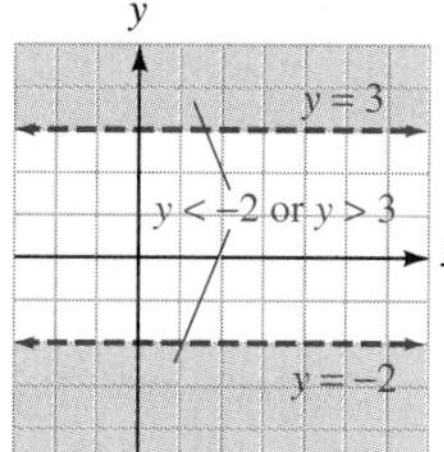

25.

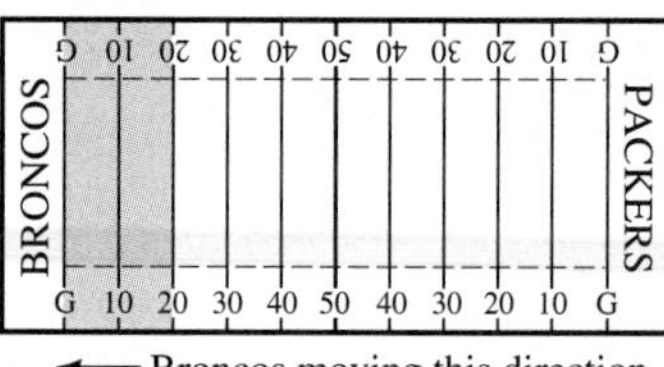

27.

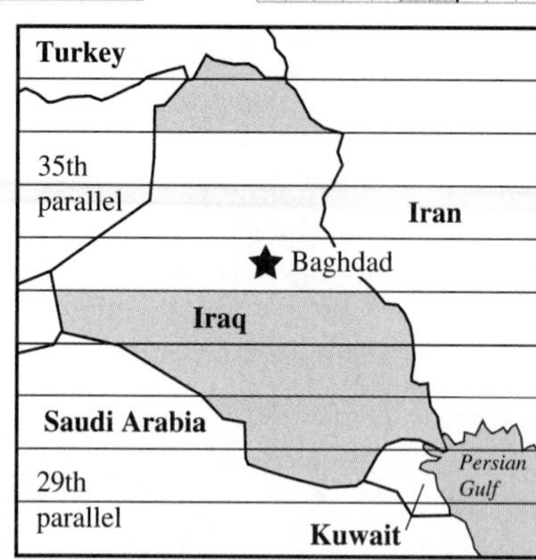

29. 1 \$10 CD and 2 \$15 CDs, 4 \$10 CDs and 1 \$15 CD

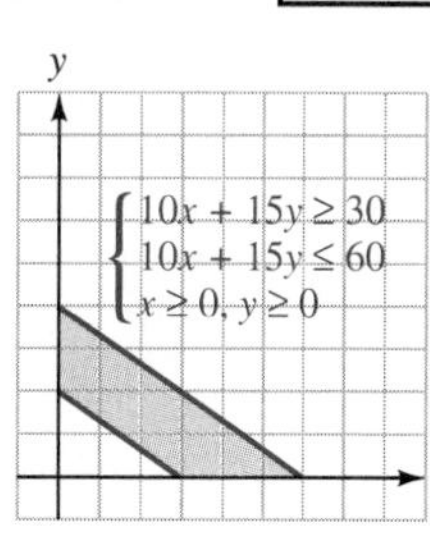

31. 2 desk chairs and 4 side chairs, 1 desk chair and 5 side chairs

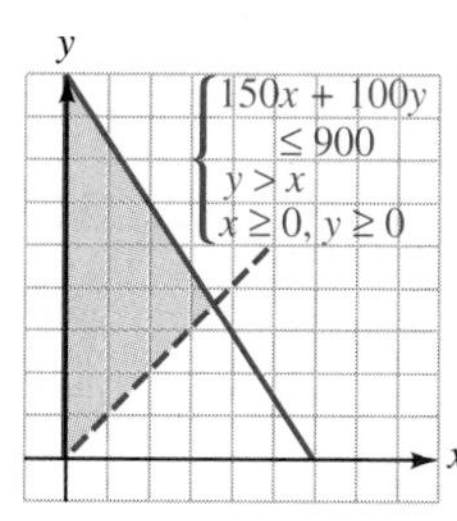

35. IV **37.** II

Key Concept (page 286)

1. a. compound inequality **b.** system of linear inequalities **c.** absolute value inequality **d.** linear inequality in two variables **e.** linear inequality in one variable **f.** double linear inequality **g.** compound inequality **h.** absolute value inequality **i.** linear inequality in two variables **2. a.** yes **b.** no **c.** yes **d.** no **e.** yes **f.** yes **g.** no **h.** no **i.** yes **3.** (number line: 3/2)

4.

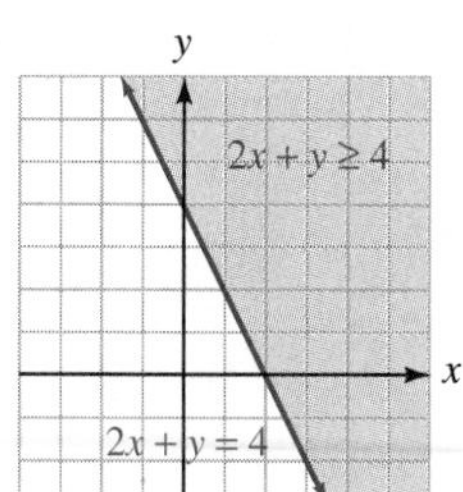

Chapter Review (page 288)

1. a. $(-\infty, 3]$ (3) **b.** $[4, \infty)$ (4) **c.** $(-\infty, 20)$ (20) **d.** $\left(-\infty, -\frac{51}{11}\right)$ (−51/11) **2.** \$20,000 or more **3. a.** yes **b.** no **4. a.** $[-10, -4)$ (−10, −4) **b.** $(-\infty, -11)$ (−11) **5. a.** $\left(-\frac{1}{3}, 2\right)$ (−1/3, 2) **b.** $[1, 9]$ (1, 9) **6. a.** yes **b.** no **7. a.** $(-\infty, -5) \cup (4, \infty)$ (−5, 0, 4) **b.** $(-\infty, \infty)$ (0) **8.** $17 \le 4l \le 25$, $4.25 \text{ ft} \le l \le 6.25 \text{ ft}$ **9. a.** 7 **b.** $\frac{5}{16}$ **c.** −71.05 **d.** −12 **10. a.** 2, −2 **b.** 3, $-\frac{11}{3}$ **c.** $\frac{26}{3}, -\frac{10}{3}$ **d.** 14, −10 **e.** $\frac{1}{5}$, −5 **f.** $\frac{13}{12}$ **11. a.** $[-3, 3]$ (−3, 3) **b.** $(-5, -2)$ (−5, −2) **c.** $\left[-3, \frac{19}{3}\right]$ (−3, 19/3) **d.** no solutions **e.** $(-\infty, -1) \cup (1, \infty)$ (−1, 0, 1) **f.** $(-\infty, -4] \cup \left[\frac{22}{5}, \infty\right)$ (−4, 22/5) **g.** $\left(-\infty, \frac{4}{3}\right) \cup (4, \infty)$ (4/3, 0, 4) **h.** $(-\infty, \infty)$ (0) **12. a.** $|w - 8| \le 2$ **b.** $[6, 10]$

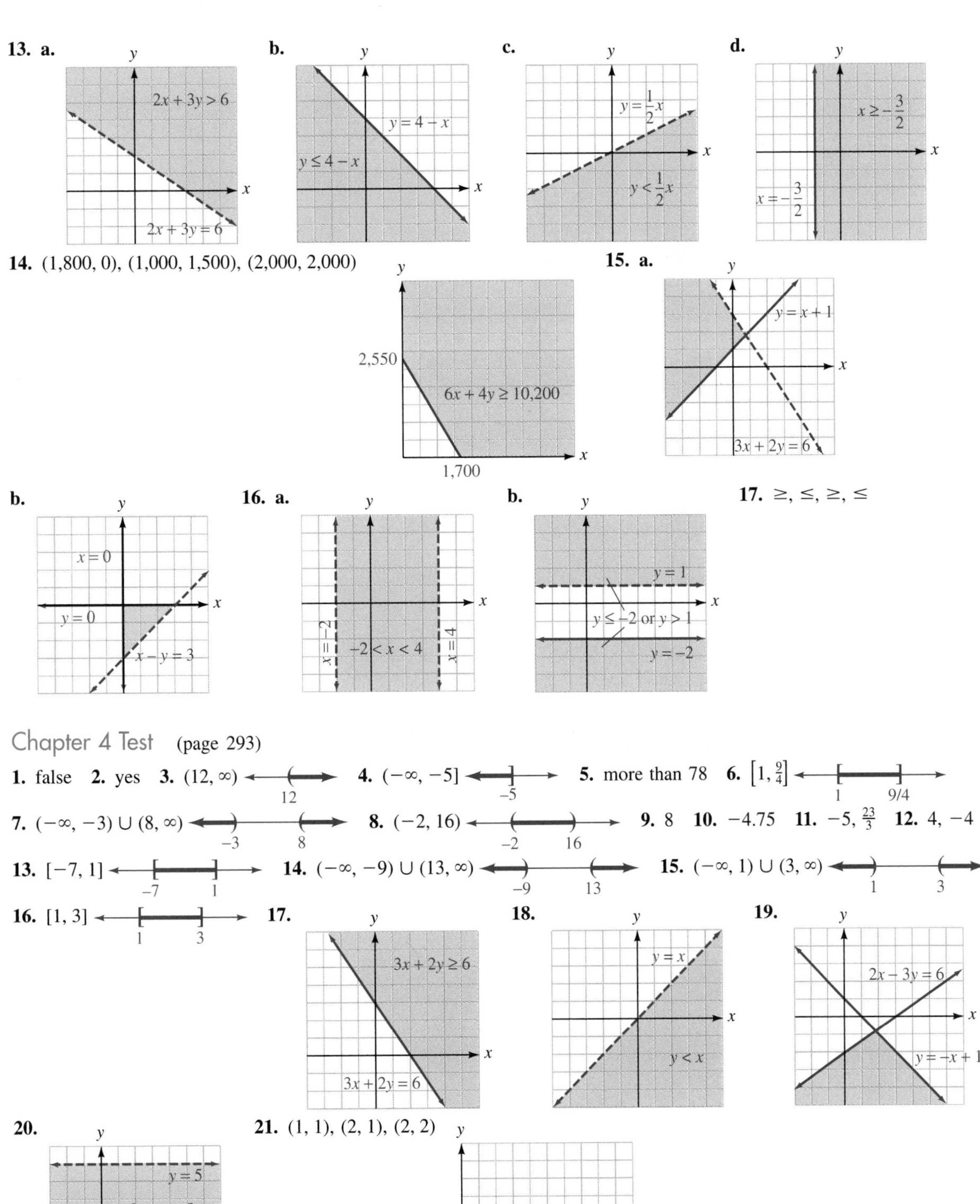

Chapter 4 Test (page 293)

1. false **2.** yes **3.** $(12, \infty)$ **4.** $(-\infty, -5]$ **5.** more than 78 **6.** $\left[1, \frac{9}{4}\right]$ **7.** $(-\infty, -3) \cup (8, \infty)$ **8.** $(-2, 16)$ **9.** 8 **10.** -4.75 **11.** $-5, \frac{23}{3}$ **12.** 4, -4 **13.** $[-7, 1]$ **14.** $(-\infty, -9) \cup (13, \infty)$ **15.** $(-\infty, 1) \cup (3, \infty)$ **16.** $[1, 3]$ **17.** **18.** **19.** **20.** **21.** (1, 1), (2, 1), (2, 2)

Cumulative Review Exercises (page 295)

1. rational numbers: terminating and repeating decimals; irrational numbers: nonterminating, nonrepeating decimals **2.** 0.125, 0.0625, 0.03125, 0.054125 **3.** 10 **4.** -6 **5.** $-26p^2 - 6p$ **6.** $-2a + b - 2$ **7.** 201 ft^2 **8.** \$20,000 **9.** $\frac{26}{3}$ **10.** 3 **11.** 6 **12.** impossible **13.** perpendicular **14.** parallel **15.** $y = \frac{1}{3}x + \frac{11}{3}$ **16.** $-\frac{8}{5}$ **17.** 9,000 prisoners/yr **18.** 1990–1995, 72,200 prisoners/yr **19.** 10 **20.** 14 **21.** (2, 1) **22.** (3, 1) **23.** (1, 1) **24.** $(-1, -1, 3)$ **25.** $(-1, -1)$ **26.** $(1, 2, -1)$ **27.** (7, 23), in 1907 the percent of U.S. workers in white-collar and farming jobs was the same (23%); (45, 42), in 1945 the percent of U.S. workers in white-collar and blue-collar jobs was the same (42%) **28.** $y = \frac{5}{9}x + 20$, about 81% **29.** 750 **30.** 250 \$5 tickets, 375 \$3 tickets,

125 \$2 tickets **31.** 3, $-\frac{3}{2}$ **32.** -5, $-\frac{3}{5}$ **33.** $(-\infty, 11]$ **34.** $(-3, 3)$ **35.** $\left[-\frac{2}{3}, 2\right]$

36. $(-\infty, -4) \cup (1, \infty)$ **37.** **38.**

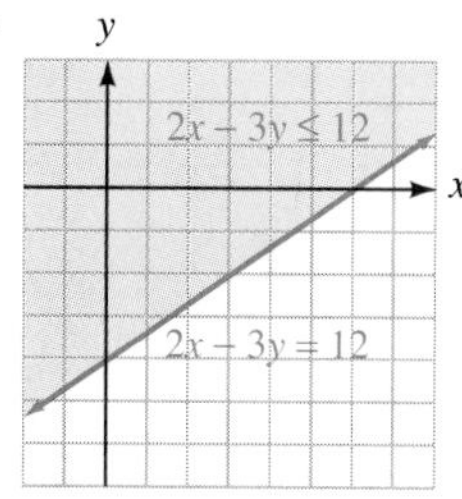

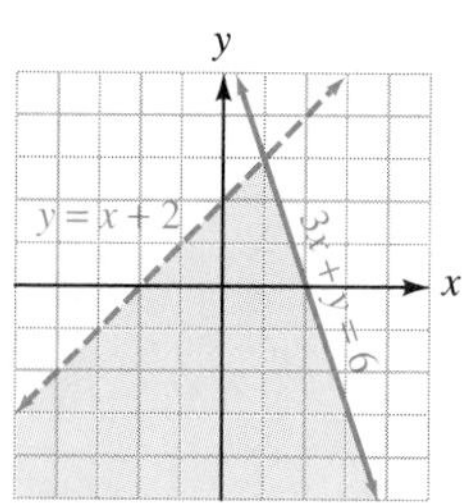

Study Set Section 5.1 (page 305)

1. base, exponent **3.** natural **5.** x^{m+n} **7.** x^ny^n **9.** 1 **11.** x^{m-n} **13.** $4^{-2} = \frac{1}{4^2}$ **15. a.** x^6 ft^2 **b.** x^9 ft^3 **17.** 9, (-2) **19.** base is 5, exponent is 3 **21.** base is x, exponent is 5 **23.** base is b, exponent is 6 **25.** base is $\frac{n}{4}$, exponent is 3 **27.** 9 **29.** -9 **31.** 9 **33.** $\frac{1}{25}$ **35.** $-\frac{1}{25}$ **37.** $\frac{1}{25}$ **39.** 1 **41.** 1 **43.** $-32x^5$ **45.** x^5 **47.** x^{10} **49.** k^7 **51.** a^4b^5 **53.** p^{10} **55.** x^5y^4 **57.** $\frac{1}{b^{72}}$ **59.** x^{28} **61.** $\frac{s^3}{r^9}$ **63.** a^{20} **65.** $-\frac{1}{d^3}$ **67.** $27x^9y^{12}$ **69.** $\frac{1}{729}m^6n^{12}$ **71.** $\frac{a^{15}}{b^{10}}$ **73.** $\frac{a^6}{b^4}$ **75.** a^5 **77.** c^7 **79.** $\frac{9}{4}$ **81.** a^4 **83.** $\frac{1}{9x^3}$ **85.** $\frac{64b^{12}}{27a^9}$ **87.** 1 **89.** $-\frac{b^3}{8a^{21}}$ **91.** 3.462825992 **93.** -244.140625 **101.** 10^{-2}, 10^{-3}, 10^{-4}, 10^{-5}, 10^{-6}, 10^{-7}, 10^{-8}, 10^{-9} **103.** $10^3 \cdot 26^3$; 17,576,000 **107.** $(-\infty, 1)$ **109.** $(-\infty, 20]$

Study Set Section 5.2 (page 312)

1. scientific **3.** 10^n **5.** left **7.** 60.22 is not between 1 and 10. **9.** 3.9×10^3 **11.** 7.8×10^{-3} **13.** 1.73×10^{14} **15.** 9.6×10^{-6} **17.** 3.23×10^7 **19.** 6.0×10^{-4} **21.** 5.27×10^3 **23.** 3.17×10^{-4} **25.** 270 **27.** 0.00323 **29.** 796,000 **31.** 0.00037 **33.** 5.23 **35.** 23,650,000 **37.** 1.817×10^{12} **39.** 5.005×10^8 **41.** 5×10^{-8} **43.** 4.005×10^{20}; 400,500,000,000,000,000,000 **45.** 4.3×10^{-3}; 0.0043 **47.** 6×10^3; 6,000 **49.** 3.6×10^{-5}; 0.000036 **51.** 2,600,000 to 1 **53.** \$$1.7 \times 10^{12}$, \$$3.9 \times 10^9$, \$$2.75 \times 10^8$, \$$3.12 \times 10^8$ **55.** 8.5×10^{-28} g **57.** about 9.5×10^{15} m **59. a.** 2.5×10^9 sec = 2,500,000,000 sec **b.** about 79 years **61.** 1.209×10^8 mi **65.** $\left[1, \frac{9}{4}\right]$ **67.** $(-\infty, 2) \cup (7, \infty)$

Study Set Section 5.3 (page 322)

1. polynomial **3.** degree **5.** cubic **7.** like **9.** monomial; 2 **11.** trinomial; 3 **13.** binomial; 2 **15.** monomial; 0 **17.** none of these; 10 **19.** binomial; 9 **21. a.** $-2x^4 - 5x^2 + 3x + 7$ **b.** $7a^3x^5 - ax^3 - 5a^3x^2 + a^2x$ **23.** like terms, $10x$ **25.** unlike terms **27.** like terms, $-5r^2t^3$ **29.** unlike terms **31.** $9x^2 + 3x - 2$ **33.** -1, -1, 1 **35.** 8, 2, 0, 2, 8

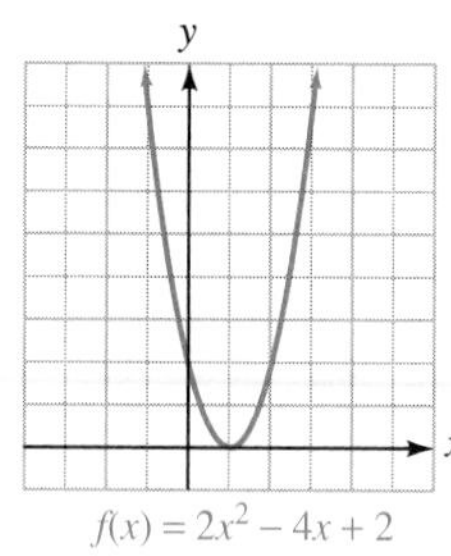

37. -42, 0, 12, 6, -6, -12, 0, 42

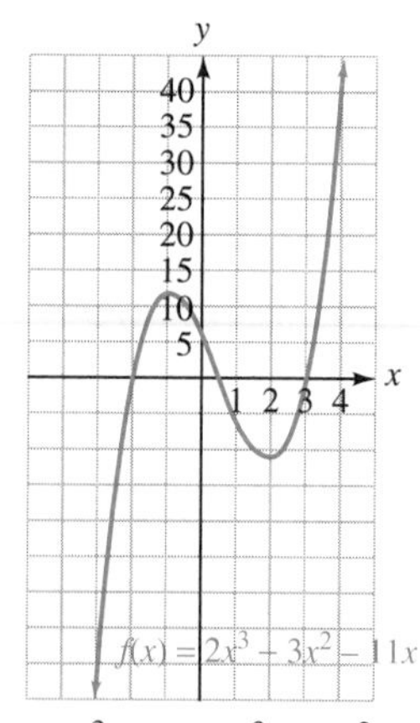

39.

41. $x^2 - 5x + 6$ **43.** $-5a^2 + 4a + 4$ **45.** $-y^3 + 4y^2 + 6$ **47.** $2x^2y^3 + 13xy + 3y^2$ **49.** $6x^3 - 6x^2 + 14x - 17$ **51.** $x^2 - 8x + 22$ **53.** $-3y^3 + 18y^2 - 28y + 35$ **55.** 20 ft **57.** 872 ft^3 **59.** 2,160 in.^{3} **61. a.** $V(x) = 2{,}500x + 275{,}000$ **b.** \$325,000 **65.** $[-5, 5]$ **67.** $(-1, 9)$

Study Set Section 5.4 (page 331)

1. product **3.** terms **5.** factors **7.** term **9.** $x^2 + 2xy + y^2$ **11.** $x^2 - y^2$ **13.** $x^2 + 2x - 8$ **15.** $16a^2 + 24a + 9$ **17. a.** $2x$, $4x$ **b.** $2x$, -3 **c.** 4, $4x$ **d.** 4, -3 **19.** $-6a^3b$ **21.** $-15a^2b^2c^3$ **23.** $-120a^9b^3$ **25.** $-405x^7y^4$ **27.** $3x + 6$ **29.** $3x^3 + 9x^2$ **31.** $-6x^3 + 6x^2 - 4x$ **33.** $7r^3st + 7rs^3t - 7rst^3$ **35.** $-12m^4n^2 - 12m^3n^3$ **37.** $x^2 + 5x + 6$ **39.** $6t^2 + 5t - 6$ **41.** $6y^2 - 5yz + z^2$ **43.** $x^2 + 4x + 4$ **45.** $a^2 - 8a + 16$ **47.** $4a^2 + 4ab + b^2$ **49.** $x^2 - 4$ **51.** $4x^2 - 9y^2$ **53.** $x^3 - y^3$ **55.** $6y^3 + 11y^2 + 9y + 2$

57. $8a^3 - b^3$ **59.** $a^3 - 3a^2b - ab^2 + 3b^3$ **61.** $3x^2 + 12x$ **63.** $-p^2 + 4pq$ **65.** $3x^2 + 3x - 11$ **67.** $5x^2 - 36x + 7$ **69.** $9.2127x^2 - 7.7956x - 36.0315$ **71.** $299.29y^2 - 150.51y + 18.9225$ **73. a.** $(x + y)(x - y)$ **b.** $x(x - y)$; $x^2 - xy$ **c.** $y(x - y)$; $xy - y^2$ **d.** They represent the same area. $(x + y)(x - y) = x^2 - y^2$ **75.** $x(12 - 2x)(12 - 2x) = 144x - 48x^2 + 4x^3$

81.

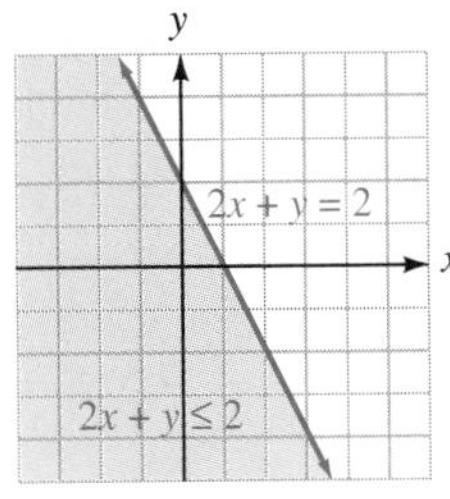

83.

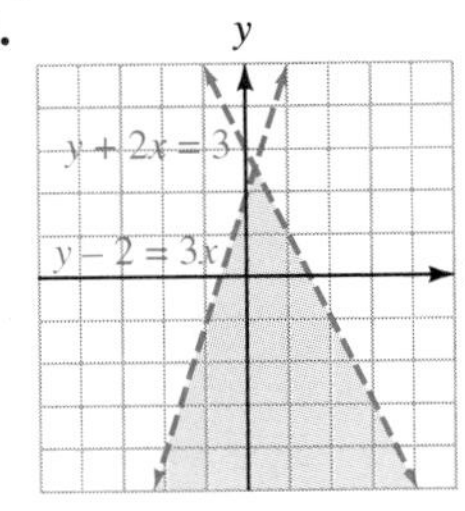

Study Set Section 5.5 (page 338)

1. factored **3.** factor **5.** greatest common factor **7.** $6xy^2$ **9. a.** The terms inside the parentheses have a common factor of 2. **b.** The terms inside the parentheses have a common factor of t. **11.** 4 **13.** x^2, 2 **15.** $2 \cdot 3$ **17.** $3^3 \cdot 5$ **19.** 2^7 **21.** $5^2 \cdot 13$ **23.** 12 **25.** 2 **27.** $4a^2$ **29.** $6xy^2z^2$ **31.** $2(x + 4)$ **33.** $2x(x - 3)$ **35.** prime **37.** $5x^2y(3 - 2y)$ **39.** prime **41.** $3z(9z^2 + 4z + 1)$ **43.** $9x^7y^3(5x^3 - 7y^4 + 9x^3y^7)$ **45.** $-3(a + 2)$ **47.** $-x(3x + 1)$ **49.** $-3x(2x + y)$ **51.** $-6ab(3a + 2b)$ **53.** $-7u^2v^3z^2(9uv^3z^7 - 4v^4 + 3uz^2)$ **55.** $(x + y)(4 + t)$ **57.** $(a - b)(r - s)$ **59.** $(m + n + p)(3 + x)$ **61.** $(u + v)(u + v - 1)$ **63.** $-(x + y)(a - b)$ **65.** $(x + y)(a + b)$ **67.** $(x + 2)(x + y)$ **69.** $(3 - c)(c + d)$ **71.** $(a + b)(a - 4)$ **73.** $(a + b)(x - 1)$ **75.** $(x + y)(x + y + z)$ **77.** $x(m + n)(p + q)$ **79.** $y(x + y)(x + y + 2z)$ **81.** $n(2n - p + 2m)(n^2p - 1)$ **83.** $r_1 = \frac{rr_2}{r_2 - r}$ **85.** $f = \frac{d_1d_2}{d_2 + d_1}$ **87.** $a^2 = \frac{b^2x^2}{b^2 - y^2}$ **89.** $r = \frac{S - a}{S - l}$ **91. a.** $\frac{1}{2}b_1h$ **b.** $\frac{1}{2}b_2h$ **c.** $\frac{1}{2}h(b_1 + b_2)$; the formula for the area of a trapezoid **93.** $r^2(4 - \pi)$ **97.** a line **99.** $(-\infty, -3)$ **101.** They are the same.

Study Set Section 5.6 (page 346)

1. squares **3.** 1, 4, 9, 16, 25, 36, 49, 64, 81, 100 **5.** $(x + 5)(x + 5) = x^2 + 10x + 25$ **7. a.** A common factor of 2 can be factored out of each binomial. **b.** $(1 - t^4)$ factors as a difference of two squares. **9. a.** $5(p^2 + 4)$ **b.** $5(p + 2)(p - 2)$ **11.** $(p^2 - pq + q^2)$ **13.** $(p - q)$ **15.** $6y$, +, $7m$ **17.** $(x + 2)(x - 2)$ **19.** $(3y + 8)(3y - 8)$ **21.** prime **23.** $(25a + 13b^2)(25a - 13b^2)$ **25.** $(9a^2 + 7b)(9a^2 - 7b)$ **27.** $(6x^2y + 7z^2)(6x^2y - 7z^2)$ **29.** $(x + y + z)(x + y - z)$ **31.** $(a - b + c)(a - b - c)$ **33.** $(x^2 + y^2)(x + y)(x - y)$ **35.** $(16x^2y^2 + z^4)(4xy + z^2)(4xy - z^2)$ **37.** $2(x + 12)(x - 12)$ **39.** $2x(x + 4)(x - 4)$ **41.** $5x(x + 5)(x - 5)$ **43.** $t^2(rs + x^2y)(rs - x^2y)$ **45.** $(a + b)(a - b + 1)$ **47.** $(a - b)(a + b + 2)$ **49.** $(2x + y)(1 + 2x - y)$ **51.** $(r + s)(r^2 - rs + s^2)$ **53.** $(x - 2y)(x^2 + 2xy + 4y^2)$ **55.** $(4a - 5b^2)(16a^2 + 20ab^2 + 25b^4)$ **57.** $(5xy^2 + 6z^3)(25x^2y^4 - 30xy^2z^3 + 36z^6)$ **59.** $(x^2 + y^2)(x^4 - x^2y^2 + y^4)$ **61.** $5(x + 5)(x^2 - 5x + 25)$ **63.** $4x^2(x - 4)(x^2 + 4x + 16)$ **65.** $2u^2(4v - t)(16v^2 + 4tv + t^2)$ **67.** $(a + b)(x + 3)(x^2 - 3x + 9)$ **69.** $\frac{4}{3}\pi(r_1 - r_2)(r_1^2 + r_1r_2 + r_2^2)$

73.

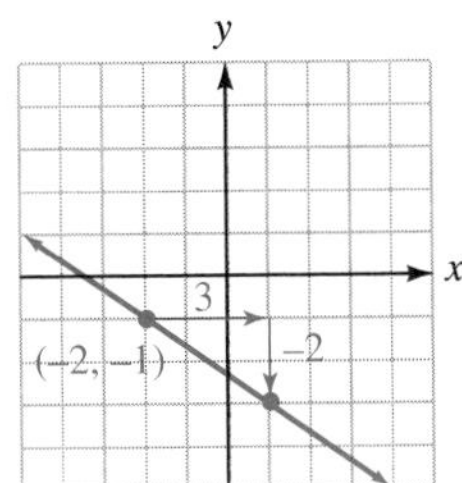

75.

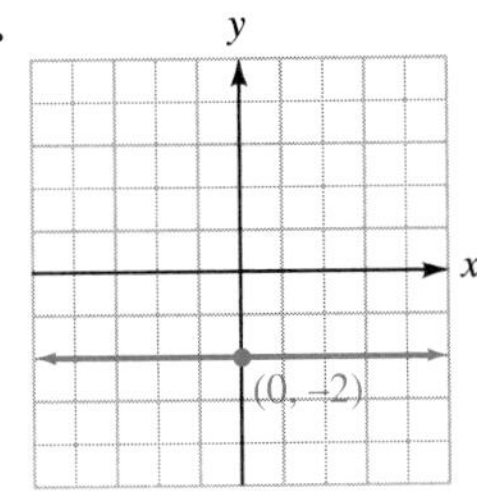

Study Set Section 5.7 (page 356)

1. trinomial **3.** lead **5. a.** positive **b.** negative **c.** positive **7.** integers **9.** $2xy + y^2$ **11.** $x^2 - y^2$ **13.** 4, −4, 1 **15.** $(x + 2)$ **17.** $(x - 3)$ **19.** $(2a + 1)$ **21.** $(x + 1)^2$ **23.** $(a - 9)^2$ **25.** $(2y + 1)^2$ **27.** $(3b - 2)^2$ **29.** $(x - 3)(x - 2)$ **31.** $(x - 2)(x - 5)$ **33.** prime **35.** $(x + 5)(x - 6)$ **37.** $(a + 10)(a - 5)$ **39.** $3(x + 7)(x - 3)$ **41.** $x^2(b - 7)(b - 5)$ **43.** $-(a - 8)(a + 4)$ **45.** $-3(x - 3)(x - 2)$ **47.** $(3y + 2)(2y + 1)$ **49.** $(4a - 3)(2a + 3)$ **51.** $(3x - 4)(2x + 1)$ **53.** prime **55.** $(4x - 3)(2x - 1)$ **57.** $(a + b)(a - 4b)$ **59.** $x(3x - 1)(x - 3)$ **61.** $-(3a + 2b)(a - b)$ **63.** $5(a - 3b)^2$ **65.** $x^2(7x - 8)(3x + 2)$ **67.** $(x^2 + 5)(x^2 + 3)$ **69.** $(y^2 - 10)(y^2 - 3)$ **71.** $(a + 3)(a - 3)(a + 2)(a - 2)$ **73.** $(x + a + 1)^2$ **75.** $(a + b + 4)(a + b - 6)$ **77.** $(x + 2 + y)(x + 2 - y)$ **79.** $(x + 1 + 3z)(x + 1 - 3z)$ **81.** $(c + 2a - b)(c - 2a + b)$ **83.** $(a - 16)(a - 1)$ **85.** $(2u + 3)(u + 1)$ **87.** $(5r + 2s)(4r - 3s)$ **89.** $x + 3$ **93.** 5 **95.** $\frac{8}{3}$ **97.** $-26p^2 - 6p$

Study Set Section 5.8 (page 361)

1. factoring **3.** cubes **5.** common **7.** trinomial **9.** Multiply the factors of $y^2z^3(x + 6)(x + 1)$ to see if the product is $x^2y^2z^3 + 7xy^2z^3 + 6y^2z^3$. **11.** $3ab$, $2b$, $2a$ **13.** $(x + 4)^2$ **15.** $(2xy - 3)(4x^2y^2 + 6xy + 9)$ **17.** $(x - t)(y + s)$ **19.** $(5x + 4y)(5x - 4y)$ **21.** $(6x + 5)(2x + 7)$ **23.** $2(3x - 4)(x - 1)$ **25.** $y^2(2x + 1)(2x + 1)$ **27.** $(x + a^2y)(x^2 - a^2xy + a^4y^2)$ **29.** $2(x - 3)(x^2 + 3x + 9)$ **31.** $(a + b)(f + e)$ **33.** $(2x + 2y + 3)(x + y - 1)$ **35.** $(25x^2 + 16y^2)(5x + 4y)(5x - 4y)$ **37.** $36(x^2 + 1)(x + 1)(x - 1)$ **39.** $(a + 3)(a - 3)(a + 2)(a - 2)$ **41.** $(x + 3 + y)(x + 3 - y)$ **43.** $(2x + 1 + 2y)(2x + 1 - 2y)$ **45.** $(x + y + 1)(x - y - 1)$ **49.** perpendicular **51.** -225

Study Set Section 5.9 (page 368)

1. quadratic **3.** At least one is 0. **5. a.** yes **b.** no **c.** yes **d.** no **7.** 3, −1 **9.** $y + 6, y - 9, -6$ **11.** 0, −2 **13.** 4, −4 **15.** 0, −1 **17.** 0, 5 **19.** −3, −5 **21.** −2, −4 **23.** $-\frac{1}{3}, -3$ **25.** $\frac{1}{2}, 2$ **27.** $1, -\frac{1}{2}$ **29.** 3, 3 **31.** $\frac{1}{4}, -\frac{3}{2}$ **33.** $2, -\frac{5}{6}$ **35.** $\frac{1}{3}, -1$ **37.** $2, \frac{1}{2}$ **39.** $\frac{1}{5}, -\frac{5}{3}$ **41.** 0, 0, −1 **43.** 0, 7, −7 **45.** 0, 7, −3 **47.** 3, −3, 2, −2 **49.** 0, −2, −3 **51.** $0, \frac{5}{6}, -7$ **53.** 16, 18 or −18, −16 **55.** 2.78, 0.72 **57.** 1 **59.** 10 in., 16 in. **61.** 3 ft **63.** 20 ft by 40 ft **65.** 11 sec and 19 sec **67.** 2 sec **69.** 6 m/sec **71.** 4 **75.** 25 ft^2

Key Concept (page 372)

1. $-5x^2 + 2x - 6$ **2.** $4s^3 - 3s^2 + 3s - 7$ **3.** $3m^2 + 5m - 12$ **4.** $3r^4st - 6r^2s^2t + 9r^2st^3$ **5.** $a^2 - 4ad + 4d^2$ **6.** $x^3 - 27y^3$ **7.** $6b^3 + 11b^2 + 9b + 2$ **8.** $-y^2 - 4y - 1$ **9.** 0; after falling for 6 seconds, the squeegee will strike the ground **10.** $V(x) = 16x^3 + 12x^2 - 4x$; 528 **11.** 9, −9 **12.** 0, 5 **13.** −3, −5 **14.** $-\frac{3}{2}, -1$ **15.** $\frac{1}{3}, -1$ **16.** 1, 0, −9

Chapter Review (page 374)

1. a. 729 **b.** −32 **c.** −64 **d.** 15 **2. a.** x^6 **b.** a^5b^6 **c.** m^{18} **d.** t^{13} **e.** $9x^4y^6$ **f.** $\frac{x^{16}}{b^4}$ **g.** −3 **h.** $\frac{1}{x^{10}}$ **i.** $-\frac{1}{625}$ **j.** $70x^4$ **k.** $\frac{x^6}{9}$ **l.** $\frac{2}{x}$ **m.** $-c^{10}$ **n.** $\frac{25}{16}$ **o.** $\frac{1}{y^8}$ **p.** $\frac{-b^3}{8a^{21}}$ **3. a.** 1.93×10^{10} **b.** 2.735×10^{-8} **4. a.** 72,770,000 **b.** 0.0000000083 **5.** 7.6×10^2 sec **6.** 1.67248×10^{-18} g **7.** 8.4×10^6 **8. a.** no **b.** yes **c.** yes **d.** no **9. a.** binomial, 2 **b.** monomial, 4 **c.** none of these, 4 **d.** trinomial, 8 **10.** 134 $in.^3$ **11. a.**

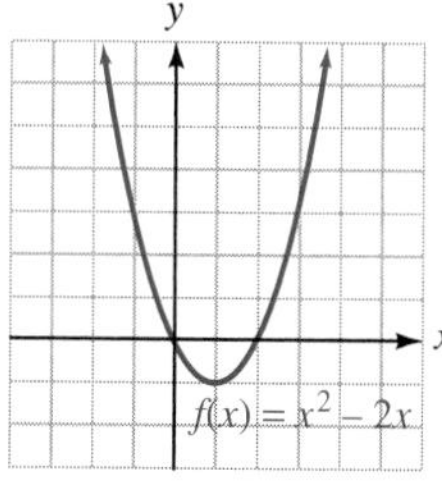

b.

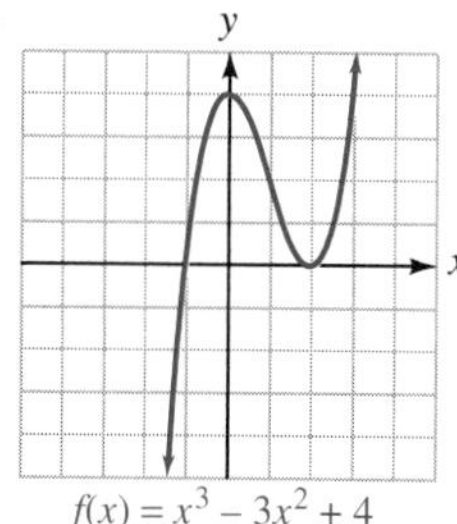

12. a. $5x^2 + 2x + 16$ **b.** $3x^2y^3 - 8x^2y + 8y$ **c.** $6x^3 + 4x^2 + x + 9$ **d.** $6k^4 - 6k^3 + 9k^2 - 2$ **13. a.** $-4a^3$ **b.** $6x^3y^2z^5$ **14. a.** $2x^4y^3 - 8x^2y^7$ **b.** $a^4b + 2a^3b^2 - a^2b^3$ **15. a.** $16x^2 + 14x - 15$ **b.** $6x^3 - 12x^2 + 4x - 8$ **c.** $25a^2 - 60a + 36$ **d.** $49c^4 - d^2$ **e.** $15x^4 - 22x^3 + 58x^2 - 40x$ **f.** $r^3 - 3r^2s - rs^2 + 3s^3$ **16.** $154x^2 + 27x + 1$ **17.** $2 \cdot 5^2 \cdot 7$ **18. a.** 6 **b.** $3xy^3$ **19. a.** $4(x + 2)$ **b.** $3x(x^2 - 2x + 3)$ **c.** prime **d.** $7a^3b(ab + 7)$ **e.** $5x^2(x + y)(1 - 3x)$ **f.** $9x^2y^3z^2(3xz + 9x^2y^2 - 10z^5)$ **20. a.** $-7(b - 2)$ **b.** $-7a^2b^2(a - b)^3(7a^2 - 7ab - 9b^2)$ **21. a.** $(x + 2)(y + 4)$ **b.** $(ry - a)(r - 1)$ **22.** $m_1 = \frac{mm_2}{m_2 - m}$ **23. a.** $(z + 4)(z - 4)$ **b.** $(y + 11)(y - 11)$ **c.** $(xy^2 + 8z^3)(xy^2 - 8z^3)$ **d.** prime **e.** $(c + a + b)(c - a - b)$ **f.** $3x^2(x^2 + 10)(x^2 - 10)$ **24. a.** $(t + 4)(t^2 - 4t + 16)$ **b.** $2y(x - 3z)(x^2 + 3xz + 9z^2)$ **25. a.** $(x + 5)^2$ **b.** $(a - 7)^2$ **26. a.** $(y + 20)(y + 1)$ **b.** $(z - 5)(z - 6)$ **c.** $-(x + 7)(x - 4)$ **d.** $(y - 8)(y + 3)$ **e.** $(4a - 1)(a - 1)$ **f.** prime **g.** $y(y + 2)(y - 1)$ **h.** $3(5x + y)(x - 4y)$ **i.** $2a^2(a + 3)(a - 1)$ **j.** $(v^2 - 7)(v^2 - 6)$ **27.** $(s + t - 1)^2$ **28.** $(k + 1 + 3m)(k + 1 - 3m)$ **29. a.** $x(x - 1)(x + 6)$ **b.** $3y(x + 3)(x - 7)$ **c.** $(z - 2)(z + x + 2)$ **d.** $(x + 1 + p)(x + 1 - p)$ **e.** $(x + 2 + 2p^2)(x + 2 - 2p^2)$ **f.** $(y + 2)(y + 1 + x)$ **g.** $4(ab + 4)(a^2b^2 - 4ab + 16)$ **h.** $36(z^2 + 1)(z + 1)(z - 1)$ **30.** $\frac{\pi}{2}h(r_1 + r_2)(r_1 - r_2)$ **31. a.** $0, \frac{3}{4}$ **b.** 6, −6 **c.** $\frac{1}{2}, -\frac{5}{6}$ **d.** $\frac{2}{7}, 5$ **e.** $0, -\frac{2}{3}, \frac{4}{5}$ **f.** $-\frac{2}{3}, 7, 0$ **32.** 17 m by 20 m

Chapter 5 Test (page 379)

1. x^9 **2.** $-8x^6y^9$ **3.** $\frac{1}{m^5}$ **4.** $\frac{m^4}{9n^{10}}$ **5.** 4.706×10^{12} **6.** 0.000245 **7.** 1.45×10^{19} **8.** 1.116×10^7 mi **9.** 5 **10.** 13 **11.** 110 ft **12.** $3x^2 + 4x - 4$ **13.**

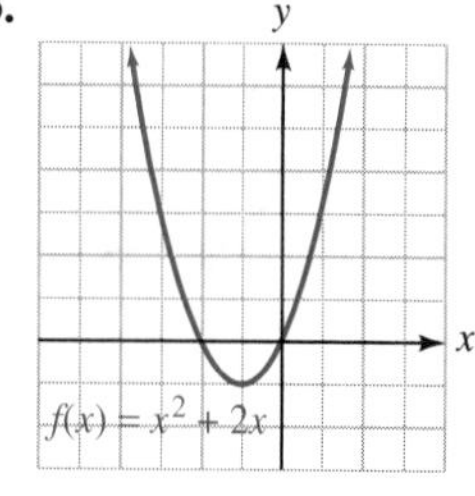

14.

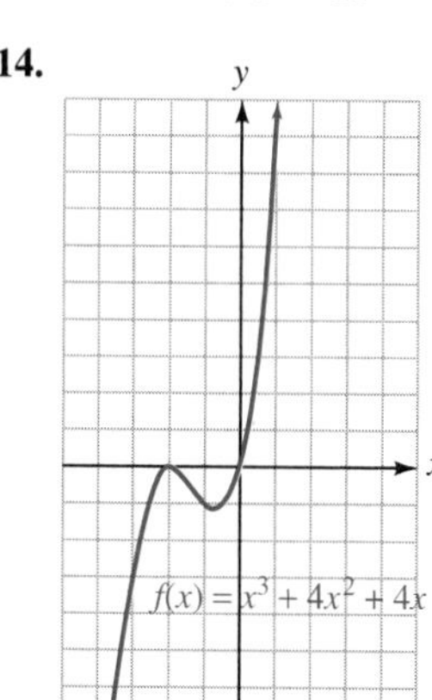

15. $5y^2 + y - 1$ **16.** $-4u^2 + 2u - 14$

17. $-6x^4yz^4$ **18.** $-15a^3b^4 + 10a^3b^5$ **19.** $z^2 - 16$ **20.** $12x^2 - 17x + 6$ **21.** $16t^2 - 72t + 81$ **22.** $2s^3 - 2st^2$ **23.** $3x(1 + 2x)$ **24.** $3abc(4a^2b - abc + 2c^2)$ **25.** $(u - v)(r + s)$ **26.** $(a - y)(x + y)$ **27.** $(x + 7)(x - 7)$ **28.** $4(y^2 + 4)(y + 2)(y - 2)$ **29.** prime **30.** $(b + 5)(b^2 - 5b + 25)$ **31.** $3(u - 2)(u^2 + 2u + 4)$ **32.** $(a - 6)(a + 1)$ **33.** $(3b + 2)(2b - 1)$ **34.** $(x + 3 + y)(x + 3 - y)$ **35.** 6, −1 **36.** $\frac{3}{4}, -\frac{3}{2}$ **37.** 0, −4 **38.** $v = \dfrac{v_1v_3}{v_3 + v_1}$ **39.** 11 ft by 22 ft

Study Set Section 6.1 (page 389)

1. rational **3.** asymptote
5. 6, 3, 1.5, 1, 0.75, 0.6, 0.5

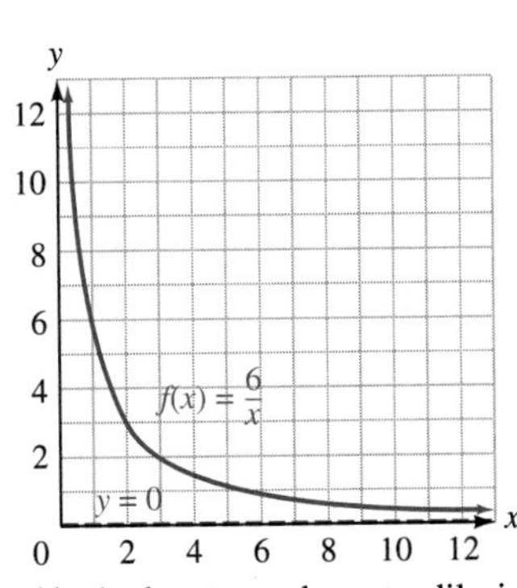

7. 3, 2, 1.5, 1.33, 1.25, 1.2, 1.17

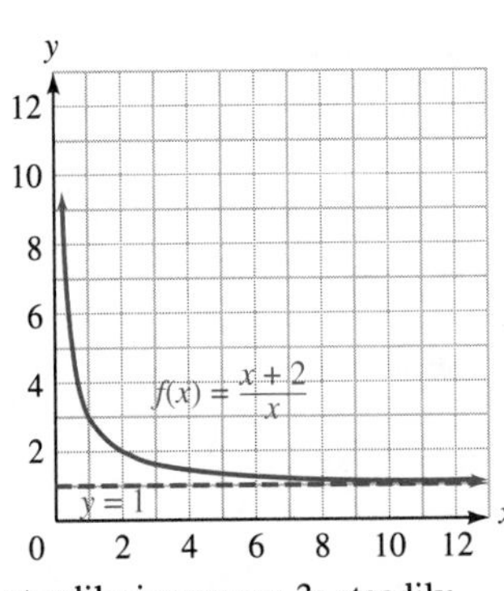

9. a. 1 **b.** 1 **c.** 1 **d.** −1 **11.** 1: decreases then steadily increases; 2: increases, decreases, then steadily increases; 3: steadily increases; 4: steadily decreases, approaching a cost of \$2.00 per unit **13. a.** true **b.** true **15.** 20 hr **17.** 12 hr **19.** $(-\infty, 0) \cup (0, \infty)$ **21.** $(-\infty, -2) \cup (-2, \infty)$ **23.** $(-\infty, 0) \cup (0, 1) \cup (1, \infty)$ **25.** $(-\infty, -7) \cup (-7, 8) \cup (8, \infty)$ **27.** $\frac{2}{3}$ **29.** $-\frac{28}{9}$ **31.** $4x^2$ **33.** $-\dfrac{4y}{3x}$ **35.** $-\dfrac{x}{2}$ **37.** −1 **39.** $\dfrac{1}{x - y}$ **41.** $\dfrac{5}{x - 2}$ **43.** $\dfrac{-3(x + 2)}{x + 1}$ **45.** in lowest terms **47.** $x + 2$ **49.** $\dfrac{x + 1}{x + 3}$ **51.** $\dfrac{m - 2n}{n - 2m}$ **53.** $\dfrac{x + 4}{2(2x - 3)}$ **55.** $\dfrac{3(x - y)}{x + 2}$ **57.** $\dfrac{2x + 1}{2 - x}$ **59.** $\dfrac{a^2 - 3a + 9}{4(a - 3)}$ **61.** in lowest terms **63.** $m + n$ **65.** $-\dfrac{m + n}{2m + n}$ or $\dfrac{-m - n}{2m + n}$ **67.** $\dfrac{x - y}{x + y}$ **69.** $\dfrac{2x - 3}{2y - 3}$ **71.** D: $(-\infty, 2) \cup (2, \infty)$; R: $(-\infty, 1) \cup (1, \infty)$ **73.** D: $(-\infty, -2) \cup (-2, 2) \cup (2, \infty)$; R: $(-\infty, \infty)$ **75. a.** \$50,000 **b.** \$200,000 **77. a.** $c(n) = 0.09n + 7.50$ **b.** $c(n) = \frac{0.09n + 7.50}{n}$ **c.** about 10¢ **79. a.** about 2.6 hr **b.** about 4.6 hr **83.** $3x(x - 3)$ **85.** $(3x^2 + 4y)(9x^4 - 12x^2y + 16y^2)$

Study Set Section 6.2 (page 401)

1. ratios **3.** extremes, means **5.** direct, inverse **7.** inverse **9.**

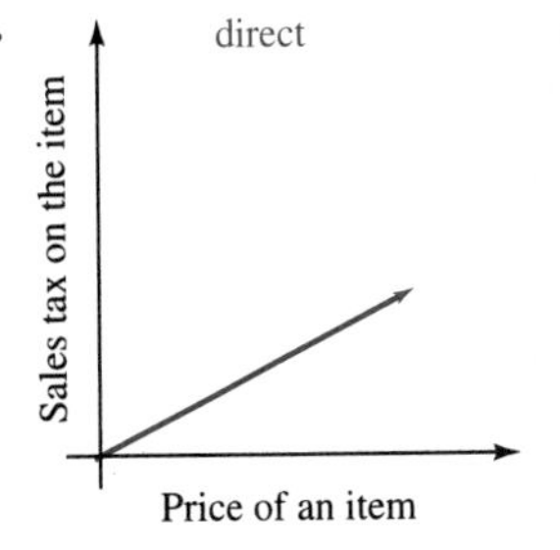

11.

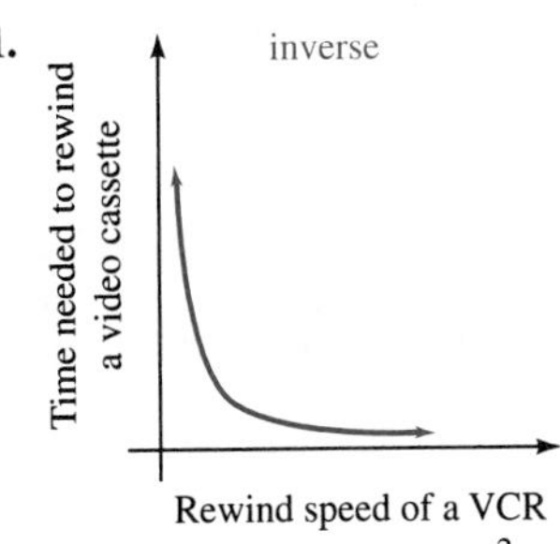

13. −7, 6, 18, −102 **15.** 3 **17.** 5 **19.** 5 **21.** 2, −2 **23.** 4, −1 **25.** $-\frac{5}{2}$, −1 **27.** $A = kp^2$ **29.** $v = \dfrac{k}{r^2}$ **31.** $P = \dfrac{ka^2}{j^3}$ **33.** L varies jointly with m and n. **35.** R varies directly with L and inversely with d^2. **37.** 202, 172, 136 **39.** about 2 gal **41.** eye: 49.9 in.; seat: 17.6 in.; elbow; 27.8 in. **43.** 7.5 in. × 12.5 in. **45.** 555 ft **47.** $46\frac{7}{8}$ ft **49.** 880 ft **51.** 1,600 ft **53.** 25 days **55.** 12 in.^{3} **57.** \$9,000 **59.** 3 ohms **61.** 0.275 in. **63.** 12.8 lb **67.** x^{10} **69.** −1

Study Set Section 6.3 (page 411)

1. divisor **3.** numerator **5.** multiply, $\frac{ac}{bd}$ **7.** 0 **9.** $(x - 5)$, $(5x - 25)$, $x(x + 3)$, $5(x - 5)$ **11.** yes, no, yes **13.** $\dfrac{5}{4}$ **15.** $-\dfrac{5}{6}$ **17.** $\dfrac{xy^2d}{c^3}$ **19.** $-\dfrac{x^{10}}{y^2}$ **21.** $x + 1$ **23.** $\dfrac{x - 4}{x + 5}$ **25.** $\dfrac{t - 1}{t + 1}$ **27.** $\dfrac{n + 2}{n + 1}$ **29.** $\dfrac{1}{x + 1}$ **31.** $(x + 1)^2$ **33.** $\dfrac{(a + 7)^2(a - 5)}{12x^2}$ **35.** $\dfrac{x + y}{x - y}$ **37.** $\dfrac{a + b}{(x - 3)(c + d)}$ **39.** $-\dfrac{x + 1}{x + 3}$ or $\dfrac{-x - 1}{x + 3}$ **41.** $x^2(x + 3)$ **43.** $\dfrac{x + 2}{x - 2}$ **45.** $\dfrac{3x}{2}$ **47.** $\dfrac{x^2 - 6x + 9}{x^6 + 8x^3 + 16}$ **49.** $\dfrac{4m^4 - 4m^3 - 11m^2 + 6m + 9}{x^4 - 2x^2 + 1}$ **51.** $k_1(k_1 + 2)$, $k_2 + 6$ **55.** $-6a^5 + 2a^4$ **57.** $6g^2 - 5gn + n^2$

Study Set Section 6.4 (page 420)

1. divisible **3.** subtract, keep, $\frac{a-c}{b}$ **5.** same **7.** LCD **9. a.** ii **b.** adding or subtracting rational expressions **c.** simplifying a rational expression **11.** $3x - 2, 3, 3x - 1$ **13.** $\frac{5}{2}$ **15.** $\frac{11}{4y}$ **17.** 2 **19.** 3 **21.** 3 **23.** $\frac{6x}{(x-3)(x-2)}$ **25.** $36x^2$ **27.** $x(x+3)(x-3)$ **29.** $(x+3)^2(x^2-3x+9)$ **31.** $(2x+3)^2(x+1)^2$ **33.** $\frac{5}{6}$ **35.** $\frac{21a-8b}{14}$ **37.** $\frac{17}{12x}$ **39.** $\frac{9a^2-4b^2}{6ab}$ **41.** $\frac{3a-5b}{a^2b^2}$ **43.** $\frac{3r+2bs}{12b^2}$ **45.** $\frac{10a+4b}{21}$ **47.** $\frac{8x-2}{(x+2)(x-4)}$ **49.** $\frac{7x+29}{(x+5)(x+7)}$ **51.** $\frac{4x+1}{x}$ **53.** 2 **55.** $\frac{3a-3}{3a-2}$ **57.** $\frac{2x^2+x}{(x+3)(x+2)(x-2)}$ **59.** $\frac{-x^2+11x+8}{(3x+2)(x+1)(x-3)}$ **61.** $\frac{-4x^2+14x+54}{x(x+3)(x-3)}$ **63.** $\frac{x^2-5x-5}{x-5}$ **65.** $\frac{11x^2+7x+1}{(2x-1)(3x+2)}$ **67.** $\frac{2}{x+1}$ **69.** $\frac{2b}{a+b}$ **71.** $\frac{2}{m-1}$ **73.** $\frac{10r+20}{r}; \frac{3t+9}{t}$ **79.** 3, 3 **81.** 0, 0, -1

Study Set Section 6.5 (page 428)

1. complex **3.** $\frac{t^2}{t^2}$ **5.** $y, y, \div, \frac{1}{6}$ **7.** $\div$ **9.** $\frac{2}{3}$ **11.** $-\frac{1}{7}$ **13.** $\frac{2y}{3z}$ **15.** $125b$ **17.** $-\frac{1}{y}$ **19.** $\frac{y-x}{x^2y^2}$ **21.** $\frac{b+a}{b}$ **23.** $\frac{y+x}{y-x}$ **25.** $y - x$ **27.** $-\frac{1}{a+b}$ **29.** $x^2 + x - 6$ **31.** $\frac{5x^2y^2}{xy+1}$ **33.** -1 **35.** $\frac{x+2}{x-3}$ **37.** $\frac{a-1}{a+1}$ **39.** $\frac{2x^2+5x}{x^3+x^2+x+1}$ **41.** $\frac{xy^2}{y-x}$ **43.** $\frac{x^2(xy^2-1)}{y^2(x^2y-1)}$ **45.** $\frac{1}{x-y}$ **47.** $\frac{x-9}{3}$ **49.** $-\frac{h}{4}$ **51.** $\frac{k_1k_2}{k_2+k_1}$ **53.** $\frac{4d-1}{3d-1}$ **57.** 8 **59.** 2, -2, 3, -3

Study Set Section 6.6 (page 438)

1. rational **3.** common denominator **5. a.** yes **b.** no **7.** $\frac{12}{x}; \frac{12}{x+15}$ **9. a.** 7 **b.** 7 **11.** $30y, \frac{9}{2y}, \frac{10}{3y}, 7y, 35$ **13.** 12 **15.** 40 **17.** $\frac{1}{2}$ **19.** no solution **21.** $\frac{17}{25}$ **23.** 0 **25.** 1 **27.** 2 **29.** 2 **31.** $\frac{1}{3}$ **33.** 0 **35.** 2, -5 **37.** -4, 3 **39.** 6, $\frac{17}{3}$ **41.** $r = \frac{E - IR_L}{I}$ **43.** $r = \frac{S-a}{S-l}$ **45.** $Q_1 = \frac{PQ_2}{1+P}$ **47.** $R = \frac{R_1R_2R_3}{R_2R_3 + R_1R_3 + R_1R_2}$ **49.** $4\frac{8}{13}$ in. **51.** $L = \frac{SN - CN}{V - C}$, 8 years **53. a.** $1\frac{7}{8}$ days **b.** Santos: $412.50, Mays: $375 **55.** about 110 sec **57.** $2\frac{4}{13}$ weeks **59.** 6 mph **61.** 35 mph and 45 mph **63.** 5 mph **65.** 9 **69.** 9.0×10^9 **71.** 4.4×10^{-22}

Study Set Section 6.7 (page 447)

1. dividend, divisor, quotient **3.** algorithm **5.** $\frac{1}{b}$ **7.** $(2x-1)(x^2+3x-4) = 2x^3+5x^2-11x+4$ **9.** $2x^2, x, 4$ **11.** $3a^2 + 5 + \frac{6}{3a-2}$ **13.** $\frac{x^2-x-12}{x-4}$, $x-4\overline{)x^2-x-12}$, $(x^2-x-12) \div (x-4)$ **15.** $\frac{y}{2x^3}$ **17.** $-\frac{3}{4a^2}$ **19.** $2x + 3$ **21.** $-\frac{2x}{3} + \frac{x^2}{6}$ **23.** $2xy^2 + \frac{x^2y}{6}$ **25.** $\frac{x^4y^4}{2} - \frac{x^3y^9}{4} + \frac{3}{4xy^2}$ **27.** $x + 2$ **29.** $3x - 5 + \frac{3}{2x+3}$ **31.** $3x^2 + x + 2 - \frac{4}{x-1}$ **33.** $3x^2 + 4x + 3$ **35.** $a + 1$ **37.** $2y + 2$ **39.** $6x - 12$ **41.** $4x^3 - 3x^2 + 3x + 1$ **43.** $a^2 + a + 1 + \frac{2}{a-1}$ **45.** $5a^2 - 3a - 4$ **47.** $6y - 12$ **49.** $x^4 + x^2 + 4$ **51.** $x^2 + x + 1$ **53.** $x^2 + x + 2$ **55.** $9.8x + 16.4 - \frac{36.5}{x-2}$ **57.** $x^2 - 5x + 6$ **59.** $3x - 2, x + 5$ **63.** $8x^2 + 2x + 4$ **65.** $-2y^3 - y^2 + 10y - 14$

Key Concept (page 450)

1. a. $\frac{3x+2}{4x+3}$ **b.** $2x - 1$ **2. a.** $\frac{d-1}{d+1}$ **b.** $3d + 2, 2d - 3, 2d + 3$ **3. a.** $\frac{8x-2}{(x+2)(x-4)}$ **b.** $\frac{x-4}{x-4}, \frac{x+2}{x+2}$ **4. a.** $\frac{3(n^2-n-2)}{n^2}$ **b.** $3n$ **5. a.** 4 **b.** $t(t-2)$ **6. a.** $\frac{1}{6}$ **b.** $(x-1)(2x+3)$ **7. a.** 1 **b.** $(x+2)(x-4)$ **8. a.** $a^2 = \frac{b^2x^2}{b^2+y^2}$ **b.** a^2b^2

Chapter Review (page 452)

1. 8, 4, 2, 1.33, 1, 0.8, 0.67, 0.57, $\frac{1}{2}$

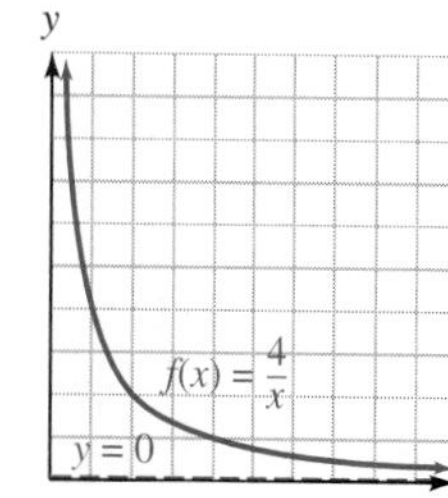

2. $y = 3, x = 0$; D: $(-\infty, 0) \cup (0, \infty)$, R: $(-\infty, 3) \cup (3, \infty)$ **3. a.** $\frac{31x}{72y}$ **b.** $\frac{x-7}{x+7}$ **c.** $\frac{1}{2(x+2)}$ **d.** $\frac{1}{x-6}$ **e.** $\frac{-a-b}{c+d}$ **f.** $\frac{-m-2n}{n+2m}$ **g.** -1 **h.** -2 **4. a.** 5 **b.** -4, -12 **5.** 70.4 ft **6.** $5,460 **7.** 1.25 amps **8.** 126.72 lb **9.** inverse variation **10.** 0.2 **11. a.** -1 **b.** $\frac{2a-1}{a+2}$ **c.** $\frac{h^2-4h+4}{h^6+8h^3+16}$ **d.** $\frac{t-2}{t}$ **e.** $\frac{3x(x-1)}{(x-3)(x+1)}$ **12. a.** $\frac{5y-3}{x-y}$ **b.** $\frac{6x-7}{x^2+2}$ **c.** $-\frac{2}{t-3}$ **d.** $-\frac{1}{p+12}$ **13. a.** $60a^2h^3$ **b.** $ab^2(b-1)$ **c.** $(x-5)(x+5)(x+1)$ **d.** $(m^2+2m+4)(m-2)^2$ **14. a.** $\frac{9a+8}{a+1}$ **b.** $\frac{3x+2yz}{12z^2}$ **c.** $\frac{4x^2+9x+12}{(x-4)(x+3)}$ **d.** $\frac{20+12y}{15y(x-2)}$ **e.** $\frac{x^2+26x+3}{(x+3)(x-3)^2}$ **15. a.** $\frac{p-3}{2(p+2)}$ **b.** $\frac{b+2a}{2b-a}$ **c.** $\frac{x-2}{x+3}$

d. $\dfrac{x^2y^2}{(x-y)^2(y^2-x^2)}$ **16. a.** 5 **b.** 2 **c.** 8, −9 **d.** −1, −2 **e.** 2 and −2 are extraneous. **f.** 0 **17. a.** $y^2 = \dfrac{x^2b^2 - a^2b^2}{a^2}$ **b.** $b = \dfrac{Ha}{2a - H}$ **18.** 5 mph **19.** 50 mph **20.** $14\frac{2}{5}$ hr **21.** $18\frac{2}{3}$ days **22.** about 3,500,000 in. lb/rad **23. a.** $\dfrac{5h^3}{k^2}$ **b.** $-\dfrac{x^3}{2y^3}$ **24. a.** $6a + \dfrac{16}{3}$ **b.** $-3x^2y + \dfrac{3x}{2} + y$ **25. a.** $b + 4$ **b.** $v^2 - 3v - 10$ **c.** $x^2 - 2x + 4$ **d.** $x^2 + 2x - 1 + \dfrac{6}{2x+3}$

Chapter 6 Test (page 457)

1. $-\dfrac{2}{3xy}$ **2.** $\dfrac{2}{x-2}$ **3.** −3 **4.** $\dfrac{2x+1}{4}$ **5.** 74 **6.** 25 decibels **7.**

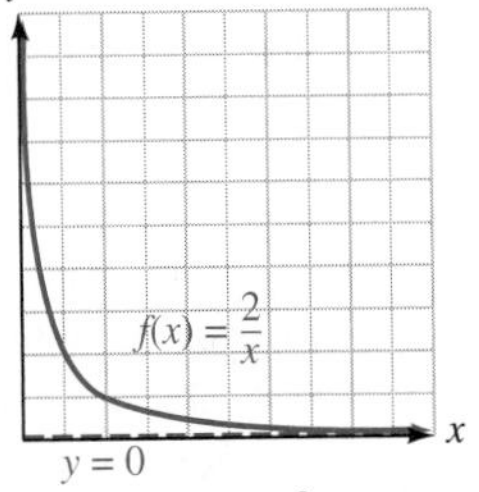

8.

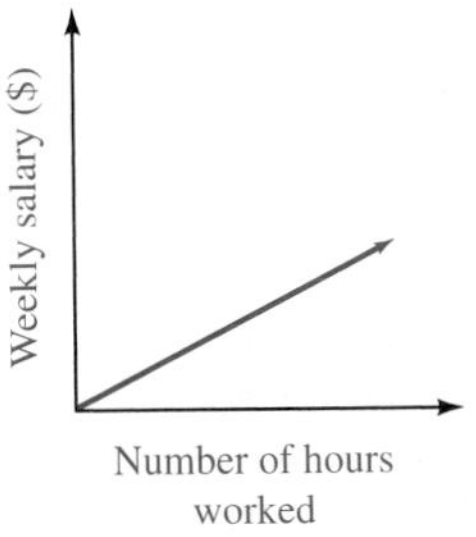

9. $\dfrac{xz}{y^4}$

10. $\dfrac{x+1}{2}$ **11.** 1 **12.** $\dfrac{(x+y)^2}{2}$ **13.** $\dfrac{2}{x+1}$ **14.** −1 **15.** $\dfrac{2}{t-4}$ **16.** 2 **17.** $\dfrac{24b^2 - 2b - 1}{3b+1}$ **18.** $\dfrac{2x+3}{(x+1)(x+2)}$ **19.** $\dfrac{u^2}{2vw}$ **20.** $\dfrac{3k^2+4k+4}{3k^2-9k-9}$ **21.** 40 **22.** 5; 3 is extraneous **23.** 6, −1 **24.** 26 **25.** $a^2 = \dfrac{x^2b^2}{b^2 - y^2}$ **26.** $r_2 = \dfrac{rr_1}{r_1 - r}$ **27.** no, $\dfrac{5}{11}$ of an hour **28.** 3 mph **29.** $\dfrac{-6x}{y} + \dfrac{4x^2}{y^2} - \dfrac{3}{y^3}$ **30.** $y^2 - 2y + 4 - \dfrac{56}{y+2}$

Cumulative Review Exercises (page 459)

1. $\frac{8}{3}$ **2.** −2 **3.** 14 **4.** Life expectancy will increase 0.1 year each year during this period. **5.** $y = -7x + 54$ **6.** $y = -\frac{11}{6}x - \frac{7}{3}$ **7.** D: $(-\infty, \infty)$; R: $(-\infty, \infty)$

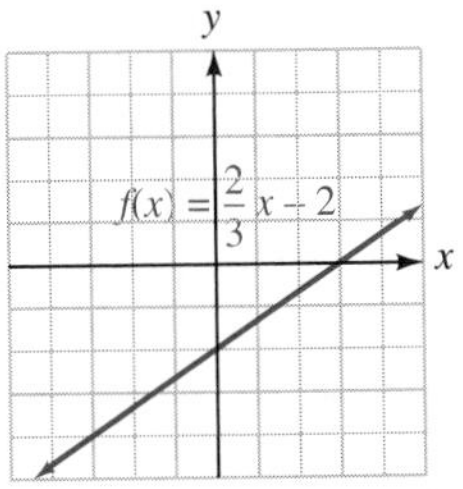

8. $T_1 = 80$, $T_2 = 60$ **9.**

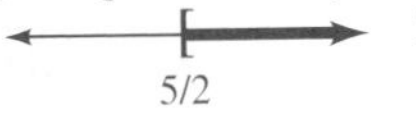

10.

11.

12.

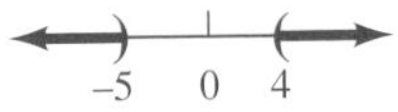

13. a^8b^4 **14.** $\dfrac{b^4}{a^4}$ **15.** 81 **16.** $\dfrac{x^{21}y^3}{8}$ **17.** 42,500 **18.** 0.000712 **19.** $y = \dfrac{kxz}{r}$ **20.** 8

21.

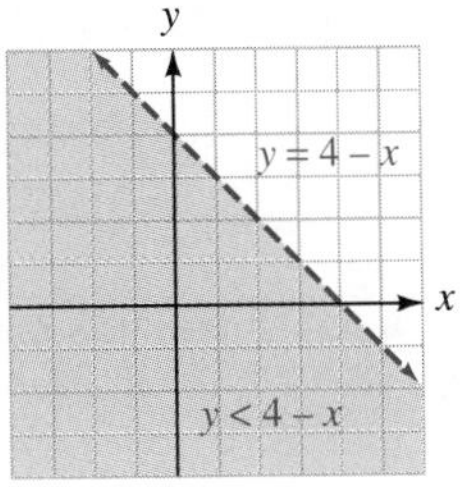

22. D: $(-\infty, \infty)$; R: $[-3, \infty)$

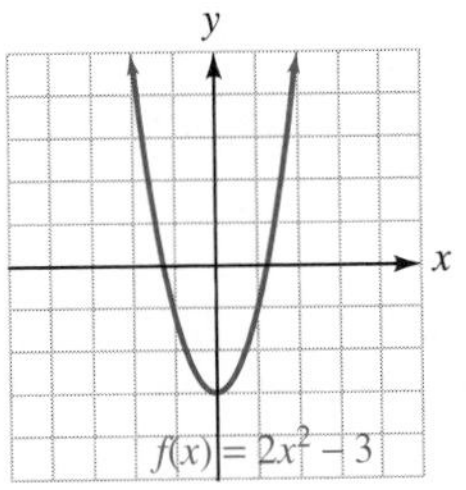

23. 18 **24.** 7 **25.** $2x^3 + x^2 + 12$

26. $-3x^2 - 3$ **27.** $6x^2 - 7x - 20$ **28.** $4x^6 - 4x^3 + 1$ **29.** yes **30.** about 17 **31.** about 14 **32.** 4 **33.** $3rs^3(r - 2s)$ **34.** $(x - y)(5 - a)$ **35.** $(x + y)(u + v)$ **36.** $(9x^2 + 4y^2)(3x + 2y)(3x - 2y)$ **37.** $(2x - 3y^2)(4x^2 + 6xy^2 + 9y^4)$ **38.** $(2x + 3)(3x - 2)$ **39.** $(3x - 5)^2$ **40.** $(5x + 3)(3x - 2)$ **41.** $-\frac{1}{3}, -\frac{7}{2}$ **42.** 0, 2, −2 **43.** $b^2 = \dfrac{a^2y^2}{a^2 - x^2}$ **44.** 9 in. by 12 in. **45.** $\dfrac{2x-3}{3x-1}$ **46.** $-\dfrac{q}{p}$ **47.** $\dfrac{4}{x-y}$ **48.** $\dfrac{a^2+ab^2}{a^2b-b^2}$ **49.** 0 **50.** −17 **51.** $x + 4$ **52.** $-x^2 + x + 5 + \dfrac{8}{x-1}$

Study Set Section 7.1 (page 471)

1. square root **3.** radical **5.** odd **7.** simplified **9.** $b^2 = a$ **11.** two, positive **13.** x **15.** 3, up **17.** 0, −1, −2, −3, −4

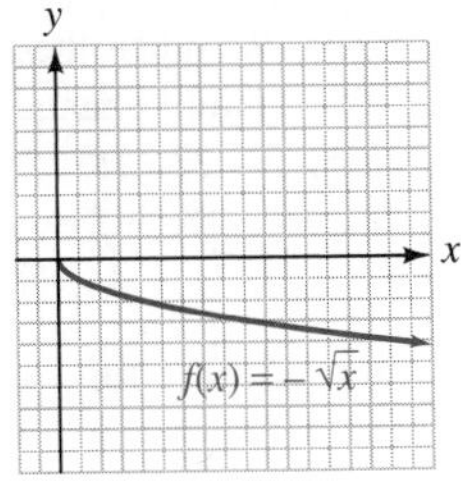

19. $\sqrt{x^2} = |x|$ **21.** $f(x) = \sqrt{x - 5}$ **23.** 11 **25.** −8 **27.** $\frac{1}{3}$ **29.** 0.5 **31.** not real **33.** 4 **35.** 3.4641 **37.** 26.0624 **39.** $2|x|$ **41.** $|t + 5|$ **43.** $5|b|$ **45.** $|a + 3|$ **47.** 0 **49.** 4 **51.** 2 **53.** −10 **55.** 4.1231 **57.** 3.3322

59. D: $[-4, \infty)$, R: $[0, \infty)$

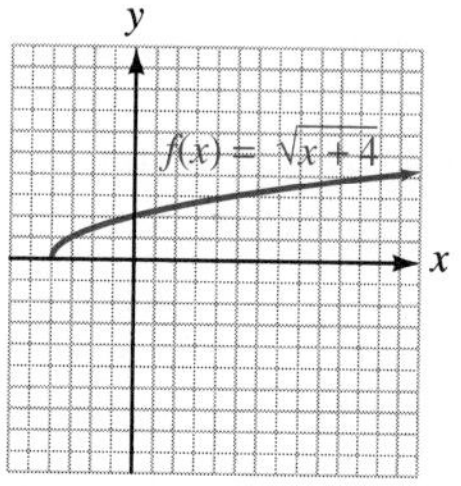

61. D: $(-\infty, \infty)$, R: $(-\infty, \infty)$

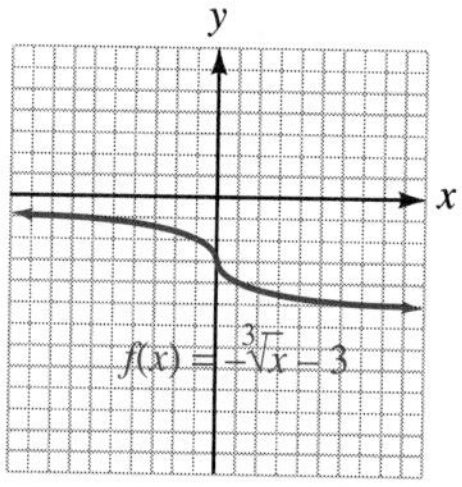

63. 1 **65.** $-\frac{2}{3}$ **67.** 0.4 **69.** $2a$ **71.** $-10pq$ **73.** $-0.4s^3t^2$ **75.** 3 **77.** -3 **79.** not real **81.** $\frac{2}{5}$ **83.** $\frac{1}{2}$ **85.** $2a$ **87.** $2a$ **89.** k^3 **91.** $\frac{1}{2}m$ **93.** $x + 2$ **95.** 7.0 in. **97.** about 61.3 beats/min **99.** 13.4 ft **101.** 3.5% **105.** 1 **107.** $\frac{3(m^2 + 2m - 1)}{(m + 1)(m - 1)}$

Study Set Section 7.2 (page 480)

1. radical **3.** extraneous **5. a.** Square both sides. **b.** Cube both sides. **7. a.** x **b.** $x - 5$ **c.** $32x$ **d.** $x + 3$ **9.** Only one side of the equation was squared. **11.** If we raise two equal quantities to the same power, the results are equal quantities. **13.** $2\sqrt{x - 2}$, 4, 16, 8, 24 **15.** 2 **17.** 4 **19.** 0 **21.** 4 **23.** 8 **25.** $\frac{5}{2}, \frac{1}{2}$ **27.** 1 **29.** 16 **31.** 14, ~~6~~ **33.** 4, 3 **35.** 2, ~~7~~ **37.** 2, -1 **39.** -1, ~~1~~ **41.** ~~1~~, no solutions **43.** 0, ~~4~~ **45.** $-$~~3~~, no solutions **47.** 1, 9 **49.** 4, ~~0~~ **51.** 2, 142 **53.** ~~6~~, no solutions **55.** $h = \frac{v^2}{2g}$ **57.** $l = \frac{8T^2}{\pi^2}$ **59.** $A = P(r + 1)^3$ **61.** $v^2 = c^2\left(1 - \frac{L_A^2}{L_B^2}\right)$ **63.** 178 ft **65.** about 488 watts **67.** 10 lb **69.** \$5 **73.** 2.5 foot-candles **75.** 0.41511 in.

Study Set Section 7.3 (page 489)

1. rational (or fractional) **3.** index, radicand **5.** $25^{1/5}$, $(\sqrt[3]{-27})^2$, $16^{-3/4}$, $(\sqrt{81})^3$, $-\left(\frac{9}{64}\right)^{1/2}$

7.

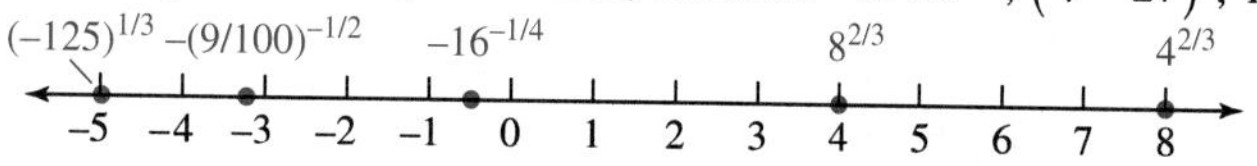

9. x^{m+n} **11.** x^{m-n} **13.** $\sqrt[n]{x}$

15. 0, 1, 2, 3, 4

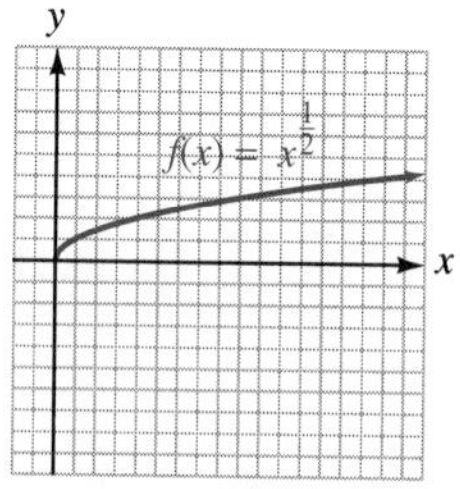

17. $\sqrt{100a^4}$, $10a^2$ **19.** $\sqrt[3]{x}$ **21.** $\sqrt[4]{3x}$ **23.** $\sqrt[4]{\frac{1}{2}x^3y}$ **25.** $\sqrt{x^2 + y^2}$ **27.** $m^{1/2}$ **29.** $(3a)^{1/4}$ **31.** $\left(\frac{1}{7}abc\right)^{1/6}$ **33.** $(a^2 - b^2)^{1/3}$ **35.** 2 **37.** 2 **39.** 2 **41.** 2 **43.** $\frac{1}{2}$ **45.** -2 **47.** not real **49.** -3 **51.** $5|y|$ **53.** $2|x|$ **55.** $3x$ **57.** not real **59.** 216 **61.** 27 **63.** 1,728 **65.** $\frac{1}{4}$ **67.** $125x^6$ **69.** $\frac{4x^2}{9}$ **71.** $\frac{1}{2}$ **73.** $\frac{1}{8}$ **75.** $\frac{1}{64x^3}$ **77.** $\frac{1}{9y^2}$ **79.** $\frac{16}{81}$ **81.** $-\frac{3}{2x}$ **83.** 2.47 **85.** 1.01 **87.** $5^{5/7}$ **89.** $4^{3/5}$ **91.** $9^{1/5}$ **93.** $\frac{1}{36}$ **95.** 3 **97.** a **99.** $a^{2/9}$ **101.** $a^{3/4}b^{1/2}$ **103.** $\frac{2x}{3}$ **105.** $\frac{1}{3}x$ **107.** $y + y^2$ **109.** $x^2 - x + x^{3/5}$ **111.** $\sqrt{p}$ **113.** $\sqrt{5b}$ **115.** 736 ft/sec **117.** 1.96 units **119.** 4,608 in.^{2}, 32 ft^2 **123.** $x < 3$ **125.** $r > 28$

Study Set Section 7.4 (page 499)

1. like **3.** factor **5.** $\sqrt[n]{a}\sqrt[n]{b}$, product, roots **7. a.** $\sqrt{4 \cdot 5}$ **b.** $\sqrt{4} \cdot \sqrt{5}$ **c.** $\sqrt{4 \cdot 5} = \sqrt{4} \cdot \sqrt{5}$ **9. a.** $\sqrt{5}$, $\sqrt[3]{5}$ (answers may vary); no **b.** $\sqrt{5}$, $\sqrt{6}$ (answers may vary); no **11.** $8k^3$, $8k^3$ **13.** 6 **15.** t **17.** $5x$ **19.** 10 **21.** $7x$ **23.** $6b$ **25.** 2 **27.** $3a$ **29.** $2\sqrt{5}$ **31.** $-10\sqrt{2}$ **33.** $2\sqrt[3]{10}$ **35.** $-3\sqrt[3]{3}$ **37.** $2\sqrt[4]{2}$ **39.** $2\sqrt[5]{3}$ **41.** $\frac{\sqrt{7}}{3}$ **43.** $\frac{\sqrt[3]{7}}{4}$ **45.** $\frac{\sqrt[4]{3}}{10}$ **47.** $\frac{\sqrt[5]{3}}{2}$ **49.** $5x\sqrt{2}$ **51.** $4\sqrt{2b}$ **53.** $-4a\sqrt{7a}$ **55.** $5ab\sqrt{7b}$ **57.** $-10\sqrt{3xy}$ **59.** $-3x^2\sqrt[3]{2}$ **61.** $2x^4y\sqrt[3]{2}$ **63.** $2x^3y\sqrt[4]{2}$ **65.** $\frac{z}{4x}$ **67.** $\frac{\sqrt[4]{5x}}{2z}$ **69.** $10\sqrt{2x}$ **71.** $\sqrt[5]{7a^2}$ **73.** $-\sqrt{2}$ **75.** $2\sqrt{2}$ **77.** $9\sqrt{6}$ **79.** $3\sqrt[3]{3}$ **81.** $-\sqrt[3]{4}$ **83.** -10 **85.** $-17\sqrt[4]{2}$ **87.** $16\sqrt[4]{2}$ **89.** $-4\sqrt{2}$ **91.** $3\sqrt{2} + \sqrt{3}$ **93.** $-11\sqrt[3]{2}$ **95.** $y\sqrt{z}$ **97.** $13y\sqrt{x}$ **99.** $12\sqrt[3]{a}$ **101.** $-7y^2\sqrt{y}$ **103.** $4x\sqrt[5]{xy^2}$ **105.** $8\pi\sqrt{5}$ ft^2; 56.2 ft^2 **107.** $5\sqrt{3}$ amps; 8.7 amps **109.** $(26\sqrt{5} + 10\sqrt{3})$ in.; 75.5 in. **113.** $\frac{-15x^5}{y}$ **115.** $3p + 4 - \frac{5}{2p - 5}$

Study Set Section 7.5 (page 508)

1. FOIL **3.** irrational **5.** perfect **7. a.** $6\sqrt{6}$ **b.** 48 **c.** can't be simplified **d.** $-6\sqrt{6}$ **9.** When the numerator and the denominator of a fraction are multiplied by the same nonzero number, the value of the fraction is not changed. **11.** A radical appears in the denominator. **13.** $\sqrt{6}$, 48, 16, 4 **15.** 11 **17.** 7 **19.** 4 **21.** $5\sqrt{2}$ **23.** $6\sqrt{2}$ **25.** 5 **27.** 18 **29.** 8 **31.** 18 **33.** $2\sqrt[3]{3}$ **35.** ab^2 **37.** $5a\sqrt{b}$ **39.** $-20r\sqrt[3]{10s}$ **41.** $x^2(x + 3)$ **43.** $12\sqrt{5} - 15$ **45.** $24\sqrt{3} + 6\sqrt{14}$ **47.** $-8x\sqrt{10} + 6\sqrt{15x}$ **49.** $-1 - 2\sqrt{2}$ **51.** $5z + 2\sqrt{15z} + 3$ **53.** $3x - 2y$ **55.** $6a + 5\sqrt{3ab} - 3b$ **57.** $18r - 12\sqrt{2r} + 4$ **59.** $-6x - 12\sqrt{x} - 6$ **61.** $\frac{\sqrt{7}}{7}$ **63.** $\frac{\sqrt{30}}{5}$ **65.** $\frac{\sqrt{10}}{4}$ **67.** 2 **69.** $\frac{\sqrt[3]{4}}{2}$ **71.** $\sqrt[3]{3}$ **73.** $\frac{\sqrt[3]{6}}{3}$ **75.** $\frac{2\sqrt{2xy}}{xy}$ **77.** $\frac{\sqrt{5y}}{y}$ **79.** $\frac{\sqrt[3]{2ab^2}}{b}$ **81.** $\frac{\sqrt[4]{4}}{2}$

83. $\frac{\sqrt[5]{2}}{2}$ **85.** $\sqrt{2} + 1$ **87.** $\frac{3\sqrt{2} - \sqrt{10}}{4}$ **89.** $2 + \sqrt{3}$ **91.** $\frac{9 - 2\sqrt{14}}{5}$ **93.** $\frac{2(\sqrt{x} - 1)}{x - 1}$ **95.** $\sqrt{2z} + 1$ **97.** $\frac{x - 2\sqrt{xy} + y}{x - y}$
99. $\frac{\sqrt{2\pi}}{2\pi\sigma}$ **101.** $\frac{\sqrt{2}}{2}$ **103.** $f/4$ **107.** 1 **109.** $\frac{1}{3}$

Study Set Section 7.6 (page 517)

1. hypotenuse **3.** Pythagorean **5.** $a^2 + b^2 = c^2$ **7.** $\sqrt{2}$ **9.** $\sqrt{3}$ **11.** 30°, 60° **13. a.** Take the positive square root of both sides. **b.** Subtract 25 from both sides. **15.** 6, 52, 4, 2 **17.** 10 ft **19.** 80 m **21.** $h = 2\sqrt{2} \approx 2.83$, $x = 2$ **23.** $x = 5\sqrt{3} \approx 8.66$, $h = 10$ **25.** $x = 4.69$, $y = 8.11$ **27.** $x = 12.11$, $y = 12.11$ **29.** $7\sqrt{2}$ cm **31.** 5 **33.** 13 **35.** 10 **37.** $2\sqrt{26}$ **39.** (3, 4) **41.** (6, 1) **43.** $\left(\frac{1}{2}, -9\right)$ **45.** $\left(\frac{\sqrt{2}}{2}, -1\right)$ **49.** $(5\sqrt{2}, 0)$, $(0, 5\sqrt{2})$, $(-5\sqrt{2}, 0)$, $(0, -5\sqrt{2})$; (7.07, 0), (0, 7.07), (−7.07, 0), (0, −7.07) **51.** $10\sqrt{3}$ mm, 17.32 mm **53.** $10\sqrt{181}$ ft, 134.54 ft **55.** about 0.13 ft **57. a.** 21.21 units **b.** 8.25 units **c.** 13.00 units **59.** yes **63.** 7

Key Concept (page 522)

1. $-3h^2\sqrt[3]{2}$ **2.** $14\sqrt[3]{e}$ **3.** $-4\sqrt{2}$ **4.** $-20r\sqrt[3]{10s}$ **5.** $3s - 2t$ **6.** $-\sqrt{21} + \sqrt{15}$ **7.** $18n - 12\sqrt{2n} + 4$ **8.** $\frac{\sqrt[3]{3k^2}}{k}$ **9.** −3 **10.** 5 **11.** 2; 7 is extraneous **12.** no solutions **13.** $3^{1/3}$ **14.** $5\sqrt[5]{a^2}$

Chapter Review (page 524)

1. a. 7 **b.** −11 **c.** $\frac{15}{7}$ **d.** not real **e.** 0.1 **f.** $5|x|$ **g.** x^4 **h.** $|x + 2|$ **2. a.** −3 **b.** −6 **c.** $4a^3b$ **d.** $\frac{s^3}{5}$ **3. a.** 5 **b.** −2 **c.** $4x^2|y|$ **d.** $22|y|$ **e.** $-\frac{1}{2}$ **f.** not real **g.** 0 **h.** 0 **4.** 48 ft **5.** 24 cm^2
6. a. D = [−2, ∞), R = [0, ∞)

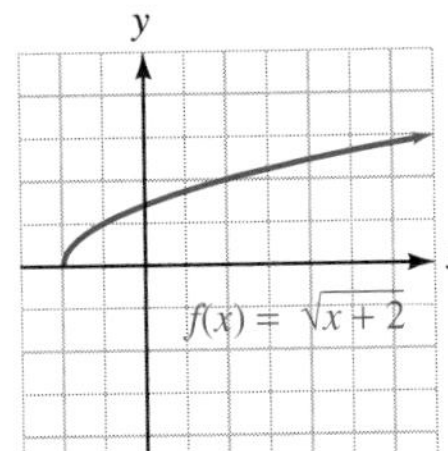

b. D = (−∞, ∞), R = (−∞, ∞)

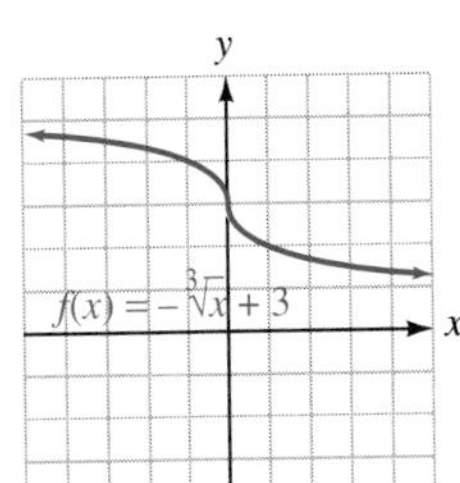

7. a. 22 **b.** 16, 9 **c.** $\frac{13}{2}$ **d.** $\frac{9}{16}$ **e.** 2, −4 **f.** 3, no solutions **8. a.** $P = \frac{A}{(r + 1)^2}$ **b.** $I = \frac{h^3b}{12}$ **9.** about 17.9 ohms **10. a.** $\sqrt{t}$ **b.** $\sqrt[4]{5xy^3}$ **11. a.** 5 **b.** −6 **c.** not real **d.** 1 **e.** $\frac{3}{x}$ **f.** $\frac{1}{3}$ **g.** −2 **h.** 5 **i.** $3ab^{1/3}$ **j.** $3cd$ **12. a.** 27 **b.** $\frac{1}{4}$ **c.** −16,807 **d.** 10 **e.** $\frac{27}{8}$ **f.** $\frac{1}{3,125}$ **g.** $125x^3y^6$ **h.** $\frac{1}{4u^4v^2}$ **13. a.** $5^{3/4}$ **b.** $a^{1/7}$ **c.** k^8 **d.** $\frac{2gh}{3}$ **14. a.** $u - 1$ **b.** $v + v^2$ **15.** $\sqrt{\frac{a}{5b}}$ **16.** Two true statements result: 16 = 16. **17. a.** $4\sqrt{15}$ **b.** $3\sqrt[3]{2}$ **c.** $2\sqrt[4]{2}$ **d.** $4\sqrt[5]{3}$ **e.** $2x^2\sqrt{2x}$ **f.** $r^5\sqrt[3]{r^2}$ **g.** $2xy\sqrt[3]{2x^2y}$ **h.** $9j^2\sqrt[3]{jk}$ **i.** $4x$ **j.** $\frac{\sqrt{17xy}}{8a^2}$ **18. a.** $3\sqrt{2}$ **b.** $11\sqrt{5}$ **c.** 0 **d.** $-8\sqrt[4]{2}$ **e.** $29x\sqrt{2}$ **f.** $13\sqrt[3]{2}$ **19.** $(6\sqrt{2} + 2\sqrt{10})$ in., 14.8 in. **20. a.** 7 **b.** $6\sqrt{10}$ **c.** 32 **d.** 72 **e.** $3x$ **f.** 33 **g.** $-2x$ **h.** $12\sqrt[3]{3}$ **i.** $4 - 3\sqrt{2}$ **j.** $-20x^3y^3\sqrt{xy}$ **k.** $3b + 6\sqrt{b} + 3$ **l.** $6u - 12 + \sqrt{u}$ **21. a.** $\frac{10\sqrt{3}}{3}$ **b.** $\frac{\sqrt{15}}{5}$ **c.** $\frac{\sqrt{xy}}{y}$ **d.** $\frac{\sqrt[3]{u^2}}{u^2v^2}$ **e.** $2(\sqrt{2} + 1)$ **f.** $\frac{a + 2\sqrt{a} + 1}{a - 1}$ **22.** $r = \frac{\sqrt[3]{6\pi^2v}}{2\pi}$ **23.** 17 ft **24.** 88 yd **25.** $7\sqrt{2}$ m **26.** $6\sqrt{3}$ cm, 18 cm **27. a.** 7.07 in. **b.** 8.66 cm **28. a.** 13; $M\left(\frac{5}{2}, -6\right)$ **b.** $2\sqrt{2}$; $M(-3, 7)$

Chapter 7 Test (page 529)

1. 0, 1, 1.41, 2, 3, 3.32, 4

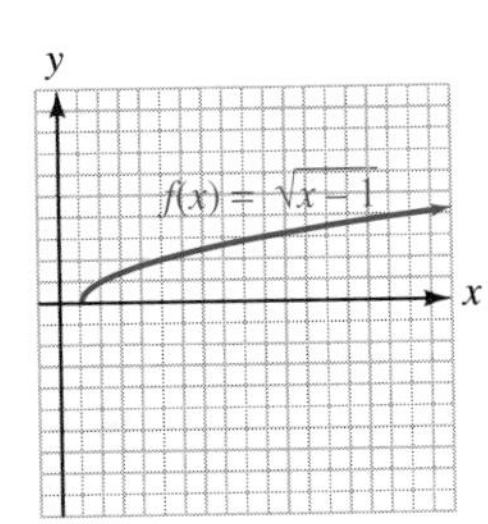

2. 3.0 mi **3.** $\frac{1}{3}$ **4.** 10 **5.** 4 is extraneous **6.** $G = \frac{4\pi^2r^3}{Mt^2}$ **7.** 2 **8.** 9

9. $\frac{1}{216}$ **10.** $\frac{9}{4}$ **11.** $2^{4/3}$ **12.** $8xy$ **13.** $|x|$ **14.** $2|x|\sqrt{2}$ **15.** $3x\sqrt[3]{2x^2}$ **16.** $3x^2y^4\sqrt{2}$ **17.** $-4xy^2$ **18.** $\frac{2}{3}a$ **19.** not real **20.** $t + 8$ **21.** $4\sqrt{3}$ **22.** $5xy^2\sqrt{10xy}$ **23.** $2x^5y\sqrt[3]{3}$ **24.** $\frac{1}{4a}$ **25.** $-\sqrt{3}$ **26.** $14\sqrt[3]{5}$ **27.** $2y^2\sqrt{3y}$ **28.** $6z\sqrt[4]{3z}$ **29.** $-6x\sqrt{y} - 2xy^2$ **30.** $3 - 7\sqrt{6}$ **31.** $\dfrac{\sqrt{5}}{5}$ **32.** $2\sqrt[3]{3}$ **33.** $\sqrt{2}(\sqrt{5} - 3)$ **34.** $\sqrt{3t} + 1$ **35.** 9.24 cm **36.** 8.67 cm **37.** 25, $\left(10, \frac{17}{2}\right)$ **38.** $\dfrac{\pi\sqrt{2}}{2}$ sec, 2.22 sec **39.** 28 in.

Study Set Section 8.1 (page 538)

1. quadratic **3.** perfect **5.** $x = \sqrt{c}, x = -\sqrt{c}$ **7.** 3, −2, 5 **9.** It is a solution. **11. a.** 36 **b.** $\frac{25}{4}$ **c.** $\frac{1}{16}$ **13. a.** Subtract 35 from both sides. **b.** Add 36 to both sides. **15.** 4 is not a factor of the numerator. It cannot be divided out. **17.** $x^2 - 8x + 15 = 0$ **19. a.** 2; $2\sqrt{5}, -2\sqrt{5}$ **b.** ± 4.47 **21.** 0, −2 **23.** 5, −5 **25.** −2, −4 **27.** 2, $\frac{1}{2}$ **29.** ± 6 **31.** $\pm\sqrt{5}$ **33.** $\pm\dfrac{4\sqrt{3}}{3}$ **35.** 0, −2 **37.** 4, 10 **39.** $-5 \pm \sqrt{3}$ **41.** $d = \dfrac{\sqrt{6h}}{2}$ **43.** $c = \dfrac{\sqrt{Em}}{m}$ **45.** 2, −4 **47.** 1, $-\dfrac{1}{2}$ **49.** $\dfrac{1}{2}, -\dfrac{2}{3}$ **51.** $-4 \pm \sqrt{10}$ **53.** $1 \pm 3\sqrt{2}$ **55.** $\dfrac{3 \pm 2\sqrt{3}}{3}$ **57.** $\dfrac{1 \pm 2\sqrt{2}}{2}$ **59.** $\dfrac{-5 \pm \sqrt{41}}{4}$ **61.** $\dfrac{-7 \pm \sqrt{29}}{10}$ **63.** width: $7\frac{1}{4}$ ft; length: $13\frac{3}{4}$ ft **65.** 1.6 sec **67.** 1.70 in. **69.** 0.92 in. **71.** 2.9 ft, 6.9 ft **75.** $2ab^2\sqrt[3]{5}$ **77.** x^3 **79.** $5ab\sqrt{7b}$

Study Set Section 8.2 (page 547)

1. quadratic **3. a.** 0, $\frac{1}{2}$ **b.** 4, −2, 0; 0, $\frac{1}{2}$ **5. a.** true **b.** true **7.** $c = \dfrac{-s \pm \sqrt{s^2 - 4rt}}{2r}$ **9.** $2 \pm \sqrt{3}$ **11. a.** The fraction bar wasn't drawn under both parts of the numerator. **b.** A ± sign wasn't written between b and the radical. **13.** −1, −2 **15.** −6, −6 **17.** $\dfrac{-5 \pm \sqrt{5}}{10}$ **19.** $-\dfrac{3}{2}, -\dfrac{1}{2}$ **21.** $\dfrac{1}{4}, -\dfrac{3}{4}$ **23.** $\dfrac{-5 \pm \sqrt{17}}{2}$ **25.** $\dfrac{-3 \pm \sqrt{17}}{4}$ **27.** $5 \pm \sqrt{7}$ **29.** 23, −17 **31.** 8.98, −3.98 **33.** 97 ft by 117 ft **35.** 0.5 mi by 2.5 mi **37.** 34 in. **39.** \$4.80 or \$5.20 **41.** 4,000 **43.** 9% **45. a.** 83% **b.** early 1976 **49.** $n^{1/2}$ **51.** $(3b)^{1/4}$ **53.** $\sqrt[3]{t}$ **55.** $\sqrt[4]{3t}$

Study Set Section 8.3 (page 558)

1. quadratic **3.** vertex **5.**

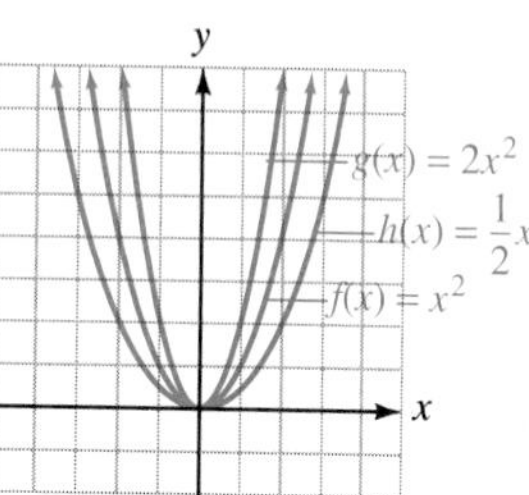

7.

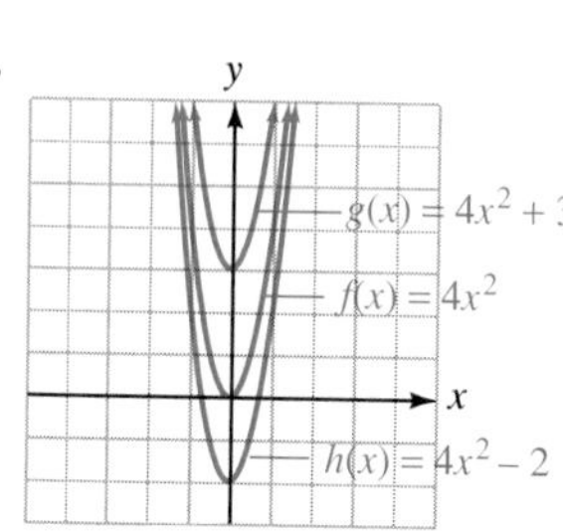

9.

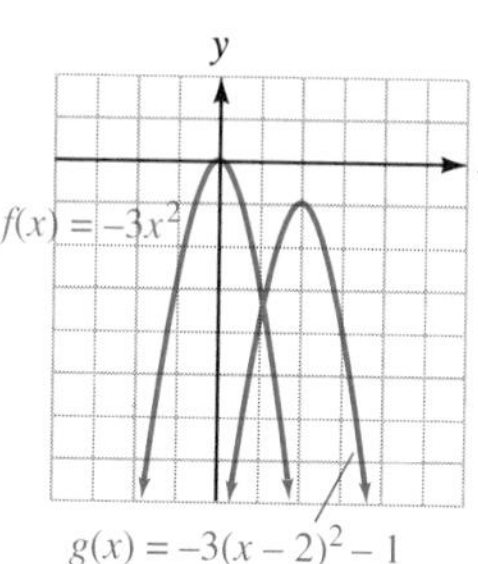

11. $h = -1$; $f(x) = 2[x - (-1)]^2 + 6$ **13.** (1, 2), $x = 1$, upward **15.** (−3, −4), $x = -3$, upward **17.** (1, −2), $x = 1$, upward **19.** (2, 21), $x = 2$, downward **21.** $\left(-\frac{2}{3}, \frac{2}{3}\right)$, $x = -\frac{2}{3}$, upward

23.

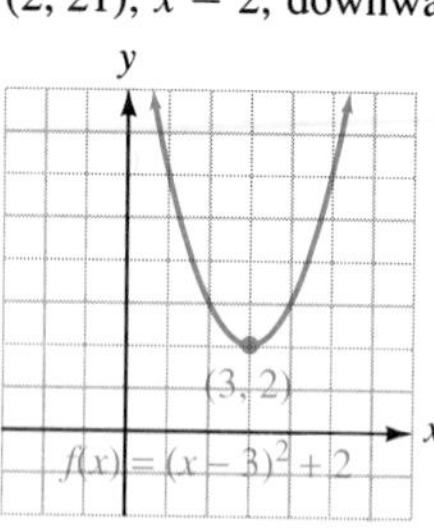

25.

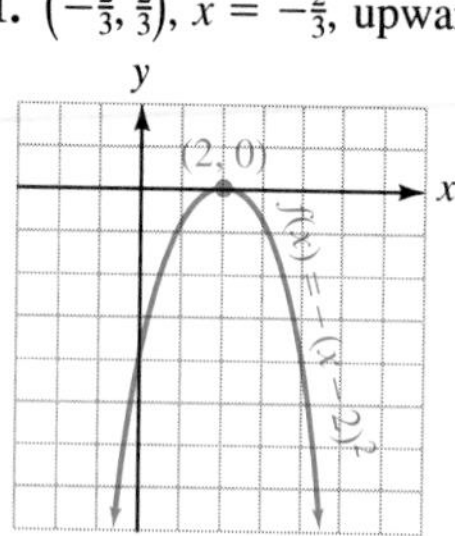

27.

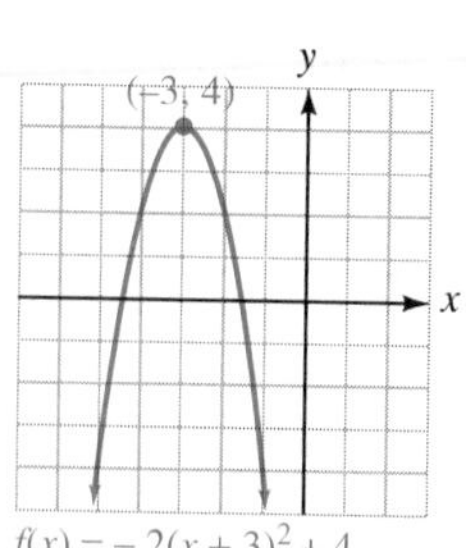

29.

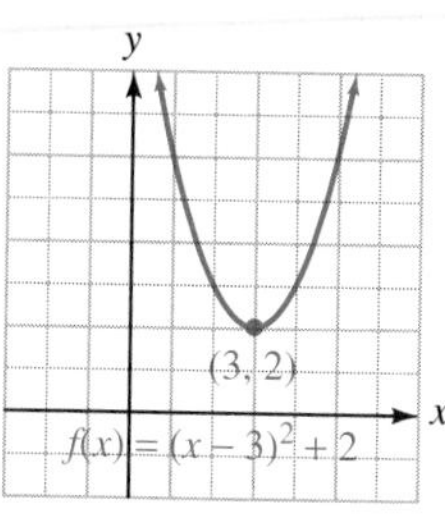

31.

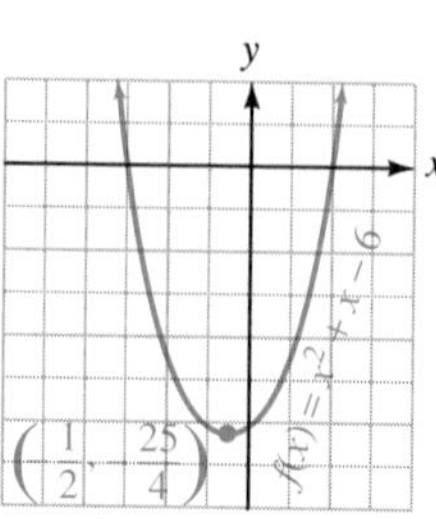

33.

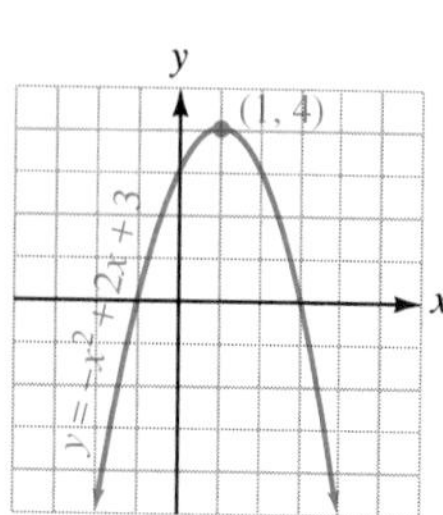

35.

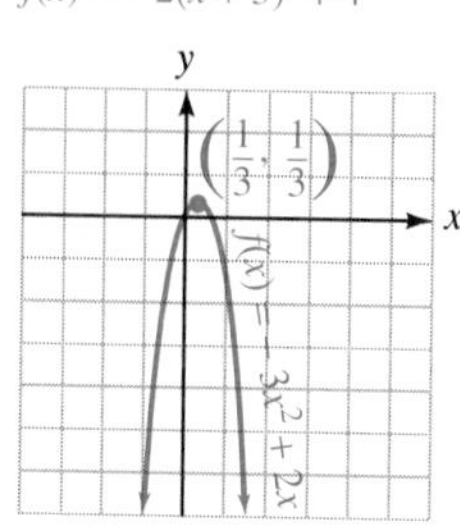

37.

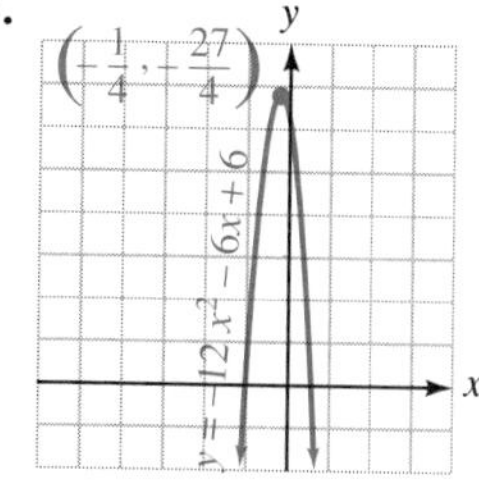

39. (0.25, 0.88) **41.** (0.50, 7.25) **43.** 2, −3 **45.** −1.85, 3.25 **47.** 3.75 sec, 225 ft **49.** 250 ft by 500 ft **51.** 15 min, $160 **53.** 1968, 1.5 million; the U.S. involvement in the war in Vietnam was at its peak **55.** 200, $7,000 **59.** $4a^2\sqrt{b}$ **61.** $\frac{\sqrt{6}}{10}$ **63.** $15b - 6\sqrt{15b} + 9$

Study Set Section 8.4 (page 569)

1. imaginary **3.** real, imaginary **5.** i or $\sqrt{-1}$ **7.** $\sqrt{-1}$ **9.** $-i$ **11.** FOIL **13.** It is a solution. **15.**

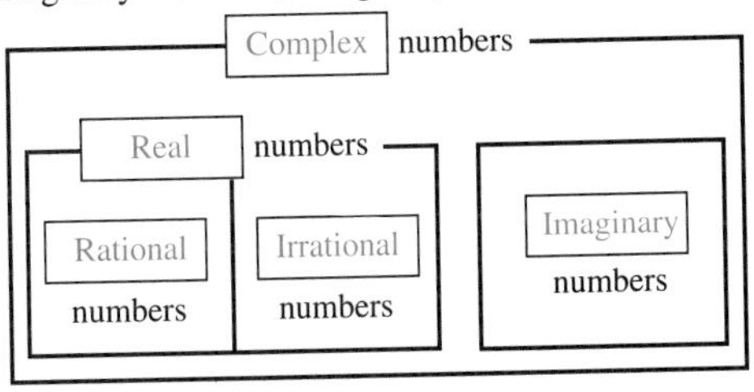

17. a. ± 1 **b.** $\pm i$ **19.** 9, $6i$, 2 **21. a.** true **b.** false **c.** false **d.** false

23. $3i$ **25.** $\sqrt{7}i$ **27.** $2\sqrt{6}i$ **29.** $-2\sqrt{6}i$ **31.** $45i$ **33.** $\frac{5}{3}i$ **35.** −6 **37.** $-2\sqrt{3}$ **39.** $\frac{5}{8}$ **41.** −20 **43.** $\pm 3i$ **45.** $\pm\frac{4\sqrt{3}}{3}i$ **47.** $-1 \pm i$ **49.** $-\frac{1}{4} \pm \frac{\sqrt{7}}{4}i$ **51.** $\frac{2}{3} \pm \frac{\sqrt{2}}{3}i$ **53.** $\frac{1}{3} \pm \frac{2\sqrt{2}}{3}i$ **55.** $8 - 2i$ **57.** $3 - 5i$ **59.** $15 + 7i$ **61.** $6 - 3i$ **63.** $-25 - 25i$ **65.** $7 + i$ **67.** $14 - 8i$ **69.** $8 + \sqrt{2}i$ **71.** $3 + 4i$ **73.** $0 - i$ **75.** $0 + \frac{4}{5}i$ **77.** $\frac{1}{8} - 0i$ **79.** $0 + \frac{3}{5}i$ **81.** $2 + i$ **83.** $-\frac{42}{25} - \frac{6}{25}i$ **85.** $\frac{1}{4} + \frac{3}{4}i$ **87.** $\frac{5}{13} - \frac{12}{13}i$ **89.** $\frac{11}{10} + \frac{3}{10}i$ **91.** $\frac{1}{4} - \frac{\sqrt{15}}{4}i$ **93.** i **95.** $-i$ **97.** 1 **99.** i **101.** $-1 + i$ **105.** 15 units, $15\sqrt{3}$ units **107.** 20 mph

Study Set Section 8.5 (page 576)

1. $b^2 - 4ac$ **3.** conjugates **5.** rational, unequal **7. a.** $x^4 = (x^2)^2$ **b.** yes **9.** 5, 6, 24, rational **11.** rational, equal **13.** complex conjugates **15.** irrational, unequal **17.** rational, unequal **19.** yes **21.** 1, −1, 4, −4 **23.** 1, −1, $\sqrt{5}$, $-\sqrt{5}$ **25.** 2, −2, $\sqrt{7}i$, $-\sqrt{7}i$ **27.** 1 **29.** no solution **31.** 16, 4 **33.** −8, −27 **35.** −1, 27 **37.** $\frac{1}{4}, \frac{1}{2}$ **39.** $1 - 2\sqrt{3}$, $1 + 2\sqrt{3}$ **41.** −1, −4 **43.** $-1, -\frac{27}{13}$ **45.** $\frac{3 + \sqrt{57}}{6}, \frac{3 - \sqrt{57}}{6}$ **47.** 1, 1, −1, −1 **49.** $1 \pm i$ **51.** 12.1 in. **53.** 30 mph **55.** 14.3 min **59.** −2 **61.** $\frac{9}{5}$

Study Set Section 8.6 (page 584)

1. quadratic **3.** interval **5.** greater **7.** undefined **9. a.** −2 **b.** −2 **c.** −2 **d.** 3 **e.** 3 **f.** 3 **11.** (1, 4),

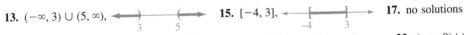

13. $(-\infty, 3) \cup (5, \infty)$,

15. $[-4, 3]$,

17. no solutions

19. $(-\infty, -3] \cup [3, \infty)$,

21. $(-5, 5)$,

23. $(-\infty, 0) \cup \left(\frac{1}{2}, \infty\right)$,

25. $\left(-\infty, -\frac{5}{3}\right) \cup (0, \infty)$,

27. $(-\infty, -3) \cup (1, 4)$,

29. $\left(-\frac{1}{2}, \frac{1}{3}\right) \cup \left(\frac{1}{2}, \infty\right)$,

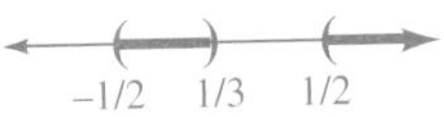

31. $(0, 2) \cup (8, \infty)$,

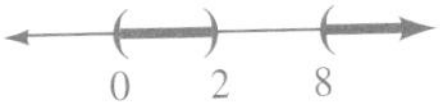

33. $\left[-\frac{34}{5}, -4\right) \cup (3, \infty)$,

35. $(-4, -2] \cup (-1, 2]$,

37. $(-\infty, -2) \cup (-2, \infty)$,

39. $(-1, 3)$. **41.** $(-\infty, -3) \cup (2, \infty)$

43.

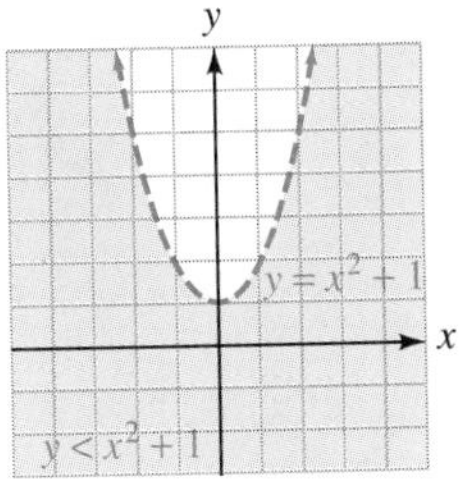

45.

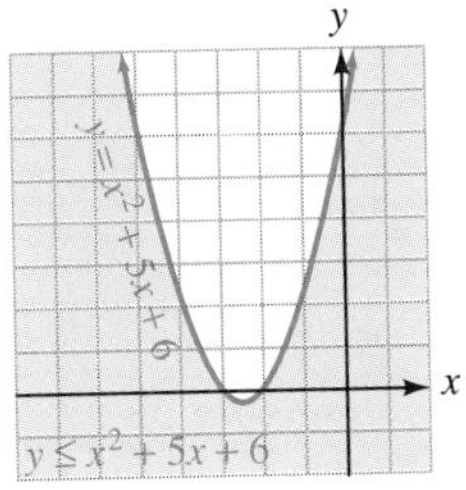

47.

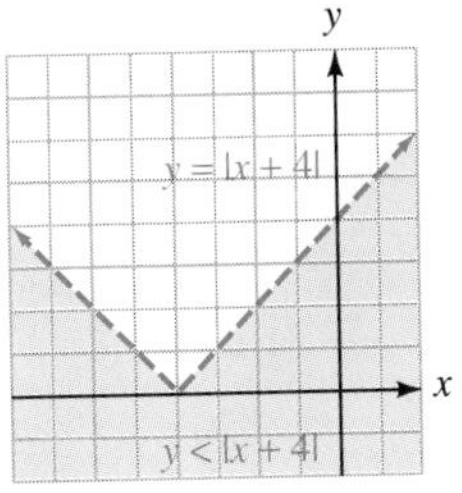

49.

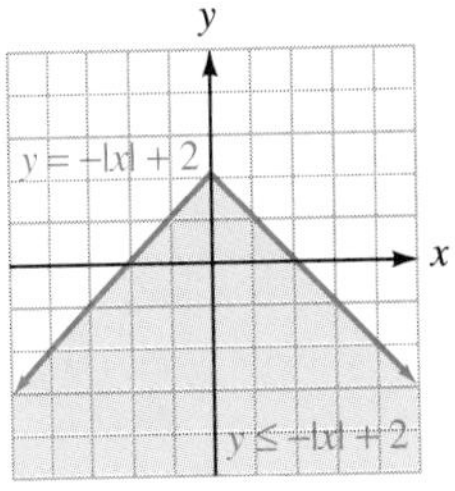

51. $(-2{,}100, -900) \cup (900, 2{,}100)$ **55.** $x = ky$ **57.** $t = kxy$ **59.** 3

Key Concept (page 587)

1. 0, −2 **2.** −3, −5 **3.** $-\frac{3}{2}$, −1 **4.** $\pm 2\sqrt{6}$ **5.** 4, 10 **6.** $\pm\frac{4\sqrt{3}}{3}$ **7.** $-5 \pm 4\sqrt{2}$ **8.** $\frac{1 \pm \sqrt{2}}{2}$ **9.** $-1 \pm i$ **10.** $\frac{3 \pm \sqrt{17}}{4}$ **11.** 23, −17 **12.** $-\frac{1}{3} \pm \frac{\sqrt{2}}{3}i$ **13.** 1, −2

Chapter Review (page 589)

1. a. −5, −4 **b.** $-\frac{1}{3}, -\frac{5}{2}$ **c.** $\pm 2\sqrt{7}$ **d.** 4, −8 **e.** 0, $-\frac{11}{5}$ **f.** $\pm\frac{7\sqrt{5}}{5}$ **2.** $\frac{1}{4}$ **3. a.** −4, −2 **b.** $\frac{3 \pm \sqrt{3}}{2}$ **4.** 6 seconds before midnight **5. a.** $5 \pm \sqrt{7}$ **b.** 0, 10 **c.** $\frac{1}{2}$, −7 **d.** $\frac{13 \pm \sqrt{163}}{3}$ **6.** sides: 1.25 in. wide; top/bottom: 2.5 in. wide **7.** 0.7 sec, 1.8 sec **8.** \$46.8 billion

9. a.

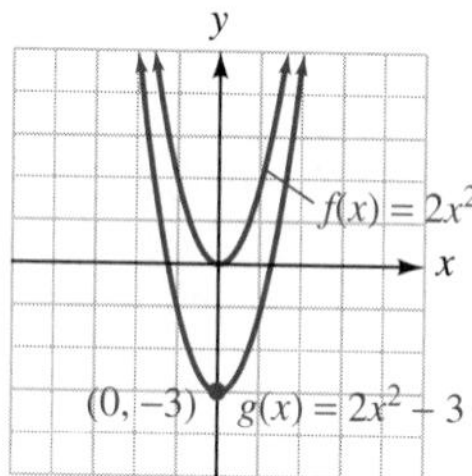

b.

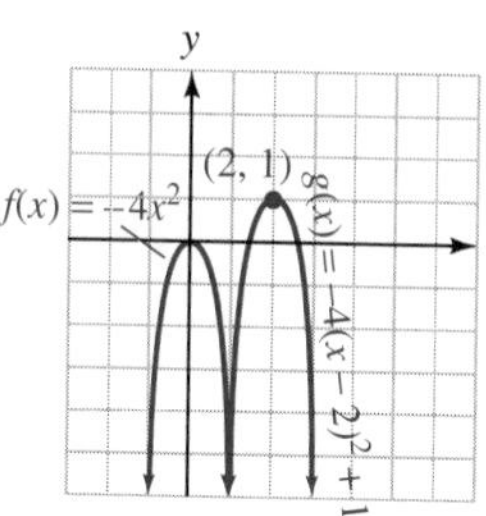

10. a.

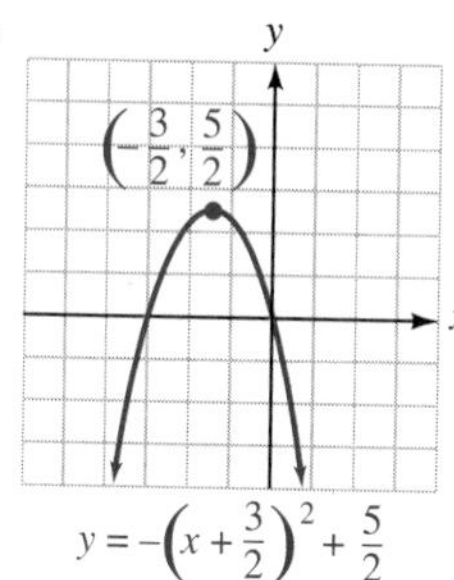

b.

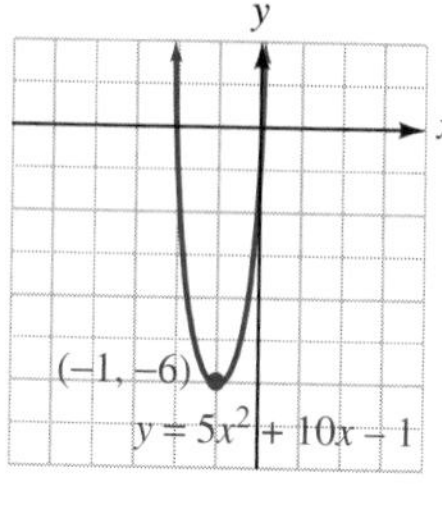

11. 1921, about 6,452 **12. a.** $2i$ **b.** $\sqrt{7}i$ **c.** $-6\sqrt{5}i$ **d.** $\frac{6}{7}i$ **e.** −3 **f.** $\frac{8}{3}$ **13. a.** $\pm 5i$ **b.** $1 \pm 2\sqrt{3}i$ **14. a.** $12 - 8i$ **b.** $2 - 68i$ **c.** $-2 - 2\sqrt{2}i$ **d.** $-18 + 128i$ **e.** $22 + 29i$ **f.** $-12 + 28\sqrt{3}i$ **15. a.** $\frac{12}{5} - \frac{6}{5}i$ **b.** $\frac{15}{17} + \frac{8}{17}i$ **c.** $-\frac{1}{7} + \frac{4\sqrt{3}}{7}i$ **d.** $0 - \frac{2}{5}i$ **16. a.** i **b.** 1 **17. a.** irrational, unequal **b.** complex conjugates **c.** equal rational numbers **18. a.** 1, 144 **b.** 8, −27 **c.** $\sqrt{3}, -\sqrt{3}, \frac{\sqrt{6}}{6}, -\frac{\sqrt{6}}{6}$ **d.** 1, $-\frac{8}{5}$ **e.** 10, 4 **19.** about 81 min **20. a.** $(-\infty, -7) \cup (5, \infty)$, −7 5 **b.** $[-9, 9]$, −9 9 **c.** $(-\infty, 0) \cup [3/5, \infty)$, 0 3/5 **d.** $(-7/2, 1) \cup (4, \infty)$, −7/2 1 4

21. a.

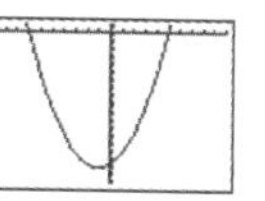

b.

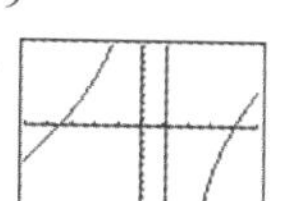

22. a.

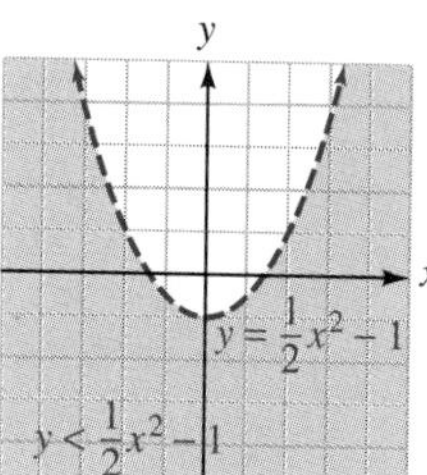

b.

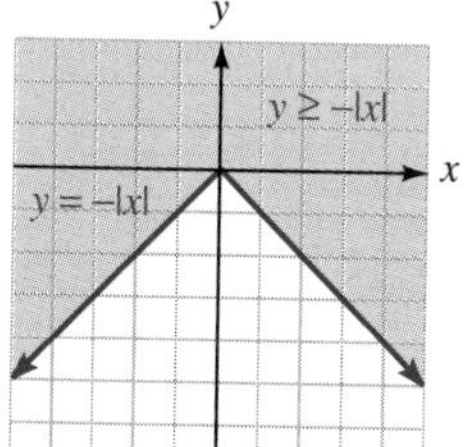

Chapter 8 Test (page 593)

1. 0, −6 **2.** $-\frac{3}{2}, -\frac{5}{3}$ **3.** 144 **4.** $\frac{1 \pm \sqrt{5}}{2}$ **5.** $\frac{4 \pm \sqrt{6}}{2}$ **6.** 4.5 ft by 1,502 ft **7.** $4\sqrt{3}i$ **8.** −1 **9.** $-1 + 11i$ **10.** $4 - 7i$ **11.** $8 + 6i$ **12.** $-10 - 11i$ **13.** $0 + i$ **14.** $\frac{1}{2} + \frac{1}{2}i$ **15.** nonreal **16.** $\pm 2\sqrt{3}i$ **17.** $2 \pm 3i$ **18.** 1, $\frac{1}{4}$

19.

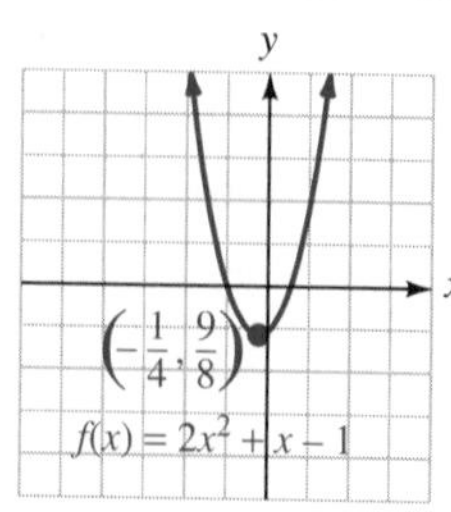

20.

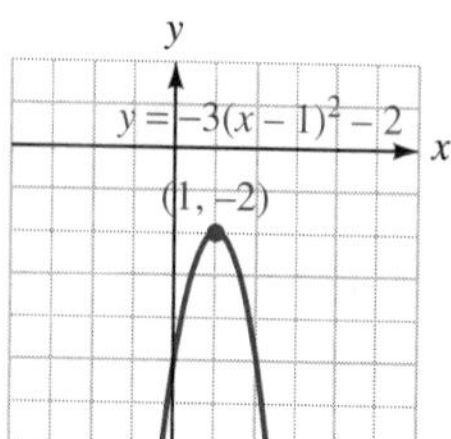

21. $(-\infty, -2) \cup (4, \infty)$, −2 4

22. $(-3, 2]$,  **23.** 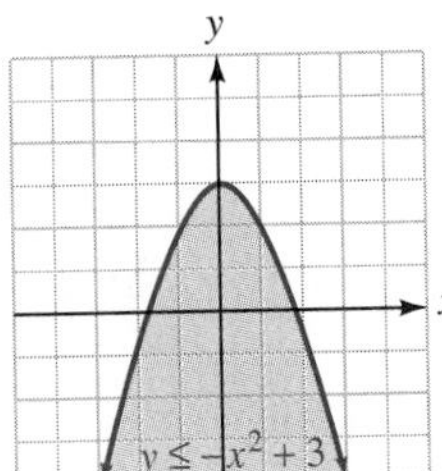

24. 20 in. **25.** 211 ft

Cumulative Review Exercises (page 595)

1. $y = 3x + 2$ **2.** $y = -\frac{2}{3}x - 2$ **3.** an increase of about 13,333 a year
4. 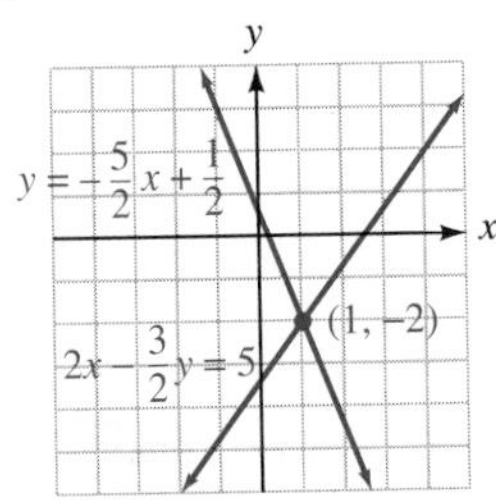

5. $(2, -1, 1)$ **6.** 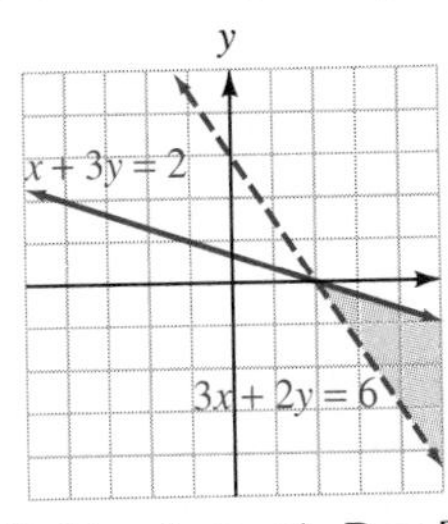

7. $\left(-\infty, -\frac{10}{9}\right)$, 
8. $(-\infty, -10] \cup [15, \infty)$, 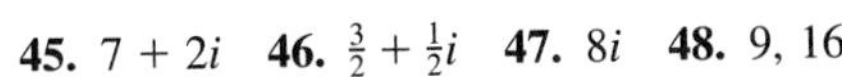 **9.** D = $(-\infty, \infty)$, R = $[-3, \infty)$ **10.** D = $(-\infty, \infty)$, R = $(-\infty, 0]$ **11.** $-4a^2 + 12a - 7$
12. $6x^2 - 5x - 6$ **13.** $(x^2 + 4y^2)(x + 2y)(x - 2y)$ **14.** $(3x + 2)(5x - 4)$ **15.** $(x - y)(x - 4)$ **16.** $(2x^2 + 5y)(4x^4 - 10x^2y + 25y^2)$
17. 6, -1 **18.** 0, $\frac{2}{3}$, $-\frac{1}{2}$ **19.** 5; 3 is extraneous **20.** $b = \dfrac{-RTV^2 + aV + PV^3}{PV^2 + a}$ **21.** $\frac{x + y}{x - y}$ **22.** 0
23. D = $(-\infty, \infty)$, R = $(-\infty, \infty)$ **24.** D = $(0, \infty)$, R = $(0, \infty)$ **25.** 9 **26.** D = $[2, \infty)$, R = $[0, \infty)$

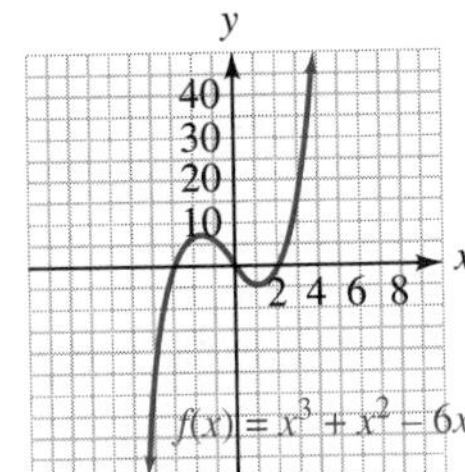

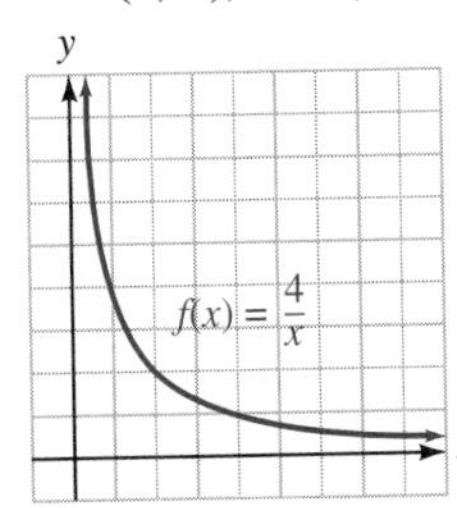

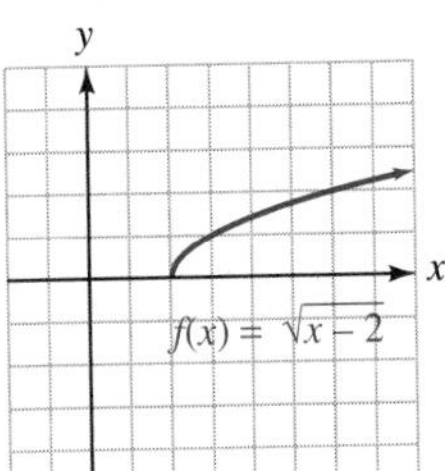

27. $-3x$ **28.** $4t\sqrt{3t}$ **29.** $\frac{1}{16}$ **30.** $x^{17/12}$ **31.** $-12\sqrt[4]{2} + 10\sqrt[4]{3}$ **32.** $-18\sqrt{6}$ **33.** $\dfrac{x + 3\sqrt{x} + 2}{x - 1}$ **34.** $\dfrac{5\sqrt[3]{x^2}}{x}$ **35.** 2, 7 **36.** $\frac{1}{4}$
37. $3\sqrt{2}$ in. **38.** $2\sqrt{3}$ in. **39.** 10 **40.** 9 **41.** 1, $-\frac{3}{2}$ **42.** $\dfrac{-2 \pm \sqrt{7}}{3}$ **43.** 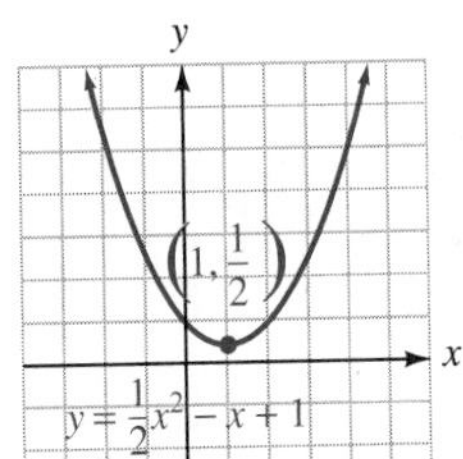

44. 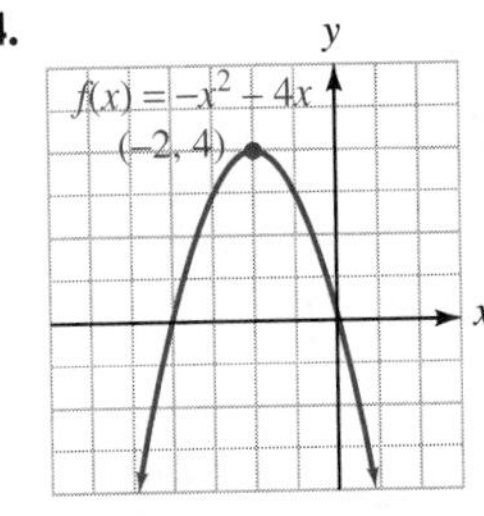

45. $7 + 2i$ **46.** $\frac{3}{2} + \frac{1}{2}i$ **47.** $8i$ **48.** 9, 16

Study Set Section 9.1 (page 604)

1. sum, $f(x) + g(x)$ **3.** product, $f(x)g(x)$ **5.** domain **7.** identity **9. a.** $g(x)$ **b.** $g(x)$
11. $-3, -2, -1, 0, 1, 2, 3$ 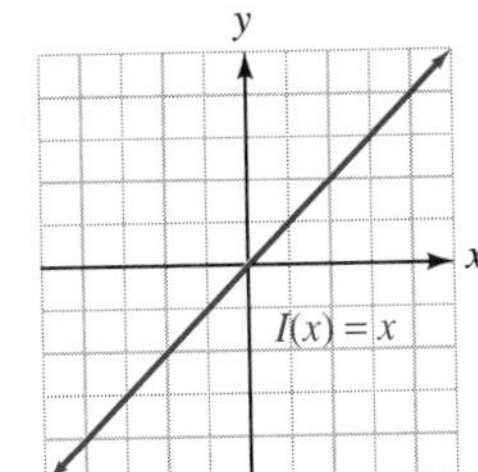

13. $(3x - 1)$, $9x$, $2x$ **15.** $7x$, $(-\infty, \infty)$ **17.** $12x^2$, $(-\infty, \infty)$ **19.** x, $(-\infty, \infty)$ **21.** $\frac{4}{3}$, $(-\infty, 0) \cup (0, \infty)$ **23.** $3x - 2$, $(-\infty, \infty)$ **25.** $2x^2 - 5x - 3$, $(-\infty, \infty)$ **27.** $-x - 4$, $(-\infty, \infty)$ **29.** $\frac{x - 3}{2x + 1}$, $(-\infty, -1/2) \cup (-1/2, \infty)$ **31.** $-2x^2 + 3x - 3$, $(-\infty, \infty)$ **33.** $(3x - 2)/(2x^2 + 1)$, $(-\infty, \infty)$ **35.** 3, $(-\infty, \infty)$ **37.** $(x^2 - 4)/(x^2 - 1)$, $(-\infty, -1) \cup (-1, 1) \cup (1, \infty)$ **39.** 7 **41.** 24 **43.** -1 **45.** $-\frac{1}{2}$ **47.** $2x^2 - 1$ **49.** $16x^2 + 8x$ **51.** 58 **53.** 110 **55.** 2 **57.** $9x^2 - 9x + 2$ **61.** $C(t) = \frac{5}{9}(2{,}668 - 200t)$ **63. a.** about \$37.50 **b.** $C(m) = \frac{1.50m}{20}$ **67.** $-\frac{3x + 7}{x + 2}$ **69.** $\frac{x - 4}{3x^2 - x - 12}$

Study Set Section 9.2 (page 612)

1. one-to-one **3.** inverses **5.** once **7.** 2 **9.** x **11.** $-6, -4, 0, 2, 8$ **13.** no **15.** 2 **17.** $2x$, $2y$, 3, y, $f^{-1}(x)$ **19.** the inverse of, inverse **21.** yes **23.** no **25.**

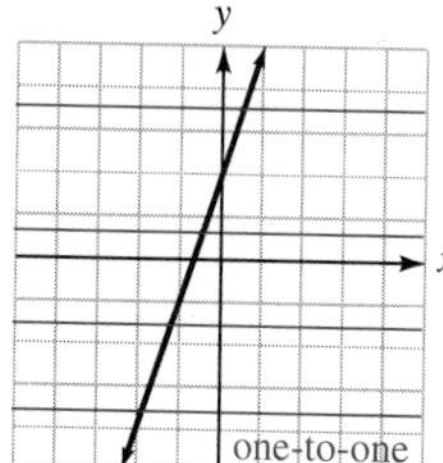
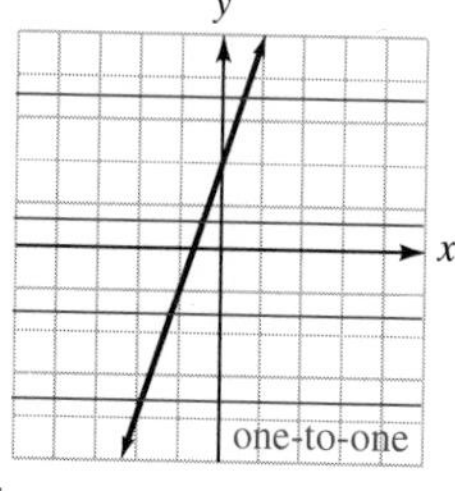

27.

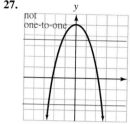

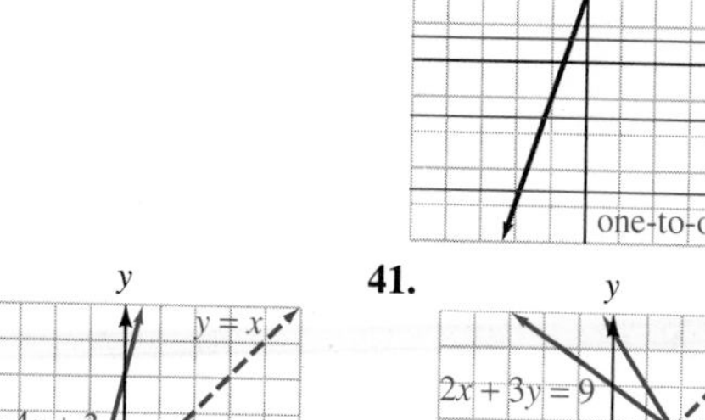

29. yes, {(2, 3), (1, 2), (0, 1)}, yes **31.** yes, {(1, 1), (1, 2), (1, 3), (1, 4)}, no **33.** $f^{-1}(x) = \frac{5}{4}x + 5$ **35.** $f^{-1}(x) = 5x - 4$ **37.** $f^{-1}(x) = 5x + 4$

39. **41.**

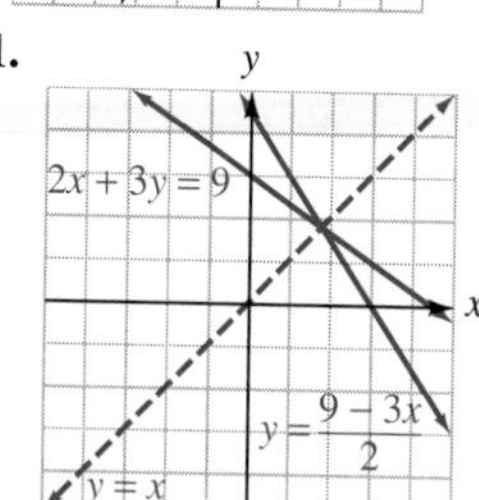

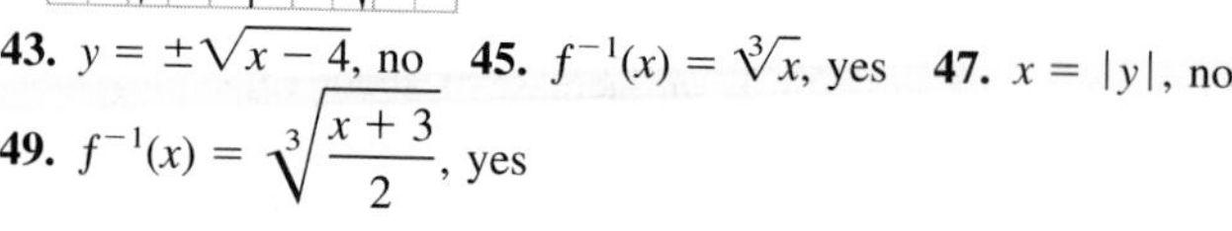

43. $y = \pm\sqrt{x - 4}$, no **45.** $f^{-1}(x) = \sqrt[3]{x}$, yes **47.** $x = |y|$, no **49.** $f^{-1}(x) = \sqrt[3]{\frac{x + 3}{2}}$, yes

51.

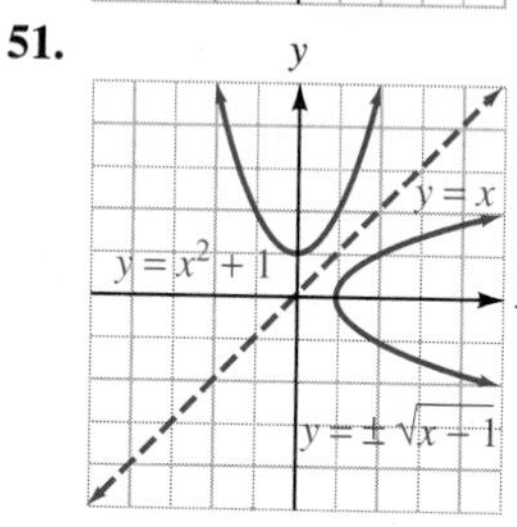

53.

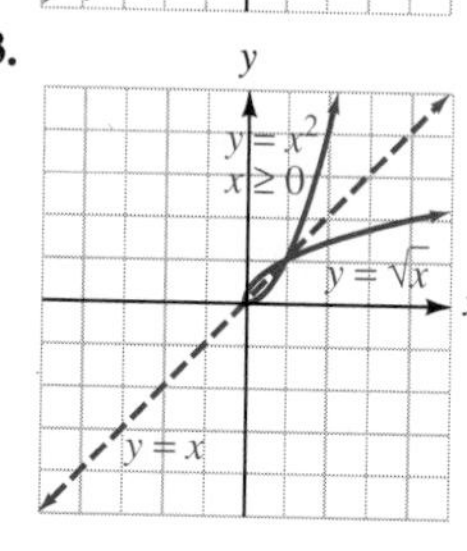

55. a. yes, no **b.** No. Twice during this period, the person's anxiety level was at the maximum threshold value. **59.** $3 - 8i$ **61.** $18 - i$ **63.** $-28 - 96i$

Study Set Section 9.3 (page 624)

1. exponential **3.** $(0, \infty)$ **5.** none **7.** increasing **9.**

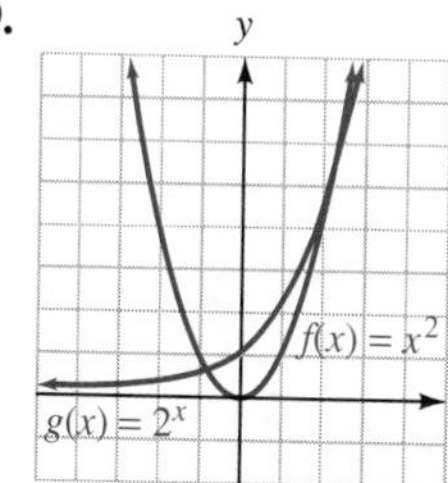

11. $A = P\left(1 + \frac{r}{k}\right)^{kt}$, $FV = PV(1 + i)^n$ **13.** $\left(1 + \frac{r}{k}\right)$, kt **15.** 2.6651 **17.** 36.5548 **19.** 8 **21.** $7^{3\sqrt{3}}$

23.

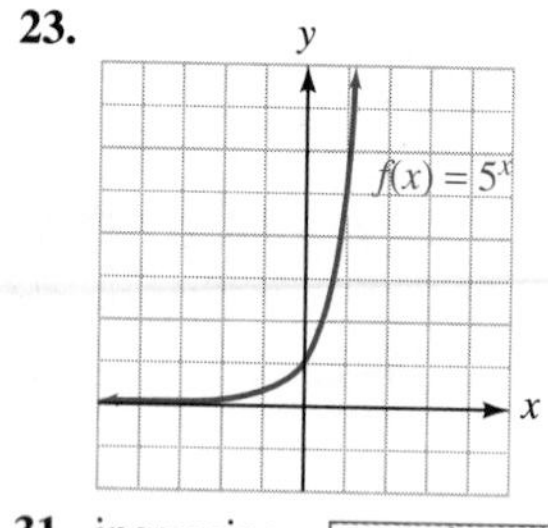

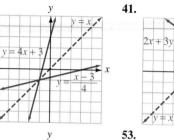

25.

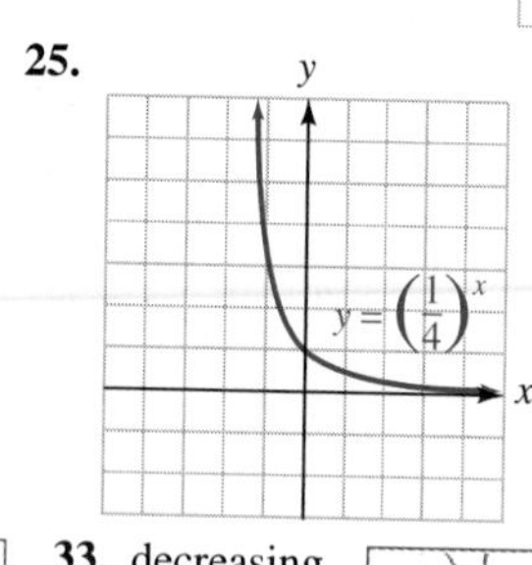

27.

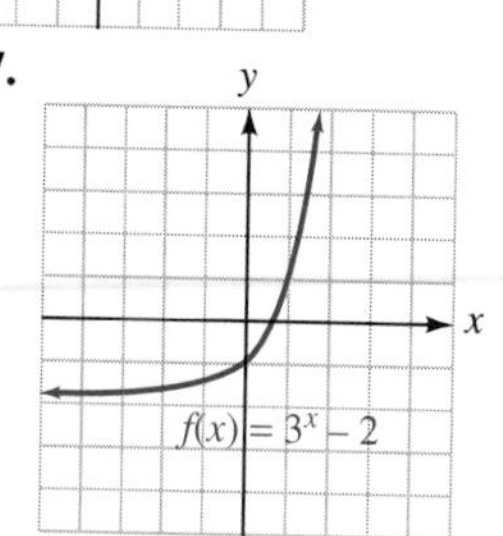

29.

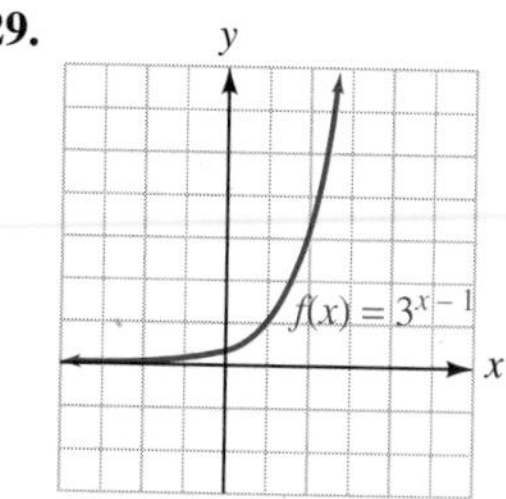

31. increasing

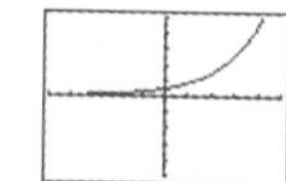

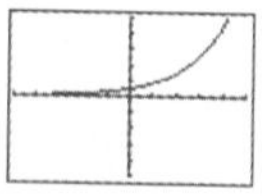

33. decreasing

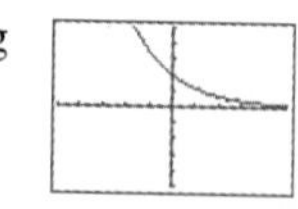

35. \$22,080.40 **37.** \$32.03 **39.** \$2,273,996.13 **41. a.** about 1500, about 1825 **b.** 6.5 billion **c.** exponential **43. a.** at the end of the 2nd year **b.** at the end of the 4th year **c.** during the 7th year **45.** 1.679046×10^8 **47.** 5.0421×10^{-5} coulombs **49.** \$1,115.33 **53.** 40 **55.** 120°

Study Set Section 9.4 (page 633)

1. the natural exponential function **3.** $(0, \infty)$ **5.** none **7.** increasing **9.** continuous **11.** 2.72

13. (number line from 0 to 4 with points marked at $\sqrt{3}/2$, $\sqrt{2}$, e, π)

15. an exponential function

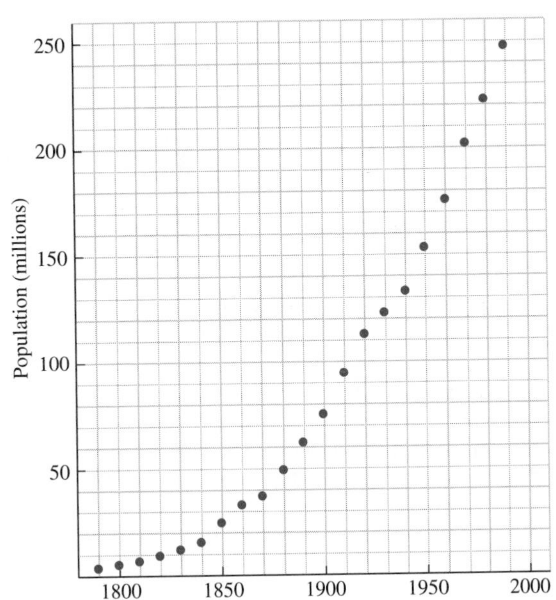

17. 1,000, 10, 0.9, 1,000

19.

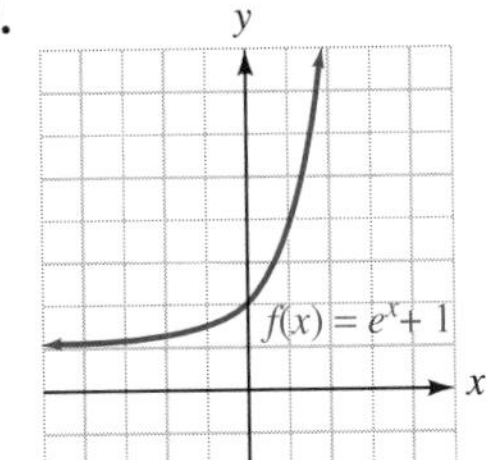

21.

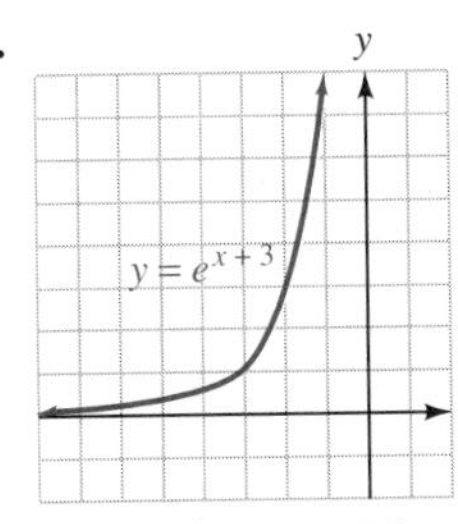

23.

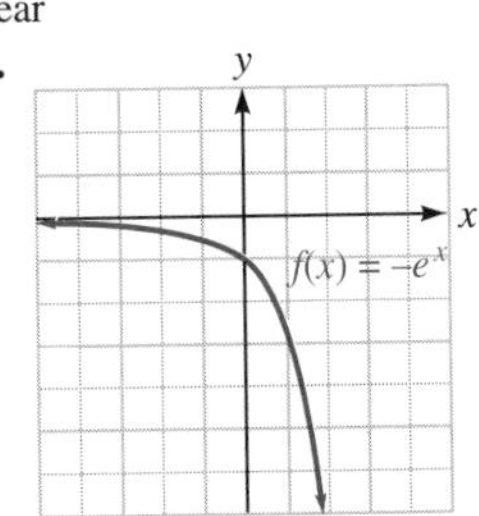

25.

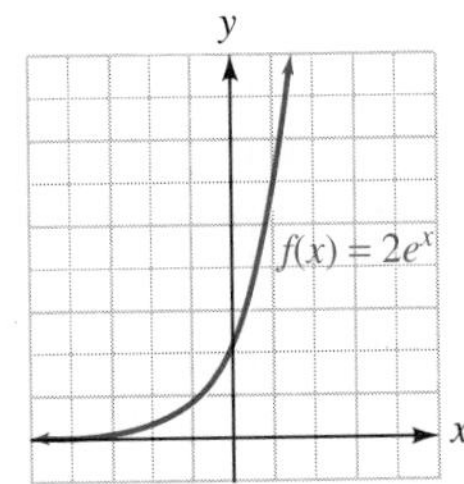

27. \$13,375.68

29. \$7,518.28 from annual compounding, \$7,647.95 from continuous compounding **31.** \$6,849.16 **33.** 10.6 billion **35.** 315 **37.** 51 **39.** 49 mps **41.** about 72 yr **45.** $4x^2\sqrt{15x}$ **47.** $10y\sqrt{3y}$

Study Set Section 9.5 (page 643)

1. logarithmic **3.** $(-\infty, \infty)$ **5.** (1, 0) **7.** increasing **9.** $x = b^y$ **11.** inverse **13.** −2, −1, 0, 1, 2 **15.** none, none, $-3, \frac{1}{2}$, 8 **17.** −0.30, 0, 0.30, 0.48, 0.60, 0.70, 0.78, 0.85, 0.90, 0.95, 1 **19.** 10 **21.** $3^4 = 81$ **23.** $12^1 = 12$ **25.** $4^{-3} = \frac{1}{64}$ **27.** $10^{-3} = 0.001$ **29.** $\log_8 64 = 2$ **31.** $\log_4 \frac{1}{16} = -2$ **33.** $\log_{1/2} 32 = -5$ **35.** $\log_x z = y$ **37.** 3 **39.** 3 **41.** 3 **43.** $\frac{1}{2}$ **45.** 64 **47.** 5 **49.** $\frac{1}{25}$ **51.** $\frac{1}{6}$ **53.** $-\frac{3}{2}$ **55.** $\frac{2}{3}$ **57.** 5 **59.** $\frac{3}{2}$ **61.** 4 **63.** 1 **65.** 0.5119 **67.** −2.3307 **69.** 25.25 **71.** 0.02

73. increasing

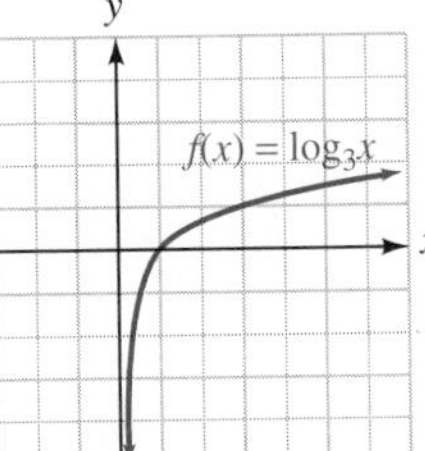

75. decreasing

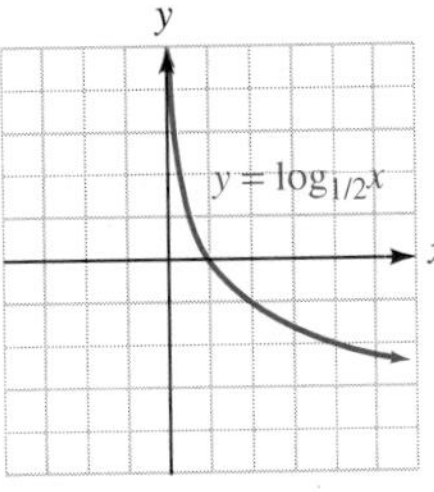

77.

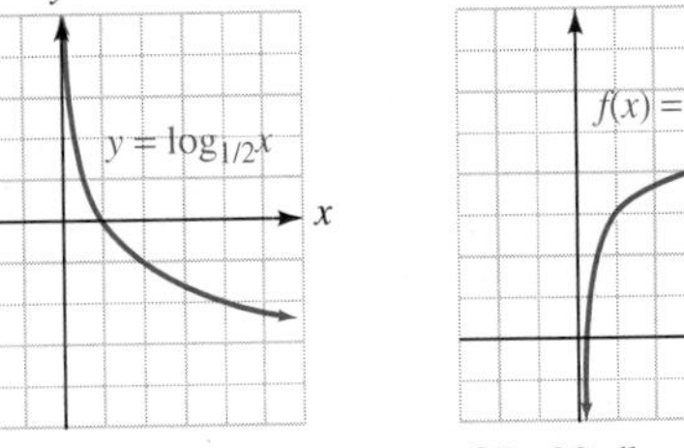

79.

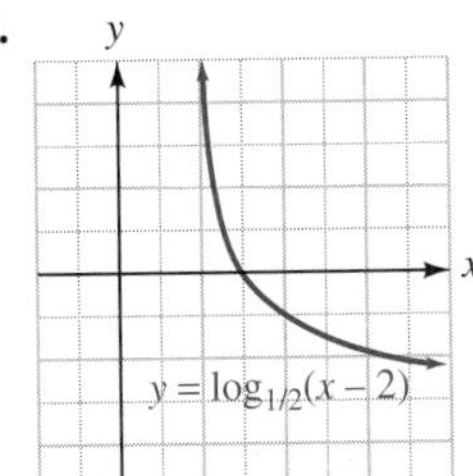

81.

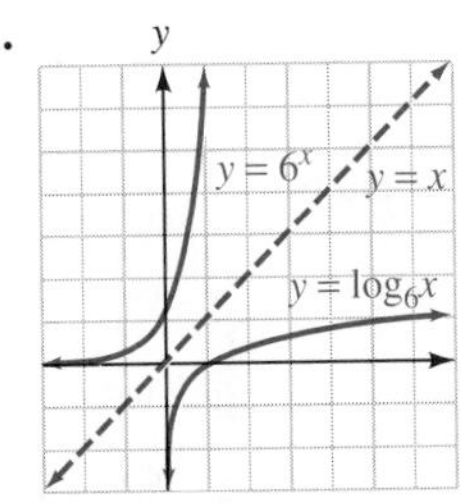

83.

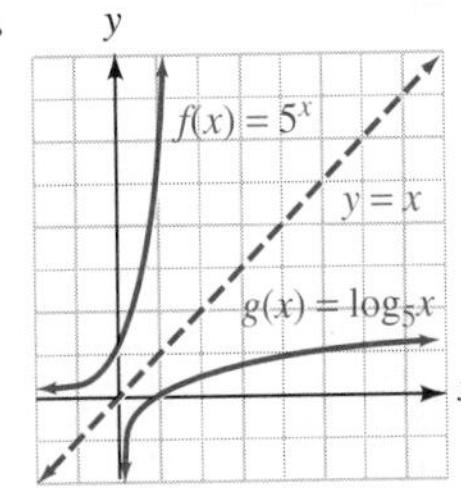

85. 29 db **87.** 49.5 db **89.** 4.4 **91.** 2,500 micrometers **93.** 3 yr old **95.** 10.8 yr **99.** 10 **101.** 0; $-\frac{5}{9}$ does not check

Study Set Section 9.6 (page 650)

1. natural **3.** −0.69, 0, 0.69, 1.10, 1.39, 1.61, 1.79, 1.95, 2.08, 2.20, 2.30 **5.** y-axis **7.** $(-\infty, \infty)$ **9.** $e^y = x$ **11.** e **13.** $\frac{\ln 2}{r}$ **15.** 3.5596 **17.** −5.3709 **19.** 0.5423 **21.** undefined **23.** 4.0645 **25.** 69.4079 **27.** 0.0245 **29.** 2.7210

31.

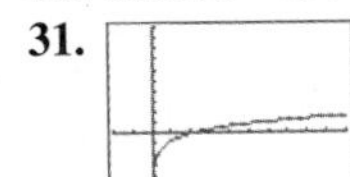

33.

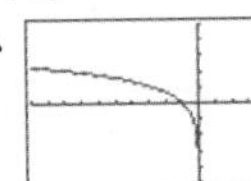

35. 5.8 yr **37.** 9.2 yr **39.** about 3.5 hr **43.** $y = 5x$ **45.** $y = -\frac{3}{2}x + \frac{13}{2}$ **47.** $x = 2$

Study Set Section 9.7 (page 658)

1. product **3.** power **5.** 0 **7.** M, N **9.** x, y **11.** x **13.** $\neq$ **15.** $\log_a b$ **17.** 0 **19.** 7 **21.** 10 **23.** 2 **25.** 1 **27.** 7 **29.** rs, t, r, s **37.** $2 + \log_2 5$ **39.** $\log_6 x - 2$ **41.** $4 \ln y$ **43.** $\frac{1}{2}\log 5$ **45.** $\log x + \log y + \log z$ **47.** $1 + \log_2 x - \log_2 y$ **49.** $3 \log x + 2 \log y$ **51.** $\frac{1}{2}(\log_b x + \log_b y)$ **53.** $\ln x + \frac{1}{2} \ln z$ **55.** $\frac{1}{3} \log_b x - \frac{1}{4} \log_b y - \frac{1}{4} \log_b z$ **57.** $\log_2 \frac{x+1}{x}$ **59.** $\log x^2y^{1/2}$ **61.** $\log_b \frac{z^{1/2}}{x^3y^2}$ **63.** $\log_b \frac{\frac{x}{z} + x}{\frac{y}{z} + y} = \log_b \frac{x}{y}$ **65.** false **67.** false **69.** true **71.** 1.4472 **73.** -1.1972 **75.** 1.1972 **77.** 1.8063 **79.** 1.7712 **81.** -1.0000 **83.** 1.8928 **85.** 2.3219 **87.** 4.8 **89.** from 2.5×10^{-8} to 1.6×10^{-7} **93.** $-\frac{7}{6}$ **95.** $\left(1, -\frac{1}{2}\right)$

Study Set Section 9.8 (page 667)

1. exponential **3.** $A_0 2^{-t/h}$ **5.** about 1.8 **7. a.** ± 3.46 **b.** 3.58 **9.** not a solution **11.** log, log 2 **13.** 1.1610 **15.** 1.2702 **17.** 1.7095 **19.** 0 **21.** ± 1.0878 **23.** 0, 1.0566 **25.** 8 **27.** $-\frac{3}{4}$ **29.** 3, -1 **31.** -2, -2 **33.** 1.8 **35.** 8.8, 0.2 **37.** 2 **39.** 3 **41.** -7 **43.** 4 **45.** 10, -10 **47.** 50 **49.** 20 **51.** 10 **53.** 10 **55.** no solution **57.** 6 **59.** 9 **61.** 4 **63.** 1, 7 **65.** 20 **67.** 8 **69.** 5.1 yr **71.** 42.7 days **73.** about 4,200 yr **75.** 5.6 yr **77.** 5.4 yr **79.** because $\ln 2 \approx 0.7$ **81.** 25.3 yr **83.** 2.828 times larger **85.** 13.3 **87.** $\frac{1}{3} \ln 0.75$ **91.** 0, 5 **93.** $\frac{2}{3}$, -4

Key Concept (page 670)

1. no **2.** no **3.** yes **4.** $f^{-1}(x) = \frac{-x-1}{2}$ **5.** $-2, 1, 3$ **6.** $\log 1{,}000 = 3$ **7.** $2^{-3} = \frac{1}{8}$ **8.** 2 **9.** $\frac{3}{2}$ **10.** 2.71828 **11.** e **12.** 0.7558 **13.**

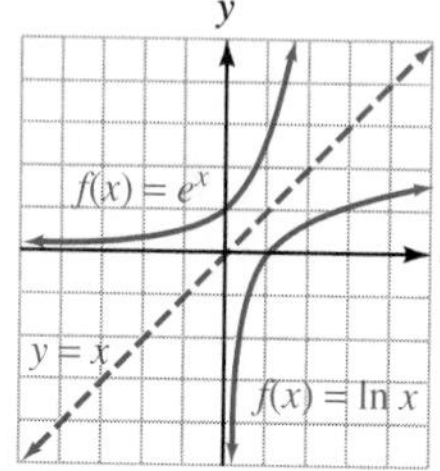

Chapter Review (page 672)

1. a. $(f + g)(x) = 3x + 1$ **b.** $(f - g)(x) = x - 1$ **c.** $(f \cdot g)(x) = 2x^2 + 2x$ **d.** $(f/g)(x) = \frac{2x}{x+1}$ **e.** 4 **f.** 3 **g.** $(f \circ g)(x) = 2(x + 1)$ **h.** $(g \circ f)(x) = 2x + 1$ **2. a.** no **b.** yes **3. a.** yes **b.** no **4.**

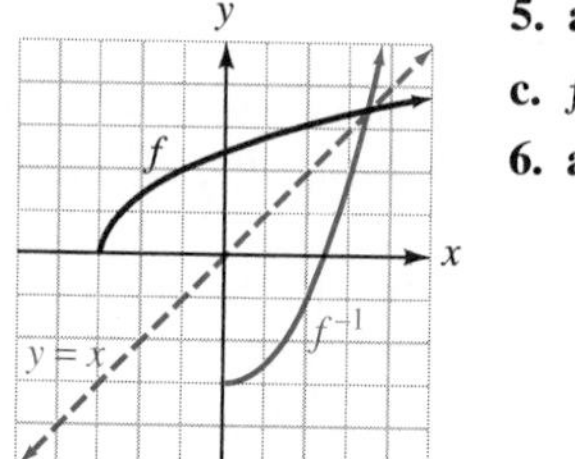

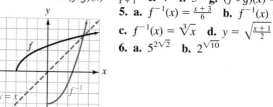

5. a. $f^{-1}(x) = \frac{x+3}{6}$ **b.** $f^{-1}(x) = \frac{x-5}{4}$ **c.** $f^{-1}(x) = \sqrt[3]{x}$ **d.** $y = \sqrt{\frac{x+1}{2}}$ **6. a.** $5^{2\sqrt{2}}$ **b.** $2^{\sqrt{10}}$

7. a. D: $(-\infty, \infty)$, R: $(0, \infty)$

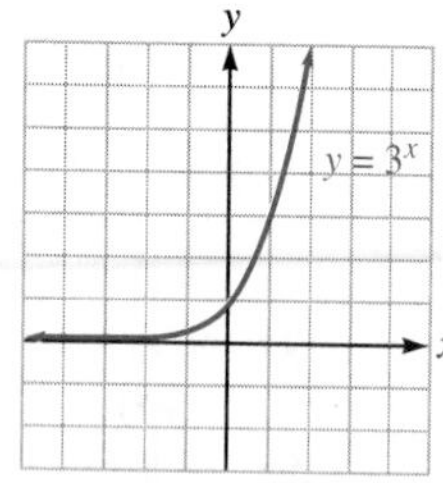

b. D: $(-\infty, \infty)$, R: $(0, \infty)$

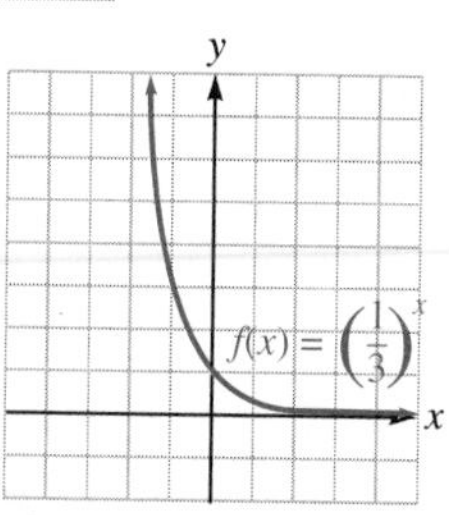

8. a.

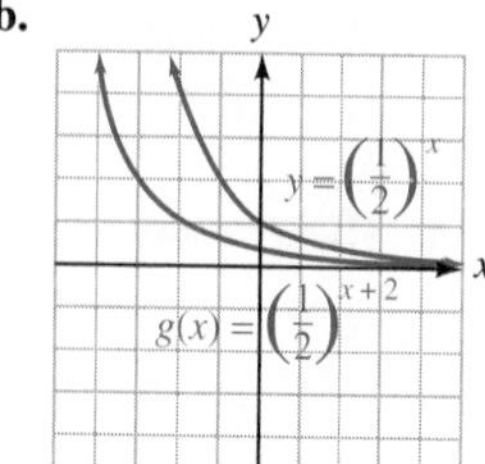

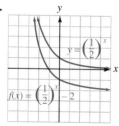

b.

9. an exponential function

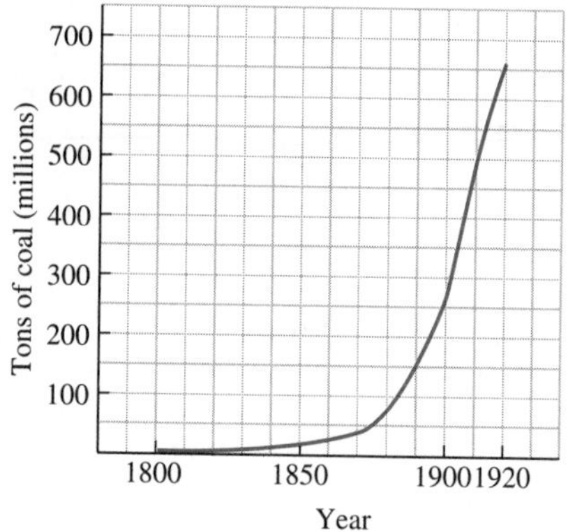

10. \$2,189,703.45 **11.** 8.22% **12.** \$2,324,767.37

13. a. D: $(-\infty, \infty)$, R: $(1, \infty)$ 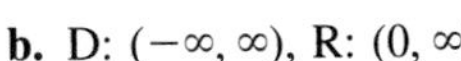 **b.** D: $(-\infty, \infty)$, R: $(0, \infty)$ 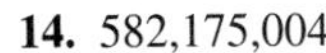 **14.** 582,175,004

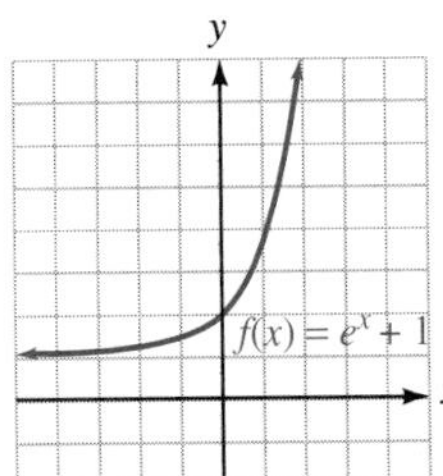

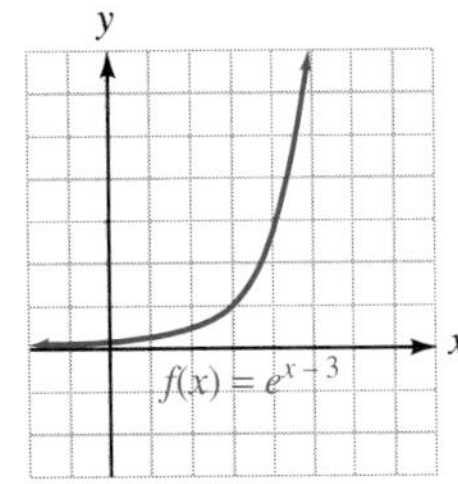

15. D: $(0, \infty)$, R: $(-\infty, \infty)$ **16. a.** $4^3 = 64$ **b.** $\log_7 \frac{1}{7} = -1$ **17. a.** 2 **b.** -2 **c.** 0 **d.** not possible **e.** $\frac{1}{2}$ **f.** 3 **18. a.** 32 **b.** $\frac{1}{81}$ **c.** 4 **d.** 10 **e.** $\frac{1}{2}$ **f.** 9 **19. a.** 0.6542 **b.** 26.9153

20. a.

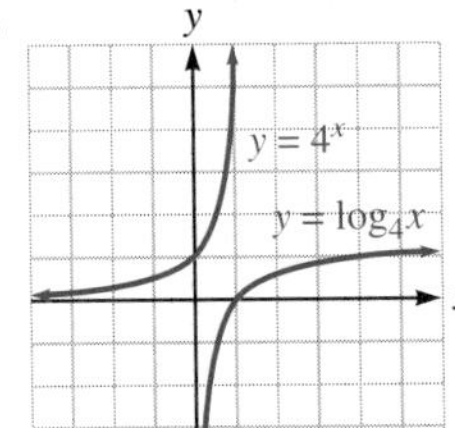

b.

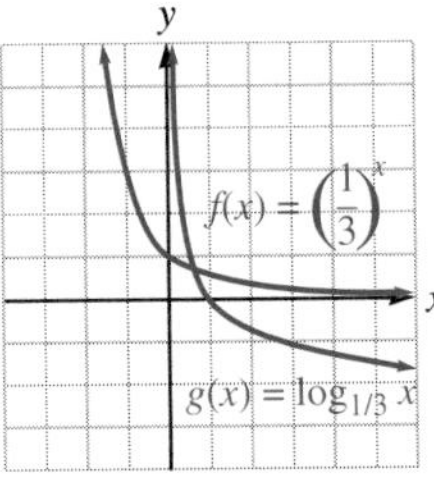

21. a.

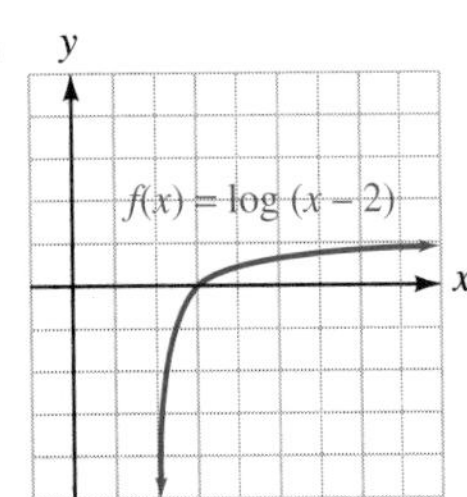

b.

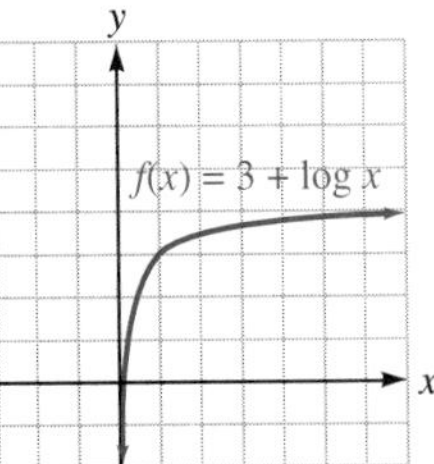

22. about 53 **23.** about 4.4 **24. a.** 6.1137 **b.** -0.1625 **25. a.** 10.3398 **b.** 0.0002 **26.** $f(x) = e^x$

27. a.

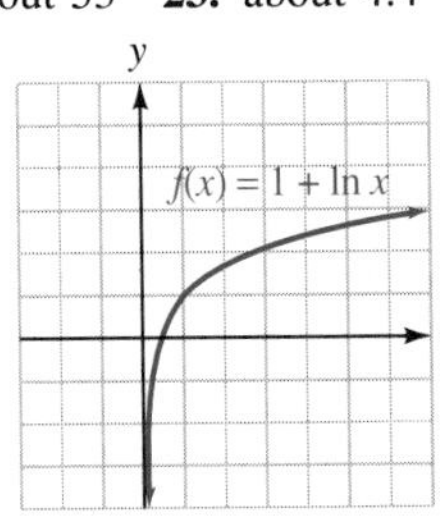

b.

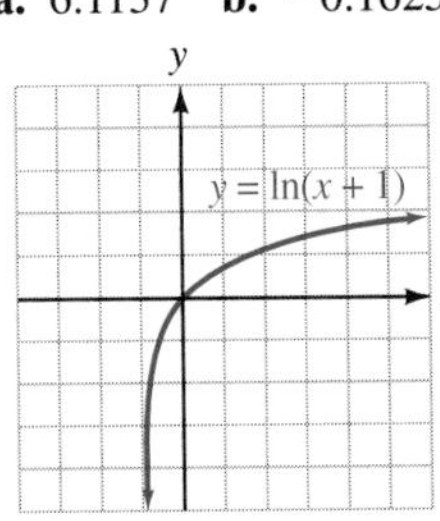

28. about 23 yr **29. a.** 0 **b.** 1 **c.** 3 **d.** 4 **e.** 4 **f.** 1 **30. a.** $3 + \log_3 x$ **b.** $2 - \log x$ **c.** $\frac{1}{2} \log_5 27$ **d.** $1 + \log a + \log b$ **31. a.** $2 \log_b x + 3 \log_b y - \log_b z$ **b.** $\frac{1}{2}(\log x - \log y - 2 \log z)$ **32. a.** $\log_2 \frac{x^3 z^7}{y^5}$ **b.** $\log_b \frac{\sqrt{x+2}}{y^3 z^7}$ **33. a.** 2.6609 **b.** 3.0000 **34.** 1.7604 **35.** about 7.9×10^{-4} gram-ions/liter **36. a.** 1.7712 **b.** 2 **c.** 2.7095 **d.** $-3, -1$ **37. a.** 25, 4 **b.** 4 **c.** 4, 3 **d.** 2 **e.** 6 **f.** 31 **38.** about 3,300 yr

Chapter 9 Test (page 679)

1. $(g + f)(x) = 5x - 1$ **2.** $(g \cdot f)(x) = 4x^2 - 4x$ **3.** 3 **4.** $4(x - 1)$ **5.** $y = \frac{12 - 2x}{3}$ **6.** $f^{-1}(x) = \sqrt{\frac{x - 4}{3}}$

7.

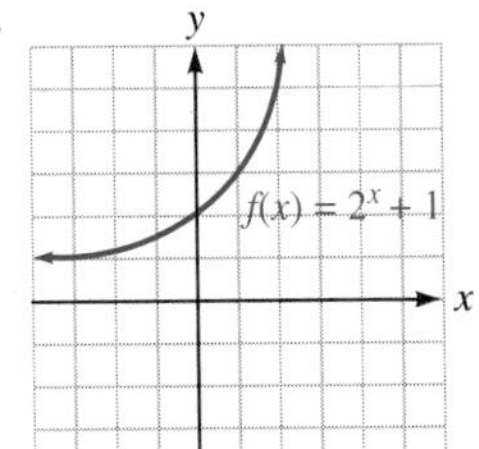

8.

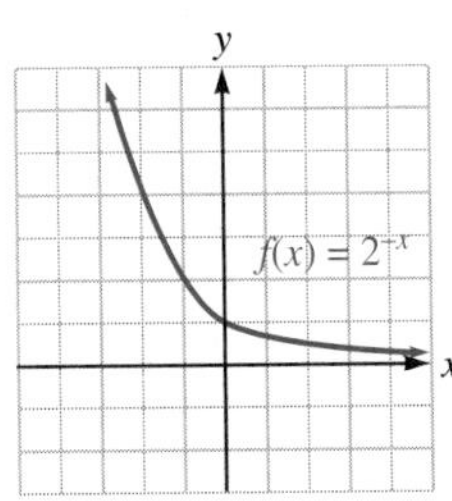

9. $\frac{3}{64}$ g = 0.046875 g **10.** \$1,060.90 **11.**

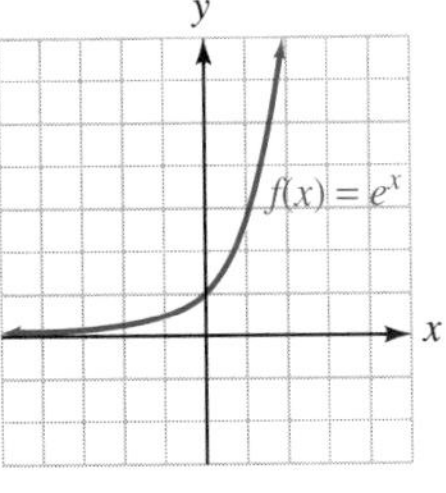

12. \$4,451.08 **13.** 2 **14.** 3 **15.** $\frac{1}{27}$ **16.** e **17.** $6^{-2} = \frac{1}{36}$ **18.** D: $(0, \infty)$, R: $(-\infty, \infty)$

19.

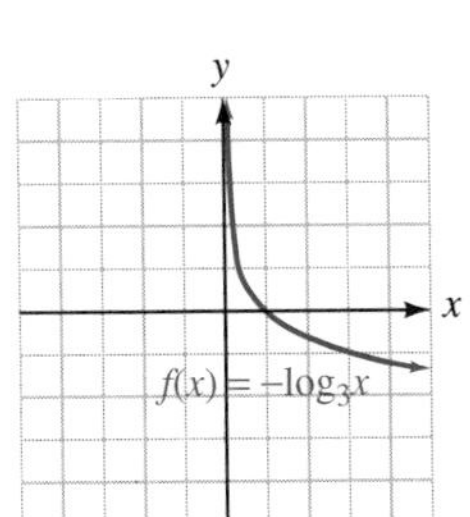

20.

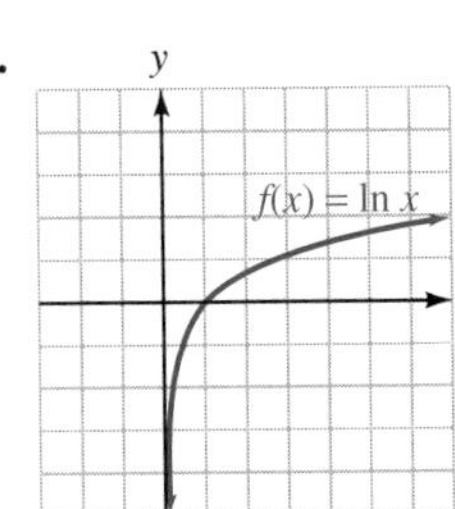

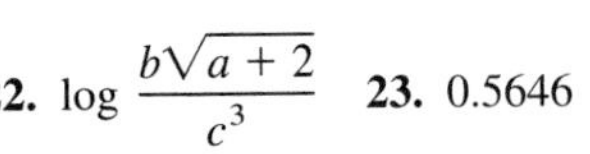

21. $2 \log a + \log b + 3 \log c$ **22.** $\log \frac{b\sqrt{a + 2}}{c^3}$ **23.** 0.5646 **24.** $y = \log x$ **25.** 6.4 **26.** 46 **27.** 0.6826 **28.** 4 **29.** 1 **30.** 10 **32. a.** yes **b.** yes **c.** 80; when the temperature of the tire tread is 260°, the vehicle is traveling 80 mph

Cumulative Review Exercises (page 681)

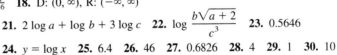

1. 0 **2.** $-\frac{4}{3}$, 5.6, 0, -23 **3.** π, $\sqrt{2}$, e **4.** $-\frac{4}{3}$, π, 5.6, $\sqrt{2}$, 0, -23, e **5.** \$8,250 **6.** $\frac{1}{120}$ db/rpm **7.** parallel **8.** perpendicular **9.** $y = -2x + 5$ **10.** $y = -\frac{9}{13}x + \frac{7}{13}$ **11.** (1, 1) **12.** (2, −2) **13.** (−1, −1) **14.** (−1, −1, 3) **15.** 85°, 80°, 15°

16.

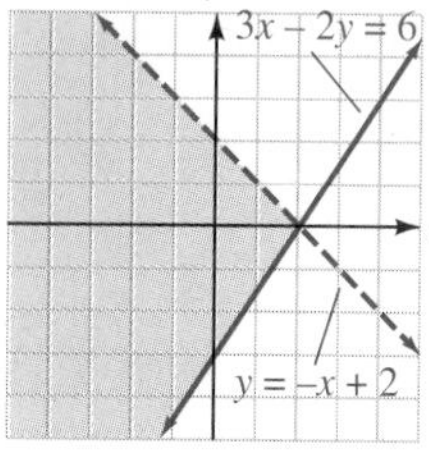

17. $\left[-3, \frac{19}{3}\right]$,

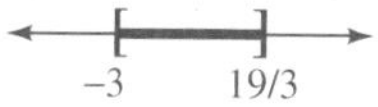

18. $(-\infty, -2)$,

19. $12x^2 - 5xy - 3y^2$ **20.** $2x^2y^3 + 13xy + 3y^2$ **21.** $a^2 - 4ab + 4b^2$ **22.** $3a^3 + 10a^2 + 6a - 4$ **23.** $xy(3x - 4y)(x - 2)$ **24.** $(16x^2y^2 + z^4)(4xy + z^2)(4xy - z^2)$ **25.** $\frac{64b^{12}}{27a^9}$ **26.** $-\frac{3x+2}{3x-2}$ **27.** $-\frac{q}{p}$ **28.** $\frac{4a-1}{(a+2)(a-2)}$ **29.** 5; 3 is extraneous **30.** $R = \frac{R_1R_2R_3}{R_2R_3 + R_1R_3 + R_1R_2}$ **31. a.** a quadratic function **b.** at about 85% and 120% of the suggested inflation **32.** about $21\frac{1}{2}$ in. **33.** $5\sqrt{2}$ **34.** $81x\sqrt[3]{3x}$ **35.** $\frac{343}{125}$ **36.** $\sqrt{3t} - 1$ **37.** $-5 + 17i$ **38.** $-\frac{21}{29} - \frac{20}{29}i$ **39.** 5, 0 does not check **40.** 0 **41.** $\frac{2}{3}, -\frac{3}{2}$ **42.** $\frac{-3 \pm \sqrt{5}}{4}$ **43.** $\frac{1}{4}, \frac{1}{2}$ **44.** $\frac{2 \pm \sqrt{2}i}{3}$

45. $4x^2 + 4x - 1$ **46.** $f^{-1}(x) = \sqrt[3]{\frac{x+1}{2}}$ **47.**

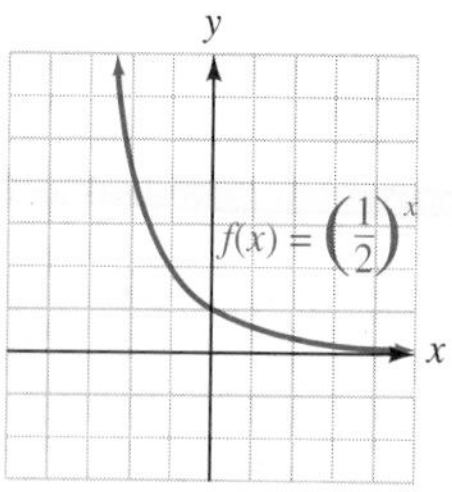

48.

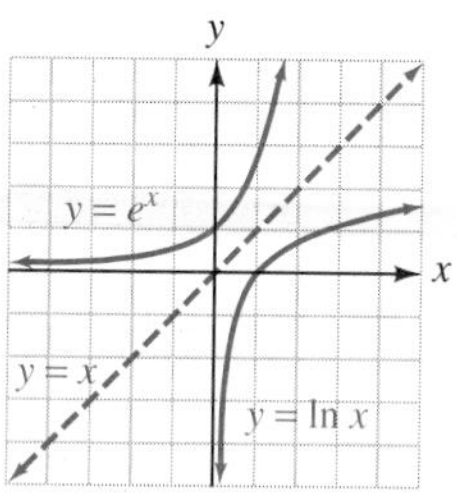

49. $2^y = x$ **50.** $\log_6 x - 2$
51. 3.4190
52. 1, −10 does not check
53. $-\frac{3}{4}$ **54.** 9

Study Set Section I.1 (page A-7)

1. circle, center **3.** parabola **5.** $x^2 + y^2 = r^2$ **7.** upward **9.** parabola, downward **11.**

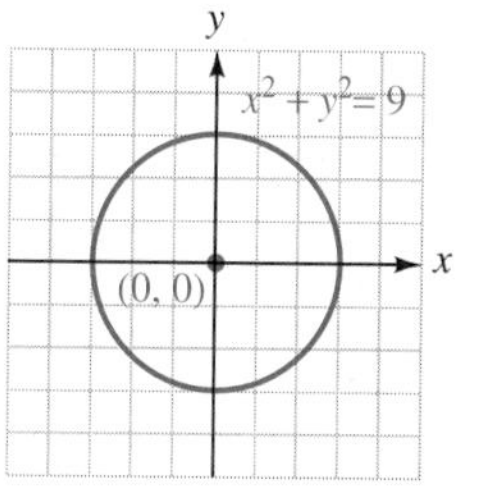

13.

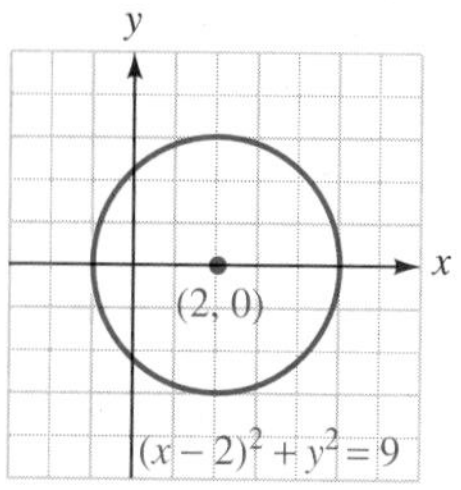

15.

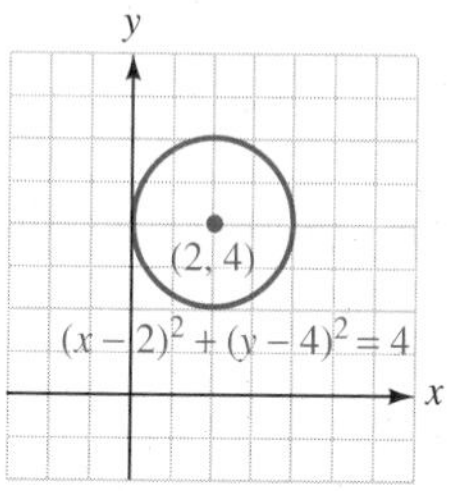

17.

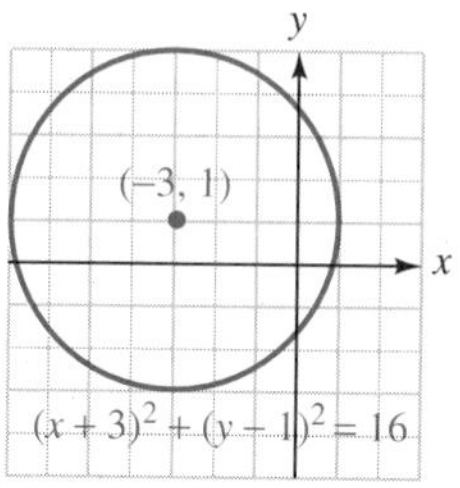

19.

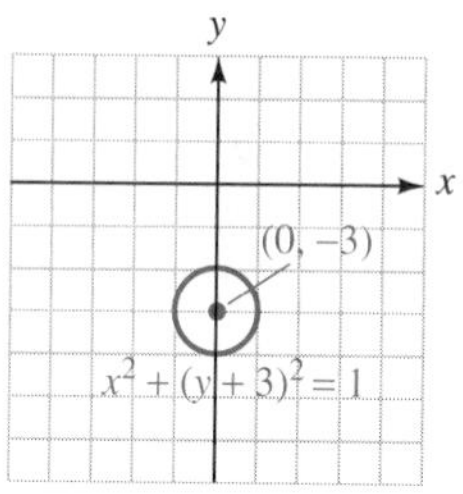

21. $x^2 + y^2 = r^2$ **23.** $(x - 6)^2 + (y - 8)^2 = 25$ **25.** $(x + 2)^2 + (y - 6)^2 = 144$ **27.** $x^2 + y^2 = 2$

29.

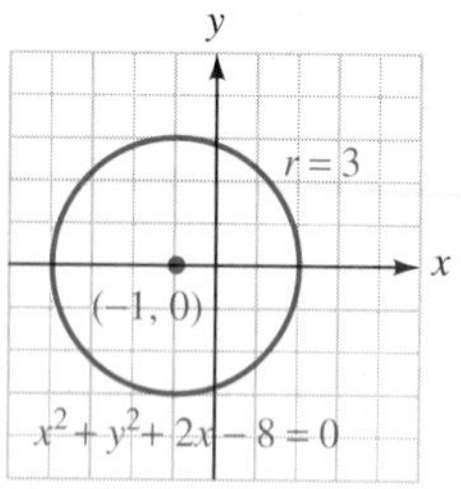

31.

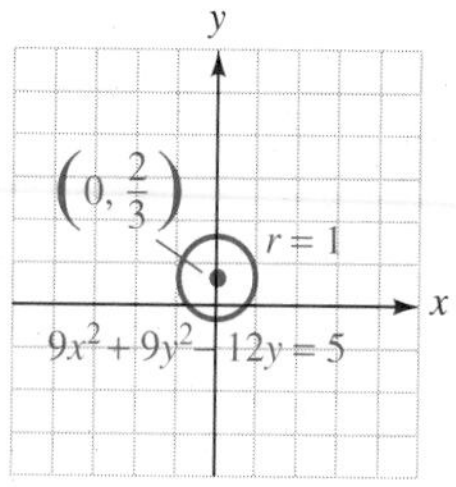

33.

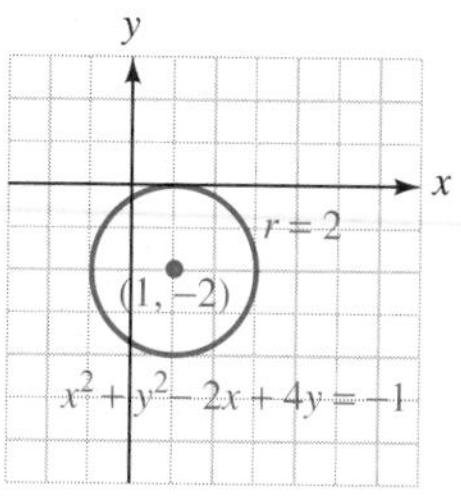

35.

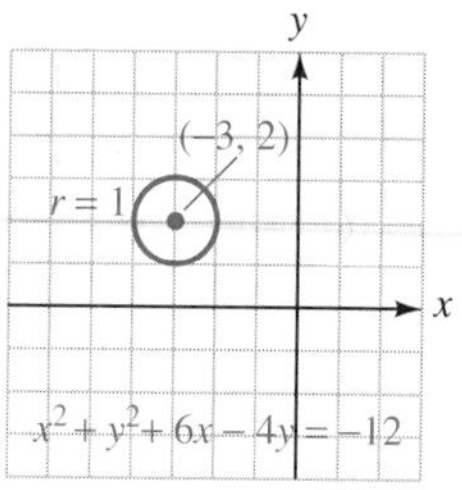

37.

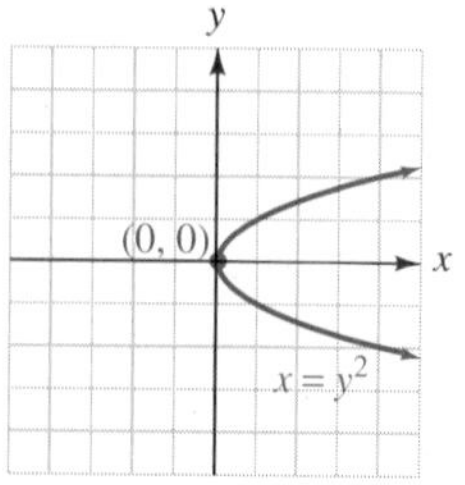

39.

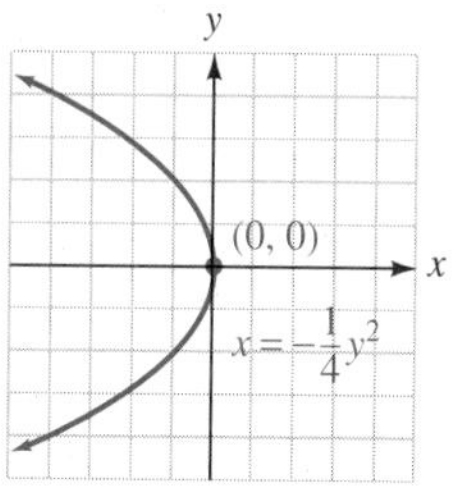

41.

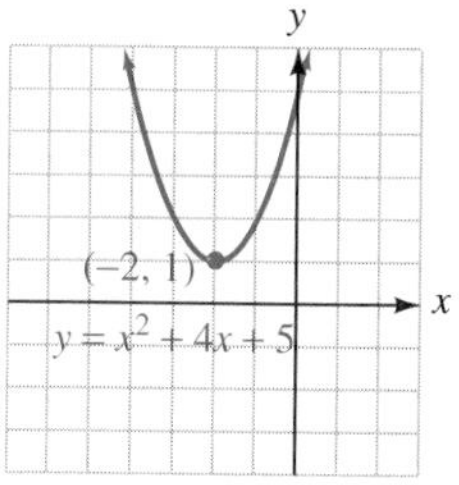

43.

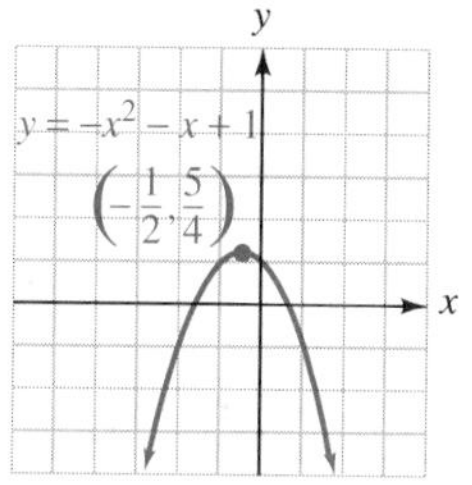

45.
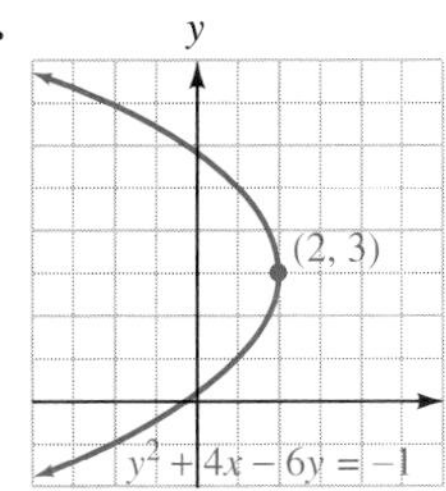

47.
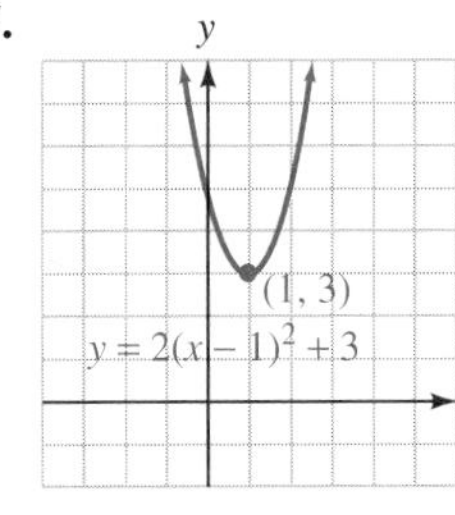

49.
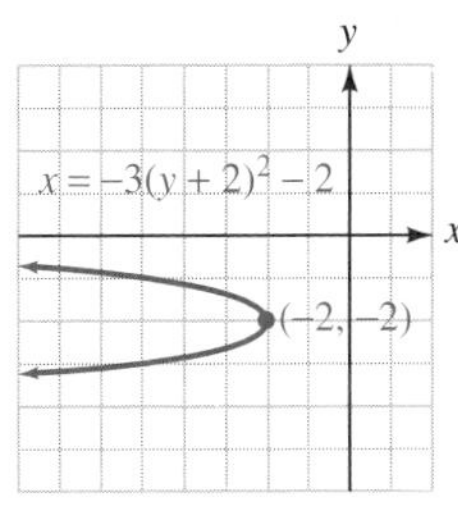

51. $(x - 7)^2 + y^2 = 9$ **53.** no
55. 30 ft away **57.** 2 AU
61. $5, -\frac{7}{3}$ **63.** $3, -\frac{1}{4}$

Study Set Section I.2 (page A-19)

1. sum, constant **3.** difference, constant **5.** $(a, 0), (-a, 0)$ **7.** $\dfrac{x^2}{a^2} + \dfrac{y^2}{b^2} = 1$ **9.** hyperbola; $(0, a), (0, -a)$

11.
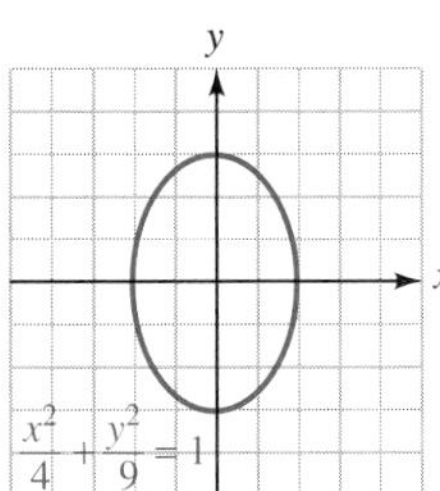

13.
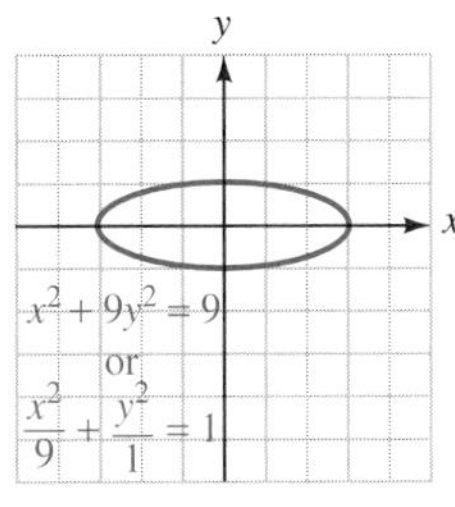

15.
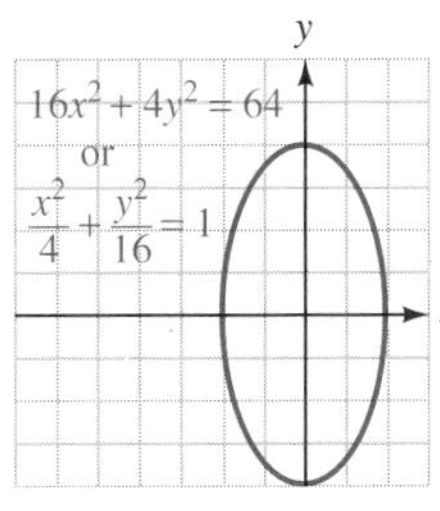

17.
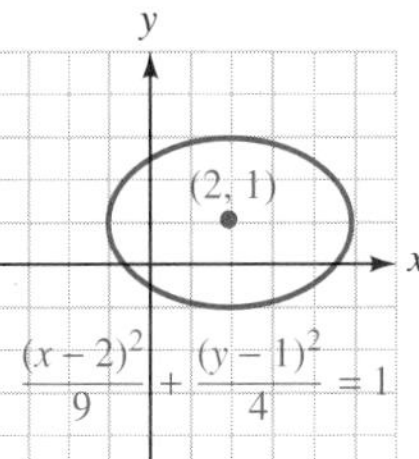

19.
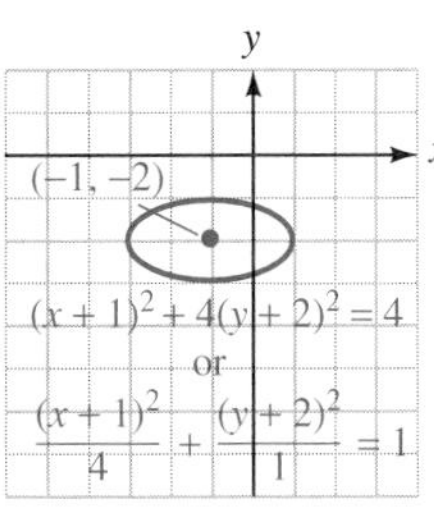

21.
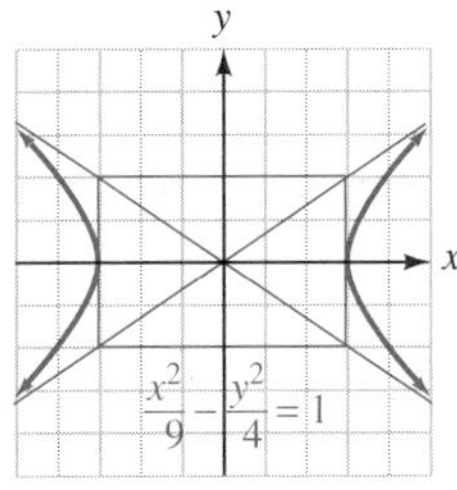

23.
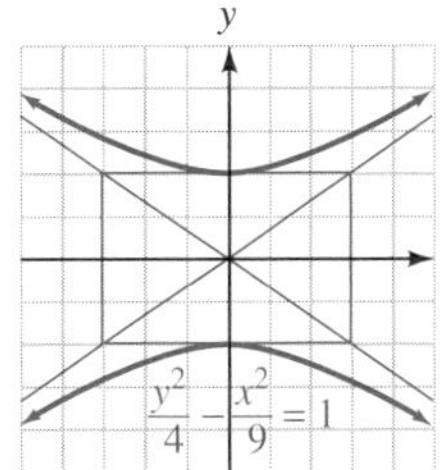

25.
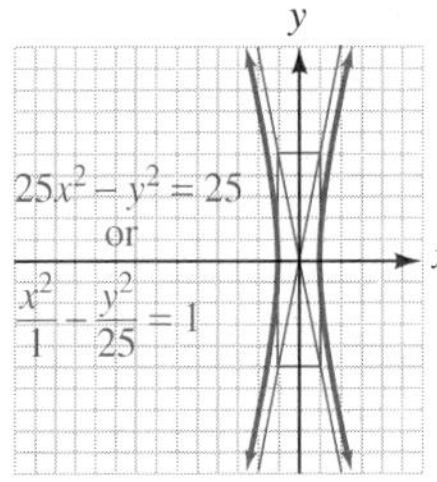

27.
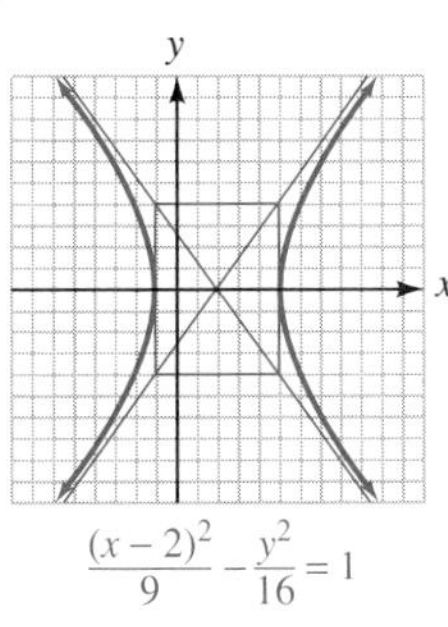

29.
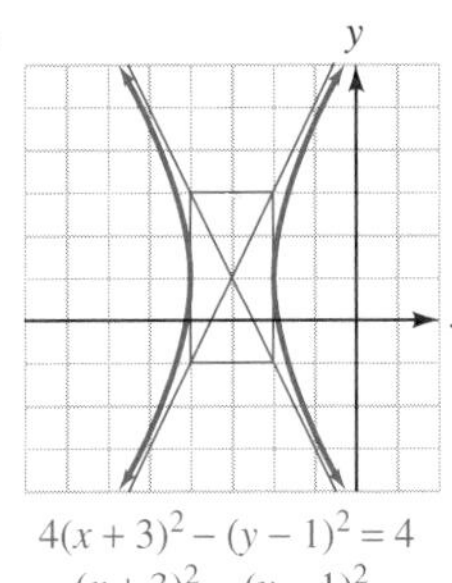

$4(x + 3)^2 - (y - 1)^2 = 4$
or $\frac{(x+3)^2}{1} - \frac{(y-1)^2}{4} = 1$

31.
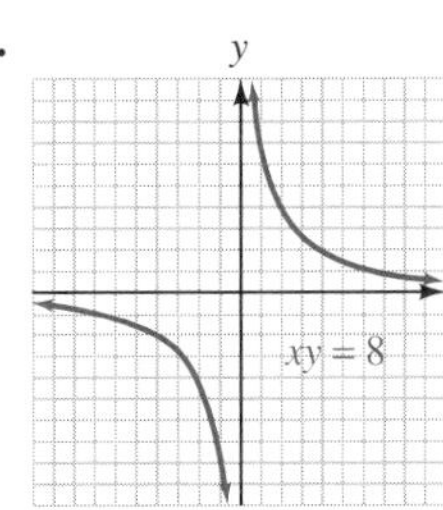

33. $\dfrac{x^2}{400} + \dfrac{y^2}{100} = 1$
35. 12π sq. units ≈ 37.7 sq. units
37. 3 units **39.** $10\sqrt{3}$ mi ≈ 17.3 mi
43. $12y^2 + \dfrac{9}{x^2}$ **45.** $\dfrac{y^2 + x^2}{y^2 - x^2}$

Study Set Section II.1 (page A-28)

1. Pascal's **3.** $3 \cdot 2 \cdot 1$ **5.** 7 **7.** 6 **9.** -120 **11.** 30 **13.** 144 **15.** 40,320 **17.** $\frac{1}{110}$ **19.** 2,352 **21.** 10 **23.** 21 **25.** $\frac{1}{168}$ **27.** $x^2 + 2xy + y^2$ **29.** $x^4 - 4x^3y + 6x^2y^2 - 4xy^3 + y^4$ **31.** $8x^3 + 12x^2y + 6xy^2 + y^3$ **33.** $x^3 - 6x^2y + 12xy^2 - 8y^3$ **35.** $8x^3 + 36x^2y + 54xy^2 + 27y^3$ **37.** $\dfrac{x^3}{8} - \dfrac{x^2y}{4} + \dfrac{xy^2}{6} - \dfrac{y^3}{27}$ **39.** $81 + 216y + 216y^2 + 96y^3 + 16y^4$ **41.** $\dfrac{x^4}{81} - \dfrac{2x^3y}{27} + \dfrac{x^2y^2}{6} - \dfrac{xy^3}{6} + \dfrac{y^4}{16}$ **43.** $3a^2b$ **45.** $-4xy^3$ **47.** $15x^2y^4$ **49.** $28x^6y^2$ **51.** $90x^3$ **53.** $640x^3y^2$ **55.** $-12x^3y$ **57.** $-70{,}000x^4$ **59.** $810xy^4$ **65.** (2, 3) **67.** 8

INDEX